Magnetic Resonance and Its Applications

Vladimir I. Chizhik · Yuri S. Chernyshev
Alexey V. Donets · Vyacheslav V. Frolov
Andrei V. Komolkin · Marina G. Shelyapina

Magnetic Resonance and Its Applications

Springer

Vladimir I. Chizhik
Yuri S. Chernyshev
Alexey V. Donets
Vyacheslav V. Frolov
Andrei V. Komolkin
Marina G. Shelyapina
Saint-Petersburg State University
Peterhof, Saint Petersburg
Russia

ISBN 978-3-319-05298-4 ISBN 978-3-319-05299-1 (eBook)
DOI 10.1007/978-3-319-05299-1
Springer Cham Heidelberg New York Dordrecht London

Library of Congress Control Number: 2014933117

Printed on acid-free paper

Springer is part of Springer Science+Business Media (www.springer.com)

Preface

The phenomena that referred to the notion of *magnetic resonance* are related to *quantum radiophysics*. Quantum Radiophysics as a branch of science was formed in the beginning of the 1960s. In this field of physics, one studies the phenomena accompanied with the emission or absorption of electromagnetic waves of radio-frequency diapason by quantum systems (nuclei, electrons, atoms, molecules, etc.). These phenomena, the physical nature of which is of independent interest, provided the basis of radiospectroscopic methods for studying the structure of matter and physicochemical processes in it. They are also used for the creation of quantum generators, amplifiers, and magnetometers. Thus, the quantum radiophysics consists of *radiospectroscopy* and *radioelectronics*.

For the first time, the selective absorption of radio waves due to the magnetic properties of electron shells was observed in 1913 by B. K. Arkad'ev, Professor of Moscow University. In 1934, C. E. Cleeton and N. H. Williams observed inversion spectra of ammonia in the range of radio frequencies (Phys. Rev., 1934, v. 45, p. 234). In 1938, I. Rabi registered the first nuclear magnetic resonance (NMR) spectra in experiments with atomic beams (Phys. Rev., 1939, v. 55, p. 526).

The particularly intensive development of radiospectroscopic methods began after 1944, when E. K. Zavoiskii, Professor of Kazan State University (Russian Federation), discovered the phenomenon of electron paramagnetic resonance—EPR (Thesis for Doctor of Science, 1944). In 1946, two scientific teams of American scientists, headed by F. Bloch and E. M. Purcell, published articles on the observation of nuclear magnetic resonance in condensed matter (Phys. Rev., 1946, v. 69, p. 127 and ibid., p. 37, consequently). In 1950, H. G. Dehmelt and G. Krueger obtained spectra of nuclear quadrupole resonance—NQR (Naturwissenschaften, 1950, Bd. 37, S. 111).

At present the latest technical achievements, including in the areas of processing and cryogenic technologies, are used in radiospectroscopy. Devices developed on the basis of these achievements have extremely high sensitivity and resolution. For example, in NMR one records quantum transitions with energies $10^{-24} \div 10^{-25}$ erg ($10^{-31} \div 10^{-32}$J) and resolves two neighboring spectral lines separated by less than 0.1 Hz. Thus, the radiospectroscopy allows the investigation of processes that lead to the finest changes in the electronic structure of atoms and molecules.

Magnetic relaxation—the achievement of the equilibrium state of nuclear or electronic macroscopic magnetization in a static magnetic field—has no direct analogs in physical phenomena which determine spectra in infrared, visible, and more shortwave diapasons, when an equilibrium state is achieved mainly due to the spontaneous transitions from excited states. The magnetic relaxation processes are determined by the intensity of fluctuating electromagnetic fields in matter. Thus, the investigation of these processes is the study of interactions of nuclei and electrons with the fluctuating electromagnetic fields in close interweaving with the study of the nature and velocity of thermal molecular motion which produces those fluctuating fields. It should be noted that magnetic relaxation is sensitive to molecular motions in a very wide range: $10^3 \div 10^{11}$ Hz. Most often, the study of spectral and relaxation characteristics of magnetic resonance have been held separately, and only in relatively rare original works attempts were made to carry out their complex analysis. However, the latter trend is increasing and no doubt in the nearest future these characteristics will be widely considered using common programs for the interpretation of responses of magnetic resonances at certain impacts on substances (or spin systems). The confirmation of this can be found in developing applications of double, ternary, and more complex resonances (2-, 3-, multidimensional spectroscopies), in which the relaxation processes are responsible for the transfer of spin polarization from one spin system to another.

The radiospectroscopic methods are widely used in molecular physics, chemistry, biology, medicine, and other sciences. In recent years, the methods of quantum radiophysics are used in engineering and industry, as well as for control of technological processes. Thus, in Russia, the NMR method is used for oil well logging, for the laboratory analysis of productivity oil collectors, and for the analysis of oil content and moisture in seeds; the ESR methods were developed for geological research, nondestructive testing of precious stones, etc.

An important step in the formation of quantum radiophysics is connected with the publication of the articles of American physicists (J. P. Gordon, H. J. Zeiger, C. H. Townes. Phys. Rev., 1955, v. 99, p. 1264) and Soviet physicists (N. G. Basov, A. M. Prokhorov. Physics-Uspekhi (Advances in Physical Sciences), 1955, v. 57, No 3, p. 481), who had independently created the first quantum generators with a beam of ammonia molecules. Both groups noted that the initial purpose of their work was the creation of high-resolution radiospectroscopes on the basis of molecular radiation, but much greater influence on the development of quantum radiophysics produced a "collateral" effect associated with the oscillating mode of the device, i.e., with its conversion in the quantum generator. Quantum generators with beams of atoms and molecules immediately attracted attention due to their unique properties: even the first devices provided the relative instability of the frequency of the order 10^{-8} (approximate level of the precision of astronomical methods of timing) and soon achieved indicators of $10^{-12} \div 10^{-13}$. At present, the generator on a beam of hydrogen atoms is characterized by the frequency instability $2 \cdot 10^{-14}$ per day and the frequency reproducibility of $\pm 5 \cdot 10^{-13}$. The high frequency stability of quantum generators has determined their use as devices of storage precise time:time and frequency standards.

Quantum amplifiers were created almost simultaneously with quantum generators. The main advantage of quantum amplifiers is an extremely low level of set (equivalent input) noises. The most perspective of these were devices created on the basis of electron paramagnetic resonance in solids. In 1956, N. Bloemdergen showed that in systems with several energy levels it was possible to create a steady-state quantum amplifier if one used an auxiliary radiation for the polarization of electron spins. It should be noted that for the first time the method of inversion of level populations with pumping radiation was proposed by N. G. Basov and A. M. Prokhorov in 1955. Quantum amplifiers have provided new possibilities in many areas of science and technology. For example, in 1963, the first in the history of television images were transmitted from America to Europe using an artificial Earth satellite (a metallic sphere of 30 m in diameter) as a passive reflector of radio waves and using a quantum amplifier as a receiver (the first translation was devoted to the funeral ceremony for American President John Fitzgerald Kennedy). Now, using quantum amplifiers, the radiolocation of all inner planets of the Solar System has been made (in order that a reader can imagine the difficulties in receiving signals reflected from planets, we indicate that the signal which passed the way to Venus and back, is weakened at 10^{27} times). Based on the revised estimations of the relative positions of planets the new theory of planetary motion in the Solar System was developed in the late 1960s. Quantum paramagnetic amplifiers are widely applied in systems of Deep Space communications and radio astronomy, where they are used as input stages of low noise receivers.

On the basis of quantum electronics devices magnetometers have been created, which have extremely high sensitivity and high accuracy measurements of magnetic fields. Using quantum magnetometers, variations of the Earth's magnetic field can be registered in limits of $10^{-7} \div 10^{-8}$, i. e., approximately $0.5 \cdot 10^{-7}$ G ($0.5 \cdot 10^{-11}$ T). Quantum magnetometers are widely used for the study of fluctuations of the Earth's magnetic field due to the change in conditions in the ionosphere and to internal processes in the globe depths, for registration of spatial variations of the geomagnetic field in geological survey, archeology, and military applications.

An important event that brought general recognition to quantum radiophysics was the invention of lasers. In 1958, for the development of ideas of quantum radioelectronics, A. L. Schawlow and C. H. Townes proposed a quantum generator of light and considered general problems related to the generation and amplification of electromagnetic waves in the optical range (Phys. Rev., 1958, v. 112, p. 940). The first pulsed optical generator was created in 1960 (T. H. Maiman. Nature, 1960, v. 187, p. 493) and the first steady-state one was realized in 1961 (A. Javan, W. R. Bennett, D. R. Herriott. Phys. Rev. Lett., 1961, v. 6, p. 106). The appearance of coherent high power light sources resulted in a colossal leap in researches on nonlinear effects in physical optics, as well as in many technical applications.

In different scientific schools, the development of the theory and experimental methods occurred with certain peculiarities that derive from the historical and economical conditions of research teams. This circumstance may lend some peculiarities of our book but we tried to be objective. In this book, we try to give a circumstantial account of the principles of magnetic resonance and its applications in accordance with our vision of the subject, our possibilities, and abilities. We tried to create a universal book but certainly some particular questions have remained out of the book. In accordance with the chosen style of the book, we have decided to follow the opinion of the great A. Abragam, who reckoned that as opposed to an article it is not necessary in a book to cite initial publications on questions. Now there are many hundreds and even thousands of important articles on each question and it is difficult to give preference to only some of them. After each chapter we suggest the list of books and reviews recommended for further reading (excuse us if we did not indicate some excellent ones).

This book is written on the basis of a few dozens of lecture courses, which have been prepared and delivered by the authors at the Faculty of Physics of Saint-Petersburg State University (Department of Quantum Magnetic Phenomena: BULLETIN DU GROUPEMENT AMPERE, October–December 2010, No. 241, http://www.ampere.ethz.ch/) for students who are taught at the directions "Physics" and "Applied Mathematics and Physics," according to the undergraduate and Master's programs "Magnetic Resonance, Physical aspects and applications," "Tomographic technologies in modern medical diagnostics," and "Quantum Radiophysics." The material is disposed so that the book would be of interest to scientists and students of physical, chemical, biological, and geological specialities.

The book contains 5 parts and 21 chapters: Alexey Donets wrote Chap. 1; Yurii Chernyshev wrote Chap. 2; Vladimir Chizhik wrote Sects. 2.7 and 2.8, Chap. 3, Sect. 4.4 (partly), Chaps. 15, 19–21; Andrey Komolkin wrote Chaps. 4 and 6; Marina Shelyapina wrote Sects. 1.9–1.11, Chaps. 5, 7–9, 17; Vyacheslav Frolov wrote Sect. 2.9, Chaps. 10–14,16. Andrei Komolkin made the key contribution to the preparation of the book as a whole, Alexey Donets edited all the figures in the book.

We would like to mention kindly and gratefully our colleagues who worked earlier with us at the Department of Quantum Magnetic Phenomena. Together, we fruitfully discussed and solved scientific problems and elaborated courses of lectures but they could not participate in the creation of the book. They are Piotr Borodin, Valentina Kasperovich, Arkadii Mel'nikov, Viktor Moskalev, and Nikolai Vecherukhin.

2013 Vladimir I. Chizhik

Contents

Chapter 1
Interaction Between Nuclei and Electrons and Their Interaction with External Electromagnetic Fields

Absorption or emission of energy by quantum systems as a result of their interaction with an electromagnetic field is the basis of all of the methods of quantum radio physics. This phenomenon in the nuclear magnetic and electron paramagnetic resonances (NMR and EPR) is caused by interaction of nuclear or electron magnetic momenta with constant and variable magnetic fields. Magnetic interaction is used in the implementation of double resonances; in some cases it is also manifested in the spectra of the nuclear quadrupole resonance (NQR). Therefore, a clear understanding of the interaction pattern of magnetic and electric momenta of microparticles with constant and variable magnetic fields allows us to understand the physical nature of the methods of quantum physics, especially NMR, EPR and NQR.

1.1 Basic Information on Electronic and Nuclear Momenta

1.1.1 The Quantum-Mechanical Description of the Properties of Angular Momentum

In accordance with the postulates of classical mechanics, moment of momentum, or angular momentum ($\vec{J}$) of an isolated system is a conserved quantity, i.e. integral of motion. Conservation law for the angular momentum holds in quantum mechanics as well. In the following, we shall use the notion of the *angular momentum operator*, denoted as $\hat{\vec{J}}$, since any physical quantity corresponds to a quantum mechanical operator. It is proved in quantum mechanics, that the operators of conserved and, therefore, simultaneously defined quantities commute with each other. In particular, the operator $\hat{J}_z$ of projection of the angular momentum of an isolated system onto an arbitrary direction (for example, axis z) commutes with the energy operator $\widehat{\mathcal{H}}$ as well as with the square of the momentum $\hat{\vec{J}}^2$. Hamiltonian $\widehat{\mathcal{H}}$ and operators $\hat{\vec{J}}^2$ and $\hat{J}_z$ form a complete set of mutually commuting operators and, therefore, have common

V. I. Chizhik et al., *Magnetic Resonance and Its Applications*,
DOI: 10.1007/978-3-319-05299-1_1,

eigenfunctions. Well-defined numerical values (*eigenvalues*) $E_{n\,J\,M_J}$, $J(J+1)$ and M_J correspond to each of these operators in the states described by these functions. Eigenfunctions are put into correspondence (according to Dirac) with *eigenvectors* of state in an abstract space designated by means of special brackets: $|\,k\,\rangle$ (*ket*-vector) or $\langle\,k\,|$ (*bra*-vector). Here, k is an index (a set of indices) numbering the states. With this notation, the assertion that the values $\widehat{\mathcal{H}}$, $\hat{\vec{J}}^2$ and $\hat{J}_z$ are simultaneously defined can be mathematically written in the following form (each state being numbered by a set of indices n, J and M_J):

$$\widehat{\mathcal{H}}\,|\,n\;J\;M_J\,\rangle = E_{n\,J\,M_J}|\,n\;J\;M_J\,\rangle, \tag{1.1}$$

$$\hat{\vec{J}}^2\,|\,n\;J\;M_J\,\rangle = \hbar^2 J(J+1)|\,n\;J\;M_J\,\rangle, \tag{1.2}$$

$$\hat{J}_z\,|\,n\;J\;M_J\,\rangle = \hbar M_J|\,n\;J\;M_J\,\rangle, \tag{1.3}$$

where $\hbar = h/(2\pi)$, h is Planck's constant, J takes any integer or half-integer values; M_J takes the values $-J$, $-J+1,\ldots,$ $J-1$, J. In quantum mechanics, momentum is typically measured in units of $\hbar$. It will be hereinafter assumed that all angular momenta are expressed in these units. So, instead of (1.2) and (1.3), we should write

$$\hat{\vec{J}}^2\,|\,n\;J\;M_J\,\rangle = J(J+1)|\,n\;J\;M_J\,\rangle, \tag{1.4}$$

$$\hat{J}_z\,|\,n\;J\;M_J\,\rangle = M_J|\,n\;J\;M_J\,\rangle. \tag{1.5}$$

The quantity M is value (in units of $\hbar$) of the angular momentum projection onto the axis z in the corresponding state, and the quantity J is the maximum possible value of this projection. Namely this value, and not $\sqrt{J(J+1)}$, as would follow from (1.2), is typically referred to as the value of the angular momentum or the quantum number for the state with the angular momentum $\hat{\vec{J}}$.

All three components ($\hat{J}_x$, $\hat{J}_y$ and $\hat{J}_z$) of the angular momentum operator do not commute with each other and, therefore, can not be defined simultaneously. However, if the $\hat{J}_z$ component is determined in accordance with (1.5), then one has relations reflecting the result of the action of operators $\hat{J}_x$ and $\hat{J}_y$ on eigenvectors of operator $\hat{J}_z$. These relations are easier to write with the help of the *shift operators*: *raising* ($\hat{J}_+$) and *lowering* ($\hat{J}_-$) operators:

$$\hat{J}_\pm = \hat{J}_x \pm i\hat{J}_y. \tag{1.6}$$

Then we have the equality

$$\hat{J}_\pm\,|\,n\;J\;M_J\,\rangle = \sqrt{(J \mp M_J)(J \pm M_J + 1)}|\,n\;J\;M_J \pm 1\,\rangle. \tag{1.7}$$

The angular momentum, which a particle has while moving along an orbit in the central field, is called orbital momentum. Its quantum mechanics counterpart is the orbital angular momentum which is denoted by $\hat{\vec{l}}$. The eigenvalue of this operator (the quantum number of l) takes integer values, starting from zero. In this book, we shall always consider only orbital angular momenta of electrons. In an atom or ion, orbital moments of electrons are added vectorially to form the total orbital angular momentum of the atom (ion) $\hat{\vec{L}}$. The eigenvalues of the operators of the projection of orbital angular momentum of a single electron ($\hat{l}_z$) and electron shell of an atom or ion ($\hat{L}_z$) are commonly denoted by letters m and M without special indices.

Electrons and nuclei also have an internal, or their own, angular momentum called *spin*. Spin is a specifically quantum mechanics quantity. The idea of spin should not be only related to rotation of a particle around its axis or displacement of the particle as a whole, which fundamentally distinguishes spin from orbital angular momentum. However, all formal properties of the operator of angular momentum given by (1.1)–(1.7) remain valid for the spin angular momentum operator, too. The operator of electron spin is denoted by $\hat{\vec{s}}$, and the operator of nuclear spin, by $\hat{\vec{I}}$. The electron spin s is $1/2$ and the nuclear spin I can be either integer or half-integer. Eigenvalues $\hat{s}_z$ and $\hat{I}_z$ are designated by m_s and m_I, respectively. The spin momenta are added (vectorially) not only with each other but also with the orbital angular momenta.

To describe the physical properties of atoms or ions, the following operators are used: *the total electron spin angular momentum* of an atom is expressed as $\hat{\vec{S}} = \sum_i^N \hat{\vec{s}}_i$ (the word "electron" is usually omitted); the total electron momentum of an atom as $\hat{\vec{J}} = \hat{\vec{L}} + \hat{\vec{S}}$; and the total momentum of an atom (including electronic and nuclear momenta) as $\hat{\vec{F}} = \hat{\vec{J}} + \hat{\vec{I}}$. Eigenvalues of the operators $\hat{\vec{J}}$ and $\hat{\vec{F}}$ can be both integer and half-integral.

A molecule is characterized by the angular momentum due to its rotation as a whole. This momentum is also designated as $\hat{\vec{J}}$. The eigenvalue of $\hat{J}_z$ (this is usually the projection onto the axis of symmetry of the molecule, if it exists) is designated by the letter K. Other combinations of angular momenta and their projections, with their own symbols, can be introduced for a molecule.

For an isolated system in a spherically symmetric field, the values of the energy states $|\, n\ J\ M_J\, \rangle$, differing only by the value M_J, coincide, that is, the energy levels degenerate $2J + 1$ times. If an electric or magnetic field of axial symmetry affects on a system then the degeneracy along $\hat{J}_z$ (projection on the direction of the field vector) is removed. In this case, the energy depends on the value of M_J.

1.1.2 The Magnetic Dipole Moments of Electrons and Nuclei

Magnetic dipole moment $\vec{\mu}$ is always associated with the angular momentum of particle $\vec{J}$:

$$\vec{\mu} = \gamma \hbar \vec{J}, \tag{1.8}$$

where γ is ratio of the magnetic moment to the angular momentum, *gyromagnetic ratio*.

According to the classical theory, a particle with charge e and mass m, which moves with angular momentum $\hbar\vec{J}$ in the central field, has a magnetic dipole moment $\vec{\mu}_l = e\hbar\vec{J}/(2m)$. If we consider an orbital motion of electron then

$$\gamma_l = -e/(2m_e). \tag{1.9}$$

The index l indicates that (1.9) is applied only in the case of orbital motion of the electron; e is the absolute value of the electron charge; m_e is the electron mass. It is suitable here to introduce the unit of measurement of the magnetic moment of an electron called *Bohr magneton* (μ_B):

$$\mu_B = e\hbar/(2m_e). \tag{1.10}$$

Then, juxtaposing the classic quantity magnetic moment $\vec{\mu}_l$, caused by orbital motion of an electron, with the operator $\hat{\vec{\mu}}_l$, we can express it in terms of the operator of orbital momentum of an electron $\hat{\vec{l}}$:

$$\hat{\vec{\mu}}_l = \mu_B\hat{\vec{l}}.$$

The presence of the spin of a particle results in a *spin magnetic moment* ($\hat{\vec{\mu}}_s$), but its value can not be calculated as simply as the orbital magnetic moment of an electron.

It follows from relativistic quantum mechanics that the gyromagnetic ratio for an electron with spin moment $\hat{\vec{s}}$ is abnormal and it can be determined with a satisfactory accuracy by the formula

$$\gamma_s = -e/m_e, \tag{1.11}$$

that is, for the spin moments the gyromagnetic ratio is twice as large as for the orbital ones.

Bohr magneton is often used as a natural unit of measurement of magnetic moment of microparticles (like the Planck's constant ($\hbar$) for an angular momentum). Therefore, the magnetic moment of microparticles can be represented by the g-factor in the form

$$\hat{\vec{\mu}}_J = g\mu_B\hat{\vec{l}}.$$

It follows from (1.8)–(1.11) that the g-factor for electronic orbital angular momentum is equal to unity ($g = g_l = 1$), and for spin momentum it is equal to two ($g = g_s = 2$). Calculations based on relativistic quantum electrodynamics, taking into account the interaction of an electron with fluctuations of the radiation field, and precise experiments have established that the value of g-factor for the spin magnetic moment of an electron is $g_s = 2.0023$. There is a correction for the orbital g-factor, but it does not exceed 10^{-6}, and it can be almost always neglected. In the case of formation of the vector sum of the spin and orbital momenta ($\hat{\vec{J}}$), the g-factor is defined by the Landé formula:

$$g_J = \frac{3}{2} + \frac{S(S+1) - L(L+1)}{2J(J+1)}. \tag{1.12}$$

In nuclear physics, by analogy with (1.10), magnetic momenta are measured in nuclear magnetons (μ_n): substituting the proton mass in (1.10) instead of the electron mass, we obtain $\mu_n = 5.505 \cdot 10^{-27}$ J T^{-1}. Then the operator of the nuclear magnetic moment is expressed as

$$\hat{\vec{\mu}}_I = g_n \mu_n \hat{\vec{I}},$$

where g_n is the *nuclear g-factor*, which, however, cannot be calculated with the help of some simple expression similar to (1.12). Magnetic moment $\hat{\vec{\mu}}_I$ is often expressed in terms of the gyromagnetic ratio γ_I for a nucleus:

$$\hat{\vec{\mu}}_I = \gamma_I \hbar \hat{\vec{I}}, \tag{1.13}$$

where γ_I, in contrast to γ_l and γ_s, is not expressed in terms of e and nuclear mass m_n.

It should be noted that since the mass of a proton is 1836 times greater than mass of an electron, the magnetic moment of an electron is by about three orders of magnitude greater than the magnetic moment of a proton. The difference in the values of the gyromagnetic ratios of an electron and nucleus is just as great, too. Due to it, the energy of interaction of electrons with a magnetic field is much higher that the energy of nuclear interactions with the same field strength.

Note that a neutron, though having no electric charge, has a magnetic moment directed against spin, as in the case of an electron.

The values of the characteristics considered for the same nuclear isotopes are given in Table 1.1, which also contains natural abundance of the corresponding isotopes and values of nuclear quadrupole momenta (eQ, see Sect. 1.1.3), as well as intensity of NMR signals relative to protons with the same number of nuclei, in the same field induction or at the same resonance frequency.

1.1.3 The Quadrupole Moment of a Nucleus

In accordance with the classical notions, a nucleus is a charged particle with a spatially distributed charge. In order to simplify the classical description of electrical properties of such a particle, the particle charge distribution can be presented in the form of an expansion by multipole moments of order k. Thus, the zero-order moment ($k = 0$) corresponds to a spherically symmetric uniform[1] charge distribution inside the particle.

---

[1] The charge density is the same throughout a body.

Table 1.1 A selection of nuclear isotopes and their properties

Element	Natural abundance (%)	I	μ/μ_n	$\gamma/(2\pi)$ (Hz/G)	$Q \cdot 10^{-24}$ (cm^2)	Relative intensity of NMR signals for the same number of nuclei	
						B_0 = const	ν_0 = const
^{1}H	99.985	1/2	2.7927	4258	–	1.000	1.00
^{2}H	0.015	1	0.8574	654	$2.8 \cdot 10^{-3}$	$9.6 \cdot 10^{-2}$	0.41
^{6}Li	7.43	1	0.0219	626	$6.4 \cdot 10^{-4}$	$8.5 \cdot 10^{-3}$	0.39
^{7}Li	92.57	3/2	3.2560	1655	$-4.2 \cdot 10^{-2}$	0.294	1.94
^{9}Be	100	3/2	1.1774	−598	$5.3 \cdot 10^{-2}$	$1.4 \cdot 10^{-2}$	0.70
^{11}B	81.17	3/2	2.6880	1366	$3.5 \cdot 10^{-2}$	0.165	1.60
^{13}C	1.11	1/2	0.7022	1070	–	$1.5 \cdot 10^{-2}$	0.25
^{17}O	0.037	5/2	1.8930	−577	$-2.6 \cdot 10^{-2}$	$2.9 \cdot 10^{-2}$	1.58
^{19}F	100	1/2	2.6273	4005	–	0.834	0.94
^{23}Na	100	3/2	2.2161	1126	0.11	$9.3 \cdot 10^{-2}$	2.22
^{27}Al	100	5/2	3.6385	1109	0.149	0.207	3.04
^{31}P	100	1/2	1.1305	1723	–	$6.6 \cdot 10^{-2}$	0.40
^{35}Cl	75.4	3/2	0.8209	417	$-8.2 \cdot 10^{-2}$	$4.7 \cdot 10^{-3}$	0.49
^{79}Br	50.6	3/2	2.0990	1067	0.33	$7.9 \cdot 10^{-2}$	1.26
^{81}Br	49.4	3/2	2.2696	1153	0.28	$9.8 \cdot 10^{-2}$	1.35
^{85}Rb	72.8	5/2	1.3483	411	0.31	$1 \cdot 10^{-2}$	1.35
^{87}Rb	27.2	3/2	2.7505	1397	0.15	0.177	1.64
^{113}Cd	12.3	1/2	0.6195	−944	–	$1.1 \cdot 10^{-2}$	0.22
^{127}I	100	5/2	2.7939	852	−0.75	$9.4 \cdot 10^{-2}$	2.33
^{133}Cs	100	7/2	2.5785	562	−0.003	$4.8 \cdot 10^{-2}$	2.80
^{205}Tl	70.5	1/2	1.6276	2480	–	0.20	0.58
^{209}Bi	100	9/2	4.0801	684	−0.46	0.91	0.14

The multipole moment of the first order ($k = 1$) is an electric dipole moment $\vec{p}$, being a vector which components can be defined in the following way in the reference frame x', y', z', with the origin located in the centre of mass of the charge distribution:

$$p_{x'} = \iiint_V \rho(x', y', z')x' \, \mathrm{d}V,$$
$$p_{y'} = \iiint_V \rho(x', y', z')y' \, \mathrm{d}V,$$
$$p_{z'} = \iiint_V \rho(x', y', z')z' \, \mathrm{d}V.$$

Integration here is carried out through the whole space.

The electric multipole moment of the second order ($k = 2$) is a tensor of the electric quadrupole moment (eQ^*):

$$eQ^* = \begin{pmatrix} eQ^*_{x'x'} & eQ^*_{x'y'} & eQ^*_{x'z'} \\ eQ^*_{y'x'} & eQ^*_{y'y'} & eQ^*_{y'z'} \\ eQ^*_{z'x'} & eQ^*_{z'y'} & eQ^*_{z'z'} \end{pmatrix},$$

where

$$eQ^*_{x'x'} = \iiint_V \rho(x', y', z')(3x'^2 - r^2)\,\mathrm{d}V\,, \quad eQ^*_{x'y'} = 3\iiint_V \rho(x', y', z')x'y'\,\mathrm{d}V\,,$$
$$eQ^*_{x'z'} = 3\iiint_V \rho(x', y', z')x'z'\,\mathrm{d}V\,;$$
$$eQ^*_{y'x'} = 3\iiint_V \rho(x', y', z')y'x'\,\mathrm{d}V\,, \quad eQ^*_{y'y'} = \iiint_V \rho(x', y', z')(3y'^2 - r^2)\,\mathrm{d}V\,,$$
$$eQ^*_{y'z'} = 3\iiint_V \rho(x', y', z')y'z'\,\mathrm{d}V\,;$$
$$eQ^*_{z'x'} = 3\iiint_V \rho(x', y', z')z'x'\,\mathrm{d}V\,, \quad eQ^*_{z'y'} = 3\iiint_V \rho(x', y', z')z'y'\,\mathrm{d}V\,,$$
$$eQ^*_{z'z'} = \iiint_V \rho(x', y', z')(3z'^2 - r^2)\,\mathrm{d}V\,; \tag{1.14}$$

r is distance to the center of mass of the charge distribution; V is volume of the body.

According to modern concepts, atomic nuclei in the ground state, while interacting with particles (e.g. electrons) located outside the nucleus, manifest themselves as systems belonging to the symmetry group $\mathcal{D}_{\infty h}$. The elements of this point group are the symmetry axis of infinite order C_∞, the mirror-reflection plane perpendicular to it, the center of inversion, and infinitely many second-order axes whose directions form a fan of vectors in the mirror-reflection plane. An ellipsoid of rotation (Fig. 1.1) is an example of a body with this symmetry. Therefore, in classical representation the

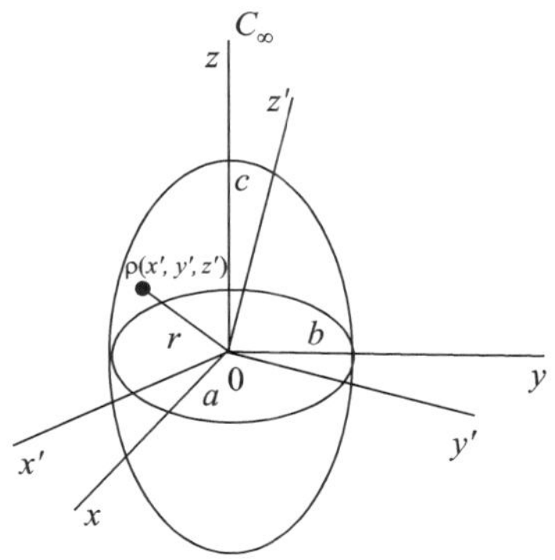

Fig. 1.1 An ellipsoid of revolution of the charge density distribution

charge density distribution in the nucleus is associated with an ellipsoid of revolution with axes $c, a = b$, with a uniform distribution of the charge density (ρ). Nuclei with spin $I = 1/2$ have a spherically symmetric charge distribution. If all three axes of an ellipsoid of revolution are equal, it degenerates into a sphere. Such nuclei can be considered as point charges at distances much greater than their size.

Nuclei with spin $I > 1/2$ are not spherically symmetric, so the charge density distribution in them should be described using multipole momenta of different ranks. Since it is known that distribution of charge within a nucleus is characterized by an inversion centre, it cannot have vector properties and, therefore, a nucleus cannot have an electric dipole moment and other odd momenta of higher order, that is confirmed by the experiment.

An electric moment of a nucleus coming next after a charge is *the quadrupole moment*. It is well known that when a particle or system of particles has axes of symmetry, the principal axes of any tensor describing properties of the particles will be directed along the axes of symmetry, and the tensor itself will be of a diagonal form. Consequently, the tensor of the nuclear quadrupole moment in the coordinate system where the z axis coincides with the direction of the axis of symmetry of infinite order C_∞ (see Fig. 1.1) and the other two axes coincide with the direction of two mutually perpendicular axes of symmetry of the second order, is axisymmetric ($eQ^*_{xx} = eQ^*_{yy}$) and diagonal:

$$eQ^* = \begin{pmatrix} eQ^*_{xx} & 0 & 0 \\ 0 & eQ^*_{xx} & 0 \\ 0 & 0 & eQ^*_{zz} \end{pmatrix}.$$

Since the sum of the diagonal terms of the quadrupole moment tensor is zero:

$$eQ^*_{xx} + eQ^*_{yy} + eQ^*_{zz} = 0,$$

then in order to characterize the quadrupole moment tensor it is sufficient to only specify eQ^*_{zz}:

$$eQ^* = eQ^*_{zz} = \iiint\limits_V \rho(x, y, z)(3z^2 - r^2)\, \mathrm{d}V. \tag{1.15}$$

This value describes the deviation of the charge density distribution inside a nucleus from spherical symmetry.

A nucleus is known to consist of protons with an electric charge e and neutrons, so that the charge density (ρ) in a nucleus is not actually continuous:

$$\rho(\vec{r}) = e \sum_k \delta(\vec{r} - \vec{r}_k), \tag{1.16}$$

where k is summation index which labels the protons that make up a nucleus. Then integration in (1.15) is replaced by summation:

$$eQ^* = e \sum_k \left(3z_k^2 - r_k^2\right). \tag{1.17}$$

All the components of the quadrupole moment and its scalar characteristic—the value of eQ^*—were determined with classical positions. However, properties of nuclei are not to be described from the standpoint of classical physics. To get down to a quantum mechanics description, it is necessary to assign an operator to the classical value eQ^*. To do it, we have to replace the coordinates in (1.17) by the operators of coordinates:

$$e\widehat{Q}^* = e \sum_k \left(3\widehat{z}_k^2 - \widehat{r}_k^2\right).$$

Then the quantity measured in experiments, described by the operator $e\widehat{Q}^*$ is a quantum-mechanical average, determined by a matrix element calculated with the help of its own wave function $| I, m, \eta \rangle$, which corresponds to the maximum projection of the nuclear spin ($m = I$). Here I is the nuclear spin, m is the magnetic quantum number, η is a set of quantum numbers characterizing the nucleon energy[2] forming a specific nucleus. In the ground state, all the lower energy levels are filled with nucleons according to the Pauli principle, so that a nucleus can be considered as a whole and its energy is defined as the total energy of individual nucleons. When external electric or magnetic fields used in radiospectroscopy are applied, only the value of the projection of the nuclear spin ($m = -I, \ldots, I$) changes; all remaining quantum numbers, characterizing the energy of the nucleus, do not change, so they are omitted in all subsequent formulas. The quantity

---

[2] In the nuclear shell model, each nucleon is characterized by a principal quantum number, an orbital number and spin quantum numbers.

$$eQ = \langle I, m | e\widehat{Q}^* | I, m \rangle_{m=I} \tag{1.18}$$

characterizes the deviation of the distribution of the charge density in a nucleus from the spherically symmetric one and is called *quadrupole moment of a nucleus*.

Reference books usually cite not the values of the quadrupole moment eQ but the value of Q, which is measured in units of area (see Eq. 1.17). Quadrupole momenta of various nuclei take the values of $-1.5 \cdot 10^{-24}$ to $+6.0 \cdot 10^{-24}$ cm^2. The positive values correspond to an along ellipsoid of revolution located along the axis $0z'$ (see Fig. 1.1), characterizing the distribution of charge density in a nucleus, and the negative ones, to an oblate ellipsoid (not shown).

1.2 The Phenomenon of Magnetic Resonance

In its simplest form, magnetic resonance can be observed when an ensemble of particles with magnetic momenta, located in a constant magnetic field are influenced by an alternating magnetic field. In classical physics, the interaction energy of a magnetic dipole ($\vec{\mu}$) with a magnetic field ($\vec{B}$) is expressed by the formula

$$E = -(\vec{\mu}\vec{B}).$$

To obtain the interaction energy operator (Hamiltonian), the classical physical quantities E and $\vec{\mu}$ should be replaced by corresponding operators $\widehat{\mathcal{H}}$ and $\hat{\vec{\mu}}$:

$$\widehat{\mathcal{H}} = -(\hat{\vec{\mu}}, \vec{B}). \tag{1.19}$$

Assuming that $\vec{B}$ is a static field $\vec{B}_0$, which vector is directed along the axis z, we can easily obtain from (1.19) allowing for (1.8)

$$\widehat{\mathcal{H}} = -\gamma\hbar\hat{J}_z B_0. \tag{1.20}$$

To find the values of the energy, it is enough to substitute the eigenvalues (M_J) of the operator $\hat{J}_z$ into (1.20):

$$E_{M_J} = -\gamma\hbar B_0 M_J. \tag{1.21}$$

This expression describing the interaction energy of the magnetic moment of an electron with the magnetic field $\vec{B}_0$ is more commonly written as $E_{M_J} = -g\mu_B B_0 M_J$. As seen from (1.21), the energy levels E_{M_J} are equidistant (Fig. 1.2), with the energy difference between adjacent ones being equal to

$$\Delta E = |E_{M_J} - E_{M_J - 1}| = |\gamma|\hbar B_0. \tag{1.22}$$

If an atom is subjected to an additional alternating magnetic field with a frequency close to $\nu = \Delta E/h$, then there emerge induced (forced) transitions between levels.

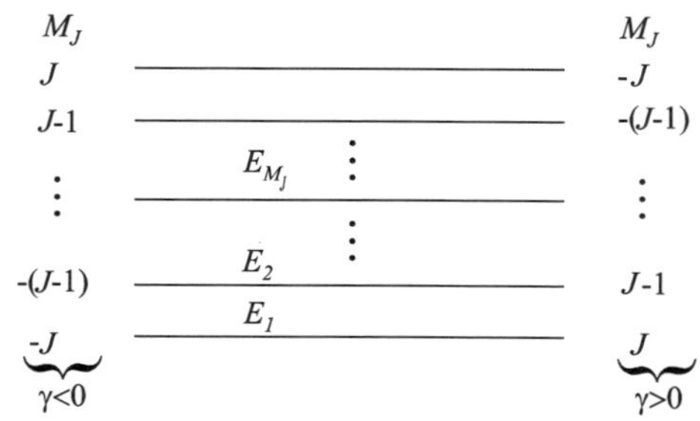

Fig. 1.2 Nuclear Zeeman sublevels

This phenomenon is called *magnetic resonance*. The probability of transition between states M_J and M'_J is proportional to the square of the modulus of the matrix element of the operator $\hat{\mu}_h$, which is the projection of the magnetic moment operator on the vector direction of the alternating magnetic field:

$$W_{M_J M'_J} \sim |\langle M_J \mid \hat{\mu}_h \mid M'_J\rangle|^2.$$

Since $\hat{\mu}_h = \gamma\hbar\hat{J}_h$ ($\hat{J}_h$ is projection of momentum on a vector of the alternating magnetic field), the transition probability is determined by the matrix element $\langle M_J \mid \hat{J}_h \mid M'_J\rangle$.

Consider two orientations of a vector of the alternating field relative to the vector of the static field $\vec{B}_0$. If the field vectors are parallel, then $\hat{J}_h = \hat{J}_z$. In accordance with (1.5), $\hat{J}_z \mid M'_J\rangle = M'_J \mid M'_J\rangle$, therefore, $\langle M_J \mid \hat{J}_l \mid M'_J\rangle = M'_J\langle M_J \mid M'_J\rangle$. The expression $\langle M_J \mid M'_J\rangle$ denotes a scalar product of the state vectors $\langle M_J \mid$ and $\mid M'_J\rangle$. Since state vectors form an orthonormal system

$$\langle M_J \mid M'_J\rangle = \delta_{M_J M'_J}$$

($\delta_{M_J M'_J}$ is the Kronecker symbol), then $W_{M_J M'_J} = 0$. If vectors of the alternating and static magnetic fields are perpendicular, we can assume, for example, $\hat{J}_h = \hat{J}_x$. Using (1.6) and (1.7), we obtain

$$\langle M_J \mid \hat{J}_x \mid M'_J\rangle = c_1\langle M_J \mid M'_J + 1\rangle + c_2\langle M_J \mid M'_J - 1\rangle \qquad (1.23)$$

(c_1, c_2 are certain constants). Expression (1.23) is not zero in case if

$$\Delta M_J = M_J - M'_J = \pm 1. \qquad (1.24)$$

Condition (1.24) determines *the selection rules for transitions between levels* in the system of magnetic momenta in a static magnetic field, which are induced by an alternating magnetic field perpendicular to it. The frequency at which a transition between adjacent levels occurs is determined, according to (1.22), by the formula

$$\nu_0 = \Delta E/h = (|\gamma|/(2\pi))B_0. \tag{1.25}$$

It follows from (1.25) that ν_0 is independent of the quantum number M, and hence, regardless of the number of energy levels, there exists a unique frequency at which magnetic resonance occurs.

Three types of transitions can occur between two levels: *spontaneous, induced and nonradiative relaxation ones*. The probability of spontaneous transitions in the radio frequency bandwidth is negligible, since it is proportional to ν_0^3 (see details in Sect. 17.1). The key role in quantum radiophysics is played by induced transitions which are the result of exposure to radio frequency fields. Nonradiative relaxation transitions occur due to exchange of energy quanta between the spin system and the lattice (see details in Sect. 1.4).

All the phenomena registering energy of quantum transitions that occur due to resonant forcings of alternating radiofrequency field are called *resonances: electron paramagnetic resonance* (EPR), *nuclear magnetic resonance* (NMR), *nuclear quadrupole resonance* (NQR) and others. Thus, electron paramagnetic resonance is a special case of magnetic resonance: the resonance occurs at the sublevels formed by interaction of magnetic moments of electrons with a constant magnetic field. In this case it is assumed that the magnetic interaction between the electron magnetic momenta of individual paramagnetic centers (atoms, ions, molecules) is much weaker than their interaction with an external magnetic field. In other words, the term electronic paramagnetic resonance is applied to the magnetic resonance in magnetically disordered systems. Electronic magnetic resonance in magnetically ordered systems (in systems with strong magnetic interactions) is commonly called, in different cases, *ferromagnetic, ferrimagnetic or antiferromagnetic resonance*.

If the angular momentum is only caused by a single electron ($s = 1/2$), in accordance with (1.25) the frequency of pure spin resonance is equal to

$$\nu_0 = \frac{\gamma_s}{2\pi} B_0 = \frac{g_s \mu_\mathrm{B}}{h} B_0. \tag{1.26}$$

Gyromagnetic ratio γ_s differs from the expression (1.11) due to relativistic corrections which leads to to the following numerical value:

$$\gamma_s/(2\pi) = 2.802 \cdot 10^{10}\ \mathrm{Hz/T}. \tag{1.27}$$

Magnetic resonance of free atoms or ions is a phenomenon which is almost such as simple, when the magnetic moment of a particle is the vector sum of spin and orbital magnetic momenta of individual electrons. It can be calculated from the known formulas of the composition of momenta theory: $\vec{\mu}_J = g_J \mu_B \vec{J}$, while resonance frequency can be obtained using formula (1.12)

$$\nu_J = \frac{g_J}{g_s} \nu_0.$$

In general, when atoms or ions are part of a crystal or molecule, electrons are involved in various interactions with their environment, which perturbs the orbital motion and leads to changes in the orbital magnetic moment. The perturbation of the orbital motion via spin–orbit interactions influences the spin moments, too. In addition, there is a direct interaction of the spins with the environment (see Sect. 1.11). For these reasons, the EPR frequency of atom or ion in a crystal differs (in some cases quite strongly) from the EPR frequency of free particles and cannot be determined by a simple formula similar to (1.26).

1.3 Macroscopic Spin Magnetization

In order to produce a strict description of system of interacting spins, quantum theory must be used. However another approach to this problem was proposed by F. Bloch in his phenomenological theory of nuclear magnetic resonance. In the basis of his theory is the fact that for a system of microparticles we can define macroscopic characteristics which represent the average values of certain physical parameters of the system. These macroscopic characteristics can be governed by the laws of classical mechanics, which allows us to use the classical theory for description of a complex system of interacting spins.

F. Bloch suggested to use the magnetization vector ($\vec{M}$), which is the total average magnetic moment of nuclei or electrons per unit volume of material, as the macroscopic characteristic of the interacting spins. In Bloch's theory, based on the classical views, behavior of the vector of nuclear or electron magnetization in constant and variable magnetic fields is considered accounting for interaction of spins with each other and with the environment.

In general, the statistical mean value, characterized by operator $\hat{A}$, can be obtained from the formula

$$\langle\hat{A}\rangle = \frac{\mathrm{Tr}\left(\hat{A}\exp[-\widehat{\mathcal{H}}/(kT)]\right)}{\mathrm{Tr}\left(\exp[-\widehat{\mathcal{H}}/(kT)]\right)}. \tag{1.28}$$

If a system of magnetic microparticles is in a static magnetic field $\vec{B}_0$, which vector is directed along the axis z, the Hamiltonian of the system is described by the formula (1.20). The components of the macroscopic magnetization (M_α, where $\alpha = x, y, z$) are found by the formula

$$M_\alpha = \sum_{i=1}^{N}\langle\mu_{\alpha i}\rangle = N\gamma\hbar\langle\hat{J}_\alpha\rangle, \tag{1.29}$$

where N is the number of microparticles per unit volume. The average values of the operators $\hat{J}_x$, $\hat{J}_y$, $\hat{J}_z$, in accordance with (1.28), are calculated as

$$\langle \hat{J}_\alpha \rangle = \frac{\mathrm{Tr}\left(\hat{J}_\alpha \exp[\gamma\hbar\hat{J}_z B_0/(kT)]\right)}{\mathrm{Tr}\left(\exp[-\widehat{\mathcal{H}}/(kT)]\right)}. \tag{1.30}$$

Since the operators $\hat{J}_x$, $\hat{J}_y$ do not commute with the operator $\hat{J}_z$, then

$$\mathrm{Tr}\left(\hat{J}_{x \text{ or } y} \exp\left[\frac{\gamma\hbar\hat{J}_z B_0}{kT}\right]\right) = 0,$$

and, consequently, the average values of the operators $\hat{\mathrm{J}}_x$ and $\hat{\mathrm{J}}_y$ and related components of macroscopic magnetization are equal to zero. Thus, the only component M_z of the macroscopic magnetization which is not zero. Designating it by M_0 and using (1.28), (1.30), we obtain:

$$M_0 = N\gamma\hbar \frac{\sum_{m=-J}^{J} m \exp\left[\gamma\hbar m B_0/(kT)\right]}{\sum_{m=-J}^{J} \exp\left[\gamma\hbar m B_0/(kT)\right]}.$$

Due to the fact that the interaction energy of magnetic moments of electrons or nuclei with external fields and with each other is much weaker than the energy of thermal motion ($E_m \ll kT$), we can expand exponentials in power series and restrict it to the first two terms:

$$M_0 = N\gamma\hbar \frac{\sum_{m=-J}^{J} m \left[1 + \gamma\hbar m B_0/(kT)\right]}{\sum_{m=-J}^{J} \left[1 + \gamma\hbar m B_0/(kT)\right]}. \tag{1.31}$$

The sum of all m (from $-J$ to J) is zero, and the summation of m^2 within the specified limits can be performed by the well-known formula:

$$\sum_{i=1}^{q} i^2 = \frac{1}{6} q(q+1)(2q+1).$$

Then we finally get

$$M_0 = \frac{N\gamma^2\hbar^2 J(J+1)B_0}{3kT}. \tag{1.32}$$

The formula (1.32) can be used to determine the static magnetic susceptibility of the ensemble of magnetic microparticles (χ_0):

$$M_0 = \chi_0 B_0. \tag{1.33}$$

Comparing (1.33) with (1.32), we obtain

$$\chi_0 = \frac{N\gamma^2\hbar^2 J(J+1)}{3kT}. \tag{1.34}$$

Equation (1.34) is the Curie law, according to which the paramagnetic susceptibility is independent of the applied field and inversely proportional to the absolute temperature: $\chi = C/T$ (C is the Curie constant, $C = N\gamma^2\hbar^2 J(J+1)/(3k)$).

Nuclear susceptibility (χ_0), in contrast to electronic susceptibility, is very small. Because of this, nuclear magnetism was not detected experimentally for a long time.

1.4 Achievement of Equilibrium Spin Magnetization: The Relaxation Times

In experiments on nuclear magnetic and electron paramagnetic, resonances the magnetization is driven from the equilibrium state, applying a radio frequency (rf) field or changing rapidly the direction of the external magnetic field $\vec{B}_0$. After such changes, a new equilibrium state of the magnetization $\vec{M}$ is achieved or it returns to its previous state. This phenomenon is called *magnetic relaxation*. This process is not instantaneous, but takes some time.

Let us consider what kind of laws the longitudinal (M_z) and transverse ($M_\perp$) components of the nuclear magnetization will obey in time during the process of achievement of the stationary state.[3]

1.4.1 Achievement of Stationary State of the Longitudinal Component of Magnetization

In condensed phases of matter, electrons and nuclei are in the complex environment of neighbouring microparticles, creating magnetic and electric fields, called local fields. In this case, the surrounding particles, electrons themselves and nuclei are involved in various forms of thermal motion, so that the local fields fluctuate and they should be considered as random functions. The spectral density of these random functions is distributed in a certain wide range of frequencies, which may include the frequency ν_0 of the magnetic resonance of the spin system . Therefore, the interaction between electrons and nuclei with neighbouring particles can cause transitions between energy levels of the spin system, and the probability of the transition will depend on factors determining the thermal motion. As a result, there is an exchange of energy between the spin system and the *lattice* . The "lattice" implies a set of

[3] There is no principle difference between nuclear and electron spin magnetizations. In this chapter we shall often refer in specific examples to nuclear magnetization.

all degrees of freedom associated with the motion of particles in matter, including the movement of electrons and nuclei. The degree of freedom, that characterizes the spin system in the simplest case is the orientation of the magnetic moments of the particles in an external magnetic field. The energy exchange between the spin system and the lattice can be caused not only by magnetic interaction of microparticles, but also by electrical interaction of their quadrupole moments with fluctuating local inhomogeneous electric fields, as well as by interactions having a more complex mechanism.

The transitions, that the spin–lattice interaction are accompanied with, are nonradiative and ensure achievement of the equilibrium state of the spin system, i.e., its relaxation. The probabilities of such transitions upward (w_{12}) and downward (w_{21}) are not equal, as opposed to the transitions due to the impact of an external RF field. Indeed, under the conditions of Boltzmann thermal equilibrium the equality holds

$$n_1^0 w_{12} = n_2^0 w_{21}\,, \tag{1.35}$$

where n_1^0, n_2^0 are equilibrium level populations. It is known also that under these conditions $n_2^0/n_1^0 = \exp[-h\nu/(kT)]$, where $h\nu$ is the energy levels difference. Consequently, we can write, using (1.35):

$$w_{12} = w_{21}\exp[-h\nu/(kT)].$$

As we can see, the equilibrium state can exist only if the probability of relaxation transitions downwards (w_{21}) is greater than the probability of transitions upwards (w_{12}).

For simplicity, we shall consider a spin system of particles with a positive gyromagnetic ratio, which is located in a magnetic field $\vec{B}_0$. Taking into account the Boltzmann distribution of nuclei at energy levels, the number of nuclei (n_m) at the level with the quantum number m can be expressed in terms of the total number of nuclei (N) per unit volume of the sample, as well as the energy of this level ($E_m = -\gamma\hbar m B_0$), as follows:

$$n_m = N\frac{\exp[-E_m/(kT)]}{\sum_{m=-I}^{I}\exp[-E_m/(kT)]} = N\frac{\exp[\gamma\hbar B_0 m/(kT)]}{\sum_{m=-I}^{I}\exp[\gamma\hbar B_0 m/(kT)]}. \tag{1.36}$$

As in derivation of (1.31), we expand the exponents in the numerator and denominator of the formula (1.36) in power series and take only the first two terms:

$$n_m = N\frac{1+\gamma\hbar B_0 m/(kT)+\ldots}{\sum_{m=-I}^{I}(1+\mu B_0 m/(kT)+\ldots)} \simeq \frac{N}{2I+1}\left(1+\frac{\gamma\hbar B_0 m}{kT}\right). \tag{1.37}$$

For $I = 1/2$, using (1.37) and assuming that populations n_1^0 and n_2^0 match the levels $m = +1/2$ and $m = -1/2$, we obtain

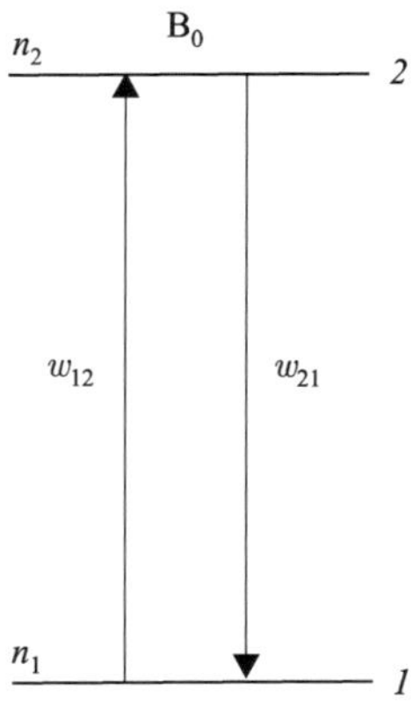

Fig. 1.3 Probability of relaxation transitions in two-levels system (I = 1/2)

$$n_1^0 = \frac{N}{2}\left(1 + \frac{\gamma\hbar B_0}{2kT}\right),$$
$$n_2^0 = \frac{N}{2}\left(1 - \frac{\gamma\hbar B_0}{2kT}\right). \tag{1.38}$$

It follows from (1.38) that the static equilibrium difference in level populations is equal to

$$\Delta n^0 = n_1^0 - n_2^0 = \frac{N\gamma\hbar B_0}{2kT}. \tag{1.39}$$

This population difference determines the equilibrium, i.e., static, magnetization M_0 in the form (1.39). Without going into details of the relaxation mechanism, we can describe the process, on the basis of a simple kinetic model of the process of achieving equilibrium.

We make a natural assumption that the probability of relaxation transitions depends neither on time nor on the state populations (see details Chap. 3). We can write an expression for the rate of change in the population of one of the levels, for example level 1 (Fig. 1.3):

$$\frac{d}{dt}n_1(t) = -n_1(t)w_{12} + n_2(t)w_{21}. \tag{1.40}$$

The term with the minus sign in (1.40) corresponds to the number of particles in the unit time leaving level 1 for level 2; the second term represents the number of particles that have arrived at level 1 from level 2. Given that $n_1(t) + n_2(t) = N$, expression (1.40) can be written as

$$\frac{d}{dt}n_1(t) = -n_1(t)(w_{12} + w_{21}) + Nw_{21}.$$

Subtracting a similar equation for $\frac{d}{dt}n_2(t)$ from this equation, we obtain the equation for the rate of change in the population difference:

$$\frac{d}{dt}\Delta n_{12}(t) = -\Delta n_{12}(t)(w_{12} + w_{21}) + N(w_{21} - w_{12}).$$

Since the second term is time-independent, it can be expressed trough the equilibrium population difference,

$$N(w_{21} - w_{12}) = \Delta n_{12}^0(w_{12} + w_{21}).$$

Finally, by introducing average transition probability $\overline{w} = (w_{12} + w_{21})/2$, we obtain the differential equation for population difference:

$$\frac{d}{dt}\Delta n_{12}(t) = -2\overline{w}(\Delta n_{12}(t) - \Delta n_{12}^0). \tag{1.41}$$

Since the static nuclear magnetization M_0 and the current value of its longitudinal component $M_z(t)$ are proportional to the corresponding differences of population levels, we can write according to (1.41):

$$\frac{d}{dt}M_z(t) = -2\overline{w}[M_z(t) - M_0]. \tag{1.42}$$

Thus, the new value of the field is matched with a new equilibrium state with a greater difference in the level populations and with a higher magnetization M_0 corresponding to it.

The characteristic time T_1, determined by the ratio $2\overline{\omega} = T_1^{-1}$, is more often used instead of transition probability. By introducing notation T_1 and integrating Eq. (1.42), we obtain

$$M_z(t) = M_0 - [M_0 - M_z(0)]\exp(-t/T_1). \tag{1.43}$$

As we see from (1.43), the approach of the longitudinal vector component (i.e., directed along the vector of the constant magnetic field $\vec{B}_0$) of nuclear magnetization M_z to its equilibrium value obeys the exponential law with time constant T_1. This is valid for the two-level system. Other cases considered in Chap. 3.

Time constant T_1 is referred to as *longitudinal relaxation time*, as it determines the course of the process of achieving the equilibrium value of the longitudinal component of nuclear magnetization. Characteristic time T_1 is also referred to as the *spin–lattice relaxation time*, because it depends on probability of the transition caused by interactions of the spin system and the lattice.

1.4.2 Achievement of Equilibrium Transverse Component of Magnetization

Under conditions of thermodynamic equilibrium, as shown in Sect. 1.3, $\vec{M} = \vec{M}_0$, and the transverse component of the magnetization vector is equal to zero ($M_\perp = 0$). A magnetization vector can be rotated at a certain angle θ, for example, with the help of a RF pulse. In this case, there emerges a non-zero transverse component of the magnetization on the plane perpendicular to the vector of the constant magnetic field. When the action of the RF pulse is over, $M_\perp \to 0$.

Let us list the most important factors causing a decrease of $M_\perp$:

1. Spin–lattice interactions, resulting in incoherent relaxation transitions between the levels. Interactions of this kind not only affect the rate of achieving the equilibrium value of the component M_z, but also reduce the component $M_\perp$.
2. Spin–spin interactions also cause relaxation transitions. Since these transitions occur as a result of the exchange of energy between interacting spins, while maintaining the overall energy balance, each transition from one level to another is matched with an reverse transition, and the population difference in bound levels remains unchanged. This means that spin–spin interactions do not change M_z, and only lead to a decrease of $M_\perp$.
3. The static magnetic field $\vec{B}_0$ can only cause a reduction of the component $M_\perp$ only in the case when it is not uniform. Variation in space of the magnitude of the static magnetic field $\vec{B}_0$ causes frequency spread in the vector precession of magnetization, inherent to various elements in the sample volume, which also leads to accelerating the decrease of $M_\perp$ down to zero.

By analogy with the variation of the longitudinal component M_z (see 1.43), Bloch proposed formulas reflecting the tend of the transverse component of the magnetization vector to its equilibrium value:

$$\frac{d}{dt} M_\perp(t) = -\frac{M_\perp(t)}{T_2^*}, \tag{1.44}$$

$$M_\perp = M_\perp(0) \exp(-t/T_2^*). \tag{1.45}$$

In these formulas, the time constant T_2^* is *transverse relaxation time*, as it determines decay of the transverse component $M_\perp$. We also introduce constant T_2, which excludes the influence of magnetic field inhomogeneity (δB), named *spin–spin relaxation time*. The time T_2^* is associated with T_2 and with inhomogeneity of magnetic field δB by the following relation:

$$1/T_2^* = 1/T_2 + \gamma \delta B/2, \tag{1.46}$$

where γ is gyromagnetic ratio.

Later experiments and rigorous quantum-mechanical calculations showed that the exponential law of change in component $M_\perp$ in the process $M_\perp \to 0$ is valid for low-viscosity liquids with narrow single NMR spectrum lines, as well as in some other cases. For solids, the law of achieving an equilibrium value for component $M_\perp$ acquires another form.

1.5 Bloch's Phenomenological Theory

The behaviour of the macroscopic nuclear magnetization can be found using classical physics laws.

It is known that motion of a "classical" magnetic moment $\vec{\mu}$ in a static magnetic field $\vec{B}$ is described by the equation

$$\frac{d\vec{\mu}}{dt} = \gamma\vec{\mu} \times \vec{B}, \tag{1.47}$$

and, following (1.47), the individual components of the vector $\vec{\mu}$ are obeyed equations

$$\begin{aligned}
\frac{d\mu_x}{dt} &= \gamma(\mu_y B_z - \mu_z B_y),\\
\frac{d\mu_y}{dt} &= \gamma(\mu_z B_x - \mu_x B_z),\\
\frac{d\mu_z}{dt} &= \gamma(\mu_x B_y - \mu_y B_x).
\end{aligned} \tag{1.48}$$

Let us direct the $0z$ axis of the coordinate frame (Fig. 1.4) along the vector of the constant magnetic field $\vec{B}_0$. Then $B_x = B_y = 0$, $B_z = B_0$, and Eqs. (1.48) take the form

$$\frac{d\mu_x}{dt} = \gamma\mu_y B_0\,, \tag{1.49}$$

$$\frac{d\mu_y}{dt} = -\gamma\mu_x B_0\,, \tag{1.50}$$

$$\frac{d\mu_z}{dt} = 0. \tag{1.51}$$

It follows from (1.51) that with vector $\vec{\mu}$ moving μ_z, and hence the angle θ (the angle between the vectors $\vec{\mu}$ and $\vec{B}_0$) remain constant. By differentiating both sides of Eq. (1.49) with respect to time and substituting $d\mu_y/dt$ in it in the form (1.50), we obtain an equation for μ_x:

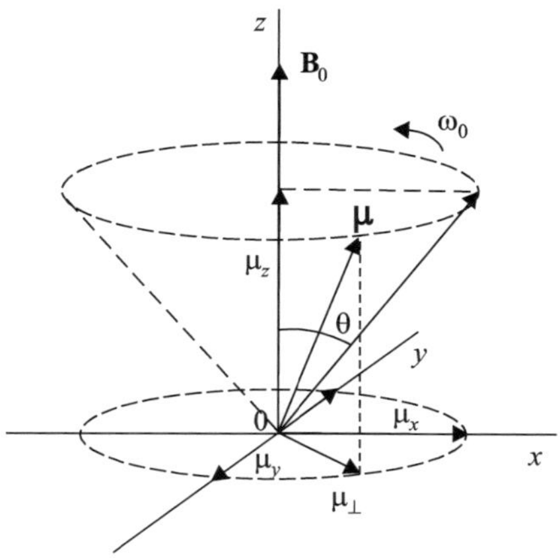

Fig. 1.4 Spin precession

$$\frac{d^2\mu_x}{dt} + (\gamma B_0)^2 \mu_x = 0. \tag{1.52}$$

Equation (1.52) is a differential equation of harmonic vibrations, its solution has the form

$$\mu_x = A\cos(\gamma B_0 t + \varphi), \tag{1.53}$$

where A is amplitude; φ is initial phase of the oscillations. Similarly, we obtain

$$\mu_y = -A\sin(\gamma B_0 t + \varphi). \tag{1.54}$$

In (1.53) and (1.54), the constant $A = \mu_\perp = \sqrt{(\mu_x)^2 + (\mu_y)^2}$ is the projection of $\vec{\mu}$ on the plane (xy), and the product γB_0 means rotation frequency (ω_0) of projection of $\mu_\perp$, i.e., $\omega_0 = \gamma B_0$.

Thus, it can be concluded that in the external magnetic field $\vec{B}_0$ magnetic moment $\vec{\mu}$ precesses around the field vector with a frequency of $\omega_0 = \gamma B_0$. This phenomenon is called *Larmor precession*.

To determine the direction of rotation of the vector of magnetic moment $\vec{\mu}$, let us assume that at some initial time t_0 the component $\mu_\perp$ was oriented along the axis $0x$, i.e., $\mu_\perp = \mu_x$, and $\mu_y = 0$ (see Fig. 1.4). Then, as it follows from (1.54), after a short time interval Δt, the component μ_y will get an increment of $\Delta\mu_y = -A\gamma B_0 \Delta t$. It means that when viewed along the direction of the field vector $\vec{B}_0$, in the case of $\gamma > 0$ the vector of magnetic moment $\vec{\mu}$ will precess in the counterclockwise direction, and when $\gamma < 0$, in the clockwise direction (see Fig. 1.4).

If the magnetic moment is simultaneously influenced by a constant ($\vec{B}_0$) and alternating ($\vec{B}^{\approx}(t)$) magnetic fields, its movement is more complicated. Let the field vector $\vec{B}^{\approx}(t)$ be oriented along the axis $0x$. The linearly polarized alternating magnetic field $B_x^{\approx}(t) = 2B_1 \cos\omega t$ can be decomposed into two components,

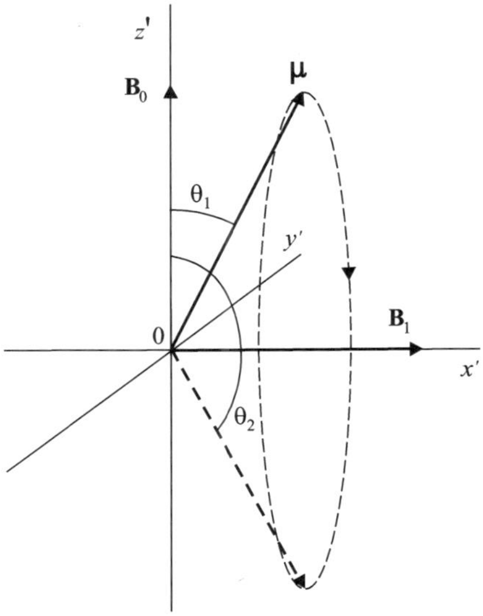

Fig. 1.5 Nutation of the magnetic moment

rotating in different directions with the same frequency ω. One of the components of the field $\vec{B}^{\approx}$ will rotate in same direction as the precessing magnetic moment (vector $\vec{\mu}$ in the field $\vec{B}_0$). To consider the interaction of this field component with the precessing magnetic moment, it is convenient to employ a coordinate frame x', y', z' rotating with the frequency of the alternating magnetic field. In this coordinate system, the magnetic field vector $\vec{B}_1$ will be motionless, i.e., it does not depend on time. When the frequency of the alternating field ω coincides with the frequency ω_0 of precession of the magnetic moment vector in the rotating system of coordinates x', y', z', the magnetic moment vector will interact with the actually constant field $\vec{B}_1$. Consequently, in the rotating frame the vector of the magnetic magnetic moment $\vec{\mu}$ will revolve around the field $\vec{B}_1$, changing the angle θ (Fig. 1.5). In mechanics, this movement is called *nutation*. Since in practice the amplitude of alternating magnetic field is small, $B_1 \ll B_0$, the frequency of the nutation is small as well, $\omega_1 = \gamma B_1 \ll \omega_0$. As a result of nutation, the angle θ will continuously change its value from θ_1 to θ_2, and then from θ_2 to θ_1, etc. Under such complex movement in the laboratory frame x, y, z, vector $\vec{\mu}$ will perform a certain curve (spiral) on the sphere surface.

Description of the behavior of the classical magnetic moment in external magnetic fields can be also used when considering the behavior of macroscopic magnetization caused by magnetic moments of nuclei and electrons.

The equation of movement for the macroscopic magnetization vector $\vec{M}$ in a magnetic field can be written as

$$\frac{\mathrm{d}\vec{M}}{\mathrm{d}t} = \gamma[\vec{M} \times \vec{B}]. \tag{1.55}$$

However, it is only valid for a system of isolated spins in the same external magnetic field $\vec{B}$. In most real cases, the spins under consideration are in the complex environment of other particles, and their magnetic moments are affected not only by external constant ($\vec{B}_0$) and alternating ($\vec{B}^{\approx}(t)$) magnetic fields, but also local fields ($\vec{B}_{loc}$), produced by the magnetic dipoles of neighbouring particles.

F. Bloch, creating a macroscopic theory of nuclear magnetic resonance, used Eq. (1.55), where $\vec{B} = \vec{B}_0 + \vec{B}^{\approx}(t)$, as the basic equation for the magnetization vector in a constant and alternating magnetic fields.

To take into account the influence of the local magnetic fields on the behaviour of the vector $\vec{M}$, F. Bloch included in the equation the additional relaxation terms that fenomenologilly characterize the interaction of the spin system with the environment and each spins with other. In general, Bloch's theory is based on three Eqs. (1.42), (1.44) and (1.55), which make up the system

$$\begin{aligned}\frac{dM_x}{dt} &= \gamma[M_y B_z - M_z B_y] - \frac{M_x}{T_2},\\ \frac{dM_y}{dt} &= \gamma[M_z B_x - M_x B_z] - \frac{M_y}{T_2},\\ \frac{dM_z}{dt} &= \gamma[M_x B_y - M_y B_x] - \frac{M_z - M_0}{T_1}.\end{aligned} \tag{1.56}$$

If an alternating field, as well as spin–spin and spin–lattice relaxations are absent ($\vec{B}^{\approx} = 0$, $T_2 = \infty$ and $T_1 = \infty$), then the Bloch equation (1.56) can be reduced to a simple equation of motion for a magnetization vector in a constant magnetic field. The analysis of this equation made at the beginning of the section, suggests that, in this case, the magnetization vector will perform precession around the vector of the static magnetic field at a frequency $\omega_0 = \gamma B_0$. If we consider the relaxation processes ($T_1 \neq \infty$ and $T_2 \neq \infty$), then, taking into account the exponential change $M_z(t)$ and $M_\perp(t)$ (see 1.43–1.45), we can conclude that precession of the magnetization vector around the field vector $\vec{B}_0$ occurs with a change in both magnetization M and angle θ.

We now consider the motion of the magnetization when exposed to constant and alternating magnetic fields without allowance for relaxation in a rotating coordinate system x', y', z'. It is known from theoretical mechanics that the derivatives of time from any vector $\vec{k}$ in a coordinate system rotating with angular velocity $\vec{\omega}'$ and in the stationary laboratory coordinate system are related by the relation

$$\left(\frac{\mathrm{d}\vec{k}}{\mathrm{d}t}\right)_{rot} = \frac{\mathrm{d}\vec{k}}{\mathrm{d}t} + [\vec{k} \times \vec{\omega}'].$$

Then Eq. (1.55) in the rotating coordinate system x', y', z' becomes

$$\left(\frac{\mathrm{d}\vec{M}}{\mathrm{d}t}\right)_{rot} = \gamma[\vec{M} \times \vec{B}_{z'}], \tag{1.57}$$

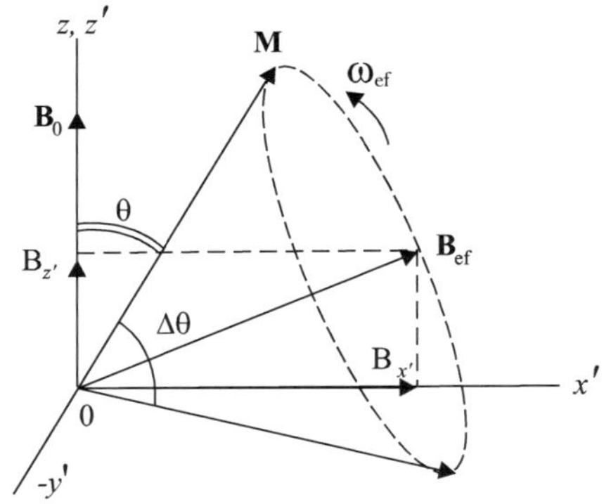

Fig. 1.6 Motion for a magnetization vector in a rotating coordinate system

where $\vec{B}_{z'} = \vec{B}_0 + \vec{\omega}'/\gamma$. As seen from (1.57), in the rotating frame the movement of the vector $\vec{M}$ is determined by magnetic field $\vec{B}_{z'}$. The angular velocity vector $\vec{\omega}'$ is directed along the axis $0z$, oppositely to the field vector $\vec{B}_0$ (Fig. 1.6). Then the absolute value is $B_{z'} = B_0 - \omega'/\gamma$.

We introduce a variable magnetic field $\vec{B}^{\approx}(t)$ with frequency ω', oriented along the axis $0x$ of the laboratory coordinate system, and expand it into two circularly polarized components. Consider how the vector $\vec{M}$ is influenced by the field component $\vec{B}^{\approx}(t)$, which rotates in the same direction as the vector $\vec{M}$ itself (we assume that it is directed along the axis $0x'$ in Fig. 1.6). As already noted, the vector $\vec{M}$ will then be involved in movement on two cones of precession. In a rotating coordinate system, such movement is reduced to the precession around the direction of the vector of the effective magnetic field: $\vec{B}_{ef} = B_z\vec{e}_z + B_{x'}\vec{e}_{x'}$, at an angular velocity $\vec{\omega}_{ef}$ (see Fig. 1.6).

We now introduce the notation for the components of the magnetization vector in a rotating coordinate system:

$$M_{x'} = u, \quad M_{y'} = -\frac{\gamma}{\mid\gamma\mid}v, \quad M_{z'} = M_z. \tag{1.58}$$

The transition from the magnetization components u, v to the components M_x and M_y in laboratory coordinate system is calculated as follows

$$M_x = u\cos\omega t - v\sin\omega t, \tag{1.59}$$

$$M_y = \pm(u\sin\omega t + v\cos\omega t). \tag{1.60}$$

The sign in Eq. (1.60) is determined by the gyromagnetic ratio γ (see 1.58). Writing the components of the vector field $\vec{B}_{ef}$

$$B_{z'} = \frac{\Delta\omega}{\gamma}, \quad B_{x'} = B_1, \quad B_{y'} = 0 \tag{1.61}$$

and substituting (1.58) and (1.61) into (1.56), we get a system of the Bloch equations in a rotating coordinate system:

$$\frac{du}{dt} + \frac{u}{T_2} + \Delta\omega v = 0, \tag{1.62}$$

$$\frac{dv}{dt} + \frac{v}{T_2} - \Delta\omega u + \mid \gamma \mid B_1 M_z = 0, \tag{1.63}$$

$$\frac{dM_z}{dt} + \frac{M_z}{T_1} - \mid \gamma \mid B_1 v - \frac{M_0}{T_1} = 0. \tag{1.64}$$

By solving Bloch's equations (1.62)–(1.64), we can obtain expressions for u, v, and M_z in the form of function of parameters $\Delta\omega$, B_0, B_1, T_1 and T_2, which are determined by experimental conditions or properties of the substance under study.

It is possible to introduce the complex dynamic susceptibility

$$\chi = \chi' - i\chi''$$

associated with the transverse magnetization. As the alternating magnetic field is directed along the axis X. We can write

$$M_x = \chi B_x = 2B_1(\chi' - i\chi'')(\cos\omega t + i\sin\omega t). \tag{1.65}$$

Hence,

$$\mathrm{Re}M_x = 2B_1(\chi'\cos\omega t + \chi''\sin\omega t). \tag{1.66}$$

Using (1.59) and (1.65), we can write:

$$\chi' = u/(2B_1), \quad \chi'' = -v/(2B_1).$$

These expressions are valid for $\gamma > 0$. In general, the value of χ'' should be defined by the formula

$$\chi'' = -\frac{\gamma}{\mid \gamma \mid}\frac{v}{2B_1}.$$

Given this formula, the expression for the energy absorbed per unit volume of substance per unit time, takes the form

$$P(\nu) = -\frac{\gamma}{\mid \gamma \mid} 2\omega B_1^2 \chi''.$$

Therefore, the v-component of magnetization and the associated magnetic susceptibility χ'' determine the energy absorbed by a spin system as a result of interaction with an RF field; this is registered in the form of an absorption signal (Fig. 1.7, curve 1). The u-component of the vector $\vec{M}$ and the real part of dynamic magnetic susceptibility (χ') are recorded as a dispersion signal (curve 2).

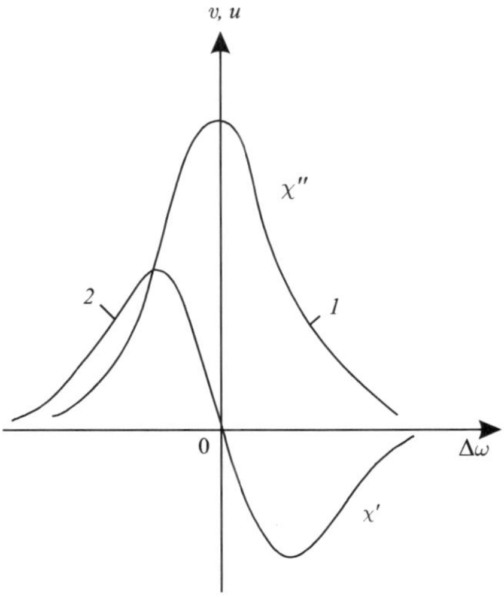

Fig. 1.7 Components of a dynamic magnetic susceptibility

1.6 Absorption and Dispersion as Parameters of Magnetization Components: The Phenomenon of Saturation

The Bloch equations can be solved for some special cases: for example, for the situation of slow passage of a field or frequency through the resonance conditions, that allows one to obtain an undistorted shape of the resonance line. The criterion for the slow passage is the rate of change of frequency ($d\omega/dt$), or the passage time through the resonance region (τ) as compared to the rate of spin–spin relaxation:

$$\frac{d\omega}{dt} \ll \frac{1}{T_2^2} \quad \text{and} \quad \tau \gg T_2. \tag{1.67}$$

Under the conditions (1.67), the components of magnetization u and v slowly change over time, and we can assume at any given time that du/dt, dv/dt and dM_z/dt will be zero, and the frequency detuning ($\Delta\omega$) will have a constant value. In this approximation, the Bloch equations (1.62)–(1.64) take the form

$$\frac{1}{T_2}u + \Delta\omega v = 0,$$

$$-\Delta\omega u + \frac{1}{T_2}v + \mid \gamma \mid B_1 M_z = 0,$$

$$-\mid \gamma \mid B_1 v + \frac{1}{T_1}M_z - \frac{1}{T_1}M_0 = 0.$$

These equations are a system of simple algebraic equations; solving them, we can obtain

$$u = \mid \gamma \mid B_1 T_2 M_0 \frac{\Delta\omega T_2}{1 + (\Delta\omega T_2)^2 + \gamma^2 B_1^2 T_1 T_2}, \tag{1.68}$$

$$v = - \mid \gamma \mid B_1 T_2 M_0 \frac{1}{1 + (\Delta\omega T_2)^2 + \gamma^2 B_1^2 T_1 T_2}, \tag{1.69}$$

$$M_z = M_0 \frac{1 + (\Delta\omega T_2)^2}{1 + (\Delta\omega T_2)^2 + \gamma^2 B_1^2 T_1 T_2}. \tag{1.70}$$

Analyzing the stationary solutions of the Bloch equations (1.68)–(1.70), we can make conclusions on the intensity, width and shape of the resonance signals of absorption and dispersion.

Amplitude and intensity. Amplitudes of absorption and dispersion signals are often defined by the maximum values of functions $v(t)$ and $u(t)$. For the component of absorption, v_{max} equals $v(t)$ at $\Delta\omega = 0$:

$$v_{max} = - \mid \gamma \mid B_1 T_2 M_0 \frac{1}{1 + \gamma^2 B_1^2 T_1 T_2}. \tag{1.71}$$

Under this condition, as can be seen from Eq. (1.68), $u(t) = 0$. With frequency detuning

$$\Delta\omega = \pm \frac{1}{T_2} \sqrt{1 + \gamma^2 B_1^2 T_1 T_2} \tag{1.72}$$

is achieved

$$u_{max} = \pm \frac{1}{2} \mid \gamma \mid B_1 T_2 M_0 \frac{1}{\sqrt{1 + \gamma^2 B_1^2 T_1 T_2}}. \tag{1.73}$$

As seen from expressions (1.71) and (1.73), dependences v_{max} and u_{max}, and hence the intensities of the components of absorption and dispersion of the field B_0 are linear: $v_{max} \sim M_0$, $u_{max} \sim M_0$. The dependence of v_{max} and u_{max} on the field B_1 is more complicated. At weak fields B_1, when the member $\gamma^2 B_1^2 T_1 T_2$ is negligibly small as compared with unity and it can be omitted, the components of absorption and dispersion are proportional to the magnitude of the field B_1. In the case of greater values of field induction B_1, as is evident from (1.71), the intensity of the component v_{max} will decrease with increasing field B_1; the impact of this field on u_{max} in such conditions gradually decreases (see 1.73).

An optimal RF field B_1, at which the maximum absorption intensity of the absorption component is achieved, will be

$$B_{1opt} = \frac{1}{\mid \gamma \mid \sqrt{T_1 T_2}}. \tag{1.74}$$

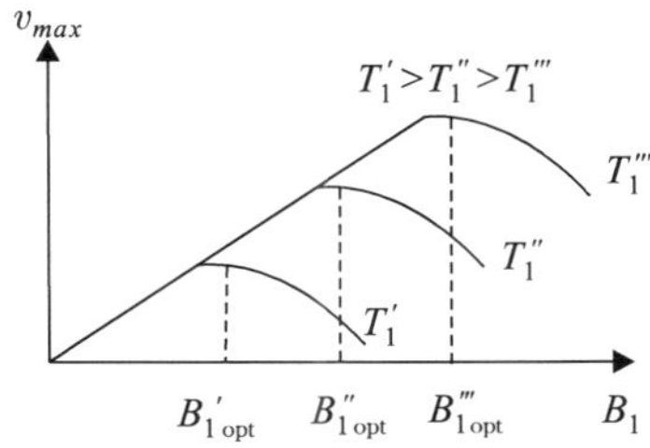

Fig. 1.8 Amplitude of an absorption in external fields for different relaxation times

Then the limiting value of the absorption component is

$$v_{\max\ \max} = \frac{1}{2}\sqrt{\frac{T_2}{T_1}}M_0.$$

The limiting value of the dispersion component is achieved at $B_1 \longrightarrow \infty$. In this case, as seen from (1.73),

$$u_{\max\ \max} = \frac{1}{2}\sqrt{\frac{T_2}{T_1}}M_0. \qquad (1.75)$$

Thus, the greatest values of absorption and dispersion signals are identical, but they are obtained at different values of the fields B_1. It is also natural that it is impossible to obtain the limiting value of a dispersion signal (1.75) as one cannot create an infinitely large field B_1. Besides in this case the line width will be infinitely large (see below).

One should also pay attention to the fact that the optimal value of the RF field in the form (1.74) is inversely proportional to the square root of the product of relaxation times T_1 and T_2. This means that in order to obtain the most intense signal for absorption of substances with long relaxation times (low-viscosity fluids, e.g., alcohols, benzene, etc.) it is necessary to influence them with an RF field $B_{1\,opt}$ having a smaller amplitude than in the case of obtaining the same signal from substances with short relaxation times (for example, glycerol, paramagnetic solutions, etc.). Dependence $v_{\max}(B_1)$ for different relaxation times T_1 is shown in Fig. 1.8 (excluding changes in T_2).

A decrease in the absorption signal intensity with an increasing field B_1 is termed *phenomenon of saturation*. Formally, this phenomenon is explained by the presence of the factor $(1+\gamma^2 B_1^2 T_1 T_2)^{-1}$ in the expression (1.71) which, in case of large fields, causes a decrease $v_{\max}$ with an increasing field B_1. This factor is called *saturation factor* and is denoted by the letter s. The physical cause of magnetic resonance saturation is alignment of the level populations. As seen from (1.68) and (1.69), magnetization components v and u are proportional to M_0, the magnitude of which is determined by the difference in populations $\Delta n^0 = n_1^0 - n_2^0$ of two adjacent levels (see also Sect. 1.4). Under thermodynamic equilibrium conditions, the relation between

n_1^0 and n_2^0 is defined by the Boltzmann factor: $n_1^0/n_2^0 = \exp(\Delta E/(kT))$, where T is the temperature of the ambient environment around the spin system (lattice). This ratio is provided by the presence of energy exchange between the spin system and the lattice, that is, due to spin–lattice relaxation. The more effective the mechanism of the spin–lattice relaxation, the sooner the thermodynamic equilibrium is established with a well-defined temperature difference in the populations of levels Δn^0 for the given temperature.

RF field B_1, causing transitions between levels, tends to level off their populations n_1^0 and n_2^0, i.e., to reduce Δn^0. Since in the presence of the field B_1 a spin system is simultaneously affected by two opposing factors: the spin–lattice relaxation, creating an equilibrium difference in level population Δn^0 in accord with (1.39), and an RF field B_1, reducing Δn^0; then, due to their impact, a new equilibrium is established with its own difference in the level populations. Formally, this new equilibrium state can also be described by the expression (1.39) by replacing the lattice temperature T with a certain temperature T_S, corresponding to the new difference in the level populations of the spin system. This temperature T_S is referred to as *spin temperature* (see also Sect. 3.2).

Thus, for the cases of the influence of a RF field B_1, the level population ratio can be expressed via spin temperature:

$$\frac{n_1}{n_2} = \exp\left(\frac{h\nu}{kT_S}\right). \tag{1.76}$$

Assuming $n_1 - n_2 = \Delta n$, $n_1 + n_2 = N$ and due to the smallness of the exponent index ($h\nu \ll kT_S$), we obtain

$$\frac{n_1}{n_2} - 1 \simeq \frac{2\Delta n}{N} \simeq \frac{h\nu}{kT_S}.$$

Figure 1.9 shows the dependence $\Delta n/N$ of $h\nu/(kT_S)$ obtained from (1.76). The graph shows that with a decreasing difference in level population Δn the spin temperature increases to infinity (at $\Delta n = 0$), and at $\Delta n < 0$ it becomes negative, i.e., at $n_2 > n_1$. Consequently, the phenomenon of complete saturation ($n_1 = n_2$) is matched with such a state of a quantum system, in which the spin temperature increases to infinity.

We can find the relation between spin temperature and saturation factor (s). Let us denote the difference in spin populations at the lower and upper levels, with the Boltzmann thermodynamic equilibrium, as Δn_0, and under the influence of the field B_1, as Δn_s. Then the saturation factor $s = \Delta n_s/\Delta n_0$. At $h\nu \ll kT$ we have an equality $\Delta n_0/\Delta n_s = T_S/T$, hence:

$$T_s = \frac{\Delta n_0}{\Delta n_s}T = \frac{1}{s}T = (1 + \gamma^2 B_1^2 T_1 T_2)T. \tag{1.77}$$

It follows from (1.77) that the T_S increases with an increasing amplitude of the field B_1.

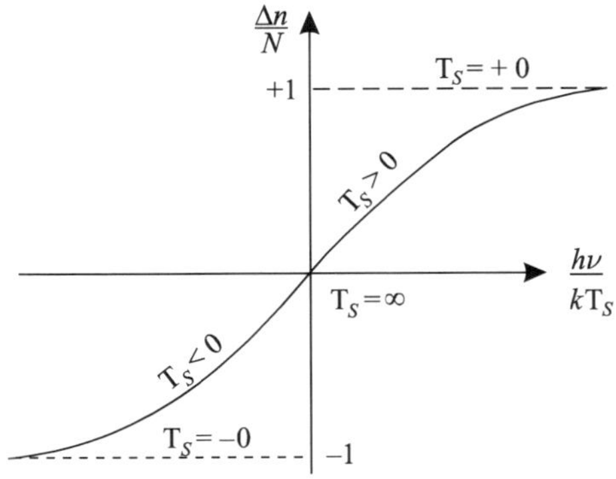

Fig. 1.9 Relation between the relative difference in the level population and the spin temperature

Line width. In spectroscopy, the line width (b) often associated with the double detuning of frequency ($\Delta\omega_{1/2}$) or field ($\Delta B_{1/2}$) at half the height of a resonance absorption line, i.e., $b = 2\Delta\omega_{1/2}$. The half-width of the absorption line can be easily found if, using the above definition, we write the equality $v(\Delta\omega) = v_{\max}/2$, substitute $v(\Delta\omega)$ and $v_{\max}$ in it by the formulas (1.69) and (1.71), and then solve this equality with respect to $\Delta\omega$:

$$\Delta\omega_{1/2} = \frac{1}{T_2}\sqrt{1 + \gamma^2 B_1^2 T_1 T_2}. \tag{1.78}$$

In (1.78), there are two factors: T_2^{-1} and $(1 + \gamma^2 B_1^2 T_1 T_2)^{1/2}$. The first one expresses inverse proportionality of the line width of spin–spin relaxation time. The second describes the line broadening due to the impact of a RF field B_1 with a high amplitude.

The form of absorption lines described by the expression (1.69) is called *Lorentzian* shape. In general, above all in crystals, the shape of the NMR lines can significantly differ from the Lorentzian shape. Some information on the spectral line shape of signals of magnetic resonance will be presented in the next section.

1.7 The Shape and Moments of the Spectral Line of Magnetic Resonance

1.7.1 A Formal Description of the Line Shape in Terms of Moments

The shape of lines of magnetic resonance described by Eqs. (1.68) and (1.69) is a consequence of the phenomenological theory of Bloch, which is of limited application. In general, the shape of a spectral line depends on the interactions in which resonating nuclei or electron are involved and which are defined by molecular dynamics and structure of matter. Therefore, the shape of a line of magnetic resonance contains important information about the microstructure of matter. Experimentally, the shape of a line can be obtained either by registering a spectrum with stationary methods, or by the Fourier transform of the envelope of free induction decay or spin echo (see Sect. 1.8.2, 1.8.3 and Chap. 2).

Mathematically, the shape of a line is described by the form factor $g(\nu)$, representing the dependence of the intensity of magnetic resonance on the frequency normalized so that

$$\int_0^\infty g(\nu)\,d\nu = 1.$$

Free induction decay is described by the function of time $f(t)$, related with $g(\nu)$ by the Fourier transform, provided that the function $g(\nu)$ is obtained in the absence of saturation. Unfortunately, functions $f(t)$ and $g(\nu)$ are not always suitable for a comparison of theoretical models and experimental data, as their analytical expression can only be obtained in the simplest cases (which does not exclude their effective use in numerical simulation). However, one can introduce such parameters of the line shape that can be quite simply calculated theoretically on the basis of information about the structure of matter. These parameters are *moments of a spectral line* (M_n). The moment of a line of the order n (or simply n-moment of line) is the quantity

$$M_n = \int (\nu - \nu_0)^n\, g(\nu)\,d\nu, \tag{1.79}$$

where ν_0 is the resonance frequency.

We shall show that the set of all moments of the line completely determines its shape. To this end, we obtain the expression of the n-th time derivative of $f(t)$. Since $f(t)$ is a Fourier preimage of the form factor $g(\nu)$, then

$$f(t) = \frac{1}{2\pi i}\int_0^\infty g(\alpha + \nu_0)\, e^{2\pi i\alpha t}\,d\alpha$$

(here we have introduced an integration variable $\alpha = \nu - \nu_0$). Consequently,

$$\frac{d^n f(t)}{dt^n} = (2\pi i)^{n-1}\int_0^\infty \alpha^n g(\alpha + \nu_0)\, e^{2\pi i\alpha t}\,d\alpha.$$

Since the integral obtained at $t = 0$ coincides with (1.79), thus the n-th moment of line is equal to within a factor to the value of the n-th derivative of the envelope of free induction decay at the initial time:

$$(2\pi i)^{n-1} M_n = \left. \frac{d^n f(t)}{dt^n} \right|_{t=0}.$$

Simultaneously, the values of derivatives $d^n f(t)/dt_n$ at $t = 0$ determine the expansion coefficients $f(t)$ into a Taylor series. Replacing the derivatives by their expressions, we obtain

$$f(t) = \sum_0^\infty \frac{(2\pi i)^{k-1}}{k!} M_k t^k.$$

Thus, the function of the free induction decay $f(t)$, and consequently, the function of the form $g(\nu)$, uniquely associated with it by the Fourier transform, are fully determined by moments M_n. In some cases, the given relationship between M_n and $d^n f/dt^n$ is used to calculate the moments of the line instead of the expression (1.79).

1.7.2 Homogeneous and Inhomogeneous Broadening of Magnetic Resonance Lines

Various mechanisms lead to broadening of the magnetic resonance lines in the aspects of two qualitatively different types: *homogeneous* and *inhomogeneous* broadening. These types can be related, for example, with different types of distribution of absorbed RF power in the spin system. If internal interactions in the system distribute this power to all the spins, the corresponding resonance line is called *homogeneously broadened.* If RF power with fixed frequency radiation is only received by a group of spins, which in this case can be ascribed their own resonant line, narrower than the total line of the magnetic resonance, then this line is *inhomogeneously broadened.* This difference is clearly seen when observing saturation of a magnetic resonance line (Fig. 1.10). In the case of a inhomogeneously broadened line, the effect of saturation (solid curves in Fig. 1.10) leads to a decrease in its intensity only near the frequency of the saturating alternating magnetic field (see Fig. 1.10a), while a homogeneously broadened line is saturated along its total width (see Fig. 1.10b). A group of spins with the same resonance frequency, linked by an interaction leading to homogeneous broadening of their magnetic resonance, is called a *spin packet*, or *isochromate*.

A good, albeit trivial, example of a cause for inhomogeneous broadening is heterogeneity of a static magnetic field (instrumentation broadening). Obviously, in this case only those spins will resonate that are in the part of the sample in which the magnetic field magnitude satisfies the resonance condition. A similar mechanism acts in the case of the magnetic dipole–dipole interaction of non-equivalent spins, differing by Larmor frequencies. In this case, the role of magnetic inhomogeneity is played by distribution of the local magnetic fields produced by the neighbouring spins. The same type of broadening occurs if the resonance line is the envelope of a number of overlapping homogeneously broadened components. Uniform broadening

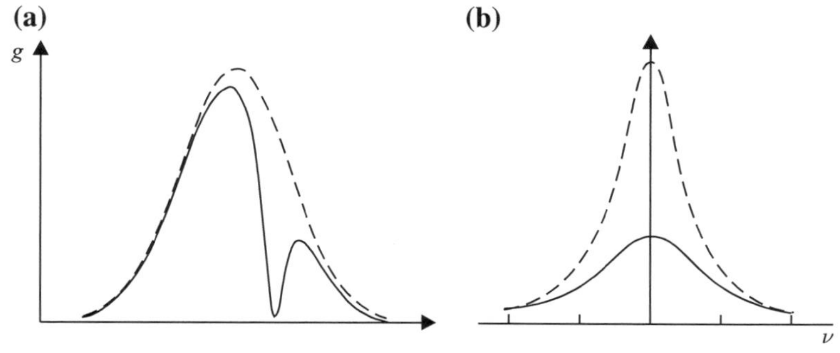

Fig. 1.10 Inhomogeneous (**a**) and homogeneous (**b**) broadening of magnetic resonance lines

of magnetic resonance is produced by a dipole–dipole interaction in an ensemble of identical spins, exchange interaction, diffusion, interaction with a radiation field.

1.8 Calculation of Response of a Spin System to Impact of Pulsed Radiofrequency Fields

1.8.1 The Motion of a Nuclear Magnetization Vector in the Absence of Relaxation

A system of the Bloch equations, which includes relaxation terms, does not have an exact solution in the general case. Therefore, in this section we shall assume that relaxation times T_1 and T_2 are sufficiently large in comparison with the considered time intervals. We shall also assume that the amplitude and phase of the rotating alternating field are constant. Taking into account these assumptions and neglecting the relaxation times, we obtain the following system of equations based on the formulas (1.62)–(1.64):

$$\frac{du}{dt} = -\Delta\omega v,$$
$$\frac{dv}{dt} = \Delta\omega u - \gamma B_1 M_z,$$
$$\frac{dM_z}{dt} = \gamma v B_1,$$

where $\Delta\omega = \omega_0 - \omega$; u, v, M_z are the components of the nuclear magnetization in a coordinate system rotating together with the vector of the alternating magnetic field: the axis z directed along the vector of the static magnetic field $\vec{B}_0$, the axis x along the vector of the rotational magnetic field with induction B_1 and frequency ω; the component u is directed along the axis x, component v is by $\pi/2$ ahead of the phase of the rotating magnetic field vector.

Under arbitrary initial conditions for u_0, v_0 and M_{z0}, it is more convenient to deal with this system using the operator method. Replacing $u(t)$ with $u(p)$, and du/dt with $pu(p) - pu_0$, and acting similarly with $v(t)$ and $M_z(t)$, we obtain

$$\begin{aligned}
&pu(p) + \Delta\omega v(p) = pu_0,\\
&-\Delta\omega u(p) + pv(p) + \gamma B_1 M_z(p) = pv_0,\\
&-\gamma B_1 v(p) + pM_z(p) = pM_{z0}.
\end{aligned}$$

Solution of this system is not difficult:

$$u(p) = \frac{p^2 u_0 - p\Delta\omega v_0 + \gamma^2 B_1^2 u_0 + \gamma B_1 \Delta\omega M_{z0}}{p^2 + \omega_{ef}^2},$$

$$v(p) = \frac{p^2 v_0 - p(\gamma B_1 M_{z0} - \Delta\omega u_0)}{p^2 + \omega_{ef}^2},$$

$$M_z(p) = \frac{p^2 M_{z0} + p\gamma B_1 v_0 + \Delta\omega(\Delta\omega M_{z0} + \gamma B_1 u_0)}{p^2 + \omega_{ef}^2},$$

where $\omega_{ef} = \gamma B_{ef}$, $B_{ef}^2 = B_1^2 + (\Delta\omega)^2/\gamma^2$. These expressions include terms of three types: $1/(p^2 + \omega_{ef}^2)$, $p/(p^2 + \omega_{ef}^2)$, $p^2/(p^2 + \omega_{ef}^2)$. Treating them as images of certain functions of time on a complex plane, we see that the last two terms are matched with $(\sin\omega_{ef}t)/\omega_{ef}$ and $\cos\omega_{ef}t$, and the first one can be presented as

$$\frac{1}{p^2 + \omega_{ef}^2} = \frac{1}{\omega_{ef}^2} - \frac{p^2}{\omega_{ef}^2(p^2 + \omega_{ef}^2)}.$$

Then the first term is matched with the expression $(1 - \cos\omega_{ef}t)/\omega_{ef}^2$. Now the functions $u(t)$, $v(t)$ and $M_z(t)$ are easily calculated:

$$u(t) = u_0 \cos\omega_{ef}t - \frac{\Delta\omega v_0 \sin\omega_{ef}t}{\omega_{ef}} + \frac{(\gamma^2 B_1^2 u_0 + \gamma B_1 \Delta\omega M_{z0})(1 - \cos\omega_{ef}t)}{\omega_{ef}^2},$$

$$v(t) = v_0 \cos\omega_{ef}t - \frac{(\gamma B_1 M_{z0} - \Delta\omega u_0)\sin\omega_{ef}t}{\omega_{ef}},$$

$$M_z(t) = M_{z0}\cos\omega_{ef}t + \frac{\gamma B_1 v_0 \sin\omega_{ef}t}{\omega_{ef}} + \frac{\Delta\omega(\Delta\omega M_{z0} + \gamma B_1 u_0)(1 - \cos\omega_{ef}t)}{\omega_{ef}^2}.$$

These expressions can conveniently be written in the matrix form:

$$\begin{pmatrix} u \\ v \\ M_z \end{pmatrix} = \boldsymbol{A} \begin{pmatrix} u_0 \\ v_0 \\ M_{z0} \end{pmatrix}, \tag{1.80}$$

where

$$\boldsymbol{A} = \begin{pmatrix} \dfrac{(\Delta\omega)^2 \cos\omega_{ef}t + (\gamma B_1)^2}{\omega_{ef}^2} & -\dfrac{\Delta\omega}{\omega_{ef}} \sin\omega_{ef}t & \dfrac{\Delta\omega\gamma B_1}{\omega_{ef}^2}(1 - \cos\omega_{ef}t) \\ \dfrac{\Delta\omega}{\omega_{ef}} \sin\omega_{ef}t & \cos\omega_{ef}t & -\dfrac{\gamma B_1}{\omega_{ef}} \sin\omega_{ef}t \\ \dfrac{\Delta\omega\gamma B_1}{\omega_{ef}^2}(1 - \cos\omega_{ef}t) & \dfrac{\gamma B_1}{\omega_{ef}} \sin\omega_{ef}t & \dfrac{(\gamma B_1)^2 \cos\omega_{ef}t + (\Delta\omega)^2}{\omega_{ef}^2} \end{pmatrix}. \tag{1.81}$$

If $\gamma B_1 \gg \Delta\omega$, which corresponds to the case of a fairly high power pulse and a rather narrow spectrum, the matrix (1.81) can be simplified. This obtained matrix, which is a simpler one, is denoted as $\boldsymbol{A}_0$:

$$\boldsymbol{A}_0 = \begin{pmatrix} 1 & 0 & 0 \\ 0 & \cos\gamma B_1 t & -\sin\gamma B_1 t \\ 0 & \sin\gamma B_1 t & \cos\gamma B_1 t \end{pmatrix}.$$

This matrix describes nutation of a nuclear magnetization vector in the rotating coordinate system around the x-axis of this system with an angular velocity γB_1.

If $B_1 = 0$, the matrix (1.81) can also be simplified. This matrix is designated as $\boldsymbol{B}$:

$$\boldsymbol{B} = \begin{pmatrix} \cos\Delta\omega t & -\sin\Delta\omega t & 0 \\ \sin\Delta\omega t & \cos\Delta\omega t & 0 \\ 0 & 0 & 1 \end{pmatrix}.$$

This matrix describes movement of a nuclear magnetization vector, which occurs in the absence of a RF field, in a rotating coordinate system around the z axis with an angular velocity $\Delta\omega$.

The general matrix (1.81) describes rotation of a nuclear magnetization vector in a rotating coordinate system with an angular velocity γB_{ef} around a certain axis located on the plane (zx) of this system and making an angle α with the axis z, while $\tan\alpha = \gamma B_1/\Delta\omega$.

It is useful to note that at $\Delta\omega < \gamma B_1$ the nuclear magnetization vector can always be turned to the plane (xy), if one selects such duration of a radio frequency pulse t_p that the component M_z is zero. Since

$$M_z = M_{z0} \frac{(\gamma B_1)^2 \cos \omega_{ef} t_p + \Delta\omega^2}{\omega_{ef}^2},$$

then for the case cosidered

$$\cos \omega_{ef} t_p = -\frac{\Delta\omega^2}{(\gamma B_1)^2}.$$

In particular, if $\Delta\omega = 0$, the magnetization vector of a spin packet rotates into the plane (xy) at $\gamma B_1 t_p = \pi/2$, and if $\Delta\omega = \gamma B_1$, when the following condition is satisfied:

$$\omega_{ef} t_p = \sqrt{\Delta\omega^2 + \gamma^2 B_1^2}\, t_p = \pi.$$

1.8.2 Free Spin Precession

Let us consider a small macroscopic volume in a sample, within which all nuclear spins are equivalent. The totality of these nuclei is can be called *spin packet*. In this case, the equivalence means identical NMR frequencies, i.e., equality of the values of $\Delta\omega$. Then the calculation of the response of a nuclear system to a sequence of RF pulses is reduced to determining the behavior of nuclear magnetization of an arbitrary spin packet and summation of signals from all the spin packets.

To excite free precession, only one RF pulse is enough, and to calculate its shape, one should first calculate the direction of the vector of the nuclear magnetization of a single spin packet by the end of a RF impulse with a length of t_p. In order to perform it, multiply the column-vector describing the initial conditions in the form (1.80) by the matrix (1.81) in which, instead of the current time t, there must be the pulse duration t_p. Then one should consider the behavior of the nuclear magnetization vector in the absence of the RF field, starting the time reading t from the end of the RF pulse, i.e., multiply the obtained result by the matrix $\boldsymbol{B}$. As to the initial column-vector, one usually adopts $u_0 = 0,\ v_0 = 0,\ M_{z0} = M_0$, that corresponds to the equilibrium initial conditions. Then

$$\begin{pmatrix} u \\ v \\ M_z \end{pmatrix} = \begin{pmatrix} \cos \Delta\omega t & -\sin \Delta\omega t & 0 \\ \sin \Delta\omega t & \cos \Delta\omega t & 0 \\ 0 & 0 & 1 \end{pmatrix}$$

$$\times \begin{pmatrix} \dfrac{\Delta\omega^2 \cos \omega_{ef} t_p + (\gamma B_1)^2}{\omega_{ef}^2} & -\dfrac{\Delta\omega}{\omega_{ef}} \sin \omega_{ef} t_p & \dfrac{\Delta\omega \gamma B_1}{\omega_{ef}^2}(1 - \cos \omega_{ef} t_p) \\ \dfrac{\Delta\omega}{\omega_{ef}} \sin \omega_{ef} t_p & \cos \omega_{ef} t_p & -\dfrac{\gamma B_1}{\omega_{ef}} \sin \omega_{ef} t_p \\ \dfrac{\Delta\omega \gamma B_1}{\omega_{ef}^2}(1 - \cos \omega_{ef} t_p) & \dfrac{\gamma B_1}{\omega_{ef}} \sin \omega_{ef} t_p & \dfrac{(\gamma B_1)^2 \cos \omega_{ef} t_p + (\Delta\omega)^2}{\omega_{ef}^2} \end{pmatrix} \begin{pmatrix} 0 \\ 0 \\ M_0 \end{pmatrix}.$$

Multiplying the matrices, we obtain

$$\begin{pmatrix} u \\ v \\ M_z \end{pmatrix} = M_0 \begin{pmatrix} \dfrac{\Delta\omega\gamma B_1}{\omega_{ef}^2}(1-\cos\omega_{ef}t_p)\cos\Delta\omega t + \dfrac{\gamma B_1}{\omega_{ef}}\sin\omega_{ef}t_p\sin\Delta\omega t \\ \dfrac{\Delta\omega\gamma B_1}{\omega_{ef}^2}(1-\cos\omega_{ef}t_p)\sin\Delta\omega t - \dfrac{\gamma B_1}{\omega_{ef}}\sin\omega_{ef}t_p\cos\Delta\omega t \\ \dfrac{(\gamma B_1)^2\cos\omega_{ef}t_p + (\Delta\omega)^2}{\omega_{ef}^2} \end{pmatrix}.$$

Thus, the components u and v of the selected spin packet are expressed by the following formulas:

$$u(t) = M_0\gamma B_1\left[\frac{\Delta\omega}{\omega_{ef}^2}(1-\cos\omega_{ef}t_p)\cos\Delta\omega t + \frac{1}{\omega_{ef}}\sin\omega_{ef}t_p\sin\Delta\omega t\right], \tag{1.82}$$

$$v(t) = M_0\gamma B_1\left[\frac{\Delta\omega}{\omega_{ef}^2}(1-\cos\omega_{ef}t_p)\sin\Delta\omega t - \frac{1}{\omega_{ef}}\sin\omega_{ef}t_p\cos\Delta\omega t\right].$$

where time t is measured from the end of the RF pulse.

To obtain the precession signal, we should sum up the components $u(t)$ and $v(t)$ of rotating magnetizations in all spin packets. If the spin packets are small enough and their parameters change gradually, the summation can be replaced by integration. In the case under study, the spin packets differ only by values $\Delta\omega$, therefore, we should only integrate with respect to $\Delta\omega$. The magnetization of individual spin packets can be different because the number of nuclei with different $\Delta\omega$ is unequal, and distribution of of spin packages by $\Delta\omega$ is described by a certain normaliz function $g(\Delta\omega)$ (form-factor). Considering this function as a weighting function for the signals from components of magnetization of individual spin packets $u(t)$ and $v(t)$, we obtain

$$U(t) = M_0\gamma B_1\int_{-\infty}^{\infty} g(\Delta\omega)\left[\frac{\Delta\omega(1-\cos\omega_{ef}t_p)}{\omega_{ef}^2}\cos\Delta\omega t + \frac{\sin\omega_{ef}t_p}{\omega_{ef}}\sin\Delta\omega t\right]d\Delta\omega,$$

$$V(t) = M_0\gamma B_1\int_{-\infty}^{\infty} g(\Delta\omega)\left[\frac{\Delta\omega(1-\cos\omega_{ef}t_p)}{\omega_{ef}^2}\sin\Delta\omega t - \frac{\sin\omega_{ef}t_p}{\omega_{ef}}\cos\Delta\omega t\right]d\Delta\omega. \tag{1.83}$$

Here $U(t)$, $V(t)$ are signals of the components of rotating magnetizations of a unit volume sample. Note that in the first formula (1.83), the integrand is an odd valued function, if $g(\Delta\omega)$ is an even function. The latter will be true if the spectrum is symmetric and the frequency of the RF field coincides with the central frequency of the spectrum; in this case $U(t) = 0$.

If the spectrum is narrow, and the amplitude of the RF pulse is large, that is, if the condition is satisfied

$$\Delta\omega \ll \gamma B_1, \tag{1.84}$$

then $\omega_{ef} = \gamma B_1$ and the formula for the free precession amplitude becomes much simpler:

$$\begin{aligned} V(t) &= -M_0 \int_{-\infty}^{\infty} g(\Delta\omega) \sin \gamma B_1 t_p \cos \Delta\omega t \, d\Delta\omega \\ &= -M_0 \sin \gamma B_1 t_p \int_{-\infty}^{\infty} g(\Delta\omega) \cos \Delta\omega t \, d\Delta\omega. \end{aligned} \tag{1.85}$$

It is clearly seen that $V(t)$ is maximum if

$$\gamma B_1 t_p = \pi/2 \quad \text{or} \quad \gamma B_1 t_p = (2m + 1)\pi/2 \tag{1.86}$$

(where m is integer), i.e., in this case the free precession amplitude reaches a maximum when the nuclear magnetization vector is rotated into a plane (xy). The pulse, for which the conditions (1.84) and (1.86) are satisfied, turns the nuclear magnetization vectors of all spin packets by 90°, and is therefore called 90°-it degree pulse.

It follows from (1.85) that the time-dependent function $V(t)$ is a cosine Fourier transform of $g(\Delta\omega)$. Assume that a spectrum has a Lorentzian shape:

$$g(\delta\omega) = \frac{1}{\pi\,\delta\omega_0[1 + (\delta\omega)^2/(\delta\omega_0)^2]}.$$

In this form, the spectrum is normalized, i.e., $\int_{-\infty}^{\infty} g(\delta\omega) d\delta\omega = 1$; $\delta\omega_0$ is the half-width of the NMR spectrum at a half height; $\delta\omega = \omega_0 - \omega_{00}$; ω_0 is still NMR frequency, ω_{00} is the central NMR frequency. If $\omega = \omega_{00}$, i.e., the frequency of the external RF field coincides with the central frequency of the NMR spectrum, $\delta\omega = \Delta\omega$. Then the time dependence of the free precession is described by the formula:

$$V(t) = -M_0 \sin \gamma B_1 t_p \int_{-\infty}^{\infty} \frac{1}{\pi\,\delta\omega_0[1 + (\Delta\omega)^2/(\delta\omega_0)^2]} \cos \Delta\omega t \, d\delta\omega.$$

The integral in this formula is easy to calculate:

$$V(t) = -M_0 \sin \gamma B_1 t_p \, e^{-\delta\omega_0 t}.$$

The larger the width of the spectrum, the faster the precession is reduced. If the spectrum has the Gaussian form, i.e.

$$g(\delta\omega) = \frac{1}{\sqrt{\pi}\,\delta\omega_0} \exp \frac{-(\delta\omega)^2}{(\delta\omega_0)^2}, \tag{1.87}$$

then the curve of the time dependence of the free precession amplitude has a Gaussian form, too (as before, $\delta\omega = \Delta\omega$):

$$V(t) = -M_0 \sin\gamma B_1 t_p \int_{-\infty}^{\infty} \frac{1}{\sqrt{\pi}\,\delta\omega_0} \exp\frac{-(\Delta\omega)^2}{(\delta\omega_0)^2} \cos\Delta\omega t \, d\Delta\omega$$
$$= -M_0 \sin\gamma B_1 t_p \exp\left(-\frac{(\delta\omega_0 t)^2}{4}\right).$$

It is helpful to consider a rectangular shape, which occurs, for example, when a sample of a cylindrical or rectangular shape (it is important that the butts of the sample are flat) is affected by an additional field when induction varies linearly, with a coordinate directed perpendicular to the butts of the sample:

$$V(t) = -M_0 \sin\gamma B_1 t_p \int_{-\delta\omega_0}^{\delta\omega_0} \frac{1}{2\delta\omega_0} \cos\Delta\omega t \, d\Delta\omega$$
$$= -M_0 \sin\gamma B_1 t_p \int_{0}^{\delta\omega_0} \frac{1}{\delta\omega_0} \cos\Delta\omega t \, d\Delta\omega$$
$$= -\frac{M_0 \sin\gamma B_1 t_p}{\delta\omega_0 t} \sin\delta\omega_0 t.$$

In this case, the precession is oscillatory. If the precession signal $V(t)$ is known an experiment, one can calculate the spectrum $g(\delta\omega)$, using the inverse Fourier transform:

$$g(\Delta\omega) = \frac{1}{2\pi M_0 \sin\gamma B_1 t_p} \int_{-\infty}^{\infty} V(t) \cos\Delta\omega t \, dt.$$

1.8.3 Formation of Echo Signals

When nuclear magnetization is exposed by two RF pulses separated with a time interval τ, there arises a signal called *spin echo*. In the simplest cases, the process of echo formation allows a very intuitive interpretation (see Chap. 2). But first, let us calculate its shapes and amplitudes. To perform it, the column-vector describing the equilibrium state of the spin packet should be multiplied on the left by the matrix $\boldsymbol{A}$, in which the current time should be replaced with the parameters of the first pulse B_1 and t_{p1}, then multiplied by the matrix $\boldsymbol{B}$, with t replaced by τ, and then again multiplied by the matrix $\boldsymbol{A}$, with the second pulse parameters B_2 and t_{p2} and then again multiplied by the matrix $\boldsymbol{B}$ with the current time t, measured from the end of the second impulse. Because of unhandiness of the calculations, we limit ourselves to the case where $\gamma B_1 \gg \Delta\omega$ and $\gamma B_2 \gg \Delta\omega$. Then, instead of the matrix $\boldsymbol{A}$, we can use a simpler matrix $\boldsymbol{A}_0$:

$$\begin{pmatrix} u \\ v \\ M_z \end{pmatrix} = \begin{pmatrix} \cos\Delta\omega t & -\sin\Delta\omega t & 0 \\ \sin\Delta\omega t & \cos\Delta\omega t & 0 \\ 0 & 0 & 1 \end{pmatrix} \begin{pmatrix} 1 & 0 & 0 \\ 0 & \cos\gamma B_2 t_{p2} & -\sin\gamma B_2 t_{p2} \\ 0 & \sin\gamma B_2 t_{p2} & \cos\gamma B_2 t_{p2} \end{pmatrix}$$
$$\times \begin{pmatrix} \cos\Delta\omega\tau & -\sin\Delta\omega\tau & 0 \\ \sin\Delta\omega\tau & \cos\Delta\omega\tau & 0 \\ 0 & 0 & 1 \end{pmatrix} \begin{pmatrix} 1 & 0 & 0 \\ 0 & \cos\gamma B_1 t_{p1} & -\sin\gamma B_1 t_{p1} \\ 0 & \sin\gamma B_1 t_{p1} & \cos\gamma B_1 t_{p1} \end{pmatrix} \begin{pmatrix} 0 \\ 0 \\ M_0 \end{pmatrix}.$$

By multiplying matrices, we obtain finally:

$$\begin{pmatrix} u \\ v \\ M_z \end{pmatrix} = \begin{pmatrix} S_1 \sin\Delta\omega\tau\cos\Delta\omega t + S_1C_2\cos\Delta\omega\tau\sin\Delta\omega t + S_2C_1\sin\Delta\omega t \\ S_1 \sin\Delta\omega\tau\sin\Delta\omega t - S_1C_2\cos\Delta\omega\tau\cos\Delta\omega t - S_2C_1\cos\Delta\omega t \\ -S_2C_1\cos\Delta\omega\tau - C_2C_1 \end{pmatrix},$$

where $S_i = \sin\gamma B_i t_{pi}$; $C_i = \cos\gamma B_i t_{pi}$. Thus, after the impact of the two RF pulses, the nuclear magnetization vector of the spin packet in a rotating coordinate system is described by the following formulas:

$$\begin{aligned} u(t) &= M_0[\sin\gamma B_1 t_{p1}\sin\Delta\omega\tau\cos\Delta\omega t \\ &\quad + \sin\gamma B_1 t_{p1}\cos\gamma B_2 t_{p2}\cos\Delta\omega\tau\sin\Delta\omega t + \sin\gamma B_2 t_{p2}\cos\gamma B_1 t_{p1}\sin\Delta\omega t], \\ v(t) &= M_0[\sin\gamma B_1 t_{p1}\sin\Delta\omega\tau\sin\Delta\omega t \\ &\quad - \sin\gamma B_1 t_{p1}\cos\gamma B_2 t_{p2}\cos\Delta\omega\tau\cos\Delta\omega t - \sin\gamma B_2 t_{p2}\cos\gamma B_1 t_{p1}\cos\Delta\omega t], \\ M_z(t) &= M_0[-\sin\gamma B_2 t_{p2}\cos\gamma B_1 t_{p1}\cos\Delta\omega\tau - \cos\gamma B_2 t_{p2}\cos\gamma B_1 t_{p1}]. \end{aligned}$$

To calculate the total signal after the second pulse, the signals must be summarized over all spin packets. If expedient to replace the summation by integration, but it should be borne in mind that the integrals of the rapidly changing integrand function are small and can be neglected. Therefore, if calculation of the free precession signal is made after the second pulse (i.e., we consider the small time interval immediately after the second RF pulse), then only those members should be chosen that do not contain τ:

$$U_2(t) = M_0 \int_{-\infty}^{\infty} g(\Delta\omega)\sin\gamma B_2 t_{p2}\cos\gamma B_1 t_{p1}\sin\Delta\omega t \, d\Delta\omega,$$

$$V_2(t) = -M_0 \int_{-\infty}^{\infty} g(\Delta\omega)\sin\gamma B_2 t_{p2}\cos\gamma B_1 t_{p1}\cos\Delta\omega t \, d\Delta\omega. \quad (1.88)$$

Other members contribute to the signal which is called "echo":

$$U_e(t) = M_0 \int_{-\infty}^{\infty} g(\Delta\omega)[\sin\gamma B_1 t_{p1} \sin\Delta\omega\tau \cos\Delta\omega t$$
$$+ \sin\gamma B_1 t_{p1} \cos\gamma B_2 t_{p2} \cos\Delta\omega\tau \sin\Delta\omega t + \sin\gamma B_2 t_{p2} \cos\gamma B_1 t_{p1} \sin\Delta\omega t]d\Delta\omega,$$
$$V_e(t) = M_0 \int_{-\infty}^{\infty} g(\Delta\omega)[\sin\gamma B_1 t_{p1} \sin\Delta\omega\tau \sin\Delta\omega t$$
$$- \sin\gamma B_1 t_{p1} \cos\gamma B_2 t_{p2} \cos\Delta\omega\tau \cos\Delta\omega t - \sin\gamma B_2 t_{p2} \cos\gamma B_1 t_{p1} \cos\Delta\omega t]d\Delta\omega. \quad (1.89)$$

If worth nothing that, the magnetization component along the z axis is not important, because it is constant and no emf in a receiver coil is induced. The integrand expressions in (1.89) contain rapidly changing and slowly changing components. The integrals of the rapidly changing components are small and can be neglected. But they are not seen explicitly and to select them, we need to perform some conversions. First products of the sines and cosines should be provided in the form of sums and differences of these functions. Using the known trigonometric formulas

$$\cos\Delta\omega\tau \cos\Delta\omega t = \frac{1}{2}[\cos\Delta\omega(t-\tau) + \cos\Delta\omega(t+\tau)],$$

$$\sin\Delta\omega\tau \sin\Delta\omega t = \frac{1}{2}[\cos\Delta\omega(t+\tau) - \cos\Delta\omega(t-\tau)]$$

substituting them into the integrals (1.89), and neglecting the integrals containing $\cos\Delta\omega(t+\tau)$, we get

$$V_e(t) = \frac{M_0}{2} \int_{-\infty}^{\infty} g(\Delta\omega) \sin\gamma B_1 t_{p1}(1 - \cos\gamma B_2 t_{p2}) \cos\Delta\omega(t-\tau)d\Delta\omega. \quad (1.90)$$

The formula for the component signal $U_e(t)$ is written similarly, but it should be borne in mind that, like in the case with the precession, $U_e = 0$, if the spectrum is symmetric and the frequency of the external RF field coincides with the central frequency of the spectrum. Taking all the terms that are independent of $\Delta\omega$, outside the integral sign in (1.90), we write:

$$V_e(t) = \frac{M_0}{2} \sin\gamma B_1 t_{p1}(1 - \cos\gamma B_2 t_{p2}) \int_{-\infty}^{\infty} g(\Delta\omega) \cos\Delta\omega(t-\tau)d\Delta\omega, \quad (1.91)$$

Recall that the time t is measured from the end of the second pulse.

The integral in (1.91) is the cosine Fourier transform of the spectrum $g(\Delta\omega)$. Obviously, in this case the maximum amplitude of the spin echo is obtained when the first pulse is the 90° pulse and the second is 180° one, i.e., the following conditions are satisfied:

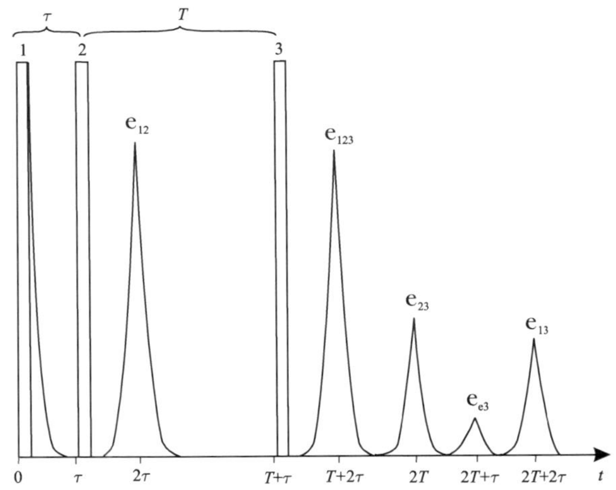

Fig. 1.11 Pulse responses after three RF pulses

$$\gamma B_1 t_{p1} = \pi/2, \quad \gamma B_2 t_{p2} = \pi.$$

Similarly, in the same approximation, we can calculate the pulse responses after three RF pulses. After each pulse, free precession emerges. If the time interval (τ, see Fig. 1.11) between the impacts of the first two impulses (*1* and *2*) is smaller than the time interval (T) between the impacts of the second and third pulses, then after the second pulse there occurs a normal signal (E_{12}) of a two-pulse spin echo. In this case, after the third pulse, there are four signals of the spin-echo. Two of them are the usual spin echo signals (E_{13} and E_{23}), arising from the first and third pulses, and from the second and third ones. Another echo is of the echo secondarily phased by the third impulse (E_{e3}), arising from the first two pulses at the moment $2T$. The fourth echo is referred to as *stimulated echo* (E_{123}). It occurs after exposure to the third pulse after time τ, equals to the time interval between the first and second pulses.

The echoes from the first and second pulses are described by the formula

$$V_{E_{12}}(2\tau) = \frac{M_0}{2} \sin \gamma B_1 t_{p1} (1 - \cos \gamma B_2 t_{p2}), \tag{1.92}$$

and from the first and third impulses, by the formula

$$V_{E_{13}}(2T + 2\tau) = \frac{M_0}{4} \sin \gamma B_1 t_{p1} (1 + \cos \gamma B_2 t_{p2})(1 - \cos \gamma B_3 t_{p3}),$$

from the second and third ones, by the formula

$$V_{E_{23}}(2T) = \frac{M_0}{2} \cos \gamma B_1 t_{p1} \sin \gamma B_2 t_{p2} (1 - \cos \gamma B_3 t_{p3}),$$

The second phase-locked echo is described by

$$V_{\mathrm{E}_{e3}}(2T+\tau)=\frac{M_0}{4}\sin\gamma B_1 t_{\mathrm{p}1}(1-\cos\gamma B_2 t_{\mathrm{p}2})(1-\cos\gamma B_3 t_{\mathrm{p}3}),$$

and the stimulated echo, by

$$V_{\mathrm{E}_{123}}(T+2\tau)=\frac{M_0}{2}\sin\gamma B_1 t_{\mathrm{p}1}\sin\gamma B_2 t_{\mathrm{p}2}\sin\gamma B_3 t_{\mathrm{p}3}.$$

If the condition $\gamma B_1 \gg \Delta\omega$ is not satisfied and it is necessary to use the matrix (1.81), the calculations are more cumbersome, but they are also possible to complete, and one can get the following expressions for the two-pulse spin-echo signals:

$$U_{\mathrm{e}}(t)=M_0\frac{\gamma B_1(\gamma B_2)^2}{2}\int\limits_{-\infty}^{\infty} g(\Delta\omega)\frac{1-\cos\omega_{\mathrm{ef}2}t_{\mathrm{p}2}}{\omega_{\mathrm{ef}1}\omega_{\mathrm{ef}2}^2}\times\left[-\frac{\Delta\omega}{\omega_{\mathrm{ef}1}}(1-\cos\omega_{\mathrm{ef}1}t_{\mathrm{p}1})\cos\Delta\omega(t-\tau)+\sin\omega_{\mathrm{ef}1}t_{\mathrm{p}1}\sin\Delta\omega(t-\tau)\right]\mathrm{d}\Delta\omega, \quad (1.93)$$

$$V_{\mathrm{e}}(t)=M_0\frac{\gamma B_1(\gamma B_2)^2}{2}\int\limits_{-\infty}^{\infty} g(\Delta\omega)\frac{1-\cos\omega_{\mathrm{ef}2}t_{\mathrm{p}2}}{\omega_{\mathrm{ef}1}\omega_{\mathrm{ef}2}^2}\times\left[\frac{\Delta\omega}{\omega_{\mathrm{ef}1}}(1-\cos\omega_{\mathrm{ef}1}t_{\mathrm{p}1})\sin\Delta\omega(t-\tau)+\sin\omega_{\mathrm{ef}1}t_{\mathrm{p}1}\cos\Delta\omega(t-\tau)\right]\mathrm{d}\Delta\omega, \quad (1.94)$$

where $\omega_{\mathrm{ef}1}=\gamma B_{\mathrm{ef}1}$, $B_{\mathrm{ef}1}^2=B_1^2+(\Delta\omega)^2/\gamma^2$, $\omega_{\mathrm{ef}2}=\gamma B_{\mathrm{ef}2}$, $B_{\mathrm{ef}2}^2=B_2^2+(\Delta\omega)^2/\gamma^2$.

It is impossible to analytically calculate the integrals (1.93), (1.94) and the integrals describing free precession after both RF impulses (1.88), even for simple forms of the spectrum such as Lorentzian or Gaussian ones. We have to only use numerical methods for the spectra of intermediate width when γB_1 and γB_2 are comparable with the spectrum width. But one can use an approximation of the infinitely wide spectrum for very wide spectra. In this approach, the above integrals can be reduced to more convenient for numerical calculation, and sometimes it is possible to get analytical formulas without integrals.

1.8.4 Free Precession and Spin Echo with an Infinitely Wide Spectrum

In practice, there can be such broad NMR spectra that it is impossible to create a desirable short pulse to satisfy the condition (1.83). To assess the consequences of a breach of this condition, it is advisable to analyze the extreme case: an infinitely wide spectrum. This spectrum is always symmetric, so the components $U(t)$, $U_2(t)$ and $U_e(t)$ in (1.83), (1.88), and (1.93) are equal to zero, but an infinitely wide spectrum cannot be normalized, and in the subsequent discussion we shall assume that $g(\Delta\omega) = 1$.

Let us consider the behaviour of free precession in detail. It is described by the formula (1.83), where $g(\Delta\omega)$ should be replaced with unity:

$$V(t) = -M_0\omega_1 \int_{-\infty}^{\infty} \left[\frac{\sin\left(\sqrt{\Delta\omega^2 + \omega_1^2}\, t_p\right) \cos \Delta\omega t}{\sqrt{\Delta\omega^2 + \omega_1^2}} + \frac{\Delta\omega\left(\cos\left(\sqrt{\Delta\omega^2 + \omega_1^2}\, t_p\right) - 1\right) \sin \Delta\omega t}{\Delta\omega^2 + \omega_1^2} \right] d\Delta\omega, \tag{1.95}$$

where $\omega_1 = \gamma B_1$. Since the integrand function is even, then the limits from $-\infty$ to $+\infty$ can be replaced with 0 and $+\infty$, and $V(t)$ (in the form 1.95) can be doubled:

$$V(t) = -2M_0\omega_1[F_1 + F_2 + F_3], \tag{1.96}$$

$$F_1 = \int_0^{\infty} \frac{\sin\left(\sqrt{\Delta\omega^2 + \omega_1^2}\, t_p\right) \cos \Delta\omega t}{\sqrt{\Delta\omega^2 + \omega_1^2}} d\Delta\omega;$$

$$F_2 = \int_0^{\infty} \frac{\Delta\omega \cos\left(\sqrt{\Delta\omega^2 + \omega_1^2}\, t_p\right) \sin \Delta\omega t}{\Delta\omega^2 + \omega_1^2} d\Delta\omega;$$

$$F_3 = -\int_0^{\infty} \frac{\Delta\omega \sin \Delta\omega t}{\Delta\omega^2 + \omega_1^2} d\Delta\omega.$$

We assume t_p to be variable, from which, according to the Laplace transform, we go to the variable p:

$$F_1(p) = \int_0^\infty \int_0^\infty \frac{\sin\left(\sqrt{\Delta\omega^2 + \omega_1^2}\, t_p\right) \cos \Delta\omega t}{\sqrt{\Delta\omega^2 + \omega_1^2}} e^{-pt_p} d\Delta\omega\, dt_p;$$

$$F_2(p) = \int_0^\infty \int_0^\infty \frac{\Delta\omega \cos\left(\sqrt{\Delta\omega^2 + \omega_1^2}\, t_p\right) \sin \Delta\omega t}{\Delta\omega^2 + \omega_1^2} e^{-pt_p} d\Delta\omega\, dt_p;$$

$$F_3(p) = -\int_0^\infty \int_0^\infty \frac{\Delta\omega \sin \Delta\omega t}{\Delta\omega^2 + \omega_1^2} e^{-pt_p} d\Delta\omega\, dt_p.$$

Changing the order of integration, we can first calculate integrals over dt_p:

$$\int_0^\infty \sin\left(\sqrt{\Delta\omega^2 + \omega_1^2}\, t_p\right) e^{-pt_p} dt_p = \frac{\sqrt{\Delta\omega^2 + \omega_1^2}}{p^2 + \Delta\omega^2 + \omega_1^2},$$

$$\int_0^\infty \cos\left(\sqrt{\Delta\omega^2 + \omega_1^2}\, t_p\right) e^{-pt_p} dt_p = \frac{p}{p^2 + \Delta\omega^2 + \omega_1^2},$$

$$\int_0^\infty e^{-pt_p} dt_p = \frac{1}{p}.$$

Note that the integrals like $\int_0^\infty \sin ax\, e^{-bx} dx$ and $\int_0^\infty \cos ax\, e^{-bx} dx$ are taken in the standard way: substitute $\sin ax$ with $(e^{iax} - e^{-iax})/2i$, and $\cos ax$ with $(e^{iax} + e^{-iax})/2$. As a result, the integrals $F_1(p)$, $F_2(p)$ and $F_3(p)$ now take the form:

$$F_1(p) = \int_0^\infty \frac{\cos \Delta\omega t}{p^2 + \Delta\omega^2 + \omega_1^2} d\Delta\omega;$$

$$F_2(p) = \int_0^\infty \frac{p\Delta\omega \sin \Delta\omega t}{(\Delta\omega^2 + \omega_1^2)(p^2 + \Delta\omega^2 + \omega_1^2)} d\Delta\omega;$$

$$F_3(p) = -\int_0^\infty \frac{\Delta\omega \sin \Delta\omega t}{p(\Delta\omega^2 + \omega_1^2)} d\Delta\omega.$$

All these integrals are easily calculated:

$$F_1(p) = \frac{\pi}{2\sqrt{p^2+\omega_1^2}} e^{-t\sqrt{p^2+\omega_1^2}},$$

$$F_2(p) = \frac{\pi}{2p}\left[e^{-t\omega_1} - e^{-t\sqrt{p^2+\omega_1^2}}\right],$$

$$F_3(p) = -\frac{\pi}{2p} e^{-t\omega_1}.$$

The Laplace transform of the function $V(t)$ in the form (1.96) is proportional to the sum $F_1(p)$, $F_2(p)$ and $F_3(p)$:

$$V(p) = \pi\frac{\omega_1 M_0}{p}\left[1 - \frac{p}{\sqrt{p^2+\omega_1^2}}\right] e^{-\sqrt{p^2+\omega_1^2}\,t}.$$

Let us write this result in the following form:

$$V(p) = \pi\frac{\omega_1 M_0}{p} G(p),$$

where

$$G(p) = \left(1 - \frac{p}{\sqrt{p^2+\omega_1^2}}\right) e^{-\sqrt{p^2+\omega_1^2}\,t}.$$

Performing the inverse Laplace transform of the function $G(p)$ and taking into consideration the fact that in the time domain the variable p is matched with the variable t_p , we obtain

$$G(t_\mathrm{p}) = \omega_1\sqrt{\frac{t_\mathrm{p}-t}{t_\mathrm{p}+t}} J_1\left(\omega_1\sqrt{t_\mathrm{p}^2-t^2}\right), \qquad t_\mathrm{p} > t,$$

$$G(t_\mathrm{p}) = 0, \qquad t_\mathrm{p} \le t$$

(J_1 is the Bessel function).

It follows from the properties of the Laplace transform that if $f(t_\mathrm{p})$ corresponds to $g(p)$, then $\int_0^{t_\mathrm{p}} f(t_\mathrm{p})dt_\mathrm{p}$ corresponds to $g(p)/p$. For $t \le t_\mathrm{p}$ we have

$$V(t_\mathrm{p}) = \pi M_0\omega_1^2 \int_0^{t_\mathrm{p}} \sqrt{\frac{t_\mathrm{p}-t}{t_\mathrm{p}+t}} J_1\left(\omega_1\sqrt{t_\mathrm{p}^2-t^2}\right) dt_\mathrm{p},$$

and if $t \geq t_p$, then $V(t_p) = 0$. The latter integral can be split into two:

$$V(t_p) = \pi M_0 \omega_1^2 \int_0^t \sqrt{\frac{t_p - t}{t_p + t}} J_1\left(\omega_1 \sqrt{t_p^2 - t^2}\right) dt_p$$

$$+ \pi M_0 \omega_1^2 \int_t^{t_p} \sqrt{\frac{t_p - t}{t_p + t}} J_1\left(\omega_1 \sqrt{t_p^2 - t^2}\right) dt_p. \quad (1.97)$$

The first integral in (1.97) is equal to zero, because within such limits t_p is always smaller than t. Now we can consider t_p as a parameter, and it is convenient to simply rename the integration variable, denoting it by x. Changing the limits of the second integral, we have for $t \leq t_p$

$$V(t) = -\pi M_0 \omega_1^2 \int_{t_p}^t \sqrt{\frac{x - t}{x + t}} J_1\left(\omega_1 \sqrt{x^2 - t^2}\right) dx. \quad (1.98)$$

It can readily be seen from this formula, that with $t = t_p$ $V(t) = 0$, since the limits of the integral become identical.

We can also get some of the implications that are found with greater difficulty from the initial formula (1.95).

Consider free precession after the weak pulse, when the condition $\omega_1 t_p \ll 1$ is satisfied. We expand the Bessel function into a series, and consider only the first member of the series:

$$J_1\left(\omega_1 \sqrt{x^2 - t^2}\right) = \frac{\omega_1 \sqrt{x^2 - t^2}}{2}.$$

Then

$$V(t) = -\frac{\pi M_0 \omega_1^3}{2} \int_{t_p}^t (x - t) dx = -\frac{\pi M_0 \omega_1^3}{4} (t_p - t)^2.$$

It follows that the initial amplitude of free precession ($t = 0$) for an infinitely wide spectrum increases in proportion to ω_1^3 and t_p^2, while under the condition $\omega_1 \gg \Delta\omega$ with growing ω_1 and t_p, it increases linearly.

In the case under study, the decay of the free precession amplitude for small $\omega_1 t_p$ is monotone. Note that numerical calculations by the formula (1.98) showed that with increasing amplitude of the RF impulse, the amplitude decay of free precession is faster, and it is imposed with oscilations which are of the greater frequency the greater the amplitude of the RF pulse.

Based on (1.98), the dependence of the initial amplitude of free precession on the RF pulse amplitude for any its duration can also be obtained. Setting $t = 0$, we obtain

$$V(0) = -\pi M_0\omega_1^2 \int_{t_p}^{0} J_1(\omega_1 x)dx = \pi V_0 M_0\omega_1[1 - J_0(\omega_1 t_p)].$$

If $\omega_1 t_p \gg 1$, the Bessel function in the square brackets can be neglected, and then one can clearly see that for large values $\omega_1 t_p$. The value $V(0)$ increases linearly and is independent of the impulse duration with an increasing amplitude of RF frequency pulse. Similarly, we can consider the two-pulse spin-echo. However, no general formulas have been received up to now. Only three special cases were studied:

(1) the long first RF pulse and the short (δ-impulse) second one;
(2) the short first and the long second RF pulses;
(3) both pulses of any duration, but of equal amplitudes.

In the first case, for $-t_{p1} < t < 0$ [time t is now counted from the moment after the end of the second pulse (τ is the period of time between the end of the first pulse and beginning of the second one)], we get a formula that is only different from that for free precession (1.98) by the sign and the presence of the factor $1 - \cos\theta_2$, where θ_2 is the angle by which the nuclear magnetization vector is rotated due to the second pulse. Since the second pulse is δ-impulse, the nuclear magnetization vectors of all excited spin packets are rotated by the same angle. Actually, the nuclear magnetization vectors of those spin packets are to rotate by the same angle that play an essential role in the formation of the spin echo. Therefore, for the validity of the formulas given below, the second pulse is to be much shorter than the first one. So, for $-t_{p1} < t < 0$ we have

$$V(t) = \frac{\pi M_0\omega_1^2(1 - \cos\theta_2)}{2} \int_{|t|}^{t_p} \sqrt{\frac{x - |t|}{x + |t|}} J_1\left(\omega_1\sqrt{x^2 - t^2}\right)dx. \tag{1.99}$$

For $0 < t < \tau$, the formula becomes more complicated, but the sign of the module of t can be omitted, because in this case the time is always positive:

$$V(t) = -\frac{\pi M_0\omega_1^2(1 - \cos\theta_2)}{2} \int_{t}^{t_{p1}} \sqrt{\frac{x - t}{x + t}} J_1\left(\omega_1\sqrt{x^2 - t^2}\right)dx$$
$$+ \pi M_0\omega_1(1 - \cos\theta_2) J_0\left(\omega_1\sqrt{t_{p1}^2 - t^2}\right). \tag{1.100}$$

If the first impulse is short and the second one is long, then for $-t_{p2} < t < t_{p2}$ one obtain:

$$V(t) = \frac{\pi M_0\omega_1^2 \sin\theta_1}{2} \int_{|t|}^{t_{p2}} J_0\left(\omega_1\sqrt{x^2 - t^2}\right)dx.$$

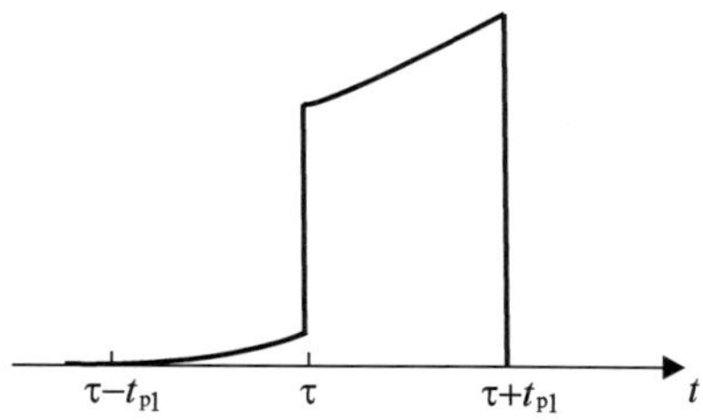

Fig. 1.12 Characteristic shape of the signal, after a long first RF pulse and a short second one

Here θ_1 is the angle by which the nuclear magnetization vector rotates under the action of the first pulse. For impulses of arbitrary length but of equal amplitudes:

$$V(t) = \frac{\pi M_0}{2}[-2S_1 + S_2 - \operatorname{sgn}(t_{p2} - t_{p1})S_3 + \operatorname{sgn}(t)(-2S_4 - 2S_5 + S_6 + S_7)], \quad (1.101)$$

$$S_1 = \omega_1^3 \int_{|t|}^{t_{p1}} (t_{p1} - x) J_0\left(\omega_1 \sqrt{x^2 - t^2}\right) dx, \quad (1.102)$$

$$S_4 = S_1 + \frac{1}{2}\omega_1^4 \int_{|t|}^{t_{p1}} (t_{p1} - x)^2 \sqrt{\frac{x - t}{x + t}} J_1\left(\omega_1 \sqrt{x - t^2}\right) dx. \quad (1.103)$$

The remaining S_i are defined as follows: S_2 and S_6 are obtained by replacing t_{p1} with $t_{p1} + t_{p2}$ in (1.102) and (1.103); S_3 and S_7 from (1.102) and (1.103) by replacing t_{p1} with $|t_{p2} - t_{p1}|$; finally, S_5—from (1.103) by replacing t_{p1} with t_{p2}. Using expressions (1.99)–(1.103), one can obtain simple formulas for calculation of the form of the spin-echo signal, under the condition that the pulses are so weak that $\omega_1 t_{p1} \ll 1$ and $\omega_2 t_{p2} \ll 1$. In the case of a long first RF pulse and the short second one in zero approximation, when only the biggest term is important (the first term of expansion of the Bessel function with a zero index which is equal to unit),—the spin echo signal is a rectangular pulse of the same duration as the first pulse, and with an amplitude of $\pi M_0 \omega_1^2 (1 - \cos\theta_2)/2$. If we take into account the terms of the next order of smallness, we obtain the characteristic shape of the signal shown in Fig. 1.12.

For $-t_{p1} < t < 0$

$$V(t) = \frac{\pi M_0 \omega_1 (1 - \cos\theta_2)(t_{p1} - |t|)^2}{8},$$

for $0 < t < t_{p1}$

$$V(t) = \frac{\pi M_0 \omega_1 (1 - \cos\theta_2)}{2} \left[1 - \frac{\omega_1^2 (t_{p1}^2 - t^2)}{2} + \frac{\omega_1^2 (t_{p1} - t)^2}{4} \right]. \tag{1.104}$$

The first two terms in the square brackets of this expression are obtained from the second term in (1.101), and the third one, from the first term in (1.101), since $J_0(x) = 1 - x^2/4$. Combining fraction in the square brackets of (1.104), we can write:

$$V(t) = \frac{\pi M_0 \omega_1 (1 - \cos\theta_2)}{2} \left[1 - \frac{\omega_1^2}{8} (t_{p1}^2 - 3t^2 + 2tt_{p1}) \right];$$

$V(t) = 0$ for other values of t.

For the short first and long second RF pulses, the echo signal acquires the characteristic triangular shape. Replacing $J_0(\omega_1 \sqrt{x^2 - t^2})$ with 1, we write:

$$V(t) = \pi M_0 \omega_1^2 \sin\theta_1 (t_{p2} - |t|).$$

When pulses are of arbitrary length, but with equal amplitudes, application of these formulas is much more complicated. For small-amplitude RF pulses, ($\omega_1 \ll 1$) $S_4 = S_1$. Replacing J_0 with 1 and calculating the remaining integral, we obtain

$$S_1 = \frac{\omega_1^3 (t_{p1} - |t|)^2}{2}. \tag{1.105}$$

The formula is valid for $|t| \le t_{p1}$. Outside this interval, $S_1 = S_4 = 0$. The remaining S_i are obtained from (1.105), by using substitution of the variables, as it was described earlier:

$$S_2 = S_6 = \frac{\omega_1^3 (t_{p1} + t_{p2} - |t|)^2}{2}, \quad |t| \le t_{p1} + t_{p2},$$

$$S_3 = S_7 = \frac{\omega_1^3 (|t_{p2} - t_{p1}| - |t|)^2}{2}, \quad |t| \le |t_{p2} - t_{p1}|,$$

$$S_5 = \frac{\omega_1^3 (t_{p2} - |t|)^2}{2}, \quad |t| \le t_{p2}.$$

Outside these intervals, all $S_i = 0$. It is advisable to write the formula (1.101) for different time periods: for $t > 0$, $t_{p1} > t_{p2}$

$$S(t) = 2(-2S_1 + S_2 + S_3 - S_5),$$

since $S_4 = S_1$, $S_6 = S_2$, $S_7 = S_3$; for $t > 0$, $t_{p2} > t_{p1}$

$$S(t) = 2(-2S_1 + S_2 - S_5), \tag{1.106}$$

since S_3 and S_7 are equal in magnitude, but they have different signs. For $t < 0$, the sign before the last parenthesis in (1.101) will change, and for $t < 0$, $t_{p1} > t_{p2}$ we get

$$S(t) = 2S_5,$$

and for $t < 0$, $t_{p2} > t_{p1}$

$$S(t) = 2(S_5 - S_3). \tag{1.107}$$

Substituting S_i in these formulas, we can get all the possible special cases. If duration of the first pulse is more than twice the duration of the second one, i.e., the condition $t_{p1} > 2t_{p2}$, is satisfied, then for $-t_{p2} \leq t \leq 0$

$$S(t) = \omega^3 (t_{p2} - |t|)^2, \tag{1.108}$$

for $0 \leq t \leq t_{p2}$

$$S(t) = \omega^3 (t_{p2}^2 - t^2 + 2tt_{p2}), \tag{1.109}$$

for $t_{p2} \leq t \leq t_{p1} - t_{p2}$

$$S(t) = 2\omega^3 t_{p2}^2, \tag{1.110}$$

for $t_{p1} - t_{p2} \leq t \leq t_{p1}$

$$S(t) = \omega^3 (t_{p2}^2 - t_{p1}^2 + 2t_{p1}t_{p2} + 2t(t_{p1} - t_{p2}) - t^2), \tag{1.111}$$

and for $t_{p1} \leq t \leq t_{p1} + t_{p2}$

$$S(t) = \omega (t_{p1} + t_{p2} - t)^2. \tag{1.112}$$

Based on the relations (1.108)–(1.112), we can conclude that the form of the spin echo turned out to be similar to a trapezium with smooth corners. The flat top of the trapezium is determined by (1.110). When $t_{p2} < t_{p1} < 2t_{p2}$, the flat top of the trapezium disappears, and the echo becomes bell-shaped (Fig. 1.13, τ is measured from the end of the second pulse). Formulas (1.108) and (1.112) are valid, but the following formulas are used instead of the rest ones: for $0 \leq t \leq t_{p1} - t_{p2}$

$$S(t) = \omega^3 (t_{p2}^2 - t^2 + 2tt_{p2}),$$

for $t_{p1} - t_{p2} \leq t \leq t_{p2}$

$$S(t) = \omega^3 (2t_{p1}t_{p2} + 2t_{p1}t - t_{p1}^2 - 2t^2),$$

for $t_{p2} \leq t \leq t_{p1}$

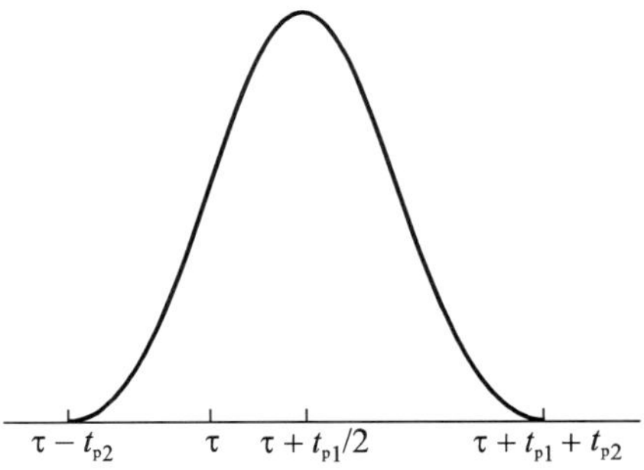

Fig. 1.13 Bell-shaped echo, after two RF pulses if $t_{p2} < t_{p1} < 2t_{p2}$

$$S(t) = \omega^3 (t_{p2}^2 - t_{p1}^2 + 2t_{p1}t_{p2} + 2tt_{p1} - 2tt_{p2} - t^2).$$

The maximum echo value is reached at $t = t_{p1}/2$:

$$S(t_{p1}/2) = \omega^3 t_{p1}(4t_{p2} - t_{p1})/2. \tag{1.113}$$

If $t_{p2} > t_{p1}$, then, substituting S_i in (1.106) and (1.107), we obtain the following expressions: for $-t_{p2} \le t \le -(t_{p2} - t_{p1})$

$$S(t) = \omega^3 (t_{p2} - |t|)^2,$$

for $-(t_{p2} - t_{p1}) \le t \le 0$

$$S(t) = \omega^3 t_{p1}(2t_{p2} - t_{p1} - 2|t|),$$

for $0 \le t \le t_{p1}$

$$S(t) = \omega^3 (2t_{p1}t_{p2} + 2t_{p1}t - t_{p1}^2 - 2t^2),$$

for $t_{p1} \le t \le t_{p2}$

$$S(t) = \omega^3 t_{p1}(t_{p1} + 2t_{p2} - 2t),$$

for $t_{p2} \le t \le t_{p1} + t_{p2}$

$$S(t) = \omega^3 (t_{p1} + t_{p2} - t)^2.$$

In the time intervals $(t_{p1} - t_{p2}) < t < 0$ and $t_{p1} < t < t_{p2}$ the dependence *S(t)* is linear, and with increase of t_{p2}, the interval increases too, i.e., with t_{p2} considerably exceeding t_{p1}, the form of the spin echo becomes a triangle with smooth angles. The size and position of the maximum echo are still determined by the formula (1.113).

1.8.5 The Single-Pulse Echo

The most common single-pulse echo arises when a nuclear system is exposed to a powerful RF pulse, the filling frequency of which is different from the NMR frequency by an amount exceeding half-width of the spectrum. In fact, a single-pulse echo is part of precession, but visually it looks like an echo signal occurring after a time interval which a close to a radio-frequency pulse duration. For calculation of a single-pulse echo we can use the general formulas (1.82) and (1.83), but it is necessary to specify the shape of the NMR spectrum. If we consider a Gaussian shape, in the presence of detuning between the frequency ω of a RF field and the central NMR spectrum frequency ω_{00} (ω_0 is a NMR frequency of any spin packet), $\delta\omega = \omega_0 - \omega_{00} = \omega_0 - \omega - \omega_{00} + \omega = \Delta\omega - \Delta\omega_0$, and then the formula (1.87) is written as follows:

$$g(\Delta\omega) = \frac{1}{\sqrt{\pi}\,\delta\omega_0}\exp\left[-\left(\frac{\Delta\omega - \Delta\omega_0}{\delta\omega_0}\right)^2\right].$$

Amplitudes of signals $U(t)$ and $V(t)$, in the NMR sensor are described as

$$U(t) = \frac{M_0\gamma B_1}{\sqrt{\pi}\delta\omega_0}\int\limits_{-\infty}^{\infty}\exp\left[-\left(\frac{\Delta\omega - \Delta\omega_0}{\delta\omega_0}\right)^2\right]$$
$$\times\left[\frac{\Delta\omega}{\omega_{\mathrm{ef}}^2}(1 - \cos\omega_{\mathrm{ef}}t_{\mathrm{p}})\cos\Delta\omega t + \frac{1}{\omega_{\mathrm{ef}}}\sin\omega_{\mathrm{ef}}t_{\mathrm{p}}\sin\Delta\omega t\right]d\Delta\omega,$$

$$V(t) = \frac{M_0\gamma B_1}{\sqrt{\pi}\,\delta\omega_0}\int\limits_{-\infty}^{\infty}\exp\left[-\left(\frac{\Delta\omega - \Delta\omega_0}{\delta\omega_0}\right)^2\right]$$
$$\times\left[\frac{\Delta\omega}{\omega_{\mathrm{ef}}^2}(1 - \cos\omega_{\mathrm{ef}}t_{\mathrm{p}})\sin\Delta\omega t - \frac{1}{\omega_{\mathrm{ef}}}\sin\omega_{\mathrm{ef}}t_{\mathrm{p}}\cos\Delta\omega t\right]d\Delta\omega.$$

Suppose that $\Delta\omega_0 t_{\mathrm{p}} \gg 1$ and introduce the dimensionless variables:

$$Z = \frac{\Delta\omega}{\Delta\omega_0},\quad A = \frac{\gamma B_1}{\Delta\omega_0},\quad \Delta_0 = \frac{\delta\omega_0}{\Delta\omega_0},\quad x_0 = \Delta\omega_0\,t_{\mathrm{p}},\quad a = \frac{t}{t_{\mathrm{p}}}.$$

Using them, we can write:

$$U(t) = \frac{M_0 A}{\sqrt{\pi}\,\Delta_0} \int_{-\infty}^{\infty} \exp\left[-\left(\frac{Z-1}{\Delta_0}\right)^2\right] \times \left[\frac{Z}{Z^2+A^2}\left(1-\cos x_0\sqrt{Z^2+A^2}\right)\cos a x_0 Z + \frac{\sin x_0\sqrt{Z^2+A^2}\sin a x_0 Z}{\sqrt{Z^2+A^2}}\right] dZ,$$

$$V(t) = \frac{M_0 A}{\sqrt{\pi}\,\Delta_0} \int_{-\infty}^{\infty} \exp\left[-\left(\frac{Z-1}{\Delta_0}\right)^2\right] \times \left[\frac{Z}{Z^2+A^2}\left(1-\cos x_0\sqrt{Z^2+A^2}\right)\sin a x_0 Z - \frac{\sin x_0\sqrt{Z^2+A^2}\cos a x_0 Z}{\sqrt{Z^2+A^2}}\right] dZ.$$

Expanding the products of trigonometric functions and introducing $V_{\text{single}}(t) = U(t) + iV(t)$ for a single-pulse echo, we obtain

$$V_{\text{single}}(t) = \frac{M_0 A}{\sqrt{\pi}\,\Delta_0} \int_{-\infty}^{\infty} \exp\left[-\left(\frac{Z-1}{\Delta_0}\right)^2\right] \times \left[\frac{Z}{Z^2+A^2}\left(e^{i x_0 a Z} - \frac{1}{2}e^{i x_0\left(aZ-\sqrt{Z^2+A^2}\right)} - \frac{1}{2}e^{i x_0\left(aZ+\sqrt{Z^2+A^2}\right)}\right)\right] + \left[\frac{1}{2\sqrt{Z^2+A^2}}\left(e^{i x_0\left(aZ-\sqrt{Z^2+A^2}\right)} - e^{i x_0\left(aZ+\sqrt{Z^2+A^2}\right)}\right)\right] dZ.$$

Among all the exponents, there is only one, $\exp\left[i x_0\left(aZ-\sqrt{Z^2+A^2}\right)\right]$, which corresponds to oscillations with low or even zero frequency for values of a close to unity. For small values of a, low-frequency vibrations are matched with the exponent $\exp(i x_0 a Z)$. The exponent $\exp\left[i x_0\left(aZ+\sqrt{Z^2+A^2}\right)\right]$ does not describe low-frequency vibrations at all, and the integral of it is small.

Thus, the single-pulse echo is calculated by the formula

$$V_{\text{single}}(t) = \frac{M_0 A}{2\sqrt{\pi}\,\Delta_0} \int_{-\infty}^{\infty} \exp\left[-\left(\frac{Z-1}{\Delta_0}\right)^2\right]\left[\frac{-Z}{Z^2+A^2} + \frac{1}{\sqrt{Z^2+A^2}}\right] e^{i x_0\left(aZ-\sqrt{Z^2+A^2}\right)} dZ.$$

The integrand expression before the second exponent can be considered to be slowly changing only if the value of Δ_0 is not too small, i.e., for a sufficiently wide spectrum. Then the expression for the single-pulse echo signal can be written as

$$V_{\text{single}}(t) = \int_{-\infty}^{\infty} F(Z) e^{i x_0 h(Z)} dZ, \tag{1.114}$$

where

$$F(Z) = \frac{M_0 A}{2\sqrt{\pi}\Delta_0} \exp\left[-\left(\frac{Z-1}{\Delta_0}\right)^2\right]\left[\frac{1}{\sqrt{Z^2+A^2}} - \frac{Z}{Z^2+A^2}\right];$$

$$h(Z) = aZ - \sqrt{Z^2+A^2}.$$

To calculate the integral in (1.114), one can use the method of stationary phase. The point in which $dh(Z)/dZ = 0$ is termed *a saddle point*. We denote it by the letter Z_0. In our case, $Z_0 = aA/\sqrt{1-a^2}$. And we can write:

$$\int_{-\infty}^{\infty} F(Z)e^{ix_0 h(Z)}dZ = F(Z_0)\exp\left[i\left(h(Z_0)x_0 + \frac{\pi}{4}\right)\right]\sqrt{2\pi\left[\left|\frac{d^2h(Z)}{dZ^2}\right|_{Z=Z_0}\right| x_0\right]^{-1}}. \tag{1.115}$$

Calculating the quantities $F(Z_0)$, $h(Z_0)$, $d^2h(Z)/dZ^2$ for $Z = Z_0$ and substituting them in the formula (1.115), we obtain

$$V_{\text{single}}(t) = \frac{M_0\sqrt{2\pi}\,(1-a)\sqrt{A}\sqrt{1-a^2}}{2\sqrt{\pi}\,\Delta_0\sqrt{x_0}\,(1-a^2)^{3/4}} \times \exp\left[-\left(\frac{aA}{\sqrt{1-a^2}} - 1\right)^2(\Delta_0)^2\right]\exp\left[i\left(-Ax_0\sqrt{1-a^2} + \frac{\pi}{4}\right)\right],$$

or

$$V_{\text{single}}(t) = \frac{M_0(1-a)^{3/4}\sqrt{A}}{\sqrt{2x_0}\Delta_0(1+a)^{1/4}} \exp\left[-\frac{\left(aA/\sqrt{1-a^2}-1\right)^2}{\Delta_0^{-2}}\right] \times \exp\left[i\left(-Ax_0\sqrt{1-a^2} + \frac{\pi}{4}\right)\right].$$

The modulus of this expression looks simpler:

$$|V_{\text{single}}(t)| = \frac{M_0(1-a)^{3/4}\sqrt{A}}{\sqrt{2x_0}\,\Delta_0(1+a^2)^{1/4}} \exp\left[-\frac{\left(aA/\sqrt{1-a^2}-1\right)^2}{\Delta_0^2}\right]. \tag{1.116}$$

If we assume that the factor before the exponent in (1.116) changes slowly enough (it is possible if the value of a is not too close to unity), then it can be approximately assumed that the maximum magnitude of a single-pulse echo signal is observed when the index in the exponent is zero. Denoting the position of the single-pulse echo signal as a_0, we find it by using the condition $a_0A/\sqrt{1-a_0^2} = 1$. Then the maximal value of the single-pulse echo signal will be reached at $a_0 = (A^2+1)^{-1}$

or at

$$t_{\max} = t_{\text{p}} \frac{\Delta\omega_0^2}{(\gamma B_1)^2 + \Delta\omega_0^2}$$

where $\Delta\omega_0$ is the difference between the frequency of the external RF field and central frequency of the NMR spectrum). With increase of the amplitude of the RF pulse B_1, $t_{\max}$ decreases, i.e., the faster the single-pulse echo signal follows the RF pulse. Let us determine the maximum amplitude of the single-pulse echo signal. To do this, we substitute the obtained value of a_0 in (1.116):

$$V_{\text{single}_{\max}}(t) = \frac{M_0 A^{1/2}\left(1-(A^2+1)^{-1}\right)^{3/4}}{\sqrt{2x_0}\,\Delta_0\left(1+(A^2+1)^{-1}\right)^{1/4}} = \frac{M_0 A^2}{\sqrt{2x_0}\Delta_0(2+A^2)^{1/4}\sqrt{1+A^2}}.$$

When $A \ll 1$ and $A \gg 1$, we obtain, respectively:

$$V_{\text{single}_{\max}} = \frac{M_0 A^2}{2\sqrt{2x_0}\,\Delta_0}, \quad V_{\text{single}_{\max}} = \frac{M_0 A^{1/2}}{\sqrt{2x_0}\,\Delta_0}.$$

However, in the case of large values of A (mind that $A = \gamma B_1/\Delta\omega_0$), the single-pulse echo signal, will rise, but simultaneously will shift to the RF pulse and therefor disappear in the much stronger signal of precession.

The obtained formulas give a very good description of the position of the single-pulse echo signal, but its amplitude and, especially, width are correct by deseribed only for rather broad NMR spectra, when the condition of slow change of the function F is satisfied.

1.9 Spin Hamiltonian: Density Matrix Formalism

As it was shown in Sect. 1.5, the phenomena of magnetic resonance within the first-order approximation can be described within the framework of the motion of classical magnetization vectors. This representation is relevant if nuclear (or electron) magnetic moments do not interact with each other. To describe interacting spins a quantum-mechanical treatment is necessary to apply.

1.9.1 Spin Hamiltonian

In the most of cases the full Hamiltonian, which describes the whole set of interactions in an atom, molecule or crystal, is extremely complex and it is impossible to obtain the exact solution of motion equations. One of the advantages of magnetic resonance is that the full description of this phenomenon can be made using a rather simple

Hamiltonian $\widehat{\mathcal{H}}$ that depends only on the spin operators and several phenomenological constants, which can be calculated using quantum-chemistry methods. This is the so-called *spin Hamiltonian*, which is widely used for description of nuclear magnetic, electron paramagnetic and nuclear quadrupole resonances.

As it is known, the behavior of any quantum-mechanical system can be described by the nonstationary Schrödinger equation, which defines the eigenfunctions Ψ:

$$\mathrm{i}\hbar\frac{\partial\Psi}{\partial t} = \widehat{\mathcal{H}}\Psi, \qquad (1.117)$$

here $\widehat{\mathcal{H}}$ is the Hamiltonian, which depends on both spacial ($\vec{r}_i$) and spin (S_i) variables of all nuclei and electrons that form the systems, and it can be both time-independent and time-dependent. Let us represent that system as two subsystems, one of which is described by spatial variables and another one by spin variables. As interactions between the subsystems are much weaker than interactions inside each subsystem, the basis functions Ψ of the full system can be represented as products of spatial-coordinate dependent functions $\Psi(\vec{r}_i, t)$ and spin-coordinate dependent functions $\Psi(S_i, t)$:

$$\Psi(\vec{r}_i, S_i, t) = \Psi(\vec{r}_i, t)\Psi(S_i, t).$$

This representation is always valid for nuclei and often can be applied to electrons. If the state is described by a wave function, which depends on spatial coordinates only and does not change at any changes occurring in the spin system, Eq. (1.117) can be averaged over these wave functions multiplying it (on the left) by the conjugate wave function $\Psi^*(\vec{r}_i, t)$ and then integrating over all coordinate space. Then instead of Eq. (1.117) we have

$$i\hbar\frac{\partial\Psi(S_i, t)}{\partial t} = \langle\Psi^*(\vec{r}_i, t)|\widehat{\mathcal{H}}|\Psi(\vec{r}_i, t)\rangle\Psi(S_i, t) = \widehat{\mathcal{H}}^{\mathrm{s}}\Psi(S_i, t), \qquad (1.118)$$

where

$$\widehat{\mathcal{H}}^{\mathrm{s}} = \langle\Psi^*(\vec{r}_i)|\widehat{\mathcal{H}}|\Psi(\vec{r}_i)\rangle. \qquad (1.119)$$

At such a transformation, the terms of the Hamiltonian, which depend on spatial variables only, give an additional term (in energy) that does not depend on the spin subsystem state. The spin operators do not change at all, however, if they are multiplied by some space-depending factors, the latter are averaged, and coefficients in the spin operators can change. Hence, for each spin Hamiltonian $\widehat{\mathcal{H}}^{\mathrm{s}}$, which describes any interaction, this question should be considered separately.

If the Hamiltonian $\widehat{\mathcal{H}}^{\mathrm{s}}$ is time-independent the eigenfunctions satisfying expression (1.118) depend on time as follows:

$$\Psi(S_i, t) = \Psi(S_i)\exp\left(-\frac{\mathrm{i}Et}{\hbar}\right). \qquad (1.120)$$

Substituting Eq. (1.120) into Eq. (1.118), we obtain the stationary Schrödinger equation to define the eigenvalues (energy levels E) and eigenvectors (the time-independent wave functions):

$$\widehat{\mathcal{H}}^s \Psi(S_i) = E\Psi(S_i). \tag{1.121}$$

Equation (1.121) is used to calculate energy levels and transition frequencies for nuclear or electron subsystems. For consideration dynamical changes in these subsystems one has to use Eq. (1.118). In order to find the spin Hamiltonian in a general view, it is necessary to analyze its components.

1.9.2 Density Matrix

Density matrix definition. In certain cases (the relaxation theory, pulse techniques in NQR, multiquantum NMR spectroscopy) the description of system by means of macroscopic magnetization is insufficient (for example, when the concept of a coherence of states is used). As it is known, if a system is in a state described by a wave function Ψ or state vector $|\Psi\rangle$ ("a pure state") then a measured value of some physical quantity Q is determined by the quantum mechanical average values $\langle Q\rangle$:

$$\langle Q\rangle = \langle\Psi|Q|\Psi\rangle.$$

Here $\hat{Q}$ is the operator corresponding to the classical quantity Q. Generally the wave function Ψ is not an eigenfunction of the operator $\hat{Q}$, but it is convenient to present it in the form of the superposition of N eigenfunctions (vectors of states) of a Hamiltonian $\widehat{\mathcal{H}}$ that does not depend on time:

$$|\Psi\rangle = \sum_{n=1}^{N} c_n|n\rangle. \tag{1.122}$$

For magnetic resonance the situation when $\widehat{\mathcal{H}}$ is a Hamiltonian of a spin system in a constant magnetic field is typical, and $\hat{Q}$ is, for example, the operator of traverse component of the spin vector entering into studied system. Thus,

$$\langle Q\rangle = \sum_{n=1}^{N}\sum_{m=1}^{N} c_n c_m^* \langle m|\hat{Q}|n\rangle.$$

Here the matrix elements $\langle m|\hat{Q}|n\rangle$ depend only on a chosen basis $|n\rangle$ and do not depend on the quantum-mechanical system state whereas coefficients c_n determine this state. In a real experiment there are "mixed" states: the system represents ensemble of subsystems, which are in their certain quantum-mechanical states. The result of measuring of the quantity Q is a statistical average on the ensemble, denoted, unlike quantum-mechanical one, by a line above averaged quantity:

$$\overline{\langle Q\rangle} = \sum_{n=1}^{N}\sum_{m=1}^{N} \overline{c_n c_m^*} \langle m|\hat{Q}|n\rangle.$$

The density matrix is defined as a matrix of elements ρ_{nm}:

$$\rho_{nm} = \overline{c_n c_m^*}.$$

The density matrix is considered as a matrix representation of an operator $\hat{\rho}$ which is called as the *operator of a probability density* (the word *probability* is usually omitted). Thus, the density matrix allows us to calculate "a classical" (experimentally measured) average of the quantity Q, if the representation of its operator in the basis of the Hamiltonian $\widehat{\mathcal{H}}$ eigenstates is known:

$$\overline{\langle Q\rangle} = \sum_{n=1}^{N}\sum_{m=1}^{N} \rho_{nm} \langle m|\hat{Q}|n\rangle$$

or, in more compact form

$$\overline{\langle Q\rangle} = Tr(\hat{\rho}\hat{Q}) \tag{1.123}$$

The variation of density matrix in time. To obtain the time dependence of Q, it is necessary to know the law of time variation of its matrix elements. It is easy to find that by substituting a state vector $|\Psi\rangle$ in the Schrödinger equation, expressed by Eq. (1.122). As a result of the transformations the following equation is:

$$\mathrm{i}h\frac{\partial\hat{\rho}(t)}{\partial t} = \widehat{\mathcal{H}}_1(t)\hat{\rho}(t) - \hat{\rho}(t)\widehat{\mathcal{H}}_1(t), \tag{1.124}$$

that is the analogue of the classical Liouville equation. In Eq. (1.124) $\widehat{\mathcal{H}}_1(t)$ is the full time-dependent Hamiltonian. For a time-independent Hamiltonian $\widehat{\mathcal{H}}$ from Schrödinger equation (1.117) it follows

$$\rho_{nm}(t) = \exp\left[-\mathrm{i}\frac{(E_m - E_n)t}{h}\right]\rho_{nm}(0),$$

that can be presented in more formal view as

$$\hat{\rho}(t) = \exp\left(\frac{-\mathrm{i}\widehat{\mathcal{H}}t}{h}\right)\hat{\rho}(0)\exp\left(\frac{\mathrm{i}\widehat{\mathcal{H}}t}{h}\right). \tag{1.125}$$

Density matrix in a rotating frame. As well as at the description of a magnetic resonance by a vector of macroscopic magnetization, it is expedient to go to a frame rotating with a frequency ω_R. With that the density matrix in a laboratory frame $\hat{\rho}(t)$ will be transformed by means of the operator of rotational displacement $\hat{R}$. The rotation operator around an axis ξ on an angle φ for a system with the total spin $\hat{F}$ looks as

$$\hat{R}_{\xi} = \exp(-\mathrm{i}\varphi\hat{F}_{\xi}). \tag{1.126}$$

For transformations in the rotating frame in magnetic resonance, the rotation around the axis z on the angle $\varphi = \omega_R t$ is carried out. Therefore a transformation of $\hat{\rho}(t)$ in a matrix in the rotating frame $\hat{\rho}_R(t)$ looks as

$$\hat{\rho}_R(t) = \exp(\mathrm{i}\omega\hat{F}_z t)\hat{\rho}(t)\exp(-\mathrm{i}\omega\hat{F}_z t). \tag{1.127}$$

Density matrix variation as a result of action of a radio-frequency pulse. The action of a radio-frequency pulse turning the magnetisation vector on the angle α around the x'-axis of the rotation frame is presented likewise:

$$\hat{\rho}_{R_+} = \exp(\mathrm{i}\alpha\hat{F}_{x'})\hat{\rho}_{R\_}\exp(-\mathrm{i}\alpha\hat{F}_{x'}). \tag{1.128}$$

Here $\hat{\rho}_{R\_}$ and $\hat{\rho}_{R_+}$ are the density matrices in the rotating frame before and after the radio-frequency pulse respectively.

Coherence of states. The density matrix element can be presented as

$$\rho_{nm} = \overline{|c_n||c_m^*|}\ \overline{\exp\left[\mathrm{i}\left(\alpha_n - \alpha_m\right)\right]}.$$

If the phases α_n and α_m are not related to each other then the exponential factor after averaging to be converted in zero and the element $\rho_{nm} = 0$. If ρ_{nm} is distinct from zero, it means a presence of a phase coherence of wave functions of states n and m, or, simply, a *coherence of states*.

The main properties of the density matrix can be derived from Eqs. (1.123), (1.124) and (1.125):

- the density matrix is a Hermite matrix ($\rho_{nm}^* = \rho_{nm}$);
- the sum of diagonal elements of the density matrix is equal to unity;
- in a stationary state the operator of the density matrix commutates with the Hamiltonian $\widehat{\mathcal{H}}$.

Its other properties can be easily obtained by decomposing the density matrix elements on the orthonormalized basis of wave functions, e.g. wave functions, which are the solutions of the stationary Schrödinger equation:

- raising the density matrix operator to the second power does not change it: $\widehat{\rho}^2 = \widehat{\rho}$;
- in the stationary state the density matrix is diagonal.

The explicit form of the density matrix operator for a spin system in the equilibrium state, characterized by the temperature T, is determined by the following way:

$$\widehat{\rho} = \frac{e^{-\widehat{\mathcal{H}}/(kT)}}{\mathrm{Tr}\, e^{-\widehat{\mathcal{H}}/(kT)}}.$$

Usually, to find the energy levels, frequencies and intensities of transitions one uses the wave function formalism, whereas when discussing various dynamical changes that occur in an ensemble of spin subsystems, the density matrix formalism is expedient.

1.10 Magnetic Dipole–Dipole Interactions

1.10.1 Hamiltonian of Dipole–Dipole Interactions

In quantum mechanics each mechanical quantity is associated with a linear self-conjugate operator. For example, the energy in quantum mechanics is associated with the Hamiltonian operator $\widehat{\mathcal{H}}$. The classical expression of the energy in the operator form is the *interaction Hamiltonian.* To find the Hamiltonian of any interaction it is necessary to find the classical expression describing it and to replace all classical quantities by corresponding operators.

If one considers particles (nuclei, electrons or ions) as classical magnetic dipoles, the energy of the dipole–dipole interaction between magnetic moments of particles i and j can be written as

$$E_{ij} = \frac{\vec{\mu}_i \cdot \vec{\mu}_j}{r_{ij}^3} - \frac{3(\mu_i \cdot \vec{r}_{ij})(\vec{\mu}_j \cdot \vec{r}_{ij})}{r_{ij}^5}, \tag{1.129}$$

where $\vec{\mu}_i$ and $\vec{\mu}_j$ are the magnetic moments of i-th and j-th particles; r_{ij} is the distance between i-th and j-th particles; $\vec{r}_{ij}$ is the radius-vector connecting two particles. To define the energy of N particles, one has to sum Eq. (1.129) over all pairs taking into account that, first, when summing every pair i and j is counted twice, and second, a particle does not interact with itself, that means $(i \neq j)$:

$$E = \frac{1}{2} \sum_{i=1}^{N} \sum_{j=1}^{N} \frac{\vec{\mu}_i \cdot \vec{\mu}_j}{r_{ij}^3} - \frac{3(\vec{\mu}_i \cdot \vec{r}_{ij})(\vec{\mu}_j \cdot \vec{r}_{ij})}{r_{ij}^5}. \tag{1.130}$$

Replacing the vectors $\vec{\mu}_i$, $\vec{\mu}_j$ in Eq. (1.130) by the corresponding operators in the form $\gamma_{i(j)}\hbar\hat{\vec{I}}_{i(j)}$ (see 1.13)[4], and energy E by the energy operator $\widehat{\mathcal{H}}$, one obtains the dipole–dipole interaction Hamiltonian:

[4] Operator $\hat{\vec{I}}$ is a nuclear or electron spin operator.

$$\widehat{\mathcal{H}} = \frac{\hbar^2}{2} \sum_{i=1}^{N} \sum_{j=1}^{N} \gamma_i \gamma_j \left(\frac{\hat{\vec{I}}_i \cdot \hat{\vec{I}}_j}{r_{ij}^3} - \frac{3(\hat{\vec{I}}_i \cdot \vec{r}_{ij})(\hat{\vec{I}}_j \cdot \vec{r}_{ij})}{r_{ij}^5} \right). \tag{1.131}$$

In the Cartesian coordinate system x, y, z scalar products of operators in Eq. (1.131) can be written as

$$\begin{aligned}
\hat{\vec{I}}_i \cdot \hat{\vec{I}}_j &= \hat{I}_{ix}\hat{I}_{jx} + \hat{I}_{iy}\hat{I}_{jy} + \hat{I}_{iz}\hat{I}_{jz}, \\
\hat{\vec{I}}_i \cdot \vec{r}_{ij} &= \hat{I}_{ix}x_{ij} + \hat{I}_{iy}y_{ij} + \hat{I}_{iz}z_{ij}, \\
\hat{\vec{I}}_j \cdot \vec{r}_{ij} &= \hat{I}_{jx}x_{ij} + \hat{I}_{jy}y_{ij} + \hat{I}_{jz}z_{ij}.
\end{aligned} \tag{1.132}$$

Substituting Eq. (1.132) into Eq. (1.131), one has:

$$\begin{aligned}
\widehat{\mathcal{H}} = \frac{\hbar^2}{2} \sum_{i=1}^{N} \sum_{j=1}^{N} \frac{\gamma_i \gamma_j}{r_{ij}^5} \Big(&\hat{I}_{ix}\hat{I}_{jx}(r_{ij}^2 - 3x_{ij}^2) + \hat{I}_{iy}\hat{I}_{jy}(r_{ij}^2 - 3y_{ij}^2) + \hat{I}_{iz}\hat{I}_{jz}(r_{ij}^2 - 3z_{ij}^2) \\
&- 3(\hat{I}_{ix}\hat{I}_{jy} + \hat{I}_{iy}\hat{I}_{jx})x_{ij}y_{ij} - 3(\hat{I}_{ix}\hat{I}_{jz} + \hat{I}_{iz}\hat{I}_{jx})x_{ij}z_{ij} - 3(\hat{I}_{iy}\hat{I}_{jz} + \hat{I}_{iz}\hat{I}_{jy})y_{ij}z_{ij} \Big).
\end{aligned} \tag{1.133}$$

Equation (1.133) can be rewritten in the matrix form:

$$\begin{aligned}
\widehat{\mathcal{H}} &= \frac{\hbar^2}{2} \sum_{i=1}^{N} \sum_{j=1}^{N} \frac{\gamma_i \gamma_j}{r_{ij}^5} (\hat{I}_{ix}, \hat{I}_{iy}, \hat{I}_{iz}) \begin{pmatrix} r_{ij}^2 - 3x_{ij}^2 & -3x_{ij}y_{ij} & -3x_{ij}z_{ij} \\ -3x_{ij}y_{ij} & r_{ij}^2 - 3y_{ij}^2 & -3y_{ij}z_{ij} \\ -3x_{ij}z_{ij} & -3y_{ij}z_{ij} & r_{ij}^2 - 3z_{ij}^2 \end{pmatrix} \begin{pmatrix} \hat{I}_{jx} \\ \hat{I}_{jy} \\ \hat{I}_{jz} \end{pmatrix} \\
&= \frac{\hbar^2}{2} \sum_{i=1}^{N} \sum_{j=1}^{N} \frac{\gamma_i \gamma_j}{r_{ij}^5} \cdot \hat{\vec{I}}_i \cdot \mathsf{D}_{ij} \cdot \hat{\vec{I}}_j.
\end{aligned}$$

Here D_{ij} are the second-rank tensors:

$$\mathsf{D}_{ij} = \begin{pmatrix} r_{ij}^2 - 3x_{ij}^2 & -3x_{ij}y_{ij} & -3x_{ij}z_{ij} \\ -3x_{ij}y_{ij} & r_{ij}^2 - 3y_{ij}^2 & -3y_{ij}z_{ij} \\ -3x_{ij}z_{ij} & -3y_{ij}z_{ij} & r_{ij}^2 - 3z_{ij}^2 \end{pmatrix}. \tag{1.134}$$

In the more compact form the components of the tensor of the dipole–dipole interactions can be written as

$$D_{ij}^{\alpha,\beta} = \delta_{\alpha,\beta} r_{ij}^2 - 3\alpha_i \beta_j,$$

where α and β mean x, y and z. As it is seen from Eq. (1.134), D_{ij} are the symmetric second-rank tensors with a zero trace. Such tensors characterize anisotropic interactions, which do not contain isotropic contributions. Thus, the dipole–dipole interaction Hamiltonian depends on the space coordinates of interacting particles $D_{ij}^{\alpha,\beta}$ and the product of the operators $\hat{I}_{i\alpha}\hat{I}_{j\beta}$.

In the spherical coordinate system $x_{ij} = r_{ij} \sin\theta_{ij} \cos\phi_{ij}$, $y_{ij} = r_{ij} \sin\theta_{ij} \sin\phi_{ij}$, $z_{ij} = r_{ij} \cos\theta_{ij}$, and the tensors D_{ij} are written as

$$\mathsf{D}_{ij} = r_{ij}^2 \begin{pmatrix} 1 - 3\sin^2\theta_{ij}\cos^2\phi_{ij} & -\frac{3}{2}\sin^2\theta_{ij}\sin 2\phi_{ij} & -\frac{3}{2}\sin 2\theta_{ij}\cos\phi_{ij} \\ -\frac{3}{2}\sin^2\theta_{ij}\sin 2\phi_{ij} & 1 - 3\sin^2\theta_{ij}\sin^2\phi_{ij} & -\frac{3}{2}\sin 2\theta_{ij}\sin\phi_{ij} \\ -\frac{3}{2}\sin 2\theta_{ij}\cos\phi_{ij} & -\frac{3}{2}\sin 2\theta_{ij}\sin\phi_{ij} & 1 - 3\cos^2\theta_{ij} \end{pmatrix}.$$

If one takes into account that $\cos\phi = (e^{\mathrm{i}\phi} + e^{-\mathrm{i}\phi})/2$ and $\sin\phi = (e^{\mathrm{i}\phi} - e^{-\mathrm{i}\phi})/(2\mathrm{i})$, whereas $\hat{I}_x = (\hat{I}_+ + \hat{I}_-)/2$ and $\hat{I}_y = (\hat{I}_+ - \hat{I}_-)/(2\mathrm{i})$, the dipole–dipole interaction Hamiltonian (1.133) in the spherical coordinate system can be written in the following form:

$$\widehat{\mathcal{H}}_{\mathrm{d}} = \frac{\hbar^2}{2} \sum_{i=1}^{N} \sum_{j=1}^{N} \frac{\gamma_i \gamma_j}{r_{ij}^3} (\hat{A}_{ij} + \hat{B}_{ij} + \hat{C}_{ij} + \hat{D}_{ij} + \hat{E}_{ij} + \hat{F}_{ij}), \tag{1.135}$$

where

$$\begin{aligned} \hat{A}_{ij} &= \hat{I}_{iz}\hat{I}_{jz}(1 - 3\cos^2\theta_{ij}); \\ \hat{B}_{ij} &= -\frac{1}{4}(\hat{I}_{i+}\hat{I}_{j-} + \hat{I}_{i-}\hat{I}_{j+})(1 - 3\cos^2\theta_{ij}); \\ \hat{C}_{ij} &= -\frac{3}{2}(\hat{I}_{i+}\hat{I}_{jz} + \hat{I}_{iz}\hat{I}_{j+}) \sin\theta_{ij}\cos\theta_{ij} \exp(-\mathrm{i}\phi_{ij}); \\ \hat{D}_{ij} &= -\frac{3}{2}(\hat{I}_{i-}\hat{I}_{jz} + \hat{I}_{iz}\hat{I}_{j-}) \sin\theta_{ij}\cos\theta_{ij} \exp(\mathrm{i}\phi) = \hat{C}_{ij}^*; \\ \hat{E}_{ij} &= -\frac{3}{4}(\hat{I}_{i+}\hat{I}_{j+}) \sin^2\theta_{ij} \exp(-2\mathrm{i}\phi_{ij}); \\ \hat{F}_{ij} &= -\frac{3}{4}(\hat{I}_{i-}\hat{I}_{j-}) \sin^2\theta_{ij} \exp(2\mathrm{i}\phi_{ij}) = \hat{E}_{ij}^*. \end{aligned} \tag{1.136}$$

Let us note that the operator $\hat{A}_{ij} \sim \hat{I}_{zi}\hat{I}_{zj}$ commutates with the Zeeman Hamiltonian $\widehat{\mathcal{H}}_{Z,ij} \sim \gamma_i \hat{I}_{zi} + \gamma_j \hat{I}_{zj}$, whereas the operators $\hat{C}_{ij}$, $\hat{D}_{ij}$, $\hat{E}_{ij}$, $\hat{F}_{ij}$ do not commutate with it; the operator $\hat{B}_{ij}$ commutes with the Zeeman Hamiltonian if the spins are equivalent and does not if the spins are different. Operators, which commute with the Zeeman Hamiltonian, are called *secular*. For such an operator one can always find a set of wave functions (representation), which is simultaneously a set of the eigenfunction of the Zeeman Hamiltonian. Then, the only non-zero matrix elements of these operators are those ones that correspond to the same energy level of the Zeeman Hamiltonian (diagonal matrix elements). In the case of equivalent spins, the secular terms $(\hat{A}_{ij} + \hat{B}_{ij})$ of the dipole–dipole interaction Hamiltonian can be written as

$$\hat{A}_{ij} + \hat{B}_{ij} = \frac{1}{2}\left(3\hat{I}_{iz}\hat{I}_{jz} - \hat{\vec{I}}_i \cdot \hat{\vec{I}}_j\right)(1 - 3\cos^2\theta),$$

taking into account that

$$\hat{\vec{I}}_i \cdot \hat{\vec{I}}_j = \hat{I}_{ix}\hat{I}_{jx} + \hat{I}_{iy}\hat{I}_{jy} + \hat{I}_{iz}\hat{I}_{jz} = \hat{I}_{iz}\hat{I}_{jz} + \frac{1}{2}\left(\hat{I}_{i+}\hat{I}_{j-} + \hat{I}_{i-}\hat{I}_{j+}\right).$$

Non-commuting operators are called *non-secular*. Their diagonal matrix elements, which can be calculated using the eigenfunctions of the Zeeman Hamiltonian, are equal to zero. Secular and non-secular terms of the Hamiltonian $\widehat{\mathcal{H}}_{\mathrm{d}}$, as we will see further, impact on the NMR (EPR or NQR) spectrum formation in different ways.

It is convenient to write $\widehat{\mathcal{H}}_{\mathrm{d}}$ as a sum of products $F_{ij}^q \hat{K}_{ij}^q$, where $\hat{K}_{ij}^q$ depend only on spin operators, whereas F_{ij}^q depend on angles θ and ϕ, which determine the orientation of the radius-vector $\vec{r}_{ij}$ relative to the external magnetic field direction:

$$\widehat{\mathcal{H}}_{\mathrm{d}} = \sum_i \sum_{j \neq i} \hbar^2 \frac{\gamma_i \gamma_j}{r_{ij}^3} \sum_{q=-2}^{2} F_{ij}^q \hat{K}_{ij}^q, \tag{1.137}$$

where

$$\begin{aligned}
\hat{K}_{ij}^0 &= \frac{1}{2}\left(\hat{I}_{iz}\hat{I}_{jz} - \frac{1}{4}\left(\hat{I}_{i+}\hat{I}_{j-} + \hat{I}_{i-}\hat{I}_{j+}\right)\right), \; F_{ij}^0 = 1 - 3\cos^2\theta_{ij}, \\
\hat{K}_{ij}^{\pm 1} &= -\frac{3}{4}\left(\hat{I}_{i\pm}\hat{I}_{jz} + \hat{I}_{iz}\hat{I}_{j\pm}\right), \; F_{ij}^{\pm 1} = \sin\theta_{ij}\cos\theta_{ij}\exp(\mp \mathrm{i}\phi_{ij}), \\
\hat{K}_{ij}^{\pm 2} &= -\frac{3}{8}\left(\hat{I}_{i\pm}\hat{I}_{j\pm}\right), \; F_{ij}^{\pm 2} \sin^2\theta_{ij}\exp(\mp 2\mathrm{i}\phi_{ij}).
\end{aligned} \tag{1.138}$$

Equations (1.136) and (1.138) correspond to each other by the following way:

$$\hat{A}_{ij} + \hat{B}_{ij} = F_{ij}^0 \hat{K}_{ij}^0,$$

$$\hat{C}_{ij} = F_{ij}^1 \hat{K}_{ij}^1, \quad \hat{D}_{ij} = F_{ij}^{-1} \hat{K}_{ij}^{-1},$$

$$\hat{E}_{ij} = F_{ij}^2 \hat{K}_{ij}^2, \quad \hat{F}_{ij} = F_{ij}^{-2} \hat{K}_{ij}^{-2}.$$

The Hamiltonian $\widehat{\mathcal{H}}_{\mathrm{d}}$ is multiparticle, and there is no analytic solution of the Schrödinger equation with such a Hamiltonian. To define the energy levels it is necessary to apply approximate methods. The energy of dipole–dipole interactions can be estimated from the relation $\gamma_i \gamma_j \hbar^2 / r_{ij}^3$ (see Eq. 1.131), that corresponds to the interaction energy of the spin with an external field. The latter is characterized by the induction value of about several tenth of mT. Hence, if the applied field is much higher than 1 mT, then to calculate corrections to the energy levels due to the dipole–dipole interactions of nuclear or electron magnetic moments, the perturbation theory can be applied. However, even in this case, and even applying modern computer technologies, it is possible to find energy levels only for a limited number of interacting particles.

1.10.2 Energy Levels for an Isolated Pair of Nuclei

For an isolated pair of nuclei it is possible to write the double-particles Hamiltonian of the dipole–dipole interactions within the zero-order approximation, neglecting interactions of this pair with other nuclei:

$$\widehat{\mathcal{H}}_{\mathrm{d},ij} = \frac{\gamma_i\gamma_j\hbar^2}{r_{ij}^3}(\hat{A}_{ij} + \hat{B}_{ij} + \hat{C}_{ij} + \hat{D}_{ij} + \hat{E}_{ij} + \hat{F}_{ij}),$$

here $\hat{A}_{ij}$, $\hat{B}_{ij}$, $\hat{C}_{ij}$, $\hat{D}_{ij}$, $\hat{E}_{ij}$ and $\hat{F}_{ij}$ are defined by Eq. (1.136). Corrections to the energy levels of the Zeeman Hamiltonian due to dipole–dipole interactions can be calculated using following formulas: within the first-order approximation

$$\Delta E_k^{(1)} = \langle k|\widehat{\mathcal{H}}_{\mathrm{d},ij}|k\rangle, \qquad (1.139)$$

within second-order approximation

$$\Delta E_k^{(2)} = \sum_{k'} \frac{|\langle k|\widehat{\mathcal{H}}_{\mathrm{d},ij}|k'\rangle|^2}{E_k - E_k'}. \qquad (1.140)$$

Here $\langle k|\widehat{\mathcal{H}}_{\mathrm{d},ij}|k\rangle$ and $\langle k|\widehat{\mathcal{H}}_{\mathrm{d},ij}|k'\rangle$ are the matrix elements of the dipole–dipole interactions Hamiltonian calculated using the eigenfunctions of the non-perturbed Zeeman Hamiltonian. These wave functions can be easily found for a two-spin system. In this case the Zeeman Hamiltonian is

$$\widehat{\mathcal{H}}_{\mathrm{Z},ij} = -\hbar B_0(\gamma_i\hat{I}_{iz} + \gamma_j\hat{I}_{jz}). \qquad (1.141)$$

Solving the Schrödinger equation with the Hamiltonian (1.141), one obtains

$$E_{ij,k} = -\hbar B_0(\gamma_i m_i + \gamma_j m_j).$$

Here k runs over the energy levels; m_i and m_j are magnetic quantum numbers for i-th and j-th nuclei, respectively.

If $\gamma_i \neq \gamma_j$ (different nuclear species), then all the $(2I_i + 1)(2I_j + 1)$ energy levels that correspond to the Hamiltonian (1.141) are non-degenerate. The eigenfunctions of such a system are the product of the wave functions of individual particles.

In particular, if the spins of the nuclei i and j are equal to 1/2, then for four non-degenerate energy levels (Fig. 1.14) the eigenfunctions are:

$$\begin{array}{ll} -E_k & \Psi_k \\ \hbar B_0 \dfrac{\gamma_i + \gamma_j}{2}, & \varphi^{(i)}_{+1/2}\varphi^{(j)}_{+1/2}, \\ -\hbar B_0 \dfrac{\gamma_i - \gamma_j}{2}, & \varphi^{(i)}_{+1/2}\varphi^{(j)}_{-1/2}, \\ -\hbar B_0 \dfrac{-\gamma_i + \gamma_j}{2}, & \varphi^{(i)}_{-1/2}\varphi^{(j)}_{+1/2}, \\ \hbar B_0 \dfrac{-\gamma_i - \gamma_j}{2}; & \varphi^{(i)}_{-1/2}\varphi^{(j)}_{-1/2}. \end{array}$$

If $\gamma_i = \gamma_j = \gamma$ (the equivalent nuclear species), then the two-spin Hamiltonian, which corresponds to the interaction of the magnetic moments with an external magnetic field, can be written as follows

$$\widehat{\mathcal{H}}_{Z,ij} = -\hbar B_0 \gamma (I_{iz} + I_{jz}). \tag{1.142}$$

In this case only two energy levels, which correspond to the minimal and maximal values of the projection of the total spin $M_{max} = \pm(I_i + I_j)$, are non-degenerate, whereas the other levels are degenerate. The level corresponding to $M = 0$ has the greatest degeneration equal to $(I_i + I_j + 1)$. In order to find the eigenfunctions of such a two-spin system it is necessary to compose a product of one-particle functions with maximal value of the total spin projection, and acting by the operator $\hat{I}_- = \hat{I}_{i-} + \hat{I}_{j-}$ to find the $(2M_{max} + 1)$ wave functions. Since the energy level, which corresponds to $(M_{max} - 1)$, is double degenerate, the second wave function can be found from the orthogonality condition. After that, one acts on the found by such a way wave function by the operator $\hat{I}_-$, and so on.

For example, if the spins are equivalent and $I_1 = I_2 = 1/2$, the maximal projection of the total spin $M_{max} = 1$, and $\Psi_1 = \varphi^{(i)}_{+1/2}\varphi^{(j)}_{+1/2}$ is the wave function, corresponding to this projection. Then

$$\begin{array}{rl} \hat{I}_-\Psi_1 & = (\hat{I}_{i-} + \hat{I}_{j-}) \left(\varphi^{(i)}_{+1/2}\varphi^{(j)}_{+1/2}\right) \\ \Downarrow & \qquad \Downarrow \\ \sqrt{2}\Psi_0^{\text{sym}} & = \varphi^{(i)}_{+1/2}\varphi^{(j)}_{-1/2} + \varphi^{(i)}_{-1/2}\varphi^{(j)}_{+1/2}. \end{array}$$

Let us remind that

$$\hat{I}_-\psi_m = \sqrt{I(I+1) - m(m-1)}\,\psi_{m-1}.$$

Thereby,

$$\Psi_0^{\text{sym}} = \frac{1}{\sqrt{2}} \left(\varphi^{(i)}_{+1/2}\varphi^{(j)}_{-1/2} + \varphi^{(i)}_{-1/2}\varphi^{(j)}_{+1/2}\right). \tag{1.143}$$

The superscript of the wave function indicates that the obtained function is symmetric relative to permutations of particles. By the similar way one can obtain the wave function Ψ_{-1}.

There is another wave function, which corresponds to the energy level with $M = 0$ and is orthogonal to Ψ_0^{sym} in the form (1.143):

$$\Psi_0^{\mathrm{asym}} = \frac{1}{\sqrt{2}}\left(\varphi_{+1/2}^{(i)}\varphi_{-1/2}^{(j)} - \varphi_{-1/2}^{(i)}\varphi_{+1/2}^{(j)}\right).$$

The superscript of the wave function indicates that the function is antisymmetric relative to permutations of particles.

Hereby, the wave functions and the corresponding energy levels (see Fig. 1.15), which describe the interaction of two identical particles with the spin number equal to 1/2, are of the following form:

$$\begin{array}{ll}
-E_k & \Psi_k \\
\gamma\hbar B_0, & \varphi_{+1/2}^{(i)}\varphi_{+1/2}^{(j)}, \\
0, & \frac{1}{\sqrt{2}}\left(\varphi_{+1/2}^{(i)}\varphi_{-1/2}^{(j)} + \varphi_{-1/2}^{(i)}\varphi_{+1/2}^{(j)}\right), \\
0, & \frac{1}{\sqrt{2}}\left(\varphi_{+1/2}^{(i)}\varphi_{-1/2}^{(j)} - \varphi_{-1/2}^{(i)}\varphi_{+1/2}^{(j)}\right), \\
-\gamma\hbar B_0, & \varphi_{-1/2}^{(i)}\varphi_{-1/2}^{(j)}.
\end{array}$$

Now it is possible to define matrix elements of the dipole–dipole interaction Hamiltonian and corrections to the Zeeman energy levels. Within the first-order approximation of the perturbation theory (1.139) only secular terms of the Hamiltonian $\widehat{\mathcal{H}}_{\mathrm{d}}$ contribute to ΔE_k, because matrix elements of these operators, determined by using the wave functions belonging to the same energy level, are non-zero.

If the spins belong to *different species*, then only the operator $\hat{A}_{ij} \sim \hat{I}_{iz}\hat{I}_{jz}$ in the form of (1.136) is secular and has non-zero matrix elements between the states with the same energy, and hence:

$$\Delta E_k = \frac{\gamma_i\gamma_j\hbar^2}{r_{ij}^3}\langle k|\hat{A}_{ij}|k\rangle = \frac{\gamma_i\gamma_j\hbar^2}{r_{ij}^3}\langle k|\hat{I}_{iz}\hat{I}_{jz}|k\rangle(1 - 3\cos^2\theta_{ij})$$
$$= \frac{\gamma_i\gamma_j\hbar^2}{r_{ij}^3}m_i m_j(1 - 3\cos^2\theta_{ij}).$$

It is worth noting that if m_i and m_j for some k-th level are of the same sign (positive or negative), the corresponding corrections are positive (at $\gamma_i\gamma_j > 0$). If m_i and m_j are of different sign, the corrections are negative. For example, for particles with spin 1/2 the corrections to all energy levels are of the same value of $(\gamma_i\gamma_j\hbar^2(1-3\cos^2\theta_{ij})/(4r_{ij}^3))$, but for the upper and lower levels they are positive, whereas for the middle levels they are negative (see Fig. 1.14). The terms $\hat{B}_{ij}$, $\hat{C}_{ij}$, $\hat{D}_{ij}$, $\hat{E}_{ij}$ and $\hat{F}_{ij}$ make contributions to the corrections to the Zeeman energy levels within the second-order approximation of the perturbation theory. The matrix elements of these operators, determined by the wave functions, corresponding to the different energy levels, are non-zero. So, for the operator $\hat{B}$, the matrix element is non-zero if the difference between mag-

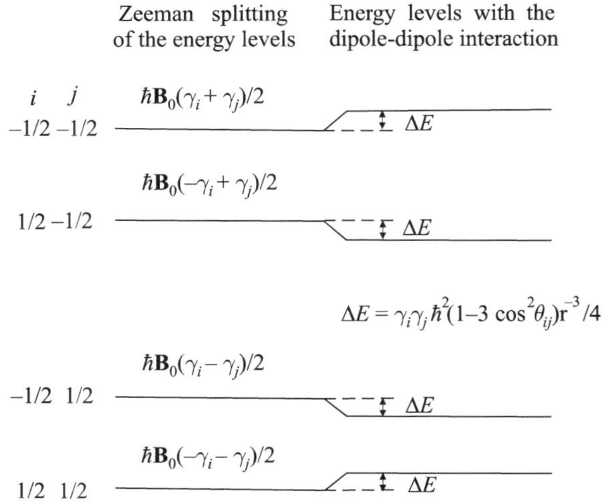

Fig. 1.14 The energy level diagram in presence of dipole–dipole interactions: different nuclear species

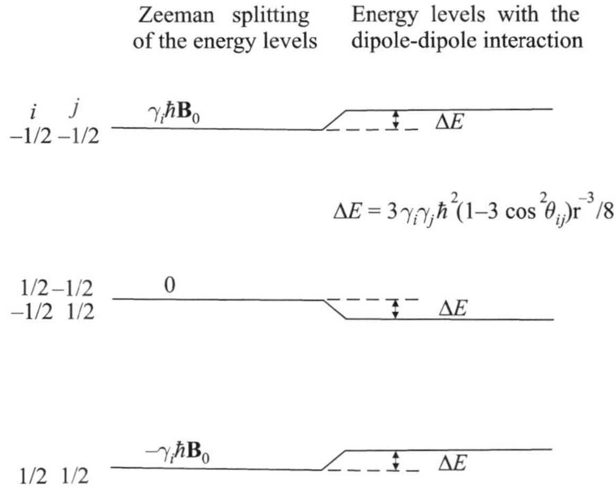

Fig. 1.15 The energy level diagram in presence of dipole–dipole interactions: equivalent nuclear species

netic quantum numbers of individual spins is ± 1: $\langle m_i, m_j|\hat{B}|m_i \pm 1, m_j \mp 1\rangle$ (see Eq. 1.136). Each of operators $\hat{C}$ and $\hat{D}$ has non-zero matrix elements between states if the difference between quantum numbers of the spin, corresponding to these states, is unity: $\langle m_i, m_j|\hat{C}|m_i - 1, m_j\rangle$, $\langle m_i, m_j|\hat{C}|m_i, m_j - 1\rangle$, $\langle m_i, m_j|\hat{D}|m_i + 1, m_j\rangle$ and $\langle m_i, m_j|\hat{D}|m_i, m_j + 1\rangle$. Operators $\hat{E}$ and $\hat{F}$ have non-zero matrix elements only between the states, which differ by simultaneous change of m_i and m_j by $+1$ (for $\hat{E}$) or by -1 (for $\hat{F}$): $\langle m_i, m_j|\hat{E}|m_i - 1, m_j - 1\rangle$, $\langle m_i, m_j|\hat{F}|m_i + 1, m_j + 1\rangle$.

However, as a rule, the corrections to the energy levels calculated within the second-order approximation of the perturbation theory are much smaller than those ones calculated within the first-order approximation. Indeed, the order of magnitude of non-diagonal matrix elements of the dipole–dipole interaction Hamiltonian, as well as diagonal, is $(\gamma_i \gamma_j)\hbar^2/r_{ij}^3$, however, when calculating the second-order corrections they are multiplied by the coefficient

$$\frac{(\gamma_i \gamma_j)\hbar^2/r_{ij}^3}{E_{im} - E_{i\,m-1}},$$

where $E_{im} - E_{i\,m-1} = \gamma_i \hbar B_0$ is a difference between energies corresponding to the non-perturbed Hamiltonian of the Zeeman interaction (see Eq. 1.140). For example, in a magnetic field of 1 T the second-order corrections are in $10^{-3} \div 10^{-4}$ times smaller than the first-order ones. That is why the contribution of the second-order corrections into the energy level values usually is not taken into account.

This analysis demonstrates that in the case of different nuclear species in order to define the energy levels it is sufficient to consider a truncated part of the dipole–dipole interaction Hamiltonian:

$$\widehat{\mathcal{H}}_{ij} = \widehat{\mathcal{H}}_{\mathrm{Z},ij} + \widehat{\mathcal{H}}_{\mathrm{d},ij} = -\hbar \left(\gamma_i \hat{I}_{zi} + \gamma_j \hat{I}_{zj} \right) B_0 + \hbar^2 \gamma_i \gamma_j \frac{\hat{I}_{zi}\, \hat{I}_{zj} (1 - 3\cos\theta_{ij})}{r_{ij}^3}. \tag{1.144}$$

Here $\widehat{\mathcal{H}}_{\mathrm{Z},ij}$ is the Zeeman Hamiltonian; $\widehat{\mathcal{H}}_{\mathrm{d},ij}$ is the truncated part of the dipole–dipole interaction Hamiltonian.

If the spins belong to the *equivalent nuclear species* the non-zero matrix elements correspond to the operators $\hat{A}_{ij} \sim \hat{I}_{iz}\hat{I}_{jz}$ and $\hat{B}_{ij} \sim \hat{I}_{i+}\hat{I}_{j-} + \hat{I}_{i-}\hat{I}_{j+}$ between the states with the same energy (see Eq. 1.136). Now the operator $\hat{B}_{ij}$ does not have non-diagonal matrix elements. The diagonal ones calculated using the wave functions of the non-perturbed Zeeman Hamiltonian are equal to

$$\Delta E_k = \frac{\gamma_i \gamma_j \hbar^2}{r_{ij}^3} \langle k | \hat{B}_{ij} | k \rangle = -\frac{\gamma_i \gamma_j \hbar^2}{4 r_{ij}^3} \left(\langle k | \hat{I}_{i+} \hat{I}_{j-} | k \rangle + \langle k | \hat{I}_{i-} \hat{I}_{j+} | k \rangle \right) (1 - 3\cos^2\theta_{ij}).$$

In other words, within the first-order approximation of the perturbation theory for the spins of the same nuclear species there are additional corrections to the Zeeman energy levels due to the operator $\hat{B}_{ij}$. Let us note that for spins equal to 1/2 the corrections to the energy level corresponding to the total spin 0 and the wave function Ψ^{asym}, are equal to zero, i.e. this level remains unshifted in the presence of dipole–dipole interactions (see Fig. 1.15). Now the terms $\hat{C}_{ij}$, $\hat{D}_{ij}$, $\hat{E}_{ij}$ and $\hat{F}_{ij}$ make contribution to corrections to the Zeeman energy levels in the second-order perturbation theory.

Matrix elements of the operators $\hat{A}_{ij}$ and $\hat{B}_{ij}$, $\hat{C}_{ij}$, $\hat{D}_{ij}$, $\hat{E}_{ij}$ and $\hat{F}_{ij}$, calculated using the wave functions of the zero-order approximation, are listed in Table 1.2. In the upper row and the left column there are wave functions of the Zeeman Hamiltonian, which are used to calculate the matrix elements. The second-order corrections to the energy level that corresponds to the wave function Ψ^{asym} are equal to zero.

Table 1.2 Matrix elements of the $\hat{A}_{ij}$, $\hat{B}_{ij}$, $\hat{C}_{ij}$, $\hat{D}_{ij}$, $\hat{E}_{ij}$ and $\hat{F}_{ij}$ operators calculated within the zero-order approximation of the perturbation theory for equivalent nuclear species

Wave function	Ψ_1	Ψ_0^{sym}	Ψ_{-1}	Ψ_0^{asym}
Ψ_1	$\frac{1}{4}(1-3\cos^2\theta)$	$-\frac{3}{2\sqrt{2}}(e^{i\Psi}\sin\theta\cos\theta)$	$-\frac{3}{4}(e^{2i\Psi}\sin^2\theta)$	0
Ψ_0^{sym}	$-\frac{3}{2\sqrt{2}}(e^{-i\Psi}\sin\theta\cos\theta)$	$-\frac{1}{2}(1-3\cos^2\theta)$	$\frac{3}{2\sqrt{2}}(e^{i\Psi}\sin\theta\cos\theta)$	0
Ψ_{-1}	$-\frac{3}{4}(e^{-2i\Psi}\sin^2\theta)$	$\frac{3}{2\sqrt{2}}(e^{-i\Psi}\sin\theta\cos\theta)$	$\frac{1}{4}(1-3\cos^2\theta)$	0
Ψ_0^{asym}	0	0	0	0

The same result can be obtained by applying the group theory. One of the main postulates of this theory is that matrix elements calculated using wave functions of different symmetry are equal to zero. As the wave functions Ψ_1, Ψ_0^{sym}, Ψ_{-1} are symmetrical relative to the particle permutation, whereas the wave function Ψ_0^{asym} is antisymmetrical, the matrix elements calculated using these functions are equal to zero within any-order approximation of the perturbation theory. Like in the case of different spins, corrections to other energy levels are negligible.

Let us introduce a concept of *local field operator* $\hat{B}_{\text{loc}}$ created by the spin i (or j) at the nucleus j (or i) site, i.e.

$$\hat{B}_{i\,\text{loc}} = -\frac{\hbar\gamma_j \hat{I}_{zj}(1-3\cos\theta_{ij})}{r_{ij}^3}, \quad \hat{B}_{j\,\text{loc}} = -\frac{\hbar\gamma_i \hat{I}_{zi}(1-3\cos\theta_{ij})}{r_{ij}^3}. \tag{1.145}$$

Then the two-particle Hamiltonian (1.144) can be represented as two one-particle ones:

$$\widehat{\mathcal{H}}_i = -\hbar\gamma_i \hat{I}_{zi}(B_0 + \hat{B}_{i\,\text{loc}}),$$

$$\widehat{\mathcal{H}}_j = -\hbar\gamma_j \hat{I}_{zj}(B_0 + \hat{B}_{j\,\text{loc}}).$$

If one replaces $\hat{I}_z$ by its eigenvalue, then from Eq. (1.145) one obtains the eigenvalue of the $\hat{B}_{i\text{ or }j\text{ loc}}$ operator:

$$B_{i\text{ or }j\text{ loc}} = -\frac{\hbar\gamma_{j\text{ or }i}m_{j\text{ or }i}(1-3\cos\theta_{ij})}{r_{ij}^3}. \tag{1.146}$$

Magnetic quantum number m may be equal to various values from $-I$ to $+I$, and hence the local field may be directed both along and against the $\vec{B}_0$ field vector. Depending on the m sign in (1.146) the total field can be both greater and less than B_0. Hence, resonance conditions are satisfied at several frequency values.

Of course, introduction of the local field concept does not add any new information concerning NMR spectra. However, it is rather useful for visual physical interpretations of the influence of dipole–dipole interactions on NMR spectra and allows us to reduce the many-particle Hamiltonian to the single-particle one.

1.11 Electron–Nucleus Interactions

In magnetic resonance experiments electron–nucleus interactions (or hyperfine interactions) are of primary importance in such phenomena as dynamic nucleus polarization, paramagnetic and nuclear magnetic relaxation, pseudo-contact chemical shift, scalar nucleus–nucleus interactions. Hyperfine interactions are used for recording the hyperfine structure of electron paramagnetic and double electron–nucleus resonances, rotatory and optical spectra (hyperfine structure of spectral lines).

Let us consider a semiclassical model of magnetic interactions between an electron and a nucleus. According to classical representations the orbital motion of an electron is equivalent to a current circuit. The current creates in the space, and at the nucleus site as well, a magnetic field and the nucleus magnetic moment interacts with this field. This interaction between nuclear and electron spins provides the contribution to electron–nucleus interactions. At long distances the electron and the nucleus interact as point magnetic dipoles, whereas at short distances there is a specific so-called *contact interaction*. The short distance means that it is comparable to the electrodynamic electron radius, which is about $e^2/(\epsilon_0 m_e) = 3 \cdot 10^{-15}$ m. So, the Hamiltonian of the electron–nucleus interaction or the *hyperfine interaction Hamiltonian* ($\widehat{\mathcal{H}}_{hf}$) can be represented as a sum of three terms:

$$\widehat{\mathcal{H}}_{hf} = \widehat{\mathcal{H}}_{orb} + \widehat{\mathcal{H}}_{dd} + \widehat{\mathcal{H}}_{cont}.$$

Let us analyze each term.

1.11.1 Interaction of the Nucleus with the Orbital Motion of the Electron

An electron moving along the orbit with the velocity $\vec{v}$ creates at the point $\vec{r}\,'$ a magnetic field $\vec{B}$:

$$\vec{B} = -\frac{\mu_0}{4\pi}\frac{e\vec{v} \times \vec{r}\,'}{r'^3}, \tag{1.147}$$

where $-e$ is the electron charge. In the coordinate system related to the nucleus, the radius vector $\vec{r}$ directs from the nucleus to the electron (Fig. 1.16). That is why taking into account that the electron momentum $\vec{p} = m_e\vec{v}$ (m_e is the electron mass) and $\vec{r}\,' = -\vec{r}$, instead of Eq. (1.147) one ought to write:

$$\vec{B}_{orb} = \frac{\mu_0}{4\pi}\frac{e\vec{p} \times \vec{r}}{m_e r^3}. \tag{1.148}$$

Finally, it is necessary to take into account the quantum mechanical nature of the momentum $\vec{p}$ and its moment of momentum (in units $\hbar$):

$$\hbar\hat{\vec{l}} = \vec{r} \times \hat{\vec{p}}. \tag{1.149}$$

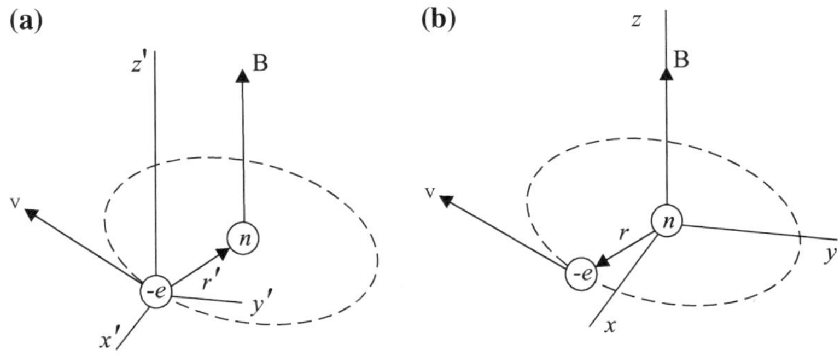

Fig. 1.16 A nucleus–electron pair in the coordinate system related to the electron (**a**) and to the nucleus (**b**)

After substitution Eq. (1.149) into Eq. (1.148) one obtains the following expression for the magnetic field created at the nucleus site by an electron with an orbital angular momentum $\vec{l}$:

$$\hat{\vec{B}}_{\text{orb}} = -\frac{\mu_0}{4\pi}\frac{e\hbar}{m_e r^3}\hat{\vec{l}}, \tag{1.150}$$

or taking into account the Bohr magneton $\mu_B = e\hbar/(2m)$,

$$\hat{\vec{B}}_{\text{orb}} = -\frac{\mu_0}{2\pi}\frac{\mu_B}{r^3}\hat{\vec{l}}. \tag{1.151}$$

Here, to make both left and right parts of Eqs. (1.150) and (1.151) equivalent the magnetic field $\hat{\vec{B}}_{\text{orb}}$ is also considered as an operator. As the Hamiltonian of the interaction of the nucleus magnetic moment $\hat{\vec{\mu}}_I$ with the magnetic field $\hat{\vec{B}}_{\text{orb}}$ is $\widehat{\mathcal{H}}_{\text{orb}} = -\hat{\vec{\mu}}_I \cdot \hat{\vec{B}}_{\text{orb}}$, then

$$\widehat{\mathcal{H}}_{\text{orb}} = \frac{\mu_0}{2\pi}\hbar\gamma_I\mu_B\frac{\hat{\vec{l}}\cdot\hat{\vec{I}}}{r^3}. \tag{1.152}$$

Here it is taken into account that $\hat{\vec{\mu}}_I = \gamma_I\hat{\vec{I}}$. In reality the electron is distributed in space according to its wave function $\psi(\vec{r})$, which is a solution of the stationary Schrödinger equation. The form of the Hamiltonian in the Schrödinger equation depends on the accepted model of the atom or molecule. In the simplest case when the electron moves in a potential field, the Hamiltonian of which, as it is known,[5] looks like

$$\widehat{\mathcal{H}} = \frac{\hat{p}^2}{2m_e} + V, \tag{1.153}$$

where V is the potential energy.

[5] See M. Goldman. Spin temperature and nuclear magnetic resonance in solids (Clarendon Press, 1970), §15.

To calculate the magnetic field at the nucleus site one usually uses the model, according to which this field is created by currents of the electron shell of the atom or molecule. As it is known from the quantum theory, the Schrödinger equation for an electron wave function in a potential field leads to the continuity equation for the probability density $\rho = \psi^*\psi$:

$$\frac{\partial \rho}{\partial t} + \operatorname{div} \vec{j} = 0, \tag{1.154}$$

here $\vec{j}$ is the current density, which is defined as $(\psi^* \hat{p}\psi - \psi \hat{p}\psi^*)$.

The current density defined accurate within a constant factor $-e$ by Eq. (1.154) is a quantum mechanical analog of the electric current density created by a moving electron charge, so to calculate the magnetic field one can use the Bio–Savare law for volumetric currents[6]:

$$\vec{B} = -\frac{\mu_0}{4\pi} \int\limits_V \frac{\vec{j} \times \vec{r}}{r^3} \mathrm{d}V.$$

Quantum mechanical calculations by applying the one-electron functions result to the Hamiltonian of the same form (1.152). However, in a many-electron atom or molecule, the magnetic field from one electron is screened by others, which can be in some moments closer to the nucleus than the electron under consideration. So Eq. (1.152) is the first-order approximation only.

1.11.2 Direct Dipole–Dipole Interactions

Description of such kind of interactions is based on the well known from the electromagnetism theory expression (1.129), which was obtained in Sect. 1.10 for the interaction energy of two magnetic dipoles. In this case one of the dipoles is an electron spin magnetic moment whereas another is a nuclear one. After substitution expressions for $\hat{\vec{\mu}}_I$ and $\hat{\vec{\mu}}_s$ into Eq. (1.129) (by analogy with Eq. (1.131)) one obtains the Hamiltonian for the dipole–dipole contribution into the electron–nucleus interaction:

$$\widehat{\mathcal{H}}_\mathrm{d} = \frac{\mu_0}{2\pi} \hbar \mu_\mathrm{B} g_\mathrm{s} \gamma_I \left[\frac{3(\hat{\vec{I}} \cdot \vec{r}) \cdot (\hat{\vec{s}} \cdot \vec{r})}{r^5} - \frac{\hat{\vec{I}} \cdot \hat{\vec{s}}}{r^3} \right].$$

[6] A. N. Matveev. Electricity and Magnetism (Mir Publishers, Moscow, 1986), § 10.

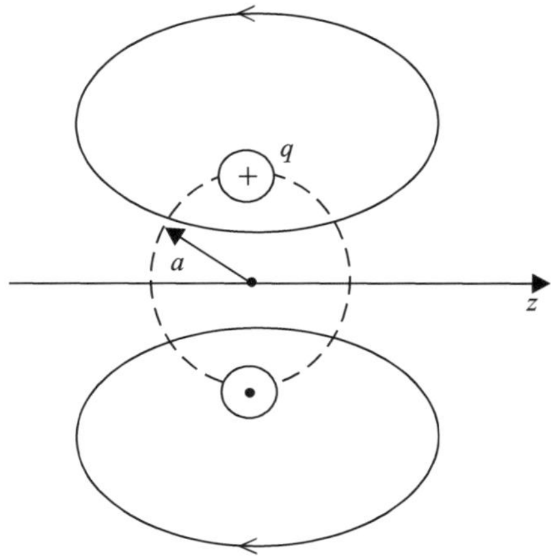

Fig. 1.17 Semiclassical representation of the Fermi contact interaction

1.11.3 Contact Interaction

The contact or *Fermi contact interaction* (following the name of the scientist who introduced it for consideration), is a specifically quantum effect, because it is supposed that both an electron and a nucleus can be situated at the same point of space. However, it can be explained from the semiclassical point of view, and it is even possible to obtain correct quantitative estimations.

First of all, let us allow that the nucleus magnetic moment is caused by the q charge motion along the circular orbit of radius a. Then, the expected value of components of the magnetic field, in which the electron stays, e.g. an s-electron, can be expressed through diagonal matrix elements on the form $\langle s|\hat{B}_z|s\rangle$. Due to the s-electron wave functions are spherically symmetric one can write

$$\langle s|\hat{B}_x|s\rangle = \langle s|\hat{B}_y|s\rangle = 0, \ \langle s|\hat{B}_z|s\rangle = \iiint\limits_R \hat{B}_z|\psi_s(r)|^2 \mathrm{d}\tau, \tag{1.155}$$

the integrals are taken over the all R space.

To estimate $\langle s|\hat{B}_z|s\rangle$ let us divide the area of integration into two parts: within the sphere of a radius a and outside it (see Fig. 1.17). The integral over the internal area can be estimated from following considerations. In the area, where currents are absent, one can introduce a magnetostatic potential Φ, such as $\vec{B} = -\vec{\nabla}\Phi$. It is known that any potential can be expanded in a series:

$$\Phi = \sum_{p=0}^{\infty} C_p r^p P_p(\cos\theta). \tag{1.156}$$

Let us differentiate the expression in the right part of Eq. (1.156) with respect to z:

$$\frac{\mathrm{d}}{\mathrm{d}z}\left(\sum_{p=0} C_p r^p P_p(\cos\theta)\right) = \sum_{p=0} C_p' r^p P_p(\cos\theta).$$

As one can see, both the potential and its derivative (1.156) are expressed by the same formulas, up to a constant factor. Hence, for the B_z component the following expansion is valid:

$$B_z = \sum_{p=0}^{\infty} C_p' r^p P_p(\cos\theta) \tag{1.157}$$

(coefficients C_p' differ from C_p). As the integration area is a sphere, one of the product terms, namely r^p, is invariant under rotations. As an integral over a spherical area can differ from zero only when integrand is invariant under rotations, the same invariance is required from the function $P_p(\cos\theta)$, which makes part of the series (1.157). However, from all the classes of $P_p(\cos\theta)$ functions only the function with $p = 0$ is invariant under rotations. Because of $r_0 P_0(\cos\theta) = 1$, the function $B_z(r)$ in the integrand (1.155) can be replaced by its own value at the reference point $B_z(0)$ and taken out of the integral sign. Besides, in assumption that at the distance a the function $\psi_s(r)$ does not change noticeably, it can be replaced by the function $\psi_s(0)$ and taken out of the integral sign. Then,

$$\langle s|\hat{B}_x|s\rangle = \langle s|\hat{B}_y|s\rangle = 0, \quad \langle s|\hat{B}_z|s\rangle = \iiint_R \hat{B}_z|\psi_s(r)|^2 \mathrm{d}\tau = \frac{4\pi}{3}a^3 B_z(0)|\psi_s(0)|^2. \tag{1.158}$$

The integral over the external area can be represented by a series

$$\Phi = \sum_{p=0} C_p r^{-p+1} P_p(\cos\theta). \tag{1.159}$$

In contrast to Eq. (1.157) all the series terms of Eq. (1.159) have an angular dependence and are not invariant under rotations, hence, for a s-electron the integral over the external area is equal to zero.

In that way, the expected value of the magnetic field acting from the nucleus side to a s-electron is defined by a product of the probability to find the s-electron at the nucleus site and the value of this field at the center of the equivalent nucleus current.

The magnetic field value at the point $r = 0$ can be expressed by means of the spin and the nucleus gyromagnetic ratio. It is known that the field at the center of a circular current i of radius a is

$$B_z(0) = -\frac{\mu_0 i}{2a}.$$

The circular current is equivalent to a magnetic dipole $\mu_I = \pi a^2$. Hence,

$$B_z(0) = -\frac{\mu_0 \mu_I}{2\pi a^3}.$$

Substitution of $B_z(0)$ into Eq. (1.158) results to the following expression for the average nucleus magnetic field:

$$\langle s|\hat{B}_z|s\rangle = -\frac{2}{3}\mu_0 \mu_I |\psi_s(0)|^2.$$

Then the energy of the interaction between a nucleus and an electron can be described as

$$E = -\mu_s B_z = \frac{2}{3}\mu_0(\vec{\mu}_s \cdot \vec{\mu}_I)|\psi_s|^2.$$

The same expression has to be obtained as a result of averaging of the corresponding Hamiltonian

$$E = \langle|\widehat{\mathcal{H}}|\rangle.$$

Hence, the contact interaction Hamiltonian can be written using a delta function in the following form:

$$\widehat{\mathcal{H}}_{\text{cont}} = \frac{2}{3}\mu_0(\vec{\mu}_I \cdot \vec{\mu}_s)\delta(\vec{r})$$

or

$$\widehat{\mathcal{H}}_{\text{cont}} = \frac{2}{3}\mu_0 \hbar \mu_{\text{B}} \gamma_I (\hat{\vec{I}} \cdot \hat{\vec{s}})\delta(\vec{r}).$$

The contact interaction of the nucleus with the electron shells characterized by the orbital number $l > 0$ does not contribute to the energy, because their electron density at the nucleus site is equal to zero. In a molecule the probability to find an electron at the nucleus site is defined by the molecular orbital and can be differ from zero even if the electron belongs to another atom or is at the state with $l > 0$.

Apart from the contact interaction there are two other interactions between nucleus and electron spins: the dipole interaction between these spins and the interaction of the nucleus spin with the orbital magnetic moment of the electron. However, they are much weaker (by an order of magnitude) than the Fermi contact interaction.

1.11.4 Electronic Magnetic Shielding

Magnetic shielding of a nucleus by electron shells is one of the most important effects of electron–nucleus interactions. Nuclear magnetic resonance frequencies in atoms and molecules slightly differ from the frequencies defined by the formula $\omega = \gamma B_0$ (where B_0 is the constant external field). Based on semiclassical presentations one can expect that this field can produce inductive currents in electron shells that cause at the nucleus site an additional magnetic field, which is proportional to B_0 and directed oppositely to the $\vec{B}_0$ vector. As a result, the magnetic field acting on the nucleus is smaller than the applied external field. Experiments show (and this is confirmed by a stricter theory) that this additional field can be of both signs. Such a phenomenon is called *magnetic shielding of the nucleus by the electron shell* (regardless of the sign of the additional field). The screen constant σ, determined from the expression

$$B_{\mathrm{n}} = (1 - \sigma)B_0,$$

is the quantitative characteristic of this effect. Here B_{n} is the field at the nucleus site. Within the quasi-relativistic approximation the influence of the magnetic field can be taken into account by replacing in the initial Hamiltonian (1.153) the momentum $\hat{\vec{p}}$ by $\hat{\vec{p}} + e\hat{\vec{A}}$, where $(-e)$ is the electron charge and $\hat{\vec{A}}$ is the vector potential of the magnetic field. It is easy to see that additional terms appear in the Hamiltonian $\widehat{\mathcal{H}}'$, which can be considered as a perturbation:

$$\widehat{\mathcal{H}}' = -\frac{e}{2m}(\vec{p} \cdot \vec{A}_0 + \vec{A}_0 \cdot \vec{p}) + \frac{e^2}{m}A_0^2. \tag{1.160}$$

In the case of a homogeneous magnetic field the vector potential may be written as

$$\vec{A} = \frac{1}{2}[\vec{B}_0 \times \vec{r}], \tag{1.161}$$

where $\vec{B}_0$ is the constant magnetic field. After substitution Eq. (1.161) into Eq. (1.160) and taking into account that $\vec{r} \times \vec{p} = \hbar\vec{l}$ and that the Bohr magneton $\mu_{\mathrm{B}} = e\hbar/(2m)$, it is easy to obtain

$$\widehat{\mathcal{H}}' = -\mu_{\mathrm{B}}\vec{l} \cdot \vec{B}_0 + \frac{e^2[\vec{B}_0 \times \vec{r}]^2}{8m}. \tag{1.162}$$

The first term in Eq. (1.162) describes the Zeeman interaction of the electron orbital moment, whereas the second one describes the magnetic susceptibility.

In the expression for current density an additional term appears [7]:

$$\vec{j}(\vec{r}) = \frac{ie\hbar}{2m}(\psi^*\vec{\nabla}\psi - \psi\vec{\nabla}\psi^*) - \frac{e^2}{m}\vec{A}_0\psi^*\psi, \tag{1.163}$$

here ψ is a function of an electron state. Within the first-order approximation of the perturbation theory

$$\psi = \psi_0 + \psi', \tag{1.164}$$

where ψ' is the first-order correction to the wave function, which can be expressed by the following way:

$$\psi' = \sum_n a_n\psi_n, \tag{1.165}$$

[7] See, for example, A. S. Davydov. Quantum Mechanics (Pergamon Press, Oxford 1965), § 54.

$$a_n = -\frac{\langle n \,|\, \widehat{\mathcal{H}}' \,|\, 0 \rangle}{E_n - E_0}.$$

As the magnetic field directs along the z-axis, within the first-order approximation of the perturbation theory the new electron wave function can be written as follows:

$$\psi = \psi_0 - \mu_{\mathrm{B}} B_0 \sum_n \frac{\langle \psi_n | \hat{l}_z | \psi_0 \rangle}{E_n - E_0} \psi_n, \quad (1.166)$$

where $\hat{l}_z$ is the z-projection of the angular momentum operator. The second term of the operator (1.162) does not contribute to (1.166), because it is diagonal within the basis of non-perturbed wave functions.

Another issue of the current density change is the appearance of an additional term in Eq. (1.160) related to the change of the electron movement direction in the presence of the magnetic field: the electron shell starts gyrating around the magnetic field vector by analogy with a free electron moving perpendicular to the magnetic field vector.

It is worth taking into account that the perturbation due to the external magnetic field is rather small relative to the energy of the electron in electrostatic field of the nucleus, even in the highest magnetic fields, which can be achieved experimentally. So when substituting Eq. (1.165) into Eq. (1.163), the terms, which are proportional to the product of A_0 and the ψ' correction in the form of (1.165), as well as the terms of higher order infinitesimal can be neglected. Keeping only the terms of the first-order infinitesimal, one has

$$j = j_0 + j_{\mathrm{para}} + j_{\mathrm{dia}},$$

where j_{para} and j_{dia} make positive and negative contributions, respectively, into the main current (j_0), and consequently to the nucleus field. From Eq. (1.163) it is possible to determine the magnetic field intensity induced at the nucleus site. Let us give an expression for z-component of the induced field:

$$B_z' = -\sigma_{zz} B_0 = -\frac{\mu_0 e \hbar}{2m} B_0 \langle \psi | \frac{2\hat{l}_z}{r^3} | \psi \rangle - \frac{\mu_0 e^2}{2m} B_0 \langle \psi | \frac{x^2 + y^2}{r^3} | \psi \rangle. \quad (1.167)$$

Substituting the expression for the wave function from (1.166) into (1.163) and keeping only terms linear in B_0, one obtains the Ramsey formula for the chemical shift caused by an electron:

$$\sigma_{zz} = \frac{\mu_0 e^2}{2m} \langle \psi_0 | \frac{x^2 + y^2}{r^3} | \psi_0 \rangle$$

$$-\frac{\mu_0 e\hbar}{2m}\sum_n\left(\frac{\langle\psi_0|\hat{l}_z|\psi_n\rangle\langle\psi_n|2\hat{l}_z/r^3|\psi_0\rangle}{E_n-E_0}+\frac{\langle\psi_0|2\hat{l}_z/r^3|\psi_n\rangle\langle\psi_n|\hat{l}_z|\psi_0\rangle}{E_n-E_0}\right). \tag{1.168}$$

The components σ_{xx}, σ_{yy} of the chemical shift tensor ($\boldsymbol{\sigma}$) are determined by the similar equations. In molecules, where there are more than one electron, the quantities $\hat{l}_z$ and $(x^2+y^2)/r^3$ must be summed up over all electrons. The first term in Eq. (1.168) is usually called as *diamagnetic*, whereas the second one as *paramagnetic*. The diamagnetic term depends only on the electron density in the ground state and it can be easily calculated theoretically. The calculation of the paramagnetic term is a more difficult task because it depends on exited states.

In the case of s-electrons only the first term in Eq. (1.168) differs from zero. Hence, only the diamagnetic shielding takes place. In other cases the both terms differ from zero and there is a competition between the paramagnetic and diamagnetic contributions. It is to be specially noted that such a situation can be realized when the shielding paramagnetism does not coincide with the magnetic susceptibility sign, i. e. the substance can be diamagnetic, whereas the shift is paramagnetic. This is related to the fact that electrons situated closer to the nucleus have a more important influence on the shielding, whereas distant electrons affect mainly the susceptibility.

1.11.5 Indirect Spin–Spin Interactions

If one supposes that the source of a disturbing field is neighbor nucleus, belonging to the same molecule, then one can define another contribution to the current density, which does not depend on the external field, but depends on the value and state of the nuclear magnetic moment and the electron shell structure. In this case one should consider a disturbed Hamiltonian, where the disturbance is the interaction of an electron, which moves in an electrostatic field, with an external magnetic field. Under such conditions two interaction mechanisms are possible: the interaction of spins through the molecule orbitals (see Sect. 1.11.3) and the interaction through the electron spin. In the latter case the main mechanism is the contact interaction considered above. Indirect spin–spin interactions between nuclei of the same molecule are, in principle, anisotropic and are described by a two-rank spin–spin interaction tensor. However in NMR spectra of liquids the anisotropy does not appear because of the molecule rotation, which averages the tensor components over all directions. So the effective is the Hamiltonian, in which the interaction is described by a scalar product of spins of interacting nuclei. The proportionality constant, called the *constant of indirect* (or *scalar*) *interaction,* contains the useful information about the relative position of atom groups in a molecule and about the intermolecular dynamics as well.

Let us consider the situation in more details. The nucleus spin disturbs valence electrons (relatively to their state in a system of charged nuclei without spins) and then the disturbed electrons create a weak magnetic field at the sites of other nuclei. The most essential interaction mechanisms between electron and nucleus spins is the contact interaction. As a result of this interaction the electron spins in the atom A tend to be antiparallel to the nucleus spin. Since electron spins interact through the chemical bonds, the electron spin polarization at the nucleus A is transmitted to the nucleus B, that orients its spin parallel or antiparallel to the spin of the nucleus A. The impact of the nucleus A on the nucleus B is rather small and proportional to $\hat{I}_z$. The Hamiltonian of such interactions is

$$\widehat{\mathcal{H}}_{\rm sc} = J\, \hat{I}_{z\rm A}\hat{I}_{z\rm B}. \tag{1.169}$$

The constant of spin–spin interaction can be calculated within the first-order approximation of the perturbation theory (using any *ab initio* or semiempirical method to calculate the chemical shift contribution) by formula

$$J = \frac{8\pi}{3} g_{\rm s}\mu_{\rm B} g_I \mu_{\rm n} \sum_n \left(\frac{\langle\psi_0|\hat{S}_{z\rm A}|\psi_{\rm n}\rangle\langle\psi_{\rm n}|\hat{S}_{z\rm B}|\psi_0\rangle}{E_n - E_0} + \frac{\langle\psi_0|\hat{S}_{z\rm B}|\psi_{\rm n}\rangle\langle\psi_{\rm n}|\hat{S}_{z\rm A}|\psi_0\rangle}{E_n - E_0} \right), \tag{1.170}$$

where $g_{\rm s}$ and g_I are the electron and nucleus g-factors, respectively; $\mu_{\rm B}$ and $\mu_{\rm n}$ are the Bohr and nucleus magnetons, respectively.

1.12 Electric Quadrupole Interactions

If the nucleus spin $I > 1/2$, then the nucleus in addition to a dipole magnetic moment has an electric quadrupole moment. At the nucleus site a non-zero electric field gradient (EFG) can exist, created, for example, by the electron shell or surrounding ions. And then, there is an interaction between the nucleus quadrupole moment and the electric field gradient. This interaction may result to noticeably changes in the NMR spectrum, and even makes it possible to observe the nuclear quadrupole resonance. The Hamiltonian of such an interaction can be deduced from an expression that determines the energy of interaction between the classical quadrupole moment and the electric field gradient, in which classical quantities are replaced by the corresponding operators. The classical energy of the interaction between a distributed charge, characterized by the density $\rho(\vec{r})$, and an electric field with a potential $U(\vec{r})$, can be represented in the following form

$$E = \iiint\limits_{\Omega_{\rm n}} \rho(\vec{r})\, U(\vec{r}) {\rm d}\Omega_{\rm n}, \tag{1.171}$$

integration in Eq. (1.171) runs over the all nucleus volume $\Omega_{\rm n}$.

Let us expand the function $U(\vec{r})$ in a Taylor series relatively to the origin of the Cartesian coordinates system associated to the nucleus:

$$U = U(0) + \sum_{\alpha} \left. \frac{\partial U}{\partial x'_\alpha} \right|_0 x'_\alpha + \frac{1}{2} \sum_{\alpha\beta} \left. \frac{\partial^2 U}{\partial x'_\alpha x'_\beta} \right|_0 x'_\alpha x'_\beta + \ldots, \tag{1.172}$$

here $x'_\alpha,\ x'_\beta$ $(\alpha, \beta = 1, 2, 3)$ correspond to the coordinates $x',\ y'$ or z'. After substitution Eq. (1.172) into Eq. (1.171) one has

$$E = U_0 \iiint\limits_{\Omega_n} \rho(\vec{r}) \mathrm{d}\Omega_n + \sum_{\alpha} \left(\frac{\partial U}{\partial x'_\alpha} \right)_{r=0} \iiint\limits_{\Omega_n} \rho(x'y'z') x'_\alpha \mathrm{d}\Omega_n$$
$$+ \frac{1}{2} \sum_{\alpha\beta} \left(\frac{\partial^2 U}{\partial x'_\alpha x'_\beta} \right)_{r=0} \iiint\limits_{\Omega_n} \rho(x'y'z') x'_\alpha x'_\beta \mathrm{d}\Omega_n + \ldots \tag{1.173}$$

The first term in Eq. (1.173) describes the energy of the Coulomb interaction with a point nucleus charge. This interaction does not depend on the nucleus orientation, and hence does not change NMR, NQR or EPR frequencies. The terms containing the first derivative of the potential are equal to zero, because they describe the interaction of the electric field with the electric dipole moment, which is equal to zero for a nucleus due to its symmetry. The third term in Eq. (1.173) defines the quadrupole interaction energy:

$$E_Q = \frac{1}{2} \sum_{\alpha\beta} V'_{\alpha\beta} \iiint\limits_{\Omega_n} x_\alpha x_\beta\, \rho(r) \mathrm{d}\Omega_n. \tag{1.174}$$

The components $V'_{\alpha\beta}$ are determined as

$$V'_{\alpha\beta} = \left. \frac{\partial^2 U}{\partial x'_\alpha \partial x'_\beta} \right|_{\vec{r}=0}$$

and form the 3×3 matrix, which characterizes the EFG tensor:

$$\mathsf{V}' = \begin{pmatrix} V_{x'x'} & V_{x'y'} & V_{x'z'} \\ V_{y'x'} & V_{y'y'} & V_{y'z'} \\ V_{z'x'} & V_{z'y'} & V_{z'z'} \end{pmatrix}. \tag{1.175}$$

As differentiation is a linear operation, the result does not depend on the order of differentiation, and hence the matrix (1.175) is symmetric and the tensor corresponding to it is the symmetric second-rank tensor. Obviously it is always possible to choose such a coordinate system $(x,\ y,\ z)$, where the matrix (1.175) is diagonal, and the tensor is defined by the three principal components: V_{xx}, V_{yy} and V_{zz}. This coordinate system is named the *principal axes frame of the EFG tensor*. As the potential satisfies the Laplace equation $\Delta U = 0$, then

$$V_{xx} + V_{yy} + V_{zz} = 0, \tag{1.176}$$

i. e. the sum of the diagonal elements of the matrix, which characterizes the EFG tensor, is equal to zero, independently on the coordinate system choice. From Eq. (1.176) it directly follows that if the nucleus is situated at the site of cubic symmetry the quadrupole interaction is equal to zero.

The point symmetry of the site, in which the nucleus under study is situated, plays an important role in defining of the EFG tensor. So, if the point symmetry is characterized by a symmetry axis, then one of the axes of the coordinate system, in which the matrix (1.175) is diagonal, is this symmetry axis. Three orthogonally related second order symmetry axes completely define direction of the principal axes of the EFG tensor.

For a more convenient form of Eq. (1.174) let us use the classical definition for the quadrupole moment tensor. Its components $eQ^*_{\alpha\beta}$ are defined by the following way (see (1.14)):

$$eQ^*_{\alpha\beta} = \iiint\limits_{\Omega_\mathrm{n}} \left(3x'_\alpha x'_\beta - \delta_{\alpha\beta} r^2\right) \rho \mathrm{d}\Omega_\mathrm{n}. \tag{1.177}$$

Then

$$\iiint\limits_{\Omega_\mathrm{n}} x'_\alpha x'_\beta \, \rho \mathrm{d}\Omega_\mathrm{n} = \frac{1}{3}\left(eQ^*_{\alpha\beta} + \iiint\limits_{\Omega_\mathrm{n}} \delta_{\alpha\beta} r^2 \rho \mathrm{d}\Omega_\mathrm{n}\right),$$

and hence

$$\begin{aligned} E_\mathrm{Q} &= \frac{1}{2}\sum_{\alpha\beta} V'_{\alpha\beta} \iiint\limits_{\Omega_\mathrm{n}} x'_\alpha x'_\beta \rho \mathrm{d}\Omega_\mathrm{n} \\ &= \sum_{\alpha\beta} \frac{1}{6}\left(V'_{\alpha\beta} eQ^*_{\alpha\beta} + V'_{\alpha\beta}\delta_{\alpha\beta} \iiint\limits_{\Omega_\mathrm{n}} r^2 \, \rho \mathrm{d}\Omega_\mathrm{n}\right). \end{aligned} \tag{1.178}$$

As the potential U satisfies the Laplace equation, the second term in Eq. (1.178) becomes zero. As a result one obtains a classical expression for the energy of quadrupole interaction of a nucleus with EFG in any coordinate system:

$$E_\mathrm{Q} = \frac{1}{6}\sum_{\alpha\beta} V'_{\alpha\beta} Q^*_{\alpha\beta}. \tag{1.179}$$

Using (similar to Sect. 1.1.3) the fact that the charge density (ρ) in the nucleus is continuous (see (1.16)), integration in Eq. (1.177) can be replaced by summation, and hence

$$eQ^*_{\alpha\beta} = e\sum_k \left(3x'_{\alpha k} x'_{\beta k} - \delta_{\alpha\beta} r_k^2\right). \tag{1.180}$$

In order to find the quantum mechanical expression for the quadrupole interaction one should replace in Eq. (1.179) the energy by the Hamiltonian operator, and in the classical expression for $Q^*_{\alpha\beta}$, Eq. (1.180), the coordinates x'_α, x'_β by the corresponding operators:

$$\widehat{\mathcal{H}}_Q = \frac{e}{6}\sum_{\alpha\beta} V'_{\alpha\beta} \sum_k \left(3\hat{x}'_{\alpha k}\hat{x}'_{\beta k} - \delta_{\alpha\beta}\hat{r}_k^2\right) = \frac{e}{6}\sum_{\alpha\beta} V'_{\alpha\beta}\hat{Q}_{\alpha\beta}, \tag{1.181}$$

where

$$\hat{Q}_{\alpha\beta} = \sum_k \left(3\hat{x}'_{\alpha k}\hat{x}'_{\beta k} - \delta_{\alpha\beta}\hat{r}_k^2\right). \tag{1.182}$$

It is rather difficult to use Eq. (1.181) for various calculations, because, first, it contains summation over all protons, and the Hamiltonian becomes many-particle. Second, calculations of the matrix elements of the operators of proton coordinates is a nontrivial task. Therefore, it is desirable to replace the coordinate operators in Eq. (1.181) by others, that do not change the matrix elements of the quadrupole interaction Hamiltonian. On this purpose let us apply the Wigner–Eckart theorem, one of the most important theorems in quantum mechanics. To formulate this theorem it is necessary to introduce notions of the Clebsch–Gordan coefficients and irreducible tensor operators.

The Clebsch–Gordan coefficients are used to determine the wave functions corresponding to the moment of momentum of a system, which consists of several subsystems. Each subsystem is characterized by its own moment of momentum and its projection. Let the system consists of two nuclei with the spins I and I'. Denote the spin projection onto the external magnetic field as m and m'. Then, the operator $\hat{\vec{I}}^{\Sigma}_z$ of such a system and its projection M can be determined due to the vector summation rule:

$$\hat{I^\Sigma}_z = \hat{I}_z + \hat{I}'_z, \qquad M = m + m'.$$

The eigenfunctions of the operators $\hat{I}_z$ and $\hat{I}'_z$ ($\psi_{I,m}$ and $\psi_{I',m'}$, respectively) are simultaneously eigenfunctions of the operators of the moment of momentum ($\hat{\vec{I}}$, $\hat{\vec{I}}'$) for each subsystem. It is obvious that the wave functions ($\Psi_{I,M}$) of the operator $\hat{I^\Sigma}_z$ can be represented as a linear combination of products of the wave function of the operators $\hat{\vec{I}}$, $\hat{\vec{I}}'$ for individual nuclei:

$$\Psi_{I^\Sigma,M} = \sum_{I,m,I',m'} C(I, I', I^\Sigma; m, m', M)\psi_{I,m}\psi_{I',m'}. \tag{1.183}$$

As it is shown in a number of quantum mechanics textbooks, such a representation is unique with an accuracy of a phase factor, which is insignificant.[8] The $C(I, I', I^{\Sigma}; m, m', M)$ coefficients in Eq. (1.183) are called the *Clebsch–Gordan coefficients*. Their values do not depend on the operator choice when determining the total wave function of the system under study. So, instead of the nuclear spin operators one can use the operators of the spin, orbital momentum or total angular momentum of electron. The Clebsch–Gordan coefficients differ from zero only if $M = m + m'$, and are equal to zero if $I^{\Sigma} \neq |I - I'|, \ldots, |I + I'|$.

Now let us proceed to definition of irreducible tensor operators. Any operator completely describing the system state can be represented as tensor components. So, the nuclear spin projection operators ($\hat{I}_x, \hat{I}_y, \hat{I}_z$) or their linear combinations form a first-rank tensor. The components, which are bilinear in these operators, form a second-rank tensor. In a general case, linear combinations of products of n operators are components of a n-rank tensor. The number of components determining the n-rank tensor is equal to $2n+1$. Usually tensor components are numbered by two indexes (n and k): n is the tensor rank, and $k = n, n-1, \ldots, -n$. Calculations of the matrix elements of these operators can be significantly simplified. For example, one can chose such linear combinations of the operators or their products that the following commutation relation between the angular momentum operators and operators, which are k-th components of the n-rank tensor ($\hat{T}_{nk}$), are satisfied:

$$[\hat{J}_{\pm}, \hat{T}_{nk}] = \sqrt{n(n+1) - k(k \pm 1)}\, \hat{T}_{nk\pm 1}, \quad [\hat{J}_z, \hat{T}_{nk}] = k\hat{T}_{nk}. \tag{1.184}$$

Here, $\hat{J}_z, \hat{J}_+, \hat{J}_-$ denote the components of the angular momenta of different nature: it can be the total angular momentum of an electron or of an ion, or intrinsic angular momentum of an electron or of a nucleus etc.

The operators, for which Eq. (1.184) is satisfied, are called the *irreducible tensor operators*. This choice of the operators allows us to draw an analogy between the tensor operators and the spherical functions Y_l^m, which obey the same commutation relations (1.184). At any coordinate transformations both spherical functions and irreducible tensor operators are transformed by the same way. In a Cartesian coordinate system the spherical functions for $l = 1, 2$ can be written as

$$Y_1^{-1} = \frac{1}{\sqrt{2}}\frac{x - \mathrm{i}y}{r}, \qquad Y_1^0 = \frac{z}{r}, \qquad Y_1^1 = -\frac{1}{\sqrt{2}}\frac{x + \mathrm{i}y}{r},$$

$$Y_2^{\pm 2} = \frac{(x \pm \mathrm{i}y)^2}{r^2}, \quad Y_2^{\pm 1} = \mp 2\frac{z(x \pm \mathrm{i}y)}{r^2}, \quad Y_2^0 = \sqrt{\frac{2}{3}}\frac{3z^2 - r^2}{r^2}. \tag{1.185}$$

Let us consider Y_1^1 to show that for the spherical functions Eqs. (1.184) are valid:

---

[8] Wave functions that describe the state of any system are always determined up to a phase factor that can be chosen arbitrarily. Because it is not the wave function but its module square, which is the probability to find the system in one or another state, has the physical meaning.

$$[l_-, Y_1^1] = \left[l_x - \mathrm{i}l_y, -\frac{1}{\sqrt{2}} \frac{x + \mathrm{i}y}{r} \right]$$
$$= -\frac{1}{\sqrt{2}r} \left([l_x, x] + \mathrm{i}[l_x, y] - \mathrm{i}[l_y, x] + [l_y, y] \right) = \sqrt{2}\frac{z}{r} = \sqrt{2}Y_1^0.$$

By the similar way it is possible to check the fulfillment of conditions (1.184) for other spherical functions as well. It is obvious that the products $r^l Y_l^m$ also satisfy to the commutation relations (1.184), as the angular momentum operators, which are in fact rotation operators, do not change the radius vector $\vec{r}$. So, it is possible to chose the product $T_{lm}(\vec{r}) = r^l Y_l^m$ as the components of l-rank tensor operators in the coordinate representation. For each tensor operator in the coordinate space there is an operator of the same rank in the spin space, which corresponds to it. So, for example, the components of the irreducible first- and second-rank tensor operators can be written in the spin and coordinate spaces as follows:

$$\begin{array}{lcc}
 & \text{Spin space} & \text{Coordinate space} \\
\text{for } T_{11}: & -\frac{1}{\sqrt{2}}J_+ = -\frac{1}{\sqrt{2}}(J_x + \mathrm{i}J_y), & -\frac{1}{\sqrt{2}}(x + \mathrm{i}y); \\
\text{for } T_{10}: & J_z, & z; \\
\text{for } T_{1-1}: & \frac{1}{\sqrt{2}}J_- = \frac{1}{\sqrt{2}}(J_x - \mathrm{i}J_y), & \frac{1}{\sqrt{2}}(x - \mathrm{i}y); \\
\text{for } T_{22}: & J_+^2, & (x + \mathrm{i}y)^2; \\
\text{for } T_{21}: & -(J_z J_+ + J_+ J_z), & -2z(x + \mathrm{i}y); \\
\text{for } T_{20}: & \sqrt{\frac{2}{3}}(3J_z^2 - J^2), & \sqrt{\frac{2}{3}}(3z^2 - r^2); \\
\text{for } T_{2-1}: & J_z J_- + J_- J_z, & 2z(x - \mathrm{i}y); \\
\text{for } T_{2-2}: & J_-^2, & (x - \mathrm{i}y)^2.
\end{array} \qquad (1.186)$$

Introduction of tensor operators is convenient due to the fact that matrix elements of individual components of tensor operators are not independent, i. e. the matrix element of any component of the tensor operator can be expressed through the matrix element of another component of the tensor operator. Relation between matrix elements of the individual components of the tensor operators can be represented as

$$\left\langle im\eta | \hat{T}_{nk} | i'm'\eta' \right\rangle = C(i'ni; m'km) \langle im\eta | \hat{T}_{nk} | i'm'\eta' \rangle \Big|_{m=i,\ k=n,\ m'=i'}, \qquad (1.187)$$

where η and η' are quantum numbers, which characterize other unvarying quantum numbers. Eq. (1.187), which is in fact the Wigner–Eckart theorem, does not depend on the representation in which components of the tensor operators are written. However, the matrix element values depend on the operator representation used for their calculation. The Wigner–Eckart theorem proving can be found in a number of textbook on quantum mechanics or group theory.[9] Equation (1.187) can be used

[9] See, for example, V. Heine. Group Theory of Quantum Mechanics (Pergamon, New York, 1960).

to express the matrix elements of the tensor operators in one representation, via the matrix elements of the operators of the same rank in another representation, if the matrix elements of the tensor operators corresponding to $m = i$, $m' = i'$ and $k = n$ in both representations are known. This statement follows from the fact that the Clebsch–Gordan coefficients do not depend on the representation choice. Indeed, if one writes Eq. (1.187) for two different representations of tensor operators, for example, for the coordinate ($\vec{r}$) and spin ($\hat{\vec{I}}$) representations

$$\left\langle im\eta|\hat{T}_{nk}(\vec{r})|i'm'\eta'\right\rangle = C(i'ni;m'km)\left\langle im\eta|\hat{T}_{nk}(\vec{r})|i'm'\eta'\right\rangle\Big|_{m=i,\ k=n,\ m'=i'}, \tag{1.188}$$

$$\left\langle im\eta|\hat{T}_{nk}(\hat{\vec{I}})|i'm'\eta'\right\rangle = C(i'ni;m'km)\left\langle im\eta|\hat{T}_{nk}(\hat{\vec{I}})|i'm'\eta'\right\rangle\Big|_{m=i,\ k=n,\ m'=i'} \tag{1.189}$$

and divides the representation (1.189) into (1.188), then one obtains

$$\frac{\left\langle im\eta|\hat{T}_{nk}(\hat{\vec{I}})|i'm'\eta'\right\rangle}{\left\langle im\eta|\hat{T}_{nk}(\vec{r})|i'm'\eta'\right\rangle} = \frac{\left\langle im\eta|\hat{T}_{nk}(\hat{\vec{I}})|i'm'\eta'\right\rangle|_{m=i,\ k=n,\ m'=i'}}{\left\langle im\eta|\hat{T}_{nk}(\vec{r})|i'm'\eta'\right\rangle|_{m=i,\ k=n,\ m'=i'}}. \tag{1.190}$$

If the matrix elements corresponding to the maximal projection of the operators

$$\left\langle im\eta|\hat{T}_{nk}(\vec{r})|i'm'\eta'\right\rangle\Big|_{m=i,\ k=n,\ m'=i'},\quad \left\langle im\eta|\hat{T}_{nk}(\hat{\vec{I}})|i'm'\eta'\right\rangle\Big|_{m=i,\ k=n,\ m'=i'},$$

are determined in some way, one can find out all the other matrix elements of the operators, written in the coordinate representation, by calculating the matrix elements of the operators in the spin representation, that is much easier. It is obvious that the operators, which have the same matrix elements when describing one or another physical phenomenon, are equivalent. In our example the operators

$$\hat{T}_{nk}(\vec{r}) \quad \text{and} \quad \frac{\left\langle im\eta|\hat{T}_{nk}(\vec{r})|i'm'\eta'\right\rangle|_{m=i,\ k=n,\ m'=i'}}{\left\langle im\eta|\hat{T}_{nk}(\hat{\vec{I}})|i'm'\eta'\right\rangle|_{m=i,\ k=n,\ m'=i'}}\hat{T}_{nk}(\hat{\vec{I}})$$

are equivalent as they have the same matrix elements.

When the description of any interaction between particles is specified by the coordinate operators, which are rather labor-consuming to calculate directly, they can be replaced by the equivalent spin operators. Then further calculations should be carried out within this representation. The operators obtained after such a replacement are called the *effective operators*. However, it is necessary to keep in mind that calculations carried out using the effective operators are correct only if the other quantum numbers (η) describing the system under study remain fixed.

The considered approach can be applied to describe quadrupole interactions. To determine the effective Hamiltonian of quadrupole interactions it is necessary to substitute the operator $\hat{Q}_{\alpha\beta}$ in form of (1.182) by an equivalent operator with the matrix elements similar to the operator $\hat{Q}_{\alpha\beta}$. Using Eq. (1.190) and taking into account that it is also valid for linear combinations of T_{nk} tensor components with a given n value, after such a substitution one has

$$\left\langle IM\eta\middle|e\sum_{k}^{\text{protons}} 3\hat{x}_{\alpha k}\hat{x}_{\beta k} - \delta_{\alpha\beta}\hat{r}_k^2\middle|IM'\eta\right\rangle = \frac{\left\langle IM\eta\middle|e\sum_k 3\hat{z}_k^2 - \hat{r}_k^2\middle|IM'\eta\right\rangle|_{M=I,\ M'=I}}{\left\langle IM\eta\middle|3\hat{I}_z{}^2 - \hat{\vec{I}}^2\middle|IM'\eta\right\rangle|_{M=I,\ M'=I}} \times \left\langle IM\eta\middle|\frac{3}{2}\left(\hat{I}_\alpha\hat{I}_\beta + \hat{I}_\beta\hat{I}_\alpha\right) - \delta_{\alpha\beta}\hat{\vec{I}}^2\middle|IM'\eta\right\rangle. \tag{1.191}$$

The value

$$\left\langle IM\eta\middle|e\sum_{k}^{\text{protons}} 3\hat{z}_k^2 - \hat{r}_k^2\middle|IM'\eta\right\rangle\Big|_{M=I,\ M'=I} \tag{1.192}$$

is the quadrupole moment of a nucleus (see Eq. 1.18). The matrix element in the denominator of Eq. (1.191) can be easily calculated. It is equal to

$$\left\langle IM\eta\middle|3\hat{I}_z{}^2 - \hat{\vec{I}}^2\middle|IM'\eta\right\rangle\Big|_{M=I,\ M'=I} = I(2I-1). \tag{1.193}$$

As it follows from Eqs. (1.191)–(1.193), the matrix elements of the operator $\hat{Q}_{\alpha\beta}$ in the form of (1.182) can be calculated using the equivalent operators:

$$\hat{Q}^{\text{eq}}_{\alpha\beta} = \frac{eQ}{I(2I-1)}\left[\frac{3}{2}\left(\hat{I}_\alpha\hat{I}_\beta + \hat{I}_\beta\hat{I}_\alpha\right) - \delta_{\alpha\beta}\hat{\vec{I}}^2\right].$$

Then the Hamiltonian of quadrupole interactions (1.181) can be replaced by the effective Hamiltonian:

$$\widehat{\mathcal{H}}_{\text{Q}}{}^{\text{eff}} = \frac{eQ}{6I(2I-1)}\sum_{\alpha\beta} V'_{\alpha\beta}\left[\frac{3}{2}\left(\hat{I}_\alpha\hat{I}_\beta + \hat{I}_\beta\hat{I}_\alpha\right) - \delta_{\alpha\beta}\hat{\vec{I}}^2\right]. \tag{1.194}$$

Further let us drop the index ‘eff’ at the effective Hamiltonian of quadrupole interaction.

It is worth noting that the effective quadrupole interaction Hamiltonian is obtained for an arbitrary orientation of the Cartesian coordinate system ($\alpha, \beta = x', y', z'$). In the principal axes frame of the EFG tensor it can be written as

$$\widehat{\mathcal{H}}_{\mathrm{Q}} = \frac{eQ}{6I(2I-1)}\left[V_{xx}\left(\hat{I}_x{}^2 - \hat{\vec{I}}^2\right) + V_{yy}\left(3\hat{I}_y{}^2 - \hat{\vec{I}}^2\right) + V_{zz}\left(3\hat{I}_z{}^2 - \hat{\vec{I}}^2\right)\right]. \tag{1.195}$$

Taking into account that $V_{xx}+V_{yy}+V_{zz}=0$, the expression (1.195) can be rewritten by the following way:

$$\widehat{\mathcal{H}}_{\mathrm{Q}} = \frac{eQV_{zz}}{4I(2I-1)}\left[(3\hat{I}_z{}^2 - \hat{\vec{I}}^2 + \frac{V_{yy}-V_{xx}}{V_{zz}}\left(\hat{I}_x{}^2 - \hat{I}_y{}^2\right)\right].$$

The value V_{zz} is called the *principal component of the EFG tensor* and is usually denoted as eq, where e is the electron charge. The expression $(V_{yy}-V_{xx})/V_{zz}$ is the *asymmetry parameter of the EFG tensor* and is denoted by η. Then the quadrupole interaction Hamiltonian in the principal axes frame of the EFG tensor can be written as follows:

$$\widehat{\mathcal{H}}_{\mathrm{Q}} = \frac{e^2qQ}{4I(2I-1)}\left[(3\hat{I}_z{}^2 - \hat{\vec{I}}^2 + \eta\left(\hat{I}_x{}^2 - \hat{I}_y{}^2\right)\right]. \tag{1.196}$$

In order to calculate the matrix elements it is convenient to use the operators $\hat{I}_+$ and $\hat{I}_-$ instead of the operators $\hat{I}_x$ and $\hat{I}_y$. Then Eq. (1.196) transforms into

$$\widehat{\mathcal{H}}_{\mathrm{Q}} = \frac{e^2qQ}{4I(2I-1)}\left[(3\hat{I}_z{}^2 - \hat{\vec{I}}^2 + \frac{\eta}{2}\left(\hat{I}_+^2 + \hat{I}_-^2\right)\right]. \tag{1.197}$$

When calculating NMR or NQR spectra of nuclei, the quadrupole moments of which interact with electric field gradients, one usually uses Eqs. (1.196) and (1.197), whereas to calculate dynamical processes in a nuclear spin system one applies the Hamiltonian of quadrupole interactions (1.194) in an arbitrary coordinate system written in a slightly modified form: the operators $\hat{I}_x$ and $\hat{I}_y$ are replaced by the operators $\hat{I}_+$ and $\hat{I}_-$, the components of the EFG tensor are replaced by their linear combinations

$$V_{20} = V_{z'z'}, \quad V_{2\pm1} = V_{z'x'} \pm V_{z'y'}, \quad V_{2\pm2} = \frac{1}{2}(V_{x'x'} - V_{y'y'} \pm iV_{x'y'}).$$

Then the quadrupole interaction Hamiltonian in an arbitrary Cartesian coordinate system is

$$\widehat{\mathcal{H}}_{\mathrm{Q}} = \frac{eQ}{4I(2I-1)}\sum_{k=-2}^{2} V_{2k}\hat{O}_k, \tag{1.198}$$

where

$$\begin{aligned}
\hat{O}_0 &= 3\hat{I}_z - \hat{\vec{I}}^2, \\
\hat{O}_1 &= \hat{I}_+\hat{I}_z + \hat{I}_z\hat{I}_+, \\
\hat{O}_{-1} &= \hat{I}_-\hat{I}_z + \hat{I}_z\hat{I}_-, \\
\hat{O}_2 &= \hat{I}_+^2, \quad \hat{O}_{-2} = \hat{I}_-^2.
\end{aligned} \tag{1.199}$$

As it is seen from Eqs. (1.198), (1.199) and (1.136), (1.138), the operator parts of the quadrupole interaction Hamiltonian and the dipole–dipole interaction Hamiltonian for the spins of equivalent nucleus species are written using the similar operators: the operators $\hat{O}_{ij}^q$ describing dipole–dipole interactions of identical spins turn into operators $\hat{O}_k$ when replacing $\hat{I}_{zi}$, $\hat{I}_{zj}$ by $\hat{I}_z$, $\hat{I}_{+i}$, $\hat{I}_{+j}$ by $\hat{I}_+$ and $\hat{I}_{-i}$, $\hat{I}_{-j}$ by $\hat{I}_-$.

Part I
Nuclear Magnetic Resonance (NMR)

Chapter 2
Basic Principles of Detection of Nuclear Magnetic Resonance

As we might expect, for the observation of NMR it is required permanent magnetic field $\vec{B}_0$, in which the Zeeman splitting of energy levels of the spin system is realized, and a alternating radio frequency field $\vec{B}_1$, stimulating transitions between the Zeeman sublevels. As a consequence of this interaction the spin system can emit or absorb the energy of the RF field, if the frequency of the field (ν) is close to the frequency of transitions (ν_0). Thus, for the observation of the nuclear magnetic resonance it is necessary to have the following main instruments: (i) sources of static and alternating magnetic fields, (ii) a device that converts the energy of the quantum transitions in the RF signals, (iii) amplifiers and hardware for registration of these signals.

Observation of NMR can be performed during continuous irradiation of alternating magnetic field $\vec{B}_1$, well as between its pulses. In this regard, the principles of observation NMR in condensed media are divided into two groups, differing from each other both experimental methods and classes of solved tasks. In the first years after its discovery the nuclear magnetic resonance observed mainly with stationary RF exposure field $\vec{B}_1$, but soon the technique of its observations in the presence of pulsed RF field $\vec{B}_1$ was developed. Until the 70-ties, both techniques developed in parallels, solving the characteristic for their class specific tasks (the study of spectra on the basis of a stationary radiation field $\vec{B}_1$ and relaxation processes by pulsed irradiation). The development of experimental technologies and the emergence of powerful and high-speed PC led to the formation of the Fourier transform NMR spectroscopy, which united both directions.

2.1 Nuclear Magnetic Resonance in Molecular and Atomic Beams

The phenomenon of nuclear magnetic resonance for the first time was observed by I. Rabi and co-workers (1938), in experiments with atomic beams. By this time, experiments of O. Stern and W. Gerlach were already well known, which allowed to

V. I. Chizhik et al., *Magnetic Resonance and Its Applications*,
DOI: 10.1007/978-3-319-05299-1_2,

estimate the magnetic moment of an atom, as well as works I. Estermann, O. Stern, and R. Frisch to determine the proton magnetic moment. All of them used the fact, that in an inhomogeneous magnetic field a particle, which has a magnetic moment ($\vec{\mu}$), is subjected by the force (see the details in Sect. 18.1)

$$\vec{F} = \mu_z \frac{\partial \vec{B}}{\partial z}, \tag{2.1}$$

where $\frac{\partial \vec{B}}{\partial z}$ is the magnetic field gradient; μ_z is the component of the magnetic moment directed along the axis z. Under the influence of ($\vec{F}$) the particle will be deflected in the direction of the axis z. The magnitude of this deviation (dz) will be determined by the kinetic energy of the particle (E), its magnetic moment ($\vec{\mu}$) and magnetic field gradient:

$$\mathrm{d}z = \vec{F}\frac{A}{E} = \mu_z \frac{\partial \vec{B}}{\partial z}\frac{A}{E} \tag{2.2}$$

(A is a constant depending on the geometric parameters of the device).

Knowing instrument parameters, the value of $\frac{\partial \vec{B}}{\partial z}$ and the average kinetic energy of particles in a beam, as seen from (2.1), one can define μ_z from the value of dz. However, even in the most advanced experiments of I. Estermann and O. Stern, the accuracy of the measurments of magnetic moments was not high: for the proton not better than 10 %. For determining the magnetic moment of microparticles by the Rabi method, on a certain part of the particle trajectory the conditions for magnetic resonance is created. The interaction of nuclear (or electronic) magnetic dipole with alternating field has been brought to its reorientation. The method consists in the following. The low-pressure chamber O (Fig. 2.1a) ($P \simeq 10^{-6}$ mm Hg $\simeq 10^{-8}$ Pa) creates a beam of particles (atoms or molecules) which trajectories sequentially pass through the gaps of magnets A and B. These magnets are identical: inhomogeneous magnetic field in their gaps is equal in magnitude $|B_\mathrm{A}| = |B_\mathrm{B}|$, but with opposite direction of the vectors of field gradients, and $|\ \partial B/\partial z\ | \simeq 800$ T/m. In this regard, the particle, flying through the gaps between magnets A and B, will be subjected to the same deflection, first, in one direction and then in the opposite direction and get on sensor detector D (Fig. 2.1b).

If the investigated particles are molecules with compensated electron spins (molecules of $^1\Sigma_0$-state) or atoms then the magnetic properties of these particles and their interaction with external magnetic fields are determined only by moments of the nuclei. Between the magnets A and B there is the magnet C, in which gap a highly-homogeneous magnetic field $\vec{B}_0$ is created. In addition, in the magnet gap the loop L of length l is set, which created an alternating magnetic field $\vec{B}_1(t)$, and which oriented perpendicular to the field $\vec{B}_0$. The principle of operation of the loop is illustrated in Figs. 2.1b and 2.2a.

As a result of effect of the static $\vec{B}_0$ and alternating $\vec{B}_1$ magnetic fields on magnetic dipole $\vec{\mu}$, this dipole will be re-oriented (see Sect. 1.2). The length of the loop l and

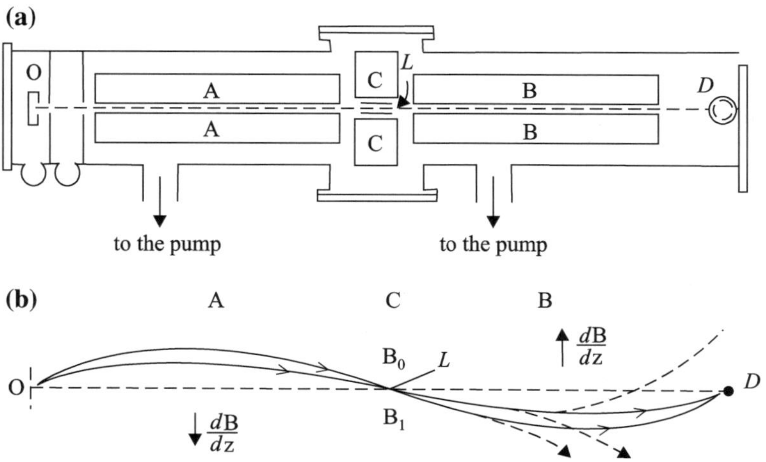

Fig. 2.1 The functional diagram of the detection of NMR by Rabi method

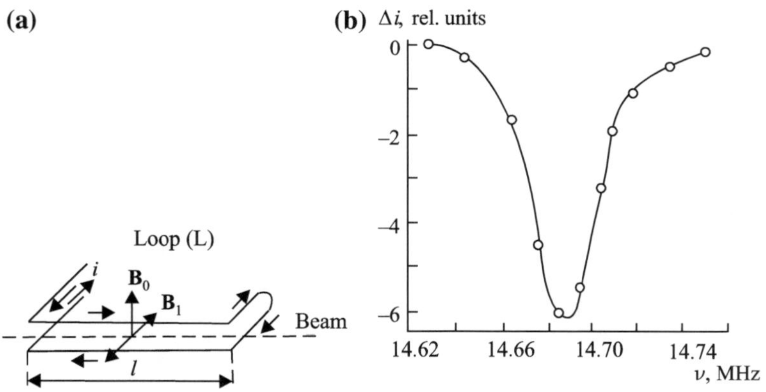

Fig. 2.2 **a** The scheme of a loop for creating of the alternating magnetic field; **b** The resonance curve of 1H nuclei in the beam of the KOH molecules

the field strength B_1 is matched so, that for the span time (t) of particles through the magnet gap, all the dipoles $\vec{\mu}$ were reoriented (sign μ_z is changed to the opposite one). Therefore, in accordance with (2.1), in the magnet gap B they will be deflected in the same direction as in the magnet gap A, and will not reach the detector D. Since the intensity of the reorientation process of magnetic dipoles increases with the frequency detuning ($\Delta\nu = \nu_0 - \nu$) decrease, then the flow particles falling into the slot of detector D must decrease. Consequently, the current i in the detector circuit will change simultaneously with the change $\Delta\nu$. This change of $i(\Delta\nu)$ has a resonant character. As an example, Fig. 2.2b shows the NMR signal from nuclei ^{1}H in the beam of the KOH molecules, obtained in the $B_0 = 0.3453$ T.

The observed resonance lines in such a way allow us to determine the frequency and line width, to investigate the shape of the signal and other characteristics of the spectrum of nuclear magnetic resonance in atomic and molecular beams.

If one studies hydrogen molecules, the NMR signal is observed only from the ortho-hydrogen (which possesses the total nuclear spin $I = 1$). In molecules of parahydrogen ($I = 0$) spins and magnetic moments of the proton pair are antiparallel, and the reorientation of the magnetic dipoles of individual protons in the magnet C does not change the magnetic properties of the molecule as a whole and the trajectory of its movement.

Using NMR on beams, gyromagnetic ratios and magnetic moments of many nuclei were measured with greater accuracy than it was done previously by other techniques.

2.2 The Sources of Magnetic Fields

The range of used fields values is approximately 0–20 T. As a rule in the region 0.1–2 T electromagnets or permanent magnets are used and fields with the induction of more than 2 T for the nuclear magnetic resonance can be obtained only in solenoids with a superconductive winding coil. At low fields (less than 0.1 T) one uses conventional solenoids or even Earth magnetic field (0.00005 T).

The main requirements to the static magnetic field are (i) sufficiently high homogeneity within the scope of an investigated sample and (ii) field stability over time. The value of the average deviation of the field ΔB within a sample divided to the magnitude of the resonance field B_0 is taken as a measure of the field homogeneity. For measurements of relaxation times and self-diffusion coefficient in many cases it is sufficient to achieve the value $\Delta B/B_0$ of order of 10^{-5} to 10^{-6}, but to obtain high resolution NMR spectra a value $\Delta B/B_0$ is required not worse than 10^{-8} to 10^{-10}. A high degree of homogeneity in the gap of an electro-or permanent magnet can only be achieved through the set of design measures to gradually increasing the field homogeneity. The yoke of magnet is usually of H-shaped form that provides sufficient rigidity and symmetry of design in conditions of significant static compression loads between the poles (mutual force attraction of the pole pieces may be several tons).

To reduce edge effects (decreasing of the induction between poles at the edges of a gap relative to the center), the diameter of the pole pieces (D) is chosen much larger than the gap between poles (d): $D/d \geq 10$. With the same aim one can use the so-called *shimming*—a decrease of the d at the periphery of the gap by introducing additional annular rings (Fig. 2.3a). By special design of the rings it is possible to expand significantly the volume of homogeneous field (see in Fig. 2.3b, dashed curve). The shimming efficiency drops rapidly with an increase of the level of B_0 due to saturation of iron at $B_0 > 1$ T and in this connection this type of shimming is almost never used in modern instruments.

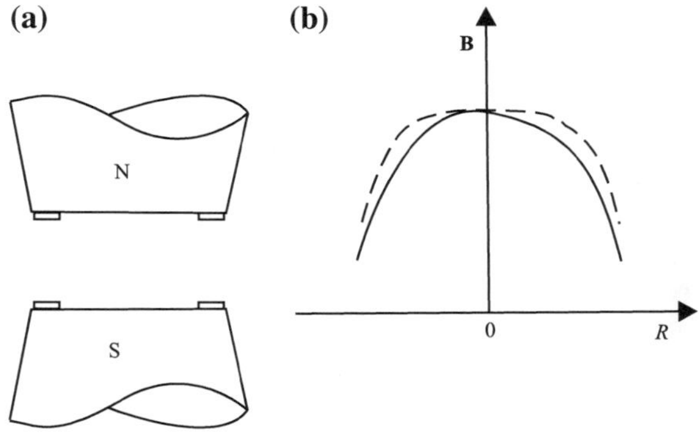

Fig. 2.3 The influence of the shimming on the magnetic field in a magnetic gap

Pole pieces are made using soft (without carbon impurities) iron, followed by a thorough multi-stage machining, polishing and annealing in hydrogen atmosphere. As a result of processing, mechanical irregularities do not exceed 0.25 μm. Parallelism of surfaces of the pole pieces also significantly effects on the resulting homogeneity. Thus, the deviation from parallelism in 2–3 μm at $D \simeq 30$ cm leads to inhomogeneity $\Delta B/B_0 \simeq 10^{-6}$. Careful adjustment of poles with control by NMR signal allows us to obtain $\Delta B/B_0 \simeq 10^{-7}$. The next step towards improving the field uniformity is the compensation of inhomogeneities using the current-carrying conductors—*current shims*.

The magnetic field in a certain volume, which has no sources, can be described analytically using the scalar potential Φ, satisfying the Laplace equation $\nabla^2\Phi = 0$. Then the field in this volume is $\vec{B} = -\vec{\nabla}\Phi$. The potential Φ can be represented as a series of three-dimensional spherical harmonics:

$$\Phi = \sum_{n=1}^{\infty}\sum_{m=0}^{n} r^n P_n^m(\cos\theta)[A_n^m \sin m\varphi + B_n^m \cos m\varphi].$$

Here $P_n^m(\cos\theta)$ is the adjoint Legendre polynomial. Coefficients A_n^m and B_n^m do not depend on coordinates and are completely determined by the sources of the field. The terms with $n = 1$ describe three components of the field (uniform field), with $n = 2$—field gradients. The problem is in the design of current loops which magnetic fields will reproduce separately each harmonic of the general expansion of the field. To compensate the inhomogeneity we need to choose the electric current in the loops so that their magnetic fields will be equal in magnitude to the corresponding harmonics and opposite in sign. It is easy to achieve experimentally, setting sequentially in each compensator such a current at which the highest resolution is obtained for the NMR spectrum. A sample should fill that volume, in which the used spatial harmonics form a closed and orthogonal system of functions, i.e. for a spherical sample centered at the origin are suitable spherical harmonics.

Due to the mutual orthogonality of spherical harmonics the configuration of the field created by one of the compensators, can not be obtained through any combination of fields of all other compensators, so the current value in each of the compensators, corresponding to the best uniformity, does not depend on currents in other compensators. Strictly speaking, the orthogonality holds only for a spherical sample centered at the origin. However, almost independent adjustment is saved for other configurations (for example, cylindrical ones), but not too elongated or flattened. In order to achieve full compensation, in principle, it is necessary to create an infinite number of harmonics, but the contribution of higher harmonics is small, thus one uses components with $n = 2, 3, 4$.

In calculations of the current configuration it can be assumed, that the component of the field, which is tangential to the surface, is perpendicular to the vector current density surface, i.e. current vector is directed along the lines of equal potential: $\Phi =$ const. Obtained configuration of conductors for planar surfaces of electromagnet pole pieces are complex and is not practically feasible to manufacture.

B. Anderson (1960) proposed a simpler system of currents based on expansion of the field in the series:

$$B_{0z} = B_0 + \left(\frac{dB_z}{dx}\right)x + \left(\frac{dB_z}{dy}\right)y + \left(\frac{dB_z}{dz}\right)z + \dots \tag{2.3}$$

Here, the first derivatives with respect to three coordinates are linear field gradients. Although it is an infinite power series, only a small number of gradients is practically significant. For a rough compensation three linear and three quadratic derivatives can be used. Linear gradient is easily obtained using a system of two parallel conductors with a current. As can be seen from Fig. 2.4a, two long conductors with equal currents create at the axis R a linearly varying field. The real configuration of the conductors, located on each of the planes pole pieces, is shown in Fig. 2.4b. Since the system antisymmetric, then the expansion (2.3) contains only odd powers decomposition. Variables h_1 and h_2 can be chosen so, that factor with third-degree expansions were equal to zero. Then the first noncompensated coefficient would be only one of the fifth degree, the value of which is negligible. The calculation gives the following relationship:

$$h_{1,2} = (3 \mp 2\sqrt{2})\delta.$$

(δ is the distance between the planes, where conductors are placed).

The system, shown in Fig. 2.4b, creates a gradient $\mathrm{d}\vec{B}_{0z}/\mathrm{d}x$. It is obvious, that for the formation of the gradient $\mathrm{d}\vec{B}_{0z}/\mathrm{d}y$, we can use a similar system, but turned on 90° around the axis z. To create the gradient $\mathrm{d}\vec{B}_{0z}/\mathrm{d}z$ it is more convenient to use two rings with opposite direction of current (see Fig. 2.4c). The optimum ratio between the radius of the ring R and value δ is given by

$$R = \frac{2}{\sqrt{3}}\delta.$$

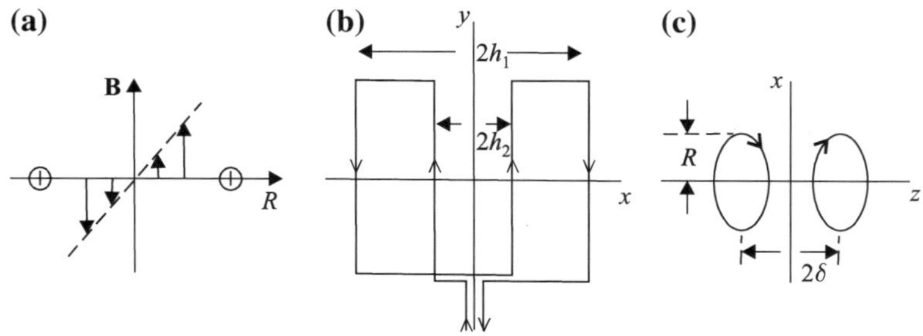

Fig. 2.4 The principles of creating of the field gradients. **a** The creating of the field gradient by two linear conductors. **b** The practical scheme for creating the field gradient $\mathrm{d}\vec{B}_{0z}/\mathrm{d}x$. **c** The scheme of two rings for creating the field gradient $\mathrm{d}\vec{B}_{0z}/\mathrm{d}z$

Fig. 2.5 The effect of sample rotation on a spectral line

Similarly the gradients of higher orders are formed.

The system of the current compensator allows us to improve the homogeneity, at least an order of magnitude (roughly to 5×10^{-8}). Further improvement of the homogeneity is achieved by spinning of a sample. If the magnitude of inhomogeneity in the sample volume is ΔB, the rotation of an ampoule with a sample averages the inhomogeneity of field to some effective value:

$$\Delta B_{\mathrm{eff}} = \gamma \langle (\Delta B)^2 \rangle \tau,$$

where $\langle (\Delta B)^2 \rangle$ is the standard deviation of the field within a sample, τ is the averaging time (time of one revolution of an ampoule). The rotation of an ampoule performed by means of a gas turbine with compressed the air. The frequency of rotation is chosen sufficiently high (up to 300 Hz) in order to reduce the value of ΔB_{eff}. In addition, rotation of a sample at low frequencies leads to satellites in a spectrum—additional signals located symmetrically with respect to the main ones (Fig. 2.5a, b). The increase of rotation speed reduces them and moves beyond the recorded range of the spectrum (Fig. 2.5c).

Requirements to the stability of the field strength must be matched with degree of homogeneity, at least for the time needed for the NMR signal registration. Stabilization of the field $\vec{B}_0$, generated by electromagnets or coils, is usually provided by a multi-stage system containing an electronic current stabilizer, stabilizer of induction flux in the magnet gap and stabilizer of the resonance conditions on the basis of the NMR phenomenon (lock-stabilization). In the superconducting solenoids required stability is ensured without any additional hardware, only by lock.

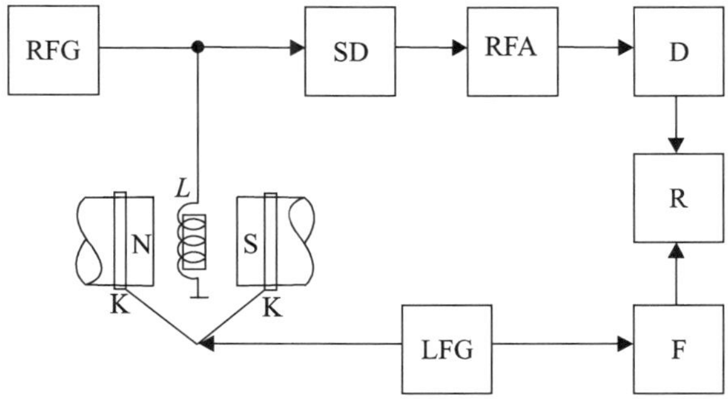

Fig. 2.6 The simplest block-scheme for the NMR observation by the stationary method

2.3 The Stationary Methods of Observation of Nuclear Magnetic Resonance

Let us first consider the basic principles of the NMR observation using stationary radiofrequency field $\vec{B}_1$. Now these methods are used in NMR in very rare cases but they are usual in EPR. So, it is reasonable to consider them in historical and methodological aspects.

The simplest functional diagram for the observation of NMR signals is shown in Fig. 2.6. The coil (L) with the investigated sample is placed in the gap of a magnet or other source of a static magnetic field $\vec{B}_0$ and oriented so that its axis and, hence, the vector of the alternating field $\vec{B}_1$ are perpendicular to the direction of the field vector $\vec{B}_0$. The radiofrequency field $\vec{B}_1$ is created in the coil L connected to the radio frequency generator (RFG). In order to perform periodic observation of the NMR the static magnetic field is modulated by additional coils (K), connected to a generator of sinusoidal or sawtooth voltage of low frequency (LFG). The horizontal sweep signal on the oscilloscope screen is realized also with the LFG. Since the NMR signal reaches a maximum twice for a period of a sine sweep, then in this case to coincide the two signals in the screen an adjustable link phase shifter (F) may be used.

The main part of the device is a spin detector (SD), which converts the energy of the quantum transitions between Zeeman sublevels in the radio frequency signals. A variety of devices containing resonant circuits (RF bridge, RF-generator of weak oscillations, superregenerator and etc.) is used as a spin detector. Obtaining in a spin detector RF signals are amplified (RFA), detected (D), and fed to the registration device (R) (oscilloscope, computer, etc.). There are four main elements of the scheme: (1) a spin detector; (2) a source of the magnetic field $\vec{B}_0$; (3) a system of amplification, detection, and registration; (4) a system for passing through the resonance (modulation of the field $\vec{B}_0$).

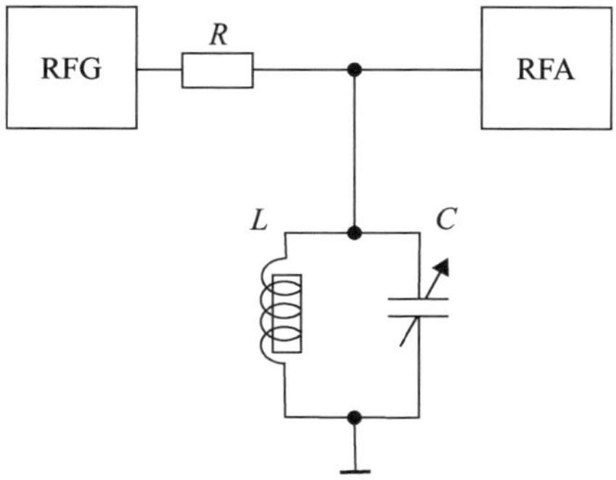

Fig. 2.7 The functional scheme of the spin detector of Rollin

2.3.1 Spin Detector

This term will denote a certain electrical circuit in which there are signals due to the precession of the vector macroscopic nuclear magnetization ($\vec{M}$). In the most cases, the basis of such a scheme is the coil (L), included in the resonant circuit tuned to the precession frequency of the $\omega_0 = \gamma B_0$. The simplest scheme of the spin detector was used by B. Rollin (see Fig. 2.7).

The RF generator with a frequency ω is connected to the parallel circuit LC, in the coil of which a sample is placed, through the ballast resistor R. The value of R is much larger than the resonant resistance of the LC circuit Z_0, so there is a mode of current generator (current, consumed from the RFG, does not depend on Z_0). We know that under NMR conditions the maximum absorption of the RF energy in a spin system takes place. The energy loss in the resonant circuit caused by this absorption leads to a decrease of quality factor Q of LC- circuit, hence to decrease of its total resistance Z_0 and, respectively, the RF voltage on it falls. In fact, this scheme is similar to the well-known in radio engineering circuit of a Q-factor meter. If, in addition, periodically to realize the resonance condition $\omega = \omega_0 = \gamma B_0$, modulating, for example, the magnetic field B_0, then RF voltage on the LC-circuit will be modulated in its amplitude. After appropriate amplification and detection the signal of nuclear resonance absorption can be observed on an oscilloscope screen or another registrator.

Let us obtain the formula for the quantitative estimates of this phenomenon. The conductivity of a parallel circuit (Y) is the sum of the conductivities of its branches:

$$Y = \frac{1}{R} + i\left(\omega C - \frac{1}{\omega L}\right). \tag{2.4}$$

The rotating component of the nuclear magnetization ($M_\perp = (M_x^2 + M_y^2)^{\frac{1}{2}}$) generates in the coil the additional magnetic flux (Φ_{add}):

$$\Phi_{\text{add}} = \xi B_1 S = \xi \cdot 4\pi M_\perp S = 4\pi \chi \xi B_1 S,$$

Here $\chi = M_{\perp}/B_1$ is the dynamic nuclear susceptibility; ξ is the filling factor of the coil by a sample, which takes into account that not the whole flow Φ_{add} crosses the transverse section of the coil. The appearance of Φ_{add} can be regarded as change of the inductance of the coil on the value of L_{add}:

$$L = \frac{\Phi + \Phi_{\mathrm{add}}}{I} = \frac{B_1 S}{I} + \frac{4\pi\chi\xi B_1 S}{I} = L_0 + L_{\mathrm{add}}.$$

Here I is the current in the coil; S is its cross section. Thus, in formula (2.4) instead of L one should write

$$L = L_0(1 + 4\pi\chi\xi),$$

and then instead of (2.4) one obtains

$$Y = \frac{\omega L_0 + \mathrm{i}\cdot 4\pi\chi\xi R}{R\omega L_0}.$$

The voltage at the resonant circuit (U) is calculated as

$$U = \frac{I}{Y} = \frac{IR}{1 + i\cdot 4\pi\chi\xi Q},$$

where $Q = R/(\omega L)$ is the Q-factor of the circuit. In the absence of the signal of nuclear resonance $U = IR = U_0$. Since the value χ is of order 10^{-5} to 10^{-7}, then $4\pi\chi\xi Q \ll 1$, even if Q-factor $Q > 100$. Therefore we can put

$$U \simeq U_0(1 - i\cdot 4\pi\chi\xi Q). \tag{2.5}$$

Nuclear susceptibility is a complex value (see Sect. 1.5): $\chi = \chi' - \mathrm{i}\chi''$, and we have

$$U = U_0[1 - 4\pi\xi Q(\chi'' - \mathrm{i}\chi')] = U_0 + U_{\chi}, \tag{2.6}$$

$$U_{\chi} = U_{\chi''} - \mathrm{i}U_{\chi'} = -4\pi\xi Q U_0(\chi'' + \mathrm{i}\chi'). \tag{2.7}$$

It is seen from (2.4), (2.5), that the voltage on the resonant circuit is changed on the value of $U_{\chi''} = -4\pi\xi Q U_0\chi''$, which is the *measure of the nuclear resonance absorption*.

The vector diagram illustrating the relationship component of $\vec{U}_0, \vec{U}_{\chi}, \vec{U}_{\chi'}$ and $\vec{U}_{\chi''}$ is shown in Fig. 2.8. Under the periodically coming nuclear resonance, the voltage on the resonant circuit changes by the value $U_{\chi''}$, causing its amplitude modulation, and the component $\vec{U}_{\chi'}$ determines phase change $\vec{U}_0$ at the angle ϕ. The modulation depth is small (about 10^{-2} to 10^{-4}), therefore, pre-amplification of the signal quickly leads to overload an amplifier, that is one of the main drawbacks of this spin detector. However, the simplicity of the scheme, the observation of the only absorption signal, the absence of complicated settings and relatively broad resonance curve of the

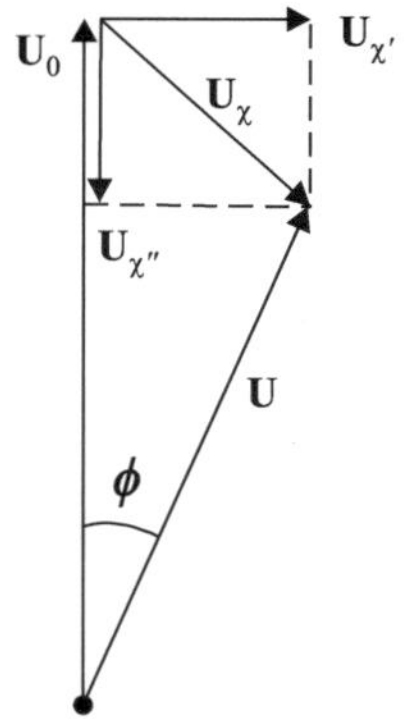

Fig. 2.8 The vector diagram of NMR signal components

resonant circuit, reducing the influence on the detector microphonics and temperature fluctuations, are advantages of this simple spin detector. To increase the modulation depth of U_0 by the NMR signal different versions of bridge circuits, one of which shown in Fig. 2.9 can be used. It should be kept in mind that the total compensation the voltage U_0 at the output of the bridge leads to the fact that the resulting signal is a mixture of dispersion and absorption signals: $|U_\chi| = (U_{\chi'}^2 + U_{\chi''}^2)^{\frac{1}{2}}$, that is inconvenient for the analysis of the spectra. Therefore, using the elements of C_1' and C_2 a certain imbalance of the bridge can be achieved, so that the output voltage is $U = U_z + U_\chi$ (U_z is the voltage unbalance, $U_z \ll U_0$). If $|U_z| \gg |U_\chi|$, then $|U| = |U_z| + |U_\chi|$ is approximately equal to the projection of U on the direction of the vector U_z. Then we have

$$U \simeq U_z - U_{\chi''} \cos\phi - U_{\chi'} \sin\phi.$$

There are two extreme cases:

$$\text{(a) } \phi = 0, \pi;\ U = U_z \pm U_{\chi''};$$
$$\text{(b) } \phi = \pi/2, 3\pi/2;\ U = U_z \pm U_{\chi'}.$$

Thus, depending on the type of the phase-balancing bridge, the amplitude modulation of the output voltage signal is caused by absorption or dispersion signals. The converse is obvious: if we measure some way phase shift of the voltage, than in the case (a) we obtain the dispersion signal, and in the case (b) the absorption signal is registered.

One of the original species of the bridge spin detector is a system of crossed coils (Fig. 2.10), which were used by F. Bloch. The axes of the transmitter (L_x) and receiving (L_y) coils are located at the angle 90°, so that, ideally, at such arrangement there is not a direct signal of RF generator from the transmitter coil in the receiving one. The field $\vec{B}_1$,which is produced by the coilL_x connected to the generator, excites

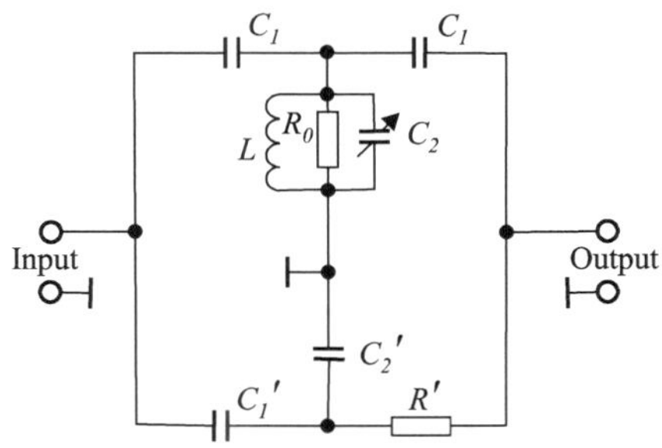

Fig. 2.9 The scheme of the RF-bridge circuits

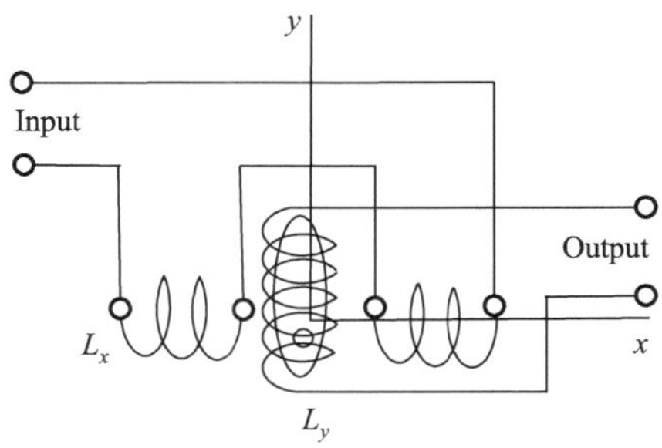

Fig. 2.10 The system of crossed coils by Bloch

in the sample the transverse component of the nuclear magnetization $M_{\perp} = M_x + iM_y$. A rotating component $M_{\perp}$ creates in the receiving coil L_y (with a cross section S and the number of turns of N) the alternating magnetic induction $\vec{B}_y = 4\pi \vec{M}_{\perp}$, which is induces an emf at its ends:

$$U_y \simeq -4\pi NS(dM_y/dt) = \pm 4\pi NS(u \cos \omega t - v \sin \omega t). \qquad (2.8)$$

Plus or minus signs depend on the sign of the gyromagnetic ratio of nuclei (see Sect. 1.5). It is seen from (2.8), that the induced emf in the coil consists of two component: $U_y(u)$ and $U_y(v)$. Thus, similar to bridge spin detectors, with the full compensation of the generator voltage at the output of the system with crossed coils there is a mixture of absorption and dispersion signals. To receive unambiguous signals of $U_{\chi'}$ or $U_{\chi''}$ it is necessary to submit a voltage from the generator to the coil L_y, that is realized by the amplitude or phase unbalance the system.

Let us consider spin detector of another type—weak oscillation generator *(autodyne)*. For the first time it was used by E. K. Zavoiskii for the observation of electron paramagnetic resonance (in 1944), and in the following to record signals caused by nuclear magnetism. As it has been established, the appearance of the NMR signal changes the impedance and resonance frequency of parallel tuned LC-circuit. As a result the change of amplitude and frequency of the generated voltage is involved, that after appropriate detection can give signals absorption. One of the simplest schemes

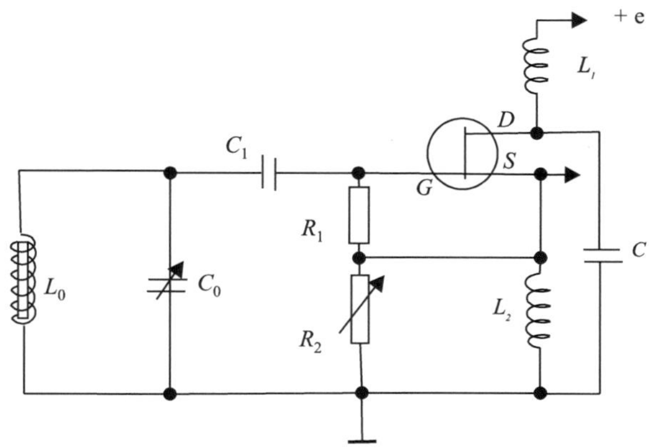

Fig. 2.11 The electrical schematic diagram of the autodyne

of this generator is shown in Fig. 2.11, where L_0, C_0 are the inductance and capacity the resonant circuit; R_1, R_2 are the resistances of feedback connection; L_1, L_2 is the inductors; C is the capacitor of lower frequencies filter.

As such, the weak oscillation generator—*autodyne*—is a conventional RF generator of Hartley type, which theory is well-developed. An autodyne feature is that it generates a voltage with a small amplitude (tens or hundreds millivolts), and in this mode even very small changes of the parameters of the resonant circuit of the generator result in significant changes in the amplitude of the voltage generated by it. From the theory of the Hartley generator it is known that stable free oscillations occur when the values of feedback coefficient(k) exceeds a certain value (k_{cr}), i.e. at

$$k > k_{cr} = \frac{R_0 C}{SL} = \frac{1}{SZ_k}, \tag{2.9}$$

here R_0 and C are the parameters of parallel resonant circuit; Z_k is its equivalent resistance; S is the slope of characteristics of amplifier element (transistor). The dependence of the RF voltage U in the circuit from the coefficient feedback at the soft mode of generation is shown in Fig. 2.12. We have already noted that equivalent resistance of the parallel circuit decreases at nuclear magnetic resonance. According to (2.9), this leads to the increase of critical values of the feedback on Δk_{cr}. The entire plot of dependence of the voltage in the circuit on the feedback coefficient is shifted to the right (dashed curve), and for the established value of feedback coefficient k_0 the amplitude of the generator reduces on the value ΔU. The figure also shows that the change ΔU under the influence of the nuclear magnetic resonance will be the greater, the closer the feedback factor of the autodyne k_0 to k_{cr}. Consequently, the maximum sensitivity is achieved at the threshold of oscillations failure.

Due to their simplicity and relatively high sensitivity autodyne spin detectors were widely used by researchers in the early years of RF spectroscopy.

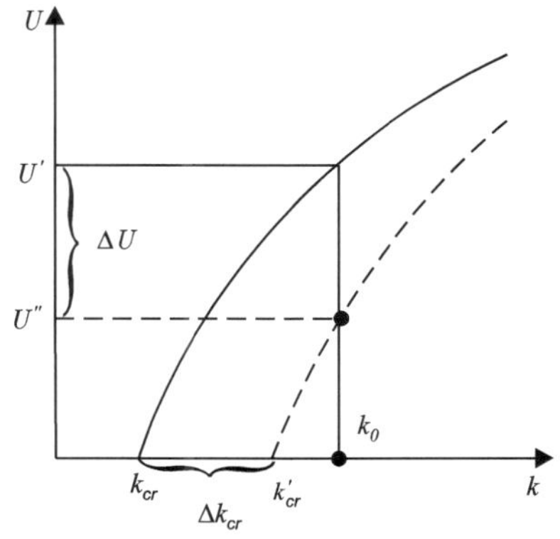

Fig. 2.12 The plot of voltage dependence of an autodyne on the values of feedback coefficients

2.3.2 Passage Through Resonance

In order to observe the spectral line of NMR it is necessary to organize the passage through resonance. This situation can be realized both with the periodic modulation of the static field B_0 near the resonant value $B_0 = \omega_0/\gamma$, and with the help of the modulation of the generator frequency. The latter method is less convenient, because it requires a large hardware efforts and has a number of shortcomings (for example, parasitic amplitude modulation is comparable to the magnitude of the signal). Recall that for the correct reproduction of the spectral shape of a line (see Sect. 1.6) it is necessary to fulfil two conditions: first, the RF amplitude of B_1 should be small enough to avoid saturation effects, and secondly, the rate of passage through the resonance should also be small, to avoid the distortion caused transient processes in the system of nuclear moments. As any physical system, the nuclear magnetization has some inertia, which can be described time constants T_1 and T_2. The previous expressions for the components of the magnetization u and v describe a stationary state of the system, i.e. processes of formation $M_\perp$ and M_z are balanced by relaxation processes (the value of the field B_1 is small enough). If at any time to stop the passage through resonance, the values component u and v from that moment can be saved as long as one wishes. In other words, transverse component of the nuclear magnetization ($M_\perp$) "should follow" the direction of the field vector $\vec{B}_1$.

In the case of fast passage through the resonance, the "detachment" $\vec{M}_\perp$ from the vector $\vec{B}_1$ occurs and then $\vec{M}_\perp$ continues to precess independently with the variable angular frequency ω_0. Since the frequency of the generator ω are the same, then in the spin detector there is the interference of two oscillations with frequencies ω and ω_0. The result is beats with increasing frequency ("wiggles"). The phase difference between two signals in the case of a linear variation of the field $B_0(t) = B_0 + \frac{\mathrm{d}B_0}{\mathrm{d}t}t$ is

$$\phi(t) = \int_0^t \gamma[B_0(t) + B_0]\mathrm{d}t = \frac{1}{2}\gamma\frac{dB_0}{dt}t^2 + \omega_0 t. \tag{2.10}$$

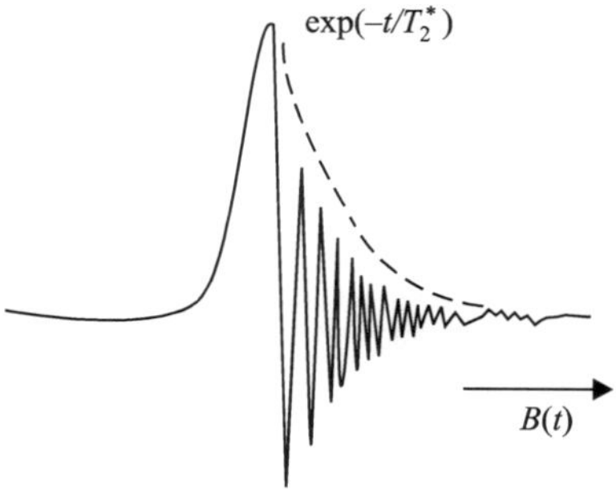

Fig. 2.13 The beats of NMR signal at the fast passage (wiggles)

The signal with beats is in the form, described by the expression (see also Fig. 2.13):

$$U = U_0 \exp\Big(-\frac{t}{T_2^*}\Big)\cos\Big(\omega_0 t + \frac{1}{2}\gamma\frac{dB_0}{dt}t^2\Big).$$

The exponential factor is introduced to take into account the decay the precession components $M_\perp$.

It was mentioned in Sect. 1.7.2 that the experimentally observed spectral line width of the magnetic resonance is characterized by the value T_2^*, which is determined by both the nature of the magnetic interactions in the matter, and inhomogeneity of the magnetic field B_0. If the nature of field inhomogeneity B_0 is similar to the Lorentz distribution, the additional broadening of the signal can be described by a value $T_2^{*'}$, and the total width of the line—by the value T_2^*:

$$\frac{1}{T_2^*} = \frac{1}{T_2} + \frac{1}{T_2^{*'}}. \tag{2.11}$$

It must be kept in mind that such a character of the broadening is not always valid and then (2.11) becomes incorrect. A detailed analysis of the solutions of the Bloch equations taking into account the passage speed of the line leads to the following criterion of the correct reproduction of the line shape:

$$\frac{dB_0}{dt} \le \frac{1}{4\gamma T_2^{*2}}. \tag{2.12}$$

If NMR signals are observed in solids, characterized by very small T_2, the criterion (2.12) is not determinative, but in this case the large width of the spectral lines $\Delta B \sim (1/T_2)$ allows one to use another approach for the choice of the amplitude of the sinusoidal modulation of the field B_m. As it can be seen from Fig. 2.14, if $B_m \ll \Delta B$, the signal at the output of the spin detector (in that situation is most often used an autodyne) is proportional to the slope of the line at the given point.

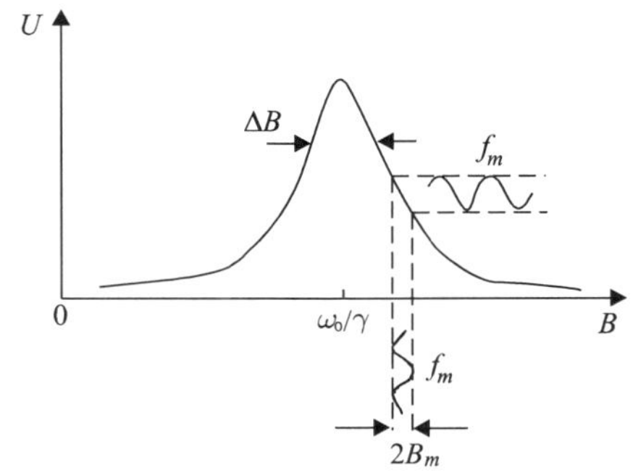

Fig. 2.14 The principle of the differential NMR passage

At the slow passage through the whole line (for example, due to changes in ω_0), the NMR signal U is proportional to the derivative of the line shape (*differential passage*). The exact correspondence to the first derivative is possible only if $B_m \to 0$, but in this case the amplitude $U \to 0$, so in practice, a distortion of the spectral line is allowed, which in principle can be taken into account with known modulation parameters. An additional advantage of the differential passage is the possibility of increasing the sensitivity of the device via a strong narrowing of a bandwidth of a receiving path, because the detected signal at the output receiver is close to sinusoidal modulation frequency f_m ($f_m \simeq$ 20–100 Hz). Receiver, recorder and modulator are not different from appropriate series of radio equipment, so that blocks are not described here.

2.4 Pulse Methods of NMR Observation

2.4.1 The Action of the 90-degree Pulse

In Sect. 1.8 the result of RF pulse action on a system of nuclear moments is discussed in detail. For a visual representation of the majority of experimental methods based on pulse action, it is convenient to use the vector model of the interaction of the macroscopic nuclear magnetization $\vec{M}_0$ and field $\vec{B}_1$, created by the coil of a spin detector.

We introduce a rectangular coordinate system x, y, z, rotating around the axis of z, which is directed along the vector of the static field $\vec{B}_0$. We assume that frequency rotation ω is equal to the frequency of RF pulse. The result of the pulse action can be obtained from the Bloch equations. Let components of the field $\vec{B}_1$ in (1.56) during the time $0 \le t \le t_{imp}$ can be expressed as follows: $B_x = 2B_1 \sin \omega t$, $B_y = 0$, $B_z = B_0$. Then, neglecting relaxation processes during the pulse (pulse duration is chosen such that $t_{imp} \ll T_1, T_2^*$), one can write (1.56) in coordinate system rotating with the frequency ω:

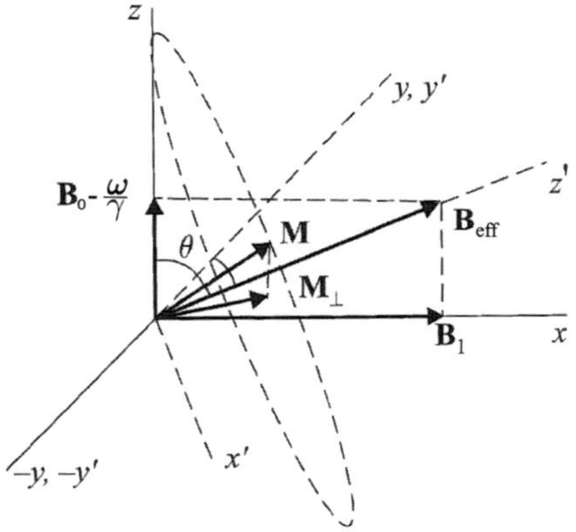

Fig. 2.15 The scheme of nutation of the magnetisation vector around the effective field B_{eff}

$$\begin{aligned} \mathrm{d}M_x/\mathrm{d}t &= \gamma(B_0 - \omega/\gamma)M_y, \\ \mathrm{d}M_y/\mathrm{d}t &= \gamma M_z B_1 - \gamma(B_0 - \omega/\gamma)M_x, \\ \mathrm{d}M_z/\mathrm{d}t &= -\gamma M_y B_1. \end{aligned} \tag{2.13}$$

The solution of (2.13) with the initial conditions $M_y(0) = 0, M_x(0) = 0, M_z(0) = M_0$ is:

$$\begin{aligned} M_x(t) &= M_{z0}\frac{\omega_1 \Delta\omega}{\omega_{\text{eff}}^2}(1 - \cos\omega_{\text{eff}}t), \\ M_y(t) &= M_{z0}\frac{\omega_1 \sin\omega_{\text{eff}}t}{\omega_{\text{eff}}}, \\ M_z(t) &= M_{z0}\left(\frac{\Delta\omega^2}{\omega_{\text{eff}}^2} + \frac{\omega_1^2\cos\omega_{\text{eff}}t}{\omega_{\text{eff}}^2}\right), \end{aligned} \tag{2.14}$$

here $\omega_{\text{eff}}^2 = \omega_1^2 + \Delta\omega^2 = \gamma^2 B_1^2 + \gamma^2(B_0 - \omega/\gamma)^2 = \gamma^2 B_{\text{eff}}^2$. Equations (2.14) describe the precession of the vector $\vec{M}$ around effective vector field $\vec{B}_{\text{eff}}$—the so-called *nutation*. It is easily seen if we go to a system of coordinates $x'y'z'$, rotated relatively to the initial system on such an angle in the plane $(z0x)$, that the axis $0z'$ coincides with the direction of the vector field $\vec{B}_{\text{eff}}$ (Fig. 2.15). In the new system the components of $\vec{M}$ will be written as:

$$M_{y'}(t) = M_0\frac{\omega_1 \sin\omega_{\text{eff}}t}{\omega_{\text{eff}}}, \quad M_{x'}(t) = M_0\frac{\omega_1 \cos\omega_{\text{eff}}t}{\omega_{\text{eff}}}, \quad M_{z'}(t) = M_0\frac{\Delta\omega}{\omega_{\text{eff}}}.$$

The vector $\vec{M}$ moves along the cone with angle θ determined by the deviation of frequency precession from the frequency of the RF pulse: $\cos\theta = \Delta\omega/\omega_{\text{eff}}$. For the small deviation $(B_1 \gg B_0 - \omega/\gamma)$ $\omega_{\text{eff}} \simeq \omega_1$, i.e. the cone degenerates into a plane $(z0y)$. If in addition the pulse duration

$$t_{\text{imp}} = \frac{\pi}{2\omega_1},$$

then at the time t_{imp} vector $\vec{M}$ will be in the plane $(x0y)$ (90-degree pulse). It is evident that nuclear magnetization vector can be rotated to the plane $(x0y)$ for sufficiently large detuning $\Delta\omega$. This requires only the condition $\Delta\omega \leq \omega_1$, i.e. the angle of the cone $\theta \geq 45°$. The duration of the pulse, rotating the vector $\vec{M}$ into plane $(x0y)$, can be found from the condition $M_z(t_{\text{imp}}) = 0$:

$$t_{\text{imp}} = \frac{1}{\omega_{\text{eff}}} \arccos \frac{\Delta\omega^2}{\omega_1^2}.$$

2.4.2 The Amplitude of Signal of the Free Induction

After the end of the 90-degree pulse we have the nonstationary state for nuclear magnetization (the vector $\vec{M}$ is perpendicular to vector field $\vec{B}_0$, the component $M_z = 0$). As it follows from Bloch equations (1.56), the behavior of the transverse components is described by functions

$$M_x(t) = M_0 \cos(\gamma B_0 t) \exp(-t/T_2^*),$$

$$M_y(t) = M_0 \sin(\gamma B_0 t) \exp(-t/T_2^*).$$

This behavior can be interpreted as a *free* (in the absence of the field $\vec{B}_1$) *precession* of the transverse component of the nuclear magnetization $M_\perp$ around the vector field $\vec{B}_0$.

Alternating magnetic flux in the coil of spin detector due to the precessing magnetization, in accordance with classical laws of electromagnetic induction, gives the emf of nuclear induction at the ends of the coil. For the correct estimation of the magnitude of this emf we use the well-known theory of electromagnetism: the reciprocity theorem for two circuits with a current. Represent an arbitrary element of a sample volume dV with nuclear magnetization $\vec{M}_0 dV$ in the form of an elementary contour ΔS with the current I: $\vec{M}_0 dV \equiv I\Delta S$. Another contour with the current I_1 is the coil of a spin detector. If we denote the flow of induction, created in the coil by the element $I\Delta S$ as Φ_{m} and the flow generated by the coil with the current I_1 through the element ΔS as Φ_1, then, by the reciprocity theorem, $n\Phi_{\text{m}}/I = \Phi_1/I_1 = M$, (where M is the coefficient of mutual inductance of two circuits; n is the number of turns in the coil).

Since $\Phi_1 = \vec{B}_1 \Delta S$, we can write for $\Phi_m = \vec{B}_1 \Delta S I/(n I_1) = \vec{B}_1 \cdot \vec{M}_0 dV/(n I_1)$. Then the flow, generated in the coil from the whole sample, is:

$$\Phi = \frac{1}{nI} \int_{V_0} \vec{B}_1 \cdot \vec{M}_0 dV,$$

and for the nuclear induction emf we have

$$E = -n\frac{d\Phi}{dt} = -\frac{1}{I_1}\frac{d}{dt}\int_{V_0} \vec{B}_1 \cdot \vec{M}_0 dV \tag{2.15}$$

(where integration is performed over the whole sample volume V_0). For further we note that the field $\vec{B}_1$, created by the coil of spin detector, depends on the coordinates of the element in the sample and has, generally speaking, three components: $\vec{B}_{1z}$, $\vec{B}_{1x}$, $\vec{B}_{1y}$. Nuclear induction signal is generated only by precessing component of $\vec{M}_\perp$, perpendicular to the vector field $\vec{B}_0$, therefore in scalar product (2.15) only a component of the field $\vec{B}_\perp = \vec{B}_x + \vec{B}_y$ should be considered, since $\vec{B}_{1z} \cdot \vec{M}_\perp = 0$. The value of $\vec{M}_\perp$ is also dependent on the coordinates, since the field $\vec{B}_1$ of the 90-degree pulse is usually created by the same coil. Therefore, $\vec{M}_\perp = \vec{M}_0 \sin(\gamma B_1 t_{90})$ (t_{90} is the duration of a 90-degree pulse.) As a result for the initial signal amplitude, taking into account that the $\vec{B}_\perp(t) = \vec{B}_{\perp\max} \cos \omega_0 t$, we can write:

$$E_0 = -\frac{\omega_0 \vec{M}_0}{I_1} \int_{V_0} \vec{B}_{\perp\max} \sin(\gamma B_{\perp\max} t_{90}) dV. \tag{2.16}$$

Equation (2.16) is suitable for numerical calculations for arbitrary configuration of a sample and a coil, but for rough estimations the simplified expression can be obtained.

Suppose that a sample is only in the middle of a coil, where the field $\vec{B}_1$ is sufficiently homogeneous and almost independent of coordinates, and parameters of the 90-degree pulse is chosen so that $\sin(\gamma B_{\perp\max} t_{90}) = 1$. Then in (2.16) $\vec{B}_{\perp\max}$ can be taken outside the integral, using for it the well-known approximate expression: $B_{\perp\max} = \mu_0 \frac{n}{l} I_1$ (l is the length of the coil; $\mu_0 = 4\pi \cdot 10^{-7}$ H / m is the magnetic permeability of the vacuum). Note also that the spin –detector coil is a component of an resonant circuit with a quality factor Q, which is tuned to the frequency resonance ω_0. As a result, we obtain an approximate formula:

$$E_0 \simeq \omega_0 M_0 \mu_0 \frac{n}{l} Q V_0, \tag{2.17}$$

here V_0 is the sample volume. For a typical case $B_0 = 1$ T, $V_0 = 10^{-6}$ m^3, $Q = 50$, $n = 10$, $l = 10^{-2}$ m and from (2.17) it follows for the protons of water that the value of $E_0 \simeq 50$ mV.

Radiotechnical registration of such a signal does not present large difficulties, however, in many cases the signal can be significantly less. Thereby the quantitative assessment of the real ratio of voltages of a signal and thermal noises, always presented in the electronic equipment, is very important. The main source of the noise component is the thermal noise of a coil of spin detector. An amplifier makes worse the initial S/N ratio, that is usually taken into account by introducing the so-called *amplifier noise factor* (F). The rms noise voltage of the parallel resonance circuit is given by the well-known Nyquist formula:

$$E_{\mathrm{N}} = \sqrt{4kT_{\mathrm{k}}R_0 \cdot \Delta f},$$

where T_{k} is the temperature of the coil; R_0 is the resonance circuit resistance at the operating frequency; Δf is the bandwidth of the receiving channel. Then the ratio of the rms voltages of signal and noise can be written in the form:

$$\frac{E_0}{E_{\mathrm{N}}} \simeq \frac{\mu_0 \omega_0 M_0 (n/l) Q V_0}{(8kT_{\mathrm{k}} F R_0 \cdot \Delta f)^{1/2}}.$$

Let us use for the calculation of R_0 the simplified expression:

$$R_0 = \omega_0 L Q \simeq \omega_0 Q \cdot 4\pi n^2 \frac{S}{l},$$

where S is the coil cross-sectional area of the coil, and substitute instead of M_0 its explicit expression (see (1.32)):

$$M_0 = \frac{N\gamma\hbar^2 I(I+1)\omega_0}{3kT}, \tag{2.18}$$

where N is the number of nuclei per unit volume. We assume that the sample temperature (T) is equal to the temperature of the spin detectors coil (T_{k}), then the S/N ratio can be calculated by the formula

$$\frac{E_0}{E_{\mathrm{N}}} \simeq \frac{N\gamma\hbar^2 I(I+1)\mu_0}{12(kTF)^{3/2}(2\pi\Delta f)^{1/2}} \frac{\omega_0^{3/2} Q^{1/2} V_0 n^{1/2}}{V_{\mathrm{L}}^{1/2}} \tag{2.19}$$

Here $V_{\mathrm{L}} = lS$ is the volume of a coil. Formula (2.19) describes the basic dependence of the S/N ratio related to the NMR frequency (ω_0), the content of the nuclei in a sample $(V_0 N)$ and the quality factor of the spin detectors coil (Q). It should be keep in mind that with the increase of the resonance frequency the ratio E_0/E_{N} increases less than $\omega^{3/2}$, as this reduce necessarily the number of turns (n) in a coil.

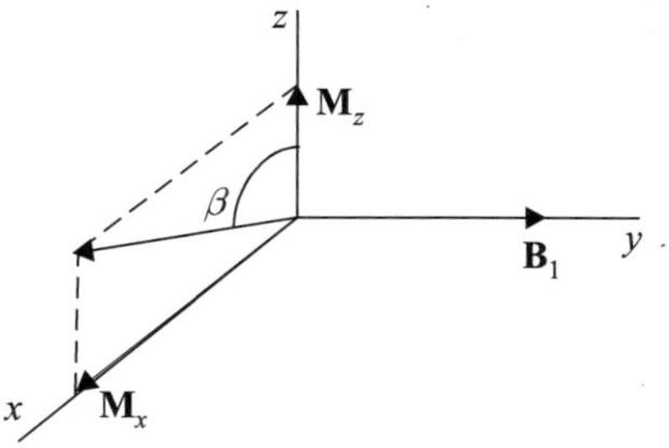

Fig. 2.16 The scheme of nutation of the magnetisation vector around $\vec{B}_1$ field

2.4.3 Principles of Fourier Transform in Nuclear Magnetic Resonance

It was noted in Sects. 1.7.1 and 1.8.2 that the Lorentzian shape of the absorption signal is the Fourier transform of the exponentially decaying signal of free induction (FID) after the of 90-degree pulse.

Consider once again the result of the exposure on the nuclear magnetization of a short pulse field $\vec{B}_1$, along the axis y of a frame rotating with the frequency of the external field ω. Let us assume, that the amplitude of the field $\vec{B}_1$ is large enough so by all possible detunings ($\Delta\omega = \omega_0 - \omega$) the rotation of vector $\vec{M}$ is almost in the plane (zx): $B_1 \gg \Delta\omega/\gamma$. If the angle of the nutation of the magnetization vector due to the pulse of duration t_{imp} (Fig. 2.16) is

$$\beta = \gamma B_1 t_{\text{imp}}, \tag{2.20}$$

then after the pulse we have:

$$\vec{M}_x(0) = \vec{M}_0 \sin\beta, \quad \vec{M}_y = 0, \quad \vec{M}_z(0) = \vec{M}_0 \cos\beta.$$

Further behavior of the components is described by obvious formulae:

$$\vec{M}_x(t) = \vec{M}_0 \sin\beta \cos\Delta\omega t \exp\left(-\frac{t}{T_2^*}\right),$$

$$\vec{M}_y(t) = \vec{M}_0 \sin\beta \sin\Delta\omega t \exp\left(-\frac{t}{T_2^*}\right)$$

or in the complex form:

$$\vec{M}_\perp(t) = \vec{M}_x(t) + \mathrm{i}\vec{M}_y(t) = \vec{M}_0 \sin\beta \exp\left[\mathrm{i}\Delta\omega t - \frac{t}{T_2^*}\right].$$

The free-induction signal $U(t)$, arising due to the precessing magnetization $\vec{M}_{\perp}(t)$, is also available in the complex form by using the principle of synchronous quadrature detection (see the following): $U = U_0 \sin\beta \exp[i\Delta\omega t - t/T_2^*]$. The Fourier transform of the signal

$$S(\omega) = \int_0^{\infty} U(t)\exp(-i\omega t)dt,$$

gives a complex spectrum:

$$S(\omega) = a(\omega) + ib(\omega),$$

where a and b are the already known (see Sect. 1.6) dispersion and absorption signals:

$$a = U_0 \sin\beta \frac{1/T_2^*}{(1/T_2^*)^2 + (\Delta\omega)^2}, \quad b = -U_0 \sin\beta \frac{\Delta\omega}{(1/T_2^*)^2 + (\Delta\omega)^2}.$$

It is clear that these signals are maximum if $\beta = \pi/2$ (90-degree pulse).

This correspondence has a deep physical meaning, since for any linear system, its frequency characteristic and the response to a single exciting pulse are connected by Fourier transform. Although both of these characteristics contain the same information, however, quite complex multiplet systems are serviceable and clear in the form of frequency characteristics of spectra. Note that the first publication about the obtaining of a NMR spectrum by Fourier transform of the free induction signal of nuclei was in 1958: A. A. Morozov, A. V. Melnikov, F. I. Skripov: "The method of free nuclear induction at low magnetic fields applied to some problems rf spectroscopy with high resolving power", Izv. AN USSR. vol. 22, p. 1141–1144. In accordance to the level of computer technology the Fourier transform was performed with a mechanical arithmometer, so the calculation of a single spectrum was very time consuming. Therefore, the article of the authors of the Leningrad (now St. Petersburg) State University did not attract the attention of contemporaries, and the method was extended only with the appearance of the third generation of computers. The analysis shows that the method of the obtaining of spectra on the basis of Fourier transform of free-induction signals in the most cases gives significant advantages over the steady-state method of registration. Note the main ones:

1. Reducing the time of obtaining of the spectrum is greater, the greater the frequency domain spectrum compared with the width of a single line. Saving of time can be used for the accumulation of signals and, thus, for the improvement of S/N ratio.

2. Relatively rapid registration of a spectrum provides new possibilities for studying spectra of short-lived compounds or intermediate stages of chemical reactions.

3. Use of complex pulse sequences combined with Fourier-analysis allowed us to open new areas of research, such as two-dimensional NMR-spectroscopy, analysis of individual self-diffusion coefficients of spectral components using a pulsed gradient, MRI, etc.

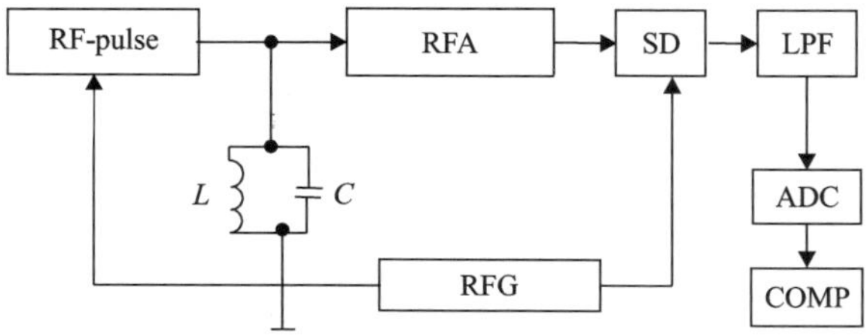

Fig. 2.17 The block-scheme of the simplest pulse Fourier transform spectrometer

Let us consider some aspects of the experiment on registration of a spectrum based on pulsed excitation of a spin system.

Functional diagram of the simplest pulse Fourier transform spectrometer is shown in Fig. 2.17. RF pulses with controlled amplitude and duration periodically excite free-induction signals in the spin detector LC. After the preliminary amplification, the signal is mixed in synchronous detector (SD) with the reference signal of the frequency ω of RF generator. The signal from SD outputs with frequency $\omega_0 - \omega$ after pre-filtering (LPF) is digitized in the analog-to-digital converter (ADC) and recorded in the computer memory.

Pulse amplitude. It was shown in Sect. 2.4.1 that the motion of the magnetization vector $\vec{M}$ under the influence of field $\vec{B}_1$ in the case of frequency detuning ($\Delta\omega$), which is comparable to the value $\gamma\vec{B}_1$, occurs over the cone around the vector field $\vec{B}_{\text{eff}}$. The initial phase of component $M_\perp$ at the end of the pulse depends on the ratio of B_1 and $\Delta\omega/\gamma$. Thus, in the case of the NMR spectrum consisting of several lines with resonance frequencies ω_{0i}, generally speaking, it is impossible to provide the accurate turn of all vectors $\vec{M}_i$ at the same angle. Besides there will be different the initial phase of the precession of each component $M_\perp$. Synchronous—phase detection of signals of all components with following the Fourier transform will distort the shape of spectral lines, depending on their location in the spectrum, as well as the distortion of the correct ratio of intensities of components in a spectrum. These effects can be minimized if we fulfill the condition

$$B_1 \gg \frac{2\pi \cdot \Delta F}{\gamma}, \tag{2.21}$$

(ΔF is the full width of a spectrum). Thus the nutation of all vectors $\vec{M}_i$ will occur around the vector $\vec{B}_{\text{eff}} \simeq \vec{B}_1$ at the same angle. For example, the observation of the proton spectrum of ethanol with the width of about 750 Hz (at a frequency of 300 MHz) the field $\vec{B}_1$ must satisfy $B_1 \gg 1 \cdot 10^{-4}$ T, which is much less commonly used fields in experiments on measurements of T_1 and T_2. But, for example, in the field of about 12 T the width of the ^{13}C spectrum can be ~30 kHz, and then $B_1 \gg 4 \cdot 10^{-3}$ T, which is not achieved in each spectrometer. The distortion of the spectral lines can be adjusted during the processing of the spectra in the computer.

The oscillator frequency can be set both outside of the limits of a spectrum (ΔF), and inside it. In the first case (Fig. 2.18a) all frequency differences $\omega_{0i} - \omega$ have one

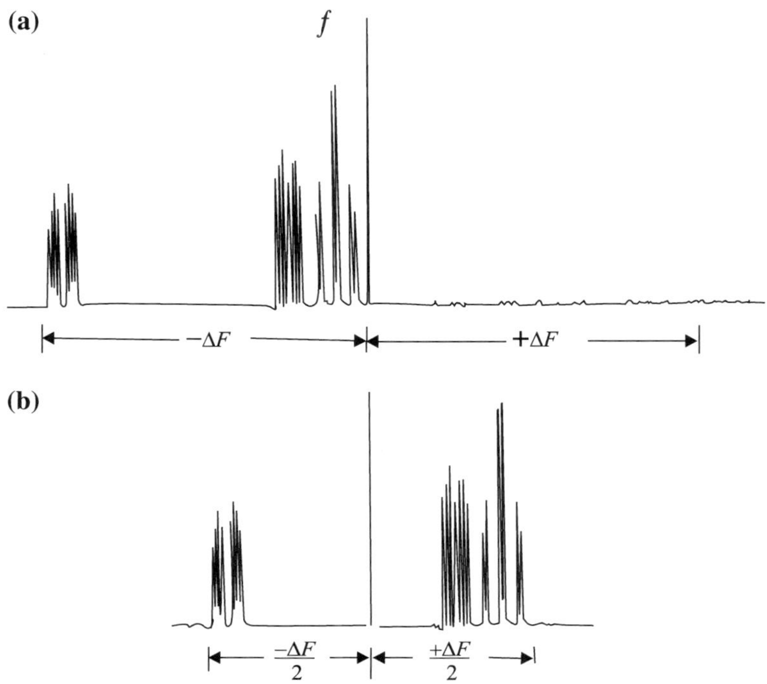

Fig. 2.18 The layout of the oscillator frequency to the frequency of a spectrum

Fig. 2.19 The block-scheme of a two-channel Fourier NMR spectrometer

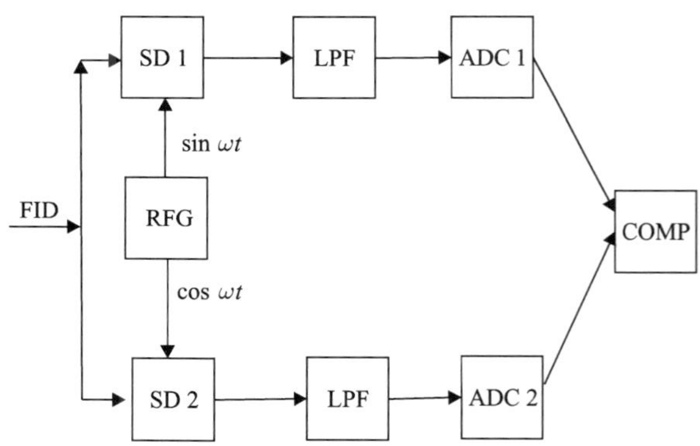

sign, which allows to obtain a spectrum on a single-channel device, the block scheme of which was shown in Fig. 2.17. The disadvantage of a single-channel system is the decrease of the S/N ratio in $\sqrt{2}$ times, as after Fourier transform the noise is summed up in the band $\pm\Delta F$.

In modern instruments the two-channel schemes are generally used (see Fig. 2.19) with two reference voltages of the generator, shifted in phase by 90°. If we place the frequency ω in the middle of the studied spectrum, then signals at the output of the synchronous detector will correspond to both positive and negative differences of $\omega_{0i} - \omega$ (see Fig. 2.18b), but the presence of two channels allows the correction of the places of all lines in a spectrum after complex Fourier transform. Additional advantage of the dual-channel scheme is the decrease of the required rate of the signal digitizing (see formula (2.22) below).

Coherent accumulation of pulse signals makes it necessary to maintain the values of ω at a fixed point of the spectrum (see Fig. 2.18), that can only be achieved by binding to the frequency ω_{0i} of certain line of NMR spectrum used as an internal or external reference. Without examining in details here the principles of nuclear resonance stabilization, we note only the obvious for these purposes: we could use the signal of dispersion, or a derivative of the absorption NMR signal. As an external standard one uses a separate ampoule with a substance which resonance line is located outside of an investigated spectrum. It is also possible to use an internal standard: the enough strong line of the spectrum or the test line of the specially selected inert substance which is added into an ampoule with the investigated solution. It is convenient to use deuterated solvents, because the resonance of deuterium can be a reference signal for field stabilization system.

Duration of the pulse. The product of the pulse duration and its amplitude defines the angle of rotation of the magnetization (β) in accordance with (2.20). The maximum signal is provided at $\beta = \pi/2$, but it is valid only for a single (rarely repeated) pulse. In cases, when it is necessary to average the signals for a certain number of repetitions during optimal time of the experiment, it must be considered the value of relaxation time T_1. Suppose that $T_2^* \ll T_1$. If the period of repetition (T_0) is comparable to the value of T_1, after a certain time the dynamic balance is established in a system. For the component $M_z(0_+)$ after the pulse, rotating it at the angle β, the following relation holds:

$$M_z(0_+) = M_z(0_-)\cos\beta,$$

where $M_z(0_-)$ is the component value before the pulse. For the components $M_z(T_0)$ at the end of the time T_0 one can write:

$$M_z(T_0) = M_z(0_+) + [M_0 - M_z(0_+)][1 - \exp(-T_0/T_1)].$$

Since at the dynamic equilibrium $M_z(T_0) = M_z(0_-)$, then we have

$$M_z(0_-) = M_0 \frac{1 - \exp(-T_0/T_1)}{1 - \cos\beta\exp(-T_0/T_1)}.$$

The initial amplitude of the signal after the pulse depends on the magnitude of the transverse component:

$$M_x(0_+) = M_0 \frac{1 - \exp(-T_0/T_1)}{1 - \cos\beta\exp(-T_0/T_1)}\sin\beta.$$

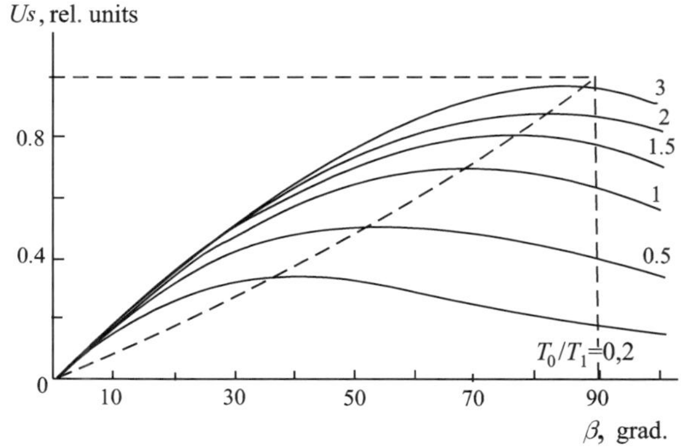

Fig. 2.20 The dependences of the NMR signal on the rotation angle of magnetisation (β) at different relations T_0/T_1

The last formula implies that the signal amplitude is maximal at some β_{opt}:

$$\cos \beta_{\text{opt}} = \exp(-T_0/T_1).$$

The dependencies of the relative amplitude of the signal (U_s) on the rotation angle (β) at different relations T_0/T_1 are shown in Fig. 2.20. The value β_{opt} is close to 90° only at $T_0/T_1 \geq 3$.

Signal to noise ratio. The possible improvement of the sensitivity of the method is determined by broadband pulse excitation of the spin system. In the stationary NMR spectroscopy for the spectra recording the successive passage through the resonance of each spectral lines is used. In this case, at any time the generator frequency corresponds to one point of the spectrum. A volume of information of a stationary method, in principle, can be improved if multiple generators and receivers tuned to different frequencies of a spectrum would be used, i.e. if we use a multi-channel spectrometer, but such a device in real situation is cumbersome and inconvenient. The excitation of a spin system with short pulse having a fairly wide spectrum is also equivalent to a multichannel spectrometer. If we exclude the time needed for Fourier transformation, it can be expected that the gain in the time, which is possible to obtain the spectrum, will depend on the ratio of the width of the whole spectrum to the width of a single line. The saved time may be used for coherent accumulation of signals. It is well known that at coherent accumulation the signal intensity is proportional to the number of repetitions (n), while the noise component increases proportionally $\sqrt{n}$. Let us consider the ratio of the signal amplitude to the rms amplitude of noises as a measure of sensitivity of the method. R. Ernst and G. Anderson showed that in the case of matched filters the ratio of the sensitivity of pulsed and stationary methods for the same measurement time is given by:

$$\frac{(S/N)_{\text{pulse}}}{(S/N)_{\text{stat}}} = 0.8\sqrt{\frac{\Delta F}{\Delta}}\, G(T_0/T_1),$$

where ΔF is the width of a spectrum; $\varDelta$ is the width of a single line, $\varDelta = 1/(\pi T_2^*)$; T_0 is the pulse repetition period.

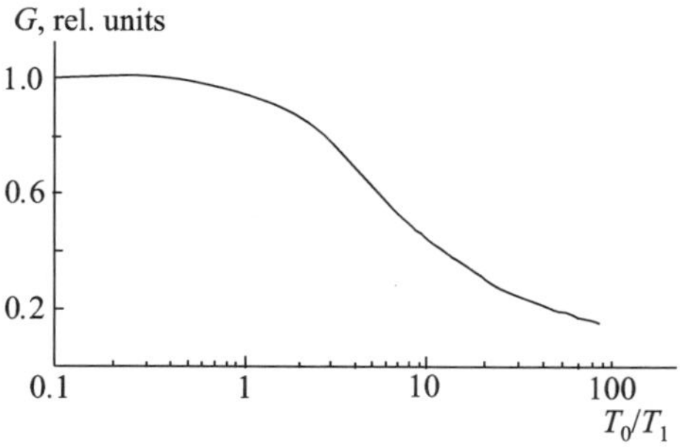

Fig. 2.21 The dependence of parameter G on T_0/T_1 (see text). Reproduced with permission from Rev. Sci. Instr., 37, 93 (1966). Copyright 1966, American Institute of Physics

The dependence of the function G on T_0/T_1 is shown in Fig. 2.21. It is evident that to increase the sensitivity the value of T_0 should be reduced. However, as we shall see below, the resolution of spectra in the method will be reduced. Therefore, the choice of the optimal duration of the pulse repetition depends on specific objectives: what is more important, to maximize the sensitivity or resolution ability in an experiment.

Sampling of NMR signal. To enter data on the NMR signal in the computer for the subsequent processing, the conversion of a continuous (analog) signal into digital form is required. This is done by the analog-to-digital converter (ADC). This procedure is a conversion of a continuous function $U(t)$, which describes the original signal, in a sequence of numbers (samplings) $\{U(t_n)\}$, $n = 0, 1, 2, \ldots$, characterizing the amplitude of the signal in some equally spaced points in time. From the information theory it is known that for the correct registration of a sinusoidal signal in the digital form, its values should be obtained at least twice per a period. This means that for a signal of complex spectrum with the maximum frequency $F_{\max}$, the interval between samplings (Δt) must be

$$\Delta t \leq 1/(2F_{\max}).$$

The obtained value of sampling at the ADC output is usually presented as a binary code with a finite number of digits (N). The value of N depends on the design of the ADC and determines its resolution and dynamic range. Minimal value of the analog signal ($U_{\min}$), that can be displayed in digital form (quantization step) is $U_{\max}/2^N$, where $U_{\max}$ is the maximum allowed for the given ADC voltage. So, if $U_{\max} = 1$ V, $N = 12$, then $U_{\min} \simeq 120\,\mu$V. Accordingly, the dynamic range of the ADC, defined as $D = U_{\max}/U_{\min}$, is the important characteristic of a system, because it determines the ability to register the signals of low intensity in the presence of a signal of high intensity. From this viewpoint it is desirable to apply ADC with maximal number of bits. Increasing this number usually leads to reduction of ADC speed, which is commonly characterized by the time conversion (t_{con}), during which occurs the setting of binary code corresponding to the input voltage. The value t_{con} must satisfy the condition:

$$t_{\text{con}} < \Delta t \leq 1/(2F_{\max}), \tag{2.22}$$

i.e. the dynamic range of the ADC (D) and the conversion time (t_{con}) must be matched with the expected width of a investigated spectrum. Note also that for two-channel device (see Fig. 2.19) $F_{max} = \Delta F/2$ in Eq. (2.22), so ADC speed can be reduced by factor of two with appropriate increasing the number of digits.

Optimization of parameters of spectral analysis. Possibility of separating the two close lines in the spectral range depends on the duration of the observation of the signal in the time domain. Thus, two neighboring lines with a minimum frequency difference ($\Delta\nu_{min}$) give the period of beats (T_{max}) in a free-induction signal:

$$T_{max} = 1/\Delta\nu_{min}.$$

This period can be observed if, at least, $T_{max} \leq T$ (T is the time of observation of the signal), that is $\Delta\nu_{min} \geq 1/T$. The end of the signal observation time (cut-off time) depends on the T_2^*, and with very good signal-to-noise ratio can be as high as $(2 \div 3)T_2^*$. Although the maximum resolution of the method is achieved at $T \gg T_2^*$, it should be kept in mind that the recorded time-domain signal is damped in time. The more we try to collect information, the more "noisy" it becomes. Here, as in any area of physical experiment, a contradiction between the desired sensitivity and resolution of the device is manifested. In practice we always must restrict the observation of a signal in limits of the order of T_2^*, and this means that, in fact, the Fourier transform is performed for the product of the nontruncated signal ($U(t)$) and the rectangular weighting function ($\Pi(t)$), which has the form

$$\Pi(t) = 1, \;\; for\; 0 \leq t \leq T; \quad \Pi(t) = 0, \;\; |t| > T.$$

Corresponding spectrum $S_n(f)$ is the convolution of the original (undistorted) spectrum $S_0(f)$ with the Fourier transform of the rectangular function $\Pi(t)$:

$$S_n(f) = A\frac{\sin \pi f T}{\pi f T}.$$

Functions of the form $\sin x/x$ correspond to characteristic oscillations on the wings of spectral lines, which can mask more weak neighboring spectral lines. Figure 2.22 demonstrates an example of the dependence of oscillations on the ratio of cutoff time T and T_2^*. The amplitude of the oscillations (A) can be significantly reduced by multiplying an original signal in the time domain by a certain weight function ("window") rather smoothly decreasing to the time $t = T$. This process is called *apodization* (the term came from optical spectroscopy) and in NMR spectroscopy in addition to reducing the oscillation, it can, as will be seen below, significantly affect the sensitivity and resolution of the Fourier spectrum. Among the most widely used apodization functions are the following (Fig. 2.23):

(1) cosine window (curve *1*): $H(t) = \cos \pi t/(2T)$;
(2) Hanning window (curve *3*): $H(t) = 0{,}5 + 0{,}5 \cos \pi t/T$;
(3) Hamming window (curve *2*): $H(t) = 0{,}54 + 0{,}46 \cos \pi t/T$.

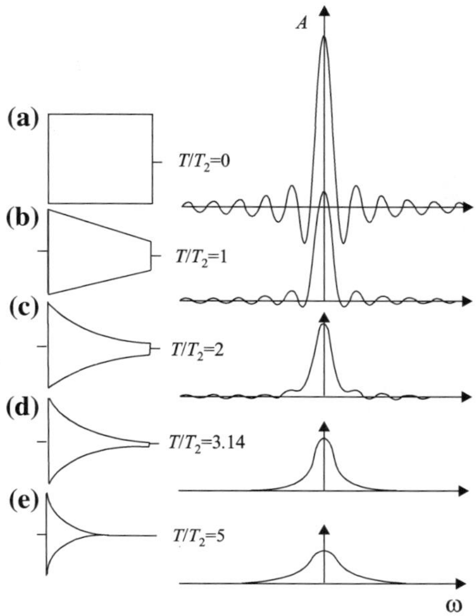

Fig. 2.22 The effect of FID truncation on the amplitude of the oscillation in the wings of spectral lines

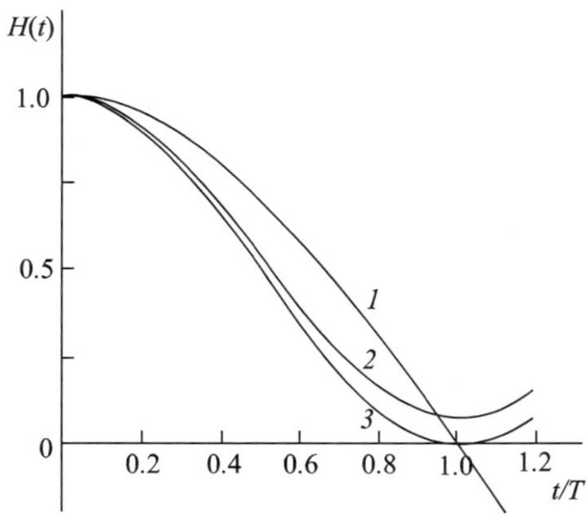

Fig. 2.23 The plot of apodization functions. *1*—cosine window, *2*—Hamming window, *3*—Hanning window

Multiplication of the signal in the time domain on a decreasing apodization function leads to some broadening of spectral lines in the frequency domain (in 1.54–2 times), but also improves the ratio of signal-to-noise because the contribution of the noise voltage from the later parts of the signal progressively decreases. In this sense, the multiplication by an exponential-weighting function of form $\exp(-t/T_2^*)$ corresponds to the so-called *optimal* (or *matched*) *filter* in the time domain. Multiplication by $\exp(-t/T_2^*)$ keeps Lorentzian line shape, but broadens it twice.

Let us evaluate quantitatively the influence of the observation time (T) and the exponential filtering on the value of $(S/N)_f$ in the frequency domain for a single spectral line. The signal in the time domain is:

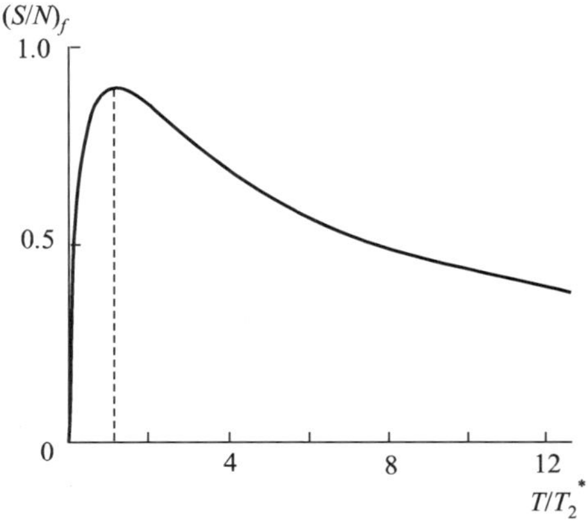

Fig. 2.24 The dependence of signal-to-noise value in the frequency domain for a single line on the ratio of the observation time (T) and relaxation time (T_2^*). Reproduced with permission from Prog. Nucl. Magn. Reson. Spec., v. 14, p. 27 (1980). Copyright 1980, Elsevier

$$U(t) = M_0 \exp(-t/T_2^*),$$

Then in the spectral domain

$$S(f) = M_0 \int_0^T \exp(-t/T_2^*)dt = M_0 T_2^*[1 - \exp(-T/T_2^*)].$$

Rms noise voltage can be written as

$$N(f) = \sqrt{\int_0^T [n(t)]^2 dt} = \bar{n}\sqrt{T},$$

where $n(t)$ is the current value of noise amplitude; $\bar{n}$ is the rms. For the ratio signal-to-noise in the frequency domain, we have:

$$\left(\frac{S}{N}\right)_f = \frac{M_0}{\bar{n}} \frac{T_2^*}{\sqrt{T}}\left[1 - \exp\left(-\frac{T}{T_2^*}\right)\right]. \tag{2.23}$$

It follows from (2.23) the clear link of the value $(S/N)_f$ and T_2^*: the more T_2^*, the narrower lines in the spectrum, and respectively greater the ratio $(S/N)_f$. At the same time, the dependence of $(S/N)_f$ on the ratio T/T_2^* has a maximum at $T/T_2^* \simeq 1.2$ (Fig. 2.24), that is also clear: if $T \gg T_2^*$, the signal information is virtually lost, but at the same time there is an accumulation of noise.

An apodizing function, in principle, can increase the resolution in spectrum if in the time domain this function prevents "decay" of a signal. In the simplest case, this function can be exponential with positive index of the form $\exp(bt/T_2^*)$. If $bt \simeq T_2^*$, then we obtain an undamped signal,that in the spectrum domain corresponds to the

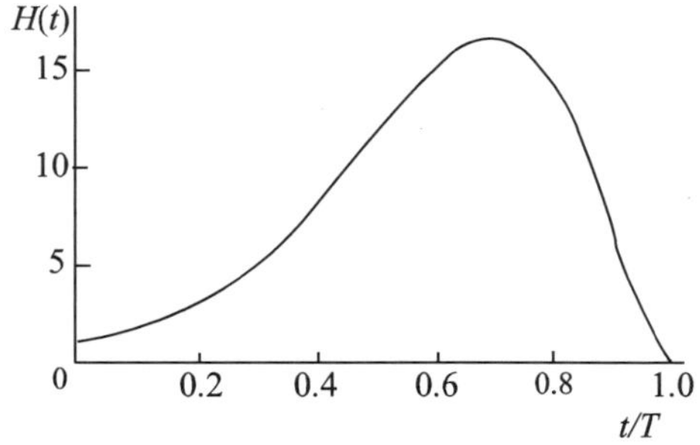

Fig. 2.25 The plot of Lorentz-Gauss apodization function

infinitely narrow δ-line. However, a line distortion and increasing noise make this procedure unreal. In literature there is a large number of functions that increase the resolution ability of the Fourier spectrum with an acceptable deterioration of the signal-to-noise ratio. The most optimum among them is Lorentz-Gauss transformation of the form $\exp(at/T - b(t/T)^2)$ (Fig. 2.25). At the initial stage of a signal a major role plays an increasing exhibitor, which improves the resolution, at the last stage the Gaussian character of a line with quickly damped wings is formed. Parameters a and b can be analytically optimized for a single line, but for a complex spectrum we must match them empirically.

Let us note one more possibility improving the character of the view of a spectrum. Because of the discrete Fourier transform the resulting spectrum is described by a set of equidistant points. The number of discrete points in the time domain is $N = T/\Delta t$, and with the $\Delta t = 1/(2F_{\max})$

$$N = 2F_{\max}T.$$

Fourier transform is a complex operation, i.e. $N/2$ points give real part of a spectrum and $N/2$ gives an imaginary one. Therefore the frequency interval between points in a spectrum is $\Delta = 2F_{\max}/N$. If the width of the observed lines is comparable with the Δ, then the spectrum may look like broken lines (Fig. 2.26). Of course, we can improve the view of a spectrum, if we increase the number N (preserving the value of $F_{\max}$, or Δt), but this is not always possible. Other possibility is the addition to data in the time domain of K zeros. This is equivalent to the prolongation of the signal observation, although this procedure does not give more information. If, for example, $K = N$, then the interval between points in the frequency domain will be halved. In principle, this addition can be infinitely big. In the limit we obtain quasicontinuous spectrum, but the restriction on this way mostly depends on the speed and memory of computers.

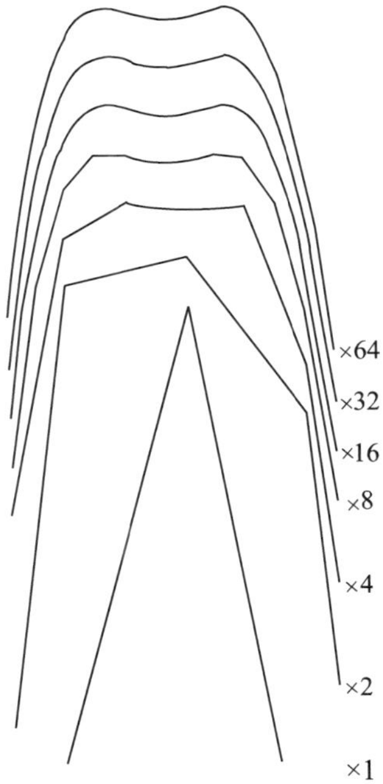

Fig. 2.26 The influence of the number of points in the time domain on the view of doublet spectrum

2.5 Measurement of Relaxation Times

The action of pulsed RF field on the system of magnetic moments is the most convenient way of removing the system from the equilibrium state. Monitoring the behavior of the system after pulses provides an analysis of the processes of establishing equilibrium, characterized by the relaxation time T_1 and T_2. From the Bloch equations (1.56) it is easy to obtain expressions for component $M_z(t)$ and $M_\perp(t)$:

$$M_z(t) - M_0 = \left[M_z(0) - M_0\right]e^{-t/T_1}, \tag{2.24}$$

$$M_\perp(t) = M_\perp(0)\, e^{-t/T_2^*}. \tag{2.25}$$

Equations (2.24) and (2.25) give a general framework to formulate concrete ways to measure the times of longitudinal (spin–lattice–T_1) and transverse (spin–spin – T_2) relaxation. Let us remind that the evolution of the longitudinal and transverse components of the nuclear magnetization occurs independently and that the relaxation time T_2^* in the Bloch equation takes into account all kinds of the broadening of the spectral line of the nuclear magnetic resonance, including the influence of inhomogeneity of the static field ΔB_0:

$$\frac{1}{T_2^*} = \frac{1}{T_2} + \frac{\gamma \Delta B_0}{2}. \tag{2.26}$$

The consequences of this circumstance will be discussed in Sect. 2.5.2. As to the Eq. (2.24), then, taking the logarithm of it, we obtain:

$$\ln \frac{M_0 - M_z(t)}{M_0 - M_z(0)} = -\frac{t}{T_1}. \tag{2.27}$$

This expression reflects the fact that the rate of the relaxation process of the longitudinal component of the vector $\vec{M}$ is independent of the initial conditions. Pulse methods of the registration of nuclear magnetic resonance give the most universal way of measuring the relaxation times in a wide range of values: for T_2—from units microseconds to hundreds of seconds, and for T_1—from a few dozen microseconds to hundreds of minutes. Concluding the general comments, we note that all experimental effects in the study of nuclear magnetic resonance are associated with the registration of the circulating transverse component of the magnetization. So in order to get an information about the components of M_z, it is necessary to turn it into a component $M_\perp$ and the signal amplitude of nuclear magnetic resonance allow us to judge about the magnitude of components M_z.

2.5.1 The Measurement of the Spin–Lattice Relaxation Time

An initial non-equilibrium state at the nuclear magnetization can be created using the pulse, turning the magnetization at any angle. The restoration of the equilibrium longitudinal magnetization, according to (2.24), is described by one of the curves shown in Fig. 2.27. Curve *1* corresponds to the rotation of the magnetization vector at a small angle. It is most convenient to use the pulse rotating of the vector $\vec{M}$ on 90 or 180 degrees. In the first case (curve *2* in Fig. 2.27), the initial values of the components of magnetization is: $M_z(0) = 0$, $M_\perp(0) = M_0$. The precessing magnetization creates in the receiving coil a signal with an initial amplitude which is proportional to the value of the longitudinal magnetization at the moment preceding the pulse. If after the time τ we initiate the second 90-degree pulse, after it the signal occurs, an initial amplitude of which is proportional to the new value of the longitudinal components $M_z(\tau)$. From (2.24) with the initial conditions indicated, we obtain:

$$M_z(\tau) = M_0\left(1 - e^{-\tau/T_1}\right),$$

and the expression (2.25) takes the form:

$$\ln \frac{M_0 - M_z(\tau)}{M_0} = -\frac{\tau}{T_1}. \tag{2.28}$$

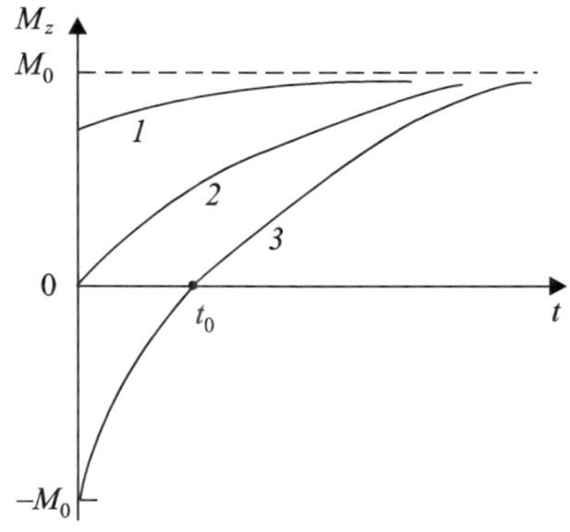

Fig. 2.27 The plot of restoration of the equilibrium of longitudinal magnetisation after pulse action: *1*—The small angle of rotation. *2*—The angle of rotation is 90°. *3*—The angle of rotation is 180°

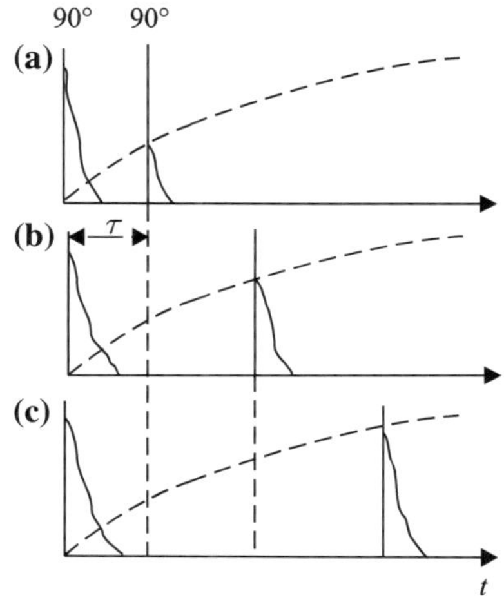

Fig. 2.28 The scheme of the E. L. Hahn experiment of T_1 measurement with pulse sequence $T - 90° - \tau - 90° - T - -$

Usually one builds function (2.28), recording the signals at different values of τ, and from the slope of the linear dependence determines T_1. The scheme of the experiment is shown in Fig. 2.28. This method of measurements has been proposed by E.L.Hahn. If about the value of the M_z component corresponding to M_0 one judges by the amplitude of the free-induction signal after the first pulse, then between consequent series of $90° - \tau - 90°$ (see Fig. 2.28a) it is necessary to keep the interval $T \simeq 5T_1$ in order to the nuclear magnetization adopted the equilibrium value (at this condition, $M_z(T) = 0.993M_0$). Therefore, the total time of the experiment is rather large if $T_1 > 1$ s (to obtain the dependence (2.28) it is desirable to fix of about (20–30) M_z values for different values of τ). But at the stable operating hardware signal,

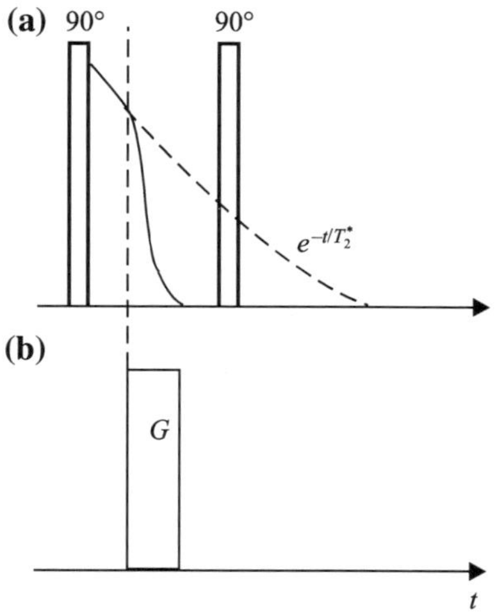

Fig. 2.29 A possible method to reduce the interval between the two pulses in the case of $T_1 \simeq T_2 \simeq T_2^*$. G is the pulse gradient

proportional to the equilibrium magnetization M_0, can be determined in a separate experiment, and then interval between pairs of pulses can be significantly reduced, because after the first pulse the longitudinal magnetization disappears and its previous state does not matter. The applicability of the described method for measuring the time T_1 is limited by the case $T_1 \simeq T_2 \simeq T_2^*$, when the free-induction signal after first pulse prevents the signal registration after the second one at small intervals between pulses (it is impossible to set $\tau < T_2^*$). To neutralize this difficulty it is possible to create for a short time after the first 90-degree the pulse of magnetic field gradient B_0 with enough large value G (Fig. 2.29). The appearance of the gradient will result in a drastic reduction of the parameter T_2^* (see (2.24)). As a result, the second 90-degree pulse can be placed close to the first one (see Fig. 2.29).

If the first pulse rotates the vector of the nuclear magnetization at the angle 180° (i.e. it becomes against the vector field $\vec{B}_0$), then $M_z(0) = -M_0$ and $M_\perp(0) = 0$. Then the evolution of the components of $M_z(\tau)$ will correspond the curve *3* in Fig. 2.27. In this case, the relaxation function has doubled in magnitude, in comparison with the expression (2.28), that is significant for a small signal-to-noise ratio. From Fig. 2.27 it is clear that in this method of measurements (it is usually denoted as $180° - \tau - 90°$ or *inversion-recovery*) the longitudinal magnetization at some time t_0 is equal to zero. From (2.27) with $M_z(0) = -M_0$ we have $t_0 = T_1 \ln 2$, i.e. in this case it is possible to identify quickly T_1 without construction of the relaxation curve (*zero*-method). For a good signal-to-noise ratio (S/N) *zero*-method provides accuracy even higher than the sequence $90° - \tau - 90°$. For example, if $S/N > 100$ the error of T_1 measurements for protons of water is realized in the limits 1–2 %, the reproducibility of measurements—to 0.5 %. In addition, when working with amplitude detector in a receiver, the zero-method eliminates errors due to non-linearity of

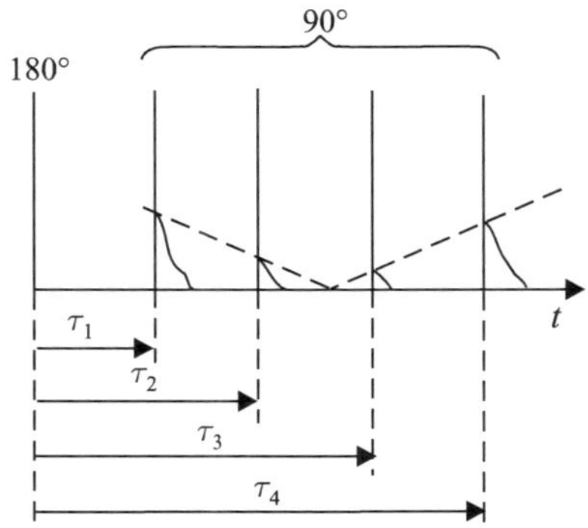

Fig. 2.30 The time diagram of NMR signals in T_1 measurments by pulse sequence $180° - \tau - 90°$

the detector at low signals. In Fig. 2.30 pulse diagram $180° - \tau - 90°$ with the peak amplitude detection of the signals is shown. For realization of the zero-method, it is necessary to have always $T > 5T_1$. Both described method are associated with the considerable time consuming in the case of large relaxation times T_1, since it is necessary to obtain a certain number of points on the relaxation curve, and it takes some time between pairs of pulses to bring the spin system in equilibrium state. To accelerate the process of measurments A. Csaki and G. Bene (1960) proposed after the first pulse to repeat periodically a group of three pulses: $90° - \tau_0 - 180° - \tau_0 - 90°$. Within of this triple pulse sequence the nuclear magnetization is registered during the interval $2\tau_0$ and then it is returned to the initial state (relative to the beginning of the triple pulses), since the total rotation of the nuclear magnetization vector is 360°. In Sect. 2.5.2 the explanation will be given why this rotation should be composed with three stages: $90° - 180° - 90°$. Here, we emphasize that the radio pulses used in this procedure must be coherent (i.e. with the same phase in the rotating frame). A necessary condition for the application of this sequence: $2\tau_0 \ll T_2$, otherwise irreversible decay of the magnetization due to the spin–spin relaxation leads to errors of measurements of T_1. The oscillogram of signal series received for "one passage" using the triples pulses is shown in Fig. 2.31. In this oscillogram each point corresponds to the NMR signal from a triple pulses, which due to the receiver channel gating are not visible. Before the series of triple pulses in at time $t = 0$ the 180-degree pulse is applied, that inverts the equilibrium nuclear magnetization (the signals are obtained with amplitude detection and therefore have a unipolar view).

There is quite a lot of modifications of pulse methods of measurements of the longitudinal relaxation time T_1. The described methods can considered as basic. Here we mention only two variants of the experiment in which one uses a continuous stream of RF pulses and which are appropriate in the case when $T_2 \ll T_1$. The first of these is known as the *progressive saturation*. In this version the continuous sequence of 90-degree pulses separated by an interval τ is applied. After three–five pulses a stationary amplitude ($A(\tau)$) of free induction signals following after each pulse is

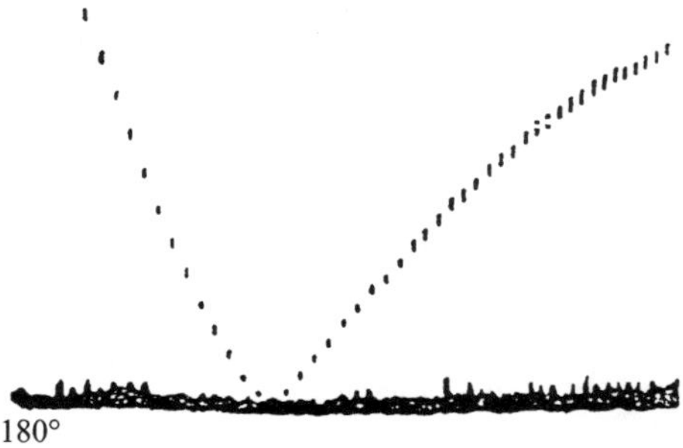

Fig. 2.31 An oscillogram of signal series using the triples pulses

established. If $\tau \gg T_1$, then $A(\tau) = A_0 \sim M_0$. For an arbitrary τ, we have:

$$A(\tau) = A_0\left[1 - e^{-\tau/T_1}\right]. \tag{2.29}$$

The application of sequence progressive saturation can significantly save the time of the experiment, especially in those cases when it is required accumulation of weak signals. The method is simple to implement, as it only uses one kind of pulse, and in fact it is usual method of accumulation of signals in the Fourier transform nuclear magnetic resonance spectroscopy. Depending on the required accuracy of the experiment time T_1 can be calculated from (2.29) for some $\tau_1 \sim T_1$ with defined beforehand A_0, or to plot the function (2.27).

Usually, all the methods measuring T_1 is based on the registration of amplitudes of nuclear magnetic resonance signals varying time intervals between pulses. However, there is an original method, in which one varies the angle of rotation of the nuclear magnetization vector by the RF pulse. Formally, this method is realized in the same way as well as method "progressive saturation", but the registration of the amplitudes of nuclear magnetic resonance signals is performed at several angles of rotation of the nuclear magnetization vector.The M_z dependence on the rotation angle β is following (see also Sect. 2.4.3):

$$M_z(\beta) = M_0 \sin\beta \frac{1 - \exp(-\tau/T_1)}{1 - \cos\beta \exp(-\tau/T_1)}. \tag{2.30}$$

The function $M_z(\beta)$ in the form (2.30) is maximized when $\cos\beta = \exp(-\tau/T_1)$. From this equation it is easy to determine T_1. In practice, the value of τ sets in the range (0.3–0.8) T_1 and a maximum amplitude of the signal is monitored at $\beta = 30$–$90°$. The main advantage of this method is the quickness of the measurement performance, that determines perspectives of its application for systems which has a dependence T_1 on the time (for example, chemical reactions).

2.5.2 The Measurement of Spin–Spin Relaxation Time

The easiest way to measure the spin-spin relaxation time (T_2) is based on the registration of free induction decay signal after a single pulse, according to (2.25). Rotation of the vector of nuclear magnetization can be done at any angle (it changes only the initial value of the transverse component of the magnetization $M_\perp(0)$), but to obtain maximum signal-to-noise ratio is appropriate to apply the 90-degree pulse. Taking logarithm of (2.25), we obtain:

$$\ln \frac{M_\perp(t)}{M_0} = -\frac{t}{T_2^*}. \tag{2.31}$$

This method can be applied in the case when the magnetic field inhomogeneity ΔB_0 does not make a significant contribution to the line width (see (2.26)), i.e. when $T_2^* = T_2$. It is easy to estimate that, if the error of the measuring time T_2 is admissible within N %, then the maximum value of T_2 in samples should not exceed a certain limit in accordance with the formula:

$$T_{2\,\max} < \frac{N}{50}\frac{R}{\omega_0}, \tag{2.32}$$

where R is the resolution magnetic system $(B_0/\Delta B_0)$; $\omega_0 = 2\pi\nu_0$. Formula (2.32) is obtained under the supposition that $N \ll 100$, i.e. the inhomogeneity of the magnetic field makes a small contribution to the total width of a line of the nuclear magnetic resonance, because otherwise measurements are meaningless. To illustrate the importance of the problem, we estimate $T_{2\,\max}$ for spectrometer of the middle class: $R = 10^9$, $\nu_0 = 300$ MHz. If an error in the measurement of T_2 is admissible within 5 %, then we obtain from (2.32) $T_{2\,\max} < 0.05$ s, that means the impossibility of measurements for a wide range of objects (for example, at room temperature for water protons $T_2 \simeq 3$ s).

It was found, however, that the dephasing of the precession of the nuclear magnetization of a sample because of the inhomogeneity of the magnetic field is reversible. E. Hahn showed that if to act on a sample with the two RF pulses at an interval of τ, which rotate the vector $\vec{M}$, respectively, on the angles β_1 and β_2, then after the time τ after the second pulse the nuclear induction signal appears in the shape composed of two mirror signals of free induction decay (Fig. 2.32). This signal is called a *spin echo*. Analytical description of this phenomenon holds in Sect. 1.3. We now carry out more visual analysis based on the vector model, taking into account the relaxation processes.

We can in mind divide a sample on so small parts, within which the field B_0 can be considered as homogeneous, i.e. within each parts is fulfiled the condition $T_2 \ll 2/(\gamma\Delta B_{0i})$, where ΔB_{0i} is the field inhomogeneity i-th part of the sample. Let the nuclear magnetization of i-th part is characterized by the value of $\vec{m}_i$. Then $\vec{M} = \sum_i \vec{m}_i$. For each magnetization $\vec{m}_i$ the Bloch equations are valid, which now contains the true time of the spin-spin relaxation of nuclei. However, the frequencies

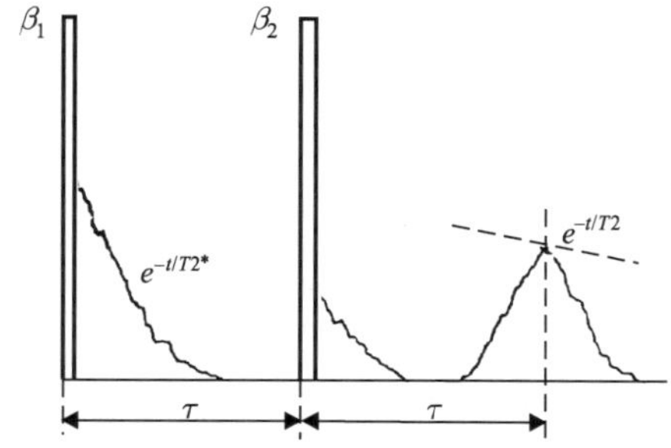

Fig. 2.32 The time diagram of NMR signals in the case of two pulse sequence (the angles of rotation approximately 90°)

of nuclear magnetic resonance $\omega_i(x, y, z)$ in each i-th part will be somewhat different from the average rate ω_0. Then the value of $M(t)$ will parametrically depend on the angles of rotation of the magnetization vector under exposed pulses and on $\Delta\omega_{0i} = \omega_0 - \omega_i(x, y, z)$. After summing we can get that at the time of 2τ the signal of nuclear magnetic resonance arises again, the maximum amplitude of which is given by (actually after integrating over all $\omega_i(x, y, z)$):

$$A(2\tau) = A(0)\, e^{-2\tau/T_2}, \tag{2.33}$$

where $A(0)$ is the value which depends on the angles of rotation of the nuclear magnetization vector under the influence of the first and second pulses.

The relaxation decay does not depend on the angles rotation of the vector, but in practice one uses the sequence of pulses $90° - \tau - 180°$, after which arises an echo signal with the greatest amplitude. A diagram of the appearance of echo signals after two pulses: $90° - \tau - 180°$ is shown in Fig. 2.33. After the 90-degree pulse the orientation of the vectors $\vec{m}_i$ is perpendicular to the vector of the external field $\vec{B}_0$ (see Fig. 2.33a), and these vectors will precess around the vector $\vec{B}_0(x, y, z)$, each with its own frequency $\omega_{0i} = \gamma B_0(x, y, z)$. The vectors $\vec{m}_i$, for which $\omega_i = \omega_0$, will be fixed in the rotating frame (position *2* in Fig. 2.33b), others will overtake the system coordinates (position *1*), or lag behind (position *3*). After some time the vectors $\vec{m}_i$ will unfold in a fan, which symbolized in Fig. 2.33b by three discrete positions. The total magnetization will decrease with time constant T_2^*, but the magnitude of each vector $\vec{m}_i$ decays with a time constant of T_2. If at the time τ we apply a RF pulse in order to turn all the vectors around the vector field $\vec{B}_1$ at 180°, then their relative position is reversed (see Fig. 2.33c). The parameters of such a pulse satisfy:

$$\gamma B_1 t_i = \pi.$$

The vectors $\vec{m}_i$ will continue to precess at the same frequency, and through the time τ after the second pulse, they will be in-phase (see Fig. 2.33d). At this moment the signal induced in the receiver coil will be maximal (spin echo). After $t = 2\tau$ the

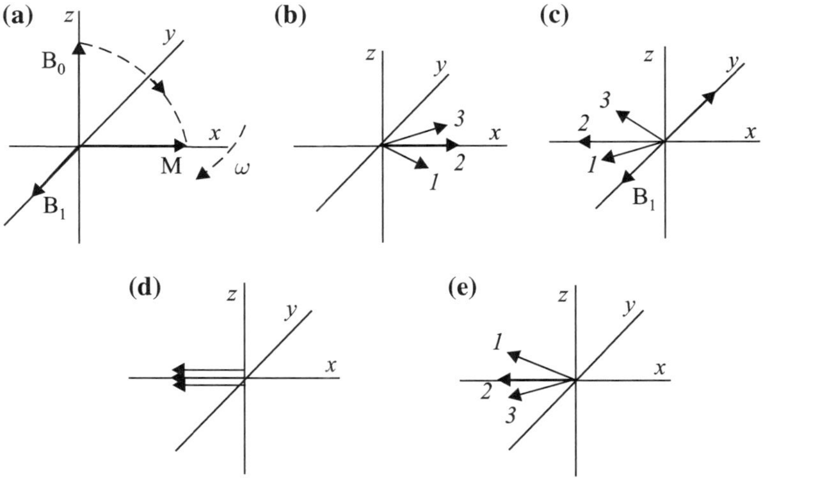

Fig. 2.33 The vector diagram of the appearence of echo signals after two pulses $90^\circ - \tau - 180^\circ$

echo signal decreases due continuing dephasing $\vec{m}_i$ (see Fig. 2.33e). Since during the time 2τ irreversible decrease of the value of vectors $\vec{m}_i$ because of the spin–spin relaxation takes place, the amplitude of the echo signals always less than the free induction signal after the first pulse and depends on time according to (2.33). Repeating the sequence of $90^\circ - \tau - 180^\circ$ with different τ, it is easy to obtain the dependence

$$\ln \frac{A(2\tau)}{A(0)} = -\frac{2\tau}{T_2}$$

and to compute T_2 from it. Naturally, that between pairs of pulses it is necessary to wait time $T = 5T_1$, in order to the spin system comes to equilibrium.

Scheme of the echo signal (see Fig. 2.33a–e) allows us to understand the need to split a 360-degree rotation of the vector magnetization into three steps ($90^\circ - 180^\circ - 90^\circ$) when measuring T_1 by method of A. Csaki and G. Bene. Indeed, the third (90-degree) pulse turns the phased magnetization (see Fig. 2.33d) along the vector field $\vec{B}_0$, i.e. the magnetization takes the orientation, which it had before the first 90-degree pulse.

This description of the process echo signal valid only in the case when the spatial coordinates of each element $\vec{m}_i$ do not depend on time, at least in the interval from the first pulse to the echo, that is for the absence in a sample of noticeable translational diffusion or macroscopic flows of the magnetic moments. Due to fast self-diffusion of molecules in low-viscosity liquids the precession phase shift of the vectors $\vec{m}_i(x, y, z)$ is different for time intervals between pulses and between the second pulse and the peak of an echo signal, i.e. frequency $\omega_i(x, y, z)$ are, to some degree, random functions of time. Analysis of the Bloch equations with taking into account the diffusion of nuclear magnetization in the presence of linear changes of the field B_0 in a sample:

$$B_z = B_0 + zG_0$$

(G_0 is the magnetic field gradient in a sample) results the following dependence of the amplitude of the echo signal on the time:

$$A(2\tau) = A(0)\, e^{-2\tau/T_2}\, e^{-\frac{2}{3}D\gamma^2 G_0^2 \tau^3}. \tag{2.34}$$

where D is the diffusion coefficient. Analysis of the relation of exponents in (2.34) allows us to estimate the possible limit of values T_2, which can be measured without any significant errors in the presence of fast molecular motion:

$$T_2 \ll 4\left(\frac{R^2 d^2}{\omega_0^2 D}\right)^{\frac{1}{3}} \approx \left(\frac{16 d^2 T^{*2}}{D}\right)^{\frac{1}{3}}, \tag{2.35}$$

Here are $R = B_0/(G_0 d)$ is the resolving power of a magnet; d is the diameter of a sample. At the derivation of (2.33) it was assumed $\tau = T_2/\,4$. If $d = 1$ cm, $\nu_0 = 300$ MHz and $R = 10^9$, then for substances having a viscosity of water ($D = 2 \cdot 10^{-5}$ cm ^{2}/s), we get $T_2 \ll 100$ s, i.e. the condition significantly better than for a single-pulse technique (see (2.32)). However, the formula (2.35) makes quite strict limits of applicability double-pulse technique for measuring time T_2 using devices equipped by magnets with a mean resolving power (for example, even at $R = 10^7$ it is impossible to measure the relaxation time of water protons, for which $T_2 \simeq 3$ s). H. Carr and E. Purcell (1958) showed that the echo signal can be repeated many times, if instead of a single 180-degree pulse to form a package of n pulses at an interval of 2τ. In this case, at the intervals $(2n)\tau = t$ echo signals will occur, the amplitudes of which (A_n) decrease in accordance with the following law:

$$A_n(t) = A(0) e^{-t/T_2} e^{-D\gamma^2 G_0^2 \tau^2 t/3}. \tag{2.36}$$

By reducing repetition period of 180° pulses we can make the diffusion attenuation as small as wished.

The amplitude of the spin echo can change in the presence of convection currents in the fluid, which often occur because of the gradient temperature within a sample under high temperatures. In this case, macroscopic parts of a sample are moved through areas of the magnetic fields with different values of B_0. If one uses a sequence of coherent pulses 90° – 180° – 180° – ... (sequence Carr-Purcell), the amplitudes of the odd echo signals are less than even, and a situation can be realized when every odd echo disappears altogether. This effect should be taken into account when measuring the coefficients diffusion, but it can be the subject of a separate study.

2.5.3 The Selective Measurement of the Relaxation Time T_1 and T_2

It was assumed in Sects. 2.5.1 and 2.5.2 that the nuclear magnetic resonance spectrum consists of a single line. The presence of nuclei electron shielding, spin–spin interactions or physicochemical heterogeneity in a substance results in the appearance of different lines in NMR spectra , which can be attributed to certain nuclear macroscopic magnetization $\vec{M}_k$, characterized by a relaxation time T_{1k} and T_{2k} (in the case of unresolved spectral lines the possible difference relaxation characteristics also needs to be considered). The most common approach to the task of determining the relaxation characteristics relating to individual spectral lines is associated with obtaining NMR spectra using the pulse technique in conjunction with the Fourier transform of free nuclear induction signals arising after pulse acting on the spin system. To do this, it is necessary to register the intensity of the spectral lines on the delay time between pulses, changing the orientation of the nuclear magnetization.

Measurements of T_{1k} for individual spectral lines are usually made using of the sequence $180° - \tau - 90°$, and the Fourier transform of the free induction signal after the second (90 degree) pulse, varying τ. Based on the dependence of the amplitude of each line of a spectrum from τ it is possible to evaluate T_{1k}.

When measuring the T_{2k} in complex spectra one uses Carr-Purcell sequence with the Fourier transform of the echo signals at different values of t.

Of course, the indicated here ways of measurements T_{1k} and T_{2k} can only be realized with the help of modern NMR-spectrometers with high-speed computers.

Another variant of measurments of relaxation characteristics of the individual components of a spectrum is to choose the pulse parameters in such a way as to affect only one line (or one group lines) in a spectrum. These pulses are called *selective*. The simplest way to implement them is to apply pulses, the spectrum (S_i) of which satisfies the expression:

$$\Delta\omega_l \ll S_l \ll \Delta\omega_{\sigma k}, \tag{2.37}$$

where $\Delta\omega_l$ is the width of an individual line in a spectrum; $\Delta\omega_{\sigma k}$ is the frequency difference between the investigated and the nearest (k) lines in a spectrum. The condition (2.37) cannot always be satisfied and then we have to use more complex pulse program or tolerate with experimental error.

As already indicated, the more difficult case is when NMR spectral lines are not resolved. In this case, the relaxation functions are complex. For example, Fig. 2.34 shows the relaxation function for the transverse component of the protons magnetization of water, saturating a porous medium. In this case, the relaxation function is well described by the sum of two exponential functions:

$$F(t) = \sum_{i=1}^{2} F_i(0)e^{-t/T_{2i}}. \tag{2.38}$$

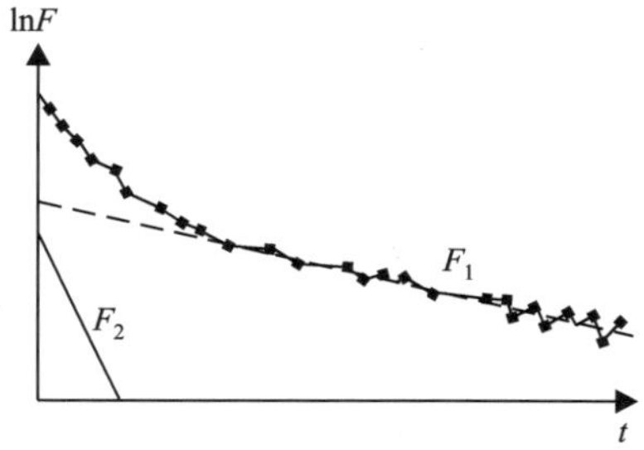

Fig. 2.34 An example of two components spin-spin relaxation (water in porous medium)

For more complex situations, the number of exponential components in (2.38) will be larger and their separation is problematic.

2.6 The Measurement of Coefficients Translational Diffusion

It follows from (2.34) that the amplitude of an echo signal in two-pulse sequence depends on both time T_2 and self-diffusion coefficient D, where the dependence on the pulse interval τ in the diffusion component significantly bigger: τ^3. Therefore with the appropriate choice of the value τ the ratio of the relaxation and diffusion components in the decay of an echo signal can be varied. Let us estimate the admissible values of the coefficient D, which can be measured in conditions of a stationary gradient. If we assume that at the time $2\tau = T_2$ the influence of self-diffusion process on the attenuation of an echo signal must be at least of the same order as due to the relaxation. We can write:

$$\exp\frac{-2\tau}{T_2} \geq \exp\left(-\frac{2}{3}D\gamma^2 G_0^2\tau^3\right).$$

Hence, it follows the evaluation the minimum measurable coefficient of self diffusion:

$$D \geq \frac{12}{T_2^3\gamma^2 G_0^2}. \tag{2.39}$$

"Apparatus" relaxation time T_2^* can be approximately expressed in terms of field inhomogeneity $G_0 d$ in a sample of a diameter d as $T_2^* = 2/(\gamma G_0 d)$, so (2.39) can be rewritten in the form:

$$D \geq 3d^2\frac{T_2^{*2}}{T_2^3}. \tag{2.40}$$

Minimum permissible value of T_2^* mainly depends on the capabilities of equipment used, because the decrease of T_2^* requires the corresponding increase of the

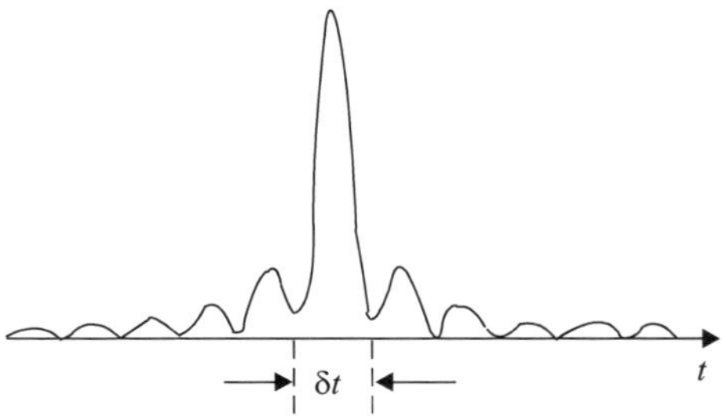

Fig. 2.35 The shape of a spin-echo signal for the cylindrical sample in the presence of the magnetic field gradient

bandwidth of receiving channel of a relaxometer. We should also bear in mind the inevitable worsening of signal/noise ratio. For a typical example, we can put $T_2^* = 10^{-4}$ c, $d = 10^{-2}$ m, then we obtain $D \geq 3 \cdot 10^{-12}$ m ^{2} c $^{-1}$. As seen from (2.40), to extend the limits of measurable values of D it may be achieved by reducing the diameter of sample ampoules. The theoretical limit for the minimum value of an ampoule diameter can be estimated from the diffusion relations $\langle r^2 \rangle = 6D\tau$. To eliminate the effect of the walls (restricted diffusion), we must have:

$$d \gg \sqrt{6D\tau}.$$

Let's do the numerical estimation for the water at 20°C ($D_{H_2O} = 2.03 \cdot 10^{-9}\,m^2s^{-1}$). If the time between pulses is 10 ms, then the diameter of ampoules should be more than 10^{-4} m, which is always performed in the experiment. Therefore, the minimum diameter of ampoules is defined by the necessity of realization of the desired level of sensitivity or some other technical reasons.

In practice, the self-diffusion coefficient is measured on the basis of the Eq. (2.34), from dependence of the echo amplitude $A(2\tau, G_0)$ on the value magnetic field gradient G_0 for fixed τ. The gradient G_0 is typically realized with the help of current-carrying conductors located at the periphery of a sample. The value of G_0, in principle, can be calculated knowing the geometry of the conductors or from the shape of an echo signal. In the presence of the magnetic field gradient in the echo envelope appear characteristic beats with a period that is associated with the size of the sample and the value G_0 (Fig. 2.35). Thus, if the sample is cylindrical, and the gradient vector G_0 is directed perpendicular to its axis, the interval between the two minima in the echo signal:

$$\delta t = \frac{15,3}{G_0 \gamma d} \tag{2.41}$$

(where d is the diameter of a sample.) Keep in mind that calculated values of the G_0 are insufficiently correct, since formula (2.41) holds for some idealized conditions (a sample is the infinitely long cylinder, a gradient coils do not create inhomogeneities of higher order (see Sect. 2.2). Corrections can be made, for example, from the measured value D of water at a fixed temperature ($D = 2.299 \cdot 10^{-9}\,m^2s^{-1}$ at $t = 25\,°C$).

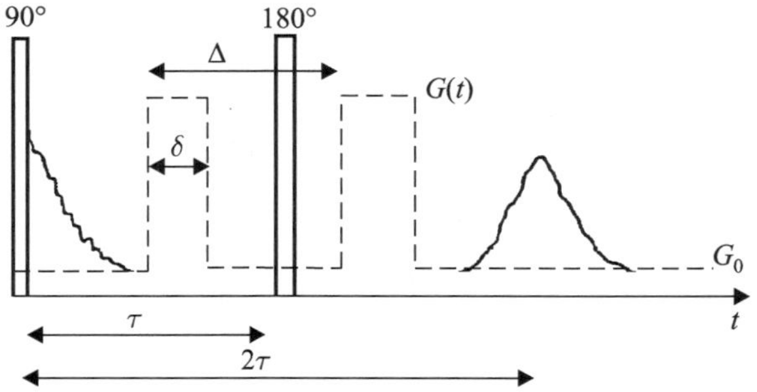

Fig. 2.36 The time diagram of spin-echo signals at 90° $-\tau-180°$ RF pulse sequence with the pulse magnetic field gradient G(t)

The basic restriction of the method with a stationary gradient, considered in previous cases, is the inability of registering of NMR signals in the presence of large gradients due to the excessive shortening of the duration of these signals: $T_2^* = 2/(\gamma G_0 d)$. It should also be taken in to account a requirement, imposed on the amplitude of the RF field B_1 when applying the 90-degree pulse:

$$B_1 \geq G_0 d. \tag{2.42}$$

The violation of this condition leads to an additional decrease of the amplitude due to unequal angles of rotation of nuclear magnetization vectors which are inherent to the individual elements of a sample. These conditions lead to the fact that the real boundary of a minimum measurable quantity D is near 10^{-11} m^{2}s^{-1}, which is only of two orders of magnitude less than the self-diffusion coefficient for water.

Significant expansion of these boundaries can be achieved if, as it was first proposed by E. Stejskal and I. Tanner (1965), magnetic field gradients are created in the form of pulses, located in the intervals between the RF pulse and the echo signals. In this case, in the condition (2.42) instead of $G_0 d$ the natural inhomogeneity of the magnet should only be considered, which can be very small. In addition, the echo signal is also observed in a relatively uniform field, and therefore limitations on the bandwidth of a receiver become softer.

The scheme of the experiment is shown in Fig. 2.36, where besides a standard pulse sequence $90° - \tau - 180°$ the dotted line shows the dependence of the magnetic field gradient on the time. Analysis yields the following dependence of the echo amplitude $A(2\tau, G)$ on the parameters of the pulse sequence ($G_0 \ll G$):

$$A(2\tau, G) = A(0)\exp\left[-\gamma^2 G^2 D\delta^2\left(\Delta - \frac{\delta}{3}\right)\right],$$

Here $A(0)$ is the amplitude of the echo signal in the absence of the gradient; δ and Δ are, respectively, the gradient pulse duration and the interval between gradient pulses, with the condition $\Delta \gg \delta$. In typical cases the value of δ can vary from tens of microseconds to tens of milliseconds, the value of Δ—from a few milliseconds to a few seconds. The value of the gradient G_{max}, is mainly determined by the specific features of the equipment and can be up to 50 T/m. The minimum measurable value of D can reach the order of $10^{-15}\,\mathrm{m^2 s^{-1}}$.

To determine the self diffusion coefficient it has been registered the dependence $\ln A(2\tau, G)/A(0)$ on G^2, δ^2 or Δ in accordance with the goals of experiment. For example, changing parametrically the interval Δ, we can study the diffusion in the presence of confinements when the mean diffusion path of particles is comparable with the size, limiting the scope (for example, a fluid in the porous environment or biological membrane).

Apart from the obvious advantages the method of pulsed gradient is rather difficult to implement for many reasons. We point out the basic ones. The analysis shows that for the correct reproduction of the signal amplitude the area of two gradient pulses must differ not more than 10^{-3} to 10^{-4} %, which is a rather complex radio engineering task. Furthermore, powerful pulses of current in the gradient coils cause the appearance of eddy currents in the magnet pole pieces and other metal parts and strongly distort the spatial and temporal gradient configuration that eventually also distort the measurement results. Solution of this problem may lie in the design of more complex configuration of gradient coils without a field outside a sample, as well as in the formation of more complex pulse sequences.

2.7 Spin Echo in a Three-Pulse Sequence

There is quite a simple pulse sequence, consisting of three 90-degree pulses, the consideration of which is appropriate in a separate section due to historical and methodological points of view. This sequence was described in the first paper of E. Hahn devoted to the phenomenon of spin-echo. In general (i.e., apart from the some specific relations T_1 and T_2) as a result of three 90-degree pulses action, three induction signals arise (after each pulse), and five signals of spin echo (see Sect. 1.8.3). The greatest interest for the relaxation and diffusion measurements are the echo signals (see Fig. 2.37), the maxima which comes at a time $2\tau_1 + \tau_2$ (stimulated echo) and $2\tau_2$ (reflected echo). The dependence of the stimulated echo amplitude on the intervals τ_1 and τ_2 is given by:

$$\exp\left(-\frac{2\tau_1}{T_2} - \frac{\tau_2}{T_1}\right) \exp\left[-\gamma^2 G_0^2 D\left(\frac{2}{3}\tau_1^3 + \tau_1^2\tau_2\right)\right], \tag{2.43}$$

where the first factor describes the relaxation contribution, and the second—the contribution of diffusion to the attenuation of the echo. Obviously, making τ_1 enough small, we can reduce the dependence of (2.43) to the expression:

$$\exp\left(-2\tau_1/T_2 - \tau_2/T_1\right).$$

If at a constant τ_1 one varies only τ_2, then the signal decrease is determined by the dependence $\exp(-\tau_2/T_1)$, and this opens the possibility of measuring the longitudinal (spin–lattice) relaxation time. This method can be recommended if the transient

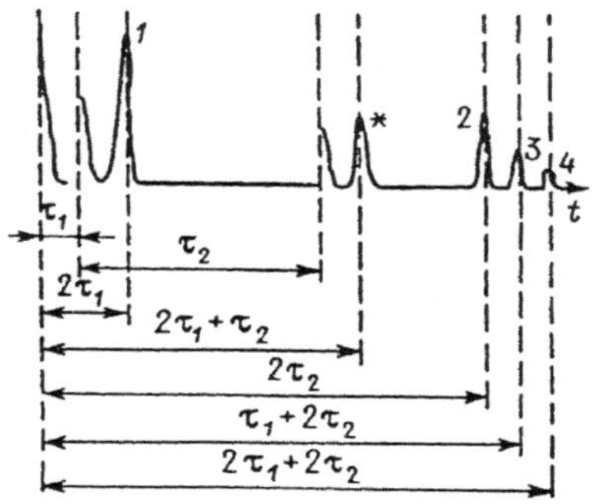

Fig. 2.37 The time diagram of NMR signals in the case of three RF pulse sequence

process in the receiver hinders the observation of free induction signal immediately after RF pulses. On the other hand, the stimulated echo expands possibilities of the spin echo method for measuring the diffusion, since the expression depending on the diffusion includes the product $\tau_1^2\tau_2$, which can be made large by τ_2 with small τ_1, that essential in the case $T_2 \ll T_1$.

2.8 Processing of Information by Means of Spin Echo

From the material of Sect. 2.5 it follows that the phenomenon of spin-echo can be used for delay and operational memory of signals. Indeed, assuming in a simple two-pulse technique the first pulse as informational and the second as control (or reader), we find that settling the interval between pulses it is possible to cause the signal (echo) in the required time. In this case the only necessary restriction is the condition

$$T_2^* \ll T < T_2,$$

where T is the time of the desirable appearance of the echo signal. The left side of the inequality defines the minimum delay between pulses for the appearance of the echo and the right one restricts the time of an experiment. Thus, the spin system operates as a storage device. Such devices, used spin echo signals, is called *spin processors*.

The cases of particular interest are when the informational pulse is weak (weak-signal approximation). Then the spin system "works" in the linear regime with respect to the information signal and the spin echo reproduces its shape (but like in a mirror, i. e. images are in the reverse order, see Fig. 2.38a). Naturally, instead of a single informational pulse one can apply a few "weak" pulses. For the direct (not inverted) sequence of signals in time the diagram, which is shown in Fig. 2.38b, can be applied. In this case, we use two strong pulses, and the formation of echo signal is the same as the formation of the stimulated echo (see Sect. 2.7).

Fig. 2.38 The scheme of spin-echoes as a random access memory for the inverted (**a**) and direct (**b**) order of information

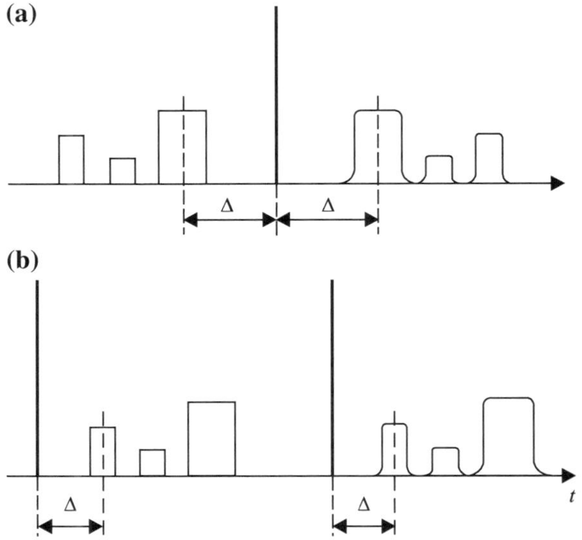

The first communications about possibilities of such use of the spin-echoes are contained in the works of E. L. Hahn with colleagues which were published in 50-ties. In particular, they showed that for the realization of the weak-signal approximation it was necessary to fulfil the following conditions:

(1) the angle of rotation of the vector of nuclear magnetization under the influence informational pulse (signal) must be weak, i. e. $\gamma B_1 t \ll 1$;

(2) the spectrum of the informational pulse (Δf) must be less than the line width of the NMR spectrum, that is equivalent to the condition $\Delta f \ll (1/T_2^*)$, and therefore pulse excites only a small portion of the spins;

(3) the spectrum of the control pulse must be considerably wider the spectrum of the information signal;

(4) the duration of the informational pulse must be much smaller than the spin-spin relaxation time T_2.

Under these conditions the echo signal at the output of the processor spin retains all the amplitude-frequency correlation of the input signal. These requirements are easily satisfied by the spin-echo technique with registration of nuclear magnetic resonance in magnetically ordered materials (see Chap. 7). This technique provides obtaining sufficiently intensive signals using compact devices (in this case the nuclear magnetic resonance is observed in internal magnetic fields without of magnets). Therefore hereinafter we assume that practical realizations of the described methods of information processing are based on the nuclear magnetic resonance in magnetically ordered substances.

Processing of amplitude-frequency modulated signals using spin processors consists of addressed signal delay, delay a signal without distortion of its shape, frequency filtering, integration, and differentiation of functions. The phenomenon of

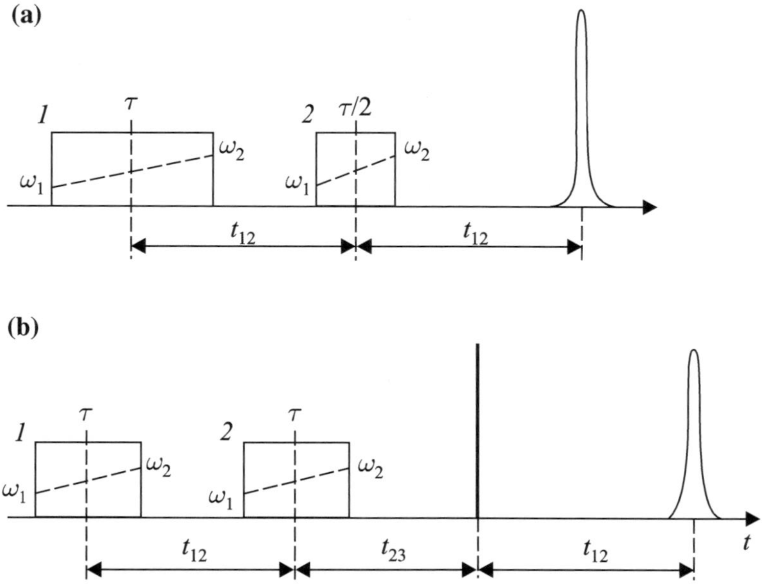

Fig. 2.39 The scheme of the "compressed" spin-echo without (**a**) and with "strong" pulse (**b**)

spin-echo can be used for matched filtering or compression of pulses with frequency modulation. In the case of the compression, the energy of a pulse of duration τ is concentrated in a time interval τ/K (K is the compression ratio, $K \gg 1$). Figure 2.39a presents the diagram of the pulse sequence, which consists of a informational pulse (*1*) and control pulse(*2*) with linear frequency modulation. The control pulse (*2*) is in two times shorter than informational one (*1*) but the change of frequency in both cases occurs in the same range: from ω_1 to ω_2. Readers are invited to explain the mechanism of the formation of the compressed echo (as an exercise) but it is evident that another coding of pulses does not yield the compressed echo. In the compression mode with two control pulses, one of which is short (δ-pulse with a wide spectrum), an informational signal can be supplied at the first, second or third place in a row (in Fig. 2.39b shows one of the options).

2.9 Radiation Damping: The Interaction of Precessing Nuclear Macroscopic Magnetization with Radio-Frequency Circuit

So far, the behavior of nuclear macroscopic magnetization was considered without taking into account the possible reverse effect of the recording equipment on it.

The main element of the radiospectrometer of nuclear magnetic resonance is the inductor (coil) which is included in the resonant circuit tuned on the frequency of the precession of the vector nuclear macroscopic magnetization $\vec{M}$. The e.m.f.

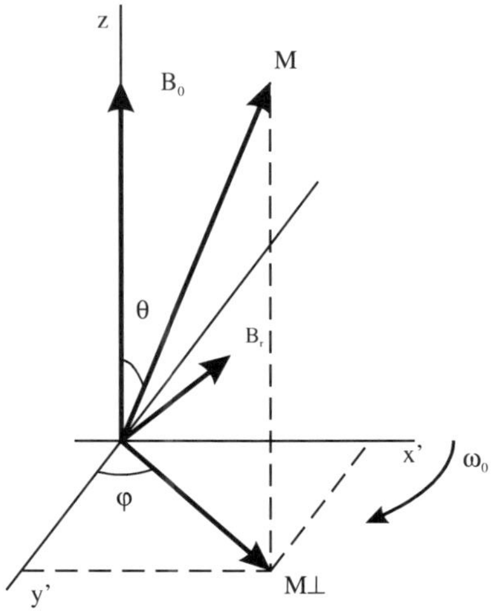

Fig. 2.40 The vector $\vec{M}$ in rotating frame

is induced in this inductor according to the formulas (2.15) and (2.16), that leads to an alternating voltage at its ends with an amplitude given by (2.17). Naturally, the alternating voltage creates an electric current in the coil, which generates an alternating magnetic field of $\vec{B}_r$ at the resonance frequency ν_0. According to the laws of electromagnetic induction, this field lags in phase on 90° from the direction of the transverse components of the vector $\vec{M}$. The physical picture of the rise of the secondary field $\vec{B}_r$ can be illustrated in Fig. 2.40, which shows the corresponding vectors in a rotating with angular frequency $\vec{\omega}_0$ coordinate system. The amplitude of the field $\vec{B}_r$ is proportional to the transverse component of the vector $\vec{M}$:

$$B_r = kM \sin\theta, \tag{2.44}$$

where k is the proportionality constant that depends on the parameters of the resonant circuit of the NMR signal sensor.

It is obvious that the emerged field $\vec{B}_r$ will affect the nuclear magnetization as well as any external alternating field with the only difference that its value depends on the orientation of $\vec{M}$. According to Fig. 2.40 under the influence of the field $\vec{B}_r$ the vector $\vec{M}$ will rotate around it in the clockwise direction and, as a result, the nuclear magnetization vector is set along the vector field $\vec{B}_0$. It is easy to check that for any initial orientation of the vector $\vec{M}$ in the space the described process develops in this way. It is worth to be emphasized that for the rise of the field $\vec{B}_r$ is connected with the existence of the circulating transverse component of the magnetization $\vec{M}_\perp$, i. e. the effect is absent for the exact orientation of the vector $\vec{M}$ along or against the vector field $\vec{B}_0$. However, as already mentioned, all experiments in the nuclear magnetic resonance are based on the registration of namely circulating transverse components of macroscopic nuclear magnetization.

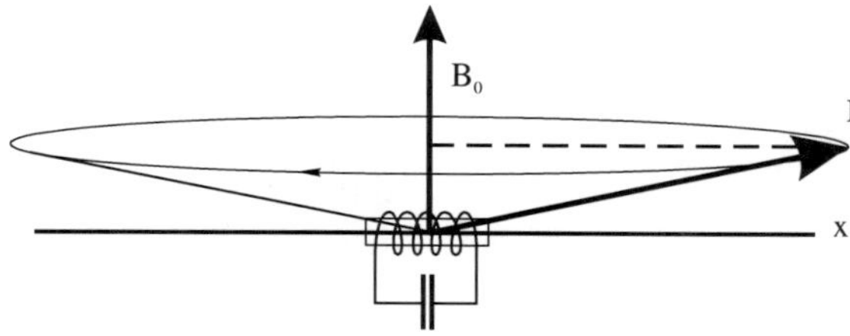

Fig. 2.41 The RF coil and the precession of the nuclear magnetization around the static magnetic field

Thus, the change in the orientation of the vector $\vec{M}$ in our case is determined by the interaction of the nuclear magnetization with a receiver circuit. Therefore, this effect is called *reaction of resonant circuit*. The Bloch equations (1.56) are insufficient to describe the effect because they do not take into account the existence of the secondary field $\vec{B}_r$, i.e. instead of three variables M_x, M_y, M_z now we have to deal with four ones: M_x, M_y, M_z and B_r. Therefore, it is necessary to complete the Bloch equations by one more equation relating the field $\vec{B}_r$ and the transverse component of the vector $\vec{M}$.

Let the receiving coil is oriented along the axis x of the laboratory system coordinates (Fig. 2.41). Then the e.m.f. will be induced only because of oscillations of the component M_x. From Eqs. (1.59) and (1.60), taking into account (1.68) and (1.69) one can obtain $M_x = -v \sin \omega_0 t$, since $u = 0$ for $\omega = \omega_0$. Then, repeating the calculations in Sect. 2.4, we obtain for the e.m.f. in the receiving coil:

$$E = 4\pi n S \omega_0 v \, \cos \omega_0 t. \tag{2.45}$$

Voltage u_s on the inductor included in the resonant circuit is more than E in Q times (Q is the quality of the circuit) and its phase is shifted on 90° with respect to the e.m.f.:

$$u_s = 4\pi n S \omega_0 Q v \, \sin \omega_0 t. \tag{2.46}$$

Taking into account the wellknown relation between the current i and the voltage u on an inductor L: $i = -L^{-1} \int u dt$ one can obtain using (2.46)

$$i = 4\pi n S L^{-1} Q v \cos \omega_0 t.$$

For simplicity we assume that the receiving coil on a design is similar to a long solenoid (i. e., its length is much larger than its diameter). Then

$$B_x \simeq 4\pi n_0 i = (4\pi)^2 n n_0 S L^{-1} Q v \, \cos \omega_0 t, \tag{2.47}$$

where n_0 is the number of turns per 1 cm. The inductance of a coil is given by $L = 4\pi n^2 S l^{-1}$, where l is the length of the coil, and we finally obtain

$$B_x = -\frac{4\pi Q}{\omega_0}\frac{\mathrm{d}M_x}{\mathrm{d}t}, \tag{2.48}$$

because

$$\frac{\mathrm{d}M_x}{\mathrm{d}t} = -\omega_0 v \cos\omega_0 t.$$

Strictly speaking, formula (2.48) should contain the attenuation coefficient η, which takes into account the incomplete filling of the receiving coil by a sample and the finite size of that coil (i.e., the field B_x less than given by (2.47) and (2.48). The coefficient η can be calculated for each case but for the average experimental conditions it amounts to about 0.3.

Thus, the system of equations for the components of the nuclear magnetization in the laboratory coordinate system is now supplemented by the Eq. (2.48):

$$\begin{aligned}
\frac{\mathrm{d}M_x}{\mathrm{d}t} &= \gamma M_y B_z - \gamma M_z B_y - \frac{M_x}{T_2},\\
\frac{\mathrm{d}M_y}{\mathrm{d}t} &= \gamma M_z B_x - \gamma M_x B_z - \frac{M_y}{T_2},\\
\frac{\mathrm{d}M_z}{\mathrm{d}t} &= \gamma M_x B_y - \gamma M_y B_x - \frac{M_z - M_0}{T_1},\\
B_x &= -\frac{4\pi Q\eta}{\omega_0}\frac{\mathrm{d}M_x}{\mathrm{d}t}.
\end{aligned} \tag{2.49}$$

It is impossible to find the general analytical solution of the system of nonlinear Eq. (2.49), however, there are a number of solutions for particular cases. To clarify the basic features of the phenomenon, consider the approximation assuming infinitely large relaxation times T_1 and $T_2(T_2^*)$, when the only reason of the damping of the precession of the vector $\vec{M}$ is the energy transfer to the receiver circuit. In this case the magnitude of the nuclear magnetization remains unchanged, and the angle θ between the vectors $\vec{M}$ and $\vec{B}_0$ gradually decreases (see Fig. 2.40). To describe the behavior of the vector $\vec{M}$ it is enough to use the last two equations in (2.49). Let us write the components of the nuclear magnetization in a spherical coordinate system and substitute the expression for B_x in the third equation of (2.49):

$$\begin{aligned}
-M\sin\theta\frac{\mathrm{d}\theta}{\mathrm{d}t} = &-\gamma M\sin\theta\cos\omega_0 t\left(-\frac{4\pi Q\eta}{\omega_0}\right)M\\
&\times\left[\cos\theta\frac{\mathrm{d}\theta}{\mathrm{d}t}\sin\omega_0 t + \omega_0\sin\theta\cos\omega_0 t\right].
\end{aligned}$$

Since the change in of the angle θ can not be fast because of the small amplitude of the field B_x, the first term in square brackets can be neglected ($\mathrm{d}\theta/\mathrm{d}t \ll \omega_0$). Making the necessary reductions, we obtain

$$\frac{d\theta}{dt} = -4\pi\gamma M Q\eta \ \sin\theta \ \cos^2\omega_0 t$$
$$= -4\pi\gamma M Q \ \sin\theta \left[\frac{1}{2} + \frac{1}{2}\cos 2\omega_0 t\right]. \tag{2.50}$$

We are interested in the change of angle θ in the time intervals, significantly exceeding the period of the nuclear magnetization precession. Then the influence of the second term in square brackets in (2.50) can be neglected, since its inclusion would only lead to very small fast changes in of the angle θ with frequency $2\omega_0$. Finally, we obtain:

$$\frac{d\theta}{\sin\theta} = -\frac{dt}{\tau_R}, \tag{2.51}$$

where

$$\tau_R = \frac{1}{2\pi\gamma M Q\eta}. \tag{2.52}$$

The solution of (2.51) is easily obtained if it is converted to a form

$$\frac{d\theta}{\sin\theta} = \frac{d(\tan\frac{\theta}{2})}{\tan\frac{\theta}{2}} = -\frac{dt}{\tau_R}.$$

Then

$$\tan\frac{\theta}{2} = Ce^{-t/\tau_R} = e^{(t_0-t)/\tau_R}, \tag{2.53}$$

where C and t_0 are the constants, and t_0 is the point in time when $\theta = \pi/2$.

The purpose of this consideration is to describe the behavior of the transverse relative field $\vec{B}_0$ components of the nuclear magnetization $M_\perp$, which recorded in NMR experiments. Taking into account that

$$M_\perp = M \ \sin\theta,$$

and expressing $\sin\theta$ with the help of (2.53), we obtain

$$M_\perp = M\frac{2}{(\tan\frac{\theta}{2})^{-1} + \tan\frac{\theta}{2}}$$
$$= M\frac{2}{e^{-(t_0-t)/\tau_R} + e^{(t_0-t)/\tau_R}} = M \ \sinh\frac{t-t_0}{\tau_R}. \tag{2.54}$$

The function (2.54) for different initial conditions is presented in Fig. 2.42.

Thus, the behavior of the transverse component does not have the character of a damped exponent, as in the case of relaxation, and is described by a more complex function. However, at small values of $\theta(0)$, when $exp[-(t_0 - t)/\tau_R] \ll 1$ Eq. (2.54) becomes

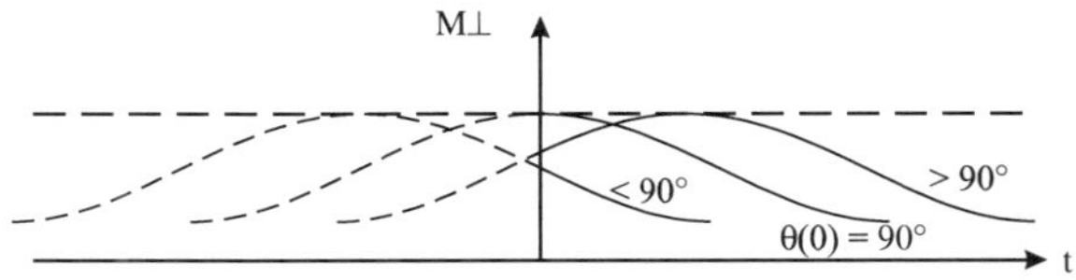

Fig. 2.42 The behavior of $M_{\perp}$ at different initial conditions

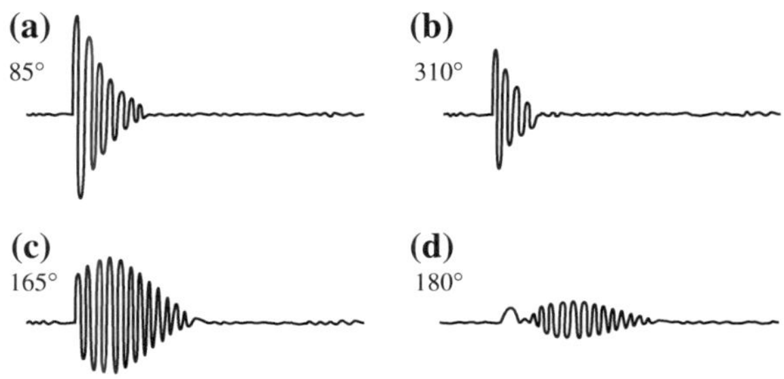

Fig. 2.43 The free induction decay signals of the ^{1}H nuclei after a single pulse at various angles of the nutation of the nuclear magnetization (to the left of each signal the initial meaning of $\theta(0)$ is indicated): 85° (**a**); 310° (**b**); 165° (**c**); 180° (**d**)

$$M_{\perp} = M(0)e^{-t/\tau_R},$$

i. e. value of τ_R, defined by (2.52), takes the sense of a damping constant. Mainly because of this, the discussed process is also called *radiation damping*.

This analysis gives the correct description of the effect of radiation damping in the case of relatively high values T_1 and T_2^*, that is confirmed by direct experiments. The free induction decay signals of the ^{1}H nuclei after a single pulse at various angles of the nutation of the nuclear magnetization are presented in Fig. 2.43 (to the left of each signal the initial meaning of $\theta(0)$ is indicated). It is obvious that the view of the signals is in accordance with the function (2.54) and Fig. 2.42. In these experiments, the the signals were registered with $T_2^* \simeq 2$ s and $\tau_R \simeq 0.3$ s.

An interesting feature of the effect is the possibility of self-excitation (see two last traces in Fig. 2.43) in two coupled resonance systems (nuclear magnetization — rf circuit). If the magnetization is directed in the opposite direction in respect to the vector field $\vec{B}_0$ (it is possible, for example, in the case of the Overhauser effect, See Chap. 16) and $\tau_R \ll T_2^*$, so-called *radiation instability* arises. On the basis of this effect a NMR generator was even created, which completely correspond to masers (see, Chap. 18).

In conclusion, it is worth noting that the additional damping of the transverse component of the nuclear magnetization leads to the broadening of spectral lines. This, in its turn, results in the deterioration of resolution of a radiospectrometer, that one should take into account in the high resolution spectroscopy. Of course, if there are many lines in a spectrum this the effect is less noticeable, because the magnetization of each line precesses at its own frequency, i. e. for the estimation of possible influence of the radiation damping using the formula (2.52) one must substitute a "partial" magnetization M_i for each spectral line with $\Sigma M_i = M$.

In spite of the main works, in which the radiation damping was discussed, published in the 50–60-ties, the interest to this problem keeps till now.

2.10 Magnetic Resonance Imaging

The term *magnetic resonance imaging* (MRI), or else *magnetic resonance tomography*, involves a lot of techniques to get the image in a certain section of the spatial distribution of any characteristic of an object to which the magnetic resonance is sensitive. In a broader sense, the magnetic resonance imaging means getting any localized information using magnetic resonance.

The idea to obtain localized nuclear magnetic resonance spectra and images with NMR was proposed a long time ago (Ivanov V. A. The method of determining the internal structure of objects. Auth. Certificate USSR, No 1112266. Priority from 21.03.60. Bulletin of discoveries and inventions. 1984. No 33. P. 125), but the first publication of the implementation of the method based on nuclear magnetic resonance appeared only in 1973. Shortly after this EPR images were obtained too. EPR imaging is still used as a laboratory research method (for example, to obtain information about the spatial distribution of free radicals), whereas NMR imaging (NMRI) for a long time went out beyond laboratories and it is widely used first of all for the purpose of medical diagnosis. Furthermore the NMRI is used in plant and animal physiology, geophysics, chemical technology, food industry etc. Therefore, the physical principles of the localization of magnetic resonance signals will be considered in case of NMR. Let us consider the principles of establishing of the mutual correspondence between the NMR signal and the position of a volume element of an object (voxel) that produces this signal. There are many ways to establish such correspondence, but the basis of all of them is the proportionality between the magnetic resonance frequency and the magnitude of the magnetic field (1.25).

2.10.1 The Obtaining of Spin Images from NMR Induction Signals in Inhomogeneous Field

A common feature of all the methods of localization is that the experiment is performed in a inhomogeneous magnetic field with a specially organized character of inhomogeneity. If we observe a magnetic resonance in an inhomogeneous static magnetic field, different points in space correspond to the different resonance frequencies in an inhomogeneous field, hence space points are mapped by the frequency. However, the mapping because of the topological properties of the magnetic field is not biunique: the same frequency corresponds not to one, but to a lot of points in a space, located on certain surface. To overcome this difficulty the character of inhomogeneity is changed successively in the course of the experiment. The obtained thereby NMR signal from object contains the information on a spatial distribution of some parameter determining this signal, but this information appears in the implicit (encoded) appearance. Process and way of entering such an information in a magnetic resonance signal are termed as spatial encoding. Decoding (deciphering) of a signal is produced by mathematical processing of the acquired data. Depending on the character of actions on spin system in the course of encoding, the visualization of

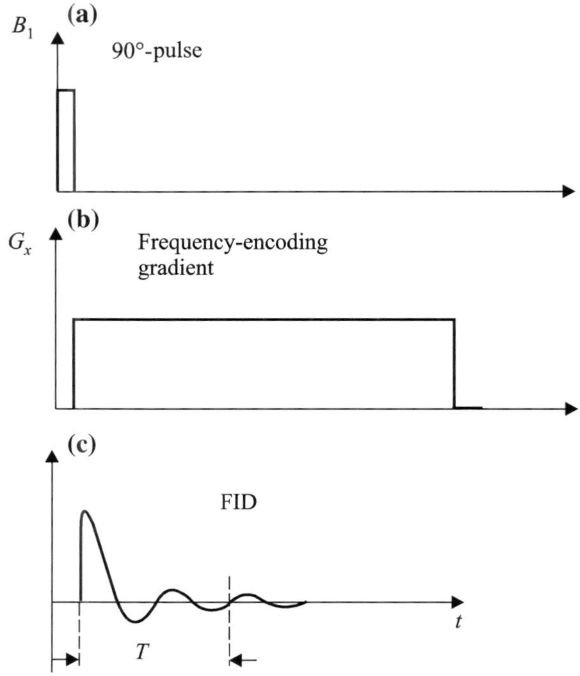

Fig. 2.44 Diagram of frequency-encoding pulse sequence with FID acquisition: **a** RF pulse; **b** frequency-encoding magnetic field gradient; **c** signal of free induction decay

spatial distribution of various parameters can be mapped, namely: concentrations of nuclei of the certain types or unpaired electrons, magnetic relaxation rates, chemical shifts, molecular diffusion, parameters of a macroscopic motion of a sample etc. The EPR imaging is still typical stationary methods of magnetic resonance detection, but for obtaining NMR images, as well as in modern NMR spectroscopy, almost exclusively the pulse method is used. A variety of methods for producing spin images are divided into two main groups: the first one is *detection of free nuclear induction signal* in a nonuniform magnetic field, the second one is *excitation of nuclear magnetic resonance signals* in such a field.

To understand the essence of the first method of obtaining the spatial information, let us consider a simple experiment. Just after the action of 90-degree pulse (B_1 in Fig. 2.44a), the magnetic field with a constant gradient (G_x in Fig. 2.44b) is created in the region where the explored object is located. In this situation in each volume element termed *voxel*, there is the precession of nuclear magnetization with the frequency spotted by a voxel position, more exactly by value of its coordinate x. The observed signal of the free induction decay signal (FID) from the whole object is the sum of signals from particular voxels in the object volume (Fig. 2.44c). Hence, the Fourier spectrum of this signal with an appropriate choice of the scale represents the distribution of density of resonating spins along the coordinate x. In other words, the frequency spectrum of the signal of the magnetic resonance is the projection of density distribution of resonating spins on the direction of gradient of a magnetic field.

Now we will give the formal description of the technique of obtaining a projection of spin density spatial distribution from a FID signal. The induction from particular voxel is determined by transversal components of nuclear magnetization u and v in a rotating frame ($u = M_{x'}$ and $v = -M_{y'}$).Their motion is described by the Bloch equations (1.62) and (1.63) under conditions $B_1 = 0$ and $\delta\omega = \gamma\vec{G} \cdot \vec{r}$, where $\vec{G}$ is the vector of gradient of a magnetic field with components G_x, G_y, G_z; $\vec{r}$ is the radius vector, defining the position of the voxel. The solution of the system (1.62), (1.63) can be presented in the form of a complex function $m_1(\vec{r}, t) = u - \mathrm{i}v$:

$$m_1(\vec{r}, t) = m_1(\vec{r}, 0) \exp\left(\mathrm{i}\vec{k} \cdot \vec{r} - \frac{t}{T_2}\right), \tag{2.55}$$

here for convenience we introduce the vector

$$\vec{k} = -\gamma \int_0^t \vec{G}(t')\mathrm{d}t' \tag{2.56}$$

Besides it is assumed that, in general, the gradient is the function of time; $m_1(\vec{r}, 0)$ is the value of complex transversal magnetization of the voxel at the time $t = 0$ (let us accept, for example, this is the moment just after a pulse of the radio-frequency field directed along an axis x' in the rotating frame). In this case $|m_1(\vec{r}, 0)|$ is proportional to spin density $\rho(\vec{r})$.

For the simplicity we assume that $t < T_2$. Then, according to the above and the Eq. (2.55), the free nuclear induction signal from whole object is expressed, up to a constant factor, by the function $s(t)$ defined by summation on the volume of the sample V:

$$s(t) = \iiint_V \rho(\vec{r}) \exp\left(\mathrm{i}\vec{k} \cdot \vec{r}\right) \mathrm{d}V. \tag{2.57}$$

Here $\mathrm{d}V$ is the volume differential. In this example $k_x = -\gamma G_x t$, $k_y = k_z = 0$. Then the signal from the whole sample can be written as

$$s(t) = \int_{x_1}^{x_1} \rho_1(x) \exp\left(\mathrm{i}k_x x\right) \mathrm{d}x \tag{2.58}$$

Here $\rho_1(x) = \iint_{S_x} \rho(\vec{r})\mathrm{d}y\mathrm{d}z$ is the projection of density of spins to the axis x, S_x is the section of object at the plane $x = \text{const}$; x_1 and x_2 are the limits of x values between which the object is located. Dependence of the integrand in (2.58) from t is contained in a variable k_x according to (2.56). The signal $s(t)$ could also be regarded as function of k_x too. It is convenient instead of $s(t)$ to enter a new function $s_1(k_x)$

by means of a relation $s\left(-\frac{k_x}{\gamma G_x}\right) = 2\pi s_1(k_x)$. Using this definition and considering $\rho_1(x) = 0$ out of the gap $[x_1, x_2]$, it is possible to present the function $s_1(k_x)$ in the form

$$s_1(k_x) = \frac{1}{2\pi} \int_{-\infty}^{+\infty} \rho_1(x) \exp(\mathrm{i}k_x x)\,\mathrm{d}x. \tag{2.59}$$

Thus, the information on the spatial distribution of spins expressed by function $\rho(x)$ is already contained in the $s(t)$ (or $s_1(k_x)$ that is equivalent) in the implicit (encoded) appearance. This technique of the input of the spatial information in a NMR signal was called *the frequency encoding*. The explicit expression of spatial distribution is easy to find, as expression (2.59) is the inverse Fourier transformation of $\rho_1(x)$. Hence, to get $\rho_1(x)$, it is sufficient to perform the direct Fourier transform of the function $s_1(k_x)$.

Certainly, in a real experiment the values of function $s(t)$ are measured only in finite number of points N and in a finite interval of time T (See Fig. 2.44c). Values $s(t)$ in these N points form the one-dimensional array of data $\{s_i\}$ to which discrete Fourier transform is applied. One obtains as the result N values ρ_j in the interval

$$-\frac{\pi N}{\gamma G_x T} \div \frac{\pi(N-2)}{\gamma G_x T}.$$

One eliminates a remained uncertainty in two other coordinates with various techniques. Let us confine ourselves to the two possibilities. One of them consists in the frequency encoding which is carried out repeatedly at different directions of a magnetic field gradient. As a result, a set of projections on different directions is obtained. The image of two-dimensional (2D) distribution of density of spins is reconstructed by means of special algorithm (the Radon transformation). Such a technique, termed as *the projection method*, is applied to obtain a pattern of distribution of spin density in certain plane, i.e. *the magnetic resonance tomogram*. The algorithm of reconstruction was applied in other types of a tomography and it does not use the peculiarities of a magnetic resonance, therefore it is not considered here.

Another technique uses so-called *the phase encoding* of a voxel position. For obtaining *an one-dimensional* distribution of density of spins by this technique, the first experiment (Fig. 2.44) should be modify as follows. A gradient pulse is applied after the action of the 90-degree pulse, as above, but the FID signal is registrated at fixed moment τ immediately after turning off the gradient G_x (Fig. 2.45b, c). At time τ the precession phase $\phi = \gamma G_x x \tau$ accumulates in every voxel with coordinate x. Thus, the voxel coordinate at the given value G_x is marked by the precession phase at the moment τ. The summed signal from the whole object at the moment τ depends on the distribution function $\rho_1(x)$ and on values G_x. Measurement of the free induction signal is repeated N times at the same interval τ after each following radio-frequency pulse, but each time with a new value of the gradient G_x: $G_{x2}, G_{x3}, \ldots, G_{xl}, \ldots, G_{xN}$. In Fig. 2.45b the arrow shows conventionally that

Fig. 2.45 Diagram of phase-encoding pulse sequence: **a** RF pulse; **b** phase-encoding gradient; **c** signal of FID free induction decay in homogeneous magnetic field

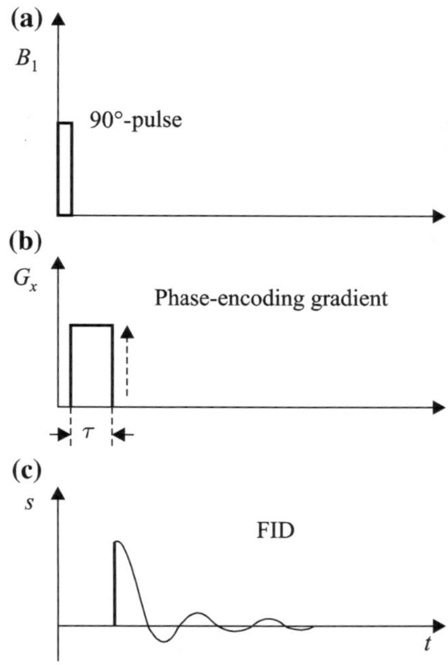

phase-encoding gradient varies at each new measuring. As well as in the previous case, the array $\{s_l\}$ from N values of signals measured at the moment τ (on Fig. 2.45c the vertical segment represents the result of one of measurements) is gotten. Unlike the previous technique, to each value l there corresponds the same value τ, but different values G_{xl}. In this case the variable is not a time, but a gradient. However the form of Eq. (2.59) remains former as it includes only quantity k_x in an explicit form. In this case $k_x = -\gamma G_x \tau$ equally dependes on τ and G_x. Hence, a spatial concentration distribution of spins $\rho_1(x)$ is also got by means of Fourier transform. As mentioned above in a real experiment the variable gradient does not change continuously, but takes discrete values. The result of discrete Fourier transform of an array $\{s_l\}$ is the array of values of spin concentration of $\{\rho_m\}$. Each value ρ_m corresponds to a certain value x. The physical difference of this method from the frequency encoding is that the voxel coordinate x is marked not by the frequency of magnetic resonance, but by the difference in phases of the precession which has run at the time τ.

Certainly, excepting some special cases, it is unsuitable to use the phase encoding for obtainment of a two-dimensional image of distribution of spin concentration because in this case the duration of image formation is N times more than at the frequency encoding. Nevertheless, the combination of both methods of encoding allows us to get the two-dimensional spatial information.

The method of obtaining the two-dimensional image is realized as follows. The transversal nuclear magnetization, as usual, is created by means of a radio-frequency pulse. As well as at obtaining the one-dimensional projection of density

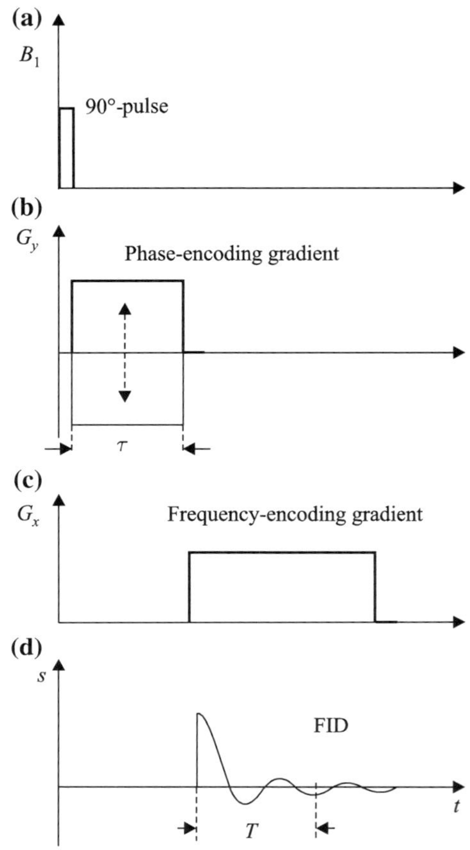

Fig. 2.46 Diagram of pulse sequence for obtaining 2D-image by phase-frequency encoding with FID acquisition: **a** RF pulse; **b** phase-encoding gradient; **c** frequency-encoding gradient; **d** signal of free induction decay

by the method of the frequency encoding, during the registration of FID, the object is in a magnetic field with a constant gradient G_x along the axis x. The difference lies in the fact that the gradient G_x turns on not just after a radio-frequency pulse B_1 (see Fig. 2.46a), but after time τ during of which the magnetic field with a gradient perpendicular to G_x, for example G_y, operates (Fig. 2.46b). At the end of the action of the gradient pulse G_y the magnetization precession for every voxel acquires a phase shift which encodes the position of the voxel along the axis y. This shift determines the initial phase of the precession in every voxel during the action of the frequency encoding gradient G_x (Fig. 2.46c). This experiment is repeated N times with the changing value G_y to provide phase encoding on the coordinate y. Fourier transform of each signal of the free induction (Fig. 2.46d) gives, as result, the spectrum which is a function of x and relates to the certain value G_y. To decode the dependence from y, it is necessary for each value x to execute the second Fourier transform, the integration variable for which is k_y, proportional to G_y. It follows immediately from Eq. (2.57) if to substitute in it $k_x = -\gamma G_x t$, $k_y = -\gamma G_y \tau$, $k_z = 0$. As a result the function of two coordinates (x and y) which represents two-dimensional distribution

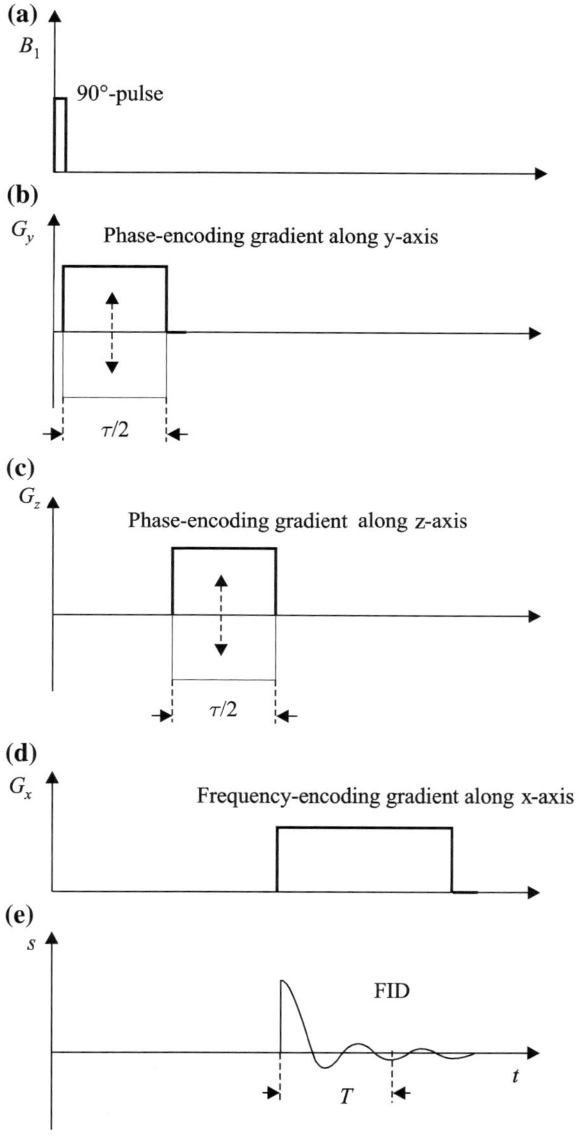

Fig. 2.47 Diagram of pulse sequence for obtaining 3D-image by phase-frequency encoding with FID acquisition: **a** RF pulse; **b** the first phase-encoding gradient (along y); **d** the second phase-encoding gradient(along z); **c** frequency-encoding gradient; **e** signal of free induction decay

of spin density in explored object is gained:

$$\rho_2(x, y) = \int_{z_1}^{z_2} \rho(\vec{r}) dz.$$

The above reasoning can easily be extended to three dimensions (the abbreviation 3D-imaging is used). In this case after the action of the 90-degree pulse B_1 (Fig. 2.47a) the phase-encoding gradient G_y operates during the time $\frac{\tau}{2}$ (Fig. 2.47b), and during the following interval $\frac{\tau}{2}$ the phase-encoding gradient G_z (Fig. 2.47c) is applied. Thus for each value of one of phase-encoding gradients, for example G_y, it is necessary to repeat the single experiment N times with the changing value G_z. Thus, on coordinates y and z phase encoding, and on coordinate x a frequency encoding (Fig. 2.47d) are carried out. The formal justification of the method follows immediately from the Eq. 2.57 in which it is necessary to substitute

$$k_x = -\gamma G_x t, \quad k_y = -\frac{\gamma G_y \tau}{2}, \quad k_z = -\frac{\gamma G_z \tau}{2}.$$

In this case the signal of the free induction $s_{G_y G_z}(t)$ depends not only on time, but also on two gradients: G_y and G_z. As well as in the one-dimensional case, instead of $s_{G_y G_z}(t)$ it is more convenient to pass to a function of arguments k_x, k_y and k_z: $s_2(k_x, k_y, k_z)$ or, more compactly, $s_2(\vec{k})$. The required distribution is represented with three-dimensional Fourier transform:

$$\rho(x, y, z) = \iiint\limits_{k_x, k_y, k_z} s_2(k_x, k_y, k_z) \exp(-\mathrm{i}k_x x) \exp(-\mathrm{i}k_y y) \exp(-\mathrm{i}k_z z) \mathrm{d}k_x \mathrm{d}k_y \mathrm{d}dk_z,$$

that can be written down in more compact form:

$$\rho(\vec{r}) = \iiint\limits_{\vec{k}} s_2(\vec{k}) \exp(-\mathrm{i}\vec{k} \cdot \vec{r}) \mathrm{d}\vec{k}.$$

An advantage of the 3D-imaging is the possibility to reflect a spatial information in an arbitrarily oriented section of an object using a corresponding mathematical processing of the three-dimensional array of data obtained in experiment.

The technique of magnetic resonance imaging reconstruction, which uses multidimensional Fourier transform, can be extended to obtain either local NMR spectra, or imaging of a spatial distribution of chemical components of an explored object which are characterized by defined chemical shift. To this purpose an acquisition of FID should be executed, as in usual NMR spectroscopy, in the homogeneous magnetic field, and a phase encoding of all three coordinates should be used. In this case the repetition of a single experiment N^3 times at different values of gradients G_x G_y and G_z is carried out. It is clear from the above that one has to execute quadruple Fourier transform for the three-dimensional localization of spectra.

2.10.2 Methods of Selective Irradiation

The receiving of data from the whole volume of an object is not always required. For saving of time during data acquisition it is more expedient to excite not all volume, but only an interesting for a researcher part. Practically the matter is about the spatial distribution in certain flat slice from which it is required to get a NMR signal. The problem is solved using *the selective irradiation* idea of which consists in simultaneous action of a radiofrequency pulse and a pulse of gradient, vector of which is perpendicular to the selected slice. It is necessary to choose the width and the shape of the radio-frequency pulse so that the rotation of equilibrium nuclear magnetization on 90 degrees would be carried out only in a narrow frequency band, and consequently, only in the narrow slice adjoining to the plane for which the resonance condition is strictly satisfied. Required parameters of a pulse can be determined by calculation on a base of Bloch equations. It should to note that the use of Fourier transform is not correct in this case because in the presence of a radio-frequency field the Bloch equations are not linear equations with constant coefficients. Therefore, the superposition principle for their solution is not valid. Nevertheless, in the case of a 90-degree pulse the order of magnitude of a covered frequency band ($\delta\omega$) corresponds to the width of spectrum of a rectangular RF pulse duration t_{sq}:

$$\delta\omega \simeq \frac{1}{t_{\mathrm{sq}}}.$$

A combination of the selective irradiation and the 2D-imaging technique allows us to obtain a map of two-dimensional distribution of spin density in the selected slice averaged on its thickness. The diagram of the pulse sequence to realize this, fairly widespread, technique is given in Fig. 2.48. Simultaneously with a RF pulse B_1 (Fig. 2.48a) the magnetic field with a gradient G_z (Fig. 2.48b) acts, thanks to which the RF pulse takes the property of spatial selection. The effects of gradient G_x and G_y (Fig. 2.48c, d) is similar to the described one for the two-dimensional imaging. The peculiarity of the given method is that a signal $s_{G_y}(t)$ (Fig. 2.48e) is received not from the whole object, but only from the selected slice. As a rule, the selective RF pulses are longer and less intensive, than usually applied in NMR spectroscopy, because they are termed *as soft*, unlike broadband, short and intensive *hard* pulses.

For more accurate fixing of boundaries of a slice, instead of square pulses one uses pulses with specially fitted shape. The 2D-imaging in the selected slices requires less time, but spatial resolution along the coordinate which is perpendicular to the selected slice plane is much worse than in the 3D-imaging.

Fig. 2.48 Diagram of pulse sequence for obtaining 2D-image in selected slice by phase-frequency encoding with FID acquisition: **a** RF pulse; **b** selecting gradient; **c** phase-encoding gradient; **d** frequency-encoding gradient; **e** signal of free induction decay

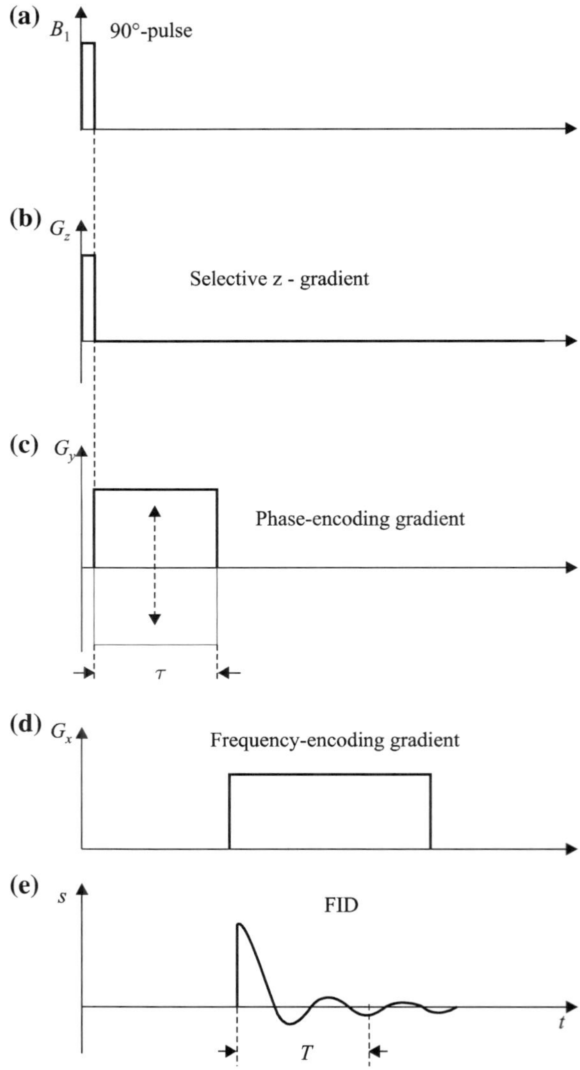

2.10.3 Versions of the Basic Methods of Formation of NMR Imaging

There are various versions of the above basic techniques. One of the most widespread techniques consists in that at a stage of the frequency encoding the signal of spin echo is used instead of free induction.

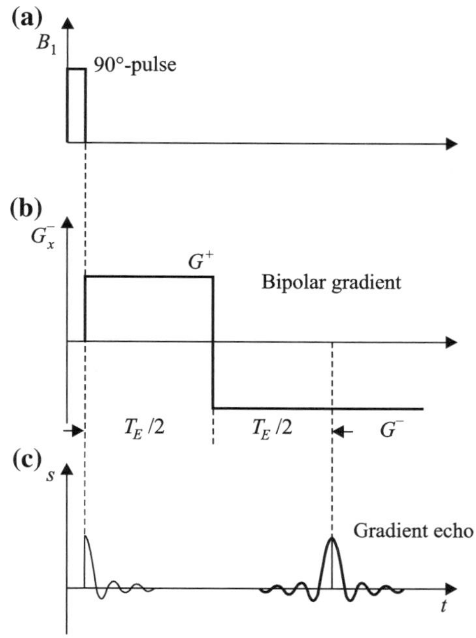

Fig. 2.49 Pulse diagram for obtaining of a gradient echo. **a** RF pulse; **b** bipolar frequency-encoding gradient; **c** gradient echo

The echo can be created either by introducing an inverting pulse in the certain time interval after the 90-degree pulse (the Hahn echo, see Sect. 2.5.2) or by using of a bipolar gradient (Fig. 2.49).

As an example, the diagram of a pulse sequence for data acquisition for 2D-imaging in the selected slice with bipolar gradient is shown in Fig. 2.50. It is very similar to the diagram in Fig. 2.48, but the gradient which carries out encoding on frequency is made bipolar (G_x in Fig. 2.50d). This leads to appearance of gradient echo with a maximum at the moment T_E after insert G_x ($\frac{T_E}{2}$ is the duration of the positive part: G_x^+) (Fig. 2.50e). The bipolar gradient allows us to get an echo without a 180-degree pulse that removes the problems related to a deviation of its parameters from nominal values. However if Hahn echo decay have a time constant T_2 (a true relaxation time), the time constant of gradient echo decay is T_2^* (an apparent relaxation time), reflecting local inhomogeneity of a magnetic field.

In other widespread updating of the technique, RF pulses, rotating the nuclear magnetization vector by an angle less than 90 degree, are used. In this way, it is possible to reduce considerably the time of data accumulation to construct an image without considerable signal intensity loss. Such a performance uses the fact that the the longitudinal magnetization is proportional to cosine, and the transversal one to sine of the nutation angle. Therefore there is an insignificant deviation from an equilibrium at the small angles that allows the essential reduction of time of equilibrium restoring.

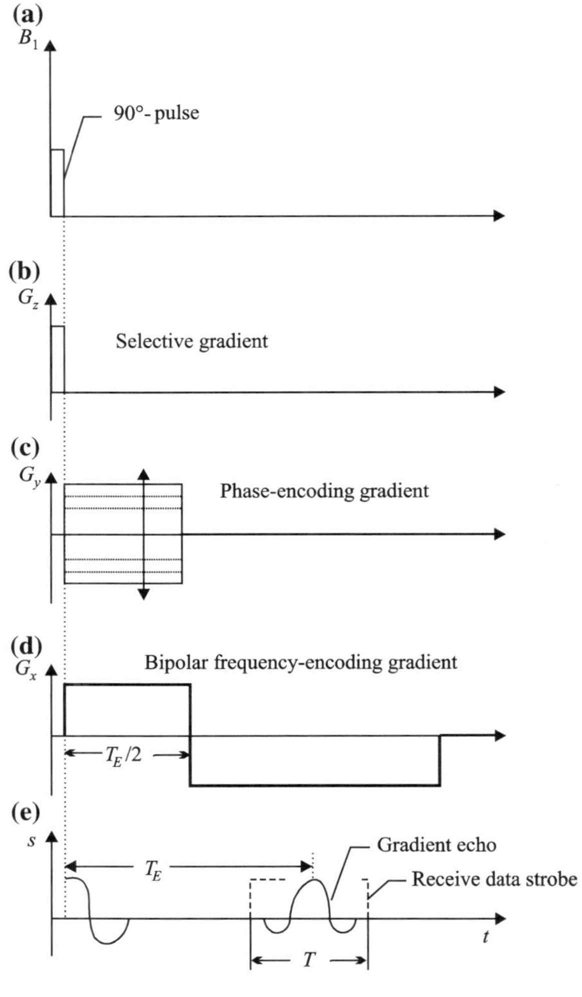

Fig. 2.50 Pulse diagram for obtaining 2D-image in selected slice with gradient echo. **a** RF pulse; **b** selective gradient; **c** phase-encoding gradient; **d** bipolar frequency-encoding gradient; **e** gradient echo

2.10.4 Contrast Enhancement in MRI

The specific feature of magnetic resonance imaging in comparison with other methods of visualization is the possibility to change a local intensity of the received signals in the dependence on various parameters characterizing nuclear magnetic resonance. The change in the appearance of nuclear magnetic resonance images depending on sensitivity to certain parameter is named as parameter contrast. They distinguish so-called *true images* of spatial distribution of a parameter, and *parameter weighted images*. In the first case the values of a parameter as a function of coordinates are calculated from the obtained data and mapped on the display. In the second case, the spatial distribution of transversal magnetization values is mapped. In this case the brightness of image reflects not only the parameter of interest, but also all other

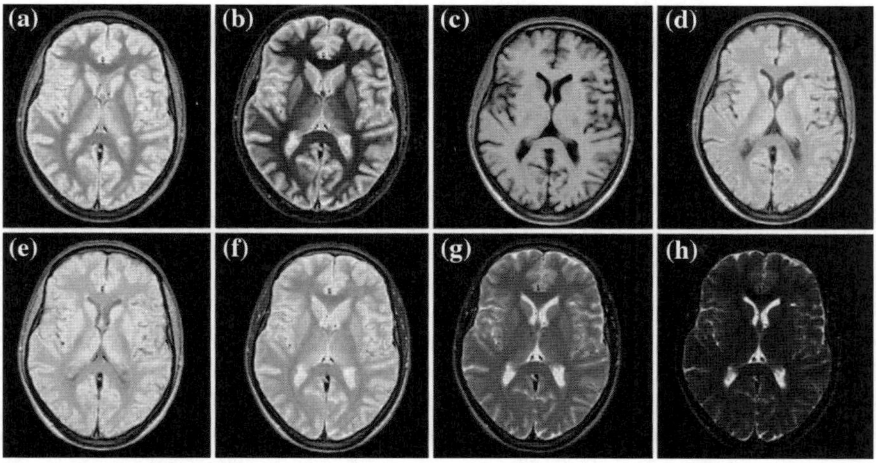

Fig. 2.51 Brain NMR-images by inversion-recovery technique (**a–d**) and by Hahn spin echo technique (**e–h**) with repetition time 3400 ms. Time recovery was: **a** 90 ms; **b** 200 ms; **c** 1000 ms; **d** 2000 ms and time echo was: **e** 5 ms; **f** 50 ms; **g** 120 ms; **h** 250 ms (The images was kindly provided by Ass. V. Parshina)

factors influencing intensity of NMR signal. Nevertheless, in many applications just the parameter weighted images provide obtaining the necessary information on an object, especially in medical diagnostic. In this cases it is possible to use the simplified techniques that essentially saves a time. Local values of relaxation times, self-diffusion coefficients and quantities characterizing a macroscopic motion of a fluid, magnetization transfer etc can be the parameters to map. Techniques of the obtaining of the true images or parameter weighted ones consist in a combination of standard methods of measurement of corresponding parameters with imaging techniques.

T_1**-relaxation contrast**. To obtain local values of T_1, 180-degree or 90-degree pulse is applied before a 90-degree pulse, as well as in usual time relaxation measurements (see 2.5.1, Fig. 2.30). All other procedures remain the same, as in the case of the imaging of spatial distribution of spin concentration. As a rule, one uses the spin echo with short echo interval T_E to data acquisition instead of FID. The whole cycle of measuring is iterated necessary number of times at different recovery times T_I. For obtaining T_1-weighted images it is enough to construct the image at one of values T_I, but its optimum value for obtaining the best contrast depends on nature of an explored object. The view of the contrasted image essentially depends on the chosen method and time interval. For example, if one uses the 180-degree pulse (inversion-recovery technique, IR), the areas characterized by longer times T_1, than pulse-to-pulse spacing, and the areas characterized by very short T_1, can look equally bright, the magnification of brightness is more typical for areas with short times of longitudinal relaxation. Figure 2.51 represents IR NMR-images of brain at different recovery times. One can see the very short T_I relative T_1 as well as the very long

(Fig. 2.51a, d) lead to a bad contrast. Intermediate values of T_I give more contrast images (Fig. 2.51b, c). For each kind of tissue the optimal value of T_I exists.

The simplest technique of T_1-weighting uses standard pulse imaging sequences with the shortened T_R-interval. Then the areas with long T_1 will be reflected with smaller brightness.

T_2-relaxation contrast. To construct T_2-weighted images, signals of Hahn echo or echo signals in Carr-Purcell sequence are used (see Sect. 2.5.1). T_2-weighted image is constructed from signals of Hahn echo with enough long delay of an echo T_E or from enough large number an echo signal in Carr-Purcell sequence (in the last case averaging signal group is often used). Contrary to T_1-weighted images, T_2-contrast leads to a reduction of brightness of the areas characterized by small values T_2 (see Fig. 2.51h, where the areas with cerebrospinal fluid are viewed only). At short T_E the contrast is determined only by water content (Fig. 2.51e), and the intermediate T_E-values permit to discriminate between white and gray matter of brain (Fig. 2.51f, g).

T_2^* relaxation contrast. According to aforesaid about the gradient echo, the images constructed under the data from echo measuring at $T_E \geq T_2^*$ reflect a spatial distribution of local inhomogeneities of the magnetic field. These techniques of contrast enhancement are used to map active zones in the cerebral cortex. The method, named the functional magnetic resonance imaging (fMRI) or the blood oxygen level dependent (BOLD) contrast, is based on an effect of augmentation of paramagnetic desoxihemoglobin in the blood feeding active cortex sites. The fMRI is used to investigate cognitive processing of motion, texture, color, sound, memory etc.

Diffusion contrast. To discriminate areas with different self-diffusion motion of particles containing resonating nuclei on NMR images, as well as in usual measurement of self-diffusion coefficients using NMR, additional magnetic field gradient pulses are introduced into the pulse sequence. It is the most convenient to use a bipolar gradient before a stage of data acquisition. As a result the smaller brightness of an image corresponds to areas with rapid self-diffusion.

Mapping of a macroscopic motion. Manifestations of effects of macroscopic motion of medium within an explored object on NMR images are very manifold and depend on the direction of motion relative the vector of magnetic field gradient and orientation of an imaged plane, the technique of obtaining and processing NMR signals etc. If the motion of medium is perpendicular to the selected slice, saturation of spins in the slice or sounding of nuclear magnetization by the second radio-frequency pulse are used. In the first case the motion leads to magnification of brightness due to inflow of nonsaturated spins, in the second case—to dimness of the image because of ablation of excited spins from the selected slice.

It is widely used so-called *phase contrast* (not to confuse with phase encoding) arising because of NMR frequency change in the voxel moving in the gradient direction. The pulse bipolar gradient is also used, but in this case the positive and negative parts have the equal areas. In case of immobile medium the dephasing of precessions under the influence of the gradient of one sign it is completely compensated by the effect of the gradient of the opposite sign. If there is a motion in the gradient direction then an additional difference in precession phase takes place after a bipolar gradient pulse. This dephasing leads to the decrease of NMR signal from moving area and, hence, to the decrease in image brightness.

2.10.5 Application of MRI

The basic area of applications of MRI is the medical diagnostics. Expansion and efficiency of the method are determined, first, by the fact that MRI uses intense signal of the protons of water, which is the main component of living systems, and, second, by the variety of parameters that can be visualized in this way.

Relaxation contrast allows us to distinguish the tissues with the identical water content, but having different water states. This is due to the fact that the relaxation is very sensitive to the mobility of water molecules (see Sects. 3.7–3.14). The last, in turn, depends on a type and a biological state of tissue. The contrast-enhanced NMR images are used for early diagnostics of pathological changes of tissues of an organism. In certain cases it is useful to use the self-diffusion contrast enhancement, in particular in the investigation of diffusion anisotropy of brain tissues. Sensitivity of the nuclear magnetic resonance to macroscopical motion in fluids allows us to visualise blood vessels and to estimate their state.

For brain diagnostics, the local spectra of the NMR ^{13}C give an important information. The course of oxidizing processes in muscles can be tracked by means of local spectroscopy ^{31}P. For physiological and cognitive investigations of brain the contrast enhancement by local inhomogeneities of magnetic field inside of a brain is used.

As well as in other methods of visualisation in medicine, the introducing of contrasting reagents in organism is applicable, however the principle of their action is specific for MRI. One uses either the paramagnetic substances providing the strong relaxation contrast, or the ferromagnetic suspensions giving contrast via T_2^*. The prospective technique of molecular visualization uses dynamic polarization (Sects. 15.2, 21.4). In this case the complexes of free radicals or ferromagnetic nanoparticles with biomarkers are used to visualize areas of pathology. In some physiological neurologic experimentes, the possibility of visualisation of trajectories of feeble current is used because of the local inhomogeneities of a magnetic field due to that current.

In non-medical areas of the NMRI, the investigation of porous media are known. The method is based on the possibility of observation of penetration of a liquids in pores and flaws of a solid sample. This direction gave practical results in the investigation of oil-bearing beds. The NMR imaging can be applied as a control device in manufacture of high-quality ceramics at a stage of preparation of ceramic mass, in the course of polymerization of plastic and etc. A special application the NMRI is a monitoring of kinetics of chemical reactions.

An investigation of a spatial distribution of paramagnetic centres, for example, of radiation induced defects and of free radicals, is carried out by the EPR imaging.

Further Readings

Berger S, Braun S (1998) 200 and more NMR experiments, 2th expanded edn. ISBN 30527-31067-3, WILEY-VCH

Blümich B (2005) Essential NMR for scientists and engineers. Springer, New York

Derome A (1987) Modern NMR techniques for chemistry research. ISBN 0-08-032513-0, Pergamon Books Ltd

Ernst RR (1994) NMR fourier transform spectroscopy. Bull Magn Reson 16:5

Friebolin H (2004) Basic one and two-dimensional NMR spectroscopy, 4th updated and revised edn. ISBN 3-522-31233-1, WILEY-VCH

Fukushima E, Roeder S (1981) Experimental pulse NMR. A nuts and bolts approach. Addison-Wesley, Redwood City

Günther H (1995) NMR spectroscopy: basic principles, concepts, and applications in chemistry, 2nd edn. Wiley, New York

Haacke EM, Brown RW, Thompson MR, Venkatesan R (1999) Magnetic resonance imaging, physical principles and sequence design. Wiley, New York. ISSN 0192–6133

Keeler J (2010) Understanding NMR spectroscopy, 2nd edn. Wiley, New York. ISBN 978-0-470-74609-7

Lenk R (1975) Diffusion and spin relaxation. Adv Mol Relax Process 6:287

Price W (1998) Pulsed-field gradient NMR as a tool for studying translational diffusion: Part 2. Exp Aspect Concepts Magn Res 10(4):197–237

Freeman R (2003) Magnetic resonance in chemistry and medicine. Oxford University Press, Oxford 336 p. ISSN 0–19-926061-3

Rinck PA (2001) Magnetic resonance in medicine, 4th completely revised edn. Blackwell, Berlin Vienna, 2001. 252 p. ISBN 0-632-05986-9; doi:10.1007/s00330-001-1154-8

Weishaupt D, Koechli VD, Marinceck B (2006) How does MRI work? An introduction to the physics and function of magnetic resonance imaging. Springer, Berlin Heidelberg, 170 p. ISSN 978–3-540-30067-0; doi:10.1007/978-3-540-378445-7

Chapter 3
Nuclear Magnetic Relaxation

In Chap. 1 the concepts of the nuclear magnetic relaxation times (rates) were introduced on the basis of phenomenological considerations, as exemplified by the theory of Bloch. However, this approach can not clarify the nature of relaxation processes and quantify the relaxation times of various nuclei in different environments. The answers to these questions are given by quantum theory, which allows the calculation of the probability of relaxation transitions. In this chapter the problems of the relaxation in two- and multi-spin systems will be considered using the theory of time-dependent perturbations.

3.1 Probability Relaxation Transitions

The interaction of nuclei with the environment (neighboring nuclei, unpaired electrons and etc.) can lead, firstly, to non-equidistant energy levels and, consequently, the splitting of the spectrum of the nuclear magnetic resonance, secondly, to non-radiative transitions of nuclei between different energy states (exchange of energy between different degrees of freedom inside a substance). These non-radiative transitions have usually chaotic character and are a fundamental mechanism of the achievement of an equilibrium (Boltzmann) distribution of nuclei among energy states. These processes, which are the main object of consideration in this chapter, are called as *relaxational*. In general, the Hamiltonian of the nuclei system can be written as

$$\widehat{\mathcal{H}} = \widehat{\mathcal{H}}_0 + \widehat{\mathcal{H}}', \tag{3.1}$$

where $\widehat{\mathcal{H}}_0$ describes the interaction of nuclei with a constant magnetic field $\vec{B}_0$, $\widehat{\mathcal{H}}'$ describes their interaction with the environment. In the investigations of nuclear magnetic resonance in the majority of cases one deals with weakly interacting nuclei. Therefore, the Hamiltonians $\widehat{\mathcal{H}}_0$ and $\widehat{\mathcal{H}}'$ contain terms, taking into account the interaction of an isolated nucleus or a small number of nuclei with the external field $\vec{B}_0$

V. I. Chizhik et al., *Magnetic Resonance and Its Applications*,
DOI: 10.1007/978-3-319-05299-1_3,

and the interaction between nuclei. Usually the interaction between nuclei sharply decreases with a distance, and that is why one can restrict the consideration of a relatively small number of partners. From the expression (1.20) it is easily to obtain for i nuclei

$$\widehat{\mathcal{H}}_0 = -\sum_i \gamma_i \hbar \hat{I}_{zi} B_0, \tag{3.2}$$

taking into account that, traditionally, the vector of the field $\vec{B}_0$ is directed along the axis z of the laboratory coordinate system. Generally speaking, the Hamiltonian $\widehat{\mathcal{H}}'$ can describe both static and time-dependent interactions: $\widehat{\mathcal{H}}' = \widehat{\mathcal{H}}'_{\text{st}} + \widehat{\mathcal{H}}'(t)$. If the term which reflects the static interaction is present, it is namely responsible for the change of energy levels, which are defined by the interaction (3.2). The problem of the calculation of new energy states can be formulated as a separate task, considering that the term $\widehat{\mathcal{H}}_0$ in (3.1) is the sum: the Hamiltonian in the form (3.2) and $\widehat{\mathcal{H}}'_{\text{st}}$.

Thus, in the consideration of the relaxation processes one can consider

$$\widehat{\mathcal{H}}' = \widehat{\mathcal{H}}'(t), \tag{3.3}$$

that practically does not decrease the generality but simplifies the description and understanding of the nature of the relaxation phenomenon.

Under the influence of the perturbation (3.3), which is random because of chaotic molecular motion in a substance, nuclei jump with certain probability w_{ik} per unit of time from a state i to a state k. The value of the probability w_{ik} depends on the type of an interaction and the "structure" of energy levels of a spin system which is determined by the energy operator of the static interaction $\widehat{\mathcal{H}}_0$. In contrast to the probability of stimulated emission or absorption, which are the same for transitions from the upper or lower levels energy (this was proved already in the works of A. Einstein), the probability relaxation transitions up and down are different. Indeed, in the case of thermodynamic equilibrium in the absence of external radiation one can write:

$$n_i^0 w_{ik} = n_k^0 w_{ki}, \tag{3.4}$$

where n_i^0 and n_k^0 are the equilibrium populations of states with the indices i and k. The Eq. (3.4) means that the number of transitions per unit time in both directions is equal. Let the index i belongs to the lower energy state. Then, in accordance with Boltzmann statistics, we have $n_i^0 > n_k^0$ and, hence, $w_{ik} < w_{ki}$ (i.e., the probability of the transition down is bigger than the one to the upper energy state, that is not surprising because for the transition up it is necessary "to take" an energy quantum from outside if it exists, and for the transition down it is enough "to give" the energy to the surrounding). However, for the transitions in the radio frequency range (small quanta of electromagnetic energy), the inequality $h\nu \ll kT$ is valid, and hence $n_i^0 \simeq n_k^0$. This yields the approximate equality of the w_{ik} and w_{ki}. Therefore, in those expressions where the probability w_{ik} and w_{ki} are added, they can be considered as equal and one can introduce the average probability of transitions: $w_n = (w_{ik} + w_{ki})/2$ (n is the conditional number of a transition connecting the states i and k). If, however, some expressions contain the difference between w_{ik} and w_{ki}, one must take into account even a small difference to get physically correct conclusions (such situations arise, for example, during the description of processes in devices of quantum radio electronics, see Chaps. 17–21).

In general, the spin–lattice (longitudinal) relaxation rate can be written for the k-th relaxation mechanism as:

$$\frac{1}{T_{1k}} = \sum_n a_{kn} w_{kn}, \tag{3.5}$$

where a_{kn} are the weight coefficients before average probabilities w_{kn}; the summation is over all allowed transitions. When several relaxation mechanisms act simultaneously, the transition probabilities are added and the total relaxation rate is calculated as follows (under the condition of the independence of relaxation processes, that is often realized):

$$\frac{1}{T_1} = \sum_k \frac{1}{T_{1k}}. \tag{3.6}$$

It must be taken into account that the formula (3.6) is valid if relaxation processes are described by simple exponential function and, in the most cases, this is true.

For the spin–spin (transverse) relaxation rate one can write the similar expressions introducing transition probabilities u_{kn}:

$$\frac{1}{T_{2k}} = \sum_n b_{kn} u_{kn}, \tag{3.7}$$

$$\frac{1}{T_2} = \sum_k \frac{1}{T_{2k}}. \tag{3.8}$$

In the case when the spectral line width $\Delta\nu_l$ of the nuclear magnetic resonance is determined by the relaxation processes only (i.e. instrumental broadening factors are excluded), it can be expressed as

$$\Delta\nu_l = \frac{1}{\pi T_2}. \tag{3.9}$$

One can see from (3.9) that the physical meaning of T_2 is the mean time life of nuclei in energy states. Indeed, from the uncertainty relation (the fundamental relation of the quantum mechanics) which reads: $\Delta E \Delta t \sim \hbar$ (ΔE is the uncertainty of an energy value, if its definition took time Δt), we obtain

$$h\Delta\nu_l \Delta t \sim 2\hbar,$$

because $h\Delta\nu = \Delta E_i + \Delta E_k$, or finally,

$$\Delta\nu_l \sim \frac{1}{\pi\Delta t}. \tag{3.10}$$

Information on energy values of nuclei can be obtained after their transitions to some other states, that is after their participation in certain process. Consequently, physical meaning of the value Δt in (3.10) is: after this interval time one can judge about the energy values of nuclei before and after the transition. Therefore Δt is the average lifetime of nuclei in states. The comparison of (3.9) and (3.10) helps us to realize that the time constant of T_2 depends on all processes determining the lifetime of nuclei in a particular energy state.

There are many methods for calculating the relaxation rates, some of them are connected with a complex mathematical apparatus. However, in the case of the calculation of the nuclear magnetic relaxation rate in gases, liquids or solids with high molecular mobility the results of all methods come to the same formulas which can be obtained in terms of the simple theory of time-dependent perturbations. Recall that to calculate, for example, the spin–lattice relaxation rate it is necessary to reckon the probabilities of transitions under the influence of a disturbance (see 3.5), a similar problem is solved for the spin–spin relaxation.

According to the perturbation theory the probability per unit time between two states is given by

$$w_{mn} = \frac{1}{t}\frac{1}{\hbar^2}\left|\int_0^t \langle m|\widehat{\mathcal{H}}'(t')|n\rangle e^{\mathrm{i}\omega_{mn}t'}\mathrm{d}t'\right|^2, \tag{3.11}$$

where $\langle m|\widehat{\mathcal{H}}'(t')|n\rangle \equiv \int\int\int \psi_m^* \widehat{\mathcal{H}}'(t')\psi_n \mathrm{d}x\mathrm{d}y\mathrm{d}z$ is the matrix element of the Hamiltonian $\widehat{\mathcal{H}}'$ (see 3.3) between states with the indices m and n of the non-perturbed system described by Hamiltonian $\widehat{\mathcal{H}}_0$.

In the standard presentation of the perturbation theory one usually implies the full Hamiltonian of a system $\widehat{\mathcal{H}}_{0\Sigma}$, which determines wave functions that depend on spatial coordinates and time:

$$\Psi_{k\Sigma}(x, y, z, t) = \psi_{k\Sigma}(x, y, z)e^{-\frac{\mathrm{i}}{\hbar}E_{k\Sigma}t}.$$

The function $\psi_{k\Sigma}$, which describes the state of a nucleus taking into account the presence of a spin, must be considered as a functions of four coordinates: three of them belong to the center of gravity of a nucleus, and the fourth one to its spin. In quantum mechanics a spin state must be characterized by two quantities: the absolute value of a spin and a spin projection at any direction, for example I_z. Since the spin of a certain type of nuclei is constant, then in fact we can talk about a single spin variable:

$$\psi_{k\Sigma} = \psi_{k\Sigma}(x, y, z, I_z).$$

For the problems considered in the nuclear magnetic resonance, the connection between spatial and spin motion can be neglected and the function $\psi_{k\Sigma}$ can be represented as

$$\psi_{k\Sigma} = \psi_{k\Sigma}(x, y, z)\psi_{\mathrm{s}}(I_z),$$

where $\psi_{\mathrm{s}}(I_z)$ is the function describing a spin state. In the following we will take into account only those interactions that determine the state of a nuclear spin, and for the Hamiltonians describing the main interaction and disturbance, we will use the notation, respectively, $\widehat{\mathcal{H}}_0$ and $\widehat{\mathcal{H}}'$.

In ordinary theory of quantum transitions the calculation of the formula (3.11) is performed for the harmonic perturbation (i.e. for the effects of a coherent electromagnetic wave). In relaxation processes the interaction has the nature of random fluctuations, and therefore it is necessary to average w_{mn} over all possible realizations (in considered relaxation processes the average values over time and across the ensemble are the same):

$$\overline{w}_{mn} = \frac{1}{t}\frac{1}{\hbar^2}\int_0^t\int_0^t \overline{\langle m\,|\,\widehat{\mathcal{H}}'(t')\,|\,n\,\rangle\langle n\,|\,\widehat{\mathcal{H}}'(t'')\,|\,m\,\rangle} e^{\mathrm{i}\omega_{mn}(t'-t'')}\mathrm{d}t'\mathrm{d}t''. \quad (3.12)$$

Taking into account (3.3), we assume that $\overline{\langle m|\widehat{\mathcal{H}}'|n\rangle} = 0$. Besides, present $\langle m|\widehat{\mathcal{H}}'(t)|n\rangle$ as

$$\langle m|\widehat{\mathcal{H}}'(t)|n\rangle = u(t)\sqrt{\overline{\left|\langle m|\widehat{\mathcal{H}}'(t)|n\rangle\right|^2}} = C_{mn}u(t),$$

here $u(t)$ satisfies the condition $\overline{|u(t)|^2} = 1$, and the denotation C_{mn} has been introduced for the shortiness. Then, omitting the bar over w_{mn}, because in the following only average probabilities will be considered, one has

$$w_{mn} = \frac{C_{mn}^2}{t\hbar^2}\int_0^t\int_0^t \overline{u(t')u^*(t'')} e^{\mathrm{i}\omega_{mn}(t'-t'')}\mathrm{d}t'\mathrm{d}t''.$$

Let us make replacement of variables $t' - t'' = \tau$:

$$w_{mn} = \frac{C_{mn}^2}{t\hbar^2}\int_0^t\int_{-t''}^{t-t''} \overline{u(t''+\tau)u^*(t'')} e^{\mathrm{i}\omega_{mn}\tau}\mathrm{d}\tau\mathrm{d}t''. \quad (3.13)$$

Now pay attention to the fact that

$$\overline{u(t''+\tau)u^*(t'')} = K(\tau) \quad (3.14)$$

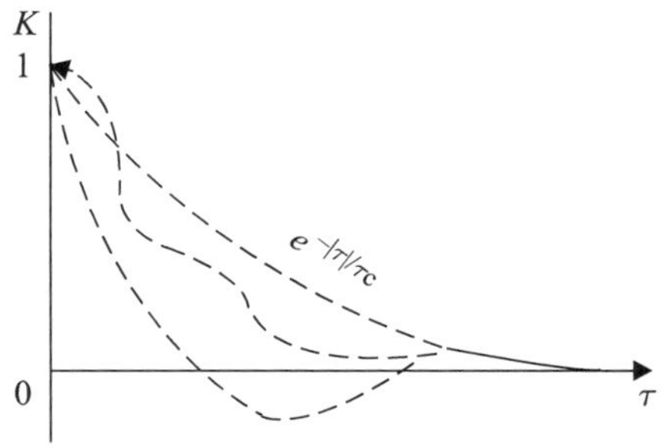

Fig. 3.1 The possible view of a correlation function

is the correlation function of the random function $u(t)$. It is easy to formulate some general properties $K(\tau)$:

1. $K(0) = \overline{|u(t)|^2}$.
2. $K(\infty) = 0$, i.e. values of a random function at large time intervals can be positive or negative with equal probability (uncorrelated).
3. $K(\tau) = K(-\tau)$, if the random process is stationary.

In other words, the correlation function must have the form shown in Fig. 3.1.

In many practically important cases, the correlation function is described by an exponential dependence

$$K(\tau) = e^{-|\tau|/\tau_c}, \tag{3.15}$$

which, of course, has the above properties. The parameter τ_c is called *correlation time* and represents the time interval, during of which a random function changes considerably its value. In the relaxation processes the fluctuations of interaction are defined by velocity of reorientations of molecules.

For liquids characteristic times of molecular motions are 10^{-9} to 10^{-12} s, for solids with high molecular mobility—10^{-4} to 10^{-6} s. Consequently, in many cases the function (3.15) can be considered as fast-damped. Thus, after the replacement of variables expression, (3.13) becomes

$$w_{mn} = \frac{C_{mn}^2}{t\hbar^2} \int_0^t \int_{-t''}^{t-t''} K(\tau) e^{\mathrm{i}\omega_{mn}\tau} \mathrm{d}\tau \mathrm{d}t''.$$

Taking into account the fast decay of the function $K(\tau)$, the limits of integration in the inner integral can be extended to $\pm\infty$ and the dependence from t'' disappears. This is equivalent to the condition: $\tau_c \ll T_2,\ T_1$. Then one can obtain:

$$w_{mn} = \frac{\overline{|\langle m|\widehat{\mathcal{H}}'(t)|n\rangle|^2}}{\hbar^2} \int_{-\infty}^{\infty} K(\tau) e^{\mathrm{i}\omega_{mn}\tau} \mathrm{d}\tau. \tag{3.16}$$

The integral in (3.16) is *the spectral density of power of a random process* at the frequency ω_{mn}, which is connected with the correlation function via Fourier transform:

$$J(\omega) = \int_{-\infty}^{\infty} K(\tau)e^{\mathrm{i}\omega\tau}\mathrm{d}\tau, \tag{3.17}$$

$$K(\tau) = \frac{1}{2\pi}\int_{-\infty}^{\infty} J(\omega)e^{-\mathrm{i}\omega\tau}\mathrm{d}\omega. \tag{3.18}$$

Thus, summarizing the results of calculations from (3.5) to (3.16), we can receive for multilevel system:

$$\frac{1}{T_{1k}} = \sum_n A_k B_{kn} J_n(\omega_n), \tag{3.19}$$

where the value A_k is proportional to the interaction energy for the k relaxation mechanism; B_{kn} are the weight coefficients to spectral densities J_n frequencies ω_n (some of B_{kn} can be equal to zero).

For the spin–spin relaxation one can write similarly to (3.7):

$$\frac{1}{T_{2k}} = \sum_n A_k B'_{kn} J_n(\omega_n). \tag{3.20}$$

3.2 Two-Level System: Relaxation, the Spin Temperature

To begin, consider the simplest system of isolated nuclei with spin $I = 1/2$ (for example, nuclei ^{1}H, ^{13}C, ^{31}P, ...). The energies of nuclei in a system, subjected to field $\vec{B}_0$, can be calculated using the formula (1.21), substituting in it the possible values of the magnetic quantum number m (eigenvalues of the operator $\hat{I}_z$): $+1/2$ and $-1/2$). Thus, the energy spectrum of the system consists of two levels, separated by an interval $\gamma\hbar B_0$. Let us construct the equations describing the "movement" of nuclei due to all possible transitions between these levels. The change of the number of nuclei at the lower level will be described by a simple equation:

$$\mathrm{d}n_1/\mathrm{d}t = -n_1 w_{12} + n_2 w_{21} + (n_2 - n_1)W_{12}, \tag{3.21}$$

where n_1 and n_2 are the populations of the energy levels; w_{12} and w_{21} are the probabilities of relaxation transitions per time unit; W_{12} is the probability of induced transitions per time unit under the influence of an external alternating magnetic field. The equation analogous to (3.21) can be formed for the population of the upper state, however, it is more convenient to replace it by the condition of the boundedness of a system:

$$n_1 + n_2 = N, \tag{3.22}$$

where N is the number of nuclei investigated in a sample. Since in the majority of experiments the observed effects are related to the difference of level populations: $\Delta n_{12} = n_1 - n_2$, construct the new equation, using (3.21) and (3.22):

$$\frac{\mathrm{d}\Delta n_{12}}{\mathrm{d}t} = -(w_{12} + w_{21} + 2W_{12})\Delta n_{12} + N(w_{21} - w_{12}). \tag{3.23}$$

It is easy to show that $N(w_{21} - w_{12}) = \Delta n_{12}^0(w_{12} + w_{21})$, where Δn_{12}^0 is the equilibrium value Δn_{12}. Therefore, the Eq. (3.23) becomes:

$$\frac{\mathrm{d}\Delta n_{12}}{\mathrm{d}t} = -(w_{12} + w_{21} + 2W_{12})\Delta n_{12} + (w_{12} + w_{21})\Delta n_{12}^0, \tag{3.24}$$

and its solution is given by

$$\Delta n_{12}(t) - \Delta n_{12}^{\mathrm{s}} = \left[\Delta n_{12}(0) - \Delta n_{12}^{\mathrm{s}}\right] e^{-t/T_1'}, \tag{3.25}$$

where $T_1' = 1/(w_{12} + w_{21} + 2W_{12})$, and the stationary solution of the Eq. (3.24) is

$$\Delta n_{12}^{\mathrm{s}} = \Delta n_{12}^0 \left(1 + \frac{2W_{12}}{w_{12} + w_{21}}\right)^{-1}.$$

If $W_{12} = 0$, i.e. external radiation is absent, $\Delta n_{12}^{\mathrm{st}} = \Delta n_{12}^0$ and

$$\frac{1}{T_1'} = \frac{1}{T_1} = w_{12} + w_{21} \simeq 2w, \tag{3.26}$$

where w is the average probability. The time constant T_1 describes the rate of the establishment of equilibrium state when there is the initial deviation of the population difference $\Delta n_{12}(0)$ from Δn_{12}^0. Figure 2.27 shows the variation of $M_z(t)$ (which is proportional to $\Delta n_{12}(t)$) from (3.25) for different initial conditions: $0 \leq \Delta n_{12}(0) \leq \Delta n_{12}^0$ (curve 1); $\Delta n_{12}(0) = 0$ (curve 2); $\Delta n_{12}(0) = -\Delta n_{12}^0$ (curve 3, the inversion of level populations).

The expression (3.26) is a particular case of the expression (3.5), in the following other realizations for more complex multilevel systems will be considered.

It should be noted that the performed analysis is valid when it is possible to use time-independent transition probabilities W_{12} (per unit of time), i.e. when the action of an external alternating magnetic field is relatively weak. If an alternating field is of very high amplitude, the probability of transition between two states is described by a sine function. This question is discussed narrowly in the chapters devoted to problems of quantum electronics. In the macroscopic approach the behavior of the spin system (or rather its macroscopic magnetization) under the influence of a strong alternating magnetic field is described by Eq. (1.56). The general solution of the problem on the basis of Bloch equations was given in by H.C. Torrey (Phys. Rev., 1949, vol. 76, p. 1059; see also the exposition of this work in the monograph: A. Lösche, Kerninduktion, Berlin, 1957).

For the transverse (spin–spin) relaxation the expression similar to (3.26) can be obtained. However, the procedure is somewhat more complicated, and appropriate consideration will be done for a concrete example (see Sect. 3.6).

Note that the main logical difficulty lies in the fact that the existence of the circulating transverse component of the nuclear magnetization means a situation when the spin system is in a non-stationary state, which cannot be described as a time-dependent superposition of stationary states (the description of the longitudinal relaxation were performed using the probability of transitions between stationary states).

The ratio of the equilibrium populations in the two-level system is

$$\frac{n_1^0}{n_2^0} = e^{\gamma \hbar B_0/(kT_L)} \tag{3.27}$$

(T_L is the lattice temperature). The equilibrium (3.27) can be disturbed by different ways (for example, it is possible to apply radiofrequency pulse, to change rapidly the magnitude of the field B_0 and so on). An arbitrary state of the spin system can be described by introducing a formal effective *spin temperature* T_S, defined by the relation

$$\frac{n_1}{n_2} = e^{\gamma \hbar B_0/(kT_S)}. \tag{3.28}$$

The spin temperature was introduced into the theory as a concept, which allowed the application of the thermodynamic approach for the description of the nuclear magnetism. The idea on the temperature of a system of nuclear spins was put forward initially as a hypothesis (without any strict proof), its validity was confirmed only by successful interpretations of experimental data.

To test the applicability of the concept of the spin temperature many experiments were carried out in different physical situations. The analysis of the limits of applicability of this concept led to the creation of new experimental methods, greatly expanded the possibilities for studying the nuclear magnetism.

As a result, the concept of spin temperature allowed the prediction of the number of new phenomena and formed the basis of the general theory of nuclear magnetism in solids. This theory results in a system a few simple equations, similar Bloch equations which are more suitable for the description of nuclear magnetic resonance in liquids.

The possibility of such an approach can be illustrated by a simple example, which history goes back to the ideas expressed by Van Vleck in 1937 concerning the generalized interpretation of Curie law Eqs. (1.32) and (1.34).

Indeed, the formula (1.32) can be rewritten for an arbitrary value of the macroscopic magnetization along the field $\vec{B}_0$ in the form

$$M_z(t) = \text{const}\,\frac{B_0}{T_S(t)}. \tag{3.29}$$

The expression (3.29) together with (1.42) yields the differential equation for T_S:

$$\frac{\mathrm{d}}{\mathrm{d}t}\frac{1}{T_S} = -\frac{1}{T_1}\left(\frac{1}{T_S} - \frac{1}{T_L}\right), \tag{3.30}$$

which, of course, must change the appearance in more complex situations, for example when a strong RF field is applied or when the field B_0 is of an order of local fields in a substance. The Eq. (3.30) shows that spin–lattice relaxation tends to equalize the spin temperature with the lattice temperature (here, as well as earlier, the lattice acts as a thermostat for a spin system).

In all these considerations we assumed that the state of a spin system can be described by the populations of energy states (only in this case the definition (3.28) of the spin temperature makes sense). However, it is possible to create a condition when the $M_z = 0$ but there is a non-zero transverse precessing component of the nuclear magnetization, which decreases during the time interval of the order of the spin–spin relaxation time T_2. Therefore, one can consider the spin temperature of a system only after the time of the order of T_2, after the disappearance of the precessing magnetization. In other words, it is correct to use the spin temperature only for systems for which $T_2 \ll T_1$.

This condition, as will be seen later, is usually fulfilled in solids, where the spin–spin interaction is much stronger than the spin–lattice one.

3.3 Multi-level System: Spin–Lattice Relaxation

If a spin system has many energy states, the question on the character of the relaxation becomes quite complex. In Sect. 3.10 it is shown that in some cases the relaxation process is described with more than a single exponent. Therefore, each particular situation must be undergone to a special study. In this section the spin–lattice relaxation in multilevel systems will be described on the basis of the concept of the spin temperature.

For a system of nuclei with spin $I > 1/2$ the existence of the spin temperature as the only parameter that describes the distribution of nuclei among all energy states are not obvious. In the case of magnetic resonance (without the influence, for example, of the quadrupole interaction on the arrangement of energy levels), the energy levels are equidistant, and the spin temperature can be entered using the relation

$$\frac{p_m}{p_{m+1}} = \exp\frac{\gamma\hbar B_0}{kT_S}, \tag{3.31}$$

where p_m is the relative population of energy states ($\sum_m p_m = 1$), and under the condition that the ratio of the populations does not depend on the index. Such a situation can occur due to the processes of spin–spin relaxation (mainly, due to "mutual upset" of spins), which proceeds with rate of $1/T_2 \gg 1/T_1$. In the presence of mutual upsets of spins the total magnetic energy of a system is conserved, but the transitions of a spin from the state m to the state $m+1$ and another one from the state m' to the state $m'-1$ change the population of individual states. Thus, for the arbitrary initial level populations this process leads to the fact that during the time of the order of T_2 the equilibrium state in a spin system is achieved, that corresponds to the most probable distribution of nuclei at the energy levels under the condition of fixed energy of a spin system in a magnetic field $\vec{B}_0$. This is Boltzmann distribution, which can be characterized by a certain effective spin temperature T_S according to (3.31). One of the interesting features of spin systems is the possibility of the inversion of energy state population when the number of nuclei at higher levels becomes larger than at lower ones. This can be done, for example, using the adiabatic fast passage through a resonance or acting on a sample with RF pulse which rotates the vector nuclear magnetization on the 180°. It follows from the expression (3.31) that the above situation corresponds to the negative spin temperature. Such a state is created in devices of quantum electronics to provide a quantum amplification or generation of electromagnetic fields. "Appearance" of the negative temperature, expressed in units of absolute temperature scale, indicates the conventional character of the conception of the spin temperature. Indeed, a spin system with $T_S = -|T_S|$, in fact, is "warmer" than any system with the positive spin temperature, because in order to achieve such a state it is necessary to bring energy from the outside (we are talking about closed system with no transportation substances; as it will be shown in Chap. 18 in experiments with atomic or molecular beams the state with the negative T_S can be achieved by sorting particles during their movement in inhomogeneous static magnetic fields). The state with the negative spin temperature can be realized only for systems with limited energy spectrum, because in this case, only a finite change of system energy is required. In conventional thermodynamic systems the energy spectrum is not limited and the effective temperature, which is introduced to them by analogy with the T_S, can never be negative.

Let us consider a spin system with non-degenerate energy spectrum: E_1, E_2, ..., E_n. Under the condition of the thermodynamic equilibrium the relative populations of these energy states are given by the formula

$$p_n = \frac{\exp[-E_n/(kT)]}{\sum_m \exp[-E_m/(kT)]},$$

where $\sum_m \exp[-E_m/(kT)] \equiv Z$ is the so-called statistical sum. Generally speaking, the energy E_n can belong to a single spin or to a set of interacting spins. We assume that any system can be described by the relation (3.31), even if it is not in equilibrium with any thermostat (the thermostat for the nuclear spins is the "lattice"). The average energy of the system can be expressed as

$$\overline{E} = \sum_n p_n E_n. \tag{3.32}$$

Let us choose the reference level of energy, so that

$$\sum_n E_n = 0, \tag{3.33}$$

and introduce for convenience the parameter $\beta = 1/(kT_S)$. Then, in principle, one can write:

$$\frac{\mathrm{d}\overline{E}}{\mathrm{d}t} = \frac{\mathrm{d}\overline{E}}{\mathrm{d}\beta}\frac{\mathrm{d}\beta}{\mathrm{d}t}. \tag{3.34}$$

From the expression (3.32) we have

$$\frac{\mathrm{d}\overline{E}}{\mathrm{d}t} = \frac{\mathrm{d}}{\mathrm{d}t}\sum_n p_n E_n = \sum_n E_n \frac{\mathrm{d}p_n}{\mathrm{d}t}. \tag{3.35}$$

Because the quantities p_n characterize the populations of energy states, then it is possible to write simple linear equations by introducing the relaxation probability (w_{mn}) of transitions from a state m to a state n:

$$\frac{\mathrm{d}p_n}{\mathrm{d}t} = \sum_m (p_m w_{mn} - p_n w_{nm}). \tag{3.36}$$

The Eq. (3.36), similar to (3.21), is often called the *basic kinetic equation*. With the help of (3.36) we rewrite (3.35) in the form

$$\frac{\mathrm{d}\overline{E}}{\mathrm{d}t} = \sum_{m,n} E_n (p_m w_{mn} - p_n w_{nm}). \tag{3.37}$$

The result of the Eq. (3.37) should not be depended on the order of summation indices and it is expedient to rewrite (3.37) in a more symmetrical form (relatively of the indices m and n):

$$\frac{\mathrm{d}\overline{E}}{\mathrm{d}t} = \frac{1}{2}\sum_{m,n} (p_m w_{mn} - p_n w_{nm})(E_n - E_m). \tag{3.38}$$

From (3.34) and (3.38) one can obtain the equation for β. For this we first compute $d\overline{E}/d\beta$:

$$\frac{\mathrm{d}\overline{E}}{\mathrm{d}\beta} = \frac{\mathrm{d}}{\mathrm{d}\beta}\sum_n p_n E_n = \frac{\mathrm{d}}{\mathrm{d}\beta}\sum_n \frac{e^{-\beta E_n}}{Z} E_n. \tag{3.39}$$

We assume that the temperature of the sample (i.e. the material carrier of a spin system) quite high: $\beta E_n \ll 1$ (the high-temperature approximation). Then, using (3.33) and taking into account only the first leading term in the expansion of the exponent in a series, we have

$$Z = \sum_n e^{-\beta E_n} = \sum_n \left(1 - \beta E_n + \frac{\beta^2 E_n^2}{2} - \ldots\right) \simeq N, \tag{3.40}$$

Here N is the number of energy states. As the result of (3.40) coincides with the value of the statistical sum for $T = \infty$, then introduce the denotation $Z = Z_\infty$. Let us return again to (3.39):

$$\frac{\mathrm{d}\overline{E}}{\mathrm{d}\beta} \simeq -\frac{1}{Z_\infty}\sum_n E_n^2. \tag{3.41}$$

Combining (3.34) and (3.41), one can obtain

$$\frac{\mathrm{d}\overline{E}}{\mathrm{d}t} = -\left(\frac{1}{Z_\infty}\sum_n E_n^2\right)\frac{\mathrm{d}\beta}{\mathrm{d}t}. \tag{3.42}$$

Rewrite (3.31) in a more general form:

$$\frac{p_n}{p_m} = e^{\beta(E_m - E_n)} \tag{3.43}$$

and assume that for any m and n the ratio p_n/p_m is expressed in the form (3.43) with the single value β. This statement is *the basic postulate on the spin temperature*.

For the state of thermodynamic equilibrium in the absence of external influences the condition

$$p_m^0 w_{mn} = p_n^0 w_{nm} \tag{3.44}$$

is always satisfied (see also 3.4) On the basis of the principle of the detailed balance (3.44) we have

$$w_{nm} = w_{mn} p_m^0 / p_n^0 = w_{mn} e^{\beta_L (E_n - E_m)}, \tag{3.45}$$

where the $\beta_L = 1/(kT_L)$; T_L is the lattice temperature. Let us use (3.45) to transform (3.38):

$$\frac{\mathrm{d}\overline{E}}{\mathrm{d}t} = \frac{1}{2}\sum_{m,n} p_m w_{mn}\left[1 - e^{(E_m - E_n)(\beta - \beta_L)}\right](E_n - E_m)$$

$$\simeq \frac{1}{2}\sum_{m,n} p_m w_{mn}(E_n - E_m)^2 \beta - \beta_L. \tag{3.46}$$

Taking into account $p_m = [\exp(-\beta E_m)]/Z_\infty \simeq 1/Z_\infty$ and using (3.42) and (3.46), one can obtain

$$\frac{d\beta}{dt} = -\frac{1}{2}\frac{\sum_{m,n} w_{mn}(E_n - E_m)^2}{\sum_n E_n^2}(\beta - \beta_L)$$

and finally

$$\frac{d\beta}{dt} = \frac{\beta_L - \beta}{T_1} \tag{3.47}$$

with

$$\frac{1}{T_1} = \frac{1}{2}\frac{\sum_{m,n} w_{mn}(E_n - E_m)^2}{\sum_n E_n^2} = \frac{\sum_k w_k(\hbar\omega_k)^2}{\sum_n E_n^2}, \tag{3.48}$$

where the index k denotes the conditional number of transitions between states of m and n.

The expressions (3.47) and (3.48) show that, when the concept of the spin temperature is valid, the spin–lattice relaxation in a multi-level system is described by an exponential function with a time constant T_1, which is determined by the structure energy levels (E_m, E_n) and the probabilities of relaxation transitions w_{mn}. The expressions (3.47) and (3.48) hold for $T_2 \ll T_1$, but in some cases they give the correct result and for $T_2 \sim T_1$. One of such examples is two-level system. If denote $(\gamma\hbar B_0/2) = E_0$, then from (3.48) we have

$$\frac{1}{T_1} = \frac{1}{2}\frac{2w(2E_0)^2}{2E_0^2} = 2w,$$

that coincides with (3.26).

3.4 Preliminary Remarks to the Theory of Redfield

Let us consider the method which is useful in cases when the concept of spin temperature is not applicable. Initially, F. Bloch and R.K. Wangsness developed the dynamical theory of nuclear magnetic resonance, and then F. Bloch (Phys. Rev., 1957, vol. 105, p. 1206) and especially A.G. Redfield (Adv. Magn. Reson., 1966, vol. 1, pp. 1–32) developed the general theory of relaxation.

The method is based on the mathematical formalism of density matrix (see Sect. 1.9) and possesses such an advantage that it allows us to describe processes both the longitudinal (spin–lattice) and transverse (spin–spin) relaxation. It is especially convenient to apply this method when a spectral line of nuclear magnetic resonance is narrowed due to molecular motion movement (fluids and gases), but one can use it to describe the nuclear magnetic resonance with broad spectral lines (solids, liquid crystals). In studying the application of the density matrix to the description of nuclear magnetic resonance it is helpful to use the book: M. Goldman. Quantum description of high-resolution NMR in liquids. Oxford, 1988, 288 pp.

Consider the case when the Hamiltonian of the system can be identified with separate independent and dependent on time contributions (see 3.1–3.3). Then the equation for the matrix density can be written as

$$\frac{\mathrm{d}\hat{\rho}}{\mathrm{d}t} = \frac{\mathrm{i}}{\hbar}\left[\hat{\rho}, \widehat{\mathcal{H}}_0 + \widehat{\mathcal{H}}'(t)\right]. \tag{3.49}$$

Let us seek a solution of (3.49) as

$$\hat{\rho}(t) = e^{-\frac{\mathrm{i}}{\hbar}\widehat{\mathcal{H}}_0 t}\hat{\rho}^*(t)e^{\frac{\mathrm{i}}{\hbar}\widehat{\mathcal{H}}_0 t}, \tag{3.50}$$

an asterisk in (3.50) and in the following does not mean the complex conjugate function, but this is only a tribute to the traditional notation. It is obvious in the absence of a time-dependent part of the Hamiltonian $\rho^*(t)$ should be constant and coincides with the density matrix for $t = 0$. If the Hamiltonian $\widehat{\mathcal{H}}'(t)$ describes small perturbations then ρ^* will change slowly over time.

Inserting (3.50) into the left side of (3.49), one can obtain

$$-\frac{\mathrm{i}}{\hbar}\left[\widehat{\mathcal{H}}_0, \hat{\rho}\right] + e^{-\frac{\mathrm{i}}{\hbar}\widehat{\mathcal{H}}_0 t}\frac{\mathrm{d}\hat{\rho}^*(t)}{\mathrm{d}t}e^{\frac{\mathrm{i}}{\hbar}\widehat{\mathcal{H}}_0 t} = \frac{\mathrm{i}}{\hbar}\left[\hat{\rho}, \widehat{\mathcal{H}}_0 + \widehat{\mathcal{H}}'(t)\right]. \tag{3.51}$$

Taking into account equal terms in (3.51) and multiplying this expression leftwards by $\exp(\mathrm{i}\widehat{\mathcal{H}}_0 t/\hbar)$ and rightwards by $\exp(-\mathrm{i}\widehat{\mathcal{H}}_0 t/\hbar)$, we will receive the equation for $\hat{\rho}^*(t)$:

$$\frac{\mathrm{d}\hat{\rho}^*(t)}{\mathrm{d}t} = \frac{\mathrm{i}}{\hbar}\left[\hat{\rho}^*(t), \widehat{\mathcal{H}}_1^*(t)\right] \tag{3.52}$$

with

$$\widehat{\mathcal{H}}_1^*(t) = e^{\frac{\mathrm{i}}{\hbar}\widehat{\mathcal{H}}_0 t}\widehat{\mathcal{H}}'(t)e^{-\frac{\mathrm{i}}{\hbar}\widehat{\mathcal{H}}_0 t}. \tag{3.53}$$

Obviously, the operator $\hat{\rho}^*$ is independent of time, if the Hamiltonian $\widehat{\mathcal{H}}'$ is equal to zero. The new expression of the density matrix is called *representation of interaction* or *Dirac representation*. This transformation corresponds to the conversion to a rotating coordinate system.

Let us try to integrate (3.52):

$$\hat{\rho}^*(t) = \hat{\rho}^*(0) + \frac{\mathrm{i}}{\hbar}\int_0^t \left[\hat{\rho}^*(t')\widehat{\mathcal{H}}_1^*(t')\right]\mathrm{d}t'. \tag{3.54}$$

The expression (3.54) is not a solution of (3.52), since the integral contains an unknown quantity, but if $\hat{\rho}^*$ changes little, it is possible to construct an approximate solution by substituting in the integrand $\hat{\rho}^*(t') = \hat{\rho}^*(0)$:

$$\hat{\rho}^*(t) = \hat{\rho}^*(0) + \frac{\mathrm{i}}{\hbar}\int_0^t \left[\hat{\rho}^*(0), \widehat{\mathcal{H}}_1^*(t')\right]\mathrm{d}t'. \tag{3.55}$$

The formula for the second approximation can be obtained in a natural way substituting (3.55) into (3.54):

$$\hat{\rho}^*(t) = \hat{\rho}^*(0) + \frac{\mathrm{i}}{\hbar}\int_0^t \left[\hat{\rho}^*(0), \widehat{\mathcal{H}}_1^*(t')\right]\mathrm{d}t' + \left(\frac{\mathrm{i}}{\hbar}\right)^2 \int_0^t\int_0^t \left[\left[\hat{\rho}^*(0), \widehat{\mathcal{H}}_1^*(t'')\right], \widehat{\mathcal{H}}_1^*(t')\right]\mathrm{d}t''\mathrm{d}t'. \tag{3.56}$$

In some tasks it is more convenient to make calculations, knowing the derivative $\mathrm{d}\hat{\rho}^*(t)/\mathrm{d}t$. From (3.56) one has

$$\frac{\mathrm{d}\hat{\rho}^*(t)}{\mathrm{d}t} = \frac{\mathrm{i}}{\hbar}\left[\hat{\rho}^*(0), \widehat{\mathcal{H}}_1^*(t)\right] + \left(\frac{\mathrm{i}}{\hbar}\right)^2 \int_0^t \left[\left[\hat{\rho}^*(0), \widehat{\mathcal{H}}_1^*(t')\right], \widehat{\mathcal{H}}_1^*(t)\right]\mathrm{d}t'. \tag{3.57}$$

The expression (3.57) is equivalent to conventional formulas of the non-stationary perturbation theory in the second approach.

3.5 Basic Equation in the Theory of Redfield

Let us start from formula (3.57) and calculate the matrix element between states with indices α and α'. It is convenient for further to "mark" states with the value of energy: the symbol α will mean the value of $E/\hbar$.

At the beginning consider the first term in the right side of the expression (3.57):

$$\langle\alpha|\left[\hat{\rho}^*(0), \widehat{\mathcal{H}}_1^*(t)\right]|\alpha'\rangle = \sum_\beta \left[\langle\alpha|\hat{\rho}^*(0)|\beta\rangle\langle\beta|\widehat{\mathcal{H}}_1^*(t)|\alpha'\rangle - \langle\alpha|\widehat{\mathcal{H}}_1^*(t)|\beta\rangle\langle\beta|\hat{\rho}^*(0)|\alpha'\rangle\right].$$

We introduce the concept of an *ensemble of ensembles*, the density matrix of which are the same for $t = 0$ but the realization of Hamiltonians $\widehat{\mathcal{H}}'(t)$ are different in each of them. Let us assume that after the averaging over an ensemble $\widehat{\mathcal{H}}'(t)$ vanishes, i.e. $\widehat{\mathcal{H}}'(t)$ does not contribute to the frequency shift (similar to (3.1–3.3) we include the static part of $\widehat{\mathcal{H}}'$ in $\widehat{\mathcal{H}}_0$).

In general, one can write (see Sects. 1.10–1.12):

$$\widehat{\mathcal{H}}'(t) = \sum_q B_q(t)\hat{K}_q, \tag{3.58}$$

here $\hat{K}_q$ is the function of spin coordinates and the multiplier $B_q(t)$ does not depend on spin. If, for example, the Hamiltonian $\widehat{\mathcal{H}}'(t)$ describes the interaction of a nuclear magnetic moment with a magnetic field, then (3.58) is the sum of three terms (the product x-, y- and z-component of a spin and relevant components of a magnetic field); in the case of the dipole–dipole interaction the formula (3.58) contains six terms (see Sect. 3.8), and so on.

As before, assume that the existing disturbance is stationary. Hence, the averaging over the ensemble and time is equivalent. Since the static term in the Hamiltonian $\widehat{\mathcal{H}}'$ is absent then we reckon $\overline{B_q(t)} = 0$, and therefore the averaging over an ensemble leads to a zero result for the operator $\widehat{\mathcal{H}}_1^*(t)$, i.e. $\overline{\langle\alpha|\widehat{\mathcal{H}}_1^*(t)|\beta\rangle} = 0$. This means that the expected time dependence is different than one in the case of the effect of an oscillating external field.

Thus, the first term in the right part of the Eq. (3.57) after averaging will be equal to zero.

Now consider the matrix element between the states α and α' for the second term in (3.57). As a result, we obtain

$$\begin{aligned}\frac{\mathrm{d}\rho_{\alpha\alpha'}}{\mathrm{d}t} = -\frac{1}{\hbar^2}\int_0^t \sum_{\beta\beta'} \Big[&\rho^*_{\alpha\beta}(0)\langle\beta|\widehat{\mathcal{H}}_1^*(t')|\beta'\rangle\langle\beta'|\widehat{\mathcal{H}}_1^*(t)|\alpha'\rangle \\ &- \langle\alpha|\widehat{\mathcal{H}}_1^*(t')|\beta\rangle\rho^*_{\beta\beta'}(0)\langle\beta'|\widehat{\mathcal{H}}_1^*(t)|\alpha'\rangle \\ &- \langle\alpha|\widehat{\mathcal{H}}_1^*(t)|\beta\rangle\rho^*_{\beta\beta'}(0)\langle\beta'|\widehat{\mathcal{H}}_1^*(t')|\alpha'\rangle \\ &+ \langle\alpha|\widehat{\mathcal{H}}_1^*(t)|\beta\rangle\langle\beta|\widehat{\mathcal{H}}_1^*(t')|\beta'\rangle\rho^*_{\beta'\alpha'}(0)\Big]\mathrm{d}t'.\end{aligned} \tag{3.59}$$

We will show that

$$\langle\beta|\widehat{\mathcal{H}}_1^*|\beta'\rangle = e^{\mathrm{i}(\beta-\beta')t}\langle\beta|\widehat{\mathcal{H}}'(t)|\beta'\rangle. \tag{3.60}$$

First, we obtain the relation between matrix elements of ρ and ρ^*. Let

$$\Psi = \sum_n c_n\psi_n = \sum_n a_n e^{-\frac{\mathrm{i}}{\hbar}E_n t}\psi_n,$$

where ψ_n and E_n are the eigenfunctions and eigenvalues of the Hamiltonian $\widehat{\mathcal{H}}_0$. If $\widehat{\mathcal{H}}'(t) = 0$, the coefficients a_n are constant over time. Using the expansion

$$e^{\frac{\mathrm{i}}{\hbar}\widehat{\mathcal{H}}_0 t}\psi_m = \left[1 + \frac{\mathrm{i}}{\hbar}\widehat{\mathcal{H}}_0 t + \ldots\right]\psi_m = \left[1 + \frac{\mathrm{i}}{\hbar}E_m t + \ldots\right]\psi_m = e^{\frac{\mathrm{i}}{\hbar}E_m t}\psi_m,$$

one can obtain

$$\langle n|\hat{\rho}(t)|m\rangle = \int \psi_n^* e^{-\frac{i}{\hbar}\widehat{\mathcal{H}}_0 t}\hat{\rho}^*(t)e^{\frac{i}{\hbar}\widehat{\mathcal{H}}_0 t}\psi_m \mathrm{d}v$$
$$= e^{\frac{i}{\hbar}(E_m - E_n)t}\int \psi_n^* \hat{\rho}^*(t)\psi_m \mathrm{d}v = e^{\frac{i}{\hbar}(E_m - E_n)t}\langle n|\hat{\rho}^*(t)|n\rangle, \quad (3.61)$$

because there is the relation $\overline{a_n a_m^*} = \overline{c_n c_m^*}\exp[\mathrm{i}(E_n - E_m)t/\hbar]$. Similarly,

$$\langle n|\widehat{\mathcal{H}}_1^*(t)|m\rangle = \int \psi_n^* e^{\frac{i}{\hbar}\widehat{\mathcal{H}}_0 t}\widehat{\mathcal{H}}'(t)e^{-\frac{i}{\hbar}\widehat{\mathcal{H}}_0 t}\psi_m \mathrm{d}v = e^{\frac{i}{\hbar}(E_n - E_m)t}\langle n|\widehat{\mathcal{H}}'(t)|n\rangle,$$

as required (see 3.60). Further, by introducing a new variable $\tau = t - t'$, transform (3.59):

$$\frac{\mathrm{d}\rho_{\alpha\alpha'}}{\mathrm{d}t} = -\frac{1}{\hbar^2}\int_0^t \sum_{\beta\beta'}\Big[\rho_{\alpha\beta}^*(0)\langle\beta|\widehat{\mathcal{H}}'(t-\tau)|\beta'\rangle\langle\beta'|\widehat{\mathcal{H}}'(t)|\alpha'\rangle e^{\mathrm{i}(\beta-\beta')(t-\tau)}e^{\mathrm{i}(\beta'-\alpha')t}$$
$$- \langle\alpha|\widehat{\mathcal{H}}'(t-\tau)|\beta\rangle\rho_{\beta\beta'}^*(0)\langle\beta'|\widehat{\mathcal{H}}'(t)|\alpha'\rangle e^{\mathrm{i}(\alpha-\beta)(t-\tau)}e^{\mathrm{i}(\beta'-\alpha')t}$$
$$- \langle\alpha|\widehat{\mathcal{H}}'(t)|\beta\rangle\rho_{\beta\beta'}^*(0)\langle\beta'|\widehat{\mathcal{H}}'(t-\tau)|\alpha'\rangle e^{\mathrm{i}(\alpha-\beta)}e^{\mathrm{i}(\beta'-\alpha')(t-\tau)}$$
$$+ \langle\alpha|\widehat{\mathcal{H}}'(t)|\beta\rangle\langle\beta|\widehat{\mathcal{H}}'(t-\tau)|\beta'\rangle\rho_{\beta'\alpha'}^*(0)e^{\mathrm{i}(\alpha-\beta)t}e^{\mathrm{i}(\beta-\beta')(t-\tau)}\Big]\mathrm{d}\tau. \quad (3.62)$$

Mind that the value of $\rho_{\alpha\beta}^*(0)$, $\rho_{\beta\beta}^*(0)$ and $\rho_{\beta'\alpha'}^*(0)$ do not depend on time.

Now let us carry out the averaging in (3.62) for ensembles with different Hamiltonians $\widehat{\mathcal{H}}'(t)$, i.e. the line over all terms of (3.62), which are randomly time-dependent, will appear, that means an averaging:

$$\frac{\mathrm{d}\rho_{\alpha\alpha'}}{\mathrm{d}t} = -\frac{1}{\hbar^2}\int_0^t \sum_{\beta\beta'}\Big[\rho_{\alpha\beta}^*(0)\overline{\langle\beta|\widehat{\mathcal{H}}'(t-\tau)|\beta'\rangle\langle\beta'|\widehat{\mathcal{H}}'(t)|\alpha'\rangle}e^{\mathrm{i}(\beta-\alpha')t+i(\beta'-\beta)\tau}$$
$$- \rho_{\beta\beta'}^*(0)\overline{\langle\alpha|\widehat{\mathcal{H}}'(t-\tau)|\beta\rangle\langle\beta'|\widehat{\mathcal{H}}'(t)|\alpha'\rangle}e^{\mathrm{i}(\beta-\alpha)\tau+i(\alpha-\beta+\beta'-\alpha')t}$$
$$- \rho_{\beta\beta'}^*(0)\overline{\langle\alpha|\widehat{\mathcal{H}}'(t)|\beta\rangle\langle\beta'|\widehat{\mathcal{H}}'(t-\tau)|\alpha'\rangle}e^{\mathrm{i}(\alpha'-\beta')\tau+i(\alpha-\beta+\beta'-\alpha')t}$$
$$+ \rho_{\beta'\alpha'}^*(0)\overline{\langle\alpha|\widehat{\mathcal{H}}'(t)|\beta\rangle\langle\beta|\widehat{\mathcal{H}}'(t-\tau)|\beta'\rangle}e^{\mathrm{i}(\beta-\beta')\tau+i(\alpha-\beta')t}\Big]\mathrm{d}\tau. \quad (3.63)$$

If products in (3.63) of a type $\overline{\langle\alpha|\widehat{\mathcal{H}}'(t)|\beta\rangle\langle\beta'|\widehat{\mathcal{H}}'(t-\tau)|\alpha'\rangle} \to 0$, when τ is greater than a certain τ_c, then the limits of integration can be expanded from 0 to ∞ (as it was done for (3.16)).

Let us introduce the correlation function (similar to 3.14):

$$G_{\alpha\beta\alpha'\beta'}(\tau) \equiv \overline{\langle\alpha|\widehat{\mathcal{H}}'(t)|\beta\rangle\langle\beta'|\widehat{\mathcal{H}}'(t+\tau)|\alpha'\rangle}. \quad (3.64)$$

Using (3.58), one can obtain

$$G_{\alpha\beta\alpha'\beta'} = \sum_{qq'} \langle\alpha|\hat{K}_q|\beta\rangle\langle\beta'|\hat{K}_{q'}|\alpha'\rangle \overline{B_q(t)B_{q'}(t+\tau)}.$$

We also introduce the spectral density $L_{qq'}(\omega)$ for the description of fluctuations of the function $B_q(t)$:

$$L_{qq'}(\omega) = \int_0^\infty \overline{B_q(t)B_{q'}(t+\tau)} e^{i\omega\tau} dt. \tag{3.65}$$

Since the product $\overline{B_q(t)B_{q'}(t+\tau)}$ is a real and even function of τ, then for $\tau_c \ll \omega^{-1}$ (the condition of fast fluctuations) the spectral densities (3.65) are real quantities and can be replaced with

$$k_{qq'}(\omega) = \mathrm{Re}L_{qq'}(\omega) = \frac{1}{2}\int_{-\infty}^{\infty} \overline{B_q(t)B_{q'}(t+\tau)} \cos\omega\tau \, d\tau.$$

Noticeable contributions to the expression (3.63) are mostly made by the terms for which $\alpha - \alpha' - \beta + \beta' = 0$ (otherwise, the integration of oscillating functions will give zero), so the second and third terms can be expressed as follows:

$$\frac{1}{\hbar^2}\sum_{\beta\beta'}\sum_{qq'}\left[\langle\alpha|\hat{K}_q|\beta\rangle\langle\beta'|\hat{K}_{q'}|\alpha'\rangle[k_{qq'}(\alpha-\beta) + k_{qq'}(\alpha'-\beta')]\rho^*_{\beta\beta'}(0)\right], \tag{3.66}$$

and the remaining terms in (3.63) have the form

$$\frac{1}{\hbar^2}\sum_{\beta\beta'}\sum_{qq'}\Big[\rho^*_{\alpha\beta}(0)[\langle\beta|\hat{K}_q|\beta'\rangle\langle\beta'|\hat{K}_{q'}|\alpha'\rangle k_{qq'}(\beta-\beta')]e^{i(\beta-\alpha')t}$$
$$+ \rho^*_{\beta'\alpha'}(0)[\langle\beta|\hat{K}_q|\beta'\rangle\langle\alpha|\hat{K}_{q'}|\beta\rangle k_{qq'}(\beta-\beta')]e^{i(\alpha-\beta')t}\Big]. \tag{3.67}$$

If now we introduce the spectral density of interaction, corresponding to the function correlation (3.64), then

$$J_{\alpha\alpha'\beta\beta'}(\omega) = \int_{-\infty}^{\infty} \overline{\langle\alpha|\widehat{\mathcal{H}}'(t)|\alpha'\rangle\langle\beta'|\widehat{\mathcal{H}}'(t+\tau)|\beta\rangle} e^{i\omega\tau} d\tau, \tag{3.68}$$

and summarizing the expressions (3.66–3.68), one obtains:

$$\frac{\mathrm{d}\rho^*_{\alpha\alpha'}}{\mathrm{d}t} = \sum_{\beta\beta'} R_{\alpha\alpha'\beta\beta'}\rho^*_{\beta\beta'}(0)e^{\mathrm{i}(\alpha-\alpha'-\beta+\beta')t}, \tag{3.69}$$

where

$$R_{\alpha\alpha'\beta\beta'} = \frac{1}{2\hbar^2}\left[J_{\alpha\beta\alpha'\beta'}(\alpha-\beta) + J_{\alpha\beta\alpha'\beta'}(\alpha'-\beta')\right] - \frac{1}{2\hbar^2}\left[\delta_{\alpha'\beta'}\sum_{\gamma} J_{\gamma\beta\gamma\alpha}(\gamma-\beta) - \delta_{\alpha\beta}\sum_{\gamma} J_{\gamma\alpha'\gamma\beta'}(\gamma-\beta')\right]. \tag{3.70}$$

In order the first order of approximations in the perturbation theory give a good convergence to the exact solution, it is necessary to have a small change $\rho^*_{\beta\beta'}(t)$ compared with $\rho^*_{\beta\beta'}(0)$ in the time interval of our interest. This means that at the necessary requirement $t \gg \tau_c$ must be a time interval for which

$$R^{-1}_{\alpha\alpha'\beta\beta'} \gg t.$$

Then in the right side of Eq. (3.69) the value of $\rho^*_{\beta\beta'}(0)$ can be replaced by $\rho^*_{\beta\beta'}(t)$ and thereby one can obtain of the differential equation Redfield for $\rho^*(t)$:

$$\frac{\mathrm{d}\rho^*_{\alpha\alpha'}(t)}{\mathrm{d}t} = \sum_{\beta\beta'} R_{\alpha\alpha'\beta\beta'}\rho^*_{\beta\beta'}(t). \tag{3.71}$$

Combined analysis of the conditions, under which the Eq. (3.71) was obtained, leads to the conclusion that the boundary of its applicability is

$$T_1, T_2 \gg \tau_c,$$

that means the assumption on the fast motions that cause fluctuations of interactions of nuclei with the environment and among themselves.

Since $R_{\alpha\alpha\beta\beta} = R_{\beta\beta\alpha\alpha}$, then the solution of Redfield Eq. (3.71) should lead to equal populations of all states, that is typical for an infinitely large temperature.

To describe the relaxation to the real equilibrium state, it is necessary to consider the contact with a thermal bath of spins (in our case, with the "lattice"). Thus, in the more strict consideration of the original Eq. (3.52) the density matrix for the overall system (including both spins and lattice) should be used. In other words, instead of the Hamiltonian $\widehat{\mathcal{H}}_0$ the sum of the "lattice" and "spin" Hamiltonians must be used. The disturbance $\widehat{\mathcal{H}}'(t)$ causes simultaneous transitions in lattice and spin system. Since in the absence of the Hamiltonian $\widehat{\mathcal{H}}'(t)$ the lattice and spins are not bound, then the required matrix density can be represented as the product of the spin ($\hat{\rho}^{*(S)}$) and lattice ($\hat{\rho}^{*(L)}$) density matrixes:

$$\hat{\rho}^* = \hat{\rho}^{*(S)}\hat{\rho}^{*(L)}.$$

As before, we assume that the lattice specific heat is infinite compared with the specific heat of the spin system. Then we have

$$\frac{\mathrm{d}\hat{\rho}^*}{\mathrm{d}t} = \hat{\rho}^{*(L)}\frac{\mathrm{d}\hat{\rho}^{*(S)}}{\mathrm{d}t},$$

since the lattice always remains in equilibrium, regardless of the state the spin system. The density matrix describing the lattice can be represented as

$$\hat{\rho}^{*(L)} = \frac{\exp[-\hat{F}/(kT)]}{\mathrm{Tr}(\exp[-\hat{F}/(kT)])}, \tag{3.72}$$

Here $\hat{F}$ is the Hamiltonian of the lattice. Assuming the temperature as sufficiently high, we can expand the exponent in the numerator of (3.72) in the series and restrict the consideration by the linear term:

$$\hat{\rho}^{*(L)} \simeq \frac{1-[\hat{F}/(kT)]}{\mathrm{Tr}(\exp[-\hat{F}/(kT)])}.$$

In all experiments on nuclear magnetic resonance the state populations of the spin system are not strongly deviate from a uniform distribution, then

$$\hat{\rho}^{*(L)}\hat{\rho}^{*(S)} \simeq \frac{1}{Z}\hat{\rho}^{*(S)} - \frac{\hat{F}}{kT}\frac{1}{ZA},$$

here Z is the number of lattice degrees of freedom; A is the number of degrees of freedom of a spin system. As a result, in the right side of the Eq. (3.71) an additional term will appear (for convenience, we have rearranged the terms under the double commutators):

$$-\frac{1}{\hbar^2}\int_0^\infty \left[\widehat{\mathcal{H}}_1^*(t), \left[\widehat{\mathcal{H}}_1^*(t-\tau), -\frac{\hat{F}}{kT}\frac{1}{ZA}\right]\right]\mathrm{d}\tau.$$

As a result the Eq. (3.71) takes the form

$$\frac{\mathrm{d}\rho_{\alpha\alpha'}^{*(S)}}{\mathrm{d}t} = \sum_{\beta\beta'} R_{\alpha\alpha'\beta\beta'}\left[\rho_{\beta\beta'}^{*(S)} - \rho_{\beta\beta'}^{*(S)}(T)\right], \tag{3.73}$$

where $\rho_{\beta\beta'}^{*(S)}(T) = \delta_{\beta\beta'}A^{-1}\exp[-\hbar\beta/(kT)]$ are the matrix elements at the thermal equilibrium. Indeed, using the relation

$$\int_{-\infty}^{\infty}\left[\widehat{\mathcal{H}}_0+\hat{F},\widehat{\mathcal{H}}_1^*(t-\tau)\right]\mathrm{d}\tau=\frac{\mathrm{i}}{\hbar}\int_{-\infty}^{\infty}\frac{\mathrm{d}}{\mathrm{d}\tau}\widehat{\mathcal{H}}_1^*(t-\tau)\mathrm{d}\tau=0,$$

which is the result of the law of conservation of total energy $\hat{F}+\widehat{\mathcal{H}}_0$, we obtain

$$\int_{-\infty}^{\infty}[\widehat{\mathcal{H}}_0,\widehat{\mathcal{H}}_1^*]d\tau=-\int_{-\infty}^{\infty}\left[\hat{F},\widehat{\mathcal{H}}_1^*\right]\mathrm{d}\tau,$$

i.e. the Hamiltonian $\hat{F}$ can be replaced by the Hamiltonian $-\widehat{\mathcal{H}}_0$, where once we arrive at (3.73), taking into account that

$$\hat{\rho}^{*(S)}(T)=\frac{\exp[-\widehat{\mathcal{H}}_0/(kT)]}{\mathrm{Tr}(\exp[-\widehat{\mathcal{H}}_0/(kT)])}\simeq\frac{\hat{E}-[\widehat{\mathcal{H}}_0/(kT)]}{A},$$

here $\hat{E}$ is the unit operator, which commutes with any operators.

Further some relaxation mechanisms will be considered, which most often can be met in the investigations of the nuclear magnetic resonance of different nuclei in the most variety of objects. Often the consideration of NMR-relaxation is started from the dipole–dipole interaction, which exists in all materials. However, methodologically it is more correct to begin the consideration with a simple case of the relaxation in a two-level system under the influence of fluctuating magnetic fields (such fields can arise due to the anisotropy electronic shielding of nuclei, or as a result of the spin-rotational interaction). On this example, it is easier to master the way of obtaining the concrete formulas for the relaxation rates T_1^{-1} and T_2^{-1}.

3.6 Fluctuations of the Magnetic Field (General Consideration)

Consider a simple two-level system, i.e. an ensemble of noninteracting nuclei with spin $I=1/2$. Let us follow the process of relaxation in this system when of fluctuating magnetic field $\vec{B}(t)$ affects on a nucleus. The Hamiltonian of nuclei, which interact with a constant magnetic field $\vec{B}_0$ and field $\vec{B}(t)$, is expressed as

$$\widehat{\mathcal{H}}=\widehat{\mathcal{H}}_0+\widehat{\mathcal{H}}'(t)=-\gamma\hbar\hat{I}_zB_0-\left(\hat{\vec{\mu}},\vec{B}(t)\right). \quad (3.74)$$

The second term in formula (3.74) can be presented as

$$\widehat{\mathcal{H}}'(t)=-\gamma\hbar\hat{I}_xB_x(t)-\gamma\hbar\hat{I}_yB_y(t)-\gamma\hbar\hat{I}_zB_z(t). \quad (3.75)$$

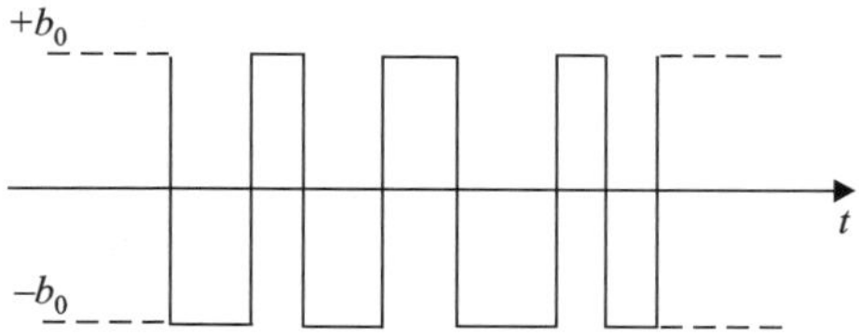

Fig. 3.2 A model of a random process

According to (3.16) and (3.26) one has:

$$\frac{1}{T_1} = 2\frac{\overline{|\langle 1|\widehat{\mathcal{H}}'(t)|2\rangle|^2}}{\hbar^2} J(\omega_0). \tag{3.76}$$

It is obvious that the matrix elements of the Hamiltonian (3.76) are determined as follows:

$$\langle 1|\widehat{\mathcal{H}}'(t)|2\rangle = -\sum_i \gamma\hbar B_i(t)\langle 1|\hat{\vec{I}}_i|2\rangle,$$

where $i = x, y, z$. If the principal basis is adopted as the eigenfunctions of $\hat{I}_z$, the eigenvalues of which are equal to $\pm 1/2$, then

$$|\langle 1|\hat{I}_x|2\rangle|^2 = |\langle 1|\hat{I}_y|2\rangle|^2 = \frac{1}{4},$$
$$|\langle 1|\hat{I}_z|2\rangle|^2 = 0.$$

If the field $\vec{B}(t)$ varies in a random way, it can be characterized by an exponential correlation function of the type (3.15) and then the spectral density interaction is given by (3.17). Let us consider this question in detail.

For simplicity, we assume that the magnetic field fluctuations are of the form shown schematically in Fig. 3.2. The slope of pulse fronts is determined by high-frequency components of the spectral density of a random function and the nuclear magnetic resonance is a relatively low frequency phenomenon. Therefore this approximation will not be too rough. According to the definition the correlation function for a given case is expressed as

$$K_i(\tau) = \overline{B_i(t)B_i^*(t+\tau)},$$

where B_i is one of the components of the fluctuating field (B_x, B_y or B_z). In accordance with the assumption the nature of the fluctuations, we assume that the magnetic field can take two possible values: $B_{i1} = b_0$ and $B_{i2} = -b_0$ (see Fig. 3.2). Let, for example, at $t = 0$ the value of the field B_{i1} is realized. Then for time $t = \tau$ can be written:

$$B_i(\tau) = P_1(\tau)B_{i1} + P_2(\tau)B_{i2}, \tag{3.77}$$

here P_1 and P_2 are the probabilities of the possible values of the B_i at the time τ. It is natural that P_1 and P_2 assume the values of 0 or 1 depending on the concrete realization (of cause, $P_1 + P_2 = 1$). Now, we express the correlation function taking into account the specific realization of $B_i(t)$:

$$K_i(\tau) = B_{i1}[p_1(\tau)B_{i1} + p_2(\tau)B_{i2}], \tag{3.78}$$

here $p_1(\tau)$ and $p_2(\tau)$ are the averaged over the ensemble probabilities that the magnetic field after a time τ will be B_{i1} and B_{i2}. The probabilities $p_1(\tau)$ and $p_2(\tau)$ can take any value within the range from 0 to 1, but $p_1(\tau) + p_2(\tau) = 1$. If we carry out the consideration for $B_i(0) = B_{i2}$, then all conclusions remain the same, but the indices in (3.77) and (3.78) are exchanged. In analogy with the process of changes in populations of energy states in a two-level system one can write the equations for the probabilities p_1 and p_2:

$$\frac{\mathrm{d}p_1}{\mathrm{d}t} = W(p_2 - p_1),$$
$$\frac{\mathrm{d}p_2}{\mathrm{d}t} = W(p_1 - p_2).$$

Hence, we obtain the equation

$$\frac{\mathrm{d}}{\mathrm{d}t}(p_1 - p_2) = -2W(p_1 - p_2)$$

with the solution

$$p_1(\tau) - p_2(\tau) = Ce^{-2W\tau}. \tag{3.79}$$

If $\tau \to 0$ the random function maintains its initial value and $p_1(\tau) \to 1$, $p_2(\tau) \to 0$. Based on these initial conditions, one can obtain $C = 1$. Taking into account formula (3.79) and the explicit view of B_{i1} and B_{i2}, the expression (3.78) is transformed to

$$K_i(\tau) = \overline{B_i(0)B_i(\tau)} = b_0^2[p_1(\tau) - p_2(\tau)] = b_0^2 e^{-\tau/\tau_c},$$

where $\tau_c = 1/(2W)$. For a stationary random process of fluctuations of the magnetic field, we have

$$K_i(\tau) = \overline{B_i^2(\tau)}e^{-|\tau|/\tau_c}, \tag{3.80}$$

that coincides up to normalization with the formula (3.15).

Strictly speaking, to describe fluctuations of the magnetic field components it may be required a few parameters, but in many cases it is sufficient to use of a single correlation time τ_c.

Substituting (3.80) into (3.17), one obtains the expression for the spectral density:

$$
\begin{aligned}
J_i(\omega) &= \overline{B_i^2(\tau)} \int_{-\infty}^{\infty} \exp\left[-\frac{|\tau|}{\tau_c} + i\omega\tau\right] d\tau \\
&= \overline{B_i^2(\tau)} \int_{-\infty}^{0} \exp\left[\tau\left(\frac{1}{\tau_c} + i\omega\right)\right] d\tau + \overline{B_i^2(\tau)} \int_{0}^{\infty} \exp\left[-\tau\left(\frac{1}{\tau_c} - i\omega\right)\right] d\tau \\
&= \overline{B_i^2(\tau)} \frac{2\tau_c}{1+\omega^2\tau_c^2}. \qquad (3.81)
\end{aligned}
$$

Using the results of the previous calculations, we finally obtain from (3.76) under the assumption isotropic nature of the magnetic field fluctuations, i.e. $\overline{B_x^2(t)} = \overline{B_y^2(t)} = \overline{B_z^2(t)} = \overline{B^2(t)}$:

$$
\frac{1}{T_1} = \gamma^2 \overline{B^2(t)} \frac{2\tau_c}{1+\omega_0^2\tau_c^2}. \qquad (3.82)
$$

The calculation for the spin–spin relaxation can not be done so simply, because when $M_\perp \neq 0$ a spin system is not in a stationary state. Before turning to the description of the calculation procedure, we obtain the expression for the spin–spin relaxation time on the basis of simple physical considerations. Based on the analysis made in Chap. 1, and also taking into account formula (3.9) one can write:

$$
\frac{1}{T_2} = w + \frac{1}{T_2'},
$$

here w is the probability of spin–lattice relaxation transitions determining one of the reasons for decrease of the lifetime in energy states; the term $1/T_2'$ describes the loss of the coherence due to frequency fluctuations of the nuclear magnetic resonance in the presence of field $\vec{B}(t)$. It is natural that probability of $w = 1/(2T_1)$ and can be expressed by the formula (3.82). The existence of the components of $\pm B_z$ causes a change in resonance frequency $\delta\omega = \pm\gamma B_z$. We assume that the fluctuation field B_z exists during the time τ_c. Then, an additional phase shift of the precessing transverse component of magnetization $M_\perp$ will be $\delta\phi = \pm\gamma B_z \tau_c$. Taking into account the random character of fluctuations, we can assume that the phase shift during the time interval Δt will be characterized by the value $\Delta\phi$, and $(\Delta\phi)^2 = n(\delta\phi)^2$, where $n = \Delta t/\tau_c$ is the number of fluctuations in time Δt. Let T_2' is an interval of time during which the significant loss of the phase memory occurs, for example, about one radian. Then

$$
\frac{1}{T_2'} = \gamma^2 B_z^2 \tau_c.
$$

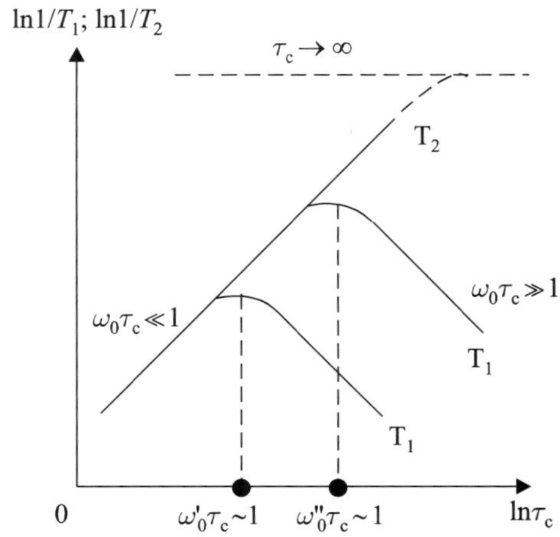

Fig. 3.3 The possible view of the dependences of relaxation rates on a correlation time

Assuming still the isotropic field fluctuations $\vec{B}(t)$, we obtain

$$\frac{1}{T_2} = \gamma^2 \overline{B^2(t)} \left[\tau_c + \frac{\tau_c}{1 + \omega_0^2 \tau_c^2} \right]. \tag{3.83}$$

It is easy to see that the first term in square brackets in (3.83) is proportional to the spectral density at zero frequency (see 3.81).

If the fluctuations of the field $\vec{B}(t)$ are very fast, i.e. $\omega_0^2\tau_c^2 \ll 1$, then

$$\frac{1}{T_1} = \frac{1}{T_2} = 2\gamma^2 \overline{B^2(t)} \tau_c, \tag{3.84}$$

and the dependence of the relaxation rates on the frequency ω_0 disappears if $B(t)$ does not depend on the magnetic field B_0. The correlation $\omega_0^2\tau_c^2 \ll 1$ is called *the condition of extreme narrowing of the nuclear magnetic resonance* (it is a typical situation for liquids and gases).

In Fig. 3.3 the dependences of $1/T_1$ and $1/T_2$, which have been constructed in accordance with formulas (3.82) and (3.83), are shown. The horizontal dashed line corresponds to the static case (no movement) when the dependence of the field $\vec{B}(t)$ on the time disappears, but there is a random distribution of local field values in the range of $\pm B$. This case is not described by the perturbation theory because the main assumption about the smallness of τ_c is violated (we used the relation $\tau_c \ll T_2$, T_1 when obtained the formula (3.16)). The behavior of the functions $1/T_1$ and $1/T_2$ is completely determined by the spectral density $J_i(\omega)$ (3.81), the approximate (but real) view of which for different values τ_c is shown in Fig. 3.4.

For a more accurate calculation of $1/T_2$ it is necessary to create new wave functions which would describe the behavior of the transverse component magnetization. The equation for the transverse magnetization component should include the opera-

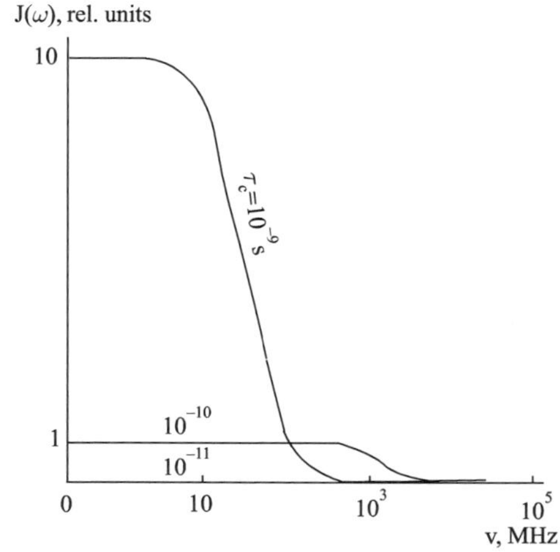

Fig. 3.4 The possible view of spectral densities

tor $\hat{I}_x$ or $\hat{I}_y$. For the considered case of a spin $I = 1/2$ we can define states, which are described, for example, by the operator $\hat{I}_x$, using the following equations:

$$\hat{I}_x\varphi_1 = \frac{1}{2}\varphi_1, \quad \hat{I}_x\varphi_2 = -\frac{1}{2}\varphi_2. \tag{3.85}$$

These two states are not eigenstates of the energy operator, however, they are orthogonal and we can talk about their populations and the transition probabilities between them u_{12} and u_{21} (similarly to w_{12} and w_{21}). For these states, we can write the kinetic equations, similar to (3.21) (or generally (3.36)) that will have the solution of the type (3.25), but instead of (3.26) we obtain

$$\frac{1}{T_2} = u_{12} + u_{21} \simeq 2u.$$

The wave functions in the expression (3.85) can always be expressed via the eigenfunctions ψ_i of the operator $\hat{I}_z$:

$$\varphi_k = \sum_i a_{ki}\psi_i. \tag{3.86}$$

In our case ($I = 1/2$)

$$\varphi_1 = \frac{1}{\sqrt{2}}(\psi_{1/2} + \psi_{-1/2}); \quad \varphi_2 = \frac{1}{\sqrt{2}}(\psi_{1/2} - \psi_{-1/2}). \tag{3.87}$$

It easy to proved by substituting (3.86) into (3.85). Using (3.86) and the orthogonality of the functions (3.87), one can obtain the generalization of (3.11) for the probabilities of transitions between states described by the functions φ_k, in the form

$$u_{kl} = \frac{1}{t}\frac{1}{\hbar^2}\left|\int_0^t \sum_{i,j}\langle i|\widehat{\mathcal{H}}'(t')|j\rangle a_{ki}a_{lj}e^{\mathrm{i}\omega_{ij}t'}\mathrm{d}t'\right|^2. \tag{3.88}$$

These probabilities are calculated in the same way as described in Sect. 3.1 for w_{mn}. The calculation of $1/T_2 \simeq 2u$, using (3.87) and (3.88) leads to expression (3.83). Now it becomes more clear the appearance in (3.83) of the spectral density at zero frequency (the first term in square brackets), which is obtained from (3.88) for the coincidental indexes i and j.

Now describe the relaxation using the mathematical apparatus of Sects. 3.4 and 3.5. There are two ways to use the Redfield Equation:

1. It is possible to find the solution for the density matrix and then compute $\langle M_i\rangle = \sum_{\alpha\alpha'}\rho_{\alpha\alpha'}\langle\alpha'|\hat{M}_i|\alpha\rangle$, from where the time constants (T_1 and T_2) for the component $M_i = M_x, M_y, M_z$ are determined.

2. Another way, which is more convenient in this case, is to derive differential equations for the components M_i:

$$\frac{\mathrm{d}}{\mathrm{d}t}\langle M_i\rangle = \frac{\mathrm{d}}{\mathrm{d}t}\sum_{\alpha\alpha'}\rho_{\alpha\alpha'}\langle\alpha'|\hat{M}_i|\alpha\rangle = \sum_{\alpha\alpha'}\frac{\mathrm{d}\rho_{\alpha\alpha'}}{\mathrm{d}t}\langle\alpha'|\hat{M}_i|\alpha\rangle, \tag{3.89}$$

where $\mathrm{d}\rho_{\alpha\alpha'}/\mathrm{d}t$ are calculated using the theory of Redfield.

Taking into account the relation between the matrix elements ρ^* and ρ, which is given by (3.61), one obtains

$$\frac{\mathrm{d}\rho^*_{\alpha\alpha'}}{\mathrm{d}t} = \mathrm{i}(\alpha-\alpha')\rho^*_{\alpha\alpha'} + e^{\mathrm{i}(\alpha-\alpha')t}\frac{\mathrm{d}\rho_{\alpha\alpha'}}{\mathrm{d}t}. \tag{3.90}$$

Substituting the Redfield Eq. (3.71) for $\rho(t)$ into (3.90) we obtain:

$$\frac{\mathrm{d}\rho_{\alpha\alpha'}}{\mathrm{d}t} = \mathrm{i}(\alpha'-\alpha)\rho_{\alpha\alpha'} + \sum_{\beta\beta'}R_{\alpha\alpha'\beta\beta'}\rho_{\beta\beta'} = \frac{\mathrm{i}}{\hbar}\left[\hat{\rho},\widehat{\mathcal{H}}_0\right]_{\alpha\alpha'} + \sum_{\beta\beta'}R_{\alpha\alpha'\beta\beta'}\rho_{\beta\beta'}.$$

Now we can express (3.89) in the form:

$$\frac{\mathrm{d}}{\mathrm{d}t}\langle M_i\rangle = \sum_{\alpha\alpha'\beta\beta'}\left(\frac{\mathrm{i}}{\hbar}\left[\hat{\rho},\widehat{\mathcal{H}}_0\right]_{\alpha\alpha'} + R_{\alpha\alpha'\beta\beta'}\rho_{\beta\beta'}\right)\langle\alpha'|\hat{M}_i|\alpha\rangle. \tag{3.91}$$

The expression (3.91) is similar to the Bloch equations for the components of the nuclear macroscopic magnetization (see Sect. 1.5). It should be noted that the number of Eq. (3.91) can be less than for elements of the density matrix.

In this case $\widehat{\mathcal{H}}_0 = -\gamma\hbar B_0\hat{I}_z = -\hbar\omega_0\hat{I}_z$, and the expression (3.58) takes the form (see also 3.75):

$$\widehat{\mathcal{H}}'(t) = -\gamma\hbar\sum_q B_q(t)\hat{I}_q, \quad q = x, y, z.$$

Keeping the notation of Sect. 3.5, we can write:

$$\langle\alpha'|\widehat{\mathcal{H}}'(t)|\alpha\rangle = -\gamma\hbar\sum_q B_q(t)\langle\alpha'|\hat{I}_q|\alpha\rangle.$$

Let us rewrite (3.91) for the particular task:

$$\frac{\mathrm{d}}{\mathrm{d}t}\langle I_i\rangle = \sum_{\alpha\alpha'}\frac{\mathrm{i}}{\hbar}\left[\hat{\rho},\widehat{\mathcal{H}}_0\right]_{\alpha\alpha'}\langle\alpha'|\hat{\vec{I}}_i|\alpha\rangle + \sum_{\alpha\alpha'\beta\beta'} R'_{\alpha\beta}\rho_{\beta\beta'}\langle\alpha'|\hat{\vec{I}}_i|\alpha\rangle. \tag{3.92}$$

Transform the first term on the right-hand side of (3.92):

$$\begin{aligned}\frac{\mathrm{i}}{\hbar}\sum_\alpha\langle\alpha|(\hat{\rho}\widehat{\mathcal{H}}_0-\widehat{\mathcal{H}}_0\hat{\rho})\hat{\vec{I}}_i|\alpha\rangle &= \frac{\mathrm{i}}{\hbar}\mathrm{Tr}(\hat{\rho}\widehat{\mathcal{H}}_0\hat{\vec{I}}_i-\widehat{\mathcal{H}}_0\hat{\rho}\hat{\vec{I}}_i) = \frac{\mathrm{i}}{\hbar}\mathrm{Tr}(\hat{\rho}\widehat{\mathcal{H}}_0\hat{\vec{I}}_i-\hat{\rho}\hat{\vec{I}}_i\widehat{\mathcal{H}}_0)\\ &= \frac{\mathrm{i}}{\hbar}\mathrm{Tr}(\hat{\rho}[\widehat{\mathcal{H}}_0,\hat{\vec{I}}_i]) = -\mathrm{i}\gamma B_0\,\mathrm{Tr}(\hat{\rho}[\hat{I}_z,\hat{\vec{I}}_i]).\end{aligned} \tag{3.93}$$

If $i = z$, then the expression (3.93) vanishes. Taking into account the usual relations between the components of the spin, we obtain for $i = x$:

$$-\mathrm{i}\gamma B_0\,\mathrm{Tr}(\hat{\rho}[\hat{I}_z,\hat{I}_x]) = -\mathrm{i}\gamma B_0\,\mathrm{Tr}(i\hat{I}_y\hat{\rho}) = \gamma B_0\langle I_y\rangle.$$

Similarly for $i = y$:

$$-\mathrm{i}\gamma B_0\,\mathrm{Tr}(\hat{\rho}[\hat{I}_z,\hat{I}_y]) = -\gamma B_0\langle I_x\rangle.$$

Finally, we obtain

$$\sum_{\alpha\alpha'}\frac{\mathrm{i}}{\hbar}[\hat{\rho},\widehat{\mathcal{H}}_0]_{\alpha\alpha'}\langle\alpha'|\hat{\vec{I}}_i|\alpha\rangle = \gamma\left[\langle\hat{\vec{I}}\rangle\times\vec{B}_0\right]_i. \tag{3.94}$$

Let us now turn to the second (relaxation) term in Eq. (3.92), which includes the values $R_{\alpha\alpha'\beta\beta'}$, consisting in their turn from the sum of the spectral densities (see (3.70)) like this:

$$\begin{aligned}J_{\alpha\beta\alpha'\beta'}(\omega) &= \int_{-\infty}^{\infty}\overline{\langle\alpha|\widehat{\mathcal{H}}'(t)|\beta\rangle\langle\beta'|\widehat{\mathcal{H}}'(t+\tau)|\alpha'\rangle}e^{\mathrm{i}\omega\tau}\,\mathrm{d}\tau\\ &= \gamma^2\hbar^2\sum_{qq'}\langle\alpha|\hat{I}_q|\beta\rangle\langle\beta'|\hat{I}_{q'}|\alpha'\rangle\int_{-\infty}^{\infty}\overline{B_q(t)B_{q'}(t+\tau)}e^{\mathrm{i}\omega\tau}\,\mathrm{d}\tau\\ &= 2\gamma^2\hbar^2\sum_{qq'}\langle\alpha|\hat{I}_q|\beta\rangle\langle\beta'|\hat{I}_{q'}|\alpha'\rangle k_{qq'}(\omega).\end{aligned} \tag{3.95}$$

If, for simplicity, we assume that the fluctuations of the field components $\vec{B}(t)$ are independent, then

$$\overline{B_q(t)B_{q'}(t+\tau)} = 0 \quad \text{for} \quad q \neq q',$$

the summation over the q' in (3.95) vanishes and the values k_{qq} define the spectral density for the magnetic field components with index q.

Let us calculate, for example, the term containing the spectral density $J_{\alpha\beta\alpha'\beta'}(\alpha' - \beta')$. Using (3.95), we will find that

$$\begin{aligned}
&\sum_{\alpha\alpha'\beta\beta'} J_{\alpha\beta\alpha'\beta'}(\alpha' - \beta')\rho_{\beta\beta'}\langle\alpha'|\hat{\vec{I}}_i|\alpha\rangle \\
&\quad = 2\gamma^2\hbar^2 \sum_{\alpha\alpha'\beta\beta'} \sum_q \langle\alpha|\hat{I}_q|\beta\rangle\langle\beta'|\hat{I}_q|\alpha'\rangle\langle\beta|\hat{\rho}|\beta'\rangle\langle\alpha'|\hat{\vec{I}}_i|\alpha\rangle k_{qq}(\alpha' - \beta') \\
&\quad = 2\gamma^2\hbar^2 \sum_{\alpha'\beta'} \sum_q \langle\beta'|\hat{I}_q|\alpha'\rangle\langle\alpha'|\hat{\vec{I}}_i\hat{I}_q\hat{\rho}|\beta'\rangle k_{qq}(\alpha' - \beta').
\end{aligned}$$

It is impossible to produce the summation over α' and β', because these indexes are included in the spectral density k_{qq}.

Carrying out similar calculations with other members, we obtain

$$\sum_{\alpha\alpha'\beta\beta'} R_{\alpha\alpha'\beta\beta'}\rho_{\beta\beta'}\langle\alpha'|\hat{\vec{I}}_i|\alpha\rangle = \gamma^2 \sum_{\alpha\beta} \sum_q \langle\beta|\hat{I}_q|\alpha\rangle\langle\alpha|\left[\left[\hat{\vec{I}}_i, \hat{I}_q\right], \hat{\rho}\right]|\beta\rangle k_{qq}(\alpha - \beta).$$

Looking over $q = x, y, z$, one can obtain for $i = z$ (the term with $q = z$ does not brings a contribution):

$$\sum_{\alpha\alpha'\beta\beta'} R_{\alpha\alpha'\beta\beta'}\rho_{\beta\beta'}\langle\alpha'|\hat{I}_z|\alpha\rangle = \gamma^2\left[k_{xx}(\omega_0) + k_{yy}(\omega_0)\right]\langle I_z\rangle. \tag{3.96}$$

Joining formulas (3.94) and (3.96) we will obtain the Eq. (3.92) in the form

$$\frac{\mathrm{d}}{\mathrm{d}t}\langle I_z\rangle = \gamma\left[\langle\hat{\vec{I}}\rangle \times \vec{B}_0\right]_z - \frac{\langle I_z\rangle - I_0}{T_1} \tag{3.97}$$

with

$$\frac{1}{T_1} = \gamma^2\left[k_{xx}(\omega_0) + k_{yy}(\omega_0)\right].$$

The appearance of I_0 in (3.97) is determined by considerations in accordance with the scheme given at the end of Sect. 3.5.

Similarly, the equations for the expected values x- and y-component of spin $\hat{\vec{I}}$ can be obtained. For example, if $i = x$, the term with $q = x$ does not contribute; for $q = y$ we will obtain the spectral density at frequency ω_0; if $q = z$ the states with indexes α and β are the same (in the selected representation the operator $\hat{I}_z$ is diagonal) and the spectral density at zero frequency ($\alpha - \beta = 0$) appears. As a result we obtain the equation

$$\frac{\mathrm{d}}{\mathrm{d}t}\langle I_x\rangle = \gamma\left[\langle\vec{\hat{I}}\rangle \times \vec{B}_0\right]_x - \frac{\langle I_x\rangle}{T_2}, \tag{3.98}$$

where

$$\frac{1}{T_2} = \gamma^2\left[k_{zz}(0) + k_{yy}(\omega_0)\right].$$

Equations (3.97) and (3.98) in combination with the similar formula for $\langle I_y\rangle$ are presented the Bloch equations (see Sect. 1.5). They were obtained for the simplest spin system, and it should not be assumed that this result is of general importance. In each case the special consideration is required to carry out (similar to that shown in this section). There are situations when the relaxation process can not be described by simple exponential dependence (see Sect. 3.8).

To calculate the spectral density $k_{qq}(\omega_0)$ it is necessary to specify a certain type of correlation function. If we assume the exponential function (3.15), for isotropic nature of the fluctuations of the field $\vec{B}(t)$ it is easy to get the results that coincide with (3.82) and (3.83).

3.7 The Influence of Fluctuations of the Magnetic Field on the NMR-Relaxation

3.7.1 *Effect of Anisotropy of the Electron Shielding*

Using the results of the previous section, we can describe the relaxation process due to *the anisotropy of electron shielding*. If the shielding of a nucleus depends on the orientation of a molecule (i.e. it is anisotropic) that often takes place, it appears that the vector of the secondary magnetic field generated in a molecule (see Sect. 1.11) does not parallel to the vector field $\vec{B}_0$. Under the influence of a chaotic rotation of molecules the secondary field will fluctuates and can cause transitions between states of a nuclear spin in the magnetic field $\vec{B}_0$.

The Hamiltonian describing the electronic shielding of a nucleus can be represented in the form

$$\widehat{\mathcal{H}}'_\sigma = \gamma\hbar\left(\sigma_x B_x \hat{I}_x + \sigma_y B_y \hat{I}_y + \sigma_z B_z \hat{I}_z\right), \tag{3.99}$$

where B_x, B_y, B_z are the components of the field $\vec{B}_0$ in the coordinate frame associated with the molecule; σ_x, σ_y, σ_z are the so-called principal values of electron shielding. For simplicity, we assume that two of the components are equal:

$$\sigma_x = \sigma_y = \sigma_\perp; \quad \sigma_z = \sigma_\parallel.$$

Then the Hamiltonian (3.99) can be rewritten as

$$\begin{aligned}\widehat{\mathcal{H}}'_\sigma &= \gamma\hbar\Big[\sigma_\parallel B_z \hat{I}_z + \sigma_\perp (B_x \hat{I}_x + B_y \hat{I}_y)\Big] \\ &= \gamma\hbar\left[\frac{\sigma_\parallel + 2\sigma_\perp}{3}\left(\vec{B}_0, \hat{\vec{I}}\right) + \frac{\sigma_\parallel - \sigma_\perp}{3}\left(2B_z\hat{I}_z - B_x\hat{I}_x - B_y\hat{I}_y\right)\right].\end{aligned}$$

The first term in square brackets in this expression contains the scalar product, which is invariant to rotation. Therefore, this term can not cause fluctuations of the magnetic field at the site of a nucleus (this term defines a constant electron screening $\sigma = (1/3)(\sigma_\parallel + 2\sigma_\perp)$ in liquids and gases). The second term is modulated by random thermal motion of molecules and it is responsible for the relaxation process. Omitting the details of the further transformation of the Hamiltonian $\widehat{\mathcal{H}}'_\sigma$, which to a great extent are presented in Sect. 1.10 for the Hamiltonian of the dipole–dipole interactions, we give the final result:

$$\begin{aligned}\frac{1}{T_{1\sigma}} &= \frac{1}{15}\gamma^2 B_0^2 (\Delta\sigma)^2 \frac{2\tau_c}{1+\omega_0^2\tau_c^2}, \\ \frac{1}{T_{2\sigma}} &= \frac{1}{45}\gamma^2 B_0^2 (\Delta\sigma)^2 \left[4\tau_c + \frac{3\tau_c}{1+\omega_0^2\tau_c^2}\right],\end{aligned} \tag{3.100}$$

where $\Delta\sigma = \sigma_\parallel - \sigma_\perp$. When $\omega_0^2\tau_c^2 \ll 1$ one can obtain $T_1/T_2 = 7/6$, that in contrast to the result (3.84) is somewhat more unit. This and other minor differences of the formulas (3.100) from (3.82) and (3.83) are due to the presence of the anisotropy of the secondary magnetic field of the electron shielding (in the previous section, was considered the isotropic case).

The most important point for the discussed relaxation mechanism is the dependence of the relaxation rate on the magnitude of the magnetic field B_0, that allows us to distinguish its contribution in the total relaxation rate in the presence of several simultaneously acting mechanisms of relaxation. It is necessary for this to carry out experiments at different magnetic fields. The contribution to the observed relaxation rate due to the anisotropy of the electron shielding is usually small, and it can be detected only in sufficiently strong magnetic fields and for nuclei with relatively large electron shells (^{13}C, ^{19}F, ^{113}Cd, ^{199}Hg, ^{205}Tl, ^{207}Pb, …). Note that the correlation time for this relaxation mechanism is $\tau_c = \tau_R$—the correlation time for rotational motion (usually rotational diffusion). The simplest expression of τ_R in terms of macroscopic parameters of medium is the Stokes-Einstein-Debye relation:

$$\tau_R = \frac{4\pi\eta a^3}{3kT}, \tag{3.101}$$

where η is the viscosity of the medium; a is the radius of a molecule. The absence, in some cases, of a good agreement between data on the relaxation rates and viscosity of the medium is determined not by the approximate character the of NMR-relaxation theory but by the lack of the proper theoretical description of the liquid state.

3.7.2 *Effect of Spin-Rotational Interaction*

Fluctuating magnetic fields can arise due to the rotation of a molecule in the absence of anisotropy of the electron shielding. Such fields appear not due to the rotational diffusion but due to the rapid quantum rotation which are typical for gases. It turns out that such a rotation is possible to exist in liquids.

If the motion of a molecule can be characterized by angular momentum:

$$\vec{p} = \hbar \vec{J} \tag{3.102}$$

(J is the rotational quantum number, which takes a value of $0, 1, 2, \ldots$), then the motion of the molecular electron shell as a whole will cause a circular current, which, in its turn, creates a magnetic field:

$$\vec{B}_{\mathrm{sr}} \sim \sum_i \frac{\vec{p}}{R_i^3}, \tag{3.103}$$

where R_i is the radius of i-th orbit of the electrons (a circular current is equivalent to a magnetic dipole). Induction of the field B_{sr} can reach 10 G (index sr is from the term *spin-rotational interaction*).

The interaction energy of the nuclear magnetic dipole $\vec{\mu} = \gamma \hbar \vec{I}$, located in a rotating molecule, with the field, which is described by (3.103), can be expressed as a scalar product

$$E_{\mathrm{sr}} = -\left(\vec{\mu}, \vec{B}_{\mathrm{sr}}(t)\right).$$

Using (3.102) and passing to the operators $\hat{\vec{I}}$ and $\hat{J}$, we can write the Hamiltonian of the spin-rotational interaction in the form of

$$\widehat{\mathcal{H}}'_{\mathrm{sr}} = -\hbar \hat{\vec{I}} \cdot \mathsf{C} \cdot \hat{J},$$

where C is the tensor of the spin-rotational interaction (in the isotropic case the tensor becomes a scalar).

Collisions of molecules in the process of thermal motion permanently change the magnitude and direction of the angular momentum $\hat{J}$, resulting in the fluctuations of the field $\vec{B}_{\mathrm{sr}}$. These fluctuations are characterized by certain correlation time $\tau_{\mathrm{c}} = \tau_{\mathrm{sr}}$ and they determine the process of *spin-rotational relaxation*.

In liquids the quantum rotation occurs within a very short time intervals and it is manifested usually at high temperatures, when thermal motion is very fast. So one can always count $\omega_0^2\tau_{sr}^2 \ll 1$. Then the relaxation rate is given by

$$\frac{1}{T_{1sr}} = \frac{1}{T_{2sr}} = \frac{8\pi^2 kT}{\hbar^2} I_0 C^2 \tau_{sr}, \tag{3.104}$$

where I_0 is the inertia moment of a molecule. If the temperature increases, the spin-rotational relaxation rate becomes higher in contrast to the trends for other relaxation mechanisms.

There are two approaches to describe the spectral density of fluctuations spin-rotational interaction. In the theory of P.S. Hubbard (Phys. Rev., 1963, vol. 131, p. 1155) the correlation time, entering into the formula (3.104), was obtained as follows:

$$\tau_{sr} = \frac{3I_0 D}{4a^2 kT},$$

where a is the radius of a spherical molecule; D is the rotational diffusion coefficient. Taking into account the relation $\tau_R = 2a^2/(9D)$ (for the rotational diffusion), one obtains

$$\tau_R \tau_{sr} = \frac{I_0}{6kT}, \tag{3.105}$$

which implies $\tau_{sr} \sim \tau_R^{-1}$, that leads to a particular temperature dependence of the spin–lattice relaxation rate (Eq. (3.105) is valid for $\tau_{sr} \ll \tau_R$).

The second approach, developed by H.S. Gutowsky and co-workers (J. Chem. Phys., 1963, vol. 38, p. 78), consists in the fact that the time dependence of the field $\vec{B}_{sr}$ (see 3.103) is approximated by a sequence of short pulses Δ, which follow with a frequency $1/\tau_R$. In this case, the correlation function has the form

$$K(\tau) = b^2 \frac{\Delta - |\tau|}{\tau_R}, \quad \text{when} \quad \tau \leq \Delta,$$

$$K(\tau) = 0, \quad \text{when} \quad \tau > \Delta,$$

where b^2 is the root-mean-square amplitude of the pulses. The spectral density of the process according to (3.17) with $\omega_0^2\tau_R^2 \ll 1$ is

$$J(\omega) \sim \frac{\Delta^2}{\tau_R},$$

that in principle is not in contradiction with formula (3.105).

Figure 3.5a presents data on the rates of the spin–lattice relaxation nuclei ^{57}Fe in the complexes [Fe(CO)$_5$] in deuterated benzene. The measurements were performed at different magnetic fields (their magnitude is indicated in the figure). In

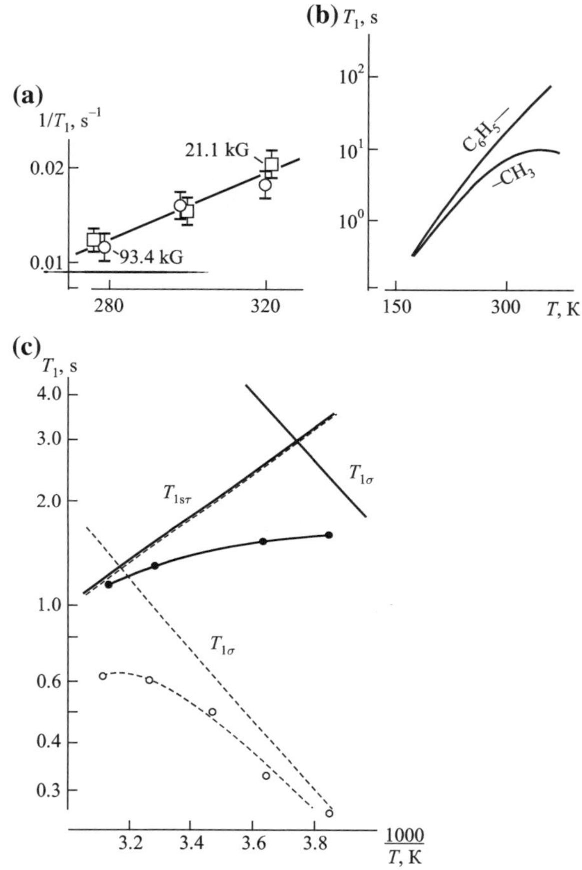

Fig. 3.5 Experimental examples of relaxation due to fluctuations of magnetic fields. See the explanations of the plots **a**, **b**, and **c** in the text. Reproduced with permission from J. Magn. Reson., 74, 433 (1987). Copyright 1987, Academic Press

the experiment the relaxation rate is completely determined by the spin-rotational interaction.

It is interesting that the spin-rotational mechanism can be significant not only in the case of a rotating molecule as a whole but also in case of the rotation of separate molecular groups (often CH_3, which exhibit high mobility—random rotation around the symmetry axis—even in solids up to the temperature of liquid nitrogen). For example, it was discovered the contribution of the spin-rotation interaction in the methyl-proton relaxation of toluene (see Fig. 3.5b). Protons belonging to the benzene ring do not exhibit this behavior, because the motion of the massive group C_6H_5 is much slower than one of CH_3. The description the intramolecular spin-rotational relaxation is very complex, and the special difficulties arise due to the absence of theoretical explanation temperature dependence of the correlation time for internal motion of the rotator in a molecule.

The spin-rotational interaction is the dominant relaxation mechanism in gases, where the rotational degrees of freedom of molecules and their fragments are mostly manifested. The modulation of the field $\vec{B}_{sr}$ occurs as a result of random collisions of molecules and the correlation time is inversely proportional to the density of the gas. On the basis of the study of the dependences of T_{1sr} and T_{2sr} on gas temperature and pressure it is possible, for example, to determine the spin-rotational interaction constant, the cross sections of collisions of molecules in gas.

It is worth noting in conclusion, that the spin-rotational relaxation mechanism must be the most important for the nuclei of heavy atoms with rich electron shells (see 3.103). For such atoms the big values of the constants of electron shielding are typical. Therefore it is not surprising that there is a relationship between the constants of spin-rotational interaction and electronic shielding. Naturally, the anisotropy of electronic shielding is stronger for heavy atoms. So the spin-rotational relaxation mechanism is often present simultaneously with the relaxation due to anisotropy electron screening, for example, that observed for nuclei ^{207}Pb, ^{13}C, ^{77}Se. In these cases, the separation of the contributions to the total relaxation rate is based on the dependence one of the mechanisms on the magnetic field B_0 (see 3.100). An example of the separation contributions to the relaxation of the nuclei ^{207}Pb for solutions of compounds $PbMe_3Cl$ is shown in Fig. 3.5c. Solid lines correspond to experiments in the magnetic field of 2.35 kG, dotted ones—of 7.05 kG, points are the experimental data.

3.8 Magnetic Dipole–Dipole Relaxation

One of the universal types of the interaction nuclei with the environment is the interaction of their magnetic moments through space: *direct dipole–dipole interaction*, which exists in the vast majority of experiments, associated with the phenomenon of the nuclear magnetic resonance. This interaction determines many features of NMR spectra and contributes to relaxation processes. Indeed, because the fluctuating magnetic fields cause the relaxation, the interaction of the magnetic moments (dipoles) should also lead to similar effects, since dipoles interact through magnetic fields, which they create in the space. The magnetic field of a dipole at the distance r is

$$B_{loc} = \frac{\mu}{r^3} f(\theta),$$

where $f(\theta)$ is the function depending on the orientation of the dipole. In the presence of thermal motion the values r and θ become random functions of time that causes fluctuations of the field B_{loc}. If the spectrum of the fluctuating B_{loc} contains components at the NMR frequency, then this field will cause the relaxation transitions when the magnetic energy of spins is exchanged with the thermal energy of molecular motion. This is, in rough features, a general mechanism of *dipole–dipole relaxation*.

The question about the possible energy states of two interacting nuclei in a static magnetic field $\vec{B}_0$ will narrowly be discussed in Sect. 5.1. Even in the simplest case of the interaction between two nuclei with spin $I_1 = I_2 = 1/2$ the scheme of energy levels is much more complex than considered in Sect. 3.6 the two-level system. For two nuclei, interacting through the dipolar mechanism, the Hamiltonian can be written as (see Sect. 1.10)

$$\widehat{\mathcal{H}}_{\mathrm{d}} = \gamma_1\gamma_2\hbar^2 r^{-3}[A + B + C + D + E + F], \tag{3.106}$$

where

$$A = \hat{I}_{1z}\hat{I}_{2z}(1 - 3\cos^2\theta),$$

$$B = -\frac{1}{4}\left[\hat{I}_{1+}\hat{I}_{2-} + \hat{I}_{1-}\hat{I}_{2+}\right](1 - 3\cos^2\theta),$$

$$C = D^* = -\frac{3}{2}\left[\hat{I}_{1z}\hat{I}_{2+} + \hat{I}_{2z}\hat{I}_{1+}\right]\sin\theta\cos\theta e^{-\mathrm{i}\phi},$$

$$E = F^* = -\frac{3}{4}\hat{I}_{1+}\hat{I}_{2+}\sin^2\theta e^{-2\mathrm{i}\phi}.$$

To clarify the role of different terms in the Hamiltonian of dipole–dipole interaction (3.106), we consider a simple case: $I_1 = I_2 = 1/2$. It appears that it is necessary to distinguish the situations when $\gamma_1 \neq \gamma_2$(different spins) and $\gamma_1 = \gamma_2$ (equivalent spins).

3.9 Relaxation due to Dipole–Dipole Interaction of Two Equivalent Spins

The theoretical description of relaxation processes due to the dipole–dipole interaction is of the most simplicity for the case of two nuclei of the same species with spin $I = 1/2$. However, even this simple case will be a much more complicated than described in Sects. 3.6 and 3.7 due to the fact that now the system of energy states includes three levels according to the Hamiltonian $\widehat{\mathcal{H}}_0$:

$$\widehat{\mathcal{H}}_0 = -\gamma_1\hbar\hat{I}_{1z}B_0 - \gamma_2\hbar\hat{I}_{2z}B_0$$

with $\gamma_1 = \gamma_2 = \gamma$.

Construct the kinetic equations for the populations of the levels (indexes correspond to the projection of total spin with respect to vector field $\vec{B}_0$: $+1,\ 0,\ -1$):

$$\frac{\mathrm{d}n_+}{\mathrm{d}t} = -n_+(w_{+0} + w_{+-}) + n_0 w_{0+} + n_- w_{-+},$$
$$\frac{\mathrm{d}n_0}{\mathrm{d}t} = -n_0(w_{0+} + w_{0-}) + n_+ w_{+0} + n_- w_{-0}, \tag{3.107}$$

$$\frac{\mathrm{d}n_-}{\mathrm{d}t} = -n_-(w_{-0} + w_{-+}) + n_0 w_{0-} + n_+ w_{+-}.$$

The longitudinal magnetization, which is proportional to an assembly average of spin $\langle I_z \rangle$, is determined by the difference of level populations:

$$M_z \sim \langle I_z \rangle \sim \left[(n_+ - n_0) + (n_0 - n_-)\right] = n_+ - n_-.$$

Therefore, it is necessary to obtain an equation for the quantity $(n_+ - n_-)$. By introducing the average probabilities w_1 and w_2 for single quantum (at frequency ω_0) and two-quantum (at a frequency of $2\omega_0$) relaxation transitions, respectively, one can obtain from Eq. (3.107) under the assumption that n_0 does not considerably differ from $(n_+ + n_-)/2$:

$$\frac{\mathrm{d}}{\mathrm{d}t}(n_+ - n_-) = -(w_1 + 2w_2)(n_+ - n_-) + C. \tag{3.108}$$

The origin of the constant C on the right side of this expression is the same as the right term in Eq. (3.24), and it is easy to obtain its value, based on the conditions of equilibrium when the derivative with respect to time is equal to zero:

$$C = (w_1 + w_2)(n_+^0 - n_-^0)$$

where n_+^0 and n_-^0 are the equilibrium populations of the levels.

It follows from the solution of Eq. (3.108) that in the case of the interaction of two identical spins the equilibrium of the longitudinal magnetization is reached in accordance with an exponential law with time constant

$$T_1 = \frac{1}{w_1 + 2w_2}, \tag{3.109}$$

that is a particular case of (3.5) and an analogue of (3.26) for a two-level system. Generally speaking, the assumption that $n_0 \simeq (n_+ + n_-)/2$ corresponds to the adoption of the postulate about the spin temperature (see Sects. 3.2 and 3.3). Therefore it is not surprising that result (3.109) can be obtained from (3.48).

The expression (3.109) is worth to be commented in the connection with the following circumstances. One of the excellent works in the area of the theory of the nuclear magnetic relaxation was published by I. Solomon in 1955 (Phys. Rev., 1955, vol. 99, p. 559). The article was devoted to the description of the NMR-relaxation in a system of two interacting spins and, in particular, so-called Solomon equations were derived namely in it. There is a logical error in this article and it is expedient to turn the attention of the NMR community to this fact because that error is still rewritten in many articles, reviews and books (moreover, many Internet materials including "Encyclopedia of NMR" contain it).

Let us consider the problem in details. First, I. Solomon wrote equations for the populations n_i of energy levels of two spins ($I = S = 1/2$) of different kinds

($\gamma_1 \neq \gamma_2$, placed in a static magnetic field. In this case there are four energy levels (see more details in Sects. 1.10 and 3.10) which correspond to the following orientations of the spins ("+" along the magnetic field, "−" in the opposite direction to the magnetic field): $++\,(n_1)$, $+-\,(n_2)$, $-+\,(n_3)$, $--\,(n_4)$. The magnetization of the spin systems is proportional to $\langle I_z \rangle$ and $\langle S_z \rangle$. For this case all results of the Solomon's approach is quite correct. But then I. Solomon proceeded to the consideration of the relaxation in a system of two interacting equivalent spins ($\gamma_1 = \gamma_2$) and made an error. He supposed, that in this case one could put $\langle I_z \rangle = \langle S_z \rangle$, and obtained the spin–lattice relaxation time (T_1):

$$T_1 = \frac{1}{2w_1 + 2w_2}.$$

This approach is not correct because it does not take into account the fact that, although any interaction eliminates the degeneration of the middle energy level, one of the new states is forbidden for the transitions from upper and lower energy levels. If we denote the wave functions for the state $+-$ as ψ_1 and for the state $-+$ as ψ_2 (undisturbed situation for the equivalent spins) the new wave functions are

$$\psi' = \frac{1}{\sqrt{2}}(\psi_1 + \psi_2);$$

$$\psi'' = \frac{1}{\sqrt{2}}(\psi_1 - \psi_2).$$

The symmetry of the first function is the same as for the upper ($--$) and lower ($++$) energy levels but the second function has another symmetry and the nuclei which got into that state cannot participate in the NMR processes including the relaxation, i.e. in the case of two equivalent spins one deals with the three-level system as it has been considered above. It is worth noting that the probabilities w_1 in both expressions are different because they should be calculated on the basis of different wave functions (incorrect or proper ones for the middle level) but w_2 is the same. The "interference" of two incorrectnesses led to the situation when the final Solomon's expression for the relaxation rate $1/T_1$ is correct! This is the reason why the error did not attract the attention of investigators during long time.

To calculate the probability of relaxation transitions of w_1 and w_2 in the expression (3.109) it is necessary to find the mean square of the matrix elements (see 3.16). Let us use the Hamiltonian of the dipole–dipole interaction in form (3.106). The calculations are more simple, if we assume that the distance between the nuclei is constant. In this case the result reflects the intramolecular contribution to the dipole–dipole relaxation (for example, the interaction of hydrogen nuclei in molecules H_2, H_2O, in isolated groups such as NH_2 and so on). Substituting (3.106) in Eq. (3.16) for the transition probabilities, we find that the terms C and D determine the probability w_1, the terms E and F dertermine the probability of w_2 and the terms A and B do not contribute to the relaxation transitions (for the static case namely these terms determine the spectrum of a system of two identical nuclei). Indeed, to obtain the probabilities w_1 and w_2 it is necessary to calculate the matrix elements between different states. Then we obtain for the term A (see the comments about the wave functions after the formula 3.11):

$$\psi_{1+}^*\psi_{2+}^*\hat{I}_{1z}\hat{I}_{2z}\begin{Bmatrix}\psi_{1-}\psi_{2-}\\ \frac{1}{\sqrt{2}}(\psi_{1+}\psi_{2-}+\psi_{1-}\psi_{2+})\end{Bmatrix}\mathrm{d}v=\begin{Bmatrix}0\\0\end{Bmatrix}, \tag{3.110}$$

for the term B

$$\psi_{1+}^*\psi_{2+}^*\left(\widehat{I}_{1+}\widehat{I}_{2-}+\widehat{I}_{1-}\widehat{I}_{2+}\right)\begin{Bmatrix}\psi_{1-}\psi_{2-}\\ \frac{1}{\sqrt{2}}(\psi_{1+}\psi_{2-}+\psi_{1-}\psi_{2+})\end{Bmatrix}\mathrm{d}v=\begin{Bmatrix}0\\0\end{Bmatrix}, \tag{3.111}$$

for the term C

$$\psi_{1+}^*\psi_{2+}^*\left(\widehat{I}_{1+}\hat{I}_{2z}+\hat{I}_{1z}\widehat{I}_{2+}\right)\begin{Bmatrix}\psi_{1-}\psi_{2-}\\ \frac{1}{\sqrt{2}}(\psi_{1+}\psi_{2-}+\psi_{1-}\psi_{2+})\end{Bmatrix}\mathrm{d}v=\begin{Bmatrix}0\\ \frac{1}{\sqrt{2}}\end{Bmatrix}, \tag{3.112}$$

for the term E

$$\psi_{1+}^*\psi_{2+}^*\widehat{I}_{1+}\widehat{I}_{2+}\begin{Bmatrix}\psi_{1-}\psi_{2-}\\ \frac{1}{\sqrt{2}}(\psi_{1+}\psi_{2-}+\psi_{1-}\psi_{2+})\end{Bmatrix}\mathrm{d}v=\begin{Bmatrix}1\\0\end{Bmatrix}. \tag{3.113}$$

The calculations for the terms D and F are similar (they describe the opposite transitions with respect to (3.112) and (3.113)).

If the molecular motion is isotropic, then the assumption (3.3) is valid and it is necessary to average the mean squares of matrix elements over the corners θ and ϕ, contained in spherical functions

$$\begin{aligned}Y_0&=1-3\cos^2\theta,\\ Y_{\pm1}&=\sin\theta\cos\theta e^{\pm i\phi},\\ Y_{\pm2}&=\sin^2\theta e^{\pm 2i\phi}.\end{aligned} \tag{3.114}$$

Perform this averaging for the function $\overline{|Y_0|^2}$:

$$\frac{1}{4\pi}\int_0^{2\pi}\int_0^{\pi}(1-3\cos^2\theta)^2\sin\theta\mathrm{d}\theta\mathrm{d}\phi=\int_0^1(1-6x^2+9x^4)\mathrm{d}x=\frac{4}{5}.$$

The calculations for the other two functions can be made using tabulated integrals with the result:

$$\overline{|Y_{\pm1}|^2} = \frac{2}{15}, \quad \overline{|Y_{\pm2}|^2} = \frac{8}{15}. \tag{3.115}$$

Taking into account the results of the calculations (3.110–3.115) and all the coefficients included in the (3.16) and (3.106), as well as the explicit form spectral density (3.81), we finally obtain the formula (3.109) for the intramolecular contribution to the relaxation rate ($r = \text{const}$):

$$\left(\frac{1}{T_{1\text{dd}}}\right)_{\text{intra}} = \frac{3}{10}\gamma^4\hbar^2 r^{-6}\left[\frac{\tau_\text{c}}{1+\omega_0^2\tau_\text{c}^2} + \frac{4\tau_\text{c}}{1+4\omega_0^2\tau_\text{c}^2}\right]. \tag{3.116}$$

We give without the derivation (which is roughly the same as presented above) the formula for the relaxation rate of nuclei with an arbitrary spin:

$$\left(\frac{1}{T_{1\text{dd}}^I}\right)_{\text{intra}} = \frac{2}{5}\gamma^4\hbar^2 I(I+1) r^{-6}\left[\frac{\tau_\text{c}}{1+\omega_0^2\tau_\text{c}^2} + \frac{4\tau_\text{c}}{1+4\omega_0^2\tau_\text{c}^2}\right]. \tag{3.117}$$

Equation (3.117) differs from (3.116) in accordance with the factor $I(I+1)$, the appearance of which is quite natural because the local magnetic field is proportional to a magnetic moment and the formula for the relaxation rate contains the square of the local magnetic field.

The dependence of $(1/T_1)_{\text{intra}}$ on the correlation time in the formulas (3.116) and (3.117) has roughly the same form as it is shown in Fig. 3.3, but the extremum of the dependence somewhat shifted (with $\omega_0\tau_\text{c} = 0.62$, but not with $\omega_0\tau_\text{c} = 1$, that was realized for the formula (3.83)).

In this section the simplest case of the interaction of two nuclei is discussed. The calculations for many interacting nuclei even with minimal spin ($I = 1/2$) is very complicated. However, investigations show that the satisfactory description of the relaxation in multispin systems can be performed using Eqs. (3.116) and (3.117) with the summation of pair interactions. Figure 3.6 shows the experimental temperature dependence of the proton relaxation in a mixture of CH_3OD–deuterated dimethyl sulfoxide at different resonance frequencies. The temperature dependence of $1/T_1$ is determined by the temperature dependence of τ_c.

To calculate the spin–spin relaxation time $1/T_2$ we need to construct new wave functions (like in Sect. 3.6), which will describe the state of the transverse component of the total spin. As a result, using the formula (3.88), we arrive at the expression

$$\left(\frac{1}{T_{2\text{dd}}}\right)_{\text{intra}} = \frac{3}{20}\gamma^4\hbar^2 r^{-6}\left[3\tau_\text{c} + \frac{5\tau_\text{c}}{1+\omega_0^2\tau_\text{c}^2} + \frac{2\tau_\text{c}}{1+4\omega_0^2\tau_\text{c}^2}\right]. \tag{3.118}$$

Under the condition of extreme narrowing of the nuclear magnetic resonance, i.e. with $\omega_0^2\tau_\text{c}^2 \ll 1$, we obtain together (3.116):

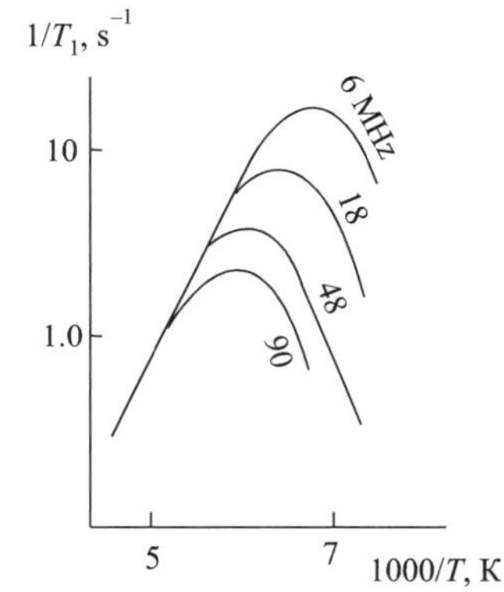

Fig. 3.6 An example of temperature dependence of proton relaxation. Reproduced with permission from J. Magn. Reson. 45, 102 (1981). Copyright 1981, Academic Press

$$\left(\frac{1}{T_{1\text{dd}}}\right)_{\text{intra}} = \left(\frac{1}{T_{2\text{dd}}}\right)_{\text{intra}} = \frac{3}{2}\gamma^4\hbar^2 r^{-6}\tau_c. \quad (3.119)$$

A significant growth in the number of works on the theoretical study of the processes of NMR-relaxation was observed in the mid-seventies of last century in connection with the development of Fourier-spectroscopy of the nuclear magnetic resonance, in particular the behaviour of multiplets in the high resolution spectra was widely discussed. These investigations are especially actual for the description the relaxation of ^{13}C nuclei. The calculations are based on the theory of Bloch-Redfield using the apparatus of the density matrix under the assumption that the dominant relaxation mechanism is the intramolecular dipole–dipole interaction.

The calculation of the intermolecular contribution to the dipole–dipole relaxation in the general case involves great mathematical difficulties. First the question of the description of intermolecular relaxation was made in the work of H.C. Torrey (Phys. Rev., 1953, vol. 92, p. 962). H.C. Torrey gave a precise mathematical formulation of the problem and proposed a method of its solution, leading to the results of the analytical or quasi-analytical form, besides a sufficiently correct model of molecular motion was used.

Low-frequency asymptotics (low frequencies or fast motion) regardless of the model of molecular motion is described by

$$1/T_{1\text{inter}} \simeq a - b\omega^{1/2},$$

and the type of high-frequency asymptotics depends on the model chosen.

Let us discuss the contribution of the intermolecular dipole–dipole interaction to the relaxation in the simplest case, when $\omega_0^2\tau_{\text{tr}}^2 \ll 1$ (τ_{tr} is the correlation time for the translational diffusion), for which there are a number of analytical expressions. We assume that there is a single nucleus in the center of a spherical molecule. Consider the interaction of the chosen nucleus with all nuclei in a spherical layer of thickness dr and of radius r. As a first approximation, we sum the pairwise interaction. If $\omega_0^2\tau_{\text{tr}}^2 \ll 1$, the contribution to the relaxation of each pairwise interaction according to (3.119) will be presented in the form

$$\frac{3\gamma^4\hbar^2}{2r^6}\tau_{\text{tr}},$$

where τ_{tr} is equal to the interval of time required for molecules to diffuse at a distance r (only in this case the interaction between two nuclei is changed to a significant amount). Taking into account the motion of both molecules one can assume $\tau_{\text{tr}} = r^2/(12D)$, where D is the diffusion coefficient. The final result is obtained by the summation of all interactions:

$$\left(\frac{1}{T_{1\text{dd}}}\right)_{\text{inter}} = \left(\frac{1}{T_{2\text{dd}}}\right)_{\text{inter}} = N_0 \int\limits_{2a}^{\infty} \frac{3\gamma^4\hbar^2}{2r^6}\frac{r^2}{12D}4\pi r^2 \text{d}r = \frac{\pi N_0\gamma^4\hbar^2}{4aD}, \quad (3.120)$$

where N_0 is the number of nuclei per unit volume; a is the radius of a molecule.

The more strict approach to the calculation of the intermolecular contribution is based on the solution of the equation describing the diffusion process:

$$\frac{\partial\Psi(\vec{r},t)}{\partial t} = D\Delta\Psi(\vec{r},t),$$

from which one can obtain the probability to find a molecule after a time t at a distance r, if at $t = 0$ it was at the point $\vec{r} = \vec{r}_0$:

$$\Psi(\vec{r},\vec{r}_0,t) = (4\pi Dt)^{-3/2}\exp\left[-\frac{(\vec{r}-\vec{r}_0)^2}{4Dt}\right] \quad (3.121)$$

with initial conditions $\Psi(\vec{r},0) = \delta(\vec{r}-\vec{r}_0)$. If two molecules (in our case, each containing one nucleus with spin $I = 1/2$) move in space independently, then from (3.121) we obtain the probability of their mutual disposition:

$$P(\vec{r},\vec{r}_0,t) = (8\pi Dt)^{-3/2}\exp\left[-\frac{(\vec{r}-\vec{r}_0)^2}{8Dt}\right].$$

The correlation function for any of the functions (3.114) is

$$K_m(\tau) = \alpha_m(8\pi D\tau)^{-3/2}N_0\int\int\frac{Y_m^*}{r_0^3}\frac{Y_m}{r^3}\exp\left[-\frac{|\vec{r}-\vec{r}_0|^2}{8D\tau}\right]\text{d}\vec{r}_0\text{d}\vec{r}, \quad (3.122)$$

here $\alpha_0 = 48\pi/15$; $\alpha_1 = 8\pi/15$; $\alpha_2 = 32\pi/15$. When integrating (3.122) it is necessary to take into account that r and r_0 can not be less than a limiting value of d, corresponding to the closest distance between molecules. If molecules are modeled by spheres of radius a, then $d = 2a$.

Finally, for the intermolecular contribution to the relaxation one can obtain (N_0 is the number of spins in 1 cm^3):

$$\left(\frac{1}{T_{1\mathrm{dd}}}\right)_{\mathrm{inter}} = \left(\frac{1}{T_{2\mathrm{dd}}}\right)_{\mathrm{inter}} = \frac{\pi N_0 \gamma^4 \hbar^2}{5aD} = \frac{6\pi^2 \gamma^4 \hbar^2 N_0 \eta}{5kT}. \tag{3.123}$$

The comparison of (3.120) and (3.123) shows that they differ only slightly (only the numerical coefficients).

If we accept the independence of the processes of intra- and intermolecular relaxation, one can obtain according to (3.6) and (3.8):

$$\frac{1}{T_{1,2}} = \left(\frac{1}{T_{1,2}}\right)_{\mathrm{intra}} + \left(\frac{1}{T_{1,2}}\right)_{\mathrm{inter}}. \tag{3.124}$$

Experiments show that the formulas (3.116–3.119), (3.123) and (3.124) give the correct functional dependence on the basic parameters (resonance frequency, temperature, distance between spins).

The comparison of the absolute values of T_1 and T_2 derived from the theory and obtained in experiment is expedient only to compare them in order of magnitude due to some uncertainties in the knowledge of parameters (for example, the distance between spins in formulas (3.116–3.119) is in the sixth degree) and discussed theory approximation. As an example, let us estimate the relaxation time of ^{1}H nuclei (protons) in liquid water, for which the condition of extreme narrowing of the NMR line ($\omega_0^2 \tau_R^2 \ll 1, \omega_0^2 \tau_{\mathrm{tr}}^2 \ll 1$) is fulfiled with a large overstock. It is possible to use the well-known information from other physical methods: $r_{\mathrm{H-H}} = 1.58\,\text{Å}$, $a = 1.74\,\text{Å}$, $N_0 = 6.75 \times 10^{22}$, $D = 2 \times 10^{-5}\,\mathrm{cm^2/s}$, $\tau \simeq 3 \times 10^{-12}\,\mathrm{s}$ (at 20 °C). As a result one can find: $T_{1\mathrm{intra}} \simeq 7\,\mathrm{s}$ and $T_{1\mathrm{inter}} \simeq 12\,\mathrm{s}$ that yields $T_1 \simeq 4.5\,\mathrm{s}$. This value is in good agreement with the experimental one: $T_1 = 3.5\,\mathrm{s}$.

As already mentioned, a difficult point in the of the theory of NMR relaxation is the description of the character of molecular motion, which modulates the interaction of nuclei with the environment. In particular, it is interesting to find out how the relaxation process proceeds in the presence of a few correlation times (which can arise due to the anisotropic motion of molecules). One of the first attempts in this direction was the work of D.W. McCall, D.C. Douglas and E.W. Anderson (J. Chem. Phys., 1959. vol. 25, p. 1272), who suggested to take the simple distribution of correlation times for calculations:

$$\begin{aligned} I(\tau_c) &= \frac{1}{\tau_c} \ln \frac{b}{a} \quad \text{if} \quad b \geq \tau_c \geq a; \\ I(\tau_c) &= 0 \quad \text{if } b < \tau_c,\ a > \tau_c. \end{aligned}$$

Such a view of the distribution of τ_c represents the most significant features of the physical picture and does not bring great mathematical difficulties. The dependences of T_1 and T_2 from the average correlation time ($\overline{\tau}_c$) are shown in Fig. 3.7. The curve 1 for $a/b = 1$ corresponds to the case of a single correlation time, as discussed here and

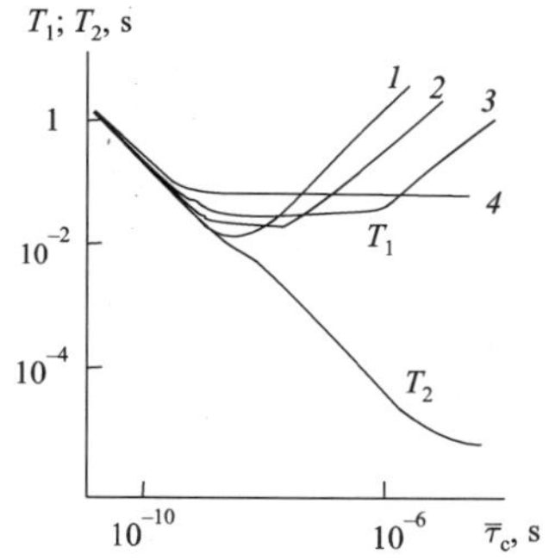

Fig. 3.7 The influence of the distribution of correlation times on the behavior of relaxation times. Reproduced with permission from J. Chem. Phys., 30, 1272 (1959). Copyright 1959, American Institute of Physics

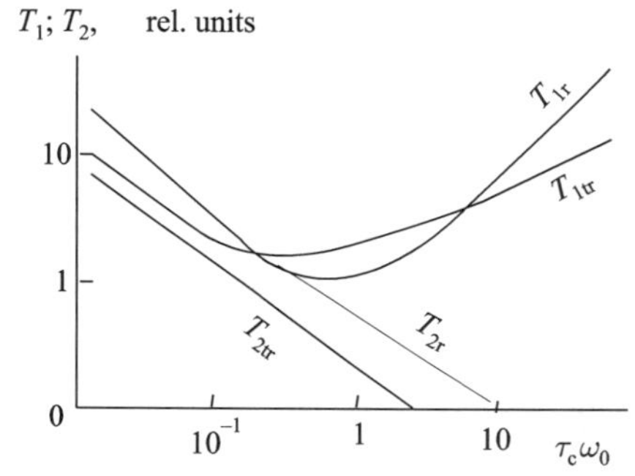

Fig. 3.8 The comparison of the behavior of the intra- and intermolecular interactions

in Sect. 3.6; the remaining curves correspond to the $a/b = 10^2$ (2), $a/b = 10^4$ (3), $a/b = 10^8$ (4).

The relaxation times were calculated also for the case of the ratio of two discrete correlation times: $\tau_{c1} = n\tau_{c2}$. The character of dependencies T_1 and T_2 was similar to that shown in Fig. 3.7: T_2 slightly depends on the difference in the correlation times, T_1 changes so that with an enhancement of the range of variation of τ_c the ratio of T_1/T_2 increases at the minimum of T_1 and this minimum becomes flatter. The case $\omega_0^2\tau_c^2 \ll 1$ is not very different from those when there is the single correlation time (the left part of the curves in Fig. 3.7).

Generally speaking, the problem of the influence of the distribution of correlation times is important and for the description of the intermolecular (translational) contribution to the dipole–dipole relaxation. Mind that the formulas (3.120) and (3.123) were derived for the special case of rapid molecular motion, when the $\omega_0^2\tau_{tr}^2 \ll 1$. If we consider the interaction of the spins, separated by large distances, this ratio for them may be violated: the fluctuations of local magnetic fields will be essential if spins shift in space relative to each other at a distance comparable to their initial separation, that implies the dependence of τ_{tr} on a distance, i.e. $\tau_{tr} = \langle r^2\rangle/(12D)$. Due to this circumstance the behavior of $(1/T_1)_{inter}$ becomes somewhat similar to that shown in Fig. 3.7. Results of numerical calculations are presented in a graphical form in Fig. 3.8. Obviously, the behavior of plots for the intermolecular contribution to the relaxation (T_{1tr} and T_{2tr}) compared to the intramolecular one (T_{1r} and T_{2r}) is in accordance with Figs. 3.3 and 3.7.

3.10 Relaxation due to Dipole–Dipole Interaction of Two Non-equivalent Spins

Let us consider again the simplest case: the interaction of two nuclei with spins equal to 1/2 but with different gyromagnetic ratios. For the convenience put in the Hamiltonian (3.106) $I_1 = I$ and $I_2 = S$. It follows from the Hamiltonian (3.2) that in this case the two-spin system in a constant magnetic field has four energy states (four levels of energy). Construct the kinetic equations describing the "motion" of nuclei between these states:

$$\begin{aligned}
\frac{\mathrm{d}n_1}{\mathrm{d}t} &= -n_1(w_{12}+w_{13}+w_{14})+n_2w_{21}+n_3w_{31}+n_4w_{41},\\
\frac{\mathrm{d}n_2}{\mathrm{d}t} &= n_1w_{12}-n_2(w_{21}+w_{23}+w_{24})+n_3w_{32}+n_4w_{42},\\
\frac{\mathrm{d}n_3}{\mathrm{d}t} &= n_1w_{13}+n_2w_{23}-n_3(w_{31}+w_{32}+w_{34})+n_4w_{43},\\
\frac{dn_4}{\mathrm{d}t} &= n_1w_{14}+n_2w_{24}+n_3w_{34}-n_4(w_{41}+w_{42}+w_{43}).
\end{aligned} \tag{3.125}$$

With the help of (3.125) one can obtain an equation for the longitudinal component of the magnetization of the spin $\langle I_z\rangle$, which is proportional to the corresponding population difference:

$$\langle I_z\rangle \sim [(n_1+n_2)-(n_3+n_4)]. \tag{3.126}$$

Differentiating (3.126) over time, expressing the derivatives of the number of nuclei at levels from (3.125) and replacing, as before, the probability of upward and downward transitions through the mean values (w_0—for transitions at the frequency $\omega_{0I} - \omega_{0S}$; w_1—for transitions at the frequency ω_{0I}; w_1'—for transitions at the frequency ω_{0S}; w_2—for transitions at the frequency $\omega_{0I}+\omega_{0S}$), we obtain the equation

$$\begin{aligned}
&\frac{\mathrm{d}}{\mathrm{d}t}[(n_1+n_2)-(n_3+n_4)]\\
&\quad= -2w_1[(n_1+n_2)-(n_3+n_4)]-2w_2(n_1-n_4)-2w_0(n_2-n_3)+C.
\end{aligned} \tag{3.127}$$

The constant C in (3.127) has the same origin as the in Eqs. (3.24) and (3.108). Adding to the second term in (3.127) $\pm w_2(n_2-n_3)$ and to the third $\pm w_0(n_1-n_4)$ one can obtain the required equation for $\langle I_z\rangle$:

$$\frac{\mathrm{d}\langle I_z\rangle}{\mathrm{d}t} = -\rho_1(\langle I_z\rangle - I_0) - \sigma_1(\langle S_z\rangle - S_0), \tag{3.128}$$

where I_0 and S_0 are the equilibrium values of $\langle I_z\rangle$ and $\langle S_z\rangle$, respectively, with

$$\langle S_z \rangle \sim [(n_1 + n_3) - (n_2 + n_4)],$$

and the constant coefficients in (3.128) are given by

$$\rho_1 = w_0 + 2w_1 + w_2; \quad \sigma_1 = w_2 - w_0. \tag{3.129}$$

The equation analogous to (3.128) can be obtained for $\langle S_z \rangle$ by the symmetrical replacing of the symbols with $\rho_1' = w_0 + 2w_1' + w_2$. The process of the calculation of probabilities w_0, w_1, w_1' and w_2 is absolutely the same as in the case of identical spins. However, in contrast to the formulas (3.110–3.113) in this case ($I_1 = I_2 = 1/2$, $\gamma_1 \neq \gamma_2$) the term B Hamiltonian $\widehat{\mathcal{H}}_\mathrm{d}$ (3.106), which describes the mutual reorientation of spins I and S, yields the contribution to the relaxation. Mind that the spectrum of two interacting nuclei of different types is determined only by the term A of the Hamiltonian $\widehat{\mathcal{H}}_\mathrm{d}$.

It follows from Eq. (3.128) that the process of establishing the longitudinal magnetization is described by the function, which is the sum of two exponents, in contrast to all previously considered cases where the relaxation function was presented by a single exponent. However, note that under the condition of rapid relaxation of, for example, spins S the establishment of the magnetization of spins I will be in accordance with a simple exponential law because in the scale of the relaxation time of spins I it can be assumed $\langle S_z \rangle \simeq S_0$ (see 3.128). This situation is usually realized if spins S are paramagnetic particles or quadrupole nuclei (see Sect. 3.12). Besides if we consider the interaction of nuclear spins I and S, the time constants

$$T_1^{II} = \frac{1}{\rho_1}, \quad T_1^{IS} = \frac{1}{\sigma_1} \tag{3.130}$$

are of the same order and the experimental relaxation function is difficult to distinguish from a single exponent in conditions of real signal to noise ratios.

The rate of the spin–lattice relaxation for the case of the interaction of spins of different types for the intramolecular contribution is given by

$$\left(\frac{1}{T_1^{II}}\right)_\mathrm{intra} = \frac{2}{15}\gamma_I^2\gamma_S^2\hbar^2 S(S+1)r^{-6} \times \left[\frac{\tau_\mathrm{c}}{1+(\omega_{0I}-\omega_{0S})^2\tau_\mathrm{c}^2} + \frac{3\tau_\mathrm{c}}{1+\omega_{0I}^2\tau_\mathrm{c}^2} + \frac{6\tau_\mathrm{c}}{1+(\omega_{0I}+\omega_{0S})^2\tau_\mathrm{c}^2}\right]. \tag{3.131}$$

The calculation of the spin–spin relaxation rates is associated with the problems, mentioned in Sects. 3.6 and 3.7. As shown by I. Solomon, the spin–spin relaxation in this case is described by a single exponent:

$$\frac{\mathrm{d}}{\mathrm{d}t}|\langle I_{x,y} \rangle| = -\rho_2 |\langle I_{x,y} \rangle|, \quad \frac{\mathrm{d}}{\mathrm{d}t}|\langle S_{x,y} \rangle| = -\rho_2' |\langle S_{x,y} \rangle|,$$

here ρ_2 and ρ_2' have the same structure as the ρ_1 and ρ_1' from (3.129) by replacing the probabilities w_i with u_i. After calculation the probabilities of u_i with (3.88) one obtains:

$$\left(\frac{1}{T_2}\right)_{\text{intra}} = \frac{1}{15}\gamma_I^2\gamma_S^2\hbar^2 S(S+1)r^{-6}\left[4\tau_c + \frac{\tau_c}{1+(\omega_{0I}-\omega_{0S})^2\tau_c^2} + \frac{3\tau_c}{1+\omega_{0I}^2\tau_c^2} + \frac{6\tau_c}{1+\omega_{0S}^2\tau_c^2} + \frac{6\tau_c}{1+(\omega_{0I}+\omega_{0S})^2\tau_c^2}\right]. \tag{3.132}$$

It is obvious that for $S = 1/2$ and in the case of the condition of extreme narrowing of a nuclear magnetic resonance line ($\omega_0^2\tau_c^2 \ll 1$) we have (compare with 3.119):

$$\left(\frac{1}{T_1}\right)_{\text{intra}} = \left(\frac{1}{T_2}\right)_{\text{intra}} = \gamma_I^2\gamma_S^2\hbar^2 r^{-6}\tau_c. \tag{3.133}$$

Problems of the description of the relaxation in a situation when there is a distribution of correlation times or in the case of the intermolecular interactions are the same as described in Sect. 3.9 for the relaxation of equivalent spins.

3.11 Scalar Relaxation

Scalar interaction (named so due to the form of Hamiltonian) occurs between the nuclei and electrons (as, for example, in hydrogen atom) or in diamagnetic substances between different nuclear spins through electronic environment (then it is called *indirect spin–spin interaction*). The Hamiltonian of the scalar interaction can be written as

$$\widehat{\mathcal{H}}_{\text{sc}} = A\left(\hat{\vec{I}}, \hat{\vec{S}}\right). \tag{3.134}$$

The coupling constant A is proportional to the square of the modulus of electron wave functions at the location of a nucleus. It follows that this interaction exists only in the presence of s-orbits of electrons (if a nucleus and electrons belong to the same atom). In diamagnetic substances magnetic moments of nuclei provide orienting effect on neighboring electrons even when electrons are paired in accordance with the principle of Pauli. This orientation is partially transferred to electrons belonging to other nuclei, and if the latter also have magnetic moments, the interaction of the type (3.134) is realized between the nuclei. Naturally, in the case of the interaction of nuclei with an unpaired electron, the constant A on a lot orders of magnitude larger than for the interaction of nuclei via an electron shell of a diamagnetic molecule.

The presence of a static part in the interaction (3.134) leads to a splitting NMR spectrum (see Chap. 4). To begin the relaxation process, it is necessary to have fluctuations of this interaction. If we consider the relaxation of the spins I, then the change of the scalar interaction can occur either as a result of its interruption, i.e. in some time intervals $A = 0$, or due to the reorientation (relaxation) of the spin S. Naturally, reorientations of the spin I also affects on the scalar product in the Hamiltonian (3.134), but this process is namely the relaxation in the system of spins I.

The disturbance of the interaction can randomly occur with a time constant τ_M due to chemical exchange of nuclei between states with different coupling constants. In this case one speaks about *scalar relaxation of the first type*. For *scalar relaxation of the second type* the correlation time is the relaxation time of spins S: T_S. If both processes are independent, the total correlation time is determined by formula

$$\frac{1}{\tau_c} = \frac{1}{\tau_M} + \frac{1}{T_S}. \tag{3.135}$$

If $\tau_M \gg T_S$ or $\tau_M \ll T_S$ both situations is formally described by identical expressions.

We note that the scalar product is invariant to rotations (it depends only on the mutual orientation of the vectors), so that the scalar interaction (3.134) does not change with the rotational diffusion of molecules or complexes.

Writing (3.134) via the components of the spin I and S and introducing the raising and lowering operators, one can obtain:

$$\widehat{\mathcal{H}}_{sc} = A\left[\hat{I}_z\widehat{S}_z + \frac{1}{2}(\widehat{I}_+\widehat{S}_- + \widehat{I}_-\widehat{S}_+)\right]. \tag{3.136}$$

Obviously, in the case of spin–lattice relaxation, in the calculation of the matrix elements in (3.16) the first term of (3.136) does not give contribution, since the matrix elements between different states are equal to zero. It is also easy to see that the terms in parentheses, "connect" only the transition at the frequency $|\omega_{0I} - \omega_{0S}|$, i.e. in the case of mutual overturnings of the spins (as usual the calculations are performed for the simplest case, when $I = S = 1/2$):

$$\psi_{I+}^*\psi_{S+}^*\hat{I}_z\widehat{S}_z \begin{Bmatrix} \psi_{I-}\psi_{S+} \\ \psi_{I+}\psi_{S-} \\ \psi_{I-}\psi_{S-} \end{Bmatrix} \mathrm{d}v = \begin{Bmatrix} 0 \\ 0 \\ 0 \end{Bmatrix},$$

$$\psi_{I-}^*\psi_{S+}^*\hat{I}_z\widehat{S}_z\psi_{I+}\psi_{S-}\mathrm{d}v = 0,$$

$$\psi_{I+}^*\psi_{S+}^*\left(\widehat{I}_+\widehat{S}_- + \widehat{I}_-\widehat{S}_+\right) \begin{Bmatrix} \psi_{I-}\psi_{S+} \\ \psi_{I+}\psi_{S-} \\ \psi_{I-}\psi_{S-} \end{Bmatrix} \mathrm{d}v = \begin{Bmatrix} 0 \\ 0 \\ 0 \end{Bmatrix},$$

$$\psi_{I-}^*\psi_{S+}^*\left(\widehat{I}_+\widehat{S}_- + \widehat{I}_-\widehat{S}_+\right)\psi_{I+}\psi_{S-}\mathrm{d}v = 1.$$

Thus, for the spin–lattice relaxation in the spin system I one obtains from (3.128) to (3.130) under the condition of the fast relaxation of the spin S:

$$\frac{1}{T_{1\mathrm{sc}}} = w_0,$$

and

$$\frac{1}{T_{1\mathrm{sc}}} = \frac{1}{2}\hbar^{-2}A^2\frac{\tau_\mathrm{c}}{1+(\omega_{0I}-\omega_{0S})^2\tau_\mathrm{c}^2}.$$

Here, as usual, before the corresponding spectral density there is the quantity which is proportional to the square of the energy of interaction causing the relaxation process.

For arbitrary spins I and S calculations give

$$\frac{1}{T_{1\mathrm{sc}}} = \frac{2}{3}S(S+1)\hbar^{-2}A^2\frac{\tau_\mathrm{c}}{1+(\omega_{0I}-\omega_{0S})^2\tau_\mathrm{c}^2}. \tag{3.137}$$

It is known that the indirect spin–spin interaction of equivalent nuclei does not lead to a splitting of NMR spectra. From these calculations we can conclude that, the relaxation effects also disappear in this case since the transition at the frequency $\omega_{0I}-\omega_{0S}$ in the system of equivalent spins is absent.

In the spin–spin relaxation of spins I the contributions are made by the spin–lattice relaxation and the effect nonaveraged z-components of the local field arising from the existence of components of S_z. For an isotropic case one has

$$\frac{1}{T_{2\mathrm{sc}}} = \frac{1}{3}S(S+1)\hbar^{-2}A^2\left[\tau_\mathrm{c}+\frac{\tau_\mathrm{c}}{1+(\omega_{0I}-\omega_{0S})^2\tau_\mathrm{c}^2}\right]. \tag{3.138}$$

Due to the circumstance that in the considered interaction only two levels participate in the relaxation process the structure of (3.137) and (3.138) is similar to one of (3.82) and (3.83). It is also seen from the expressions (3.137) and (3.138) that in the case of the fast motion (small τ_c and $(\omega_{0I}-\omega_{0S})^2\tau_\mathrm{c}^2 \ll 1$) we obtain $T_{1\mathrm{sc}} = T_{2\mathrm{sc}}$, but in the case of the slow motion (big τ_c and $(\omega_{0I}-\omega_{0S})^2\tau_\mathrm{c}^2 \gg 1$) the scalar interaction makes a significant contribution only to the spin–spin relaxation, that leads to a situation when $T_1 > T_2$ (in the presence of other relaxation mechanisms).

3.12 The Quadrupole Relaxation

Up to now we have examined the relaxation mechanisms which are connected with the fluctuations in the magnetic field or, more precisely, with the fluctuations of the magnetic interactions. If nuclei have spin $I > 1/2$, then they can participate in the principally different mechanism of relaxation through the *electric quadrupole interaction* (see Sects. 1.12 and 5.4). The fact is that for these nuclei the distribution of their charge is not spherically symmetric (in contrast to nuclei with spin $I = 0$ or $I = 1/2$), and they are oblate or prolate ellipsoids. The orientation of such—quadrupolar—nuclei in inhomogeneous electric fields is not longer arbitrary. If an inhomogeneous electric is static like as, for example, in solids, liquid crystals or glasses, then the system of energy levels arises and the transitions between them (in the absence of a magnetic field B_0) are known as nuclear quadrupole resonance (NQR). In liquids, the electric field gradient fluctuates over direction and magnitude, the quadrupole splitting of energy levels disappears, but the quadrupole interaction causes relaxation transitions in the system of energy levels, determined by magnetic interactions with $\vec{B}_0$.

Physically the process of magnetic relaxation in this case proceeds as follows. Electric fields via the quadrupole interaction "provoke" random reorientations of the electric nuclear quadrupole, i.e. nucleus is undergone by random rotatations. But it also changes the orientation of the magnetic moment of a nucleus in space and this means transition between different energy states of the magnetic moment of a nucleus in a static magnetic field $\vec{B}_0$. Consequently, *the magnetic relaxation in a spin system can occur as a result of electrical interactions* and its intensity depends on the structure of electric fields at the site of nuclei. The quadrupole relaxation mechanism is one of the most effective for the diamagnetic compounds due to the large value of the quadrupole interaction and it is widely spread because among 116 magnetic isotopes 87 have spin $I > 1/2$.

In the considered case, formula (3.1) can be written as

$$\widehat{\mathcal{H}} = \widehat{\mathcal{H}}_0 + \widehat{\mathcal{H}}_Q(t),$$

where $\widehat{\mathcal{H}}_Q(t)$ is the Hamiltonian of the quadrupole interaction.

In general, the representation of the Hamiltonian of the quadrupole interaction in a convenient form for the further is connected with cumbersome calculations. If electric field gradients are $q_{xx} \neq q_{yy} \neq q_{zz}$, then the coordinate system in which the $q_{\alpha\beta} = 0$ for $\alpha \neq \beta$ (i.e. the orientation of the coordinate axes is determined by the structure of the electric field), the Hamiltonian $\widehat{\mathcal{H}}_Q$ takes the form (see the details in Chaps. 1 and 8):

$$\widehat{\mathcal{H}}_Q = \frac{eQ}{4I(2I-1)}\left[q_{zz}(3\hat{I}_z^{\,2} - \hat{I}^2) + (q_{xx} - q_{yy})\left(\hat{I}_x^{\,2} - \hat{I}_y^{\,2}\right)\right]. \tag{3.139}$$

As it has repeatedly been pointed out, it is expedient instead of the operators $\hat{I}_x$ and $\hat{I}_y$ to use the operators $\hat{I}_+$ and $\hat{I}_-$. Then the expression (3.139) at an arbitrary coordinate system is transformed as follows:

$$\widehat{\mathcal{H}}_Q = \frac{eQ}{4I(2I-1)}\Big[V_0\left(3\hat{I}_z^{\,2} - \hat{I}^2\right) + V_{+1}\left(\hat{I}_-\hat{I}_z + \hat{I}_z\hat{I}_-\right) + V_{-1}\left(\hat{I}_+\hat{I}_z + \hat{I}_z\hat{I}_+\right) + V_{+2}\hat{I}_-^2 + V_{-2}\hat{I}_+^2\Big], \tag{3.140}$$

where $V_0 = q_{zz}$; $V_{\pm 1} = q_{xz} \pm iq_{yz}$; $V_{\pm 2} = (q_{xx} - q_{yy})/2 \pm iq_{xy}$.

In the following we assume that the interaction $\widehat{\mathcal{H}}_Q$ does not contain a static part and causes only relaxation effects in the system of energy levels belonging to the magnetic moment of a nucleus in a magnetic field $\vec{B}_0$, i.e. let us consider a fundamentally new mechanism of relaxation in nuclear magnetic resonance (due to not magnetic but electrical interactions). The energy of the quadrupole interaction is often much greater than the energy of magnetic interactions, discussed earlier. Therefore the quadrupole relaxation mechanism is usually dominant. We would like to emphasize that *fluctuations of electric field gradients cause the reorientations of nuclei via the quadrupole interaction* (3.140) *and, as a result, the reorientations of the magnetic moment of nuclei, i.e. there are transitions between different states of the magnetic moment in a static magnetic field* (*magnetic relaxation*).

In investigations of the microstructure of materials the interesting aspect is the comparison of temperature and concentration dependencies of relaxation rates of different isotopes with $I = 1/2$ and $I > 1/2$.

As before, let us consider in the details the simplest case: $I = 1$. In accordance with the Hamiltonian $\widehat{\mathcal{H}}_0 = -\gamma\hbar\hat{I}_z B_0$ the energy levels are determined by the possible values of the magnetic quantum number $m = +1, 0, -1$. Therefore, the system of kinetic equations for the populations of levels coincides with (3.107). Consequently, the relaxation process is described by a single exponential function with the time constant $T_1 = (w_1 + 2w_2)^{-1}$ (the denotations of the probabilities of relaxation transitions retain the same). To determine the probabilities of w_1 and w_2, one must calculate the matrix elements. Taking into account the effect of the operators $\widehat{I}_\pm$ on wave functions

$$\widehat{I}_\pm \psi_m = [(I \mp m)(I \pm m + 1)]^{1/2} \psi_{m\pm 1},$$

it is possible easily to obtain

$$\begin{aligned}
\langle m|\widehat{\mathcal{H}}_Q|m\rangle &= A\Big[3m^2 - I(I+1)\Big]V_0,\\
\langle m\pm 1|\widehat{\mathcal{H}}_Q|m\rangle &= A(2m\pm 1)[(I\mp m)(I\pm m+1)]^{1/2}V_{\mp 1},\\
\langle m\pm 2|\widehat{\mathcal{H}}_Q|m\rangle &= A[(I\mp m)(I\mp m-1)(I\pm m+1)(I\pm m+2)]^{1/2}V_{\mp 2},\\
\langle m\pm n|\widehat{\mathcal{H}}_Q|m\rangle &= 0 \quad \text{with} \quad n > 2,
\end{aligned} \tag{3.141}$$

with $A = eQ/[4I(2I-1)]$.

If a nucleus enters in a molecule, the electric field gradient $q_{\alpha\beta}$ $(\alpha, \beta = x, y, z)$ defined by structure of the molecular electron shell. Therefore, the calculation of $q_{\alpha\beta}$ is a complex theoretical problem, since for this we need to know accurately the configuration of electronic shell of molecules in the nearest vicinity of the nucleus chosen. Quantum chemical calculations are now attracted for the interpretation of quadrupole relaxation data. However, in some cases, the calculations can be made very simply. For example, a nucleus located in the center of a spherically symmetric atom or ion does not participate in the quadrupole interaction with its own electronic shell. Then it is easy to calculate the electric field gradients, arising from any external charge e_k (Fig. 3.9).

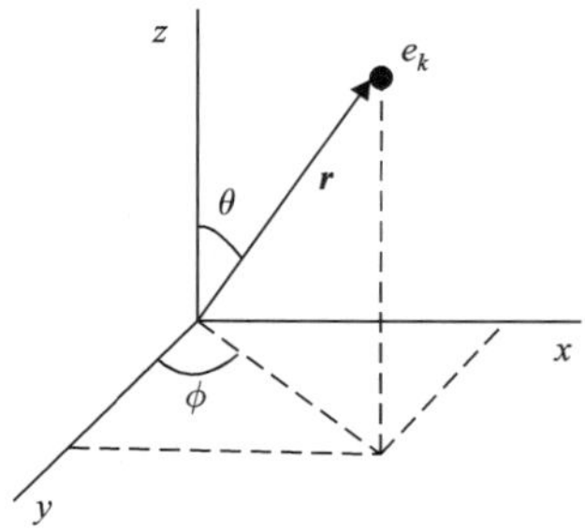

Fig. 3.9 To the question of the influence of a single charge on quadrupole relaxation

The potential of the electric field of a charge e_k is given by

$$V = \frac{e_k}{r} = \frac{e_k}{\sqrt{x^2 + y^2 + z^2}}.$$

Calculating successively by the first and second derivatives with respect to coordinates, we can get

$$\frac{d^2 V}{dx_\alpha^2} = e_k \frac{3x_\alpha^2 - r^2}{r^5}, \tag{3.142}$$

where $x_\alpha = x, y, z$. If $x_\alpha = z$ then

$$V_0 = \frac{e_k}{r^3}(3\cos^2\theta - 1), \tag{3.143}$$

and for $x_\alpha \neq x_\beta$

$$\frac{d^2 V}{dx_\alpha dx_\beta} = e_k \frac{3x_\alpha x_\beta}{r^5}. \tag{3.144}$$

Using (3.142) and (3.144), and taking into account that

$$\cos\phi \pm i\sin\phi = \exp(\pm i\phi),$$
$$(\cos\phi \pm i\sin\phi)^2 = \cos^2\phi \pm 2i\sin\phi\cos\phi - \sin^2\phi = \exp(\pm 2i\phi),$$

we obtain, in addition to (3.143)

$$V_{\pm 1} = \frac{e_k}{r^5}(3xz \pm 3iyz) = \frac{3e_k}{r^3}\sin\theta\cos\theta e^{\pm i\phi},$$
$$V_{\pm 2} = \frac{3}{2}\frac{e_k}{r^5}(y^2 \pm 2ixy - x^2) = \frac{3}{2}\frac{e_k}{r^3}\sin^2\theta e^{\pm 2i\phi}. \tag{3.145}$$

Summarizing the results of the formulas (3.109), (3.141), (3.145) and averaging the spherical functions for isotropic diffusion of the charge e_k around the nucleus at a constant distance r (in the same way as it was done for the intramolecular contribution to the dipole–dipole relaxation), we obtain in the case of spin $I = 1$:

$$\frac{1}{T_{1Q}} = \frac{3}{10}\left(\frac{e_k eQ}{\hbar r^3}\right)^2\left[\frac{\tau_c}{1+\omega_0^2\tau_c^2} + \frac{4\tau_c}{1+4\omega_0^2\tau_c^2}\right]. \tag{3.146}$$

Without going into details of calculation, which in many respects are similar to ones discussed in Sects. 3.7–3.9, we present the formula for the spin–spin relaxation rate:

$$\frac{1}{T_{2Q}} = \frac{3}{20}\left(\frac{e_k eQ}{\hbar r^3}\right)^2\left[3\tau_c + \frac{5\tau_c}{1+\omega_0^2\tau_c^2} + \frac{2\tau_c}{1+4\omega_0^2\tau_c^2}\right]. \tag{3.147}$$

If $\omega_0^2\tau_c^2 \ll 1$, then

$$\left(\frac{1}{T_{1,2}}\right)_Q = \frac{3}{2}\left(\frac{e_k eQ}{\hbar r^3}\right)^2 \tau_c. \tag{3.148}$$

Formulas (3.146–3.148) do not take into account the fact that external charges distort (polarize) a spherical electron shell. As a result the gradients at the nucleus will increase. This effect is called the *antishielding*. So one can write:

$$q_{zz} = q_{zz}^0 + q_{zz}' = q_{zz}^0[1 - \gamma(r)],$$

where q_{zz}^0 is the electric field gradient calculated using the formulas (3.142) and (3.143); q_{zz}' is the additional effect due to the antishielding; $\gamma(r)$ is the antishielding coefficient, which is also called as the *Sternheimer antishielding factor* (named after the scientist, who considered the effect for the first time in 1951–1954).

If the charge e_k is inside of a spherically symmetric shell, then $\gamma(r) \ll 1$. If the charge is outside of an electron shell, then theoretical calculations predict, for example, the following antishielding factors: Al^{3+} – 2.5; Rb^+ – 51; Cs^+ – 99; Cl^- – 48. Since the antishielding factor can not be determined from independent experiments, there are certain difficulties in the interpretation of absolute values of the quadrupole relaxation rates.

For nuclei with spin $I > 1$, the relaxation function is generally not described by a single exponential function (similar to the case of the relaxation due to the dipole–dipole mechanism for a couple of different spins). Indeed, it has been confirmed as theoretically and experimentally, but for spins $I = 3/2$, 5/2, 7/2 the deviations from a single exponent is not big. Therefore, some researchers suggested in similar situations to approximate the relaxation function by a single exponent with a time constant of T_{1ef}.

If the condition of the extreme narrowing of a NMR-line is valid, i.e. $\omega_0^2\tau_c^2 \ll 1$, then the relaxation function is described by a single exponent for all values of the spin I. A. Abraham obtained the corresponding formula for the rates of the quadrupole relaxation (The Principles of Nuclear Magnetism. 1961, Oxford):

$$\frac{1}{T_{1Q}} = \frac{1}{T_{2Q}} = \frac{3}{40} \frac{2I+3}{I^2(2I-1)} \left(1 + \frac{\eta^2}{3}\right) \left(\frac{eq_{zz}Q}{\hbar}\right)^2 \tau_c, \tag{3.149}$$

where $\eta = (q_{xx} - q_{yy})/q_{zz}$ is the so-called *asymmetry parameter*, which characterizes the structure of the inhomogeneity electric field (the deviation from the axial symmetry).

3.13 Relaxation in Conditions of Heterogeneity of a System and Superposition of Different Interactions

If in a substance under investigation the nuclei are in chemically non-equivalent positions, spectra of the nuclear magnetic resonance are split (see Sects. 4.1, 4.2). In the absence of a instrumental broadening (due to inhomogeneity of the field $\vec{B}_0$ or other reasons), the spectrum consists of a number of resolved lines, each of which the macroscopic nuclear magnetization $\vec{M}_k$ can be assigned. The magnetization $\vec{M}_k$ can be described by (1.32), if N means the number of nuclei belonging to the k-th spectral line. This magnetization accepts all the arguments given in previous sections. Naturally, being in different molecules or molecular groups, nuclei relax in different ways, depending on the type and energy of interaction, as well as on the type of spectral density of the interaction fluctuations determined by velocity and type of molecular motion. For the well-resolved spectra there are developed ways to measure T_1 and T_2 for each spectral line (with Fourier transform, see Sect. 2.4). However, in magnetic fields with middle or low homogeneity the splitting of the spectrum does not occurs. In this case, information about the relaxation can be obtained if the number of components with different relaxation characteristics is not great. Thus, it may be said, the "spectral heterogeneity" of the relaxation characteristics manifested (in the conditions of homogeneous structure sample: pure liquids or ideal solutions).

It is obvious that heterogeneity of medium should also lead to the increase of the number of magnetizations $\vec{M}_k$, differing by rates of relaxation processes. These objects include, for example, systems: fluids in porous solids, gels, biological objects, etc. The difference in relaxation characteristics of heterogeneous objects is determined mainly by differences in molecular mobility for different structural domains. Interpretation of results for heterogeneous (dispersed) systems are usually complicated due to the uncertainty of phase distributions and the presence of exchange of molecules (or nuclei) between different phases. In the simplest case of two phases (with indexes a and b) and in the absence of the exchange of nuclei between them is obvious that

$$F(t) = F_a(0)e^{-t/T_a} + F_b(0)e^{-t/T_b}, \tag{3.150}$$

where $F(t)$ stands for the relaxation function of longitudinal or transverse component of the nuclear macroscopic magnetization.

If between phases a and b the relatively slow exchange is realized then simultaneously with the relaxation process the magnetization is transferred between the phases. Therefore, the nuclear magnetization of each phase "carries" the reflex of another phase, and the expression (3.150) is transformed into

$$F(t) = F(0)\left[p'_a e^{-t/T'_a} + p'_b e^{-t/T'_b}\right], \tag{3.151}$$

where p'_a, p'_b, T'_a, T'_b are "apparent" phase populations and "apparent" relaxation times, which are functions of the lifetimes of nuclei in phases a and b.

In the case of the rapid exchange there is a complete averaging of relaxation process and it is impossible to separate the individual exponent in (3.151), while the average relaxation rate is given by the formula (regardless the nature of the exchange and the number of phases):

$$\frac{1}{T_{1,2}} = \sum_i \frac{p_i}{(T_{1,2})_i}, \tag{3.152}$$

where p_i is the relative concentration of the i-th phase; it is natural that $\sum_i p_i = 1$. The exchange is rapid, if the lifetime in the i-th state is much less than the $(T_{1,2})_i$ in this state.

Let us briefly consider the effects that occur if there is the simultaneous existence of several relaxation mechanisms. If the processes of relaxation transitions of nuclei under the influence of different interactions are independent, then it is obvious that the transition probabilities are added and the total rate relaxation is given by (3.6) and (3.8):

$$\frac{1}{T_{1,2}} = \sum_k \left(\frac{1}{T_{1,2}}\right)_k, \tag{3.153}$$

and the values $(T_{1,2})_k$ are defined by the general expressions (3.19) and (3.20). The problem of the selection of each term in (3.153) is a complex methodological challenge. Usually the measurements are carried out by varying in the broadest possible range of temperature or some other parameters of an experiment. The existence of several relaxation mechanisms, on the one hand, complicates the interpretation of experimental data, but on the other hand, creates conditions for more information about the processes taking place in the object investigated.

In the case when the relaxation rate is determined only by dipole–dipole ($T_{1\mathrm{dd}}$) and the spin-rotational ($T_{1\mathrm{sr}}$) interactions, the curve of the dependence $1/T_1$ on temperature has a minimum where $1/T_{1\mathrm{dd}} = 1/T_{1\mathrm{sr}}$. Then

$$\frac{1}{T_1} = \frac{1}{T_{1\mathrm{dd}}} + \frac{1}{T_{1\mathrm{sr}}} = \frac{2}{T_{1\mathrm{dd}}} = \frac{2}{T_{1\mathrm{sr}}},$$

and the separation of contributions can be made easily.

A similar procedure can be made if instead of the dipole interaction the significant relaxation mechanism is due to the anisotropy electronic shielding. In this case it is expedient to measure the relaxation rate at different values of the magnetic field B_0, because it is the only relaxation mechanism in which $1/T_1$ increases in proportion to B_0^2 (see 3.100). That dependence allows the selection of this relaxation mechanism in other cases.

In the presence of the scalar interaction the situation is often realized when T_2 is much smaller than the T_1. This is especially valid in the case of fast spin relaxation of spins S, interacting with investigated spins I (i.e. spins S are the quadrupole nuclei or paramagnetic particles). This effect must be taken into account when analyzing the spectra of high resolution.

The problem of separation of the dipole–dipole and quadrupole contributions to the total relaxation rate arises rarely because of the large intensity of the quadrupole interaction (dominant mechanism). However, for nuclei with small quadrupole moments (for example, ^{7}Li) such a problem exists, and it is usually solved with the help of isotopic substitution of neighboring magnetic nuclei (usually ^{1}H and ^{2}H).

3.14 Relaxation in a Rotating Coordinate Frame

It was shown in Sect. 1.5 that in the rotating coordinate frame the nuclear magnetization is affected by the effective field, which is equal to the field $\vec{B}_1$ under the condition of the sharp resonance ($\omega_0 - \omega = 0$). Initially the method of the rotating coordinate system was considered only as a convenient mathematical method to describe the movement of the magnetization vector, but in 1959, the I. Solomon carried out the measurements of the transverse relaxation time (T_2) in the presence of the rf field (Compt. Rend., 1959, vol. 249, p. 1631). It was found that the results depend on the orientation of the magnetization vector with respect to the vector the rf field. In particular, it was found that if the magnetization vector is directed along the vector of the rotating field $\vec{B}_1$, then relaxation time in some cases depends strongly on the field $\vec{B}_1$. Since the relaxing magnetization vector remains parallel to the vector field $\vec{B}_1$, which is the only active field in the rotating coordinate system, the relaxation time in this case is denoted by symbol of $T_{1\rho}$. Thus, the measurement of the relaxation rate in the rotating coordinate system implies the registration of the evolution of the transverse (!) with respect to the vector field $\vec{B}_0$ component of the nuclear magnetization in the presence of rf field $\vec{B}_1$.

The interest to measurements of the relaxation time in the rotating system coordinates is due the fact that, as a rule, $T_{1\rho}$ differs from the T_2 if there is a dependence of the relaxation times on the frequency of nuclear magnetic resonance. In this case, the role of the resonance frequency plays the nutation frequency $\omega_1 = \gamma B_1$, i.e. it is available to measure correlation times of the order of ω_1^{-1}. Since $\omega_1 \ll \omega_0$, the method of measurement relaxation time in the rotating coordinate system allows us to study very slow molecular motions in materials, in particular, in solids.

The theory of the influence of molecular motion on T_ρ was constructed by G.P. Jones, and M. Goldman in 1960–1970s, who used the transformed Hamiltonian of dipole–dipole interaction.

Thus, in the rotating with a frequency ω coordinate system the effective field is

$$\vec{B}_{\mathrm{ef}} = \left(\vec{B}_0 + \frac{\vec{\omega}}{\gamma}\right) + \vec{B}_1 = \frac{\vec{\omega}_{\mathrm{e}}}{\gamma},$$

which at the exact resonance ($\omega = \omega_0$) is

$$\vec{B}_{\mathrm{ef}} = \vec{B}_1 = -\frac{\vec{\omega}_1}{\gamma}.$$

The vector of the effective field $\vec{B}_{\mathrm{ef}}$ is at the angle α in respect to the vector of the external static field $\vec{B}_0$:

$$\tan\alpha = \frac{B_1}{B_0 - \omega/\gamma}.$$

The relaxation time $T_{1\rho}$ characterizes the process of the establishing of the equilibrium magnetization in the rotating coordinate system along the vector field $\vec{B}_{\mathrm{ef}}$.

Taking into account the dipole–dipole interaction in the presence of the rf field, the Hamiltonian of a system in laboratory frame can be written as

$$\widehat{\mathcal{H}} = \widehat{\mathcal{H}}_0 + \widehat{\mathcal{H}}_{\mathrm{rf}} + \widehat{\mathcal{H}}_{\mathrm{d}}. \tag{3.154}$$

The expression (3.58) for the dipole part of the Hamiltonian of a system of two identical spins becomes

$$\widehat{\mathcal{H}}_{\mathrm{d}} = \sum_{q=-2}^{2} F_q \hat{K}_q,$$

where

$$\hat{K}_0 = \hat{I}_{1z}\hat{I}_{2z} - \frac{1}{4}\left(\hat{I}_{1+}\hat{I}_{2-} + \hat{I}_{1-}\hat{I}_{2+}\right), \quad \hat{K}_{\pm 1} = \hat{I}_{1z}\hat{I}_{2\pm} + \hat{I}_{2z}\hat{I}_{1\pm}, \quad \hat{K}_{\pm 2} = \hat{I}_{1\pm}\hat{I}_{2\pm},$$

$$F_0 = \gamma^2\hbar^2(1 - 3\cos^2\theta)r^{-3}, \quad F_{\pm 1} = -\frac{3}{2}\gamma^2\hbar^2\sin\theta\cos\theta e^{\mp \mathrm{i}\phi}r^{-3},$$

$$F_{\pm 2} = -\frac{3}{4}\gamma^2\hbar^2\sin^2\theta e^{\mp 2\mathrm{i}\phi}r^{-3}.$$

The calculation of the relaxation time in the rotating coordinate system requires the knowledge of the appropriate Hamiltonian, which can be obtained from (3.154) as a result of three transformations. At the first step the transition to the rotating coordinate system is realized with help of the operator $\exp(\mathrm{i}\hat{I}_z\omega t)$. Then the new coordinate system is oriented as follows: the axis of z' directed along the vector of the field $\vec{B}_{\mathrm{ef}}$, and the axis x' lies in plane in which the vectors of the fields $\vec{B}_0$ and $\vec{B}_1$ are disposed. Finally, the transition to the rotating with the frequency ω_{e} coordinate frame is performed using the operator $\exp(\mathrm{i}\hat{I}_{z'}\omega_e t)$, i.e. we made the transition to the double rotating coordinate system. As a result, the operator dipole–dipole interaction in a rotating system coordinates is obtained in a view of quite cumbersome expression, in which there are new components of the K'_q, representing the spin operators in the axes x', y', z' but the components of F_q remain the same:

$$\begin{aligned}
\hat{\mathcal{H}}'_d = \frac{1}{2}F_0\Big[&(1-3\cos^2\alpha)\hat{K}'_0 - \frac{3}{2}\sin\alpha\cos\alpha\left(\hat{K}'_1e^{\mathrm{i}\omega_e t} + \hat{K}'_{-1}e^{-\mathrm{i}\omega_e t}\right)\\
&-\frac{3}{4}\sin^2\alpha\left(\hat{K}'_2e^{\mathrm{i}2\omega_e t} + \hat{K}'_{-2}e^{-\mathrm{i}2\omega_e t}\right)\Big]\\
&+F_1e^{\mathrm{i}\omega_0 t}\Big[\sin\alpha\hat{K}'_0 + \cos\frac{3}{2}\alpha\cos\frac{1}{2}\alpha\hat{K}'_1e^{\mathrm{i}\omega_e t} - \sin\frac{3}{2}\alpha\sin\frac{1}{2}\alpha\hat{K}'_{-1}e^{-\mathrm{i}\omega_e t}\\
&\qquad -\sin\alpha\cos^2\frac{1}{2}\alpha\hat{K}'_2e^{\mathrm{i}2\omega_e t} + \sin\alpha\sin^2\frac{1}{2}\alpha\hat{K}'_{-2}e^{-\mathrm{i}2\omega_e t}\Big]\\
&+F_{-1}e^{-\mathrm{i}\omega_0 t}\Big[\sin 2\alpha\hat{K}'_0 - \sin\frac{3}{2}\alpha\sin\frac{1}{2}\alpha\hat{K}'_1e^{\mathrm{i}\omega_e t}\\
&\qquad +\cos\frac{3}{2}\alpha\cos\frac{1}{2}\alpha\hat{K}'_{-1}e^{-\mathrm{i}\omega_e t} + \sin\alpha\sin^2\frac{1}{2}\alpha\hat{K}_2e^{\mathrm{i}2\omega_e t}\\
&\qquad -\sin\alpha\cos^2\frac{1}{2}\alpha\hat{K}'_{-2}e^{-\mathrm{i}2\omega_e t}\Big]\\
&+F_2e^{\mathrm{i}2\omega_0 t}\Big[\sin^2\alpha\hat{K}'_0 + \sin\alpha\cos^2\frac{1}{2}\alpha\hat{K}'_1e^{\mathrm{i}\omega_e t} - \sin\alpha\sin^2\frac{1}{2}\alpha\hat{K}'_{-1}e^{-\mathrm{i}\omega_e t}\\
&\qquad +\cos^4\frac{1}{2}\alpha\hat{K}'_2e^{\mathrm{i}2\omega_e t} + \sin^4\frac{1}{2}\alpha\hat{K}'_{-2}e^{-\mathrm{i}2\omega_e t}\Big]\\
&+F_{-2}e^{-\mathrm{i}2\omega_0 t}\Big[\sin^2\alpha\hat{K}'_0 - \sin\alpha\sin^2\frac{1}{2}\alpha\hat{K}'_1e^{\mathrm{i}\omega_e t}\\
&\qquad +\sin\alpha\cos^2\frac{1}{2}\alpha\hat{K}'_{-1}e^{-\mathrm{i}\omega_e t} + \sin^4\frac{1}{2}\alpha\hat{K}'_2e^{\mathrm{i}2\omega_e t}\\
&\qquad +\cos^4\frac{1}{2}\alpha\hat{K}'_{-2}e^{-\mathrm{i}2\omega_e t}\Big].
\end{aligned}$$

The analysis of this expression shows that for the spin–lattice relaxation one should take into account the probability of transitions at frequencies $\pm\omega_e$, $\pm 2\omega_e$, $\pm(\omega_0+\omega_e)$, $\pm(\omega_0-\omega_e)$, $\pm(\omega_0+2\omega_e)$, $\pm(\omega_0-2\omega_e)$, $\pm(2\omega_0+\omega_e)$, $\pm(2\omega_0-\omega_e)$, $\pm(2\omega_0+2\omega_e)$ $\pm(2\omega_0-2\omega_e)$ (see 3.48 and the comments to that formula).

Assuming the isotropic motion of the two spins ($I_1 = I_2 = 1/2$) in the space and using the standard scheme of calculations of the spin–lattice relaxation with an exponential correlation function, one can obtain:

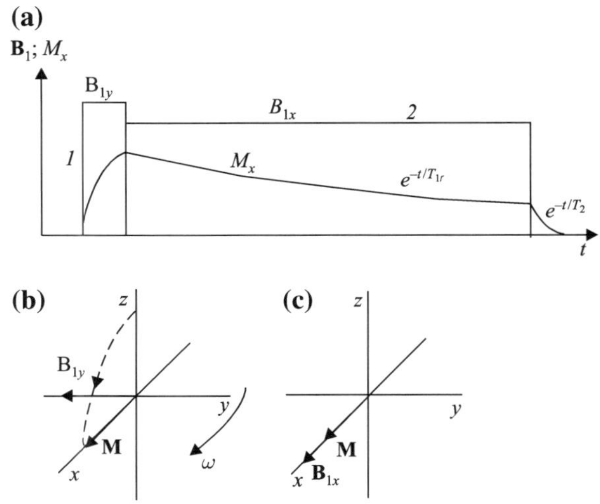

Fig. 3.10 To the explanation of the relaxation in the rotary frame; the pulse sequence (**a**) and the mutual orientation of rf fields and magnetization (**b** and **c**)

$$
\begin{aligned}
\frac{1}{T_{1\rho}} = \frac{3}{10}\gamma^4\hbar^2 r^{-6}\Bigg\{\frac{3}{2}\bigg[&\sin^2\alpha\cos^2\alpha\frac{\tau_c}{1+\omega_e^2\tau_c^2}+\sin^4\alpha\frac{\tau_c}{1+4\omega_e^2\tau_c^2}\bigg]\\
+\bigg[&\cos^2\frac{3}{2}\alpha\cos^2\frac{\alpha}{2}\frac{\tau_c}{1+(\omega_0+\omega_e)^2\tau_c^2}+\sin^2\alpha\sin^2\frac{\alpha}{2}\frac{\tau_c}{1+(\omega_0-\omega_e)^2\tau_c^2}\\
&+\sin^2\alpha\cos^4\frac{\alpha}{2}\frac{\tau_c}{1+(2\omega_0+\omega_e)^2\tau_c^2}+\sin^2\alpha\sin^4\frac{\alpha}{2}\frac{\tau_c}{1+(2\omega_0-\omega_e)^2\tau_c^2}\bigg]\\
+4\bigg[&\sin^2\alpha\cos^4\frac{\alpha}{2}\frac{\tau_c}{1+(\omega_0+2\omega_e)^2\tau_c^2}+\sin^2\alpha\sin^4\frac{\alpha}{2}\frac{\tau_c}{1+(\omega_0-2\omega_e)^2\tau_c^2}\\
&+\cos^8\frac{\alpha}{2}\frac{\tau_c}{1+(2\omega_0+2\omega_e)^2\tau_c^2}+\sin^8\frac{\alpha}{2}\frac{\tau_c}{1+(2\omega_0-2\omega_e)^2\tau_c^2}\bigg]\Bigg\}. \quad (3.155)
\end{aligned}
$$

If $\alpha = 0$, then (3.155) turns into (3.116), i.e. $T_{1\rho} = T_1$. Further, for the rapid movement of ($\omega^2\tau_c^2 \ll 1$) the expression (3.155) and (3.116) become (3.119), i.e. $T_1 = T_2 = T_{1\rho}$. Usually $T_{1\rho}$ is defined in conditions of the exact resonance ($\alpha = 90°$) and $\omega_1 \ll \omega_0$. Then (3.155) becomes

$$
\frac{1}{T_{1\rho}} = \frac{3}{10}\gamma^4\hbar^2 r^{-6}\left[\frac{3}{2}\frac{\tau_c}{1+4\omega_1^2\tau_c^2}+\frac{5}{2}\frac{\tau_c}{1+\omega_0^2\tau_c^2}+\frac{\tau_c}{1+4\omega_0^2\tau_c^2}\right]. \quad (3.156)
$$

In the case of slow motions ($\omega_0^2\tau_c^2 \gg 1$) the first term in (3.156) becomes a dominant and the position of the minimum $T_{1\rho}$ corresponds to ($\omega_1^2\tau_c^2 \simeq 1$). As always, $\omega_1 \ll \omega_0$, and the frequency ω_1 can be changed by an experimentalist changing the magnitude of the field B_1 with a constant operating frequency of a spectrometer.

An experiment on the investigation of the relaxation in the rotating frame involves three phases: (1) the orientation of the nuclear magnetization along the vector of the alternating field $\vec{B}_1$; (2) the relaxation of the magnetization in the field $\vec{B}_1$ during a certain time interval t; (3) the measurement of the magnetization observing the free induction decay after switching off the field $\vec{B}_1$. It is suggested several methods to implement the first phase of the experiment, and the adiabatic rapid passage through the resonance is among them (more exactly, a half cycle of rapid adiabatic passage through the resonance). But the main method to obtain the desirable orientation of the nuclear magnetization is to apply the 90° pulse, after which a long pulse is included with the required amplitude of the field $\vec{B}_1$ and the filling phase of which differs by 90° from the filling phase of the first pulse (Fig. 3.10a). Thus, after the first pulse the magnetization vector $\vec{M}$ is directed to the axis of x (Fig. 3.10b) and along the vector of the field $\vec{B}_1$, which is also oriented along the x during the second (2) pulse (Fig. 3.10a) due to its phase shift.

The required combination of pulses can be obtained in a standard pulsed spectrometer with a possibility to change filling phases of individual pulses. To determine the $T_{1\rho}$ the series of measurements with different durations of the second pulse is required. The pulse sequence, shown in Fig. 3.10, is called *spin–locking*.

Further Readings

Abragam A (1961) The principles of nuclear magnetism. Oxford University Press, New York

Bakhmutov V (2004) Practical NMR relaxation for chemists. Wiley, New York

Banci L, Bertini I, Luchinat C (1991) Nuclear and electron relaxation. VCH Publishers, Deerfield Beach

Bertini I, Luchinat C (1996) NMR of paramagnetic substances. Coord Chem Rev 151:1

Boere RT, Kidd RG (1982) Rotational correlation times in nuclear magnetic relaxation. Annu Rep NMR Spectrosc 13:319

Chizhik VI (2004) Nuclear magnetic relaxation. Publishing House SPbSU, Saint Petersburg (in Russian)

Ernst RR, Bodenhausen G, Wokaun A (1987) Principles of NMR in one and two dimensions. Oxford University Press, New York

Harris RK, Wasylishen RE (eds) (2012) Encyclopedia of NMR. Wiley, New York (ten volumes; ISBN 978-0-470-05821-3)

Kowalewski J (1990) Nuclear spin relaxation in diamagnetic fluids. 1. General aspects and inorganic applications. Annu Rep NMR Spectrosc 22:308

Kowalewski J (1991) Nuclear spin relaxation in diamagnetic fluids. 2. Organic systems and solutions of macromolecules and aggregates. Annu Rep NMR Spectrosc 23:289

Kowalewski J, Kruk D, Parigi G (2005) NMR relaxation in solution of paramagnetic complexes: recent theoretical progress for S 3 1. Adv Inorg Chem 57:42

Kowalewski J, Maler L (2006) Nuclear spin relaxation in liquids: theory, experiments and applications. Taylor & Francis, New York (Series in, Chemical Physics, Vol. 2)

Kruk D (2007) Theory of evolution and relaxation in multi-spin systems. Arima Publishing (Abramis), New York

Levitt MH (2008) Spin dynamics: basics of NMR, 2nd edn. Wiley, New York

Mackor EL, MacLean C (1967) Relaxation processes in system of two non-identical spins. Prog Nucl Magn Reson Spectrosc 3:129

McConnell JR (1987) The theory of nuclear magnetic relaxation in liquids. Cambridge University Press, New York

Redfield AG (1965) The theory of relaxation processes. Adv Magn Reson 1:1
Slichter CP (1990) Principles of magnetic resonance, 3rd edn. Springer, New York
Vold RL, Vold RR (1978) Nuclear magnetic relaxation in coupled spin systems. Prog Nucl Magn Reson Spectrosc 12:79
Webb GA (ed) (2006) Modern magnetic resonance. Springer, New York (three volumes)
Werbelow LG, Grant DM (1977) Intramolecular dipolar relaxation in multispin systems. Adv Magn Reson 9:190

Chapter 4
Nuclear Magnetic Resonance in Liquids

In isotropic liquids there are relatively fast rotational and translational movements of molecules. Most of molecules are subjected to *conformational transitions* (see Sect. 4.1.1). The term "isotropic" in the case of liquids means that any orientation of the axes of the molecular frame are equiprobable in the space (i.e. in laboratory frame) and reorientation of the molecular frames are faster than the characteristic time of the nuclear magnetic resonance (relaxation time T_2^*). The axes of the molecular frame may either coincide with the principal axes of the inertia tensor, or connect to any configurational stable group of atoms. Rotation of molecules leads to the situation when values of the intramolecular dipole–dipole and quadrupole interactions are averaged to zero. Translational motion also reduces intermolecular dipole–dipole interactions to zero. Both tensors of chemical shift and J-coupling become scalar:

$$\sigma_i = \frac{1}{3}\operatorname{Tr}\boldsymbol{\sigma}_i = \frac{1}{3}(\boldsymbol{\sigma}_{xx,i} + \boldsymbol{\sigma}_{yy,i} + \boldsymbol{\sigma}_{zz,i})$$
$$J_{ij} = \frac{1}{3}\operatorname{Tr}\mathbf{J}_{ij} = \frac{1}{3}(\mathbf{J}_{xx,ij} + \mathbf{J}_{yy,ij} + \mathbf{J}_{zz,ij})$$

As a result, NMR spectra in liquids are mainly determined by the three Hamiltonians in the isotropic form (hereinafter in this Chapter (except Sect. 4.3.1) the Hamiltonians are in the units of frequency, Hz or cycles per second):

1. Zeeman:

$$\widehat{\mathcal{H}}_Z = -\sum_i \frac{\gamma_i}{2\pi} B_0 \hat{I}_{zi};$$

2. Chemical shift:

$$\widehat{\mathcal{H}}_\sigma = \sum_i \sigma_i \frac{\gamma_i}{2\pi} B_0 \hat{I}_{zi};$$

V. I. Chizhik et al., *Magnetic Resonance and Its Applications*,
DOI: 10.1007/978-3-319-05299-1_4,

3. J-coupling:

$$\widehat{\mathcal{H}}_\mathrm{J} = \sum_{i<j} J_{ij}\, \hat{\vec{I}}_i \cdot \hat{\vec{I}}_j.$$

The values of interactions of protons (nuclei of light Hydrogen) in the $B_0 \sim 1$ Tl are the following: Zeeman interactions ~42 MHz, chemical shifts are in the range from 0 to 1 kHz (i.e. the values of σ_i in different substances vary in the range of 25 ppm of value of Zeeman interactions), J-couplings are in the range from 0 to 20 Hz (does not depend on external magnetic field). For the heavy atoms, chemical shift may be of 10 000 ppm, and J-coupling of hundreds Hertz.

4.1 Preliminary Remarks on Calculation of NMR Spectra

4.1.1 Calculation of Chemical Shift Tensor and Constants of J-Couplings

The physical model of interactions of nuclei and electrons in atoms and molecules, which results in chemical shift and J-coupling, was described in Sects. 1.11.4 and 1.11.5, respectively. Therefore in this section we shall discuss the methods of calculation of chemical shifts (σ_i) and constants of indirect spin–spin coupling J_{ij}.

To calculate the tensors of chemical shift and J-coupling it is necessary to make quantum chemical calculation of the electronic structure of a molecule, i.e. to calculate all wave functions of electrons in the molecule. This is a very important but complex task by itself. The complexity of the calculation grows rapidly with the increasing of the number of atoms (and, consequently, number of electrons) in the molecule. So, the analytical calculations of NMR properties are possible for relatively small molecules. As in all tasks of quantum chemistry, the results of calculations strongly depend on the positions of nuclei in space, the basis set for the electron orbitals, and the method of calculations. While the spatial structure of the molecule is more or less known for a researcher, the parameters of quantum chemical calculations are the "voluntaristic" (or of state-of-art) choice of researcher. The latest is not a subject of this book, and we refer the reader to the proper literature on quantum chemistry. Also, we just state here that there are a plenty of commercial and free of charge programs which make quantum chemistry calculations and are able to estimate the NMR properties of atoms.

Let us consider here the problems with spatial arrangement of atoms. Most of molecules in liquid phase (both in pure liquids and solutions) have intramolecular mobility. For instance, molecular fragments can rotate around covalent single bond. Such a rotation called *conformational mobility*. Rotation is usually not uniform: there are several (one, two, three) minima of the potential energy, which means that a molecule spends more time in these *conformations*, and changes one conformation

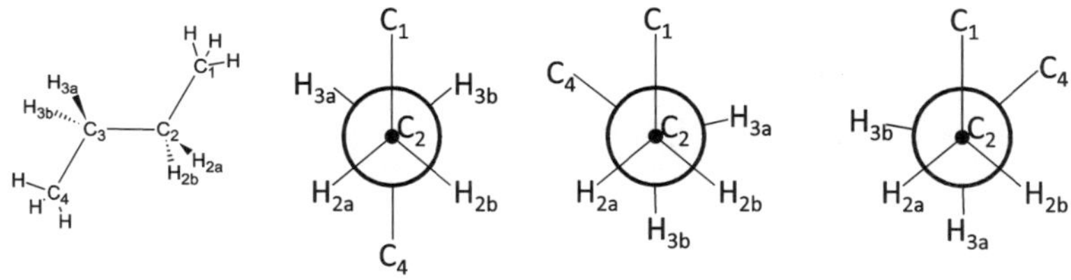

Fig. 4.1 Molecule of butane and its three conformations: trans, gauche^{-}, and gauche^{+} (*left* to *right*)

to another with fast jumps. The probabilities of these conformations are not equal to each other but rather they are distributed according to the Boltzmann statistics. If conformational changes are fast (see criterion in Sect. 4.3) then spectrum depends on average values of spin interactions (let us denote with ξ one of σ_i or J_{ij} in the following formulae):

$$\overline{\xi} = \sum_k P_k\, \xi_k,$$

where k numerates the conformations and P_k is the statistical weight of the kth conformation.

As an example, let us consider the model of Rotational Isomeric States (RIS) of the chain molecules. According to the model, each of the link $-CH_2-CH_2-$ of the chain molecule can be in the one of three confomations: plain *trans* and two twisted $gauche^{+}$ and $gauche^{-}$. The probability of gauche-conformations is less than the probability of trans-conformation. In the Fig. 4.1 the butane molecule is shown in the three conformations. In order to calculate the spin properties of both ^{13}C and ^{1}H nuclei we should perform two different calculations of electronic structure of the molecule—for trans and for one of the gauche conformations (the third calculation is not necessary due to reflection (mirror) symmetry of both gauche conformations). During the calculation of the electronic structure of the molecule in the each conformation we need to optimize geometry of the molecule. After these calculations it is possible to obtain the average value of the interaction of spins:

$$\overline{\xi} = \frac{1}{Z}\left(\xi_{\text{trans}} \exp\left(-\frac{E_{\text{trans}}}{k_B T}\right) + 2\xi_{\text{gauche}} \exp\left(-\frac{E_{\text{gauche}}}{k_B T}\right)\right),$$

where exponential factor gives us the Boltzmann distribution and Z is the partition function $Z = \exp(-E_{\text{trans}}/k_B T) + 2\exp(-E_{\text{gauche}}/k_B T)$.

Both number of conformations of a molecule and time of quantum chemical calculation for each conformation increase with the increasing of the complexity of molecule. This leads to longer calculations and to decreasing of the accuracy of the calculations.

4.1.2 Spin Hamiltonian in Liquids

It is useful to join Zeeman and chemical shift Hamiltonians in one single term:

$$\widehat{\mathcal{H}}_0 = \widehat{\mathcal{H}}_Z + \widehat{\mathcal{H}}_\sigma = -\sum_i \frac{\gamma_i}{2\pi}(1-\sigma_i)B_0\hat{I}_{zi}.$$

Then the total Hamiltonian of spin system can be written as the sum of two terms:

$$\widehat{\mathcal{H}} = \widehat{\mathcal{H}}_0 + \widehat{\mathcal{H}}_J. \tag{4.1}$$

To calculate NMR spectra in liquids it is necessary to solve the time-independent Schrödinger equation with total Hamiltonian of spin system:

$$\widehat{\mathcal{H}} \mid \varphi_n \rangle = E_n \mid \varphi_n \rangle. \tag{4.2}$$

After we found all eigenvalues of energy E_n (remind once again that they are in the units of frequency, Hz) and eigenfunctions (eigenstates)[1] $\mid \varphi_n \rangle$, the frequencies (ν_{nm}) and intensities (more precisely, probabilities P_{nm}) of NMR resonance lines (between energy levels n and m with $\Delta M = \pm 1$) can be calculated. We shall consider the case of a strong external magnetic field B_0. Resonance transitions are induced by weak electromagnetic harmonic field $B_x(t) = B_x \sin(2\pi\nu_{nm}t)$ along the axis X (orthogonal to the direction of $\vec{B}_0$) which acts at the resonance frequency ν_{nm}:

$$\nu_{nm} = E_m - E_n, \quad P_{nm} \sim \left|\langle \varphi_m \mid \sum_i \hat{I}_{xi} \mid \varphi_n \rangle\right|^2. \tag{4.3}$$

As it was stated above, all intermolecular spin–spin interactions average to zero, so Hamiltonian—i.e. energy of spin–spin interactions—describes a spin system of each single molecule. In the case of pure liquid the spectrum of whole sample is like the spectrum of a single molecule, so we can solve the Schrödinger equation for the N-spin system of one molecule. In the mixture, the spectrum is a superposition of the individual spectra of the components taken proportionally to the mixture's composition.

Dimension of Hilbert space is $n = \prod_{i=1}^{N}(2I_i + 1)$, where I_i—spin of the ith nucleus. If a molecule contains N spins $I_i = 1/2$ then dimension of Hilbert space is 2^N. For instance, dimension of space for the proton system of ethanol molecule CH_3CH_2OH is 64 if the molecule contains ^{12}C and ^{16}O species with zero spin; that

[1] We should note that terms "eigenvalue of energy" (quantum mechanics) and "eigenvalue" (linear algebra), as well as "eigenfunctions" ("eigenstates") (quantum mechanics) and "eigenvectors" (linear algebra) are about the same in the context of this book. This is because of function space is a Hilbert space and Hamiltonians are linear operators, so they are isomorphic to vectors and matrices in $\mathbb{C}^n$, respectively, and *time-independent Schrödinger equation* corresponds to the *eigenvalue equation* in linear algebra.

means the vectors of states have length 64 and the matrices of Hamiltonian have dimension 64 × 64.

To calculate a proton NMR spectrum of organic molecules in liquids or solutions, a researcher can neglect the interactions of protons with Carbon and Oxygen nuclei because of the most abundant isotopes ^{12}C and ^{16}O have a zero spin. Natural abundance of ^{13}C nuclei is about 1 %, so the signal of proton NMR of the molecules which contains nucleus of ^{13}C in the certain position is two order of magnitude weaker than signal of proton NMR of the whole sample. Such a signal is practically below the level of noise. In the Carbon-13 NMR spectroscopy *vice versa* the researcher must consider all of the proton–Carbon-13 spin–spin interactions, but the whole sample of ethanol can be considered as a mixture 1:1 of two "substances" labelled with Carbon-13 in one of the positions: $^{13}CH_3{}^{12}CH_2OH$ and $^{12}CH_3{}^{13}CH_2OH$. Take into account that the molecules with two Carbon-13 nuclei ($^{13}CH_3{}^{13}CH_2OH$) give 50 times weaker signal than each of the single-labelled molecules.

It is useful to calculate the matrix of the Hamiltonian in the basis of eigenfunctions of Z-projection of the total spin operator $\hat{I}_z = \sum_{i=1}^{N} \hat{I}_{zi}$. Such functions are called *Zeeman eigenfunctions*, and the basis is called *Zeeman eigenbasis*. For the simplicity here the only spins $I = 1/2$ will be considered. In this case the eigenfunctions of $\hat{I}_{zi}$ are usually denoted as $|\,\alpha_i\,\rangle$ and $|\,\beta_i\,\rangle$, their eigenvalues are $m_i = +1/2$ and $-1/2$, respectively, so $\hat{I}_{zi}\,|\,\alpha_i\,\rangle = +1/2\,|\,\alpha_i\,\rangle$, $\hat{I}_{zi}\,|\,\beta_i\,\rangle = -1/2\,|\,\beta_i\,\rangle$. The total Zeeman functions have a form $|\,\alpha_1\beta_2\ldots\beta_N\,\rangle$. Often, indices are skipped and in the function $|\,\alpha\beta\ldots\beta\,\rangle$ left symbol denotes state of the first spin and right symbol denotes state of the Nth spin.

While numbering the Zeeman eigenfunctions, it is useful to sort them in according to their eigenvalues ($M_k = \sum_i m_i$). Let $M_1 = N/2$, $M_2 = N/2 - 1,\ldots,$ $M_{N+1} = -N/2$. In this Zeeman eigenbasis, a matrix of the Hamiltonian of spin–spin interactions has a block-diagonal form. Off-diagonal elements of the matrix are equal to zero for the vectors with $M_n \neq M_k$: $\langle\,\varphi_n\,|\,\widehat{\mathcal{H}}\,|\,\varphi_k\,\rangle = 0$.

To proof this we remind the rules of commutation: if two operators acts on states of different particles then they commute ($\left[\hat{A}_i,\ \hat{A}_k\right] = 0,\ k \neq i$); square of spin-operator $\hat{\vec{I}}^2$ commute with all projections of $\hat{\vec{I}}$, i.e. $\hat{I}_z$, $\hat{I}_x$, and $\hat{I}_y$.

Let us show that the commutator $\left[\hat{I}_z,\ \widehat{\mathcal{H}}\right] = 0$. Obviously each term of $\hat{I}_z$ commutes with each term of $\widehat{\mathcal{H}}_0$:

$$\left[\hat{I}_z,\ \widehat{\mathcal{H}}_0\right] = -\left[\sum_i \hat{I}_{zi},\ \sum_k \frac{\gamma_k}{2\pi}\,(1-\sigma_k)\,B_0\,\hat{I}_{zk}\right]$$
$$= -\sum_i\sum_k \frac{\gamma_k}{2\pi}\,(1-\sigma_k)\,B_0\,\left[\hat{I}_{zi},\ \hat{I}_{zk}\right] = 0.$$

If $i \neq k$ then operators act on the eigenfunctions of different spins, if $i = k$ then these operators are the same, so, all terms of double sum are equal to zero.

Let us expand the commutator $\left[\hat{I}_z,\ \widehat{\mathcal{H}}_{\mathrm{J}}\right]$:

$$\left[\sum_k \hat{I}_{zk},\ \sum_{i<j} J_{ij}\,\hat{\vec{I}}_i\cdot\hat{\vec{I}}_j\right]=\sum_{i<j}\left(\left[\sum_k \hat{I}_{zk},\ J_{ij}\,\hat{\vec{I}}_i\cdot\hat{\vec{I}}_j\right]\right). \tag{4.4}$$

The commutator in the parentheses may be written as a sum of three terms:

$$\left[\sum_k \hat{I}_{zk},\ J_{ij}\,\hat{\vec{I}}_i\cdot\hat{\vec{I}}_j\right]=\left[\sum_{k\neq i,\,j} \hat{I}_{zk},\ J_{ij}\,\hat{\vec{I}}_i\cdot\hat{\vec{I}}_j\right]+\left[\hat{I}_{zi},\ J_{ij}\,\hat{\vec{I}}_i\cdot\hat{\vec{I}}_j\right]+\left[\hat{I}_{zj},\ J_{ij}\,\hat{\vec{I}}_i\cdot\hat{\vec{I}}_j\right]$$

Indices k, i, and j in the first term denotes different spins, and consequently, this commutator is equal to zero. The right two terms are also equal to zero. Because $(\hat{\vec{I}}_i+\hat{\vec{I}}_j)^2$ commute with its projection $(\hat{I}_i+\hat{I}_j)_z=\hat{I}_{zi}+\hat{I}_{zj}$, then:

$$\begin{aligned}0\equiv\left[(\hat{I}_i+\hat{I}_j)_z,\ (\hat{\vec{I}}_i+\hat{\vec{I}}_j)^2\right]&=\left[\hat{I}_{zi}+\hat{I}_{zj},\ \hat{\vec{I}}_i^{\,2}+\hat{\vec{I}}_j^{\,2}+2\hat{\vec{I}}_i\cdot\hat{\vec{I}}_j\right]\\ &=\left[\hat{I}_{zi},\ \hat{\vec{I}}_i^{\,2}\right]+\left[\hat{I}_{zi},\ \hat{\vec{I}}_j^{\,2}\right]+\left[\hat{I}_{zj},\ \hat{\vec{I}}_i^{\,2}\right]+\left[\hat{I}_{zj},\ \hat{\vec{I}}_j^{\,2}\right]+2\left[\hat{I}_{zi}+\hat{I}_{zj},\ \hat{\vec{I}}_i\cdot\hat{\vec{I}}_j\right]\equiv 0.\end{aligned}$$

Here, the commutator of the sum of operators is written as the sum of commutators. Obviously four left commutators are equal to zero. That means that the fifth term, $\left[\hat{I}_{zi}+\hat{I}_{zj},\ \hat{\vec{I}}_i\cdot\hat{\vec{I}}_j\right]$ is equal to zero too. We have proved both $\widehat{\mathcal{H}}_0$ and $\widehat{\mathcal{H}}_{\mathrm{J}}$ commute with $\hat{I}_z$, i.e. $\left[\widehat{\mathcal{H}},\ \hat{I}_z\right]=0$.

Let us write matrix element of this commutator, which is identical to zero:

$$0\equiv\langle\varphi_n\,|\left[\widehat{\mathcal{H}},\ \hat{I}_z\right]|\,\varphi_k\rangle \tag{4.5}$$

$$\begin{aligned}&=\langle\varphi_n\,|\,\widehat{\mathcal{H}}\,\hat{I}_z\,|\,\varphi_k\rangle-\langle\varphi_n\,|\,\hat{I}_z\,\widehat{\mathcal{H}}\,|\,\varphi_k\rangle\\ &=\langle\varphi_n\,|\,\widehat{\mathcal{H}}\Big(\hat{I}_z\,|\,\varphi_k\rangle\Big)-\Big(\langle\varphi_n\,|\,\hat{I}_z\Big)\widehat{\mathcal{H}}\,|\,\varphi_k\rangle\\ &=M_k\langle\varphi_n\,|\,\widehat{\mathcal{H}}\,|\,\varphi_k\rangle-M_n\langle\varphi_n\,|\,\widehat{\mathcal{H}}\,|\,\varphi_k\rangle\\ &=(M_k-M_n)\langle\varphi_n\,|\,\widehat{\mathcal{H}}\,|\,\varphi_k\rangle\equiv 0.\end{aligned} \tag{4.6}$$

In order to satisfy the identity (4.6) when $M_n\neq M_k$, the matrix element $\langle\varphi_n\,|\,\widehat{\mathcal{H}}\,|\,\varphi_k\rangle$ must be equal to zero.

In the following example, the matrix of Hamiltonian of two-spin system, basis functions, and eigenvalues M_k are shown:

$$(H_{ij}) = \begin{pmatrix} H_{11} & 0 & 0 & 0 \\ 0 & H_{22} & H_{23} & 0 \\ 0 & H_{32} & H_{33} & 0 \\ 0 & 0 & 0 & H_{44} \end{pmatrix}, \quad \begin{matrix} |\,\alpha\alpha\,\rangle \\ |\,\alpha\beta\,\rangle \\ |\,\beta\alpha\,\rangle \\ |\,\beta\beta\,\rangle \end{matrix}, \quad \begin{matrix} +1 \\ 0 \\ 0 \\ -1 \end{matrix}. \tag{4.7}$$

During the diagonalization of such a matrix, the vectors with different eigenvalues are not mixed. It allows us to diagonalize each submatrix individually. Dimension of submatrix is equal to binomial coefficients C_N^k. The maximum dimension is $C_N^{[N/2]}$. For instance, the Hamiltonian of the 6-spin system of protons of ethanol molecule has the following submatrices: $1\times 1, 6\times 6, 15\times 15, 20\times 20, 15\times 15, 6\times 6$, and 1×1. Two eigenvalues and eigenfunctions are known: they are $H_{1,1}$ and $|\,\varphi_1\,\rangle = |\,\alpha\alpha\alpha\alpha\alpha\alpha\,\rangle$, $H_{64,64}$ and $|\,\varphi_{64}\,\rangle = |\,\beta\beta\beta\beta\beta\beta\,\rangle$.

Consider a system of N nuclei with spin $I = 1/2$ (different or the same species). The Hamiltonian $\widehat{\mathcal{H}}_0$ contributes to the only diagonal elements of the matrix:

$$\langle\,\varphi_m\,|\,\widehat{\mathcal{H}}_0\,|\,\varphi_m\,\rangle = -\frac{1}{2\pi}\sum_i \gamma_i(1-\sigma_i)\,B_0\,\langle\,\varphi_m\,|\,\hat{I}_{zi}\,|\,\varphi_m\,\rangle = \frac{1}{4\pi}\sum_i \mp\gamma_i(1-\sigma_i)\,B_0. \tag{4.8}$$

Here, the sign + or – before gyromagnetic ratio depends on the state of the ith spin in the basis vector $|\,\varphi_m\,\rangle$: $|\,\alpha_i\,\rangle$ leads to sign minus, and $|\,\beta_i\,\rangle$—to sign plus.

Consider one of the term in the operator $\widehat{\mathcal{H}}_J$:

$$J_{ij}\,\hat{\vec{I}}_i\cdot\hat{\vec{I}}_j = J_{ij}\left(\hat{I}_{zi}\hat{I}_{zj} + \tfrac{1}{2}(\hat{I}_{+i}\hat{I}_{-j} + \hat{I}_{-i}\hat{I}_{+j})\right).$$

Its diagonal elements are dependent on Z-projection of $\hat{\vec{I}}$:

$$\begin{aligned}
\langle \ldots\alpha_i\ldots\alpha_j\ldots \,|\, J_{ij}\hat{I}_{zi}\hat{I}_{zj} \,|\, \ldots\alpha_i\ldots\alpha_j\ldots\rangle &= +\tfrac{1}{4}J_{ij};\\
\langle \ldots\alpha_i\ldots\beta_j\ldots \,|\, J_{ij}\hat{I}_{zi}\hat{I}_{zj} \,|\, \ldots\alpha_i\ldots\beta_j\ldots\rangle &= -\tfrac{1}{4}J_{ij};\\
\langle \ldots\beta_i\ldots\alpha_j\ldots \,|\, J_{ij}\hat{I}_{zi}\hat{I}_{zj} \,|\, \ldots\beta_i\ldots\alpha_j\ldots\rangle &= -\tfrac{1}{4}J_{ij};\\
\langle \ldots\beta_i\ldots\beta_j\ldots \,|\, J_{ij}\hat{I}_{zi}\hat{I}_{zj} \,|\, \ldots\beta_i\ldots\beta_j\ldots\rangle &= +\tfrac{1}{4}J_{ij}.
\end{aligned}$$

States of other spins are denoted with dots. Remember that in each of the formulae above the vectors bra and ket are the same, $\langle\,\varphi_m\,|$ and $|\,\varphi_m\,\rangle$. Sign plus before J_{ij} appears if spins i and j are collinear and sign minus if they are anti collinear.

Off-diagonal elements of the matrix of Hamiltonian are dependent on the shift operators $\hat{I}_{+i}\hat{I}_{-j}$:

$$H_{mk} = \langle\,\varphi_m\,|\,\tfrac{1}{2}J_{ij}\big(\hat{I}_{+i}\hat{I}_{-j} + \hat{I}_{-i}\hat{I}_{+j}\big)\,|\,\varphi_k\,\rangle.$$

Let us write the states of spin i and j in the vectors $\langle\,\varphi_m\,|$ and $|\,\varphi_k\,\rangle$ explicitly and states of other spins denote with dots:

$$\begin{aligned}H_{mk} =&\tfrac{1}{2} J_{ij} \, \langle \ldots \alpha_i \ldots \beta_j \ldots \mid \hat{I}_{+i} \hat{I}_{-j} \mid \ldots \beta_i \ldots \alpha_j \ldots \rangle \\ &+ \tfrac{1}{2} J_{ij} \langle \ldots \alpha_i \ldots \beta_j \ldots \mid \hat{I}_{-i} \hat{I}_{+j} \mid \ldots \beta_i \ldots \alpha_j \ldots \rangle \\ =&\tfrac{1}{2} J_{ij} \langle \ldots \alpha_i \ldots \beta_j \ldots \mid \ldots \alpha_i \ldots \beta_j \ldots \rangle + \tfrac{1}{2} J_{ij} \langle \ldots \alpha_i \ldots \beta_j \ldots \mid 0 \rangle = \tfrac{1}{2} J_{ij}.\end{aligned}$$

This matrix element is not equal to zero only if vectors $\langle \varphi_m \mid$ and $\mid \varphi_k \rangle$ differ with orientation of spins i and j but states of all other spins are the same. The second non-zero matrix element is H_{km}. Other matrix elements are zero.

Each term in the double sum in the Hamiltonian $\widehat{\mathcal{H}}_{\mathrm{J}}$ contributes to the only one off-diagonal element of the matrix and each element of the matrix are defined by the only one term in the sum. Some elements may be equal to zero.

The general formula for the matrix element is $H_{km} = 1/2 \, J_{ij}$, if states of all the spins in vectors $\langle \varphi_m \mid$ and $\mid \varphi_k \rangle$ are the same except the spins i and j, which are anti collinear and "swapped". In the following examples for the 4-spin system $I = 1/2$ states of "swapped" spins are underlined:

$$\begin{aligned}\langle \alpha \underline{\beta} \underline{\alpha} \beta \mid \widehat{\mathcal{H}}_{\mathrm{J}} \mid \alpha \underline{\alpha} \underline{\beta} \beta \rangle &= \tfrac{1}{2} J_{23}, \\ \langle \underline{\beta} \alpha \beta \underline{\alpha} \mid \widehat{\mathcal{H}}_{\mathrm{J}} \mid \underline{\alpha} \alpha \beta \underline{\beta} \rangle &= \tfrac{1}{2} J_{14}, \\ \langle \alpha \beta \alpha \beta \mid \widehat{\mathcal{H}}_{\mathrm{J}} \mid \beta \alpha \beta \alpha \rangle &= 0.\end{aligned}$$

In the last sample two pairs of spins are "swapped", that means this matrix element is equal to zero.

Consider now general case of a spin system of p arbitrary spins $I_i, i = 1, 2, \ldots, p$. Eigenvectors of $\hat{I}_{zi}$ are denoted in accordance to their eigenvalues: $\hat{I}_{zi} \mid m_i \rangle = m_i \mid m_i \rangle$. Diagonal elements of the matrix of the spin Hamiltonian depend on the term with inclusion of single operator $\hat{I}_{zi}$ and with product of two operators $\hat{I}_{zi} \hat{I}_{zj}$:

$$\begin{aligned}\langle m_1 m_2 \ldots m_p \mid \hat{I}_{zi} \mid m_1 m_2 \ldots m_p \rangle &= m_i, \\ \langle m_1 m_2 \ldots m_p \mid \hat{I}_{zi} \hat{I}_{zj} \mid m_1 m_2 \ldots m_p \rangle &= m_i \, m_j;\end{aligned} \tag{4.9}$$

while off-diagonal elements depend on the terms with shift operators $\hat{I}_{+i} \hat{I}_{-j}$:

$$\begin{aligned}&\langle \ldots, m_i + 1, \ldots, m_j - 1 \ldots \mid \hat{I}_{+i} \hat{I}_{-j} + \hat{I}_{-i} \hat{I}_{+j} \mid \ldots, m_i, \ldots, m_j \ldots \rangle \\ &\qquad = \sqrt{I_i(I_i + 1) - m_i(m_i + 1)} \sqrt{I_j(I_j + 1) - m_j(m_j - 1)},\end{aligned} \tag{4.10}$$

$$\begin{aligned}&\langle \ldots, m_i - 1, \ldots, m_j + 1 \ldots \mid \hat{I}_{+i} \hat{I}_{-j} + \hat{I}_{-i} \hat{I}_{+j} \mid \ldots, m_i, \ldots, m_j \ldots \rangle \\ &\qquad = \sqrt{I_i(I_i + 1) - m_i(m_i - 1)} \sqrt{I_j(I_j + 1) - m_j(m_j + 1)}.\end{aligned} \tag{4.11}$$

Remember that states of all other spins except mentioned i and j are the same in bra and ket. The formulae (4.9–4.11) contain all the combinations of spin operators in the total Hamiltonin $\widehat{\mathcal{H}}$.

4.1.3 Spectral Line Shape

In calculation of NMR spectra with time-independent Schrödinger equation (4.2) it was assumed that all molecules are in the equivalent external conditions and the width of resonance lines is infinitely small. In experiments, many factors lead to the broadening of resonance lines. There are both natural causes such as thermal motion of molecules, intermolecular spin–spin interactions and others, and instrumental factors such as inhomogeneity of the magnetic field of a spectrometer, noise in its electronic blocks, finite time of registration of free induction decay, and others. Each spectral line is not infinitely narrow rather occupies some range of frequencies with the certain distribution of the amplitude in this range. Two close lines even can overlap each other and one partially overlapped or partially resolved line appears in a spectrum. According to this, we should rewrite (4.3) in another manner: probability P_{nm} should be changed with probability density $p_{nm}(\nu)$:

$$\nu_{nm} = E_n - E_m, \quad p_{nm}(\nu) \sim \left| \langle \varphi_n \mid \sum \gamma_i \hat{I}_{xi} \mid \varphi_m \rangle \right|^2 g(\nu - \nu_{nm}), \quad \int_{-\infty}^{\infty} g(\nu)\,\mathrm{d}\nu = 1. \tag{4.12}$$

The function $g(\nu)$ is a line shape (cf. with form factor in Sect. 1.7.1). Notation $g(\nu - \nu_{nm})$ means intensity of the line on frequency ν close to the frequency of resonance transition ν_{nm}. The function is normalized, so $P_{nm} = \int p_{nm}(\nu)\,\mathrm{d}\nu$. In the idealistic case the function $g(\nu)$ is the Dirac delta-function $\delta(\nu)$. In liquids, the line shape is Lorentzian function

$$g(\nu) = \frac{1}{\pi\,\delta\nu} \frac{1}{(1 + (\nu/\delta\nu)^2)},$$

where $\delta\nu$ is a half-width of the line on the half-height. Amplitude (height) of the line on the central frequency is $1/(\pi\,\delta\nu)$.

In the NMR spectroscopy, *intensity* of a line means its *integral* intensity which is proportional to the probability of transition P_{nm}. *Amplitude* (height) of a line on the resonance frequency depends on spectrometer settings while *intensity* (probability) does not. Strictly speaking, relations of intensities of different lines in a spectrum do not depend on instrument settings.[2]

In order to compare calculated NMR spectra with experimental one we should take into account line shape. The function $g(\nu)$ is another variable parameter to NMR spectra calculation.

[2] The special case is numerous of correlation experiments, when experimental conditions are specially chosen in order to affect some spins and, so, to distort a system, to break equilibrium conditions. In such an experiment a researcher knowingly changes *intensity* of some resonance lines.

4.1.4 Spectra Designation. Magnetically Equivalent Spins

In strong magnetic field, where Zeeman interactions are significantly stronger than indirect spin–spin couplings ($|(2\pi)^{-1}\gamma_i B_0| \gg |J_{ij}|$) and chemical shifts are weak ($\sigma_i \ll 1$), the shape of NMR spectrum depends on the relation of $|J_{ij}|$ and difference of resonance frequencies $|\nu_i - \nu_j| = (2\pi)^{-1}(|\gamma_i|\sigma_i - |\gamma_j|\sigma_j)B_0$. Here, it is necessary to introduce the definition of *magnetically equivalent spins*.

Definition: *magnetically equivalent spins are those of one species which in each time moment have the same electronic environment, i.e. equal value of σ_i. Those spins, which have equivalent electronic environment in average during fast intramolecular (conformational) transitions, also become magnetically equivalent.*

Example 1. In the molecule of methane CH_4 all four protons (nuclei of Hydrogen) are magnetically equivalent because of symmetry of the molecule (all Hydrogen atoms are indistinguishable).

Example 2. In the molecule of fluoroethane CH_2F–CH_3 three protons in the methyl group (–CH_3) in the each time moment are non-equivalent because of different positions relative to Fluorine atom—one of them in *trans*- and two other in *gauche*-positions. However, because of fast rotation around C–C single bond they become magnetically equivalent. This molecule contains three sets of magnetically equivalent spins: three protons of –CH_3 group, two protons of –CH_2– group and spin of Fluorine nucleus. The system have three resonance frequencies ν_{H_3}, ν_{H_2}, and ν_F.

Example 3. In the molecule of 1,1-di-fluoroethylene

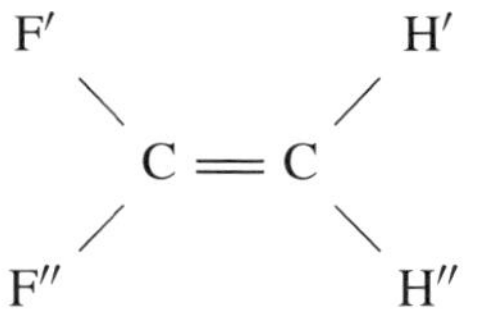

,

rotation around double chemical bond C=C is prohibited. Both protons and both Fluorine nuclei are magnetically equivalent because of symmetry of the molecule. However, constants of indirect spin–spin interactions of proton H′ with F′ and with F″ are different: $J_{H'F'} = J_{H''F''}$, $J_{H'F''} = J_{H''F'}$, but $J_{H'F'} \neq J_{H'F''}$. The cause of this inequality is *cis*- and *trans*- positions of these atoms relative to the bond between two Carbon atoms.

NMR spectrum in liquids in general case consists of several groups of lines (multiplets). Each group of lines belongs to one of the groups of magnetically equivalent spins. The multiplets can be separated, may overlaps, or may join into more complex groups of lines.

To name NMR spectrum, the following rules are applied:

1. Each group of magnetically equivalent spins is denoted by Roman capital Latin letter, usually starting with A. Number of spins in the group is shown as index. See item 3 for exception.

2. If the difference between resonance frequencies of spins in two groups much greater than the constant of J-coupling of spins between these groups (for instance, these are spins of different species or their chemical shifts differ significantly) then these groups are denoted by letters from different parts of the alphabet. As an example, AX. If $|\nu_i - \nu_j|$ is comparable with $|J_{ij}|$ then next letters are used, AB or XY.
3. If J-coupling of spins in two groups are different then instead of the indexing, the letter is repeated with increasing number of apostrophes: AA′A″A‴.

Examples. A spectrum of methane CH_4 is denoted A_4 assuming Carbon-12 isotope, and either A_4X or AX_4 assuming Carbon-13 atom in the molecule; ethane CH_3–CH_3 is A_6; trifluoroethane CH_3–CF_3 is A_3X_3; fluoroethane CH_2F–CH_3 is either A_3B_2X or A_3N_2X depending on strength of $\vec{B}_0$; 1-chloro-1,1-difluoroethane $CClF_2$–CH_3 is AN_2X_3; 1,1-difluoroethylene $CF_2{=}CH_2$ is AA′XX′.

4.2 Calculation of NMR Spectra

4.2.1 Analytical and Numerical Calculations, Symmetry of Spin System

Exact calculations of NMR spectra can only be carried out for such systems, which has the Hamiltonian matrix of small dimension. To calculate eigenvalues and eigenvectors of the time-independent Schrödinger equation (4.2) one should solve the secular equation for the Hamiltonian matrix:

$$\det |H - \lambda_k \mathbb{E}| = 0, \tag{4.13}$$

where $\mathbb{E}$ is unit matrix, and λ_k are eigenvalues (i.e. eigenvalues of energy E_k in (4.2)). We mentioned earlier that the matrix has diagonal-block structure, so we can solve this secular equation for each submatrix separately. Based on this requirement and using the basis of Zeeman functions, we can analytically calculate the spectra of systems containing not more than three nuclei with spin $I = 1/2$, because of equations of the fifth degree have no general analytical solution.

To reduce the size of the submatrices in the matrix of the Hamiltonian, we can use the *symmetry-adapted basis vectors*. To find out symmetry-adapted basis vectors, we should refer each spin of the group of magnetically equivalent spins as point in space. Then, we can use irreducible representations of *point symmetry groups*. Symmetry operations lead to permutation of indices of spins. Symmetry-adapted vector changes direction to opposite (changes its sign but not length) or preserves direction (and length) after such permutation of indices of spins. If a vector preserves its direction then the vector is called *symmetric*. If a vector changes direction to opposite then the vector is called *antisymmetric*. Symmetric vectors correspond to operations of rotation, but antisymmetric correspond to operations of reflection. Consider some examples.

Example 1. A system of two magnetically equivalent spins (A_2) can be represented as two point in space. In this system there are two symmetry operations: rotation around the axis of symmetry C_2 and reflection in the plane. Both of these operations lead to the same result—permutation of indices of spins $1 \longleftrightarrow 2$. Make such permutation for the vector $|\,\alpha\alpha\,\rangle$: $|\,\alpha\alpha\,\rangle = |\,\alpha_1\alpha_2\,\rangle \longrightarrow |\,\alpha_2\alpha_1\,\rangle = |\,\alpha\alpha\,\rangle$. This is a symmetric vector. Vector $|\,\beta\beta\,\rangle$ is also symmetric. However, Zeeman vector $|\,\alpha\beta\,\rangle$ changes to $|\,\beta\alpha\,\rangle$ after permutation of indices of spins. Both of them are not symmetry-adapted vectors. Their linear combinations $(|\,\alpha\beta\,\rangle + |\,\beta\alpha\,\rangle)/\sqrt{2}$ and $(|\,\alpha\beta\,\rangle - |\,\beta\alpha\,\rangle)/\sqrt{2}$ are symmetric and antisymmetric vectors, respectively.

Example 2. Three-spin system A_3 we can imagine as the vertices of an equilateral triangle. This system has a point group C_{3v} and three symmetry operation: (1) rotation around the axis C_3 which correspond to cyclic permutation of indices of spins ($123 \longrightarrow 312$), (2) rotation around the axis C_2, and (3) reflection in the plane σ_v. Let both the C_2 axis and the σ_v plane contain point 1. Both of the operations (2) and (3) lead to permutation of indices of the spins 2 and 3 ($1\,23 \longrightarrow 1\,32$), but the operation (3) is antisymmetric and operation (2) is symmetric. Once again, vector $|\,\alpha\alpha\alpha\,\rangle$ is symmetric. Three vectors with eigenvalue $M_k = +1/2$, namely $|\,\alpha\alpha\beta\,\rangle$, $|\,\alpha\beta\alpha\,\rangle$, and $|\,\beta\alpha\alpha\,\rangle$ should be changed with $(|\,\alpha\alpha\beta\,\rangle + |\,\alpha\beta\alpha\,\rangle + |\,\beta\alpha\alpha\,\rangle)/\sqrt{3}$ (symmetric), $(|\,\alpha\alpha\beta\,\rangle + |\,\alpha\beta\alpha\,\rangle - 2|\,\beta\alpha\alpha\,\rangle)/\sqrt{6}$ (symmetric), and $(|\,\alpha\alpha\beta\,\rangle - |\,\alpha\beta\alpha\,\rangle)/\sqrt{2}$ (antisymmetric).

The primary property of the symmetry-adapted basis vectors is that the Hamiltonians of spin–spin interactions $\widehat{\mathcal{H}}$ and radio frequency $\widehat{\mathcal{H}}_{\text{rf}}$ are completely symmetric under the interchange of magnetically equivalent spins. This leads to the fact that the matrix elements between the symmetry-adapted functions of different symmetry are zero. Consequently, the probability of transition between eigenstates with different symmetry is zero. This condition significantly reduces the number of possible transitions in systems that include groups of equivalent spins. Furthermore, the matrix elements of the Hamiltonian between the states belonging to different irreducible representations are also equal to zero. In the example 1, all off-diagonal matrix elements of the Hamiltonian are equal to zero.

The use of the symmetry-adapted basis functions reduces the size of linear algebra problems to solve. However, for the modern computers and for general purpose NMR computer programs, the Zeeman basis function are preferable. It is easy to generate Zeeman basis function and calculate matrix elements of the Hamiltonian with this basis.

The restriction of the size of a spin system is based on the computer precision but not on the computer's memory size and processor performance. The problem of eigenvalues and eigenvectors of matrix of size greater than 1000×1000 may be unstable. This has a nature of finite computer precision and rounding of results of arithmetic float-point operation. The number of such operations is big enough to the problem to be unstable. For instance, maximum size of matrix of Hamiltonian of 12-spin system ($I = 1/2$) is 924×924. So, number of spins in the system of about 12–14 is the limit for numerical calculation of NMR spectra.

4.2.2 Spectrum A_p of a System of p Magnetically Equivalent Spins

The total Hamiltonian $\widehat{\mathcal{H}}$ of the system A_p is of a simple form:

$$\widehat{\mathcal{H}} = \widehat{\mathcal{H}}_0 + \widehat{\mathcal{H}}_J, \quad \widehat{\mathcal{H}}_0 = \nu_0 \sum_i^p \hat{I}_{zi}, \quad \widehat{\mathcal{H}}_J = \sum_{i<j}^p J_{ij}\, \hat{\vec{I}}_i \cdot \hat{\vec{I}}_j.$$

NMR spectrum of such a system contains the only line with frequency ν_0. To show this, we state the following theorem:

Theorem 1. *Let the Hamiltonian $\widehat{\mathcal{H}}$ consists of such two components $\widehat{\mathcal{H}}_0 + \widehat{\mathcal{H}}_1$ that $\left[\widehat{\mathcal{H}}_0, \widehat{\mathcal{H}}_1\right] = 0$ and $\left[\hat{I}_x, \widehat{\mathcal{H}}_1\right] = 0$. Then the NMR spectrum determined by the Hamiltonian $\widehat{\mathcal{H}}$, coincides with the NMR spectrum determined by the Hamiltonian $\widehat{\mathcal{H}}_0$.*

The proof is reduced to proving the two statements:

1. all frequencies of allowed transitions (that is of non-zero intensity) between the eigenenergies of a system with the Hamiltonian $\widehat{\mathcal{H}}$, coincide with the frequencies of allowed transitions between eigenenergies of a system with the Hamiltonian $\widehat{\mathcal{H}}_0$;
2. for each of the allowed frequencies combined (total) probability of transitions between the energy eigenstates of the Hamiltonian $\widehat{\mathcal{H}}$, and the total probability of transitions between the energy eigenstates of the Hamiltonian $\widehat{\mathcal{H}}_0$, are equal.

It follows from the first statement that NMR spectra, calculated using both Hamiltonians, contain the same frequencies despite the fact that the sets of eigenvalues of each Hamiltonians are different. From the second statement one can conclude that the intensities of each of the spectral lines (the sum of the intensities of all the transitions at a given frequency) given by both Hamiltonians are equal. But this means that the NMR spectra, determined by both Hamiltonians, are the same. We shall not proof the theorem here and we refer readers to, for example Pople et al. (1959).

Both Hamiltonians $\widehat{\mathcal{H}}_0$ and $\widehat{\mathcal{H}}_J$ satisfy Theorem 1. To show this one can expand sum in terms as in Eq. (4.4) and proof that each terms of $\widehat{\mathcal{H}}_J$ commute with terms of $\widehat{\mathcal{H}}_0$ and $\hat{I}_x$. On the basis of Theorem 1 we can reduce the problem of calculation of the spectrum of A_p to finding the NMR spectrum, described by the Hamiltonian $\widehat{\mathcal{H}}_0$. The matrix of Zeeman Hamiltonian has diagonal form in the basis of Zeeman functions. The allowed transitions are between the energy eigenstates with $\Delta M = \pm 1$. Energies of such states differ by ν_0, that means the single line with frequency ν_o is observed in the NMR spectrum of the A_p system.

4.2.3 First-Order Spectrum of a System A_pX_q

First-order spectra are called so because it is enough use the first-order time-independent perturbation theory to calculate these spectra.

There are p nuclei with resonance frequency ν_A and q nuclei with resonance frequency ν_X in a system A_pX_q. Values of J-coupling between each nucleus of group A with each nucleus of group X are equal to each other. Let it be J_{AX}. The Hamiltonian of the system is:

$$\widehat{\mathcal{H}} = \widehat{\mathcal{H}}_0 + \widehat{\mathcal{H}}_{1A} + \widehat{\mathcal{H}}_{1X} + \widehat{\mathcal{H}}_{1AX},$$

where

$$\widehat{\mathcal{H}}_0 = \nu_A \hat{I}_{zA} + \nu_X \hat{I}_{zX};$$

$$\hat{I}_{zA} = \sum_{i=1}^{p} \hat{I}_{zi}^{A}, \quad \hat{I}_{zX} = \sum_{i=1}^{q} \hat{I}_{zi}^{X};$$

$$\widehat{\mathcal{H}}_{1A} = \sum_{i<j}^{p} J_{ij}^{A} \vec{\hat{I}}_i^{A} \cdot \vec{\hat{I}}_j^{A}, \quad \widehat{\mathcal{H}}_{1X} = \sum_{i<j}^{q} J_{ij}^{X} \vec{\hat{I}}_i^{X} \cdot \vec{\hat{I}}_j^{X};$$

$$\widehat{\mathcal{H}}_{1AX} = \sum_{i}^{p} \sum_{j}^{q} J_{AX} \vec{\hat{I}}_i^{A} \cdot \vec{\hat{I}}_j^{X} = J_{AX} \vec{\hat{I}}^{A} \cdot \vec{\hat{I}}^{X}.$$

It can be easily shown that Hamiltonians $\widehat{\mathcal{H}}' = \widehat{\mathcal{H}}_0 + \widehat{\mathcal{H}}_{1AX}$ and $\widehat{\mathcal{H}}'' = \widehat{\mathcal{H}}_{1A} + \widehat{\mathcal{H}}_{1X}$ satisfy Theorem 1, i.e. $\left[\widehat{\mathcal{H}}', \widehat{\mathcal{H}}''\right] = 0$ and $\left[\hat{I}_x{}^{A} + \hat{I}_x{}^{X}, \widehat{\mathcal{H}}''\right] = 0$. Note that we did not assume any relation of $|\nu_A - \nu_X|$ and $|J_{AX}|$. An important consequence of this: the NMR spectrum of any system does not depend on the energy of the indirect spin–spin interactions within each group of magnetically equivalent nuclei.

In the system A_pX_q, the difference between the resonance frequencies of the nuclei always substantially exceed the constant of indirect spin–spin coupling: $|\nu_A - \nu_X| \gg |J_{AX}|$. Let us compute the spectrum in accordance with the first-order perturbation theory:

$$\widehat{\mathcal{H}}_{unp} = -\nu_A \hat{I}_{zA} - \nu_X \hat{I}_{zX},$$

$$\widehat{\mathcal{H}}_{pert} = J_{AX} \vec{\hat{I}}^{A} \cdot \vec{\hat{I}}^{X}.$$

Here, $\widehat{\mathcal{H}}_{unp}$ is the Hamiltonian of the unperturbed system, and $\widehat{\mathcal{H}}_{pert}$ describes the perturbation.

Consider the spectrum of the unperturbed system. We introduce the notation for the eigenvalues and eigenvectors of operators $\hat{I}_{zA}$ and $\hat{I}_{zX}$:

$$\hat{I}_{zA} \mid M_{Ai} \rangle = M_{Ai} \mid M_{Ai} \rangle, \qquad \hat{I}_{zX} \mid M_{Xk} \rangle = M_{Xk} \mid M_{Xk} \rangle$$

and write:

$$\widehat{\mathcal{H}}_{unp} \mid M_{Ai}, M_{Xk} \rangle = (\nu_A M_{Ai} + \nu_X M_{Xk}) \mid M_{Ai}, M_{Xk} \rangle.$$

Since the minimum difference between the eigenvalues of energy of this Hamiltonian is $|\nu_A - \nu_X|$, and by assumption, this value is much larger than $|J_{AX}|$, then the applicability of the first-order perturbation theory in this case is justified. Correction to the energy levels in the first-order is equal to the diagonal matrix element of the perturbation operator $\widehat{\mathcal{H}}_{pert}$:

$$\Delta E = \langle M_{Ai}, M_{Xk} \mid J_{AX} \hat{\vec{I}}^{A} \cdot \hat{\vec{I}}^{X} \mid M_{Ai}, M_{Xk} \rangle = J_{AX} M_{Ai} M_{Xk}.$$

The eigenenergies of the full Hamiltonian is written like this:

$$E = \nu_A M_{Ai} + \nu_X M_{Xk} + J_{AX} M_{Ai} M_{Xk}.$$

Taking into account the selection rules for the allowed transitions, the lines with either $\Delta M_{Ai} = \pm 1$ while $M_{Xk} = \text{const}$ or $\Delta M_{Xi} = \pm 1$ while $M_{Ak} = \text{const}$ appear in the NMR spectrum. Their frequencies are determined by:

$$\Delta E_{Ak} = \nu_A + J_{AX} M_{Xk}, \qquad \Delta E_{Xi} = \nu_X + J_{AX} M_{Ai}. \tag{4.14}$$

Values of M_{Ai} and M_{Xk} are in ranges form $-M_{A,max}$ to $+M_{A,max}$ and from $-M_{X,max}$ to $+M_{X,max}$, respectively. Here, $M_{A,max} = I^A + 1$, where I^A is the total (cumulative) spin of nuclei A. For example, in the system A_3 of spins $I = 1/2$ the total spin $I^A = 3I = 3/2$. The number of lines in the part of spectrum of nuclei X is $2I^A + 1$. For example, spectrum of the AX_2 system with both spins $I = 1/2$ consists of a triplet close to the resonance frequency ν_A (total spin of nuclei X is 1), and a doublet around the frequency ν_X (total spin of nuclei A is 1/2).

Total (cumulative) probability (P_k) of transition with frequency ΔE_{Ak} depends on the combination number of orientations of spins X, which gives the same value M_{Xk}. Consider once again the AX_2 system with both spins $I = 1/2$. Eigenfunctions of the X subsystem are $|\varphi_1\rangle = |\alpha\alpha\rangle$ with $M_{X,1} = +1$, $|\varphi_2\rangle = |\alpha\beta\rangle$ with $M_{X,2} = 0$, $|\varphi_3\rangle = |\beta\alpha\rangle$ with $M_{X,3} = 0$, and $|\varphi_4\rangle = |\beta\beta\rangle$ with $M_{X,4} = -1$. The number of states with $M_X = 0$ is two, so the probability of transition with ν_A is twice of probabilities of transitions with frequencies $\nu_A + J_{AX}$ and $\nu_A - J_{AX}$, in other words the intensities of three lines in spectrum of the subsystem A are 1 : 2 : 1. In the system AX with spin $I_X = 1$ intensities of triplet lines of spin A are 1 : 1 : 1.

This rule can be applied to spectra of systems consisting of a large number of groups of magnetically equivalent nuclei. For example, the $A_pN_qX_r$ system has the following energy levels:

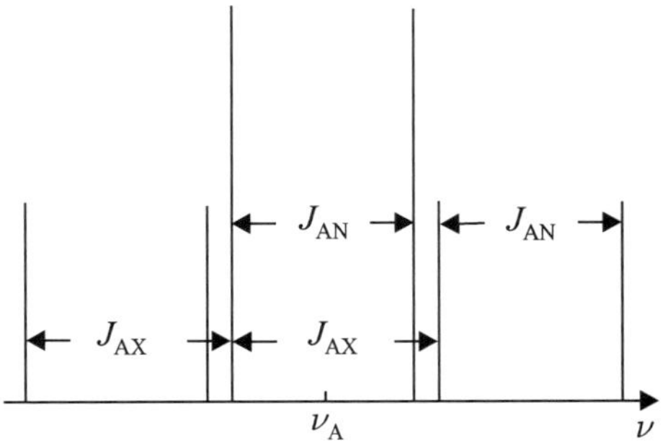

Fig. 4.2 First-order spectrum of nuclei A of the system ANX_2 with all spins $I = 1/2$

$$E = \nu_A M_A + \nu_N M_N + \nu_X M_X + J_{AX} M_A M_X + J_{AN} M_A M_N + J_{NX} M_N M_X$$

and transition frequencies:

$$\begin{aligned}\Delta E_A &= \nu_A + J_{AN} M_N + J_{AX} M_X,\\ \Delta E_N &= \nu_N + J_{AN} M_A + J_{NX} M_X,\\ \Delta E_X &= \nu_X + J_{AX} M_A + J_{NX} M_N.\end{aligned}$$

Line of subsystem A is splitted in a multiplet with lines spaced at J_{AN}, each of them, in turn, is splitted into a multiplet with distance between lines J_{AX}. Line intensities are expressed as a product of the degeneracy degree of states of M_N and M_X. For instance, in Fig. 4.2 the part of spectrum of ANX_2 system is shown. All spins are $I = 1/2$.

In a system $A_p N_q X_r \ldots$, the total integral intensities of all lines in parts of a spectra belonging to the magnetically non-equivalent spins relate as $p : q : r : \ldots$

4.2.4 Spectrum of a System AB

The Hamiltonian of a system AB (both $I = 1/2$) is as following:

$$\widehat{\mathcal{H}} = \underbrace{\tfrac{1}{2\pi}\gamma B_0(1-\sigma_A)}_{\nu_A} \hat{I}_{zA} + \underbrace{\tfrac{1}{2\pi}\gamma B_0(1-\sigma_B)}_{\nu_B} \hat{I}_{zB} + J\, \hat{\vec{I}}^{A} \cdot \hat{\vec{I}}^{B}.$$

Let us enumerate the basic vectors as it is done in (4.7). The matrix of the Hamiltonian has four diagonal elements, which are defined by the formula (4.9), and two off-diagonal, which are the same ($H_{23} \equiv H_{32}$) and defined by the formula (4.10) or (4.11). The matrix elements are:

$$\begin{aligned}
H_{11} &= \tfrac{1}{2}(+\nu_A + \nu_B) + \tfrac{1}{4}J \quad &\text{for} \quad |1\rangle = |\alpha\alpha\rangle,\\
H_{22} &= \tfrac{1}{2}(+\nu_A - \nu_B) - \tfrac{1}{4}J \quad &\text{for} \quad |2\rangle = |\alpha\beta\rangle,\\
H_{33} &= \tfrac{1}{2}(-\nu_A + \nu_B) - \tfrac{1}{4}J \quad &\text{for} \quad |3\rangle = |\beta\alpha\rangle,\\
H_{44} &= \tfrac{1}{2}(-\nu_A - \nu_B) + \tfrac{1}{4}J \quad &\text{for} \quad |4\rangle = |\beta\beta\rangle,\\
H_{23} &= H_{32} = \tfrac{1}{2}J.
\end{aligned}$$

To diagonalize the matrix of the Hamiltonian we need to diagonalize the only its central submatrix of 2×2:

$$\begin{pmatrix} H_{22} & H_{23} \\ H_{32} & H_{33} \end{pmatrix} = \begin{pmatrix} \tfrac{1}{2}(+\nu_A - \nu_B) - \tfrac{1}{4}J & \tfrac{1}{2}J \\ \tfrac{1}{2}J & \tfrac{1}{2}(-\nu_A + \nu_B) - \tfrac{1}{4}J \end{pmatrix}.$$

The secular equation for this submatrix is

$$\left(\tfrac{1}{2}(+\nu_A - \nu_B) - \tfrac{1}{4}J - E_n\right)\left(\tfrac{1}{2}(-\nu_A + \nu_B) - \tfrac{1}{4}J - E_n\right) - \tfrac{1}{4}J^2 = 0.$$

Expand the parenthesis and reduce similar terms:

$$\begin{aligned}
E_n^2 - E_n &\left(\tfrac{1}{2}(+\nu_A - \nu_B) - \tfrac{1}{4}J + \tfrac{1}{2}(-\nu_A + \nu_B) - \tfrac{1}{4}J\right)\\
&+ \left(\tfrac{1}{2}(+\nu_A - \nu_B) - \tfrac{1}{4}J\right)\left(\tfrac{1}{2}(-\nu_A + \nu_B) - \tfrac{1}{4}J\right) - \tfrac{1}{4}J^2\\
&= E_n^2 + \tfrac{1}{2}JE_n - \tfrac{1}{4}(\nu_A - \nu_B)^2 - \tfrac{3}{16}J^2 = 0.
\end{aligned}$$

The roots of this quadratic equation are

$$\begin{aligned}
E_2 &= +C - J/4,\\
E_3 &= -C - J/4,
\end{aligned}$$

where $C = +\sqrt{(\nu_A - \nu_B)^2 + J^2}\,/2$.

Eigenvectors $|\varphi_2\rangle$ and $|\varphi_3\rangle$ can be written as a linear combination of $|2\rangle$ and $|3\rangle$: $|\varphi_2\rangle = a_2|2\rangle + b_2|3\rangle$ and $|\varphi_3\rangle = a_3|2\rangle + b_3|3\rangle$. The coefficients a_2, b_2, a_3, and b_3 are the solution of the set of equations:

$$\begin{cases} a_2(H_{22} - E_2) + b_2H_{23} = 0 \\ a_2^2 + b_2^2 = 1 \\ a_3^2 + b_3^2 = 1 \\ a_2a_3 + b_2b_3 = 0 \end{cases} . \tag{4.15}$$

The second and third equations of the Set of equations (4.15) are the normalization of $|\varphi_2\rangle$ and $|\varphi_3\rangle$, the fourth line is the condition of their orthogonality. These are not linear equations, so we can find the solutions with the coefficient ±1. One of the solutions is:

Table 4.1 The frequencies and probabilities of transitions in the system AB ($I = 1/2$)

Transition	Frequency	Probability
$1 \leftrightarrow 2$	$\nu_c - C + J/2$	$(1 + \sin 2\theta)/4$
$1 \leftrightarrow 3$	$\nu_c + C + J/2$	$(1 - \sin 2\theta)/4$
$2 \leftrightarrow 4$	$\nu_c + C - J/2$	$(1 + \sin 2\theta)/4$
$3 \leftrightarrow 4$	$\nu_c - C - J/2$	$(1 - \sin 2\theta)/4$

Here, value $\nu_c = (\nu_A + \nu_B)/2$ is the central frequency of the quadruplet

$$| \varphi_2 \rangle = +\cos\theta \,| 2 \rangle + \sin\theta \,| 3 \rangle,$$
$$| \varphi_3 \rangle = -\sin\theta \,| 2 \rangle + \cos\theta \,| 3 \rangle.$$

Here, the value of θ can be find from the equation

$$\sin 2\theta = J/C. \tag{4.16}$$

All transitions in the system AB ($1 \leftrightarrow 2$, $1 \leftrightarrow 3$, $2 \leftrightarrow 4$, $3 \leftrightarrow 4$) are allowed. Their frequencies and probabilities are calculated using Formula (4.3). For example, let us calculate probability of transition $1 \leftrightarrow 2$:

$$P_{12} \sim \left| \langle \alpha\alpha \,|\, I_{xA} + I_{xB} \,|\, \varphi_2 \rangle \right|^2$$
$$= \left| \langle \alpha\alpha \,|\, I_{xA} + I_{xB} \,(\cos\theta | \alpha\beta \rangle + \sin\theta | \beta\alpha \rangle) \right|^2$$
$$= (\cos\theta + \sin\theta)^2/4 = (1 + \sin 2\theta)/4.$$

It is obviously from the example, why the Eq. (4.16) is written for $\sin 2\theta$. The frequencies and probabilities of all transitions in the system AB ($I = 1/2$) are presented in Table 4.1.

Form of the spectrum depends on the only relation $|J|/|(\nu_B - \nu_A)|$ but not the signs of these values. The spectrum consists of four lines: two inner and two outer. Inner lines are more intensive then outer ones. Distance (in frequency scale) between outer and inner lines is equal to modulus J, and the distance between centres of the left and right pairs of lines is $2C$. It is impossible to find the sign of J and to determine which resonance frequency higher, ν_B or ν_A.

If we take the limit $\nu_B \to \nu_A$, then a system AB becomes a system A_2. Value of $\sin 2\theta$ is equal to 1 in this system, and $C = J/2$. Probability of transitions $1 \leftrightarrow 3$ and $3 \leftrightarrow 4$ becomes zero, the frequencies of inner lines becomes $\nu_c = \nu_B = \nu_A$. So, the spectrum has a form of singlet with probability of the line being equal to 1. Another limit $|J| \ll |\nu_A - \nu_B|$ turns the system AB into AX. Here, $\sin 2\theta \to 0$, $C \to (\nu_B - \nu_A)/2$. Probabilities of all transitions will be equal to each other. Centres of the left and right pairs of lines become ν_B and ν_A, the splitting of the lines is J. Both limits confirm the correctness of the solution presented in Table 4.1.

Table 4.2 Basis functions for calculation of spectrum of ABX system

Number	Basis function	Total spin
$\vert 1\rangle$	$\vert\alpha\alpha\ \alpha\rangle$	+3/2
$\vert 2\rangle$	$\vert\alpha\alpha\ \beta\rangle$	+1/2
$\vert 3\rangle$	$\vert\alpha\beta\ \alpha\rangle$	+1/2
$\vert 4\rangle$	$\vert\beta\alpha\ \alpha\rangle$	+1/2
$\vert 5\rangle$	$\vert\alpha\beta\ \beta\rangle$	−1/2
$\vert 6\rangle$	$\vert\beta\alpha\ \beta\rangle$	−1/2
$\vert 7\rangle$	$\vert\beta\beta\ \alpha\rangle$	−1/2
$\vert 8\rangle$	$\vert\beta\beta\ \beta\rangle$	−3/2

The first and second symbols correspond to the states of the spins A and B, respectively; third symbol is the state of the spin X. *Horizontal lines* separate states which do not mix to each other

4.2.5 Spectrum of a System ABX

In this subsection we briefly describe the spectrum of a system ABX (all spins $I = 1/2$). The size of the Hamiltonian matrix of the system is 8×8. In general, the matrix will be composed of four submatrices with dimensions $1 \times 1, 3 \times 3,\ 3 \times 3$ and 1×1. In the Table 4.2, the Zeeman basis functions of the system is written. Taking into account the fact that eigenstates with opposite orientation of spin X do not mix, we can state that vectors $|\,2\,\rangle$ and $|\,7\,\rangle$ are the eigenvectors of the total Hamiltonian. So, each of submatrix 3×3 split into two parts 2×2 and 1×1.

A quadratic equation for a system ABX is similar to that of a system AB, but to solve it we have to introduce a pair of values C_+ and angle θ_+ for one orientation of spin X, and C_- and angle θ_- for another. Maximum number of transitions in the system is 15, but transition $7 \leftrightarrow 2$ is forbidden. So, fourteen transitions are allowed: eight transitions close to frequency $(\nu_A + \nu_B)/2$, and six transitions close to frequency ν_X. Absolute value of all constants of J-coupling and sign of the relation J_{AX}/J_{BX} can be found from the spectrum.

The above examples of the analytical calculation of NMR spectra of the systems AB and ABX are the simplest. There are methods for analytical calculation of such complex systems as AB_qX_r, A_2B_q, and others. However, in practice NMR spectra are calculated numerically on a computer. To simplify the spectra and to facilitate the finding of energies of spin interactions, double resonances and two-dimensional spectroscopy are used (see Chap. 15).

4.3 Dynamic Effects in the NMR Spectra

The shape and width of the NMR spectrum is sensitive to the processes with characteristic frequencies of the same order of magnitude as the width of the spectrum, measured in frequency units (Hz). If characteristic frequencies of a process less than a spectrum width, a spectrum consists of separate parts corresponding to the each state of nuclei. If characteristic frequencies of a process are much greater than a spectrum width, a spectrum is defined by the average state of nuclei. This case satisfies the definition of magnetically equivalent nuclei (see Sect. 4.1.4). Therefore, NMR spectroscopy is a very powerful method to study various dynamic processes in materials, when characteristic frequencies of processes are close to a spectrum width.

The time-depending nuclear interactions, which changes due to the random motion of molecules in liquids (such as dipole-dipole or quadrupole), influence the shape and width of a spectral line. Other types of molecular motion, such as chemical exchange of atoms or ions between the molecules of liquid, internal rotation of molecular groups in molecules (conformational transitions), formation and collapse of molecular associations, etc. also cause a variety of dynamic effects in NMR spectra. A characteristic feature of some forms of motion is that a nucleus changes its electronic environment, and therefore, it changes its chemical shift (σ). In some cases the constants of indirect spin-spin interaction (J-coupling) of a moving nucleus with other nuclei in a molecule are subject to change in time. For example, in a molecule of ethanol $CH_3 - CH_2 - OH$, the proton of the hydroxyl group –OH is involved in the chemical exchange with the hydroxyl protons of other molecules of alcohol, as well as the protons of water molecules.

In general, the calculation of the NMR spectrum of molecular systems with chemical exchange, hindered rotation, and other forms of movement is a difficult task.

4.3.1 Exchange Between Two States

In simplified methods of spectra calculation, a system of nuclei with chemical exchange between two magnetically inequivalent states A and B is considered. The resonance frequencies[3] of a spin system in states A and B are ω_A and ω_B. That means these frequencies is due to different chemical shift in the two states (σ_A and σ_B). It is assumed also that there are no other spin–spin interactions between spins in both states (J-couplings $J_{AB} = 0$). If chemical exchange is absent in such a system, we can assume the system consists of two independent parts. Magnetization of both parts are described with the Bloch equations (1.62–1.64). In the case of weak radio

[3] In this Section, an angular frequency $\omega = 2\pi\nu$ in units of radians per second is used for the simplification of equations.

frequency magnetic field B_x, the magnetization ($\vec{M}$) is slightly different from its equilibrium value $\vec{M}_0$. The equations for the orthogonal components of magnetization u and v in the rotational frame are:

$$\frac{du}{dt} + \frac{u}{T_2} + \Delta\omega v = 0, \tag{4.17}$$

$$\frac{dv}{dt} + \frac{v}{T_2} - \Delta\omega u = -\gamma B_x M_0. \tag{4.18}$$

Using the complex magnetization $M = u + iv$ we can rewrite the Eqs. (4.17) and (4.18) in the complex form:

$$\frac{dM}{dt} + \alpha M = -i\gamma B_x M_0, \tag{4.19}$$

where $\alpha = T_2^{-1} - i\Delta\omega$.

In the case of the absence of chemical exchange between states A and B, the Eq. (4.19) can be used for the each state independently:

$$\frac{dM_A}{dt} + \alpha_A M_A = -i\gamma B_x M_{0A}, \quad \frac{dM_B}{dt} + \alpha_B M_B = -i\gamma B_x M_{0B},$$

where $\alpha_A = T_{2A}^{-1} - i\Delta\omega_A$, $\Delta\omega_A = \omega_A - \omega$; $\alpha_B = T_{2B}^{-1} - i\Delta\omega_B$, $\Delta\omega_B = \omega_B - \omega$.

Let us consider that transitions of nuclei A to state B and back are characterized with random jumps. The average lifetimes of nuclei in states A and B between jumps are τ_A and τ_B, respectively. The jumps are instant, so we can neglect the jumps between the same states (A↔A) and the existence of some intermediate states between A and B. In this case we can write the equations for M_A and M_B in the following form:

$$\frac{dM_A}{dt} + \alpha_A M_A = -i\gamma B_x M_{0A} - \tau_A^{-1} M_A + \tau_B^{-1} M_B, \tag{4.20}$$

$$\frac{dM_B}{dt} + \alpha_B M_B = -i\gamma B_x M_{0B} - \tau_B^{-1} M_B + \tau_A^{-1} M_A, \tag{4.21}$$

where $M_{0A} = P_A M_0$, $P_A = \tau_A(\tau_A + \tau_B)^{-1}$; $M_{0B} = P_B M_0$, $P_B = \tau_B(\tau_A + \tau_B)^{-1}$.

If we use CW (continues wave) method for the spectrum recording than

$$\frac{dM_A}{dt} = 0, \quad \frac{dM_B}{dt} = 0,$$

and Eqs. (4.20) and (4.21) have a form

$$\begin{aligned} (\alpha_A + \tau_A^{-1})M_A - \tau_B^{-1} M_B &= -i\gamma B_x M_0 P_A, \\ -\tau_B^{-1} M_A + (\alpha_B + \tau_B^{-1})M_B &= -i\gamma B_x M_0 P_B. \end{aligned}$$

To find a line shape we should calculate the imagine part v of total magnetization M:

$$M = M_{\mathrm{A}} + M_{\mathrm{B}} = u + \mathrm{i}v.$$

Let us write solutions for M_{A}, M_{B}, and M in the form

$$M_{\mathrm{A}} = -\mathrm{i}\gamma B_{\mathrm{x}} M_0 \frac{P_{\mathrm{A}}(\alpha_{\mathrm{B}} + \tau_{\mathrm{B}}^{-1}) + P_{\mathrm{B}}\tau_{\mathrm{B}}^{-1}}{(\alpha_{\mathrm{A}} + \tau_{\mathrm{A}}^{-1})(\alpha_{\mathrm{B}} + \tau_{\mathrm{B}}^{-1}) - (\tau_{\mathrm{A}}\tau_{\mathrm{B}})^{-1}}, \tag{4.22}$$

$$M_{\mathrm{B}} = -\mathrm{i}\gamma B_{\mathrm{x}} M_0 \frac{P_{\mathrm{B}}(\alpha_{\mathrm{A}} + \tau_{\mathrm{A}}^{-1}) + P_{\mathrm{A}}\tau_{\mathrm{A}}^{-1}}{(\alpha_{\mathrm{A}} + \tau_{\mathrm{A}}^{-1})(\alpha_{\mathrm{B}} + \tau_{\mathrm{B}}^{-1}) - (\tau_{\mathrm{A}}\tau_{\mathrm{B}})^{-1}}, \tag{4.23}$$

$$M = -\mathrm{i}\gamma B_{\mathrm{x}} M_0 \frac{\tau_{\mathrm{A}} + \tau_{\mathrm{B}} + \tau_{\mathrm{A}}\tau_{\mathrm{B}}(P_{\mathrm{A}}\alpha_{\mathrm{B}} + P_{\mathrm{B}}\alpha_{\mathrm{A}})}{(1 + \alpha_{\mathrm{A}}\tau_{\mathrm{A}})(1 + \alpha_{\mathrm{B}}\tau_{\mathrm{B}}) - 1}. \tag{4.24}$$

The above expression (for the component v) is rather complex, so we consider special cases.

Slow exchange: the values of τ_{A} and τ_{B} satisfies the inequalities

$$\tau_{\mathrm{A}}^{-1} \ll |\omega_{\mathrm{A}} - \omega_{\mathrm{B}}|, \qquad \tau_{\mathrm{B}}^{-1} \ll |\omega_{\mathrm{A}} - \omega_{\mathrm{B}}|.$$

In addition, let the resonance lines of nuclei in the states A and B does not overlap (in absence of exchange). From (4.22) and (4.23) it follows that in the frequency range close to the ω_{A} value M_{A} can be written as

$$M_{\mathrm{A}} = -\mathrm{i}\gamma B_{\mathrm{x}} M_0 \frac{P_{\mathrm{A}}}{(\alpha_{\mathrm{A}} + \tau_{\mathrm{A}}^{-1})} = \mathrm{i}\gamma B_{\mathrm{x}} M_0 \frac{P_{\mathrm{A}}}{(T'_{2\mathrm{A}})^{-1} - \mathrm{i}\Delta\omega_{\mathrm{A}}},$$
$$M_{\mathrm{B}} = 0,$$

here $(T'_{2\mathrm{A}})^{-1} = T_{2\mathrm{A}}^{-1} + \tau_{\mathrm{A}}^{-1}$. Consequently,

$$v \simeq v_{\mathrm{A}} = -\gamma B_{\mathrm{x}} M_0 P_{\mathrm{A}} \frac{T'_{2\mathrm{A}}}{1 + (T'_{2\mathrm{A}}\Delta\omega_{\mathrm{A}})^2},$$

and in the frequency range close to ω_{B}

$$v \simeq v_{\mathrm{B}} = -\gamma B_{\mathrm{x}} M_0 P_{\mathrm{B}} \frac{T'_{2\mathrm{B}}}{1 + (T'_{2\mathrm{B}}\Delta\omega_{\mathrm{B}})^2},$$

here $(T'_{2\mathrm{B}})^{-1} = T_{2\mathrm{B}}^{-1} + \tau_{\mathrm{B}}^{-1}$.

So, in the case of slow exchange, two lines with the resonance frequencies ω_{A} and ω_{B} does not overlap. Their widths depend on τ_{A} and τ_{B} and they are greater than their widths in the case on absence of exchange.

Fast exchange: the values of τ_A and τ_B are very small, i.e.

$$\tau_A^{-1} \gg (\omega_A - \omega_B), \qquad \tau_B^{-1} \gg (\omega_A - \omega_B).$$

In this case we can neglect the terms with $\tau_A\tau_B$ in the Eq. (4.24). Magnetization M has form

$$M = -i\gamma B_x M_0 \frac{\tau_A + \tau_B}{\alpha_A\tau_A + \alpha_B\tau_B} = -i\gamma B_x M_0 \frac{1}{(T'_2)^{-1} - i(\omega_c - \omega)},$$

here $(T'_2)^{-1} = P_A(T_{2A})^{-1} + P_B(T'_{2B})^{-1}$, $\omega_c = P_A\omega_A + P_B\omega_B$ is the average frequency. The component v can be expressed as:

$$v = -\gamma B_x M_0 \frac{T'_2}{1 + [T'_2(\omega_c - \omega)]^2}.$$

In the case of fast exchange, the single line exists with the following parameters: resonance frequency is ω_c, the width of the line is $2(T'_2)^{-1}$. The position of the line and line width depend on the probabilities (P_A and P_B) of states A and B.

Intermediate exchange: its conditions are

$$\tau_A^{-1} \simeq (\omega_A - \omega_B), \qquad \tau_B^{-1} \simeq (\omega_A - \omega_B).$$

The general formula (4.24) is used in this case. Expression for the value of v is very complex, so we consider the special case:

$$P_A = P_B = 1/2, \qquad \tau_A = \tau_B = 2\tau, \qquad T_{2A}^{-1} = T_{2B}^{-1} \simeq 0. \qquad (4.25)$$

The last condition means the line width is vanishingly small compare with distance between lines. From the equation (4.24) it follows:

$$M = -i\gamma B_x M_0 \frac{4\tau - i2\tau^2(\Delta\omega_A - \Delta\omega_B)}{(1 - i2\tau\Delta\omega_A)(1 - i2\tau\Delta\omega_B) - 1},$$
$$v = -\frac{1}{4}\gamma B_x M_0 \frac{\tau(\omega_A - \omega_B)^2}{(\omega_c - \omega)^2 + \tau^2(\omega_A - \omega_B)^2(\omega_B - \omega)^2}, \qquad (4.26)$$

where $\omega_c = (\omega_A - \omega_B)/2$. If the condition (4.25) is satisfied, then the Eq. (4.26) is valid for any exchange rates. It is suitable for a wide range of τ from high values (slow exchange) to low values (fast exchange). Equation (4.26) predicts two non-overlapping lines at ω_A and ω_B if τ is high, and one line with average frequency $\omega_c = (\omega_A - \omega_B)/2$. One or two wide lines are observed in a spectrum if τ has intermediate value. Calculating positions of extremum $v(\omega)$ according to (4.26), the distance δ between them (in frequency units) as function of τ $(\omega_A - \omega_B)$ can be estimated:

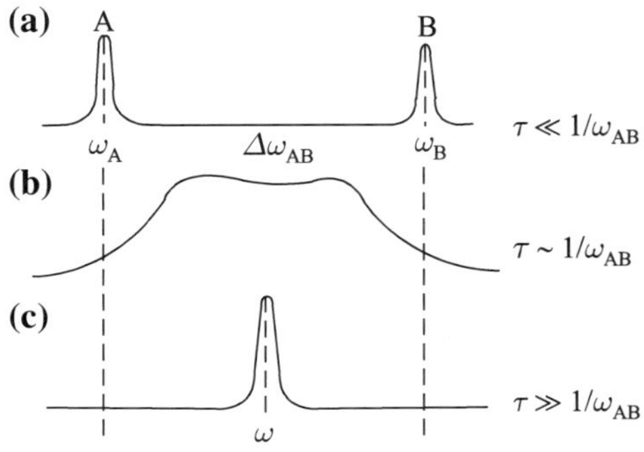

Fig. 4.3 Dependence of line shape in the case of slow (**a**) intermediate (**b**), and fast exchange (**c**)

$$\delta = \sqrt{1 - \frac{2}{\tau^2(\omega_A - \omega_B)^2}}(\omega_A - \omega_B).$$

This formula predicts two lines in spectrum if $\tau(\omega_A - \omega_B) \geq \sqrt{2}$, and one line if $\tau(\omega_A - \omega_B) \leq \sqrt{2}$.

Line shapes in the Fig. 4.3 are shown for some values of parameter $k = \tau(\omega_A - \omega_B)$. If τ is high, two separate lines become broad (Fig. 4.3a). With changing τ to lower values the broadening is accompanied with decreasing splitting between the lines. If $k = \sqrt{2}$ then lines are united (Fig. 4.3b). The following decrease of value τ leads to narrowing the single line at the frequency $\omega_{AB} = \omega_c$ (Fig. 4.3c).

4.3.2 Exchange Between Multiple States

If there are n stable states, then we can generalize the above result in the following manner. Denote ω_j and T_{2j} resonance frequency and transverse relaxation time of spins in the state j, respectively, $j = 1, 2, \ldots, n$. Introduce the value

$$\alpha_j = T_{2j}^{-1} - i(\omega_j - \omega).$$

Assume the probability of transition of nucleus from the state i to k per unit time is τ_{jk}^{-1}. In the equilibrium state, probabilities P_j of finding a nucleus in the state j satisfies the equation

$$P_j \sum_k \tau_{jk}^{-1} = \sum_k P_k \tau_{kj}^{-1}.$$

Because of transition from state i to the same state i does not affect the spectrum, we assume $\tau_{ii}^{-1} = 0$.

The differential equation for a complex magnetization M_j, which corresponds to state j, can be written as generalization of Eq. (4.21):

$$\frac{dM_j}{dt} + \alpha_j M_j = -i\gamma B_x M_{0j} + \sum_k (\tau_{kj}^{-1} M_k - \tau_{jk}^{-1} M_j). \tag{4.27}$$

In order to obtain the steady-state solution of the differential equation we should set $dM_j/dt = 0$, than solve the linear algebraic equation and calculate linear part

$$M = \sum_j M_j.$$

In the case of slow exchange, assume the transitions lead to the only broadening the line. The solution of Eq. 4.27 is

$$M_j = -i\gamma B_x M_0 P_j \tau_j (1 + \alpha_j \tau_j)^{-1}.$$

We can calculate the width of a line:

$$(T'_{2j})^{-1} = T_{2j}^{-1} + \tau_j^{-1},$$

here $\tau_j^{-1} = \sum_k \tau_{jk}^{-1}$ is the probability of transition of nucleus from the state j to any other state per unit time. So, if the natural line width (connected to T_{2j}^{-1}) is known or we can neglect this value, then we can estimate the average life time of nucleus in the state j independently for each spectral line.

4.3.3 *Effect of Chemical Exchange on J-Coupling*

The previous sections deal with effect of chemical exchange on the spectra of spins without J-coupling, with the only different chemical shifts. Discuss now how the exchange influences on the multiplet structure of spectra. To simplify the formulae, assume the system consists of two spins I_1 and I_2 with J-coupling:

$$\hat{\mathcal{H}} = -\omega_1 \hat{I}_{z1} - \omega_2 \hat{I}_{z2} + J_{12} \hat{\vec{I}}_1 \cdot \hat{\vec{I}}_2,$$

where $\omega_1 = \gamma_1(1-\sigma_1)B_0$, $\omega_2 = \gamma_2(1-\sigma_2)B_0$. Both spins are in the same molecule but exchange is intermolecular and breaks the coupling.

Assume $J_{12} \ll |\omega_1 - \omega_2|$, that means the spectrum is AX. According to formula (4.14), the spectrum consists of two multiplets with frequencies

$$\omega_1 - J_{12} I_{z2}, \qquad \omega_2 - J_{12} I_{z1},$$

where $I_{z2} = -I_2, -I_2 + 1, \ldots, +I_2$, $I_{z1} = -I_1, -I_1 + 1, \ldots, +I_1$. Assume that the nucleus 1 permanently resides in a molecule but the nucleus 2 takes part in the intermolecular exchange. So, for the simplest case of two spins $I_1 = I_2 = 1/2$ in absence of exchange the spectrum consists of two doublets with the resonance frequencies

$$\omega_1' = \omega_1 + J_{12}/2, \quad \omega_1'' = \omega_1 - J_{12}/2,$$
$$\omega_2' = \omega_2 + J_{12}/2, \quad \omega_2'' = \omega_2 - J_{12}/2.$$

The splitting between the lines in the doublets is J_{12}. When the nucleus 2 performs random jumps from one molecule to another, its resonance frequency changes abruptly, taking the values of ω_2' or ω_2'', depending on the value of I_{z1} of the spin 1 in that molecule to which currently attached nucleus 2. The resonance frequency of the spin 1, which remains in the molecule, also changes randomly from ω_1' to ω_1'', since new spin 2, which arrives to the molecule during the exchange act, has different value of I_{z2}. Thus, if at least one nucleus is involved in the exchange, the resonance frequencies of both nuclei undergo random intermittent changes, and therefore the exchange process affects both doublets $\omega_1'\omega_1''$ and $\omega_2'\omega_2''$ simultaneously.

The general dependence of the multiplet structure on the exchange rate is similar to that obtained in the previous sections. The condition of slow exchange is

$$\tau^{-1} \ll J_{12},$$

while for fast exchange it is

$$\tau^{-1} \gg J_{12},$$

where τ is the mean lifetime of spin 2 in the molecule between jumps. Slow exchange leads to widening of the multiplet's lines. Fast exchange leads to the loosing of the multiplet structure and to the appearance of two singlets on the central frequencies ω_1 and ω_2.

These results for two-positional exchange can be generalized for the case of several nuclei with arbitrary spins as well as groups of magnetically equivalent nuclei. The dependence of the structure of NMR spectra on the rate of chemical exchange is widely used for investigation of dynamical processes in matter.

For example, consider proton NMR spectra of two samples of ethanol: pure dry alcohol and alcohol with addition of hydrochloric acid. The spectra are shown in the Fig. 4.4. The spectrum of pure ethanol is A_3N_2X, i.e. it can be calculated using the perturbation theory. The values of J-couplings are: $J_{A_3-N_2} \simeq 7.1$ Hz, $J_{N_2-X} \simeq 4.8$ Hz. Spin–spin interactions between spins of –CH_3 and –OH groups are very small and do not lead to splitting the corresponding lines in the spectrum. The spectrum of –CH_2– group contains 8 lines which origin is the splitting of the line into 4 lines because of the interaction with –CH_3 group (total spin $I_{CH_3} = 3/2$) and the splitting of each line into 2 lines because of interaction with –OH. Both spectra of –CH_3 and

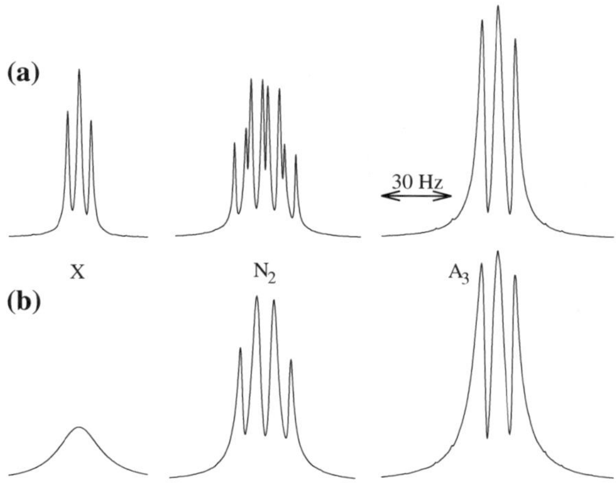

Fig. 4.4 Proton NMR spectra of ethanol: **a** pure dry alcohol, **b** alcohol with addition of hydrochloric acid. Spectrum type is A_3N_2X. Separated parts (scale is the same for all parts) are shown: X– group –OH, N_2– group –CH_2–, A_3– group –CH_3

–OH groups are triplets because of total spin of –CH_2– group is $I_{CH_2} = 1$. Addition of hydrochloric acid leads to the fast exchange of hydroxyl protons with protons of the acid and with similar protons in other molecules. As a result, J-coupling of this proton with protons of –CH_2– group is broken. Its spectrum becomes singlet, the spectrum of –CH_2– group is simplified to quadruplet, but the spectrum of –CH_3 group remains untouched.

In addition to the intermolecular exchange, other time-dependent processes, which lead to random reorientation one of the spin (changing the projection of spin $\hat{I}_{zi}$), influence on the multiplet structure of spectra. For the quadrupolar nucleus, quadrupolar relaxation is such a process. Fast reorientation of spin of quadrupolar nucleus breaks spin–spin interactions of this nucleus with other spins and spectra becomes more simple.

4.4 NMR-Relaxation in Liquids

In this section we present some examples of the study of the microstructure of liquids by measurements of the nuclear magnetic relaxation times. Key achievements in this area are related to the determination of parameters of molecular motion (reorientation time, activation energy, etc.). Because of such motion depends on intermolecular interactions, in many cases it is possible to draw conclusions on the character of the interactions and their variations under different external conditions (composition of a solution, temperature, pressure, etc.).

4.4.1 Separation of the Inter- and Intramolecular Contributions to the Dipole–Dipole Relaxation

In the Chap. 3, the problems of separation of contributions of different kinds of interactions to the relaxation rate are discussed. As for dipole–dipole relaxation, it is necessary to separate the contributions of intramolecular interactions (spins residue in one molecule) and intermolecular (spins belong to different molecules), see Sect. 3.9. To reach the goal one can use the isotopic substitution of atoms of interest if the magnetic moment of another isotope is zero or negligibly small. Thus, in the studying of magnetic resonance of protons, the deuterium (heavy Hydrogen) is used. Another method for the studying of inter- and intramolecular spin–spin interactions is a dilution of solute with inert magnetically weak substance like CCl_4 or CS_2. Using the both methods, one can not allow chemical exchange of nuclei of different species, as it is realized in water. Because of exchange, soon after preparing of mixture of H_2O and D_2O in proportion of 1:1, three kind of molecules will exist in the substance: 25 % of H_2O, 25 % of D_2O, and 50 % of HDO. Intramolecular interaction, which are usually the subject of investigation, are changed in the case of intermolecular chemical exchange.

A classic example of the correct separation of intra- and intermolecular contributions to the dipole relaxation of the nuclei ^{1}H was realized for benzene (J. C. Duplan, A. Briguet, J. Delman. *J. Chem. Phys.*, **54**(8), 3648-3649 (1971)). Using the formulae of Sect. 3.9 and introducing the concentration of protonated benzene (α) in the mixture, one obtains the following equation:

$$\frac{1}{T_1} = \left(\frac{1}{T_1^{\mathrm{H}}}\right)_{\mathrm{intra}} + \alpha\left(\frac{1}{T_1^{\mathrm{H}}}\right)_{\mathrm{inter}} + (1-\alpha)\left(\frac{1}{T_1^{\mathrm{D}}}\right)_{\mathrm{inter}}, \tag{4.28}$$

where the subscript H denotes the interaction between the protons of the same or different molecules, the index D refers to the interaction of protons with deuterons in the molecule C_6D_6. Further, considering the concentration of spins and gyromagnetic ratio of the proton and deuteron, the formula (4.28) can be transformed to the expression

$$\frac{1}{T_1} = \left[\left(\frac{1}{T_1^{\mathrm{H}}}\right)_{\mathrm{intra}} + 0.042\left(\frac{1}{T_1^{\mathrm{H}}}\right)_{\mathrm{inter}}\right] + 0.958\alpha\left(\frac{1}{T_1^{\mathrm{H}}}\right)_{\mathrm{inter}},$$

The dependence of $T_1^{-1}(\alpha)$ in benzene for the temperature 20 °C is shown in Fig. 4.5. Experimental data allows us to obtain separately values of $(1/T_1)_{\mathrm{inter}} = 0.036\,\mathrm{s}^{-1}$ and $(1/T_1)_{\mathrm{intra}} = 0.014\,\mathrm{s}^{-1}$. Knowing the value of the intramolecular contribution to the spin-lattice relaxation of protons, one can calculate the rotational reorientation time of the benzene molecule: $\tau_R = 1.7 \times 10^{-12}$ s.

The correctness of the method of isotopic substitution is confirmed, for example, with experiment for a mixture of normal and deuterated dioxane (Fig. 4.6). It is

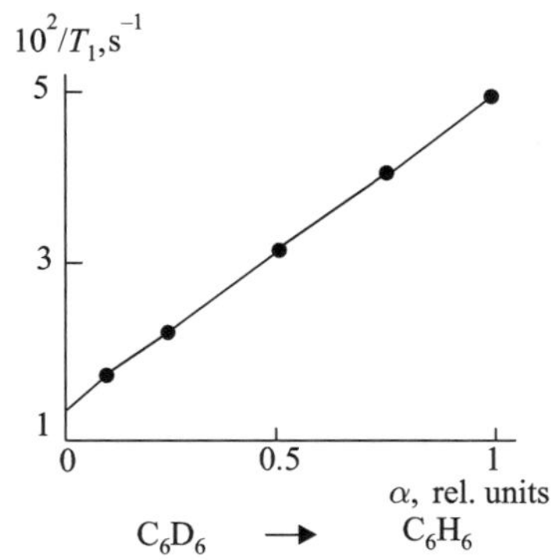

Fig. 4.5 Dependence of relaxation rate of proton NMR-relaxation at 20 °C on the concentration of deuterated benzene. Reproduced with permission from Duplan et al. (1971). Copyright 1971, American Institute of Physics

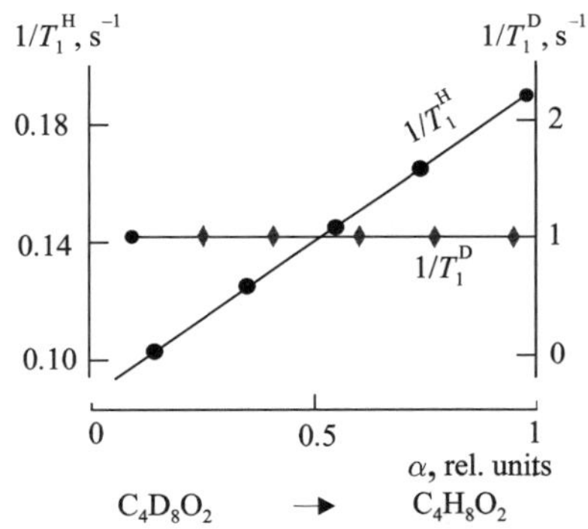

Fig. 4.6 Dependence of relaxation rates of proton and deuteron NMR-relaxation on the concentration of deuterated dioxane

followed from the experiment that the mobility of deuterated molecules is nearly the same as protonated ones in the whole range of concentrations. This conclusion is based on the formula (3.149) and on the fact that the quadrupole coupling constants in the dioxane molecule remain the same, which is determined by the structure of the chemical bonds.

If the material used for the "magnetic dilution" is not isotopically substituted analogs, but other liquids, then we should take into account the change of viscosity of a solution.

4.4.2 *Pure Liquids and Their Mixtures*

Due to contactless and nondestructive character of investigation by means of NMR-relaxation, it is powerful method for the studying molecular motion in liquids under variable conditions of temperature, pressure, composition of liquids, and other parameters influencing on the molecular mobility.

Before turning to the analysis of experimental results, we note that in the investigation of the resonance of nuclei with spin $I = 1/2$ (for example, ^{1}H), with the purely magnetic mechanism of relaxation (see Sects. 3.7–3.10), it necessary to degas the samples to eliminate the effect of dissolved molecular oxygen on the relaxation rate. Molecules O_2 are paramagnetic because of high electronic magnetic moment. They force significant relaxation of the neighbour spins (see Sect. 3.10). For example, relaxation rate of protons of water in degassed and nondegassed samples differs by 30 % while solubility of gas oxygen is 5 mL in 1 L of water at 20 °C. The amount of dissolved oxygen can vary with temperature and with changing of the composition of the medium. This can greatly complicate the interpretation of results. In the future, we mean that the dissolved oxygen removed from the samples, where it is necessary. Relaxation of nuclei with spin $I > 1/2$ is usually governed by quadrupole mechanism, which is much more intense than the magnetic relaxation channels, and degassing of samples is not necessary (except to experiments with nuclei ^{7}Li with small quadrupole moment).

Dependence the relaxation rates on the temperature. The main parameter, that determines the dependence of the nuclear magnetic relaxation on the temperature, is the correlation time (τ_c) in the formula in Sects. 3.7–3.12. Other values do less affect on the temperature dependence of $1/T_1$ and $1/T_2$. The interpretation of experimental data is usually based on activation model of motion:

$$\tau_c = \tau_c^0 \exp\left[E_a/(RT)\right], \tag{4.29}$$

where E_a is the activation energy. If motion of particles is fast ($\omega^2\tau_c^2 \ll 1$) then for the most of the processes the following formula is valid:

$$\frac{1}{T_{1,2}} \sim \tau_c.$$

In the ln $1/T_{1,2}$–$(1/T)$ (or ln $T_{1,2}$–$(1/T)$) coordinates, the experimental data should lie on the straight line (see, for example, part of the graph 3.3 for $\omega^2\tau_c^2 \ll 1$). In the study of liquids which are relatively simple in nature of intermolecular interactions, the experiment fully confirmed the theoretical predictions. From the slope of the line, it is easy to calculate the activation energy. The absolute value of τ_c is usually not calculated from the data of relaxation measurements due to the fact that the values of the quantities involved in the previously mentioned formulae are known with insufficient accuracy. The exception is when at low temperatures it is possible to study the so-called dispersion part of the temperature dependence of $1/T_1$, i.e. for $\omega\tau_c \simeq 1$. Since the resonance frequency is known, the correlation time at the extremum of temperature dependence is determined accurately. Then using Formula (4.29) we can calculate τ_c for different temperature ranges. However, in the most cases we can not reach a sufficient slow motion before the liquid freezing (there is a liquid to solid phase transition).

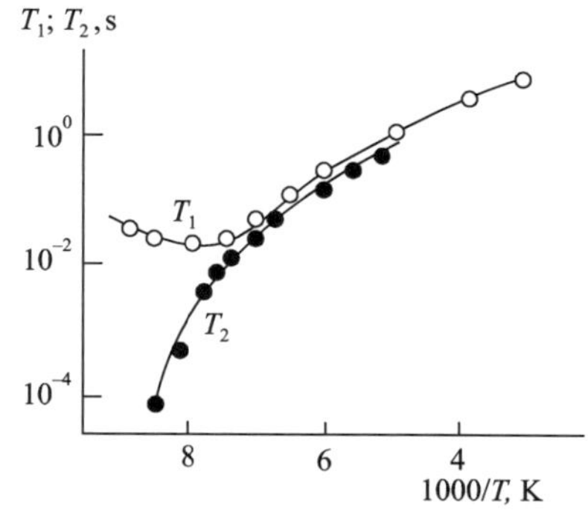

Fig. 4.7 The temperature dependence of the spin–lattice relaxation time of protons in liquid isobutyl bromide ($(CH_3)_2CHCH_2Br$)

A good example is the study of the temperature dependence of the spin–lattice relaxation time of protons in liquid isobutyl bromide ($(CH_3)_2CHCH_2Br$). This substance is good for testing the theory of relaxation, since, first, thermal motion in its liquid phase varies widely in the accessible range of the temperature (including supercooled state), and secondly, the molecules have a simple geometry and do not form associates, and thirdly, the NMR relaxation is determined by relatively simple interactions (dipole–dipole). The experimental results is depicted in Fig. 4.7, and they in general coincide with the prediction of the theory.

Similar results were obtained for glycerol and some other simple liquids. Strictly speaking, of course, there are some deviations of the experimental results from the predictions of the considered theory. They are expressed, in particular, in the deviation of the ratio T_1 / T_2 from the calculated value of 2.3 in the minimum of T_1. For example, this ratio for glycerol is about 4. In addition, the slopes of the straight lines in high and low temperatures are different. These circumstances are probably due to the presence of multiple correlation times (or due to the distribution of the correlation times—see Fig. 3.7), which can occur for the bulky anisotropic molecules.

Modern Fourier-transform radio spectrometers allow one to make the relaxation measurements for the nuclei of all chemical elements and to characterise the molecular motion in a variety of pure liquids and solutions. For example, a series of works of J. Kowalewski and co-workers devoted to studying the relaxation of ^{13}C in organic liquids. Based on relaxation data they studied the rotational diffusion, the internal dynamics in organic molecules, evaluated the activation energy of motion of different species. Data on the relaxation rates of the nuclei ^{33}S are used as a source of information about the motion of the molecules of dimethyl sulfoxide (DMSO) in aqueous solutions. The study of the relaxation of the quadrupole nuclei (nuclear spin of ^{33}S $I = 3/2$) situated in the molecular structures with low symmetry, gave information on the correlation times of rotational diffusion of molecules. The value of τ_R for the molecules of DMSO is found: at 303 K in pure DMSO $\tau_R = 3.7 \cdot 10^{-12}$ s.

Magnetic relaxation in liquid water. Let us now consider a more complicated case of the temperature dependence of the relaxation rate, which is realized in such an important object of study, like water. Modern experimental instruments allows one to register signal of magnetic resonance of all of the nuclei in water molecules in the conditions of their natural abundance: ^{1}H – 100 %, ^{2}H – 1.6 × 10^{-2} %, ^{17}O – 3.7 × 10^{-2} % (basic oxygen isotope ^{16}O has no magnetic moment). Actual research is usually governed with samples enriched in deuterium (up to 100 %) and Oxygen-17 (up to (5–10) %). Comparison of relaxation data of these nuclei, the mechanisms of relaxation of which are different, allows us to obtain reliable information about the molecular motion in the water. The results of measurements of the relaxation rates of different nuclei in the water have been published in a large number of articles, but the most complete data are contained in works of J. C. Hindman and colleagues (in the beginning of 1970s). They have shown that the temperature dependence of the relaxation rates of all these nuclei is not described by a simple single exponential function. Within experimental error at temperatures up to 408 K the data are satisfactorily approximated by two exponential functions:

$$\frac{1}{T_1} = A' \exp(E'_{a1}/(RT)) + B' \exp(E'_{a2}/(RT)).$$

The values of E'_{a1} and E'_{a2} are close to each other for any nuclei: $E'_{a1} \simeq 40\,\mathrm{kJ\,mol^{-1}}$ for low temperatures (<30 °C), but $E'_{a2} \simeq 14\,\mathrm{kJ\,mol^{-1}}$ for high temperatures (50–150 °C).

These two exponents can be associated with the double-structure model of the water, according to which some of the water molecules are associated with neighboring molecules with large number of hydrogen bonds (two to four), and other water molecules are relatively free (weak-bonded, which are connected to neighbors with just one H-bond or even absolutely free). It is known that number of the weak-bonded molecules enlarges with the increasing of temperature. In this case parameters A' and B' should depends on temperature. So, the values of E'_{a1} and E'_{a2} should be changed. The temperature dependence of A' and B' can be adequately interpolated with functions $\exp[\pm\alpha/(RT)]$. In this case we obtain

$$\frac{1}{T_1} = A \exp(\alpha_1/(RT)) \exp(E_{a1}/(RT)) + B \exp(-\alpha_2/(RT)) \exp(E_{a2}/(RT)),$$

where A and B is constants, and α_1 and α_2 are characteristics of temperature dependence of numbers of strong and weak H-bonded molecules. As a result, E'_{a1} and E'_{a2} have values of $-2\,\mathrm{kJ\,mol^{-1}}$ and $+6\,\mathrm{kJ\,mol^{-1}}$, respectively. It is interesting that energy $E_a \simeq 40\,\mathrm{kJ\,mol^{-1}}$ is close to activation energy of molecular motion in ice, where tetrahedral network of H-bonds is realised. The value $E_a \simeq 20\,\mathrm{kJ\,mol^{-1}}$ is about the energy of single H-bond in water.

Investigation of multi component systems. The study of binary and more complex systems by measuring the magnetic relaxation rates of different nuclei can track changes in the molecular structure and the nature of the thermal motion of the molecules in the mixture, depending on its composition and temperature. These dependencies are based on changes of the correlation times for rotational and translational diffusion. Since the correlation time can not be determined in an independent experiment, in the most studies, the NMR relaxation rates are compared with the viscosity of the substance (η) or with the diffusion coefficients (D) of molecules containing the resonating nuclei. Therefore, it is important to analyze the theory that relates the macroscopic characteristics of η and D with reorientation times. This analysis is beyond the scope of this book, but we refer to the three models which are actively used in the interpretation of experimental data:

1. *Stokes–Einstein hydrodynamic model*, which considers the motion of a molecule as movements of solid sphere in a medium with viscosity η;
2. *model of microviscosity*, which takes into account the finite size of the neighbouring particles (molecules);
3. *model of mutual viscosity*, which connects the reorientation times to moments of inertia and sizes of molecules of a mixture.

The latest is most convenient for the interpretation of data for the molecules without specific intermolecular interactions.

In the first years of application of relaxation methods to study the microstructure of solutions, researches did not put much attention on the homogeneity of the magnetic field B_0. The homogeneity of the magnets that time was about 10^{-5}. As a result, it was impossible to resolve the NMR spectra of each substance of a mixture. The "average" relaxation rate of all components are studied. This complicates interpretation of the data. This was the reason why substances with simple spectra were used in the early experiments.

Due to the complexity of handling multi-exponential curves, it is easy to study objects with single lines of nuclear magnetic resonance. This can be done by using isotopic substitution or by registering resonance of nuclei which are contained in one component of a mixture. For example, the investigation of the relaxation of ^{1}H and ^{2}H in mixtures of either protonated or fully deuterated dioxane with, respectively, deuterated or normal water is shown in Fig. 4.8. If we consider the average slope of the curves, we can determine that the maxima on the curves correspond to the concentration of dioxane is about 0.2. Deviation from linearity (increasing the relaxation rate) is usually associated with the formation of molecular associates. So in the case of dioxane–water we can assume the presence of labile complexes $C_4H_8O_2$—$4H_2O$. Similar extremums are observed in solutions of alcohols in water, in the system DMSO–H_2O, and in many other binary systems. Such behavior is typical for these objects and for the diffusion coefficients, which means symbate changes of rates of both rotational and translational motion.

Kowalewski and co-workers investigated the relaxation rates of ^{2}H, ^{13}C, ^{14}N, ^{35}Cl and measured diffusion coefficients (with spin–echo method with pulsed gradient, see Sect. 2.6) in binary systems acetonitrile–chloroform (H. Kovacs and other, *J. Phys. Chem.*, **93**(2), 962-969 (1989)). They determined the reorientation times of acetonitrile and chloroform at different temperatures and compositions of solutions, making it possible to discuss the applicability of various diffusion models to describe the motion of the molecules in these liquids.

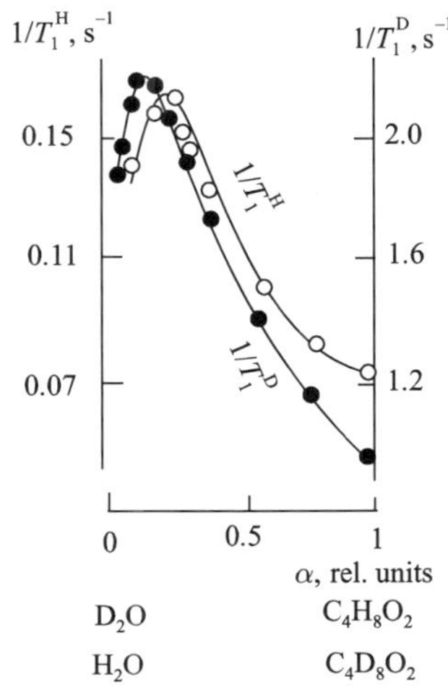

Fig. 4.8 The relaxation of ^1H and ^2H in mixtures of either protonated or fully deuterated dioxane with, respectively, deuterated or normal water

4.4.3 Diamagnetic Electrolyte Solutions

The study of NMR relaxation in electrolyte solutions is very promising, since changing the composition and temperature of the solution leads to a wide range of the conditions of the experiment, remaining however the same class of substances. A very important fact is that the data on resonances of different nuclei (located in both solvent and solute) can be compared to each other. The electrolyte solutions are interesting objects in view of their important role in the physical, chemical, biological processes, and in industry. This section will focus on the solutions of simple electrolytes, containing relatively small ions. Thus, we do not touch the study of ionized macromolecules and complex organic ions in solution. However, using the specified above range of objects, we show the problems and prospects of NMR relaxation in the study of the microstructure of electrolyte solutions.

In these solutions, almost all the basic mechanisms of the nuclear magnetic relaxation appear:

1. dipole–dipole;
2. scalar;
3. quadrupole;
4. spin–rotational;
5. chemical shift anisotropy.

Interaction of these types cause magnetic relaxation because of they are modulated by the thermal molecular motion. The first works in this area appeared in 1953–1955, and now their number exceeds many thousands. In particular, the interesting features of pulsed nuclear magnetic resonance revealed in the study of transport processes

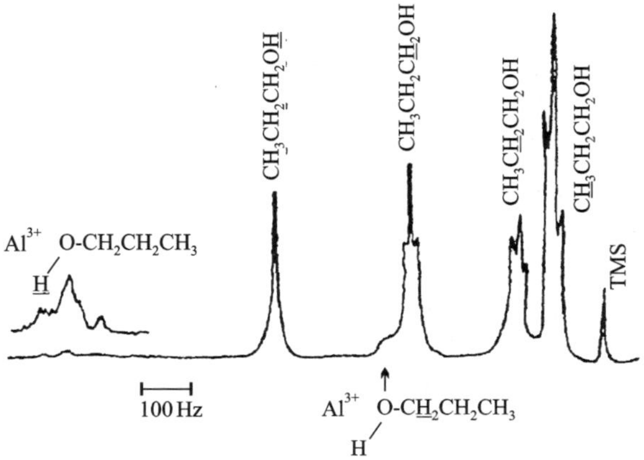

Fig. 4.9 The proton spectrum in n-propanol–0.44 moles of aluminum chloride at −25°C. Resonance frequency is 60 MHz

in electrolyte solutions under the conditions of flow of electric current through the sample.

Relaxation of magnetization of nuclei of solvent. If the nuclei in the substance are in structures with different electronic surroundings, the nuclear magnetic resonance spectrum should consist of several lines (see Sects. 4.1, 4.2). The ratio of the integral intensities of spectral lines is in strict proportion to the number of nuclei in each magnetically equivalent group of atoms. The position of the spectral lines shows the structure of the electronic shell of molecules (or molecular groups). The chemical shift may be sensitive to the environment of the molecule, on its solvation shell. Integral intensity of the line shows the number of molecules (number of atoms) in different groups of molecules. The relaxation rates of each spectral line give information on mobility of these molecules. These properties of NMR spectra are used, in particular, to study the structure of the solvation shells of the ions.

First the individual signals from molecules of solvation shells were obtained in the 1960's. Currently, resolved spectral lines of solvation shells of certain ions and the bulk solvent are recorded for all species of nuclei of the solvent molecules. An example of the proton spectrum in n-propanol–0.44 moles of aluminum chloride at −25 °C is shown in Fig. 4.9. Character of the lines splitting of the spectrum allows us to conclude on the positions and orientations of the molecule of alcohol with respect to the aluminum ion.

It is known that the process of chemical exchange prevents the registration of the individual signals from nuclei in different states. If the lifetime of a nucleus in the particular state is small, i.e. $1/\tau \simeq \pi\Delta\nu/2$ ($\Delta\nu$—is the difference in the resonance frequencies of the nuclei in different states), then spectrum consists of single line. Therefore, all the experimental results in a slow exchange conditions are obtained for multiply-charged small ions, which form stable solvate shells at low temperatures. Most of such studies are related to the solutions of aluminum. It is shown that the first solvation shell of the ion Al^{3+} consists of six molecules of the solvent in most cases. This is true for solutions of various aluminum salts in methanol, dimethyl sulfoxide, trimethyl phosphate, acetone, etc. From these data it can be concluded that the activation energy of the rotational motion of water molecules in the hydration shell of Al^{3+} is much smaller than the activation energy of the exchange of water molecules between hydration shell and the free bulk water. Similar studies was made for the ions Ga^{3+}, In^{3+}, Be^{2+}, Mg^{2+}, and so.

However, most often in electrolyte solutions, it is realised the condition of fast exchange of solvent molecules between all states. These states are first and other solvation (hydration) shells of cations and anions, bulk (free) solvent. Their spectrum consist of one single line. In this case the measured relaxation rate is described with Formula (3.152). Thus, one target value depends on many parameters (relative population of different states, mobility of nuclei in these states, etc.). So, the correctness of the assumptions about the structure of the solution as a whole or its individual structures determines the success of the investigations.

To increase the information obtained from the experiments we should measure the temperature and concentration dependence of the relaxation rates. It turns out that the relaxation characteristics are more sensitive to the restructuring of the solution than the chemical shifts of the spectral lines. The fact that the electronic screening of nuclei depends on more factors than the process of nuclear magnetic relaxation. From one hand, it "smoothes" the concentration or temperature dependence of the chemical shifts. On the other hand, it introduces difficulties to interpret the experimental results.

We start from the well-known and trusted expression (3.152) for the spin-lattice relaxation time in the conditions of fast exchange. The arguments given below are valid for the resonances of the nuclei ^1H, ^2H and ^{17}O of water molecules (in other solvents the nuclei are ^{13}C, ^{14}N). Generally speaking, the relaxation time T_1 and T_2 in low-viscosity fluids are equal, and therefore the time T_2 does not require a separate discussion. However, in the water, the spin–spin relaxation time of protons is smaller than the spin–lattice relaxation time by 30 %. This difference is due to the presence of scalar interaction of protons with oxygen isotope ^{17}O in the presence of proton exchange and the difference may vary depending on the concentration of the electrolyte. This greatly complicates the interpretation of data on the T_2. Because it effects on the spin-spin relaxation time of protons only, hereafter we consider only the spin-lattice relaxation, for which we can write:

$$\frac{1}{T_1} = \sum_{i=1}^{N-1} \frac{mn_i\alpha^{\pm}}{M}\frac{1}{T_{1i}} + \left[1 - \sum_{i=1}^{N-1} \frac{mn_i\alpha^{\pm}}{M}\right]\frac{1}{T_{10}} \tag{4.30}$$

or

$$\frac{1}{T_1} = \frac{1}{T_{10}} + \sum_{i=1}^{N-1} \frac{mn_i\alpha^{\pm}}{M}\left[\frac{1}{T_{1i}} - \frac{1}{T_{10}}\right], \tag{4.31}$$

here m is molality of the solution; M is number of moles of solute in 1 kg of solvent; T_{1i} is the relaxation time of nuclei of solvent in the ith state (index 0 correspond to

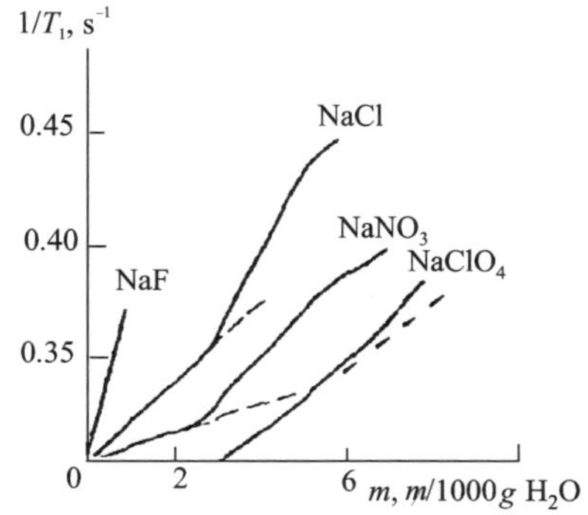

Fig. 4.10 The concentration dependence of the proton spin–lattice relaxation rates in aqueous solutions of Sodium salts (23 °C)

bulk solvent); n_i is the number of solvent molecules in the ith state per one ion; $\alpha^{\pm}$ is number of ions obtained after dissociation of solute. Formulae (4.30) and (4.31) predict a linear dependence of relaxation rate on the concentration of the electrolyte. Such dependence, however, is only observed up to concentrations of 2–5 moles of salt per 1 kg of H_2O (see, e.g., Fig. 4.10). Deviation from linearity can be explained either by changes in T_{1i}, or by the replacing of some states by other states with increasing concentration of the electrolyte ($n_i \to 0$ while n_j increases). Simultaneously with the elucidation of this question it was possible to study the influence of anions and cations on the measured value $(1/T_1)$.

Studies have shown the following:

1. In a certain combination of ions, the states of solvent molecules are determined by one selected species of ions over a wide concentration range. For instance, they are Na^+, Ca^{2+}, Ba^{2+} and so long, with anion ClO_4^-, which weak interacts with water molecules and does not form ion pairs. The analysis of the concentration dependence of relaxation rates of deuterons and protons in aqueous $NaClO_4$ solution allows us to reliably calculate coordination ions Na^+. In Fig. 4.11 the concentration dependence of $1/T_1$ for deuterons at 0 °C is shown. Additional confirmation of the conclusions was obtained from the analysis of the resonance of the nuclei ^{23}Na. After that, using the additive scheme (4.31), which validity is tested experimentally, we calculated value n_i for other states.
2. Study of resonances of the nuclei of other ions (^{19}F, ^{23}Na and others) revealed that the structure of the hydration shells of these ions is unchanged up to high concentrations. For example, the relaxation rate of the nuclei ^{19}F in a solution of KF–H_2O changes by approximately 15 % in the concentration range of KF from 1 to 7 mole per 1 kg water, while the relaxation of the protons accelerated by a factor of 2.5. This fact allowed us to estimate the mobility of water molecules in different states in the solution.
3. Based on the measurement of the spin–lattice relaxation time of protons and deuterons, and on the developed method of processing these data, it was possible

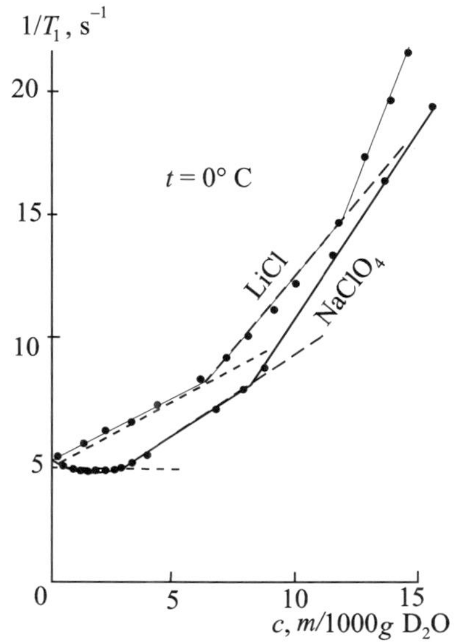

Fig. 4.11 The concentration dependence of the deuteron spin–lattice relaxation rates in some aqueous solutions

to calculate the essential characteristics of the microstructure of electrolyte solutions: coordination numbers of ions (n_i); relative values $\lambda_i = \tau_i/\tau_0 = T_{10}/T_{1i}$, which describe relative changing of mobility of the water molecules in the ith state with respect to bulk water; activation energies of thermal motion of molecules in different states in the solutions. These data for the temperature of 23°C is shown in Table 4.3.

Since the developed method is based on the study of the peculiarities of the concentration and temperature dependence of the nuclear magnetic relaxation, it is most suitable for studying the structure of solutions of highly soluble salts. However, if the information about the microstructure of hydration (solvation) shells of ions in the solute is known, we can set the general rules and the structure of the solution for poorly soluble compounds.

It should be noted that the method is justified only in cases if high precision (better than 2 %) of the measurements of the relaxation rates is realized. In other case the singularities of the concentration dependencies of $1/T_1$ become fuzzy, positions of bends of linear functions become uncertain.

Relaxation of nuclei of solute. Rate of relaxation of magnetization of ions determined, first of all, the structure of the first solvation shell, the closest to the ion. In particular, the solving of the problem of separation of the impact of anions and cations on the structure of the solution is impossible without the investigation of the resonance of the nuclei of dissolved salts.

Relatively small number of nuclei in the ions have spin $I = 1/2$. They are, for instance, ^{1}H, ^{13}C, ^{19}F, ^{31}P, ^{113}Cd, ^{207}Pb and others. Nuclei of Fluorine and

Table 4.3 Coordination numbers of ions in the aqueous solutions

Ion	n_1	λ_1	n_2	λ_2	n_Σ	λ_Σ
Li^+	4	2.3	—	—	4	2.3
Na^+	6	1.6	12	0.8	18	1.1
K^+	8	1.0	16	0.8	24	0.9
Mg^{2+}	6	4.5	12	1.3	54	1.4
Ca^{2+}	6	3.3	12	0.8	18	1.6
Sr^{2+}	8	2.7	16	0.8	24	1.3
Ba^{2+}	12	2.1	24	0.8	36	1.1
F^-	4	3.3	—	—	4	3.3
Cl^-	4	1.3	—	—	4	1.3
Br^-	4	0.9	—	—	4	0.9
I^-	4	0.8	—	—	4	0.8
H_3O^+	3	1.9	9	1.0	12	1.2
NH_4^+	4	0.8	—	—	4	0.8
NO_3^-	6	0.9	—	—	6	0.9
SO_4^{2-}	8	3.2	—	—	8	3.2
N_3^-	6	0.8	—	—	6	0.8
SCN^-	4	0.8				
CO_3^{2-}	9	2.5	—	—	9	2.5

Phosphorus are part of negative ions, and relaxation times of their magnetization are very sensitive to small impurities of transition metal cations, which leads to additional difficulties in the experiments and comparing the results of different investigators.

Most nuclei of ions have spin $I > 1/2$, and therefore the primary mechanism of relaxation for them is the quadrupole interactions. It is advisable to consider separately the quadrupole relaxation of nuclei in polyatomic and monatomic ions.

Electric field gradient at the location of the nucleus is determined by the electron shell of the ion, the neighboring solvent molecules, and oppositely charged ions. If the fluctuations of these species (and their impact on the electric field gradient) are independent, we can write, for example, the spin-lattice relaxation time:

$$\frac{1}{T_{1Q}} = \left(\frac{1}{T_{1Q}}\right)_{\text{internal}} + \left(\frac{1}{T_{1Q}}\right)_{\text{solution}} + \left(\frac{1}{T_{1Q}}\right)_{\text{ions}}, \tag{4.32}$$

which can be undoubtedly considered as acceptable approximation to reality.

In the case of polyatomic ions, electric field gradient at the location of the nucleus is determined (at least up to certain electrolyte concentrations), basically, by the electrons of the chemical bonds between atoms (which is corresponding to the first term in (4.32)). The fluctuations of the gradient in this case are mostly due to the reorientation of ions in space. Thus, the relaxation process is similar to that occurs in the molecules. With the increasing the concentration of the electrolyte, the role of the third term in (4.32) increases. This is true especially in the case when internal electric field has a high symmetry and quadrupole coupling constant is small (for example, for nuclei ^{35}Cl in the anion ClO_4^-).

In monatomic ions their own (spherically symmetric) electron shell does not create the electric field gradient at the position of the nucleus. The first term in (4.32) is explicitly absent. The effect of antiscreening should be considered separately, see Sect. 3.12. Fluctuating electric field in the position of the nucleus of a monatomic ion is generated by the solvent molecules and ions of opposite charge. The second term in (4.32) describes the relaxation rate of the ion in the condition of "infinite" dilution. The third term reflects the interaction with other ions, which may occur by direct contact or through layers of solvent molecules (indirectly). Generally speaking, because of the correlation in the motion of ions and solvent molecules, the terms in the right side of (4.32) are interdependent. At low concentrations of dissolved ions, it can be assumed that the experimentally measured relaxation rates reflect only the interaction of ions with the solvent molecules. Therefore two cases are considered separately: the relaxation processes at "infinite" dilution and at a real concentration of dissolved substances.

The theory of the quadrupole relaxation of ions on the basis of electrostatic model was developed in the 1960s by K. A. Valiev and H. G. Hertz. Despite the natural in general, and similar assumptions, the Valiev–Hertz theory could not describe a number of relaxation features in simple systems. In particular, the values of the relaxation rates, calculated on the basis of this theory, sometimes differ by an order of magnitude from the experimental values. This difference can not be caused by the inaccurate knowledge of the values of the parameters in the formulae (for example, the distance between ion center and the location of the charge in the molecule of solvent or counterions). Therefore, K. A. Valiev suggested in an article the influence of symmetry of hydration shells of the rate of quadrupole relaxation of monatomic ions. The development of these ideas was presented in works of V. I. Chizhik ("Quadrupole relaxation of hydrated ion nuclei", *Journal of Structural Chemistry*, **8**(2), 303-304 (1967)).

If some of the ion formed symmetrical hydration (solvation) shell, the gradient of the electric field at the position of the nucleus disappears, and the intensity of the quadrupole relaxation decreases. If under the influence of thermal motion one or more of the ligands are changed (or strong distortion of the complex takes place), the gradient appears (Fig. 4.12 b). Relaxation in this case is slower than in the case of Brownian motion (Fig. 4.12a). Calculation of the rate of the quadrupole relaxation of monatomic ions leads to the formula

$$\frac{1}{T_{1,2}} = \frac{9}{20}\,\frac{2I+3}{I^2(2I-1)}\,n_1\left[\frac{eQ}{\hbar}\,\frac{(1-\gamma)pZ}{r^3}\right]^2\left(\frac{\Delta}{\tau_1}\right)^2\tau_1, \tag{4.33}$$

where n_1 is the number of the solvent molecules in the first solvation (hydration) shell, τ_1 is the reorientation time of molecules in this shell (n_1 and τ_1 are listed in Table 4.3), Δ is the time interval when symmetry of the complex is broken. To obtain Formula (4.33), it was assumed that the solvent molecules around an ion can be approximated with set of effective charges of Z (at a distance r from the center of the ion), rather than with point dipoles, as it was done in the Valiev–Hertz theory.

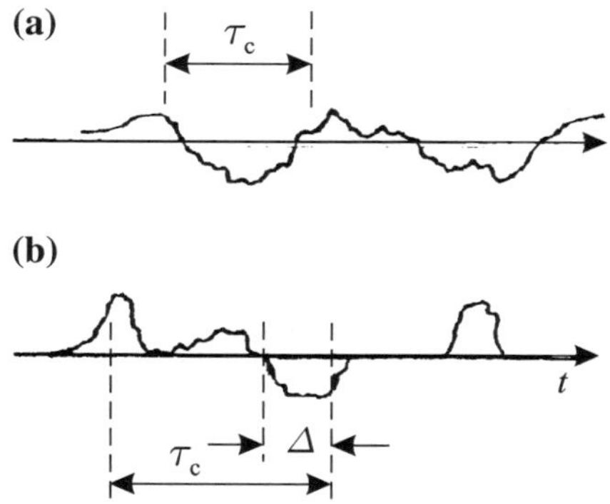

Fig. 4.12 The dependence of electric field gradient in accordance with the models of Valiev–Hertz (**a**) and Chizhik (**b**)

Interpreting the experimental data with the help of Formula (4.33), one can estimate the relative time of symmetry breaking of the complex if one of the molecules reorients (Δ/τ_1) and if of all molecules move independently in the complex ($n_1\Delta/\tau_1$).

4.4.4 Paramagnetic Electrolyte Solutions

Electron–nucleus spin–spin interaction is usually much stronger than other interaction of the nuclei spins in the matter. Therefore, it leads to so large contribution to the relaxation rate that the nuclear magnetic resonance at the nucleus of the paramagnetic ion, as a rule, is not observed due to excessive broadening of the spectral line. In solutions of paramagnetic substances, a magnetic relaxation rate of the *solvent molecules* is significantly higher than in absence of paramagnetic species. This is due to dipole–dipole and scalar interactions (see Sects. 3.10, 3.11) with unpaired electrons of the paramagnetic particles. The gyromagnetic ratio of electron is about 1000 times greater than that of any of the nuclei. Therefore, even in low concentrated solutions of paramagnetic substances, the relaxation time of the magnetic nuclei in the molecules of the solvent is determined mainly by the concentration and properties of paramagnetic particles. With the nuclear magnetic relaxation method one can measure the concentration of the solute paramagnetic substances, kinetic and thermodynamical parameters of solvation, can study the processes of complexation, kinetics of ligand exchange, oxidation-reduction reactions, microstructure of paramagnetic complexes and the electron density distribution in them, the formation of ionic pairs, etc.

If paramagnetic solutions are characterized by the high rate of exchange of solvent molecules between the solvation shell of the paramagnetic ion and the rest of the solution, then the relaxation rates are described by Eq. (3.152). The equation in this case can be rewritten as

$$\frac{1}{T_{1,2}} = \frac{p_{\mathrm{M}}}{(T_{1,2})_{\mathrm{M}}} + \frac{p_{\mathrm{s}}}{(T_{1,2})_{\mathrm{s}}}, \tag{4.34}$$

where p_{M} and p_{s} are the relative parts of water molecules in the solvate shell (index M) and the rest of solution (index s). Because of $(T_{1,2})_{\mathrm{M}} \ll (T_{1,2})_{\mathrm{s}}$, it is not necessary to distinguish different structures in the rest of solution, i.e. to divide p_{s} into subcategories. In many cases except noted below, the first term in Eq. (4.34) is much higher then the second term starting with the very low concentrations of the paramagnetic solute (of order of 10^{-3}–10^{-2} mole per 1000 cm^3 of water). Equation (4.34) transfers to

$$\frac{1}{T_{1,2}} = \frac{p_{\mathrm{M}}}{(T_{1,2})_{\mathrm{M}}}. \tag{4.35}$$

It is following from Eq. (4.35) that the problem of separation of impacts of cations and anions on the relaxation rate is not sufficient in the study of paramagnetic solutions unlike in the study of the diamagnetic solutions. This increases the validity of the interpretation of the experimental data, but reduces the amount of information on the structure of the solution in general.

Usually the data on the nuclear magnetic relaxation in the presence of paramagnetic ions are interpreted in terms of Solomon–Bloembergen equations, which, in general, are the combination of Eqs. (3.10), (3.137) and (3.138) with the assumption that the resonance frequency of electrons (ω_{0S}) is much higher than the frequency of nuclear magnetic resonance (ω_{0I}). The Solomon–Bloembergen equations are valid if two conditions are met:

1. electronic relaxation (motion of the electron spins) are not correlated with the molecular motion;
2. the main (on a scale of energy) interaction of the electrons is the interaction with the external static magnetic field $\vec{B}_0$ (Zeeman interaction).

For the nuclear spin I we can write (taking into account $\omega_{0I}^2\tau_{\mathrm{c}}^2 \ll 1$):

$$\begin{aligned}
\frac{1}{T_{1\mathrm{M}}} &= \frac{2}{15}\gamma_I^2\gamma_S^2\hbar^2 S(S+1)r^{-6}\left[3\tau_{\mathrm{c}1} + \frac{7\tau_{\mathrm{c}1}}{1+\omega_{0S}^2\tau_{\mathrm{c}1}^2}\right] \\
&\quad + \frac{2}{3}S(S+1)\hbar^{-2}A^2\frac{\tau_{\mathrm{c}2}}{1+\omega_{0S}^2\tau_{\mathrm{c}2}}, \\
\frac{1}{T_{2\mathrm{M}}} &= \frac{1}{15}\gamma_I^2\gamma_S^2\hbar^2 S(S+1)r^{-6}\left[7\tau_{\mathrm{c}1} + \frac{13\tau_{\mathrm{c}1}}{1+\omega_{0S}^2\tau_{\mathrm{c}1}^2}\right] \\
&\quad + \frac{1}{3}S(S+1)\hbar^{-2}A^2\left[\tau_{\mathrm{c}2} + \frac{\tau_{\mathrm{c}2}}{1+\omega_{0S}^2\tau_{\mathrm{c}2}^2}\right].
\end{aligned} \tag{4.36}$$

Depending on the value of $\omega_{0S}\tau_c$, Formulae (4.36) predict different relations of relaxation rates $1/T_1$ and $1/T_2$. Note that under the condition $\omega_{0I}^2\tau_c^2 \ll 1$, the relation $T_1 \simeq T_2$ is always true for the dipolar mechanism of relaxation independently on the value of $\omega_{0S}\tau_c$. If $\omega_{0S}^2\tau_c^2 \ll 1$ (that means that $\omega_{0I}^2\tau_c^2$ much less than unity), then strict equality $T_1 = T_2$ is valid for both channels of magnetic relaxation.

The question on the values of correlation times τ_{c1} and τ_{c2} (which is used in Formulae (4.36)) is rather complex in the theory of magnetic relaxation in the paramagnetic systems. Changing of the values of dipole–dipole and scalar interactions can be influenced by the motion of different types. First of all, these motions may be rotations. For instance, rotation of a complex as whole, rotation of a molecule around the axis paramagnetic ion–molecule, etc. In the case of independent rotations around two axes we can write

$$\frac{1}{\tau_c} = \frac{1}{\tau_r} + \frac{1}{\tau_R}.$$

The general formula for the dipole–dipole part in the relaxation can be obtained if we assume chemical exchange of the atoms (with resonant nuclei) between the molecules in the solution with characteristic time τ_M, the influence of reorientation of the neighboring nucleus of the paramagnetic particle with the relaxation time T_S, independence of fluctuation of all magnetic fields which lead to the relaxation:

$$\frac{1}{\tau_{c1}} = \frac{1}{\tau_r} + \frac{1}{\tau_R} + \frac{1}{\tau_M} + \frac{1}{T_S}. \tag{4.37}$$

As for the scalar part, Formula (3.135) is valid but we should rewrite it here taking the notation of the Formulae (4.36):

$$\frac{1}{\tau_{c2}} = \frac{1}{\tau_M} + \frac{1}{T_S}. \tag{4.38}$$

Note that scalar interactions are independent on the rotations.

The mentioned at the beginning of this Section methods of studying the structure of paramagnetic electrolyte solutions do only work if the paramagnetic particle (ion) is characterized by long time of magnetic relaxation of electrons. In this case we can neglect the value $1/T_S$ in Formulae (4.37) and (4.38). That means relaxation rates depend on correlation times of thermal motions of species in the matter. In other words, the relaxation rates reflects the molecular structure of the solution.

Such ions as Ni^{2+}, Co^{2+}, rare earths (except Gd^{3+}) and some others show opposite behaviour. They are characterized with fast electronic relaxation. Consequently, it is $\tau_c \simeq T_S$ in Formulae (4.37) and (4.38). Usually electron relaxation in these systems is so fast that the lines of electron paramagnetic resonance are very broad and become unobservable. Thus, on the one hand, nuclear magnetic resonance is the only way to study the dynamic behaviour of the electrons, but on the other hand, the study of nuclear relaxation, which is determined by the dipole–dipole and a scalar interactions, does not allow one to determine the parameters of molecular motion around the paramagnetic ions.

We describe a method based on a comparative study of the concentration and temperature dependencies of the relaxation rates of protons and deuterons, which allows us bypass this difficulty and obtain information on the microstructure of hydration shells of paramagnetic ions with short electronic relaxation times. Proton relaxation in these systems is completely determined by magnetic mechanism of relaxation and is described by Formulae (4.35) and (4.36). In some cases, if the exchange of the solvent molecules in the first hydration shell is not too fast, instead of (4.34) we use the expression

$$\frac{1}{T_1^{\rm H}} = \frac{p_M}{T_{1M}^{\rm H} + \tau_M}. \tag{4.39}$$

This occurs, for example, at low temperatures for a solution containing ions Ni^{2+}. The relaxation of the deuterium nuclei is determined not only by magnetic interactions, but also by electric quadrupole relaxation mechanism, and therefore, according to (3.6) and (4.39), we have

$$\frac{1}{T_1^{\rm D}} = \frac{p_M}{T_{1M}^{\rm D} + \tau_M} + \frac{1}{T_{1{\rm Q}}}, \tag{4.40}$$

where $1/T_{1\mathrm{Q}}$ is relaxation rate caused by the quadrupole interactions (see Sect. 3.12), which is described by Formulae (3.152) and, the same, (4.31) in the case of fast chemical exchange between all possible states. Relaxation rates of neclei ^{2}H in the water molecules are expressed with (3.149) because of $\omega_0^2\tau_c^2 \ll 1$ (spin of deuterium nucleus is $I = 1$).

The term in Formula (4.40) which correspond to the magnetic mechanism of relaxation can be calculated from the data on the proton resonance. According to Formulae (4.36), the relation $T_{1M}^{\rm D}/T_{1M}^{\rm H} = \gamma_{\rm H}^2/\gamma_{\rm D}^2 = 42.4$ for both dipole–dipole and scalar channels of relaxation.

So, based on the ideas of the Sect. 4.4.3, one can determine the coordination numbers of ions and the parameters which characterise mobility of the water molecules in the hydration shell. It was discovered that ions Co^{2+}, Ni^{2+}, Sm^{3+}, and Eu^{3+} have double–layered hydration shell. For instance, the hydration shells of both ions Europium and Samarium have hydration numbers $n_1 = 9$, $n_2 = 18$, and $\lambda_1 \simeq 3$.

Note that the theory of magnetic relaxation in the solutions with the fast electron relaxation is not completely developed up to now. Particulary, Solomon–Bloembergen equations (4.36) can not describe the number of features of relaxation in such systems. This is the reason why scientists offer new approaches to the description of the phenomenon of NMR-relaxation in the systems with fast electron relaxation.

A contribution to the theory is made by the groups of J. Kowalewski (*Progr. NMR Spectr.*, **17**, 141-185, (1985)) and I. Bertini (L. Banci, I. Bertini, C. Luchinat. "Nuclear

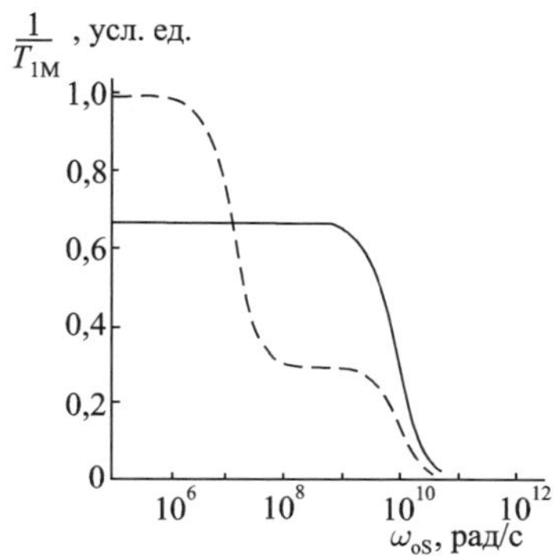

Fig. 4.13 The dependence of T_{1M}^{-1} on the electron frequency ω_{0S} for protons interacting with paramagnetic center $S = 1$ (the *dotted line* corresponds to the Solomon–Bloembergen theory)

and electron relaxation", Weinheim, 1991). They developed the theories of relaxation, which account slow motion ($\tau_R, \tau_M \gg T_S$). Solomon–Bloembergen equations can be obtained in these theories as a special case, the restrictions in the applicability of these equations is discussed. In these theories, the correlation functions account motion of both the lattice and the electron system. It means, electrons are treated as part of the lattice and relaxation time of electron magnetization become unnecessary. New theoretical conclusions predict other, than in the Solomon–Bloembergen theory, dependencies of relaxation rates on the parameters of electron spins (see Fig. 4.13).

4.4.5 Heterogeneous Systems

There are a lot of heterogeneous systems which worth to be considered in this Chapter. That means the magnetic relaxation in these systems are similar to relaxation in liquids. These include porous solids saturated (or partially saturated) with liquid, colloid solutions, sols and gels, ion-exchange resins (polymers), synthetic and biological membranes, layered silicates, etc. Systematic investigations of such systems by the NMR methods were started in the 1960's after works of J. R. Zimmerman, W. E. Brittin, D. E.Woessner (*J. Chem. Phys.*, **35**(1), 41-48 (1961); *J. Phys. Chem.*, **61**(9), 1328-1333 (1957)). Progress of NMR that time promises to obtain the important information on the structure of heterogeneous systems, correlation times, orientational order parameters, thickness of the boundary layer, etc. However, the following studies have shown that these problems have no easy solution. Therefore, to interpret the experimental results, it was necessary to significantly improve the theory of NMR relaxation for liquids that are included in the heterogeneous systems. This process was particularly intense in the 1980's. Features of NMR relaxation in heterogeneous systems are summarized in Sect. 3.13. Here we mention some of them:

1. Spin–lattice relaxation time (T_1) much greater than spin–spin relaxation time (T_2). The ratio T_1/T_2 at the minimum value of T_1 is much greater than value which is predicted by the Formulae in the Sects. 3.6–3.12. Minimum of the temperature

dependence of T_1 is flatter than it should be, for example, according to Formula (3.117).

2. Sometimes there are several minima in the temperature dependence of $\log T_1$ (or $\log 1/T_1$).
3. Often, multiexponential relaxation dependencies are observed in heterogeneous systems.

To explain these features one should consider the distribution of the correlation times, the anisotropy of the rotational and translational diffusion, the conditions of chemical and spin exchange, the possible large gradients of the magnetic field on the boundary of solid–liquid and some other circumstances.

In Sect. 3.13 it was stated that the heterogeneity leads to certain features of the relaxation due to the nuclear spins are in different states (phases). We should emphasize the difference of the term "phase" in the NMR spectroscopy of heterogeneous systems and in the thermodynamics. In NMR, under "phase" one should understand the part of atoms or molecules with the nuclei under investigation, which can be characterized by their own relaxation times T_{1i}, T_{2i} and relative concentration p_i. That means there is no fast exchange between these states. Multi-phase, for example, may be due to the existence of active sites of adsorption on a solid surface, distribution of sizes of pores of the adsorbent, etc. During the measurement time nuclei can be in certain states (phases), and then the relaxation curves are characterised with a number of relaxation times. In a certain conditions (temperature, pressure and other), chemical exchange of the nuclei or molecules between different phases can take place up to fast exchange, and the relaxation parameters become average.

With great care a researcher should interpret experimental results on the relation T_1/T_2. From the simple theory (see Sects. 3.6–3.12), it follows that $T_1/T_2 > 1$ in the systems with slow molecular mobility. Therefore, at first sight, the difference in the values of T_1 and T_2 can be explained by the fact that the molecules of the liquid lose mobility on the solid surface or close to a macromolecules. This leads to the relation $T_1 > T_2$ or even $T_1 \gg T_2$. If the exchange between surface and bulk phase exists, then the relaxation rate is average. The origin of the difference of T_1 and T_2 can be the variety of correlation times. In Fig. 4.14, the temperature dependencies of T_1 and T_2 of protons of water adsorbed in the NaX zeolite is shown. Zeolites are aluminosilicates with highly ordered pores and channels of nanometer scale. The internal structure of pores are characterized by variety of chemical compositions and values of surface activity. These properties distinguish zeolites among other microporous materials. The temperature dependence of T_2 can be used to determine the residence time of protons of the OH-groups linked to the structure of the zeolite.

There is another reason for the difference in the relaxation times T_1 and T_2, which is not directly connected to the process of relaxation. It is an inhomogeneous broadening of the NMR spectral lines in the presence of heterogeneity of the samples due to the difference in the magnetic susceptibility of liquid and solid components. The fact is that near the solid surface there are strong gradients of magnetic field. A simple calculation shows that in the magnetic field $B_0 \simeq 1$–$2\,\mathrm{T}$, gradient of magnetic field near the surface of a glass balls of radius $10^{-6}\,\mathrm{m}$ is of the order of $10^{-1}\,\mathrm{T\,cm^{-1}}$.

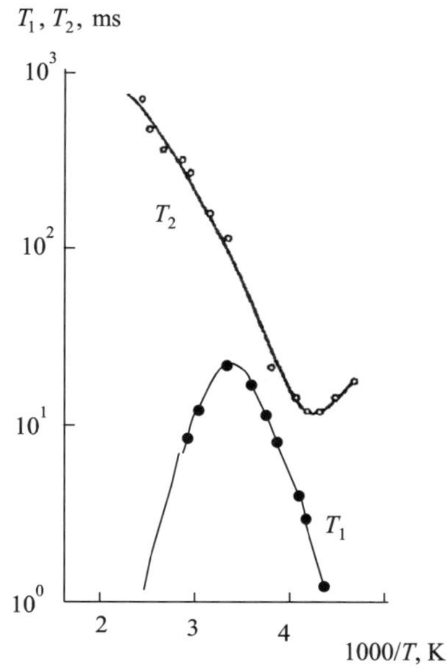

Fig. 4.14 The temperature dependencies of T_1 and T_2 of protons of water adsorbed in the NaX zeolite

In most cases, the structure of heterogeneous systems can be modelled by the spherical formations with certain size distribution. In heterogeneous systems with an anisotropy of magnetic inhomogeneities, orientational factor of line broadening of nuclear magnetic resonance is important. The inhomogeneous broadening should lead to a reduction of the spin-spin relaxation time (T_2), i.e. according to the Formula (1.46) T_2 become T_2^*. Application of multipulse methods for measuring the transverse relaxation time do not eliminate errors in the measurement of T_2 in such systems because of diffusion decay of spin echo signals. In such gradients, the diffusion decay of spin echo signals is significant even at very high pulse repetition rates, for instance, up to 20 kHz. If researcher neglect this decay, it can be concluded on a new channel of relaxation in the system instead of relaxation in the inhomogeneous magnetic field B_0.

Influence of geometrical factors on the relaxation in heterogeneous systems can be manifested in the form of non-exponential nature of the relaxation curves, which is due to the spatial distribution of pores, bulk and surface runoff.

In some cases, the inconsistency of the theory with experiment due to the existence of unaccounted relaxation mechanisms. In inorganic heterogeneous systems a researcher should carefully examine the effect of paramagnetic impurities and inclusions in a hard surface on the relaxation rates of nuclei in liquid which are close to the surface. Considering the great efficiency of paramagnetic channel of magnetic relaxation (see Sect. 3.10), it is necessary to perform specific experiments to detect this effect in the systems under investigation. In this context it may be interesting one of the early works of D. Michel, who measured the relaxation times of protons in mixture of light and heavy (D_2O) water adsorbed on different samples of silica gel. The total amount of water remained unchanged. If the relaxation of protons is completely determined by the interaction with paramagnetic ions, we can expect the independence of the proton relaxation time on the concentration of D_2O. Otherwise, time T_1 should growth with increasing concentration of heavy water in accordance with a formula similar to (4.28). A similar relationship holds for the time T_2. The results of the experiment confirm the assumption on existence of paramagnetic impurities in the silica gel: the relaxation times in the several samples weakly depended or even did not depend at all on the concentration of D_2O. It is interesting that paramagnetic species can not be detected with electron paramagnetic resonance method even with instrument with sensitivity of 10^{11} spins g^{-1}.

Further Readings

Günther H (1995) NMR spectroscopy: basic principles, concepts, and applications in chemistry, 2nd edn. Wiley, New York

Keeler J (2010) Understanding NMR spectroscopy, 2nd edn. Wiley, New York (ISBN 978-0-470-74609-7)

Levitt MH (2008) Spin dynamics: basics of NMR, 2nd edn. Wiley/Interscience, New York

Pople JA, Schneider WG, Bernstein HJ (1959) High-resolution nuclear magnetic resonance. McGraw-Hill Book Company, New York, Toronto, London

Chapter 5
Nuclear Magnetic Resonance in Diamagnetic Solids

Application of the NMR methods to solids provides much more information than for liquids or gases. The matter is that interactions, which determine NMR spectra of solids, in contrast to interactions in liquids, keep anisotropic properties. The main interactions are dipole–dipole interactions, quadrupole interactions, and interactions of nuclear magnetic moments with magnetic moments of electrons or paramagnetic particles.

Dipole–dipole interactions between nuclear magnetic moments create additional local magnetic fields at the nucleus sites that results in splitting and broadening of NMR lines. These interactions usually are the most important in solids and play the key role in structural studies. However, they also can mask other interactions that prevents extraction of additional information. If this is a case, *high resolution NMR* techniques should be applied.

Quadrupole interactions, which are the interactions between nucleus quadrupole moments and electric field gradients, are only for nuclei with spins $I > 1/2$ situated in noncubic lattice sites. These interactions can be both greater and smaller the dipole–dipole ones. Moreover, in several cases they can be much greater the Zeeman interactions of nucleus magnetic moments with an external magnetic field. In these cases one deals with another magnetoresonance method, namely, the *nuclear quadrupole resonance* (NQR).

Quadrupole interactions lead to resonance NMR lines split and shift. The latter is the most important difference between the quadrupole and dipole–dipole interactions: the dipole–dipole interactions never shift the "center of gravity" of the NMR spectrum.

In liquids and gases both the dipole–dipole and quadrupole interactions are averaged to zero due to high mobility of atoms and molecules, that leads to the loss of information, which can be obtained from such interactions.

Interactions of nucleus magnetic moments with magnetic moments of electrons or paramagnetic particles can be classified by the following way:

V. I. Chizhik et al., *Magnetic Resonance and Its Applications*, DOI: 10.1007/978-3-319-05299-1_5,

- If the compound under study is paramagnetic, that means it contains paramagnetic ions, there are dipole–dipole interactions between nuclear magnetic moments and magnetic moments of neighboring ions. It should be emphasized that despite the dipole–dipole interactions between nuclei in this case the nucleus magnetic moments interact with average magnetic moments of electrons or ions, as the latter change their directions much faster (in thousands times) than the nucleus reorientation velocity. So the nuclei "feel" only an average field created by magnetic moments of electrons or ions. These interactions are of the same order than the dipole–dipole ones and shift NMR lines.
- If the nuclei belong to paramagnetic ions or atoms, then we deal with *superfine interactions* between magnetic moments of nuclei and electrons. If it is not the case, it is *indirect superfine interaction* of nucleus magnetic moments with electron shells of paramagnetic ions. These kinds of interactions can not be described as dipole–dipole ones, they should be interpreted only within the quantum mechanics formalism. Superfine and indirect superfine interactions occur only if distances between nuclei and unpaired electrons are comparable to the orbit radii, in which the unpaired electrons are located. They can be both much greater and smaller than dipole–dipole interactions and even greater Zeeman ones. In the latter case such methods as *electronic paramagnetic resonance* (EPR) or *ferromagnetic resonance* (FMR) should be used. Superfine and indirect superfine interactions result in the shift of NMR lines.
- Interactions between nuclear magnetic moments and electron shells of diamagnetic atoms or ions also shift the nuclear magnetic resonance frequency. These so-called *chemical shifts* of either paramagnetic or diamagnetic nature are described within the quantum mechanics. In liquids the anisotropic part of these interactions is averaged to zero and only the isotropic one is observed. In solids they are always anisotropic and contain more information about the properties of studied materials, however, they are usually masked by dipole–dipole interactions and can be detected only by high resolution NMR methods, such as special multipulse sequences or magic angle spinning.
- *Indirect spin–spin interactions* are interactions between magnetic moments of nuclei by means of electron shells of atoms or ions. They shift and split NMR lines and are much smaller than dipole–dipole interactions.

Thus we have listed above the main interactions, which define NMR spectra in solids. The first three (except the isotopic superfine interactions, which also shift spectral lines in liquids) are specific for solids.

However, despite NMR spectra contain a lot of information about various properties of solids, it is rather difficult to extract it, as it can be rather problematic to separate contributions from interactions of different kinds. So it is necessary to reveal how all these interactions affect resonance frequencies and line shapes. To solve this task one has to know how to calculate NMR spectra in the presence of various interactions.

5.1 Effect of Dipole–Dipole Interactions on NMR Spectra in Solids

In Sect. 1.10 we have considered corrections to energy levels of the Zeeman Hamiltonian due to the pair dipole–dipole interactions between spins of the same magnetic species. It has been shown that dipole–dipole interactions remove the degeneracy of Zeeman energy levels and result in the appearance of a non-equidistant energy level system.

5.1.1 NMR Spectrum of an Isolated Spin Pair of Different Magnetic Species

Let us consider a system of two spins i and j of different magnetic species in the presence of an external magnetic field. Such a system is described by the Hamiltonian $\widehat{\mathcal{H}}_Z + \widehat{\mathcal{H}}_d$. In order to define NMR frequencies and intensities of transitions between energy levels in this system one has to calculate the transition probabilities arising under the action of a rf magnetic field. In the laboratory frame the interaction with this alternate magnetic field

$$B^{\approx} = 2B_1 \cos \omega t, \tag{5.1}$$

whose direction is perpendicular to the constant external magnetic field, is described by the following Hamiltonian[1]:

$$\widehat{\mathcal{H}}_{ij}^{\approx} = -\hbar B_1 \left(\gamma_i \left(\hat{\vec{I}}_{i+} + \hat{\vec{I}}_{i-} \right) + \gamma_j \left(\hat{\vec{I}}_{j+} + \hat{\vec{I}}_{j-} \right) \right) \cos \omega t. \tag{5.2}$$

The transition probabilities from the level a to the level b under a harmonic action like (5.1) is defined as follows [2]:

$$W_{ab} = \frac{\pi}{\hbar^2} \left| \langle a | \hat{V}_{ab} | b \rangle \right|^2 \delta(\omega_{ab} - \omega), \tag{5.3}$$

where $\langle a | \hat{V}_{ab} | b \rangle$ is the matrix element calculated from the time independent part of the Hamiltonian (5.2) using the wave functions corresponding to the levels a and b; δ is the delta-function. Hence, the probability of a transition from a level k to a level k' due to the perturbation (5.2) can be calculated according to the following expression

$$W_{kk'} \sim \left| \langle k | \hat{\vec{I}}_{i(j)+} + \hat{\vec{I}}_{i(j)+} | k' \rangle \right|^2 \delta(\omega_{ab} - \omega). \tag{5.4}$$

---

[1] C. P. Slichter. Principle of Magnetic Resonance, 2nd ed. Spring, Berlin, 1980.

[2] A. Carrington and A. D. McLachlan. Introduction to magnetic resonance with application to chemistry and chemical physics. Harper & Row, New York, 1967.

Non-zero matrix elements of the operators $\hat{\vec{I}}_+$ and $\hat{\vec{I}}_-$ appear only between states with $\Delta m_i = \pm 1$ or $\Delta m_j = \pm 1$. However, within the first-order approximation of the perturbation theory, under the action of operators $\hat{B}_{ij}$, $\hat{C}_{ij}$, $\hat{D}_{ij}$, $\hat{E}_{ij}$ and $\hat{F}_{ij}$, functions, which correspond to the k' levels, are admixed to the wave functions Ψ_k:

$$\Psi_k^{(1)} = \Psi_k^{(0)} - \sum_{k' \neq k} \frac{\left|\langle k|\widehat{\mathcal{H}}_{ij}|k'\rangle\right|}{E_k - E_{k'}} \Psi_{k'}^{(0)}. \tag{5.5}$$

Particularly, for a selected nucleus pair with the spins equal to 1/2 but with different gyromagnetic ratios, in the presence of dipole–dipole interactions the wave functions can be written as follows

$$\begin{aligned}
\Psi_1^{(1)} &= \varphi_{+1/2}^{(i)}\varphi_{+1/2}^{(j)} - \frac{\hbar^2\gamma_i\gamma_j}{r_{ij}^3} \\
&\times \left(\frac{\langle 1|\hat{C}_{ij}|2\rangle}{E_1 - E_2}\varphi_{+1/2}^{(i)}\varphi_{-1/2}^{(j)} + \frac{\langle 1|\hat{C}_{ij}|3\rangle}{E_1 - E_3}\varphi_{-1/2}^{(i)}\varphi_{+1/2}^{(j)} + \frac{\langle 1|\hat{E}_{ij}|4\rangle}{E_1 - E_4}\varphi_{-1/2}^{(i)}\varphi_{-1/2}^{(j)} \right), \\
\Psi_2^{(1)} &= \varphi_{+1/2}^{(i)}\varphi_{-1/2}^{(j)} - \frac{\hbar^2\gamma_i\gamma_j}{r_{ij}^3} \\
&\times \left(\frac{\langle 2|\hat{D}_{ij}|1\rangle}{E_2 - E_1}\varphi_{+1/2}^{(i)}\varphi_{+1/2}^{(j)} + \frac{\langle 2|\hat{B}_{ij}|3\rangle}{E_2 - E_3}\varphi_{-1/2}^{(i)}\varphi_{+1/2}^{(j)} + \frac{\langle 2|\hat{C}_{ij}|4\rangle}{E_2 - E_4}\varphi_{-1/2}^{(i)}\varphi_{-1/2}^{(j)} \right), \\
\Psi_3^{(1)} &= \varphi_{-1/2}^{(i)}\varphi_{+1/2}^{(j)} - \frac{\hbar^2\gamma_i\gamma_j}{r_{ij}^3} \\
&\times \left(\frac{\langle 3|\hat{D}_{ij}|1\rangle}{E_3 - E_1}\varphi_{+1/2}^{(i)}\varphi_{+1/2}^{(j)} + \frac{\langle 3|\hat{B}_{ij}|2\rangle}{E_1 - E_2}\varphi_{+1/2}^{(i)}\varphi_{-1/2}^{(j)} + \frac{\langle 3|\hat{C}_{ij}|4\rangle}{E_3 - E_4}\varphi_{-1/2}^{(i)}\varphi_{-1/2}^{(j)} \right), \\
\Psi_4^{(1)} &= \varphi_{-1/2}^{(i)}\varphi_{-1/2}^{(j)} + \frac{\hbar^2\gamma_i\gamma_j}{r_{ij}^3} \\
&\times \left(\frac{\langle 4|\hat{F}_{ij}|1\rangle}{E_4 - E_1}\varphi_{+1/2}^{(i)}\varphi_{+1/2}^{(j)} + \frac{\langle 4|\hat{D}_{ij}|3\rangle}{E_4 - E_3}\varphi_{-1/2}^{(i)}\varphi_{+1/2}^{(j)} + \frac{\langle 4|\hat{D}_{ij}|2\rangle}{E_4 - E_2}\varphi_{+1/2}^{(i)}\varphi_{-1/2}^{(j)} \right).
\end{aligned} \tag{5.6}$$

Here γ_i, γ_j are the gyromagnetic ratios for the nuclei i and j, respectively; $\psi_{\pm m}^{(i,j)}$ are the wave functions of the Zeeman Hamiltonian for the spin i or j; the definitions for operators $\hat{C}_{ij}$, $\hat{D}_{ij}$, $\hat{E}_{ij}$ and $\hat{F}_{ij}$ can be found in Sect. 1.10 (see Eq. (1.136)). Corrections to the energy levels of the Zeeman interaction, which arise if one takes

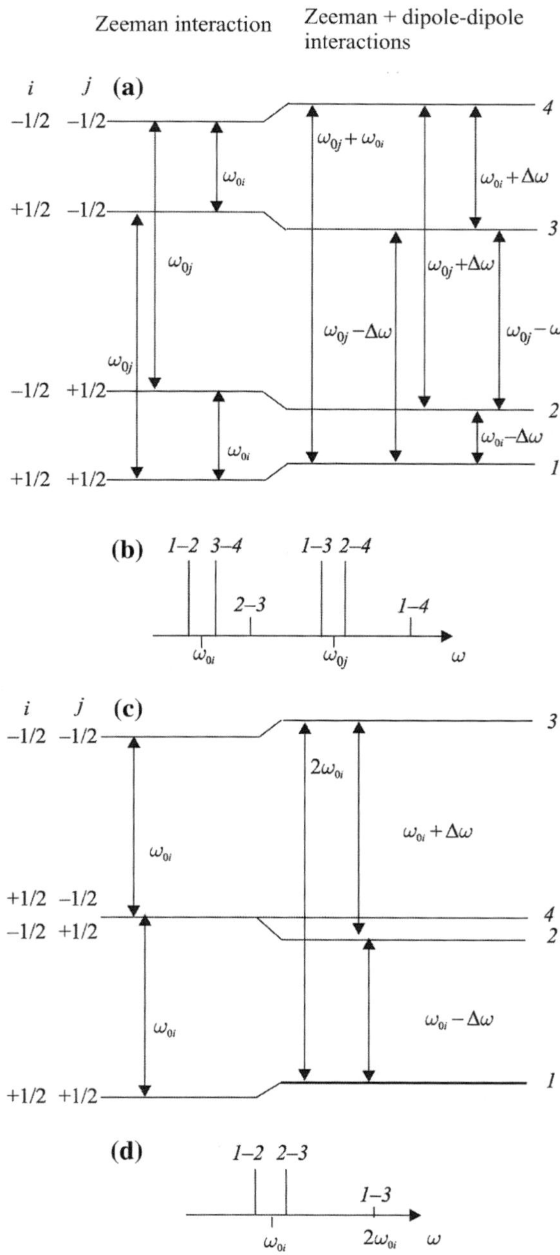

Fig. 5.1 The energy level schema (the Zeeman and dipole–dipole interactions) and the corresponding NMR spectra for two nuclei with $I = 1/2$ of different (**a**, **b**) or equal (**c**, **d**) species

into account the dipole–dipole interactions of two different particles with spins equal to 1/2, can be calculated using Eq. (1.139). The changes in the Zeeman energy level schema due to dipole–dipole interactions are illustrated in Fig. 5.1a.

As it is seen from Eq. (5.6), the wave functions $\Psi_k^{(1)}$, which correspond to any of four energy levels, contain wave functions $\varphi_{+1/2}^{(i)}\varphi_{+1/2}^{(j)}$, $\varphi_{+1/2}^{(i)}\varphi_{-1/2}^{(j)}$, $\varphi_{-1/2}^{(i)}\varphi_{+1/2}^{(j)}$ and $\varphi_{-1/2}^{(i)}\varphi_{-1/2}^{(j)}$ with different coefficients. That is why the probabilities of all possible transitions between levels (1–4) (Fig. 5.1a) calculated using the wave functions $\Psi_k^{(1)}$ are not equal to zero, though for transitions prohibited within the zero-order approximation they are too small, and within the first-order approximation are proportional to $\left|\frac{\langle 3|\hat{C}_{ij}-\hat{F}_{ij}|4\rangle}{E_3-E_4}\right|^2$. In this case, there are six lines in the spectrum (see Fig. 5.1b), which correspond to the following transition frequencies:

$$\omega_{1\text{-}2} = \frac{E_1 - E_2}{\hbar} = \gamma_i\left(B_0 - \frac{\gamma_j \hbar (1 - 3\cos^2\theta_{ij})}{2r_{ij}^3}\right) = \omega_{0i} - \Delta\omega,$$

$$\omega_{1\text{-}3} = \omega_{0j} - \Delta\omega, \quad \omega_{2\text{-}4} = \omega_{0j} + \Delta\omega, \quad \omega_{3\text{-}4} = \omega_{0i} + \Delta\omega,$$

$$\omega_{1\text{-}4} = \omega_{0i} + \omega_{0j}, \quad \omega_{2\text{-}3} = \omega_{0j} - \omega_{0i}.$$

Distances between the lines (1–2)–(3–4) and (1–3)–(2–4) are similar and equal to $2\Delta\omega$. Here

$$\Delta\omega = \frac{\gamma_i \gamma_j \hbar (1 - 3\cos^2\theta_{ij})}{2r_{ij}^3}, \tag{5.7}$$

where θ_{ij} is the angle between the external magnetic field vector and the radius vector connecting i and j nuclei; r_{ij} is the distance between them.

Hence, the spectral line, which corresponds to the nucleus spin i (or j) in the Zeeman field, due to the dipole–dipole interactions is split into two lines placed symmetrically relative to the frequency ω_{0i} (or ω_{0j}) and of the equal intensities. The two lines that correspond to transitions (1–4) and (2–3) are of much lower intensity (about several orders) than other spectral lines (see Fig. 5.1b). In order to observe all the spectrum at any fixed frequency of the alternate magnetic field $B_\approx$ it is necessary to vary the constant field B_0 within a rather broad range. As a rule in experiment the value B_0 is varied in such a manner that to observe the lines of only one doublet corresponding to the nucleus i or j.

5.1.2 NMR Spectrum of an Isolated Spin Pair of Equivalent Magnetic Species

In the case of an isolated spin pair of equivalent magnetic species (with the same gyromagnetic ratio) the probabilities of transitions between energy levels described by the Zeeman and dipole–dipole interactions can be calculated in the similar way as for the case of different nuclei. The appearance of additional lines in the NMR spectrum is caused by the operators $\hat{C}_{ij}$, $\hat{D}_{ij}$, $\hat{E}_{ij}$ and $\hat{F}_{ij}$. Under their action, within the first-order approximation of the perturbation theory, the wave functions corresponding to the k' levels are admixed to the wave functions Ψ_k. These functions can be calculated using Eq. (5.5). However, the calculations prove that only function of the same symmetry can be mixed with themselves. Particularly, if the nucleus spin is equal to 1/2 then, as it follows from Table 1.2 no function is admixed to the wave function Ψ_0^{asym} in any-order approximation of the perturbation theory. That means that under the action of an alternate magnetic field the probability of a transition from one level to another for function Ψ_0^{asym} is equal to zero. Hence, in the case of equivalent nuclei within the second-order approximation of the perturbation theory the only allowed are the transitions between levels (1–2), (2–3) and (1–3) (see Fig. 5.1c), and the NMR spectrum consists of three lines (see Fig. 5.1d), which correspond to the following transition frequencies:

$$\begin{gathered}\omega_{1\text{–}2} = \frac{E_1 - E_2}{\hbar} = \gamma_i B_0 - \frac{3}{4}\gamma_i^2\hbar\frac{1-3\cos^2\theta}{r^3} = \omega_{0i} - \Delta\omega_i,\\ \omega_{2\text{–}3} = \omega_{0i} + \Delta\omega_i, \quad \omega_{1\text{–}3} = 2\omega_{0i},\end{gathered} \tag{5.8}$$

where

$$\Delta\omega_i = \frac{3}{4}\gamma_i^2\hbar\frac{1-3\cos^2\theta}{r^3}.$$

The line intensities corresponding to the first two transitions are several order greater than the line intensities corresponding to the transition (1–3) (see Fig. 5.1d), and as for the case of different nuclei only two first lines are experimentally observed. The distance between these lines is defined by

$$\Delta\omega = \omega_{2\text{–}3} - \omega_{1\text{–}2} = \frac{3}{2}\gamma_i^2\hbar\frac{1-3\cos^2\theta}{r^3}. \tag{5.9}$$

It is evident that for two isolated nuclei the lines are infinitely narrow, however, in reality, there are always interactions with other surrounding nuclei that broaden the lines. If the interaction between magnetic moments of the isolated pair is comparable to that one with external magnetic moments, experimentally one observes one rather broadened line.

Summing up the results of Sects. 5.1.2 and 1.10 the conclusions are following:

- In the case of nuclei of different magnetic species only the operator $\hat{A}_{ij}$ makes contribution to the energy levels, whereas for the equivalent nuclei the both $\hat{A}_{ij}$ and $\hat{B}_{ij}$ operators contribute.
- Nonsecular terms of the dipolar Hamiltonian ($\hat{B}_{ij}$, $\hat{C}_{ij}$, $\hat{D}_{ij}$, $\hat{E}_{ij}$, $\hat{F}_{ij}$ for different nuclei and $\hat{C}_{ij}$, $\hat{D}_{ij}$, $\hat{E}_{ij}$, $\hat{F}_{ij}$ for equivalent nuclei) give extremely small contributions to the energy levels, but admix the wave functions of the unperturbed (Zeeman) Hamiltonian that results in the appearance of additional transitions at frequencies $n\omega_{0i} \pm k\omega_{0j}$ (n and k are even numbers) for different nuclei and at zero

frequency and frequencies $n\omega_{0i}$ for equivalent nuclei. The intensities of additional lines are much lower than the intensities of the main spectral lines.

- The dipole–dipole interaction does not depend on the applied field value.
- If in a crystal there are isolated pairs of different nuclei with spin equal to 1/2 (for example, the ^{1}H and ^{19}F nuclei) then the NMR spectrum for the both nuclei i and j is a doublet with the same splitting, which is described by

$$\Delta\omega_{j(i)} = \frac{\gamma_i \gamma_j \hbar (3\cos^2\theta_{ij} - 1)}{r_{ij}^3}.$$

- If the nuclei i and j are of the same magnetic species the spectrum is also a doublet but the line splitting is one and a half times greater than for different nuclei.
- The value of the doublet splitting is defined by the internuclear distance and the angle between the radius vector, which connects the nuclei, and the external magnetic field direction. Hence, by recording the orientation dependence of the NMR lines splitting it is possible to determine unambiguously both the distance between atoms and the radius vector direction (relative to the magnetic field vector). In other words, the dipole–dipole interactions helps to determine some structural parameters of crystals.

5.2 Determination of Structural Parameters by NMR

The NMR method is widely used to determine coordinates of light atoms, usually hydrogen atoms, which are invisible for X-ray methods. The accuracy of determining hydrogen positions in a crystal lattice by NMR is comparable to the neutron diffraction methods, but in contrast to the neutrons NMR does not destroy the sample under study. However, sometimes NMR fails to decrypt completely the crystal structure as the crystal may include non-magnetic nuclei. Nevertheless, it can be applied in order to precise the crystal symmetry, or, as it was mentioned above, to determine positions of light nuclei. If the crystal structure in general is determined by any method (X-ray or neutron diffraction), but the location of light atoms is unknown, the application of the NMR method is very promising. As it can be seen from Eqs. (5.7) and (5.9), if there is an isolated pair, within the first-order approximation of the perturbation theory the distance between the NMR lines is determined only by the positions of nuclei relative to each other and to the external magnetic field. This fact also allows us to study by NMR objects that contain isolated pairs of magnetic nuclei but do not have a long-range order, such as polymers, liquid crystals, disordered solid solutions and glasses. Studying of such objects by X-ray or neutron diffraction is a rather challenging task, whereas for NMR it is almost the routine one.

The first work on determining the coordinates of hydrogen atoms by NMR, which has become a classic, is the work of Pake,[3] who studied a gypsum crystal

[3] G. E. Pake. Nuclear Resonance Absorption in Hydrated Crystals: Fine Structure of the Proton Line. J. Chem. Phys. 16 (1948) 327.

$CaSO_4 \cdot 2H_2O$. Since then a lot of crystal hydrates, solids containing the groups CH_2, NH_2 or groups of light nuclei such as ^7Li, ^{10}B, ^{19}F etc. have been studied by NMR.

Let us consider an example of the $K_2C_2O_4 \cdot H_2O$ crystal, for which the proton NMR spectrum is a doublet. It points out that there is an isolated pair of proton nuclei, so let us find their positions. The $K_2C_2O_4 \cdot H_2O$ crystal has a monoclinic structure with tree independent lattice parameters a, b, c, and angles $\alpha, \gamma = 90°$, $\beta \neq 90°$. The unit cell contains two formula units. The proton pairs occupy two nonequivalent sites, whose coordinates are related to each other by the ac mirror plane reflection. However, X-ray fails to determine unambiguously the proton positions and the point group of the crystal. The X-ray pattern refinement proposes two possible symmetry groups: the first one contains the only symmetry element, which is the ac mirror plane, whereas the second one contains two symmetry elements, a two-fold axis and a plane, which is perpendicular to that axis. The NMR method helped to make a right choice between them.

The most complete and accurate information about the distances between the protons and the direction of interproton radius vector can be obtained by studying the orientation dependence of the line splitting of the single crystal NMR spectrum. To do this, the crystal is rotated in a magnetic field around an axis, which can be arbitrary or coincide with one of the crystallographic axes. The latter simplifies NMR data handling.

Following the work of Pedersen let us consider the more common case when the sample rotation occurs around an arbitrary axis.[4] It would demonstrate how to describe an orientation dependence of a NMR spectrum, taking into account not only the dipole–dipole interactions, but also other types of interactions, such as quadrupole, hyperfine, indirect spin–spin, etc. Orientation dependencies of NMR spectra are usually measured when the crystal is rotated about an axis perpendicular to the external magnetic field direction. The direction of the field $\vec{B}_0$ in the plane, which is perpendicular to the rotation axis can be defined relative to an arbitrary zero line by an angle ϕ (Fig. 5.2). The radius vector direction ($\vec{r}_{ij}$) connecting two protons can be defined by an angle δ between $\vec{r}_{ij}$ and its projection on the plane, which is perpendicular to the rotation axis, and by an angle ϕ_1, which is the angle between the projection $\vec{r}_{ij}$ on the plane perpendicular to the rotation axis and the zero line, which lies within this plane, and relative to which the magnetic field $\vec{B}_0$ direction is defined.

In order to describe the orientation dependence of the distance between the doublet lines (in frequency units) one has to apply Eq. (5.9) after transforming it in such a manner that to extract the dependence $\Delta\omega$ on the angle ϕ, which characterizes the crystal rotation in the external magnetic field. As it follows from simple trigonometric equations (see Fig. 5.2) the angle θ between the magnetic field and the radius vector can be defined from the following equation:

[4] B. Pedersen and D. F. Holsomb. NMR in hydrate crystals: Structural information from broadened fine-structure lines. J. Chem. Phys. 38 (1962) 61.

Fig. 5.2 The proton pair orientation in an external magnetic field in the laboratory frame

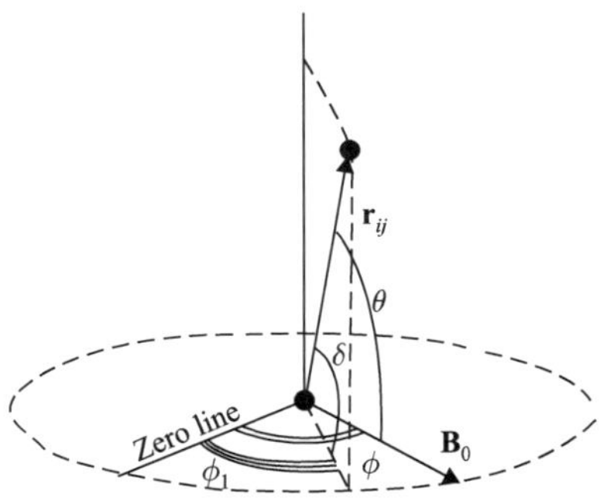

$$\cos\theta = \cos\delta\cos(\phi_1 - \phi). \tag{5.10}$$

By substituting (5.10) into (5.9) and taking into account that

$$\cos^2(\phi_1 - \phi) = \frac{1}{2}(1 + \cos 2(\phi_1 - \phi)),$$

one has:

$$\begin{aligned}\Delta\omega &= \frac{3\gamma_i^2\hbar}{2r_{ij}^3}(3\cos^2\theta - 1)\\ &= \frac{3\gamma_i^2\hbar}{4r_{ij}^3}(3\cos^2\delta\cos 2(\phi_1 - \phi) + 3\cos^2\delta - 2) = A + B\cos 2(\phi_1 - \phi),\end{aligned} \tag{5.11}$$

where

$$A = \frac{3\gamma_i^2\hbar}{4r_{ij}^3}(3\cos^2\delta - 2), \quad B = \frac{9\gamma_i^2\hbar}{4r_{ij}^3}\cos^2\delta.$$

Thus, as it follows from Eq. (5.11) the orientation dependence of the distance between the lines of the doublet is described by a cosine function with the amplitude B and the constant component A, which can be easily found by fitting the experimental orientational dependence. Knowing A and B parameters, one can determine the interproton distance and the angle δ.

According to X-ray data there are two water molecules in the $K_2C_2O_4{\cdot}H_2O$ unit cell. Therefore, if there are two symmetry elements the water molecules must be located either in the mirror plane, or on the symmetry axis. All the interproton vectors in these cases are parallel to each other, i.e. all the proton pairs are magnetically equivalent, and hence for any orientation of the rotation axis one should observe only one doublet in the NMR spectrum. If there is a symmetry plane only, the proton pairs are not magnetically equivalent, and at an arbitrary orientation of the magnetic field there should be two doublets. It has been found experimentally that for any orientation of the magnetic field the ^{1}H NMR spectrum consists of only one doublet, and, consequently, the symmetry group of the crystal contains two elements of symmetry.

After determining the symmetry group it is possible to find the hydrogen atoms coordinates from the NMR spectra. To do this it is necessary

- to determine the distance and the direction of the radius vector $\vec{r}_{ij}$ connecting the propons of the isolated pair (relative to the crystal axes);
- if the oxygen positions are known from X-ray data, knowing the interproton distance and the bond angle ∠H–O–H, which is 108° for the water molecule and does not change noticeably from one compound to another, it is possible to determine the hydrogen coordinates.

The determination accuracy of the hydrogen positions depends on the measurement precision of the distance between the lines in the doublet. Dipole–dipole interactions of the isolated proton pair with other magnetic nuclei lead to a broadening of the doublet. As a result, they may overlap each other and the position of the maxima of the doublet lines changes making it difficult to measure the distance between the lines. Sometimes it is more efficient to measure the distance between the lines of the doublet on the position of the "center of gravity" of each line, which forms a doublet. This method provides much more accurate values for the line splitting.

To separate the lines of the doublet one approximates each line by a model function. The quantitative calculation of line shapes for a large number of interacting nuclei is a complex task which is solved only numerically within an approximation of given shapes for individual lines. It was found that for an isolated nucleus with spin number 1/2 if one excludes a small number (~10) of the nearest neighbors the interaction with other nuclei with spin 1/2 leads to a Gaussian shape of an individual line $g(\omega)$:

$$g(\omega) = \frac{1}{\Delta\omega\sqrt{\pi}} \exp\left(-\frac{(\omega - \omega_0)^2}{\Delta\omega^2}\right), \tag{5.12}$$

here $\Delta\omega$ is the half-width of the NMR spectral line, defined at a distance $1/e$ of its maximum value. However, most often in solids the line shape is described by a function, which is intermediate between Gaussian and Lorentzian. In most crystallohydrates the splitting of the doublet lines is significantly greater than the doublet lines width, and one can find a function that accurately fits the NMR spectrum, and therefore with a high degree of accuracy to determine interproton distances. But if the splitting of the doublet lines is comparable with their width, the approximation accuracy is greatly reduced. That leads to a large error in the determination of the interproton distances.

Experimental NMR spectra have a form of a doublet only in solids containing isolated pairs of nuclei (the distance between the nuclei within a pair is essentially less than distances to other magnetic nuclei). In the most cases, the NMR spectrum is a bell-shaped line, as usually the local field at each nucleus is due to its interaction with a large number of neighbors. The local field created by all surrounding nuclei depends on the relative positions of nuclei and their distribution over the energy levels. This total local field is of different value for nuclei of the same magnetic species that leads to different frequencies of transitions between energy levels of nucleus spins, so in whole one obtains a symmetric, with respect to the frequency ω_0, spectral line of a finite width.

5.3 The van Vleck Method

If in a crystal there is no isolated pairs of nuclei dipole–dipole interactions lead to a significant broadening of NMR lines and information about the crystal structure can be obtained by a numerical approximation of the spectral line shape on the basis of model assumptions about possible crystal structures or by using the so-called *method of moments* developed by van Vleck.[5]

5.3.1 Calculation of the Second Moment of a Spectral Line

The statistical determination of moments of a random variable, which obeys a distribution law (commonly it is the normal distribution law), is well known. If the random variable is the number of nuclei resonating at a certain frequency, the NMR spectral line shape can be represented as a function of distribution of these nuclei.

By definition, the n-order moment S_n is:

$$S_n = \overline{\Delta\omega^n} = \overline{(\omega - \omega_0)^n} = \frac{\int_{-\infty}^{\infty} f(\omega)(\omega - \omega_0)^n \mathrm{d}\omega}{\int_{-\infty}^{\infty} f(\omega)\mathrm{d}\omega}.$$

Here $f(\omega)$ is the function describing the line shape. If $f(\omega)$ is an even function, then all its odd moments are equal to zero. Since dipole–dipole interactions do not shift the center of gravity of the NMR spectrum, the line shape due to these interactions is an even function.

The most important characteristics of the NMR spectral line are the second and fourth moments.

Let us consider the second moment of the adsorption line and how it is related to the crystal structure parameters. The second moment is

$$S_2 = \frac{\int_{-\infty}^{\infty} f(\omega)(\omega - \omega_0)^2 \mathrm{d}\omega}{\int_{-\infty}^{\infty} f(\omega)\mathrm{d}\omega} \tag{5.13}$$

[5] J. H. Van Vleck. The Dipolar Broadening of Magnetic Resonance Lines in Crystals. Phys. Rev. 74 (1948) 1168.

or

$$S_2 = \frac{\int_{-\infty}^{\infty} f(\omega)\omega^2 \mathrm{d}\omega}{\int_{-\infty}^{\infty} f(\omega)\mathrm{d}\omega} - 2\omega_0 \underbrace{\frac{\int_{-\infty}^{\infty} f(\omega)\omega \mathrm{d}\omega}{\int_{-\infty}^{\infty} f(\omega)\mathrm{d}\omega}}_{\omega_0} + \underbrace{\frac{\int_{-\infty}^{\infty} f(\omega)\omega_0^2 \mathrm{d}\omega}{\int_{-\infty}^{\infty} f(\omega)\mathrm{d}\omega}}_{\omega_0^2}$$

$$= \frac{\int_{-\infty}^{\infty} f(\omega)\omega^2 \mathrm{d}\omega}{\int_{-\infty}^{\infty} f(\omega)\mathrm{d}\omega} - \omega_0^2. \tag{5.14}$$

The function $f(\omega)$ is proportional to the probability of transitions at the frequency ω, which is determined by Eq. (5.4),

$$f(\omega) \sim W_{k\,k'} \sim \left| \langle k | \hat{\vec{I}}_x | k' \rangle \right|^2 \delta(\omega_{k\,k'} - \omega),$$

if the external magnetic field is homogeneous within the sample and relaxation processes provide the Boltzmann equilibrium in the spin system. The probabilities of transitions are nonzero only for a discrete number of frequencies and, hence, the integration in Eq. (5.14) can be replaced by the summing:

$$S_2 = \frac{\sum_{k,k'}^{\infty} \left| \langle k | \hat{\vec{I}}_x | k' \rangle \right|^2 \omega_{k\,k'}^2}{\sum_{k,k'}^{\infty} \left| \langle k | \hat{\vec{I}}_x | k' \rangle \right|^2} - \omega_0^2. \tag{5.15}$$

It is to note that

$$\omega_{k\,k'} = \frac{E_k - E_{k'}}{\hbar}, \tag{5.16}$$

and according to quantum mechanics

$$\sum_{k,k'}^{\infty} \left| \langle k | \hat{\vec{I}}_x | k' \rangle \right|^2 (E_k - E_{k'})^2 = -\operatorname{Tr}[\widehat{\mathcal{H}}, \hat{\vec{I}}_x]^2, \tag{5.17}$$

$$\sum_{k,k'}^{\infty} \left| \langle k | \hat{\vec{I}}_x | k' \rangle \right|^2 = \sum_{k,k'}^{\infty} \langle k | \hat{\vec{I}}_x | k' \rangle \langle k' | \hat{\vec{I}}_x | k \rangle = \sum_{k}^{\infty} \langle k | \hat{\vec{I}}_x^2 | k \rangle = \operatorname{Tr} \hat{\vec{I}}_x^2. \tag{5.18}$$

After substitution Eqs. (5.16)–(5.18) into (5.15) one has

$$S_2 = -\frac{\operatorname{Tr}[\widehat{\mathcal{H}}, \hat{\vec{I}}_x]^2}{\hbar^2 \operatorname{Tr} \hat{\vec{I}}_x^2} - \omega_0^2. \tag{5.19}$$

The Hamiltonian in Eq. (5.19) is the sum of the Zeeman and dipole–dipole interactions: $\widehat{\mathcal{H}} = \widehat{\mathcal{H}}_Z + \widehat{\mathcal{H}}_d$. Taking into account

$$\frac{\mathrm{Tr}[\widehat{\mathcal{H}}_Z, \hat{\vec{I}}_x]^2}{\hbar^2\ \mathrm{Tr}\,\hat{\vec{I}}_x^2} = \omega_0^2,$$

instead of Eq. (5.19) one has

$$S_2 = -\frac{\mathrm{Tr}[\widehat{\mathcal{H}}_d, \hat{\vec{I}}_x]^2}{\hbar^2\ \mathrm{Tr}\,\hat{\vec{I}}_x^2}. \tag{5.20}$$

Such a recording is more convenient as the Tr value depends neither on the choice of the set of orthogonal wave functions, which are calculated using the matrix elements, nor on the coordinate system. In general, in Eq. (5.19) it is necessary to consider the Hamiltonian describing only those types of interactions that occur in solids, and which affect the observed NMR spectral line.

When taking into account dipole–dipole interactions the additional lines, which arise at frequencies 0, $2\omega_0$, $3\omega_0, \ldots$ due to nonsecular part of the corresponding Hamiltonian, are not observed experimentally. Hence, in Eq. (5.20) only the secular part of the Hamiltonian $\widehat{\mathcal{H}}_d$ (the operator $\hat{A}$ for different spins and the operators $\hat{A}$, $\hat{B}$ for identical spins) can be taken into account. If one considers the total dipole–dipole Hamiltonian it results in disagreement between the calculated and experimental values of the second moment and does not provide the correct information about the crystal structure parameters. As Eq. (5.20) contains the Hamiltonian of dipole–dipole interactions, which depends on the mutual location of nuclei, the second moment calculated using not difficult but tedious calculations will depend on the crystal lattice parameters.[6]

Originally the expression for the second moment was found by van Vleck. For a crystal containing nuclei of equivalent magnetic species ($\mu_i = \mu_j = \mu$), Van Vleck obtained the following expression for the second moment of the adsorption line:

$$S_2 = \frac{3\mu^2}{4N} \sum_{i=1}^{N} \sum_{j \neq i} \frac{(3\cos^2\theta_{ij} - 1)^2}{r_{ij}^6}, \tag{5.21}$$

where $\mu = \gamma\hbar\sqrt{I(I+1)}$ is the nucleus magnetic moment (the total nucleus magnetic moment of the nucleus with the spin I, not its projection). After choosing a nucleus i one has to sum $(3\cos^2\theta_{ij} - 1)^2 r_{ij}^{-6}$ over all nuclei j of the crystal. Usually it is sufficiently to consider the nuclei within the radius of 5–10 Å. Then one chooses another nucleus i and repeats the procedure. It is clear that only nuclei with

---

[6] The details can be found in A. L?sche. Kerninduktion. Berlin, Deutscher Verlag der Wissenschaften, 1957.

different summations over j should be considered. Their number is N and can not be greater than the number of nuclei in the unit cell, but can be less than it if there are crystallographically equivalent nuclei in the unit cell. For example, if all the nuclei in the unit cell are equivalent, $N = 1$, and there is only one summing in Eq. (5.21).

The contributions to the second moment from the nuclei of other species (k) is $(3/2)^2 = 9/4$ times less than the contribution from the nuclei of the same species:

$$S_2 = \frac{1}{3N}\sum_{i=1}^{N}\sum_{k}\mu_k^2\frac{(3\cos^2\theta_{ik}-1)^2}{r_{ik}^6}. \tag{5.22}$$

The total second moment of the resonance line is the sum of the contributions from the same (see Eq. (5.21)) and different (see Eq. (5.22)) magnetic species:

$$S_2 = \frac{3\mu^2}{4N}\sum_{i=1}^{N}\sum_{j\neq i}\frac{(3\cos^2\theta_{ij}-1)^2}{r_{ij}^6} + \frac{1}{3N}\sum_{i=1}^{N}\sum_{k}\mu_k^2\frac{(3\cos^2\theta_{ik}-1)^2}{r_{ik}^6}. \tag{5.23}$$

It is worth noting that for a monocrystal an additional information issue is the orientation dependence of the second magnetic moment, i.e. its dependence on the orientation of the magnetic field $\vec{B}_0$ direction relative to the crystallographic axes.

If one deals with polycrystals, i.e. solids that are composed of many crystallites with a random variation in direction relative to the external magnetic field, the Eq. (5.23) can be simplified by averaging over θ:

$$S_2 = \frac{3\mu^2}{5N}\sum_{i=1}^{N}\sum_{j\neq i}r_{ij}^{-6} + \frac{4}{15N}\sum_{i=1}^{N}\sum_{k}\mu_k^2 r_{ik}^{-6}. \tag{5.24}$$

Then, by comparing the experimental value of the second moment with that one calculated using Eq. (5.24) or (5.23) for various structural models of the studied crystal one can make a conclusion about the crystal structure.

5.3.2 Experimental Determination of the Second Moment

To determine the second moment value experimentally one should use Eq. (5.13) where instead of the function $f(\omega)$ one substitutes the digitized experimental values of the NMR line shape, and hence the integrals in Eq. (5.13) are to be found numerically. If one obtains the NMR absorption line in a differential form, that is usual for the continuous wave (cw) technics, to calculate the second moment in the differential form Eq. (5.13) should be transformed as

$$S_{2\exp} = \frac{\int_{-\infty}^{\infty} f'(\omega-\omega_0)(\omega-\omega_0)^3 d\omega}{\int_{-\infty}^{\infty} 3f'(\omega-\omega_0)(\omega-\omega_0) d\omega}. \tag{5.25}$$

One should keep in mind that the second moment calculated using Eq. (5.25) contains both systematic and statistic errors. Systematic errors are often due to (i) an insufficient slow rate of the spectrum recording; (ii) an insufficient small modulation amplitude; (iii) saturation; (iv) an uniform drift of the magnetic field. Static errors, in general, are due to (i) a thermal noise of the electronics; (ii) instabilities of the magnetic field; (iii) oscillation generator instabilities. Moreover, when determining experimentally the second moment, the integrals in Eq. (5.25) are calculated numerically, the infinite integration limits are substituted by the finite ones, as outside a certain frequency range the NMR signal intensity is equal to zero. However, very often the noise impedes to define correctly the frequency range, masking a part of the NMR spectrum. An incorrect choice of the frequency range may result in a huge error in the second moment value, as the terms that correspond to the large frequency mismatch value give a rather noticeable contribution to S_2. Sometimes reducing the influence of some error sources leads to increasing in the influence of others: for example, decreasing the modulation amplitude reduces the corresponding systematic error, but it decreases the signal-to-noise ratio as well; reducing the recording rate not only extends the recording time, but also increases the role of the magnetic field drift. That is why usually one intentionally allows significant systematic errors due to recording conditions, and then corrects them:

$$S_{2\text{real}} = S_{2\text{exp}} - (\upsilon\tau_0)^2 - h_{\text{m}}^2/4. \quad (5.26)$$

Here υ is the spectrum recording rate (in G/s); τ is the time constant of the synchronous detector; h_{m} is the modulation amplitude (in G).

When deriving Eq. (5.26) it was supposed that correction for the resonance passage time and finite modulation amplitude value are not too great (less than 10–20% of the real second moment value).

The saturation (as a source of systematic error) to some extent may be eliminated, if one plots the dependence of the second moment, calculated from the first half of the absorption line, of the generation amplitude and extrapolate the result to an infinitesimal generation amplitude. The effect of the magnetic field drift (another source of systematic errors) can be eliminated by writing each line twice up and down the drift.

The statistic error, as usual, can be reduced by making multiple recording of spectral lines under the same conditions.

5.3.3 Shape and Second Moment of a Spectral Line in the Presence of Molecular Motion

Experimental studies of the NMR line shape in solids showed that very often the observed lines are much narrower than the calculated ones (in effect, the second moment is the root-mean-square line width). The main reason for such a discrepancy is the molecular motion. It turns out that more or less symmetrical molecule can participate in rotational or translational motions without disturbing the regular crystal structure. This movement is usually in the form of fast jumps between states corresponding to the potential energy minima. Sometimes only a part of molecules participates in the motion. For example, the group $-CH_3$, which makes part of a number of molecules, easy rotates around the three-fold symmetry axis, even at liquid nitrogen temperature. In polymers the molecular motion can be very diverse: parts of a long complex molecule can easily move relative to each other.

If there is any kind of molecular motion the second moment can also be calculated theoretically. All these calculations are based on the assumption that the nuclei "feel" only an average local field (as a rule, the molecular motion is a very fast process compared to the NMR frequency). Usually the average local field is much lower than its static value and, therefore, the motion decreases the average local fields, and thus the line width. By the degree of the spectral line narrowing one can judge about the type of the molecular motion, as well as about its speed. That is why NMR soon after its discover, became a powerful tool to study the atomic and molecular motion in solids.

It is known that in solids the rotational and translational motion are not totally free. The dependence of the potential energy on the molecule orientation is quite complex, and usually there is a number of local minima, whose positions are determined primarily by the symmetry. At sufficiently low temperature and relatively high potential barriers the molecule is in one of its positions corresponding to the minimum of the potential energy, and the only allowed movement is the rotational oscillation. At higher temperatures there is a probability that the molecule jumps over a potential barrier. In the simplest case, this process is described by one correlation time τ_c, which characterizes the average lifetime of a molecule in one particular orientation (sometimes one considers the correlation frequency $\omega_c = 2\pi/\tau_c$). The temperature dependence of the correlation time is usually expressed by

$$\tau_c = \tau_0 \exp\left(\frac{V}{kT}\right),$$

where V is the potential barrier height; T is the temperature (in K); k is the Boltzmann constant; τ_0 is a pre-exponential factor. However, this formula is not always true. Sometimes reorientation processes can not be characterized by the only one correlation time, for example, if the tunnel transitions should be taken into account. However, in any case one can consider that the rotational motion of the molecule is its reorientation between positions that correspond to the potential energy minima.

If magnetic nuclei are in motion, the NMR spectral line changes, because the interaction between nuclear dipoles changes as well. From the classical theory point of view, if there is a motion, the local fields become time dependent. At a rather fast nuclear motion, the NMR spectrum is defined by an average field, and the nature of the averaging depends of the motion type. For example, if there is a reorientation of a nuclear pair about the two-fold axis the average local field does not change, if the characteristic time of jumping is essentially small. It is due to the fact that the local field is proportional to $(3\cos^2\theta - 1)$ and, hence, it is the same for the both orientations. If there is a higher-fold axis, one can consider that the reorientation and uniform rotation make the similar contributions to the local fields. That is why the model of uniform rotation is usually applied to treat experimental NMR spectra.

As it was shown by Andrew and Newing on a simple example, the rotation produces weak side-spectra, whereas random reorientations leads to associated sidebands.[7] Their character depends on the motion rate. For slow thermal motions, when $\Delta\omega > \omega_c$ ($\Delta\omega$ is the line width), the spectrum shape changes, whereas the second moment value S_2 remains almost constant and the Van Vleck formulas can be applied. If the motion is noticeably fast, $\Delta\omega \gg \omega_c$, the sidebands intensity is much less than the central line one. As a result only the central part of the spectrum is observed, and its second moment is defined by the time averaged local field. In other words, the time averaging of the local magnetic field corresponds to neglecting the sidebands and wings. In this case the second moment means the second moment of the central (experimentally observed) line. Generally speaking, here the resonance line is narrower than for immobile nuclei, i.e. the second moment is lower. If $\Delta\omega \equiv \omega_c$ then S_2 has an intermediate value.

In order to calculate the second moment of a NMR line when $\Delta\omega \ll \omega_c$, it is necessary to make a time averaging of $(3\cos^2\theta(t) - 1)r^{-3}(t)$, then to square the obtained quantity and substitute into the van Vleck formula (5.23). For a polycrystalline sample the result must be averaged over θ.

In solids there is a number of different motion types: (i) intra- and intermolecular translational and rotational oscillations; (ii) molecule rotations as a whole around the gravity center and rotations of isolated groups of nuclei inside the molecule (that means not only the almost free rotations, but the uneven reorientations as well); (iii) self-diffusion. For $\Delta\omega \ll \omega_c$ all these various motion types lead to different changes in the second moment value and require individual analysis.

The simplest case is the self-diffusion. Usually it is observed in solids where molecules can rotate. Thus, the self-diffusion was found in the methane, methanol, cyclohexane, and several other compounds. In this case, the local field $\vec{B}_{loc}$ vanishes after averaging and, therefore, $S_2 \to 0$.

If there are weak nuclear vibrations the second moment varies a little. Translational vibrations corrects the second moment, in such a manner that its value increases, whereas the rotational oscillations decrease it.

Another interesting case occurs when molecules, ions or groups of atoms rotate as a whole. Then it is convenient to represent the expression for the second moment as a sum of two terms, which reflect inter- and intramolecular contributions. It is clear that when a molecule rotates as a whole the distances between nuclei belonging to this molecule do not change, and to calculate the intramolecular contribution S_2^{intra} it is only necessary to average $(3\cos^2\theta - 1)$, that is much easier. Gutowski and Pake

[7] E. R. Andrew and R.A. Newing. The narrowing of nuclear magnetic resonance spectra by molecular rotation in solids. Proc. Phys. Soc. 72 (1958) 959.

calculated S_2^{inter} for a pair of nuclei freely rotating around an arbitrary axis.[8] The similar result can be obtained if one considers S_2^{intra} for a pair of nuclei reorientating about the three- and higher-fold axis. For these cases

$$\overline{(3\cos^2\theta - 1)r^{-3}} = \frac{1}{2}(3\cos^2\theta' - 1)(3\cos^2\gamma - 1)r^{-3}.$$

Here θ is the angle between the external magnetic field direction and the internuclear vector; θ' is the angle between the rotation axis and the external magnetic field direction; γ is the angle between the rotation axis and the internuclear vector.

When calculating the intramolecular contribution using Eq. (5.23) one has to put $(3\cos^2\theta' - 1)^2(3\cos^2\gamma - 1)^2 r^{-6}/4$ under the summation sign instead of $(3\cos^2\theta_{ij} - 1)^2 r_{ij}^{-6}$. For polycrystalline samples, in addition, it is necessary to average the term $(3\cos^2\theta' - 1)^2$ over the unit sphere. As a result, in Eq. (5.24) one has to put $(3\cos^2\gamma\ 1)^2 r^{-6}/4)$ instead of r^{-6}.

Thus, by rotating a pair of nuclei about an axis the second moment depends only on the orientation of the rotation axis relative to the external field and the internuclear vector directions. If $\gamma = 90°$, then the intramolecular contribution to the second moment is reduced by four times, and if $\theta' = 54.7°$, that corresponds to $(3\cos^2\theta' - 1 = 0)$, the second moment is zero, regardless of the direction of the internuclear vector. The latter is applied in high-resolution NMR methods in solids: to suppress dipole–dipole interactions the crystal is rotated about an axis, which makes an angle 54.7° with the magnetic field directions, the so-called magic angle spinning (MAS) method.

Isotropic reorientations of a pair of nuclei, i.e. such reorientations for which all the directions of the internuclear vector are equiprobable, give $S_2 = 0$.

Calculation of the intermolecular contribution to the second moment is a more difficult task and have been exactly solved only in a few specific cases.

5.4 Influence of Quadrupole Interactions on NMR Spectra

Let us consider a system of nuclear spins with $I > 1/2$ in an external magnetic field. In the presence of quadrupole interactions the total Hamiltonian of such a system is the sum of the Zeeman and quadrupole interaction Hamiltonians (see (1.197)). In general, it is not possible to solve the Schrödinger equation for such a Hamiltonian, and in order to find energy levels and transition frequencies the perturbation theory should be applied. However, for some specific cases the exact solution of the Schrödinger equation is possible. For example, if the external magnetic field direction coincides with the principal z axis of the axisymmetric EFG tensor, $\eta = 0$, then the total Hamiltonian of the system is:

---

[8] H. S. Gutowsky and G. E. Pake. Structural investigations by means of nuclear magnetism. II. Hindered rotation in solids. J. Chem. Phys. 18 (1950) 162.

$$\widehat{\mathcal{H}} = -\gamma\hbar\hat{I}_z B_0 + \frac{e^2 q Q}{4I(2I-1)}\left(3\hat{I}_z^2 - \hat{I}^2\right).$$

Here eQ is the nucleus quadrupole moment; eq is the zz-component of the EFG tensor at the nucleus site.

In this case the Zeeman Hamiltonian $\widehat{\mathcal{H}}_Z$ commutes with the quadrupole interaction Hamiltonian $\widehat{\mathcal{H}}_Q$. Hence, the eigenfunction of the Zeeman Hamiltonian are simultaneously the eigenfunctions of the quadrupole interaction Hamiltonian, and the energy levels E_m can be easily calculated, as it is well known how the $\hat{I}_z$ and $\hat{I}$ operators act on the eigenfunctions of the Zeeman Hamiltonian:

$$E_m = -\gamma\hbar m B_0 + \frac{e^2 q Q}{4I(2I-1)}(3m^2 - I(I+1)). \tag{5.27}$$

As it is seen from Eq. (5.27), the quadrupole interaction energy does not depend on the m sign, therefore, the distance between the energy levels with quantum number $m = 1/2$ and $m = -1/2$ does not change. However, the energy levels are not equidistant anymore. The intensities of transitions between these levels (I_m) (let us remind that they are proportional to the matrix elements of the Zeeman Hamiltonian (5.2)) correlate with each other as follows

$$\frac{I_{m\longleftrightarrow(m-1)} = I_{-(m-1)\longleftrightarrow -m}}{I_{m-1\longleftrightarrow m-2} = I_{-(m-1)\longleftrightarrow -(m-2)}} = \frac{I(I+1) - m(m-1)}{I(I+1) - (m-1)(m-2)}. \tag{5.28}$$

For example, for $I = 5/2$ according to Eq. (5.28)

$$I_{5/2\longleftrightarrow 3/2} : I_{3/2\longleftrightarrow 1/2} : I_{1/2\longleftrightarrow -1/2} : I_{-3/2\longleftrightarrow -1/2} : I_{-5/2\longleftrightarrow -3/2} \\ = 5 : 8 : 9 : 8 : 5.$$

The intensities of the spectral lines, which correspond to transitions between energy levels with $\Delta m \neq 1$ are zero. Thereby, in the presence of quadrupole interactions there are $2I$ lines in the NMR spectrum, with frequencies defined by

$$\omega_{m\longleftrightarrow m-1} = \frac{E_m - E_{m-1}}{\hbar} = \omega_0 + \frac{3e^2 q Q(2m-1)}{4I(2I-1)\hbar}. \tag{5.29}$$

In other words, it leads to the appearance of additional lines in the NMR spectrum, the so-called *satellites*, which are situated symmetrically relative to the resonance frequency ω_0 without quadrupole interactions. For half-integer spins at the frequency ω_0 there is a line, which corresponds to the transition $1/2 \longleftrightarrow -1/2$, the so-called *central* line. It is to be noted that to calculate the components of the EFG tensor one usually uses the frequency difference corresponding to the $3/2 \longleftrightarrow 1/2, -1/2 \longleftrightarrow -3/2$ transitions for half-integer spins and to the $1 \longleftrightarrow 0, 0 \longleftrightarrow -1$ transitions for integer spins, as these spectral lines have the highest intensities. They are the

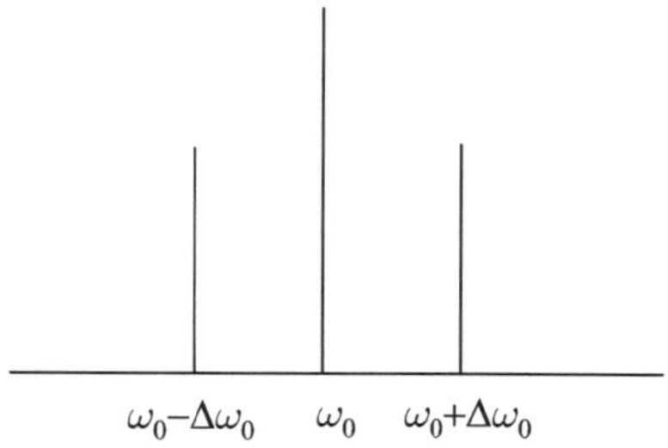

Fig. 5.3 NMR spectrum for a nucleus with $I = 3/2$ in the presence of quadrupole interactions

so-called *fist satellites*. The spectral lines corresponding to $5/2(2) \longleftrightarrow 3/2(1)$, $-3/2(-1) \longleftrightarrow -5/2(-2)$ transitions are called the *second satellites* an so on.

For example for the spin $I = 3/2$ the corrections to the Zeeman energy levels due to the quadrupole interaction are following:

$$\Delta E_{\pm 1/2} = -\frac{e^2 q Q}{4}, \qquad \Delta E_{\pm 3/2} = \frac{e^2 q Q}{4}.$$

The rf field $B_1 \cos \omega t$, which is perpendicular to the constant external magnetic field, causes transitions between these levels. The transition probabilities are proportional to the matrix element square:

$$W_{m \longleftrightarrow m'} \sim \left| \langle m | \hat{I}_x | m' \rangle \right|^2 . \tag{5.30}$$

As it follows from Eq. (5.30)

$$W_{3/2 \longleftrightarrow 1/2} : W_{1/2 \longleftrightarrow -1/2} : W_{-1/2 \longleftrightarrow -3/2} = 3 : 4 : 3.$$

The transition frequencies corresponding to the non-zero transition probabilities are

$$\omega_{3/2 \longleftrightarrow 1/2} = \omega_0 + \frac{e^2 q Q}{2\hbar}, \quad \omega_{1/2 \longleftrightarrow -1/2} = \omega_0, \quad \omega_{-1/2 \longleftrightarrow -3/2} = \omega_0 - \frac{e^2 q Q}{2\hbar},$$

i.e. in the NMR spectrum there are three spectral lines (instead of one): the central, its frequency coincides with ω_0, and two symmetric satellites. The distance between these satellites is exactly equal to the quadrupole coupling constant in frequency units $e^2 q Q/\hbar$.

Thereby, if the external field direction coincides with the principal z-axis of the EFG tensor and if $\eta = 0$ the central line frequency $1/2 \longleftrightarrow -1/2$ remains the same for any ratio between the energies of the Zeeman and quadrupole interactions, and the NMR spectrum is symmetric relative to the ω_0 frequency. Figure 5.3 represents the NMR spectrum for a nucleus with $I = 3/2$ in the presence of quadrupole interactions with inhomogeneous electric field.

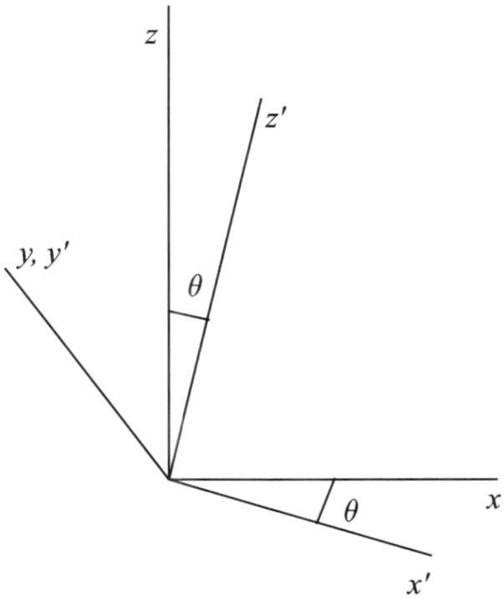

Fig. 5.4 Relations between the laboratory frame and the EFG principal axes frame

If the direction of the external magnetic field does not coincide with the direction of the principal EFG axis, the quadrupole interaction Hamiltonian does not commute to the Zeeman Hamiltonian any more, and to calculate the energy levels one applies the perturbation theory. As in NMR the Zeeman interaction is much stronger the other ones, its Hamiltonian is considered an unperturbed one, whereas the quadrupole interaction Hamiltonian is a small parameter. To calculate the first- and higher-order corrections to the Zeeman energy it is reasonable to write the quadrupole interaction Hamiltonian in the laboratory frame, in which the z-axis is along the magnetic field $\vec{B}_0$.

It can be done by two ways:

- in Eq. (1.198) written in an arbitrary frame (for example, in the laboratory one) to express the components of the EFG tensor trough the principal component of the EFG tensor eq, the asymmetry parameter η and the Euler angles;
- in the Hamiltonian (1.197) to rewrite the operator part, which is written in the coordinate system related to the directions of the principal components of the EFG tensor, in the laboratory frame.

If $\eta = 0$ the second way is easier. In this case the operator part of the quadrupole interaction Hamiltonian in the principal axes frame of the EFG tensor consists of only one operator:

$$\hat{O}_0 = 3\hat{I}_{z'}{}^2 - \hat{\vec{I}}^2, \tag{5.31}$$

and the only operator to be transformed is $\hat{I}_{z'}{}^2$ because the operator $\hat{\vec{I}}^2$ does not depend on the frame choice. Without loss of generality it is possible to chose such a frame where the z, x axes of the laboratory frame and axes z', x', which define the directions of the principal axes of the EFG tensor, lie in the same plane, whereas axes y and y' coincide (Fig. 5.4). Then

$$\hat{I}_{z'} = \hat{I}_z \cos\theta + \hat{I}_x \sin\theta. \tag{5.32}$$

After substituting Eq. (5.32) into (5.31) one has

$$\widehat{\mathcal{H}}_{\mathrm{Q}} = \frac{e^2qQ}{4I(2I-1)}\left[\left(3\hat{I}_z^2 - \hat{\vec{I}}^2\right)\frac{3\cos^2\theta - 1}{2} + \frac{3}{2}\left(\hat{\vec{I}}_+\hat{I}_z + \hat{I}_z\hat{\vec{I}}_+\right)\sin\theta\cos\theta \right.$$
$$\left. + \frac{3}{2}\left(\hat{\vec{I}}_-\hat{I}_z + \hat{I}_z\hat{\vec{I}}_-\right)\sin\theta\cos\theta + \frac{3}{4}\hat{\vec{I}}_+^2\sin^2\theta + \frac{3}{4}\hat{\vec{I}}_-^2\sin^2\theta\right]. \tag{5.33}$$

Thereby, even if $\eta = 0$ but the orientation of the external magnetic field relative to the principal axes of the EFG tensor is arbitrary, in the quadrupole interaction Hamiltonian nonsecular terms appear. That makes impossible to solve exactly the Schrödinger equation with such a Hamiltonian, and hence the perturbation theory should be applied. If the Zeeman energy is much greater than the quadrupole interaction energy ($|\gamma\hbar B_0| \gg |e^2qQ|$), the first-order approximation is sufficient. In this case only the secular terms of the quadrupole interaction Hamiltonian make contribution to the Zeeman energy levels (the first term in Eq. (5.33)):

$$\Delta E_m^{(1)} = \frac{e^2qQ}{8I(2I-1)}(3m^2 - I(I+1))(3\cos^2\theta - 1). \tag{5.34}$$

Equation (5.34) differs from the quadrupole interaction energy for the exact solution of the Schrödinger equation (5.27) only by the factor $(3\cos^2\theta - 1)/2$, and all the aforesaid may be applied to this case as wall. It is worth noting that neither for the exact solution of Eq. (5.27) nor for the first-order approximation the corrections to the Zeeman energy levels do not depend on the applied magnetic field value.

In the second-order approximation of the perturbation theory the nonsecular terms of the quadrupole interaction Hamiltonian (5.33) (the last three terms) give additional corrections to the energy levels:

$$\Delta E_m^{(2)} = -m\frac{1}{\hbar\gamma B_0}\left(\frac{e^2qQ}{4I(2I-1)}\right)^2$$
$$\times\left(\frac{3}{8}\sin^2 2\theta(8m^2 - 4I(I+1)+1) + \frac{3}{8}\sin^4\theta(2I(I+1) - 2m^2 - 1)\right). \tag{5.35}$$

A qualitative analysis of the positions of the energy levels within the first- and second-order approximations of the perturbation theory can be done by substituting the first-order corrections, calculated using Eq. (5.34), as the function of m^2, and the second-order corrections, calculated using Eq. (5.35), as the product of m and a function, which depends on m^2 and several constants:

$$\Delta E_m^{(2)} = -mf(m^2). \tag{5.36}$$

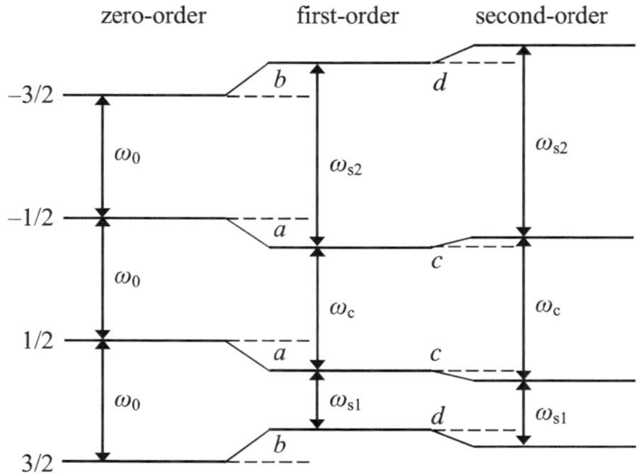

Fig. 5.5 Corrections to the Zeeman energy levels due to quadrupole interactions within the zero-, first- and second-order approximations of the perturbation theory

As it was mentioned above the first-order corrections do not depend on the m sign, whereas the second-order corrections are of the similar value but different sign for the positive and negative values of the magnetic quantum number m (see Eq. (5.36)). Let us denote the corrections to the energy levels with $m = \pm 1/2$ and $m = \pm 3/2$ calculated within the first-order approximation, as a and b, and those ones calculated within the second-order approximation as $\mp c$ and $\mp d$, respectively. Then the expressions for the energy levels of the total Hamiltonian corresponding to the quantum numbers $m = \pm 1/2$ and $m = \pm 3/2$ can be written as follows

$$\begin{aligned} E_{3/2} &= (-3/2)\gamma\hbar B_0 + b - d, \\ E_{1/2} &= (-1/2)\gamma\hbar B_0 - a - c, \\ E_{-1/2} &= (1/2)\gamma\hbar B_0 - a + c, \\ E_{-3/2} &= (3/2)\gamma\hbar B_0 + b + d. \end{aligned} \tag{5.37}$$

The energy levels for the Zeeman and quadrupole interactions within the zero-, first- and second-order approximations of the perturbation theory are represented in Fig. 5.5.

As it follows from Eq. (5.37) the frequency transitions between the levels with $\Delta m = \pm 1$ are described by

$$\begin{aligned} \omega_{s1} &= \omega_{3/2 \longleftrightarrow 1/2} = \omega_0 - (a + b + c - d)/\hbar, \\ \omega_c &= \omega_{1/2 \longleftrightarrow -1/2} = \omega_0 + 2c/\hbar, \\ \omega_{s2} &= \omega_{-3/2 \longleftrightarrow -1/2} = \omega_0 + (a + b - c + d)/\hbar. \end{aligned}$$

Here ω_{s1}, ω_{s2}, ω_c are the frequencies of two corresponding satellites and the central line of the NMR spectrum. Therefore, within the second-order approximation of the perturbation theory: (i) the central line shifts from the ω_0 position by the value of $2c/\hbar$; (ii) the satellites are not symmetrically placed anymore relative to the central line, and their center of symmetry shifts relative to ω_0; (iii) the distance between the satellites remains the same as within the first-order approximation and is $2(a+b)/\hbar$.

The nonsecular terms of the quadrupole interaction Hamiltonian not only shift the energy levels but mix the wave functions corresponding to different values of the magnetic quantum number m. Since the operator parts of the quadrupole interaction Hamiltonian and the dipole–dipole interaction Hamiltonian are similar, all the reasoning about the relative changing of the wave functions within the first-order approximation of the perturbation theory described in Sects. 5.1.1 and 5.1.2 are applicable to the quadrupole interactions. So, the operators $\hat{\vec{I}}_+\hat{I}_z$ and $\hat{\vec{I}}_-\hat{I}_z$ admix Ψ_{m+1} and Ψ_{m-1} functions, respectively, to the wave function Ψ_m of the "pure" state that corresponds to the Zeeman energy E_m (the "pure" state means that it is not disturbed by any internal interaction). That means the transitions with $\Delta m = \pm 2, \pm 3$ are not forbidden anymore and their probabilities can be calculated using these mixed wave functions. Moreover, the transitions between the energy levels may occur even if the rf field is parallel to the constant magnetic field. In the case of the quadrupole interactions, as opposed to the dipole–dipole interactions, these probabilities can be rather noticeable, because often the energy of the quadrupole interactions is much greater than the energy of the dipole–dipole interactions. For example, for the spin $I = 3/2$ instead of three spectral lines (one central and two satellite lines) in the NMR spectra there are additional lines at the frequencies

$$\begin{aligned}
\omega_{d1} &= \omega_{3/2\longleftrightarrow -1/2} = 2\omega_0 + (-a - b + c + d)/\hbar,\\
\omega_{d2} &= \omega_{1/2\longleftrightarrow -3/2} = 2\omega_0 + (a + b + c + d)/\hbar,\\
\omega_{d3} &= \omega_{3/2\longleftrightarrow -3/2} = 3\omega_0 + 2d/\hbar.
\end{aligned}$$

It is to be noted that the frequencies of the additional lines in the NMR spectrum are not strictly aliquot to ω_0. Within the first-order approximation of the perturbation theory the lines corresponding to the frequencies $\omega_{3/2\longleftrightarrow -1/2}$ and $\omega_{-3/2\longleftrightarrow 1/2}$ are symmetrically placed relative to the frequency $2\omega_0$, whereas within the second-order approximation they are shifted relative to it by the value of $(c+d)/(2\hbar)$. It can be shown that the transition $\omega_{3/2\longleftrightarrow -3/2}$ is of the lowest intensity. The exact values of the transition frequencies can be calculated using Eqs. (5.33) and (5.35). The total NMR spectrum for the spin 3/2 is illustrated in Fig. 5.6.

These conclusions are true even if $\eta \neq 0$. In this case within the first-order approximation the transition frequencies are determined by the following expression:

$$\omega_{m\longleftrightarrow m-1} = \omega_0 + \frac{3e^2qQ(2m-1)}{8I(2I-1)\hbar}(3\cos^2\theta - 1 + \eta\sin^2\theta\cos 2\phi), \qquad (5.38)$$

and within the second-order approximation by

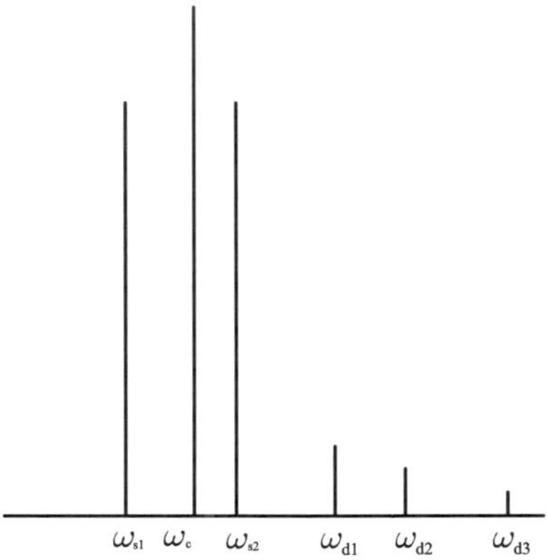

Fig. 5.6 An NMR spectrum of a nucleus with $I = 3/2$ at the presence of the quadrupole interaction within the second-order approximation of the perturbation theory

$$\begin{aligned}\omega_{m\longleftrightarrow m-1} = &\left(\frac{3e^2qQ}{4I(2I-1)\hbar}\right)^2 \frac{1}{3\omega_0}\left\{2\left[\left(I+\frac{3}{2}\right)\left(I-\frac{1}{2}\right)-3\left(m-\frac{1}{2}\right)^2\right]\right.\\ &\times\left[\cos^4\theta\left(\frac{3}{8}-\frac{\eta}{4}\cos 2\phi+\frac{\eta^2}{24}\cos^2 2\phi\right)\right.\\ &\left.+\cos^2\theta\left(-\frac{3}{4}+\frac{\eta^2}{6}-\frac{\eta^2}{12}\cos^2 2\phi\right)+\left(\frac{3}{8}+\frac{\eta}{4}\cos 2\phi+\frac{\eta^2}{24}\cos^2 2\phi\right)\right]\\ &-4\left[\left(I+\frac{3}{2}\right)\left(I-\frac{1}{2}\right)-6\left(m-\frac{1}{2}\right)^2\right]\left[\cos^4\theta\left(-\frac{3}{2}+\eta\cos 2\phi-\frac{\eta^2}{6}\cos^2 2\phi\right)\right.\\ &\left.\left.+\cos^2\theta\left(\frac{3}{2}-\frac{\eta^2}{6}-\eta\cos 2\phi+\frac{\eta^2}{3}\cos^2 2\phi\right)+\frac{\eta^2}{6}\sin^2 2\phi\right]\right\}.\end{aligned}$$

Here ϕ is the angle between the principal x-axis of the EFG tensor and the intersection line of the plane perpendicular to the external magnetic field vector and the plane perpendicular to the principal z-axis of the EFG tensor.

The frequency of the central line within the second-order approximation is defined by following:

$$\omega_{1/2\longleftrightarrow -1/2} = \omega_0 - \frac{\omega_Q^2\,(I(I+1)-3/4)}{6\omega_0}\left(A(\phi)\cos^4\theta + B(\phi)\cos^2\theta + C(\phi)\right),$$

where

$$\begin{aligned}A(\phi) &= -\frac{27}{8}+\frac{9}{4}\eta\cos 2\phi-\frac{3}{8}\eta^2\cos^2 2\phi;\\ B(\phi) &= \frac{30}{8}-2\eta\cos 2\phi-\frac{1}{2}\eta^2+\frac{3}{4}\eta^2\cos^2 2\phi;\\ C(\phi) &= -\frac{3}{8}+\frac{1}{3}\eta^2-\frac{1}{4}\eta\cos 2\phi-\frac{3}{8}\eta^2\cos^2 2\phi;\end{aligned}$$

$$\omega_Q = \frac{3e^2 q Q}{2I(2I-1)\hbar}.$$

Thus, the quadrupole interaction leads to the splitting of the NMR spectral lines. Within the first-order approximation of the perturbation theory the satellites are placed symmetrically relative to the initial resonance frequency, whereas within the second-order approximation they are shifted relative to the both central spectral line and the initial frequency ω_0.

Since the NMR lines number and their frequencies depend on the nucleus quadrupole moment value and the value of the EFG tensor at the nucleus site, the NMR method is widely used for measuring the quadrupole interaction tensor, the components of which $D_{\alpha,\beta}$ are related to the components of the EFG tensor by the following way:

$$D_{\alpha,\beta} = \frac{3eQ}{2I(2I-1)} V_{\alpha,\beta} \qquad \alpha, \beta \equiv x, y, z.$$

Here $V_{\alpha,\beta}$ are the components of the EFG tensor (see Eq. (1.175)).

For the most nuclei their quadrupole moments are known with a rather high precision. Hence by determining the quadrupole interaction tensor one determines the EFG tensor as well. And it is the latter that contains the information about the crystal structure, electron density distribution, phase transitions, local ordering for solid solutions, dynamics of atoms or groups of atoms in solids etc. Even small changing in atomic positions lead to a rather noticeable changing in the EFG values. This fact allows us to define the structural parameters of the sample under study with a very high accuracy. That is why let us discuss methods of determining the EFG tensor from NMR spectra. In the general case this problem is rather complicated and to solve it two classical methods can be applied: the universal method proposed by Volkoff[9] and the single-rotation method.

5.5 Determination of the EFG Tensor by NMR

The electric field gradient is a symmetric second-rank tensor with the zero trace and has only five independent components, which are to be determined by NMR. Experimentally the components of the EFG tensor are obtained in an arbitrary frame usually determined by the crystal faceting, and they must be transformed into the principal axes frame of the EFG tensor. In this coordinate system the five independent parameters are eq, η and three direction cosines.

[9] M. Volkoff, H. E. Petch, and D. W. L. Smellie. Nuclear electric quadrupole interactions in single crystals. Can. J. Phys. 30 (1952) 270.

5.5.1 The Volkoff Method

An universal method for determining the components of the EFG tensor proposed by Volkoff is based on the measuring of three orientation dependencies of the difference between the satellite frequencies ω_m and ω_{-m} corresponding to the transitions $m \longleftrightarrow m-1$, $-m \longleftrightarrow -m+1$ by rotating the crystal about three mutually transverse axes x, y and z, which may not coincide to the crystallographic axes. This method is applied if there is no information about the crystal symmetry. It allows us to determine the both quadrupole coupling constant and asymmetry parameter even if the directions of the crystallographic axes are unknown. For half-integer spin numbers one usually measures the splitting between the first satellites, which correspond to the $3/2 \longleftrightarrow 1/2$ and $-1/2 \longleftrightarrow -3/2$ transitions. To extract the components of the EFG tensor it is necessary to transform Eq. (5.38) to such a form where the splitting between the first satellites $\Delta\omega = \omega_{3/2\longleftrightarrow 1/2} - \omega_{-3/2\longleftrightarrow -1/2}$ explicitly depends on the angle ϕ, which determines the rotation angle of the crystal in the external magnetic field. To achieve the aim let us in Eq. (5.38) pass from the principal axes frame of the EFG tensor to the x, y, z frame, which coincides with the set of the mutually transverse axes of crystal rotations. After that, depending on about which axis α ($\alpha = x, y, z$) the crystal is rotated, the satellite splitting $\Delta\omega$ can be represented as follows

$$2\Delta\omega = \frac{3eQ}{I(2I-1)\hbar}(A_\alpha + B_\alpha \cos 2\phi_\alpha + C_\alpha \sin 2\phi_\alpha). \tag{5.39}$$

If the crystal rotates about the z-axis:

$$A_z = -\frac{1}{2}V_{zz}, \qquad B_z = \frac{1}{2}(V_{xx} - V_{yy}), \qquad C_z = -V_{xy}; \tag{5.40}$$

about the x-axis:

$$A_x = -\frac{1}{2}V_{xx}, \qquad B_x = \frac{1}{2}(V_{yy} - V_{zz}), \qquad C_x = -V_{zy}; \tag{5.41}$$

and about the y-axis:

$$A_y = -\frac{1}{2}V_{yy}, \qquad B_y = \frac{1}{2}(V_{zz} - V_{xx}), \qquad C_y = -V_{xz}. \tag{5.42}$$

Eq. (5.39) can be transformed to

$$2\Delta\omega = \frac{3eQ}{I(2I-1)\hbar}(A_\alpha + D_\alpha \cos(2\phi_\alpha + \phi_{\alpha 0})), \tag{5.43}$$

$$D_\alpha = \sqrt{B_\alpha^2 + C_\alpha^2}; \qquad \operatorname{tg}\phi_{\alpha 0} = -C_\alpha / B_\alpha. \tag{5.44}$$

As it is seen from Eq. (5.43) the orientation dependence of the satellite splitting is a sinusoidal function. Thus by measuring the $2\Delta\omega(\phi_\alpha)$ it is easy to determine both the sine curve amplitude D_α and constant component A_α for each chosen rotation axis. However, as it follows from Eqs. (5.40)–(5.42) the determination of the constant components A_x, A_y, A_z is equivalent to the determination of all diagonal elements of the EFG tensor in the x, y, z frame. As the trace of the EFG tensor is equal to zero the sum $(A_z + A_x + A_y)$ is to be zero as well. This condition is very useful to verify experimentally the orthogonality of the rotation axes. After the diagonal components of the EFG tensor are found one can determine the off-diagonal components (see Eqs. (5.40) – (5.42) and (5.44)):

$$
\begin{aligned}
C_z &= -V_{xy} = \pm\sqrt{D_z^2 - B_z^2} = \pm\sqrt{D_z^2 - (A_y - A_x)^2},\\
C_x &= -V_{yz} = \pm\sqrt{D_x^2 - B_x^2} = \pm\sqrt{D_x^2 - (A_z - A_y)^2},\\
C_y &= -V_{xz} = \pm\sqrt{D_y^2 - B_y^2} = \pm\sqrt{D_y^2 - (A_x - A_z)^2}.
\end{aligned}
\qquad (5.45)
$$

The off-diagonal components of the EFG tensor are defined ambiguously by Eq. (5.45) (they can be both positive and negative). The unique solution choice is possible only due to other physical reasons. Such an ambiguity is the most important disadvantage of the Volkoff method.

It is to be mentioned that by NMR or NQR methods the EFG tensor can be determined up to a sign only (+ or −), as the energy of the quadrupole interaction (and hence the NMR and NQR spectra) does not change if one inverses the direction of the quadrupole moment relative to the magnetic field vector. However, an incorrect choice of the sing of individual components of the tensor, off-diagonal ones, for example, leads to changing of the orientation of the principal axes of the EFG tensor, and consequently makes it impossible to use the obtained results.

Hereby, with reservations, the Volkoff method allows us to determine all the components of the EFG tensor $\mathsf{V}_{x,y,z}$ in the frame, whose axes coincide with the directions of the three mutually perpendicular crystal rotation axes. Often these axes are chosen on the basis of the crystal faceting (directions of crystallographic axes are determined by the crystal growing) or by the X-ray diffraction analysis. Sometimes the directions of the crystallographic axes can be detected from NMR spectra (it will be discussed further). If the direction cosines of the crystal rotation axes relative to the crystallographic frame are known, it is possible to determine the EFG tensor in this frame (V_{cr}):

$$
\mathsf{V}_{\mathrm{cr}} = \mathsf{T}^{-1}\mathsf{V}_{x,y,z}\mathsf{T},
$$

here T is the direction cosines matrix that transforms one coordinate frame to another; T^{-1} is the transposed matrix. By diagonalizing the EFG tensor V_{cr} obtained in such a manner it is possible to calculate the value of the principal z-component of the EFG tensor at the nucleus site, the asymmetry parameter η and the direction cosines of the principal axes of the EFG tensor with respect to the crystal frame. If the direction cosines are unknown, the V_{zz} and η values can be obtained by diagonalizing the

$\mathsf{V}_{x,y,z}$ tensor. In this case the directions of the principal axes of the EFG tensor are determined with respect to the crystal rotation axes.

The Volkoff method due to its generality has a number of disadvantages:

- the off-diagonal components of the EFG tensor are defined ambiguously (it has been discussed above);
- the rotation axes should be mutually perpendicular, non-adherence to this condition would result in lack of accuracy of the EFG tensor determination;
- measurement of three orientation dependencies of the satellite splitting is a rather time consuming task;
- the information about the crystal symmetry is not used.

The latter, if taken into account, may reduce the number of parameters to be found. For example, if there is a symmetry axis passing through a nucleus site, in which the EFG tensor should be determined, one of the principal axis of the tensor coincides with this axis, and therefore, with the direction of one of the crystallographic axes.[10] If this is the three- or higher-fold axis, then $\eta = 0$. This is due to the fact that the EFG tensor does not change after symmetry transformations.

Let us consider such an example of the four-fold axis passing through the nucleus site, at which the EFG tensor is to be determined. Then after rotation of the frame by $\pi/2$ about the symmetry axis, which coincides, for example, with the z-axis of the crystal, one passes to another coordinate system: $\tilde{x}$, $\tilde{y}$, $\tilde{z}$ related to the former by the following way: the axis $\tilde{x}$ coincides with y, the axis $\tilde{y}$ is directed along $-x$, and the axis $\tilde{z}$ coincides with z. In this case the component $V_{\tilde{x}\tilde{y}}$ of the EFG tensor can be expressed trough the components in the x, y, z frame:

$$V_{\tilde{x}\tilde{y}} = \frac{\partial^2 U}{\partial \tilde{x} \partial \tilde{y}} = -\frac{\partial^2 U}{\partial x \partial y} = -V_{xy}.$$

Other components of the EFG tensor can be found by the similar way. The matrix representation has the following form:

$$\mathsf{V} = \begin{pmatrix} V_{yy} & -V_{xy} & V_{yz} \\ -V_{xy} & V_{xx} & -V_{xz} \\ V_{yz} & -V_{xz} & V_{zz} \end{pmatrix}.$$

But such a transformation can not change the tensor value, and hence

$$V_{xx} = V_{yy}; \quad V_{xy} = -V_{xy} = 0; \quad V_{xz} = V_{yz}; \quad V_{yz} = -V_{xz}.$$

After rotation of the coordinate system by the angle π about the z-axis, the $\tilde{x}$-axis coincides with x, the $\tilde{y}$-axis coincides with y, and the $\tilde{z}$-axis is along the z-axis as before. After transformations, similar to described above, one has the following tensor representation:

[10] One of the crystallographic axes is always directed along the highest-fold symmetry axis.

$$V = \begin{pmatrix} V_{xx} & V_{xy} & -V_{xz} \\ V_{yx} & V_{yy} & -V_{yz} \\ -V_{zx} & -V_{zy} & V_{zz} \end{pmatrix}$$

and then $V_{xz} = -V_{xz} = 0$; $V_{yz} = -V_{yz} = 0$. Finally

$$V_{xx} = V_{yy}, \quad V_{xy} = V_{xz} = V_{yz} = 0.$$

That is the principal axes frame of the EFG tensor coincides with the x, y, z frame, which is crystallographic, as a rule; $\eta = 0$ and the tensor is determined by the only component eq, that was to be proved.

Now, the only one component of the tensor should be found experimentally. In this way, we can always determine which components of the EFG tensor must be zero.

If both a second-fold axis and a mirror plane passe trough the site of the interest, the direction of one of the EFG principal axes coincides with the symmetry axis. However, the asymmetry parameter in this case is not equal to zero and there are three unknown tensor component.

For positions located in the center of inversion, none of the components of the EFG tensor goes to zero, and there are still five components to be found experimentally.

Of course, sometimes it turns out that one of the components is zero, but it does not follow from the crystal symmetry and does not reduce the number of unknown components. The presence of two uncoinciding three- and higher-fold axes leads to zero values for all tensor components. Such an analysis makes easier determining the EFG tensor.

Very often there are symmetry axes or planes that do not passe through the nucleus sites under study but relate between each other the crystallographically equivalent sites, in which the EFG tensor are not equal but related by a frame transformation. In this case the main components of the EFG tensor eq are equal, and only the direction cosines differ. The presence of magnetically inequivalent sites in the crystal is used to determine the EFG tensor from NMR spectra in the method of the single rotation.

5.5.2 The Method of the Single Rotation

This method can be applied only if in a crystal there are inequivalent sites occupied by quadrupole nuclei. This does not reduce significantly the range of studied materials, as almost all crystals has symmetry elements. In the method of the single rotation, as it follows from its name, the only one orientation dependence, when rotating a crystal about a single axis, is measured. That makes the experiment much easier, as it does not require precise adjusting of the crystal in the external magnetic field. Nevertheless, there are some limits on the rotation axis direction. As it will be shown further, the rotation axis can not be perpendicular to a symmetry plane (or parallel to a symmetry axis), reflections in which (or rotations about which) transform one

Fig. 5.7 The geometry of the single rotation experiment

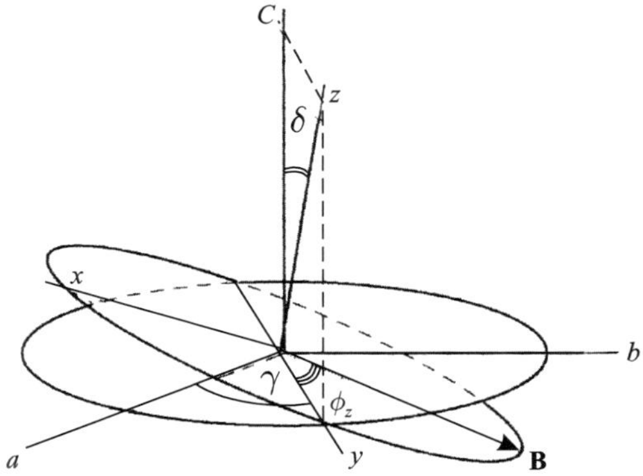

magnetically inequivalent site to others. The method of the single rotation reduces the time of the experiment, but increases the number of calculations required to determine the components of the EFG tensor.

Let us consider how to apply this method if there is a crystal with two magnetically inequivalent nucleus sites, that can be transformed one to another by the mirror reflection in the crystal plane *ab*. In other words, the coordinates of the one site are transformed to the coordinates of the other by changing the z-axis direction (crystallographic axis c). The EFG tensors in these sites are also passe from one to another by this transformation. Let us write the EFG tensors, which correspond to these magnetically inequivalent sites, in general form in the a, b, c frame using the crystal symmetry:

$$\mathsf{V}_{\mathrm{cr}}^{(1)} = \begin{pmatrix} V_{aa} & V_{ab} & V_{ac} \\ V_{ab} & V_{bb} & V_{bc} \\ V_{ac} & V_{bc} & V_{cc} \end{pmatrix} \iff \mathsf{V}_{\mathrm{cr}}^{(2)} = \begin{pmatrix} V_{aa} & V_{ab} & -V_{ac} \\ V_{ab} & V_{bb} & -V_{bc} \\ -V_{ac} & -V_{bc} & V_{cc} \end{pmatrix}. \tag{5.46}$$

Let us choose an arbitrary frame x, y, z. Let z is the rotation axis, and it can be defined by the Euler angle δ, γ with respect to the crystallographic frame (Fig. 5.7). Without losing generality, assign the y-axis is the line of intersection between two planes: the crystallographic plane *ab* and the plane perpendicular to the rotation axis z.

As it follows from Eqs. (5.40), (5.43) and (5.44), when rotating the crystal about the z-axis for each inequivalent nucleus site the orientation dependence of the splitting between the first satellites $\Delta\omega$ can be written as

$$2\Delta\omega^{(1,2)} = \frac{3eQ}{I(2I-1)\hbar}\left(A_z^{(1,2)} + D_z^{(1,2)}\cos\left(2\phi_z + \phi_{z0}^{(1,2)}\right)\right), \tag{5.47}$$

where

$$A_z^{(1,2)} = -\frac{1}{2}V_{zz}^{(1,2)}; \quad D_z^{(1,2)} = \sqrt{\left(B_z^{(1,2)}\right)^2 + \left(C_z^{(1,2)}\right)^2};$$

$$\operatorname{tg}\phi_{z0}^{(1,2)} = -C_z^{(1,2)}/B_z^{(1,2)}; \tag{5.48}$$

$$B_z^{(1,2)} = \frac{1}{2}\left(V_{xx}^{(1,2)} - V_{yy}^{(1,2)}\right); \quad C_z^{(1,2)} = -V_{xy}^{(1,2)};$$

the angle ϕ_z is counted from the y-axis. Usually the direction of the y-axis is unknown and the rotation angle of the crystal in the magnetic field is defined by ϕ'_z, which is counted from an arbitrary chosen direction. If there is a symmetry plane, which usually coincides or is parallel to one of the crystallographic planes, it is possible to determine using NMR spectra the direction of the y-axis, and to count the angle from it. In our case at an arbitrary orientation of the crystal in the magnetic field, the NMR spectrum consists of two groups of lines, corresponding to different magnetically inequivalent nucleus sites. If the magnetic field direction is parallel to the symmetry plane, which coincides with the ab-plane, this two groups interflow. It is obvious if one considers the crystal rotation about the c-axis. In this case in Eqs. (5.47) and (5.48) it is necessary to substitute indexes x, y, z by a, b, c, respectively. Then as it follows from the general form of the tensors $\mathbf{V}_{cr}^{(1,2)}$ (see Eq. (5.46))

$$A_c^{(1)} = A_c^{(2)} = -\frac{1}{2}V_{cc}; \quad B_c^{(1)} = B_c^{(2)} = \frac{1}{2}(V_{aa} - V_{bb}); \quad C_c^{(1)} = C_c^{(2)} = -V_{ab}.$$

Therefore, the splitting between the satellites for both magnetically inequivalent nucleus sites is the same. Such a confluence of the spectral lines may occur not only if the magnetic field vector is parallel to the symmetry plane, but in a random way as well. However, if the direction of the rotation axis with respect to the crystallographic ones is known at least approximately, it is not so difficult to determine at which crystal orientation the magnetic field vector is parallel to the symmetry plane.

Thereby, by measuring the orientation dependence of the splitting between satellites for each magnetically inequivalent nucleus site and by determining the angle ϕ' value, at which the lines corresponding to different inequivalent sites interflow, it is possible to find the direction of the y-axis. In Fig. 5.8 we plot the orientation dependencies of the splitting between satellites for two inequivalent nucleus sites when rotating the crystal about an arbitrary axis. The confluence of satellites corresponding to different inequivalent sites occurs when $\phi'_0 = 50°$ and $\phi'_0 = 140°$. Let the axis rotation is chosen in such a manner that the magnetic field vector is parallel to the symmetry plane at $\phi'_0 = 50°$. This relative angle corresponds to the angle $\phi_z = 0°$. At $\phi'_{\max} = 26°$ there is a maximal splitting between satellites for one of the inequivalent sites $(\cos(2\phi_z^{(1)} + \phi_{z0}^{(1)}) = 1)$. It corresponds to the angle $\phi_z = -24°$. Therefore, $\phi_{z0}^{(1)} = 48°$. By the similar way one can find for the second site that $\phi_{z0}^{(2)} = -28°$.

Fitting the experimental data by a function like (5.47) it is possible to find the values $A_z^{(1,2)}$, $D_z^{(1,2)}$ and $\phi_{z0}^{(1,2)}$, and hence to calculate, using Eq. (5.48), the values $V_{zz}^{(1,2)}$, $(V_{xx}^{(1,2)} - V_{yy}^{(1,2)})^2 + (V_{xy}^{(1,2)})^2$ and $V_{xy}^{(1,2)}/(V_{xx}^{(1,2)} - V_{yy}^{(1,2)})$ in the frame related

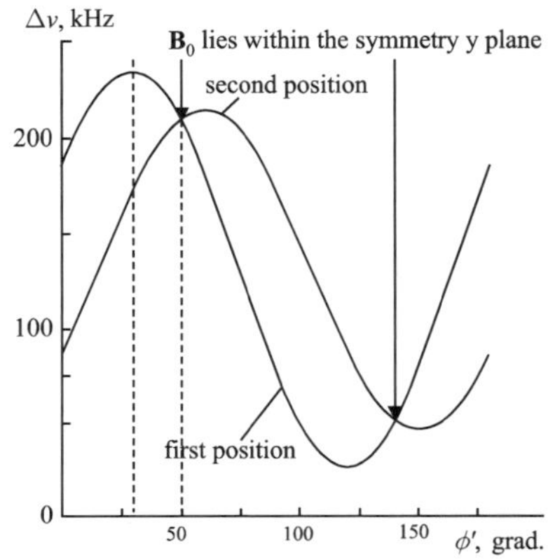

Fig. 5.8 Orientation dependencies of the splitting between satellites for two inequivalent nucleus sites

to the crystal rotation. Solving the system of six equations, one can find six unknown quantities: $V_{zz}^{(1,2)}$, $V_{xx}^{(1,2)}$, $V_{xy}^{(1,2)}$.

The following step in the method of the single rotation is to transform the tensors $\mathsf{V}_{\text{cr}}^{(1,2)}$ into the x, y, z frame. Such a transformation can be done if the direction cosines l, m, n of the x, y, z coordinate system with respect to the crystallographic a, b, c frame are known:

$$\mathsf{T} = \begin{pmatrix} l_1 & l_2 & l_3 \\ m_1 & m_2 & m_3 \\ n_1 & n_2 & n_3 \end{pmatrix}.$$

Then

$$\mathsf{V}_{x,y,z}^{(1,2)} = \mathsf{T}^{-1}\mathsf{V}_{\text{cr}}\mathsf{T}.$$

As from the orientation dependence of the splitting between the first satellites in the x, y, z frame only three components of the tensor $V_{zz,xx,xy}^{(1,2)}$ are determined, it is necessary to ascertain how these components are related to the components of the V_{cr} tensor. Without loss of generality let $l_3 = 0$. Then for one of the nucleus site one has:

$$\begin{aligned} V_{zz}^{(1)} &= n_1^2 V_{aa} + n_2^2 V_{bb} + n_3^2 V_{cc} + 2n_1 n_2 V_{ab} + 2n_1 n_2 V_{ac} + 2n_2 n_3 V_{bc}, \\ V_{xx}^{(1)} - V_{yy}^{(1)} &= (l_1^2 - m_1^2) V_{aa} + (l_2^2 - m_2^2) V_{bb} - m_3^2 V_{cc} \\ &\quad + 2(l_1 l_2 - m_2 m_1) V_{ab} - 2m_3 m_1 V_{ac} - 2m_2 m_3 V_{bc}, \\ V_{xy}^{(1)} &= l_1 m_1 V_{aa} + l_2 m_2 V_{bb} + (l_2 m_1 + l_1 m_2) V_{ab} + l_1 m_3 V_{ac} + l_2 m_3 V_{bc}. \end{aligned}$$

For another site, taking into account Eq. (5.46) these relations can be rewritten as follows:

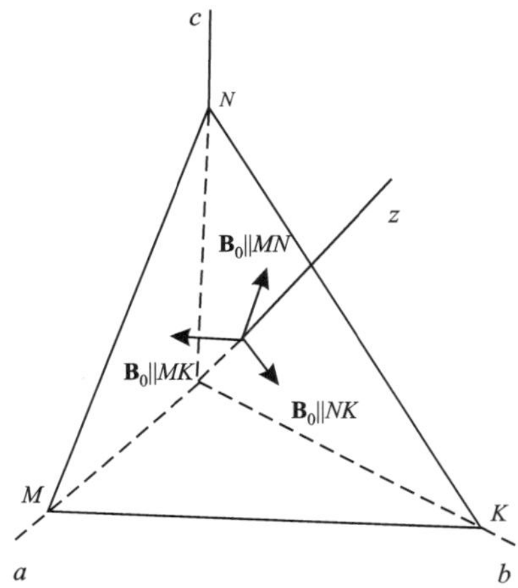

Fig. 5.9 The experiment geometry to determine directions of the principal axes of the EFG tensor

$$
\begin{aligned}
V_{zz}^{(2)} &= n_1^2 V_{aa} + n_2^2 V_{bb} + n_3^2 V_{cc} + 2n_1 n_2 V_{ab} - 2n_2 n_3 V_{ac} - 2n_2 n_3 V_{bc}, \\
V_{xx}^{(2)} - V_{yy}^{(2)} &= (l_1^2 - m_1^2) V_{aa} + (l_2^2 - m_2^2) V_{bb} - m_3^2 V_{cc} \\
&\quad + 2(l_1 l_2 - m_2 m_1) V_{ab} + 2m_3 m_1 V_{ac} + 2m_2 m_3 V_{bc}, \\
V_{xy}^{(2)} &= l_1 m_1 V_{aa} + l_2 m_2 V_{bb} + (l_2 m_1 + l_1 m_2) V_{ab} - l_1 m_3 V_{ac} - l_2 m_3 V_{bc}.
\end{aligned}
$$

Thereby, one obtained six equations to determine five unknown quantities (the sixth equation serves as a verification). After determining all the components of the EFG tensor it has to be diagonalized in order to find its principal components and directions of the principal axes.

It is to be mentioned that in several cases it is possible to determine from NMR spectra the direction of the arbitrary chosen rotation axis as well. For example, if in the crystal there are three mutually perpendicular symmetry planes and their intersections define the directions of the crystallographic axes, then the plane perpendicular to the rotation axis crosses these planes along the lines NK, MK and MN, which form a triangle (see Fig. 5.9). During the crystal rotation at some values of the angle ϕ' the direction of the magnetic field vector is parallel to each of these symmetry planes. As it was mentioned above, in this case the NMR spectra corresponding to inequivalent nucleus sites, related between each other by a mirror reflection transformation, parallel to which the magnetic field is directed, interflow. Measuring the values of the ϕ' angle for such orientations of the magnetic field it is possible to determine the angles of the triangle formed by the intersection lines of the symmetry planes with the plane perpendicular to the rotation axis. After, it is easy to find the direction of the rotation axis.

5.6 NMR Line Shape in Polycrystals

As a polycrystalline sample consists of a large number of single crystals randomly oriented in space, the orientation of the principal axes of a tensor, which describes an interaction, changes also randomly. The line shape for a polycrystal is a superposition of spectral lines of single crystals, and it can be easily calculated if the tensors describing internal interactions are axial symmetric. The principal axes of these tensors are equiprobable distributed in the space. Their directions can be represented as vectors drawn from one point in different directions. Then the points, which are the intersection points of these vectors with a unit sphere, uniformly fill the sphere surface. The direction of the external magnetic field vector also can be represented as a point on the sphere. Then the angle θ between the magnetic field vector $\vec{B}_0$ and the principal axis of the tensor is equal to the arc length, passing through two points corresponding to the magnetic field direction and the principal axis of the tensor in one of the single crystals. It is obvious that the angle θ is the same for all directions of the principal axes lying on the circle, drawing on the sphere surface and whose plane is perpendicular to the magnetic field vector. Frequencies of the NMR spectral lines depend on the angle θ. For any axially symmetric interaction the orientation dependence of the spectral line frequency can be represented in the following way:

$$\omega = \omega_0 + \text{const} \cdot (3\cos^2\theta - 1)/2,$$

where const $= 3\gamma_i^2\hbar/(2r^3)$ for dipole–dipole interactions of nuclei of the same species (see Eq. (5.8)), and const $= 3e^2qQ(2m-1)/4I(2I-1)\hbar$ for quadrupole interactions (see Eq. (5.34)).[11] It is reasonable to introduce a dimensionless quantity $x = (\omega - \omega_0)/\,\text{const}$, which means a relative shift of the NMR frequency due to nuclei interact with intercrystalline fields of different nature:

$$x = (3\cos^2\theta - 1)/2. \tag{5.49}$$

The superposition intensity of spectral lines corresponding to individual single crystals $I(x)\mathrm{d}x$ at a dimensionless frequency x is proportional to the number of nuclei with the frequency lying within the range $\mathrm{d}x$, and therefore it is proportional to a spherical segment between θ and $\theta + \mathrm{d}\theta$:

$$I(x)\mathrm{d}x = \frac{1}{2\pi}\int_0^{2\pi}\int_0^{\pi} \sin\theta \mathrm{d}\theta \mathrm{d}\phi. \tag{5.50}$$

After integrating Eq. (5.50) over ϕ and dividing both parts of the obtained expression by $\mathrm{d}x$, one has

$$I(x) = \frac{\mathrm{d}\theta}{\mathrm{d}x}\sin\theta. \tag{5.51}$$

[11] For other internal interactions the dependence of the NMR line frequency on the tensor orientation is similar but with other constats.

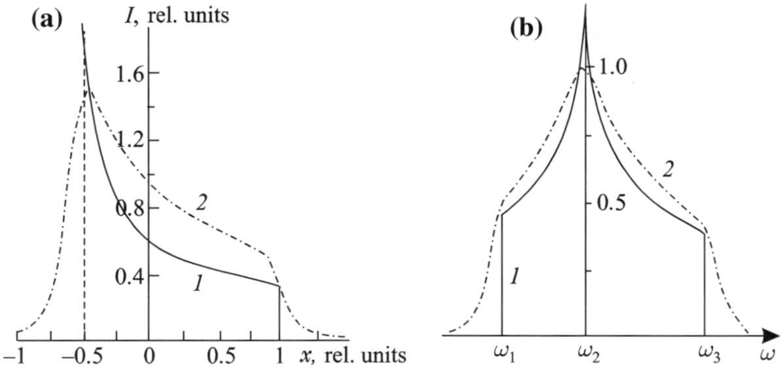

Fig. 5.10 Powder patterns: **a** — for dipole–dipole (or quadrupole with $\eta = 0$) interactions; **b** — for chemical shift anisotropy or quadrupole interactions with $\eta \neq 0$. Curves *1* and *2* correspond to the cases when the line shapes of single crystals are described by δ-functions or by Gaussian lines, respectively

It is to be noted that $\sin\theta d\theta = -d(\cos\theta)$. Let us find from Eq. (5.49) an expression that defines $\cos\theta$ as a function of relative frequency x, and put it into Eq. (5.51). Then

$$I(x) = \sqrt{\frac{1}{3(2x+1)}} = \sqrt{\frac{1}{3}\left(\frac{\text{const}}{2(\omega - \omega_0) + \text{const}}\right)}. \tag{5.52}$$

Thereby one obtains how the NMR spectral line intensity depends on the frequency x. Let us analyze this expression. As $\cos^2\theta$ can change within the range from 1 to 0, the x value, as it follows from Eq. (5.49), can change from $-1/2$ to 1. The spectral line intensity corresponding to $x = 1$ is equal to 1/3, and if $x = -1/2$, then $I(x) \to \infty$. Therefore, the maximal intensity of the NMR line corresponds to the frequency

$$\omega = \omega_0 - \text{const}/2,$$

and measuring the frequency, which corresponds to the maximum of the NMR line, it is possible to determine the constant. It means that if there is a selected pair of nuclei in the crystal it is possible to determine the distance between these nuclei, and if there is a quadrupole interaction, it is possible to determine the quadrupole coupling constant.

Obviously it is much easy to treat polycrystalline spectra as compared to single crystal ones. However, when one deals with polycrystals a considerable part of information is lost. For example, it is impossible to determine the direction of the radius vector connecting two nuclei or directions of the principal axes of the EFG tensor. Moreover, the accuracy in determining one or another parameter decreases. The latter is due to the fact that when deriving Eq. (5.52) it was supposed that the NMR line for a single crystal is a δ-function. In this case the line shape for a polycrystal is described by the curve 1 in Fig. 5.10a. In real materials the line shape significantly differs from it. To take into account broadening effects, it is necessary to calculate the convolution product of the function $I(x)$ and the function describing the line shape of an individual component $g(x - x')$:

$$I_t(x) = \int_{-\infty}^{\infty} I(x')g(x - x')\mathrm{d}x'.$$

Here x' is the frequency corresponding to the maximum intensity of the individual spectral line in a single crystal. Such calculations are made in assumption that the line shape of the single crystal can be approximated, for example, by Lorentzian or Gaussian function. It is rather difficult to treat analytically the obtained formula, and the computational modeling is usually applied. In Fig. 5.10a, the curve 2 corresponds to the line shape in a polycrystalline sample in the case when the NMR line of a single crystal can be approximated by the Gaussian.

If the interaction that determines the line shape in a polycrystal is characterized by a tensor with the non-zero asymmetry parameter, then it can be represented as two sectionally continuous functions, each of which has the following form:

$$I(\omega) = \frac{1}{\pi\sqrt{[(\omega_3 - \omega_2)(\omega - \omega_1)]}} K\left\{\arcsin\sqrt{\frac{(\omega_3 - \omega)(\omega_2 - \omega_1)}{(\omega_3 - \omega_2)(\omega - \omega_1)}}\right\} \tag{5.53}$$

for $\omega_2 \le \omega \le \omega_3$ and

$$I(\omega) = \frac{1}{\pi\sqrt{[(\omega_3 - \omega)(\omega_2 - \omega_1)]}} K\left\{\arcsin\sqrt{\frac{(\omega_3 - \omega_2)(\omega - \omega_1)}{(\omega_3 - \omega)(\omega_2 - \omega_1)}}\right\} \tag{5.54}$$

for $\omega_1 \le \omega \le \omega_2$. Here $K(\arcsin k)$ is the complete first-order elliptical integral, which is tabulated and can be computed; $\omega_1, \omega_2, \omega_3$ are the main components of the tensor. Eqs. (5.53) and (5.54) are obtained under the assumption that the NMR line of a single crystal is a δ-function. The calculated spectral line is plotted in Fig. 5.10b (curve 1). As it is seen the maximum intensity of the NMR line corresponds to the frequency ω_2. The line shape has two bends, which correspond to the frequency values ω_1 and ω_3. Therefore, on the basis of the analysis of the NMR line shape in polycrystals it is possible to determine all three principal components of the tensor, however, the accuracy essentially depends on the line width of the individual single crystal. The real line shape in a polycrystalline sample for $\eta \neq 0$ and when the individual single crystal line is fitted by the Gaussian, is shown in Fig. 5.10b (curve 2). As it is seen, the curve is rather smoothed, that reduces the determination accuracy of the principal components of the tensor.

5.7 High Resolution NMR in Solids

As it was shown in Sects. 1.10 and 1.12 the NMR spectral lines in solids are broaden mainly due to dipole–dipole interactions, and if a nucleus spin is greater than 1/2 there are quadrupole interactions of nuclei with electric field gradients statistically distributed over a crystal, which make an important contribution to the NMR line width. In paramagnetic crystals superfine interactions also contribute into the spectral line width. Large NMR line widths do not allows us to fix line shifts caused by electron shielding, i.e. chemical shifts, as well as shifts due to indirect spin–spin interactions. However, those interactions provide extensive knowledge about properties of liquids, where both dipole–dipole and quadrupole interactions are averaged to zero, and there are only isotropic contributions of superfine interactions. That is why one of the most important problem of the solid state NMR is to suppress or at least to reduce the impact of these interactions (essentially the dipole–dipole one) on the NMR spectra. The solution has been proposed by Waugh.[12] It is obvious that to suppress the dipole–dipole interactions in solids it is necessary to change experimental conditions in such a manner that to average them, like in liquids. As soon as all quantities measured experimentally are always time averaged it is necessary to make the internal Hamiltonian time depending. It can be done by two ways:

- by a physical displacement of solids, that leads to an explicit dependence of parameters of the Hamiltonian on time; an example of such an approach is the rotation of a sample about an axis, which makes a definite angle with the magnetic field vector;
- by such an external action on a sample that results in the time dependence of the operators, which determine internal interactions in solids; an example of such an approach is the application of special pulse sequences.

If during the life time of a nucleus in a definite state the Hamiltonian operators can change many times then similar to liquids the NMR spectrum observed in experiment is determined by the time averaged Hamiltonian operators. The averaging may occur in the coordinate space, if the Hamiltonian parameters are time depending, or in the spin space, if the spin operators are time depending.

5.7.1 NMR Line Narrowing by Magic Angle Spinning

In solids in the absence of molecular motion the line width $\Delta\omega$, caused by direct dipole–dipole interactions, is so large that it makes difficult to study electron shielding and indirect spin–spin interactions. To measure chemical shifts in solids it is necessary to narrow NMR lines by averaging dipole–dipole interactions. One of such methods is the macroscopic *magic angle spinning* (MAS).

[12] J.S Waugh, M.M Maricq, R Cantor. Rotational spin echoes in solids. J. Mag. Res. 29 (1969) 183.

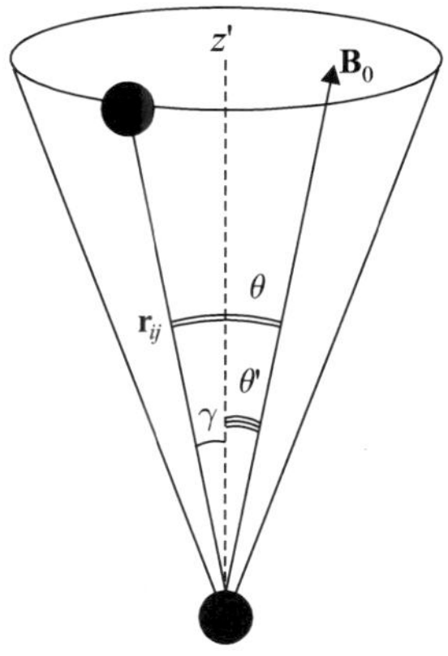

Fig. 5.11 Reorientations of the internuclear vector $\vec{r}_{ij}$ when rotating a sample about a fixed axis z' in an external magnetic field

Let us consider how macroscopic spinning of a sample affects its NMR spectrum. In the presence of dipole–dipole interactions between nucleus magnetic moments the NMR spectrum is determined by the secular part of the dipole–dipole Hamiltonian, which for the nuclei of the same magnetic species can be written as follows

$$\widehat{\mathcal{H}}_{\mathrm{d}} = \frac{\gamma^2\hbar^2}{2}\sum_{ji} r_{ij}^{-3}(3\cos^2\theta_{ij} - 1)\left(\hat{\vec{I}}_i\hat{\vec{I}}_j - 3\hat{I}_{zi}\hat{I}_{zj}\right), \tag{5.55}$$

and for the nuclei of different magnetic species as

$$\widehat{\mathcal{H}}_{\mathrm{d}} = \frac{1}{2}\sum_{ij}\gamma_i\gamma_j\hbar^2 r^{-3}(1 - 3\cos^2\theta_{ij})\left(\hat{I}_{zi}\hat{I}_{zj}\right).$$

When rotating a sample with the angular velocity ω_{rot} each internuclear vector $\vec{r}_{ij}$ describes a cone about the z'-axis (Fig. 5.11). As it is seen from the figure, the value of $\cos\theta_{ij}$ can be found by the following way:

$$\cos(\theta_{ij}) = \cos\theta'\cos\gamma_{ij} + \sin\theta'\sin\gamma_{ij}\cos(\omega_{\mathrm{rot}}t + \phi_{0ij}), \tag{5.56}$$

and after substitution Eq. (5.56) into Eq. (5.55), one can find that

$$\begin{aligned}\widehat{\mathcal{H}}_{\mathrm{d}} = \frac{1}{2}\sum_{ij}\gamma^2\hbar^2 r_{ij}^{-3}\Bigg[&\frac{1}{2}(3\cos^2\theta' - 1)(3\cos^2\gamma_{ij} - 1)\\ &+\frac{3}{2}\sin 2\theta'\sin 2\gamma_{ij}\cos(\omega_{\mathrm{rot}}t + \phi_{0ij})\\ &+\frac{3}{2}\sin^2\theta'\sin^2\gamma_{ij}\cos(2\omega_{\mathrm{rot}}t + \phi_{0ij})\Bigg]\left(\hat{\vec{I}}_i\hat{\vec{I}}_j - 3\hat{I}_{zi}\hat{I}_{zj}\right).\end{aligned}$$

It is obvious that the first term in the square brackets is constant, whereas the other two are periodic with the average equal to zero. Thereby, the first term of the Hamiltonian

$$\widehat{\mathcal{H}}_{\rm d} = \frac{1}{2}(3\cos^2\theta' - 1)\frac{\gamma^2\hbar^2}{2}\sum_{ij} r_{ij}^{-3}(3\cos^2\gamma_{ij} - 1)\left(\hat{\vec{I}}_i\hat{\vec{I}}_j - 3\hat{I}_{zi}\hat{I}_{zj}\right) \quad (5.57)$$

determines the contribution of the dipole–dipole interactions into the reduced NMR spectrum, whereas the second and third ones with frequencies $\omega_{\rm rot}$ and $2\omega_{\rm rot}$ respectively, correspond to sidebands at frequencies $\omega_0 \pm n\omega_{\rm rot}$. At a high rotation speed in the Hamiltonian (5.57), which describes the central part of the spectrum, a "narrowing" factor appears. As it is seen from Eq. (5.57), if the angle θ' between the rotation axis and the magnetic field vector $\vec{B}_0$ can satisfy the condition

$$(3\cos^2\theta' - 1) = 0,$$

i.e. $\theta = 54°44'$ ("magic" angle), then at a sufficiently high rotation speed $\omega_{\rm rot} > \Delta\omega$ (usually several kHz) the theoretical value of the line width caused by dipole–dipole interactions is zero. However, the sidebands appearing due to the rotation make difficult the quantitative analysis, because they must be taken into account in calculations of the total second moment of spectral lines.

In MAS experiments not only the dipole–dipole interactions are averaged to zero, but also the magnetic shielding anisotropy. The components of the anisotropic shielding tensor can be measured either by cw NMR (by analyzing the spectral line shape) or by applying special pulse sequences.

5.7.2 NMR Line Narrowing by Pulse Sequences

Since the first attempts to suppress dipole–dipole interactions a variety of pulse sequences, which allow us to modify intentionally the Hamiltonian describing the internal interactions in solids, have been developed. It makes possible to improve the resolution of NMR spectra by eliminating or scaling selected interactions (such as, for example, homonuclear and heteronuclear dipole–dipole interactions), and to increase significantly the amount of information retrieved from such spectra.

Currently, a coherent averaging of spin Hamiltonian operators by acting on a sample by periodic multipulse sequences or by rotating a sample is widespread, because it essentially simplifies NMR spectra. The resulting spectra are described by modified Hamiltonian operators, which do not contain time-dependent terms. In the case when the effects of periodic disturbances can be completely described by a time independent Hamiltonian it is convenient to apply the average Hamiltonian theory. This approach leads to simple analytical results and is especially useful to analyze how periodic multipulse sequences affect the spin Hamiltonian. However, such an approach is not universal. In many cases, there are more resonance lines than the theory predicts. In particular, if one rotates a sample at an angle to the magnetic field vector additional lines (sidebands) appear, and they can not be described any more by this time independent Hamiltonian.

Before we start to expose the average Hamiltonian theory and the methods, which are used to suppress various interactions, let us consider what is similar and different in the Hamiltonian operators describing various types of interactions that determine a NMR spectrum.

As it was already shown, all interactions of nuclear spins (between each other, with external or internal electric and magnetic fields) can be described, with respect to constituting operators, by three kinds of Hamiltonian operators, which contain terms linear, bilinear and quadratic in spin operators. Let us discuss them.

The linear terms of the Hamiltonian include the Zeeman interaction with a static magnetic field $\vec{B}_0$, the interaction with a radiofrequency field $\vec{B}_1(t)$, and the interactions resulting in chemical shifts: paramagnetic and diamagnetic shielding, hyperfine interactions between magnetic moments of nuclei and electrons (both direct and indirect). All these interactions can be described by the following Hamiltonian:

$$\widehat{\mathcal{H}} = -\sum_i^N \gamma_i \hbar \hat{\vec{I}}_i \, \mathsf{T}_i \, \vec{B}_0. \tag{5.58}$$

Depending on the interaction type the tensor T_i is defined by the following way:

- $\mathsf{T}_i \equiv 1$, if the Hamiltonian (5.58) describes the Zeeman interaction of nuclear magnetic moments with the static ($\vec{B}_0$) or alternating ($\vec{B}_1$) external field;
- $\mathsf{T}_i \equiv -\boldsymbol{\sigma}$, if the Hamiltonian (5.58) describes effects of magnetic shielding of nuclei by the electron shell ($\boldsymbol{\sigma}$ is the magnetic shielding tensor);
- $\mathsf{T}_i \equiv -\mathsf{A}$, if the Hamiltonian (5.58) describes the direct or indirect superfine interaction of nuclei and electrons (A is the tensor of the superfine interaction).

For the most studied materials tensors of the magnetic shielding and hyperfine interactions are symmetric second rank tensors with non-zero traces. Such tensors can always be represented as a sum of an isotropic component t_{iso} independent on the crystal orientation and a second rank tensor with a non-zero trace described by the following matrix

$$\mathsf{T} = \begin{pmatrix} t_{xx} & t_{xy} & t_{xz} \\ t_{yx} & t_{yy} & t_{yz} \\ t_{zx} & t_{zy} & -(t_{xx}+t_{yy}) \end{pmatrix}. \tag{5.59}$$

If the matrix (5.59) defines the tensor in the laboratory frame, then the Hamiltonian (5.58) can be represented as

$$\widehat{\mathcal{H}} = -\sum_i^N \widehat{\mathcal{H}}_i, \tag{5.60}$$

where

$$\widehat{\mathcal{H}}_i = -\gamma_i \hbar B_0 \left(\frac{\hat{\vec{I}}_{i+}}{2}(t_{xz} - \mathrm{i}t_{yz}) + \frac{\hat{\vec{I}}_{i-}}{2}(t_{xz} + \mathrm{i}t_{yz}) + \hat{\vec{I}}_{iz}(t_{zz} + t_{\mathrm{iso}}) \right).$$

The bilinear terms in the Hamiltonian describe dipole–dipole and indirect spin–spin interactions between nuclear magnetic moments. In this case

$$\widehat{\mathcal{H}} = \sum_{i=1}^{N} \sum_{j=1}^{N} \hat{\vec{I}}_i \mathsf{T}_{ij} \hat{\vec{I}}_j, \tag{5.61}$$

here depending on the interaction type the tensor T_{ij} is defined by the following way:

- $\mathsf{T}_{ij} = (\gamma_i \gamma_j \hbar^2/(2r_{ij}^5))\mathsf{D}$, if the Hamiltonian (5.61) describes dipole–dipole interactions between magnetic moments of nuclei i and j (D is the tensor of dipole–dipole interactions (1.134));
- $\mathsf{T}_i \equiv \mathsf{J}$, if the Hamiltonian (5.61) describes the indirect spin–spin interactions of nuclei through the electron shell (J is the tensor of indirect spin–spin interactions).

The tensor of dipole–dipole interactions is a symmetric second rank tensor with a zero trace, whereas the tensor of indirect spin–spin interactions in the most cases is a symmetric second rank tensor with a non-zero trace, which can be represented as a sum of the isotropic component t_{iso} and a 3×3 matrix with a zero trace (see Eq. (5.59)). Using the matrix representation of tensors, the Hamiltonian (5.61) after multiplication of the corresponding matrices is

$$\widehat{\mathcal{H}} = \sum_{i=1}^{N} \sum_{j=1}^{N} \left(\hat{A}_{ij} + \hat{B}_{ij} + \hat{C}_{ij} + \hat{D}_{ij} + \hat{E}_{ij} + \hat{F}_{ij} \right), \tag{5.62}$$

where

$$\begin{aligned}
\hat{A}_{ij} &= \hat{\vec{I}}_{iz} \hat{\vec{I}}_{jz} (t_{zz} + t_{\mathrm{iso}}); \\
\hat{B}_{ij} &= -\tfrac{1}{4} \left(\hat{\vec{I}}_{i+} \hat{\vec{I}}_{j-} + \hat{\vec{I}}_{i-} \hat{\vec{I}}_{j+} \right) (t_{zz} + t_{\mathrm{iso}}); \\
\hat{C}_{ij} &= -\tfrac{3}{2} \left(\hat{\vec{I}}_{i+} \hat{\vec{I}}_{jz} + \hat{\vec{I}}_{iz} \hat{\vec{I}}_{j+} \right) (t_{zx} - \mathrm{i}t_{zy}); \\
\hat{D}_{ij} &= -\tfrac{3}{2} \left(\hat{\vec{I}}_{i-} \hat{\vec{I}}_{jz} + \hat{\vec{I}}_{iz} \hat{\vec{I}}_{j-} \right) (t_{zx} + \mathrm{i}t_{zy}); \\
\hat{E}_{ij} &= -\tfrac{3}{4} \left(\hat{\vec{I}}_{i+} \hat{\vec{I}}_{j+} \right) \left(\tfrac{t_{xx} - t_{yy}}{2} - \mathrm{i}t_{xy} \right); \\
\hat{F}_{ij} &= -\tfrac{3}{4} \left(\hat{\vec{I}}_{i-} \hat{\vec{I}}_{j-} \right) \left(\tfrac{t_{xx} - t_{yy}}{2} + \mathrm{i}t_{xy} \right).
\end{aligned} \tag{5.63}$$

Expressions in Eq. (5.63) are the sum of the products of terms, which contain the spin operators, and terms, which depend only on space variables and are determined by tensors. The expressions that contain only the spin variables are exactly identical to those that describe the dipole–dipole interactions (1.136). It is possible to demonstrate that in a spherical frame for axially symmetric tensors the space parts of both types of interactions are the same up to a constant.

The quadratic terms in the Hamiltonian describe the quadrupole interactions of nuclear magnetic moments with electric field gradients. The Hamiltonian of such an interaction is

$$\widehat{\mathcal{H}} = \sum_{i=1}^{N} \frac{eQ_i}{4I_i(2I_i - 1)} \hat{\vec{I}}_i \mathsf{T}_i \hat{\vec{I}}_i, \tag{5.64}$$

where $\mathsf{T} = \mathsf{V}_i$ is the EFG tensor. It is a symmetric second rank tensor with a zero trace. The Hamiltonian (5.64) can be also represented as a sum of the operators (5.63):

$$\widehat{\mathcal{H}} = \sum_{i=1}^{N} \left(\hat{A}_i + \hat{B}_i + \hat{C}_i + \hat{D}_i + \hat{E}_i + \hat{F}_i \right),$$

where

$$\begin{aligned}
\hat{A}_i + \hat{B}_i &= \left(3\hat{\vec{I}}^2_{iz} - \hat{\vec{I}}^2 \right) t_{zz}; \\
\hat{C}_i &= \left(\hat{\vec{I}}_{i+}\hat{\vec{I}}_{iz} + \hat{\vec{I}}_{iz}\hat{\vec{I}}_{i+} \right) (t_{zx} - \mathrm{i}t_{zy}); \\
\hat{D}_i &= \left(\hat{\vec{I}}_{i-}\hat{\vec{I}}_{iz} + \hat{\vec{I}}_{iz}\hat{\vec{I}}_{i-} \right) (t_{zx} + \mathrm{i}t_{zy}); \\
\hat{E}_i &= \hat{\vec{I}}^2_{i+} \left(\tfrac{t_{xx} - t_{yy}}{2} - \mathrm{i}t_{xy} \right); \\
\hat{F}_i &= \hat{\vec{I}}^2_{i-} \left(\tfrac{t_{xx} - t_{yy}}{2} + \mathrm{i}t_{xy} \right).
\end{aligned} \tag{5.65}$$

The first equation in (5.65) is a transformed expression for the sum of the operators $(\hat{A}_i + \hat{B}_i)$. By the similar way the sum of these operators can be represented in (5.63), if the interacting spins are of the same species. As it is seen from Eq. (5.65), the spin parts of the operators are similar to those that were considered above when describing dipole–dipole and indirect spin–spin interactions.

Within the first approximation all the Hamiltonian operators that describe interactions in the system of nuclear spins are time independent: the dipole–dipole interaction Hamiltonian may depend on time if the orientation of the radius vector connecting nuclei or the internuclear distance depends on time; the Hamiltonian operators of the quadrupole, superfine, indirect spin–spin interactions or electron shielding depend on time only if the tensors corresponding to these operators are time dependent. In solids at low temperature these time dependencies can be neglected if we are not interested in the study of dynamic processes. Moreover, these processes are internal and can not be controlled by an experimentalist. In this Section we will consider time dependencies that are created by an experimentalist to suppress or scale one or another interaction. The description of these external actions is done within the formalism of exponential operators.

5.7.3 Exponential Operators

There is a certain analogy in the consideration of exponential operators and exponential functions of complex variables. Thus, the exponential function of a complex variable z can be represented as a Taylor series:

$$e^z = 1 + z + \frac{z^2}{2!} + \frac{z^3}{3!} + \frac{z^4}{4!} + \dots.$$

This power series converges to e^z for all values of z. In such a manner it is possible to represent the exponential operators $e^{\hat{P}}$ and $e^{i\hat{P}}$:

$$\begin{aligned} e^{\hat{P}} &= 1 + \hat{P} + \tfrac{\hat{P}^2}{2!} + \tfrac{\hat{P}^3}{3!} + \dots, \\ e^{i\hat{P}} &= 1 + i\hat{P} + \tfrac{(i\hat{P})^2}{2!} + \tfrac{(i\hat{P})^3}{3!} + \dots. \end{aligned} \tag{5.66}$$

Let us consider the properties of exponential operators. It is easy to verify the validity of the relations given further by applying these relations to the right part of Eq. (5.66): if $\hat{A}$ and $\hat{B}$ commute between each other, $[\hat{A}, \hat{B}] = [\hat{B}, \hat{A}]$, then all the relations, that are true for usual exponential function, are true for exponential operators as well:

$$\hat{A}e^{\hat{B}} = e^{\hat{B}}\hat{A}, \quad e^{(\hat{A}+\hat{B})} = e^{\hat{B}}e^{\hat{A}}; \tag{5.67}$$

if $\hat{A}$ and $\hat{B}$ do not commute between each other then Eq. (5.67) should be replaced by the following ones:

$$\begin{gathered} e^{(\hat{A}+\hat{B})} = e^{\hat{A}}e^{\hat{B}}e^{-[\hat{A},\hat{B}]/2} = e^{[\hat{A},\hat{B}]/2}e^{\hat{B}}e^{\hat{A}}, \\ e^{\hat{B}}e^{\hat{A}} = \exp\left(\hat{A} + \hat{B} + \frac{1}{2}[\hat{B},\hat{A}] + \frac{1}{12}\left([\hat{B},[\hat{B},\hat{A}]] + [[\hat{B},\hat{A}],\hat{A}] + \dots\right) + \dots\right), \\ \left(e^{\hat{B}}e^{\hat{A}}\right)^{-1} = \left(e^{\hat{A}}\right)^{-1}\left(e^{\hat{B}}\right)^{-1}. \end{gathered} \tag{5.68}$$

Application of the exponential operators allows us to present in a compact form the transformation from the laboratory frame[13] to a noninertial one, for example, rotating about an axis of the laboratory frame. Let us show that the operator $\hat{I}_{x^z}$ in the frame rotating about the z-axis of the laboratory frame is related to the $\hat{I}_x$ operator

[13] Mind that traditionally the z-axis of the laboratory frame is directed along the external magnetic field vector $\vec{B}_0$.

of the laboratory frame by the following transformation:

$$\hat{I}_{x^z} = \exp(-\mathrm{i}\omega \hat{I}_z t)\hat{I}_x \exp(\mathrm{i}\omega \hat{I}_z t). \tag{5.69}$$

Let us express the right part of Eq. (5.69) in the form of a function depending on the angle $\phi = \omega t$:

$$f(\phi) = \exp(-\mathrm{i}\phi \hat{I}_z)\hat{I}_x \exp(\mathrm{i}\phi \hat{I}_z), \tag{5.70}$$

and differentiate Eq. (5.70) with respect to ϕ:

$$\frac{\partial f}{\partial \phi} = \exp(-\mathrm{i}\phi \hat{I}_z)(-\mathrm{i}\hat{I}_z\hat{I}_x + \mathrm{i}\hat{I}_x\hat{I}_z)\exp(\mathrm{i}\phi \hat{I}_z). \tag{5.71}$$

Taking into account that $[\hat{I}_z\hat{I}_x] = \hat{I}_z\hat{I}_x - \hat{I}_x\hat{I}_z = \mathrm{i}\hat{I}_y$, one has, instead of Eq. (5.71):

$$\frac{\partial f}{\partial \phi} = \exp(-\mathrm{i}\phi \hat{I}_z)\hat{I}_y \exp(\mathrm{i}\phi \hat{I}_z). \tag{5.72}$$

By analogy, differentiating Eq. (5.72) and considering $[\hat{I}_z\hat{I}_y] = -\mathrm{i}\hat{I}_x$ one obtains the differential equation for the function $f(\phi)$:

$$\frac{\partial^2 f}{\partial \phi^2} = \exp(-\mathrm{i}\phi \hat{I}_z)(-\mathrm{i}\hat{I}_z\hat{I}_y + \mathrm{i}\hat{I}_y\hat{I}_z)\exp(\mathrm{i}\phi \hat{I}_z) = -\exp(-\mathrm{i}\phi \hat{I}_z)\hat{I}_x \exp(\mathrm{i}\phi \hat{I}_z) = -f$$

or

$$\frac{\partial^2 f}{\partial \phi^2} + f = 0, \tag{5.73}$$

After solving the differential equation (5.73) one has

$$f(\phi) = A\cos\phi + B\sin\phi. \tag{5.74}$$

The constants A and B in Eq. (5.74) are easy to find from the initial conditions. As it follows from Eq. (5.70), $f(0) = \hat{I}_x$, $f'(0) = \hat{I}_y$ and, therefore, $\hat{A} = \hat{I}_x$, $\hat{B} = \hat{I}_y$.

By the similar way one can obtain other relations:

$$\begin{aligned}
\hat{I}_{x^z} &= \exp(-\mathrm{i}\phi \hat{I}_z)\hat{I}_x \exp(\mathrm{i}\phi \hat{I}_z) \equiv \hat{I}_x\cos\phi + \hat{I}_y\sin\phi,\\
\hat{I}_{y^z} &= \exp(-\mathrm{i}\phi \hat{I}_z)\hat{I}_y \exp(\mathrm{i}\phi \hat{I}_z) \equiv -\hat{I}_x\sin\phi + \hat{I}_y\cos\phi,\\
\hat{I}_{z^z} &= \exp(-\mathrm{i}\phi \hat{I}_z)\hat{I}_z \exp(\mathrm{i}\phi \hat{I}_z) \equiv \hat{I}_z.
\end{aligned} \tag{5.75}$$

If we consider the angular momentum operators as ordinary vectors, then from Eq. (5.75) it follows that $\hat{I}_{x^z}$, $\hat{I}_{y^z}$ and $\hat{I}_{z^z}$ are the components of the angular momentum vector along the axes x^z, y^z and z^z, which rotates about the z-axis of the laboratory frame on the angle $\phi = \omega t$. Thereby, the exponential operator $\exp(\mathrm{i}\phi \hat{I}_z)$ is a rotation operator, and Eq. (5.75) is the transformation to a frame rotating with the angular velocity ω.

By analogy it is possible to show that the operators $\exp(\mathrm{i}\phi\hat{I}_x)$ and $\exp(\mathrm{i}\phi\hat{I}_y)$ are the rotation operators about the axes x and y, respectively:

$$\begin{aligned}
\hat{I}_{x^x} &= \exp(-\mathrm{i}\omega\hat{I}_x t)\hat{I}_x \exp(\mathrm{i}\omega\hat{I}_x t) \equiv \hat{I}_x,\\
\hat{I}_{y^x} &= \exp(-\mathrm{i}\omega\hat{I}_x t)\hat{I}_y \exp(\mathrm{i}\omega\hat{I}_x t) \equiv \hat{I}_z \sin\omega t + \hat{I}_y \cos\omega t,\\
\hat{I}_{z^x} &= \exp(-\mathrm{i}\omega\hat{I}_x t)\hat{I}_z \exp(\mathrm{i}\omega\hat{I}_x t) \equiv \hat{I}_z \cos\omega t - \hat{I}_y \sin\omega t;\\
\hat{I}_{x^y} &= \exp(-\mathrm{i}\omega\hat{I}_y t)\hat{I}_x \exp(\mathrm{i}\omega\hat{I}_y t) \equiv \hat{I}_x \cos\omega t - \hat{I}_z \sin\omega t,\\
\hat{I}_{y^y} &= \exp(-\mathrm{i}\omega\hat{I}_y t)\hat{I}_y \exp(\mathrm{i}\omega\hat{I}_y t) \equiv \hat{I}_y,\\
\hat{I}_{z^y} &= \exp(-\mathrm{i}\omega\hat{I}_y t)\hat{I}_z \exp(\mathrm{i}\omega\hat{I}_y t) \equiv \hat{I}_x \sin\omega t + \hat{I}_z \cos\omega t.
\end{aligned} \tag{5.76}$$

Rotation about an arbitrary axis can be represented as three successive rotations of the laboratory frame.

Now it is possible to use exponential operators to make the quantum-mechanical transformation similar to the classical transformation to a rotating frame. The consideration of the quantum-mechanical analogue of the rotating frame is important and widely used in the average Hamiltonian theory. Moreover, as it was shown earlier (see Eqs. (1.62)–(1.64)), all the quantities recorded by NMR are observed in the rotating frame.

5.7.4 Quantum-Mechanical Analogue of the Rotating Frame

Let a static magnetic field $\vec{B}_0$ and an alternating magnetic field $\vec{B}_1$, which is perpendicular to $\vec{B}_0$ and rotating with the angular velocity ω, affect a system of noninteracting spins, and

$$\vec{B}_1(t) = \vec{i}B_1 \cos\omega t - \vec{j}B_1 \sin\omega t,$$

where $\vec{i}$ and $\vec{j}$ are the unit vectors directed along the x and y axes of the laboratory frame. The Hamiltonian of the interaction of a spin with the external magnetic fields can be written as follows

$$\widehat{\mathcal{H}}_i = -\gamma\hbar(\hat{\vec{I}}_i \cdot \vec{B}) = -\gamma_i\hbar\left(B_0\hat{I}_{zi} + B_1(\hat{I}_{xi}\cos\omega t - \hat{I}_{yi}\sin\omega t)\right)$$

or taking into account the first line in (5.75):

$$\widehat{\mathcal{H}}_i = -\gamma_i\hbar\left(B_0\hat{I}_{zi} + B_1(\exp(\mathrm{i}\omega\hat{I}_{zi}t)\hat{I}_{xi}\exp(-\mathrm{i}\omega\hat{I}_{zi}t))\right).$$

Then the Schrödinger equation is written as

$$-\frac{\hbar}{\mathrm{i}}\frac{\partial\Psi}{\partial t} = -\gamma_i\hbar\left(B_0\hat{I}_{zi} + B_1(\exp(\mathrm{i}\omega\hat{I}_{zi}t)\hat{I}_{xi}\exp(-\mathrm{i}\omega\hat{I}_{zi}t))\right)\Psi. \tag{5.77}$$

By multiplying Eq. (5.77) on the left by $\exp(-\mathrm{i}\omega\hat{I}_{zi}t)$ and using the expressions

$$\frac{\partial \exp(-\mathrm{i}\omega\hat{I}_{zi}t)\Psi(t)}{\partial t} = -\mathrm{i}\omega\hat{I}_{zi}\exp(-\mathrm{i}\omega\hat{I}_{zi}t)\Psi + \exp(-\mathrm{i}\omega\hat{I}_{zi}t)\frac{\partial\Psi}{\partial t}, \tag{5.78}$$

one has, instead of Eq. (5.77),

$$-\frac{\hbar}{\mathrm{i}}\frac{\partial(\exp(-\mathrm{i}\omega\hat{I}_{zi}t)\Psi)}{\partial t} = -\left(\hbar(-\omega+\gamma_i B_0)\hat{I}_{zi} + \gamma_i\hbar B_1\hat{I}_{xi}\right)\exp(-\mathrm{i}\omega\hat{I}_{zi}t)\Psi.$$

Let us assume

$$\Psi' = \exp(-\mathrm{i}\omega\hat{I}_{zi}t)\Psi \qquad \text{or} \qquad \Psi = \exp(\mathrm{i}\omega\hat{I}_{zi}t)\Psi'. \tag{5.79}$$

The physical meaning of Eq. (5.79) is that the functions Ψ and Ψ' transform from one to another when the coordinate axes rotate on the angle ωt. Then the Schrödinger equation for the wave function Ψ' in the rotating frame can be written as follows:

$$-\frac{\hbar}{\mathrm{i}}\frac{\partial\Psi'}{\partial t} = -\left(\hbar(-\omega+\gamma_i B_0)\hat{I}_{zi} + \gamma_i\hbar B_1\hat{I}_{xi}\right)\Psi'. \tag{5.80}$$

The Hamiltonian in Eq. (5.80) is the Hamiltonian in the rotating frame:

$$\widehat{\mathcal{H}}_{\mathrm{Z}}^{\mathrm{rot}} = -\hbar(-\omega+\gamma_i B_0)\hat{I}_{zi} - \gamma_i\hbar B_1\hat{I}_{xi}. \tag{5.81}$$

It is time independent and describes the interactions of a nuclear spin with an effective magnetic field $\vec{B}_{\mathrm{eff}}$, which coincides with the effective field in the classical Bloch equations in the rotating frame:

$$\vec{B}_{\mathrm{eff}} = \left(B_0 - \frac{\omega}{\gamma}\right)\vec{k} + B_1\vec{i},$$

here $\vec{k}$ and $\vec{i}$ are unit vectors in the rotating frame directed along the z'- and x'-axes, respectively. At the resonance conditions, $\omega = \gamma B_0$, the effective field coincides (in both the direction and the value) with the $\vec{B}_1$ field. Thereby, transformation to the rotating frame leads to such a modification of the Hamiltonian describing interactions of spins with external magnetic fields, that the term describing the interaction with the constant magnetic field disappears and the interaction with the alternating magnetic field becomes time independent.

Now it is possible to find out the wave function Ψ' from Eq. (5.80). As the spin Hamiltonian $\widehat{\mathcal{H}}_{\mathrm{Z}}^{\mathrm{rot}}$ in the rotating frame is time independent, the wave function $\Psi'(t)$ can be written as

$$\Psi'(t) = \exp((-\mathrm{i}/\hbar)\widehat{\mathcal{H}}_{\mathrm{Z}}^{\mathrm{rot}}t)\Psi'(0), \tag{5.82}$$

and it is the solution of Eq. (5.80), that is easy to prove by direct substitution Eq. (5.82) into Eq. (5.80). Substituting the explicit form of the operator (5.81) in Eq. (5.82) at resonance conditions one has:

$$\Psi'(t) = \exp\left(\mathrm{i}\gamma_i B_1 \hat{I}_{xi} t\right)\Psi'(0) = \exp\left(\mathrm{i}\omega_1 \hat{I}_{xi} t\right)\Psi'(0), \tag{5.83}$$

here $\omega_1 = \gamma B_1$. It is to be noted that the wave function Ψ' is a slowly varying function of time.

The wave function Ψ in the laboratory frame can be obtained by substituting Ψ' from Eq. (5.82) or (5.83) into Eq. (5.79):

$$\Psi(t) = \exp(\mathrm{i}\omega \hat{I}_{zi} t)\exp((-\mathrm{i}/\hbar)\widehat{\mathcal{H}}_{\mathrm{Z}}^{\mathrm{rot}} t)\Psi'(0), \tag{5.84}$$

and as it follows from Eq. (5.79) at $t = 0$, we have $\Psi(0) = \Psi'(0)$. Hence, Eq. (5.84) is a solution of Eq. (5.77). The wave function $\Psi(t)$, in contrary to $\Psi'(t)$, is a rapidly oscillating function of time. Thereby, transformation to the rotating frame not only changes the Hamiltonian, but suppresses rapid oscillations of the wave function as well.

It is worth noting that the exponential operators in Eq. (5.84) cannot be permuted, as the operators $\hat{I}_{zi}$ and $\widehat{\mathcal{H}}_{\mathrm{Z}}^{\mathrm{rot}}$ do not commute between each other. Of cause, transformation to the rotating frame does not change physical properties of the system, but it significantly simplify the solving of the Schrödinger equation, as in the rotating frame the evolution of a system is governed by the modified Hamiltonian, in which at resonance conditions the term describing the interaction of the spin with the static magnetic field completely disappears. The wave function (5.84) can be used to calculate average values of various operators. Let us note that namely the average value of an operator represents the measured physical quantity corresponding to a given operator.

The described above transformations of the spin Hamiltonian are a particular case of transformations, which one performs when describing the results of pulse sequences affecting a system.

5.7.5 Hamiltonian of Internal Interactions in the Rotating Frame: The Average Hamiltonian Method

The ability to create such a mathematical formalism that allows us to get the general solution of the nonstationary Schrödinger equation for any periodic pulse sequence is the main advantage of the average Hamiltonian method. It is no longer necessary to solve the Schrodinger equation in each particular case once again, and one can use the general solution obtained before.

As it was shown in Sect. 5.7.4, the solving of the Schrödinger equation with the Hamiltonian, which takes into account only interactions of nuclear spins with the external magnetic field, is simplified by transformation from the laboratory frame to the rotating one. A similar approach can be used if one considers a system of nuclei interacting between each other and affected by various pulse sequences. The only problem is the correct choice of the frame.

The total spin Hamiltonian consists of two parts: the time independent Hamiltonian $\widehat{\mathcal{H}}_{\text{int}}$ describing various internal interactions in the system and the Hamiltonian $\widehat{\mathcal{H}}_{\text{ext}}$ describing various external actions on the system. The latter may depend on time and can be changed by an experimentalist:

$$\widehat{\mathcal{H}} = \widehat{\mathcal{H}}_{\text{ext}}(t) + \widehat{\mathcal{H}}_{\text{int}}. \tag{5.85}$$

It is clear that whatever it would be the external action, it can not change the Hamiltonian $\widehat{\mathcal{H}}_{\text{int}}$ in Eq. (5.85) written in the laboratory frame. However, external actions changes the wave functions. Experimentally one observes the average, in quantum-mechanical terms, value of the corresponding operator, then the fact that $\widehat{\mathcal{H}}_{\text{int}}$ is time independent does not contradict the possibility of time averaging in principle. The Hamiltonian $\widehat{\mathcal{H}}_{\text{ext}}$, in its turn, consists of two parts: the time independent Hamiltonian describing interactions of nuclear magnetic moments with a constant external magnetic field and the time dependent Hamiltonian that describes interactions of nuclear magnetic moments with an alternating external field. The alternating magnetic field affects the system of interacting spins only at definite time moments, between the pulses the system evolves under the action of the constant magnetic field. In this case in the laboratory frame the Hamiltonian $\widehat{\mathcal{H}}$ describing both internal and external interactions has the form

$$\widehat{\mathcal{H}} = -\gamma\hbar\hat{I}_{zi}B_0 + \widehat{\mathcal{H}}_{\text{int}}. \tag{5.86}$$

Let us find the solution of the non-stationary Schrödinger equation in the frame rotating with the angular velocity $\omega = \omega_0$. It was shown in Sect. 5.7.4 how the Zeeman Hamiltonian changes when transforming to the rotating frame. Now let us consider how such a transformation affects the Hamiltonian of internal interactions. For this as before we introduce the wave function

$$\Psi' = \exp(-\mathrm{i}\omega_0\hat{I}_{zi}t)\Psi. \tag{5.87}$$

The Schrödinger equation with the Hamiltonian (5.86) is

$$-\frac{\hbar}{\mathrm{i}}\frac{\partial\Psi}{\partial t} = (-\gamma_i\hbar B_0 + \widehat{\mathcal{H}}_{\text{int}})\Psi. \tag{5.88}$$

After multiplying Eq. (5.88) on the left by $\exp(-\mathrm{i}\omega_0\hat{I}_{zi}t)$ and considering Eq. (5.78) we have

$$-\frac{\hbar}{\mathrm{i}}\frac{\partial(\exp(-\mathrm{i}\omega_0\hat{I}_{zi}t)\Psi)}{\partial t}$$
$$= -\hbar(-\omega_0+\gamma_i B_0)\hat{I}_{zi}\exp(-\mathrm{i}\omega_0\hat{I}_{zi}t)\Psi + \exp(-\mathrm{i}\omega_0\hat{I}_{zi}t)\widehat{\mathcal{H}}_{\mathrm{int}}\Psi. \tag{5.89}$$

As a rule the operator $\exp(-\mathrm{i}\omega_0\hat{I}_{zi}t)$ does not commute with the Hamiltonian of internal interactions, so one can not permute the product terms in the last term of Eq. (5.89). In order to obtain the equation to determine the wave function in the rotating frame, it is necessary to divide and then multiply the last term in Eq. (5.89) by $\exp(-\mathrm{i}\omega_0\hat{I}_{zi}t)$. Then taking into account that $(-\omega_0 + \gamma_i B_0 = 0)$, Eq. (5.89) transforms to

$$-\frac{\hbar}{\mathrm{i}}\frac{\partial\Psi'}{\partial t} = \exp(-\mathrm{i}\omega_0\hat{I}_{zi}t)\widehat{\mathcal{H}}_{\mathrm{int}}\exp(\mathrm{i}\omega_0\hat{I}_{zi}t)\Psi'. \tag{5.90}$$

As it is seen from Eq. (5.90) the wave function evolution in the rotating frame is defined by the only Hamiltonian of internal interactions, that is why sometimes the rotating frame is called the *interaction representation*. It is to be noted that not always transformations of the form (5.87) result in the interaction representation.

The Hamiltonian

$$\widetilde{\widehat{\mathcal{H}}}_{\mathrm{int}} = \exp(-\mathrm{i}\omega_0\hat{I}_{zi}t)\widehat{\mathcal{H}}_{\mathrm{int}}\exp(\mathrm{i}\omega_0\hat{I}_{zi}t), \tag{5.91}$$

which defines the evolution of the wave function Ψ' in Eq. (5.90), is a periodic function of time with a period $2\pi/\omega_0$. In the rotating frame in the absence of the magnetic field the wave function Ψ' slowly varies in time, whereas the Hamiltonian of internal interactions varies rapidly, i.e. the transformation like Eq. (5.87) transfers the time dependence from the wave function to the Hamiltonian of internal interactions. But it is the external interactions that determine the time dependence of the wave function, so such a transfer allows us to handle the form of the Hamiltonian.

In this case it is also possible to find a formal solution of Eq. (5.90). If the chosen time intervals δt are sufficiently small then the Hamiltonian $\widetilde{\widehat{\mathcal{H}}}_{\mathrm{int}}(t)$ within the time range from t to $t+\delta t$ can be replaced by its value averaged over this time range:

$$\overline{\widehat{\mathcal{H}}}_{\mathrm{int}} = \int_t^{t+\delta t}\widetilde{\widehat{\mathcal{H}}}_{\mathrm{int}}(t')\,\mathrm{d}t'.$$

Then during the time interval δt the wave function changes as follows:

$$\Psi'(t+\delta t) = \left(\exp\left(-\frac{\mathrm{i}}{\hbar}\int_t^{t+\delta t}\widetilde{\widehat{\mathcal{H}}}_{\mathrm{int}}(t')\,\mathrm{d}t'\right)\right)\Psi'(t),$$

i.e. the wave function at the moment $t' = 0 + \delta t$ is equal to

$$\Psi'(\delta t) = \left(\exp \left(-\frac{\mathrm{i}}{\hbar} \int\limits_0^{\delta t} \widetilde{\widetilde{\mathcal{H}}}_{\mathrm{int}}(t')\,\mathrm{d}t' \right) \right) \Psi'(0),$$

and at the moment $2\delta t$ it is

$$\Psi'(2\delta t) = \left(\exp \left(-\frac{\mathrm{i}}{\hbar} \int\limits_{\delta t}^{2\delta t} \widetilde{\widetilde{\mathcal{H}}}_{\mathrm{int}}(t')\,\mathrm{d}t' \right) \exp \left(-\frac{\mathrm{i}}{\hbar} \int\limits_0^{\delta t} \widetilde{\widetilde{\mathcal{H}}}_{\mathrm{int}}(t'')\,\mathrm{d}t'' \right) \right) \Psi'(0).$$

To find the wave function describing the system in the rotating frame at the moment t_n one has to multiply n exponential operators of the form $\exp(-\mathrm{i}/\hbar) \int_t^{t+\delta t} \widetilde{\widetilde{\mathcal{H}}}_{\mathrm{int}}(t)\,\mathrm{d}t$ acting in a strict order, because the Hamiltonian operators $\widetilde{\widetilde{\mathcal{H}}}_{\mathrm{int}}(t)$ in different time moments t do not commute between each other:

$$\Psi'(t_n) = \left(\prod_{k=0}^{n} \exp \left(-\frac{\mathrm{i}}{\hbar} \int\limits_{k\delta t}^{(k+1)\delta t} \widetilde{\widetilde{\mathcal{H}}}_{\mathrm{int}}(t)\,\mathrm{d}t \right) \right) \Psi'(0). \tag{5.92}$$

The action of the exponential operators in Eq. (5.92) on the wave function $\Psi'(0)$ can be calculated by expanding into Taylor series:

$$\Psi'(t_n) = \left(\prod_{k=0}^{n} \left(1 - \frac{\mathrm{i}}{\hbar} \int\limits_{k\delta t}^{(k+1)\delta t} \widetilde{\widetilde{\mathcal{H}}}_{\mathrm{int}}(t)\,\mathrm{d}t + \ldots \right) \right) \Psi'(0).$$

If the exponent in Eq. (5.92) were the usual scalar functions, the product of exponential functions could be replaced by a single exponential function, the exponent of which is a sum of functions that are exponents of the individual exponential functions. To make the writing of the wave function (5.92) more compact the product of exponential non-commuting functions formally can be written as:

$$\prod_{k=0}^{n} \exp \left(-\frac{\mathrm{i}}{\hbar} \int\limits_{k\delta t}^{(k+1)\delta t} \widetilde{\widetilde{\mathcal{H}}}_{\mathrm{int}}(t)\,\mathrm{d}t \right) = \exp \left(-\frac{\mathrm{i}}{\hbar} \int\limits_0^{t_n} \widetilde{\widetilde{\mathcal{H}}}_{\mathrm{int}}(t)\,\mathrm{d}t \right),$$

taking into account that the sum of the integrals $\sum_{k=0}^{n} \int_{k\delta t}^{(k+1)\delta t} \widetilde{\widetilde{\mathcal{H}}}_{\mathrm{int}}(t)\,\mathrm{d}t$ is equal to $\int_0^{t_n} \widetilde{\widetilde{\mathcal{H}}}_{\mathrm{int}}(t)\,\mathrm{d}t$. To underline that in this case it concerns the product of exponential functions written in a strict time order one introduces the Dyson time ordering operator $\hat{T}$:

$$\Psi'(t_n) = \hat{T} \exp \left(-\frac{\mathrm{i}}{\hbar} \int\limits_0^{t_n} \widetilde{\widetilde{\mathcal{H}}}_{\mathrm{int}}(t)\,\mathrm{d}t \right) \Psi'(0). \tag{5.93}$$

As it follows from Eq. (5.87) in the rotating frame the wave function Ψ' at the moment $t = 0$ coincides with the wave function $\Psi(0)$ in the laboratory frame. Then, after substituting Eq. (5.87) into Eq. (5.93) one has the solution of the Schrödinger equation in the laboratory frame:

$$\Psi(t_n) = \exp(\mathrm{i}\omega_0 \hat{I}_{zi} t_n) \hat{T} \exp\left(-\frac{\mathrm{i}}{\hbar} \int_0^{t_n} \widehat{\widetilde{\widetilde{\mathcal{H}}}}_{\mathrm{int}}(t)\, \mathrm{d}t \right) \Psi(0).$$

One can write the time dependence of the wave function Ψ in a more compact form by means of a *propagator*, the concept of which is widely used in quantum mechanics. In our case the propagator for external actions is

$$\exp(\mathrm{i}\omega_0 \hat{I}_{zi} t) = \hat{U}_{\mathrm{ext}}, \tag{5.94}$$

and the propagator for internal interactions is

$$\hat{T} \exp\left(-\frac{\mathrm{i}}{\hbar} \int_0^{t} \widehat{\widetilde{\widetilde{\mathcal{H}}}}_{\mathrm{int}}(t)\, \mathrm{d}t \right) = \hat{U}_{\mathrm{int}}.$$

The propagator for external actions determines the system evolution in the absence of internal interactions and is the quantity reciprocal to the operator transforming the wave function from the laboratory frame to the rotating one.

The method of solving the Schrödinger equation described above is classical and is used not only in the average Hamiltonian theory.

The average Hamiltonian theory was developed by Waugh. He noticed that if the Hamiltonian and the propagator of external actions are periodic functions of time such that

$$\widehat{\mathcal{H}}_{\mathrm{ext}}(t + t_{\mathrm{m}}) = \widehat{\mathcal{H}}_{\mathrm{ext}}(t), \qquad \hat{U}_{\mathrm{ext}}(t + t_{\mathrm{c}}) = \hat{U}_{\mathrm{ext}}(t), \tag{5.95}$$

then at the time moments multiple of the cycle period t_{c} the propagator of external actions is equal to the unit operator:

$$\hat{U}_{\mathrm{ext}}(N\, t_{\mathrm{c}}) = \exp(\mathrm{i}\omega_0 \hat{I}_{zi} N\, t_{\mathrm{c}}) = \hat{U}_{\mathrm{ext}}(0) = 1.$$

Then the wave function evolution in the laboratory frame at the time moments multiple of t_{c} is determined by the modified Hamiltonian of internal interactions $\widehat{\widetilde{\widetilde{\mathcal{H}}}}_{\mathrm{int}}$, and therefore all average (observed) quantities determined by this wave function depend only on $\widehat{\widetilde{\widetilde{\mathcal{H}}}}_{\mathrm{int}}$. Of cause in this case the observation must be stroboscopic: only at the time moments multiple of t_{c}.

Despite the Hamiltonian of external interactions $\widehat{\mathcal{H}}_{\mathrm{ext}} = -\mathrm{i}\omega_0 \hat{I}_{zi}$ is not strictly speaking a periodic function (its period $t_{\mathrm{m}} = 0$, and in this regard it is an exception to the common rule), the propagator $\hat{U}_{\text{ext}}$ in the form of (5.94) is a periodic function with period $t_c = 2\pi/\omega_0$.

The next step made by Waugh was substitution of the Hamiltonian $\widetilde{\widetilde{\mathcal{H}}}_{\text{int}}$ by its average value. Let us consider in which cases and how it can be done.

The Hamiltonian $\widetilde{\widetilde{\mathcal{H}}}_{\text{int}}$ is a periodic function of time (see (5.91)) with period $t_c = 2\pi/\omega_0$, therefore:

$$\widetilde{\widetilde{\mathcal{H}}}_{\text{int}}(t + Nt_c) = \widetilde{\widetilde{\mathcal{H}}}_{\text{int}}(t) \quad \text{for integer } N. \tag{5.96}$$

From Eq. (5.96) it follows that

$$\hat{T}\exp\left(-\frac{\mathrm{i}}{\hbar}\int\limits_0^{Nt_c}\widetilde{\widetilde{\mathcal{H}}}_{\text{int}}(t)\,\mathrm{d}t\right) = \left(\hat{T}\exp\left(-\frac{\mathrm{i}}{\hbar}\int\limits_0^{t_c}\widetilde{\widetilde{\mathcal{H}}}_{\text{int}}(t)\,\mathrm{d}t\right)\right)^N,$$

i.e. to describe the state of the system at any time moment multiple of t_c it is sufficient to determine the system evolution within the only one cycle.

Now it is possible to find such an "average" Hamiltonian $\overline{\widetilde{\mathcal{H}}}_{\text{int}}$, in order to fulfill the equation

$$\Psi(nt_c) = \hat{T}\exp\left(-\frac{\mathrm{i}}{\hbar}\int_0^{t_c}\widetilde{\widetilde{\mathcal{H}}}_{\text{int}}(t)\mathrm{d}t\right)\Psi(0) = \exp\left(-\frac{\mathrm{i}}{\hbar}\overline{\widetilde{\mathcal{H}}}_{\text{int}}t_c\right)\Psi(0).$$

It is obvious that one can not just write

$$\overline{\widetilde{\mathcal{H}}}_{\text{int}}t_c = \int_0^{t_c}\widetilde{\widetilde{\mathcal{H}}}_{\text{int}}\mathrm{d}t,$$

because Eq. (5.93) is only the compact writing form for (5.92), and in the real calculations one must use Eq. (5.92) instead of Eq. (5.93). That is why let us consider the product of operators in Eq. (5.92) grouping them in such a way that it becomes possible to write in the explicit form the terms linear in the operators $\int_{k\delta t}^{(k+1)\delta t}\widetilde{\widetilde{\mathcal{H}}}_{\text{int}}(t)\mathrm{d}t$:

$$\begin{aligned}
\exp\left(-\frac{\mathrm{i}}{\hbar}\overline{\widetilde{\mathcal{H}}}_{\text{int}}t_c\right) &= \hat{T}\exp\left(-\frac{\mathrm{i}}{\hbar}\int\limits_0^{t_c}\widetilde{\widetilde{\mathcal{H}}}_{\text{int}}(t)\mathrm{d}t\right) = \prod_{k=0}^{n}\left(1 - \frac{\mathrm{i}}{\hbar}\int\limits_{k\delta t}^{(k+1)\delta t}\widetilde{\widetilde{\mathcal{H}}}_{\text{int}}(t)\mathrm{d}t + \ldots\right)\\
&= 1 - \frac{\mathrm{i}}{\hbar}\left(\int\limits_{(n-1)\delta t}^{n\delta t}\widetilde{\widetilde{\mathcal{H}}}_{\text{int}}(t)\mathrm{d}t + \int\limits_{(n-2)\delta t}^{(n-1)\delta t}\widetilde{\widetilde{\mathcal{H}}}_{\text{int}}(t)\mathrm{d}t + \cdots + \int\limits_{0\cdot\delta t}^{\delta t}\widetilde{\widetilde{\mathcal{H}}}_{\text{int}}(t)\mathrm{d}t\right)\\
&\quad + \ldots\\
&= 1 - \frac{\mathrm{i}}{\hbar}\int\limits_0^{t_c}\widetilde{\widetilde{\mathcal{H}}}_{\text{int}}(t)\mathrm{d}t + \text{higher-order terms.}
\end{aligned}$$

On the other hand the average Hamiltonian $\overline{\widetilde{\mathcal{H}}}_{\text{int}}$ always can be represented as a sum of the operators $\overline{\widehat{\mathcal{H}^n}}_{\text{int}}$ $(n = 0, 1, \ldots)$ each of which contains only the operators of the n-th exponent:

$$\exp\left(-\frac{\mathrm{i}}{\hbar}\overline{\widetilde{\mathcal{H}}}_{\text{int}}t_c\right) = \exp\left(-\frac{\mathrm{i}}{\hbar}\left(\overline{\widehat{\mathcal{H}^0}}_{\text{int}} + \overline{\widehat{\mathcal{H}^1}}_{\text{int}} + \overline{\widehat{\mathcal{H}^2}}_{\text{int}} + \ldots\right)t_c\right)$$
$$= 1 - \frac{\mathrm{i}}{\hbar}\left(\overline{\widehat{\mathcal{H}^0}}_{\text{int}} + \overline{\widehat{\mathcal{H}^1}}_{\text{int}} + \overline{\widehat{\mathcal{H}^2}}_{\text{int}} + \ldots\right)t_c + \ldots. \tag{5.97}$$

Then

$$\overline{\widehat{\mathcal{H}^0}}_{\text{int}} = \frac{1}{t_c}\int_0^{t_c}\widetilde{\widehat{\mathcal{H}}}_{\text{int}}(t')\mathrm{d}t', \tag{5.98}$$

$$\overline{\widehat{\mathcal{H}^1}}_{\text{int}} = -\frac{\mathrm{i}}{2t_c}\int_0^{t_c}\mathrm{d}t''\int_0^{t''}\mathrm{d}t'\left[\widetilde{\widehat{\mathcal{H}}}_{\text{int}}(t''), \widetilde{\widehat{\mathcal{H}}}_{\text{int}}(t')\right]. \tag{5.99}$$

Let us note that Eqs. (5.98), (5.99)[14] (the so-called *Magnus expansions*), were obtained rigorously and independently of the $\widetilde{\widehat{\mathcal{H}}}_{\text{int}}$ value. However, for a successful application of the average Hamiltonian theory it is advisable to consider only the first few terms in the series expansion (5.97). For this it is necessary that $\overline{\widetilde{\mathcal{H}}}_{\text{int}}$ has such a small value that $\exp\left((-\mathrm{i}/\hbar)\overline{\widetilde{\mathcal{H}}}_{\text{int}}t_c\right) \simeq 1$, i.e. the wave function describing the system state has to change only a little within the period t_c. It is correct if interactions of the spins with external magnetic field are considerably greater than all the other interactions. In our case this means that $\omega_0 = \gamma B_0 \gg \langle n|\widehat{\mathcal{H}}_{\text{int}}|m\rangle$, where $\langle n|\widehat{\mathcal{H}}_{\text{int}}|m\rangle$ is the matrix element of the Hamiltonian of internal interactions for any states n and m. Really, $t_c = 2\pi/\omega_0$, therefore:

$$\exp\left(-\mathrm{i}\overline{\widetilde{\mathcal{H}}}_{\text{int}}t_c\right) \simeq \exp\left(-\mathrm{i}\overline{\widetilde{\mathcal{H}}}_{\text{int}}(2\pi/\omega_0)\right) \simeq 1 \quad \text{at} \quad \left|\langle n|\overline{\widetilde{\mathcal{H}}}_{\text{int}}|m\rangle\right| \ll \omega_0.$$

The fulfillment of this condition means that in high magnetic fields in the average Hamiltonian theory it is possible to consider only the first term $\overline{\widehat{\mathcal{H}^0}}_{\text{int}}$ in Eq. (5.97).

Let us obtain the Hamiltonian $\widetilde{\widehat{\mathcal{H}}}_{\text{int}}$ in the explicit form in the case when the Hamiltonian of internal interactions can be represented as a sum of operators like Eq. (5.62). As the exponential operators $\exp(\pm\mathrm{i}\omega_0\hat{I}_{zi}t)$ do not affect space coordinates it is sufficiently to consider only changing in the spin part of operators like Eq. (5.63). If the operators commute with each other then in accordance with Eq. (5.67) they can be permuted. If the spins are of different species, then the operator $\hat{I}_{zi}$ commutes with the terms $\hat{A}$, and if the spins are of the same species then it commutes with both $\hat{A}$ and $\hat{B}$, the other operators in Eq. (5.63) do not commute with the operator $\hat{I}_z$. For

[14] The other terms are too tedious and can be found anywhere, see, for example, M.Mehring. High resolution NMR spectroscopy in solids. Springer-Verlag, 1976.

spins of the same species

$$\tilde{\tilde{\mathcal{H}}}_{\text{int}}(t) = \hat{A} + \hat{B} + \exp(-\mathrm{i}\omega_0 \hat{I}_z t)(\hat{C} + \hat{D} + \hat{E} + \hat{F}) \exp(\mathrm{i}\omega_0 \hat{I}_z t).$$

The operators $\hat{C}$, $\hat{D}$, $\hat{E}$ and $\hat{F}$ contain the operators $\hat{\vec{I}}_+$, $\hat{\vec{I}}_-$, $\hat{I}_z$, which under the action of the exponential operators, as it follows from Eq. (5.75), turn into the operators $\hat{\vec{I}}_+ \exp(-\mathrm{i}\omega_0 t)$, $\hat{\vec{I}}_- \exp(\mathrm{i}\omega_0 t)$ and $\hat{I}_z$, respectively. Then

$$\tilde{\tilde{\mathcal{H}}}_{\text{int}}(t) = \hat{A} + \hat{B} + \hat{C}\exp(-\mathrm{i}\omega_0 t) + \hat{D}\exp(\mathrm{i}\omega_0 t) + \hat{E}\exp(-2\mathrm{i}\omega_0 t) + \hat{F}\exp(2\mathrm{i}\omega_0 t). \tag{5.100}$$

The Hamiltonian $\tilde{\tilde{\mathcal{H}}}_{\text{int}}$ does not commute with itself in different moments of time, therefore, to determine the average Hamiltonian it is necessary to use the Magnus expansions (5.98) and (5.99). One part of the Hamiltonian $\tilde{\tilde{\mathcal{H}}}_{\text{int}}(t)$ is time independent (for any integration limits), whereas another is a periodic function of time with period $t_c = 2\pi/\omega_0$. Therefore, if the integration limits are chosen to be multiple of the period t_c, for example from 0 to t_c, then all the time dependent terms in Eq. (5.100) are averaged to zero, and the evolution of the wave function at the time moments $t_c, 2t_c, \ldots$ is determined by the time-average Hamiltonian $\overline{\tilde{\tilde{\mathcal{H}}}}_{\text{int}} = \hat{A} + \hat{B}$.

In a real experiment NMR spectral envelopes are recorded. That means the observation is made at time moments multiple of t_c, and as it follows from the average Hamiltonian theory, within the zero-order approximation the system evolves under the action of the secular part of the Hamiltonian, which describes interactions bilinear and quadratic in spin operators. Thereby, the zero-order approximation of the average Hamiltonian theory coincides with the first-order approximation of the perturbation theory, described in Sect. 1.10. Using Eq. (5.99) It is easy to demonstrate that the first-order approximation of the average Hamiltonian theory is equal to the second-order approximation of the perturbation theory an so on.

Thus, high magnetic field results in averaging of parameters in the spin space: all nonsecular terms of the Hamiltonian of internal interactions are neglected within the first-order approximation of the average Hamiltonian theory.

As a conclusion:

- to determine the form of the average Hamiltonian it is necessary to find such a frame in which the evolution of the wave function is determined by the Hamiltonian of internal interactions only;
- an external action on the spin system leads to averaging of internal interactions only if (i) both its Hamiltonian and propagator are periodic function of time (if the external action is a constant magnetic field, it can be considered as a periodic function with zero period, and the propagator period is equal to the period of the Larmor precession of spins); (ii) the observation is made stroboscopically (at time moments multiple of period t_c); (iii) internal interactions are weaker than

interactions with external magnetic fields (in this case only the first terms of the Magnus expansions should be taken into account).

5.7.6 Arbitrary Periodic Action on a Spin System

Let us consider how an arbitrary pulse sequence modifies the Hamiltonian of internal interactions. The total Hamiltonian of a spin system can be written as a sum of the Hamiltonian $\widehat{\mathcal{H}}_{\text{ext}}$ describing interactions of spins with external magnetic fields and the Hamiltonian $\widehat{\mathcal{H}}_{\text{int}}$ describing internal interaction in the spin system (5.85). In order to find the form of the average Hamiltonian at an arbitrary external action it is necessary as before to pass to the interaction representation (sometimes it is called the toggling frame), where the Hamiltonian $\widehat{\mathcal{H}}_{\text{ext}}$ disappears and the system evolution is determined by the Hamiltonian $\widehat{\mathcal{H}}_{\text{int}}$ only. For this let us make transformations similar to those set forth in Sect. 5.7.5. The Hamiltonian of external interactions is now time dependent, hence to determine the wave function we use formal operators similar to Eq. (5.93), in which the Hamiltonian $\widehat{\widetilde{\mathcal{H}}}_{\text{int}}$ is replaced by the Hamiltonian $\widehat{\mathcal{H}}_{\text{ext}}$. Then instead of the wave function Ψ' in the form (5.87) we have

$$\Psi'(t) = \hat{U}'_{\text{ext}}\Psi(t),$$

where

$$\hat{U}'_{\text{ext}}(t) = \hat{T}\exp\left(\frac{\mathrm{i}}{\hbar}\int_0^t \widehat{\mathcal{H}}_{\text{ext}}(t')\,\mathrm{d}t'\right), \tag{5.101}$$

here $\hat{T}$ is the Dyson time ordering operator. The exponential operator in Eq. (5.101) as before is a product of exponential operators:

$$\hat{U}'_{\text{ext}}(t) = \hat{T}\exp\left(\frac{\mathrm{i}}{\hbar}\int_0^t \widehat{\mathcal{H}}_{\text{ext}}(t')\,\mathrm{d}t'\right) = \hat{T}\left(\prod_{k=0}^{n}\exp\left(\frac{\mathrm{i}}{\hbar}\int_{k\delta t}^{(k+1)\delta t}\widehat{\widetilde{\mathcal{H}}}_{\text{ext}}(t)\,\mathrm{d}t\right)\right).$$

It is obvious that if the spin system is under the action of the external magnetic field only, then

$$\hat{U}'_{\text{ext}}(t) = \exp\left(-\frac{\mathrm{i}}{\hbar}\gamma\hat{I}_z B_0 t\right). \tag{5.102}$$

We emphasize that the toggling frame coincides with the rotating frame only if the wave function transformation is determined by an operator like (5.102). In general case the rotating frame and the toggling frame are different.

The Schrödinger equation in the interaction representation can be obtained by making transformations similar to Eqs. (5.88)–(5.90). Then

$$-\frac{\hbar}{\mathrm{i}}\frac{\partial \Psi'}{\partial t} = \left(\hat{U}'\widehat{\mathcal{H}}_{\mathrm{int}}\hat{U}\right)\Psi'$$
$$= \hat{T}\exp\left(\frac{\mathrm{i}}{\hbar}\int_0^t \widehat{\mathcal{H}}_{\mathrm{ext}}(t')\mathrm{d}t'\right)\widehat{\mathcal{H}}_{\mathrm{int}}\hat{T}\exp\left(-\frac{\mathrm{i}}{\hbar}\int_0^t \widehat{\mathcal{H}}_{\mathrm{ext}}(t')\mathrm{d}t'\right)\Psi'. \tag{5.103}$$

Once again, the evolution of the wave function is defined by the modified Hamiltonian of internal interactions only:

$$\widetilde{\widetilde{\mathcal{H}}}_{\mathrm{int}} = \hat{T}\exp\left(\frac{\mathrm{i}}{\hbar}\int_0^t \widehat{\mathcal{H}}_{\mathrm{ext}}(t')\,\mathrm{d}t'\right)\widehat{\mathcal{H}}_{\mathrm{int}}\hat{T}\exp\left(-\frac{\mathrm{i}}{\hbar}\int_0^t \widehat{\mathcal{H}}_{\mathrm{ext}}(t')\,\mathrm{d}t'\right).$$

The wave function Ψ', which is the solution of the Schrödinger equation (5.103), is described by

$$\Psi'(t) = \hat{U}_{\mathrm{int}}\Psi'(0) = \hat{T}\exp\left(-\frac{\mathrm{i}}{\hbar}\int_0^t \widetilde{\widetilde{\mathcal{H}}}_{\mathrm{int}}(t')\,\mathrm{d}t'\right)\Psi'(0).$$

As at zero time $\Psi'(0) = \Psi(0)$, the wave function in the laboratory frame is

$$\Psi(t) = \hat{U}_{\mathrm{ext}}\hat{U}_{\mathrm{int}}\Psi(0),$$

here the propagators $\hat{U}_{\mathrm{ext}}$ and $\hat{U}_{\mathrm{int}}$ are determined by the following way:

$$\hat{U}_{\mathrm{ext}} = \hat{T}\exp\left(-\frac{\mathrm{i}}{\hbar}\int_0^t \widehat{\mathcal{H}}_{\mathrm{ext}}(t')\,\mathrm{d}t'\right), \quad \hat{U}_{\mathrm{int}} = \hat{T}\exp\left(-\frac{\mathrm{i}}{\hbar}\int_0^t \widetilde{\widetilde{\mathcal{H}}}_{\mathrm{int}}(t')\,\mathrm{d}t'\right).$$

As it was pointed out in Sect. 5.7.5, to make averaging of internal interactions more effective the Hamiltonian $\widehat{\mathcal{H}}_{\mathrm{ext}}$ and the propagator $\hat{U}_{\mathrm{ext}}$ must satisfy the periodic conditions (5.95). It is obvious that for an arbitrary but satisfying the conditions (5.95) external action the propagator $\hat{U}_{\mathrm{ext}}$ at time moments multiple of t_{c} is $\hat{1}$, and therefore the system evolution in these time moments is determined only by the Hamiltonian of internal interactions written in the toggling frame. As before, using the Magnus expansions one can pass to the average Hamiltonian and determine wave functions in the laboratory frame, and therefore the values of parameters measured experimentally.

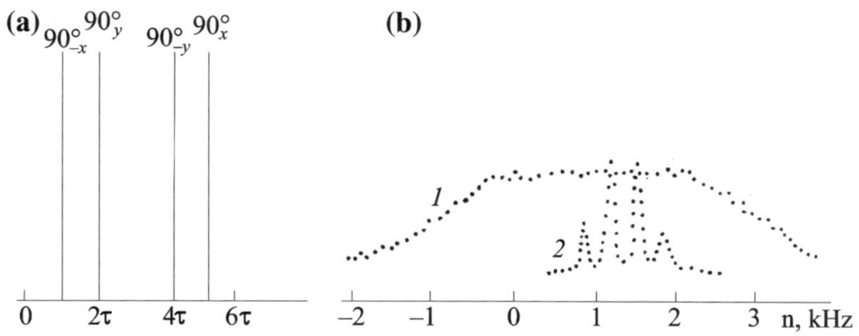

Fig. 5.12 The WHH-4 pulse sequence (**a**) and an example of narrowing NMR lines in solids by WHH-4 (**b**): a NMR spectrum without (*1*) and after applying (*2*) WHH-4

5.7.7 WHH-4 Pulse Sequence

The WHH-4 pulse sequence (or WAHUHA, named after its inventors, Waugh, Huber and Haeberlen) is the basic multipulse sequence, which averages down to zero the Hamiltonian operators of dipole–dipole and quadrupole interactions but does not affect isotropic components of indirect spin–spin interactions and interactions that cause chemical shifts of NMR lines. It consists of four pulses of different phases: $\tau - 90^\circ_{-x} - \tau - 90^\circ_y - 2\tau - 90^\circ_{-y} - \tau - 90^\circ_x - \tau$ (Fig. 5.12a). Here $90^\circ_{\pm\alpha}$ is the pulse of the rf field directed along the $\pm\alpha$-axis ($\alpha = x,\, y$) for which $\omega_1 t_\mathrm{p} = 90^\circ$; τ is the time interval between two pulses.

The cycle time t_c in this case is equal to 6τ, and for each time interval it is necessary to find the propagator of external actions and to determine $\widetilde{\widetilde{\mathcal{H}}}_\mathrm{int}$ in the rotating frame. Let us consider that (i) the pulse duration is short, so we can neglect changing in the Hamiltonian of internal interactions during the pulses; (ii) the cycle time is considerably shorter the spin relaxation time; (iii) the pulse currier frequency coincides with the resonance frequency of the studied nuclei.

In the rotating frame only the secular part of the Hamiltonian of internal interaction is important, and, as it is known, it contains only the operators $\hat{I}_z$ or their products and the operators that do not change after a transformation from one frame to another (for example, $\hat{\vec{I}}_i \hat{\vec{I}}_j$). That is why to obtain a general expression for any Hamiltonian of internal interactions let us consider how this pulse sequence affects the operator $\hat{I}_z$.

During the time interval τ before the fist pulse $\widetilde{\hat{I}}_z\big|_{0\le t\le \tau} = \hat{I}_z$. By the end of the fist 90°-pulse of duration $t_{\mathrm{p}1}$, while the magnetic field vector $\vec{B}_1$ is directed along the x-axis, the propagator in the rotating frame is already determined in Eq. (5.83). Now the vector $\vec{B}_1$ is along the $-x$ axis and the propagator differs from Eq. (5.83) only by sign in exponential exponents:

$$\hat{U}\big|_{t=\tau+t_{\mathrm{p}1}} = \exp(-\mathrm{i}\omega_1 \hat{I}_x t_{\mathrm{p}1})\big|_{\omega_1 t_{\mathrm{p}1}=\pi/2} = \exp\frac{-\mathrm{i}\hat{I}_x \pi}{2}.$$

Using this propagator and taking into account the third line in Eq. (5.76) one can find $\tilde{\hat{I}}_z$ within the time interval $\tau \le t \le 2\tau$:

$$\tilde{\hat{I}}_z\Big|_{\tau \le t \le 2\tau} = \exp\frac{-\mathrm{i}\hat{I}_x\pi}{2}\hat{I}_z\exp\frac{\mathrm{i}\hat{I}_x\pi}{2} = \hat{I}_y. \tag{5.104}$$

By the end of the second 90°-pulse applying along the y-axis of the rotating frame the propagator is

$$\hat{U}\Big|_{t=2\tau+t_{p2}} = \exp\frac{\mathrm{i}\hat{I}_y\pi}{2}\exp\frac{-\mathrm{i}\hat{I}_x\pi}{2}. \tag{5.105}$$

Considering Eq. (5.76) the operator $\tilde{\hat{I}}_z$, which does not change within the time interval 2-4τ, can be written as follows:

$$\tilde{\hat{I}}_z\Big|_{2\tau \le t \le 4\tau} = \exp\frac{\mathrm{i}\hat{I}_x\pi}{2}\exp\frac{-\mathrm{i}\hat{I}_y\pi}{2}\hat{I}_z\exp\frac{\mathrm{i}\hat{I}_y\pi}{2}\exp\frac{-\mathrm{i}\hat{I}_x\pi}{2} = \hat{I}_x.$$

Here we have used properties of noncommuting exponential operators (the last line in Eq. (5.68)).

By the similar way one can determine the propagators of external actions and the operators $\tilde{\hat{I}}_z$ within the time intervals $4\tau \le t \le 5\tau$ and $5\tau \le t \le 6\tau$:

$$\hat{U}\Big|_{t=4\tau+t_p} = \exp\frac{-\mathrm{i}\hat{I}_y\pi}{2}\exp\frac{\mathrm{i}\hat{I}_y\pi}{2}\exp\frac{-\mathrm{i}\hat{I}_x\pi}{2} = \exp\frac{-\mathrm{i}\hat{I}_x\pi}{2}, \tag{5.106}$$

$$\hat{U}\Big|_{t=5\tau+t_p} = \exp\frac{\mathrm{i}\hat{I}_x\pi}{2}\exp\frac{-\mathrm{i}\hat{I}_y\pi}{2}\exp\frac{\mathrm{i}\hat{I}_y\pi}{2}\exp\frac{-\mathrm{i}\hat{I}_x\pi}{2} = 1, \tag{5.107}$$

$$\tilde{\hat{I}}_z\Big|_{4\tau \le t \le 5\tau} = \exp\frac{\mathrm{i}\hat{I}_x\pi}{2}\hat{I}_z\exp\frac{-\mathrm{i}\hat{I}_x\pi}{2} = \hat{I}_y,$$

$$\tilde{\hat{I}}_z\Big|_{5\tau \le t \le 6\tau} = \hat{I}_z.$$

Now it is easy to obtain the average Hamiltonian describing the chemical shift within the zero-order approximation:

$$\begin{aligned}
\overline{\mathcal{H}^{(0)}}_{\mathrm{int}\sigma} &= \frac{1}{6\tau}\gamma_i\hbar B_0\sigma\Bigg(\sum_i \tau\tilde{\hat{I}}_{zi}\Big|_{0\le t\le\tau} + \tau\sum_i\tilde{\hat{I}}_{zi}\Big|_{\tau\le t\le 2\tau} + 2\tau\sum_i\tilde{\hat{I}}_{zi}\Big|_{2\tau\le t\le 4\tau} \\
&\quad + \tau\sum_i\tilde{\hat{I}}_{zi}\Big|_{4\tau\le t\le 5\tau} + \tau\sum_i\tilde{\hat{I}}_{zi}\Big|_{5\tau\le t\le 6\tau}\Bigg) \\
&= \frac{1}{6\tau}\gamma_i\hbar B_0\sigma\sum_i\left(\tau\hat{I}_{zi} + \tau\hat{I}_{yi} + 2\tau\hat{I}_{xi} + \tau\hat{I}_{yi} + \tau\hat{I}_{zi}\right) \\
&= \frac{1}{3}\gamma_i\hbar B_0\sigma\sum_i\left(\hat{I}_{zi} + \hat{I}_{yi} + \hat{I}_{zi}\right),
\end{aligned}$$

here $\sigma = \sigma_{zz} + \sigma_{\text{iso}}$ are anisotropic and isotropic components of the magnetic shielding tensor, respectively.

Thereby, the WHH-4 pulse sequence averages the chemical shift Hamiltonian but not down to zero.

To establish relationships between averaged values of the secular part of the magnetic shielding Hamiltonian and its unaveraged values let us express the sum of the operators $(\hat{I}_{zi} + \hat{I}_{yi} + \hat{I}_{zi})$ through the operator $\hat{I}_{zi}$ applying to them the vector sum rule. If one considers the operators $\hat{I}_{zi}$, $\hat{I}_{yi}$ and $\hat{I}_{xi}$ as vectors of the same length but directed along the x-, y- and z-axes, respectively, then the total vector is equal to $\sqrt{3}\hat{I}_{zi}$. Therefore, the chemical shift value measured by applying the WHH-4 pulse sequence is $\sqrt{3}$ times smaller. This is the so-called *scaling factor*.

To determine how the WHH-4 pulse sequence affects the secular part of the Hamiltonian of dipole–dipole interactions,[15] let us find the corresponding Hamiltonian $\widetilde{\widehat{\mathcal{H}}}_{\text{d,int}}$. There is no need to consider the propagators and to make a complete analysis similar to that made previously. It is sufficiently to notice that if the operator $\hat{I}_z$ under the effect of any propagator transforms into $\hat{I}_y$, the secular part of the Hamiltonian of dipole–dipole interactions

$$\widehat{\mathcal{H}}_{\text{d},z} = \gamma^2\hbar^2 \sum_{i \neq j} (3\hat{I}_{zi}\hat{I}_{zj} - \hat{\vec{I}}_i\hat{\vec{I}}_j) r_{ij}^{-3}$$

transforms into

$$\widehat{\mathcal{H}}_{\text{d},y} = \gamma^2\hbar^2 \sum_{i \neq j} (3\hat{I}_{yi}\hat{I}_{yj} - \hat{\vec{I}}_i\hat{\vec{I}}_j) r_{ij}^{-3}.$$

If $\hat{I}_z$ under the effect of any propagator transforms into $\hat{I}_x$, then the secular part of the Hamiltonian of dipole–dipole interactions transforms into

$$\widehat{\mathcal{H}}_{\text{d},x} = \gamma^2\hbar^2 \sum_{i \neq j} (3\hat{I}_{xi}\hat{I}_{xj} - \hat{\vec{I}}_i\hat{\vec{I}}_j) r_{ij}^{-3}.$$

Then within the zero-order approximation the average Hamiltonian of dipole–dipole interactions for spins of the same species can be written as

$$\frac{1}{3}\sum_{i \neq j}(\widehat{\mathcal{H}}_{\text{d},z} + \widehat{\mathcal{H}}_{\text{d},y} + \widehat{\mathcal{H}}_{\text{d},x}) = \gamma^2\hbar^2 \sum_{i \neq j} (3\hat{I}_{zi}\hat{I}_{zj} + 3\hat{I}_{yi}\hat{I}_{yj} + 3\hat{I}_{xi}\hat{I}_{xj} - 3\hat{\vec{I}}_i\hat{\vec{I}}_j) r^{-3}. \tag{5.108}$$

The first three terms in the right part of Eq. (5.108) are the scalar products $3\hat{\vec{I}}_i\hat{\vec{I}}_j$, therefore, the average Hamiltonian of dipole–dipole interactions is zero within the zero-order approximation.

[15] As it was shown in Sect. 5.7.5 its nonsecular part is averaged to zero just by applying a constant magnetic field.

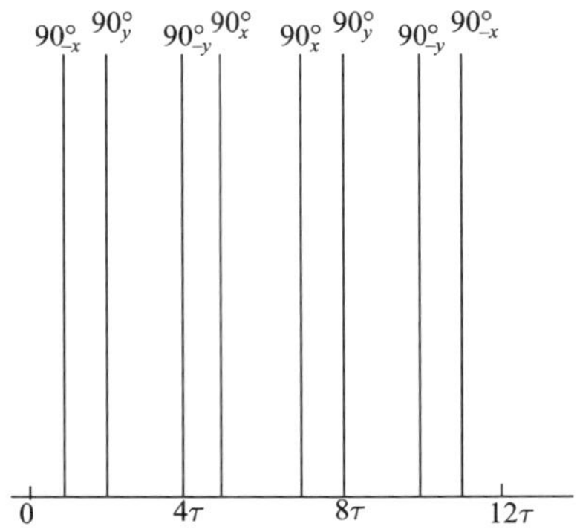

Fig. 5.13 Schematic representation of the MREV-8 pulse sequence

Thus, the WHH-4 pulse sequence is selective with respect to different interactions: within the zero-order approximation it averages to zero the Hamiltonian of dipole–dipole interactions for spins of the same species, the Hamiltonian of quadrupole interactions and anisotropic part of the Hamiltonian of indirect spin–spin interactions, but it allows us to measure the chemical shift and the shift caused by hyperfine interactions. When applying WHH-4 (as opposed to MAS experiments) it is possible to determine not only the isotropic component of the magnetic shielding tensor, but all of them.

An example of narrowing NMR lines in solids by WHH-4 is shown in Fig. 5.12b, where 1 is the NMR spectrum without using a pulse sequence, and 2 is the NMR spectrum after applying it.

The only disadvantage of this pulse sequence is that it is extremely sensitive to the pulse imperfection and inhomogeneity of both constant and alternating magnetic fields. These "instrumental" errors do not allow us to average down to zero the Hamiltonian of dipole–dipole interactions and change the scaling factors. These problems can be partly solved by applying another widely used pulse sequence, which will be discussed further.

5.7.8 MREV-8 Pulse Sequence

The MREV-8 pulse sequence, called after its authors Mansfield, Rhim, Elleman and Vaughan, is shown in Fig. 5.13. It consists of two phase-cycled WHH-4 pulse sequences (the phases of the second WHH-4 sequence are shifted at π as compared with the first one).

The effect of the first WHH-4 cycle has been discussed above. The propagators obtained after the action of the second four pulses can be obtained from Eqs. (5.104)–(5.107) by substituting plus for minus and vice versa. Then

$$\widetilde{\hat{I}_z}\big|_{6\tau\le t\le 7\tau} = \hat{I}_z, \quad \widetilde{\hat{I}_z}\big|_{7\tau\le t\le 8\tau} = -\hat{I}_y, \quad \widetilde{\hat{I}_z}\big|_{8\tau\le t\le 10\tau} = \hat{I}_x,$$

$$\widetilde{\hat{I}_z}\Big|_{10\tau\leq t\leq 11\tau} = -\hat{I}_y, \quad \widetilde{\hat{I}_z}\Big|_{11\tau\leq t\leq 12\tau} = \hat{I}_z.$$

And within the zero-order approximation the average Hamiltonian of the magnetic shielding can be defined by the following way:

$$\overline{\widehat{\mathcal{H}}_{\sigma\,\mathrm{int}}} = \frac{1}{3}\gamma_i \hbar B_0 \sigma \sum_i (\hat{I}_{zi} + \hat{I}_{xi}). \tag{5.109}$$

The vector corresponding to the sum of operators in the right part of Eq. (5.109) directs along the diagonal of the face made by the z and x axes, therefore its length is $\sqrt{2}\,\hat{I}_z$.

Thereby, the MREV-8 pulse sequence results in decreasing the measured chemical shift in $\sqrt{2}/3$ times as compared with the real one, i.e. the greater decreasing of the measured chemical shift value than when applying the WHH-4 pulse sequence. It is obvious that the MREV-8 pulse sequence, similar to the WHH-4 one, averages down to zero the Hamiltonian of dipole–dipole interactions for spins of the same species, the Hamiltonian of quadrupole interactions and anisotropic part of the Hamiltonian of indirect spin–spin interactions.

5.8 Relaxation in Diamagnetic Solids: Role of Slow Motions

Relaxation mechanisms in diamagnetic solids in general terms is similar to those in liquids, but relative contributions of different mechanisms may be essentially different.

Dipole–dipole interactions in solids can be modulated by thermal motions of various kinds: random reorientations of molecules or molecular groups about symmetry axes, thermal vibrations of molecules or atoms about their equilibrium positions, diffusive motions. But in contrast to liquids, in solids thermal motions are anisotropic, and therefore, the averaging of local fields is smaller and the result of such an averaging depends on the direction of the external magnetic field vector. It leads to orientation dependencies of relaxation times. At low temperature, in general, there are only small oscillations of particles about their equilibrium positions. As a result, the spin–lattice relaxation times become very long (see Eqs. (3.116), (3.117), (3.132) and Fig. 3.3 on the right of the dash line): in particular cases the spin–lattice relaxation times of ^{1}H or ^{19}F achieve several thousands seconds. An example of the T_1 temperature dependence of protons in the solid benzol is shown in Fig. 5.14a.

It is interesting to compare this dependence with the temperature dependence of the second moment of the NMR spectral line S_2, represented in Fig. 5.14b. As it is seen, the spin–lattice relaxation time is rather sensitive to changing of the velocity of thermal motions in greater temperature range than changing of the second moment value.

Relaxation studies provide especial benefits for studies of jump reorientation of molecules or molecular groups around two-fold symmetry axis, since such a

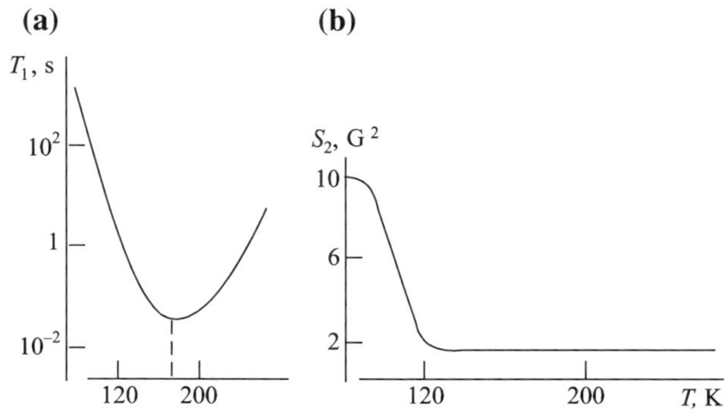

Fig. 5.14 Temperature dependence of [1]H T_1 (**a**) and S_2 (**b**) in the solid benzol

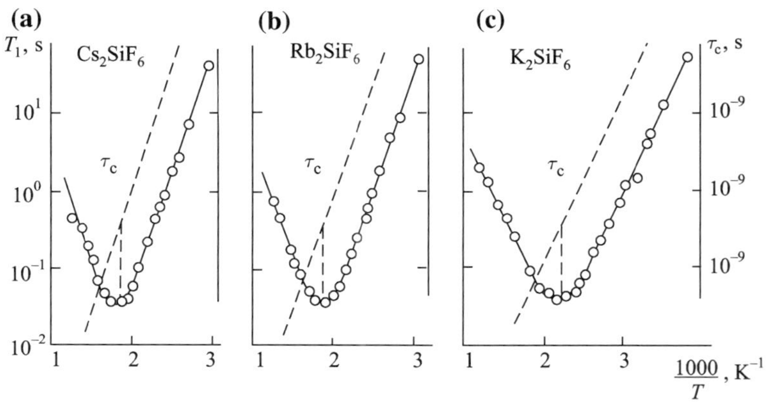

Fig. 5.15 Temperature dependencies of the spin–lattice relaxation times of ^{19}F in Cs_2SiF_6 (**a**), Rb_2SiF_6 (**b**) and K_2SiF_6 (**c**)

motion does not change the second moment (and width) of the spectral lines. Such reorientations of water molecules are rather typical for crystalline hydrates. Form Fig. 5.14a it is easy to determine the correlation time of the motion at the minimum of the $T_1(T)$ dependence and then to calculate this parameter for other temperature values. Thus, for ^{1}H nuclei in the benzol one has a linear dependence of $\ln \tau_c$ on the inverse temperature $(1/T)$, and it is possible to draw a conclusion about the activation character of the molecular motion in agreement with the formula $\tau_c = \tau_c^0 \exp[E_a/(RT)]$ and to determine the activation energy E_a (the potential barrier that blocks the molecular motion).

In Fig. 5.15 we plot the temperature dependencies of ^{19}F T_1 ($\nu_0 = 30$ MHz) in several crystal hexaphosphides: Cs_2SiF_6 (*a*), Rb_2SiF_6 (*b*) and K_2SiF_6 (*c*). It was found that in all studied hexaphosphides anions execute isotropic reorientation motions. Therefore, to treat the experimental data one can use formulas of Sects. 3.9 and 3.10,

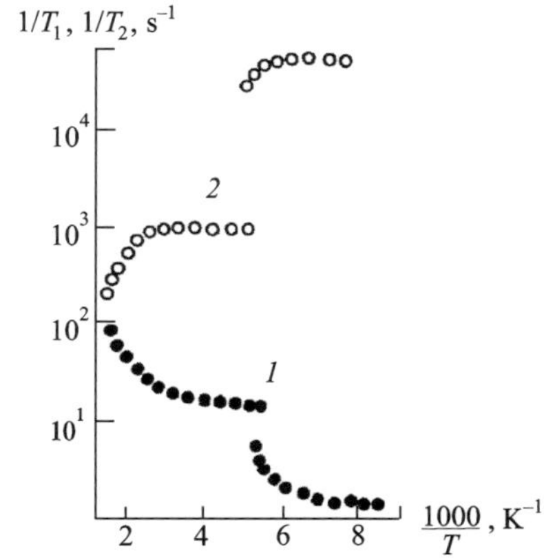

Fig. 5.16 Temperature dependence of the ^{19}F spin–lattice (*1*) and spin–spin (*2*) relaxation times at 30 MHz in the $CsPbF_3$ powder sample

including the formulas for calculations correlation times τ_c (dash lines in Fig. 5.15) and activation energies E_a (from the slope of τ_c or T_1 curves).

It is worth noting that in solids it is not always possible to register extremum of T_1 (or $1/T_1$) temperature dependencies. More frequently molecular motions are rather slow, $\omega_0^2\tau_0^2 \gg 1$, and the solid melts before one achieves the condition $\omega_0\tau_c \simeq 1$, i.e. a phase transition occurs. It should be said that NMR is a very powerful tool to study phase transitions, moreover it is more convenient (as compared to X-ray or neutron diffraction) to study melting–solidification phase transitions, as NMR allows us to observe the appearance of liquid phase in the finest capillaries, thin layers, microscopic internal areas. The liquid phase is detected by the appearance of narrow lines in the NMR spectrum.

NMR is also a rather effective method to fix other types of phase transitions, for example, structural phase transitions. In Fig. 5.16 we plot temperature dependencies of T_1 and T_2 for ^{19}F nuclei at 30 MHz in $CsPbF_3$. The jumps in the temperature dependencies at 180 K correspond to the phase transition from the cubic to tetragonal structure.

If there are only lattice vibrations theoretical estimations of the relaxation times (see Sects. 3.9 and 3.10) lead to much greater values than those obtained experimentally. That is why another relaxation mechanism, which takes into account combinative effects, should be considered. If nuclei participate in lattice vibrations (which are always more high-frequency compared to the NMR frequency) internuclear distances depend on time

$$r = r_0 + \Delta r_1 \sin \omega_1 t + \Delta r_2 \sin \omega_2 t$$

(ω_1 and ω_2 are random frequencies of lattice oscillations), and due to interactions are nonlinear ($\sim r^{-3}$) the combination frequencies $\omega_1 \pm \omega_2$ appear, which can coincide with the NMR frequency. Therefore, the role of magnetic field fluctuations is more important in relaxation processes. However, this model does not provide a satisfactory agreement with experimental data.

There are some proofs that in many cases relaxation intensity increases due to paramagnetic impurities, and even small concentrations like $\sim 10^{-6}$ may make an essential contribution. It is rather typical for ionic crystals. In organic solids, for which dipole–dipole interactions modulated by the motion of molecules or molecular groups (e. g. CH_3) are rather strong, this mechanism is not so frequent. The role of interactions between nuclear and electron spins was discussed in details for liquids. It was shown that even small impurities essentially impact nuclear magnetic relaxation rates. But in liquids one deals with a large thermal "reservoir" (thermal motion energy) with a broad spectrum, and paramagnetic particles may approach very close to all nuclei (by turn, yet for a short period of time). In solids electron spins are fixed and even strong interactions exist only for nuclei, which are nearby paramagnetic centers. However, due to the EPR relaxation of paramagnetic centers, strong dipole–dipole interactions of nuclear spins and, as a result, effective spin diffusion, the nuclear magnetic relaxation runs over all the sample volume. In solids if one of magnetic moments belongs to an electron spin whereas another belongs to a nuclear one, the distance r between them is more or less constant. That is why the dipole–dipole interaction is not averaged and fluctuates only a little. But in the Hamiltonian $\widehat{\mathcal{H}}_{\mathrm{d}ij}$ (see Eqs. (1.135), (1.136)), where i refers to the nuclear spin I and j refers to the electron spin S (further when describing dipole–dipole interactions between nuclear and electron spins the indexes i and j will be omitted), there are terms $\hat{C}$ and $\hat{D}$:

$$\hat{C} = \hat{D}^* = -\frac{3}{2}\sin\theta\cos\theta\exp(-\mathrm{i}\phi)\left(\hat{\hat{I}}_+\hat{S}_z + \hat{\hat{S}}_+\hat{I}_z\right),$$

where the first term in the operator part describes the reorientation of the nuclear spin I keeping the orientation of the electron spin S unchanged. Such a process requires an energy quant $\hbar\omega_{0I}$ that is much less than $\hbar(\omega_{0I} \pm \omega_{0S})$ or $\hbar\omega_{0S}$ but may be important for other terms of the Hamiltonian (see Eqs. (1.135), (1.136)) or if there are other kind of interactions (for example, scalar interactions). The relaxation of the nuclei closest to the paramagnetic center is transmitted over the sample by spin-flop (e. g. cross-relaxation) mechanisms. Such a spin diffusion mechanism is described by the $\hat{B}$ term, which contains the product of operators $\hat{\hat{I}}_{i\pm}\hat{\hat{I}}_{k\mp}$ (see Eq. (1.136)).

The analysis shows that the relaxation processes can be characterized by an effective time constant $T_{1\mathrm{eff}}$, which depends on the paramagnetic impurity concentration, "diffusion" barrier, spin-flop probability, crystal structure and so on. Such kind of reasoning were confirmed by many experiments, for example, by Andrew. In a pure LiF single crystal the spin–lattice relaxation times are equal to 5 min for ^{7}Li and to 2 min for ^{19}F. However, after X-ray irradiation of the sample, which creates paramagnetic centers, the spin–lattice relaxation times decrease dramatically: $T_1 = 30$ s for ^{7}Li and $T_1 = 10$ s for ^{19}F. Currently, X-ray irradiation is used if it is necessary to reduce effectively relaxation times to avoid suturation in cw NMR experiments. It is to be noted that paramagnetic centers in crystals may appear due to impurities of transition metal ions or due to exited atoms etc.

In solids relaxation times depend on the crystal orientation that confirms, in particular, the spin diffusion mechanism for nuclei (the process probability depends of the spectral line shape, which, in its turn, depends on the crystal orientation with respect to the magnetic field direction). However, this mechanism usually does not contribute into the NMR line width, as the number of nuclei close to paramagnetic impurities is a minority with respect to the total number of nuclei (local fields decreases as $1/r^3$, see Eq. (1.135)). In materials with a large number of paramagnetic ions the spectral line width may be noticeably greater that one caused by nuclear dipole–dipole interactions. In this case the main contribution is due to interactions of nuclei with fast fluctuating electron magnetic moments (enormous compared to nuclear ones). The NMR line shape is close to Lorentz line, as in liquids, and the time dependence of the free induction decay is close to exponential.

Lattice vibrations more affect the relaxation of quadrupole nuclei, as the energy of quadrupole interactions essentially exceeds the energy of dipole–dipole interactions. A theoretical description of relaxation processes in this case becomes more complex, as the energy levels are not equidistant anymore, and NMR spectra are split (the number of lines depends on the nuclear spin). Moreover, relaxation times depend on the transition frequency they are measured at. So, for example, for a nuclear spin equal to $7/2$, there are seven transition frequencies with different relaxation times, and it is very difficult to separate them. That is why at relaxation measurements the crystal should be oriented in such a manner that satellite lines interflow with the central one: it corresponds to a two-level energy system with the only spin–lattice relaxation time. The presence of the quadrupole relaxation channel can be detected by observing how the NMR spectra react on a high alternating magnetic field: the saturation of one spectral line leads to changing of the intensities of the others.[16]

It was already mentioned that in solids $\omega_0^2\tau_c^2 \gg 1$. That is why for a more reliable determination of thermal motion parameters from nuclear spin–lattice relaxation rates, particularly the correlation time τ_c, it is better to use low resonance frequency ω_0, in order to be within the T_1 dispersion range, i.e. to fulfill the relation $\omega_0\tau_c \simeq 1$.

Numerical values of the relaxation time depend on many factors, and it is practically impossible to take them all into account within any theoretical consideration, so it is not reliable to obtain parameters characterizing the molecular thermal motion on the basis of numerical values of relaxation times. More definite is the analysis, based on the positions of the minima of the temperature dependence of the spin–lattice relaxation time. In this case, it is more correct to affirm that if such a type of molecular motion has a characteristic correlation time τ_c, then the value of τ_c close to ω_0^{-1} corresponds to the minimum of T_1.

Thereby, at usual values of magnetic fields this method allows us to measure correlation times τ_c in a rather narrow range (10^{-9} to 10^{-7} s for ^{1}H NMR). It is essential that the method is low-sensitive to slow molecular motions, which are more typical for solids. Of cause, it is always possible to decrease frequency and to work in lower magnetic fields. But it reduces sensitivity of the method and ask wide-range

[16] More details can be found in A. Abraham. The Principles of Nuclear Magnetism. Oxford University Press, London, 1961

spectrometers, that complicates experiment. Moreover, it increases the dead time of receiver. That is why reducing the operating frequency of the spectrometer is not reasonable. More effective is to carry out the experiment in the rotating frame and to apply multipulse sequences (it will be consider further).

As it was shown in Sect. 1.5, in the rotating frame the nuclear magnetization is affected by an effective field (1.61), which at the exact resonance condition ($\omega_0 - \omega = 0$) is equal to B_1. Initially this method was considered only as a convenient mathematical device to describe the magnetization motion in the presence of a high-frequency field. But in 1959 Solomon found that the spin–spin relaxation time T_2 measured in the presence of a radiofrequency field depends on the orientation of the nuclear magnetization vector with respect to the radiofrequency field vector.[17] In particular, if the magnetization vector keeps the direction along the rotating field $\vec{B}_1$, the relaxation time essentially depends on the field value $\vec{B}_1$. The magnetization vector, even relaxing, keeps its direction along $\vec{B}_1$, which is the only field in the rotating frame. The time that characterizes the relaxation of the nuclear magnetization component parallel to $\vec{B}_1$ is denoted by $T_{1\rho}$ (if $\omega \neq \omega_0$, then $T_{1\rho}$ is the relaxation time for the nuclear magnetization component parallel to the $\vec{B}_{\text{eff}}$ vector in the rotating frame). Thereby, to measure the relaxation rate in the rotating frame it is necessary to record the evolution of the transversal (!) component, transversal with respect to the $\vec{B}_0$ vector, in the presence of the rf field $\vec{B}_1$.

Interest to relaxation time measurements in the rotating frame is also caused by the fact that, as a rule, $T_{1\rho}$ differs from T_2 if relaxation times depend on the NMR frequency, i.e. if there are internal motions in crystals. In these cases the nutation frequency $\omega_1 = \gamma B_1$, in a sense, acts the resonance frequency, and one can measure correlation time values of about ω_1^{-1}. As $\omega_1 \ll \omega_0$ it is clear that such a method allows us to study rather slow molecular motions in solids.

The theoretical basis of the role of molecular motions on $T_{1\rho}$ were developed by Goldman and Jones using the modified Hamiltonian of dipole–dipole interactions.[18] Usually $T_{1\rho}$ is determined at the exact resonance condition ($\omega_0 - \omega = 0$) and at $\omega_1 \ll \omega_0$

$$\frac{1}{T_{1\rho}} = \frac{3}{10}\gamma^4\hbar^2 r^{-6}\left[\frac{3}{2}\frac{\tau_c}{1+4\omega_1^2\tau_c^2} + \frac{5}{2}\frac{\tau_c}{1+\omega_0^2\tau_c^2} + \frac{\tau_c}{1+4\omega_0^2\tau_c^2}\right].$$

In the case of slow molecular motions ($\omega_0^2\tau_c^2 \gg 1$) the first term dominates, and the characteristic value, at which $T_{1\rho}$ achieves its minimum, is $\omega_1^2\tau_c^2$. The position of the $T_{1\rho}$ minimum, first, corresponds to the slower motion (as it is always $\omega_1 \ll \omega_0$), and second, can be changed by an experimenter by changing the B_1 value keeping the spectrometer adjustment frequency. This method was successfully applied to study

[17] I. Solomon. Rotary Spin Echoes. Phys. Rev. Lett. 2 (1959) 301.

[18] M. Goldman and L. Shen. Spin-spin relaxation in LaF_3. Phys. Rev. 144 (1966) 321. G. P. Jones. Spin-Lattice relaxation in the rotating frame: weak-collision case. Phys. Rev. 148 (1966) 332.

methyl group rotations in the tetramethylammonium halides ,[19] in tetramethylsilane [20] and other molecular crystals[21] including biological objects.

A relaxation study experiment in the rotating frame assumes three steps:

- to fix the nuclear magnetization vector along the vector of the alternating field $\vec{B}_1$;
- the magnetization relaxation in the field $\vec{B}_1$;
- the magnetization measurement by recording the free induction signal after the field $\vec{B}_1$ is switched off.

There are several ways to implement the first step of the experiment. The main way to obtain the desired orientation is to apply a usual 90°-pulse, after which one applies another pulse of required amplitude of the $\vec{B}_1$ field with the phase filling differs by 90° from the phase filling of the first pulse. The first pulse orients the $\vec{M}$ vector along the x-axis. Due to the phase shift $\vec{B}_1$ is also directed along the x-axis during the second pulse. Such a pulse combination can be obtained using coherent pulse spectrometer with adjustable phase filling of individual pulses. This is a so-called *spin-locking* technics. To determine $T_{1\rho}$ it is necessary to carry out a series of measurements with various durations of the second pulse. It is rather time-consuming as after each cycle it is necessary to wait till the spin–lattice system comes to a Boltzmann equilibrium, that is about several T_1 values (let us remind that in solids T_1 values may achieve few minutes). In Fig. 5.17 we give an example of T_1 and $T_{1\rho}$ temperature dependencies of protons of water molecules in the gypsum crystal for different values of the spin-locking field obtained in the laboratory (T_1) and rotating ($T_{1\rho}$) frames.

To gain the experiment time sometimes the second (long) pulse is replaced by a series of 90°-pulses separated by short time intervals: $\tau \ll T_2$, which allows us to observe the change of magnetization in a single passage, recording NMR signals in "windows" between the pulses. This is a so-called *pulse spin-locking* experiment.

Relaxation processes in the rotating frame is also studied by applying more sophisticated pulse sequences used to suppress dipole–dipole interactions in solids.

As pulse sequences are sensitive to slow molecular motion, one can use the same sequences that average dipole–dipole interactions, such as the WHH-4 sequence mentioned above, the HW-8 pulse sequence developed by Haeberlen and Waugh, the phase alternating MW-2 and MW-4 (Mansfield–Waugh) pulse sequences.

---

[19] S. Albert, H. S. Gutowsky, and J. A. Ripmeester. On a T_1 and $T_{1\rho}$ study of molecular motion and phase transitions in the tetramethylammonium halides. J. Chem. Phys. 56 (1972) 3672.

[20] S. Albert and J. A. Ripmeester. Correlation effects and molecular tumbling in NMR studies of solid β-(CH_3)4Si. J. Chem. Phys. 57 (1972) 2641.

[21] See, for example, J. Huang, L. Zhang, and H. Tang. Solid-state NMR analysis reveal the structure dependence of the molecular dynamics for ω-amino acids. J. Phys. Chem. B 116 (2012) 2096.

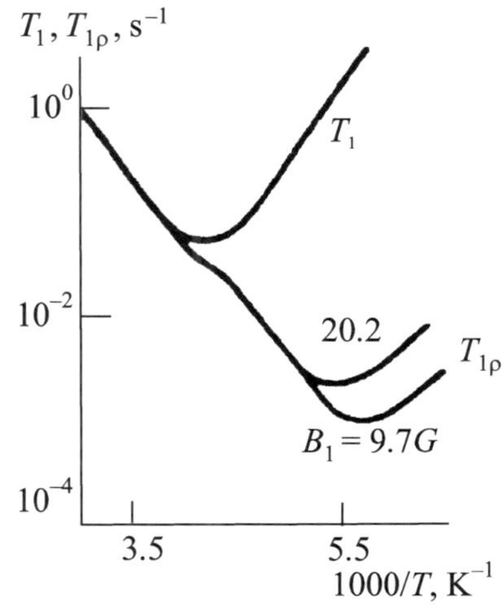

Fig. 5.17 Temperature dependencies of the proton spin–lattice relaxation times in the laboratory (T_1) and rotating ($T_{1\rho}$) frames in the gypsum crystal at B_1 equal to 20.2 and 9.7 G (curves 1 and 2, respectively). Reproduced with permission from J. Chem. Phys. 44, 2995 (1966). Copyright 1966, American Institute of Physics

All these pulse sequences consist of a number of 90°-pulses:

$$\begin{aligned}
&\text{WHH-4}: (\tau - \hat{P}_x - 2\tau - \hat{P}_{-x} - \tau - \hat{P}_{-y} - 2\tau - \hat{P}_y)_n, && t_c = 6\tau,\\
&\text{MW-2:} \quad (\tau - \hat{P}_x - 2\tau - \hat{P}_{-x} - \tau)_n, && t_c = 4\tau,\\
&\text{MW-4:} \quad (\tau - \hat{P}_x - 2\tau - \hat{P}_x - 2\tau - \hat{P}_x - 2\tau - \hat{P}_x - \tau)_n, && t_c = 8\tau,\\
&\text{HW-8:} \quad (\text{WHH-4}) - (-\text{WHH-4}), && t_c = 12\tau.
\end{aligned}$$

Here P_α denotes a 90°-pulse along the $\alpha = x, y, z$ axis.

As it was shown in Sects. 5.7.7 and 5.7.8, implementing WHH-4 and MREV-8 pulse sequences allows us to average to zero dipole–dipole interactions of nuclear spins of the same species if the NMR signal is observed stroboscopically at time moments multiple of the cycle time t_c. That means the relaxation channel due to these interactions is eliminated from relaxation processes and the time necessary to come to equilibrium state is much greater than T_2. Such an equilibration process for the nuclear magnetization component perpendicular to the $\vec{B}_1$ vector (or to $\vec{B}_{\text{eff}}$ if the alternating field frequency is not equal to the resonance frequency) is characterized by T_{2e}. It is clear that the transversal relaxation time T_{2e} depends on the molecular motion rate. It is well known that fast molecular motion ($\omega_0^2\tau_c^2 \ll 1$) averages dipole–dipole interactions that results in narrowing of NMR spectral lines and long T_1 and

T_2 relaxation time values. In this case multipulse sequences intended to average dipole–dipole interactions do not affect NMR signals and the envelope curve decay is determined by the relaxation time $T_2 = T_{2e} = T_1$. On the other hand, in the case of the rigid lattice these pulse sequences also lead to the long time magnetization decay. However, if the lattice is not quite rigid and the correlation time characterizing its motion is about the time intervals between the pulses τ, the value of dipole–dipole interactions is not constant within one pulse cycle. As a result, the dipole–dipole interactions are not averaged completely that leads to T_{2e} decreasing. Thus, considering T_{2e} as a function of τ_c one should expect a minimum near $\tau \sim \tau_c$.

Theoretical studies of how the molecular motion impacts on the spin system response after implementing a multipulse sequence are based on the same method as in the case of the rigid lattice, but in the total Hamiltonian of the system there are two time dependent terms:

$$\widehat{\mathcal{H}} = \widehat{\mathcal{H}}_Z + \widehat{\mathcal{H}}_1(t) + \widehat{\mathcal{H}}_d(t).$$

Here $\widehat{\mathcal{H}}_Z$ describes interactions of the spins with a constant magnetic field:

$$\widehat{\mathcal{H}}_Z = -\gamma\hbar\hat{\vec{I}}_z B_0;$$

$\widehat{\mathcal{H}}_1(t)$ represents interactions of the spins with an external high frequency field:

$$\widehat{\mathcal{H}}_1 = 2\omega_1 \cos\omega_0 t\left[f_x(t)\hat{\vec{I}}_x + f_y(t)\hat{\vec{I}}_y\right],$$

here $f_x(t)$ and $f_y(t)$ are pulse functions equal to ± 1 during the in-phase (anti-phase) or shifted by $\pm 90°$ pulses and equal to zero the rest of the time. These two terms do not differ from those one considered in the rigid lattice case. The third term

$$\widehat{\mathcal{H}}_d(t) = \frac{3}{2}\frac{\gamma^2\hbar^2}{r^3(t)}\left[1 - 3\cos^2\theta(t)\right]\left[3\hat{\vec{I}}_{1z}\hat{\vec{I}}_{2z} - (\hat{\vec{I}}_1, \hat{\vec{I}}_2)\right]$$

is a function of time, as both the distance $r(t)$ between nuclei and their mutual orientation determining by the angle $\theta(t)$ change in time.

Calculations carried out in the rotating frame result to a rather complex time dependence of the average value of the magnetization component:

$$\langle\hat{\vec{I}}_x(t)\rangle = \exp\left[-tR_1(\alpha) + \tau_c(1 - \exp(t/\tau_c)R_2(\alpha))\right],$$

$$R_1(\alpha) = \frac{2S_2}{3\alpha^2}\left[\alpha + \frac{\sinh\alpha(2 - 5\cosh\alpha)}{4\cosh^2\alpha - 1}\right];$$

Table 5.1 Theoretical expressions for T_{2e} for various pulse sequences

Pulse sequence	Cycle time	Expression for T_{2e} At $\tau_c \ll \tau$	At $\tau = \pi\tau_c/2$	At $\tau_c \gg \tau$
MW-2	4τ	$T_{2e} = \frac{1}{S_2\tau_c}$	$T_{2e} = \frac{3.8}{S_2\tau}$	$T_{2e} = \frac{3\tau_c}{S_2\tau^2}$
MW-4	8τ			
WHH-4	6τ	$T_{2e} = \frac{3}{2S_2\tau_c}$	$T_{2e} = \frac{6.4}{S_2\tau}$	$T_{2e} = \frac{9\tau_c}{2S_2\tau^2}$
MREV-8	12τ			

$$R_2(\alpha) = \frac{S_2}{2\alpha}\frac{[4\sinh\alpha(\cosh\alpha - 1)]^2}{\cosh 6\alpha - 1};$$

$\alpha = \tau/\tau_c$; S_2 is the second moment of the NMR spectral line for the rigid lattice. However, these expressions can be simplified for $t \gg \tau$, then the magnetization decay can be approximately described by the only exponent with the time constant T_{2e}, and

$$\frac{1}{T_{2e}} = \tau_c S_2\left\{\frac{2}{n\alpha}\left[1 - \frac{(2+n)\cosh(n-2)\alpha - 2}{\alpha\sinh n\alpha}\sinh^2\alpha\right]\right\}. \tag{5.110}$$

Here $n = 2$ for the MW-2 and MW-4 sequences, i.e. these sequences affect the T_{2e} value by the similar way. For WHH-4 and MREV-8 $n = 3$, and their action on the magnetization component perpendicular to the magnetic field $\vec{B}_1$ vector is identical.

In Table 5.1 we list values of the transversal relaxation time T_{2e} for different pulse sequences and ratios between the correlation time τ_c and time interval between pulses τ.

It should be noted that application of the MW-4 sequence is very profitably to study molecular motions. Its main advantage is that it rather effectively (in practice) averages dipole–dipole interactions (up to 30,000). Simultaneously, this sequence averages chemical shifts that limits it application in high-resolution solid state NMR experiment but does not impede molecular motion studies.

As it is seen the multipulse techniques and spin-locking method have similar abilities in studying molecular motions in the spin systems with dipole–dipole interaction. This can be explained by the fact that the effect of a pulse sequence and a high (with respect to the local dipole–dipole fields) rf field on such a spin system is quite similar: in both cases the dipole–dipole interactions are suppressed, either due to subsequent orientation of the nuclear magnetization vector along the all unit vectors of the rotating frame or due to its orientation along the effective field, contribution to which from local fields is negligibly small.

As a result $T_{1\rho}$ and T_{2e} are almost equal to each other if the average value of the B_1 field of high frequency pulses is equal to the value of the high radiofrequency field

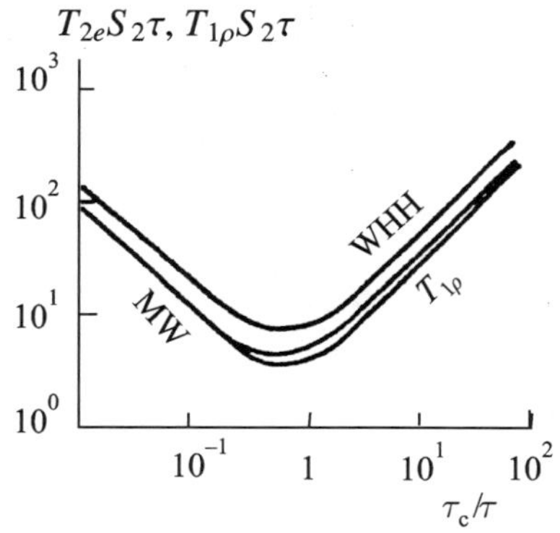

Fig. 5.18 Illustration of similar abilities of the multipulse and spin-locking methods to study molecular motions. Reproduced with permission from J. Chem. Phys. 43, 3575 (1965). Copyright 1965, American Institute of Physics

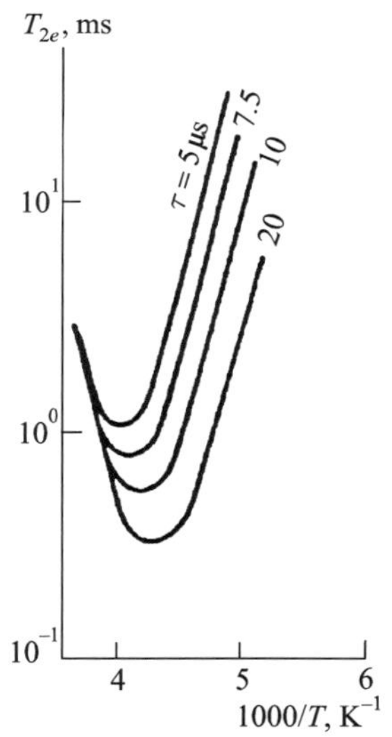

Fig. 5.19 Temperature dependence of ^{1}H T_{2e} in C_6H_6 measured by MW-4. Reproduced with permission from J. Chem. Phys. 43, 3575 (1965). Copyright 1965, American Institute of Physics

applied in the spin-locking methods. To illustrate it in Fig. 5.18 we plot dependencies of dimensionless quantities $T_{2e}S_2\tau$, $T_{1\rho}S_2\tau$ versus τ_c/τ.

Nevertheless, multipulse techniques have an advantage over the method of relaxation in the rotating frame: the T_{2e} value can be found within the only one cycle, and it is not necessary to repeat the cycle with various durations of high-frequency pulses. This leads to great time saving, as to repeat the cycle one needs about (3–5)T_1, and in solids T_1 itself can be very long.

To illustrate how such pulse sequences are applied to study molecular motions in solids we plot in Fig. 5.19 temperature dependencies of the time T_{2e} of the proton

NMR signal decay in the solid cyclohexane C_6H_6 obtained using the MW-4 pulse sequence at various time intervals between pulses. According to Eq. (5.110) and data of Table 5.1 the minimal T_{2e} values are determined as

$$T_{2e\ \mathrm{min}} = 3.8/(S_2\tau). \tag{5.111}$$

Eq. (5.111) allows us to make an independent verification of the theory without any fitting parameter. The T_{2e} values calculated using Eq. (5.111) at the second moment value equal to $7.8 \cdot 10^8\ \mathrm{s}^{-2}$ (obtained independently) are in fair agreement with experimental data.

Further Readings

Barnes RG (1997) NMR in metal hydrogen systems. Top Appl Phys 73:93–151
Callaghan PT (2011) Translational dynamics and magnetic resonance: principles of pulsed gradient spin echo NMR. Oxford University Press, Oxford
Duer MJ (ed) (2002) Solid state NMR spectroscopy: principles and applications. Wiley-Blackwell, New York
Gerstein BC, Dybowski CR (1985) Transient techniques in NMR of solids. Academic Press, San Diego
Harris RK, Wasylishen RE, Duer MJ (2009) NMR crystallography. Wiley, Chichester
Klinowski J (ed) (2010) New techniques in solid-state NMR., Topics in current chemistrySpringer, Berlin
MacKenzie KJD, Smith ME (2002) Multinuclear solid-state NMR of inorganic materials., Pergamon materials seriesElsevier, Oxford
Mehring M (1976) High resolution NMR spectroscopy in solids. Springer-Verlag, Berlin
Saito H, Ando I, Naito A (2006) Solid state NMR spectroscopy for biopolymers: principles and applications. Springer, Dordrecht
Schmidt-Rohr K, Spiess HW (1994) Multidimensional solid-state NMR and polymers. Academic Press, San Diego
Stejskal EO, Memory JD (1994) High resolution NMR in the solid state: fundamentals of CP/MAS. Oxford University Press, Oxford

Chapter 6
Nuclear Magnetic Resonance in Liquid Crystals

6.1 Structure and Properties of Liquid Crystals

Liquid crystal or *mesophase* is a thermodynamic phase of matter which differs from isotropic liquids and single crystals. However, it combines two inconsistent properties of matter: fluidity and anisotropy of physical properties. Fluidity is specific for liquids, anisotropy is attribute of crystals. Anisotropy (dependence of properties on direction in a sample) of optical, electrical, magnetic, mechanical and other properties in liquid crystals appears due to correlation of molecular orientation at long distances which much longer than the size of molecules. This is called orientational order. Translational order absent in a molecular system because of fluidity of the sample. Molecules more or less free to move through a sample. In some kind of liquid crystals molecules arrange in a layers or even in 3-dimensional lattices, but diffusion makes these layers and lattices smooth. Depending on orientational and translational order, the molecules forms different mesophases.

There are two classes of liquid crystals. *Lyotropic* liquid crystals are multicomponent systems containing solvent (usually water) and specific *amphiphilic* solute molecules having a polar, water-soluble group attached to a nonpolar, water-insoluble hydrocarbon chain. Phase transitions between the isotropic homogeneous solution and different liquid crystalline phases appear due to changing the concentration of the solute. The simple example of the lyotropic liquid crystal is a soap bubble (or a thin film of soapy water); cell membranes also show liquid crystalline properties. *Thermotropic* liquid crystals are usually single component systems consisting of the molecules of significantly anisotropic form: elongate (rod-like) or plane (disk-like). Other types of molecules are banana-shaped, brick-shaped, etc. Phase transitions appear with changing of temperature.

The examples of rod-like liquid crystalline molecules of homologous series *p*-*n*-alkyloxybenzyliden-*p'*-*n*-butylaniline (a) and disk-like molecules of series 1,2,5,6,9,10-hexa-*n*-alkoxy-triphenylene (b) are shown in Fig. 6.1.

The main distinction of liquid crystal from liquids is the *long-range orientational order* of molecules. The word "long-range" means that the molecules have preferred

V. I. Chizhik et al., *Magnetic Resonance and Its Applications*,
DOI: 10.1007/978-3-319-05299-1_6,

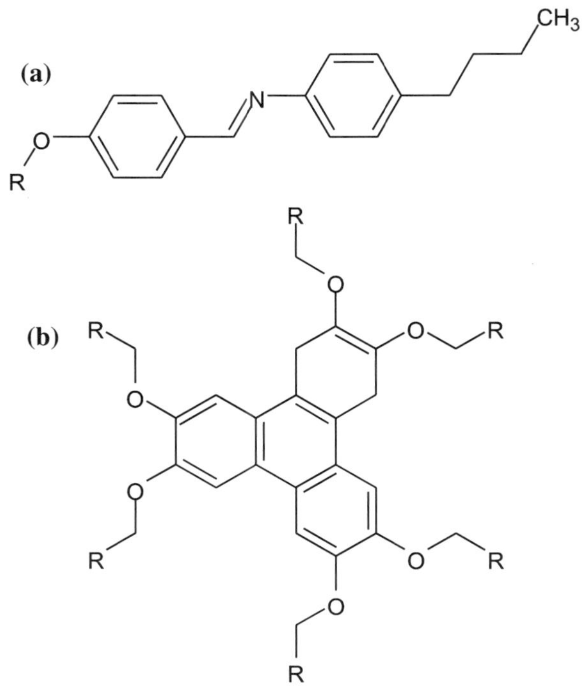

Fig. 6.1 Examples of homologous series of rod-like (**a**) and disk-like (**b**) molecules. R means *n*-alkyl chain

orientation in certain direction in a part of a sample with much greater size than the size (length) of one molecule. In a constant electric field or a magnetic field preferred orientation may be the same in the whole sample, such samples have homogeneous orientation.

Let us introduce the molecular reference frame xyz linked to one arbitrary molecule. The axes of the molecular frame coincide with the principal axes of inertia tensor of the molecule. The z axis in rod-like molecule coincides with the axis of minimal inertia, and in the disk-like molecule with the axis perpendicular to the molecule's plane. The direction in space with preferred orientation of z axis is called *director* and denotes with the symbol $\vec{n}$. Director of liquid crystals usually orients along or perpendicular to the magnetic field of NMR spectrometer. In this Chapter we do not consider the case of partial or inhomogeneous orientation of a liquid crystal during the NMR experiment.

While in the crystalline solids all the molecules are oriented strictly along one direction, in a liquid crystal the molecular motion leads to fluctuation of orientation of the molecules around director. The important numerical characteristic of the quality of orientation of the molecules is *orientational order parameter* (S_{zz} or simply S):

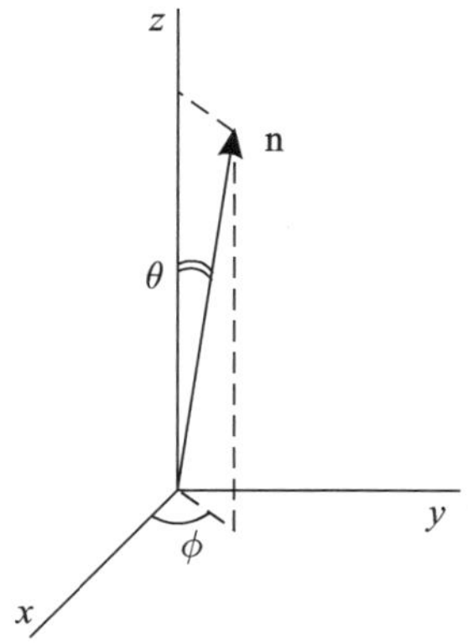

Fig. 6.2 Orientation of director $\vec{n}$ in the molecular frame xyz

$$S \equiv S_{zz} = \left\langle \frac{3\cos^2\theta - 1}{2} \right\rangle = \langle P_2(\cos\theta) \rangle. \tag{6.1}$$

Here, the angle brackets denote averaging over time, θ is the angle between the molecular z axis and the director (see Fig. 6.2). $P_2(x)$ is the second degree Legendre polynomial.

Because of the molecules has nearly but not exact axial symmetry, than their tilt in the plains (xz) and (yz) are different, which is described by the *biaxiality* of liquid crystal $S_{xx} - S_{yy}$:

$$S_{xx} - S_{yy} = \left\langle \frac{3\sin^2\theta\cos 2\phi}{2} \right\rangle.$$

In isotropic phase both parameters are zero. Order parameter is about 0.4 just below the temperature of isotropic to liquid crystal transition (so called *clearing point*) and it growth up to 0.8–0.9 with decreasing temperature. In the discotic liquid crystals order parameter may be of 0.95. Biaxiality is important for the rod-like liquid crystals and is limited by 0.04–0.05.

In addition to translational and rotational mobility of the molecules as whole, there is conformational mobility (see Sect. 4.1.1) of the molecules of liquid crystals. After the conformational transition occurred, the distance between some atoms and the orientation of the internuclear vector in the molecular frame are subject to change. Because of we choose the molecular frame coincides with the frame of the inertia tensor, than rotations of molecular groups (intramolecular, conformational rotations) do not depend on the rotation of the molecule as whole. This is important for the averaging of the dipole–dipole and quadrupolar interactions.

The average values of intramolecular dipole-dipole and quadrupolar interactions in liquid crystals are not zero because of preferential orientation of the molecules along the director. Intermolecular dipole-dipole interactions are zero because of fast translational molecular motion. The average value of gradient of electric field caused by the nearest molecules is also equal to zero. So, intermolecular interactions do not influence on the shape of the spectrum. A spectrum of liquid crystal is similar to the spectrum of one molecule. (Really, the intermolecular spin–spin interactions influence on the relaxation rate which leads to the broadening of the resonance lines). Chemical shifts are tensors in liquid crystals. The observed chemical shift of the spectral lines depends on the orientation of the director in the magnetic field and on the orientational order parameter. J-couplings have minor values relative to other mentioned above interactions and do not influence on the spectral line shape.

6.2 Spin Hamiltonian in Liquid Crystals

6.2.1 Dipole–Dipole Interactions

Direct dipole–dipole interactions determine line shape of ^{1}H and ^{19}F NMR spectra. In partially deuterated substances, dipole–dipole interactions of protons and deuterons influence on spectra of both ^{1}H and ^{2}H, but in the deuteron spectra they lead to the broadening or weak splitting of lines. In the experiments on double ^{13}C–^{1}H resonance, both spectra of protons and carbons-13 are splitted.

Secular part of the Hamiltonian of dipole–dipole interactions determines the spectrum of the spin system in strong external magnetic field. For the system of same isotopic species (homonuclear interactions) it can be written in the form

$$\widehat{\mathcal{H}}_{\mathrm{d}} = -\sum_{i<j}^{N} 2D_{ij}\left(\hat{I}_{zi}\hat{I}_{zj} - \tfrac{1}{3}\hat{\vec{I}}_i \cdot \hat{\vec{I}}_j\right). \tag{6.2}$$

Summation is limited by the number of spins N in one molecule. In the case of heteronuclear interactions (different species) the dot product of spins in Formula (6.2) should be omitted. The *constants of dipole–dipole interactions* can be written as

$$D_{ij} = \frac{3}{2}\gamma_i\gamma_j\hbar\left\langle \frac{3\cos^2\theta_{ij} - 1}{2r_{ij}^3}\right\rangle.$$

In the SI units, the coefficient $\mu_0/4\pi$ is necessary to obtain the value of D_{ij} in radian per second. Angle θ_{ij} is the angle between the internuclear vector and the direction of magnetic field $\vec{B}_0$. The average $\langle \ldots \rangle$ should be applyed over time. During the time both internuclear distance r_{ij} may change becouse of conformational transitions and angle θ_{ij} because of conformational transitions and rotation of the molecule as whole.

Note that in literature another value

$$b_{ij} = -\frac{\mu_0}{4\pi}\frac{\gamma_i\gamma_j\hbar}{r_{ij}^3}$$

is called *dipole–dipole coupling constant*, and *secular dipole–dipole coupling* d_{ij} is

$$d_{ij} = b_{ij} \frac{3\cos^2\theta_{ij} - 1}{2}.$$

The value of D_{ij} used in this Chapter relates to those as

$$D_{ij} = -\frac{3}{2}\left\langle d_{ij} \right\rangle.$$

For example, if two spins ^1H is separated by a distance of 1 Å and $\theta_{ij} = 0$ then $D_{ij}/2\pi = 180.144\,\text{kHz}$.

We can consider the angle θ_{ij} as the sum the following three angles: the angle α between director of liquid crystal $\vec{n}$ and the direction of external magnetic field $\vec{B}_0$, the angle θ between the long molecular axis z and the director, and the angle $\theta_{ij,z}$ between the internuclear vector and the axis z of the molecular frame. Because of molecular motion and conformational transitions are independent if the molecular reference frame coincides with the principal axis of inertia tensor, the average of Legendre polynomial of $\cos\theta_{ij}$ is equal to product of average polynomials of cosine of each component of the angle:

$$\left\langle P_2(\cos\theta_{ij})\right\rangle = \left\langle P_2(\cos\alpha)\right\rangle \left\langle P_2(\cos\theta)\right\rangle \left\langle P_2(\cos\theta_{ij,z})\right\rangle. \tag{6.3}$$

After the homogeneous orientation of the sample is set in the magnetic field of spectrometer, the value of $P_2(\cos\alpha)$ may be equal to either 1 or -0.5 depending on the orientation of director along or perpendicular to magnetic field. Let us consider hereinafter the first case. The second factor in the right part of Formula (6.3) is the orientational order parameter S_{zz}. The third factor has completely intramolecular nature.

Taking into account order parameter and biaxiality we can write the constants of dipole–dipole interactions in the form

$$D_{ij} = \frac{3}{2}\gamma_i\gamma_j\hbar\left[S_{zz}\left\langle\frac{3\cos^2\theta_{ij,z} - 1}{2r_{ij}^3}\right\rangle + (S_{xx} - S_{yy})\left\langle\frac{\cos^2\theta_{ij,x} - \cos^2\theta_{ij,y}}{2r_{ij}^3}\right\rangle\right]. \tag{6.4}$$

Here, $\theta_{ij,x}$ and $\theta_{ij,y}$ denote the angles between the internuclear vector and the axes x and y of molecular frame (like $\theta_{ij,z}$).

In the temperature range of liquid crystalline phase (typically from a few degrees to several tens of degrees) we can assume with a high degree of accuracy that the distribution of molecular conformations does not depend on temperature. This assumption means that the average values in the Formula (6.4) also does not depend on temperature. The only variable value is the orientational order parameter. The constants of dipole–dipole interactions are proportional to $S_{zz} = S_{zz}(T)$, so the temperature dependence of $\widehat{\mathcal{H}}_d(T)$ has the form

$$\widehat{\mathcal{H}}_d(T) = S_{zz}(T)\,\widehat{\mathcal{H}}_d\big|_{S_{zz}=1}.$$

This is confirmed by the fact that line shape of proton NMR spectra practically does not depend on temperature, the only frequency scale is changing with temperature.

6.2.2 Quadrupolar Interactions

Quadrupolar interactions appear in NMR of ^{2}H, ^{14}N, ^{35}Cl and ^{37}Cl.

To study a substance by the deuteron NMR, it is useful to synthesize a partially deuterated substances. Usually, aliphatic chains (as terminal groups) and benzene rings (as central rigid fragments of molecules) are deuterated separately. Natural abundance of heavy Hydrogen is enough to record spectra using the modern spectrometers, but it is necessary to use heteronuclear decoupling with strong radio frequency pulses while duration of the experiments may takes several hours.

In organic molecules, the electric field is caused by the electrons of the chemical bonds. In the case of deuterium, the c axis of the electric field gradient tensor in liquid crystalline molecule is directed along the covalent chemical bond C–D. In the frame of principal axes abc of the electric field gradient tensor, the secular (first-order) part of the full Hamiltonian of quadrupolar interactions has form

$$\widehat{\mathcal{H}}_{\mathrm{Q}} = \frac{e^2 q Q}{4I(2I-1)} \left[3\hat{I}_c{}^2 - I(I+1) + \frac{\eta}{2}\left(\hat{I}_+{}^2 + \hat{I}_-{}^2\right)\right].$$

Here, axis c corresponds to the largest principal value of the tensor, $V_{cc} = eq$, the order of axes corresponds to the following order of principal values of gradient: $V_{cc} > V_{aa} > V_{bb}$. Therefore, asymmetry parameter $\eta = (V_{aa} - V_{bb})/V_{cc} > 0$. The values of quadrupole moment of a nucleus Q are listed in Table 1.1. The value

$$q_{\mathrm{C-D}} = \frac{e^2 q Q}{4I(2I-1)}$$

is called the first-order quadrupolar coupling.

The first-order quadrupolar coupling $q_{\mathrm{C-D}}$, expressed in the units of frequency, is about 165–175 kHz for the deuterons in aliphatic chains and about 185–195 kHz for the deuterons in aromatic rings. The asymmetry of electric field gradient tensor η is small: it is about 0.01 in aliphatic groups and 0.04 in aromatic groups.

Transition from the frame abc to the laboratory frame XYZ we can do through the molecular frame xyz. The spectrum of nucleus with spin $I = 1$ consists of two lines with the splitting

$$\Delta\nu = \frac{3}{2} q_{\mathrm{C-D}} \left[S_{zz} \left\langle \frac{3\cos^2\theta_{cz} - 1}{2} + \eta \frac{\cos^2\theta_{az} - \cos^2\theta_{bz}}{2} \right\rangle + (S_{xx} - S_{yy}) \right.$$
$$\left. \times \left\langle \frac{\cos^2\theta_{cx} - \cos^2\theta_{cy}}{2} + \eta \frac{\cos^2\theta_{ax} - \cos^2\theta_{ay} + \cos^2\theta_{by} - \cos^2\theta_{bx}}{3} \right\rangle \right]. \quad (6.5)$$

Here, θ_{ax} and others are the angles between axes abc and xyz.

If we ignore the parameters of biaxiality of liquid crystal and asymmetry of the electric field gradient, the above Formula becomes much simply:

$$\Delta\nu = \frac{3}{2}\, q_{\mathrm{C-D}} S_{zz} \left\langle \frac{3\cos^2\theta_{cz} - 1}{2} \right\rangle. \tag{6.6}$$

Average value $\langle P_2(\cos\theta_{cz})\rangle = \langle (3\cos^2\theta_{cz} - 1)/2 \rangle$ is sometime called *the parameter of orientational order of chemical bond* C–D, and it is usually denoted $S_{\mathrm{C-D}}$. Than Formula (6.6) has form

$$\Delta\nu = \frac{3}{2}\, q_{\mathrm{C-D}} S_{zz} S_{\mathrm{C-D}}. \tag{6.7}$$

6.2.3 Chemical Shifts

Chemical shifts in spectroscopy of liquid crystals is usually observed in spectra of ^{13}C, but in strong magnetic field also in spectra ^{1}H. Chemical shift is not isotropic in liquid crystals but tensor. In each of the conformational states of the molecule, this tensor has different principal values and the orientation of the principal axes in the molecular frame of reference, since the screening depends on the spatial distribution of the electron shells of neighboring atoms. However, with rapid conformational transitions and almost free rotation of the molecule around the long molecular axis, the tensor becomes average with the principal axes coinciding with the axes of molecular frame. In the isotropic phase orientational order parameter is equal to zero, so the chemical shift is isotropic:

$$\sigma_{i,\,\mathrm{iso}} = \frac{1}{3}\,\mathrm{Tr}\,\boldsymbol{\sigma}_i = \frac{1}{3}(\boldsymbol{\sigma}_{\mathrm{xx},i} + \boldsymbol{\sigma}_{\mathrm{yy},i} + \boldsymbol{\sigma}_{\mathrm{zz},i}).$$

Anisotropic part of the shift in a liquid crystalline phase is calculated according

$$\sigma_{i,\,\mathrm{aniso}} = \sum_{\alpha=x,y,z} S_{\alpha\alpha} \boldsymbol{\sigma}_{\alpha\alpha,i}. \tag{6.8}$$

If we neglect the biaxiality of liquid crystal, then $S_{xx} = S_{yy} = -S_{zz}/2$ and Formula (6.8) becomes

$$\sigma_{i,\,\mathrm{aniso}} = S_{zz}\left(\boldsymbol{\sigma}_{\mathrm{zz},i} - \frac{\boldsymbol{\sigma}_{\mathrm{xx},i} + \boldsymbol{\sigma}_{\mathrm{yy},i}}{2}\right).$$

A researcher can determine the anisotropy of the chemical shifts if record several spectra with different orientation of the director in magnetic field (with the techniques of variable angle spinning).

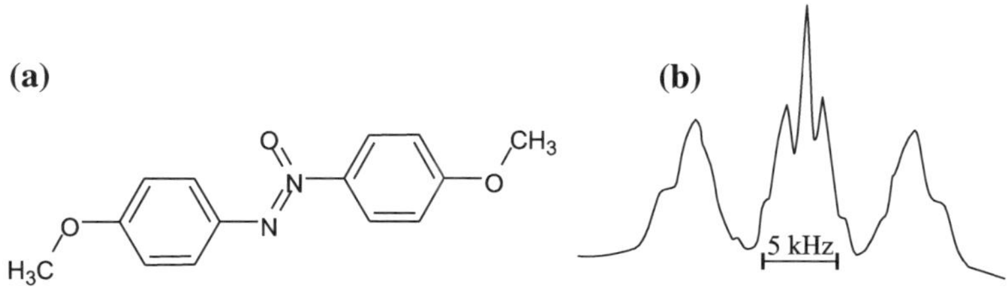

Fig. 6.3 Molecule PAA (**a**) and its proton NMR spectrum (**b**)

6.3 NMR Spectroscopy of Liquid Crystals

First proton NMR spectrum of liquid crystal was published as early as 1953 by R. D. Spence, H. A.Moses and P. C. Jain (*J. Chem. Phys.*, **21**, 380 (1953)). They recorded the temperature dependence of spectra of 4,4′-dimethoxyazoxybenzene (better known as *para*-azoxyanisole, PAA, see Fig. 6.3a) in isotropic and nematic phases. The spectra of liquid crystal were the wide triplet. Later the fine structure of these broad lines were discovered (Fig. 6.3b).

One of the basic task of the proton NMR spectroscopy of liquid crystals is to determine the spatial structure and set of possible conformations of a liquid crystalline molecule. Because of number of lines in a spectrum is very high, the individual lines join to one broad, partly resolved line. In this case we can not solve the inverse problem of NMR spectroscopy, namely to calculate the parameters of spin–spin interactions of all pairs of spins from the spectrum. The direct problem of spectroscopy, calculation of NMR spectrum based on the known (from any model) parameters of spin interactions can be solved approximately.

Molecule of PAA is the simplest liquid crystalline molecule. It contains 14 protons. Dimension of Hamiltonian matrix for such a system is $2^{14} \times 2^{14}$. As it was stated in the Sect. 4.2.1, accurate diagonalization of matrix is possible for the spin systems containing 10–12 spins. Therefore, there are two directions of investigations of liquid crystals.

First direction is the simplification of the spin system with partial substitution of light Hydrogen with heavy Hydrogen (Deuterium). In first approximation, the dipole–dipole interactions of ^{1}H–^{2}H can be neglected, as they are about 6.5 times less than interactions of protons. Selectively deuterated samples can be obtained during the chemical synthesis if the deuterated components are used. For example, the partially deuterated in the methyl groups PAA contains only 8 atoms of ^{1}H in the benzene rings. Spectrum of such a system can be easy calculated. This is useful method because of researcher can also record deuterium spectra.

Second direction in the proton spectroscopy is the use of approximate methods of spectra calculation based on the simplification of spin Hamiltonian. One of the most accurate method of approximate calculation of proton NMR spectra was developed by A. V. Komolkin and Yu. V. Moltchanov (*Liquid Crystals*, **4**(2), 117–123 (1989)). The method is based on the subdivision of the total spin system into several subsystems (blocks). Size of blocks should be less than 10–12 spins in order we can calculate the spectrum of each block. Protons of different blocks should interact much weaker than inside each subsystem. Hamiltonian of dipole–dipole interaction for each block is written in the full form while interactions of spins belonging to different blocks accounting approximately. The method assumes "action" (but not interaction) of external protons on the protons of a block. If we choose as external spins any other species (for example, Fluorine or Deuterium), this method produces accurate (not approximate) result as for the system A_pX_q (see Sect. 4.2.3). Hamiltonian of k-th block is written as sum of two terms, first is Hamiltonian of the block and second is the Hamiltonian of spins of the block in the local magnetic fields of external spins:

$$\widehat{\mathcal{H}}_{\mathrm{d}}^{k} = -\sum_{i<j} 2D_{ij}\left(\hat{I}_{zi}\hat{I}_{zj} - \tfrac{1}{3}\hat{\vec{I}}_i \cdot \hat{\vec{I}}_j\right) - \sum_{i} \hat{I}_{zi}\left(\sum_{n} 2D_{in}m_n\right).$$

Here, indices i and j denote spins of the block and n denote external spins. Value m_n is projection of the external spin on axis Z.

In the low magnetic field (resonance frequency less than 100 MHz) chemical shifts does not distort the wide spectral line. Spectra are symmetric with respect to the centre. In the high field (frequency greater than 300 MHz) differences in chemical shifts of aliphatic and aromatic protons can lead to asymmetry of the spectra.

Interpretation of proton NMR spectra can provide complete information on structure and conformational mobility of liquid crystals because the spectra depends on interactions of all pairs of spins.

Deuterium is another important nucleus for investigation of liquid crystals. The deuterium spectrum consists of resolved doublets, each doublet belongs to one group of magnetically equivalent deuterons. For instance, methyl groups $-CD_3$ and $-CD_2-$, benzene rings. In the Fig. 6.4 the molecule of 4-n-pentyl-d_{11}-4′-cyanobiphenyl (also known as 5CB-d11) partially deuterated in the pentyl chain and deuterium spectrum are shown.

Temperature dependence of deuterium spectra provides relative temperature dependence of $S_{zz}(T)$. But it is impossible to find absolute values of S_{zz} and $S_{\mathrm{C-D}}$. So, for the structural investigation, it is not enough to use the only deuterium NMR spectroscopy.

Carbon-13 NMR spectroscopy is difficult because of researcher must apply radio pulses for the heteronuclear decoupling in order to eliminate influence of ^{1}H atoms on the spectra of ^{13}C. The difficulties is not in the NMR experimental methods. They are well developed and widely used for non-liquid-crystalline samples. The problem is to stabilize the temperature of liquid crystalline sample with high accuracy (better than 1°) and without temperature gradient inside the sample. During the decoupling pulse sequence, the radio coil heats and it can affect the temperature stabilization system.

Methods of double-quantum Carbon-13 NMR spectroscopy on the samples with natural abundance were developed by group of Malcolm H. Levitt (D. Sandström, M. H. Levitt, *J. Am. Chem. Soc.*, **118**(29), 6966–6974 (1996)). Because of natural abundance of this isotope is about 1 %, the only one of 400–700 molecules of liquid

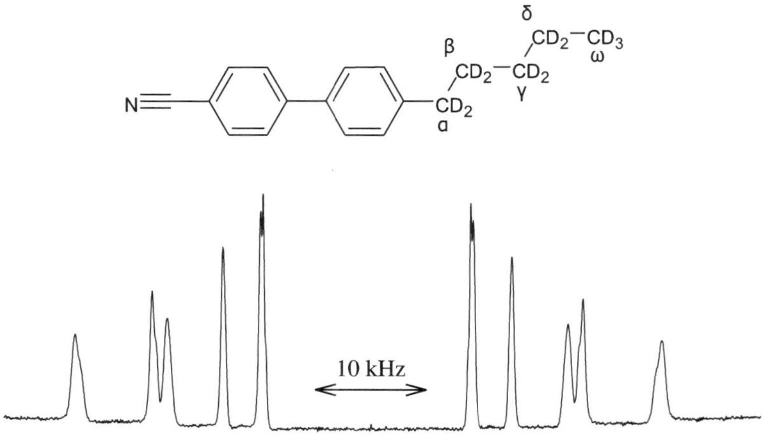

Fig. 6.4 Molecule of partly deuterated 5CB-d11 with designation of methyl groups and its deuterium NMR spectrum at 26.5 °C

crystal contains two atoms of this species. Double quantum spectrum is registered from such molecules. It consists of doublets. Additional methods of two-dimensional spectroscopy allow researcher to correlate the values of dipole–dipole interactions to chemical shifts of the atoms, that means to assign the values to pair of ^{13}C atoms. These are backbone atoms of the molecule, two nearest atoms have covalent link. So, the values of dipole couplings allow researcher to find average angle between the direction of this bond and the long molecular axis.

Methods of separated local field spectroscopy (SLF) including proton detected local fields (PDLF) were developed for the liquid crystals. They allow one to measure and to assign dipole–dipole couplings to ^{1}H and ^{13}C atoms (S. V. Dvinskikh, "Separated Local Field NMR Spectroscopy in Columnar Liquid Crystals" in: *Thermotropic Liquid Crystals. Recent Advances*, ed. by A. Ramamoorthy, Springer, 2007, p. 117–140).

6.4 Relaxation in Liquid Crystals

Molecular motions play a major role, on the one hand, in the formation of liquid crystals, and on the other, in the processes of NMR relaxation. Magnetic dipole–dipole and electric quadrupole interactions determine the relaxation processes in liquid crystals. In this book we consider dipolar mechanism of relaxation.

In the liquid crystals, there are some types of motions:

1. intramolecular (vibrations of the atoms along the chemical bonds, the conformational transitions);
2. molecular reorientations (rotations) around long and short molecular axes;
3. molecular diffusion corresponding to translational motion;
4. order director fluctuation.

The listed motions differ from each other with their rates. We can consider them as independent. The fastest motions are vibrations and conformational rotations. The correlation times of these motions are in the range of 10^{-15} to 10^{-13} s. Correlation times of molecular rotations around long and reorientations (jump-like) around short axes are of order of 10^{-11} and 10^{-8} s, respectively. Oscillation of the long molecular axis with respect to the director (rotation on the small angle around short axis which does not lead to reorientation of the molecule) has characteristic time of order of 10^{-10} s. Slowest motions are a group motions of the molecules. They include oscillation of director of small domain around the average director of the whole sample (correlation times are 10^{-8} to 10^{-6} s). The amplitude of these oscillations are not usually exceed 20°. Obtaining homogeneous orientation of the directors in the whole sample and reorientation of the director in the magnetic field of the spectrometer is the slowest processes with characteristic times from milliseconds to hundreds of seconds. However, they do not affect the relaxation process, as researcher begins the experiment only after these processes are completed.

We can consider the affect of the motions to the relaxation rate as following:

$$\frac{1}{T_1} = \frac{1}{T_{1,\mathrm{ODF}}} + \frac{1}{T_{1,\mathrm{SD}}} + \frac{1}{T_{1,\mathrm{R}}}, \tag{6.9}$$

where the terms are order director fluctuation (ODF), translational self-diffusion (SD) and rotational diffusion (R).

To estimate the individual inputs of these motions, it is necessary to carefully study the frequency and temperature dependencies of relaxation rates. The frequency range is from kilohertz to hundreds megahertz. Method of fast field cycling is suitable for this purpose. Spectrometer has air coil with low inductance. NMR experiment (preparation and free induction decay registration) is at the fixed frequency. But relaxation occurs at low frequency. The time of switching between two levels of magnetic field (and two resonance frequencies) is short relative to relaxation time. The levels of the magnetic field depend on electric current through the coil.

The basic principles of mechanisms of relaxation is described in Chap. 3. However, in liquid crystals translational motion is anisotropic, so it affect the process of relaxation.

Let us express the Hamiltonian of the dipole-dipole and quadrupole interactions (the only case of axially symmetric interactions with $\eta = 0$) in the form of irreducible tensor operators in spherical coordinates:

$$\widehat{\mathcal{H}}_\lambda = C_\lambda \sum_{p=-2}^{2} A_{20}^{p} \mathcal{D}_{p0}^{2}(\Omega) T_{2p}^{\lambda}, \tag{6.10}$$

where λ means type of interactions (dipole-dipole or quadrupole), C_λ is constant, A_{20}^{p} depends on the spacial coordinates of atoms, $\mathcal{D}_{p0}^{2}(\Omega)$ are the Wigner D-matrices, T_{2p}^{λ} is the spin part of the Hamiltonian.

If we consider the time-dependent part of the Hamiltonian, the following correlation functions can be obtained:

$$G_p(\tau) = \left\langle \left[\mathcal{D}^2_{p0}(\Omega, t) - \overline{\mathcal{D}^2_{p0}(\Omega, t)}\right]\left[\mathcal{D}^2_{p0}(\Omega, t+\tau) - \overline{\mathcal{D}^2_{p0}(\Omega, t)}\right]^*\right\rangle. \quad (6.11)$$

To obtain this equation we assume that time-dependent part of Eq. (6.10) is $\mathcal{D}^2_{p0}(\Omega)$, while conformational transitions (which affect A^p_{20}) are fast enough to influence on the relaxation.

Correlation function for the molecular reorientation depends on the model of motion (stochastic process, model of strong interactions, model of rotational diffusion etc.). Usually, to calculate $T^{-1}_{1,\mathrm{R}}$, the model of elongated ellipsoids is used. The model assume that correlation time of rotation around long axis of ellipsoid is much shorter then correlation time of rotation around short axis.

Translational diffusion in the liquid crystals is also anisotropic. Theory, initially proposed by H. C. Torrey for the intermolecular contribution to the dipole relaxation, was developed by S. Žumer and M. Vilfan. They obtained the following expression for the nematic phase:

$$\frac{1}{T_{1,\mathrm{SD}}} = \frac{9\gamma^2\hbar^2}{8}\frac{N_0\tau_\perp}{d^3} Q\left(\omega\tau_\perp, \frac{\langle r_\perp^2\rangle}{d^2}, \frac{D^0_{\parallel}}{D^0_{\perp}}\right), \quad (6.12)$$

where N_0 is the density of spins; $\langle r_\perp^2\rangle$ is average square of distance of "jump" of molecule in the perpendicular direction to the long molecular axis; d is diameter of the model cylindrical molecule; $D^0_{\parallel}$ and $D^0_{\perp}$ are diffusion coefficients in the homogeneously oriented sample in the parallel and perpendicular to the long molecular axis direction; $\tau_\perp$ is correlation time in the perpendicular direction and

$$\tau_\perp = \frac{\langle r_\perp^2\rangle}{4D^0_{\perp}}.$$

Function Q is dimensionless and has no analytical expression.

For the typical NMR frequency range ($\nu > 10\,\mathrm{MHz}$) Eq. (6.12) becomes similar to that in the Torrey's theory:

$$\frac{1}{T_{1,\mathrm{SD}}} = \frac{1}{1.4}\frac{1}{T_{1,\mathrm{inter}}}.$$

In the low frequency limit Eq. (6.12) has the form

$$\frac{1}{T_{1,\mathrm{SD}}} \sim \left[\mathrm{const} - (\omega_0\tau_\perp)^n\right],$$

and for the nematic phase $n = 1/2$ but for the smectic phase n varies between 1/2 and 1. For the extremely high frequency

$$\frac{1}{T_{1,\mathrm{SD}}} \sim \omega^{-2}.$$

Accounting the slow motion of the director leads to the following expression for the correlation function:

$$G(\tau) = AS^2 S'^2 \left\langle \delta\vec{n}(\vec{r}, t) \cdot \delta\vec{n}^*(\vec{r}, t + \tau) \right\rangle, \tag{6.13}$$

where director fluctuation $\delta\vec{n}$ is small and axial symmetric, the factor A depends on the orientation of a molecule, S and S' are the orientational order parameter (6.1) and the same parameter which describes the orientation of conformers. To calculate the correlation function (6.13), the superposition of the plane waves is used. Each wave has a mean square amplitude

$$\left\langle |\delta\vec{N}_i|^2 \right\rangle = \frac{kT}{Kl^2 V}$$

and time decay

$$\tau_l = \frac{\eta}{Kl^2}.$$

Here, K is elastic constant, η is macroscopic viscosity, l is the magnitude of wave vector, V is the volume of the sample. In this case

$$\langle \delta\vec{N}(\vec{r}, t) \cdot \delta\mathbf{N}^*(\vec{r}, t + \tau) \rangle = \frac{V}{(2\pi)^3} \int\limits_{l_1}^{l_2} \langle |\delta\vec{N}|^2 \rangle e^{-\tau/\tau_l} 4\pi l^2 dl \tag{6.14}$$

(l_1 and l_2 show the limits of wave lengths of fluctuation modes). Minimal length should be of order of molecular length. Maximal length defines by the radius of correlation of director. Fourier transform of the correlation function (6.13) with accounting of (6.14) results in spectral densities of director fluctuation. In the nematic phase usual result is

$$\frac{1}{T_{1,\mathrm{ODF}}} = \frac{\mathrm{const}}{\sqrt{\omega}}.$$

In smectic phase an expression for $T_{1,\mathrm{ODF}}$ is more complicated because we need to account different models of director fluctuation.

Another kind of fluctuations—order parameter fluctuation—appears in the temperature range of a few degrees below and above the transition point from nematic to isotropic phase.

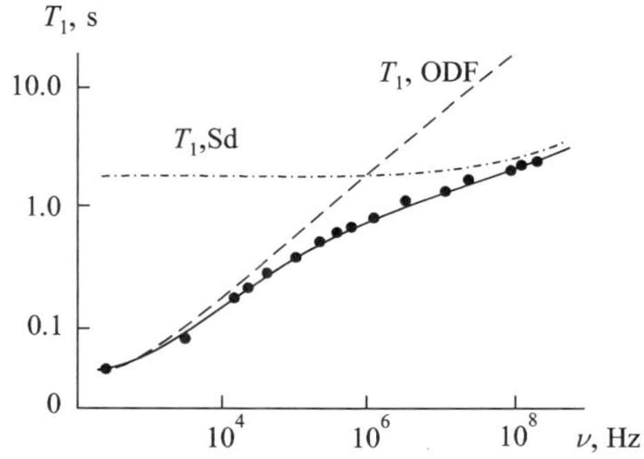

Fig. 6.5 Dispersion of relaxation time T_1 in PAA

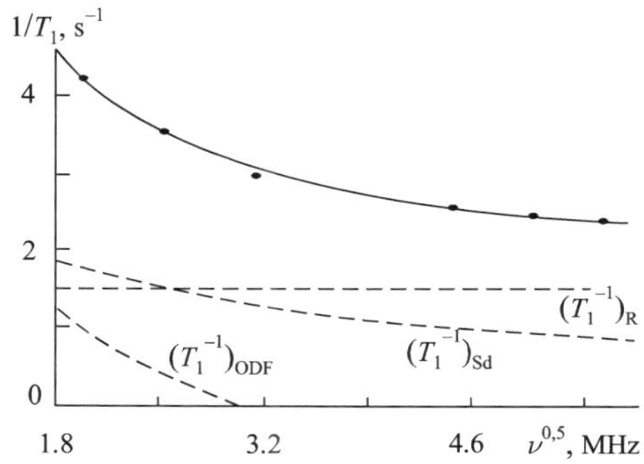

Fig. 6.6 Dispersion of relaxation rate T_1^{-1} in 4*O*.6

Examples of frequency dependencies of times and rates of proton spin–lattice relaxation are shown in Figs. 6.5 and 6.6. Data for the PAA is depicted in Fig. 6.5. On the low frequency (less than 1 MHz) dispersion is determined by the slow director fluctuation. In the range of "traditional" NMR frequencies (tens and hundreds MHz) relaxation is caused by several mechanisms. Similar results were obtained for other nematics. F. Noack and co-authors investigated in details the relaxation in liquid crystals with fast field cycling method using the partially deuterated samples. So, in the Fig. 6.6 dependence of relaxation rate T_1^{-1} is shown as function of square root of resonance frequency in the nematic phase of 4-n-butyloxybenzilidene-4' -n-hexylaniline (also known as 4*O*.6). Obviously, the director fluctuation affects on the relaxation rate on the frequencies less than 2 MHz. Rotational motion is constant in the investigated frequency rate, so characteristic frequencies of rotational motion is much higher.

In conclusion we should mention that some topics was not displayed here. They are:

1. methods of interpretation of spectra of samples with inhomogeneous orientation, for example, polymeric liquid crystals;

2. methods of investigation of small molecules dissolved in liquid crystals and, consequently, oriented by the nematic;
3. NMR of liquid crystals in zero external magnetic field.

These topics are reviewed in special issues on spectroscopy and relaxation in liquid crystals.

Further Readings

Burnell EE, de Lange CA (2003) NMR of ordered liquids. Kluwer Academic Publisher, Dordrecht ISBN 1-4020-1343-4

Chandrasekhar S (1992) Liquid crystals, 2nd edn. Cambridge University Press, Cambridge ISBN 0 521 42741 3

Dong RY (2010) Nuclear magnetic resonance spectroscopy of liquid crystals. World Scientific, Singapore ISBN 978-9814273664

Noack F, Schweikert KH (1994) Nuclear spin relaxation mechanisms in liquid crystals studied by field cycling NMR. In: Luckhurst GR, Veracini CA (eds) The molecular dynamics of liquid crystals., NATO ASI SeriesKluwer Academic, Dordrecht, pp 233–255 ISBN 978-94-010-4509-4

Chapter 7
Nuclear Magnetic Resonance in Magnetic Materials

Nuclear magnetic resonance in magnetically ordered materials was first observed by Gossard and Portis in the metallic cobalt on ^{59}Co nuclei.[1] Since then numerous studies of different magnetically ordered materials were performed. The main difference of the NMR in magnetically ordered materials from the NMR in diamagnetic materials is that it can be observed without an external magnetic field, because in magnetically ordered materials due to hyperfine interactions there are large and fairly homogeneous magnetic fields on the nuclei of paramagnetic ions.

Besides, interactions of the nuclear magnetic moments with the external alternating magnetic field are realized through electron shells of paramagnetic ions. That leads to a significant intensity increase in NMR signals. In real magnetically ordered materials the resulting alternating magnetic field at the nuclei has a very wide variation in magnitude, that, in particular, has a very significant impact on the shape of free precession and spin-echo signals.

All this leads to a significant simplification of the NMR equipment: there is no need in magnets, and very often to excite a pulse response of a system rf pulses of low amplitude are sufficient. Therefore, the NMR equipment is more compact.

Before discuss specific features of NMR in magnetics, let us remind the main properties of magnetically ordered materials and introduce necessary notions.

7.1 Magnetism and Magnetically Ordered Materials

Interpretation of magnetic phenomena in solids is based on two main conceptions: (i) ions in solids may have discrete magnetic moments; (ii) these magnetic moments interact with each other not only as usual magnetic dipoles (such interactions are too weak to be essential) but also as quantum mechanical particles, and hence are related by forces of quantum mechanical nature (exchange forces).

[1] A. C. Gossard and A. M. Portis, Observation of Nuclear Resonance in a Ferromagnet. Phys. Rev. Lett. **3** (1959) 164–166.

V. I. Chizhik et al., *Magnetic Resonance and Its Applications*,
DOI: 10.1007/978-3-319-05299-1_7,

An electron is a magnetism carrier, as besides a charge it has also a spin, and hence, a magnetic moment corresponding to it (the Bohr magneton $\mu_B = 9.274 \times 10^{-24}$ J/T). A complete description of the nature of magnetic moments of isolated atoms can be found in any textbook on atomic physics.[2] Briefly speaking, an atom has a non-zero magnetic moment if its inner d- or f-electron shells are not complete. That leads to the spin and orbital moments of electrons in the shell are not compensated. Such a situation can be realized if atoms belong to one of the following groups: the iron group ($3d$), the palladium group ($4d$), the platinum group ($5d$), the lanthanides ($4f$), and the actinides ($5f$). It is supposed that atomic magnetic moments still exist if atoms become ions in solids or form a metallic lattice. It is well known that in solids there is a strong interaction between valence electrons of atoms. However, d- and f-shells interact in different way. For example, the f-shell, which is localized and strongly bounded to the ion, is almost not affected by the influence of neighboring ions. And as a rule, the ion keeps its atomic characteristics. Hence, similar to isolated atoms, within the first-order approximation magnetic moments of ions can be considered as an integer number of Bohr magnetons. It is more or less fair for dielectrics, however in metals the situation is much more complex due to conduction electrons, which are delocalized and are an additional issue of magnetic moments. For more details we refer to any textbook on theory of magnetism.[3]

7.1.1 Exchange Interactions

A magnetic state that appears in a solid due to spontaneous magnetic moments of atoms (ions) is the result of several competitive agents. On the one hand, there are some ordering forces that are responsible for the mutual orientation of atomic magnetic moments in solids. But on the other hand, thermal motion impedes the ordering. It is possible to divide all types of magnetism into two classes: noncooperative and cooperative. In the first case, magnetic moments of individual ions do not "feel" each other and behave independently, an ordering state is created only by an external magnetic field. In the seconde case, ordering is the result of exchange interactions between magnetic moments of ions, and the external magnetic field is only a tool to detect the magnetic ordering that already exists.

Exchange interactions between magnetic moments in substances with cooperative magnetism can be classified depending on several factors, but all of them are a consequence of the Pauli exclusion principle. In Fig. 7.1 we represent a schematic classification of exchange interactions. A *direct* (or *contact*) *exchange interaction* is realized between magnetic moments of ions, which are located sufficiently close to each other (their wave functions overlap). This is a very strong interaction that rapidly

[2] See, for example, M. Born. Atomic Physics: 8th Edition. Dover Publications, New York, USA (1969).

[3] See, for example, K. Yosida. Theory of magnetism. Springer-Verlag, Berlin Heidelberg, Germany (1996).

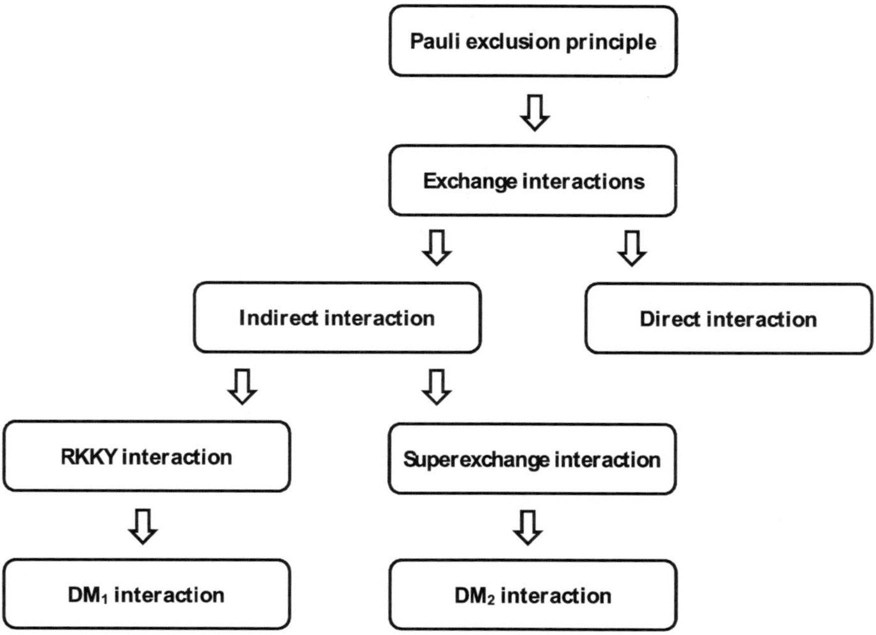

Fig. 7.1 Classification of exchange interactions. Reproduced with permission from Contemp. Phys. 23, 469 (1982). Copyright 1982, Taylor and Francis Ltd

decreases with the interionic distance. This is also the interaction that occurs between magnetic moments of different orbitals of the same ion. An *indirect exchange interaction* relates magnetic moments located relatively distant from each other, so their wave functions do not overlap. The exchange is realized through a mediator that can be either collective electrons (in metals) or nonmagnetic ions (in dielectrics). Such an exchange is called the *Ruderman–Kittel–Kasuya–Yosida* (RKKY) interaction, in the first case, or the *superexchange* (or Kramers–Anderson superexchange) interaction, in the second case.

It is easy to show that the interaction between two magnetic moments (or spins) depends on their relative orientation. It can be represented as an isotropic interaction of two spins, $\hat{\vec{S}}_i$ and $\hat{\vec{S}}_j$, which only depends on the interionic distance r_{ij}. Then, the exchange interaction energy of these two spins is

$$E_{\text{ex}}^{ij} = -J(r_{ij})\hat{\vec{S}}_i\hat{\vec{S}}_j, \tag{7.1}$$

where $J(r_{ij})$ is the so-called exchange interaction parameter. In solids it is necessary to sum over all pair interactions:

$$E_{\text{ex}} = -\sum_{i,j} J(r_{ij})\hat{\vec{S}}_i\hat{\vec{S}}_j. \tag{7.2}$$

For a direct exchange interaction that occurs between two electrons of the same atom, the J parameter is always positive. However, if it is an exchange between two different ions it can be either positive or negative. For indirect interactions it also can be of different sign (in metals, where the RKKY interaction is realized, it is an oscillating function of r_{ij}).

However, Eq. (7.2) does not take into account the anisotropy of exchange interactions. Such an anisotropy appears due to interactions of ionic magnetic moments with the crystal lattice (see Sect. 7.1.3), and there are two main issues of it: the anisotropy of geometrical locations of the neighboring ions, which creates an anisotropic crystal field, and the spin–orbit interaction. In the simplest case, for an axial magnetic and neglecting spin–orbit effects, the resulting energy of the anisotropy and exchange interactions can be written as

$$E = -\sum_i D_i(\hat{\vec{S}}_i)^2 - \sum_{i,j} J(r_{ij})\hat{\vec{S}}_i\hat{\vec{S}}_j, \tag{7.3}$$

where D is the intensity of the axial crystal field.

A strong spin–orbit interaction may lead to a so-called *Dzyaloshinskii–Moriya* (DM) interaction, which, in its turn, results to non-collinear spin configurations. The coupling can be also realized through either collective electrons or nonmagnetic ions, in Fig. 7.1 these two varieties of the DM interaction are denoted as DM_1 and DM_2, respectively. If there is a DM interaction between magnetic moments, the energy (7.1) in Eq. (7.2) must be replaced by

$$E_{\mathrm{DM}}^{ij} = d_{ij}(\hat{\vec{S}}_i \times \hat{\vec{S}}_j), \tag{7.4}$$

where d_{ij} is a coupling constant.

7.1.2 Types of Magnetism

Despite magnetism is essentially a quantum-mechanics phenomenon, and up to now the nature of several types of magnetism is still debatable, the types of magnetism can be easily classified. There are five main classes of magnetism: diamagnetism, paramagnetism, ferromagnetism, antiferromagnetism and ferrimagnetism. Let us briefly describe all of them.

Diamagnetism. The origin of the diamagnetism lie in a shielding effect due to the electron motion: the orbital electron motion or conduction electron current in metals. The latter is a source of the Landau diamagnetism.

An ideal diamagnetism is a noncooperative phenomenon, which is characterized by a small negative magnetic susceptibility χ, that does not depend on temperature. It is worth noting that a diamagnetic component exists in all materials, but in magnetic materials it is usually masked by more intensive interactions. Examples of diamagnetic substances are crystalline NaCl, Cu, Al_2O_3 or amorphous SiO_2.

Paramagnetism. It is another example of noncooperative magnetism, which appears due to ions have spontaneous magnetic moments. In an ideal paramagnetic all these moments are identical, localized in an isotropic medium ($D = 0$ in Eq. (7.3)) and are sufficiently distant one from another to be independent. Each magnetic moment is affected, on the one hand, by a disorienting thermal motion characterized by $k_B T$, and on the other hand, by an ordering action of the external magnetic field. Hence, magnetic ordering is the result of the orientation of magnetic moments under the effect of the external magnetic field despite the thermal disorder. An ideal paramagnetism is characterized by a positive magnetic susceptibility χ, which linearly depends on the inverse temperature (the Curie law): $\chi = C/T$, where C is the Curie constant.

However, in practice, in solids there are always noticeable exchange interactions between magnetic ions. The Curie law is a particular case of the more general Curie–Weiss law: $\chi = C/(T - \Theta)$, where Θ ia a constant characterizing interionic interactions. The Curie–Weiss law takes into account a cooperative magnetism, which may dominate if the thermal motion energy is sufficiently low. With temperature decreasing cooperative effects become more and more important and below a certain ordering temperature another magnetic state appears. If this state is ferromagnetic or ferrimagnetic, the ordering temperature is called the Curie temperature (T_C); for an antiferromagnetic state this is the Neel temperature (T_N) etc.

Ferromagnetism. It is a cooperative effect, which is characterized by a collinear alignment of atomic magnetic moments with a long-range order. A magnetization exists even without an external magnetic field. In an ideal ferromagnetic all ions have identical spontaneous magnetic moments and occupy identical crystallographic sites. The exchange interaction dominates the magnetic crystallographic anisotropy, $|J| \gg |D|$, and $J > 0$. To minimize the magnetostatic energy a ferromagnetic crystal usually divides into domains, each of which is magnetized along the easy magnetization axis defined by the D value. The external magnetic field may change the domain sizes, increasing those ones that are oriented along the external field however, it almost does not affect the magnetization value. In ferromagnetics the external magnetic field only helps to reveal the magnetic ordering that already exists.

In any cooperative magnetic state with temperature increasing one can achieve such a temperature value, at which the thermal motion energy, at first, becomes comparable with the exchange energy, and then exceeds it. In a ferromagnetic it occurs at the Curie temperature T_C. The spontaneous magnetization decreases with temperature increasing and disappears at $T = T_C$. Above T_C, a ferromagnetic becomes a paramagnetic and obeys the Curie–Wess law: the inverse of the susceptibility, $1/\chi$, varies linearly with temperature but does not tend to zero at zero temperature. Despite the fact that ferromagnetic materials are known for a long time, there are only few pure substances, which are ferromagnetic: three elements of the iron group (Fe, Co, Ni) and rare earth elements (Gd, Tb, Dy etc.). The majority of ferromagnetic materials is alloys with better magnetic characteristics than the pure iron. There are also non-metallic ferromagnets, such as EuS and similar compounds with sulfur or selenium, and chalcogenides.

Antiferromagnetism. Similar to ferromagnetism it is a cooperative effect, which is characterized by a long-range order. In an ideal case, magnetic ions occupy crystallographically identical sites. Exchange interactions dominate over disturbances, $|J| \gg |D|$, but $J < 0$. As a result the magnetic moments of neighboring ions are ordered antiparallel and the resulting magnetization is zero. In other words a simple antiferromagnetic can be represented as an assemblage of two interpenetrating identical ferromagnetic lattices, with antiparallel magnetizations. A simple antiferromagnetic can not be amorphous, as it is impossible to divide an amorphous material on two identical sublattices. It is worth noting that ordering of magnetic moments in an atiferromagnetic may be ambiguous. For example, in a body centered cubic lattice there are two sorts of ordering, in a face centered cubic lattice there are even more.

Antiferromagnetism was predicted by Landau in 1933 and just after Neel developed the theory of antiferromagnetism. However, experimentally, antiferromagnetism was first observed only in 1938 in MnO by Bizette, Squire and Tsai.[4] By now about several hundreds of antiferromagnetic compounds are known. The majority of them are ionic compounds like FeO, CoO, MnS etc., however there are antiferromagnetic elements, such as Cr and α-Mn, several light lanthanides like Sm, and ordered alloys (CrPt, Fe_3Mn etc.). The temperature dependence of $1/\chi$ of antiferromagnetics is rather specific. Above the ordering temperature T_N an antiferromagnetic becomes a paramagnetic and $1/\chi$ follows the Curie–Weiss law, but below the Neel temperature, the dependence is nonlinear: the susceptibility nonlinearly decreases with temperature decreasing as the long-order becomes stronger.

Ferrimagnetism. It differs from other main classes by the fact that in a ferrimagnetic material there are two and more different magnetic species. They can be ions of the same substance but with different valence (for example, Fe^{2+} and Fe^{3+}) or ions of different elements (for example, Gd and Fe). The mostly studied ferrimagnetics are ferrites with the spinel structure ($MgAl_2O_4$). Their chemical formula can be written as MFe_2O_4, here M is a two-valence metal ion (usually of the iron group: Mn^{2+}, Fe^{2+}, Co^{2+}, Ni^{2+}, Cu^{2+}, or even Mg^{2+}, Zn^{2+} and Cd^{2+}) with the ion radius $0.6 \div 0.9$ Å, the iron ions in this structure have a valence 3+.

Magnetic ions in a crystalline ferrimagnetic occupy two (or more) kinds of lattice sites, A and B, which have different magnetic environment. Within each sublattice all magnetic moments are ordered ferromagnetically, whereas magnetic moments of the sublattices A and B are antiparallel to each other. As soon as generally the number of A and B sites is not the same, and moreover, the values of the ion magnetic moments at the A and B sites are different, there is a spontaneous magnetization in the crystal.

Temperature dependencies of magnetic properties of a ferrimagnet are rather similar to those in a ferromagnet, but the spontaneous magnetization decreases more rapidly and $1/\chi$ above the Curie temperature in a paramagnetic region increases with temperature, but not linearly.

[4] H. Bizette, C. F. Squire, and B. Tsai. Le point de transition X de la susceptibilite magnetique du protoxyde de manganese, MnO. Compt. Rend. **207** (1938) 449.

Opposite to antiferromagnets, there are amorphous ferrimagnets, where ions of A and B sublattices occupy randomly distributed sites. Examples of amorphous ferrimagnetic materials are $REFe_2$ alloys (with RE = Tb or Gd) with ferromagnetic Fe–Fe and RE–RE interactions, and antiferromagnetic Fe-RE interactions.

For a long time the magnetism have been considered as a problem of the ordering of magnetic moments located in sites of regular crystal lattices. The situation was changed in 1970-th due to dramatic growth of theoretical and experimental studies of two related systems: amorphous solids, which do not contain any pair of equivalent atomic sites, and disordered solids, where atoms randomly occupy crystallographic sites. That resulted in discovering of new types of magnetic ordering, which appear in systems without any long-range order: speromagnetism, asperomagnetism, mictomagnetism, helimagnetism, spin glasses, superparamagnetism etc. A brief but rather comprehensive review of the nature of different types of magnetism and their interrelationship was made by.[5]

7.1.3 Magnetic Crystallographic Anisotropy and Domains

In a rather small monocrystal the direction of the magnetic moment of a unity volume ($\vec{M}_e$) is the same over the sample, and is absolutely defined for a given crystal. The matter is that the interaction with neighboring ions is not reduced to exchange interactions only. Due to spin–orbit interactions magnetic moments are oriented relative to the electron orbits in such a manner to minimize the electron energy. As the distribution of the electron orbits is determined by the crystal structure, magnetic moments of ions are also oriented relative to the crystallographic axes by a definite way. This orientation is not always unique. Sometimes there are several directions strictly or approximately equivalent. It is to be kept in mind that a simultaneous change in directions of both spin and orbital moments to their opposite values does not change the electron energy. If one finds the direction of the magnetic moment corresponding to the maximum and minimum energy and calculates their difference, one can get a so-called *magnetocrystalline anisotropy energy* or *anisotropy energy* (E_a). The directions that correspond to the minimal energy are the *easy magnetization axes*, whereas the directions that corresponds to the maximal energy are the *hard magnetization axes*. Crystals with one easy magnetization axis, such as Co, are called the crystals of the *easy axis* type. Cubic crystals usually have three easy magnetization axes. In α-Fe the easy magnetization axes are the directions $< 100 >$, whereas in Ni they are the $< 111 >$ axes. There are also crystals where magnetization axes form a plane or a cone, so-called *easy plane* or *easy cone* type crystals, respectively.

A uniformly magnetized crystal creates around itself a large magnetic field of a significant energy reserve. It is a so called *demagnetization energy*. Such a state is energetically unfavorable for the crystal, so in the absence of a large external magnetic field, it spontaneously breaks up into regions (domains) with different directions of the magnetic moment. Within each domain the magnetic moments

[5] C. M. Hurd. Varieties of magnetic order in solids. Contemp. Phys. **23** (1982) 469–493.

usually direct along one of the easy magnetization axes. If the crystal has only one easy magnetization axis, the magnetic moment in the domains is oriented either along or against the axis. This rule can be violated only in a few domains, and only if the increase in the anisotropy energy is compensated enough by decreasing in the demagnetization energy.

At the boundary between the domains a smooth rotation of the magnetization vector occurs: from its direction in one domain to its direction in another one. This results in reducing the exchange energy, since if the boundary is sharp, the mutual orientation of the nearest paramagnetic ions from different domains does not correspond to the minimal exchange energy. But the presence of domain walls leads to an increase in the anisotropy energy. Because of this despite reducing the size of domains helps to low the demagnetization energy, it is not energetically favorable for a crystal to break up into very small domains with a lot of domain walls. As a result, the number of domains and the thickness of domain walls are such that the total energy of the exchange interaction, anisotropy and demagnetization is minimal. If inside a domain wall the magnetic moment direction lies within the wall plane, such a domain wall is called the *Bloch wall*. There is also the *Neel wall*: the magnetization rotates not within the wall plane but perpendicular to it.

For bodies of simple geometric shapes (such as thin films) and ideal (without defects) crystals, it is possible to develop a rather simple and reliable model of domain structures. However, it should be kept in mind that if the domain size, thickness and shape of domain walls can be calculated, it is difficult to establish a specific location of boundaries between domains, as it is always defined by defects.

Domain walls have a rather great influence on NMR phenomena in magnetics. Very often despite the fact that the number of nuclei in the domains is much greater than in the domain walls, the signal from nuclei in domain walls is more intensive. In calculations, where the structure of domain walls is important, one usually uses the Bloch wall model (it will be discussed in more details in Sect. 7.5). The structure of such a wall is determined by the minimum of the exchange and anisotropy energies.

7.2 Local Fields

The *local field* is the field at the nucleus site. In magnetically ordered materials this field on the nuclei of paramagnetic ions is mainly due to hyperfine interactions, which are interactions of the nucleus magnetic moment with the magnetic moment of its electron shell. The value of the local field can be rather high. For example, for nuclei of the iron group ions it is tens of Tesla, whereas for rare-earth ions it may achieve hundreds of Tesla. Therefore, even for nuclei with a relatively small nuclear gyromagnetic ratio the NMR frequency may lie within a rather wide range from tens to thousands of megahertz.

The mechanism of the appearance of local fields is quite complex. In the absence of an external magnetic field completely filled electron shells of ions do not create magnetic fields at the nucleus sites, because both orbital and spin magnetic moments are zero. In a paramagnetic ion there is always a partially filled shell (the $3d$-shell for ions of the iron group, the $4f$-shell for rare-earth ions, etc.) and, therefore, there is a nonzero magnetic moment. In addition, in paramagnetic ions a partially filled electron shell, which has its own mechanical and magnetic moments, polarizes all other shells. It means that their wave functions are not identical for oppositely directed spins. For all shells with $l \neq 0$ (where l is the orbital spin number) it is difficult to separate this field, as it is undistinguishable from the field created by $3d$- or $4f$-electrons. Nevertheless, the s-shell polarization leads to a qualitatively different result, as s-shells are spherically symmetric, and their orbital magnetic moments are zero. However, as the wave functions of s-electrons (ψ_s) are not equal to zero at the nucleus site and have different space distribution due to their interaction with uncompensated spins of unfilled shells, a nonzero spin density at the nucleus site appears. This is the issue of the so-called *Fermi contact interaction*. The local field created by such an interaction is proportional to $|\psi_{s\uparrow}(0)|^2 - |\psi_{s\downarrow}(0)|^2$ (here, the arrows indicate the spin directions). It is to be noted that both inner and external[6] shells, with respect to the partly filled shell, are polarized. Moreover, it results in the appearance of some additional spin density on the external, earlier empty, s-shells. Local fields created by different s-shells can be of different sign and partly compensate each other. Nevertheless, due to a rather great value of the Fermi contact interaction the contribution of s-electrons into the local field usually dominates (this is the case for $3d$-elements). As a rule, the total field created by s-shells is of the sign opposite to the field created by the partly filled shell.

For ions of the iron group there is a *quenching of the orbital angular momentum*. Electric fields generated by neighboring paramagnetic ions result in splitting of the energy levels, which were degenerate. This splitting is much larger than kT, and wave functions of the populated lower levels may not have the orbital angular momentum. In this case only the spin magnetic moment is is significant. However, a complete freezing is very rare: due to the spin-orbit coupling the effective magnetic moment is still different from the pure spin one. That is why very often one introduces some *effective spin* and *g-factor*, which determines the value of the energy level splitting in an external magnetic field.

Ions of the iron group with five $3d$-electrons, Fe^{3+} and Mn^{2+}, represent a very interesting example. In these ions the spin magnetic moments are parallel, whereas the orbital magnetic moment even in a free ion is equal to zero. In this case the electron density distribution is spherically symmetric and such an electron shell does not create any local field. Hence, the local field is completely due to polarized s-shells. According to calculations, the main contributions are from the $2s$- and $3s$-shells. For the Mn^{2+} ion, the field created by $1s$-, $2s$- and $3s$-shells are -3, -140 and $+74$ T, respectively. It results in the total field of -69 T, that is in fair agreement with experimental data (-65 T). And the values of these fields are very sensitive to changes in wave functions. In rare earth ions an unquenched orbital angular momentum of the $4f$-shell creates much higher local fields. In metals conduction electrons are also polarized, and hence, local fields, as well as NMR frequencies, are very different from those ones in dielectrics.

---

[6] For the ions of the iron group this is the $4s$-shell.

The part of the hyperfine field that is created by the spherically symmetric s-shells is called the *isotropic hyperfine field*, the other shells create the *anisotropic field*.

Another part of the local field is the *dipole field* ($\vec{B}_d$), i. e. the field created by all neighboring ions, which have magnetic moments. To calculate the dipole field it is necessary to summarize contributions from all ions of the crystal. It is possible to calculate this field precisely within a sphere circled about the nucleus under study. The contribution of outer dipoles is possible to estimate supposing that the rest of the crystal can be represented as a continuum with a magnetization, and the crystal itself has a shape of an ellipsoid. For a crystal of an arbitrary form the internal field is inhomogeneous, that leads to a dispersion of both local field values and their directions. In an ellipsoid the local field is the same in all the part of the sample. Particular extreme cases of an ellipsoid are a sphere, a thin plate and a long thin cylinder. In multidomain samples to calculate the local dipole field it is necessary to know in details the domain structure of the sample. However, this field from distant parts of the sample is rather small, as the contributions from different domains are mutually compensated.

In magnetically ordered materials there are also rather important induced magnetic fields on nuclei and non-magnetic ions. They appear due to overlapping of wave functions of non-magnetic ions with those ones of paramagnetic ions that polarize electron shells. This polarization can be high enough: about 0.1 % for s-shells and a few percent for others. Therefore, local fields at nuclei of non-magnetic ions is only one order lower than those ones at magnetic ions. A similar phenomenon is observed in magnetic metals. Here the connection is via conduction electrons, and its effects can be estimated by observing NMR on nuclei of nonmagnetic ions in alloys.

The hyperfine interaction of a nucleus spin with the electron shell of a free ion can be represented by the following Hamiltonian:

$$\widehat{\mathcal{H}}_{\rm hf} = \hat{\vec{I}}\,\mathsf{A}\,\hat{\vec{J}}, \tag{7.5}$$

where A is the hyperfine interaction tensor; $\hat{\vec{J}}$ is the operator of the total moment of momentum of the electron shell; $\hat{\vec{I}}$ is the operator of the nucleus spin. For elements of the iron group due to the quenching of the orbital angular momentum $\hat{\vec{J}}$ should be replaced by the operator of the spin moment of momentum $\hat{\vec{S}}$. In many cases, when the isotropic hyperfine interaction is much greater than the anisotropic one, A can be considered as a constant. As the nucleus ($\vec{M}_n$) or electron ($\vec{M}_e$) magnetizations are proportional to the corresponding moments of momentum let us, supposing that A is a constant, rewrite Eq. (7.5) in the following form:

$$\widehat{\mathcal{H}}_{\rm hf} = A_0 \vec{M}_e \cdot \vec{M}_n, \tag{7.6}$$

where $A_0 = A/(N\gamma_e\gamma\hbar^2)$; N is the number of magnetic moments in the unit volume; γ_e is the electron gyromagnetic ratio.

As the operator of the interaction between the nuclear magnetization and the magnetic field can be written as

$$\widehat{\mathcal{H}} = -(\vec{M} \cdot \vec{B}), \tag{7.7}$$

then by comparing the Hamiltonian operators (7.6) and (7.7) for the hyperfine contribution of the local field one has

$$\vec{B}_{\mathrm{hf}} = -A_0 \vec{M}_{\mathrm{e}}, \tag{7.8}$$

and the total local field is

$$\vec{B}_{\mathrm{loc}} = \vec{B}_{\mathrm{hf}} + \vec{B}_{\mathrm{d}}.$$

If there is an external magnetic field, it is necessary to add the corresponding term.

At low temperatures thermal motions lead to consistent oscillations of the ion magnetic moments and hence the magnetization can be well described within the theory of spin waves. At temperature close to the Curie point, these fluctuations become more chaotic. But in any case, with temperature increasing the average magnetization becomes smaller. Since frequencies characterizing the magnetization motion are rather large, as compared to NMR frequencies, nuclei "feel" only the constant component of the magnetization, i.e. its average value. So in effect, only the average value of the local field, determined by the average value of the magnetization, is significant:

$$\langle \vec{B}_{\mathrm{loc}} \rangle = -A_0 \langle \vec{M}_{\mathrm{n}} \rangle + \langle \vec{B}_{\mathrm{d}} \rangle.$$

If one supposes that (i) the studied crystal is ideal (free of defects and impurities), single-domain and of an ellipsoidal shape; (ii) the nuclei, on which NMR is observed, occupy equivalent sites in the crystal, then the NMR spectrum consists of only one line, the width of which is determined by relaxation processes. Various impurities, domain structure, porosity of the sample, a deviation from the ellipsoidal shape lead to a broadening of the spectral line. Therefore, usually the line width of the NMR spectrum is much greater than the line width defined by relaxation processes. For example, in cobalt films the spectral line width is about 10 MHz, and is non-uniform, whereas the homogeneous width of the spectral line is only 15 kHz.

In ferrimagnets and antiferromagnets there are at least two sublattices. In antiferromagnets, as the two sublattices are identical, without an external magnetic field there is only one NMR spectral line. However, it is difficult to observe. Canted antiferromagnets are an exception: the intensity of the NMR signal is rather high. Application of the external magnetic field, whose vector is parallel to one of the magnetization vectors of the sublattices, leads to splitting of the NMR line. In ferrimagnets, as the sublattices are nonidentical, there are always at least two NMR lines.

By measuring local fields it is possible to determine the spin density at the nucleus site (for each sublattice) and to compare it with theoretical calculations.It is also very important to know the temperature dependence of the NMR frequencies (and hence, of the local fields), especially at low temperature and near the Curie temperature, as it helps to verify the theory of spin waves.

7.3 Mechanisms of the rf Field Enhancement in Domains

A direct impact of the external rf field on the nuclei of paramagnetic ions is relatively small. Quantum resonance transitions of nuclear spins are induced by the variable component of the local field rather than by this external rf field. The latter is a result of the influence of the external magnetic field on magnetic moments of paramagnetic ions. The ratio of the amplitude of the alternating field acting on the nucleus to the amplitude of the external rf field is called the *enhancement factor*, usually denoted by η.

There are two principle mechanisms, which define the η value: one acts in domains, another acts in domains walls.

In domains in the absence of an external influence, the electron magnetization vector ($\vec{M}_\mathrm{e}$) is settled along one of the easy magnetization axes. Let us consider the simplest case: a ferromagnetic with one easy magnetization axis. A derivation of the magnetization vector from this axis by the angle θ results in increasing the anisotropy energy. This energy can be written as follows:

$$E_\mathrm{a} = K_1 \sin^2\theta + K_2 \sin^4\theta, \tag{7.9}$$

$$K_1 + K_2 > 0.$$

If the deviation is small, the angle θ is small and $\sin\theta$ can be expanded in a series. Then, keeping only quadratic terms in θ, one has

$$\Delta E_\mathrm{a} = K_1\theta^2. \tag{7.10}$$

The same dependence can be obtained if one supposes that the electron magnetization interacts with a fictitious field, the so-called *anisotropy field* $\vec{B}_\mathrm{a}$. In a general case

$$E_\mathrm{a} = -(\vec{M}_\mathrm{e}\cdot\vec{B}_\mathrm{a}) = -M_\mathrm{e}B_\mathrm{a}\cos\theta. \tag{7.11}$$

At small θ values it is possible to expand $\cos\theta$ in a Taylor series, then

$$E_\mathrm{a} = -M_\mathrm{e}B_\mathrm{a}(1-\theta^2/2),$$

$$\Delta E_\mathrm{a} = M_\mathrm{e}B_\mathrm{a}\theta^2/2. \tag{7.12}$$

It is obvious that if

$$B_\mathrm{a} = 2K_1/M_\mathrm{e}, \tag{7.13}$$

then Eqs. (7.10) and (7.12) are equivalent.

It should be stressed that one can introduce the anisotropy field only if θ is small. Special attention should be paid to the fact that from Eq. (7.9) it follows that $E_a(0) = E_a(\pi)$. However, it can not be obtained from Eq. (7.11). In other words, the requirement of smallness of the θ angle, when introducing the anisotropy field, is not connected with neglecting the higher-order terms in the expansion of $\cos\theta$ in a Taylor series.

Thus, if in different domains magnetization vectors are in opposite directions, but along the easy axis, the anisotropy energy is the same. To be able to use the concept of the "anisotropy field" one should always assume that the direction of its vector and the equilibrium magnetization vector coincide.

Motion of the electron magnetization vector can be described by the well-known equation:

$$\frac{\mathrm{d}\vec{M}_e}{\mathrm{d}t} = -\gamma_e[\vec{M}_e \times \vec{B}]. \tag{7.14}$$

The electron gyromagnetic ratio γ_e is negative, but here it is positive and its sing is written in the explicit form. Supposing that $\vec{M}_e$ is under the influence of two fields, the anisotropy field $\vec{B}_a$ directed along the z-axis and the alternating field $\vec{b}e^{i\omega t}$ directed perpendicular to $\vec{B}_a$, let us looks for the solution of Eq. (7.14) in the form of

$$\vec{M}_e = \vec{M}_{e=} + \vec{M}_{e\sim}e^{i\omega t}.$$

Let $M_{e\sim} \ll M_e$ and $b \ll B_a$. Within the zero-order approximation supposing $b = 0$ and $M_{e\sim} = 0$, one has

$$[\vec{M}_{e=} \times \vec{B}_a] = 0.$$

It means the vectors $\vec{M}_{e=}$ and $\vec{B}_a$ are parallel. The value $\vec{M}_{e=}$ is equal to M_{e0} (where M_{e0} is the equilibrium magnetization value). Within the first-order approximation, after substituting into Eq. (7.14) the sum of the constant and alternating magnetizations instead of $\vec{M}_e$, and the sum of the anisotropy and alternating rf fields instead of $\vec{B}$, keeping terms of the first infinitesimal, one has:

$$i\omega\vec{M}_{e\sim} + \gamma_e[\vec{M}_{e\sim} \times \vec{B}_a] = -\gamma_e[\vec{M}_{e0} \times \vec{b}].$$

Assuming that the $\vec{b}$ vector directs along the x-axis and opening the vector product one writes:

$$i\omega M_{ex} + \Omega_0 M_{ey} = 0, \tag{7.15}$$

$$-\Omega_0 M_{ex} + i\omega M_{ey} = -\gamma_e M_{e0} b_x, \tag{7.16}$$

here M_{ex} and M_{ey} are the projection of the $\vec{M}_{e\sim}$ vector on the x- and y-axes, respectively; $\Omega_0 = \gamma_e B_a$ is the precession frequency of the electron magnetization vector.

Since in observing NMR signals the ω frequency is close to the resonance frequency, which is usually much less than the precession frequency of the electron magnetization vector, supposing $\omega \ll \Omega_0$, from Eq. (7.15) one obtains that $M_{ey} \ll M_{ex}$, and hence one can consider $M_{ey} = 0$. Then from Eq. (7.16) it follows that

$$M_{ex} = M_{e0} b_x / B_a \quad \text{or} \quad M_{ex}/M_{e0} = b_x / B_a.$$

Hence the summary vector of the electron magnetization is parallel to the summary vector of the magnetic field. Therefore, one can assume that under the effect of the rf field with the frequency much lower than the precession frequency of the electron magnetization this electron magnetization follows the vector of the summary magnetic field.

Under the effect of the alternating component of the electron magnetization, which acts through the hyperfine interaction, an alternating component of the local field $B_\sim$ appears (see Eq. (7.8)):

$$B_\sim = -A_0 M_{ex} = -A_0 M_{e0} \frac{b_x}{B_a}.$$

Hence, the enhancement factor, which is defined as

$$\eta = B_\sim / b_x,$$

is equal to

$$\eta = -A_0 M_{e0} / B_a.$$

The expression in the numerator is the local field. Then

$$\eta = B_{loc} / B_a. \tag{7.17}$$

Despite Eq. (7.17) is derived for ferromagnets, it can be applied to ferrimagnets and even to canted antiferromagnets.

Usually the enhancement factor in domains is not very large, $\eta = 10$–10^3. However, there are materials with very low E_a value. For example, in the canted antiferromagnet $FeBO_3$ it is only 0.5 G, and η achieves the value of about 10^6.

If the rf vector directs arbitrarily with respect to the anisotropy field vector, in a crystal with one easy magnetization axis only the project of the rf field perpendicular to $\vec{B}_a$ is essential. Such a situation inevitably appears in polycrystalline samples. In this case formally one can suppose that for "unsuccessfully" placed domains the enhancement factor is lower and can be determined as $\eta = \eta_0 \sin \alpha$, where α is the angle between the directions of the anisotropy field vector and the alternating rf field vector; η_0 is the enhancement factor for domains where $\alpha = \pi/2$.

The enhancement factor determined in such a manner turns out to be different for differently directed domains and loses its physical meaning. There is a dispersion of the enhancement factor over the sample. However, there is another more important issue of this dispersion. Values of the anisotropy field induction listed in reference books, in fact, are average values that reflect how easy it is possible to deviate the magnetization vector of all the sample from its equilibrium state, in which the magnetization vector directs along the anisotropy field. In real crystals the anisotropy fields is not the same in different areas of the sample, there is a dispersion in both values and directions. They are so-called *local anisotropy fields*. And in the expression for the enhancement factor one should put these local anisotropy fields. That is why there is always a dispersion of the η values over the sample that leads to very interesting specific features in free precession and spin echo signals (see Sect. 7.6). In cubic crystals, where in different domains magnetization vectors direct along different easy magnetization axes, the angle α is always different.

In crystals with an easy magnetization plane the anisotropy energy slightly depends on the nuclear magnetization vector orientation in this plane. To take it into account one should add to Eq. (7.9) another term. In hexagonal crystals, which has the six-fold symmetry axis (for an easy plane crystal it is the hard magnetization axis) Eq. (7.9) transforms into

$$E_a = K_1 \sin^2\theta + K_2 \sin^4\theta + K_3 \sin 6\phi$$

(ϕ is the azimuth angle in the plane perpendicular to the six-fold axis). If the coefficient K_3 is small, the effective anisotropy field, which keeps the nuclear magnetization vector in a specified direction within this plane, is very low, however, the corresponding enhancement factor is very high. It is interesting to note that in this case two essentially different anisotropy fields are realized. The high anisotropy field acts when the external magnetic field, whose vector directs, for example, along the hard magnetization axis, derives the electron magnetization vector from the easy plane, whereas the low one prevents the deviation of the electron magnetization vector within the easy plane from the direction corresponding to the minimal E_a value. That is why at an arbitrary orientation of the alternating field only the projection perpendicular to the vector of this small anisotropy field, but lying within the easy plane, is important.

In antiferromagnets, in the absence of external magnetic fields the total magnetization is equal to zero. If one applies an alternating magnetic field of the amplitude b perpendicularly to the anisotropy field vector it will create an alternating component of the magnetization. As it leads to the changing in the angle between the magnetizations of the sublattices, to calculate the alternating component of the magnetization vector it is necessary to take into account the exchange interaction between these sublattices. The exchange interaction energy, which is usually written as

$$E_{ex} = J_0(\vec{M}_{e1} \cdot \vec{M}_{e2})$$

($\vec{M}_{e1}$ and $\vec{M}_{e2}$ are the magnetization vectors of the corresponding sublattices), can be represented through an effective field $\vec{B}_{ex}$, which is

$$|\vec{B}_{ex}| = J_0 M_{e0}, \tag{7.18}$$

here M_{e0} is the module of the sublattice magnetizations; J_0 is a value proportional to the exchange interaction constant J; $\vec{B}_{ex}$ is the so-called *exchange field*. If the frequencies of antiferromagnetic resonance (AFR) are much greater than the NMR frequencies, one can suppose that the directions of the magnetization vector $\vec{M}_{e1}$ and $\vec{M}_{e2}$ in each moment coincide with the summary field vector. The alternating magnetization component of any of the sublattices is

$$M_{e\sim} = M_{e0}\frac{b}{B_{ex} + B_a},$$

and the enhancement factor is

$$\eta = \frac{B_{loc}}{B_{ex} + B_a}.$$

Since usually $B_a \ll B_{ex}$, then $\eta = B_{loc}/B_{ex}$. For an antiferromagnetic with the Neel point T_N close to the room temperature the exchange field is of about 10^3 T, whereas $\eta < 1$. Fortunately, the majority of antiferromagnetics has rather low Neel temperature, and hence the exchange field is one or two order lower, but the enhancement factor is still low.

The impact of the constant magnetic field $\vec{B}_0$, whose vector directs along the easy magnetization axis, leads to rather interesting effects, and in the presence of such a field the enhancement factor can be much larger. This problem is discussed in more details in Sect. 7.4.

In magnetically ordered materials there is not only the enhancement of the alternating magnetic field, but the enhancement of the NMR signal as well. Under the action of the rotating nuclear magnetization the electron magnetization starts to move, that induces an electromotive force in the coil of the receiver circuit containing the sample.

Reasoning as in above, one can show that in ferromagnets the enhancement factor of the signal is the same as the enhancement factor of the alternating rf field. However, in ferrites and canted antiferromagnetics it is not the case.[7]

7.4 Influence of Constant and Alternating Magnetic Fields on an Antiferromagnet

When calculating the influence of constant and alternating magnetic fields on an antiferromagnet it is necessary to take into account three contributions into the total energy: the exchange energy, the anisotropy energy and the energy of the interaction with an external static magnetic field. To simplify calculations let us suppose that the antiferromagnet is of the easy axis type and let us neglect in Eq. (7.9) the second term for the anisotropy energy. Then the total energy can be written as follows:

---

[7] For more details see A. G. Gurevich. Magnetic resonance in ferrites and antiferromagnets [in Russian]. Nauka, Moscow, 1973.

$$E = J_0(\vec{M}_{e1} \cdot \vec{M}_{e2}) - (\vec{B}_0 \cdot \vec{M}_{e1}) - (\vec{B}_0 \cdot \vec{M}_{e2}) + K(\sin^2\theta_1 + \sin^2\theta_2).$$

Here K is the constant determining the anisotropy energy; θ_1, θ_2 are the angles between the magnetization vectors of the corresponding sublattices (directed along the easy axis) and the external magnetic field vector $\vec{B}_0$. After opening the scalar products one has

$$E = JM_{e0}^2\cos(\theta_1 - \theta_2) - B_0 M_{e0}\cos\theta_1 - B_0 M_{e0}\cos\theta_2 + K(\sin^2\theta_1 + \sin^2\theta_2). \quad (7.19)$$

It is obvious that in the absence of the external magnetic field the minimal energy corresponds to $\theta_1 = 0$ and $\theta_2 = \pi$ (or vice-versa). To determine the configurations that correspond to the minimal energy in the presence of the external magnetic field it is necessary to differentiate Eq. (7.19) in θ_1 and θ_2, and then to equate the derivatives to zero:

$$\frac{dE}{d\theta_1} = -JM_{e0}^2\sin(\theta_1 - \theta_2) + B_0 M_{e0}\sin(\theta_1) + 2K\sin\theta_1\cos\theta_1 = 0,$$

$$\frac{dE}{d\theta_2} = JM_{e0}^2\sin(\theta_1 - \theta_2) + B_0 M_{e0}\sin(\theta_2) + 2K\sin\theta_2\cos\theta_2 = 0.$$

Solving these equations one can find that there are three possible configurations of the magnetization vectors of the sublattices and the magnetic field $\vec{B}_0$ (see Fig. 7.2):

(1) $\theta_1 = 0$ and $\theta_2 = \pi$;
(2) $\theta_1 = -\theta_2 = \theta$ (taking into account the symmetry axis of the system);
(3) $\theta_1 = 0$ and $\theta_2 = 0$.

To determine the angle θ corresponding to the second case one has the only equation:

$$-JM_{e0}^2\sin 2\theta + B_0 M_{e0}\sin\theta + K\sin 2\theta = 0.$$

Using Eq. (7.18) for the exchange energy and Eq. (7.13) for the anisotropy energy one obtains the following expression:

$$-B_{ex}\sin 2\theta + B_0\sin\theta + \frac{B_a}{2}\sin 2\theta = 0.$$

Using $\sin 2\theta = 2\sin\theta\cos\theta$ and canceling by $\sin\theta$, one has

$$-2B_{ex}\cos\theta + B_0 + B_a\cos\theta = 0$$

or

$$\cos\theta = \frac{B_0}{2B_{ex} - B_a}. \quad (7.20)$$

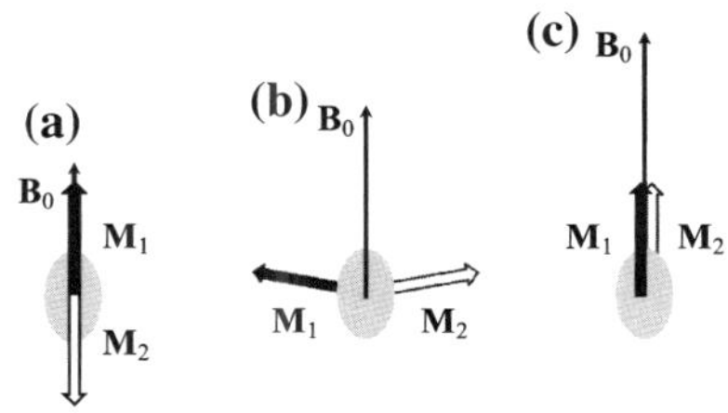

Fig. 7.2 Possible configurations of the magnetization vectors of two sublattices in an antiferromagnet under the action of an external magnetic field $\vec{B}_0$: (**a**) a low magnetic field; (**b**) an intermediate magnetic field; (**c**) a high magnetic field

Since $B_{ex} \gg B_a$ and B_0, then $\cos\theta$ is small, and θ is close to $\pi/2$.

Now it is possible to determine which energy minimum is more profound. In the first case, as it follows from Eq. (7.19):

$$E_1 = -JM_{e0}^2 = -B_{ex}M_{e0},$$

in the second case:

$$\begin{aligned} E_2 &= JM_{e0}\cos 2\theta - 2B_0M_{e0}\cos\theta + 2K\sin^2\theta \\ &= M_{e0}B_{ex}\cos 2\theta - 2B_0M_{e0}\cos\theta + M_{e0}B_a\sin^2\theta. \end{aligned}$$

and finally in the third case:

$$E_3 = JM_{e0}^2 - 2B_0M_{e0} = M_{e0}(B_{ex} - 2B_0),$$

As one can see in the first state the energy does not depend on the value of the applied field B_0, whereas in the second one it is lower the greater B_0 is. Let us find at which B_0 value $E_1 = E_2$. Denote this field value as B_{12}. Equating the energies in these two states, canceling by M_{e0}, and replacing $1 - \cos^2\theta$ for $\sin^2\theta$ and $\cos^2\theta - \sin^2\theta = 2\cos^2\theta - 1$ for $\cos 2\theta$ one obtains the following expression:

$$(2B_{ex} - B_a)\cos^2\theta - 2B_{12}\cos\theta + B_a = 0.$$

Substituting $\cos\theta$ in the form of (7.20) one has

$$\frac{B_{12}^2}{2B_{ex} - B_a} - \frac{2B_{12}^2}{2B_{ex} - B_a} + B_a = 0,$$

and finally

$$B_{12} = \sqrt{B_a(2B_{ex} - B_a)}.$$

If $B_{ex} \gg B_a$, that is almost always the case, then

$$B_{12} \cong \sqrt{2B_a B_{ex}}.$$

If the field B_a is sufficiently low, then the field $B_0 > B_{12}$ necessary to make the second state more favorable is rather low and can be realized in practice (Fig. 7.2b).

If one compares by the similar way the second and third states ($E_2 = E_3$), denoting the field of transition as B_{23} one has

$$B_{ex} - 2B_{23} = -\frac{B_{23}^2}{2B_{ex} - B_a} - B_{ex} + B_a$$

or

$$(2B_{ex} - B_a)^2 - 2B_{23}(2B_{ex} - B_a) + B_{23}^2 = 0.$$

Let us solve the quadratic equation:

$$B_{23} = 2B_{ex} - B_a + \sqrt{(2B_{ex} - B_a)^2 - (2B_{ex} - B_a)^2} = 2B_{ex} - B_a \cong 2B_{ex}.$$

Hence, if $B_0 < \sqrt{2B_a B_{ex}}$ then the state with inverse directions of the magnetization vectors of two sublattices is more favorable. For $0 < B_0 < \sqrt{2B_a B_{ex}}$ the magnetization vectors of the sublattices remain antiparallel. However, if $\sqrt{2B_a B_{ex}} < B_0 \ll 2B_{ex}$ then the second state is more profitable, and the sublattice magnetizations make the same (except for sign) angle with the external magnetic field vector, settling almost perpendicular to it (see Eq. (7.20) and Fig. 7.2c). It means that at $B_0 = \sqrt{2B_a B_{ex}}$ (a critical field) the magnetization vectors of the sublattices turn over. With following increasing of the external magnetic field, taking into account that $B_0 < 2B_{ex}$, the second state is still realized, but the angle θ becomes smaller than $\pi/2$, i.e. a small component of the magnetic moment parallel to the external magnetic field appears. If $B_0 \geq 2B_{ex}$ the third state is realized: the magnetization vectors of the sublattices are parallel to each other and to the external magnetic field vector as well. It is to be noted that this case is rarely realized in practice as it is difficult to create an external magnetic field compared with the exchange field.

In the second case if one applies an alternating magnetic field with vector perpendicular to the easy magnetization axis, the resulting magnetization can oscillate without changing the angle between the magnetization vectors of the sublattices. In every moment the directions of the total magnetization vector and of the summary field vector (formed by the external magnetic field, the anisotropy field and the exchange field) coincide. Then, assuming that the exchange field is much greater than the anisotropy one, the alternate component of any sublattice magnetization is equal to

$$M_{e\sim} = M_{e0}\frac{b}{B_a + B_0}.$$

Since at such conditions the external magnetic field B_0 is also greater than $\sqrt{2B_a B_{ex}}$, and hence is greater than B_a, the enhancement factor is

$$\eta = \frac{-A_0 M_{e\sim}}{b} = \frac{B_{loc}}{B_0}.$$

At low values of the anisotropy field the enhancement factor can be rather high. It can be even higher if one takes into account two factors: (i) at rather low anisotropy values the AFM frequencies can be close to the NMR frequencies, that leads to greater values of the alternating component of the magnetizations of the sublattices; (ii) the mobility of the magnetization can be higher if the external field is close to its critical value. More accurate calculations result in the following expression for the enhancement factor in antiferromagnets[8]:

$$\eta = \frac{B_{loc} B_0}{B_0^2 - 2B_{ex} B_a}.$$

7.5 Mechanisms of the rf Field Enhancement in Domain Walls

The second mechanism of the rf field enhancement occurs in domain walls. Let us consider a 180-degree Bloch wall. To calculate the enhancement factor it is necessary to know the law, which the orientation of the electron magnetization vector within the wall depth obeys. This problem can be solved for a rather ideal model: a single flat wall between two domains of infinite extent.

The domain wall structure is determined by the exchange and anisotropy energies. To calculate the domain wall structure means to solve a variational problem. As a result one obtains a function $\theta(y)$, where θ is the angle between the magnetization vector in a given place of the domain wall and the magnetization vector in the domain; y is the axis coming from the domain wall center and directed perpendicular to its plane (see Fig. 7.3).

The exchange energy for two ions is defined by the scalar product of their spins (see Eq. (7.1)). For a 3D continuum it can be written as follows:

$$E_{ex} = J_0 \left[(\nabla M_{ex})^2 + (\nabla M_{ey})^2 + (\nabla M_{ez})^2 \right] \frac{1}{M_{e0}^2},$$

here ∇ is the gradient operator; M_{e0} is the modulus of the electron magnetization vector; J_0 is the value proportional to the exchange interaction constant J. The total energy characterizing this domain wall is described by the following formula:

$$E = \int_{-\infty}^{\infty} (E_{ex} + E_a) \mathrm{d}y.$$

[8] M. I. Kurkin and E. A. Turov, NMR in Magnetically Ordered Substances and Its Application [in Russian], Nauka, Moscow (1990).

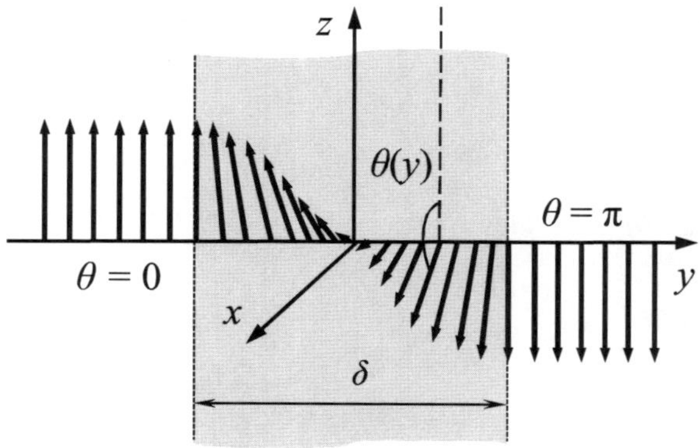

Fig. 7.3 Changing of the magnetization vector direction within the Bloch wall

Assuming that there is only one easy magnetization axis and using Eq. (7.9) for the anisotropy energy ($K_2 = 0$) one has

$$E = \int_{-\infty}^{\infty} \left[J_0 \left[(\nabla M_{ex})^2 + (\nabla M_{ey})^2 + (\nabla M_{ez})^2 \right] \frac{1}{M_{e0}^2} + K \sin^2 \theta \right] \mathrm{d}y.$$

As the domain wall considering here is of infinite extent along the x and z directions the orientation of the $\vec{M}_e$ vector depends only on the y coordinate. But for a Bloch wall by definition $M_{ey} = 0$. Then $M_{ez} = M_{e0} \cos \theta$, $M_{ex} = M_{e0} \sin \theta$ and

$$E = \int_{-\infty}^{\infty} \left[J_0 \left[\left(\frac{\mathrm{d}\theta}{\mathrm{d}y} \right)^2 \sin^2 \theta + \left(\frac{\mathrm{d}\theta}{\mathrm{d}y} \right)^2 \cos^2 \theta \right] + K \sin^2 \theta \right] \mathrm{d}y$$

or

$$E = \int_{-\infty}^{\infty} \left[J_0 \left(\frac{\mathrm{d}\theta}{\mathrm{d}y} \right)^2 + K \sin^2 \theta \right] \mathrm{d}y. \qquad (7.21)$$

To define the function $\theta(y)$ that corresponds to the minimal energy of the domain wall it is necessary that the expression in square brackets in Eq. (7.21) obeys the Euler condition. Denoting $F \equiv J_0(\mathrm{d}\theta/\mathrm{d}y)^2 + K \cos^2 \theta$ the Euler condition can be written as follows:

$$F_{\theta}^{'} - \frac{\mathrm{d}}{\mathrm{d}y} F_{\theta'}^{'} = 0. \qquad (7.22)$$

In our case

$$\frac{\mathrm{d}F}{\mathrm{d}\theta} = 2K \sin \theta \cos \theta, \qquad \frac{\mathrm{d}F}{\mathrm{d}\theta'} = 2J_0 \frac{d\theta}{\mathrm{d}y}.$$

Hence Eq. (7.22) can be rewritten as

$$-J_0\frac{\mathrm{d}^2\theta}{\mathrm{d}y^2}+K\sin\theta\cos\theta=0. \tag{7.23}$$

Multiplying both terms of Eq. (7.23) by $\mathrm{d}\theta/\mathrm{d}y$ one obtains

$$-J_0\frac{\mathrm{d}^2\theta}{\mathrm{d}y^2}\frac{\mathrm{d}\theta}{\mathrm{d}y}+K\sin\theta\cos\theta\frac{\mathrm{d}\theta}{\mathrm{d}y}=0. \tag{7.24}$$

The first term in Eq. (7.24) is the derivation of $-(\mathrm{d}\theta/\mathrm{d}y)^2$, whereas the second one is the derivation of $\sin^2\theta$ (or $-\cos^2\theta$). After integrating Eq. (7.24) one obtains

$$-J_0\left(\frac{\mathrm{d}\theta}{\mathrm{d}y}\right)^2+K\sin^2\theta=\mathrm{const}\,.$$

At the interval limits both $\sin\theta$ and $\mathrm{d}\theta/\mathrm{d}y$ are equal to zero, hence const $=0$. Then

$$\frac{\mathrm{d}\theta}{\mathrm{d}y}=\sqrt{\frac{K}{J_0}}\sin\theta \quad \text{or} \quad \frac{\mathrm{d}\theta}{\sin\theta}=\sqrt{\frac{K}{J_0}}\mathrm{d}y.$$

Integrating it one has the following equation for y:

$$y=\sqrt{\frac{J_0}{K}}\operatorname{artanh}(\cos\theta)+\mathrm{const}\,. \tag{7.25}$$

The integration constant in Eq. (7.25) is also equal to zero as at the center of the domain wall $y=0,\ \theta=\pi/2,\ \cos\theta=0$ and $\operatorname{artanh}(\cos\theta)=0$. Taking into account that const $=0$, the second equation in (7.25) means that

$$\tanh\frac{y}{\delta}=\cos\theta,$$

where $\delta=\sqrt{J_0/K}$. This can be rewritten in a generally accepted form

$$\sin\theta=\left(\cosh\frac{y}{\delta}\right)^{-1}. \tag{7.26}$$

Having the dependence $\theta(y)$ in the explicit form it is possible to calculate how the enhancement factor η depends on the y coordinate.

The motion of domain walls that occurs under the action of external fields is described by a second-order differential equation. It means the domain walls exhibit resonance characteristics and have their own resonance oscillation frequencies. However, one can suppose that forced vibrations, with a frequency noticeably lower the resonance one, are in time with oscillations of the external field. The magnetic field, whose vector coincides with the magnetization vector direction in the domains, causes such a displacement of the domain wall that the size of the energy-favorable domain (the magnetization is parallel to the magnetic field) increases, whereas the size of the energy-unfavorable domain (the magnetization is antiparallel to the magnetic field) decreases (see Fig. 7.3). The alternating magnetic field with vector parallel to the magnetization of one of the domains results in vibrations of the domain wall, this, in its turn, leads to the appearance of the alternating component of the electron magnetization in each given place, and hence to the appearance of the alternating component of the local field.

If under the action of the alternating field of the amplitude b the domain wall oscillates with an amplitude $\mathrm{d}y$ then the changing of the θ angle in a given place of the domain wall is equal to

$$\mathrm{d}\theta = \frac{\mathrm{d}\theta}{\mathrm{d}y}\mathrm{d}y.$$

Differentiation of the both parts of Eq. (7.26) leads to

$$\cos\theta \mathrm{d}\theta = -\frac{\sinh y/\delta}{\delta \cosh^2 y/\delta}\mathrm{d}y = -\frac{\tanh y/\delta}{\delta \cosh y/\delta}\mathrm{d}y = -\frac{\cos\theta \sin\theta}{\delta}\mathrm{d}y.$$

And canceling by $\cos\theta$ one has

$$\frac{\mathrm{d}\theta}{\mathrm{d}y} = -\frac{1}{\delta}\sin\theta.$$

The wall displacement can be easily related to the initial value of the magnetic susceptibility $\chi(0)$:

$$\mathrm{d}y = \chi(0)b/(2M_{e0}).$$

Then

$$\mathrm{d}\theta = -\frac{b}{2M_{e0}\delta}\chi(0)\sin\theta,$$

$$B_{10} = B_{\mathrm{loc}}\mathrm{d}\theta = \frac{bB_{\mathrm{loc}}}{2M_{e0}\delta}\chi(0)\sin\theta,$$

$$\eta = \eta_0 \sin\theta,$$

$$\eta_0 = \frac{B_{\mathrm{loc}}}{2M_{e0}\delta}\chi(0). \tag{7.27}$$

The major difference between the enhancement factor in the domain wall and in the domain is that there is a dispersion of their values over the domain wall depth.

To calculate the NMR signal induced in the receiver of the oscillating circuit it is necessary to summarize the η values over the cross sectionof the domain wall.

In polycrystalline samples, in addition, one has to summarize over orientations of domain walls, because now only the projection of the rf field parallel or antiparallel to the magnetization vector in the neighboring domain is important.

7.6 Influence of the Enhancement Factor Dispersion on NMR Signals

As it was shown in previous Section the enhancement factor changes within a domain wall. However, it can be different even in domains if there is a dispersion of local anisotropy fields or one deals with polycrystalline samples (or multidomain samples with nonparallel magnetization vectors in different domains). That dispersion of the enhancement factor over a sample results in specific features of free precession and spin echo signals.

7.6.1 Formation of Free Precession and Spin Echo Signals

First of all let us discuss the free precession signal, supposing the pulse affecting a sample is short (much shorter than relaxation times) and of high amplitude (sufficient enough to fulfill the condition $\gamma B_{10} \gg \Delta\omega$ for all spin packets). Mind that γ is the nucleus gyromagnetic ratio; B_{10} is the rf field intensity affecting a nucleus; $\Delta\omega = \omega_0 - \omega$; ω_0 is the NMR frequency; ω is the rf field frequency. It is to note that if there is a dispersion of the enhancement factor one cannot use widespread notions of 90- or 180°-pulses; for different spin packets with different η values the rotation angle of magnetization vectors is not the same, it depends on the enhancement factor.

After a rf pulse the behavior of a spin packet within the time interval much shorter than relaxation times can be described by Eq. (1.82). If $\gamma B_{10} \gg \Delta\omega$ these expressions can be simplified:

$$u(t) = M_0 \sin \gamma B_{10} t_p \sin \Delta\omega t,$$

$$v(t) = M_0 \sin \gamma B_{10} t_p \cos \Delta\omega t.$$

In magnetically ordered compounds, where an enhancement effect of the fr field takes place, B_{10} is the rf field amplitude affecting nuclei. Since $B_{10} = \eta b_1$, where b_1 is the amplitude of the external rf field, then

$$u(t) = M_0 \sin \gamma\eta b_1 t_p \sin \Delta\omega t,$$

$$v(t) = M_0 \sin \gamma\eta b_1 t_p \cos \Delta\omega t.$$

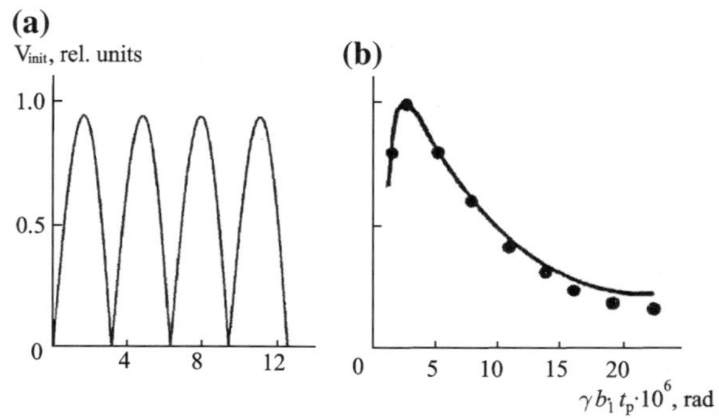

Fig. 7.4 The initial amplitude of the free precession signal as a function of the rf pulse intensity calculated using Eq. (7.28) (**a**) and measured experimentally (**b**)

To calculate the initial amplitude of the free precession signal, let us take $t = 0$ for all spin packets, then

$$u(0) = 0, \qquad v(0) = M_0 \sin \gamma \eta b_1 t_p.$$

If all spin packets had the same enhancement factor, the calculation of the precessing magnetization of the all sample at the initial time would come to multiplying the expression for v by the number of the spin packets. That would lead to the following expression for the initial amplitude of the free precession signal:

$$V_{\text{init}} = V_0 M_0 \eta \sin(\gamma \eta b_1 t_p) \tag{7.28}$$

(V_0 is a coefficient determined by instrumental factors). If $\eta = 1$, then V_{init} is a periodic function of the integral pulse intensity $\gamma b_1 t_p$ (see Fig. 7.4a). However, the experimental dependence has a considerably different form (see Fig. 7.4b). The question is: is it possible to explain such a disagreement between calculations and experimental data by the dispersion of the enhancement factor over a polycrystalline sample or over a domain wall depth?

First, let us consider a polycrystalline sample of the easy axis type, in which the signal observed experimentally is the signal from domains. And let the domains are oriented chaotically, but in such a way that any domain orientations, characterized by the easy axis direction (or by the anisotropy field vector), are equiprobable. Let the easy axis direction is characterized by an angle α, which is the angle between the easy axis and the rf field vector. Then only that component of the rf field that is perpendicular to the easy axis affects the sample. The effective enhancement factor is $\eta = \eta_0 \sin \alpha$, here η_0 is the enhancement factor for the domain, the magnetization vector of which is perpendicular to the rf field vector. But the number of domains with different α is different. Supposing an equiprobable distribution of the easy magnetization axis, the number of domains with a certain α is proportional to the surface area of a unit spherical ring: $2\pi \sin \alpha \, d\alpha$. Then

$$V_{\text{init}} = \frac{2V_0 M_0}{4\pi} \int_0^{\pi/2} 2\pi\eta_0 \sin\alpha \sin(\gamma\eta_0 b_1 t_{\text{p}} \sin\alpha) \sin\alpha \mathrm{d}\alpha;$$

4π in the denominator is the surface area of a unit sphere; the argument of the sine function ($\gamma\eta_0 b_1 t_{\text{p}} \sin\alpha$) indicates the angle, on which the rf pulse rotates the nucleus magnetization vector (in the rotating frame) in domains with the same α value. $M_0 \sin(\gamma\eta_0 b_1 t_{\text{p}} \sin\alpha)$ is the projection of the nucleus magnetization vector on the y-axis in the frame rotating with the rf field. The factor ($\eta_0 \sin\alpha$) means that the signal induced in the inductance coil by the precessing magnetization of a spin packet is proportional to the enhancement factor for this spin packet. Thus

$$V_{\text{init}} = V_0 M_0 \eta_0 \int_0^{\pi/2} \sin(\gamma\eta_0 b_1 t_{\text{p}} \sin\alpha) \sin^2\alpha \mathrm{d}\alpha.$$

Such an integral can be easy calculated.

Let us introduce $\phi = \pi/2 - \alpha$, then $\sin\alpha = \cos\phi$, and $\sin^2\alpha = 1 - \sin^2\phi$, $\mathrm{d}\alpha = -\mathrm{d}\phi$, $\phi_0 = \gamma\eta_0 b_1 t_{\text{p}}$. Then

$$V_{\text{init}} = V_0 M_0 \eta_0 \left[-\int_{\pi/2}^{0} \sin(\phi_0 \cos\phi)\mathrm{d}\phi + \int_{\pi/2}^{0} \sin^2\phi \sin(\phi_0 \cos\phi)\mathrm{d}\phi \right].$$

Changing the integration limits one obtains:

$$V_{\text{init}} = V_0 M_0 \eta_0 \left[\int_0^{\pi/2} \sin(\phi_0 \cos\phi)\mathrm{d}\phi - \int_0^{\pi/2} \sin^2\phi \sin(\phi_0 \cos\phi)\mathrm{d}\phi \right]. \quad (7.29)$$

The integrals in Eq. (7.29) can be written in terms of the well known Struve functions H_ν defined as follows:

$$H_\nu(z) = \frac{2z^\nu}{\sqrt{\pi}\, 2^\nu \Gamma(\nu + 1/2)} \int_0^{\pi/2} [\sin(z\cos t) \sin^{2\nu} t]\mathrm{d}t,$$

where Γ is the Γ-function. For $\nu = 0$ and $\nu = 1$ the Struve functions are

$$H_0(z) = \frac{2}{\pi} \int_0^{\pi/2} [\sin(z\cos t)]\mathrm{d}t, \qquad H_1(z) = \frac{2z}{\pi} \int_0^{\pi/2} [\sin(z\cos t)\sin^2 t]\mathrm{d}t,$$

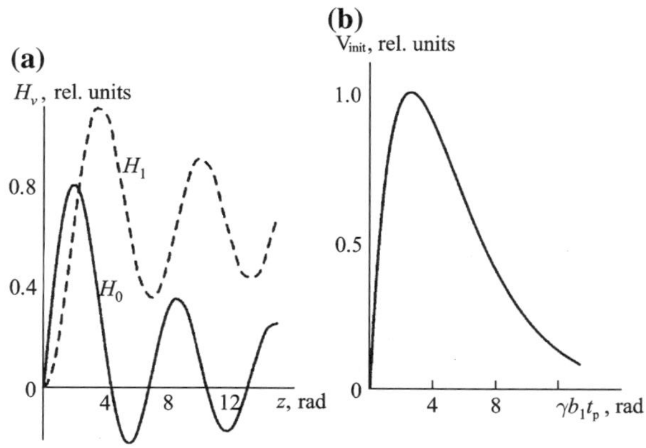

Fig. 7.5 The $H_0(z)$ and $H_1(z)$ Struve functions (**a**) and the initial amplitude of the free precession signal as a function of rf pulse intensity calculated using Eq. (7.30) (**b**)

(as $\Gamma(1/2) = \sqrt{\pi}$ and $\Gamma(3/2) = \sqrt{\pi}/2$). The Struve functions $H_0(z)$ and $H_1(z)$ are plotted in Fig. 7.5a. Taking into account the expressions for $H_0(z)$ and $H_1(z)$ one can write:

$$V_{\text{init}} = \frac{\pi V_0 M_0 \eta_0}{2}\left[H_0(\phi_0) - \frac{H_1(\phi_0)}{\phi_0}\right]. \tag{7.30}$$

The envelope of the function described by Eq. (7.30) is shown in Fig. 7.5b. As it is seen, this is a damped function. At high ϕ_0 values it decreases proportionally to $\sqrt{\phi_0}$.

By the similar way it is possible to calculate the dependence of the initial amplitude of the free precession signal on the rf pulse intensity for an assemblage of domain walls situated parallel to each other or chaotically distributed in space.

Let us analyze the first situation, when the domain walls are parallel each other and separate domains with opposite orientations. Let the rf field vector is parallel to the magnetization vector in domains.

To calculate the initial amplitude of the free precession signal it is necessary to summarize signals over all elementary planes with the same θ value. As the free precession signal corresponding to one elementary plane of a width $\mathrm{d}y$ is proportional to $\eta \sin(\gamma\eta b_1 t_p)\mathrm{d}y$, then the total signal from each boundary is proportional to $\int_{-\infty}^{\infty} \eta \sin(\gamma\eta b_1 t_p)\mathrm{d}y$. Taking into account that for a domain wall $\eta = \eta_0 \sin\theta$, and $\sin\theta = (1/\delta)(\cosh y/\delta)^{-1}$, one can write:

$$\mathrm{d}y = \frac{\delta}{\sin\theta}\mathrm{d}\theta,$$

$$V_{\text{init}} = V_0 M_0 \int_0^{\pi/2} \frac{\delta\eta_0}{\sin\theta} \sin\theta \sin(\phi_0 \cos\theta)\mathrm{d}\theta = V_0 M_0 \eta_0 \delta \int_0^{\pi/2} \sin(\phi_0 \cos\theta)\mathrm{d}\theta.$$

This can be reduced to the Struve function H_0:

$$V_{\text{init}} = V_0 M_0 \pi \eta_0 \delta H_0(\phi_0).$$

If one put in this equation the value η_0 in the form (7.27), one obtains that V_{init} does not depend on the domain wall depth:

$$V_{\text{init}} = \frac{V_0 M_0 \pi B_{\text{loc}}}{2 M_{\text{e0}}} \chi(0) H_0(\phi_0). \tag{7.31}$$

It is rather understandable as the thicker the domain wall the more nuclei it contains, but the lower enhancement factor is.

In a polycrystalline sample the rf field ($b \cos\psi$) rotates the nuclear magnetization vector. The number of the domain walls, the normal to which makes an angle ψ with the fr field vector, is proportional to $\sin\psi$. Hence

$$V_{\text{init}} = \pi V_0 M_0 \eta_0 \delta \int_0^{\pi/2} [\sin\psi \cos\psi H_0(\phi_0 \cos\psi)] \mathrm{d}\psi.$$

Introducing $z = \cos\psi$ one writes

$$V_{\text{init}} = \frac{\pi V_0 M_0 \delta}{2} \int_0^1 \eta_0 z H_0(\phi_0 z) \mathrm{d}z = \frac{V_0 \pi M_0 \delta \eta_0 H_1}{4\phi_0}. \tag{7.32}$$

Despite with ϕ_0 increasing the function $V_{\text{init}}(\phi_0)$ for polycrystals described by Eq. (7.32) decreases faster than V_{init} calculated using Eqs. (7.30) and (7.31) it keeps its oscillating character.

It is to be noted that for all considered here cases the calculated dependence of V_{init} versus the rf pulse intensity essentially differs from experimental data. An agreement can be achieved only for a very complex model, where domain walls are represented as rigidly fixed round membranes.[9]

Knowing the dependence of free precession signal on the rf pulse amplitude it is possible to calcullate the similar dependence for a spin echo signal (V_{echo}). Let us put in Eq. (1.90) instead of B_1 and B_2, ηb_1 and ηb_2, respectively:

$$\begin{aligned} V_{\text{echo}}(t) &= \frac{M_0}{2} \sin\gamma\eta b_1 t_{\text{p1}} (1 - \cos\gamma\eta b_2 t_{\text{p2}}) \\ &= \frac{M_0}{2} \left[\sin\gamma\eta b_1 t_{\text{p1}} - \frac{1}{2} \sin\gamma\eta (b_1 t_{\text{p1}} - b_2 t_{\text{p2}}) - \frac{1}{2} \sin\gamma\eta (b_1 t_{\text{p1}} + b_2 t_{\text{p2}}) \right]. \end{aligned} \tag{7.33}$$

[9] M. B. Stearns. Spin-echo and free-induction-decay measurements in pure Fe and Fe-rich ferromagnetic alloys: Domain-wall dynamics. Phys. Rev. **162** (1967) 496.

Here b_1, b_2 are the amplitudes of the first and second rf pulse, respectively; t_{p1}, t_{p2} are the durations of these pulses. Eq. (7.33) represents a sum of three terms, each of which describes the dependence of the initial amplitude of the free precession signal after a corresponding pulse. Making transformations similar to made before, one obtains:

$$V_{echo} = \frac{1}{2} V_{init}(\gamma\eta_0 b_1 t_{p1}) - \frac{1}{4}\left[V_{init}(\gamma\eta_0(b_1 t_{p1} + b_2 t_{p2})) + V_{init}(\gamma\eta_0(b_1 t_{p1} - b_2 t_{p2}))\right]. \tag{7.34}$$

It is possible to write simpler formulas for two particular cases: if $b_1 t_{p1} = b_2 t_{p2}$, then

$$V_{echo} = \frac{1}{2} V_{init}(\gamma\eta_0 b_1 t_{p1}) - \frac{1}{4} V_{init}(2\gamma\eta_0 b_1 t_{p1});$$

if $2b_1 t_{p1} = b_2 t_{p2}$, then

$$V_{echo} = \frac{3}{4} V(\gamma\eta_0 b_1 t_{p1}) - \frac{1}{4} V(3\gamma\eta_0 b_1 t_{p1}).$$

The calculations are in good agreement with experimental data if one applies the membrane model.

7.6.2 Distribution Function of Nuclei Over Enhancement Factors

There is another approach to solve the problem. It is not necessary to know the enhancement factor dispersion, but to describe its variation by a function $F(\eta)$, which indicates how much nuclei have one or another η value. I. e. $F(\eta)$ is a *distribution function* of nuclei over enhancement factors. If $\gamma\eta b_1 \gg \Delta\omega$ the calculation of the initial amplitude of the free precession signal comes to summing up over all spin packets with different η taking into account the function $F(\eta)$, in other words Eq. (7.28) should be multiplied by $F(\eta)$ and then integrated over η:

$$V(b_1 t_p) = V_0 M_0 \int_0^{\infty} \eta F(\eta) \sin(\gamma\eta b_1 t_p) \mathrm{d}\eta.$$

This is a Fourier sine transformation. Hence, if the dependence of V on $b_1 t_p$ is known applying an inverse Fourier transformation one can find the distribution function of nuclei over enhancement factors:

$$F(\eta) = \frac{1}{\pi\eta V_0 M_0} \int_0^{\infty} V(\gamma b_1 t_p) \sin(\gamma\eta b_1 t_p) \mathrm{d}(\gamma b_1 t_p). \tag{7.35}$$

Now if one approximates the experimental dependance $V_{\text{init}}(b_1 t_p)$ by an appropriate function it is possible to calculate $F(\eta)$. The simplest function that reflects the dependence plotted in Fig. 7.4b is following:

$$V_{\text{init}}(b_1 t_p) = 2V_{\max} \frac{\gamma\eta_0 b_1 t_p}{1 + (\gamma\eta_0 b_1 t_p)^2}. \tag{7.36}$$

Here $V_{\max}$ is the maximal value of $V_{\text{init}}(b_1 t_p)$. It is achieved at $\gamma\eta_0 b_{\max} t_p = 1$ ($b_{\max}$ is the amplitude of the rf field that corresponds to the maximal value of V_{init}). If γ, t_p and $b_{\max}$ are known, it is possible to determine η_0. After substituting Eq. (7.36) into Eq. (7.35) one obtains, accurate within a constant factor, that:

$$F(\eta) = \eta^{-1} \exp(-\eta/\eta_0). \tag{7.37}$$

This function due to its simplicity is very suitable in further calculations. Here η_0 is the root-mean-square enhancement factor. A very essential disadvantage of the function (7.37) is that it can not be normalized. However, despite it can not be used to determine absolute values of physical quantities, it does not prevent calculations of various dependencies. In effect, one can suppose that Eq. (7.37) is true in a rather large neighborhood of η_0, but for $\eta \ll \eta_0$ one should apply another approximation of $F(\eta)$. For example, on can suppose it to be constant. Such an approach is possible, as it is easy to show that spin packets with either too small (as it is small) or too large (as there are only a few of such spin packets) η value do not make any noticeable contribution into the free precession and spin echo signals.

However, sometimes at large $\gamma b_1 \eta_0 t_p$ values the amplitude decreases much faster than it follows from Eq. (7.36). And then one has to use other expressions for $F(\eta)$ (they can be found anywhere). The most interesting among them is

$$F(\eta) = \eta^{-1/2} \exp(-\eta/\eta_0).$$

To normalize this function one has to add to the right part of this expression a factor $1/\sqrt{\pi\eta_0}$. The function $V_{\text{init}}(\phi_0)$, which corresponds to such a $F(\eta)$ dependence, is much more complex, as compared to Eq. (7.36):

$$V_{\text{init}}(\phi_0) = \frac{V_0 M_0}{\sqrt{2}} \left[\frac{1}{2} + \frac{1}{\sqrt{1+\phi_0^2}} \right] \frac{\sqrt{\sqrt{1+\phi_0^2} - 1}}{1+\phi_0^2},$$

where $\phi_0 = \gamma\eta_0 b_1 t_p$.

Let us come back to the distribution function of nuclei over enhancement factors in the form (7.37) and write expressions for dependence of the spin echo amplitude on the rf pulse intensity. Using Eq. (7.34) one obtains:

$$V_{\text{echo}}(\phi_1, \phi_2) = V_0 M_0 \left[\frac{\phi_1}{1+\phi_1^2} + \frac{\phi_2 - \phi_1}{1+(\phi_2-\phi_1)^2} + \frac{\phi_2+\phi_1}{1+(\phi_2+\phi_1)^2} \right], \quad (7.38)$$

where $\phi_1 = \gamma\eta_0 b_1 t_{p1}$, $\phi_2 = \gamma\eta_0 b_2 t_{p2}$; b_1 and b_2 are the amplitudes of the first and second pulses, respectivelly; t_{p1} and t_{p2} are their intensities.

The dependence $V_{\text{echo}}(\phi_1, \phi_2)$ is rather interesting. To find out its several specific features let us determine extremuma of the function (7.38) (varying ϕ_2 at fixed ϕ_1). The first term in square brackets in Eq. (7.38) does not depend on ϕ_2, hence let us skip it and focus on the rest.

If $\phi_1 < \sqrt{3}$ then the spin echo amplitude as a function of ϕ_2 has the only maximum at

$$\phi_2 = \sqrt{1+\phi_1^2 + 2\sqrt{1+\phi_1^2}}. \quad (7.39)$$

If $\phi_1 > \sqrt{3}$, there are two extremuma: the first one is described by Eq. (7.39), whereas another takes place at

$$\phi_2 = \sqrt{1+\phi_1^2 - 2\sqrt{1+\phi_1^2}}.$$

If $\phi_2 = \sqrt{\phi_1^2 - 3}$ the spin echo amplitude is equal to zero.

In Fig. 7.6 we plot the spin echo amplitude as a function of the first rf pulse intensity ϕ_1 at different ratios between the integral intensities of the first and second rf pulses, $n = \phi_2/\phi_1$. As it is clearly seen, at small ϕ_1 and ϕ_2 values the maximal echo amplitude is achieved at almost equal integral intensities of the first and second pulses, however, for large ϕ_1 and ϕ_2 values even slight changes in n have an important influence of the spin echo amplitude.

These conclusions are preserved if one applies other distribution functions of nuclei over enhancement factors. Forms of the free precession and spin echo signals are still determined by the NMR spectrum shape, if there is no correlation between the NMR frequency and the enhancement factor. However, it is not always true. For example, the resonance frequency varies across the domain wall in such a manner that in the middle of the wall, where η has the highest value, the NMR frequency is minimal.

7.7 Infinitely Broad NMR Spectrum

To calculate free precession and spin echo signals for a NMR signal of an arbitrary shape it is necessary to summarize signals over all spin packets with different frequencies (ω_0) and enhancement factors (η), that can be done only numerically. If ω_0 and η do not correlate between each other to obtain the expression for the free

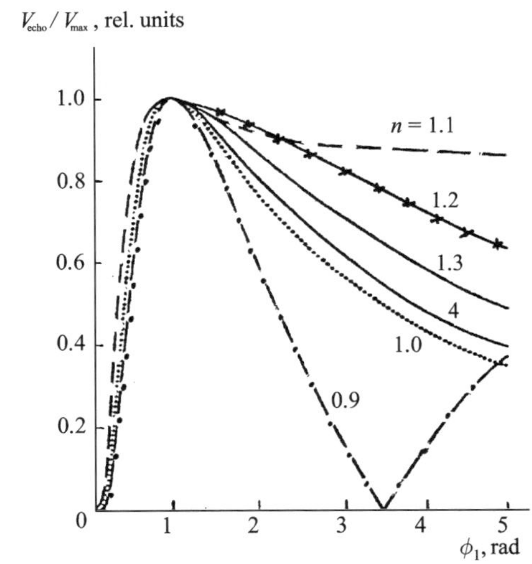

Fig. 7.6 The spin echo amplitude versus the first rf pulse intensity at different ratios $n = \phi_2/\phi_1$

precession amplitude one should in Eq. (1.85) substitute ηb_1 for B_1, multiply by $\eta F(\eta)$ and integrate over η. The final expressions have the following form:

$$V_x = V_0 M_0 \int_0^{\infty} \int_{-\infty}^{\infty} \eta u(\Delta\omega, \eta b_1) g(\Delta\omega) F(\eta)\, \mathrm{d}\Delta\omega \mathrm{d}\eta,$$

$$V_y = V_0 M_0 \int_0^{\infty} \int_{-\infty}^{\infty} \eta v(\Delta\omega, \eta b_1) g(\Delta\omega) F(\eta) \mathrm{d}\Delta\omega \mathrm{d}\eta,$$

here u and v are calculated using Eq. (1.82).

Using Eq. (7.37) for the distribution function of nuclei over the enhancement factor and the function $g(\Delta\omega) = (1 + \Delta\omega/\Delta\omega_0)\exp(-\Delta\omega/\Delta\omega_0)$ for description of the line shape,[10] one can calculate one of the integrals. The second integration can be done numerically (making an appropriate change of variables it is possible to come to finite integration limits). The expressions obtained in such a manner can be applied to an infinite spectrum as well, but for such a spectrum it is possible to obtain a rather simpler expression, wich does not contain any integrals.

7.7.1 Free Precession

If one uses the second formula from Eq. (1.83), which contains the expression for a signal component shifted on $\pi/2$ in phase with respect to the external rf field,

[10] It describes a spectral line of a shape intermediate between Lorentzian and Gaussian.

it is possible to obtain an analytical expression for the free precession amplitude. The component, which is in phase with the external rf field, is equal to zero for a symmetric spectrum if the rf frequency coincides with the central frequency of the spectrum (it is always the case for an infinitely broad spectrum). Substituting in Eq. (1.83) ηb_1 for B_1, multiplying all the expression by $\eta F(\eta) = \exp(-\eta/\eta_0)$ and integrating it over η one obtains:

$$V = V_0 \pi M_0 \int_0^\infty e^{-\frac{\eta}{\eta_0}} \eta^2 \gamma^2 b_1^2 \int_t^{t_p} \left(\frac{x-t}{x+t}\right)^{\frac{1}{2}} J_1[\eta\gamma b_1 (x^2 - t^2)^{\frac{1}{2}}] \mathrm{d}x \mathrm{d}\eta.$$

Let us permute the integration order:

$$V = V_0 \pi M_0 \gamma^2 b_1^2 \int_t^{t_p} \left(\frac{x-t}{x+t}\right)^{\frac{1}{2}} \int_0^\infty \eta^2 \exp\left(-\frac{\eta}{\eta_0}\right) J_1[\eta\gamma b_1 (x^2 - t^2)^{\frac{1}{2}}] \mathrm{d}\eta \mathrm{d}x.$$

It is known that

$$I_1 = \int_0^\infty \eta^2 \exp\left(-\frac{\eta}{\eta_0}\right) J_1(\eta, k) \mathrm{d}\eta = \frac{3k\eta_0^4}{(1 + k^2\eta_0^2)^{5/2}},$$

$$k = \gamma b_1 (x^2 - t^2)^{1/2}.$$

Then

$$V = 3V_0 \pi M_0 \eta_0^4 (\gamma b_1)^3 \int_t^{t_p} \frac{x-t}{[1 + \gamma^2 b_1^2 \eta_0^2 (x^2 - t^2)]^{5/2}} \mathrm{d}x.$$

This integral can be easily calculated. Introducing $\phi_0 = \gamma b_1 \eta_0 t_p$ and $a = t/t_p$, one has

$$V = \frac{V_0 \pi \phi_0 \eta_0 M_0}{t_p (1 - \phi_0^2 a^2)^2} \left[1 + a^2\phi_0^2 - \frac{(1 - a^2\phi_0^2)^2 + a\phi_0^2(3 + 2\phi_0^2 - 3a^2\phi_0^2)}{[1 + (1 - a^2)\phi_0^2]^{3/2}}\right]. \tag{7.40}$$

Though at $a\phi_0 = 1$ the quantity $(1 - a^2\phi_0^2)$ is zero, the expression in square brackets is zero as well. Let us single out the term

$$\frac{(1 - a^2\phi_0^2)^2}{(1 - a^2\phi_0^2)^2 [1 + (1 - a^2)\phi_0^2]^{3/2}} = \frac{1}{[1 + (1 - a^2)\phi_0^2]^{3/2}}.$$

If one put in it $a\phi_0 = 1$ one has $1/\phi_0^3$. In the rest part, evaluating an indeterminate form, finally one has:

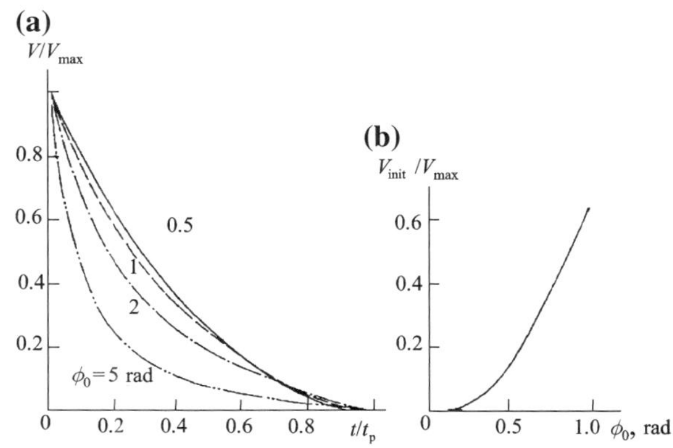

Fig. 7.7 Shapes of the free precession signal for an infinitely broad NMR spectrum in the presence of the η inhomogeneity at different values of the rf pulse amplitude: ϕ_0 = 0.5, 1, 2, and 5 rad (**a**). The initial amplitude of the free precession signal versus the rf pulse amplitude (**b**)

$$V_{\text{init}} = \frac{V_0 \pi \eta_0 M_0}{t_{\text{p}}} \left(\frac{\phi_0}{4} - \frac{1}{\phi_0^2} + \frac{3}{4\phi_0^3} \right).$$

The initial amplitude of the free precession signal, at ($t = 0$), is expressed by a very simple formula:

$$V_{\text{init}} = \frac{V_0 \pi \eta_0 M_0}{t_{\text{p}}} f(\phi_0) \left(1 - \frac{1}{(1 + \phi_0^2)^{3/2}} \right) = \frac{V_0 \pi \eta_0 m_0}{t_{\text{p}}} f(\phi_0). \qquad (7.41)$$

The results obtained by applying Eqs. (7.40) and (7.41) are plotted in Fig. 7.7a and b, respectively. If there is a dispersion of the enhancement factor, that is always the case for magnetically ordered compounds, the free precession decay is always monotonous and is faster the higher the rf field amplitude is (see the curve for $\phi_0 = 5$ rad in Fig. 7.7a). The initial amplitude of the free precession signal at low values of the rf field amplitude changes essentially nonlinearly, but then this dependence becomes linear (see Fig. 7.7b). At $t = t_{\text{p}}$ similar to diamagnetics the amplitude of the free precession signal in magnetically ordered compounds is equal to zero.

7.7.2 Spin Echo

The echo signal can be calculated by the similar way using Eqs. (1.90) and (1.91). As the calculations are rather tedious here we present only the final result.

For two pulses of arbitrary durations and equal amplitudes the formulas are following:

$$Q(t) = -2Q_1 + Q_2 - \text{sgn}(t_2 - t_1) Q_3 + \text{sgn}(t)(-2Q_4 - 2Q_5 + Q_6 + Q_7),$$

where

$$Q_1 = \frac{3\theta\phi_0^2}{5t_1}\left[(2+3\phi^2)\left(\frac{5P_0}{z^3}+\frac{P_{50}}{\phi_0}\right)+\frac{P_{33}-3P_{11}}{z^2}+3P_{51}-\frac{3P_{52}+2P_{30}}{\phi_0}\right];$$

$$Q_4 = Q_1 + \frac{15\theta\phi_0^3}{14t_1}(C_1+C_2+C_3+C_4+C_5);$$

$$C_1 = \frac{7\phi(3\phi^2+4)}{z^4}\left(P_{33}-P_{11}+\frac{P_{77}}{7}-\frac{3P_{55}}{5}\right);$$

$$C_2 = -(4+3\phi^2)(1+2a)P_{70}+\frac{3}{\phi_0^2}\left(\frac{8P_{30}}{15}+P_{74}+\frac{4P_{52}}{5}\right);$$

$$C_3 = [3\phi\phi_0-(2+a)(4+3\phi^2)]\frac{P_0/z^3-P_{71}}{\phi_0};$$

$$C_4 = \left(3+6a-3a^2-\frac{4}{\phi_0^2}\right)\left(\frac{2P_{50}}{5}+P_{72}\right);$$

$$C_5 = 3\frac{2+a}{\phi_0}\left[\frac{3P_{11}-P_{33}}{5z^2}-P_{73}-\frac{3P_{51}}{5}\right];$$

$$P_0 = P_{11}-\frac{2P_{33}}{3}+\frac{P_{55}}{5},\quad P_{nk} = \frac{\phi_0^k}{(\phi_0^2-\phi^2+1)^{n/2}}-\phi^k;$$

$$z = 1-\phi^2,\quad \phi_0 = \eta_0\omega_1 t_{\mathrm{p1}},\quad \phi = \eta_0\omega_1|t|,\quad a = \phi/\phi_0 = |t|/t_{\mathrm{p1}};$$

here the time t is counted off from the time moment τ after the second rf pulse; the other terms Q_i are determined by the similar way described in Sect. 1.8.4, Q_2 can be obtained by substituting $t_{\mathrm{p1}}+t_{\mathrm{p2}}$ for t_{p1} in the expression for Q_1; by analogy one can find Q_6 from Q_4; Q_3 and Q_7 from Q_1 and Q_4, respectively, substituting $|t_{\mathrm{p2}}-t_{\mathrm{p1}}|$ for t_{p1}, and finally Q_5 from Q_4 substituting t_{p2} for t_{p1}.

Contrary to the spin precession signal the spin echo form can be rather various and even have several peaks. The number of the peaks depends on the ratio between the durations of the pulses. A relatively simple shape is if pulses are of equal durations or if one is much shorter than another. But the simplest spin echo shape can be obtained for low values of the pulse amplitudes, when the majority of spin packets rotates to small (compared to $\pi/2$) angles. However, in such cases the intensity of the spin echo signal is rather low. If $\phi_0 \ll 1$ the dispersion of the enhancement factor does not affect the shape of the spin echo signal and one can apply Eqs. (1.104) and (1.105). The shape of the echo signal for different ratio between the pulse durations is shown in Fig. 7.8.

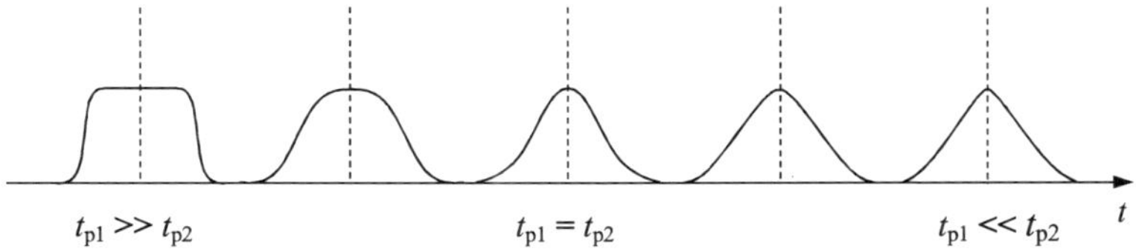

Fig. 7.8 Shapes of the spin echo signal for low values of the rf pulse amplitude but different ratio between the durations of two pulses

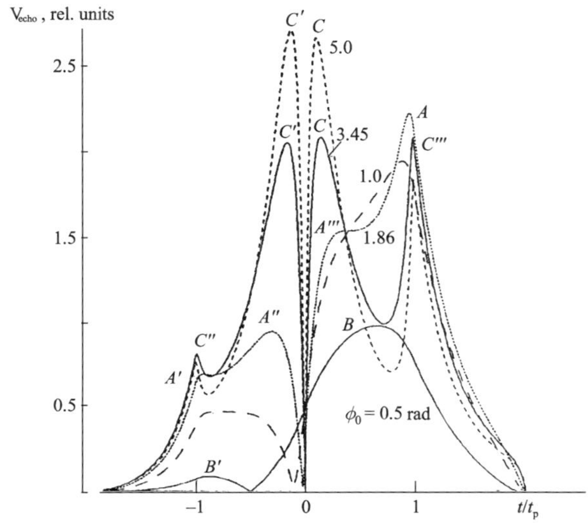

Fig. 7.9 Shapes of the spin echo signal for an infinitely broad NMR spectrum at different values of the rf pulse amplitude: $\phi_1 = \phi_2 \equiv \phi_0 =$ 0.5, 1.0, 1.86, 3.45, and 5 rad

If the integral amplitudes of the pulses are rather large, the shape of the spin echo envelope becomes more complex (see Fig. 7.9). First, let us suppose that both durations and amplitudes of the pulses are equal. At $\phi_0 = 0.5$ rad an additional peak of important intensity, B', appears on the left of the main echo signal,[11] whereas the main echo signal B becomes slightly asymmetric (the solid line in Fig. 7.9). The higher the integral amplitude of the pulses, the more intensive the left peak is, the main line becomes more and more asymmetric and its maximum shifts more and more to the right. At $\phi = 1.86$ rad a new four-peak echo shape appears (see the peaks A', A'', A''' and A at the dotted line in Fig. 7.9). For $\phi_0 = 3.45$ rad it becomes even more pronounced (the solid line in Fig. 7.9): there are two central peaks close to each other, C and C', and two peaks more distant from the center, C'' and C'''. The peaks C'' and C''' appear at time moment close to $\pm t_p$. It is clearly seen from Fig. 7.9 that the location of the peaks C and C' depends on the ϕ_0 value (compare locations of the C' and C peaks on the curves for $\phi_0 = 5.0$ and $\phi_0 = 3.45$ rad). Further increasing of the pulse amplitude results in increasing the intensity of the

[11] In fact, it always exists but at low ϕ_0 values its intensity is negligibly low.

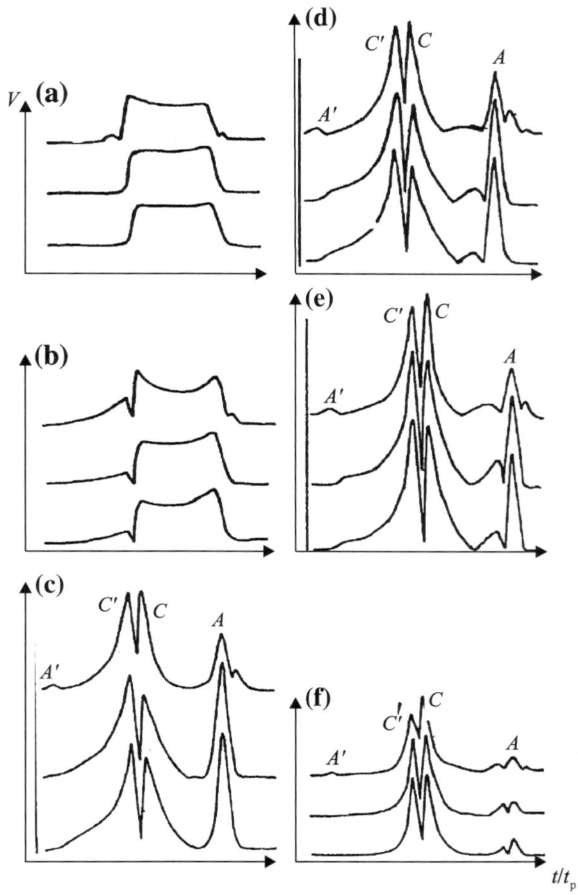

Fig. 7.10 Dependencies of the spin echo shape on the integral intensity of the first pulse if $t_{p2} \ll t_{p1}$. The intensity increases from a to f

central peaks, the intensity of outer peaks decreases. At $\phi_0 = 3.45$ rad the peaks C' and C are of equal intensity.

It is to be noted that for an infinitely broad NMR spectrum the peak intensity increases unlimitedly and linearly with ϕ_0, as more distant parts of the spectrum participate in the spin echo formation. However, according to calculations (see Fig. 7.10), at a certain value of the rf pulse amplitude for a broad but finite spectrum the increasing of the central peaks intensity becomes slower and does not change within a rather large range of the rf pulse amplitude.

At unequal durations of rf pulses the shape of the spin echo envelope is much more complex than represented in Fig. 7.9. If $t_{p1} < t_{p2}$, the left part of the spin echo becomes more complex, if $t_{p1} > t_{p2}$, the right part. The location of the peak C'' depends on the duration of the second pulse. Besides these peaks, another one peak appears, and its location is determined by the $t_{p1} - t_{p2}$ value.

An essential difference in pulse durations leads to another one characteristic shape of the spin echo envelope. If $t_{p1} \gg t_{p2}$ with the rf pulse amplitude increasing the envelope shape changes from a trapezoidal (Fig. 7.10a, c) to a certain characteristic shape (Fig. 7.10f). The main differences are following: (i) the left peak A' almost disappears; (ii) the right peak A is split, the location of the right part does not change whereas the location of the left part depends on the duration of the first pulse (see Fig. 7.10c, d, e).

7.8 Coupled Electron–Nucleus System

Hyperfine interaction leads not only to a large local magnetic fields at the nuclei, but to magnetic fields on electron shells induced by nucleus magnetic moments (it is the field that causes the hyperfine structure of EPR spectra). The constant component of this field adds with the anisotropy field and affects the precession frequency of the electron magnetization and the enhancement factor of the rf field. The variable component adds with the external alternating field vector, and the total effect is hard to predict in advance. It is clear that a significant effect occurs when fields created by nuclei are comparable in magnitude to the external magnetic field. And this is the case if the nuclei have large magnetic moments and are at temperature low enough (when the nuclear magnetization is important). High η values also enhance the effect, because in real experiments the applied external field are rather low. Therefore, a more precise statement of the problem requires to study both systems related by the hyperfine interaction.

7.8.1 Eigenfrequencies of the Electron System

Within a classic approach and neglecting relaxation effects the behavior of both electron and nucleus magnetizations is described by two vector (or six scalar) equations:

$$\frac{d\vec{M}_e}{dt} = \gamma_e[\vec{M}_e \times \vec{B}_e],$$
$$\frac{d\vec{M}}{dt} = \gamma[\vec{M} \times \vec{B}_{hf}], \tag{7.42}$$

where $\vec{M}_e$ and $\vec{M}$ are the vectors of the electron and nucleus magnetizations, respectively; $\vec{B}_e$ and $\vec{B}_{hf}$ are effective fields determined by the following way:

$$\vec{B}_e = \vec{B}_a - A_0\vec{M}, \qquad \vec{B}_{hf} = -A_0\vec{M}_e,$$

here A_0 is the hyperfine interaction constant.

Let the z-axis directs along the anisotropy field vector $\vec{B}_a$. At the equilibrium state

$$M_{ez} = M_{e0}, \quad M = M_0, \quad M_{ex} = M_{ey} = M_x = M_y = 0.$$

In this section we consider only small oscillations of the nucleus and electron magnetizations near the equilibrium (in Sect. 7.8.2 this limitation is removed) and find eigenfrequencies of these oscillations. In this case Eq. (7.42), which are nonlinear, can be linearized if one lets $M_{ez} = M_{e0}$, $M_z = M_0$, and supposes that M_{ex}, M_{ey}, M_x and M_y are small quantities.[12] Let us consider Eq. (7.42), but skip the equations for $\mathrm{d}M_{ez}/\mathrm{d}t$ and $\mathrm{d}M_z/\mathrm{d}t$, as within the chosen approximation M_{ez} and M_z do not depend on time:

$$\begin{aligned}
\frac{\mathrm{d}M_{ex}}{\mathrm{d}t} &= \gamma_e[M_{ey}(B_a - A_0 M_0) + M_{e0} A_0 M_y],\\
\frac{\mathrm{d}M_{ey}}{\mathrm{d}t} &= \gamma_e[-M_{e0} A_0 M_x - M_{ex}(B_a - A_{e0} M_0)],\\
\frac{\mathrm{d}M_x}{\mathrm{d}t} &= \gamma[-M_y A_0 M_{e0} + M_0 A_0 M_{ey}],\\
\frac{\mathrm{d}M_y}{\mathrm{d}t} &= \gamma[-M_0 A_0 M_{ex} + M_x A_0 M_{e0}].
\end{aligned} \tag{7.43}$$

Let us consider the first pair of equations of (7.43): multiply the second equation by i and summarize with the first one. After introducing new variables $M_e^+ = M_{ex} + \mathrm{i}M_{ey}$, $M^+ = M_x + \mathrm{i}M_y$, one obtains:

$$\frac{\mathrm{d}M_e^+}{\mathrm{d}t} = -\mathrm{i}\gamma_e[M_e^+(B_a - A_0 M_0) + M^+ M_{e0} A_0]. \tag{7.44}$$

By the similar way multiplying the forth equation of (7.43) by i and summarizing it with the third equation one has:

$$\frac{\mathrm{d}M^+}{\mathrm{d}t} = -\mathrm{i}\gamma[-M^+ A_0 M_{e0} + M_e^+ M_0 A_0]. \tag{7.45}$$

By analogy it is possible to multiply the second and forth equations of (7.43) by i and subtract them from the first and third equations, respectively. Then, introducing $M_e^- = M_{ex} - \mathrm{i}M_{ey}$, $M^- = M_x - \mathrm{i}M_y$, one obtains:

$$\begin{aligned}
\frac{\mathrm{d}M_e^-}{\mathrm{d}t} &= \mathrm{i}\gamma_e[M_e^-(B_a - A_0 M_0) + M^- M_{e0} A_0],\\
\frac{\mathrm{d}M^-}{\mathrm{d}t} &= \mathrm{i}\gamma[-M^- A_0 M_{e0} + M_e^- M_0 A_0].
\end{aligned} \tag{7.46}$$

[12] Indeed, M_{ez} and M_z differ from M_{e0} and M_0, but the difference is of the second order of smallness.

Supposing that both nucleus and electron magnetizations are harmonic functions of time one has:

$$M_{\mathrm{e}}^{\pm} = M_{\mathrm{e}A}^{\pm} e^{\mathrm{i}\omega t}, \quad M^{\pm} = M_{A}^{\pm} e^{\mathrm{i}\omega t}, \tag{7.47}$$

here $M_{\mathrm{e}A}^{\pm}$ and $M_{A}^{\pm}$ are the amplitudes of the precessing vectors $\vec{M}_{\mathrm{e}}$ and $\vec{M}$. Substituting Eq. (7.47) into Eqs. (7.44)–(7.46) and canceling by $\mathrm{i}e^{\mathrm{i}\omega t}$, one comes to two systems of equations, one to determine the electron and nucleus magnetizations $M_{\mathrm{e}A}^{+}$ and M_{A}^{+}, and another to determine $M_{\mathrm{e}A}^{-}$ and M_{A}^{-}:

$$\begin{aligned}
\omega M_{\mathrm{e}A}^{+} &= -\gamma_{\mathrm{e}}[M_{\mathrm{e}A}^{+}(B_{\mathrm{a}} - A_0 M_0) + M_{A}^{+} M_{\mathrm{e}0} A_0],\\
\omega M_{A}^{+} &= -\gamma[-M_{A}^{+} A_0 M_{\mathrm{e}0} + M_{\mathrm{e}A}^{+} M_0 A_0];\\
\omega M_{\mathrm{e}A}^{-} &= \gamma_{\mathrm{e}}[M_{A}^{-}(B_{\mathrm{a}} - A_0 M_0) + M_{A}^{-} M_0 A_0],\\
\omega M_{A}^{-} &= \gamma[-M_{A}^{-} A_0 M_{\mathrm{e}0} + M_{\mathrm{e}A}^{-} M_0 A_0].
\end{aligned}$$

This is the main advantage of introducing the $M_{\mathrm{e}}^{\pm}$ and $M^{\pm}$ quantities. After writing it in a standard form one has the following system of equations:

$$\begin{aligned}
&\omega M_{\mathrm{e}A}^{+} + \gamma_{\mathrm{e}} M_{\mathrm{e}A}^{+}(B_{\mathrm{a}} - A_0 M_0) + \gamma_{\mathrm{e}} M_{A}^{+} M_{\mathrm{e}0} A_0 = 0,\\
&\gamma M_{\mathrm{e}A}^{+} M_0 A_0 + \omega M_{A}^{+} - \gamma M_{A}^{+} A_0 M_{\mathrm{e}0} = 0,\\
&\omega M_{\mathrm{e}A}^{-} - \gamma_{\mathrm{e}} M_{\mathrm{e}A}^{-}(B_{\mathrm{a}} - A_0 M_0) + \gamma_{\mathrm{e}} M_{A}^{-} M_{\mathrm{e}0} A_0 = 0,\\
&\gamma M_{\mathrm{e}A}^{-} M_0 A_0 - \omega M_{A}^{-} - \gamma M_{A}^{-} A_0 M_{\mathrm{e}0} = 0.
\end{aligned} \tag{7.48}$$

The system of homogeneous equations (7.48) has a nonzero solution only if its determinant is equal to zero. For the first and second pairs of equations in (7.48) one has, respectively:

$$\begin{vmatrix} \omega + \gamma_{\mathrm{e}} B_{\mathrm{a}} + \gamma_{\mathrm{e}} A_{\mathrm{e}0} M_0 & \gamma_{\mathrm{e}} A_0 M_{\mathrm{e}0} \\ \gamma A_0 M_0 & \omega - \gamma A_0 M_{\mathrm{e}0} \end{vmatrix} = 0,$$

and

$$\begin{vmatrix} \omega - \gamma_{\mathrm{e}} B_{\mathrm{a}} - \gamma_{\mathrm{e}} A_0 M_0 & \gamma_{\mathrm{e}} A_0 M_{\mathrm{e}0} \\ \gamma A_0 M_0 & -\omega - \gamma A_0 M_{\mathrm{e}0} \end{vmatrix} = 0.$$

After expanding these determinants one obtains quadratic equations, which differ one from another only by the sign, to determine eigenfrequencies of the this coupled electron–nucleus system (the upper sign corresponds to the first pair of equations, the lower corresponds to the second pair:

$$\omega^2 \pm \omega(\gamma_{\mathrm{e}} B_{\mathrm{a}} - \gamma_{\mathrm{e}} A_0 m M_0 - \gamma A_0 M_{\mathrm{e}0}) - \gamma_{\mathrm{e}} \gamma B_{\mathrm{a}} A_0 M_{\mathrm{e}0} = 0. \tag{7.49}$$

Now, taking into account that the electron gyromagnetic ratio is negative one writes:

$$\omega^2 \mp \omega(|\gamma_e|B_a - |\gamma_e|A_0M_0 + \gamma A_0M_{e0}) + |\gamma_e|\gamma B_a A_0 M_{e0} = 0.$$

The solutions for the first and second pairs of Eq. (7.48) are following:

$$\omega_{1,2} = \frac{1}{2}(|\gamma_e|B_a - |\gamma_e|A_0M_0 + \gamma A_0M_{e0}) \pm \sqrt{\frac{1}{4}(|\gamma_e|B_a - |\gamma_e|A_0M_0 + \gamma A_0M_{e0})^2 - |\gamma_e|\gamma B_a A_0 M_{e0}};$$

$$\omega_{3,4} = -\frac{1}{2}(|\gamma_e|B_a - |\gamma_e|A_0M_0 + \gamma A_0M_{e0}) \pm \sqrt{\frac{1}{4}(|\gamma_e|B_a - |\gamma_e|A_0M_0 + \gamma A_0M_{e0})^2 - |\gamma_e|\gamma B_a A_0 M_{e0}}. \quad (7.50)$$

The expression under the square root is always greater than the expression before the root. Moreover, the frequency is an essentially positive quantity. Hence, in Eq. (7.50) one should take the sign "+" before the root for each pair of solutions.

The expression under the root can be simplified:

$$\frac{1}{4}(|\gamma_e|B_a - |\gamma_e|A_0M_0 + \gamma A_0M_{e0})^2 - |\gamma_e|\gamma B_a A_0 M_{e0} = \frac{1}{4}(|\gamma_e|B_a - |\gamma_e|A_0M_0 - \gamma A_0M_{e0})^2 + |\gamma_e|\gamma A_0^2 M_{e0} M_0.$$

Hereby

$$\omega_{1,2} = \frac{1}{2}(|\gamma_e|B_a - |\gamma_e|A_0M_0 - \gamma A_0M_{e0}) + \sqrt{\frac{1}{4}(|\gamma_e|B_a - |\gamma_e|A_0M_0 - \gamma A_0M_{e0})^2 + |\gamma_e|\gamma A_0^2 M_{e0} M_0}. \quad (7.51)$$

The largest term in Eq. (7.51) is $|\gamma_e|B_a$. Neglecting other terms, within the first-order approximation

$$\omega_1 = |\gamma_e|B_a.$$

Hence ω_1 is the frequency of ferromagnetic resonance (FMR) if one neglects interactions of the electron magnetization with the nucleus one.

Let us come back to Eq. (7.51). The first term in parentheses under the root is the FMR frequency; the second term represents how this frequency changes under the action of a maximum possible field created by the nucleus magnetization; the third term is the NMR frequency in the local field. The term beyond the parentheses is the product of the NMR frequency and the changes in the FMR frequency. Supposing that the second and third terms are much less than the first one, i. e. are the quantities of the first-order smallness, the term beyond the parentheses is of the second-order smallness and can be neglected. Then one can extract the square root:

$$\omega_1 = |\gamma_e|B_a - |\gamma_e|A_0M_0 = |\gamma_e|B_a\left(1 - \frac{A_0M_0}{B_a}\right).$$

Multiplying the both numerator and denominator of the fraction in the parentheses of the last expression by M_{e0} and taking into account that $\eta = B_{loc}/B_a$, and $B_{loc} = -A_0M_{e0}$, one obtains

$$\omega_1 = |\gamma_e|B_a\left(1 + \frac{\eta M_0}{M_{e0}}\right).$$

The NMR frequency can be calculated even easier. Taking into account that the product of the quadratic equation roots is the free term of this equation, for the equation roots of Eq. (7.49) one can write:

$$\omega_1\omega_2 = A_0M_{e0}B_a\gamma|\gamma_e|$$

or

$$\omega_2|\gamma_e|B_a\left(1 + \frac{\eta M_0}{M_{e0}}\right) = A_0M_{e0}B_a\gamma|\gamma_e|.$$

Canceling the same terms on the left and on the right and considering that at small x values $(1+x)^{-1} = 1 - x$, one has:

$$\omega_2 = \gamma B_{loc}\left(1 - \frac{\eta M_0}{M_{e0}}\right).$$

Despite the ratio M_0/M_{e0} is quite small, at high values of the enhancement factor the frequency shift can be rather noticeable. Moreover, as everywhere in our computations we replaced M_z by M_0 it is possible to consider that resonance frequencies of a strongly coupled electron–nucleus system depend on the z-component of the nucleus magnetization, M_z, but to prove this assumption on has to make such calculations, in which deviations of the nucleus magnetization vector from its equilibrium state M_0 are not small anymore. It will be discussed in the further section.

7.8.2 Behavior of the Nuclear Magnetization Vector

Let us come back to the system of Eq. (7.42). To solve it easier one assume that the NMR frequency and the frequency of the external rf field essentially differ from the eigenfrequencies of the electron system. Then one supposes that in every specific time moment the electron magnetization vector is parallel to the magnetic field vector, which is the sum of an anisotropy field, an external magnetic field and an alternating field created by the nucleus magnetization. If the anisotropy field vector directs along the z-axis, the vectors of alternating fields lye within the xy plane, and the enhancement factor has an axial symmetry along the z-axis, then it is easy to set up an equation, describing the behavior of the nucleus magnetization taking into account the feedback of the nucleus system on the electron one.

First of all one should describe in explicit form the fields created by the nucleus system. For example, the u component creates a rf field of the amplitude $B_{eu} = -A_0 u$. Under the action of this field an alternating electron magnetization appears:

$$M_{eu} = M_{e0}\frac{B_{eu}}{B_a} = -\frac{A_0 M_{e0} u}{B_a} = \eta u.$$

The alternating electron magnetization in its turn creates an alternating field at the nucleus sites:

$$B_u = -A_0 M_{eu} = \eta A_0 u.$$

By analogy v creates at the nucleus sites an alternating field proportional to v:

$$B_v = \eta A_0 v.$$

Hereby, at the presence of an external alternating magnetic field there are two components that affect nuclei: the component, which is in phase with this field

$$B_{in} = \eta b_1 + B_u = \eta b_1 + \eta A_0 u, \tag{7.52}$$

and the component, which is in advance the external field at $\pi/2$:

$$B_{ad} = B_v = \eta A_0 v. \tag{7.53}$$

Of cause in such an approach one does not take into account nonlinearities of the electron system. Moreover, one neglects the influence of the static field on the electrons (the field created by the nuclear magnetization M_z). The latter means that the anisotropy field is much greater the maximum value of the static field created by the nucleus system ($B_a \gg A_0 M_0$).

Thereby, there are two rotating fields affecting the nucleus system. A transformation to the magnetizations in the rotating frame (rotating together with the external field) was made only for one field. That is why one has to repeat it again.

Let two rf field with the amplitudes B_{10} and B_{20} affect the nucleus magnetization:

$$B_{1x} = B_{10}\cos\omega t, \qquad B_{1y} = -B_{10}\sin\omega t,$$

$$B_{2x} = B_{20}\sin\omega t, \qquad B_{2y} = B_{20}\cos\omega t.$$

In addition, of cause, there is a static magnetic field B_{loc}.

The rotating components of the magnetization can be written in the same form as before:

$$\begin{aligned} M_x &= u\cos\omega t - v\sin\omega t, \\ M_y &= -u\sin\omega t - v\cos\omega t. \end{aligned} \tag{7.54}$$

Then let us compute the derivatives $\mathrm{d}M_x/\mathrm{d}t$, $\mathrm{d}M_y/\mathrm{d}t$ and substitute them and M_x, M_y on the form (7.54) into the second equation of the system (7.42). Writing it for individual components one has a system of three differential equations. Equating coefficients at $\cos\omega t$ and $\sin\omega t$, one obtains a rather simple system of two differential equations, which does not contain terms changing with frequencies close to ω:

$$\begin{aligned} \frac{\mathrm{d}u}{\mathrm{d}t} &= -\Delta\omega v + \gamma B_{20} M_z, \\ \frac{\mathrm{d}v}{\mathrm{d}t} &= \Delta\omega u - \gamma B_{10} M_z, \end{aligned} \tag{7.55}$$

$$\frac{\mathrm{d}M_z}{\mathrm{d}t} = \gamma(uB_{20} + vB_{10}). \tag{7.56}$$

If one puts into Eqs. (7.55) and (7.56) B_{10} and B_{20} in the form of (7.52) and (7.53), respectively, one obtains the following system of equations:

$$\begin{aligned} \frac{\mathrm{d}u}{\mathrm{d}t} &= -\Delta\omega v + \gamma\eta A_0 M_z v, \\ \frac{\mathrm{d}v}{\mathrm{d}t} &= \Delta\omega u - \gamma M_z \eta b_1 - \gamma\eta A_0 M_z u, \\ \frac{\mathrm{d}M_z}{\mathrm{d}t} &= \gamma v\eta b_1, \end{aligned} \tag{7.57}$$

where, as before, $\Delta\omega = \omega_0 - \omega = \gamma B_{\mathrm{loc}} - \omega$. This system of equations can be reduced to the only differential equation of the second order for M_z, whereas u and v can be expressed trough M_z. First let us do it for $\Delta\omega = 0$. Then Eq. (7.57) has the following form

$$\begin{aligned} \frac{\mathrm{d}u}{\mathrm{d}t} &= \gamma\eta A_0 M_z v, \\ \frac{\mathrm{d}v}{\mathrm{d}t} &= -\gamma M_z \eta b_1 - \gamma\eta A_0 M_z u, \\ \frac{\mathrm{d}M_z}{\mathrm{d}t} &= \gamma v\eta b_1. \end{aligned} \tag{7.58}$$

After dividing the first equation by the third one, one obtains:

$$\frac{\mathrm{d}u}{\mathrm{d}M_z} = \frac{A_0}{b_1} M_z.$$

Then, separating variables and integrating:

$$u = \frac{A_0}{2b_1} M_z^2 + C.$$

As at $t = 0$ $M_z = M_0$ and $u = 0$, then $C = -(A_0/(2b_1))M_0^2$ and

$$u = \frac{A_0}{2b_1}(-M_0^2 + M_z^2). \tag{7.59}$$

Let us divide the second of equations in (7.58) by the first one:

$$\frac{\mathrm{d}v}{\mathrm{d}u} = -\frac{u}{v} - \frac{b_1}{A_0 v},$$

then separate variables and integrate. One has

$$A_0 v^2 = -A_0 u^2 - 2b_1 u + C.$$

The constant $C = 0$, as at $t = 0$ $u = 0$ and $v = 0$. Thereby

$$v^2 = -\frac{2b_1 u}{A_0} - u^2.$$

To this one can add an obvious relation

$$u^2 + v^2 + M_z^2 = M_0^2.$$

Substituting Eq. (7.59) into the right part of the second equation in (7.58), one writes:

$$\frac{\mathrm{d}v}{\mathrm{d}t} = -\gamma M_z \eta b_1 + \frac{\gamma\eta A_0^2 M_z}{2b_1}(M_0^2 - M_z^2).$$

Differentiating the third of equations in (7.58) and substituting into it the obtained value of $\mathrm{d}v/\mathrm{d}t$, one has

$$\frac{\mathrm{d}^2 M_z}{\mathrm{d}t^2} = -\gamma^2\eta^2 b_1^2 M_z + \frac{\gamma^2\eta^2 A_0^2 M_z}{2}(M_0^2 - M_z^2).$$

Let us denote the right part of this equation by $f(M_z)$. Then

$$\frac{\mathrm{d}^2 M_z}{\mathrm{d}t^2} = f(M_z).$$

Since

$$\frac{\mathrm{d}}{\mathrm{d}t}\left(\frac{\mathrm{d}M_z}{\mathrm{d}t}\right)^2 = 2\frac{\mathrm{d}M_z}{\mathrm{d}t}\frac{\mathrm{d}^2 M_z}{\mathrm{d}t^2} = 2\frac{\mathrm{d}M_z}{\mathrm{d}t} f(M_z),$$

$$\mathrm{d}\left(\frac{\mathrm{d}M_z}{\mathrm{d}t}\right)^2 = 2f(M_z)\mathrm{d}M_z,$$

then

$$\left(\frac{\mathrm{d}M_z}{\mathrm{d}t}\right)^2 = 2\int f(M_z)\mathrm{d}M_z + C = 2\eta^2\left[-\frac{\gamma^2 b_1^2}{2}M_z^2 + \frac{\gamma^2 A_0^2 M_0^2}{4}M_z^2 - \frac{\gamma^2 A_0^2}{8}M_z^4\right] + C.$$

Thus,

$$\frac{\mathrm{d}M_z}{\mathrm{d}t} = \eta\sqrt{-\gamma^2 b_1^2 M_z^2 + \frac{\gamma^2 A_0^2 M_0^2}{2}M_z^2 - \frac{\gamma^2 A_0^2}{4}M_z^4 + C}.$$

To determine the constant C mind that at $t = 0$, $M_z = M_0$ and $\mathrm{d}M_z/\mathrm{d}t = 0$. The last condition follows from the third equation of (7.58). Then

$$C = \gamma^2 b_1^2 M_0^2 - \frac{\gamma^2 A_0^2 M_0^4}{4},$$

$$\frac{\mathrm{d}M_z}{\mathrm{d}t} = \frac{\gamma\eta A_0}{2}\sqrt{-M_z^4 - \left(2M_0^2 + \frac{4b_1^2}{A_0^2}\right)M_z^2 + \frac{4M_0^2 b_1}{A_0^2} - M_0^4}. \tag{7.60}$$

The expression under the square root is a fourth-degree polynomial function, and its roots are $M_{z1}^2 = M_0^2$, $M_{z2}^2 = M_0^2 - 4b_1^2/A_0^2$. That is why Eq. (7.60) can be represented in the following form

$$\frac{\mathrm{d}M_z}{\mathrm{d}t} = \frac{\gamma\eta A_0}{2}\sqrt{-(M_z^2 - M_0^2)\left(M_z^2 - M_0^2 + \frac{4b_1^2}{A_0^2}\right)}.$$

Separating variables and integrating one has

$$\int_{M_0}^{M_z} \frac{\mathrm{d}M_z}{\sqrt{(M_0^2 - M_z^2)\left(M_z^2 - M_0^2 + 4b_1^2/A_0^2\right)}} = \frac{\gamma\eta A_0 t}{2}.$$

Let us introduce the following variables: a dimensionless z-component of the nucleus magnetization $Z = M_z/M_0$; an integral intensity of the alternating local field $\tau = \gamma\eta A_0 M_0 t/2$ created by the nucleus subsystem; an enhancement factor of an external linearly polarized field by the nucleus subsystem $k = A_0 M_0/(2b_1)$. Then instead of the last equation one obtains

$$\tau = M_0 \int_1^Z \frac{\mathrm{d}Z}{\sqrt{(1 - Z^2)\left(Z^2 - 1 + k^{-2}\right)}}. \tag{7.61}$$

If $2b_1 > A_0 M_0$, then making the change of variables: $\sin\phi = \sqrt{1 - M_z/M_0}$, the integral in the right part of Eq. (7.61) is a standard elliptic integral:

$$\tau = -k \int_1^{\arccos Z} \mathrm{d}\phi \sqrt{1 - k^2 \sin^2 \phi}.$$

Thus

$$\tau = -kF\left(\arccos(Z), k\right),$$

where F is an incomplete first-order elliptic integral.

If $2b_1/(A_0 M_0) < 1$, then the integral in the right part of Eq. (7.61) can be reduced to the standard elliptic integral:

$$\tau = \int_1^{\arcsin k^{-1}\sqrt{1-Z^2}} \frac{\mathrm{d}\phi}{\sqrt{1 - k^{-2}\sin^2 \phi}}.$$

Hereby,

$$\tau = F\left(\arcsin k^{-2}\sqrt{1 - Z^2}, k^{-1}\right).$$

If the local alternating field created by the nucleus subsystem is equal to the external field b_1, i. e. $k = 1$, then Eq. (7.61) is reduced to

$$\int_1^Z \frac{\mathrm{d}Z}{Z\sqrt{1 - Z^2}} = \tau.$$

The solution of this equation can be obtained in elementary functions:

$$\ln \frac{1 + \sqrt{1 - Z^2}}{Z} = -\tau.$$

Then the dimensionless z-component of the nucleus magnetization is

$$Z = \frac{2\exp(-\tau)}{1 + \exp(-2\tau)}.$$

For the dimensionless x- and y-components of the nucleus magnetization one can obtain the following expressions:

$$X = \frac{1 - \exp(-2\tau)^2}{1 + \exp(-2\tau)^2},$$

$$Y = \frac{\exp(-2\tau)\sqrt{\exp(-4\tau) + 2\exp(-2\tau) - 3}}{[1 + \exp(-2\tau)]^2}.$$

In a general case if there is a frequency mismatch $\Delta\omega$ between the frequency of the external alternating field ω and the eigenfrequency of the nucleus subsystem ($\omega_0 = \gamma B_{\mathrm{loc}}$), the development of formulas for nucleus magnetization components becomes more complex yet analogous. To the dimensionless parameters introduced above one should add another one related to the mismatch

$$q = \frac{\Delta\omega}{\gamma\eta A_0 M_0},$$

which is the ratio between the frequency mismatch and the local alternating field created by the nucleus subsystem. The dimensionless projections of the magnetization X and Y are expressed, as before, through the projection Z:

$$X = qk(Z - 1) - k(Z^2 - 1) = k(Z - 1)(q - Z - 1),$$

$$Y^2 = k^2[(1 - Z^4) + 2q(Z^3 - 1) - (q^2 + 2q - 2 + k^{-2})(Z^2 - 1) + 2q(q - 1)(Z - 1)].$$

The variable Z can be found by solving a differential equation of the first order similar to Eq. (7.60):

$$\frac{\mathrm{d}Z}{\mathrm{d}t} = -\sqrt{(1 - Z)Q(Z)}, \tag{7.62}$$

$$Q(Z) = Z^3 + (1 - 2q)Z^2 + (q^2 + k^{-2} - 1)Z + 2q - 1 + k^{-2} - q^2. \tag{7.63}$$

If one denotes the roots of Eq. (7.63) by Z_1, Z_2 and Z_3, then the solution of Eq. (7.62) can be represented in the following form:

$$\tau = \int\limits_Z^1 \frac{\mathrm{d}Z}{\sqrt{(1 - Z)(Z - Z_1)(Z - Z_2)(Z - Z_3)}}.$$

The last integral can be reduced to elliptical functions, but the exact form depends on are all the roots of the equation $Q(Z) = 0$ real or not. Let us consider several particular cases:

- if all the roots of the polynom $Q(Z)$ in the form (7.63) are real (let us enumerated them in the order of decreasing, i. e. $1 > Z_1 > Z_2 > Z_3$) and Z may change within a range Z_2–1, then the solution of the differential equation has the following form:

$$\tau = \frac{2}{\sqrt{(1 - Z_2)(Z_1 - Z_3)}} F(\mu, r),$$

where

$$\mu = \arcsin\sqrt{\frac{(Z_1 - Z_3)(1 - Z)}{(1 - Z_1)(Z - Z_3)}}, \quad r = \sqrt{\frac{(1 - Z_1)(Z_2 - Z_3)}{(1 - Z_2)(Z_1 - Z_3)}};$$

- if Z_1 is real, but Z_2 and Z_3 are complex conjugate (where m and n are their real and imaginary parts, respectively), then

$$\tau = \tau_0 - \frac{1}{\sqrt{sp}} F\left(2\mathrm{arctg}\sqrt{\frac{p(1 - Z)}{s(Z - Z_1)}}, \frac{1}{2}\sqrt{\frac{(1 - Z_1)^2 - (s - p)^2}{sp}}\right),$$

where

$$s^2 = (m - 1)^2 + n^2; \; p^2 = (m - Z_1)^2 + n^2; \; \tau_0 = \frac{2}{\sqrt{sp}} F\left(\frac{\pi}{2}, \frac{1}{2}\sqrt{\frac{(1 - Z_1)^2 - (s - p)^2}{sp}}\right).$$

The minimal value of Z is determined by the largest root of the polynom $Q(Z)$, i. e. by Z_1. According to calculations, the dependence of the minimal value of the dimensionless z-component of the nucleus magnetization $M_{z\,\min}$ on the dimensionless mismatch q has a minimum (see Fig. 7.11). At low values of the enhancement factor k this minimum coincides with the NMR frequency, but with k increasing it shifts towards positive q values.

Several extreme cases lead to very simple results. Let us discuss a case when $\Delta\omega = 0$ and the maximal field that nuclei create on electrons (it is equal to $A_0 M_0$) multiplied by the enhancement factor η, which is caused by the electron subsystem, is much greater the external field, i. e. $A_0 M_0 \eta \gg b_1$. It is easy to obtain that

$$M_z = M_0\sqrt{1 - k^{-2}\sin^2\frac{\gamma\eta A_0 M_0 t}{2}},$$

$$u = \frac{M_0}{2k}(1 - \cos\gamma\eta A_0 M_0 t), \quad v = \frac{M_0}{2k}\sin\gamma\eta A_0 M_0 t, \quad k = \frac{A_0 M_0}{2b_1}.$$

Hence, the value of M_z remains always close to the M_0 value, and the components u and v are small compared to M_0. Graphically, the nucleus magnetization vector moves over a conical surface, the cone director makes an angle β with the z-axis

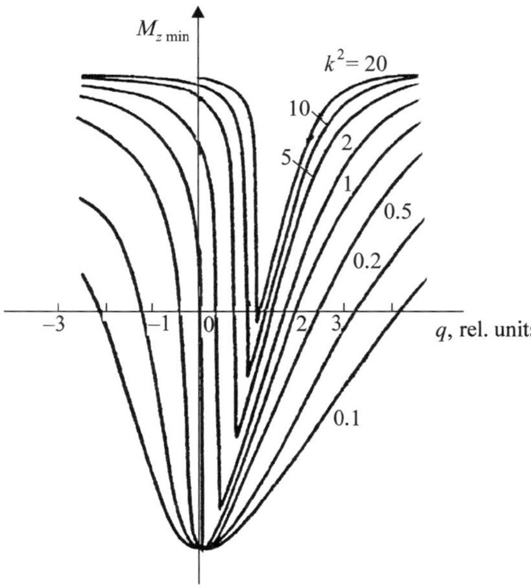

Fig. 7.11 Dependencies of the minimal value of the dimensionless z-component of the nucleus magnetization on the dimensionless mismatch q at different values of the enhancement factor k

($\text{tg}\beta = 1/(2k)$). The frequency of such a motion is determined by the value $\gamma\eta A_0 M_0$, i. e. this frequency is equal to the NMR frequency in the maximal field created by nuclei on electrons multiplied by the enhancement factor. Similar motions occur if the external field affects nuclei with a non-resonance frequency.

All the stated above is in fair agreement with measurements of eigenfrequencies of an electron-nucleus system due to hyperfine interactions. A more detail study exhibits that the amplitude of the nucleus magnetization oscillations increases with deviations of the rf field frequency from the eigenfrequency of the nucleus subsystem. However, it remains low if $\gamma\eta A_0 M_0 \gg b_1$.

Further Readings

Wohlfarth EP (1993) Ferromagnetic materials: a handbook on the properties of magnetically ordered substances. North-Holland Publishing Company, Netherlands

Carretta P, Keren A (2011) In: Lacroix C, Mendels P, Mila F (eds) NMR and μSR in highly frustrated magnets., Introduction to frustrated magnetism. Springer series in soid-state science 164Springer, Berlin, Heidelberg.

Kurkin MI, Turov EA (1990) NMR in magnetically ordered materials and its applications. Nauka, Moscow

Gurevich AG, Melkov GA (1996) Magnetization oscillations and waves. CRC Press, Boca Raton Inc

Meny C, Panissod P (2006) In: Webb G (ed) Nuclear magnetic resonance in ferromagnetic multilayers and nanocomposites: investigations of their structural and magnetic properties in modern magnetic resonance. Springer, Heidelberg

Part II
Nuclear Quadrupole Resonance (NQR)

Chapter 8
Nuclear Quadrupole Resonance

Nuclear quadrupole resonance (NQR) is one of the most sensible tools to study the local structure of materials, electronic density distributions near the nuclei under study, the nature of defects in solids, the mobility of molecules, or their parts, phase transitions and so on. It is very helpful to investigate physical properties of solids, molecular crystals and polymers especially, but can be applied to metals and glasses as well. In liquids quadrupole interactions are averaged to zero and NQR is not applicable.

In contrast to NMR, NQR can be observed in zero or weak magnetic fields. Hence, the technical equipment required to NQR is less complicated that makes wider the area of its application.

The main physical parameter that can be determined from NQR spectra is the tensor of quadrupole interactions, which are interactions between the nuclear quadrupole moment and intercrystalline electric field gradients (EFG). The information about components of the EFG tensor can be obtained by measuring the frequencies of the transitions between the energy levels of this interaction. It is evident that NQR can be observed only for nuclei with $I \geq 1$, which have quadrupole moments. However, that does not restrict the area of studied objects, because almost every nucleus has an isotope with a nonzero quadrupole moment.

NQR frequencies lie within the range from 20 kHz to several GHz. Commonly NQR is used to study objects with strong quadrupole interactions where $e^2qQ/h \geq 1\,\text{MHz}$. At lower frequencies it is rather difficult to obtain a good signal-to-noise ratio and application of NMR is more reasonable. That is why NQR is usually applied to covalent crystals where EFG values are rather high. Ionic or ionic-covalent crystals are objects for NMR study. NQR is also rarely used to study light nuclei, such as lithium, deuterium, sodium, because their quadrupole moments are too small. As a rule, bigger mass number of nucleus corresponds to bigger quadrupole moment (Table 1.1), and hence, bigger e^2qQ/h values. However, if the EFG value at the nucleus site is too small due to symmetry reasons it also leads to low NQR frequencies (see, for example, in Table 8.1 data for ^{27}Al in the yttrium aluminium garnet $Y_3Al_5O_{12}$, where Al nuclei occupy two different sites with different symmetry).

V. I. Chizhik et al., *Magnetic Resonance and Its Applications*,
DOI: 10.1007/978-3-319-05299-1_8,

Table 8.1 Selected data on NQR frequencies in crystals

Compound	Nucleus	eQ (10^{24} cm^2)	I	e^2qQ/h (MHz)
$LiNbO_3$	^{7}Li	$-4.2 \cdot 10^{-2}$	3/2	0.055
	^{93}Nb	−0.28	9/2	22
$Y_3Al_5O_{12}$	^{27}Al	0.149	5/2	6.6
	^{27}Al	0.149	5/2	0.6
CCl_4	^{35}Cl	−0.08249	5/2	35
$LiTaO_3$	^{181}Ta	3.44	7/2	14

8.1 Theory of Nuclear Quadrupole Resonance

Similar to NMR, to obtain the information about physical properties of solids from NQR spectra one has to apply a rf field and to measure frequencies of transitions between energy levels of quadrupole interactions. That means it is necessary to know the energy levels of quadrupole interactions.

8.1.1 Energy Levels, Frequencies and Intensities of NQR Transitions

If a nucleus with a non-zero quadrupole moment ($I > 1/2$) is in an inhomogeneous electric field, its interaction with this field, as it has been shown in Chap. 1, can be described by the Hamiltonian of quadrupole interactions $\widehat{\mathcal{H}}_{\mathrm{Q}}$ (see Eq. 1.197). In the principal axes frame of the EFG tensor, it can be written as

$$\widehat{\mathcal{H}}_{\mathrm{Q}} = \frac{e^2qQ}{4I(2I-1)}\left(3\hat{I}_z{}^2 - \hat{\vec{I}}^2 + \frac{\eta}{2}(\hat{I}_+^2 + \hat{I}_-^2)\right), \tag{8.1}$$

where e is the electron charge; $eq \equiv V_{zz}$ is the principal component of the EFG tensor; Q is the nuclear quadrupole moment; I is the nuclear spin; $\hat{I}_z$ is the nuclear spin projection operator on the direction of the principal axis of the EFG tensor; $\eta = |(V_{xx} - V_{yy})/V_{zz}|$ is the asymmetry parameter of the EFG tensor; $\hat{I}_+$, $\hat{I}_-$ are the ladder operators. To determine the eigenfunctions and energy levels that correspond to this Hamiltonian, one has to solve the stationary Schrödinger equation:

$$\widehat{\mathcal{H}}_{\mathrm{Q}}\Psi_n = E_n\Psi_n.$$

The index n runs over the energy levels and eigenfunctions of the Hamiltonian $\widehat{\mathcal{H}}_{\mathrm{Q}}$.

In general case it is not possible to obtain the exact solution of the Schrödinger equation with such a Hamiltonian. However, if the asymmetry parameter $\eta = 0$, the Schrödinger equation with the Hamiltonian

$$\widehat{\mathcal{H}}_{\mathrm{Q}} = \frac{e^2qQ}{4I(2I-1)}(3\hat{I}_z{}^2 - \hat{\vec{I}}^2) \tag{8.2}$$

can be solved exactly. This Hamiltonian commutes with the operator of the nuclear spin projection $\hat{I}_z$, hence, they have a common set of eigenfunctions. Therefore, energy levels of the quadrupole interaction can be determined as follows:

$$E_m = \frac{e^2qQ}{4I(2I-1)}(3m^2 - I(I+1)). \tag{8.3}$$

As it is clearly seen from Eq. (8.3), the quadrupole interaction energy does not depend on the magnetic quantum number m. Therefore, if the nucleus has a half-integer spin the energy levels of quadrupole interactions are doubly degenerate. This effect is well-known from the Kramers theorem, which states that the energy levels of a nucleus with a half-integer spin remain at least doubly degenerate in the presence of a purely electric field. For a nucleus with an integer spin there are double degenerate levels and one nondegenerate level that corresponds to the magnetic quantum number $m = 0$. Thereby, the number of energy levels for quadrupole interaction is $(I + 1)$ for integer spins and $(I + 1/2)$ for half-integer spins.[1] These energy levels are not equidistant, that is why there is always I lines for integer spin numbers and $(I - 1/2)$ lines for half-integer spin numbers at frequencies

$$\omega_{\pm m \leftrightarrow \pm(m-1)} = \frac{3e^2qQ}{4I(2I-1)\hbar}(2m-1). \tag{8.4}$$

For example, for the spin number $I = 2$ at presence of quadrupole interactions the five-fold degenerate energy level, which corresponds to the ground state of the nucleus, splits in such a manner that there is one nongenerate energy level with $m = 0$ and two doubly degenerate energy levels with $m = \pm 1$ and ± 2. According to Eq. (8.3) the energy values of these levels are

$$E_{\pm 2} = \frac{e^2qQ}{4}, \quad E_{\pm 1} = -\frac{e^2qQ}{8}, \quad E_0 = -\frac{e^2qQ}{4},$$

and the NQR spectrum consists of two lines with transition frequencies

$$\omega_1 = \frac{3e^2qQ}{8\hbar}, \quad \omega_2 = \frac{e^2qQ}{8\hbar}.$$

The ratio between the frequencies $\omega_1 : \omega_2$ is 3:1.

For a nucleus with $I = 5/2$ the energy levels of quadrupole interactions are

$$E_{\pm 5/2} = \frac{e^2qQ}{4}, \quad E_{\pm 3/2} = -\frac{e^2qQ}{20}, \quad E_{\pm 1/2} = -\frac{e^2qQ}{5}.$$

[1] It is well-known that in NMR the number of energy levels of the Zeeman interaction is $(2I + 1)$.

In the NQR spectrum one still observes two lines, but their frequencies are

$$\omega_1 = \frac{3e^2qQ}{10\hbar}, \quad \omega_2 = \frac{3e^2qQ}{20\hbar},$$

and the ratio between the frequencies $\omega_1 : \omega_2$ is $2 : 1$.

In general, the ratio of the transition frequencies for quadrupole interaction, see Eq. (8.4), is

$$\frac{\omega_{\pm m \leftrightarrow \pm(m-1)}}{\omega_{\pm(m-1) \leftrightarrow \pm(m-2)}} = \frac{2m-1}{2m-3},$$

i.e., for integer spin numbers the ratio of the transition frequencies represents a row of integer odd numbers, whereas for half-integer spin numbers it is a row of natural numbers.

If the asymmetry parameter η does not equal to zero, even an approximate solution of the Schrödinger equation with the Hamiltonian given by Eq. (8.1) cannot be obtained in a general form. The reason for that is that two non-commuted parts of the Hamiltonian (8.1) are of the same order of accuracy, and the application of the perturbation theory is not reasonable: it is necessary to take into account higher approximations of the perturbation theory. That is why the problem of finding out the energy levels and transition frequencies is to be solved for each spin number separately. For spin numbers 1, 3/2, 2 and 3 it is possible to get the exact solution of the Schrödinger equation, whereas for other spin numbers only approximate solutions are possible.

8.1.2 Exact Solution of the Secular Equation for $I = 1$ and $I = 3/2$

To determine the energy levels and transition frequencies of quadrupole interactions on can use the standard procedure for searching eigenfunctions and eigenvalues of an operator, that is the Hamiltonian operator in our case. Using this procedure, one has to calculate matrix elements of the operator within some representation. The more reasonable is to use the representation of the $\hat{I}_z$ operator, in other words to calculate matrix elements one can use wave functions ψ_m, which are eigenfunctions of the operator $\hat{I}_z$ and satisfy the equation $\hat{I}_z \psi_m = m\psi$. The matrix, calculated in such a manner, is to be diagonalized. The elements of the diagonal matrix are the eigenvalues of the Hamiltonian operator.

Indeed, eigenfunctions of an operator can be expanded into a series of orthonormal functions:

$$\Psi_k = \sum_m c_{km} \psi_m. \tag{8.5}$$

Let us put Eq. (8.5) into the stationary Schrödinger equation $\widehat{\mathcal{H}}\Psi_k = E_k\Psi_k$:

$$\sum_m c_{km}\widehat{\mathcal{H}}\psi_m = E_k \sum_m c_{km}\psi_m, \tag{8.6}$$

then multiply the both parts of Eq. (8.6) by ψ_{m_1} and integrate it. Taking into account that the wave functions of the operator $\hat{I}_z$ are orthonormal, one can write:

$$\sum_m c_{km}\langle m_1|\widehat{\mathcal{H}}|m\rangle = E_k c_{km_1}. \tag{8.7}$$

Multiplying sequentially Eq. (8.6) by wave functions with another $m_1 = m'$ value, one obtains $(2I+1)$ homogeneous equations to determine the expansion coefficients c_{km}:

$$\sum_m c_{km}\langle m'|\widehat{\mathcal{H}}|m\rangle - E_k c_{km'} = 0, \tag{8.8}$$

where m and m' run from $-I$ to I. A nonhomogeneous system of linear equations has a unique non-trivial solution if and only if its determinant is zero.

Exact solution of the secular equation for $I = 1$ From Eq. (8.8) one obtains an equation to determine the energy levels of quadrupole interactions. Hence, for $I = 1$ the matrix elements of the quadrupole interaction Hamiltonian $\langle\psi_m|\widehat{\mathcal{H}}_\mathrm{Q}|\psi_{m'}\rangle/(e^2qQ)$, calculated using the eigenfunctions of the operator $\hat{I}_z$, produce the following matrix

ψ_1	ψ_{-1}	ψ_0	
1/4	$\eta/4$	0	ψ_1
$\eta/4$	1/4	0	ψ_{-1}
0	0	−1/2	ψ_0

Subtracting from this matrix the diagonal one, which consists of the elements λ, and equating its determinant to zero one has

$$\begin{vmatrix} 1/4-\lambda & \eta/4 & 0 \\ \eta/4 & 1/4-\lambda & 0 \\ 0 & 0 & -1/2-\lambda \end{vmatrix} = 0. \tag{8.9}$$

After solving the secular equation one obtains

$$\lambda_1 = -\frac{1}{2}, \qquad \lambda_{2,3} = \frac{1}{4}(1 \pm \eta).$$

Then, the energy levels of the quadrupole interaction for spin $I = 1$ are

$$E_1 = -\frac{e^2qQ}{2}, \qquad E_{2,3} = \frac{e^2qQ}{4}(1 \pm \eta). \tag{8.10}$$

Thus, at $\eta \neq 0$ the energy levels degeneracy is broken (according to the Kramers theorem electric field can remove the degeneracy of the energy levels for integer spins).

A rf field may cause transitions between these levels. Let us calculate their probabilities. First of all it is necessary to determine eigenfunctions of the quadrupole interaction Hamiltonian, i.e. to find c_{mk} coefficients of the expansion of the wave function Ψ in eigenfunctions of the $\hat{I}_z$ operator.

According to the standard procedure the matrix, which corresponds to the determinant (8.9), where λ is sequentially replaced by λ_1, λ_2 and λ_3 related to E_1, E_2 and E_3 by Eq. (8.10), is to be multiplied by the column vector, which consists of coefficients c_{m1}, c_{m2}, c_{m3}. For $E_1 = -e^2qQ/2$ one obtains:

$$\begin{pmatrix} 1/4+1/2 & \eta/4 & 0 \\ \eta/4 & 1/4+1/2 & 0 \\ 0 & 0 & -1/2+1/2 \end{pmatrix} \begin{pmatrix} c_{11} \\ c_{-11} \\ c_{01} \end{pmatrix} = 0. \tag{8.11}$$

As a result one has a system of linear homogeneous equations to determine the coefficients c_{m1}. Solving it one obtains that c_{01} is arbitrary, and can be chosen equal to 1, taking into account the normality condition, whereas $c_{11} = c_{-11} = 0$.

Thus the eigenfunction of the quadrupole interaction Hamiltonian, which corresponds to the energy level E_1, coincides with the eigenfunction of the operator $\hat{I}_z$ at $m = 0$. If in Eq. (8.11) one substitutes another eigenvalue for E_1, e.g. E_2, one obtains the following system of equations:

$$\begin{aligned} \eta/4c_{12} + \eta/4c_{-12} &= 0, \\ \eta/4c_{12} + \eta/4c_{-12} &= 0, \\ (-3/4 - \eta/4)c_{02} &= 0. \end{aligned} \tag{8.12}$$

Thereby, it appears that $c_{02} = 0$ and $c_{12} = -c_{-12}$, however coefficients $c_{\pm12}$ are arbitrary. Taking into account the normality condition they are $c_{12} = -c_{-12} = 1/\sqrt{2}$. By the similar way one can find other coefficients that determine the eigenfunctions for E_3: $c_{13} = c_{-13} = 1/\sqrt{2}$. Hence, the wave functions that correspond to the energy levels E_1, E_2 and E_3 are following

$$\begin{aligned} \Psi_1 &= \psi_0, \\ \Psi_2 &= \frac{1}{\sqrt{2}}(\psi_1 - \psi_{-1}), \\ \Psi_3 &= \frac{1}{\sqrt{2}}(\psi_1 + \psi_{-1}). \end{aligned} \tag{8.13}$$

Let us calculate probabilities of transitions between the levels E_1, E_2, E_3 that occur under the action of a rf field ($B_{1z}\cos\omega t$), its vector is parallel to the principle z-axis of the EFG tensor. The Hamiltonian describing the interaction of the nuclear magnetic moment with the magnetic field B_{1z} is

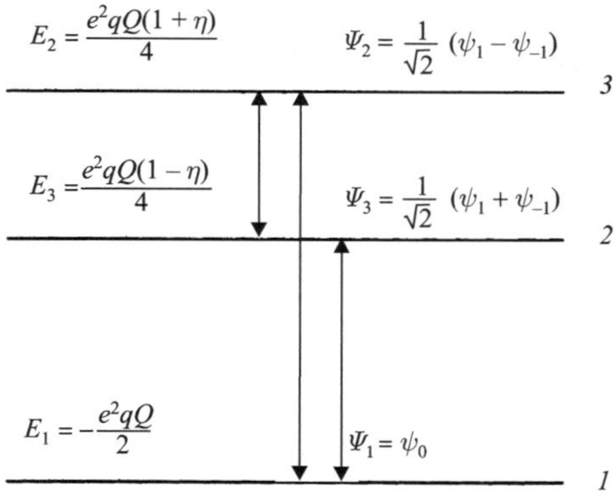

Fig. 8.1 Energy levels of the quadrupole interaction for a nucleus with $I = 1$ and $\eta \neq 0$

$$\widehat{\mathcal{H}}_1^Z = -\gamma\hbar\hat{I}_z B_{1z}\cos\omega t.$$

Thereby, the transition probability between the energy levels (2–3) (see Fig. 8.1) can be found from the following expression:

$$W_{2\leftrightarrow 3} \simeq \langle \Psi_2|\hat{I}_z|\Psi_3\rangle^2. \tag{8.14}$$

If one puts Eq. (8.13) into Eq. (8.14) one obtains that $W_{2\leftrightarrow 3} \simeq 1$. Note that in NMR if the vector of the radiofrequency magnetic field directs along the external magnetic field there is no transitions. Having done similar calculations for transitions $(1 \leftrightarrow 2)$ and $(1 \leftrightarrow 3)$, one gets the transition probability is equal to zero. Hence, under the influence of the rf field directed along the z-axis, only transitions between the levels $2 \leftrightarrow 3$ take place. The frequency that corresponds to this transition is

$$\omega_{2\leftrightarrow 3} = \frac{e^2qQ}{2\hbar}\eta. \tag{8.15}$$

If the rf field vector directs along the x or y principal axes of the EFG tensor, applying the procedure describing above one can obtain that the non-zero probabilities ($\sim$1) are those that correspond to $(1 \leftrightarrow 3)$ and $(1 \leftrightarrow 2)$ transitions, respectively. The frequencies of these transitions are

$$\omega_{1\leftrightarrow 3} = \frac{e^2qQ}{4\hbar}(3+\eta), \qquad \omega_{1\leftrightarrow 2} = \frac{e^2qQ}{4\hbar}(3-\eta).$$

The sum of frequencies $\omega_{1\leftrightarrow 3}$ and $\omega_{1\leftrightarrow 2}$ is equal to $3e^2qQ/(2\hbar)$. It means that by summing up the transition frequencies $\omega_{1\leftrightarrow 3}$ and $\omega_{1\leftrightarrow 2}$ it is possible to find out the quadrupole coupling constant, whereas from Eq. (8.15) on can estimate the asymmetry parameter. Knowing the rf field vector direction for each transition (for example, relative to the crystal faceting) one can determine the direction of the principal axes of the EFG tensor, as well. For an arbitrary direction of the rf field vector it is possible to observe all three transitions at once. This is the usual case for polycrystalline materials.

Exact solution of the secular equation for $I = 3/2$ For a nucleus with $I = 3/2$ the matrix elements of the quadrupole interaction Hamiltonian $\langle\psi_m|\widehat{\mathcal{H}}_Q|\psi_{m'}\rangle/(e^2qQ)$ calculated using the wave functions of the $\hat{I}_z$ operator form the following matrix:

$$\begin{array}{cccc|c} \psi_{3/2} & \psi_{-1/2} & \psi_{1/2} & \psi_{-3/2} & \\ \hline 1/4 & \eta/(4\sqrt{3}) & 0 & 0 & \psi_{3/2} \\ \eta/(4\sqrt{3}) & -1/4 & 0 & 0 & \psi_{-1/2} \\ 0 & 0 & -1/4 & \eta/(4\sqrt{3}) & \psi_{1/2} \\ 0 & 0 & \eta/(4\sqrt{3}) & 1/4 & \psi_{-3/2} \end{array} . \tag{8.16}$$

This is a block matrix with two equal blocks. Hence, the energy levels of quadrupole interactions for $I = 3/2$ can be obtained by equating to zero the following determinant:

$$\begin{vmatrix} 1/4 - \lambda & \eta/(4\sqrt{3}) \\ \eta/(4\sqrt{3}) & -1/4 - \lambda \end{vmatrix} = 0. \tag{8.17}$$

From Eq. (8.17) one has

$$\lambda_{1,2} = \pm\frac{1}{4}\sqrt{1 + \frac{\eta^2}{3}}, \tag{8.18}$$

and hence

$$E_{1,2} = e^2qQ\lambda = \pm\frac{e^2qQ}{4}\sqrt{1 + \frac{\eta^2}{3}}. \tag{8.19}$$

For the spin 3/2 the energy levels remain double degenerate even if $\eta \neq 0$, as it is evidenced from the Kramers theorem. For the NQR frequency from Eq. (8.19) one gets

$$\omega = \frac{e^2qQ}{2\hbar}\sqrt{1 + \frac{\eta^2}{3}}.$$

It is obvious that by measuring this frequency one can not find out both e^2qQ and η simultaneously. To determine both parameters it is necessary to complicate the experiment, for example, by applying a stationary magnetic field.

By subtracting the eigenvalues (8.18) from diagonal matrix elements of the matrix (8.16) and equating the determinant to zero one obtains an equation to determine the expansion coefficients of the eigenfunctions of the Hamiltonian on the eigenfunctions of the operator $\hat{I}_z$:

$$\begin{aligned} \left(\frac{1}{4} - \lambda_{1,2}\right) C_{\pm 3/2} + \frac{\eta}{4\sqrt{3}} C_{\mp 1/2} = 0, \\ \frac{\eta}{4\sqrt{3}} C_{\pm 3/2} - \left(\frac{1}{4} + \lambda_{1,2}\right) C_{\mp 1/2} = 0. \end{aligned} \tag{8.20}$$

The determinant of the nonhomogeneous system of linear equations (8.20) is nonzero, hence the coefficients $C_{\pm 3/2}$ and $C_{\pm 1/2}$ are not equal to zero. Applying the normality condition one gets the wave functions for the upper level ($\lambda = \lambda_1 = 1/4\sqrt{1+\eta^2/3}$):

$$\Psi_{\text{up}} = \frac{1}{\sqrt{2\lambda_1}} \left(\sqrt{\lambda_1 - 1/4}\, \psi_{\mp 1/2} + \sqrt{\lambda_1 + 1/4}\, \psi_{\pm 3/2} \right); \tag{8.21}$$

For the lower level ($\lambda = \lambda_2 = -1/4\sqrt{1+\eta^2/3} = -\lambda_1$) the wave function is

$$\Psi_{\text{dn}} = \frac{1}{\sqrt{2\lambda_1}} \left(\sqrt{\lambda_1 + 1/4}\, \psi_{\mp 1/2} + \sqrt{\lambda_1 - 1/4}\, \psi_{\pm 3/2} \right). \tag{8.22}$$

As it is evidenced from Eqs. (8.21) and (8.22) if the asymmetry parameter $\eta \neq 0$ then the wave function for the upper or lower energy level is a linear combination of all wave functions of the operator $\hat{I}_z$. It means that transitions between these levels may occur at any direction of the rf magnetic field vector. The probability of such transitions is proportional to the square of the matrix element for the Hamiltonian, which describes the interaction of the nuclear magnetic moment with the rf field:

$$\begin{aligned} W_z &= \frac{\eta^2}{3(1+\eta^2/3)}, && \text{if } \vec{B}_1 \| z; \\ W_x &= \frac{3\sqrt{3}(\eta/3+1)^2}{1+\eta^2/3}, && \text{if } \vec{B}_1 \| x; \\ W_y &= \frac{3\sqrt{3}(\eta/3-1)^2}{1+\eta^2/3}, && \text{if } \vec{B}_1 \| y. \end{aligned} \tag{8.23}$$

The probabilities of transitions (8.23), and therefore the NQR signal intensities, depend on the direction of the applied field. For $\eta = 0$ only the rf field perpendicular to the principal z-axis of the EFG tensor leads to transitions between energy levels. If it directs along the z-axis, the bigger η value the more intensive the NQR signal is.

It is possible to obtain analytic formulas for the energy levels of quadrupole interactions for the spin numbers equal to 2 and 3. For these spin numbers the matrix constructed from matrix elements of the quadrupole interaction Hamiltonian has the block character with the block size of 2×2. With the spin number increasing the polynom degree in the secular equation raises, and it becomes impossible to obtain the solution in an analytical form. For half-integral spins the matrix also is of block type, but the block size is bigger then two, and the secular equation in a general case can be solved only numerically. For example, for the spin 5/2 the matrix elements of the quadrupole interaction Hamiltonian $\langle \psi_m | \widehat{\mathcal{H}}_Q | \psi_{m'} \rangle / (e^2 q Q)$, calculated using the wave functions of the operator $\hat{I}_z$ form the matrix like

$$\begin{array}{cccccc|c}
\psi_{5/2} & \psi_{1/2} & \psi_{-3/2} & \psi_{3/2} & \psi_{-1/2} & \psi_{-5/2} & \\
\hline
10 & \sqrt{10}\,\eta & 0 & 0 & 0 & 0 & \psi_{5/2} \\
\sqrt{10}\,\eta & -8 & 3\sqrt{2}\,\eta & 0 & 0 & 0 & \psi_{1/2} \\
0 & 3\sqrt{2}\,\eta & -2 & 0 & 0 & 0 & \psi_{-3/2} \\
0 & 0 & 0 & -2 & 3\sqrt{2}\,\eta & 0 & \psi_{3/2} \\
0 & 0 & 0 & 3\sqrt{2}\,\eta & -8 & \sqrt{10}\,\eta & \psi_{-1/2} \\
0 & 0 & 0 & 0 & \sqrt{10}\,\eta & 10 & \psi_{-5/2}
\end{array} \tag{8.24}$$

Applying the above mentioned procedure one has the following secular equation:

$$\lambda^3 - 84\left(1 + \frac{\eta^2}{3}\right)\lambda - 160(1 - \eta^2) = 0. \tag{8.25}$$

Equation (8.25) can be solved exactly for a particular case, $\eta = 1$:

$$\lambda_2 = 0, \qquad \lambda_{3,1} = \pm 4\sqrt{7}.$$

In this particular case the energy levels become equidistant. It means that there is only one line in the NQR spectrum. Let us remind that if the asymmetry parameter η is equal to zero, there are two lines in the NQR spectrum with the ratio of transition frequencies 2:1. For other η values Eq. (8.25) can be solved only numerically.

Figure 8.2 represents the energy levels of the quadrupole interaction (a), the transition frequencies (b) and their ratio (c) as functions of the asymmetry parameter for the spin number equal to 5/2. As it is seen in Fig. 8.2, for $I = 5/2$ the levels ±5/2, ±3/2, ±1/2 remain degenerated at any value of the asymmetry parameter. The energy of the lower level is more sensible to changing in the η value. The energy of the upper level remains almost constant (it changes only by 4 %). It is worth noting that if $\eta \leq 0.2$, changing in transition frequencies (see Fig. 8.2b) and their ratio (see Fig. 8.2c) do not exceed 5 %. If the NQR line is broad it is difficult to register such a change. It makes the task of determining the asymmetry parameter almost impossible. At great values of the asymmetry parameter it can be found from the ratio of transition frequencies.

To determine the probabilities of transitions it is necessary to find the wave functions. As it follows from Eq. (8.24) a superposition of wave functions corresponds to each level: $\psi_{1/2}$ admixes to $\psi_{5/2}$, $\psi_{-1/2}$ admixes to $\psi_{-5/2}$; $\psi_{\pm 5/2}$ and $\psi_{\mp 3/2}$ admix to $\psi_{\pm 1/2}$, respectively. Now it is possible to observe transitions (1–2), (2–3), (1–3) (see Fig. 8.2a).

As in the case of the spin number 3/2, transitions between the energy levels appear not only under the effect of the rf field applied perpendicular to the principal axis of the EFG tensor, but under the field parallel to this axis as well.

So, for the spin number 5/2 by measuring the NQR frequencies it is possible to find both the quadrupole coupling constant and the asymmetry parameter (for $\eta \geq 0.2$). However, it is almost impossible to determine the orientation of the principal axes of the EFG tensor, as it is technically difficult to change the orientation of the rf field.

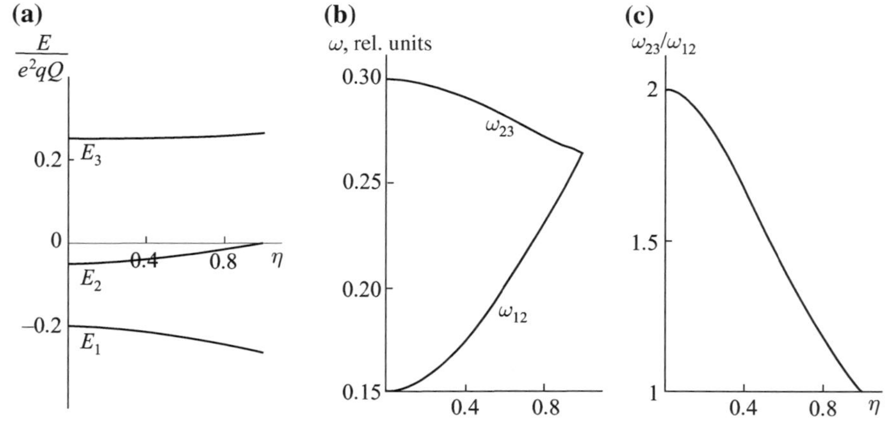

Fig. 8.2 Quadrupole interactions for $I = 5/2$ and $\eta \neq 0$: the energy levels (**a**), transition frequencies (**b**) and their ratio (**c**) as functions of η

Thus, if $\eta = 0$, then

- for any nucleus spin number the energy levels of quadrupole interactions are twofold degenerate, except the level with $m = 0$ for integer spin numbers;
- the ratio between NQR transition frequencies for integer spins is a series of odd integer numbers, whereas for half-integer spins it is a series of natural numbers;
- transitions between the energy levels occur only under the action of the rf field perpendicular to the principal z-axis of the EFG tensor;
- it is impossible to determine the orientation of the EFG tensor relative to the crystallographic axes;
- the intensities that correspond to the higher frequencies are higher than the intensities of lower frequency transitions.

If $\eta \neq 0$, then

- the degeneracy of energy levels of quadrupole interactions is removed for integer spin numbers, whereas for half-integer spin numbers the energy levels remain degenerated;
- for integer spin numbers transitions between the split (± 1) energy levels are allowed; the transitions occur under the action of an arbitrarily directed radiofrequency field;
- for integer spin numbers both the quadrupole coupling constant and the asymmetry parameter can be determined unambiguously;
- for half-integer spin numbers the main transitions ($\Delta m = \pm 1$) have higher intensity as compared with transitions $\Delta m = \pm 2$;
- for half-integer spins transitions under the action of a rf field, whose vector is parallel to the principal z-axis of the EFG tensor, are less probable than transition under the field perpendicular to it;

- the ($\pm 1/2 \leftrightarrow \pm 3/2$) transition frequency is more sensitive to the η value (as compared to other transitions);
- for $I = 3/2$ it is impossible to determine both the asymmetry parameter and the quadrupole coupling constant, but it is possible to determine if $\eta = 0$ or not; for other spin numbers the asymmetry parameter and the quadrupole coupling constant can be found only for big η values;
- from NQR spectra it is almost impossible to determine the direction of the principal axes of the EFG tensor.

8.1.3 NQR in the Presence of Magnetic Field

An external magnetic field removes the degeneracy of energy levels of quadrupole interactions. Usually, one applies magnetic fields, for which the Zeeman energies are much lower than the energy of quadrupole interactions. It is more convenient to write the total Hamiltonian of Zeeman and quadrupole interactions in the principal axes frame of the EFG tensor:

$$\widehat{\mathcal{H}} = \frac{e^2 qQ}{4I(2I-1)}\left(3\hat{I}_z^{\,2} - \hat{I}^2 + \frac{\eta}{2}(\hat{I}_+^2 + \hat{I}_-^2)\right) - \gamma\hbar B_0\left[\hat{I}_z\cos\theta + \left(\frac{\hat{I}_+}{2}\exp(-\mathrm{i}\phi) + \frac{\hat{I}_-}{2}\exp(\mathrm{i}\phi)\right)\sin\theta\right].$$

Here θ and ϕ are the Euler angles that determine orientation of the external magnetic field vector relative to the principal axes of the EFG tensor. In this case it is impossible to obtain an analytic solution of the Schrödinger equation in general form. However, if the magnetic field vector directs along one of the principal axis of the EFG tensor, the equation can be solved analytically. As the energy levels at quadrupole interactions are determined only for a specific spin value, corrections to these levels that appear when applying an external magnetic field should be also calculated for a specific spin number. Let us consider how magnetic field affects the energy levels for the spin numbers 1 and 3/2.

NQR energy levels for $I = 1$ in magnetic field For $I = 1$ if there are both quadrupole and Zeeman interactions to calculate energy levels one has to consider the determinant of the following form:

$$\begin{vmatrix} \frac{1}{4} - \frac{R}{4}\cos\theta - \lambda & \eta/4 & -\frac{R}{4\sqrt{2}}\sin\theta e^{\mathrm{i}\phi} \\ \eta/4 & \frac{1}{4} + \frac{R}{4}\cos\theta - \lambda & -\frac{R}{4\sqrt{2}}\sin\theta e^{-\mathrm{i}\phi} \\ -\frac{R}{4\sqrt{2}}\sin\theta e^{\mathrm{i}\phi} & -\frac{R}{4\sqrt{2}}\sin\theta e^{-\mathrm{i}\phi} & -1/2 - \lambda \end{vmatrix} = 0. \tag{8.26}$$

Here $R = 4\gamma\hbar B_0/(e^2 qQ)$ is the ratio between the Zeeman and quadrupole interactions.

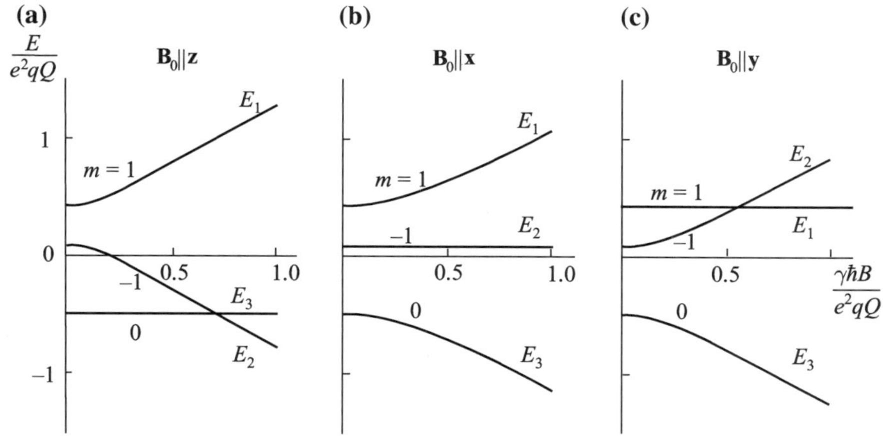

Fig. 8.3 Energy levels versus R if magnetic field vector directs along the z (**a**), x (**b**) and y (**c**) principal axis oft the EFG tensor ($I = 1, \eta = 0.5$)

Let us consider how the NQR spectrum changes if the magnetic field vector $\vec{B}_0$ is parallel to one of the principal axes of the EFG tensor.

Let $\theta = 0$ and $\phi = 0$ ($\vec{B}_0 || \vec{z}$), then nondiagonal matrix elements, which depend on the external field, disappear from the determinant (8.26) and the matrix has quasidiagonal form (similar to the case when the field is absent), that helps to obtain an analytic expression for energy levels:

$$E_{1,2} = \frac{e^2qQ}{4}\left(1 \pm \sqrt{\eta^2 + R^2}\right), \quad E_3 = -\frac{e^2qQ}{2}. \tag{8.27}$$

As it is seen from (8.27), the magnetic field completely removes the degeneracy of energy levels at any asymmetry parameter value.

In Fig. 8.3 we plot values of energy levels as a function of applied magnetic field (in R units) at various directions of the external filed. As it is seen, the energy level E_3, which corresponds to the zero projection of the operator $\hat{I}_z$ on the external magnetic field, does not depend on the applied field. The energy levels $E_{1,2}$, which correspond to the linear combinations of the wave functions (see Eq. (8.13)) with $m = \pm 1$, depend on the magnetic field value, see Fig. 8.3a. And this dependence is linear if $\eta = 0$.

If $R^2 = 9 - \eta^2$, then $E_2 = E_3$ and hence there is only one line in the NQR spectrum. By measuring the magnetic filed value, which corresponds to the NQR spectrum degeneracy, it is possible to determine the asymmetry parameter, whereas the measured transition frequency gives the value of the quadrupole coupling constant (e^2qQ). However, the intersection point of the energy levels is at rather high magnetic field and its value increases with η decreasing. For example, or $\eta = 1$, R is equal to $2\sqrt{2}$, and if quadrupole interactions are strong enough this R value corresponds to a rather high magnetic field that is not used in NQR. With following increasing of the external magnetic field the energy levels look like in the case of nuclear magnetic resonance.

If $\theta = \pi/2$ and $\phi = 0$ (the magnetic field vector directs along the x-axis, see Fig. 8.3b), then the secular equation has three roots again, which correspond to the following values of the energy levels:

$$E_{1,3} = \frac{e^2qQ}{8}\left(-(1-\eta) \pm \sqrt{(3+\eta)^2 + 4R^2}\right), \quad E_2 = \frac{e^2qQ}{4}(1-\eta).$$

Now the external magnetic field does not affect the E_2 level, and it is the level to which the zero projection of the operator $\hat{I}_z$ on the magnetic field vector corresponds. Let us note that now the direction of the magnetic field vector does not coincide with the z-axis, and in the absence of magnetic field, this energy level corresponds to a linear combination of the wave functions with $m = \pm 1$. The levels $E_{1,3}$ depend on the magnetic field, and if $\eta = 0$ this dependence is linear. The energy levels in this case do not intersect for any value of the magnetic field value (see Fig. 8.3b).

If $\theta = \pi/2$ and $\phi = \pi/2$ (the magnetic field vector directs along the y-axis, Fig. 8.3c, the secular equation has three roots that correspond to the following energy levels:

$$E_{2,3} = \frac{e^2qQ}{8}\left(-(1+\eta) \pm \sqrt{(3-\eta)^2 + 4R^2}\right), \quad E_1 = \frac{e^2qQ}{4}(1+\eta).$$

Now magnetic field does not have any influence on E_1, and the intersection of energy levels is when $R^2 = 6\eta + 2\eta^2$ ($R = 2\sqrt{2}$ if $\eta = 1$). If $\eta < 1$ the level intersection occurs at lower field as compared to the case when the magnetic field vector is parallel to the z-axis (compare Fig. 8.3a and c): the smaller η the lower the R value, which corresponds to the intersection point, is. If for some reasons, for example, due to the crystal symmetry, the direction of the principal axes is known, then it is possible to find the asymmetry parameter of the EFG tensor by applying the external magnetic field parallel to the x-axis and by determining the magnetic field value, at which the NQR spectrum consists of one line only. The strongest dependence of the transition frequencies ratio of η is observed if the magnetic field vector directs along the x-axis. However, as a rule the direction of the principal axes is unknown beforehand, and in order to determine it one has to measure an orientation dependence of NQR frequencies.

It is worth noting that for $I = 1$ and without an external stationary magnetic field it is possible to measure both the quadrupole coupling constant and η, but it is practically impossible to determine the orientation of the EFG tensor. For $I = 3/2$ using stationary NQR methods without magnetic field it is impossible to determine separately the asymmetry parameter and the quadrupole coupling constant.

NQR energy levels for $I = 3/2$ in magnetic field In this case the determinant to calculate the energy levels is

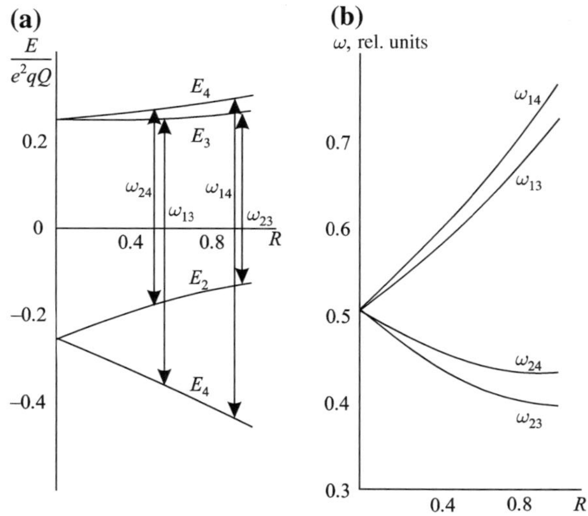

Fig. 8.4 Energy levels (**a**) and transition frequencies (**b**) versus R for $I = 3/2$ at $\eta = 0, 5, \theta = \pi/6, \phi = \pi/3$

$$\begin{vmatrix} \frac{1}{4} - \frac{R}{4}\cos\theta - \lambda & \eta/(4\sqrt{3}) & -\frac{R}{4\sqrt{3}}\sin\theta e^{\mathrm{i}\phi} & 0 \\ \eta/(4\sqrt{3}) & -\frac{1}{4} + \frac{R}{12}\cos\theta - \lambda & -\frac{R}{6}\sin\theta e^{-\mathrm{i}\phi} & -\frac{R}{4\sqrt{3}}\sin\theta e^{\mathrm{i}\phi} \\ -\frac{R}{4\sqrt{3}}\sin\theta e^{-\mathrm{i}\phi} & -\frac{R}{6}\sin\theta e^{\mathrm{i}\phi} & -\frac{1}{4} - \frac{R}{12}\cos\theta - \lambda & \eta/(4\sqrt{3}) \\ 0 & -\frac{R}{4\sqrt{3}}\sin\theta e^{-\mathrm{i}\phi} & \eta/(4\sqrt{3}) & \frac{1}{4} + \frac{R}{4}\cos\theta - \lambda \end{vmatrix} = 0,$$

here $R = 6\gamma\hbar B_0/(e^2qQ)$. The secular equation for $I = 3/2$ can be written as

$$\lambda^4 - 2a\lambda^2 - b\lambda + c = 0, \tag{8.28}$$

where

$$\begin{aligned} a &= \frac{1}{16}\left(1 + \frac{\eta^2}{3} + \frac{5R^2}{9}\right), \\ b &= \frac{R^2}{72}(3\cos^2\theta - 1 + \eta\sin^2\theta\cos 2\phi), \\ c &= \frac{1}{256}\left(1 + \frac{\eta^2}{3}\right)^2 \\ &+ \frac{R^2}{64}\left(\frac{R^2}{36} + \frac{1}{18} - \frac{1}{3}\cos^2\theta + \frac{2}{9}\eta\sin^2\theta\cos 2\phi + \frac{\eta^2}{18}\cos 2\theta\right). \end{aligned} \tag{8.29}$$

The solution of Eq. (8.28) in general form can be obtained only numerically. It has four roots. As example of the dependence of the energy levels of the external magnetic field is plotted in Fig. 8.4a. As it is seen, the magnetic field has a weak influence on the upper energy levels, but it affects noticeably the lower ones. The

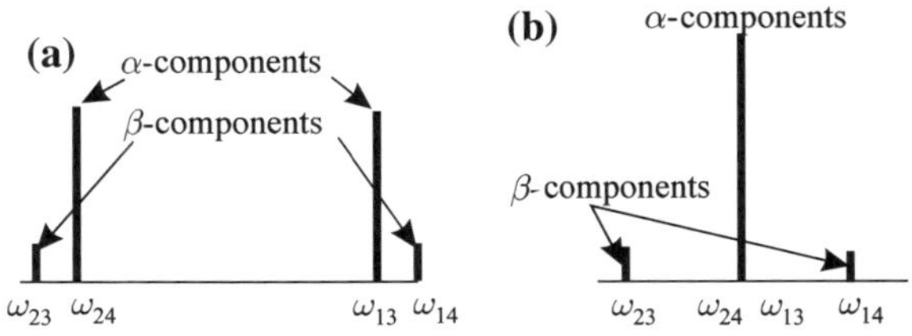

Fig. 8.5 A NQR spectrum for $I = 3/2$ with $\eta = 0.5$ at two different orientations of the external magnetic fields: $\theta = \pi/6$ and $\phi = \pi/3$ (**a**); the magnetic field vector coincides with the direction of the zero splitting cone generatrix (**b**)

magnetic field commixes the wave functions in such a manner that each energy level corresponds to a superposition of wave functions with different projection values of the operator $\hat{I}_z$. That is why transitions $1 \leftrightarrow 4$, $2 \leftrightarrow 3$, $1 \leftrightarrow 3$, $2 \leftrightarrow 4$ are not forbidden anymore, and the NQR spectrum consists of four lines. The transitions ω_{23}, ω_{24} decrease with magnetic field increasing, whereas the frequencies ω_{13}, ω_{14} increase (see Fig. 8.4b).

The NQR spectrum lines corresponding to the transitions $1 \leftrightarrow 4$ and $2 \leftrightarrow 3$ are called the *outer doublet* or *β-components*, and the lines corresponding to the transitions $1 \leftrightarrow 3$ and $2 \leftrightarrow 4$ are called the *inner doublet* or *α-components* (Fig. 8.5a). The frequencies of these transitions depend on the direction of the external magnetic field vector. Therefore, by measuring the orientation dependence of transition frequencies in a weak magnetic field it is possible to determine the quadrupole coupling constant, the asymmetry parameter and the direction of the principal axes of the EFG tensor. However, recording such orientation dependencies and their interpretation is a rather laborious task. It can be noticeably simplified if one can find such an orientation of the external magnetic field, for which the lines of the inner doublet interflow. Analysis of Eq. (8.28) shows that it is possible if $b = 0$. Then Eq. (8.28) becomes biquadratic and can be solved analytically.

Let us set the parameter b equal to zero:

$$3\cos^2\theta - 1 + \eta(1 - \cos^2\theta)\cos 2\phi = 0. \tag{8.30}$$

Equation (8.30) contains only the asymmetry parameter and the angles, which determine the direction of the external magnetic field relative to the principal axes of the EFG tensor. Let us write Eq. (8.30) as

$$\sin^2\theta = \frac{2}{3 - \eta\cos 2\phi}. \tag{8.31}$$

Equation (8.31) describes a cone surface with an elliptical section. In this case the expression for energy levels can be represented as follows

$$E_{1,2,3,4} = \pm e^2 qQ\sqrt{a \pm \sqrt{a^2 - c}}.$$

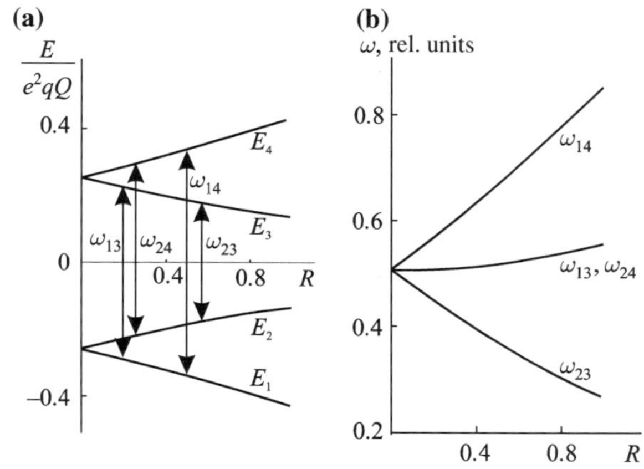

Fig. 8.6 Energy levels (**a**) and transition frequencies (**b**) versus R for $I = 3/2$ if the magnetic field vector coincides with the direction of the generatrix of the zero splitting cone

Let us consider the case when $c > 0$ and $a > \sqrt{a^2 - c}$. As it is seen from Fig. 8.6a, the magnetic field splits the upper and lower energy levels by the same value. The corresponding transition frequencies can be written as

$$\omega_{1\leftrightarrow 4} = 2e^2qQ\sqrt{a + \sqrt{a^2 - c}},\ \omega_{2\leftrightarrow 3} = 2e^2qQ\sqrt{a - \sqrt{a^2 - c}},$$
$$\omega_{2\leftrightarrow 4} = \omega_{1\leftrightarrow 3} = e^2qQ\left(\sqrt{a + \sqrt{a^2 - c}} + \sqrt{a - \sqrt{a^2 - c}}\right). \qquad (8.32)$$

As one can see from Eq. (8.32), α-lines (the transitions $2 \leftrightarrow 4$ and $1 \leftrightarrow 3$) have the same frequency, which weakly depends on the external magnetic field value (see Fig. 8.6b). So, at any arbitrary orientation of the external magnetic field vector there are four lines in the NQR spectrum. But, if the external magnetic field vector lies on the surface described by Eq. (8.31) the quartet transforms to the triplet, i.e. one observes a zero splitting of the α-components (see Fig. 8.5b). The surface described by Eq. (8.31) is called *the zero splitting cone*. By measuring these three NQR frequencies it is possible to determine the quadrupole coupling constant e^2qQ and factors a, c as well. After that from Eq. (8.29), taking into account that $b = 0$, one can find the asymmetry parameter and the orientation of the principal axes of the EFG tensor (angles θ and ϕ) relative to the external magnetic field vector. If the direction of the principal axes of the EFG tensor is known beforehand, for example, for symmetry reasons, the task becomes simpler, as in this case η can be found directly from Eq. (8.30). It is worth noting that if one measures the orientation dependence by rotating about an arbitrary axis it is possible not to find such a position when the external magnetic field vector coincides with the generatrix of the zero splitting cone.

For the spin number equal to 3/2, as well as for any other spin numbers, the exact solution of the secular equation (8.28) can be obtained only if the external magnetic field vector is parallel to one of the principal axes of the EFG tensor.

If $\theta = 0$ and $\phi = 0$ ($\vec{B}_0 \| \vec{z}$), then the secular equation has four roots, which correspond to the following energy levels:

$$\begin{aligned} E_{1,2} &= \frac{e^2qQ}{2}\left(\pm\sqrt{(R/3-1/2)^2+\eta^2/12}-R/6\right), \\ E_{3,4} &= \frac{e^2qQ}{2}\left(\pm\sqrt{(R/3+1/2)^2+\eta^2/12}+R/6\right); \end{aligned} \tag{8.33}$$

if $\theta = \pi/2$ and $\phi = 0$ ($\vec{B}_0 \| \vec{x}$), then

$$\begin{aligned} E_{1,2} &= \frac{e^2qQ}{4}\left(\pm\sqrt{(R/3+1)^2+(R-\eta)^2/3}-R/3\right), \\ E_{3,4} &= \frac{e^2qQ}{4}\left(\pm\sqrt{(R/3-1)^2+(R+\eta)^2/3}+R/3\right); \end{aligned} \tag{8.34}$$

if $\theta = \pi/2$ and $\phi = \pi/2$ ($\vec{B}_0 \| \vec{y}$), then

$$\begin{aligned} E_{1,2} &= \frac{e^2qQ}{4}\left(\pm\sqrt{(R/3+1)^2+(R+\eta)^2/3}-R/3\right), \\ E_{3,4} &= \frac{e^2qQ}{4}\left(\pm\sqrt{(R/3-1)^2+(R-\eta)^2/3}+R/3\right). \end{aligned} \tag{8.35}$$

As on can see from Eqs. (8.33)–(8.35), there are four lines in the NQR spectrum, which can degenerate to three lines at rather high fields.

Thus, in order to determine the quadrupole coupling constant and the asymmetry parameter it is necessary to apply an external magnetic field. The most simple interpretation of NQR spectra is when the the magnetic field vector is along the generatrix of the zero splitting cone.

8.2 Temperature Dependence of NQR Frequencies

In crystals at temperature higher than absolute zero ions or molecules oscillate. With temperature increasing some parts of molecules or molecules as a whole can execute rotatory swinging motions or even rotate about any symmetry axis. These oscillations do not change the form of the unperturbed Hamiltonian describing the Zeeman interaction of nuclear magnetic moments with an external magnetic field and affect only relaxation times, as well as shape, width and splitting of NMR lines. In NQR such a mobility of molecules or their individual groups modulates the most intensive interaction that is the interaction between nuclear quadrupole moments and intracrystalline electric field gradients. That affects not only resonance line shape,

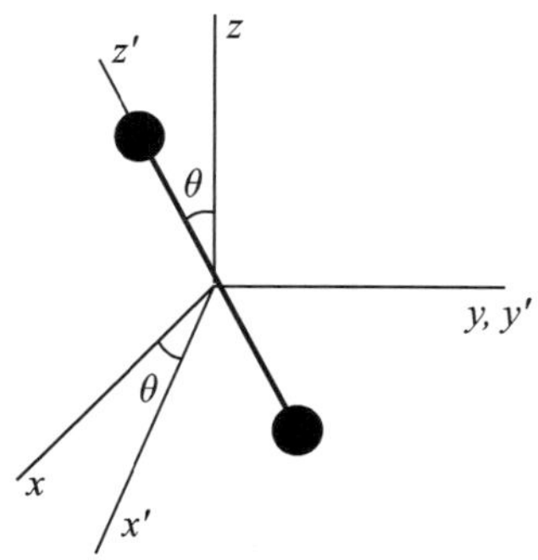

Fig. 8.7 Rotatory oscillation motion: relations between (x, y, z) and (x', y', z') frames

but NQR frequencies as well. In order to calculate NQR spectra it is necessary to average the Hamiltonian of quadrupole interactions over all kinds of motion.

8.2.1 Rotatory Oscillation Motion

This kind of motion has the most important influence on NQR spectra. The frequencies of such oscillations are of about 10^{12} Hz, that is several orders greater than NQR frequencies. Let us find the average value of the quadrupole coupling constant taking into account the rotary oscillation motion of a molecule about an axis, supposing that the asymmetry parameter of the EFG tensor, η, is equal to zero. Without loss of generality one can make an assumption that the molecule rotation axis is perpendicular to the xz-plane in the principal axes frame of the EFG tensor (x, y, z) and the molecule rotates about it by angle $\theta(t)$ (Fig. 8.7). Then the EFG tensor (V') in the coordinate system (x', y', z') is connected with the tensor (V), written in the principal axes frame, by the following transformation:

$$\mathsf{V}' = \mathsf{T}^{-1}\mathsf{V}\mathsf{T},$$

where T is the matrix connecting the (x, y, z) and (x', y', z') frames:

$$\mathsf{T} = \begin{pmatrix} \cos\theta(t) & 0 & \sin\theta(t) \\ 0 & 1 & 0 \\ -\sin\theta(t) & 0 & \cos\theta(t) \end{pmatrix}.$$

Taking into account that in the (x, y, z) frame the EFG tensor is diagonal and $V_{zz} = eq_{zz}$, $V_{xx} = V_{yy} = -eq_{zz}/2$, after multiplying the corresponding matrices one has

$$V'_{z'z'} = eq_{z'z'} = eq_{zz}\left(1 - \frac{3}{2}\sin^2\theta(t)\right).$$

Let us suppose that the q_{zz} does not change. Then the NQR frequency depends on the rotation angle by the following way:

$$\omega = \omega_0 \left(1 - \frac{3}{2}\sin^2\theta(t)\right).$$

As in solids a molecule deviates from its equilibrium state only by a small angle: $\theta(t) \ll 1$, then it is possible to expand $\sin\theta(t)$ in a Taylor series considering only the first term. Then

$$\omega = \omega_0 \left(1 - \frac{3}{2}\theta^2(t)\right). \tag{8.36}$$

If one supposes that the molecule executes harmonic oscillations, then the time dependence of the rotation angle of the molecule can be written as

$$\theta(t) = \theta_1 \sin\omega_{\text{rot}}t, \tag{8.37}$$

here ω_{rot} is the rotary oscillation frequency. Substituting Eq. (8.37) into Eq. (8.36) after time averaging one has

$$\overline{\omega} = \omega_0 \left(1 - \frac{3\theta_1^2}{2T_p}\int_{-T_p/2}^{T_p/2} \sin^2\omega_{\text{rot}}t\,\mathrm{d}t\right) = \omega_0\left(1 - \frac{3\theta_1^2}{4}\right) \tag{8.38}$$

(T_p is the rotary oscillation period, $T_p = 2\pi/\omega_{\text{rot}}$).

The rotary oscillation energy can be calculated if the moment of inertia of the molecule relative to the oscillation axis (J_1) is known:

$$E = J_1\omega_{\text{rot}}^2\theta_1^2/2. \tag{8.39}$$

As the molecule executes harmonic oscillations the energy in form of (8.39) can be equate with the energy of a harmonic oscillator:

$$E = \hbar\omega_{\text{rot}}\left(\frac{1}{2} + \frac{1}{\exp\left[\hbar\omega_{\text{rot}}/(kT)\right] - 1}\right) = \frac{J_1\omega_{\text{rot}}^2\theta_1^2}{2}. \tag{8.40}$$

After determining θ_1^2 from Eq. (8.40) one has

$$\overline{\omega} = \omega_0\left(1 - \frac{3\hbar}{2J_1\omega_{\text{rot}}}\left(\frac{1}{2} + \frac{1}{\exp[\hbar\omega_{\text{rot}}/(kT)] - 1}\right)\right). \tag{8.41}$$

If the rotary oscillation energy is much smaller than kT the exponent in Eq. (8.41) can be expanded in a series, and then

$$\overline{\omega} = \omega_0\left(1 - \frac{3\hbar}{4J_1\omega_{\text{rot}}} - \frac{3kT}{2J_1\omega_{\text{rot}}^2}\right). \tag{8.42}$$

The second term in the parentheses does not depend on the temperature and represent the frequency of zero-point oscillations. It can be neglected if $kT \gg \hbar\omega_{\text{rot}}$. Then Eq. (8.42) can be written as

$$\overline{\omega} = \omega_0 \left(1 - \frac{3kT}{2J_1\omega_{\text{rot}}^2}\right). \tag{8.43}$$

For any melting temperature known by now the thermal vibration energy is smaller than $2J_1\omega_{\text{rot}}^2$, and at high temperature the NQR frequency linearly decreases with temperature increasing. The greater the moment of inertia (or the rotary oscillation frequency) the smaller coefficients of linear regression are. It should be noted that the most important contribution to the temperature dependence of these frequencies is due to low-frequency oscillations, as the coefficient of linear regression inversely depends on the squared rotation frequency.

This formula was obtained by H. Bayer, and Eq. (8.43) is called the *Bayer law*. This law is usually used for a qualitative analysis of temperature dependencies of NQR frequencies.

If a molecule executes oscillations about several axes simultaneously, then after transformations similar to stated above and taking into account that oscillation amplitudes are small, one has

$$\overline{\omega} = \omega_0 \left(1 - \frac{3}{4}(\theta_1^2 + \theta_2^2)\right), \tag{8.44}$$

here θ_1^2 and θ_2^2 are the amplitudes of harmonic oscillations about two mutually perpendicular axes. As it is seen from Eq. (8.44) if oscillations are independent of each other, then their influences on the NQR frequency are additive. Amplitudes of rotary oscillations can be expressed through the corresponding moments of inertia. If a molecule oscillates simultaneously about n axes the formula (8.41) can be rewritten as:

$$\overline{\omega} = \omega_0 \left(1 - \sum_{i=1}^{n} \frac{3\hbar}{2J_i\omega_{\text{rot}i}} \left(\frac{1}{2} + \frac{1}{\exp\left[\hbar\omega_{\text{rot}i}/(kT)\right] - 1}\right)\right). \tag{8.45}$$

As it follows from Eqs. (8.41) and (8.45) it is possible to estimate frequencies of rotary oscillations and moments of inertia from temperature dependency of NQR frequencies.

The temperature coefficient of NQR frequency decreasing can be determined by differentiating Eq. (8.45) with respect to temperature:

$$\frac{\mathrm{d}(\overline{\omega} - \omega_0)}{\omega_0 \mathrm{d}T} = -\frac{3\hbar^2}{2kT^2} \sum_{i=1}^{n} \frac{\exp[\hbar\omega_{\text{rot}i}/(kT)]}{J_i(\exp[\hbar\omega_{\text{rot}i}/(kT)] - 1)^2}.$$

If the rotation occurs about an arbitrary axis, which makes an angle α with the principal axis of the EFG tensor, then the contribution of each rotary oscillation into

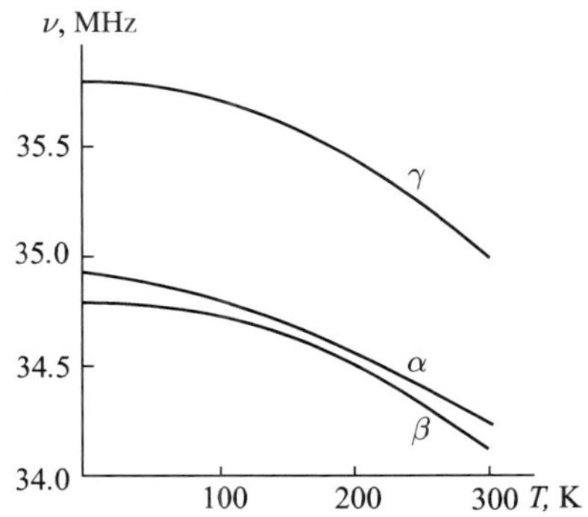

Fig. 8.8 Temperature dependence of the ^{35}Cl NQR frequency in n-$C_6H_4Cl_2$ for different Cl sites

the temperature coefficient decreases:

$$\frac{d(\overline{\omega}-\omega_0)}{\omega_0 dT} = -\frac{3\hbar^2}{2kT^2}\sum_{i=1}^{n}\frac{\exp[\hbar\omega_{\mathrm{rot}i}/(kT)]}{J_i(\exp[\hbar\omega_{\mathrm{rot}i}/(kT)]-1)^2}\sin^2\alpha_i. \quad (8.46)$$

However, often directions of rotary axes are unknown, moreover, usually a molecule executes several oscillations simultaneously (with different moments of inertia). In this case in order to describe the temperature dependence of NQR frequency one can apply a generalized formula:

$$\overline{\omega} = \omega_0(1 - AT + B/T),$$

where constants A and B are determined experimentally.

If $\eta \neq 0$ then the temperature coefficient decreases as well:

$$\frac{d(\overline{\omega}-\omega_0)}{\omega_0 dT} = -\frac{(3-\eta)\hbar^2}{2kT^2}\sum_{i=1}^{n}\frac{\exp[\hbar\omega_{\mathrm{rot}i}/(kT)]}{J_i(\exp[\hbar\omega_{\mathrm{rot}i}/(kT)]-1)^2}.$$

Let us give an example of temperature dependence of the ^{35}Cl NQR frequency in n-$C_6H_4Cl_2$ crystal (Fig. 8.8). The nucleus ^{35}Cl has spin 3/2, and in the NQR spectrum one can observe only one line from each unequivalent site. In the n-$C_6H_4Cl_2$ crystal chlorine nuclei may occupy three unequivalent sites (α, β, γ), and hence there are three lines in the NQR spectrum, whose frequencies decrease with temperature increasing.

Thus, if in crystal there are molecules, which execute rotary oscillations, then

- NQR frequencies decrease with temperature increasing;
- if rotation occurs about an axis, which makes an angle α with the direction of the principal axis of the EFG tensor, the temperature coefficient $d\overline{\omega}/(\omega_0 dT)$ decreases;
- with η increasing the temperature coefficient decreases as well;

- low-frequency rotary oscillations make greater contribution to the temperature changing of NQR frequencies than high-frequency ones;
- in high-temperature approximation NQR frequencies linearly depend on temperature.

8.2.2 Translation Vibrations

This kind of motion is of second importance that leads to the temperature dependence of the NQR frequency. There are *acoustic* and *optic* vibrations. Acoustic translation vibrations are the vibrations, for which the wave vector $\vec{k} \to 0$ when $\omega \to 0$. Optic vibrations are the vibrations, for which the wave vector $\vec{k} \neq 0$ if $\omega \to 0$ (see Fig. 8.9). In the acoustic mode (curve (*2*) in Fig. 8.9) atoms belonging to the same unit cell displace in the same directions, whereas in the optic mode (curve (*1*) in Fig. 8.9) they displace in opposite directions; vibration amplitudes are inversely proportional to their masses.

In ionic crystals the unit cell contains ions with opposite charges, that is why optic vibrations result in important changing in dipole moments of the unit cell. They determine the optical behavior of the crystal, and, being high-frequency, almost do not influence on the NQR frequency.

In three-dimensional crystals there are three mutually perpendicular acoustic modes. The propagation direction of one of the modes coincides with the wave vector direction, such vibrations are called *longitudinal vibrations*. Two other modes, whose propagation directions are perpendicular to the wave vector, are *transversal vibrations*.

In isotropic crystals vibration frequencies $\omega(\vec{k})$ do not depend on the wave vector direction. Two modes of transversal vibrations have the same frequencies $\omega(\vec{k})$ for each wave vector $\vec{k}$ value. As a rule these frequencies are lower than the longitudinal vibration frequencies, which correspond to the same $\vec{k}$ values. Therefore, with temperature increasing the main contribution to the NQR frequencies is due to transversal acoustic vibrations.

Let us remind some formulas from the theory of solid state acoustics, which will be used in forthcoming calculations. Each acoustic vibration frequency (ω) can be matched with a phonon of energy $\hbar\omega$. The total number of phonons ($3N$), which correspond to acoustic vibrations (or waves), can be calculate by means of the following integral:

$$3N = \frac{3\Omega_c n}{2\pi^2 \overline{v}^3} \int_0^{\omega_{\max}} \omega^2 \mathrm{d}\omega = \frac{\Omega_c n}{2\pi^2 \overline{v}^3} \omega_{\max}^3,$$

here Ω_c is the unit cell volume; n is the number of atoms in the unit cell; $\overline{v}$ is the average vibration propagation speed of the corresponding acoustic mode; $\omega_{\max}$ is the maximal value of the acoustic mode frequency.

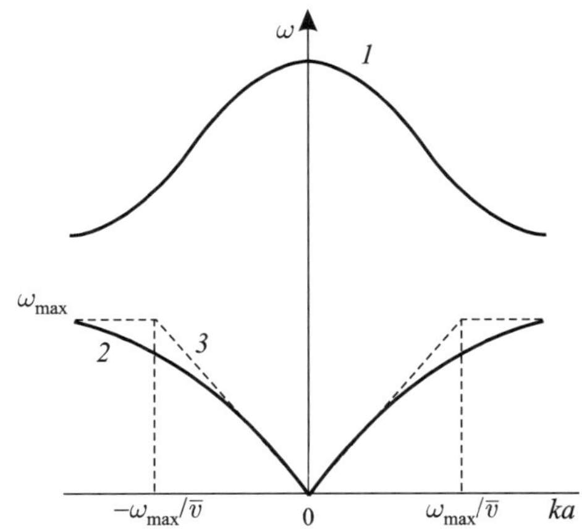

Fig. 8.9 Dispersion curves for acoustic and optic modes of translation vibrations; *a* is the lattice parameter

Table 8.2 Typical values of the acoustic transversal frequencies and Debay temperature in several crystals

Crystal	$\omega_{\max}$ (10^{13} MHz)	T_D (K)
NaCl	3.09	320
KCl	2.67	174
TlCl	1.61	124
Diamond	–	2000

Within the Debye approximation (curve 3 in Fig. 8.9) the energy of acoustic vibrations $\hbar\omega(\vec{k})$ is equal to zero if the wave vector value is greater than $\omega_{\max}/\overline{v}$ and equal to $\overline{v}\vec{k}$ if the wave vector is less or equal to $\omega_{\max}/\overline{v}$. The maximal energy value corresponds to the *Debye temperature*. In Table 8.2 we listed typical values of the acoustic transversal frequencies and Debay temperature in several crystals.

The frequency density of acoustic waves is determined from the following relation

$$\frac{\mathrm{d}N}{\Omega_{\mathrm{c}}\mathrm{d}\omega} = \frac{\omega^2}{2\pi^2 v^3}, \tag{8.47}$$

where $\mathrm{d}N$ is the number of acoustic waves in a frequency interval from ω to $\omega + \mathrm{d}\omega$.

In an isotropic elastic medium thermal disturbances of a crystal form a system of standing waves propagating with sonic speed (Fig. 8.10). The equation of a standing wave propagating along the x axis can be written as follows

$$y = A \sin\left(\omega_{\mathrm{tr}} t - 2\pi\frac{x}{\lambda}\right),$$

where y is the deviation of an atom or a molecule from its equilibrium state; A is the vibration amplitude; ω_{tr} is the frequency of translation vibrations; λ is the wave length. Then the speed of the particle displacement can be determined by the following way:

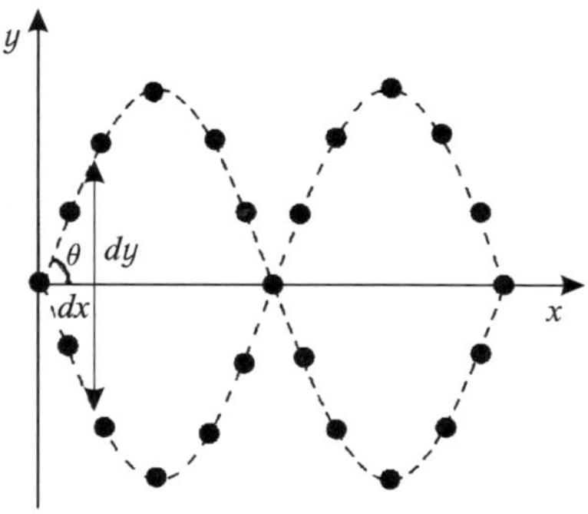

Fig. 8.10 Deviations of particles from their equilibrium state: an acoustic mode

$$c = \frac{\mathrm{d}y}{\mathrm{d}t} = \omega_{\mathrm{tr}} A \cos\left(\omega_{\mathrm{tr}} t - 2\pi \frac{x}{\lambda}\right) = c_{\max} \cos\left(\omega_{\mathrm{tr}} t - 2\pi \frac{x}{\lambda}\right).$$

After equating the kinetic energy of the wave to the energy of the harmonic oscillator one has

$$\frac{mc_{\max}^2}{2} = \frac{\omega_{\mathrm{tr}}^2 A^2 \rho \Omega_{\mathrm{c}}}{2} = \hbar\omega_{\mathrm{tr}} \left(\frac{1}{2} + \frac{1}{\exp\left[\hbar\omega_{\mathrm{tr}}/(kT)\right] - 1}\right), \tag{8.48}$$

here ρ is the crystal density and m is its mass. From Eq. (8.48) one can determine the wave amplitude:

$$A^2 = \frac{2\hbar}{\omega_{\mathrm{tr}} \rho \Omega_{\mathrm{c}}} \left(\frac{1}{2} + \frac{1}{\exp\left[\hbar\omega_{\mathrm{tr}}/(kT)\right] - 1}\right). \tag{8.49}$$

The deviation of a particle from its equilibrium state on an infinitesimal segment $\mathrm{d}y$ of the wave propagation divided by $\mathrm{d}x$ represents the angle of rotation of a particle about an axis perpendicular to the propagation direction (see Fig. 8.10). Then

$$\mathrm{tg}\theta \simeq \theta(t) = \frac{\mathrm{d}y}{\mathrm{d}x} = -\frac{2\pi A}{\lambda} \cos\left(\omega_{\mathrm{tr}} t - 2\pi \frac{x}{\lambda}\right). \tag{8.50}$$

As it follows from Eq. (8.50), in the harmonic approximation the rotation angle amplitude of a particle is

$$\theta_1(\omega_{\mathrm{tr}}) = \frac{2\pi A}{\lambda} = \frac{\omega_{\mathrm{tr}} A}{c_{\mathrm{tr}}}$$

($c_{\mathrm{tr}} = \omega_{\mathrm{tr}}\lambda/(2\pi)$ is the acoustic wave speed). Now, after finding the squaredamplitude $(\overline{\theta_1^2})$ averaged over all possible frequencies, it is possible to determine the changing in the NQR frequency using Eq. (8.38) for rotary oscillations. It can be done considering the frequency dependence of acoustic waves in the form of (8.47) and the amplitude of these waves in the form of (8.49):

$$\overline{\theta_1^2}(\omega_{\text{tr}}) = \int\limits_0^{\omega_\text{D}} \theta_1^2(\omega_{\text{tr}})\mathrm{d}N = \frac{1}{2}\int\limits_0^{\omega_\text{D}} \theta_1^2(\omega_{\text{tr}})\Omega_\text{c}\frac{\omega_{\text{tr}}^2}{\pi^2 c_{\text{tr}}^3}\mathrm{d}\omega_{\text{tr}}$$
$$= \frac{\hbar}{\pi^2\rho c_{\text{tr}}^5}\int\limits_0^{\omega_\text{D}}\left(\frac{\omega_{\text{tr}}^3}{2} + \frac{\omega_{\text{tr}}^3}{\exp\left[\hbar\omega_{\text{tr}}/(kT)\right]-1}\right)\mathrm{d}\omega_{\text{tr}}. \tag{8.51}$$

The first integral in Eq. (8.51) is easy to calculate, whereas the second one is the well known Debye function:

$$D(x_\text{D}) = \frac{3}{x_\text{D}^3}\int\limits_0^{x_\text{D}}\frac{x^3}{\exp x - 1}\mathrm{d}x,$$

$$x_\text{D} = \frac{T_\text{D}}{T} = \frac{\hbar\omega_\text{D}}{kT},\quad x = \frac{\hbar\omega_{\text{tr}}}{kT}.$$

After integration in Eq. (8.51), one has

$$\overline{\theta^2}_1 = \frac{k^4 T_\text{D}^4}{8\pi^2\rho c_{\text{tr}}^5\hbar^3}\left(1 + \frac{8}{3}\frac{T}{T_\text{D}}D\left(\frac{T_\text{D}}{T}\right)\right). \tag{8.52}$$

After substitution Eq. (8.52) into Eq. (8.38) one has an expression describing the temperature dependence of the NQR frequency:

$$\overline{\omega} = \omega_0\left[1 - \frac{3k^4 T_\text{D}^4}{32\pi^2\rho c_{\text{tr}}^5\hbar^3}\left(1 + \frac{8}{3}\frac{T}{T_\text{D}}D\left(\frac{T_\text{D}}{T}\right)\right)\right]. \tag{8.53}$$

The Debye function goes to zero at $T_\text{D}/T \to \infty$ ($T \to 0$). At $T \ll T_\text{D}$ the Debye function can be expanded in a series:

$$D(x_\text{D}) = \frac{\pi^4}{4x_\text{D}^4} + \cdots + \frac{A}{x_\text{D}^n}. \tag{8.54}$$

Substituting Eq. (8.54) into Eq. (8.53) and taking into account the first expansion term only one has

$$\overline{\omega} = \omega_0\left[1 - \frac{3k^4 T_\text{D}^4}{32\pi^2\rho c_{\text{tr}}^5\hbar^3}\left(1 + \frac{8\pi^4}{12}\left(\frac{T}{T_\text{D}}\right)^5\right)\right]. \tag{8.55}$$

As one can see from Eq. (8.55), the second term in the parentheses decreases proportionally to T^5, and at $T < 0.25\ T_{\mathrm{D}}$ it is one order less than unity and can be neglected. Then in Eq. (8.55) there is the only term that determines zero-point oscillations. These oscillations are always in solids, therefore the NQR frequency is not equal to ω_0 at any temperature. That means that when comparing theoretical and experimental data it is always necessary to take into account zero-point oscillations.

If $T \gg T_{\mathrm{D}}$ then it is possible to expand the Debye function in a series again:

$$D(x_{\mathrm{D}}) = 1 - \frac{3}{8}x_{\mathrm{D}} + \frac{1}{20}x_{\mathrm{D}}^2 + \cdots. \tag{8.56}$$

Substituting Eq. (8.56) into Eq. (8.53), then taking into account only the first expansion term and neglecting zero-point oscillations one obtains:

$$\overline{\omega} = \omega_0\left(1 - \frac{k^4 T_{\mathrm{D}}^4}{4\pi^2 \rho c_{\mathrm{tr}}^5 \hbar^3}\frac{T}{T_{\mathrm{D}}}\right).$$

Therefore, at high temperature lattice vibrations, as well as rotary oscillations, lead to linear decreasing of the NQR frequency.

8.2.3 Anomalous Temperature Dependence of the NQR Frequency

In most crystals the NQR frequency decreases with temperature increasing. However, there are such solids, for which an anomalous temperature dependence of NQR frequencies is observed. Particulary, many ferroelectrics exhibit such temperature dependencies, for example $LiNbO_3$. In order to understand such a behavior, it is necessary to decline several assumptions made by Bayer. The first issue is the influence of anharmonic oscillations on the NQR spectrum.

Let us suppose that rotary oscillations of a molecule are of such kind that the angle changing can be described by the following function

$$\begin{aligned} \theta &= \theta_1 \sin\omega t \quad &\text{for} \quad 0 \le t \le T_p/n, \\ \theta &= 0 \quad &\text{for} \quad T_p/n < t \le T_p, \end{aligned}$$

here T_p is the period of rotary oscillations of the molecule; n is a temperature dependent parameter. Then the mean-square value of θ can be found by the following way:

$$\overline{\theta^2} = \theta_1^2 \frac{1}{T_p}\int_0^{T_p/n} \sin^2\omega_{\mathrm{tr}} t\, \mathrm{d}t = \frac{\theta_1^2}{2n} - \theta_1^2 \frac{\sin(4\pi/n)}{\omega_{\mathrm{tr}}}. \tag{8.57}$$

If n rises with temperature increasing, then the frequency rises as well.

The second issue of the anomaly is the dependence of the quadrupole coupling constant on the rotation angle of the molecule. Such dependence can appear if rotation leads to changing in intramolecular distances.

The third issue of NQR frequency rising with temperature increasing can be decreasing of the unit cell volume.

Finally, in some crystals temperature changing leads to displacement of ions from their sites in the unit cell that also results in changing in intramolecular and intermolecular distances as well. In such cases the quadrupole coupling constant can either increases or decreases. The lithium niobate crystal $LiNbO_3$ provides a good illustration of such a behavior of the ^7Li quadrupole coupling constant. In this crystal with temperature increasing the ions displacement results to increasing of electric filed gradients at the lithium sites and to their decreasing at niobium sites.

8.3 Quadrupole Relaxation in Solids

Relaxation mechanisms of nuclear magnetization under NQR conditions, in general, are similar to those that were considered in NMR. However, there are some features determined by strong coupling of the spin subsystem to the lattice, presence of degenerated energy levels and their nonequidistant arrangement (for example, if $I = 1$ and $\eta \neq 0$, or for $I > 3/2$).

Quadrupole relaxation is the result of modulation of quadrupole interactions by thermal motion of different nature (see Chap. 3). Lattice vibrations directly affect the quadrupole interaction of nuclei, so NQR is more sensitive to lattice movements than NMR. In solid state the NMR theory of spin-lattice relaxation processes can be successfully described within the spin temperature concept, which is especially useful when one deals with a two-level system. In this case the spin subsystem can be characterized by a single spin temperature and hence a single spin–lattice relaxation time. In the theory of nuclear quadrupole resonance, two-level system arises when considering the interaction of a spin 3/2 with EFG tensors. In this case the only difference from NMR is that these energy levels are double degenerated. As it will be shown further in Sect. 9.2 (see Eq. (9.41)), the evolution of the spin system can be described by two magnetizations, $\vec{M}^+$ and $\vec{M}^-$, equal in magnitude and oppositely directed, each of which has its own system of non-degenerated energy levels. The magnetizations $\vec{M}^+$ and $\vec{M}^-$ are defined by the difference of the level populations $1/2 \leftrightarrow 3/2$ and $-1/2 \leftrightarrow -3/2$, respectively. If the spin subsystem is disturbed from its equilibrium state, for example, by an alternating rf field, then in the interaction representation these two vectors rotate in different directions, and since they practically do not interact with each other, equilibrium recovery processes for each magnetization can be considered separately, i.e. in each case one can consider a nondegenerate two-level system, one equivalent to another.

Thus, if $I = 3/2$, one can use a single spin temperature, and the only spin–lattice relaxation time. This conclusion can be made strictly on the basis of kinetic equations, which reflect the principle of detailed balance (changes in the populations

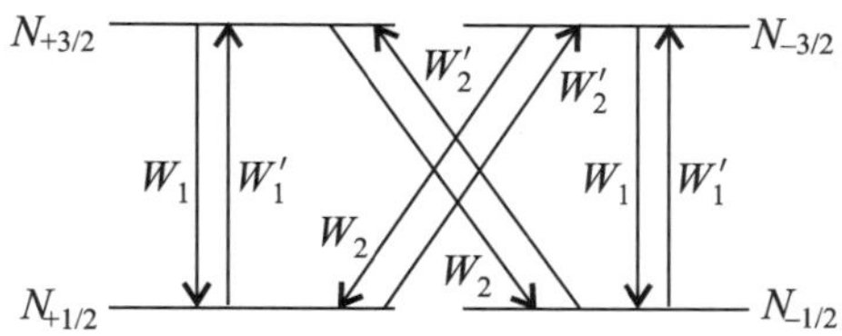

Fig. 8.11 Energy levels, populations and transitions (single- and double-quantum) in a quadrupole system with $I = 3/2$ and $\eta = 0$

of energy levels arise due to interaction between the nucleus quadrupole moment and inhomogeneous electric field gradients (see Fig. 8.11)).

Changes in the population of energy levels may occur under the influence of single-quantum (between the levels $\pm 1/2 \leftrightarrow \pm 3/2$) and double-quantum (between the levels $\mp 1/2 \leftrightarrow \pm 3/2$) transitions, that correspond to a change in the magnetic quantum number by one and two, respectively. Let us denote probabilities of upwards and downwards relaxation single-quantum transitions as $W_1^{'}$ and W_1, respectively. The probabilities of double quantum transitions are denoted by the symbols $W_2^{'}$ and W_2 (Fig. 8.11). With this notation, the change in the populations of the energy level (N) can be described by the following equations:

$$\begin{aligned}\frac{\mathrm{d}N_{\pm 3/2}}{\mathrm{d}t} &= N_{\pm 1/2}W_1^{'} - N_{\pm 3/2}W_1 + N_{\mp 1/2}W_2^{'} - N_{\pm 3/2}W_2,\\ \frac{\mathrm{d}N_{\pm 1/2}}{\mathrm{d}t} &= -N_{\pm 1/2}W_1^{'} + N_{\pm 3/2}W_1 - N_{\mp 1/2}W_2^{'} + N_{\pm 3/2}W_2.\end{aligned} \tag{8.58}$$

Of cause, the probabilities of the upward and downward relaxation transitions are not equal to each other, as they should ensure a Boltzmann equilibrium for the spin subsystem in the stationary state:

$$\frac{\overline{N}_{\pm 1/2}}{\overline{N}_{\pm 3/2}} = \frac{W^{'}}{W} = \exp\left(-\frac{\hbar\omega}{kT}\right), \tag{8.59}$$

here $\overline{N}_\alpha$ (with $\alpha = \pm 1/2, \pm 3/2$) are the equilibrium populations of the energy levels; W and $W^{'}$ are the probabilities of the downward and upward relaxation transitions, respectively. Hence,

$$W^{'} = \exp(-\hbar\omega/(kT))W. \tag{8.60}$$

Within the high-temperature approximation $\hbar\omega/(kT) \ll 1$, $\exp(-\hbar\omega/(kT)) \simeq 1$, and then $W^{'} = W$.

Since in the experiment the intensity of a NQR signal is proportional to the difference between the populations of energy levels, let us set up the kinetic equations for the difference of populations $n = N_{\pm 1/2} - N_{\pm 3/2}$. Subtracting in (8.58) the second equation from the first one we obtain:

$$\frac{\mathrm{d}n}{\mathrm{d}t} = 2(W_1 + W_2)n.$$

The solution of this equation is well known:

$$n = A\exp(-2t(W_1 + W_2)) + C. \tag{8.61}$$

Constants A and C can be found from the following conditions:

- at $t \to \infty$ the spin subsystem tends to an equilibrium state $n(\infty) = n_0$, then $C = n_0$;
- at $t = 0$ the population difference depends on how much the spin subsystem is derived from its equilibrium. If, for example, the spin subsystem in the initial time was in a saturated state ($n(0) = 0$) then $A = -C = -n_0$.

Taking into account that the NQR signal intensity $S(t)$ is proportional to $n(t)$, one obtains

$$S(t) = S_\infty[1 - \exp(-2t(W_1 + W_2))].$$

Hence, the evolution of the spin subsystem is described by one relaxation constant, which is

$$T_1 = 1/(2(W_1 + W_2)). \tag{8.62}$$

To calculate the spin–lattice relaxation time it is necessary to calculate the corresponding probabilities of relaxation transitions. This problem can be solved by modeling thermal translational vibrations, rotational fluctuations and thermal motion of different nature in order to determine the time-dependent part of the Hamiltonian, which describes quadrupole interactions, and to calculate corresponding matrix elements, as it was done in Sect. 3.1. It should be kept in mind that in contrast to NMR, it is necessary to use the density matrix of the quadrupole interactions Hamiltonian, and such an interaction representation, as will be shown in Sect. 9.2, does not coincide with the rotating frame. This density matrix was already considered in detail in this Chapter when one dealt with calculations of the energy levels. Calculations themselves are quite tedious,[2] so we will not dwell on the details of such calculations.

To date, there are only rather rough models that simulate thermal motions of different kind, so the calculation results are only in qualitative agreement with experimental data. The most common is the *Bayer model*, that takes into consideration the influence of molecule rotational oscillations in molecular crystals on spin-lattice relaxation processes. In this theory and its generalizations thermal vibrations in a crystal are supposed to be harmonic. As it follows from the Bayer theory, the probability of single-quantum relaxation transitions W_1, for $I = 3/2$ linearly depends on the temperature, whereas the probability of double quantum relaxation transitions W_2, is proportional to T^2. Usually $W_1 \ll W_2$, then, as it is seen from Eq. (8.62), the spin–lattice relaxation is proportional to T^2 as well. Such a behavior of the spin–lattice relaxation time is often a priori accepted as universal for rotational oscillations in molecular crystals. To a certain extent this point of view is supported by research

[2] They can be found, for example, in V. S. Grechishkin. Yadernye Kvadrupol'nye Vzaimodeistviya v Tverdykh Telakh (Nuclear Quadrupole Interactions in Solids, Moscow: Nauka, 1973).

practice, although it would be better to consider the similar experimental behavior of the relaxation rate as a particular case, which only reflects the fact that molecular rotational oscillations are close to the harmonic approximation.

As real solids are characterized by anharmonicity of molecular rotational oscillations, it is natural to expect a deviation from the form obtained theoretically within the framework of a pure harmonic model. It should be also kept in mind that the condition $W_1 \ll W_2$ is not universal. The ratio between W_1 and W_2 depends on the angle between the directions of the principal axes of the EFG tensor and the axis, around which rotational oscillations occur. The temperature behavior of the spin–lattice relaxation time, different from the quadratic dependence, can be predicted if one considers the existence of quasi-harmonic oscillations. This approximation is more general than the Bayer theory and is widely used. For nuclei with $I = 3/2$ and if the EFG tensor is axially symmetric expressions to calculate the probabilities of relaxation transitions are following:

$$W_1 = \frac{3\hbar\omega^2}{2J\omega_{\text{rot}}^3\tau_a}\coth\frac{\hbar\omega_{\text{rot}}}{2kT},$$

$$W_2 = 3\tau_a\left(\frac{(\hbar\omega)^2}{4J\omega_{\text{rot}}}\right)^2\left[\left(1+\exp\frac{-\hbar\omega_{\text{rot}}}{kT}\right)\sinh^2\frac{\hbar\omega_{\text{rot}}}{2kT}\right]^{-1},$$

where ω is the NQR frequency; ω_{rot} is the frequency of molecular rotational oscillations; J is the moment of inertia; τ_a is the average lifetime of discrete states of molecule oscillations. Supposing quasi-harmonic rotational oscillations, the temperature dependence of the spin–lattice relaxation is described by three terms: linear, quadratic and cubic in temperature:

$$\frac{1}{T_1} \simeq aT + bT^2 + cT^3. \tag{8.63}$$

It is clear that contributions of terms with different degrees of T in Eq. (8.63) are different for each particular substance, but within this approximation the exponent in the temperature power law dependence of the relaxation rate is always between one and three. Although the theoretical function, which describes the temperature dependence of the relaxation rate contains three terms with different degrees of T, in practice, within a limited temperature range where a molecular crystal exists (from 77 K to the melting point) it is possible with a sufficient accuracy to approximate the temperature dependence by a power function of the following form

$$T_1^{-1}(T) = aT^n, \tag{8.64}$$

where n is a non-integer value, which for each substance is determined from experimental data. If there is no special feature in the molecule behavior, the power in Eq. (8.64) is within the range from 1 to 3. For many compounds this range is even narrower, $1.5 \leq n \leq 2.5$.

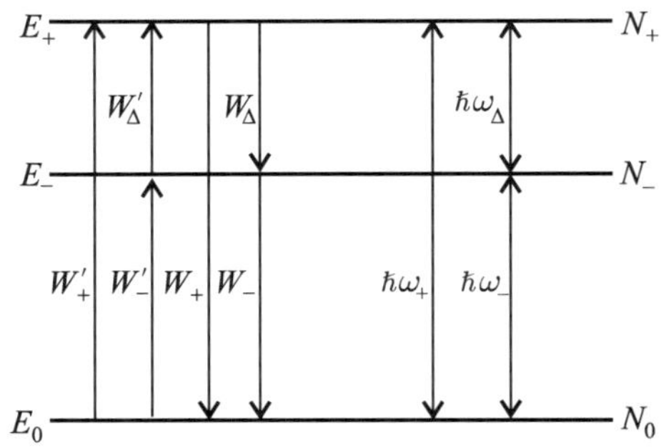

Fig. 8.12 Energy levels, populations and transitions in a quadrupole system with $I = 1$ and $\eta \neq 0$

Another mechanism of spin–lattice relaxation times is translational vibrations, which similar to rotational oscillations modulate the unperturbed Hamiltonian of quadrupole interactions. If one apply to lattice vibrations the Debye distribution with characteristic temperature T_D, for $T > T_D$ the temperature dependence of the quadrupole spin–lattice relaxation time is:

$$\frac{1}{T_1} \simeq T^2 \left(a - \frac{b}{T^2} \right),$$

and for temperatures much smaller than the Debye temperature

$$1/T_1 \simeq T^7.$$

If the spin system is characterized by more than two energy levels relaxation processes can not be described by a single spin–lattice relaxation time. For example, if $I = 1$ and $\eta \neq 0$ the quadrupole interactions lead to the three non-degenerate energy levels (Fig. 8.12). In a three-level system the spin–lattice relaxation, as will be shown further, can not be described by a single exponent. It is determined by the sum of two exponential terms, each of which is characterized by its own relaxation time.

In order to describe the process of establishing an equilibrium between the spin subsystem and the lattice one can again apply the kinetic equations for the populations, which reflect the principle of detailed equilibrium (see Chap. 3). Changes in the number of nuclei at each level can be described by simple equations:

$$\begin{aligned}
\frac{dN_+}{dt} &= W_+^{'} N_0 - W_+ N_+ + W_\Delta^{'} N_- - W_\Delta N_+,\\
\frac{dN_-}{dt} &= W_-^{'} N_0 - W_- N_- - W_\Delta^{'} N_- + W_\Delta N_+, \qquad (8.65)\\
\frac{dN_0}{dt} &= -W_+^{'} N_0 + W_+ N_+ - W_-^{'} N_0 + W_- N_-.
\end{aligned}$$

Expressing the probability of relaxation transitions from the lower to the higher level in terms of the probabilities of transitions from the higher to the lower level from Eq. (8.60) and substituting it into Eq. (8.65), one can write:

$$\begin{aligned}
\frac{\mathrm{d}N_+}{\mathrm{d}t} &= W_+ \exp\left(-\frac{\hbar\omega_+}{kT}\right) N_0 - (W_+ + W_\Delta)N_+ + W_\Delta \exp\left(-\frac{\hbar\omega_\Delta}{kT}\right) N_-,\\
\frac{\mathrm{d}N_-}{\mathrm{d}t} &= W_- \exp\left(-\frac{\hbar\omega_-}{kT}\right) N_0 - \left(W_- + W_\Delta \exp\left(-\frac{\hbar\omega_\Delta}{kT}\right)\right) N_- + W_\Delta N_+,\\
\frac{\mathrm{d}N_0}{\mathrm{d}t} &= -N_0\left(W_+ \exp\left(-\frac{\hbar\omega_+}{kT}\right) + W_- \exp\left(-\frac{\hbar\omega_-}{kT}\right)\right) + W_- N_- + W_+ N_+.
\end{aligned} \tag{8.66}$$

Since the amplitude of the NQR signal is proportional to the difference of populations of the corresponding levels, let us introduce the following notation for them:

$$n_+ = (N_+ - N_0), \quad n_- = (N_- - N_0), \quad n_\Delta = N_+ - N_-.$$

Equation (8.66) can be written in terms of n_α (with $\alpha = +, -, \Delta$). The factor $\exp(-\frac{\hbar\omega_\pm}{kT})$ for the majority of crystals is very close to unity. Thus, for the most frequently studied ^{14}N nucleus with $I = 1$, the quadrupole coupling constant is less than 5 MHz. Hence, the smallest value of the Boltzmann factor is of 0.999997 at $T = 77$ K. Then all exponents in Eq. (8.66) can be replaced by unity. It should be noted that T in Eqs. (8.59) and (8.66) is the spin temperature, which is different for different pairs of energy levels. However, as the spin temperature for all pairs of levels is usually higher or equal to the lattice temperature, when considering the kinetic equations the difference in spin temperatures is not essential. Finally, one has:

$$\begin{aligned}
\frac{\mathrm{d}n_+}{\mathrm{d}t} &= -n_+(2W_+) - n_- W_- - n_\Delta W_\Delta,\\
\frac{\mathrm{d}n_-}{\mathrm{d}t} &= -n_+ W_+ - n_-(2W) + n_\Delta W_\Delta,\\
\frac{\mathrm{d}n_\Delta}{\mathrm{d}t} &= -n_+ W_+ + n_- W_- - n_\Delta(2W_\Delta).
\end{aligned} \tag{8.67}$$

Solutions of differential equations (8.67) are sought in the form of

$$\begin{aligned}
n_+ &= \sum_i a_i \exp\left(-\frac{t}{T_{1i}}\right),\\
n_- &= \sum_i b_i \exp\left(-\frac{t}{T_{1i}}\right),\\
n_\Delta &= \sum_i c_i \exp\left(-\frac{t}{T_{1i}}\right).
\end{aligned} \tag{8.68}$$

Substituting Eq. (8.68) into Eq. (8.67) and equating coefficients of the same exponents one obtains

$$\begin{aligned}
-a_i/T_{1i} &= -2W_+a_i - W_\Delta c_i - W_-b_i,\\
-b_i/T_{1i} &= -2W_-b_i + W_\Delta c_i - W_+a_i,\\
-c_i/T_{1i} &= -W_+a_i - 2W_\Delta c_i + W_-b_i.
\end{aligned}$$

Since $n_\Delta = n_+ - n_-$, $c_i = a_i - b_i$, in the system of three equations (8.67) the latter is a linear combination of two first. Then

$$\begin{aligned}
-a_i/T_{1i} &= -2W_+a_i + W_\Delta(b_i - a_i) - W_-b_i,\\
-b_i/T_{1i} &= -2W_-b_i - W_\Delta(b_i - a_i) - W_+a_i
\end{aligned}$$

or

$$\begin{aligned}
a_i(-1/T_{1i} + 2W_+ + W_\Delta) + b_i(-W_\Delta + W_-) = 0,\\
a_i(-W_\Delta + W_+) + b_i(-1/T_{1i} + 2W_- + W_\Delta) = 0.
\end{aligned} \tag{8.69}$$

The system (8.69) has nonzero solutions for the coefficients a_i and b_i only if its determinant is equal to zero:

$$\left(-\frac{1}{T_{1i}} + 2W_+ + W_\Delta\right)\left(-\frac{1}{T_{1i}} + 2W_- + W_\Delta\right) - (-W_\Delta + W_-)(-W_\Delta + W_+) = 0. \tag{8.70}$$

The spin–lattice relaxation time can be found from Eq. (8.70):

$$1/T_{1i} = W_+ + W_- + W_\Delta \pm \sqrt{W_+^2 + W_-^2 + W_\Delta^2 - W_+W_- - W_+W_\Delta - W_-W_\Delta}. \tag{8.71}$$

As one can see from Eq. (8.71), there are two solutions for the relaxation times. Let us denote the shorter time by T_s, and the longer one by T_l. Now one can write the equations that determine the changes in the difference between the populations of particular levels:

$$\begin{aligned}
n_+ &= a_\mathrm{s}\exp(-t/T_{1\mathrm{s}}) + a_\mathrm{l}\exp(-t/T_{1\mathrm{l}}),\\
n_- &= b_\mathrm{s}\exp(-t/T_{1\mathrm{s}}) + b_\mathrm{l}\exp(-t/T_{1\mathrm{l}}).
\end{aligned} \tag{8.72}$$

As before, the coefficients of the exponents in Eq. (8.72) are determined from the initial conditions. If the spin system is affected by a rf pulse with the filling frequency equal, for example, to ω_+, then the population N_+ increases by the value of ΔN. Then the population N_0 is reduced by the same amount, whereas the population N_- remains unchanged. These processes occur in the spin subsystem in a time much shorter than the spin–lattice relaxation. As it follows from (8.68), in this case the n_+ value increases by $2\Delta N$, whereas the n_- value increases only by ΔN, i.e. $\Delta n_- = 0.5\Delta n_+$.

If before applying a rf pulse (a reading pulse) the spin subsystem was derived from an equilibrium state (e.g. saturation) and was relaxing for a time τ, then the signal amplitude, which is proportional to $n_+(0) - n_+(\tau)$, can be written as follows

$$S = S_\infty \left(a_s \left(1 - \exp\left(\frac{-\tau}{T_{1s}}\right)\right) + a_l \left(1 - \exp\left(\frac{-\tau}{T_{1l}}\right)\right)\right),$$

here S_∞ is the amplitude of the signal after all relaxation processes. Without loss of generality $n_+(0)$ can be normalized to unity. Then

$$a_s + a_l = 1$$

and taking into account that $\Delta n_- = 0.5\Delta n_+$, for the coefficients b_i one has

$$b_s + b_l = 1/2.$$

In this case the amplitude of the NQR signal after the rf pulse can be written as

$$S = S_\infty \left(\left(1 - a_s \exp\left(\frac{-\tau}{T_{1s}}\right)\right) - (1 - a_s) \exp\left(\frac{-\tau}{T_{1l}}\right)\right). \tag{8.73}$$

Thus, the relaxation process for $I = 1$ and $\eta \neq 0$ is described by two relaxation constants. The coefficient a_s is determined by the initial nonequilibrium state, from which the spin subsystem comes to the equilibrium. If the rf pulse has the filling frequency ω_-, to determine the time dependence of the NQR signal for this frequency it is necessary to make calculations similar to mentioned above. Obviously, in this case, the spin–lattice constants are also defined by Eq. (8.71), but with another factor a_s in Eq. (8.73). Sometimes this fact can allows us to determine both relaxation constants from measurements made at different transitions. But commonly, only one constant can be found from such an experiment. To find the two constants, they must be significantly different from each other, whereas the coefficients of the corresponding exponents must be approximately the same.

For an arbitrary nuclear spin the solution of the kinetic equations (8.67) has the following form:

$$n_i(t) = \sum_k a_i^{(k)} \exp\left(-\frac{t}{T_{1k}}\right) + n_i(\infty),$$

here $k = 1, 2, \ldots, I$ for integer spins and $k = 1, 2, \ldots, I - 1/2$ for half-integer ones.

Thus, the non-equidistant multilevel systems exhibit a complex relaxation behavior that can not be described by a simple single-exponential law.

8.4 Methods of Calculations of NQR Spectral Parameters

By analyzing NQR spectra one can find the components of the EFG tensor. In crystals electric field gradients are created by charged ions and electrons located in the non-spherical symmetric orbitals. Distortion of electron shells can be caused by several reasons such as their overlapping, a covalency of chemical bonds, an influence of electric fields induced by ions, and so on. Electric field gradients vary considerably even with small distortions of the electronic structure. Therefore, the measured values of the EFG tensor can be used to determine the electronic structure of the crystal. However, to extract such an information it is necessary to calculate the EFG tensor.

There is a number of methods to calculate the EFG tensor distribution in crystals. The most popular in the past methods are: the point charge model, the point multipole model, the Townes–Dailey method, the Mulliken–Wolfsberg–Helmholtz method, the method of full or partial neglecting of differential overlapping. Each of these methods has its own advantages and disadvantages. The choice of the calculation method depends on various factors: first, a research problem; second, studied objects; third, technical means available to a researcher. For example, for ionic crystals the simple point charge model is sufficient to describe the EGF tensors at the nucleus sites. The Townes–Dailey method is traditionally used to estimate components of the EFG tensor in molecular crystals. However, to reproduce correctly the electron density distribution near the nucleus under study ab initio calculations are required, and at present almost all calculations are carried out within various ab initio approaches.

Let us consider in more detail each of these methods.

8.4.1 Townes–Dailey Method

This method developed by Townes and Dailey[3] was rather widely used in the past to analyze NQR experimental data. Despite its simplicity, this method allows us to estimate quite accurately resonance frequencies that is extremely important when searching for a NQR signal. In addition, it provides qualitative information about chemical bond parameters in a molecule (or molecular crystal) such as hybridization, ionicity, double bond degree).

In the Townes–Dailey method only the valence electrons of the atom are taken into account, since the filled shell of the atom does not affect the EFG value at the nucleus site. The wave functions of the molecule are written as MO LCAO functions (Molecular Orbitals as a Linear Combination of Atomic Orbitals). To take into account distortion of electron shells created by electric fields generated by inner or outer charges (with respect to the considered shell) in this method one introduces the Sternheimer antishielding factor, which takes into account that the EFG value at

---

[3] C. H. Townes and B. P. Dailey. Determination of electronic structure of molecules from nuclear quadrupole effects. J. Chem. Phys. **17** (1949) 782–796. B. P. Dailey and C. H. Townes. The ionic character of diatomic molecules. J. Chem. Phys. **23** (1965) 118–123.

the nucleus site caused by charged particles is reinforced due to the electron shell distortion.

The Townes–Dailey method is based on the following assumptions:

- Point charges and multipoles do not give a significant contribution to the EFG tensor. In molecular crystals to which this method is applied, chemical bonds are essentially covalent. If the ionicity is less then 50 %, then the contribution of point multipoles into the EFG tensor does not exceed 10 %.
- The Sternheimer antishielding factor does not change in the extension from atom to molecule. In fact, the difference can achieve a few percent.
- It is assumed that the main contribution to EFG is due to the valence p-electrons. Inner shells are completely filled, therefore, spherically symmetric. Actually a small contribution to the EFG tensor due to a decompensation of electron shells and symmetry breaking is possible (as a rule it is less then 10 %). Since the shells of s-electrons are spherically symmetric, electric field gradients induced by them at the nucleus site are equal to zero.
- It is supposed that the EFG tensor at the nucleus site of the isolated atom is known.

In order to justify theoretically the third sentence, let us represent the nuclear wave function as a product of a radial function and a spherical harmonic:

$$\psi = \chi(r) Y_{l,m}(\theta, \phi).$$

Then zz-component of the EFG tensor has the following form:

$$V_{zz} = \int \psi^* \frac{3\cos^2\theta - 1}{r^3} \psi \, d\tau = -\frac{2\left(3m^2 - l(l+1)\right)}{(2l-1)(2l+3)} \left\langle \frac{1}{r^3} \right\rangle = A_{lm} \left\langle \frac{1}{r^3} \right\rangle,$$

here l and m are the orbital and magnetic quantum numbers, respectively. The values of the coefficient A_{lm} for various l and m are listed below:

l	m	orbital	A_{lm}
0	0	s	0
1	±1	p_x, p_y	−2/5
1	0	p_z	4/5
2	±2	$d_{xy}, d_{x^2-y^2}$	2/7
2	±1	d_{xz}, d_{yz}	−4/7
2	0	d_{z^2}	4/7

The contribution of p-electrons predominates due to the fact that the average value of $\langle 1/r^3 \rangle$ inversely proportional to l^3:

$$\left\langle \frac{1}{r^3} \right\rangle = \left(\frac{Z}{a_0 n l} \right)^3,$$

here Z is the charge number; a_0 is the radius of the first Bohr orbit; n is the principal quantum number. At the same n the ratio between the values of $\langle 1/r^3 \rangle$ for p- and d-electrons is 8:1. With n increasing the contribution to the EFG value from corresponding electrons rapidly decreases.

Let us consider several examples that exhibit how to apply the Townes–Dailey method to estimate parameters of chemical bonding in simple atoms and molecules.

Example 1. Chlorine atom (Cl) The valence configuration of the atom Cl is $1s^2 2s^2 p^6 3s^2 p^5$. The inner shells are completely filled, spherically symmetric and do not contribute to EFG. The shell $3s^2$ is also spherically symmetric and the main contribution to the EFG tensor is due the $3p^5$-shell:

$$V_{zz}^{\text{at}} = \sum_{p} (V_{zz})_p \,. \tag{8.74}$$

Each p- shell consists of three mutually perpendicular orbitals: p_x, p_y, p_z, on each of which, according to the Pauli principle, can be found two electrons. Let us choose the coordinate axes in such a manner that these three p-orbitals are directed along the corresponding coordinate axes and on the p_z-orbital there is only one electron. Let us introduce the notation $(V_{zz})_{p_z} = V_{zz}^{(0)}$. Eq. (8.74) can be rewritten as

$$V_{zz}^{\text{at}} = V_{zz}^{(0)} - \frac{1}{2} V_{zz}^{(0)} \cdot 4 = -V_{zz}^{(0)}.$$

Here it is taken into account that the contribution to the EFG tensor from the p_x- and p_y-orbitals is twice smaller than the contribution from the p_z-orbitals and has the opposite sign.

Since p-electrons are outer to the whole shell, the resulting gradient should be multiplied by the Sternheimer antishielding factor. Thus, finally one has

$$V_{zz}^{\text{at}} = -(1 - \gamma_\infty)\, V_{zz}^{(0)}.$$

Example 2. Chlorine molecule (Cl_2) We assume that only $3s$- and $3p$-orbitals are involved into the bonding formation between two chlorine atoms. Hereby hybridized sp-atomic orbitals are formed (a part of electrons from the s-orbital goes to the p-orbital):

$$\psi_{\text{b}} = \alpha\phi_{3s} + \left(1 - \alpha^2\right)^{1/2} \phi_{3p_z} \quad \rightarrow \quad \text{a bonding orbital}$$

$$\psi_{\text{ab}} = \left(1 - \alpha^2\right)^{1/2} \phi_{3s} - \alpha\phi_{3p_z} \quad \rightarrow \quad \text{an antibonding orbital}$$

Here α is the degree of hybridization, which characterizes the quantity of electrons, which go from s- to p-orbital. Orbitals p_x, p_y do pot participate in the bonding formation. A bonding orbital is such an orbital whose electron energy lies lower the energy of electrons in $3s$- and $3p_z$-orbitals; an antibonding orbital is such an orbital whose electron energy corresponds to the energy higher then the energy of electrons in $3s$- and $3p_z$-orbitals.

Let us construct the bonding molecular orbital:

$$\psi_{\mathrm{b}} = A\left(\psi_{\mathrm{b}}^{(1)} + \psi_{\mathrm{b}}^{(2)}\right),$$

here $\psi_{\mathrm{b}}^{(1)}$, $\psi_{\mathrm{b}}^{(2)}$ are normalized bonding orbitals of chlorine atoms 1 and 2, respectively; A is a normalization factor which is defined by the following normalization condition

$$\int \psi_{\mathrm{b}}^{*}\psi_{\mathrm{b}}\mathrm{d}\tau = 1.$$

Then

$$A = \frac{1}{\sqrt{2(1+S)}},$$

where

$$S = \int \psi_{\mathrm{b}}^{(1)*}\psi_{\mathrm{b}}^{(2)}\mathrm{d}\tau = \int \psi_{\mathrm{b}}^{(2)*}\psi_{\mathrm{b}}^{(1)}\mathrm{d}\tau$$

is the overlap integral. Let us find the contribution of ψ_{b} into the EFG value:

$$\begin{aligned}(V_{zz})_{\psi_{\mathrm{b}}} &= \int \psi_{\mathrm{b}}^{*}\left(\frac{3\cos^2\theta - 1}{r^3}\right)\psi_{\mathrm{b}}\mathrm{d}\tau = A^2\left[\int \psi_{\mathrm{b}}^{(1)*}\left(\frac{3\cos^2\theta - 1}{r^3}\right)\psi_{\mathrm{b}}^{(1)}\mathrm{d}\tau\right.\\ &+ \int \psi_{\mathrm{b}}^{(2)*}\left(\frac{3\cos^2\theta - 1}{r^3}\right)\psi_{\mathrm{b}}^{(2)}\mathrm{d}\tau + \int \psi_{\mathrm{b}}^{(1)*}\left(\frac{3\cos^2\theta - 1}{r^3}\right)\psi_{\mathrm{b}}^{(2)}\mathrm{d}\tau\\ &+ \left.\int \psi_{\mathrm{b}}^{(2)*}\left(\frac{3\cos^2\theta - 1}{r^3}\right)\psi_{\mathrm{b}}^{(1)}\mathrm{d}\tau\right] \equiv A^2\left[q_{11} + q_{22} + q_{12} + q_{21}\right].\end{aligned}$$

Terms q_{11} and q_{22} are the contributions to EFG from chlorine atoms 1 and 2, respectively. Since one calculates EFG at the nucleus site of the chlorine atom number 1, the contribution from the distant chorine atom number 2 can be neglected. The last two terms q_{12} and q_{21} are also small and can be ignored. The latter can be easily proved by a simple example of the hydrogen molecule. The hydrogen wave function has the following form

$$\psi = \left(2 - \frac{r}{3a_0}\right) r e^{-r/(3a_0)} \cos\theta.$$

For the angle θ, which is within the range from $\pi/2$ to $-\pi/2$, the maximum of the wave function is achieved at $r = 1.8 \cdot a_0$, the minimum corresponds to $r = 10 \cdot a_0$, and zero is at $r = 6 \cdot a_0$. The probability to find an electron at $r > 6 \cdot a_0$ is about 90 %. However, if one calculates for this function the $\langle 1/r^3 \rangle$ value, the contribution to it from the region $r > 6 \cdot a_0$ is only 4 % of the total value, and the contribution from the range $r > 1.8 \cdot a_0$ is 14 %. It is explained by the fact that $\langle 1/r^3 \rangle$ is a rapidly decreasing function. For the molecule Cl_2 a rough estimation leads to $q_{12} = -0.05 \cdot q_{11}$.

Thus, the contribution from the bonding orbital to V_{zz} is

$$(V_{zz})_{\psi_b} = A^2 \int \psi_b^{(1)*} \left(\frac{3\cos^2\theta - 1}{r^3} \right) \psi_b^{(1)} \mathrm{d}\tau$$
$$= \frac{1-\alpha^2}{2(1+S)} \int \psi_{3p_z}^* \left(\frac{3\cos^2\theta - 1}{r^3} \right) \psi_{3p_z} \mathrm{d}\tau = \frac{1-\alpha^2}{2(1+S)} V_{zz}^{(0)},$$

Here one takes into account that $3s$ does not contribute to EFG. Moreover, as the wave functions s and p are orthogonal to each other, the cross terms do not contribute to EFG as well.

By the similar way one can obtain the contribution of the antibonding orbital:

$$(V_{zz})_{\psi_{ab}} = \alpha^2 V_{zz}^{(0)}.$$

As p_x- and p_y-orbitals do not take part in the chemical bonding their contributions to EFG do not change. Finally, taking into consideration all the contributions and the Sternheimer antishielding factor one obtains the V_{zz} value in the Cl_2 molecule:

$$V_{zz}^{\mathrm{mol}} = (1-\gamma_\infty) \left[\frac{1-\alpha^2}{2(1+S)} V_{zz}^{(0)} \cdot 2 + \alpha^2 V_{zz}^{(0)} \cdot 2 - \frac{1}{2} V_{zz}^{(0)} \cdot 2 - \frac{1}{2} V_{zz}^{(0)} \cdot 2 \right]$$
$$= (1-\gamma_\infty) V_{zz}^{(0)} (1-\alpha^2) \left(\frac{1}{1+S} - 2 \right). \tag{8.75}$$

If the overlapping is too small ($S \ll 1$) this expression can be rewritten as follows

$$V_{zz}^{\mathrm{mol}} \simeq V_{zz}^{\mathrm{at}} (1-\alpha^2)(1+S).$$

Neglecting the second-order terms, one has

$$V_{zz}^{\mathrm{mol}} \simeq V_{zz}^{\mathrm{at}} (1-\alpha^2+S). \tag{8.76}$$

Thus, with the degree of hybridization increasing EFG decreases and, conversely, increases with overlap increasing. If V_{zz}^{mol}, V_{zz}^{at} and S are known one can estimate the degree of hybridization, but this estimation is rather inaccurate.

In the common case for V_{zz}^{mol} in the form of (8.75) one can write :

$$V_{zz}^{\mathrm{mol}} = \left(\frac{N_x + N_y}{2} - N_z \right) V_{zz}^{\mathrm{at}}, \tag{8.77}$$

where N_x, N_y and N_z are populations of the corresponding p_x, p_y and p_z orbitals of the chlorine atom.

If p_x- and p_y-orbitals do not participate in the chemical bonding, then $N_x = N_y = 2$. If there is a so-called π-bond, that leads to decreasing of N_x, N_y populations, and hence, to decreasing of EFG. Taking into account π-bond Eq. (8.76) can be written as

$$V_{zz}^{\text{mol}} \simeq V_{zz}^{\text{at}}(1 - \alpha^2 + S - \pi),$$

where π is a measure of π-bonding.

Next, as $3d$-orbital initially is vacant, if there is a pd-hybridization it results in decreasing of N_z, and hence in increasing of EFG:

$$V_{zz}^{\text{mol}} \simeq V_{zz}^{\text{at}}(1 - \alpha^2 + S - \pi + d).$$

Example 3. *R*–Cl molecule Let us denote by X_{Cl} and X_R the electronegativity of the Cl and the halogen R atoms, respectively. Let the chlorine atom is more electronegative than the halogen atom ($X_{\text{Cl}} > X_R$). The chemical bonding can be represented as a superposition of pure covalent and pure ionic bonds:

$$\psi = (1 - I)(R - \text{Cl}) + I(R^+\text{Cl}^-),$$

here $I(R^+\text{Cl}^-)$ characterizes the ionicity degree. The ionic contribution can be calculated within other models, for example, within the point-charge model. Finally, taking into account the ionicity degree the covalent contribution can be written as follows

$$V_{zz}^{\text{mol}} \simeq V_{zz}^{\text{at}}(1 - I)(1 - \alpha^2 + S - \pi + d) \simeq V_{zz}^{\text{at}}(1 - I - \alpha^2 + S - \pi + d). \quad (8.78)$$

If the halogen atom R has the electronegativity less than the electronegativity of the Cl atom, the expression for V_{zz}^{mol} has the form

$$V_{zz}^{\text{mol}} \simeq V_{zz}^{\text{al}}(1 + I - \alpha^2 + S - \pi + d). \quad (8.79)$$

In some cases, when studying the nature of chemical bonding, the useful information can be obtained from the asymmetry parameter of the EFG tensor:

$$\eta = \left| \frac{V_{xx} - V_{yy}}{V_{zz}} \right|.$$

Taking into account Eq. (8.77) for V_{zz} if the EFG tensor is not axially symmetric one has the following expressions for xx-, yy-component of the EFG tensor and η:

$$V_{xx}^{\text{mol}} = \left(N_x - \frac{N_y + N_z}{2}\right) V_{zz}^{\text{at}}, \qquad V_{yy}^{\text{mol}} = \left(N_y - \frac{N_x + N_z}{2}\right) V_{zz}^{\text{at}},$$

$$\eta = \frac{3/2(N_x - N_y)}{N_z - (N_x + N_y)/2}. \quad (8.80)$$

If $\eta \ll 1$, Equation (8.80) can be rewritten as follows:

$$\eta \simeq \frac{3}{4}(N_x - N_y)\frac{eQV_{zz}^{\mathrm{at}}}{\hbar\omega_{\frac{1}{2}\leftrightarrow\frac{3}{2}}} \quad \text{for } I = \frac{3}{2},$$
$$\eta \simeq \frac{2}{3}(N_x - N_y)\frac{eQV_{zz}^{\mathrm{at}}}{\hbar\omega_{\frac{1}{2}\leftrightarrow\frac{3}{2}}} \quad \text{for } I = \frac{5}{2}. \tag{8.81}$$

As it is seen from Eq. (8.81) the asymmetry parameter of the EFG tensor is defined by the occupancy difference of p_x- and p_y-orbitals.

8.4.2 Determination of Chemical Bond Parameters from NQR Spectra

The Townes–Dailey method allows us to express the basic parameters of the quadrupole interaction (namely, the quadrupole coupling constant and the asymmetry parameter of the EFG tensor) in terms of the main parameters of the chemical bonding and molecular structure. As it is seen shown in Eqs. (8.78) and (8.79), the main component of the EFG tensor depends on several parameters, and hence for a correct interpretation of the experimental data auxiliary semiempirical schemes are required. Let us consider in more details the main parameters of chemical bonding that affect NQR spectra.

Hybridization It is well known that hybridization of the orbitals, which form chemical bonds, change both bond lengths and angles between the bonds.

Let us consider how hybridization affects the C–H and C–Cl bonds. According to experimental data, the C–H bond length decreases with increasing the contribution of bonding orbitals of s-type, i.e. when coming from sp^2- (C_2H_2) to sp-orbitals (C_2H_4). However, this influence is not very important. Similar results were obtained for C–Cl bonds. Below there are the C–Cl bond length values for various hybridization type:

Hybridization	Molecule	C–Cl bond length (Å)
sp^3	$ClCH_2CH_3$	1.76
sp^2	$ClCHCH_2$	1.69
sp	ClCCH	1.63

However, the influence of the hybridization on bond angles is much more essential. If a halogen atom is bounded to several atoms (ligands), the degree of hybridization α^2 can be calculated by the following formula:

$$\alpha^2 = \frac{\cos\beta}{\cos\beta - 1}, \tag{8.82}$$

where β is the angle between bonds. It is supposed that the orbitals are directed along the lines, which connect atoms. However this expression may fail if there is a strong repulsion between ligands or in the presence of pd-hybridization. The latter concerns heavy atoms.

Clark and Grandinetti have studied the relationships between ^{17}O quadrupolar coupling parameters and the local structure around the bridging oxygen in germanates.[4] Within the semiempirical Townes–Dailey method they have obtained analytical expression for the NQR frequency (ω) and asymmetry parameter (η) as functions of the angle $\Omega = \angle$G–O–Ge:

$$\omega(\Omega) = a\left(\frac{1}{2} + \frac{\cos\Omega}{\cos\Omega - 1}\right)^{c_1},$$

$$\eta(\Omega) = b\left(\frac{1}{2} - \frac{\cos\Omega}{\cos\Omega - 1}\right)^{c_2}$$

Here a, b, c_1 and c_2 are fitting parameters. The best agreement can be achieved with c_1 values near 2. The similar relationships were obtained by Hussin et al. for amorphous GeO_2 glasses.[5]

In order to define the degree of hybridization α^2 in binary compounds one usually applies other relationships. For example, Towns and Dailey proposed the following rule to determine the hybridization degree:

$$|X_A - X_B| \leq 0.25 \Rightarrow \alpha^2 = 0,$$

$$|X_A - X_B| > 0.25 \Rightarrow \alpha^2 = 0.15.$$

Bond ionicity Based on the rules for determining the hybridization degree and using the experimental values of the quadrupole coupling constants of nuclei in different binary molecules, Townes and Dailey calculated the ionicity degree and compared it with empirical dependencies previously proposed by Gordy,[6] Pauling[7] and Wilmshurst.[8] Here are the most common ways to determine the degree of ionicity:

---

4 T. M. Clark, P. J. Grandinetti. Relationships between bridging oxygen ^{17}O quadrupolar coupling parameters and structure in germanates. J. Non-Cryst. Solids 265 (2000) 75–82.

5 R. Hussin, D. Holland, R. Dupree. Does six-coordinate germanium exist in Na_2O–GeO_2 glasses? Oxygen-17 nuclear magnetic resonance measurements. J. Non-Cryst. Solids **232-234** (1998) 440–445. R. Hussin, R. Dupree, D. Holland. The Ge–Ô–Ge bond angle distribution in GeO_2 glass: a NMR determination. J. Non-Cryst. Solids **246** (1999) 159–168.

6 W. Gordy. Dependence of bond order and of bond energy upon bond length. J. Chem. Phys. **15** (1947) 305–310.

7 L. Pauling. Nature of the chemical bond (Oxford University Press, London, 1950).

8 J. K. Wilmshurst. Orbital radii and the dependence of bond length upon ionicity, hybridization, and bond order. J. Chem. Phys. **33** (1960) 813–820.

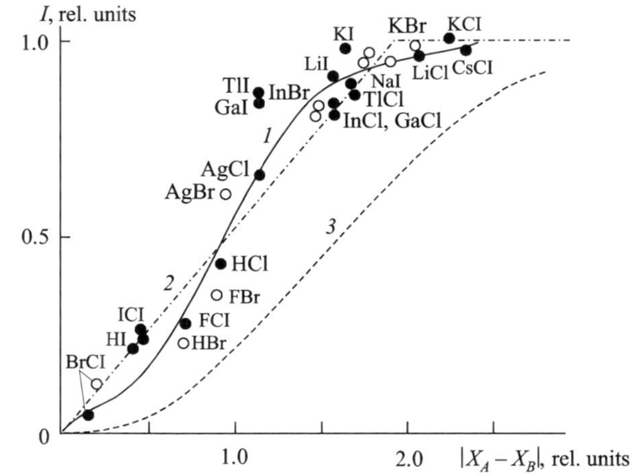

Fig. 8.13 Bond ionicity calculated within the Townes–Dailey (*curve 1*) compared to the Gordy (*curve 2*) and Pauling (*curve 3*) methods

- by Gordy:

$$I = \frac{|X_A - X_B|}{2}, \quad \text{if } |X_A - X_B| \leq 2,$$

$$I = 1, \quad \text{if } |X_A - X_B| > 2;$$

- by Pauling:

$$I = 1 - e^{-\left(\frac{X_A - X_B}{2}\right)^2};$$

- by Wilmshurst:

$$I = \frac{|X_A - X_B|}{X_A + X_B}.$$

The results obtained by these different methods are given in Fig. 8.13. As it is seen, the Gordy method predicts the greatest ionicity and is very close to the Townes–Dailey curve. The Pauling estimation leads to the essentially low ionicity degree. It is worth noting that neglecting the *sp*-hybridization in the Townes–Dailey method leads to the similar results.

It is to be noted that the Wilmshurst estimation of the ionicity results in a greater *p*-hybridization degree as compared to the Townes–Dailey method. This, in its turn, decreases the ionicity value.

It is interesting to consider a compound of two halogens, where the bond ionicity degree can be determined according to NQR parameters of both nuclei. The results, obtained by Townes and Dailey for BrCl and ICl are following:

$$\text{BrCl}: \ \text{for } ^{35}\text{Cl} \ I = 0.056, \ \text{for } ^{81}\text{Br} \ I = 0.110;$$
$$\text{ICl}: \ \text{for } ^{35}\text{Cl} \ I = 0.115, \ \text{for } ^{127}\text{I} \ I = 0.229.$$

Here the sp-hybridization is taken into account. As it is seen, the bond ionicity values obtained from the NQR parameters of different nuclei are rather different. It is interesting to note that for ICl neglecting the sp-hybridization leads to a better agreement:

$$\mathrm{ICl}: \text{ for } {}^{35}\mathrm{Cl}\ \ I = 0.248, \text{ for } {}^{127}\mathrm{I}\ \ I = 0.226.$$

For BrCl nothing has changed since $X_{\mathrm{Br}} - X_{\mathrm{Cl}} = 0.2$, hence, $\alpha^2 = 0$.

Studies on the chemical bonding in complexes of Py–ICl made by Igarashi et al.[9] have shown that as a result of the ICl complexation with Py, the ^{14}N and ^{35}Cl NQR frequencies are shifted towards lower values, whereas the ^{127}I NQR frequency is shifted towards higher values. This result can be explained within the framework of the Townes–Dailey approach. To calculate V_{zz}^{Cl} and V_{zz}^{I} one can apply Eqs. (8.78) and (8.79), from which it follows that the experimentally observed changes of nuclear quadrupole resonance frequency are caused by increasing of the I–Cl bond ionicity.

Double-bonding The relationship between the asymmetry parameter of the EFG tensor η and the double-bonding degree π was determined by Goldstein and Bragg[10]:

$$\pi = \frac{2}{3}\frac{|V_{xx} - V_{yy}|}{V_{zz}^{\mathrm{at}}}.$$

They carried out a detailed analysis of the double-bonding degree for planar carbon containing molecules. Within the framework of the molecular orbital theory a double bond in, for example, the C–Cl bond, means the chlorine atom loses its π-electron. Then the measured asymmetry parameter is related to the double-bonding degree as follows

$$\eta = \frac{2}{3}\pi\frac{e^2Qq^{\mathrm{at}}}{e^2Qq^{\mathrm{mol}}}.$$

To estimate the double-bonding degree Pauling proposed the following expression:

$$\pi = \frac{r_1 - r}{r_1 + 2r - 3r_2},$$

where r_1 is the single-bond length, r_2 is the double-bond length, r is an intermediate case.

---

9 T. Igarashi, T. Oyama, O.K. Poleshschuk, J. Koput, J.N. Latosinska, B. Nogaj. Ab inito study of the bonding and nuclear quadrupole coupling in the Py-ICl complex. J. Mol. Struct. **513** (1999) 29–34.

10 J. H. Goldstein and J. K. Bragg. Nuclear quadrupole effects in the microwave spectrum of the asymmetric top molecule,vinyl chloride. Phys. Rev. **75** (1949) 1453–1454.

8.4.3 Correlation Equations for NQR Frequencies

Another method widely used to interpret NQR spectra is the method of correlation equations, which relates changes in resonance frequency with structural changes of studied materials.

For example, it has been found a correlation between the ^{35}Cl NQR frequencies and the electronegativity of the R atom, X_R, in binary Cl–R compounds. This correlation follows from the following equation:

$$V_{zz} = (1 - I)(1 - \alpha^2).$$

Within the first approximation, the sp-hybridization degree α^2 can be considered independent on R, whereas I linearly dependent on X_R (by Gordy.) Then zz-component of the EFG tensor is linearly proportional to X_R. Consequently, with decreasing the R atom electronegativity, the ^{35}Cl NQR frequency should decrease as well, as it experimentally observed.

According to NQR studies of aromatic compounds and tetrahedral molecules of chlorine-substituted methane the resonance frequency increases with the number of halogen atoms in the molecule. A possible explanation is follows: as $X_{\mathrm{H}} < X_{\mathrm{Cl}}$, the substitution of Cl by H leads to a charge transfer from the carbon to the chlorine atom. It should result in decreasing of the C–Cl bond ionicity degree, and hence, in increasing of the ^{35}Cl NQR frequency.

Let us consider an example with chlorine-substituted methane $\mathrm{CCl}_n\mathrm{H}_{4-n}$. One supposes that: (i) the substitution does not change neither bond lengths nor bond angles; (ii) the charge redistribution is due to one of two possible mechanisms:

- if hydrogen atoms are a part of the molecule, then despite their number the change in the charge density Δx is redistributed between all chlorine atoms. Then the charge density on each chlorine atom changes by value of

$$\Delta y = \frac{(4 - n)\Delta x}{n}; \tag{8.83}$$

- each chlorine atom, despite the number of chlorine atoms in the molecule, pulls a part of the electron density from each hydrogen atom toward itself:

$$\Delta y = (4 - n)\Delta x. \tag{8.84}$$

Let us denote by ω_n the ^{35}Cl NQR frequency in a molecule, which contains n chlorine atoms. According to the Townes–Dailey theory

$$\omega_4 = uV_{zz}^{\mathrm{at}},$$

$$\omega_n = (u - \Delta y)V_{zz}^{\mathrm{at}}, \tag{8.85}$$

Table 8.3 Data for ξ and $\Delta\nu$ for selected substituted hydrocarbons

Atom, group of atoms	Cl		Br	
	$\Delta\nu$ (MHz)	ξ	$\Delta\nu$ (MHz)	ξ
Cl	2.4	0.47	22	0.5
Br	2.6	0.47	–	–
COOH	2.8	0.49	34	0.3
NO_2	3.5	0.34	36.5	0.31

here u is a proportionality constant; Δy can be defined by one of the aforementioned ways. After substitution of Eq. (8.83) or Eq. (8.84) into Eq. (8.85), in the first case one has

$$\omega_n = \omega_4 + \frac{1}{3}(\omega_4 - \omega_1) - \frac{4}{3n}(\omega_4 - \omega_1),$$

whereas in the second case one obtains

$$\omega_n = \omega_4 - (4-n)\frac{\omega_4 - \omega_1}{3}.$$

According to experimental data ω_n linearly depends on n, hence the second assumption is correct.

For $R(CH_2)_nR$ (where R is a halogen atom) an oscillating dependence of the resonance frequency of the R nucleus (ν_R) versus n was obtained. This dependence can be described by the following equation:

$$\nu_R = \nu_0 + \Delta\nu \sum_{i=1}^{n-1} \xi_i,$$

where $\Delta\nu$ and ξ_i are parameters ($\xi_i < 1$). For example in $Cl(CH_2)_nCl$ the ^{35}Cl NQR frequency oscillates around the value of 33 MHz: $\nu < 33$ MHz for even n numbers and $\nu > 33$ MHz for odd n numbers.

For saturated hydrocarbons the parameters ξ and $\Delta\nu$ are quite universal. For example, in R_mCH_{3-m}–$CH_{3-n}R_n$ compounds the NQR frequency can be described by expressions

$$\nu_m = \nu_{01} + \Delta\nu n\xi + \Delta\nu m, \quad \nu_n = \nu_{02} + \Delta\nu m\xi + \Delta\nu n$$

with the same parameters ξ and $\Delta\nu$.

In Table 8.3 one lists the values of ξ and $\Delta\nu$ for saturated hydrocarbons when a hydrogen atom is substituted by an atom or by a group of atoms. For the Cl atom: $\nu_{01} = 38$ MHz, $\nu_{02} = 32$ MHz, for the Br atom: $\nu_{01} = 250.9$ MHz, $\nu_{02} = 248.3$ MHz.

The study of a relationship between changes in NQR parameters and reaction constants is of particular interest. Let us consider the Hammett equation well-known in chemistry. Any substituent can be characterized by a constant σ (Hammett parameter), which is given by

$$\lg \frac{k}{k_0} = p\sigma,$$

where k, k_0 are reaction rates; p is the reaction constant. For example, it was found that in some compounds there is a correlation between changes in the NQR frequency and σ. For example, for the chlorbenzene and bromobenzene a linear correlation between the resonance frequency and the Hammett parameter of the substituent was obtained:

$$\nu(^{35}\mathrm{Cl}) = \left(34.826 + 1.024 \sum_i \sigma_i\right) \mathrm{MHz},$$
$$\nu(^{81}\mathrm{Br}) = \left(226.932 + 7.639 \sum_i \sigma_i\right) \mathrm{MHz},$$

summing runs over all the substituents. These correlation equations allowed us to predict the Hammett parameter values for several compounds.

In m-substituted fluorobenzene there is a correlation between the chemical shift of the ^{19}F nucleus and the Hammett parameter σ:

$$\delta(^{19}\mathrm{F}) = (0.6 - 7.1\sigma)\ \mathrm{ppm}$$

For the asymmetry parameter of the EFG tensor in chlorine aromatic compounds the following correlation equation exists:

$$\eta = \left(7.4 - 2.2 \sum_i \sigma_i\right) \%.$$

In aliphatic and aromatic compounds there is a relation between the NQR frequency ν and dissociation constant $(pK\alpha)$:

$$\nu(^{35}\mathrm{Cl}) = (40.67 - 1.56pK\alpha)\ \mathrm{MHz},$$
$$\nu(^{79}\mathrm{Br}) = (319.31 - 12.58pK\alpha)\ \mathrm{MHz}.$$

However, in the presence of conjugated bonds in the substituent this relation can be broken. This can be explained by the fact that the bond conjugation affects stronger the dissociation constant, whereas the NQR frequency does not change noticeably.

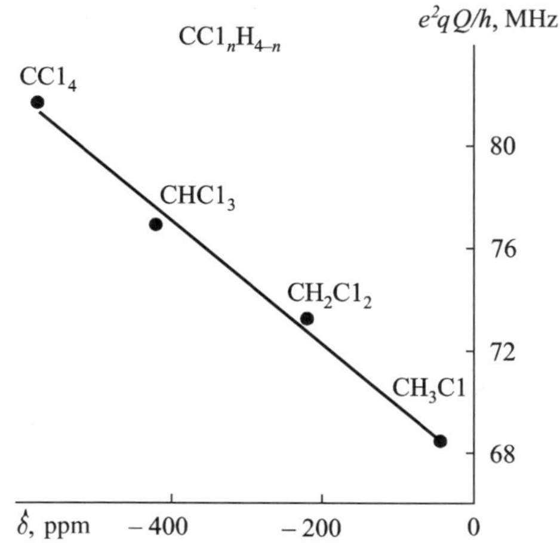

Fig. 8.14 The ^{35}Cl quadrupole coupling constant versus the chemical shift in $CH_{4-x}Cl_x$. Reproduced with permission from Can. J. Chem, 43, 2530 (1965). Copyright 1965, NRC Research Press

Let us note the work of Saito,[11] in which it was made an attempt to relate the observed chemical shifts of the NMR frequencies with parameters of chemical bonding determined by NQR. Such a relationship is not obvious. In chlorine-substituted methane it was found a direct correlation between the quadrupole coupling constant e^2qQ/h) of ^{35}Cl and the chemical shift (δ_{Cl}) (see Fig. 8.14). That means that both the paramagnetic chemical shift and quadrupole coupling constant are determined by the same electron distribution.

To summarize, one can say that in spite of the widespread use of the Townes–Dailey method the interpretation of the NQR spectra within this approach has several disadvantages: the most significant is the presence of a large number of parameters, each of which is determined from empirical relationships established only for a narrow class of compounds. In addition, this method is not quite acceptable for compounds with a high bond ionicity degree.

The Townes–Dailey method was developed in 1960s, when ab initio or even semi-empirical calculations were rather difficult. Currently there are more sophisticated methods. Nevertheless, the Townes–Dailey method can qualitatively explain the relationship between the measured parameters of the NQR spectrum and the chemical structure of the studied material.

8.4.4 Ab initio Methods: Basis and Applications to NQR

All ab initio quantum calculation methods of solids can be divided into two classes: *band* and *cluster*. In the band theory of solids, the long-range order, the periodicity of the crystal structure plays the fundamental role. Yet recently, the majority of the band theory calculations was carried out using the plane waves basis, and electrons

[11] Y. Saito. ^{35}Cl and ^{37}Cl magnetic resonance of simple chlorine compounds. Can. J. Chem. **43** (1965) 2530–2534.

were considered in a periodic field of atomic nuclei, or cores. This approach gives a rather good results in studying the bulk properties of the crystal, the heat capacity, electrical conductivity, the Fermi surface, and so on, and is justified for metals. For correct description of ionic or covalent crystals the band calculation methods, which use the atomic orbitals, are much more appropriate. However, even by now such calculations are not widespread as require a lot of computational resources. To gain the computational time it is possible to combine the both methods mentioned above using so-called "muffin tin" approximation, firstly proposed by Slater[12] and widely used in many modern electronic structure methods, such as the Augmented Plane Wave (APW) method, the Korringa-Kohn-Rostoker (KKR) method,[13] which is based on the implementation of Green's functions, etc.

When studying effects of the short-range order, electronic structure of amorphous solids, surface phenomena, various deviations from stoichiometry, such as impurity atoms, localized defects, lattice distortions etc., the cluster approach is much more suitable. Certainly it can be applied to study the ideal structures, at least to have the reference point in studying defect structures. The cluster approach is more flexible and allows us to take into account such effects as electronic correlations in more accurate way.

In cluster calculations usually only a short-range order is taken into account, i.e. the interaction of atoms with a few nearest neighbors. However, the modeling of a solid by a small fragment raises a number of serious problems. First of all this is a problem of the boundary conditions choice, or how to take into account the environment of the cluster in the crystal lattice. Consideration of the cluster as a free molecule has also some drawbacks: the atomic bonds on the cluster surface are broken, the environment of the surface atoms is different from the real one. Compensation of these effects is closely related to the correct modeling of the filling of electron states of the crystal. For the correct reproduction of the fully occupied valence band in the studied compound, the total number of electrons in the cluster should correspond to the configurations of the atoms defined by their formal valence. In this case the cluster has a rather large excess charge.

However, the calculation of charged clusters may be rather problematic. If the compound under study is not purely ionic, that is very often the case as usually the effective charges of atoms are quite different from their formal valence, then the total number of electrons in the cluster must be less due to the covalency of bonds between ligands and atoms in the next coordination spheres. Ignoring these effects can lead to unphysical energy shifts, destabilize bounded states or can result in poor convergence of the iterative procedure. One of the possible ways to saturate broken bonds is to introduce pseudoatoms (e.g. hydrogen atoms). The pseudoatoms are placed outside the cluster under study, usually on the extensions of the bonds of the central atom with ligands or ligands with atoms belonging to the next coordination sphere that are

[12] J. C. Slater. Wave Functions in a Periodic Potential. Phys. Rev. **51** (1937) 846–851.

[13] J. Korringa. On the calculation of the energy of a Bloch wave in a metal. Physica **13** (1947) 392–400. W. Kohn and N. Rostoker. Solution of the Schrödinger equation in periodic lattices with an application to metallic lithium Phys. Rev. **94** (1954) 1111–1120.

not considered in the cluster. In this case the length of the ligand–pseudoatom bond is changed in such a manner to achieve the agreement with experimental data (for example, with the EFG value). This technique allows us to reproduce more accurate the electron density distribution around the ligands.

Besides taking into account the redistribution of the electron density between the boundary atoms of the cluster and the surrounding crystal another important aspect is the correct modeling of the influence of the crystal potential on the electronic state inside the cluster, and, above all, on the characteristics of atoms located on the cluster boundary. One of the most simple way to take it into account is to use the boundary conditions in the form of a uniformly charged Watson sphere circumscribed around the cluster with the charge equal in magnitude and opposite in sign to the cluster charge. However, such boundary conditions do not reflect the anisotropy of the crystal environment.

Sometimes a crystalline environment is simulated be the Madelung potential (the potential created by point ions located at the lattice sites). When calculating electric field gradients induced by lattice, there is considerable arbitrariness in the parameters choice (see Sect. 8.4.6). However, in cluster methods to calculate the contribution to the EFG tensor from the crystal rest surrounding the cluster the most commonly used method is the point-charge model, in which the ion charge values are chosen in accordance with their formal valences. There are several ways to take into account the lattice contributions. There is a very common method when the contribution of the infinite lattice is considered separately and added to the result obtained for a single cluster. However, the question is: What is the Sternheimer antishielding factor for the cluster? Sometimes, the Madelung potential is initially introduced into the Hartree–Fock equations. Another way is to simulate the potential of an infinite lattice by a finite number (hundreds or thousands) of point charges placed on the surface of the surrounding cluster.

Thus, to study the electronic structure of solids within the cluster approach one needs to solve the following problems:

- the cluster choice (the fragment size and form, the central atom etc.);
- both adequate and cost-effective (from the computational time point of view) boundary conditions;
- the choice of the calculation method, which is suitable for description of the studied characteristics with required accuracy.

Usually in cluster calculations one applies schemes based on the Hartree–Fock–Roothaan method.[14] Therefore, before one moves on to the application of the cluster approach directly for calculating the NQR parameters, let us give the basic terms and formulas of this method.

Hartree–Fock–Roothaan method

The electronic structure of a molecule or a cluster is described by the stationary Schrödinger equation

---

[14] C.C.J. Roothaan. New developments in molecular orbital theory. Rev. Mod. Phys. **23** (1951) 69–89.

$$\widehat{\mathcal{H}}\Psi' = E\Psi'. \tag{8.86}$$

Here Ψ' is the wave function of electrons and nuclei; $\widehat{\mathcal{H}}$ is the total Hamiltonian of the system:

$$\widehat{\mathcal{H}} = \widehat{\mathcal{H}}_{\mathrm{e}} + \widehat{\mathcal{H}}_{\mathrm{n}} + \widehat{\mathcal{H}}_{\mathrm{en}},$$

$\widehat{\mathcal{H}}_{\mathrm{e}}$ is the Hamiltonian of the electron subsystem:

$$\widehat{\mathcal{H}}_{\mathrm{e}} = -\frac{1}{2}\sum_{i} \nabla^2(i) + \frac{1}{2}\sum_{i \neq j} \frac{1}{|\vec{r}_i - \vec{r}_j|},$$

the sum runs over all electrons; $\widehat{\mathcal{H}}_{\mathrm{n}}$ is the Hamiltonian of the nuclear subsystem:

$$\widehat{\mathcal{H}}_{\mathrm{n}} = -\frac{1}{2}\sum_{n} \frac{1}{m_n} \nabla^2(n),$$

the sum runs over all nuclei; $\widehat{\mathcal{H}}_{\mathrm{en}}$ is the Hamiltonian of electron–nucleus interactions:

$$\widehat{\mathcal{H}}_{\mathrm{en}} = -\sum_{n}\sum_{i} \frac{z_n}{|\vec{r}_i - \vec{R}_n|} + \frac{1}{2}\sum_{n}\sum_{n' \neq n} \frac{z_n z_{n'}}{|\vec{R}_n - \vec{R}_{n'}|},$$

z_n, m_n and $\vec{R}_n$ are the charge, mass and radius vector of the nth nucleus, $\vec{r}_i$ is the radius vector of the ith electron, the sum runs over all nuclei and electrons;

$$\vec{\nabla} \equiv \vec{e}_x \frac{\partial}{\partial x} + \vec{e}_y \frac{\partial}{\partial y} + \vec{e}_z \frac{\partial}{\partial z}.$$

Here and below the atomic units are used: $\hbar = 1, m_{\mathrm{e}} = 1, e = 1$. The wave function Ψ' depends on both space and spin coordinates of all nuclei and electrons and even for a simple system it is a very difficult task to solve Eq. (8.86) exactly, and it is necessary to use a number of approximations.

Adiabatic or Born–Oppenheimer approximation Because the nucleus mass m_{n} is much greater than the electron mass m_{e} and the average velocity of the nuclei is much less than the average velocity of electrons one can consider that the nucleus motion is sufficiently slow and electrons always have time to follow it. Then in Eq. (8.86) the motion of electrons and nuclei can be separated, and the total wave function of the system $\Psi'(q, Q)$ (q is a set of all electron coordinates, Q is a set of all nucleus coordinates) can be represented as a product of the electron and nucleus wave functions:

$$\Psi'(q, Q) = X(Q)\Psi(q, Q),$$

here $\Psi(q, Q)$ is the solution of the Schrödinger equation for electrons if the nucleus coordinates are fixed

$$\left[\widehat{\mathcal{H}}_{\mathrm{e}} + \widehat{\mathcal{H}}_{\mathrm{en}}\right]\Psi(q, Q) = E(q)\Psi(q, Q). \tag{8.87}$$

One-electron or Hartree approximation In a many-electron system all electrons interact with each other, the motion of each one is determined by the motion of all others. But according to the Hartree approximation, electrons move independently of each other and the motion of each electron is determined by its coordinates in the mean potential field generated by other nuclei and electrons. Then the motion of electrons is separated, and each electron can be introduced by one-electron function, which is a solution of the Schrödinger equation of the form (8.87), where instead of $|\vec{r}_i - \vec{r}_j|$ in the Hamiltonian $\widehat{\mathcal{H}}_{\mathrm{e}}$ there is an effective potential V_{eff}. The wave function of a many-electron system is a product of one-electron functions (orbitals):

$$\Psi = C\prod_{i=1}^{n}\psi_i,$$

where C ia a normalization factor.

Hartree–Fock approximation The Hartree approximation does not take into account the influence of the electron spin on its space coordinates. According to the Pauli exclusion principle the many-electron wave function must be antisymmetric with respect to mutual permutation of any pair of electrons. This condition can be satisfied if the wave function Ψ is represented as a Slater determinant:

$$\Psi = C\begin{vmatrix} \psi_1(1)\alpha(1) & \psi_1(1)\beta(1) & \dots & \psi_n(1)\alpha(1) & \psi_n(1)\beta(1) \\ \psi_1(2)\alpha(2) & \psi_1(2)\beta(2) & \dots & \psi_n(2)\alpha(2) & \psi_n(2)\beta(2) \\ \vdots & \vdots & \ddots & \vdots & \vdots \\ \psi_1(N)\alpha(N) & \psi_1(N)\beta(N) & \dots & \psi_n(N)\alpha(N) & \psi_n(N)\beta(N) \end{vmatrix}, \tag{8.88}$$

where N is a number of electrons, $n = N/2$ is a number of orbitals. For orthonormalized one-electron wave functions the normalization factor is $C = (N!)^{-1/2}$. In a general case Ψ should be a linear combination of determinants like Eq. (8.88). However for closed shell cases the solution of Eq. (8.87) may be represented as a single Slater determinant.

For systems containing unpaired electrons in singly occupied orbitals ψ (an open-shell case) this can not be done. In such cases, many-electron wave function must be represented as a linear combination of Slater determinants of $\widetilde{\Psi}_l$, which correspond to the various ways of the energy level occupation (electronic configurations):

$$\Psi = \sum_l \widetilde{\Psi}_l.$$

For closed-shell systems the orthonormalized orbitals $\psi_i(r)$ are the solution of the Hartree–Fock equation

$$\hat{F}\psi_i(\vec{r}) = \varepsilon_i \psi_i(\vec{r}) \tag{8.89}$$

with the Fock operator

$$\hat{F} \equiv \hat{h} + \hat{J} - \hat{K} = \hat{h} + \sum_{l=1}^{n} \left(2\hat{J}_l - \hat{K}_l\right).$$

The eigenvalues ε_i in Eq. (8.89) represent the orbital energies. The operator $\hat{h}$ describes the kinetic energy of electrons and their interaction with the nucleus core:

$$\hat{h} \equiv -\frac{1}{2}\sum_i \nabla^2(i) + \sum_i \sum_n \frac{z_n}{|\vec{r}_i - \vec{R}_n|}.$$

The Coulomb ($\hat{J}_l$, $\hat{J}$) and exchange ($\hat{K}_l$, $\hat{K}$) operators are defined by equations

$$\hat{J}\psi(\vec{r}) \equiv \left(\int \frac{\hat{\rho}(\vec{r}\,'|\vec{r}\,')}{|\vec{r} - \vec{r}\,'|} \mathrm{d}V'\right)\psi(\vec{r}), \quad \hat{K}\psi(\vec{r}) \equiv \left(\int \frac{\hat{\rho}(\vec{r}|\vec{r}\,')}{|\vec{r} - \vec{r}\,'|} \mathrm{d}V'\right)\psi_l(\vec{r}),$$

$$\hat{J}_l\psi(\vec{r}) \equiv \left(\int \frac{\psi_l(\vec{r}\,')\psi_l^*(\vec{r}\,')}{|\vec{r} - \vec{r}\,'|} \mathrm{d}V'\right)\psi(\vec{r}), \quad \hat{K}_l\psi(\vec{r}) \equiv \left(\int \frac{\psi(\vec{r}\,')\psi_l^*(\vec{r}\,')}{|\vec{r} - \vec{r}\,'|} \mathrm{d}V'\right)\psi_l(\vec{r}),$$

$$\hat{\rho}(\vec{r}|\vec{r}\,') = 2\sum_{i=1}^{n} \psi_i(\vec{r})\psi_i^*(\vec{r}\,').$$

Here $\hat{\rho}(\vec{r}|\vec{r}\,')$ is the spinless density matrix of the first order. The Coulomb operator $\hat{J}_l$ describes the mean electrostatic field created by all electrons except lth. The operator $\hat{K}_l$ describes the exchange interaction, which has no classical analogue and is a consequence of the Pauli exclusion principle. Introducing the following notation

$$J_{ij} \equiv \int\int \frac{\psi_i(\vec{r})\psi_i^*(\vec{r})\psi_j(\vec{r}\,')\psi_j^*(\vec{r}\,')}{|\vec{r} - \vec{r}\,'|} \mathrm{d}V\mathrm{d}V',$$

$$K_{ij} \equiv \int\int \frac{\psi_i(\vec{r})\psi_j^*(\vec{r})\psi_j(\vec{r}\,')\psi_i^*(\vec{r}\,')}{|\vec{r} - \vec{r}\,'|} \mathrm{d}V\mathrm{d}V'$$

one obtains an expression for the total energy of the system:

$$E = 2\sum_{i=1}^{n} \varepsilon_i - \sum_{j=1}^{n} \left(2J_{ij} - K_{ij}\right).$$

The wave functions ψ_i can be found using the *variational method* by minimizing the total energy.

The Hartree–Fock method as it is stated above is applicable only to atoms. When searching for an analytical solution for a molecule its one-electron wave function is chosen as a linear combination of orthonormalized basis functions:

$$\Psi_i = \sum_{\mu} C^i_{\mu} \psi_{\mu}. \tag{8.90}$$

The substitution of molecular orbitals in the form of Eq. (8.90) into Eq. (8.89) leads to the system of Hartree–Fock–Roothaan equations.

There are several methods for constructing a basis set. For example, in the MO LCAO method the atomic functions centered on each atomic nuclei within the molecule (or cluster) are chosen as the basis functions. The coefficients C^i_{μ} are determined by solving the secular equation

$$\sum_{\nu} \langle \mu|\hat{h}|\nu\rangle + \sum_{\lambda\sigma} P_{\lambda\sigma} \left[\langle \mu\nu|\lambda\sigma\rangle - \frac{1}{2}\langle \mu\lambda|\nu\sigma\rangle - \varepsilon_i S_{\mu\nu} \right] C^i_{\nu} = 0, \tag{8.91}$$

where

$$\langle \mu|\hat{h}|\nu\rangle = \int \psi^*_{\mu}(\vec{r})\hat{h}\psi_{\nu}(\vec{r})\mathrm{d}V$$

are the matrix elements of the Hamiltonian $\hat{h}$;

$$S_{\mu\nu} = \int \psi_{\mu}(\vec{r})\psi_{\nu}(\vec{r})\mathrm{d}V$$

are the overlap integrals;

$$\langle \mu\nu|\lambda\sigma\rangle = \int\int \frac{\psi^*_{\mu}(\vec{r})\psi_{\nu}(\vec{r}\,')\psi^*_{\lambda}(\vec{r})\psi_{\sigma}(\vec{r}\,')}{|\vec{r}-\vec{r}\,'|}\mathrm{d}V\mathrm{d}V' \tag{8.92}$$

are the Coulomb or exchange electron interaction integrals;

$$P_{\mu\nu} = 2\sum_{i} C^i_{\mu} C^i_{\nu}$$

are the density matrix elements.

The Roothaan equations in the form of Eq. (8.91) represent a system on linear equations for C^i_{ν} coefficients in the decomposition of molecular orbitals Ψ_i. This system of equations can be solved by iteration method. First, one chooses a zero-order approximation for $C^{i,(0)}_{\nu}$, second, one calculates the expression in square brackets and solves the system of equations. As a result one obtains a first-order approximation for $C^{i,(1)}_{\nu}$. The procedure should be repeated up to a self-consistent solution is achieved, namely, up to the condition $C^{i,(n)}_{\nu} - C^{i,(n-1)}_{\nu} < \delta$ (where δ is the given accuracy) is satisfied. For each eigenvalue ε_i one obtains a set of C^i_{ν} coefficients, i.e.

an approximate solution of the Hartree–Fock equation. Knowing the wave functions Ψ_i it is possible to calculate the EFG tensor at the nucleus site A:

$$V^A_{\alpha\beta} = e\langle\Psi|\sum_k \frac{(\delta_{\alpha\beta} r^2_{Ak} - 3r^{\alpha}_{Ak} r^{\beta}_{Ak})}{r^5_{Ak}}|\Psi\rangle,$$

here Ψ is a many-electron wave function; r_{Ak} is the radius vector between the A atom and kth electron.

Basis set choice Before starting the Hartree–Fock–Rutaana calculations it is necessary to solve the problem of choice of the basis set. When solving the system of equations (8.91) the main difficulty is to calculate two-center integrals (8.92). Therefore, the basis functions, on the one hand, must be such as even a small number of them can approximate well the molecular orbital, and, on the other hand, the matrix elements (8.92) should be computed fast enough. The *Slater* and *Gaussian* basis functions types are the most widely used in such calculations. Functions of both types are single-center, i.e. they are tied to a specific point in space and are symmetric with respect to this point. In molecular calculations usually one uses the basis functions centered on different points in space, at the nuclei, or less commonly at the atomic bonds. However it is possible to use functions with the "floating" enter, i.e. the center point of the basis functions is not fixed in advance, but is chosen from the optimization conditions.

The Slater orbitals have a well-known analytical form:

$$\psi^{(S)}_{klm}(\vec{r}) = R^{(S)}_k(r) Y_{lm}(\theta, \phi),$$

$$R^{(S)}(r) = N^{(S)}_k r^{n_k - 1} e^{-\xi_k r}, \quad N^{(S)}_k = \frac{(2\xi_k)^{n_k + 1/2}}{\sqrt{2n_k!}},$$

here ξ_k, n_k are parameters; l, m are the orbital and magnetic quantum numbers. The Slater basis sets are usually denoted as STO. The main advantage of the STO basis sets is that to achieve a rather good approximate solution of the Hartree–Fock equation only a few number of Slater functions is required. Calculations within the STO basis sets allow us to obtain atomic and molecular orbitals, which properly describe the electron density distribution both near the atomic nucleus and far from it. However, the calculations of multicenter integrals, Eq. (8.92), is rather problematic and time consuming, because it is difficult to reduce Slater functions centered on different points to a single coordinate system.

The Gaussian orbitals (GTO) allows us to avoid this disadvantage of the STO basis sets. Such orbitals can be written as follows:

$$\psi^{(G)}_k(\vec{r}) = N^{(G)}_k x^{l_k} y^{m_k} z^{n_k} e^{-\alpha_k r^2},$$

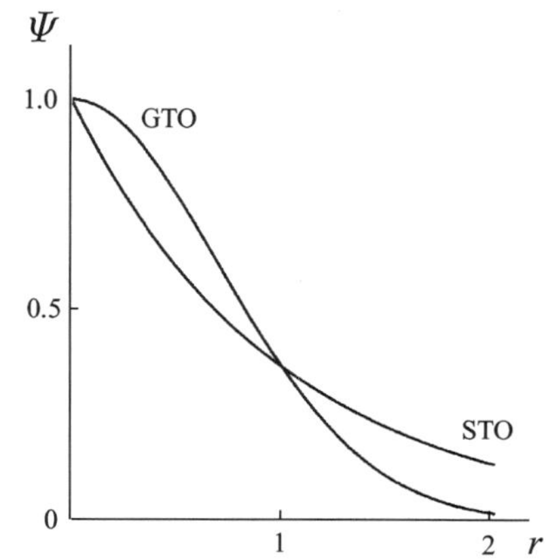

Fig. 8.15 The Slater and Gaussian 1*s* functions for a hydrogen atom

$$N_k^{(G)} = \left(\frac{2\alpha_k}{\pi}\right)^{3/4} \left[\frac{\alpha^{l_k+m_k+n_k}}{(2l_k-1)!!(2m_k-1)!!(2n_k-1)!!}\right]^{1/2},$$

here l_k, m_k, n_k are non-negative integer numbers; $\alpha_k > 0$ are parameters. The pre-exponential factors describe both radial and angle dependencies of the function $\psi_k(\vec{r})$. For $l_k + m_k + n_k = 0$ one has the 1*s*-orbital, for $l_k + m_k + n_k = 1$—three 2*p*-orbitals. If $l_k + m_k + n_k = 2$ there are six different products of x, y and z coordinates (xx, yy, zz, xz, xy, yz), from which it is possible to construct five 3*d*-orbitals and one 3*s*-orbital (because $x^2 + y^2 + z^2 = r^2$).

Due to the quadratic dependence on the distance of the Gaussian functions, it is much easier to reduce the product of two Gaussian functions centered on different nuclei to a single coordinate system, as compared to the Slater functions. This leads to fast and easy calculations of multicenter integrals (8.92). However, such a dependence also determines the main drawback of the Gaussian functions.

Figure 8.15 represents the 1*s* wave function for a hydrogen atom in both Gaussian and Slater representations. First of all, it should be noted that the GTO function provides a poor description of the electron density behavior both near the nucleus and far from it. Another disadvantage of the GTO basis is its rather poor convergence: it is necessary to take a much larger number of Gaussian functions than in the case of STO basis to obtain comparable results. Moreover, for the ground state of the hydrogen atom the Slater orbital gives the exact solution:

$$\psi_{1s}^{(S)} = \sqrt{\frac{\xi^3}{\pi}} e^{-\xi r},$$

whereas the Gaussian orbital gives

$$\psi_{1s}^{(G)} = \left(\frac{2\alpha}{\pi}\right)^{3/4} e^{-\alpha r^2}$$

that leads to the underestimated energy of the ground state. To achieve the result comparable to that one obtained with using of only one Slater orbital one needs ten Gaussian orbitals. However, if one deals with a many-electron system using the GTO basis sets becomes more profitable.

The employment of large basis sets is rather inconvenient because one has to store and process a huge number of many-electron integrals, mainly two-electron ones. To reduce the basis set size one uses not elementary Gaussian functions but their linear combinations, so-called *contracted Gaussian orbitals*:

$$\psi_\gamma = \sum_i c_{i\gamma} \psi_i^{(G)}.$$

Let us describe how to construct the contracted orbitals. Any GTO basis set is characterized by a number of primitive GTO of different type (s, p, d and so on). These numbers are written in parentheses, the number of contracted GTO is written in square brackets. For example, $(9s5p)/[4s3p]$ means that nine s-orbitals are arranged in four groups and five p-orbitals in three groups. Such a contraction, as a rule, can be effected in different ways, so sometimes it is pointed out how many primitive GTO functions each contracted GTO function contains.

The basis contraction should be implemented in such a way that each contracted GTO function describes the electron density distribution in a definite point in space. Usually, one group includes those Gaussian orbitals, which have the largest coefficients in the atomic orbitals expansion (for example, $1s$ or $2s$). By this way one obtains contracted GTO functions, which approximately correspond to these atomic orbitals. If the parameter α of a GTO function is such that the maximum of this function is in the "valence" region between neighboring nuclei, such a function forms a separate group. If a GTO function gives a significant contribution to two or more atomic orbitals, this orbital also should not be included in any group in the construction of the contracted GTO basis set.

Another way to construct a linear combination of Gaussian orbitals proposed by Hudzinaga is to fit ns Slater orbitals by $1s$ Gaussian orbitals, np Slater orbitals by $2p$ Gaussian orbitals, nd Slater orbitals by $3d$ Gaussian orbitals and so on. This basis obtained by fitting a Slater orbital by k Gaussian orbitals is denoted by STO-kG.

The simplest level of ab initio calculations includes the use of minimal basis set of functions centered on nuclei. For the minimal basis set the number of basis functions is equal to the number of orbitals. Thus, the hydrogen and helium are presented by one s-function, lithium and beryllium atoms are described by a pair of such functions. For the following elements of the first series in addition to these two s-functions one considers a full set of three p-functions, and etc. Here is a minimal basis set for the oxygen and niobium atoms:

Oxygen: $1s, 2s, 2p_x, 2p_y, 2p_z$;
Niobium: $1s, 2s, 2p_x, 2p_y, 2p_z, 3s, 3p_x, 3p_y, 3p_z, 3d_{3zz-rr}, 3d_{xx-yy}, 3d_{xy}, 3d_{xz}, 3d_{yz}, 4s, 4p_x, 4p_y, 4p_z, 4d_{3zz-rr}, 4d_{xx-yy}, 4d_{xy}, 4d_{xz}, 4d_{yz}, 5s, 5p_x, 5p_y, 5p_z$.

Here the orbitals $1s, 2s, 2p_x, 2p_y, 2p_z$ etc. can be written as either Staler or Gaussian functions.

Next, it is necessary to specify the form of basis functions. We have already mentioned that one way is to use the STO-*k*G basis sets. The GTO parameters expansion coefficients are determined by minimizing (for example, by the least squares method) deviations of the Gaussian decomposition from the exact Slater functions.

It is worth noting that the total energy of atoms and molecules, calculated within the minimal STO-*k*G basis set, is highly dependent on the *k* value, whereas other properties, such as the charge distribution, optimized geometry, electric dipole moment etc. are less sensitive to the number of the expansion terms. However, the accuracy of such calculations is not too high, that is a consequence of a number of drawbacks of this basis. For example, for the lithium atom, which has only three electrons, and for the oxygen atom, which has eight electrons, it is proposed to use the same number of functions. Therefore, description of the oxygen atom within the minimal basis is rather poorer than the description of the elements with a small number of electrons. In addition, the Gaussian functions have fixed exponents and can not be compressed or stretched out in accordance with different molecular or crystalline environment, because the minimal basis sets contain only one valence functions for each particular type of symmetry.

Let us consider a MO_n cluster, where a metallic atom M is surrounded by *n* oxygen atoms. For correct description of the electronic density distribution near the central atom, which determines the EFG value, it is necessary to define precisely the valence orbitals of the oxygen atom. The most appropriate way is to represent valence orbitals by more than one basis function (each of which can in turn be composed of a fixed linear combination of primitive Gaussian functions). Such split-valence basis sets in which there are multiple basis functions corresponding to each valence atomic orbital are called valence double, triple, quadruple-zeta, and so on, basis sets. The oxygen basis functions in the TZV (triple-zeta-valence) representation have following form: $1s, 2s', 2p'_x, 2p'_y, 2p'_z, 2s'', 2p''_x, 2p''_y, 2p''_z, 2s''', 2p'''_x, 2p'''_y, 2p'''_z$.

Another group of split-valence basis sets are sets proposed by Pople and co-workers. The Pople's basis set notation reflects the split valence nature of these sets. For example, $n - ij$G or $n - ijk$G means following: n is the number of primitive Gaussian for the inner shells; ij or ijk are the numbers of primitive Gaussian functions for contractions in the valence shell; ij describes sets of double-zeta-valence quality and ijk—sets of triple-zeta-valence quality.

A common feature of these basis sets is that functions are centered on the nucleus. In some cases, it is necessary to shift the center of the electron density distribution from the nucleus site. Then, to polarize the valence s- and p-orbitals one can add to the basis set orbitals that are not part of the ground states wave functions, the so-called *polarization* functions: p-functions for a hydrogen atom, d-functions for an oxygen atom etc.

The described basis sets have proved themselves in molecular calculations. When one deals with charged systems (anions or clusters simulating the crystal), there are specific problems due to the fact that the outer electrons are weakly bounded, i.e. a part of the electron density is shifted from the center of the cluster to the periphery. This can be taken into account by introducing of so-called *diffuse* functions, with small exponent. The optimal values of exponents for polarization and diffuse functions can be found in corresponding tables.

The Pople's basis sets can also be augmented with polarization or diffusion functions: the 6-311G* notation means that the 6-311G basis set is augmented with polarization functions, the 6-311+G* notation means addition of diffuse functions.

8.4.5 Semi-empirical Mulliken–Wolfsberg–Helmholtz method

When carrying out Hartree–Fock–Roothaan calculations the main difficulty is to calculate the $\hat{F}$ matrix elements in the form of Eq. (8.89). If there are m orbitals in the system the number of overlap integrals is equal to m^2, and the number of multi-centered integrals is m^4. Semi-empirical methods allows us to gain computational time, by using experimental data. In the semi-empirical Mulliken–Wolfsberg–Helmholtz (MWH) method the following assumptions are made:

- only the valence orbitals are considered;
- all filled orbitals are taken into account;
- exited states are taken into account;
- the Mulliken approximation is used:

$$\psi_{\mu_A}(1)\psi_{\nu_B}(1) = \frac{1}{2} S_{\mu\nu} \left[\psi_{\mu_A}(1)\psi_{\mu_A}(1) + \psi_{\nu_B}(1)\psi_{\nu_B}(1)\right], \tag{8.93}$$

  where $S_{\mu\nu}$ is the overlap integral;
- the Fock operator $\hat{F}$ is substituted by an effective operator $\hat{F}^{\text{eff}}$ and its matrix elements $F^{\text{eff}}_{\mu\nu}$ are the parameters taken from an experiment.

Within the framework of this method the following notations are adopted:

$$F^{\text{eff}}_{\mu\nu} \equiv H_{\mu\nu}.$$

With these approximations the problem of finding the expansion coefficients of MO LCAO is reduced to solving the eigenvalue problem:

$$\sum_{\omega} \left(H_{\mu\omega} - \varepsilon_i S_{\mu\omega}\right) = 0.$$

Diagonal matrix elements characterize the electron energy of the ψ_μ orbital, and in the MWH method they are supposed to be equal to the corresponding ionization potentials:

$$H_{\mu\mu} = -V_\mu.$$

It is known that V_μ depends on the charge Q and the electron configuration. For an integer electron configuration $d^m s^n p^l$ the charge dependence is described by the following expression:

$$V_\mu(m,n,l) = A_\mu(m,n,l)Q^2 + B_\mu(m,n,l)Q + C_\mu(m,n,l).$$

The coefficients $A_\mu(m,n,l)$, $B_\mu(m,n,l)$, $C_\mu(m,n,l)$ can be found in the atomic terms tables, however, their definition can be ambiguous.

Configuration obtained in calculations, can be fractional. The ionization potential for a fractional configuration can be approximated by a linear combination of V_μ for integer configurations. The coefficients are chosen in such a manner that the total configuration matches the initial fractional one.

Let us consider an atom with a valence shell, which includes s-, p- and d-electrons and represent its configuration $d^{N_d}s^{N_s}p^{N_p}$ (where N_s, N_p, N_d are populations of corresponding orbitals, and $N_s + N_p + N_d = n$) through the configurations d^n, $d^{n-1}s$, $d^{n-1}p$:

$$d^{N_d}s^{N_s}p^{N_p} = ad^n + bd^{n-1}s + cd^{n-1}p.$$

Then the fractional population of the d-orbital, N_d, is a sum of integer configurations d^n, $d^{n-1}s$ and $d^{n-1}p$ with corresponding weighting coefficients a, b and c, respectively:

$$N_d = an + b(n-1) + c(n-1) = n - N_s - N_p, \qquad N_s = b, \qquad N_p = c.$$

Similarly, one can write the fractional configuration N_s and N_p. Hence one gets $a = 1 - N_s - N_p$. Consequently, the ionization energy of the valence states with fractional electron configuration can be written as

$$-H_{\mu\mu} = -(1 - N_s - q_p)V_\mu(d^n) + N_s V_\mu(d^{n-1}s) + N_p V_\mu(d^{n-1}p).$$

Here $V_\mu(d^m p^l s^k)$ is the ionization potential of the atomic orbital ψ_m for an atom with the valence configuration $d^m p^l s^k$ ($m + l + k = n$ is a number of valence electrons). The ionization potential $V_\mu(d^m p^l s^k)$ depends quadratically on the total atomic charge:

$$Q = n - N_s - N_p - N_d.$$

Let us consider the doubly occupied molecular orbital, which consists of two atomic orbitals centered on different atoms:

$$\Psi_i = C^i_\mu \psi_{\mu A} + C^i_\nu \psi_{\nu B}.$$

Taking into account the normalization of ψ_μ, the electron density is given by

$$2\int \Psi_i^*(1)\Psi_i(1)\mathrm{d}r = 2(C_\mu^i)^2 + 2(C_\nu^i)^2 + 4C_\mu^i C_\nu^i S_{\mu\nu} = 2.$$

Within the Mulliken approximation (8.93) the overlap population is shared equally between the A and B atoms. Then the population of atomic orbitals ψ_{μ_A}, which is part of the molecular orbitals Ψ_i, is

$$n_{\mu_A}^i = 2C_\nu^{i\,2} + 2C_\mu^i C_\nu^i S_{\mu\nu}.$$

The total population of the atomic orbitals ψ_{μ_A} is the sum over all occupied molecular orbitals: $n_{\mu_A} = \sum_i n_{\mu_A}^i$, and the total population of all orbitals of the atom A is $n_A = \sum_\mu n_{\mu_A}$. Now it is easy to calculate the atomic charge value:

$$Q = n - n_A.$$

After defining the atomic charges, we can calculate more accurate the electron energy values ε_i, and consequently, the new values of the $H_{\mu\mu}$ matrix element and new MO LCAO expansion coefficients C_μ^i. Knowing these coefficients, again one calculates the $H_{\mu\mu}$ matrix element and so on. Thus, this is a self-consistent procedure.

Let us go back to finding the matrix elements of the Fock operator within the MWH approximation. It has been shown that the diagonal matrix elements $H_{\mu\mu}$ can be associated with the ionization potential of the electron. The off-diagonal matrix elements should be proportional to the overlap integral $S_{\mu\nu}$ and depend on the energy of the ψ_{μ_A} and ψ_{μ_B} orbitals. Below there is a list of the most common approximations for off-diagonal matrix elements.

- Wolfsberg–Helmholtz approximation[15]:

$$H_{\mu\nu} = \frac{1}{2} K S_{\mu\nu}(H_{\mu\mu} + H_{\nu\nu}); \qquad (8.94)$$

- Balhauzen–Gray approximation[16]:

$$H_{\mu\nu} = K S_{\mu\nu}\sqrt{H_{\mu\mu} + H_{\nu\nu}}; \qquad (8.95)$$

- Cusachs approximation[17]:

$$H_{\mu\nu} = \frac{1}{2}(H_{\mu\mu} + H_{\nu\nu}) S_{\mu\nu}(2 - |S_{\mu\nu}|); \qquad (8.96)$$

---

[15] M. Wolfsberg and L.J. Helmholz. The Spectra and Electronic Structure of the Tetrahedral Ions MnO^{4-}, CrO^{4--}, and ClO^{4-}. J. Chem. Phys. **20** (1952) 837–843.

[16] C.J. Ballhausen and H.B. Gray. The electronic structure of the vanadyl ion. Inorg. Chem. **1** (1962) 111–122.

[17] L. C. Cusachs. Semiempirical molecular orbitals for general polyatomic molecules. II. One-electron model prediction of the H–O–H angle J. Chem. Phys. **43** (1965) S157–S159.

- Morokuma–Fukui approximaton[18]:

$$H_{\mu\nu} = S_{\mu\nu}\left[\frac{H_{\mu\mu} + H_{\nu\nu}}{2} + K\right]; \tag{8.97}$$

where K is a numerical coefficient.

These approximations are equivalent to neglecting kinetic energy integrals and some three-center integrals. This makes it necessary to vary the K coefficient. Moreover, it appears that in some cases, in order to achieve an agreement with experimental data for electron transitions energies it is required to vary the coefficients even for compounds within a homologous series. This means that the MWH method can be used only as a rough approximation. For more correct description of the electronic structure of compounds ab initio methods should be used. Nevertheless, the MWH method is quite applicable to calculate the ground state characteristics.

8.4.6 Point-Multipole Model

Cluster calculation methods based on the Hartree–Fock–Ruhtaan or some semiempirical methods allow us (more or less accurate) to describe the electron density distribution near the studied nucleus. To construct a more complete picture it is necessary to expand the cluster, or go to the band calculation methods. However, in the latter case one can lose the local structure specification. However, if crystals with mainly ionic chemical bonding a rather good results for the EFG tensor can be obtained within a simple point-multipole model. This model was widely used to analyze experimental data of the solid state NMR on quadrupole nuclei and NQR. In addition, it is used for modeling boundary conditions in ab initio cluster calculations.

First, let us consider the point-charge model one of the simplest modifications of the point-multipole model. In the point-charge model the crystal is represented as a set of ions located in the lattice sites. This system of point charges q_k at points $\vec{r}_k$ creates EFG, which in the Cartesian coordinate system is written as follows

$$V^0_{j,\alpha\beta} = \sum_{k\neq j} \frac{q_k(\delta_{\alpha\beta} r^2_{jk} - 3r^{\alpha}_{jk} r^{\beta}_{jk})}{r^5_{jk}}, \tag{8.98}$$

where $r_{jk} = |\vec{r}_j - \vec{r}_k|$; $\alpha, \beta = x, y, z$; δ is the Kronecker delta; the sum runs over all lattice sites. Since the size of neighboring ions is less than $\vec{r}_k$, then they can be considered as point charges, whose values are determined by the valence of the ions. Thus, the problem is reduced to the lattice sums calculation.

---

[18] K. Fukui, Y. Yonezawa, H. Shingu. A molecular orbital theory of reactivity in aromatic hydrocarbons. J. Chem. Phys. **20** (1952) 722–725. K. Morokuma. Molecular orbital studies of hydrogen bonds. III. C=O · · ·H––O hydrogen bond in H_2CO· · ·H_2O and H_2CO· · ·$2H_2O$. J. Chem. Phys. **55** (1971) 1236–1244.

Due to its simplicity, the point-charge model has been widely used in the interpretation of NMR and NQR experimental data. However, in compounds with a significant part of covalency in chemical bonding, this model often underestimates the ion charges.

The lattice point-charges deform the wave functions of ions. Partly, it can be taken into account by considering polarization of ions. Within the point-multipole model both point charge and induced dipole contributions to the EFG tensor are calculated. As compared to the point-charge model the calculation within the point-multipole model is a bit more complicated, because first it is necessary to calculate the dipole moments ($\vec{\mu}$). In order to calculate $\vec{\mu}$ one has to multiply the electric field value $\vec{E}$, which is created on the ion site by surrounding charges, by the electronic polarizability coefficient (α). Then, as the electric field is defined by point charges and induced dipoles of the neighboring ions, $\vec{E}^0$ and $\vec{E}(\vec{\mu})$, respectively, the dipole moment $\vec{\mu}$ can be written as follows

$$\vec{\mu} = \alpha\left[\vec{E}^0 + \vec{E}(\vec{\mu})\right],$$

where

$$\vec{E}_j^0 = \vec{\nabla}\left[\sum_{k\neq j}\frac{q_k}{r_{jk}}\right], \tag{8.99}$$

$$\vec{E}_j(\vec{\mu}) = \vec{\nabla}\left[\sum_{k\neq j}\left(\vec{\mu}_k, \vec{\nabla}\left(\frac{1}{r_{kj}}\right)\right)\right]. \tag{8.100}$$

Since the dipole moments depend on the electric field, and it, in its turn, is determined by the dipole moments, this is a self-consistent problem that can be solved using an iterative procedure. The total EFG created by point charges and induced dipoles is given by

$$V_{j,\alpha\beta} = V^0_{j,\alpha\beta} + V_{j,\alpha\beta}(\vec{\mu}),$$

where the dipole contribution is described by

$$V_{j,\alpha\beta}(\vec{\mu}) = \sum_{k\neq j}\frac{3\left(\delta_{\alpha\beta}r^2_{jk} - 5r^{\alpha}_{jk}r^{\beta}_{jk}\right)}{r^7_{jk}}\left(\vec{r}_{jk}, \vec{\mu}_k\right) + \frac{3\left(r^{\alpha}_{jk}\mu^{\beta}_k + r^{\beta}_{jk}\mu^{\alpha}_k\right)}{r^5_{jk}}. \tag{8.101}$$

However, there is an uncertainty in the electronic polarizability choice, its values are not well determined. In addition, the polarizability depends on the crystal structure.

Earlier, one of the major disadvantages the EFG calculations within the point-multipole model was the slow convergence of the series (8.98)–(8.101). Practice shows that to achieve the convergence of a few percent it is necessary to choose the summation radius of 100–120 Å. There are several methods to improve the convergence of the series, such as the Ewald method,[19] the Evjen method.[20] However, with development of efficient computational techniques the slow convergence is not a problem anymore.

Besides the lattice effect it is necessary to take into account the contribution of valence electrons of the ion. The electronic shell shape of the ion is not spherical. External sources like point charges or dipoles inducing electric field gradients at the nuclei, simultaneously polarize the electronic shells of the ion. The deformed electronic shell creates a secondary gradient, which has the same symmetry as the initial one induced by external sources. Then, the resulting EFG from both external sources and the electron shell can be represented by the following expression

$$V_{\alpha\beta}^{\mathrm{tot}} = (1 - \gamma_{\infty})\, V_{\alpha\beta},$$

where $(1 - \gamma_{\infty})$ is the Sternheimer antishielding factor, for heavy ions it can be equal to several hundreds. Moreover, for anions the $(1 - \gamma_{\infty})$ value may change essentially from one compound to another. For cations it is almost constant and in calculations one can use the value obtained for the free ion. When analyzing experimental data the parameters values (ion charges and polarizabilities) are chosen in such a way that the calculated EFG tensor agrees well with the experimental one.

It is to be noted that besides the dipole contribution it is possible to take into account the quadrupole, octupole contributions and so on. However, with the multipole number n increasing, first, the contribution to the EFG tensor decreases, second, the uncertainty in the parameters choice increases. For example, if $n = 1$ the only parameter is the ion charge, whereas for the dipole term ($n = 2$) the varying parameter is the polarizability, which is anisotropic. For the quadrupole contribution, $n = 3$, the varying parameter is the anisotropic tree-rank tensor etc. That is why when calculating the EFG tensor the quadrupole and higher multipoles contributions are neglected. In oxides usually only the oxygen polarizability is taken into account, as its value is about an order greater than that one of cations.

Further Readings

Grimes RW, Catlow CRA, Shluger AL (1992) Quantum mechanical cluster calculations in solid state studies. World Scientific, Singapore

Huzinaga S (1984) Gaussian basis sets for molecular calculations. Elsevier, Amsterdam

Semin G, Babushkina T, Jakobson G (1975) Nuclear quadrupole resonance in chemistry. Wiley, New York

Slater JC (1963) Quantum theory of molecules and solids: the self-consistent field for molecules and solids. McGraw-Hill, New York

---

[19] P.P. Ewald. Die Berechnung optischer und elektrostatischer Gitterpotentiale. Ann. Phys. **64** (1921) 253–287.

[20] H.M. Evjen. On the Stability of Certain Heteropolar Crystals. Phys. Rev. **39** (1932) 675–687.

Chapter 9
Experimental Methods in NQR

In nuclear quadrupole resonance, splitting of energy levels of a spin system occurs under the influence of the interaction between the nuclear quadrupole moment and intracrystalline electric field gradients. That allows us to observe NMR signals in zero or weak static magnetic field. The absence of electromagnets essentially decreases the necessary instrument set that is a great advantage of this method over NMR.

In order to initiate transitions between sublevels of a spin system, similar to the nuclear magnetic resonance, it is necessary to apply an rf field with frequency that may change within a rather wide range, because in NQR the signal frequency is unknown beforehand.

9.1 Peculiarities of NQR Spectra Registration

For NQR signal detection both stationary and pulse methods are used, and the latter becomes more and more popular. Principles of operation of NQR spectrometers are similar to NMR, therefore, usually to record NQR signals one can use standard high-quality NMR spectrometers, see Chap. 2 of this textbook. But stationary methods have some advantages over pulse ones: (i) they are more convenient when one deals with a complex multiplet structure; (ii) they allow us to reproduce broad NQR lines without distortion[1]; (iii) they are rather simple for spectrum analysis. Stationary NQR spectrometers, as a rule, have more simple hardware for frequency tuning of transmitter and receiver that makes them irreplaceable in signal searching within a broad frequency range.

Let us consider bases principles of registration of NQR signals under the action of an rf field $\vec{B}_1$.

The functional diagram for registration of NQR signals differs from that one in a NMR spectrometer by the absence of a magnet and by the application of frequency

[1] When recording NQR signals by pulse methods it is very difficult to obtain an undistorted signal shape.

V. I. Chizhik et al., *Magnetic Resonance and Its Applications*,
DOI: 10.1007/978-3-319-05299-1_9,

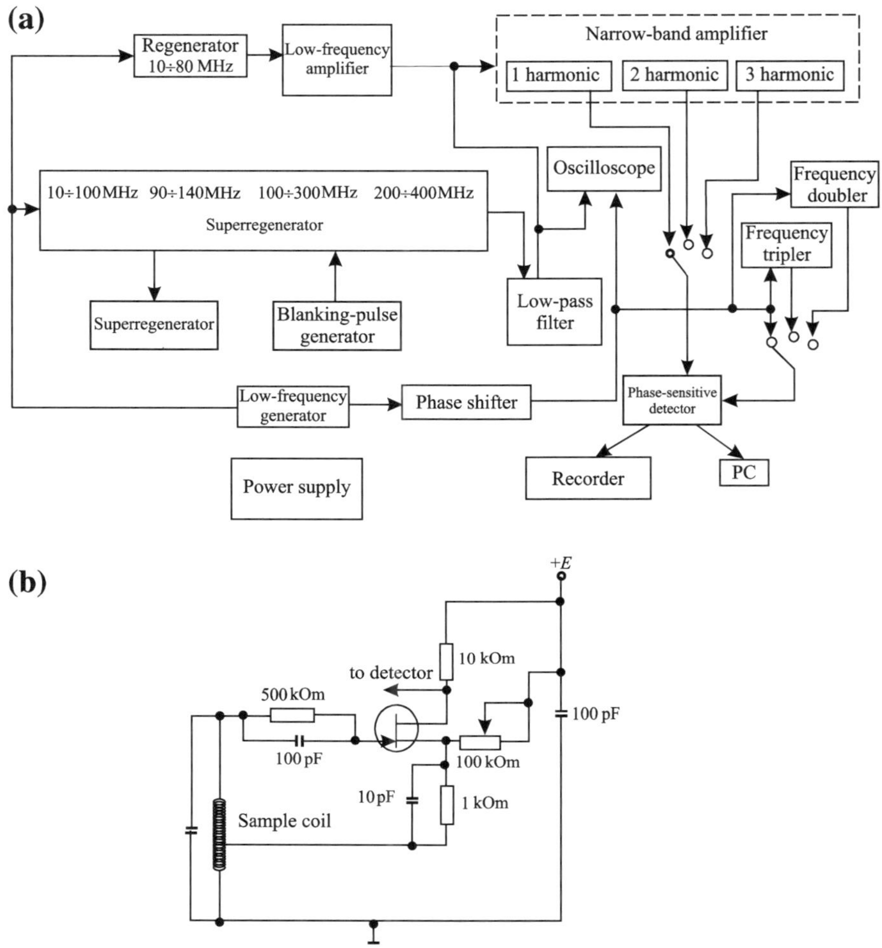

Fig. 9.1 A functional diagram for registration of NQR signals

passage trough resonance conditions. The block-scheme is shown in Fig. 9.1a. It is possible to mark out three basic elements of this diagram:

- the high-frequency generator and spin detector;
- the system of signal amplification, detection and registration;
- the system of resonance passage (frequency modulation of $\vec{B}_1$).

The high-frequency generator and spin detector, as a rule, are realized into the same functional part, which is a supergenerator, regenerator, bridge circuit or Bloch crossed coils.

The principle of operation of a supergenerator is similar to an autodyne spin detector described in Chap. 2 of this textbook. As it has been noted, the sensitivity of a self-oscillating detector noticeably increases with decreasing of oscillation

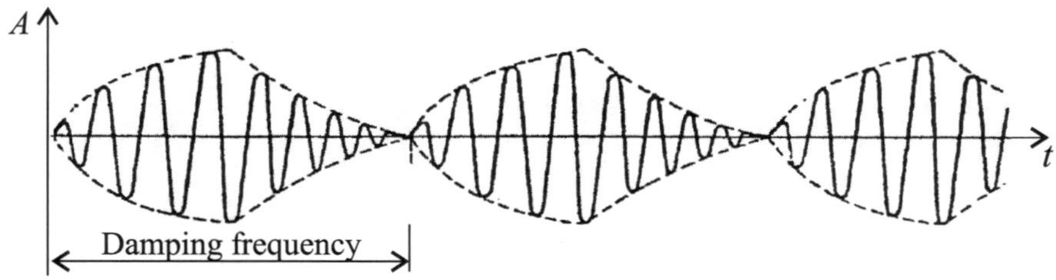

Fig. 9.2 Variations of the high-frequency signal amplitude in time in a supergenerator

amplitude. Therefore, for sensitivity enhancement it is possible to use periodical changing of amplitude generation near the zero value with the frequency Ω_d, which is called the *damping frequency*. Oscillations are damped by an external generator (*external damping*) or by internal process (*selfdamping*). A generator with periodic excitation represents a supergenerator.[2] In such a generator the high-frequency signal amplitude varies with time in a way shown in Fig. 9.2. The simplest circuit of the supergenerator is shown in Fig. 9.1b. Self-excitation conditions are ensured by a field-effect transistor hookup by three-point circuit. The unit RC in the switch circuit provides periodical interruption of oscillations. During operation at oscillation conditions the capacitor is charging, in consequence of that the oscillations end until it discharges again through the shunt resistance. The damping frequency is determined by the time constant RC. By means of resistance of $100\,\mathrm{k}\Omega$ one can change the transistor mutual conductance in operating point, where the oscillation threshold of the circuit is achieved. The frequency modulation is realized by means of a vibrating capacitor or varicaps. The vibrating capacitor allows the circuit to work within the all frequency range up to several hundreds MHz. At higher frequencies it is necessary to apply continuous systems.

To excite transitions between energy sublevels of a spin system, split due to quadrupole interaction, and to register the generator energy absorbed by the system, it is necessary that the oscillation frequency is several order greater than the damping frequency, and the modulation frequency Ω_m, used for resonance passage, is several order less than the damping frequency:

$$\omega_0 \gg \Omega_\mathrm{d} \gg \Omega_\mathrm{m}.$$

Such a method of registration is especially convenient for detection of weak signals, whose frequency is unknown beforehand, as a supergenerator can be easily tuned in a wide frequency range. That is why this method is widespread for investigation of various materials by NQR. When using a supergenerator the registered shape of the signal is a superposition of the absorption and dispersion signals, and there is an additional hardware line broadening. For these reasons and because of instability of the amplitude of the rf field $\vec{B}_1$ it is impossible to determine the shape of NQR lines.

[2] A. Lösche. Kerninduktion. Deutscher Verlag d. Wissenschaften, 1957

However, in this case the line form is not of great importance as in NMR, because the information about electric field gradients can be obtained by measuring only the NQR frequency. Another shortcoming of using a supergenerator is the fact that the rf field is not monochromatic: the frequency spectrum besides the main frequency contains additional sidebands at the distance of damping frequency (or its higher harmonics). In order to determine the central component of the spectrum one changes the damping frequency that results in shifting of sidebands but does not change the main resonance signal. In addition the sidebands can be suppressed by means of slow modulation of the damping frequency. Due to the shifts of sidebands they can be averaged by a phase detector with a large time constant. A certain weakening of the sidebands can be obtained by changing the pulse ratio. The regenerator is free of above-mentioned shortcomings. It represents a generator of continuous oscillations with positive feedback. An essential advantage of regenerative circuits is that they allow us to observe the pure absorption signal without an admixture of a dispersion signal, which causes the frequency modulation only that does not show at amplitude detection. However, the regenerator sensitivity is rather low as compared with the supergenerator. That is why standard NQR spectrometers contain a supergenerator for signal searching and a regenerator for recording of undistorted NQR lines.

Advantages and disadvantages of bridge circuits and Bloch crossed coils are described in detail in Chap. 2.

Receiver, recorder and modulator do not differ in essence from corresponding serial radiotechnical devices, so they are not discussed in this textbook.

The registration technics of NQR spectra in external magnetic field is characterized by some specifics. Usually the studied crystal is fixed on an axis and put into a metallic (diamagnetic) cylinder serving as a screen. The external magnetic field is created by the Helmholtz coils, which can rotate about one or two perpendicular axes. Rotation angle values are counted by limbs. In experiment one uses radiofrequency coils of two types. At the first stage, the coil axis is almost parallel to the rotation axis, whereas at the second stage it is perpendicular to the rotation axis. That excludes any line missing out. Since a resonance line intensity depends on squared sinus of the angle between the radiofrequency coil axis and the z-axis of EFG in the crystal, the only opportunity to miss the line is if these two axes coincide. Therefore, the magnetic field orientation is changed in one degree, and at every orientation one records the spectrum lying near the unsplit NQR line.

9.2 Physical Bases of Pulse NQR

During last decades pulse NQR techniques are becoming more and more popular, especially as a powerful tool for explosives and drugs detection. In order to provoke transitions between energy levels at quadrupole interactions it is necessary to apply a rf field. As it is shown in Sect. 8.1.2, for all nuclei, except nuclei with $I = 1$, the most intensive NQR signal is observed when the rf field vector is perpendicular to the principal axis of the EFG tensor. However, this direction is usually unknown. Therefore first of all it is necessary to find the NQR signal in powder samples and after that one can precise the results by studying monocrystals. When applying pulse NQR methods transitions are excited by one or several high-power pulses, after which, similar to NMR, a free induction or an echo signal appears. Its Fourier transform gives a NQR signal. In NMR the resonance pulse turns the nuclear magnetization vector, which initially directs along the stationary magnetic field, about the rf field. Under the impact of the 90-degree pulse the magnetization vector turns into the (xy)-plane, the phase of the rf pulse determines the magnetization direction in this plane.

However, as opposed to NMR, in NQR there is no macroscopic magnetization in the stationary state. Under the influence of electric field gradients an ordered arrangement of quadrupole (not magnetic) moments appears. The quadrupole moments represent symmetric second-rank tensors (see (1.175)), and as it is shown in Sect. 5.4, at presence of quadrupole interactions the energy levels do not depend on the direction of the nucleus quadrupole moment: along or opposite to the direction of the principal axis of the EFG tensor. The absence of magnetization in the stationary state directly follows from calculations of the average magnetic moment of the unit volume. The simplest way to calculate it is to use the density matrix (see Sect. 1.9.2):

$$\langle \hat{M}_z \rangle = \mathrm{Tr}(\hat{\rho}\hat{M}_z), \tag{9.1}$$

where $\hat{\rho}$ is the density matrix corresponding to the stationary state

$$\hat{\rho} = A \exp\left(-\widehat{\mathcal{H}}_{\mathrm{Q}}/(kT)\right), \tag{9.2}$$

here $\widehat{\mathcal{H}}_{\mathrm{Q}}$ is the Hamiltonian of quadrupole interactions in the form of (8.1); k is the Boltzmann constant; T is the temperature; A is a constant. As the energy of quadrupole interaction as a rule is much less than kT, the exponent in (9.2) can be expanded in a Taylor series. Taking into account only the first two terms one has

$$\hat{\rho} = A\left(1 - \widehat{\mathcal{H}}_{\mathrm{Q}}/(kT)\right). \tag{9.3}$$

After substitution Eqs. (9.3) and (8.1) into Eq. (9.1) and considering that $\hat{M}_z = \gamma\hbar N\hat{I}_z$, one has

$$\begin{aligned}
\langle \hat{M}_z \rangle =& \gamma\hbar A N \left\{ \mathrm{Tr}(\hat{I}_z) - \frac{e^2 q Q}{4I(2I-1)kT} \right. \\
& \left. \times \left[\mathrm{Tr}\left(\hat{I}_z(3\hat{I}_z{}^2 - \hat{\vec{I}}^2)\right) + \frac{\eta}{2}\mathrm{Tr}\left(\hat{I}_z(\hat{\vec{I}}_+^2 + \hat{\vec{I}}_-^2)\right)\right]\right\} \\
=& \gamma\hbar A N \left\{ \sum_{m=-I}^{I} m - \frac{e^2 q Q}{4I(2I-1)kT} \right. \\
& \left. \times \left[\sum_{m=-I}^{I} m(3m^2 - I(I+1)) + \frac{\eta}{2}\mathrm{Tr}\left(\hat{I}_z(\hat{\vec{I}}_+^2 + \hat{\vec{I}}_-^2)\right)\right]\right\}.
\end{aligned} \tag{9.4}$$

The first two terms in Eq. (9.4) are equal to zero, as the sum of odd powers of m is equal to zero. The last term is equal to zero as well, as the operators $\hat{I}_z$, $\hat{I}_+^2$ and $\hat{I}_-^2$ do not commute with each other, and hence the product of these operators does not have nondiagonal matrix elements.

The second difference between NQR and NMR is that in the first case the energy levels are almost equidistant, as any interaction touching nuclear spins is weaker than their interaction with an external stationary magnetic field. That is why, without special conditions, in NMR a rf pulse causes transitions between all energy levels simultaneously, that is a *nonselctive excitation*. In the second case all energy levels are unequidistant and the rf pulse excites only the transition, whose frequency is close to the radiofrequency pulse filling, that is a *selective excitation*. Even if the transition frequencies are the same ($m \to m+1$ and $-m \to -m-1$), each transition is excited by the rf field circularly polarized either clockwise or anticlockwise. In other words, in NQR there is always a selective excitation. If the rf field is linearly polarized, for example, along the x-axis, it can be decomposed into two components: B^+ and B^-, one of which is polarized clockwise, whereas the other is polarized anticlockwise:

$$\begin{aligned} B_{\approx}(t) &= 2B_{1x}\cos(\omega t - \phi) = B^+ + B^-, \\ B^+ &= B_{1x}\cos(\omega t - \phi) - B_{1y}\sin(\omega t - \phi), \\ B^- &= B_{1x}\cos(\omega t - \phi) + B_{1y}\sin(\omega t - \phi). \end{aligned} \tag{9.5}$$

The component B^+ causes transitions at frequency ω between levels $m \to m+1$, the component B^- at frequency $-\omega$ between levels $-m \to -m-1$. At NMR conditions only the B^+ component is active, the B^- component results only in a small shift of the resonance frequency, that is the *Bloch–Zeigert shift*. Usually it is not taken into account. At nuclear quadrupole resonance conditions it is always necessary to consider both polarized components of the rf field.

The absence of macroscopic magnetization in the stationary case makes unobvious the fact that after an excitation of the spin system by a radiofrequency pulse a free induction signal appears (and after several pulses there is an echo signal). It is also unclear, if one can apply (and under what conditions) for the spin system description the vector model, which is successfully used in NMR. However, free induction signals are observed in NQR as well, and as it will be shown further by quantum mechanics calculations the vector model can be applied if the asymmetry parameter of the EFG tensor, η, is equal to zero. It is necessary to mention another difference between nuclear quadrupole and magnetic resonances: in NQR the free induction decay is caused not by a nonuniformity of the external magnetic field, but by a nonuniformity of the EFG distribution.

9.3 NQR Free Induction

At present to describe the magnetization evolution one uses the density matrix apparatus description. At conditions of both NQR and NMR the recorded signals are proportional to the expected value of the magnetization $\langle \hat{M}_x \rangle$, whose vector is perpendicular to the z-axis of the laboratory frame (in NMR the z-axis directs along the external magnetic field vector, in NQR it is along the principal axis of the EFG tensor). Therefore, to calculate the free precession signal means to determine the behavior the expected magnetization of the unit volume $\langle \hat{M}_x \rangle$. In order to find $\langle \hat{M}_x \rangle$ one can use the relation (9.1), in which it is necessary to substitute $\hat{M}_z$ by $\hat{M}_x$. However, free precession signals appear at a nonstationary state of the spin system, and hence they can not be described by the stationary density matrix. To determine the density matrix one has to solve the nonstationary Liouville equation for two cases: at presence and in absence of a rf pulse. Let us note the density matrices corresponding to the first and second cases as $\hat{\rho}(t)$ and $\hat{\rho}_0(t)$, respectively. When neglecting relaxation processes the Liouville equation for these two cases[3] are

$$\mathrm{i}\hbar \frac{\partial \hat{\rho}}{\partial t} = [(\widehat{\mathcal{H}}_\mathrm{Q} + \widehat{\mathcal{H}}_1)\hat{\rho}], \tag{9.6}$$

$$\mathrm{i}\hbar \frac{\partial \hat{\rho}_0}{\partial t} = [\widehat{\mathcal{H}}_\mathrm{Q}\hat{\rho}_0], \tag{9.7}$$

here

$$\widehat{\mathcal{H}}_1 = -\gamma\hbar\hat{I}_x B_1 \cos\omega(t - t_0) = -\omega_1\hbar\hat{I}_x \cos\omega(t - t_0), \quad \omega_1 = \gamma B_1. \tag{9.8}$$

The Hamiltonian in the right part of Eq. (9.6) depends on time, therefore, to solve it let us pass to another representation, where this dependence disappears. The expression for the density matrix is

$$\hat{\rho} = \exp\left(-\mathrm{i}\frac{\widehat{\mathcal{H}}_\mathrm{Q}}{\hbar}(t - t_0)\right)\hat{\rho}^* \exp\left(\mathrm{i}\frac{\widehat{\mathcal{H}}_\mathrm{Q}}{\hbar}(t - t_0)\right). \tag{9.9}$$

After substitution Eq. (9.9) into Eq. (9.6), one has the nonstationary Liouville equation for the density matrix $\hat{\rho}^*$:

[3] During the radiofrequency pulse relaxation processes can be neglected if their duration is much less than the spin-spin and spin-lattice relaxation times, however, they should be taking account between the pulses. Nevertheless, firstly one carries out calculations without any relaxation effects, and after that one can take into account their influence phenomenologically.

$$
\begin{aligned}
&\widehat{\mathcal{H}}_{\mathrm{Q}} \exp\left(-\mathrm{i}\frac{\widehat{\mathcal{H}}_{\mathrm{Q}}}{\hbar}(t-t_0)\right)\hat{\rho}^* \exp\left(\mathrm{i}\frac{\widehat{\mathcal{H}}_{\mathrm{Q}}}{\hbar}(t-t_0)\right) \\
&\quad + \mathrm{i}\hbar \exp\left(-\mathrm{i}\frac{\widehat{\mathcal{H}}_{\mathrm{Q}}}{\hbar}(t-t_0)\right)\frac{\partial \hat{\rho}^*}{\partial t} \exp\left(\mathrm{i}\frac{\widehat{\mathcal{H}}_{\mathrm{Q}}}{\hbar}(t-t_0)\right) \\
&\quad + \mathrm{i}\hbar \exp\left(\mathrm{i}\frac{\widehat{\mathcal{H}}_{\mathrm{Q}}}{\hbar}(t-t_0)\right)\hat{\rho}^* \left(\mathrm{i}\frac{\widehat{\mathcal{H}}_{\mathrm{Q}}}{\hbar} \exp\left(\mathrm{i}\frac{\widehat{\mathcal{H}}_{\mathrm{Q}}}{\hbar}(t-t_0)\right)\right) \\
&= \widehat{\mathcal{H}}_{\mathrm{Q}} \exp\left(-\mathrm{i}\frac{\widehat{\mathcal{H}}_{\mathrm{Q}}}{\hbar}(t-t_0)\right)\hat{\rho}^* \exp\left(\mathrm{i}\frac{\widehat{\mathcal{H}}_{\mathrm{Q}}}{\hbar}(t-t_0)\right) \\
&\quad - \exp\left(-\mathrm{i}\frac{\widehat{\mathcal{H}}_{\mathrm{Q}}}{\hbar}(t-t_0)\right)\hat{\rho}^* \exp\left(\mathrm{i}\frac{\widehat{\mathcal{H}}_{\mathrm{Q}}}{\hbar}(t-t_0)\right)\widehat{\mathcal{H}}_{\mathrm{Q}} \\
&\quad + \widehat{\mathcal{H}}_1 \exp\left(-\mathrm{i}\frac{\widehat{\mathcal{H}}_{\mathrm{Q}}}{\hbar}(t-t_0)\right)\hat{\rho}^* \exp\left(\mathrm{i}\frac{\widehat{\mathcal{H}}_{\mathrm{Q}}}{\hbar}(t-t_0)\right) \\
&\quad - \exp\left(-\mathrm{i}\frac{\widehat{\mathcal{H}}_{\mathrm{Q}}}{\hbar}(t-t_0)\right)\hat{\rho}^* \exp\left(i\frac{\widehat{\mathcal{H}}_{\mathrm{Q}}}{\hbar}(t-t_0)\right)\widehat{\mathcal{H}}_1 .
\end{aligned} \tag{9.10}
$$

By multiplying Eq. (9.10) on the right by $\exp\left(\mathrm{i}\frac{\widehat{\mathcal{H}}_{\mathrm{Q}}}{\hbar}(t-t_0)\right)$ and on the left by $\exp\left(-\mathrm{i}\frac{\widehat{\mathcal{H}}_{\mathrm{Q}}}{\hbar}(t-t_0)\right)$ after cancelation of the similar terms in the left and right parts one has

$$
\mathrm{i}\hbar\frac{\partial \hat{\rho}^*}{\partial t} = [\widehat{\mathcal{H}}_1^* \hat{\rho}^*], \tag{9.11}
$$

$$
\widehat{\mathcal{H}}_1^* = \exp\left(\mathrm{i}\frac{\widehat{\mathcal{H}}_{\mathrm{Q}}}{\hbar}(t-t_0)\right)\widehat{\mathcal{H}}_1 \exp\left(-\mathrm{i}\frac{\widehat{\mathcal{H}}_{\mathrm{Q}}}{\hbar}(t-t_0)\right). \tag{9.12}
$$

The solution of the Liouville equation (9.11) in general form with the Hamiltanion (9.12) can be obtained approximately using properties of irreducible tensor operators (see Sect. 1.12) and exactly for a specific nuclear spin value using the matrix representation of operators and introducing fictitious spin operators. Currently to describe the spin subsystem behavior after the pulse action both methods are used. The first one, especially in NQR, demands a lot of cumbersome mathematical transformations, thus here we consider the second one.

9.3.1 Formation of the Free Induction for $I = 3/2$ at $\eta = 0$

The fictitious spin operators are used in the case if it is necessary to describe a selective excitation of the spin subsystem by a rf pulse. This approach is based on the matrix representation of spin operators.

The matrix representation of spin operators for the spin number 1/2 trough the Pauli matrix is well-known. Let us remind that rows and columns of these matrix contain matrix elements of the corresponding spin operators $\hat{I}_z$, $\hat{I}_+$, $\hat{I}_-$, $\hat{I}_x$ and $\hat{I}_y$ calculated by means of eigenfunctions ψ_m ($m = \pm 1/2$) of the operator $\hat{I}_z$. So the matrix

$$\begin{pmatrix} \langle\psi_{1/2}|\hat{I}_z|\psi_{1/2}\rangle & \langle\psi_{-1/2}|\hat{I}_z|\psi_{1/2}\rangle \\ \langle\psi_{1/2}|\hat{I}_z|\psi_{-1/2}\rangle & \langle\psi_{-1/2}|\hat{I}_z|\psi_{-1/2}\rangle \end{pmatrix} \to \frac{1}{2}\begin{pmatrix} 1 & 0 \\ 0 & -1 \end{pmatrix},$$

corresponds to the operator $\hat{I}_z$, the matrix

$$\begin{pmatrix} \langle\psi_{1/2}|\hat{I}_+|\psi_{1/2}\rangle & \langle\psi_{-1/2}|\hat{I}_+|\psi_{1/2}\rangle \\ \langle\psi_{1/2}|\hat{I}_+|\psi_{-1/2}\rangle & \langle\psi_{-1/2}|\hat{I}_+|\psi_{-1/2}\rangle \end{pmatrix} \to \begin{pmatrix} 0 & 0 \\ 1 & 0 \end{pmatrix},$$

corresponds to $\hat{I}_+$. Others matrices are of the following form

$$\hat{I}_- \to \begin{pmatrix} 0 & 1 \\ 0 & 0 \end{pmatrix}, \quad \hat{I}_x = \frac{\hat{I}_+ + \hat{I}_-}{2} \to \frac{1}{2}\begin{pmatrix} 0 & 1 \\ 1 & 0 \end{pmatrix}, \quad \hat{I}_y = \frac{\hat{I}_+ - \hat{I}_-}{2\mathrm{i}} \to \frac{\mathrm{i}}{2}\begin{pmatrix} 0 & -1 \\ 1 & 0 \end{pmatrix}. \tag{9.13}$$

Matrix representations of spin operators for other spin numbers can be obtained by the similar way. For example, for $I = 3/2$ the matrix corresponding to the operator $\hat{I}_\mathrm{Q} = \hat{I}_z^2 - \hat{\vec{I}}^2/3$ is

$$\hat{I}_\mathrm{Q} \to \begin{pmatrix} \langle\psi_{3/2}|\hat{I}_\mathrm{Q}|\psi_{3/2}\rangle & \langle\psi_{1/2}|\hat{I}_\mathrm{Q}|\psi_{3/2}\rangle & \langle\psi_{-1/2}|\hat{I}_\mathrm{Q}|\psi_{3/2}\rangle & \langle\psi_{-3/2}|\hat{I}_\mathrm{Q}|\psi_{3/2}\rangle \\ \langle\psi_{3/2}|\hat{I}_\mathrm{Q}|\psi_{1/2}\rangle & \langle\psi_{1/2}|\hat{I}_\mathrm{Q}|\psi_{1/2}\rangle & \langle\psi_{-1/2}|\hat{I}_\mathrm{Q}|\psi_{1/2}\rangle & \langle\psi_{-3/2}|\hat{I}_\mathrm{Q}|\psi_{1/2}\rangle \\ \langle\psi_{3/2}|\hat{I}_\mathrm{Q}|\psi_{-1/2}\rangle & \langle\psi_{1/2}|\hat{I}_\mathrm{Q}|\psi_{-1/2}\rangle & \langle\psi_{-1/2}|\hat{I}_\mathrm{Q}|\psi_{-1/2}\rangle & \langle\psi_{-3/2}|\hat{I}_\mathrm{Q}|\psi_{-1/2}\rangle \\ \langle\psi_{3/2}|\hat{I}_\mathrm{Q}|\psi_{-3/2}\rangle & \langle\psi_{1/2}|\hat{I}_\mathrm{Q}|\psi_{-3/2}\rangle & \langle\psi_{-1/2}|\hat{I}_\mathrm{Q}|\psi_{-3/2}\rangle & \langle\psi_{-3/2}|\hat{I}_\mathrm{Q}|\psi_{-3/2}\rangle \end{pmatrix}$$

$$\to \begin{pmatrix} 1 & 0 & 0 & 0 \\ 0 & -1 & 0 & 0 \\ 0 & 0 & -1 & 0 \\ 0 & 0 & 0 & 1 \end{pmatrix}, \tag{9.14}$$

and the matrix

$$\hat{I}_x = \frac{\hat{I}_+ + \hat{I}_-}{2} \to \frac{1}{2}\begin{pmatrix} 0 & \sqrt{3} & 0 & 0 \\ \sqrt{3} & 0 & 2 & 0 \\ 0 & 2 & 0 & \sqrt{3} \\ 0 & 0 & \sqrt{3} & 0 \end{pmatrix} \tag{9.15}$$

corresponds to the operator $\hat{I}_x$. By the similar way it is possible to represent various spin operators for other nucleus spin numbers.

Let us consider an example with $I = 3/2$ and $\eta = 0$. What representations of spin operators are to be introduced in order to describe the evolution of a spin subsystem at presence of quadrupole interactions and interaction with a rf field? Using the matrix representations of the operators $\hat{I}_{\mathrm{Q}}$ and $\hat{I}_x$, one can shows that $\widehat{\mathcal{H}}_1^*$ (Eq. (9.12)) contains two parts: slowly varying with time and rapidly oscillating. The latter, similar to NMR, can be neglected.

The Hamiltonian of quadrupole interactions for $I = 3/2$ can be written as

$$\widehat{\mathcal{H}}_{\mathrm{Q}} = \frac{e^2qQ}{4}\left(\hat{I}_z{}^2 - \hat{\vec{I}}^2/3\right). \tag{9.16}$$

Taking into account that the NQR frequency (ω_0) corresponding to the transition $\pm 3/2 \leftrightarrow \pm 1/2$ is equal to $e^2qQ/(2\hbar)$, let us rewrite Eq. (9.16):

$$\widehat{\mathcal{H}}_{\mathrm{Q}} = \frac{\hbar\omega_0}{2}\hat{I}_{\mathrm{Q}}. \tag{9.17}$$

After substitution Eq. (9.17) into Eq. (9.12) and taking into account Eq. (9.8), one has the following expression for the operator $\widehat{\mathcal{H}}_1^*$:

$$\begin{aligned}\widehat{\mathcal{H}}_1^* &= -\omega_1\hbar\cos\omega(t-t_0)\exp(\mathrm{i}\alpha\hat{I}_{\mathrm{Q}})\hat{I}_x\exp(-\mathrm{i}\alpha\hat{I}_{\mathrm{Q}})\\ &= -\omega_1\hbar\frac{\exp[\mathrm{i}\omega(t-t_0)]+\exp[-\mathrm{i}\omega(t-t_0)]}{2}\exp(\mathrm{i}\alpha\hat{I}_{\mathrm{Q}})\hat{I}_x\exp(-\mathrm{i}\alpha\hat{I}_{\mathrm{Q}}),\end{aligned} \tag{9.18}$$

where

$$\alpha = \frac{\omega_0}{2}(t-t_0). \tag{9.19}$$

Let us expand the operator $\exp(\mathrm{i}\alpha\hat{I}_{\mathrm{Q}})$ in a Taylor series:

$$\exp(\mathrm{i}\alpha\hat{I}_{\mathrm{Q}}) = 1 + \frac{\mathrm{i}\alpha}{1!}\hat{I}_{\mathrm{Q}} + \frac{(\mathrm{i}\alpha)^2}{2!}\hat{I}_{\mathrm{Q}}^2 + \frac{(\mathrm{i}\alpha)^3}{3!}\hat{I}_{\mathrm{Q}}^3 + \cdots. \tag{9.20}$$

It is worth noting that the matrix representation of the operator $\hat{I}_{\mathrm{Q}}^2$ is the unit matrix (it is easy to see by multiplying the matrix corresponding to the operator $\hat{I}_{\mathrm{Q}}$ by itself), and the matrix representation of all even powers of the operator $\hat{I}_{\mathrm{Q}}$ is the unit matrix as well. The odd powers of the operator $\hat{I}_{\mathrm{Q}}$ are equal to the operator itself:

$$(\hat{I}_{\mathrm{Q}})^{2n} = [1], \quad (\hat{I}_{\mathrm{Q}})^{2n+1} = \hat{I}_{\mathrm{Q}}. \tag{9.21}$$

Equation (9.20) can be rearranged in such a way:

$$\exp(\mathrm{i}\alpha\hat{I}_{\mathrm{Q}}) = \left(1 - \frac{(\alpha)^2}{2!} + \cdots\right)[1] + \mathrm{i}\left(\frac{\alpha}{1!} - \frac{(\alpha)^3}{3!} + \cdots\right)\hat{I}_{\mathrm{Q}}. \tag{9.22}$$

The first and second terms of Eq. (9.22) represent the expansion in a Taylor series of the functions $\cos\alpha$ and $\sin\alpha$, respectively. Then

$$\begin{aligned}\exp(\mathrm{i}\alpha\hat{I}_{\mathrm{Q}}) &= [1]\cos\alpha + \mathrm{i}\hat{I}_{\mathrm{Q}}\sin\alpha,\\ \exp(-\mathrm{i}\alpha\hat{I}_{\mathrm{Q}}) &= [1]\cos\alpha - \mathrm{i}\hat{I}_{\mathrm{Q}}\sin\alpha.\end{aligned} \tag{9.23}$$

Now Eq. (9.23) can be written in the matrix form as:

$$\begin{aligned}\exp(\mathrm{i}\alpha\hat{I}_{\mathrm{Q}}) \rightarrow &\begin{pmatrix} \cos\alpha & 0 & 0 & 0\\ 0 & \cos\alpha & 0 & 0\\ 0 & 0 & \cos\alpha & 0\\ 0 & 0 & 0 & \cos\alpha \end{pmatrix} + \mathrm{i}\begin{pmatrix} \sin\alpha & 0 & 0 & 0\\ 0 & -\sin\alpha & 0 & 0\\ 0 & 0 & -\sin\alpha & 0\\ 0 & 0 & 0 & \sin\alpha \end{pmatrix}\\ &= \begin{pmatrix} \exp \mathrm{i}\alpha & 0 & 0 & 0\\ 0 & \exp(-\mathrm{i}\alpha) & 0 & 0\\ 0 & 0 & \exp(-\mathrm{i}\alpha) & 0\\ 0 & 0 & 0 & \exp \mathrm{i}\alpha \end{pmatrix}\end{aligned} \tag{9.24}$$

and

$$\exp(-\mathrm{i}\alpha\hat{I}_{\mathrm{Q}}) \rightarrow \begin{pmatrix} \exp(-\mathrm{i}\alpha) & 0 & 0 & 0\\ 0 & \exp \mathrm{i}\alpha & 0 & 0\\ 0 & 0 & \exp \mathrm{i}\alpha & 0\\ 0 & 0 & 0 & \exp(-\mathrm{i}\alpha) \end{pmatrix}. \tag{9.25}$$

The matrix representation of the operator $\widehat{\mathcal{H}}_1^*$ (see (9.75)) can be obtained by multiplying the matrices (9.24), (9.15) and (9.25) and taking into account (9.19):

$$\begin{aligned}\widehat{\mathcal{H}}_1^* = -\frac{\sqrt{3}}{2}\omega_1\hbar\\ \times\begin{pmatrix} 0 & \begin{matrix} e^{\mathrm{i}(-\omega_0+\omega)(t-t_0)}\\ +e^{-\mathrm{i}(\omega_0+\omega)(t-t_0)}\end{matrix} & 0 & 0\\ \begin{matrix} e^{\mathrm{i}(\omega_0+\omega)(t-t_0)}\\ +e^{\mathrm{i}(\omega_0-\omega)(t-t_0)}\end{matrix} & 0 & \begin{matrix}\frac{2}{\sqrt{3}}(e^{\mathrm{i}\omega(t-t_0)}\\ +e^{-\mathrm{i}\omega(t-t_0)})\end{matrix} & 0\\ 0 & \begin{matrix}\frac{2}{\sqrt{3}}(e^{\mathrm{i}\omega(t-t_0)}\\ +e^{-\mathrm{i}\omega(t-t_0)})\end{matrix} & 0 & \begin{matrix} e^{\mathrm{i}(\omega_0-\omega)(t-t_0)}\\ +e^{\mathrm{i}(\omega_0+\omega)(t-t_0)}\end{matrix}\\ 0 & 0 & \begin{matrix} e^{-\mathrm{i}(\omega_0-\omega)(t-t_0)}\\ +e^{-\mathrm{i}(\omega_0+\omega)(t-t_0)}\end{matrix} & 0 \end{pmatrix}.\end{aligned} \tag{9.26}$$

The exponents containing the frequency (ω) or the sum of frequencies ($\omega_0 + \omega$), are rapidly oscillating and within the pulse duration ($t_p = t - t_0$) are averaged to zero.

The exponents with the difference of frequencies ($\Delta\omega = \omega_0 - \omega$) are slowly varying and under condition $\Delta\omega t_p < 1$ they are equal to 1. Then the matrix (9.26) can be rewritten in the following way

$$\widehat{\mathcal{H}}_1^* = -\frac{\sqrt{3}}{2}\omega_1\hbar \begin{pmatrix} 0\,1\,0\,0 \\ 1\,0\,0\,0 \\ 0\,0\,0\,1 \\ 0\,0\,1\,0 \end{pmatrix}. \tag{9.27}$$

The matrix (9.27) is the block matrix, and each submatrix coinsides with doubled matrix expression of the operator $\hat{I}_x$ from (9.13) for $I = 1/2$. Hence, the Hamiltonian $\widehat{\mathcal{H}}_1^*$ can be expressed through the fictitious spin operators $\hat{I}_x{}^*$, which are the spin operators $\hat{I}_x$ for $I = 1/2$:

$$\widehat{\mathcal{H}}_1^* \to -\omega_1\hbar\sqrt{3}\begin{pmatrix} \hat{I}_x{}^* & 0 \\ 0 & \hat{I}_x{}^* \end{pmatrix}. \tag{9.28}$$

Thereby, the Hamiltonian $\widehat{\mathcal{H}}_1^*$ does not depend on time, and one can write the solution of the Liouville equation for the density matrix $\hat{\rho}^*$ with the Hamiltonian $\widehat{\mathcal{H}}_1^*$ in the form of (9.11) as:

$$\hat{\rho}^*(t - t_0) = \exp\left(-\frac{\mathrm{i}}{\hbar}\left(\widehat{\mathcal{H}}_1^*(t - t_0)\right)\right)\hat{\rho}^*(t_0)\exp\left(\frac{\mathrm{i}}{\hbar}\left(\widehat{\mathcal{H}}_1^*(t - t_0)\right)\right). \tag{9.29}$$

It is to be noted that $\hat{\rho}^*(t_0) = \hat{\rho}(t_0)$ (see Eq. (9.9)). The density matrix $\hat{\rho}(t_0)$ represents a stationary density matrix in the form (9.2) describing the state of the spin subsystem before the rf pulse. Taking into account that the quadrupole interaction energy is much less than the thermal motion energy one can expend the stationary density matrix in a Taylor series considering only the first term of the expansion:

$$\hat{\rho}(t_0) = \frac{\exp[-\widehat{\mathcal{H}}_Q/(kT)]}{\mathrm{Tr}(\exp[-\widehat{\mathcal{H}}_Q/(kT)])} = \frac{N}{2I+1}\left(1 - \frac{e^2qQ}{4kT}\hat{I}_Q\right).$$

As it is seen from Eq. (9.14) the matrix representation of the operator $\hat{I}_Q$ for the spin number 3/2 is the diagonal matrix, which can be divided into two block matrices, and each of them is the matrix representation of the operator $\hat{I}_z$ for $I = 1/2$:

$$\hat{I}_Q \to \begin{pmatrix} 1 & 0 & 0 & 0 \\ 0 & -1 & 0 & 0 \\ 0 & 0 & -1 & 0 \\ 0 & 0 & 0 & 1 \end{pmatrix} = 2\begin{pmatrix} +\hat{I}_z{}^* & 0 \\ 0 & -\hat{I}_z{}^* \end{pmatrix}. \tag{9.30}$$

In this case the operator $\hat{I}_z$, as well as the operator $\hat{I}_x$ for $I = 1/2$, is a fictitious operator, hence in Eq. (9.30) it is marked by an asterisk.

Introduction of fictitious operators is rather convenient: the commutational relations, which are true for the common spin operators, are valid for the fictitious operators as well, as the fictitious operators are the spin operators for $I = 1/2$.

As the matrix (9.30) is the block matrix, the matrix representation for $\hat{\rho}(t_0)$ has the similar block form:

$$\hat{\rho}(t_0) \to \begin{pmatrix} \hat{\rho}^+(t-t_0) & 0 \\ 0 & \hat{\rho}^-(t-t_0) \end{pmatrix},$$

where

$$\hat{\rho}^+(t_0) = \frac{N}{4}\left(1 - \frac{e^2qQ}{2kT}\hat{I}_z{}^*\right), \quad \hat{\rho}^-(t_0) = \frac{N}{4}\left(1 + \frac{e^2qQ}{2kT}\hat{I}_z{}^*\right). \tag{9.31}$$

In order to determine the macroscopic magnetizations $\langle \hat{M}_x \rangle$, $\langle \hat{M}_y \rangle$ and $\langle \hat{M}_z \rangle$ it is necessary to express the spin operators $\hat{I}_x$, $\hat{I}_y$ and $\hat{I}_z$ for the spin number 3/2 through the fictitious spin operators. Let us show how it can be done for the operator $\hat{I}_z$:

$$\hat{I}_z = \begin{pmatrix} 3/2 & 0 & 0 & 0 \\ 0 & 1/2 & 0 & 0 \\ 0 & 0 & -1/2 & 0 \\ 0 & 0 & 0 & -3/2 \end{pmatrix} = \begin{pmatrix} 1/2 & 0 & 0 & 0 \\ 0 & -1/2 & 0 & 0 \\ 0 & 0 & 1/2 & 0 \\ 0 & 0 & 0 & -1/2 \end{pmatrix} + \begin{pmatrix} 1 & 0 & 0 & 0 \\ 0 & 1 & 0 & 0 \\ 0 & 0 & -1 & 0 \\ 0 & 0 & 0 & -1 \end{pmatrix}$$

$$= \begin{pmatrix} \hat{I}^*{}_z & 0 \\ 0 & \hat{I}^*{}_z \end{pmatrix} + \begin{pmatrix} [1] & 0 \\ 0 & -[1] \end{pmatrix}$$

(here [1] is the two-dimensional unit matrix). The mean value of the second term is equal to zero, hence the operator $\hat{I}_z = \begin{pmatrix} \hat{I}^*{}_z & 0 \\ 0 & \hat{I}^*{}_z \end{pmatrix}$.

By the same way one can obtain the representations of the operators $\hat{I}_x$ and $\hat{I}_y$:

$$\hat{I}_x = \sqrt{3}\begin{pmatrix} \hat{I}^*{}_x & 0 \\ 0 & \hat{I}^*{}_x \end{pmatrix} + \begin{pmatrix} 0 & \hat{I}^*_+ \\ \hat{I}^*_- & 0 \end{pmatrix},$$

$$\hat{I}_y = -\sqrt{3}\begin{pmatrix} \hat{I}^*{}_y & 0 \\ 0 & \hat{I}^*{}_y \end{pmatrix} + \begin{pmatrix} 0 & -\hat{I}^*_+ \\ \hat{I}^*_- & 0 \end{pmatrix}.$$

One can show that the mean values of the operators corresponding to the second term of these expressions calculated using the diagonal density matrix are equal to zero. Then

$$\hat{I}_x = \sqrt{3}\begin{pmatrix} \hat{I}^*{}_x & 0 \\ 0 & \hat{I}^*{}_x \end{pmatrix}, \quad \hat{I}_y = -\sqrt{3}\begin{pmatrix} \hat{I}^*{}_y & 0 \\ 0 & \hat{I}^*{}_y \end{pmatrix}.$$

Now the stationary macroscopic magnetizations can be calculated by the common way using single elements of the density matrix $\hat{\rho}^+(t_0)$ and $\hat{\rho}^-(t_0)$. For example, for $\langle \hat{M}_z{}^{\pm} \rangle$ one has:

$$\langle \hat{M}_z{}^{\pm} \rangle = \mathrm{Tr}\left(\hat{\rho}^{\pm}(t-t_0)M_z^{\pm}\right) = \frac{\gamma\hbar N}{4}\left(\mathrm{Tr}\,\hat{I}_z{}^* \mp \frac{e^2qQ}{2kT}\,\mathrm{Tr}(I_z^*)^2\right)$$

$$= \frac{\gamma\hbar N}{4}\left(\sum_{m=-1/2}^{1/2} m \mp \frac{e^2qQ}{2kT}\sum_{m=-1/2}^{1/2} m^2\right) = \mp M_0, \tag{9.32}$$

$$M_0 = \frac{\gamma\hbar e^2 qQN}{16kT}.$$

The total stationary magnetization of the unit volume $M_z = 0$, as it follows from general consideration of the magnetization in presence of the quadrupole interaction only. But it can be decomposed into two components. The vector of one of them is along the principal axis of the EFG tensor, whereas the other directs in the opposite side: $M_z = M_z^+ + M_z^- = 0$.

The matrix representation of the Hamiltonian $\widehat{\mathcal{H}}_1^*$ in the form of (9.28) is also the block matrix, hence a block matrix corresponds to exponential operators in Eq. (9.29) as well. Then

$$\hat{\rho}^*(t-t_0) \to \begin{pmatrix} \hat{\rho}^{*+}(t-t_0) & 0 \\ 0 & \hat{\rho}^{*-}(t-t_0) \end{pmatrix}.$$

The components of this matrix can be found by substituting into (9.29) the expression for $\hat{H}_1^*$ in the form of (9.28):

$$\hat{\rho}^{*+}(t-t_0) = \frac{N}{4} e^{\mathrm{i}\sqrt{3}\,\omega_1 \hat{I}_x{}^*(t-t_0)}\left(1 - \frac{e^2qQ}{2kT}\hat{I}_z{}^*\right) e^{-\mathrm{i}\sqrt{3}\,\omega_1 \hat{I}_x{}^*(t-t_0)},$$

$$\hat{\rho}^{*-}(t-t_0) = e^{\mathrm{i}\sqrt{3}\,\omega_1 \hat{I}_x{}^*(t-t_0)} \frac{N}{4}\left(1 + \frac{e^2qQ}{2kT}\hat{I}_z{}^*\right) e^{-\mathrm{i}\sqrt{3}\,\omega_1 \hat{I}_x{}^*(t-t_0)}. \tag{9.33}$$

As it was shown beforehand in Chap. 5 (see Eq. (5.76)), the transformation like $\exp(\mp \mathrm{i}\alpha \hat{I}_x{}^*)\hat{I}_z{}^* \exp(\pm \mathrm{i}\alpha \hat{I}_x{}^*)$ is equivalent to the rotation about the x-axis by the angle $\pm\alpha$. As a result one has

$$\hat{\rho}^{*\pm}(t-t_0) = \frac{N}{4}\left(1 \mp \frac{e^2qQ}{2kT}(\hat{I}_z{}^* \cos(\sqrt{3}\,\omega_1(t-t_0)) + \hat{I}_y{}^* \sin(\sqrt{3}\,\omega_1(t-t_0))\right). \tag{9.34}$$

The density matrix $\hat{\rho}^*$ describes the behavior of a spin system in the coordinate system where the quadrupole interactions are equal to zero and the system evolves under the effect of the Hamiltonian $\widehat{\mathcal{H}}_1^*$, which does not depend on time. This frame is similar to the rotating frame widely used when describing NMR phenomena. In essence at $\eta = 0$ each component of the density matrix $\hat{\rho}^{*\pm}$ describes the behavior of a single component of the magnetization $\pm M_0$ in two frames rotating in opposite directions. The expected values of magnetization components $\langle \hat{M}_z{}^{*\pm}(t-t_0)\rangle$ in these rotating frames can be found by the standard way:

$$\begin{aligned}\langle \hat{M}_z{}^{*\pm}(t-t_0)\rangle = \mathrm{Tr}(\hat{\rho}^{*\pm}(t-t_0)\hat{M}_z{}^{\pm}) &= \frac{\gamma\hbar N}{4}\,\mathrm{Tr}\,\hat{I}_z{}^* \\ &\mp \frac{\gamma\hbar e^2 qQN}{8kT}\Big(\mathrm{Tr}(\hat{I}_z{}^*)^2\cos(\sqrt{3}\,\omega_1(t-t_0)\Big) \\ &- \mathrm{Tr}(\hat{I}_z{}^*\hat{I}_y{}^*)\sin(\sqrt{3}\,\omega_1(t-t_0)).\end{aligned}$$

The operators $\hat{I}_z{}^*$ and $\hat{I}_y{}^*$ do not commute between each other, hence the diagonal elements corresponding to the product of these operators are equal to zero ($\mathrm{Tr}\,\hat{I}_z{}^*\hat{I}_y{}^* = 0$). Then

$$\langle \hat{M}_z{}^{*\pm}(t-t_0)\rangle = \mp\frac{\gamma\hbar e^2 qQN}{16kT}\cos(\sqrt{3}\,\omega_1 t_\mathrm{p}) = \mp M_0\cos(\sqrt{3}\,\omega_1 t_\mathrm{p}). \tag{9.35}$$

It is evident that the x-component of the expected magnetization in this frame is equal to zero as no operator determining the density matrix $\hat{\rho}^*$ in the form of (9.34) commutes with the operator $\hat{I}_x{}^*$:

$$\langle \hat{M}_x{}^{*\pm}(t-t_0)\rangle = \mathrm{Tr}(\hat{\rho}^{*\pm}(t-t_0)\hat{M}_x{}^{\pm}) = 0. \tag{9.36}$$

For the y-component of the expected magnetization one has

$$\begin{aligned}\langle \hat{M}_y{}^{\pm}(t-t_0)\rangle &= \mathrm{Tr}(\hat{\rho}^{*\pm}(t-t_0)\hat{M}_y{}^{\pm}) \\ &= \pm\frac{e^2 qQN\gamma\hbar\sqrt{3}}{16kT}\sin(\sqrt{3}\,\omega_1(t-t_0)) = \pm M_0\sin(\sqrt{3}\,\omega_1(t-t_0)).\end{aligned} \tag{9.37}$$

Thus, as it is seen from Eq. (9.32) for $I = 3/2$ and $\eta = 0$ the stationary magnetization of the unit volume can be represented by two components of the same value but of opposite directions: $\langle \hat{M}^{\pm}\rangle = \pm M_0$. Under field $B_x^{\approx}$ in the frame determined by the transformation (9.9), each component $\pm M_0$ is turning about the x-axis by the angle $\pm\sqrt{3}\,\omega_1 t_\mathrm{p}$ (see Eqs. (9.35)–(9.37)). Hence, to describe the behavior of each component it is possible to apply the vector model and Bloch equations.

The density matrix in the laboratory frame after the rf pulse can be found from Eq. (9.9):

$$\hat{\rho}(t_\mathrm{p}) = \exp\left(-\mathrm{i}\frac{\omega_0(t_\mathrm{p})}{2}\hat{I}_\mathrm{Q}\right)\hat{\rho}^*(t_\mathrm{p})\exp\left(\mathrm{i}\frac{\omega_0 t_\mathrm{p}}{2}\hat{I}_\mathrm{Q}\right). \tag{9.38}$$

After the pulse the system evolves only under the influence quadrupole interactions. Within $t_{\mathrm{p}} \leq t \leq \tau$ the density matrix, which is the solution of Eq. (9.7), can be found by the standard way:

$$\hat{\rho}(t - t_0) = \exp(-\mathrm{i}\widehat{\mathcal{H}}_{\mathrm{Q}}(t - t_0))\hat{\rho}(t_0)\exp(\mathrm{i}\widehat{\mathcal{H}}_{\mathrm{Q}}(t - t_0)).$$

However, it is to be noted that the initial time now corresponds to the termination of the rf pulse ($t_0 = t_{\mathrm{p}}$), and as the initial density matrix instead of $\rho(t_0)$ one has to use $\hat{\rho}(t - t_{\mathrm{p}})$ (see. Eq. (9.34)). Then

$$\hat{\rho}(t - t_{\mathrm{p}}) = \begin{pmatrix} \hat{\rho}^{+}(t - t_{\mathrm{p}}) & 0 \\ 0 & \hat{\rho}^{-}(t - t_{\mathrm{p}}) \end{pmatrix}, \tag{9.39}$$

where

$$\begin{aligned}
\hat{\rho}^{\pm}(t - t_{\mathrm{p}}) &= \frac{\gamma\hbar N}{4}\Bigg(1 \mp \frac{e^2qQ}{2kT} e^{\mp \mathrm{i}\omega_0 \hat{I}_z{}^*(t-t_{\mathrm{p}})}\Big(\hat{I}_z{}^* \cos(\sqrt{3}\,\omega_1 t_{\mathrm{p}}) \\
&\quad + \hat{I}_y{}^* \sin(\sqrt{3}\,\omega_1 t_{\mathrm{p}})\Big) e^{\pm \mathrm{i}\omega_0 \hat{I}_z{}^*(t-t_{\mathrm{p}})}\Bigg) \\
&= \frac{\gamma\hbar N}{4}\Bigg(1 \mp \frac{e^2qQ}{2kT}\Big(\hat{I}_z{}^* \cos(\sqrt{3}\,\omega_1 t_{\mathrm{p}}) + (\hat{I}_y{}^* \cos\omega_0(t - t_{\mathrm{p}}) \\
&\quad \mp \hat{I}_x{}^* \sin\omega_0(t - t_{\mathrm{p}}))\sin(\sqrt{3}\,\omega_1 t_{\mathrm{p}})\Big)\Bigg).
\end{aligned} \tag{9.40}$$

The density matrix (9.41) can be used to calculate components of the macroscopic magnetization, which defines intensity of the free induction signal after the first pulse:

$$\begin{aligned}
\langle \hat{M}_z{}^{\pm}(t - t_{\mathrm{p}})\rangle &= \mathrm{Tr}(\hat{\rho}^{\pm}(t - t_{\mathrm{p}})M_z^{\pm}) \\
&= \mp\frac{e^2qQ\gamma\hbar N}{16kT}\cos(\sqrt{3}\,\omega_1 t_{\mathrm{p}}) = \mp M_0 \cos(\sqrt{3}\,\omega_1 t_{\mathrm{p}}), \\
\langle \hat{M}_x{}^{\pm}(t - t_{\mathrm{p}})\rangle &= \mathrm{Tr}(\hat{\rho}^{\pm}(t - t_{\mathrm{p}})M_x^{\pm}) = \frac{e^2qQ\gamma\hbar N}{16kT}\sqrt{3}\,\sin(\sqrt{3}\,\omega_1 t_{\mathrm{p}})\sin\omega_0(t - t_{\mathrm{p}}) \\
&= M_0\sqrt{3}\,\sin(\sqrt{3}\,\omega_1 t_{\mathrm{p}})\sin\omega_0(t - t_{\mathrm{p}}), \\
\langle \hat{M}_y{}^{\pm}(t - t_{\mathrm{p}})\rangle &= \mathrm{Tr}(\hat{\rho}^{\pm}(t - t_{\mathrm{p}})M_y^{\pm}) = \pm\frac{e^2qQ\gamma\hbar N}{16kT}\sqrt{3}\,\sin(\sqrt{3}\,\omega_1 t_{\mathrm{p}})\cos\omega_0(t - t_{\mathrm{p}}) \\
&= \pm M_0\sqrt{3}\,\sin(\sqrt{3}\,\omega_1 t_{\mathrm{p}})\cos\omega_0(t - t_{\mathrm{p}}).
\end{aligned} \tag{9.41}$$

Thereby, after the rf pulse the z- and y-components of the total magnetization are equal to zero:

$$\langle \hat{M}_z(t - t_{\mathrm{p}})\rangle = \langle \hat{M}_z{}^{+}(t - t_{\mathrm{p}})\rangle + \langle \hat{M}_z{}^{-}(t - t_{\mathrm{p}})\rangle = 0,$$

$$\langle \hat{M}_y(t-t_p)\rangle = \langle \hat{M}_y{}^+(t-t_p)\rangle + \langle \hat{M}_y{}^-(t-t_p)\rangle = 0.$$

And the only expected magnetization along the x-axis, which coincides with the vector of the linearly polarized rf field (see Eq. 9.5)), differs from zero, as only for it the components $\langle \hat{M}_x{}^+(t-t_p)\rangle$ and $\langle \hat{M}_x{}^-(t-t_p)\rangle$ are of the same sign (see (9.41)). Then

$$\langle \hat{M}_x(t-t_p)\rangle = \frac{e^2qQ\gamma\hbar N}{8kT}\sqrt{3}\sin(\sqrt{3}\,\omega_1 t_p)\sin\omega_0(t-t_p). \tag{9.42}$$

It is the component $\langle M_x(t-t_p)\rangle$ which defines the free precession signal after the rf pulse. As it is seen from Eq. (9.42), the free precession amplitude does not change with time. That corresponds to the absence of relaxation precesses.

In NQR the main issue of the free induction decay is an inhomogeneity of electric field gradients due to various defects of the crystal lattice, thermal motion etc. If the EFG distribution obeys a normal law, the distribution of the NQR frequencies (ω_0) is described by the Gauss function:

$$g(\omega_0' - \omega_0) = \exp\left(-\frac{(\omega_0' - \omega_0)^2}{2\delta\omega^2}\right)(\sqrt{2\pi}\,\delta\omega)^{-1}, \tag{9.43}$$

where

$$\int_{-\infty}^{\infty} g(\omega_0' - \omega_0)\mathrm{d}\omega_0' = 1;$$

ω_0 is the mean distribution frequency; $\delta\omega$ is root mean square deviation.

The free induction decay after a pulse can be obtained by averaging the expression (9.42) over all frequencies with the distribution function (9.43):

$$\begin{aligned}\overline{\langle \hat{M}_x(t-t_p)\rangle} &= \int_{-\infty}^{\infty} g(\omega_0' - \omega_0)\langle \hat{M}_x(t-t_p)\rangle \mathrm{d}\omega_0' \\ &= \frac{e^2qQ\gamma\hbar N}{8kT}\sqrt{3}\sin(\sqrt{3}\,\omega_1 t_p)\sin(\omega_0(t-t_p))\exp\frac{-\delta\omega^2(t-t_p)^2}{2}.\end{aligned} \tag{9.44}$$

Similar to NMR the intensity of the free induction decay signal just after a rf pulse depends on the rotation angle of individual magnetization components $\langle \hat{M}_z{}^{\pm}\rangle$, before the pulse they direct along and against the principal axis of the EFG tensor. However, in NQR the turn rate of the magnetization is higher as compared to NMR, i.e. at the same value of the rf field individual components $\langle|\hat{M}_z{}^{*\pm}|\rangle$ turn on a wider angle. For the transition $\pm 3/2 \leftrightarrow \pm 1/2$ the turn rate increases $\sqrt{3}$ times as compared to NMR. According to calculations, for other transitions this increasing can be even greater. The maximal intensity of the free induction signal for the spin number 3/2 can be observed if $\sqrt{3}\,\omega_1 t_p = \pi/2$.

Thus, for $I = 3/2$ assuming that $\eta = 0$, $t_p < T_1, T_2$, $\Delta\omega t_p < 1$ one can make the following conclusions:

- The stationary macroscopic magnetization of a nuclear spin system in nonzero electric field gradients is equal to zero; however, there are two oppositely directed magnetizations and the behavior of each of them can be described using the vector model.
- It is more convenient to consider the spin subsystem evolution in two frames rotating in opposite directions, where the quadrupole interaction disappears and the system evolves under the action of a time independent disturbance described by the Hamiltonian $\widehat{\mathcal{H}}_1^*$.
- Rotation of each stationary magnetization component occurs under the action of the corresponding component of the rf field polarized in opposite directions.
- In NQR the rotation rate of each stationary magnetization component is higher as compared to NMR.
- In the laboratory frame after a rf pulse the magnetization vector directed along the x-axis appears. Its value is maximal if to excite the spin subsystem one applies a 90-degree pulse.
- The issue of the free induction decay is electric field gradient inhomogeneously distributed in the crystal.

Using the above-stated formalism it is easy to consider the evolution of the spin subsystem under the effect of two, three or more pulses.

9.3.2 Formation of the Free Induction Signal for $I = 1$ and $\eta \neq 0$

In order to determine the density matrix and the expected value of the magnetization it is reasonable to consider the matrix representation of spin operators and to introduce fictitious spin operators. In this purpose one uses eigenfunctions of the quadrupole interaction Hamiltonian. For $I = 1$ they are functions like (8.13). Following the formalism proposed by Osokin let us use the function $\mathrm{i}\Psi_2$ instead of the function Ψ_2 in the form of (8.13) and change the numeration of levels by substitution Ψ_2 for Ψ_3: $\Psi_3' = \mathrm{i}\Psi_2$, $\Psi_2' = \Psi_3$, $\Psi_1' = \Psi_1$. Matrix representations of the operators can be determined by using any set of wave functions, but the method proposed by Osokin, makes easy calculations of the free induction and echo signals. For $I = 3/2$ it is sufficient to introduce only three fictitious operators. In the case of $I = 1$ the situation is more complicated as the matrix representation, for example for the operator $\hat{I}_Q$, is of the form

$$\hat{I}_Q = 3\hat{I}_z^{\,2} - \hat{\vec{I}}^{\,2} = \begin{pmatrix} 1 & 0 & 0 \\ 0 & 1 & 0 \\ 0 & 0 & -2 \end{pmatrix}. \tag{9.45}$$

Equation (9.45) shows that squaring the operator $\hat{I}_Q$ does not result to the unit matrix, hence, the formalism stated in the previous section can not be applied. Nevertheless, introduction of fictitious operators is possible in this case as well. It is worth noting that the matrix (9.45) can be represented as a sum of two following matrices:

$$\hat{I}_{\mathrm{Q}} = 3\hat{I}_z^{\,2} - \hat{\vec{I}}^{2} = \begin{pmatrix} 1 & 0 & 0 \\ 0 & 1 & 0 \\ 0 & 0 & -2 \end{pmatrix} = \begin{pmatrix} 1 & 0 & 0 \\ 0 & 0 & 0 \\ 0 & 0 & -1 \end{pmatrix} + \begin{pmatrix} 0 & 0 & 0 \\ 0 & 1 & 0 \\ 0 & 0 & -1 \end{pmatrix}. \tag{9.46}$$

The block 2 × 2 matrices represent a matrix operator $\pm 2\hat{I}_z$ for the spin number 1/2. Let us note that operators for such a spin always describe a two-level system, i.e. relate two energy levels. The fictitious operators $\hat{I}_z{}^*$ also relate two energy levels corresponding to different values of eigenfunctions of the operator $\hat{I}_z$ for $I = 1$. In order to distinguish fictitious operators from real ones they are indicated by superscripts that point out which energy levels they relate. Thus, the second matrix in the expansion of the operator $\hat{I}_{\mathrm{Q}}$ represents a fictitious operator that relates the energy levels 1 and 2, it is designated as $2\hat{I}_z{}^{2\to 1}$, the first fictitious operator relates the energy levels 1 and 3, and is designates as $2\hat{I}_z{}^{1\to 3}$.

The matrix for the operator $\hat{I}_{\mathrm{Q}}$ is diagonal and of 3 × 3 size, that is why it is possible to represent it in the form of two matrices. In general case, if a matrix is not diagonal for its expansion over fictitious operators, nine operators are required. If the spin is equal to I, then $N = 2I + 1$ and each projection is described by a $N \times N$ matrix. To expand such an operator it is necessary to use N^2 matrix fictitious operators. Moreover, for expansion of unit operators described by the unit $N \times N$ matrix, one needs N fictitious operators more. Introduction of the latter, as a rule, is not necessary, but essentially simplifies calculations. As an example, let us write out all 12 fictitious operators for $I = 1$:

$$\begin{aligned}
&\hat{I}_x{}^{2\to 1} = \frac{1}{2}\begin{pmatrix} 0 & 0 & 0 \\ 0 & 0 & 1 \\ 0 & 1 & 0 \end{pmatrix}, \quad
\hat{I}_y{}^{2\to 1} = \frac{1}{2}\begin{pmatrix} 0 & 0 & 0 \\ 0 & 0 & -\mathrm{i} \\ 0 & \mathrm{i} & 0 \end{pmatrix}, \quad
\hat{I}_z{}^{2\to 1} = \frac{1}{2}\begin{pmatrix} 0 & 0 & 0 \\ 0 & 1 & 0 \\ 0 & 0 & -1 \end{pmatrix}; \\
&\hat{I}_x{}^{1\to 3} = \frac{1}{2}\begin{pmatrix} 0 & 0 & 1 \\ 0 & 0 & 0 \\ 1 & 0 & 0 \end{pmatrix}, \quad
\hat{I}_y{}^{1\to 3} = \frac{1}{2}\begin{pmatrix} 0 & 0 & \mathrm{i} \\ 0 & 0 & 0 \\ -\mathrm{i} & 0 & 0 \end{pmatrix}, \quad
\hat{I}_z{}^{1\to 3} = \frac{1}{2}\begin{pmatrix} -1 & 0 & 0 \\ 0 & 0 & 0 \\ 0 & 0 & 1 \end{pmatrix}; \\
&\hat{I}_x{}^{3\to 2} = \frac{1}{2}\begin{pmatrix} 0 & \mathrm{i} & 0 \\ -\mathrm{i} & 0 & 0 \\ 0 & 0 & 0 \end{pmatrix}, \quad
\hat{I}_y{}^{3\to 2} = \frac{1}{2}\begin{pmatrix} 0 & 1 & 0 \\ 1 & 0 & 0 \\ 0 & 0 & 0 \end{pmatrix}, \quad
\hat{I}_z{}^{3\to 2} = \frac{1}{2}\begin{pmatrix} 1 & 0 & 0 \\ 0 & -1 & 0 \\ 0 & 0 & 0 \end{pmatrix}; \\
&\hat{1}^{2\to 1} = \frac{1}{2}\begin{pmatrix} 0 & 0 & 0 \\ 0 & 1 & 0 \\ 0 & 0 & 1 \end{pmatrix}, \quad
\hat{1}^{1\to 3} = \frac{1}{2}\begin{pmatrix} 1 & 0 & 0 \\ 0 & 0 & 0 \\ 0 & 0 & 1 \end{pmatrix}, \quad
\hat{1}^{3\to 2} = \frac{1}{2}\begin{pmatrix} 1 & 0 & 0 \\ 0 & 1 & 0 \\ 0 & 0 & 0 \end{pmatrix}.
\end{aligned} \tag{9.47}$$

Numeration of energy levels is made in such a way that the projection of the operator $\hat{I}_z$, which is equal to zero, corresponds to the lower energy level (1), the projections equal to +1 and −1 correspond to the second and third energy levels, respectively.

Fictitious operators introduced in such a manner are called *single-transition operators*. One supposes that

$$\hat{I}_i^{r\to s} = -\hat{I}_i^{s\to r}$$
$$(i = x, y, z, r \to s = 1 \to 2, 1 \to 3 \text{ or } 3 \to 2).$$

The introduced fictitious operators possess a distinctive feature: the square of any operator is the unit matrix $\hat{1}^{2\to 1}$, $\hat{1}^{1\to 3}$ or $\hat{1}^{3\to 2}$. This fact allows us to represent any exponential operator $\exp \mathrm{i}\alpha \hat{I}_i^{r\to s}$ in the following form:

$$\exp \mathrm{i}\alpha \hat{I}_i^{r\to s} = 1 + \left(1 + \frac{(\alpha)^2}{2!} + \cdots\right)\hat{1}^{r\to s} - \hat{1}^{r\to s} + \mathrm{i}\left(\frac{\alpha}{1!} - \frac{(\alpha)^3}{3!} + \cdots\right)\hat{I}_i^{r\to s}. \tag{9.48}$$

Here terms in parentheses are the expansions in a Taylor series of the functions $\cos\alpha$ and $\sin\alpha$, respectively. Then

$$\exp(\pm \mathrm{i}\alpha \hat{I}_i^{r\to s}) = 1 + \hat{1}^{r\to s}(\cos\alpha - 1) \pm \mathrm{i}\hat{I}_i^{r\to s}\sin\alpha. \tag{9.49}$$

Single-transition operators have rather simple commutation relations. If operators relate identical energy levels ($r \to s = 1 \to 2$, or $1 \to 3$, or $3 \to 2$), then they have standard commutation relations:

$$[\hat{I}_i^{r\to s}, \hat{I}_j^{r\to s}] = \mathrm{i}\hat{I}_k^{r\to s}. \tag{9.50}$$

For operators relating different energy levels ($r \to t$ and $s \to t$), the commutation relations are following:

$$\begin{aligned}
&[\hat{I}_z{}^{r\to t}, \hat{I}_z{}^{s\to t}] = 0,\\
&[\hat{I}_x{}^{r\to t}, \hat{I}_x{}^{s\to t}] = [\hat{I}_y{}^{r\to t}, \hat{I}_y{}^{s\to t}] = \frac{\mathrm{i}\hat{I}_y{}^{r\to s}}{2},\\
&[\hat{I}_x{}^{r\to t}, \hat{I}_y{}^{s\to t}] = \mathrm{i}/2\hat{I}_x{}^{r\to s},\\
&[\hat{I}_x{}^{r\to t}, \hat{I}_z{}^{s\to t}] = -\mathrm{i}/2\hat{I}_y{}^{r\to t},\\
&[\hat{I}_y{}^{r\to t}, \hat{I}_z{}^{s\to t}] = \mathrm{i}/2\hat{I}_x{}^{r\to t}.
\end{aligned} \tag{9.51}$$

Changing in the order of the s, t and r, s indices leads to the sign change of fictitious operators, and hence to the sing change in the commutation relations (9.50) and (9.51). Operators relating different energy levels ($r \to t$ and $s \to u$)[4] always commutate with each other. Now any operator for $I = 1$ can be represented as a sum of fictitious operators with corresponding coefficients.

[4] For the spin number equal to 1 such operators do not exist.

As an example let us consider the Hamiltonian describing quadrupole interactions. If $\eta \neq 0$ then for $I = 1$ such a Hamiltonian in the principal axes frame of the EFG tensor is

$$\widehat{\mathcal{H}}_{\mathrm{Q}} = \frac{e^2qQ}{4}\left(3\hat{I}_z^2 - \hat{\vec{I}}^2 + \frac{\eta}{2}\left(I_+^2 + I_-^2\right)\right).$$

As it is seen from Eq. (9.46) the operator $\hat{I}_{\mathrm{Q}}$ can be expressed in terms of single-transition operators by the following way:

$$\hat{I}_{\mathrm{Q}} = 3\hat{I}_z{}^2 - \hat{\vec{I}}^2 = 2(\hat{I}_z{}^{2\to 1} + \hat{I}_z{}^{3\to 1}).$$

The matrix representation of the sum of the operators $\hat{I}_+^2$ and $\hat{I}_-^2$ for $I = 1$ can be written as

$$\hat{I}_+^2 = \begin{pmatrix} -1 & -\mathrm{i} & 0 \\ -\mathrm{i} & 1 & 0 \\ 0 & 0 & 0 \end{pmatrix}, \quad \hat{I}_-^2 = \begin{pmatrix} -1 & \mathrm{i} & 0 \\ \mathrm{i} & 1 & 0 \\ 0 & 0 & 0 \end{pmatrix}, \quad \hat{I}_+^2 + \hat{I}_-^2 = 2\begin{pmatrix} -1 & 0 & 0 \\ 0 & 1 & 0 \\ 0 & 0 & 0 \end{pmatrix}. \tag{9.52}$$

By comparing Eqs. (9.52) and (9.47), one has

$$\hat{I}_+^2 + \hat{I}_-^2 = 4\hat{I}_z{}^{2\to 3}.$$

Finally, in terms of one-spin operators the Hamiltonian for the quadrupole interaction is:

$$\widehat{\mathcal{H}}_{\mathrm{Q}} = \frac{e^2qQ}{2}\left(\hat{I}_z{}^{2\to 1} + \hat{I}_z{}^{3\to 1} + 2\eta\hat{I}_z{}^{2\to 3}\right). \tag{9.53}$$

By the similar way in terms of single-transition operators it is possible to express the Zeeman Hamiltonian:

$$\begin{aligned} \widehat{\mathcal{H}}_1 &= -\gamma\hbar\hat{I}_x B_1 \cos\omega(t - t_0) = -\omega_1\hbar\hat{I}_x \cos\omega(t - t_0) \\ &= 2\hbar\omega_1\hat{I}_x{}^{2\to 1}\cos\omega(t - t_0). \end{aligned} \tag{9.54}$$

It is to be noted that now both $\widehat{\mathcal{H}}_{\mathrm{Q}}$ and $\widehat{\mathcal{H}}_1$ contain only terms linear in spin operators that essentially simplify further calculations. Now it is possible to use fictitious operators to solve the Liouville equations (9.6) and (9.7) with the total Hamiltonian written as a sum of the Hamiltonian operators (9.53) and (9.54).

Following the above-stated formalism let us pass to another representation in which the Hamiltonian in the nonstationary Liouville equation does not depend on time:

$$\hat{\rho} = \exp\left(-\mathrm{i}\frac{\widehat{\mathcal{H}}_{\mathrm{Q}}}{\hbar}(t - t_0)\right)\hat{\rho}^* \exp\left(\mathrm{i}\frac{\widehat{\mathcal{H}}_{\mathrm{Q}}}{\hbar}(t - t_0)\right).$$

Then the Liouville equation can be written as

$$\mathrm{i}\hbar\frac{\partial\hat{\rho}^*}{\partial t} = [\widehat{\mathcal{H}}_1^*\hat{\rho}^*],$$

where

$$\widehat{\mathcal{H}}_1^* = \exp\left(\mathrm{i}\frac{\widehat{\mathcal{H}}_\mathrm{Q}}{\hbar}(t-t_0)\right)\widehat{\mathcal{H}}_1\exp\left(-\mathrm{i}\frac{\widehat{\mathcal{H}}_\mathrm{Q}}{\hbar}(t-t_0)\right).$$

If instead of the Hamiltonian operators $\widehat{\mathcal{H}}_1$ and $\widehat{\mathcal{H}}_\mathrm{Q}$ one uses their representation through fictitious operators, then taking into account Eq. (9.49) and omitting fast oscillating terms it is easy to obtain the following expression for the Hamiltonian $\widehat{\mathcal{H}}_1^*$:

$$\widehat{\mathcal{H}}_1^* = \omega_1\hat{I}_x{}^{2\to1},$$

and by the end of the first pulse the density matrix can be written as:

$$\hat{\rho}^*(t-t_0) = \frac{1}{9}\left(\hat{1}^{1\to2}\hat{1}^{1\to3} + \frac{e^2qQ}{4kT}(\hat{I}_z{}^{1\to3} - \hat{I}_z{}^{1\to2})\cos\omega_1 t_\mathrm{p}\right.$$
$$\left. + \hat{I}_x{}^{1\to3}\sin\omega_1 t_\mathrm{p}\cos\omega t + \hat{I}_y{}^{2\to3}\sin\omega_1 t_\mathrm{p}\sin\omega t - 2\eta\hat{I}_x{}^{2\to3}\right).$$

Then the signal intensity just after the first pulse is

$$\langle\hat{M}_y{}^{\pm}(t-t_0)\rangle = \mathrm{Tr}(\hat{\rho}^{*\pm}(t-t_0)\hat{M}_y{}^{\pm}) = \pm\frac{e^2qQN}{4kT}\frac{1+\eta}{1+\eta^2}\sin\left(\frac{1+\eta}{1+\eta^2}\sqrt{2}\,\omega_1(t-t_0)\right)$$
$$= \pm M_0\sin\left(\frac{1+\eta}{1+\eta^2}\sqrt{2}\,\omega_1(t-t_0)\right).$$

Thus, the formalism of fictitious spin operators (or single-transition operators) can be successfully used for calculation of free induction signals. It works well even in cases of more complicated Hamiltonian, for example, if one takes into account dipole–dipole interactions in the spin system and/or magnetic shielding (chemical shift). That is why this formalism is usually applied to calculate signals after complex pulse sequences.

9.4 NQR Spin Echo

Let us consider a system of quadrupole nuclei placed into an electric crystal field with nonzero gradients and let us apply two rf pulses of durations $t_{\mathrm{p}1}$ and $t_{\mathrm{p}2}$ with the same modulation frequency (ω) and with the space-time interval between them τ (see Fig. 9.3). As it was shown previously for the free induction signal, if $\Delta\omega t_p < 1$ the density matrix by the end of the first pulse is related to the density matrix before the pulse action by the transformations (9.29) and (9.9):

$$\hat{\rho}(t_{p1}) = \hat{U}(t_{p1})\hat{\rho}(t_0)\hat{U}^{-1}(t_{p1}),$$

where $\hat{U}(t_{p1})$ is the density matrix evolution operator (or propagator) during the pulse:

$$\hat{U} = \exp\left(-\frac{i}{\hbar}\mathcal{H}_Q t_{p1}\right), \quad \hat{U}^*(t_{p1}) = \exp\left(-\frac{i}{\hbar}\widehat{\mathcal{H}}_1^* t_{p1}\right).$$

The density matrix by the beginning of the next pulse is related to the density matrix before the first pulse by the following way:

$$\hat{\rho}(t - t_{p1}) = \hat{S}\hat{\rho}(t_{p1})\hat{S}^{-1} = \hat{S}\hat{U}\hat{\rho}(t_0)\hat{U}^{-1}\hat{S}^{-1},$$

where $\hat{S}$ is the density matrix evolution operator after the first pulse:

$$\hat{S} = \exp\left(-\frac{i}{\hbar}\widehat{\mathcal{H}}_Q(t - t_{p1})\right).$$

It is obvious that the total evolution operator for density matrix after the n-th pulse ($\hat{R}_n$), can be written as a product of individual evolution operators:

$$R_n = \hat{S}_n\hat{U}_n\hat{S}_{n-1}\hat{U}_{n-1}\cdots\hat{S}_1\hat{U}_1,$$

where the index n numbers the evolution operators after the n-th pulse. Let us note that it is prohibited to change the order of the evolution operators as they may not commute with each other.

Using the evolution operator $R_{n=2}$ the density matrix after the second pulse can be written by the following way:

$$\hat{\rho}(t - \tau - t_{p2} - t_{p1}) = \hat{R}_2\hat{\rho}(t_0)\hat{R}_2^{-1} = \hat{S}_2\hat{U}_2\hat{S}_1\hat{U}_1\hat{\rho}(t_0)\hat{U}_1^{-1}\hat{S}_1^{-1}\hat{U}_2^{-1}\hat{S}_2^{-1}, \quad (9.55)$$

and the expected value of any operator, e.g. of the operator $\hat{I}_x$, can be calculated by the formula

$$\langle\hat{I}_x(t - \tau - t_{p2} - t_{p1})\rangle = \mathrm{Tr}(\hat{\rho}(t - \tau - t_{p2} - t_{p1})\hat{I}_x) = \mathrm{Tr}(\hat{R}_2\hat{\rho}(t_0)\hat{R}_2^{-1}\hat{I}_x).$$

Thereby if the matrix representations of the individual evolution operators are known in order to determine the expected magnetization value it is sufficient to multiply the corresponding density matrices.

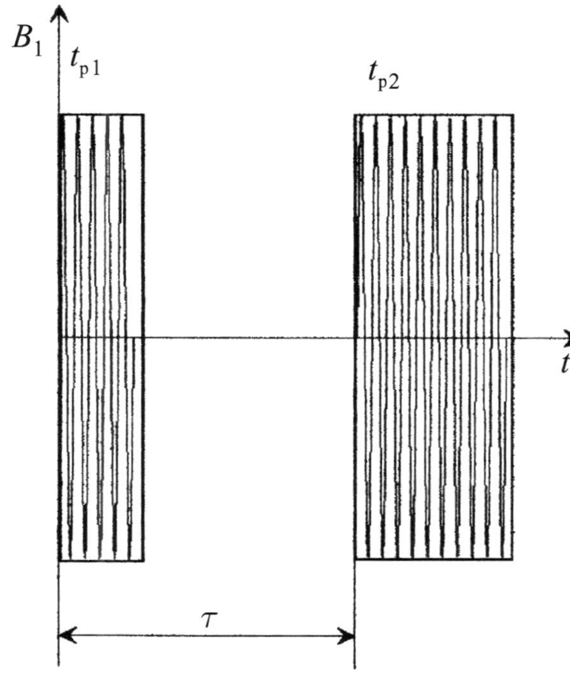

Fig. 9.3 The echo pulse sequence

9.4.1 Echo Signal Formation for $I = 3/2$ and $\eta = 0$

Let us consider the evolution of a spin subsystem for the spin number equal to 3/2 and $\eta = 0$ after two rf pulses. The matrix representations of the evolution operators have been obtained before (see Eqs. (9.33) and (9.40)). The matrix

$$\hat{U}_1^* \rightarrow \begin{pmatrix} e^{i\sqrt{3}\,\omega_1 \hat{I}_x{}^*(t-t_0)} & 0 \\ 0 & e^{i\sqrt{3}\,\omega_1 \hat{I}_x{}^*(t-t_0)} \end{pmatrix} \tag{9.56}$$

corresponds to the evolution operator $\hat{U}_1^*$ during the first pulse, i.e. within $t_0 \geq t \geq t_{p1}$, whereas the matrix

$$\hat{S}_1 \rightarrow \begin{pmatrix} e^{-i\frac{\omega_Q}{2}\hat{I}_z{}^*(t-t_{p1})} & 0 \\ 0 & e^{i\frac{\omega_Q}{2}\hat{I}_z{}^*(t-t_{p1})} \end{pmatrix}. \tag{9.57}$$

corresponds to the evolution operator $\hat{S}_1$ after the first pulse, i.e. within $t_{p1} \geq t \geq \tau$.

Let us note that in the evolution operators (9.56), (9.57) the time reference point varies with respect to what the time point this operator acts. The operator $\hat{U}_1$ acts within the time interval $t_0 \geq t \geq t_{p1}$ and for it the time reference point is the time point $t = t_0$ corresponding to the beginning of the first pulse. At $t = t_0$ exponents in

Eq. (9.56) are equal to 1 and the operator $\hat{U}_1$ does not transform the density matrix at all. By the end of the first pulse $t - t_0 = t_{p1}$. The operator $\hat{S}_1$ acts within the time interval $t_{p1} \geq t \geq \tau$ and for it the time reference point is $t = t_{p1}$, which corresponds to the end of the first pulse. At $t = t_{p1}$ exponents in Eq. (9.57) are equal to 1, hence the $\hat{S}_1$ does not transform the density matrix. By the beginning of the next pulse the time in the exponents defining the matrix of the operator $\hat{S}_1$ is equal to $t = \tau - t_{p1}$ (see Fig. 9.3, which represents the echo pulse sequence). Two other evolution operators $\hat{U}_2^*$ and $\hat{S}_2$ acting within the time intervals $\tau \geq t \geq \tau + t_{p2}$ and $\tau + t_{p2} \geq t$, respectively, are

$$\hat{U}_2^* \rightarrow \begin{pmatrix} e^{i\sqrt{3}\omega_1 \hat{I}_x{}^*(t-\tau)} & 0 \\ 0 & e^{i\sqrt{3}\omega_1 \hat{I}_x{}^*(t-\tau)} \end{pmatrix},$$

$$\hat{S}_2 \rightarrow \begin{pmatrix} e^{-i\frac{\omega_0}{2}\hat{I}_z{}^*(t-\tau-t_{p2})} & 0 \\ 0 & e^{i\frac{\omega_0}{2}\hat{I}_z{}^*(t-\tau-t_{p2})} \end{pmatrix}.$$

Now it is easy to define the density matrix (9.55) describing the spin behavior after the second rf pulse taking into account that the density matrix before the second pulse is known (see Eqs. (9.39) and (9.40)):

$$\hat{\rho}(t - \tau - t_{p2} - t_{p1}) = \hat{R}_2 \hat{\rho}(t_0) \hat{R}_2^{-1} = \hat{S}_2 \hat{U}_2 \hat{\rho}(\tau - t_{p1}) \hat{U}_2^{-1} \hat{S}_2^{-1},$$

$$\hat{\rho}(\tau - t_{p1}) = \begin{pmatrix} \hat{\rho}^+(\tau - t_{p1}) & 0 \\ 0 & \hat{\rho}^-(\tau - t_{p1}) \end{pmatrix},$$

The components of the density matrix before the second pulse can be obtained by substitution t for τ in Eq. (9.40):

$$\hat{\rho}^{\pm}(\tau - t_{p1}) = \frac{\gamma\hbar N}{4}\left(1 \mp \frac{e^2 Qq}{2kT}\left(\hat{I}_z{}^* \cos(\sqrt{3}\,\omega_1 t_{p1}) + (\hat{I}_y{}^* \cos\omega_0(\tau - t_{p1}) \right.\right.$$
$$\left.\left. \mp \hat{I}_x{}^* \sin\omega_0(\tau - t_{p1}))\sin(\sqrt{3}\,\omega_1 t_{p1})\right)\right). \quad (9.58)$$

The expected values of the $\hat{M}_x$ magnetization component after the second pulse calculated by means of the density matrix (9.55) are

$$\langle \hat{M}_x(t - t_p)\rangle = \frac{e^2 qQ\gamma\hbar N \cdot \sqrt{3}}{8kT}\left(\sin(\sqrt{3}\,\omega_1 t_{p1})\cos^2\frac{\sqrt{3}\,\omega_1 t_{p2}}{2}\sin\omega_0(t - t_{p1})\right.$$
$$- \sin(\sqrt{3}\,\omega_1 t_{p1})\sin^2\frac{\sqrt{3}\,\omega_1 t_{p2}}{2}\sin\omega_0(t - 2\tau - 2t_{p2})$$
$$\left. + \sin(\sqrt{3}\,\omega_1 t_{p1})\cos(\sqrt{3}\,\omega_1 t_{p2})\sin\omega_0(t - \tau - t_{p2})\right).$$

Table 9.1 Expressions to calculate the spin echo amplitude for various I value

Nuclear spin	Transition	Echo amplitude
1	$0 \leftrightarrow \pm 1$	$\frac{e^2qQ\gamma\hbar N\cdot\sqrt{2}}{4kT}\sin(\sqrt{2}\,\omega_1 t_{p1})\sin^2\frac{\sqrt{2}\,\omega_1 t_{p2}}{2}$
3/2	$\pm 1/2 \leftrightarrow \pm 3/2$	$\frac{e^2qQ\gamma\hbar N\cdot\sqrt{3}}{8kT}\sin(\sqrt{3}\,\omega_1 t_{p1})\sin^2\frac{\sqrt{3}\,\omega_1 t_{p2}}{2}$
2	$0 \leftrightarrow \pm 1$	$\frac{e^2qQ\gamma\hbar N\cdot\sqrt{6}}{40kT}\sin(\sqrt{6}\,\omega_1 t_{p1})\sin^2\frac{\sqrt{6}\,\omega_1 t_{p2}}{2}$
2	$\pm 1 \leftrightarrow \pm 2$	$\frac{3e^2qQ\gamma\hbar N}{20kT}\sin(2\omega_1 t_{p1})\sin^2(\omega_1 t_{p2})$
5/2	$\pm 1/2 \leftrightarrow \pm 3/2$	$\frac{e^2qQ\gamma\hbar N\cdot\sqrt{8}}{40kT}\sin(\sqrt{8}\,\omega_1 t_{p1})\sin^2\sqrt{2}\,\omega_1 t_{p2}$
5/2	$\pm 3/2 \leftrightarrow \pm 5/2$	$\frac{e^2qQ\gamma\hbar N\cdot\sqrt{5}}{20kT}\sin(\sqrt{5}\,\omega_1 t_{p1})\sin^2\frac{\sqrt{5}\,\omega_1 t_{p2}}{2}$

If one takes into account the influence of EFG inhomogeneity, as it was done to calculate the free induction signal, then

$$\begin{aligned}\overline{\langle\hat{M}_x(t)\rangle} = {} & \frac{e^2qQ\gamma\hbar N\cdot\sqrt{3}}{8kT} \\ & \times\Bigg(\sin(\sqrt{3}\,\omega_1 t_{p2})\cos^2\frac{\sqrt{3}\,\omega_1 t_{p2}}{2}\sin\omega_0(t-t_{p1})\exp(-(t-t_{p1})^2\delta\omega^2) \\ & -\sin(\sqrt{3}\,\omega_1 t_{p1})\sin^2\frac{\sqrt{3}\,\omega_1 t_{p2}}{2}\sin\omega_0(t-2\tau)\exp(-(t-2\tau)^2\delta\omega^2) \\ & +\sin(\sqrt{3}\,\omega_1 t_{p1})\cos(\sqrt{3}\,\omega_1 t_{p2})\sin\omega_0(t-\tau-t_{p1})\exp(-(t-\tau-t_{p2})^2\delta\omega^2)\Bigg).\end{aligned} \tag{9.59}$$

As it is seen from Eq. (9.59) the second and fourth lines describe the free induction signals after the first and the second pulses, whereas the third line corresponds to the echo signal at $t = 2\tau$. In NQR when forming the echo signal, similar to formation of the free induction signal, the both components of the rf field polarized in opposite directions are important, and the both magnetization components rotating in different directions in the coordinate system defined by Eq. (9.9) contribute to the echo signal. As before the vector of the macroscopic magnetization of the unit volume directs along the x-axis and there is no any macroscopic nuclear magnetization in any other direction, regardless of the amplitude and duration of the applied rf pulse. The echo shape is described by the Gaussian function, if the EFG distribution obeys a normal law.

Spin echo amplitudes for other nuclear spin values can be obtained by the similar way (see Table 9.1). As it is seen the rotation rate of the magnetization increases with the nuclear spin value, and it is greater the lower the transition frequency is. So, for example, for $I = 2$ the transition $\pm 1 \leftrightarrow \pm 2$ corresponds to the higher frequency and the rotation rate of magnetization for this transition is lower than for the transition $0 \leftrightarrow \pm 1$. Increasing the rotation rate means that in order to turn the magnetization vector by angle 90 or 180° at the same value of the radiofrequency field it is necessary to decrease the pulse duration.

9.4.2 Free Induction and Echo in a Weak Magnetic Field

Let us suppose that a system of quadrupole nuclei is in a weak stationary magnetic field. Let EFG at the nucleus sites differs from zero. If the Zeeman interaction of the nuclear spin with magnetic fields is much weaker than the quadrupole interactions such a field is called a "weak magnetic field".

Let us consider a case when $\eta = 0$ and the stationary magnetic field vector makes the angle θ with the principal z axis of the EFG tensor. The Hamiltonian of such a system is

$$\begin{aligned} \widehat{\mathcal{H}}_{\mathrm{s}} &= \widehat{\mathcal{H}}_{\mathrm{Q}} + \widehat{\mathcal{H}}_{\mathrm{Z}} && \text{without a rf pulse,} \\ \widehat{\mathcal{H}}_{\mathrm{p}} &= \widehat{\mathcal{H}}_{\mathrm{Q}} + \widehat{\mathcal{H}}_{\mathrm{Z}} + \widehat{\mathcal{H}}_1 && \text{with a rf pulse.} \end{aligned} \tag{9.60}$$

The Hamiltonians $\widehat{\mathcal{H}}_{\mathrm{Q}}$, $\widehat{\mathcal{H}}_{\mathrm{Z}}$ and $\widehat{\mathcal{H}}_1$ in the principal axes frame of the EFG tensor can by written as

$$\begin{aligned} \widehat{\mathcal{H}}_{\mathrm{Q}} &= \frac{e^2 q Q}{4I(2I-1)}(3\hat{I}_z^2 - \hat{I}^2), \\ \widehat{\mathcal{H}}_{\mathrm{Z}} &= -\gamma B_0\left((\hat{I}_z)\cos\theta + \frac{1}{2}(\hat{I}_+ \exp(-\mathrm{i}\phi) + \hat{I}_- \exp(\mathrm{i}\phi))\sin\theta\right), \\ \widehat{\mathcal{H}}_1 &= -\gamma\hbar B_1 \hat{I}_x \cos\omega(t - t_0). \end{aligned}$$

As it was shown before, magnetic field removes the magnetic quantum number degeneracy and commixes wave functions with different m number. In order to define the density matrices in presence of rf pulses ($\hat{\rho}$) and without them ($\hat{\rho}_0$) it is possible to use fictitious operators again. For the transition $\pm 3/2 \leftrightarrow \pm 1/2$ and any spin number the expected magnetization value ($\langle M_x(t - t_{\mathrm{p}})\rangle$) after the first pulse is written as follows:

$$\begin{aligned} \overline{\langle \hat{M}_x(t - t_{\mathrm{p}1})\rangle} &= \frac{3e^2 q Q \gamma\hbar N a}{2I(2I-1)(2I+1)kT} \sin\left(\sqrt{I(I+1) - m(m-1)}\,\omega_1 t_{\mathrm{p}1}\right) \\ &\quad \times \sin\omega_0(t - t_{\mathrm{p}}) \exp(-(t - t_{\mathrm{p}})^2 \delta\omega^2)\left(\frac{f+1}{f}\cos\left(\frac{3-f}{2}(\omega_z\cos\theta)t\right)\right. \\ &\quad \left. + \frac{f-1}{f}\cos\left(\frac{3+f}{2}(\omega_z\cos\theta)t\right)\right), \end{aligned} \tag{9.61}$$

$$a = \sqrt{I(I+1) - m(m-1)}, \quad f = \sqrt{1 + \left(\frac{2I-1}{2}\right)^2 \mathrm{tg}^2\theta}.$$

By comparing Eqs. (9.44) and (9.61) for $I = 3/2$ one can see that the maximal intensity of the free induction signal depends on the applied magnetic field and its direction with respect to the principal axis of the EFG tensor. The time evolution of the free induction signal has oscillations with amplitude and frequency defined by the value and direction of the applied field if $\delta\omega > \omega_z$. If the magnetic field vector directs along the principal axis of the EFG tensor ($\theta = 0$) or is perpendicular to it ($\theta = 90°$), then $f = 1$ and the expression (9.61) for the spin equal to 3/2 can be rewritten as follows:

$$\overline{\langle \hat{M}_x(t - t_{p1})\rangle} = \frac{\sqrt{3}e^2 qQ\gamma\hbar N}{8kT} \sin\left(\sqrt{I(I+1) - m(m-1)}\,\omega_1 t_{p1}\right) \times \sin\omega_0(t - t_{p1})\exp(-(t - t_{p1})^2\delta\omega^2) 2\cos\omega_z t.$$

After the second pulse the echo signal is formed. For the transition $\pm 3/2 \leftrightarrow \pm 1/2$ the expected magnetization value $\overline{\langle \hat{M}_x(t - t_p)\rangle}$, which defines the form and intensity of the echo signal, can be calculated using the following formula:

$$\begin{aligned}\overline{\langle \hat{M}_x(t - t_p)\rangle} &= \frac{3e^2 qQ\gamma\hbar Na}{2I(2I-1)(2I+1)kT}\exp(-(t - 2\tau)^2\delta\omega^2)\\ &\times \sin\left(\sqrt{I(I+1) - m(m-1)}\,\omega_1 t_{p1}\right)\sin^2\left(\sqrt{I(I+1) - m(m-1)}\frac{\omega_1 t_{p2}}{2}\right)\\ &\times \sin\omega_0(t - 2\tau)\left[\left(\frac{f+1}{2f}\right)^2\cos\left(\frac{3-f}{2}(\omega_z\cos\theta)(t - 2\tau)\right)\right.\\ &+ \left(\frac{f-1}{2f}\right)^2\cos\left(\frac{3+f}{2}(\omega_z\cos\theta)(t - 2\tau)\right)\\ &+ \frac{f^2-1}{2f^2}\left(\cos\left(\frac{3\omega_z\cos\theta}{2}(t - 2\tau)\right)\cos\frac{f\omega_z\cos\theta}{2}t\right.\\ &\left.\left.+ \cos\frac{3\omega_z\cos\theta}{2}t\cos\left(\frac{f\omega_z\cos\theta}{2}(t - 2\tau)\right) - \cos\frac{3\omega_z\cos\theta}{2}t\cos\frac{f\omega_z\cos\theta}{2}t\right)\right].\end{aligned} \tag{9.62}$$

By comparing Eqs. (9.59) and (9.62) for $I = 3/2$ one can see that the maximal echo intensity also depends on the applied external magnetic field value and its direction with respect to the principal axis of the EFG tensor. If $\delta\omega > \omega_z$ the shape of the echo signal also exhibits oscillations, whose amplitude and frequency are governed by both value and direction of the applied field. In Fig. 9.4 one plots the shape of the echo signal at different orientations of the external magnetic field vector.

If the magnetic field vector directs along the EFG tensor principal axis ($\theta = 0$) or is perpendicular to it ($\theta = 90°$) then $f = 1$ and the expression (9.62) for the spin equal to 3/2 becomes essentially simpler:

$$\begin{aligned}\overline{\langle \hat{M}_x(t - t_p)\rangle} &= \frac{\sqrt{3}e^2 qQ\gamma\hbar N}{8kT}\exp(-(t - 2\tau)^2\delta\omega^2)\\ &\times \sin\left(\sqrt{I(I+1) - m(m-1)}\,\omega_1 t_{p1}\right)\sin^2\left(\sqrt{I(I+1) - m(m-1)}\frac{\omega_1 t_{p2}}{2}\right)\\ &\times \sin\omega_0(t - 2\tau)\cos\omega_z(t - 2\tau).\end{aligned} \tag{9.63}$$

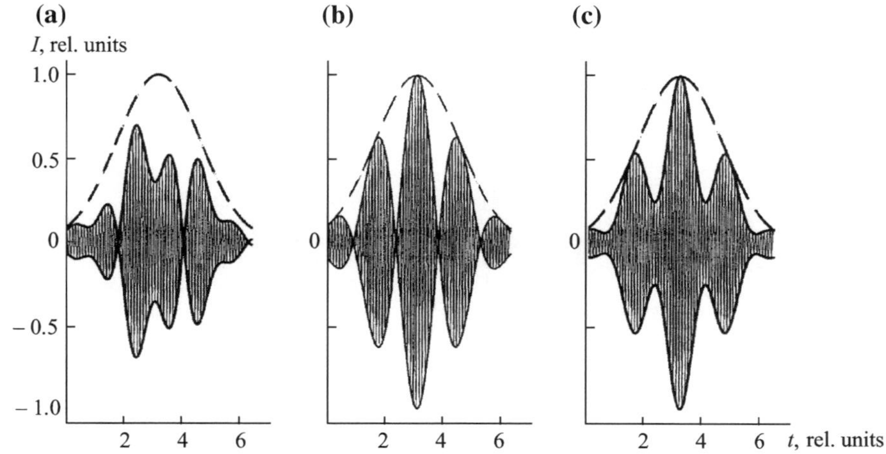

Fig. 9.4 The echo signal for $I = 3/2$ at various orientations of the external magnetic field relative to the principal z-axis of the EFG tensor: $\theta = 45°$ (**a**), $\theta = 0$ (**b**), the magnetic field vector directs along the generatrix of the zero splitting cone (**c**). The *dash lines* correspond to the echo shape without any magnetic field

As it is seen from Eq. (9.63) the form of the echo signal is modulated by the harmonic function $\cos \omega_z(t - 2\tau)$ (see Fig. 9.4b).

If the magnetic field vector directs along the generatrix of the zero splitting cone (see (8.31), then $f = 3$ and the free induction signal can be calculated by the following formula:

$$\overline{\langle \hat{M}_x(t - t_p)\rangle} = \frac{\sqrt{3}e^2qQ\gamma\hbar N}{8kT} \sin\left(\sqrt{I(I+1) - m(m-1)}\,\omega_1 t_{p1}\right) \times \sin\omega_0(t - t_{p1}) \exp(-(t - t_{p1})^2\delta\omega^2)\left(\frac{4}{3} + \frac{2\cos\omega_z t}{3}\right).$$

The echo envelope is modulated again by a harmonic function, however the modulation is not 100 % (see Fig. 9.4c). Thereby by measuring the orientation dependence of the echo shape it is possible to obtain the information about direction of the principal axes of the EFG tensor that is much easier as compared to stationary methods.

Another way to obtain this information is to record the maximal echo intensity (I) versus the pulse delay (τ). For an arbitrary orientation of applied magnetic field relative to the principal axes of the EFG tensor this dependence can be obtained from Eq. (9.62) at $t = 2\tau$:

$$I(\tau) \sim \frac{f^2 - 1}{2f^2}\left[\cos\frac{f\omega_z\cos\theta}{2}\tau + \cos\frac{3\omega_z\cos\theta}{2}\tau - \cos\frac{3\omega_z\cos\theta}{2}\tau\cos\frac{f\omega_z\cos\theta}{2}\tau\right].$$

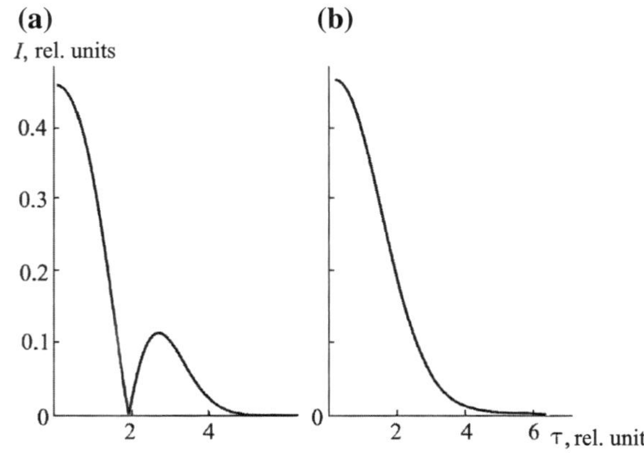

Fig. 9.5 Spin-echo amplitude as a function of the pulse delay (τ) at two different orientations of the external magnetic field: $\theta = 30°$ (**a**), $\theta = 90°$ (**b**)

Figure 9.5 represents two examples of such a dependence. As it is seen it can be nonmonotonic.

If $I > 3/2$ and $\eta = 0$ then for the transitions $\pm 3/2 \leftrightarrow \pm 5/2$ and higher neither free induction, nor echo signals are modulated. Decreasing of the echo amplitude with τ increasing is also gradual. However, if $\eta \neq 0$ the modulation of the echo signal is observed for higher transitions as well. This allows us to determine if the asymmetry parameter is equal to zero or not without any changing of the absorption frequency.

9.5 Two-Frequency NQR Spin Echo

In NQR if $I > 3/2$ the arrangement of energy levels is not equidistant. That means that to excite the system one can apply pulses with different modulation frequencies and to obtain a so-called *two-frequency echo*:

$$B^{\approx}(t) = B_1(\cos\omega_1 t + \cos\omega_2 t). \tag{9.64}$$

The frequencies ω_1 and ω_2 should be close or equal to the frequencies of one or another transition: for example ω_1 is close to the frequency of the transitions ($\pm 1/2 \leftrightarrow \pm 3/2$), whereas ω_2 is close to the frequency of the transitions $\pm 3/2 \leftrightarrow \pm 5/2$. Hence, the signals of free induction and echo are also observed at different frequencies.

To calculate the free induction and echo signals it is possible to apply the methodology stated above. Skipping the calculation details, let us discuss the final formulas describing behavior of the macroscopic magnetization after one and two pulses for $I = 5/2$.

The frequencies of the transitions $\pm 1/2 \leftrightarrow \pm 3/2$ and $\pm 3/2 \leftrightarrow \pm 5/2$ differ in two times, hence, Eq. (9.64) can be written in the following form

$$B^{\approx}(t) = B_1(\cos \omega t + \cos 2\omega t).$$

The pulse durations for different frequencies are chosen to be equivalent.

Let us consider the case when (i) the vector of radiofrequency field is perpendicular to the principal axis of the EFG tensor; (ii) the time space between two pulses is much greater the spin–spin and spin–lattice relaxation times ($\tau \ll T_1, T_2$); (iii) nuclei do not interact with each other; (iv) EFG tensors at the nuclear sites are axisymmetric ($\eta = 0$). In this case the signal of free precession $I_{\text{pr}1}$ at frequency ω_0 is

$$I_{\text{pr}1} \sim (2 \sin \alpha \cos \alpha - 15 \sin \alpha) \sin \omega_0 t, \tag{9.65}$$

and at frequency $2\omega_0$ it is proportional to

$$I_{\text{pr}2} \sim (-\sin \alpha \cos \alpha - 12 \sin \alpha) \sin 2\omega_0 t. \tag{9.66}$$

Here α is the rotation angle of the magnetization vector,

$$\alpha = \frac{\gamma B_1 \cdot \sqrt{13}}{2} t_{\text{p}1}. \tag{9.67}$$

At one-frequency excitation the rotation angle of the magnetization for each transition considered here is defined as follows:

$$\text{for } \omega_0: \qquad \alpha = \frac{\gamma B_1 \cdot \sqrt{8}}{2} t_{\text{p}1}, \tag{9.68}$$

$$\text{for } 2\omega_0: \qquad \alpha = \frac{\gamma B_1 \cdot \sqrt{5}}{2} t_{\text{p}1}. \tag{9.69}$$

Hence, at two-frequency excitation in order to turn the magnetization vector by a given angle one needs a shorter pulse for a fixed value of the rf field. As it follows from Eqs. (9.65) and (9.68), the maximum of the free induction signal is observed at the following α angles:

$$\text{for } \omega_0: \qquad \alpha = 98.5487^\circ \tag{9.70}$$

$$\text{for } 2\omega_0: \qquad \alpha = 85.9437^\circ. \tag{9.71}$$

Thereby, generally speaking for different transitions one should apply different pulse durations. However, difference in signal intensities at α values indicated in Eq. (9.70) and at $\alpha = 90^\circ$ is not greater then 1 %. Nevertheless, one should remember that the integral intensities of applied pulses ($\gamma B_1 t_{\text{p}1}$), as it is seen from Eqs. (9.67) and (9.68), are essentially different.

The intensity of the echo signals ($I_{\text{e}1}$) at frequency ω_0 can be written as follows:

$$I_{\text{e}1} \sim A \sin \omega_0 (t - 2\tau) + B \sin \omega_0 (t - 3\tau) + C \sin \omega_0 (t - 4\tau). \tag{9.72}$$

According to calculations made by using the formula (9.72) at two-frequency excitation several echo signals can be observed:

$$\begin{aligned} &\text{at} \quad t = 2\tau: \quad && \text{usual echo,} \\ &\text{at} \quad t = 3\tau: \quad && \text{additional echo,} \\ &\text{at} \quad t = 4\tau: \quad && \text{“forbidden” echo.} \end{aligned}$$

The amplitudes of these signals depend on the angles $\alpha_{1,2} = \sqrt{13}\gamma B_1 t_{p1,2}/2$ by the following way:

$$\begin{aligned} A &= (2\sin\alpha_1\cos\alpha_1 - 15\sin\alpha_1)\sin^2\alpha_2, \\ B &= (\sin\alpha_1\cos\alpha_1 + 12\sin\alpha_1)(\cos\alpha_2 - 1)\cos\alpha_2, \\ C &= (2.66\cos\alpha_1(\cos\alpha_1 - 1) + 4(\cos\alpha_1 - 1) + 0.585\sin^2\alpha_1) \\ &\quad \times (\cos\alpha_2 - 1)\sin\alpha_2. \end{aligned} \tag{9.73}$$

Here α_2 is defined by Eq. (9.67).

As it is seen from Eq. (9.73), to record the maximal spin echo signal at different time moments one needs pulses of different duration (at the same value of their amplitude). For the maximal value of the free induction signal after the first pulse the echo signal is

$$\begin{aligned} A(t = 2\tau) &\sim \sin^2\alpha_2, \\ B(t = 3\tau) &\sim \cos\alpha_2(\cos\alpha_2 - 1), \\ C(t = 4\tau) &\sim \sin\alpha_2(\cos\alpha_2 - 1). \end{aligned}$$

Thereby in order to obtain the maximal intensity of the usual echo at 2τ the angle α_2 should be equal to 90°, i.e. the duration of the second pulse is about the duration of the first one. At this condition the additional echo signal (at $t = 3\tau$) is not observed, the intensity of the "forbidden" echo signal decreases in 1.3 times as compared to its maximal value.

The maximal intensity of the additional echo signal can be obtained at $\alpha_2 = 180°$, i.e. the duration of the second pulse is about two times greater the first one. This ratio between the pulses is similar to that one at one-frequency excitation of the spin echo in the cases of NMR and NQR. For such a duration of the second pulse neither usual nor forbidden echo signals are observed.

The maximal intensity of the forbidden echo signal can be obtained at $\alpha_1 = 180°$ or $\alpha_1 = 69.7°$ and $\alpha_2 = 120°$, i.e. the ratio between the durations of the first and second pulses should be equal to 0.66 or 1.74, respectively. If the duration of the first pulse corresponds to the angle of 180° neither usual nor additional echo signals are observed. The optimal conditions to observe all three kinds of echo simultaneously are following:

$$\alpha_1 = \pi/2; \quad \alpha_2 = \pi/3.$$

The echo signals (I_{e2}) at frequency $2\omega_0$ can be written as

$$I_{e2} \sim D\sin 2\omega_0(t-2\tau) + E\sin 2\omega_0(t-3\tau/2) + F\sin(2\omega_0(t-5\tau/2)).$$

At frequency $2\omega_0$ at two-frequency excitation one can also observe a number of echo signals:

$$\begin{aligned} &\text{at}\quad t=2\tau: && \text{usual echo,}\\ &\text{at}\quad t=(3/4)\tau: && \text{additional echo,}\\ &\text{at}\quad t=(5/2)\tau: && \text{"forbidden" echo.}\end{aligned}$$

The amplitudes of these echo signals depend on the angles $\alpha_{1,2} = \sqrt{13}\gamma B_1 t_{p1,2}/2$ by the following way:

$$\begin{aligned} D &= (\sin\alpha_1\cos\alpha_1 + 12\sin\alpha_1)\sin^2\alpha_2,\\ E &= (2\sin\alpha_1\cos\alpha_1 - 15\sin\alpha_1)(\cos\alpha_2 - 1)\cos\alpha_2,\\ F &= (2.66\cos\alpha_1(\cos\alpha_1 - 1) + 4(\cos\alpha_1 - 1) + 0.585\sin^2\alpha)\\ &\quad\times(\cos\alpha_2 - 1)\sin\alpha_2. \end{aligned} \tag{9.74}$$

Again, as it is seen from Eq. (9.74) to record the maximal echo signal at different time moments one needs pulses of different duration (at the same pulse amplitude). Conditions imposed on the pulse durations are almost the same as for the echo recording at ω_0.

If $\eta \neq 0$, the conditions for the two-frequency echo formation do not change, i.e. the angles α_1 and α_2, which allow us to obtain the maximal echo intensity, can be found from Eqs. (9.65), (9.66) and (9.73) as before. However, the integral intensities of the rf pulses depend on η:

$$\alpha_{1,2} = \frac{\gamma B_1\cdot\sqrt{13}}{2}\left(1 + \frac{15}{36}\eta + \frac{361}{648}\eta^2\right)t_{p1,2}.$$

Thereby, if the EFG tensor is not axisymmetric then in order to obtain specified values of the $\alpha_{1,2}$ angles at the same amplitude of rf pulse it is necessary to decrease the pulse duration as compared with the case for $\eta = 0$. At $\eta \neq 0$ the usual echo signal is observed at the moment $t = 2\tau$ as before, whereas the signals of the additional and forbidden echoes shift:

$$\begin{aligned} t &= \left(3 - \frac{70}{27}\eta^2\right)\tau: && \text{additional echo at frequency } \omega_0,\\ t &= \left(4 - \frac{70}{27}\eta^2\right)\tau: && \text{forbidden echo at frequency } \omega_0, \end{aligned} \tag{9.75}$$

$$t = (\frac{3}{2} + \frac{35}{54}\eta^2)\tau : \qquad \text{additional echo at frequency } 2\omega_0,$$
$$t = (\frac{5}{2} + \frac{35}{54}\eta^2)\tau : \qquad \text{forbidden echo at frequency } 2\omega_0. \qquad (9.76)$$

As it is seen from Eqs. (9.75) and (9.76) the influence of the EFG asymmetry parameter is stronger for lower transitions than for higher ones.

Let us note some other special features of the two-frequency echo which are not obvious from calculations but are observed experimentally: (i) to describe the echo amplitude attenuation with increasing the time intervals between pulses, τ, it is necessary to use different relaxation times for different echo types that is conditioned by different nature of relaxation processes recorded by means of usual, additional and "forbidden" echoes; (ii) crystal defects affect more the intensity of the additional echo than the usual one (with impurity concentration increasing the amplitude of the usual echo decreases slower, as compared to the additional one).

Thus, by recording the two-frequency echo signals it is possible

- to obtain the values of the quadrupole coupling constant and the asymmetry parameter by measuring frequencies of both transitions and time of the echo appearance (the determination accuracy of the asymmetry parameter is higher if η is determined by measuring the time corresponding to the maximal value of the echo amplitude than by measuring the signal frequency);
- to study relaxation processes of different nature;
- to determine the concentration of impurities and their distribution over the studied crystal;
- to identify lines corresponding to different nonequivalent nuclear sites.

9.6 Multipulse Line Narrowing Sequences in NQR

By now in NMR there is a lot of multipulse sequences specially developed to study various properties of solids. For example they allow us to obtain high resolution NMR spectra, from which it is possible to determine chemical shifts that provide a lot of information about the structure of studied compounds, its physical properties and chemical bonding. In NQR it is impossible to observe chemical shifts as there is no external magnetic field. However, application of multipulse sequences in NQR allows us to increase the sensitivity of pulse spectrometers, helps in signal searching (that is especially important for new compounds where quadrupole coupling constants are unknown beforehand).

Before discussing ways of narrowing NQR spectral lines let us underline the main differences between NMR and NQR, which specify application of different multipulse NQR techniques:

- in zero magnetic field there are, as a rule, a multilevel system with unequidistant spaces between levels;

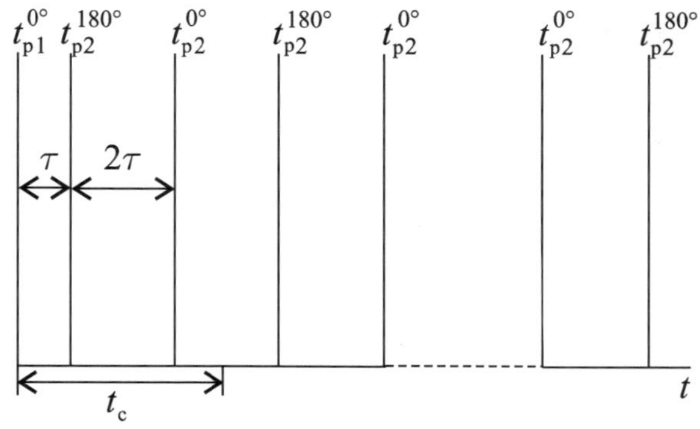

Fig. 9.6 The PAPS sequence

- in polycrystalline samples orientations of the rf field relative to the principal axes of the EFG tensor are randomly distributed that leads to different rotation angles of the nuclear magnetization vector in different regions of the sample;
- a NQR spectrum is usually broaden due to inhomogeneity of electric field gradients; in addition the main changes in linewidths are caused by heteronuclear dipole–dipole interactions, whereas the influence of homonuclear dipole–dipole interactions usually is much weaker;
- for each nuclear spin number the result of the multipulse sequence should be analyzed separately.

Due to the features mentioned above not all the pulse sequences developed for NMR are can be used in NQR. The first sequence used to obtain a high-resolution NQR spectrum was the OW sequence, proposed by Ostroff and Waugh.This sequence consists of a series of 90-degree pulses. The filling frequency of the pulses is continuously shifted in phase ($t_{p1}^{0°} - (\tau - t_{p2}^{90°})_N$). At present, there is a sufficiently large number of pulse sequences, the most popular is the PAPS sequence (Phase Alternated Pulse Sequence). It consists of N pulses with alternating phases: $t_{p1}^{0°} - (\tau - t_{p2}^{180°} - 2\tau - t_{p2}^{0°} - \tau)_N$ (Fig. 9.6). In fact, it is a modified sequence of Carr–Purcell.

In NQR spectroscopy to describe the impact of any multipulse sequence one applies the same mathematical formalism as in NMR. Let us consider how to calculate the effect of any pulse sequence for $I = 3/2$ using the PAPS sequence as an example.

Similar to NMR, calculations of high-resolution NQR spectra are carried out within the framework of the average Hamiltonian theory described in Sect. 5.7.5, but it has some special features. In particular, in the average Hamiltonian theory it is usually assumed that the pulse line narrowing can be obtained if the Hamiltonian of the internal interactions is averaged to zero during the sequence cycle. In NQR if one deals with a polycrystalline sample the Hamiltonian describing it can not be averaged to zero in principle due to a random distribution of rf field directions in the sample. However, this requirement is not necessary. To narrow NQR lines it is quite enough that the average Hamiltonian commutes with density matrix describing the spin system after a preparatory pulse before the first cycle of the pulse sequence.In this case, the Liouville equation (9.7) on the right is zero and hence the density matrix does not depend on time. Hence, any observable quantity, which is determined by the time-independent density matrix, does not depend on time as well, i.e. there is no any decay process.

The total Hamiltonian for the spin system consists of several terms:

$$\widehat{\mathcal{H}} = \widehat{\mathcal{H}}_{\mathrm{Q}} + \widehat{\mathcal{H}}_1 + \Delta\widehat{\mathcal{H}}_{\mathrm{Q}} + \widehat{\mathcal{H}}_{\mathrm{d}}. \tag{9.77}$$

Here $\widehat{\mathcal{H}}_{\mathrm{Q}}$ is the quadrupole interaction Hamiltonian; $\widehat{\mathcal{H}}_1$ is the Zeeman Hamiltonian describing the interaction with rf field; $\widehat{\mathcal{H}}_{\mathrm{d}}$ is the Hamiltonian describing the dipole interactions of different spins; $\Delta\widehat{\mathcal{H}}_{\mathrm{Q}}$ is the Hamiltonian describing the quadrupole interaction with electric field gradients inhomogeneously distributed in the sample. This Hamiltonian corresponds to the main contribution to inhomogeneous broadening of NQR lines. Let $I = 3/2$, then in the principal axes frame of the EFG tensor this Hamiltonian has the following form:

$$\Delta\widehat{\mathcal{H}}_{\mathrm{Q}} = \frac{\Delta\omega_Q\hbar}{6}\left(3\hat{I}_z{}^2 - \hat{I}^2 + \frac{\eta}{2}(\hat{I}_+^2 + \hat{I}_-^2)\right),$$

here $\Delta\omega_Q = e^2\Delta q Q/(2\hbar)$; Δq is the dispersion of the electric field gradients in the sample. Let us express all the contributions that make up the total Hamiltonian (9.77) in terms of single transition operators for $I = 3/2$:

$$\widehat{\mathcal{H}}_{\mathrm{Q}} = \frac{\omega_Q\sqrt{1+\eta^2/3}}{2}\left(\hat{I}_z{}^{1\to3} - \hat{I}_z{}^{2\to4}\right); \tag{9.78}$$

$$\begin{aligned}\widehat{\mathcal{H}}_1 = -2\omega_1\cos\omega t \Big(&(\sqrt{3}\cos\theta + \sin\theta)(\hat{I}_x{}^{1\to2} + \hat{I}_x{}^{3\to4}) \\ &+ (\sqrt{3}\sin\theta - \cos\theta)(\hat{I}_x{}^{1\to4} - \hat{I}_x{}^{3\to2}) + (\hat{I}_x{}^{1\to4} - \hat{I}_x{}^{2\to3})\Big);\end{aligned} \tag{9.79}$$

$$\mathrm{tg}\theta = \eta/\sqrt{3}.$$

$$\Delta\widehat{\mathcal{H}}_{\mathrm{Q}} = \frac{\Delta\omega_Q\sqrt{1+\eta^2/3}}{2}\left(\hat{I}_z{}^{1\to3} - \hat{I}_z{}^{2\to4}\right);$$

$$\begin{aligned}\widehat{\mathcal{H}}_{\mathrm{d}} = F_x\Big(&(\sqrt{3}\sin\theta - \cos\theta)(\hat{I}_x{}^{1\to4} - \hat{I}_x{}^{2\to3}) - (\sqrt{3}\cos\theta + \sin\theta)(\hat{I}_x{}^{1\to2} + \hat{I}_x{}^{3\to4}) \\ &+ (\hat{I}_x{}^{1\to4} - \hat{I}_x{}^{2\to3})\Big) + F_y\Big((\sqrt{3}\sin\theta + \cos\theta)(\hat{I}_y{}^{1\to4} + \hat{I}_y{}^{2\to3}) \\ &- (\sqrt{3}\cos\theta - \sin\theta)(\hat{I}_y{}^{1\to2} - \hat{I}_y{}^{3\to4}) - (\hat{I}_y{}^{1\to4} - \hat{I}_y{}^{2\to3})\Big)\end{aligned}$$

$$+ F_z\left(2\cos\theta(\hat{I}_z^{1\to4} + \hat{I}_z^{2\to3}) + 2\sin\theta(\hat{I}_x^{1\to3} - \hat{I}_x^{2\to4}) + (\hat{I}_z^{1\to4} - \hat{I}_z^{2\to3})\right);$$

here F_x, F_y and F_z are the x-, y-, z-projections of the local magnetic field created by S spins at the I spin site on the direction of the principal axes of the EFG tensor.

Following the procedure described above, when considering the formation of the free precession signal, let us pass to another representation, in which the Hamiltonian $\widehat{\mathcal{H}}_1$ in the form of (9.79) becomes independent of time, and the Hamiltonian of the main interaction $\widehat{\mathcal{H}}_Q$ in the form (9.78) is reduced to zero. Such a conversion can be made by means of the following transformation:

$$\tilde{\widehat{\mathcal{H}}} = \hat{U}^{-1}\widehat{\mathcal{H}}\hat{U},$$

$$\hat{U} = \exp\left(-\mathrm{i}\frac{\widehat{\mathcal{H}}_Q}{\hbar}(t - t_0)\right). \tag{9.80}$$

It can be done by replacing the exponential operators in (9.80) by their expansion in a Taylor series in the form of Eq. (9.49) and using the commutation relations for the fictitious spin operators. Then the Hamiltonian operators considered here are:

$$\widetilde{\widehat{\mathcal{H}}_Q} = 0,$$

$$\begin{aligned}\widetilde{\widehat{\mathcal{H}}_1} = -2\omega_1\cos\omega t\Bigg[&(\sqrt{3}\cos\theta + \sin\theta)(\hat{I}_x^{1\to2} + \hat{I}_x^{3\to4})\cos\frac{\omega_Q\sqrt{1+\eta^2/3}}{2}t\\
&+ (\hat{I}_y^{1\to2} + \hat{I}_y^{4\to3})\sin\frac{\omega_Q\sqrt{1+\eta^2/3}}{2}t + (\sqrt{3}\sin\theta - \cos\theta)(\hat{I}_x^{1\to4} - \hat{I}_x^{3\to2})\\
&+ (\hat{I}_x^{1\to4} - \hat{I}_x^{2\to3})\Bigg],\end{aligned} \tag{9.81}$$

$$\Delta\widetilde{\widehat{\mathcal{H}}_Q} = \Delta\widehat{\mathcal{H}}_Q,$$

$$\begin{aligned}\widehat{\mathcal{H}}_d = F_x&\left((\hat{I}_x^{1\to4} + \hat{I}_x^{2\to3}) + (\sqrt{3}\sin\theta - \cos\theta)(\hat{I}_x^{1\to4} - \hat{I}_x^{2\to3})\right)\\
&+ F_y\left((\sqrt{3}\sin\theta + \cos\theta)(\hat{I}_y^{1\to4} + \hat{I}_y^{2\to3}) - (\hat{I}_y^{1\to4} - \hat{I}_y^{2\to3})\right)\\
&+ F_z\left(2\cos\theta(\hat{I}_z^{1\to4} + \hat{I}_z^{2\to3}) + (\hat{I}_z^{1\to4} - \hat{I}_z^{2\to3})\right)\\
&- F_x(\sqrt{3}\cos\theta + \sin\theta)\Bigg((\hat{I}_x^{1\to2} + \hat{I}_x^{3\to4})\cos\frac{\omega_Q\sqrt{1+\eta^2/3}}{2}t\end{aligned}$$

$$- (\hat{I}_y{}^{1\to 2} - \hat{I}_y{}^{4\to 3}) \sin \frac{\omega_Q\sqrt{1+\eta^2/3}}{2}t\Bigg)$$

$$- F_y(\sqrt{3}\cos\theta - \sin\theta)\Bigg((\hat{I}_y{}^{1\to 2} + \hat{I}_y{}^{3\to 4}) \cos \frac{\omega_Q\sqrt{1+\eta^2/3}}{2}t$$

$$+ (\hat{I}_x{}^{1\to 2} - \hat{I}_x{}^{4\to 3}) \sin \frac{\omega_Q\sqrt{1+\eta^2/3}}{2}t\Bigg)$$

$$+ 2F_z \sin\theta\Bigg((\hat{I}_x{}^{1\to 3} + \hat{I}_x{}^{2\to 4}) \cos \frac{\omega_Q\sqrt{1+\eta^2/3}}{2}t$$

$$- (\hat{I}_y{}^{2\to 4} + \hat{I}_y{}^{1\to 3}) \sin \frac{\omega_Q\sqrt{1+\eta^2/3}}{2}t\Bigg). \tag{9.82}$$

Neglecting rapidly oscillating terms in Eqs. (9.81) and (9.82), one finally obtains:

$$\widetilde{\widetilde{\mathcal{H}}}_1 = -\omega_1(\sqrt{3}\cos\theta + \sin\theta)(\hat{I}_x{}^{1\to 2} + \hat{I}_x{}^{3\to 4});$$

$$\begin{aligned}\widehat{\mathcal{H}}_d = {} & F_x(\hat{I}_x{}^{1\to 4} + \hat{I}_x{}^{2\to 3} + (\sqrt{3}\sin\theta - \cos\theta)(\hat{I}_x{}^{1\to 4} - \hat{I}_x{}^{2\to 3})) \\ & + F_y\left((\sqrt{3}\sin\theta + \cos\theta)(\hat{I}_y{}^{1\to 4} + \hat{I}_y{}^{2\to 3}) - (\hat{I}_y{}^{1\to 4} - \hat{I}_y{}^{2\to 3})\right) \\ & + F_z\left(2\cos\theta(\hat{I}_z{}^{1\to 4} + \hat{I}_z{}^{2\to 3}) + (\hat{I}_z{}^{1\to 4} - \hat{I}_z{}^{2\to 3})\right).\end{aligned}$$

If the interaction of spins with the rf field is much stronger than other interactions, then during the pulse the latter can be neglected. Then the Hamiltonian describing the behavior of the spin subsystem during the pulse contains only one term, and in the absence of the pulse the system evaluates only under the influence of internal interactions:

$$\begin{aligned}\widetilde{\widetilde{\mathcal{H}}} &= \widetilde{\widetilde{\mathcal{H}}}_1 && at\ 0 < t < t_p, \\ \widetilde{\widetilde{\mathcal{H}}} &= \widetilde{\Delta\widetilde{\mathcal{H}}_Q} + \widetilde{\widetilde{\mathcal{H}}}_d && at\ t_p < t.\end{aligned}$$

Now let us consider how the Hamiltonian operators describing the internal interactions change under the influence of the PAPS pulse sequence. The first pulse of duration t_{p1} (see Fig. 9.6) is preparatory, it acts on the spin system before the pulse sequence. It prepares the initial density matrix. And a time interval τ later one applies a sequence narrowing the NQR spectral lines, which consists of two rf pulses of duration t_{p2}. The filling frequency of each successive pulse in the sequence is shifted by 180° relative to the previous pulse. The cycle time is (t_c) for this sequence is equal to 4τ. As mentioned in Sect. 5.7.5, for the effective averaging internal interactions the Hamiltonian $\widehat{\mathcal{H}}_{ext}$ and propagator of external interactions

$$\hat{U}_{\text{ext}} = \hat{T} \exp\left(-\frac{\mathrm{i}}{\hbar}\int_0^t \widehat{\mathcal{H}}_{\text{ext}}(t')\mathrm{d}t'\right)$$

($\hat{T}$ is the Dyson time ordering operator) must satisfy the periodicity condition (5.95):

$$\begin{aligned} \widehat{\mathcal{H}}_{\text{ext}}(t+t_m) &= \widehat{\mathcal{H}}_{\text{ext}}(t), \\ \hat{U}_{\text{ext}}(t+t_{\text{c}}) &= \hat{U}_{\text{ext}}(t). \end{aligned} \tag{9.83}$$

If conditions (9.83) are satisfied, at times multiples of t_{c} the propagator of external interactions is equal to the unit operator:

$$\hat{U}_{\text{ext}}(N\,t_{\text{c}}) = \hat{U}_{\text{ext}}(0) = 1.$$

It is obvious that conditions (9.83) are satisfied for the PAPS sequence, and consequently one can apply the average Hamiltonian theory.

Following the formalism set out in Sect. 5.7.5, one obtains the values of the average Hamiltonian describing the interaction considered above:

$$\begin{aligned} \overline{\Delta\widehat{\mathcal{H}}_{\text{Q}}} = {} & \frac{\Delta\omega_{\text{Q}}\sqrt{1+\eta^2/3}}{2} \\ & \times \left((\hat{I}_z^{\,1\to2} - \hat{I}_z^{\,3\to4})\cos\frac{\alpha_2}{2} + (\hat{I}_y^{\,2\to1} + \hat{I}_y^{\,3\to4})\sin\frac{\alpha_2}{2} \right)\cos\frac{\alpha_2}{2}, \end{aligned} \tag{9.84}$$

$$\alpha_2 = \omega_1(\sqrt{3}\cos\theta + \sin\theta)t_{\text{p2}};$$

$$\begin{aligned} \widehat{\mathcal{H}}_{\text{d}} = {} & F_x\Bigg[\hat{I}_x^{\,1\to4} + \hat{I}_x^{\,2\to3} + (\sqrt{3}\sin\theta - \cos\theta)\Bigg((\hat{I}_x^{\,1\to4} - \hat{I}_x^{\,2\to3})\cos\frac{\alpha_2}{2} \\ & - (\hat{I}_y^{\,2\to4} + \hat{I}_y^{\,3\to1})\sin\frac{\alpha_2}{2}\Bigg)\cos\frac{\alpha_2}{2}\Bigg] \\ & - F_y\Bigg[\Bigg((\hat{I}_y^{\,1\to4} - \hat{I}_y^{\,2\to3})\cos\frac{\alpha_2}{2} - (\hat{I}_x^{\,1\to3} - \hat{I}_x^{\,2\to4})\sin\frac{\alpha_2}{2}\Bigg)\cos\frac{\alpha_2}{2} \\ & - (\sqrt{3}\sin\theta + \cos\theta)(\hat{I}_y^{\,1\to4} + \hat{I}_y^{\,2\to3})\Bigg] \\ & + F_z\Bigg[\Bigg((\hat{I}_z^{\,1\to4} - \hat{I}_z^{\,2\to3})\cos\frac{\alpha_2}{2} + (\hat{I}_y^{\,1\to2} + \hat{I}_y^{\,3\to4})\sin\frac{\alpha_2}{2}\Bigg)\cos\frac{\alpha_2}{2} \\ & + 2\cos\theta(\hat{I}_z^{\,1\to4} + \hat{I}_z^{\,2\to3})\Bigg]. \end{aligned} \tag{9.85}$$

As it can be seen from Eqs. (9.84) and (9.85), neither the Hamiltonian describing the quadrupole interaction with inhomogeneously distributed electric field gradients nor the Hamiltonian describing the dipole–dipole interactions are not averaged to zero under the action of the PAPS sequence. Before the impact of the preparatory pulse the spin system is in a thermodynamic equilibrium, and the density matrix corresponding to it can be determined in the usual way:

$$\hat{\rho}(0) = A\left(1 - \frac{\widehat{\mathcal{H}}_{\mathrm{Q}}}{kT}\right).$$

In the interaction representation, the density matrix in terms of the single transition operators has the form

$$\hat{\rho}(0) = A\left(1 - \frac{2\Delta\omega_{\mathrm{Q}}\sqrt{1+\eta^2/3}(\hat{I}_z{}^{1\to 3} - \hat{I}_z{}^{2\to 4})}{kT}\right).$$

After application of a pulse of duration t_{p}, the density matrix is written as following

$$\hat{\rho}(t_{\mathrm{p}1}) = \exp\left(-\mathrm{i}\widetilde{\widetilde{\mathcal{H}}}_1 t_{\mathrm{p}1}\right)\hat{\rho}(0)\exp\left(\mathrm{i}\widetilde{\widetilde{\mathcal{H}}}_1 t_{\mathrm{p}1}\right). \tag{9.86}$$

Using the single transition operators, Eq. (9.86) can be transformed as follows:

$$\hat{\rho}(t_{\mathrm{p}1}) = (\hat{I}_z{}^{1\to 2} - \hat{I}_z{}^{3\to 4})\cos\alpha_1 - (\hat{I}_y{}^{2\to 1} + \hat{I}_y{}^{3\to 4})\sin\alpha_1,$$

$$\alpha_1 = \omega_1(\sqrt{3}\cos\theta + \sin\theta)t_{\mathrm{p}2}.$$

Now, it is easy to show that the density matrix obtained after the first preparatory pulse commutes with the operators $\widetilde{\widetilde{\Delta\mathcal{H}_{\mathrm{Q}}}}$ and $\widetilde{\widetilde{\mathcal{H}}}_{\mathrm{d}}$. Hence, under the influence of the PAPS pulse sequence the free induction decay or echo signals are equal to zero, and NQR lines narrow to their relaxation width.

9.7 Detection of Drags and Explosives (Illicit Substances) by NQR

As it was underlined before, one on the most important advantages of NQR technique, compared to NMR, is that a magnet is not required. Therefore, it is less expensive and more compact. Another advantages of NQR are that it is non-toxic, non-destructive, can be used to analyze rather large quantities of substances, even nonhomogeneous and mixed. The main disadvantage of NQR is that to detect a substance one needs a rather larger quantity of it (tens grammes). Moreover, due to a low intensity of NQR signals the direct NQR methods are time consuming, whereas indirect ones are too cumbersome.

In NQR a rf pulse excites transitions between nucleus states created not by an external magnetic field but internal quadrupole interactions. And the transition

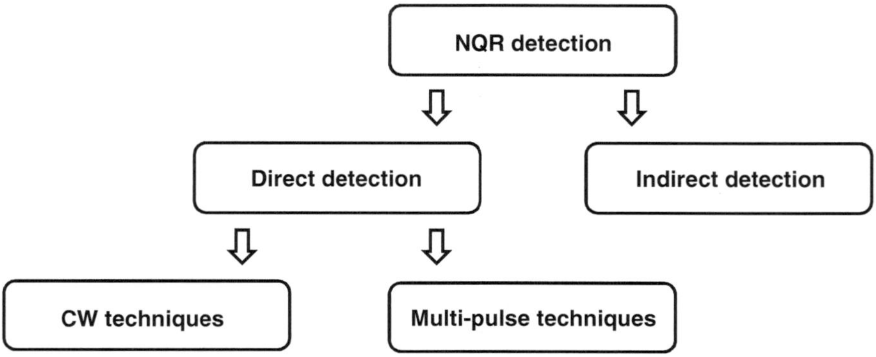

Fig. 9.7 NQR methods for explosives and drags detection

frequencies are strongly related to a specific environment of a resonating quadrupole nucleus. Since ^{14}N and ^{35}Cl are quadrupole nuclei, which enter to compositions of the most of drags and explosives, NQR techniques can be used for detection of such substances.

The isotope ^{14}N has spin $I = 1$. Its natural abundance is 93.6 %. Frequencies of NQR transitions lies within the frequency range between 0 and 6 MHz, the exact value is determined by the EFG distribution at the nucleus site created mostly by the nearest neighbors. ^{35}Cl with spins $I = 5/2$ and natural abundance of 75.4 % is another quadrupole nucleus used for drugs and explosives detection. Despite NQR frequencies of ^{14}N and ^{35}Cl nuclei are rather low (and hence sensitivity of the NQR method is low), NQR becomes more and more applied in such areas as transport security as it can provide automated detection of substances missed by imaging X-ray methods.

The more frequently used explosives contain NO_2 groups. As opposed to nitrides of alkali metals, such as $NaNO_2$, $CaNO_2$, $CsNO_2$, where sp^2-orbital of the nitrogen atom is occupied by a lone-electron pair, in organic nitrogen-containing compounds the group NO_2 forms a σ-bond N–C. This leads to a dramatic decrease of the ^{14}N NQR frequencies, and hence to a disastrous decreasing of the sensitivity. The task of the sensitivity increasing without making the apparatus part too complex is rather challenging and is actually solving by various scientific and technical groups.

There are difference experiment NQR realizations based on direct or indirect technics of NQR signal detection (see Fig. 9.7). Indirect NQR detection methods, namely double NQR, are discussed in Sect. 15.4. More details can be found in the review paper of Grechishkin and Sinyavskii.[5]

[5] V. S. Grechishkin, N. Ya. Sinyavskii. New technologies: nuclear quadrupole resonance as an explosive and narcotic detection technique. Phys. Usp. **40** (1997) 393.

Further Readings

Das TP, Hahn EL (1958) Nuclear quadrupole resonance spectroscopy. Academic Press, New York

Fraissard J, Lapina O (2009) Explosives detection using magnetic and nuclear resonance techniques, Wiley, New York (ISBN 978-9048130610)

Miller JB (2007) Nuclear quadrupole resonance detection of explosives, in Counterterrorist detection techniques of explosives. In: Yinon J (ed) Elsevier, Amsterdam, pp 131–156.

Parish RV (1990) NMR, NQR, EPR, and Mossbauer spectroscopy in inorganic chemistry. Ellis Horwood, New York

Rudakov TN (2009) Detection of explosives by NQR methods: main aspects for transport security. In: Fraissard J, Lapina O (eds) Explosives detection using magnetic and nuclear resonance techniques. NATO Science for Peace and Security Series, Springer, New York, p 111

Smith JAS (ed) (1983) Advances in nuclear quadrupole resonance, vol 5. Heyden, London

Weiss A (1972) Nuclear quadrupole resonance. In: Topics in current chemistry, vol 30. Springer, Berlin.

Part III
Electron Paramagnetic Resonance (EPR)

Chapter 10
Basic Interactions of an Electron in Solids

This part focuses the attention on the approach suitable for studying assemblies of weakly interacting paramagnetic centers. Here we will not consider magnetically ordered systems where ferromagnetic, ferrimagnetic or antiferromagnetic resonance phenomena can be observed, as well as systems with free electrons (metals and semiconductors). Any atomic or molecular system which is characterized by an uncompensated angular momentum (and, consequently, by a non-zero magnetic momentum) is called a *paramagnetic center*. Magnetic resonance in such a system of electrons is known as *electron paramagnetic resonance* (EPR).

Most stable atomic and molecular systems cannot act as paramagnetic centers since, when forming a molecule, electrons usually tend to interact in such a way that their resultant momentum is equal to zero, whereas atoms with non-zero momentum are unstable under normal conditions and tend to form chemical bonds. Nevertheless, there are many paramagnetic systems of great scientific and practical interest, which can be successfully studied by electron paramagnetic resonance, including paramagnetic molecules with non-zero angular momentum (e.g. O_2, NO); atoms and ions with incomplete d- and f-shells; stable and stabilized radicals; electron systems in crystals with a non-zero angular momentum representing impurity and radiation-induced defects.

The applicability of NMR and EPR to the study of the microscopic structure of a substance is determined by the fundamental difference between these methods. Characteristics of a nucleus in different substances and aggregate states remain stable. A nucleus simply act as a convenient natural "gauge", perceiving and transmitting information about its environment, whereas the electron state forms the structural characteristics of the material. In turn, electron paramagnetic resonance is a fairly sensitive indicator of the changing state of the electron.

V. I. Chizhik et al., *Magnetic Resonance and Its Applications*,
DOI: 10.1007/978-3-319-05299-1_10,

10.1 Crystal Electrostatic Field Model

In general, a paramagnetic center in a magnetic field does not result in a Zeeman system of energy levels (i.e. a system of equally spaced levels), that leads to the splitting of the magnetic resonance line. Sometimes in order to calculate the system of magnetic levels determining the EPR spectrum before including the magnetic field as a small perturbation one first has to analyze a complex hierarchy of non-magnetic interactions of the electron. The assumed weakness of interactions between paramagnetic centers makes it possible to determine the system of energy levels for the whole assembly by simply calculating the levels for one such center interacting with its immediate environment. For research purposes, this interaction may be conveniently divided into two types: electrostatic interaction between the center and its immediate environment (usually diamagnetic ions or molecules) and magnetic and electrostatic (quadrupole) interaction with the nearest nuclei. The first type of interaction leads to the appearance of the fine structure of the EPR spectrum whereas electron-nuclear interactions are responsible for its hyperfine structure (in this case the levels are much less splitted). In this chapter we will review the theory describing the formation of energy levels for paramagnetic centers in crystal environments. First, we will introduce the electron Hamiltonian (without derivation) in the presence of magnetic and electric fields; next, we will show how to use the quantum-mechanical perturbation theory to calculate the energy levels for paramagnetic centers in specific cases. Free atom and ion models can be directly used to calculate the energy levels of paramagnetic centers (which is the main purpose of electron paramagnetic resonance theory) only for rarefied gases and atomic beams. In the most interesting cases, the movement of electrons in an atom in condensed phases (liquid and solid) is perturbed by surrounding particles. Even if these perturbations are relatively weak, in the sense that the electrons remain link to particular atoms (ions), they may cause considerable changes in the system of magnetic sublevels. This is due to the fact that magnetic interactions are usually the weakest interactions in which the electron may be involved. A simple method to detect the interaction between electrons in an atom or ion with their nearest environment is the internal electrostatic crystal field model. This model was developed in the 1930s by van Vleck to explain magnetic properties of paramagnetic crystals and later was successively applied to treat EPR spectra. In this section we will discuss the main principles of the theory and how it can be applied to different paramagnetic systems. We will also describe a qualitative approach to the problem of energy level splitting of paramagnetic centers in the electrostatic crystal field.

10.1.1 Van Vleck Assumptions of the Model

The basic assumption of the crystal field model is that the effect of the surrounding ions and molecules on the paramagnetic center, regardless of the nature of interaction, is the result of the effective electrostatic field created by ions and molecules at the site of the paramagnetic center. This is so-called crystal field. To simplify theoretical analysis, a few additional assumptions are usually introduced:

(1) All interactions that cannot be accounted for by the crystal field (e.g. those associated with the overlapping electron clouds of neighboring paramagnetic centers or of a paramagnetic center and surrounding ions) are excluded from analysis;
(2) The fields of the surrounding ions or molecules can be calculated as the fields of point charges or dipoles.

These assumptions can simplify the calculation of energy levels for paramagnetic centers in crystals and express values that define the spectrum of electron paramagnetic resonance through the parameters of the crystal field. Within this model EPR spectrum calculations are performed in two stages. First, the energy levels and states of the paramagnetic center are determined in the presence of the crystal field. Weaker magnetic interactions at this stage are neglected. In this case the Hamiltonian is written as follows:

$$\widehat{\mathcal{H}} = \widehat{\mathcal{H}}_0 + \widehat{\mathcal{H}}_{\text{cryst}}. \tag{10.1}$$

Here, $\widehat{\mathcal{H}}_0$ includes interactions, which are much stronger compared to those between the electrons and the crystal field corresponding to $\widehat{\mathcal{H}}_{\text{cryst}}$. It is essential to our calculations that the interaction described by the Hamiltonian $\widehat{\mathcal{H}}_{\text{cryst}}$ acts only on the spatial part of the electron wave function (but not on the spin one). At the second stage one takes into account weaker magnetic interactions. The perturbation theory is re-applied, which allows us to obtain the *spin Hamiltonian* (see Sect. 1.9) containing only spin operators. At this stage we are able to find out the energy levels and states of the paramagnetic center in a magnetic field, which directly determine the EPR spectrum.

10.1.2 Perturbation of Free Atom or Ion by Crystal Field

Depending on the value of the interaction between the crystal field and the electrons relative values of electron interactions in the atom, crystal fields are classified into three types: *weak, intermediate and strong*. If $\langle\widehat{\mathcal{H}}_{\text{cryst}}\rangle$ is lower than the spin-orbit interaction $\lambda\hat{L}\hat{S}$ ($\hat{L}$ and $\hat{S}$ are orbital and spin momenta of atom, λ is spin-orbit interaction constant), than the crystal field is qualified as weak. If $\langle\widehat{\mathcal{H}}_{\text{cryst}}\rangle$ is higher than the spin-orbit interaction, but weaker than the residual, or correlation interaction (see Sect. 10.2.5) it is regarded as intermediate. The crystal field is classified as strong field if its interaction with the paramagnetic center is higher than the residual interaction. Depending on the field type the Hamiltonian $\widehat{\mathcal{H}}_0$ in (10.1) should include different interactions.

Table 10.1 Interactions at the stages of perturbation theory depending on a class of crystal field

Crystal field	Hamiltonian	Quantum numbers non-pert.states	Perturbation	
			I stage	II stage
Strong	$\widehat{\mathcal{H}}_{\text{ssscf}} = \sum_i^N \left[p_i^2/(2m) - eV_i'(r_i)\right]$	$n_i l_i (m_i m_{si})$	$\widehat{\mathcal{H}}_{\text{cryst}}$	$\widehat{\mathcal{H}}_{\text{res}} + \lambda \hat{\vec{L}} \cdot \hat{\vec{S}} + \mu_B(\hat{\vec{L}} + 2\hat{\vec{S}}) \cdot \vec{B}$
Intermediate	$\widehat{\mathcal{H}}_{\text{ssscf}} + \widehat{\mathcal{H}}_{\text{res}}$	$n_i l_i LS(M M_S)$	$\widehat{\mathcal{H}}_{\text{cryst}}$	$\lambda \hat{\vec{L}} \cdot \hat{\vec{S}} + \mu_B(\hat{\vec{L}} + 2\hat{\vec{S}}) \cdot \vec{B}$
Weak	$\widehat{\mathcal{H}}_{\text{ssscf}} + \widehat{\mathcal{H}}_{\text{res}} + \lambda \hat{\vec{L}} \cdot \hat{\vec{S}}$	$n_i l_i LSJ(M_J)$	$\widehat{\mathcal{H}}_{\text{cryst}}$	$g_L \mu_B \hat{\vec{J}} \cdot \vec{B}$

Proceeding from the given definitions, in the case of a weak crystal field a free atom is described by total momentum $\hat{\vec{J}}$ and by Hamiltonian $\widehat{\mathcal{H}}_0$, which includes the spin-orbit interaction (see Sect. 10.1.3). In an intermediate field, spin-orbit interaction is neglected at the first stage, so the total momentum of the atom (ion) $\hat{\vec{J}}$ loses its meaning. It is reasonable to suggest that the crystal field "breaks" the bond between the orbital and spin momenta. So the Hamiltonian $\widehat{\mathcal{H}}_0$ is represented as $\widehat{\mathcal{H}}_0 = \widehat{\mathcal{H}}_{\text{ssscf}} + \widehat{\mathcal{H}}_{\text{res}}$ where $\widehat{\mathcal{H}}_{\text{ssscf}}$ is the Hamiltonian of free atom in the Hartree-Fock approximation of spherically symmetric self-consistent field and $\widehat{\mathcal{H}}_{\text{res}}$ is he residual interaction. In a strong field, the total orbital momentum $\hat{\vec{L}}$ also loses its meaning as the residual interaction between electrons resulting in the orbital momentum is weaker than the effect of the crystal field on each electron. The unperturbed state in a strong field is the atomic state of Hartree-Fock approximation, whereby the atomic state is determined by the state of its electrons subject to the disturbing effect of the strong crystal field. The strong crystal field changes electron wave functions and relieves degeneration of level with the same l values. The residual perturbation is taken into account at the second stage only. This results in the further removal of degeneration and to the formation *of spectral terms* in the presence of the crystal field. However, in this case, it is impossible to assign a defined value of the orbital momentum L as due to the absence of the spherical symmetry $\hat{\vec{L}}$ is not the integral of motion. Interactions which should be included in the Hamiltonian representing the unperturbed state and which should be taken into account at the second stage in each of the three cases are presented in Table 10.1.

10.1.3 Different Realization of the Crystal Field Model

The crystal field theory is most effectively used for the interpretation of the experimental EPR data for crystals which contain paramagnetic ions or paramagnetic centers (defects of various kinds) arising due to the exposure of crystals to radiation.

Paramagnetic ions are metal ions with incomplete electron shells. Elements which are able to form paramagnetic ions are grouped in the periodic table according to the criterion of the disturbance of the formal shell occupation order. In such cases a shell

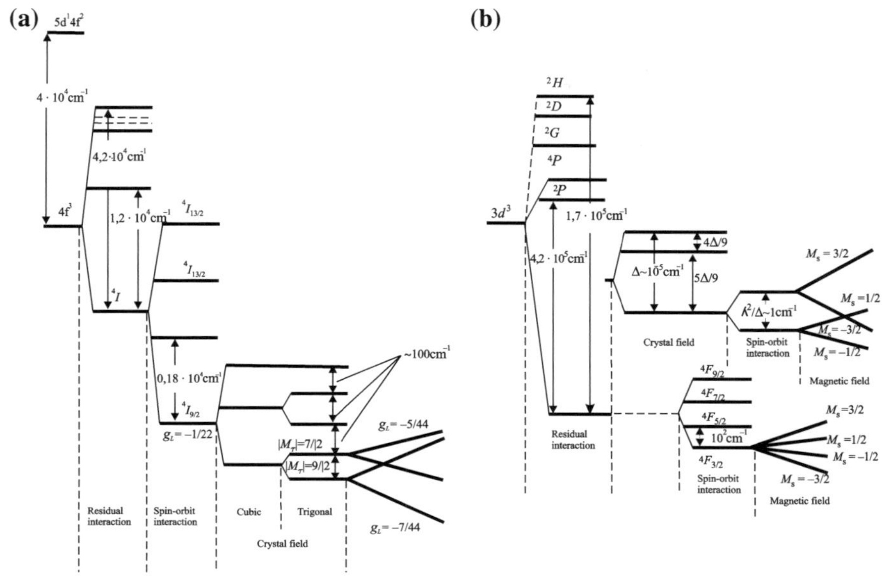

Fig. 10.1 Level diagrams for electron configurations $5d^14f^2$ of ion Nd^{3+} (**a**) and $3d^3$ of ion Cr^{3+} (**b**)

with higher n is occupied before a shell with lower n but higher l. There are five such groups in the periodic table: three transition groups (elements with $4s^23d^k$, $5s^24d^k$, $6s^25d^k$ shell configurations); lanthanoids ($6s^25d^14f^k$) and actinoids ($7s^26d^15f^k$).

Spectroscopic methods, including EPR, exclusively relies on interactions which result in the splitting of the degenerate levels of paramagnetic centers. When estimating the splitting character, it should be taken into account that the energy of the interaction of an electron with the crystal field depends on the symmetry of a disposition of charges forming the field, the distances between an ion and charges and the effects of electrostatic screening by other electron shells.

The effective potential of the crystal field is formed only by those field components that depend on coordinates in certain way. Thus, for the elements of the first transition group, the effective potential results in an interaction energy corresponding to $2 \times 10^4\,\text{cm}^{-1}$. As specified in the above classification of crystal fields, ions of the transition groups should be regarded as situated in the intermediate field as their spin-orbit interaction is characterized by the constant $\lambda/(hc) \simeq 100\,\text{cm}^{-1}$. At the same time, the residual interaction determining the distance between spectral terms measures ca. $10^5\,\text{cm}^{-1}$ and more (see intervals between terms for Cr^{3+} left in Fig. 10.1b). In the absence of the crystal field the spin-orbit interaction would lead to the further level split (of about $100\,\text{cm}^{-1}$) and in a magnetic field would occur standard Zeeman effect (bottom right in Fig. 10.1b). Crystal field splits a ground level Cr^{3+} at intervals of 10^4–$10^5\,\text{cm}^{-1}$, and the spin-orbit interaction together with a magnetic field leads to a more complex system of levels. It is the transitions between

Table 10.2 Contributions of different interactions for various ions

Group of	Electron configuration	Interaction type (order of magnitude, cm^{-1})		
		Residual	Spin-orbit	Crystal field
Lanthanoides	$4f^k$	10^5	$6 \cdot 10^2$–$3 \cdot 10^3$	10^2
Fe group	$3d^k$	10^5	(2–8) $\cdot 10^2$	10^4
Pd group, Pt group and actinoids	$4d^k, 5d^k, 5f^k$	10^5	$3 \cdot 10^2$–$3 \cdot 10^3$	10^5

them those give the EPR spectrum (top right in Fig. 10.1b). Lanthanoide ions with incomplete $4f$ shells always have the complete $5s^2p^6$ shell located outside incomplete shells. The electrostatic screening effect produced by outer shells decreases the influence of crystal field on $4f$ electrons by one or two orders. At the same time, λ grows with the increase of the element number, so lanthanide ions are impacted by a weak crystal field. This means that in the present case the effect of the crystal field may be regarded as a disturbance to the states arising from the spin-orbit interaction. An example of energy level system with a weak crystal field (ion Nd^{3+}) is presented in Fig. 10.1a.

Ions of $4d$ and $5d$ transition elements as well as some actinoids have a distinctive tendency to form covalent bond complexes. In this case, reliable data can be obtained within the strong crystal field model; to achieve better accuracy, however, one needs to reject the first assumption in the crystal field model and to use the covalent bond model (see Chap. 12). Contributions of different interactions (arranged in crystal field type) for paramagnetic ions are presented in Table 10.2.

10.2 Elements of Atom Theory

10.2.1 General Quasi-Relativistic Hamiltonian of an Electron

The quasi-relativistic Hamiltonian $\widehat{\mathcal{H}}_e$ is a generalized and convenient form of representing the diverse interactions of an electron in an atom, molecule or crystal:

$$\widehat{\mathcal{H}}_e = \frac{1}{2m_e}(\widehat{\vec{p}} + e\vec{A})^2 - eV + \frac{e\hbar}{2m_e^2c^2}\hat{\vec{s}} \cdot (\vec{\mathcal{E}} \times \widehat{\vec{p}}) + 2\mu_B\hat{\vec{s}} \cdot \vec{B}, \tag{10.2}$$

where $\vec{A}$ and V are the vector and scalar potentials of the external electromagnetic field; $\vec{\mathcal{E}}$ is the electric field, $\vec{\mathcal{E}} = -\nabla V$. The first two terms of the Hamiltonian (10.2) describe the kinetic energy of the electron and the energy of its interaction as a charged spinless particle with the external electromagnetic field. The other two terms are conditioned by the existence of the electron spin and the spin magnetic moment: the third term describes the spin-orbit interaction; the fourth expresses the interaction between the spin magnetic moment and the external static magnetic field $\vec{B}$.

According to the classical non-relativistic theory, no interaction is possible between the spin magnetic moment of the electron and the electrostatic field; however, its emergence can be explained within the framework of the relativity theory by the fact that the electric and magnetic fields are mutually transformed in coordinate systems which move relative to each other. In our case, an electromagnetic field (electrostatic in the laboratory coordinate system associated with the electron) is converted into a magnetic field. The third term of the Hamiltonian (10.2) describes the interaction between the magnetic spin moment of the electron with this effective magnetic field. An accurate derivation of the Hamiltonian (10.2) follows from the Dirac equation for an electron in an external field in the limit where $v/c \ll 1$ (v is the mean electron velocity in an atom, c is the light velocity).

10.2.2 Electron in the Central Electric Field

The Hamiltonian transformation in case of the spherical symmetry. In this case, the electrostatic field is almost spherically symmetrical, and the magnetic field (applied to a sample) is homogeneous. For a constant homogeneous magnetic field one can assume that

$$\vec{A} = \frac{1}{2}\vec{B} \times \vec{r}, \tag{10.3}$$

and the intensity of the spherically symmetrical electrostatic field can be expressed through its scalar potential, with the atomic nucleus used as the origin:

$$\vec{\mathcal{E}} = -\frac{\partial V}{\partial r}\frac{\vec{r}}{r}, \tag{10.4}$$

Taking into account (10.3), the first term of the Hamiltonian (10.2) is transformed into

$$\frac{1}{8m_{\mathrm{e}}}[4\widehat{\vec{p}}^2 + 2e\widehat{\vec{p}} \cdot (\vec{B} \times \vec{r}) + 2e(\vec{B} \times \vec{r}) \cdot \widehat{\vec{p}} + e^2(\vec{B} \times \vec{r})^2]. \tag{10.5}$$

To calculate mixed products of the $\widehat{\vec{p}} \cdot (\vec{B} \times \vec{r})$ and $(\vec{B} \times \vec{r}) \cdot \widehat{\vec{p}}$ type in Eq. (10.5) a standard cyclic permutation of vector-factors may be performed for the triple vector product as these products only contain commuting heteronymic components $\widehat{\vec{p}}$ and $\vec{r}$. Therefore, $\widehat{\vec{p}} \cdot (\vec{B} \times \vec{r}) = (\vec{r} \times \widehat{\vec{p}}) \cdot \vec{B} = (\vec{B} \times \vec{r}) \cdot \widehat{\vec{p}}$. In this case the sum of the second and third terms in (10.5) can be written as $4(\vec{r} \times \widehat{\vec{p}}) \cdot \vec{B}$. The operator of the orbital angular momentum of the electron ($\widehat{\vec{l}}$) in units $\hbar$ is associated with the linear momentum by the following known relation:

$$\widehat{\vec{l}} = \frac{1}{\hbar}\vec{r} \times \widehat{\vec{p}}, \tag{10.6}$$

that makes it possible to write Eq. (10.5) as

$$\frac{p^2}{2m_e} + \frac{e\hbar}{2m_e}\hat{\vec{l}} \cdot \vec{B} + \frac{e^2\left(\vec{B}\times\vec{r}\right)^2}{8m_e}.$$

Using the definition $\mu_B = e\hbar/(2m)$, this expression can be reduced to

$$\frac{p^2}{2m_e} + \mu_B\hat{\vec{l}} \cdot \vec{B} + \frac{e^2\left(\vec{B}\times\vec{r}\right)^2}{8m_e}. \tag{10.7}$$

The first term in (10.7) describes the non-relativistic kinetic energy of the electron.

Electron in a central field. This situation occurs in a hydrogen atom or ion with a single electron (hydrogen-like ion), where the electron moves in the field of a point charge. The solution for this case is well known and can be found in any textbook on atomic physics and quantum mechanics (See, for example, L.D. Landau, E.M. Lifshitz. Quantum Mechanics: Non-Relativistic Theory. Vol. 3 (3rded.). Pergamon Press. 1977. ISBN 978-0-08-020940-1). The energy operator can be obtained by retaining only the first term from (10.5) and the second term from (10.2) in the form $V = -\frac{eZ}{r}$ (Coulomb potential, Z is nuclear charge). The eigenvalues of this operator are $E_n = \frac{e^2Z^2}{2a_0n^2}$, where n is the integer index, known *as the principal quantum number* and $a_0 = 0.529 \times 10^{-10}$ m is the atomic unit of length (*the Bohr radius*). The principal quantum number determines the energy of an electron in a Coulomb electrostatic field. Wave functions ψ_{nl}^m in the coordinate representation can be written as

$$\psi_{nl}^m(r,\theta,\phi) = R_{nl}(r)Y_l^m(\theta,\phi), \tag{10.8}$$

where $R_{nl}(r)$ is the radial part of the wave function and $Y_l^m(\theta,\phi)$ is the angular part representing the spherical functions of spherical coordinates θ, ϕ. Table 10.3 contains the explicit forms of spherical functions $Y_l^m(\theta,\phi)$ for l from 1 to 4. Indices of spherical harmonics l and m, also called *quantum numbers*, can take the following values: $l = 0, 1, \ldots, n-1$; $m = -l, -l+1, \ldots, l$.

The state vector with quantum numbers n, l and m can be written as $|nml\rangle$. It is also the eigenvector for the squared angular momentum $(\widehat{l^2})$ and its projection on the z axis $(\widehat{l_z})$ with eigenvalues $l(l+1)$ and m, respectively. This is why l and m are known as the *orbital* and *magnetic* quantum numbers, respectively. In various areas of spectroscopy the state of the electron can be signified by a symbol consisting of a value equal to n and the subsequent character standing for the value of l: thus, s corresponds to $l = 0$; p corresponds to $l = 1$; d stands for $l = 2$; f for $l = 3$; g for $l = 4$, etc. This notation also applies to many-electron atoms or ions with the difference that an index denoting the number of electrons in a given state is added above on the right. For example, $3d^4$ means that the shell contains four electrons in the state with $n = 3, l = 2$, while $6s^25d^14f^3$ means three electrons with $n = 4$, $l = 3$, one with $n = 5$, $l = 2$ and two with $n = 6$, $l = 0$. Thus, it is assumed that we

Table 10.3 Spherical harmonics Y_l^m, relevant homogeneous polynomials U_l^m and equivalent operators $\widehat{Q}_l^m$

l	m	$Y_l^m(\theta,\phi)$	$U_l^m(x,y,z)$	$\widehat{Q}_l^m$
1	0	$\sqrt{\frac{3}{4\pi}}\cos\theta$	z	$\hat{l}_z$
	±1	$-\sqrt{\frac{3}{4\pi\cdot2}}\sin\theta\, e^{\pm i\phi}$	x	$\frac{1}{2}(\hat{l}^+ + \hat{l}^-)$
2	0	$\sqrt{\frac{5}{4\pi\cdot4}}(3\cos^2\theta-1)$	$(3z^2-r^2)$	$3\hat{l}_z{}^2-\widehat{l^2}$
	±1	$-\sqrt{\frac{5\cdot3}{4\pi\cdot2}}\cos\theta\sin\theta e^{\pm i\phi}$	xz	$\frac{1}{4}\{\hat{l}_z,(\hat{l}^+)+(\hat{l}^-)\}$
	±2	$\sqrt{\frac{5\cdot3}{4\pi\cdot8}}\sin^2\theta\, e^{\pm2i\phi}$	x^2-y^2	$\frac{1}{2}[(\hat{l}^+)^2+(\hat{l}^-)^2]$
3	0	$\sqrt{\frac{7}{4\pi\cdot4}}(5\cos^3\theta-3\cos\theta)$	$(5z^3-3zr^2)$	–
	±1	$\sqrt{\frac{7\cdot3}{4\pi\cdot16}}(5\cos^2\theta-1)\sin\theta\, e^{\pm i\phi}$	$x(5z^2-r^2)$	–
	±2	$\sqrt{\frac{7\cdot15}{4\pi\cdot8}}\cos\theta\sin^2\theta\, e^{\pm2i\phi}$	$z(x^2-y^2)$	–
	±3	$\sqrt{\frac{7\cdot5}{4\pi\cdot16}}\sin^3\theta\, e^{\pm3i\phi}$	x^3-3xy^2	–
4	0	$\sqrt{\frac{9}{4\pi\cdot64}}(35\cos^4\theta-30\cos^2\theta+3)$	$(35z^4-30z^2r^2+3r^4)$	$35\hat{l}_z{}^4-30\hat{l}_z{}^2\widehat{l^2}+25\hat{l}_z{}^2-6\widehat{l^2}+3\widehat{l^4}$
	±1	$\sqrt{\frac{9\cdot5}{4\pi\cdot16}}(7\cos^3\theta-3\cos\theta)\sin\theta\, e^{\pm i\phi}$	$xz(7z^2-3r^2)$	$\frac{1}{4}\{7\widehat{l_z^3}-3\widehat{l_z}\widehat{l^2}-\widehat{l_z},\widehat{l^+}+\widehat{l^-}\}$
	±2	$\sqrt{\frac{9\cdot5}{4\pi\cdot32}}(7\cos^2\theta-1)\sin^2\theta\, e^{\pm2i\phi}$	$(x^2-y^2)(7z^2-r^2)$	$\frac{1}{4}\{7\hat{l}_z{}^2-\widehat{l^2}-5,(\widehat{l^+})^2+(\widehat{l^-})^2\}$
	±3	$\sqrt{\frac{9\cdot35}{4\pi\cdot16}}\sin^3\theta\cos\theta\, e^{\pm3i\phi}$	$xz(x^2-3y^2)$	$\frac{1}{4}\{\widehat{l_z},(\widehat{l^+})^3+(\widehat{l^-})^3\}$
	±4	$\sqrt{\frac{9\cdot35}{4\pi\cdot128}}\sin^4\theta\, e^{\pm4i\phi}$	$x^4-6x^2y^2+y^4$	$\frac{1}{2}[(\hat{l}^+)^4+(\hat{l}^-)^4]$

can use the same quantum numbers as for one-electron atoms to describe each of the electrons in the first approximation. The set of quantum numbers n_i and l_i indicating the number of electrons in each of these states determines the *electron configuration* of the atom (or the electron shell in question). One distinctive feature of the Coulomb potential is the fact that electron energy depends only on the value of the quantum number n. In the electron paramagnetic resonance theory the Coulomb potential can be used to obtain analytical expressions for mean values of the distance between the electron and the nucleus, for example:

$$\langle r^2\rangle = \frac{\left[5n^4-3n^2l\,(l+1)\right]a_0^2}{2Z^2}. \tag{10.9}$$

For $3d$-electrons, i.e. $n = 3$, $l = 2$, Eq. (10.9) results in to the expression $\langle r^2\rangle = 126a_0^2/Z^2$. For the same electron state,

$$\langle r^4\rangle = 25{,}5a_0^4/Z^4. \tag{10.10}$$

Of course, the values obtained with the help of (10.9) or (10.10) should be regarded as approximate because the Coulomb field hardly ever occurs in actual paramagnetic centers. However it is very important to be able to make at least an approximate comparison of theoretical estimates and actual EPR spectrometric data.

Wave functions of p- and d- states. Further we will need expressions for electron wave functions in the spherically symmetric field as the real linear combinations of the wave functions in the form (10.8). Let us construct such combinations for values $l = 1$ and $l = 2$ frequently used in electron paramagnetic resonance theory. As mentioned above, the state vector with known n, l and m is generally denoted by $|nlm\rangle$. However, if for a given problem the values of n and l are fixed, for a reason of simplicity n and l indices are omitted and $|nlm\rangle \equiv |m\rangle$. For example, state vectors for $l = 1$ and any n value are denoting by $|0\rangle$, $|1\rangle$ and $|-1\rangle$. As states $|m\rangle$ and $|-m\rangle$ correspond to a pair of complex conjugates of the wave functions, they can be replaced with their real combinations. The corresponding wave functions are:

$$|m_+\rangle = \frac{1}{\sqrt{2}}\left(|m\rangle + |-m\rangle\right); \quad |m_-\rangle = \frac{1}{i\sqrt{2}}\left(|m\rangle - |-m\rangle\right). \tag{10.11}$$

Using explicit expressions for the spherical functions listed in Table 10.3, one can easily obtain real wave functions for $l = 1$

$$\begin{gathered} |0\rangle \equiv p_z = \sqrt{\frac{3}{4\pi}} R_{n1}(r)\,\frac{z}{r}, \\ |1_+\rangle \equiv p_x = \sqrt{\frac{3}{4\pi}} R_{n1}(r)\,\frac{x}{r}, \qquad |1_-\rangle \equiv p_y = \sqrt{\frac{3}{4\pi}} R_{n1}(r)\,\frac{y}{r}. \end{gathered} \tag{10.12}$$

By similar way one can construct five d-orbitals:

$$\begin{gathered} |0\rangle \equiv d_{z^2} = \sqrt{\frac{5}{4\pi}} R_{n1}(r)\,\frac{3z^2 - r^2}{r^2}, \\ |1_+\rangle \equiv d_{zx} = \sqrt{\frac{5}{4\pi}} R_{n1}(r)\,\frac{zx}{r^2}, \qquad |1_-\rangle \equiv d_{yz} = \sqrt{\frac{5}{4\pi}} R_{n1}(r)\,\frac{yz}{r^2}, \\ |2_+\rangle \equiv d_{x^2-y^2} = \sqrt{\frac{5}{4\pi}} R_{n1}(r)\,\frac{x^2 - y^2}{r^2}, \qquad |2_-\rangle \equiv d_{xy} = \sqrt{\frac{5}{4\pi}} R_{n1}(r)\,\frac{xy}{r^2}. \end{gathered} \tag{10.13}$$

For the sake of clarity, schematic representations of the spatial distribution of electron charge density for orbitals are frequently used. The construction of such distributions is based on the interpretation of the squared absolute value of the wave function as the probability density of finding an electron at a given point. Figure 10.2 presents the contours bounding the area where the electron charge is focused (the contour line corresponds to approximately 0.1 of the maximum probability density) for some p- and d-orbitals described by Eqs. (10.12) and (10.13).

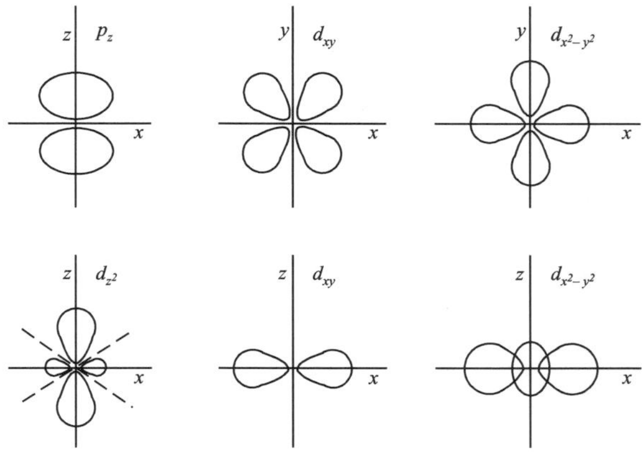

Fig. 10.2 Contours bounding the area where the electron charge is focused for some p- and d-orbitals

10.2.3 Interaction with Magnetic Field and Spin-Orbital Interaction

The second term in Eq. (10.7) corresponds to the energy of the interaction between an electron orbital angular momentum and a magnetic field. The third term determines the diamagnetic contribution to the magnetic susceptibility of the substance. Indeed, the magnetic susceptibility (χ) can be defined as

$$\chi = \mu_0 M/B \qquad (10.14)$$

where (μ_0 is the magnetic constant, M is the magnetization). Magnetization is expressed, in turn, as a derivative of energy with respect to magnetic field:

$$M = -n_0 \frac{\partial E}{\partial B} \qquad (10.15)$$

(where n_0 is the number of paramagnetic particles per a unit of volume). The last term in (10.7) is conveniently expressed in the ρ, ϕ, z cylindrical coordinate system (where the z axis is directed along the magnetic field vector $\vec{B}$): $\frac{e^2 B^2 \rho^2}{8m_e}$. After averaging this term with the wave functions of the given electron state, diamagnetic contribution to the total energy of the electron (E_{dia}) is

$$E_{dia} = \frac{e^2 B^2 \langle \rho^2 \rangle}{8m_e},$$

hence the expression of the diamagnetic contribution to the magnetization according to (10.15):

$$M_{dia} = -\frac{e^2 n_0 B \langle \rho^2 \rangle}{4m_e}.$$

Using (10.14) one easily obtains the expression for diamagnetic susceptibility:

$$\chi_{\text{dia}} = -\frac{e^2 n_0 \langle \rho^2 \rangle}{4m_e}. \tag{10.16}$$

By substituting the spherically symmetrical electrostatic field (10.4) into Eq. (10.2), one obtains the third term of the Hamiltonian in the form $\zeta(r)\hat{\vec{l}} \cdot \hat{\vec{s}}$, with

$$\zeta(r) = -\frac{e\hbar^2}{2m_e^2 c^2} \frac{1}{r} \frac{\partial V}{\partial r}. \tag{10.17}$$

Here we used (10.6) for the orbital angular momentum.

This phenomenon is called the spin-orbit interaction because the third term of the Hamiltonian is a scalar product of the orbital and spin angular momenta. It should be noted that usually $\zeta(r) < 0$ as the scalar potential appearing in (10.17) tends to decrease with increasing the distance from a nucleus.

Thus, for the interaction of an electron, with the spherically symmetric electrostatic field of the nucleus and the homogeneous external magnetic field, the Hamiltonian of the electron can be written in the following form:

$$\widehat{\mathcal{H}}_e = \frac{p^2}{2m_e} - eV + \mu_B \left(\hat{\vec{l}} + 2\hat{\vec{s}} \right) \cdot \vec{B} + \zeta(r)\hat{\vec{l}} \cdot \hat{\vec{s}} + \frac{e^2 \left(\vec{B} \times \vec{r} \right)^2}{8m_e c^2}. \tag{10.18}$$

10.2.4 Many-Electron Atom: Spherically Symmetric Self-Consistence Field

The Hamiltonian form for a many-electron case. Equation (10.18) can be extended to many electron atoms (ions). For a N-electron atom the Hamiltonian can be written in the following form:

$$\widehat{\mathcal{H}}_e = \sum_{i=1}^{N} \left[\frac{p_i^2}{2m_e} - eV_i + \sum_{k<i} \frac{e^2}{r_{ik}} \right] + \mu_B \left(\hat{\vec{L}} + 2\hat{\vec{S}} \right) \cdot \vec{B} + \lambda \hat{\vec{L}} \cdot \hat{\vec{S}} + \sum_{i=1}^{N} \frac{e^2 \left(\vec{B} \times \vec{r}_i \right)^2}{8m_e c^2}. \tag{10.19}$$

Here we have added a term describing the Coulomb interaction between electrons ($r_{ik} = |\vec{r}_i - \vec{r}_k|$ being the distance between the interacting electrons) and introduced the full atomic orbital ($\hat{\vec{L}}$) and spin ($\hat{\vec{S}}$) moments (see Sect. 1.1); the spin-orbit interaction constant λ appearing in (10.19) is no longer determined by a formula similar to (10.17). The expression for the spin-orbit interaction is written in a simplified form containing only the dependence on the total moments $\hat{\vec{L}}$, $\hat{\vec{S}}$. This simplified form will be sufficient to solve the problems discussed herein. Note that Eqs. (10.18) and (10.19) do not take into account the interaction of an atom with the magnetic fields produced by other particles (e.g. nuclei). The required changes will be made as appropriate. Thus, in this section, we have presented the electron Hamiltonian in a form convenient for calculations and clarified the physical meaning of its individual terms. The electrostatic interaction of the electrons with the nucleus and between electrons makes a major contribution to energy; however, the electron paramagnetic resonance theory is mostly interested in fine interactions such as those between spin and orbital magnetic moments with the magnetic field as well as spin-orbit interaction. Their role in the formation of the EPR spectra is a major subject of this part of the book.

Hartree-Fock approximation. If one neglects the magnetic interaction in an atom with N electrons, the Hamiltonian (10.19) takes the following form:

$$\widehat{\mathcal{H}}_{\mathrm{e}} = \sum_{i=1}^{N} \left(\frac{p_i^2}{2m} - \frac{e^2 Z}{r_i} \right) + \sum_{i=1}^{N} \sum_{k<i} \frac{e^2}{r_{ik}}. \tag{10.20}$$

Here, as before, $r_{ik} = |\vec{r}_i - \vec{r}_k|$.

The interaction between the electrons is strong enough and cannot be regarded as a perturbation. However, the total interaction can be divided into two parts: in this case, it is possible to obtain easily wave functions and energies for the main part, and consider the rest as a perturbation. This method is generally known as the *self-consistent field method (SCF)* or *Hartree-Fock approximation.*

Following this approach, one may conveniently represent the main part of the interaction as the sum interactions of each of electrons with a spherically symmetric potential $V_i'(r)$. This potential is created by the nucleus and all the electrons, except for the i-th; i.e. in this approximation, called the spherically symmetric self-consistent field approximation, the Hamiltonian can be written as

$$\widehat{\mathcal{H}}_{\mathrm{ssscf}} = \sum_{i}^{N} \left[\frac{p_i^2}{2m} - eV_i'(r_i) \right]. \tag{10.21}$$

Here r_i is the coordinate of the i-th electron. Every potential $V_i'(r)$ is calculated as an average over all directions; the averaging is performed using wave eigenfunctions of the state described by the Hamiltonian (10.21) which, in its turn, depends on the type of this potential. This field is described as a self-consistent field. Without getting into highly non-trivial approaches to calculating self-consistent field potentials and wave functions, let us highlight one important feature of the Hamiltonian (10.21) compared to (10.20). The Hamiltonian (10.21) describes the motion of an electron in a spherically symmetric field. In this case, the wave functions have the same form (10.8) as for the hydrogen-like ion, differing only in the radial part. This removes the degeneracy in l, i.e. electron energy E_{inl} now depends both on n and the angular momentum l. In this case, for fixed n, the energy grows with the l increase. Like for the hydrogen atom, radial functions have $n - 1$ nodes between 0 and ∞ and vanish at $r = 0$ if $l \neq 0$. This analogy allows us to continue using the quantum numbers n and l to classify states.

Model of the atom within a spherically symmetric SCF approximation. The energy of an atom in this approximation is completely determined by a set of quantum numbers n_i and l_i, i.e. by its electron configuration. The state of the atom, described by its wave function, is determined by the values of the quantum numbers n_i, l_i and m_i. Thus, the i-th vector of the state of the j-th electron can be written as $|n_i l_i m_i\rangle^{(j)}$. Then, following the particle identity principle, the state vector of the whole atom in this approximation is a combination of the products of state vectors for all electrons, which is anti-symmetric with respect to the permutation of any two electrons. It is known that this anti-symmetric combination $|l_1 l_2 \ldots l_N; m_1 m_2 \ldots m_N\rangle_{\text{à}}$ represents the Slater determinant:

$$|l_1 l_2 \ldots l_N; m_1 m_2 \ldots m_N\rangle_{\grave{a}} = \det\left\{|l_i m_i\rangle^{(j)}\right\}$$

In case of equal n_i and l_i values for all electrons, the Slater determinant can be written as follows (n_i and l_i indices is omitted to simplify):

$$\det\left\{|m_i\rangle^{(j)}\right\} = \frac{1}{\sqrt{N!}} \begin{vmatrix} |m_1\rangle^{(1)} & |m_1\rangle^{(2)} & \ldots & |m_1\rangle^{(N)} \\ |m_2\rangle^{(1)} & |m_2\rangle^{(2)} & \ldots & |m_2\rangle^{(N)} \\ \ldots & \ldots & \ldots & \ldots \\ |m_N\rangle^{(1)} & |m_N\rangle^{(2)} & \ldots & |m_N\rangle^{(N)} \end{vmatrix}. \tag{10.22}$$

Instead of the set of vectors $|m_i\rangle$ one can use any other set of linearly independent vectors obtained by the superposition of vectors from the initial set. In this case the atomic state vector will be determined by a formula similar to (10.22).

10.2.5 Many-Electron Atom: Correlation Interaction

The energy of the atom within the spherically-symmetric self-consistent field approximation does not depend on the total orbital moment L and spin moment S of the atom. The residual (or correlation) interaction ($\widehat{\mathcal{H}}_{\text{res}}$), which determines the formation of L and S, is a perturbation with respect to the interaction described by the Hamiltonian $\widehat{\mathcal{H}}_{\text{ssscf}}$. Unlike the above Hamiltonian, which is symmetrical to independent rotations of coordinates associated with individual electrons, the Hamiltonian (10.20) and, consequently $\widehat{\mathcal{H}}_{\text{res}}$, is invariant only with respect to *simultaneous* rotations of the coordinates associated with all the electrons. If one takes into account the residual interaction, the L and S degeneracy, which takes place for the spherically symmetric SCF approximation, is mainly removed. Now the atomic energy for the same electron configuration depend on the total orbital and spin momenta. As a result, the energy level corresponding to the given electronic configuration is split into individual sublevels (the spectral terms). The spectral terms are denoted by symbols

^{2S+1}S, where $2S + 1$ (multiplicity) determines the number of sub-levels resulting from the subsequent splitting of a term with the given L under the spin-orbit interaction. The letters $S, P, D, \ldots$ are used by analogy to describe one-electron states instead of the digitally set number $L = 0, 1, 2, \ldots$.

Electron paramagnetic resonance can only be observed in atomic and molecular systems remaining in a given state existing during considerably long time (i.e. in stable or metastable state). Therefore, for the theory of the electron paramagnetic resonance it is of great importance which of the terms corresponds to the lowest energy. The term with the lowest energy (the principal term of the atom or ion) is determined by Hund rules:

(1) among all the terms in the given configuration, select the terms with maximum S;
(2) among all the terms with maximum S, select the term with the highest L.

The terms possible in the given configuration are determined by momentum addition rules and the Pauli principle. The simplest case is the term with the highest possible values of S and L for the given configuration.

Example 1. Let us determine the ground state for the Cr^{3+} ion. The configuration of the outer shells for the Cr atom can be written as $4s3d^5$, for the Cr^{3+} ion as $3d^3$. Electrons with $l = 2$ can have $2l + 1 = 5$ orbital states differing in the values of m projections. Consequently, all three electrons can have the same spin state without violating the Pauli principle, with the maximum value $S = 3/2$ according to Hund's first rule. In this case, the only valid values for m resulting in the highest S are 2, 1 and 0 as the states of the electrons must be different. Thus, $\max L = \max M = \sum_{i=1}^{3} m_i = 3$. Consequently, the ground term of the Cr^{3+} ion is written as ^4F.

Example 2. In the Fe^{2+} ion (configuration $3d^6$) only five of the electrons can have the same spin projection, while the sixth should have the reverse spin direction, i.e. $S = 2$. Five electrons with the same spin projections will result in the zero total orbital angular momentum as all their projections m must be different, and they adopt all possible values. The total orbital angular momentum is determined by the sixth electron, i.e. $L = l = 2$. Consequently, the ground term of the Fe^{2+} ion is ^5D.

Note that in the absence of external fields, atomic energy does not depend on the values of the projections M and M_S. Therefore, each term remains degenerate $(2L + 1)(2S + 1)$ times. The state vectors in this approximation are determined in accordance with Wigner theorem, which will be discussed in Sect. 10.3.4.

10.2.6 Spin-Orbital Interaction in a Many-Electron Atom

When taking into account the spin-orbit interaction, the state of an atom must be characterized by the value of the total angular momentum $\hat{\vec{J}} = \hat{\vec{L}} + \hat{\vec{S}}$. The value of J is specified in accordance to the moment a addition rule and take the values from $L - S$ to $L + S$. Each of the terms with certain L and S is splitted under the spin-orbit interaction with accordance of a number of possible values J, i.e. $2L + 1$ or $2S + 1$. Each level characterized by a certain value of J degenerates $2J + 1$ times. This degeneracy can be completely eliminated under the action of the magnetic field (*Zeeman effect*).

The ground state is determined by a rule supplementing the Hund's rules: if the shell is filled less than half, the state with the lowest J $(J = |L - S|)$ has the lowest energy; however, if the shell is filled more than half, the state with the highest J $(J = |L + S|)$ has the lowest energy. In symbolic notation, the value of J is written as a subscript to the right from the term: ${}^{2S+1}L_J$.

To demonstrate the successive effect of these interactions on the energy level structure, a diagram of levels for typical paramagnetic ions (a lanthanide ion Nd^{3+} (*a*) and a transition group ion Cr^{3+} ion in its ground configuration $3d^3$ (*b*) are presented in Fig. 10.1. This diagram also shows the changes resulting from additional interactions with the electrostatic crystal field of the lattice and the magnetic field applied during the experiment. These interactions will be discussed in detail in the following sections; the splitting of the levels occurring in the magnetic field will be considered in Chap. 13.

10.3 Selected Information on the Group Theory

The application of perturbation theory to the crystal field model necessitates the use of numerical computation when solving secular equations, which makes it impossible to analyze the results in the general form or to draw qualitative conclusions. This difficulty can be avoided if one takes in account the properties of the crystal field symmetry with the aid of group theory. As applied to electron paramagnetic resonance, the group theory provides a simple yet strict method for solving numerous problems: determining the qualitative pattern of energy level splits in the crystal electrostatic field, unifying the calculations of crystal field matrix elements and determining the conditions at which they become zero, developing selection rules for transitions. The key principles of group theory necessary to use the properties of crystal field symmetry are outlined below.

10.3.1 Elements and Transformations of Symmetry

If an object is transformed into itself as a result of a certain transformation, we say that this object is invariant to this transformation, or that this transformation is a *symmetry transformation*. Symmetry transformations are determined by symmetry elements: a symmetry plane or reflecting plane (σ), an axis of n-fold symmetry (C_n) and a rotoreflection axis (S_n). A rotoreflection axis S_n determines a single symmetry transformation, constituting from a rotation by an angle of $2\pi k/n$ with an obligatory subsequent reflexion onto the plane perpendicular to the S_n axis. It should be mentioned that an individual rotation or reflection could not be regarded as a symmetry transformation for rotoreflection axes.

Because of symmetry transformations may be treated as some operations over a function or a vector, they are denoted similarly to quantum-mechanical operators: $\widehat{C}_n^k$, $\widehat{S}_n^k$ and correspond to pure rotations, rotation–reflections by $2\pi k/n$, and reflection operations by $\widehat{\sigma}$ respectively. The $\widehat{\sigma}$ symbols may have the lower indices. These are used to denote the symmetry plane orientation relative to the symmetry axis: $\widehat{\sigma}_h$ means that the plane is perpendicular to the axis (the "horizontal" plane), $\widehat{\sigma}_v$ means that the plane passes through the axis (the "vertical" plane). If there are the axes C_2 perpendicular to the first symmetry axis C_n, then it is possible to introduce the symmetry planes σ_d, which pass through the axis C_n and divide the angle between two axes C_2 in half. One special type of transformation is inversion (point reflection) $\hat{\imath}$. It is evident that $\hat{\imath} \equiv \widehat{S}_2$. Sometimes it is preferred to consider rotation-inversion $\hat{\imath}\widehat{C}_n$ rather than rotation-reflection. The result of two sequential transformations (first is $\widehat{G}_1$, then $\widehat{G}_2$) is called the transformation product: $\widehat{G}_3 = \widehat{G}_2\widehat{G}_1$. In general, transformations do not commute $\widehat{G}_1\widehat{G}_2 \neq \widehat{G}_2\widehat{G}_1$.

Group concept. A set of transformations (group elements) $\widehat{G}_i$ forms a group $\mathcal{G}$ if it includes products of any $\widehat{G}_i\widehat{G}_k$, an identity transformation $\widehat{E}$ and a transformation $\widehat{G}_i^{-1}$ inverse to each of the $\widehat{G}_i$ transformations. Not only transformations, but objects of any other character (such as numbers, vectors and matrices) can be *elements of group*. A group, all elements of which commute, is called an *Abelian group*. A group, all elements of which can be presented as a power of a certain element, is called a *cyclic group*. The number of elements in a group is called the *group order* and is denoted as g (the number of elements in a group may be equal to infinity).

Usually, the group of symmetry transformations, or *the symmetry group*, is denoted with the same letter as the main symmetry element, but of a different font: $\mathcal{C}_n$, $\mathcal{S}_n$. The corresponding groups include all rotations $\widehat{C}_n^k$ $(k = 0, 1, \ldots, n-1)$ or $\widehat{S}_n^k$ (n is always even for $\mathcal{S}_n$). Indices added to the group designation mean that other symmetry elements are added to the symmetry axis C_n. If a group contains a plane σ_h or σ_v, it is denoted by ($\mathcal{C}_{nh}$) or ($\mathcal{C}_{nv}$) respectively. Groups which along with C_n have two-fold symmetry axes (C_2') perpendicular to C_n are designated as $\mathcal{D}_n$. If a group contains planes σ_d yet, then it is denoted by $\mathcal{D}_{nd}$.

It is easy to see that the addition of one symmetry element often results of necessity in the appearance a number of other elements: for example, if one plane σ_v is added to the axis C_n, then n − 1 other planes σ_v appear automatically, and when a plane σ_v is added in the group $\mathcal{D}_n$ a plane σ_h appears. Groups formed by a single reflection plane and a center of inversion have special designations: C_s and C_i respectively. It is evident that $\widehat{C}_i \equiv \widehat{S}_2$ and $\widehat{C}_s \equiv \widehat{C}_{1h}$.

The so-called *cubic groups* play an important part. These groups, which could be built on the basis of the group $\mathcal{D}_2$, have four C_3' axes tilted at equal angles to three mutually perpendicular axes. When the three perpendicular axes are C_2, similarly to the group $\mathcal{D}_2$, the group $\mathcal{T}$ is formed which includes all rotations around the tetrahedron axes (Fig. 10.3a). Groups $\mathcal{T}_h$ and $\mathcal{T}_d$ are formed by adding planes σ_h or σ_d, respectively. If C_2 axes are replaced with C_4, six inclined C_{2d} axes are then added automatically (see Fig. 10.3b) and a group $\mathcal{O}$ of rotations around cubic or

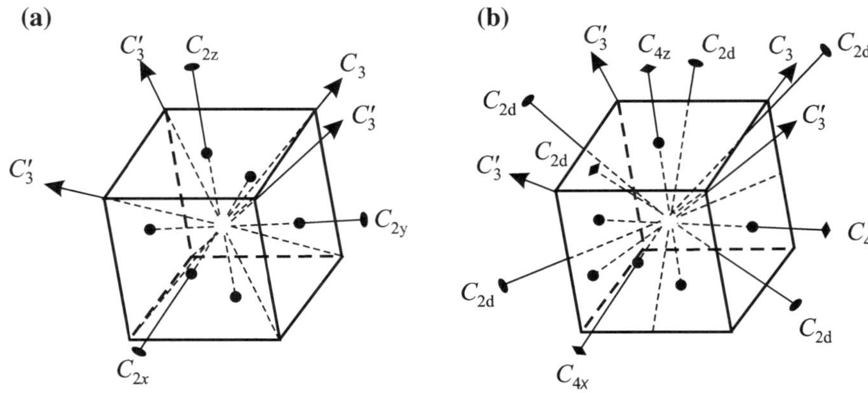

Fig. 10.3 Simmetry axes of tetrahedron (**a**) and cube (**b**) or octahedron

octahedron symmetry axes is obtained. The group contains 24 transformations: $\widehat{E}$, $4\widehat{C}_3^1$, $4\widehat{C}_3^2$, $3\widehat{C}_2$, $3\widehat{C}_4^1$, $3\widehat{C}_4^3$, $6\widehat{C}_{2d}$. Inversion adds another 24 elements: $\hat{\imath}$, $4\widehat{S}_3^1$, $4\widehat{S}_3^2$, $3\widehat{\sigma}_h$, $3\widehat{S}_4^1$, $3\widehat{S}_4^3$ è $6\widehat{\sigma}_d$. The resulting group of 48 elements is the complete symmetry group of a cube (or octahedron) $\mathcal{O}_h$.

All the finite groups are *point groups* as all the transformations they contain keep one point at its place. In crystals there are only one-, two-, three-, four- and six-fold symmetry axes. There are 32 point groups known as it crystallographic groups, each of them belongs to a specific crystallographic system: triclinic (point groups $\mathcal{C}_i$) , monoclinic ($\mathcal{C}_2$, $\mathcal{C}_{2h}$ and $\mathcal{C}_s$), trigonal ($\mathcal{C}_3$, $\mathcal{C}_{3v}$, $\mathcal{S}_6$, $\mathcal{D}_3$ and $\mathcal{D}_{3h}$), rhombic ($\mathcal{D}_2$, $\mathcal{D}_{2h}$ and $\mathcal{C}_{2v}$), tetragonal ($\mathcal{S}_4$, $\mathcal{C}_{4v}$, $\mathcal{D}_{2d}$, $\mathcal{D}_4$ and $\mathcal{D}_{4h}$), hexagonal ($\mathcal{D}_6$, $\mathcal{D}_{3h}$, $\mathcal{D}_{6h}$ and $\mathcal{C}_{6v}$) and cubic system ($\mathcal{T}$, $\mathcal{T}_d$, $\mathcal{T}_h$, $\mathcal{O}$ and $\mathcal{O}_h$). The all above listed groups contain a finite number of elements. Along with the above mentioned groups, there is an infinite point *rotation group* $\mathcal{K}$ which is of great importance for practical applications and includes arbitrary angle rotations around arbitrary axes passing through the center. With added inversions and, as a consequence, an infinite number of arbitrary reflection planes passing through the center, the group becomes a *full orthogonal group* ($\mathcal{K}_h$), i.e. a spherical symmetry group.

Conjugated elements and classes. Elements $\widehat{G}_1$ and $\widehat{G}_2$ are called *conjugated* if they are related by the requirement $\widehat{G}_2 = \widehat{G}_i\widehat{G}_1\widehat{G}_i^{-1}$, where $\widehat{G}_i$ is any element of the group. The assembly of all elements conjugated with the element in question, i.e. those generated by all elements of the $\widehat{G}$ group, forms a *class*. Element $\widehat{E}$ forms a class by itself. Each element of an Abelian group also constitutes a class. It is possible to show that conjugate transformations are rotations (reflections) around axes (in planes) which are converted into each other by means of any other group transformations. The ϕ and $2\pi - \phi$ angle rotations belong to one class if the group contains a transformation which changes the rotation direction (the rotation of an axis by 180 degrees or its reflection in the plane passing through the axis). These rules help to divide group transformations into classes. A class may be determined by a list of constituent elements in braces. Thus, $\mathcal{C}_{nv}$ groups with odd n include classes$\{\widehat{E}\}, \{\widehat{C}_n^k, \widehat{C}_n^{n-k}\}, \{\widehat{\sigma}_{v1}, \widehat{\sigma}_{v2}, \ldots, \widehat{\sigma}_{vn}\}$, groups with even n include classes $\{\widehat{E}\}, \{\widehat{C}_n^k, \widehat{C}_n^{n-k}\}, \{\widehat{\sigma}_{v1}, \widehat{\sigma}_{v3}, \ldots, \widehat{\sigma}_{v\,n-1}\}, \{\widehat{\sigma}'_{v2}, \widehat{\sigma}'_{v4}, \ldots, \widehat{\sigma}'_{vn}\}$. Rotations around the axis C_n convert σ_{vi} with different i (or σ'_{vi} with different i) into each other, but never σ_{vi} into σ'_{vi} or vica versa. 24 elements in the group $\mathcal{O}$ are divided into five classes: $\{\widehat{E}\}, \{8\widehat{C}_3\}, \{3\widehat{C}_2\}, \{6\widehat{C}_4\}, \{6\widehat{C}_{2d}\}$.

Isomorphism and homomorphism. Groups $\mathcal{G}$ and $\mathcal{H}$ are *isomorphic* if their elements $\widehat{G}_i$ and $\widehat{H}_j$ are in one-to-one correspondence so that a product $\widehat{G}_i\widehat{G}_j$ corresponds to the product of the corresponding elements $\widehat{H}_i\widehat{H}_j$. If a certain element of the group $\mathcal{H}$ corresponds to certain element of the group $\mathcal{G}$, but the same element of the group $\mathcal{H}$ may also correspond to other elements of the group $\mathcal{G}$, then the group $\mathcal{H}$ is homomorphic to the group $\mathcal{G}$. For example, groups $\mathcal{C}_2$, $\mathcal{C}_i \equiv \mathcal{S}_2$ and $\mathcal{C}_s \equiv \mathcal{C}_{1h}$; $\mathcal{C}_{4v}$, $\mathcal{D}_4$ and $\mathcal{D}_{2d}$; $\mathcal{C}_n$ and $\mathcal{S}_n$; $\mathcal{C}_{nv}$ and $\mathcal{D}_n$; $\mathcal{O}$ and $\mathcal{T}_d$ are isomorphic. The homomorphism concept is used in the definition of the group representation, the most important concept in the group theory.

10.3.2 Group Representation

The *linear representation* $\mathfrak{G}$ of the group $\mathcal{G}$ is refered as a group of square matrices $\boldsymbol{G}(\widehat{G}_i)$ of the f order homomorphic to this group. If we transform all representation matrices by the following way $\boldsymbol{S}\boldsymbol{G}(\widehat{G}_i)\boldsymbol{S}^{-1} = \boldsymbol{G}'(\widehat{G}_i)$, where $\boldsymbol{S}$ is the same matrix for all $\widehat{G}_i$ transformations, then the set of matrices $\boldsymbol{G}'(\widehat{G}_i)$ forms a representation $\mathfrak{G}'$ termed *equivalent* to the representation $\mathfrak{G}$. It is possible to match an equivalent unitary representation consisting of unitary matrices to any linear representation. Linear representation matrices may be treated as tables of coefficient $G_{\mu\nu}$ of the linear transformation in a certain linear space of the f-dimension, invariant to symmetry transformations of the $\mathcal{G}$ group. If $|\nu\rangle$ is a vector of this space, then the symmetry transformation effect on it will be expressed by the following equation:

$$\widehat{G}|\nu\rangle = \sum_{\rho=1}^{f} |\rho\rangle G_{\rho\nu}. \tag{10.23}$$

The set of vectors $|\nu\rangle$ transformed according to (10.23) is called the *representation basis*. A transfer to an equivalent representation is equipollent to a transition to new basis vectors in this space. As it is always possible to obtain an equivalent unitary representation, it is always possible to select a basis from mutually orthogonal vectors. The representation matrix order f is known as *representation dimension.* Matrix traces (spurs) are also called *representation characters* $(\chi(\widehat{G}))$. The matrix spurs are of such a nature that equivalent representations have equal characters and that the characters of all transformations pertaining to one class are also equal. A representation is called *irreducible* if neither this representation itself nor any of its possible equivalent representations consist of quasi-diagonal matrices of similar structure. A representation is *reducible* if all transformations defined by its matrices keep a certain subspace invariant (this subspace should not be identical to the whole space). As shown by the group theory, the required and sufficient evidence of the representation irreducibility is the absence of a matrix that differs from the multiple of the identity matrix and that *commutes with all representation matrices*, i.e. there should be no matrix $\boldsymbol{A}$, where $\boldsymbol{G}(\widehat{G}_i)\boldsymbol{A} - \boldsymbol{A}\boldsymbol{G}(\widehat{G}_i) \neq 0$, if only $\boldsymbol{A} \neq k\boldsymbol{I}$, where $\boldsymbol{I}$ is an identity matrix (Schur lemma).

Properties of irreducible representations of finite groups. The matrix elements of irreducible representations of finite groups satisfy the requirements of orthogonality relations proceeding from the Schur lemma:

$$\sum_{\widehat{G}} G_{\mu\nu}^{(i)}(\widehat{G}) G_{\kappa\lambda}^{(j)*}(\widehat{G}) = \frac{g}{f_i} \delta_{ij} \delta_{\mu\kappa} \delta_{\nu\lambda}. \tag{10.24}$$

Here indices i, j stand for irreducible representation numbers; f_i is the dimension (matrix order) of the irreducible representation $\mathfrak{G}^{(i)}$; μ, κ are the row numbers; ν, λ are the column numbers. The number of inequivalent irreducible representations is equal to the number of classes (n_C) in the group. Dimensions f_i of the irreducible representations satisfy the equation

$$\sum_{i=1}^{n_C} f_i^2 = g. \tag{10.25}$$

The orthogonality relation (10.24) between the matrix elements of an irreducible representation results in the orthogonal relation between the following characters:

$$\sum_{\mathcal{G}} \chi^{(i)}(\widehat{G}) \chi^{(j)}(\widehat{G})^* = g\delta_{ij}. \tag{10.26}$$

Here $\chi^{(i)}(\widehat{G})$ and $\chi^{(j)}(\widehat{G})$ are the characters of irreducible representations $\mathfrak{G}^{(i)}(\widehat{G})$ and $\mathfrak{G}^{(j)}(\widehat{G})$.

One-dimensional representations $f_i = 1$ constitute a special case of irreducible representations. In this case, the character $\chi^{(j)}(\widehat{G})$ is a multiplication factor of the basis vector used in the transformation $\widehat{G}$. There always exists an *identity representation* corresponding to a basis vector which is invariant to group transformations. One-dimensional irreducible representations are denoted by $\mathfrak{A}$ and $\mathfrak{B}$; in this case $\mathfrak{B}$ indicates the reversal of the basis vector sign when rotating by the angle $\frac{2\pi}{n}$. Irreducible representations differing in the behaviors of the basis vectors in cases of reflection are indicated with one or two prime marks. The g index is used for parity (invariancy) and u for imparity (sign reversal) relative to inversions; elsewhere numeric notation is applied. Thus, the symbol $\mathfrak{T}_{1g}$ is used for presentations of object transformations in three-dimensional space which are invariant to inversion.

The character of the identity transformation $\widehat{E}$ for any representations is equal to their dimensionality.

Table 10.4 The characters of some groups and bases of their irreducible representations

IR of the group:	The characters of the transformation					Bases transforming as per IR		
C_i	$\widehat{E}$				$\hat{\imath}$	C_i	C_2	C_s
C_2	$\widehat{E}$				$\widehat{C}_2$			
C_s	$\widehat{E}$				$\widehat{\sigma}$			
$\mathfrak{A}_g$ $\mathfrak{A}$ $\mathfrak{A}'$	1				1		z	x; y
$\mathfrak{A}_u$ $\mathfrak{B}$ $\mathfrak{A}''$	1				−1	x; y; z	x; y	z
C_{3v}	$\widehat{E}$	$2\widehat{C}_3$			$3\widehat{\sigma}_v$	—		—
$\mathcal{D}_3$	$\widehat{E}$	$2\widehat{C}_3$			$3\widehat{C}_2$	C_{3v}		$\mathcal{D}_3$
$\mathfrak{A}_1$	1	1			1	z		
$\mathfrak{A}_2$	1	1			−1	—		z
$\mathfrak{E}$	2	−1			0	x, y		x, y
$\mathfrak{E}'_1$	{ 1	−1			i	—		—
	{ 1	−1			$-i$			
$\mathfrak{E}'_2$	2	1			0			
C_{4v}	$\widehat{E}$	$\widehat{C}_2$	$2\widehat{C}_4$	$2\widehat{\sigma}_v$	$2\widehat{\sigma}'_v$	C_{4v}	$\mathcal{D}_4$	$\mathcal{D}_{2d}$
$\mathcal{D}_4$	$\widehat{E}$	$\widehat{C}_2$	$2\widehat{C}_4$	$2\widehat{C}_2$	$2\widehat{C}''_2$			
$\mathcal{D}_{2d}$	$\widehat{E}$	$\widehat{C}_2$	$2\widehat{S}_4$	$2\widehat{C}'_2$	$2\widehat{\sigma}'_d$			
$\mathfrak{A}_1$	1	1	1	1	1	z	—	—
$\mathfrak{A}_2$	1	1	1	−1	−1	—	z	—
$\mathfrak{B}_1$	1	1	−1	1	−1	—	—	—
$\mathfrak{B}_2$	1	1	−1	−1	1	—	—	z
$\mathfrak{E}$	2	−2	0	0	0	x, y	x, y	x, y
$\mathfrak{E}'_1$	2	0	$\sqrt{2}$	0	0	—	—	—
$\mathfrak{E}'_2$	2	0	$-\sqrt{2}$	0	0	—	—	—
$\mathcal{O}$	$\widehat{E}$	$8\widehat{C}_3$	$3\widehat{C}_2$	$6\widehat{C}_{2d}$	$6\widehat{C}_4$	—		—
$\mathcal{T}_d$	$\widehat{E}$	$8\widehat{C}_3$	$3\widehat{C}_2$	$6\widehat{\sigma}_d$	$6\widehat{S}_4$	$\mathcal{O}$		$\mathcal{T}_d$
$\mathfrak{A}_1$	1	1	1	1	1	—		—
$\mathfrak{A}_2$	1	1	1	−1	−1	$x\,y\,z$		—
$\mathfrak{E}$	2	−1	2	0	0	$\frac{\sqrt{3}}{2}(x^2-y^2)$, $\frac{1}{2}(3z^2-r^2)$		$\frac{\sqrt{3}}{2}(x^2-y^2)$, $\frac{1}{2}(3z^2-r^2)$
$\mathfrak{T}_1$	3	0	−1	−1	1	x, y, z		—
$\mathfrak{T}_2$	3	0	−1	1	−1	xy, yz, zx		xy, yz, zx; x, y, z
$\mathfrak{E}'_1$	2	1	0	0	$\sqrt{2}$	—		—
$\mathfrak{E}'_2$	2	1	0	0	$-\sqrt{2}$	—		—
$\mathfrak{G}'$	4	−1	0	0	0	—		—

Irreducible representations (IR) and their characters for different finite groups are presented below in Table 10.4. The table also contains irreducible representations for transformations of orthogonal coordinates themselves.

Characters determination of group $\mathcal{O}$. As an example, let us consider the characters of an irreducible representation of the group $\mathcal{O}$. It follows from the properties of the irreducible representation that $\mathcal{O}$ has five irreducible representations, whereas

Eq. (10.25) at $g = 24$ and $n_C = 5$ results in $f_i = 1, 1, 2, 3, 3$. For group $\mathcal{O}$ it is convenient to select coordinate axes in such a way that they coincide with C_4 symmetry axes. In this case, the representation matrices have the simplest form, for example:

$$\begin{matrix} \widehat{E} & \widehat{C}_3 & \widehat{C}_2 & \widehat{C}_{2d} & \widehat{C}_4 \\ \begin{pmatrix} 1 & 0 & 0 \\ 0 & 1 & 0 \\ 0 & 0 & 1 \end{pmatrix}, & \begin{pmatrix} 0 & 1 & 0 \\ 0 & 0 & 1 \\ 1 & 0 & 0 \end{pmatrix}, & \begin{pmatrix} -1 & 0 & 0 \\ 0 & -1 & 0 \\ 0 & 0 & 1 \end{pmatrix}, & \begin{pmatrix} 0 & 1 & 0 \\ 1 & 0 & 0 \\ 0 & 0 & -1 \end{pmatrix}, & \begin{pmatrix} 0 & 1 & 0 \\ -1 & 0 & 0 \\ 0 & 0 & 1 \end{pmatrix} \end{matrix}$$

etc.

From here the characters listed in Table 10.4 are immediately obtainable. It follows from the table that the set of orthogonal coordinates $x,\ y,\ z$ may serve as a basis for the irreducible representation $\mathfrak{T}_1$ of the group $\mathcal{O}$, whereas the set of products $xy,\ yz$ and zx may be used as a basis for the irreducible representation $\mathfrak{T}_2$. A set of orthogonal homogeneous polynomials $(1/2)(3z^2-r^2)$ and $(\sqrt{3}/2)(x^2-y^2)$ may serve as a basis of the two-dimensional irreducible representation $\mathfrak{E}$ (see Table 10.4). To verify this directly let us consider the transformation of polynomials for one of the rotations $\widehat{C}_3$:

$$\widehat{C}_3^1\left[\frac{1}{2}(3z^2 - r^2)\right] = \frac{1}{2}(3x^2 - r^2) = \frac{1}{4}(3z^2 - r^2) + \frac{3}{4}(x^2 - y^2),$$

$$\widehat{C}_3^1\left[\frac{\sqrt{3}}{2}(x^2 - y^2)\right] = \frac{\sqrt{3}}{2}(y^2 - z^2) = -\frac{\sqrt{3}}{4}(3z^2 - r^2) - \frac{\sqrt{3}}{4}(x^2 - y^2)$$

or in the matrix form:

$$\widehat{C}_3^1\begin{bmatrix} \frac{1}{2}(3z^2 - r^2) \\ \frac{\sqrt{3}}{2}(x^2 - y^2) \end{bmatrix} = \begin{bmatrix} -\frac{1}{2} & \frac{\sqrt{3}}{2} \\ -\frac{\sqrt{3}}{2} & -\frac{1}{2} \end{bmatrix}\begin{bmatrix} \frac{1}{2}(3z^2 - r^2) \\ \frac{\sqrt{3}}{2}(x^2 - y^2) \end{bmatrix}. \tag{10.27}$$

The matrix in (10.27) is the matrix of the irreducible representation $\mathfrak{E}$ corresponding to one of the $\widehat{C}_3$ rotations. Its character $\chi = -1$. The matrix will take a different form for other $\widehat{C}_3$ rotations, but in all cases $\chi = -1$ because all $\widehat{C}_3$ rotations belong to the same class. In case of $\widehat{C}_2$ transformations, both polynomials remain unchanged, i.e. $\chi = 2$, whereas under $\widehat{C}_4$ and $\widehat{C}_{2d}$ transformations only term $(\sqrt{3}/2)(x^2 - y^2)$ changes its sign, than gives $\chi = 0$. Thus, the resulting set of characters is identical to that given in Table 10.4 for the irreducible representation $\mathfrak{E}$.

Irreducible representations of the rotation group. The rotation group $\mathcal{K}$ has an infinite number of irreducible representations denoted by $\mathfrak{D}^{(l)}$. The basis of the irreducible representation $\mathfrak{D}^{(l)}$ is a set of $2l + 1$ linearly independent spherical harmonics $Y_l^m(\theta, \phi)$, where θ and ϕ are the polar and azimuthal angles in spherical coordinates respectively. The characters of irreducible representations are calculated according to

$$\chi(\phi) = \frac{\sin[(2l + 1)\phi/2]}{\sin(\phi/2)}. \tag{10.28}$$

10.3.3 Basic Theorems

Decomposition of a reducible representation into irreducible components. From Eq. (10.26) one easily obtains a formula for expanding reducible representations into irreducible ones if one knows the characters of the reducible representation. As the characters of equivalent representations are identical, the character of the reducible representation $\chi(\widehat{G})$ can be written as

$$\chi(\widehat{G}) = \sum_{j=1}^{n_C} n_j \chi^{(j)}(\widehat{G}), \tag{10.29}$$

where n_j is the number of irreducible representations of the $\mathfrak{G}^{(j)}$ type. By multiplying both parts of Eq. (10.29) by $\chi^{(j)}(\widehat{G})^*$ and summing by group taking into account (10.26), one obtains

$$n_j = \frac{1}{g} \sum_{\widehat{G}} \chi(\widehat{G}) \chi^{(j)}(\widehat{G})^*. \tag{10.30}$$

As the characters of the transformations are identical for the same class, it is more convenient to use a modified formula:

$$n_j = \frac{1}{g} \sum_{C=1}^{n_C} g_C \chi(\widehat{G}_C) \chi_C^{(j)}(\widehat{G})^*, \tag{10.31}$$

where $\chi(\widehat{G}_C)$, $\chi_C^{(j)*}$ are transformation characters for a class C; g_C is the number of transformations in this class, n_C is the number of classes; the summation is performed over all classes.

For basis vectors of irreducible representations, let us use the notation $|j\mu\rangle$, where j is the number of the irreducible representation, μ is the number of the basis vector for this irreducible representation. If $|a\rangle$ is one of the vectors in the vector space defined by the reducible representation, it can be decomposed into non-normalized basis vectors $|aj\rangle$ transformed according to the irreducible representation of $\mathfrak{G}^{(j)}$:

$$|a\rangle = \sum_{j=1}^{n_C} |aj\rangle, \tag{10.32}$$

The vector $|aj\rangle$ can be determined according to

$$|aj\rangle = \frac{f_i}{g} \sum_{\widehat{G}} \chi^{(j)}(\widehat{G})^* \widehat{G} |a\rangle. \tag{10.33}$$

Equation (10.33) can be proved by inserting (10.32) into it and summation by diagonal elements taking into account Eq. (10.30). In this case $|aj\rangle = \sum_{i=1}^{N} |aj\mu\rangle$, where

$|aj\mu\rangle$ is transformed to μ row of the irreducible representation $\mathfrak{G}^{(j)}$ (here the vectors are also normalized). The operation on $|a\rangle$ in Eq. (10.33) has the meaning of projecting $|a\rangle$ onto the basis vector $|aj\rangle$, that is why the operator

$$\widehat{P}_j = \frac{f_i}{g} \sum_{\widehat{G}} \chi^{(j)}(\widehat{G})^* \widehat{G}$$

is called the projection operator.

Direct group products and their representations. If a group consists of elements representing products of elements from another two groups $\mathcal{F}$ and $\mathcal{H}$, all elements of which are different except for $\mathcal{E}$, and all elements of $\mathcal{F}$ commute with elements of $\mathcal{H}$, then group $\mathcal{G}$ is the *direct product of groups* $\mathcal{F}$ and $\mathcal{H}$: $\mathcal{G} = \mathcal{F} \otimes \mathcal{H}$. The product of the irreducible representation bases $\mathfrak{F}^{(i)}$ (basis $|i\mu\rangle$) and $\mathfrak{H}^{(j)}$ (basis $|j\nu\rangle$) can be used as the basis of irreducible representations $\mathfrak{G}^{(i\otimes j)}$:

$$|ij\mu\nu\rangle = |i\mu\rangle|j\nu\rangle, \tag{10.34}$$

where μ, ν change from 1 to f_i or f_j, respectively. The resulting basis (10.34), in its turn, is irreducible with respect to group products

$$\mathfrak{G}^{(i\otimes j)}_{\mu\nu,\kappa\lambda} = \mathfrak{F}^{(i)}_{\mu\kappa}\mathfrak{H}^{(j)}_{\nu\lambda}, \tag{10.35}$$

where the sets of indices μ, ν and κ, λ respectively, number the rows and columns of the matrix $\mathcal{G}$. This matrix is *the direct product of matrices* $\boldsymbol{F}$ and $\boldsymbol{H}$: $\boldsymbol{G} = \boldsymbol{F} \otimes \boldsymbol{H}$. Representation $\mathfrak{G}^{(i\otimes j)}$ is called *the direct product of representation* $\mathfrak{F}^{(i)}$*) for group* $\mathcal{F}$ *and representation* $\mathfrak{H}^{(j)}$ *for group* $\mathcal{H}$. By summing over the diagonal elements $\mu = \kappa$ and $\nu = \lambda$, one may show that the characters of the direct group product are equal to the products of the characters of groups-multiplicands:

$$\chi^{(i\otimes j)}(\widehat{F} \otimes \widehat{H}) = \chi^{(i)}(\widehat{F})\chi^{(j)}(\widehat{H}). \tag{10.36}$$

For example, by multiplying directly the inversion group ($\mathcal{C}_i$) or reflection group ($\mathcal{C}_s$) and some finite groups (see Table 10.4), one obtains other finite groups such as $\mathcal{C}_{3h} = \mathcal{C}_3 \otimes \mathcal{C}_s$, $\mathcal{C}_{4h} = \mathcal{C}_4 \otimes \mathcal{C}_i$, $\mathcal{C}_{6h} = \mathcal{C}_6 \otimes \mathcal{C}_i$, $S_6 = \mathcal{C}_3 \otimes \mathcal{C}_i$, $\mathcal{D}_{3d} = \mathcal{D}_3 \otimes \mathcal{C}_i$, $\mathcal{D}_{4h} = \mathcal{D}_4 \otimes \mathcal{C}_i$, $\mathcal{D}_{6h} = \mathcal{D}_6 \otimes \mathcal{C}_i$, $\mathcal{O}_h = \mathcal{O} \otimes \mathcal{C}_i$, $\mathcal{T}_h = \mathcal{T} \otimes \mathcal{C}_i$. The tables of characters for these groups are constructed from tables of characters of the original groups according to (10.36). The full orthogonal group is also represented as the direct product of groups: $\mathcal{K}_h = \mathcal{K} \otimes \mathcal{C}_i$.

Direct product of irreducible representations of the same group. Formula (10.35) can also be used to construct direct products of irreducible representations of the same group: $\mathfrak{G} \sim \mathfrak{G}^{(i)} \otimes \mathfrak{G}^{(j)}$. Generally, the new representation $\mathfrak{G}^{(i\otimes j)}$, is reducible and can be decomposed into irreducible components:

$$\mathfrak{G}^{(i)} \otimes \mathfrak{G}^{(j)} \sim \sum_{k=1}^{N_C} n_k \mathfrak{G}^{(k)} \tag{10.37}$$

(The Clebsch-Gordan expansion). The n_k coefficients for finite groups can be determined from Eq. (10.31) accounting Eq. (10.36), where one sets $\widehat{F} = \widehat{H} = \widehat{G}$. In particular, it follows from Eqs. (10.31) and (10.26) that the identity representation is contained in the product of irreducible representations (and only once) only if complex conjugate irreducible representations are multiplied.

Direct product of irreducible representations of the rotation group. The product of basis vectors of the representations $|lm\rangle$ and $|l'm'\rangle$, where m and m' vary from $-l$ to l and from $-l'$ to l', respectively, can be used as the basis for a representation which is the direct product of irreducible representations $\mathfrak{D}^{(l)}$ and $\mathfrak{D}^{(l')}$. By identifying subspaces invariant to group $\mathcal{K}$ in the space of basis vectors $(|lm\rangle\ |l'm'\rangle)$, one can obtain a set of such subspaces with dimensions from $2(|l-l'|)+1$ to $2(l+l')+1$. Hence,

$$\mathfrak{D}^{(l)} \otimes \mathfrak{D}^{(l')} \sim \mathfrak{D}^{(|l-l'|)} \oplus \mathfrak{D}^{(|l-l'|+1)} \oplus \ldots \oplus \mathfrak{D}^{(l+l')}.$$

Decomposition into irreducible representations corresponds to the addition of vectors $\vec{l}$ and $\vec{l}'$. The total length of the summary vector $\vec{L}$ may range from $|l-l'|$ to $l+l'$. In this case, basis vectors $|LM\rangle$ for each of the resulting irreducible representations $\mathfrak{D}^{(L)}$ are linear combinations of products $|lm\rangle|l'm'\rangle$:

$$|LM\rangle = \sum_{m}\sum_{m'} |lm\rangle|l'm'\rangle\langle ll'mm'|LM\rangle. \tag{10.38}$$

Here, $\langle ll'mm'|LM\rangle$ are the Clebsch–Gordan coefficients. It can be shown that nonzero coefficients must satisfy the condition $m'+m = M$; therefore, Eq. (10.38) will only include the terms with all possible values of m' and m satisfying this condition. Specific values for Clebsch-Gordan coefficients are cited in Table 11.4.

10.3.4 Application of the Group Theory to Electron Paramagnetic Resonance Problems

The use of symmetry properties of physical systems is based on a number of theorems following from the group theory, namely, *Wigner theorem* on the symmetry group of the Hamiltonian, the theorem on the split of a degenerate energy level caused by perturbation, the *Wigner–Eckart theorem* on matrix elements, the *Kramers theorem* and the *Jahn–Teller theorem*.

Wigner theorem. A coordinate transformation that does not change the form of a Hamiltonian $\widehat{\mathcal{H}}$ is known as a Hamiltonian symmetry transformation. The set of all Hamiltonian symmetry transformations forms the Hamiltonian symmetry group $\mathcal{G}$. Formally, this means that if $\widehat{G}$ is one of the group operations, then

$$\widehat{G}\widehat{\mathcal{H}}\widehat{G}^{-1} = \widehat{\mathcal{H}}. \tag{10.39}$$

The group $\mathcal{G}$ satisfies the following requirement (Wigner theorem): *The eigenvectors of the Hamiltonian $\widehat{\mathcal{H}}$ belonging to the same eigenvalue E_i are transformed in the course of symmetry operations according to one of the irreducible representations of the Hamiltonian symmetry group $\mathcal{G}$.*

Indeed, if $|i\mu\rangle$ is an eigenvector belonging to an eigenvalue E_i, that is,

$$\widehat{\mathcal{H}}|i\mu\rangle = E_i|i\mu\rangle. \tag{10.40}$$

Then, acting by the $\widehat{G}$ operator on both parts of (10.40), one easily obtains

$$\widehat{G}\widehat{\mathcal{H}}|i\mu\rangle = E_i\widehat{G}|i\mu\rangle,$$

$$\widehat{G}\widehat{\mathcal{H}}\widehat{G}^{-1}\widehat{G}|i\mu\rangle = E_i\widehat{G}|i\mu\rangle,$$

hence subject to (10.39):

$$\widehat{\mathcal{H}}\widehat{G}|i\mu\rangle = E_i\widehat{G}|i\mu\rangle. \tag{10.41}$$

Equation (10.41) means that $\widehat{G}|i\mu\rangle$, similarly to $|i\mu\rangle$, is an eigenvector of the Hamiltonian $\widehat{\mathcal{H}}$ belonging to the same eigenvalue E_i.

Thus, eigenvectors belonging to the same energy value, transform through each other on the action of the symmetry operations of the group $\mathcal{G}$. This means that such vectors form the basis of a linear vector space invariant to the group $\mathcal{G}$, and, therefore, provide a certain representation $\mathfrak{G}$ of the group $\mathcal{G}$ (see Sect. 10.3.1).

One can also show that it is an irreducible representation. The opposite assertion that two irreducible representations always correspond to different energy levels, is not absolutely strict. However, the coincidence of levels for two different vectors, which do not transform through each other can be either random (accidental degeneracy) or associated with some unaccounted Hamiltonian symmetry transformation, so the group $\mathcal{G}$, in fact, is not a group of all Hamiltonian symmetry transformations. Some examples are: energy level intersection in the magnetic field and the degeneration of hydrogen atom levels over the orbital angular momentum. To sum up, each energy level corresponds to a certain irreducible representation of the the group $\mathcal{G}$. Some of these representations may be identical or equivalent.

Let $\mathfrak{G}$ stand for irreducible representations of the group $\mathcal{G}$, and E_i^{τ} for levels corresponding to the same irreducible representation. The index τ numerates the levels related to the same irreducible representation $\mathfrak{G}^{(i)}$, $\tau = 1, 2, \ldots, n_i$. The whole space of the state vectors is divided into $n = \sum_{i=1}^{n_C} n_i$ subspaces each of which is determined by the basis of the irreducible representation $\mathfrak{G}^{(i)}$. The number of linearly independent vectors determines the dimensionality f_i of the irreducible representation, which, consequently, is equal to the degeneracy order of the energy level E_i^{τ}. In summary, the following correlation can be established between concepts of the group theory and the quantum mechanics:

Abstract group	$\leftrightarrow$ Hamiltonan symmetry group $\mathcal{G}$
Group element	$\leftrightarrow$ Hamiltonan symmetry transformation $\widehat{G}$
Irreducible representation (IR)	$\leftrightarrow$ Energy level E_i^τ
Basis vectors of IR	$\leftrightarrow$ Eigenvectors $\lvert\tau i\mu\rangle$ belonging to the same energy level E_i^τ
IR dimension	$\leftrightarrow$ Order degeneracy

As an example of a Wigner theorem application, let us determine the form of the state vector for a many electron atom within the approximation, which takes in account the formation of the total orbital and spin momenta due to the residual interaction. The symmetry with respect to rotations of the atom as a whole requires that the atomic state vectors form a basis of an irreducible representation of the rotation group. Such vectors can be constructed in accordance to (10.38) as a combination of atomic state vectors within the approximation of a spherically symmetric self-consistent field in the form of the Slater determinant (10.22). Thus, in case of two electrons (without regard to the spin), the vector has the following form:

$$|LM\rangle = \frac{1}{\sqrt{2}} \sum_{m_1+m_2=M} \left[|l_1m_1\rangle^{(1)}|l_2m_2\rangle^{(2)} - |l_1m_1\rangle^{(2)}|l_2m_2\rangle^{(1)}\right] \langle l_1l_2m_1m_2|LM\rangle.$$

Here, L and M denote the total orbital angular momentum of the atom and its projection, respectively, L takes the values $|l_1 - l_2|$, $|l_1 - l_2| + 1$, $|l_1 - l_2| + 2$, $\dots, l_1 + l_2 - 1, l_1 + l_2$ and M takes the values $-L, -L+1, \dots, L-1, L$.

Theorem on energy level split under a perturbation. Let a perturbation has a lower symmetry (the group $\mathcal{G}$) than the unperturbed Hamiltonian (the group $\mathcal{G}_0$). Then the theorem can be formulated as follows: *A perturbation leads to the split of the unperturbed level E_{j0} into the number of levels equal to the number of irreducible representations of $\mathcal{G}$ contained in the representation (irreducible with respect to $\mathcal{G}_0$ and reducible with respect to $\mathcal{G}$) by which eigenvectors of unperturbed Hamiltonian belonging to the level E_{j0} are transformed. The degeneracy of the resulting levels is equal to the dimensions of the respective irreducible representations of the group $\mathcal{G}$.*

Indeed, the total Hamiltonian symmetry is described by the group of the least symmetry, i.e. the group $\mathcal{G}$. According to the Wigner theorem, the eigenvectors $|j\alpha\rangle$ of the unperturbed Hamiltonian belonging to the level E_{j0} are transformed by one of the irreducible representations of $\mathfrak{G}^{(j)}$ of the group $\mathcal{G}$. In accordance with the perturbation theory (see Sect. 11.3), a linear combination of the eigenvectors $|j\alpha\rangle$ of the unperturbed Hamiltonian with a fixed j and all possible values of α should be taken as a zero-order approximation for eigenvectors of the perturbed Hamiltonian. Consequently, the set of vectors $|j\alpha\rangle$ also provides a certain representation of $\mathcal{G}$ (generally reducible), i.e. $\mathfrak{G}_0^{(j)}$ is an irreducible representation of $\mathcal{G}_0$ and, at the same time, a certain representation of $\mathcal{G}$. However, according to the Wigner theorem, eigenvectors belonging to the same value E_{jk} of a perturbed Hamiltonian must be transformed by one of the irreducible representations $\mathfrak{G}^{(k)}$ of $\mathcal{G}$. The number n_{jk}, indicating how many times each irreducible representation $\mathfrak{G}^{(k)}$ is contained in $\mathfrak{G}_0^{(j)}$

Table 10.5 The group theory notations for perturbed and unperturbed Hamiltonians

Symbol meaning	Unperturbed Hamiltonian $\widehat{\mathcal{H}}_0$	Total Hamiltonian $\widehat{\mathcal{H}}_0 + \widehat{\mathcal{H}}'$
Symmetry group	$\mathcal{G}_0$	$\mathcal{G}$
Eigenvalue	$\widehat{E}_{j0}$	$\widehat{E}_{jk}$
Irreducible representation	$\mathfrak{G}_0^{(j)}$	$\mathfrak{G}^{(k)}$
Degeneracy order	$\sum_{k=1}^{n_C} f_k n_{jk}$	f_k
Irreducible representation basis	$\|j\alpha\rangle_0$	$\|k\mu\rangle$
Reducible representation	–	$\mathfrak{G}_0^{(j)}$
Characters corresponding to transformations from class C of the group $\mathcal{G}$	$\chi_0^{(j)}(\widehat{G}_C)$	$\chi_C^{(k)}$

is determined by decomposition into irreducible representations (10.31), in which $\chi = \chi_0^{(j)}(\widehat{G}_C)$:

$$n_{jk} = \frac{1}{g} \sum_{C=1}^{n_C} g_C \chi_0^{(j)}(\widehat{G}_C) \chi_C^{(k)*}. \tag{10.42}$$

Here the summation runs over all classes of $\mathcal{G}$; $\chi_C^{(k)}$ are the characters of the representation $\mathfrak{G}^{(k)}$ of the group $\mathcal{G}$ for class C; $\chi_0^{(j)}(\widehat{G}_C)$ are the characters of the irreducible representation $\mathfrak{G}_0^{(j)}$ of the group $\mathcal{G}_0$ corresponding to the transformations of the class C of the group $\mathcal{G}$. Thus, the total number of levels resulting from the split of the level E_{j0} equals

$$n = \sum_{k=1}^{N} n_{kj}, \tag{10.43}$$

and their degeneracies are equal to the dimensions of the respective irreducible representations $\mathcal{G}^{(k)}$. It is evident $\sum_{k=1}^{N} n_{kj} f_k = f_0$, where f_0 is the degeneracy of the initial level.

Table 10.5 contains notations corresponding to the group theory concepts for perturbed and unperturbed Hamiltonians.

The Wigner–Eckart theorem defines the properties of matrix elements of the $\langle \tau j \mu | \widehat{Q}_\kappa^\nu | \tau' j' \mu' \rangle$ type where $\widehat{Q}_\kappa^\nu$ is the ν-th component of *the irreducible* (with respect to the group $\mathcal{G}$) *tensor operator* Q of the rank κ. Components of such a tensor are defined as a set of operators transforming according to the irreducible representation $\mathfrak{G}^{(\kappa)}$ of the group $\mathcal{G}$.

If the matrix element is produced with the state vectors transformed by irreducible representations $\mathfrak{G}^{(j)}$ and $\mathfrak{G}^{(j')}$ of the same group, then, according to the Wigner–Eckart theorem, the matrix element is represented as

$$\langle \tau j \mu | \widehat{Q}^{\nu}_{\kappa} | \tau' j' \mu' \rangle = \sum_{\sigma=1}^{n_j} \langle \tau j \| \widehat{\mathsf{Q}}_{\kappa} \| \tau' j' \rangle_{\sigma} S^{j\kappa j'}_{\sigma, \mu\nu\mu'}. \tag{10.44}$$

The theorem is proved by means of the fundamental lemma of the representation theory (*the Schur lemma*) and by a decomposition of the direct product of $\mathfrak{G}^{(\kappa)} \otimes \mathfrak{G}^{(j')}$ into irreducible components according to Eq. (10.37). Index σ in (10.44) numbers the repeating irreducible representations of type $\mathfrak{G}^{(j)}$ occurring in this decomposition. The quantity $\langle \tau j || \widehat{\mathsf{Q}}_{\kappa} || \tau' j' \rangle_{\sigma}$ is called a *reduced matrix element of the irreducible tensor* $\widehat{\mathsf{Q}}_{\kappa}$. The coefficients $S^{j\kappa j'}_{\sigma, \mu\nu\mu'}$ do not depend on the particular realization of the tensor $\widehat{\mathsf{Q}}_{\kappa}$, but are defined exclusively by its transformation properties at transformations of the $\mathcal{G}$ group.

A direct consequence of the Wigner–Eckart theorem is a principle that defines the selection rules for matrix elements of the different operators: *the matrix element* $\langle \tau j \mu | \widehat{Q}^{\nu}_{\kappa} | \tau' j' \mu' \rangle$ *vanishes if the decomposition* $\mathfrak{G}^{(\kappa)} \otimes \mathfrak{G}^{(j')}$ *does not contain a identity representation* $\mathfrak{G}^{(j)}$.

The same condition can be formulated otherwise: *the matrix element is zero if the decomposition* $\mathfrak{G}^{(j)} \otimes \mathfrak{G}^{(j')}$ *does not contain representation* $\mathfrak{G}^{(\kappa)}$.

Equivalent operator method. In case the decomposition of the direct product of irreducible representation contains no repeated irreducible representations, the sum in (10.44) has one term only. This situation is always the case for the rotation group, if $\widehat{\mathsf{Q}}_p$ stands for spherical tensor operators of the $2p + 1$ rank. Therefore, the matrix elements of any such operators of the $\langle lm | \widehat{Q}^q_p | l'm' \rangle$ type (except for the constant factor which is common for the given l, p and l' and independent on m and m'), are identical:

$$\langle lm | \widehat{V}^q_p | l'm' \rangle = \langle l || p || l' \rangle \langle lm | \widehat{Q}^q_p | l'm' \rangle. \tag{10.45}$$

Here we assume that $|lm\rangle$ and $|l'm'\rangle$ are transformed by irreducible representations $\mathfrak{D}^{(l)}$ and $\mathfrak{D}^{(l')}$, respectively, and $\widehat{V}^q_p$ and $\widehat{Q}^q_p$ by representation $\mathfrak{D}^{(p)}$, where the proportionality factor $\langle l || p || l' \rangle$ is independent of m, m' or q.

The equivalent operator method is used as follows. The calculation of matrix elements of the operator $\widehat{V}^q_p$, presenting one of components of the crystal field electrostatic potential and, therefore, depending on the coordinates, requires rather complex integration. Instead, first one determines the matrix elements of the *equivalent operator* $\widehat{Q}^q_p$, obtained from $\widehat{V}^q_p$ by replacing coordinates with the corresponding angular momentum projections. The ambiguity associated with the non-commutativity of different momentum components is eliminated by averaging over their all possible combinations. For example, $\widehat{Q}^2_2 = \frac{1}{2}(\widehat{l}_x \widehat{l}_y + \widehat{l}_y \widehat{l}_x)$ is the equivalent operator for $\widehat{V}^2_2 = xy$. However, one of the elements $\langle lm | V^q_p | l'm' \rangle$ must be calculated explicitly by integration to determine the common proportionality coefficient $\langle l \| p \| l' \rangle$. Expressions for tensor operators components are given in the right column of Table 10.3. In the Table, symbols { } are used for anti-commutator: $\{\hat{a}, \hat{b}\} = \frac{1}{2}(\hat{a}\hat{b} + \hat{b}\hat{a})$.

Further Readings

Abragam A, Bleaney B (1970) Electron paramagnetic resonance of transition ions. Clarendon Press, Oxford, p 911
Heine V (1970) Group theory in quantum mechanics, 3rd edn. Pergamon Press, New York

Chapter 11
Energy Levels of Paramagnetic Center in Crystal Field

11.1 Qualitative Approach

The analysis of the nature of splitting of ion levels under the influence of the crystal field is based on the Wigner theorem and Eq. 10.30. The unperturbed Hamiltonian $\widehat{\mathcal{H}}$ is the Hamiltonian of a free ion which is invariant with respect to transformations from the full orthogonal the group $\mathcal{K}_h$. As it can be seen from Table 10.1, the perturbation is represented by the Hamiltonian of interaction between the electrons and the crystal field:

$$\widehat{\mathcal{H}}' = \widehat{\mathcal{H}}_{\text{cryst}} = -\sum_{i=1}^{N} eV_{\text{cryst}}(\vec{r}_i)\,.$$

Its symmetry is determined by the symmetry group ($\mathcal{G}$) of the arrangement of surrounding ions and molecules producing the crystal field. According to the above mentioned theorem, it is sufficient to know the angular momentum of ion J (without perturbation) and the environment symmetry group $\mathcal{G}$ to determine the character of the splitting. Following (10.43), the total number of levels n resulting from the split of the unperturbed level equals $n = \sum_{i=1}^{N_C} n_i$; summation is over all irreducible representations (IR) of the group $\mathcal{G}$, the number of which is equal to the number of classes N_C, while n_i are calculated by the formula

$$n_i = \frac{1}{g}\sum_{C=1}^{N_C} g_C \chi_J(\phi_C)\chi_C^{(i)*}, \tag{11.1}$$

representing a special case of Eq. (10.42). In (11.1) the angle ϕ_C corresponds to the rotation belonging to the class C of the group $\mathcal{G}$. Characters $\chi_C^{(i)}$ of irreducible representations of the group $\mathcal{G}$ are taken from tables of characters, while characters of the reducible representation $\chi_J(\phi_C)$ are characters of irreducible representations of $\mathcal{K}_h$, calculated as per (10.28). For convenience, this formula was tabulated for some values of J and rotations occurring in finite point symmetry groups (Table 11.1).

V. I. Chizhik et al., *Magnetic Resonance and Its Applications*,
DOI: 10.1007/978-3-319-05299-1_11,

Table 11.1 Characters of irreducible representations of $\mathcal{K}_h$

J	ϕ_C				
	0	$\pi/2$	$2\pi/3$	π	$\pi/3$
0	1	1	1	1	1
1	3	1	0	−1	2
2	5	−1	−1	1	1
3	7	−1	−1	−1	−1
4	9	1	0	1	−2
3/2	4	0	−1	0	3
5/2	6	−2	0	0	0
7/2	8	0	1	0	−3

It should be noted that the number of levels resulting from the split of the undisturbed level and the degeneracy of each level depends only on the environment symmetry and the angular momentum of the unperturbed system. The nature of the momentum does not affect the qualitative result. To illustrate this let us determine the mode of the level split for a particle with angular momentum $j = l = 3$ in a crystal field with symmetry $\mathcal{O}_h$. The calculations and results are presented below in the convenient tabular form (see Table 11.2).

Thus, the representation $\mathfrak{D}^{(3)}$ can be reduced to three irreducible representations: $\mathfrak{D}^{(3)} \sim \mathfrak{A}_2 \oplus \mathfrak{T}_1 \oplus \mathfrak{T}_2$. This form of record indicates that under the influence of the crystal field with the symmetry $\mathcal{O}$ the energy level corresponding to $l = 3$ splits into three levels: one non-degenerate (singlet) and two thrice-degenerate (triplets). This follows from the fact that $\mathfrak{A}_2$ is a one-dimensional representation whereas $\mathfrak{T}_1$ and $\mathfrak{T}_2$ are three-dimensional ones ($f_{\mathfrak{A}_2} = 1$, $f_{\mathfrak{T}_1} = f_{\mathfrak{T}_2} = 3$). In addition, it can be argued that the state vector and the wave function of the singlet level are transformed by an irreducible representation $\mathfrak{A}_2$ of the group $\mathcal{O}$, that is they inverse sign under rotations around the C_{2d} axes and under $\pi/2$ and $3\pi/2$ rotations around the C_4 axes; no sign inversion, however, takes place under other rotations (see Table 10.4). The state vectors (wave functions) for the obtained levels form bases of the corresponding representations $\mathfrak{T}_1$ and $\mathfrak{T}_2$. The matrix elements of these representations define the transformation law for state vectors on application of symmetry operations from the group $\mathcal{O}$. For the sake of simplicity, characters for $\mathcal{O}_h$ and $\mathcal{K}_h$ the groups were replaced with characters for $\mathcal{O}$ and $\mathcal{K}$. The result does not depend on the presence of inversion; however, the state parity should be additionally indicated. The latter is completely determined by the parity of the spatial wave function, i.e. by the value of $(-1)^l$.

In this example, the final representation for $l = 3$ should be written as $\mathfrak{D}^{(3)} \sim \mathfrak{A}_{2u} \oplus \mathfrak{T}_{1u} \oplus \mathfrak{T}_{2u}$. Representations $\mathfrak{A}_{2u}$, $\mathfrak{T}_{1u}$ and $\mathfrak{T}_{2u}$ now belong to the $\mathcal{O}_h$ the group.

The results shown in Table 11.3 for other values of moment l and other symmetry groups can be obtained in a similar manner. In fact, these results cover all important cases. As one can see, the highest order of the symmetry axis or the presence of the cubic axis C_3 (i.e. the crystallographic syngony of the environment) determines the overall picture of the split. Thus, for a paramagnetic center the splitting of levels is

Table 11.2 The calculations to fulfill expansions of IR of rotation group on the IR of octahedron group

l	IR	$\widehat{E}$	$8\widehat{C}_3$	$3\widehat{C}_2$	$6\widehat{C}_{2d}$	$6\widehat{C}_4$	$\sum_{C=1}^{n_C} g_C \chi_3(\phi_C)\chi_C^{(l)}$	n_j
		Characters of operations $\widehat{G}$						
1	$\mathfrak{A}_1$	1	1	1	1	1	$1\cdot1\cdot7+8\cdot1\cdot1-3\cdot1\cdot1-6\cdot1\cdot1-6\cdot1\cdot1=0$	0
2	$\mathfrak{A}_2$	1	1	1	−1	−1	$1\cdot1\cdot7+8\cdot1\cdot1-3\cdot1\cdot1-6\cdot1\cdot1+6\cdot1\cdot1=24$	1
3	$\mathfrak{E}$	2	−1	2	0	0	$1\cdot2\cdot7-8\cdot1\cdot1-3\cdot2\cdot1-6\cdot0\cdot1-6\cdot0\cdot1=0$	0
4	$\mathfrak{T}_1$	3	0	−1	−1	1	$1\cdot3\cdot7+8\cdot0\cdot1+3\cdot1\cdot1+6\cdot1\cdot1-6\cdot1\cdot1=24$	1
5	$\mathfrak{T}_2$	3	0	−1	1	−1	$1\cdot3\cdot7+8\cdot0\cdot1+3\cdot1\cdot1-6\cdot1\cdot1+6\cdot1\cdot1=24$	1
	$\mathfrak{D}^{(3)}$	7	1	−1	−1	−1	—	—

Table 11.3 The expansions of rotation group irreducible representation in the IR of octahedron group

l	$\mathcal{O}, \mathcal{O}_h, \mathcal{T}_d$	$\mathcal{D}_{4h}$	$\mathcal{D}_{3d}, \mathcal{D}_3, \mathcal{C}_{3v}$	$\mathcal{D}_{2d}, \mathcal{D}_4$
1	$\mathfrak{T}_{1u}$	$\mathfrak{A}_{2u} \oplus \mathfrak{E}_u$	$\mathfrak{A}_{2u} \oplus \mathfrak{E}_u$	$2\mathfrak{A}_{1u} \oplus \mathfrak{B}_{1u}$
2	$\mathfrak{T}_{2g} \oplus \mathfrak{E}_g$	$\mathfrak{A}_{1g} \oplus \mathfrak{B}_{1g} \oplus \mathfrak{B}_{2g} \oplus \mathfrak{E}_g$	$\mathfrak{A}_{1g} \oplus 2\mathfrak{E}_g$	$3\mathfrak{A}_{1g} \oplus 2\mathfrak{B}_{1g}$
3	$\mathfrak{A}_{2u} \oplus \mathfrak{T}_{1u} \oplus \mathfrak{T}_{2u}$	$\mathfrak{A}_{2u} \oplus \mathfrak{B}_{1u} \oplus \mathfrak{B}_{2u} \oplus 2\mathfrak{E}_u$	$\mathfrak{A}_{1u} \oplus 2\mathfrak{A}_{2u} \oplus 2\mathfrak{E}_u$	$3\mathfrak{A}_{2u} \oplus 2\mathfrak{B}_{1u} \oplus 2\mathfrak{B}_{2u}$
3/2	$\mathfrak{G}'$	$2\mathfrak{E}'$	$2\mathfrak{E}'$	$2\mathfrak{E}'$
5/2	$\mathfrak{G}' \oplus \mathfrak{E}'_1$	$3\mathfrak{E}'$	$3\mathfrak{E}'$	$3\mathfrak{E}'$
7/2	$\mathfrak{G}' \oplus \mathfrak{E}'_1 \oplus \mathfrak{E}'_2$	$4\mathfrak{E}'$	$4\mathfrak{E}'$	$4\mathfrak{E}'$

determined solely by its angular momentum in the initial state and the environment symmetry. However, this qualitative approach does not allow any conclusions with regard to either the values of the split or the sequence of energy levels. Further we will consider a method for calculating energy intervals between levels in the crystal field within the first-order approximation of the perturbation theory, which, despite of general inconsistency with experimental finding, provides us with the correct relative position of levels and the order of intervals between them. For more accurate results, higher approximations are used (adjusted for excited configurations, the spatial distribution of charges forming the crystal field, etc). To apply the method, one must first obtain some analytical expression for the effective potential of the crystal field.

11.2 Analytical Expression for the Potential of Crystal Field

Any scalar potential of the crystal field in a restricted region of space can be expanded in spherical harmonics:

$$V_{\text{cryst}} = \sum_{p=0}^{\infty} \sum_{q=-p}^{p} B_p^q r^p Y_p^q(\theta, \phi) \tag{11.2}$$

where coefficients B_p^q are the crystal field parameters; $Y_p^q(\theta, \phi)$ are the above mentioned spherical functions of the order p with index q (see Sect. 10.2.2):

$$Y_p^q(\theta, \phi) = (-1)^q \sqrt{\frac{(2p+1)(p-q)!}{4\pi(p+q)!}} P_p^{|q|}(\cos\theta)e^{iq\phi};$$

$P_p^{|q|}(\cos\theta)$ are the associated Legendre functions. The potential of V_p^q must be real that leads to the condition $B_p^q = B_p^{-q}$. The real potential may also be represented by an equivalent series of homogeneous orthogonal polynomials U_p^q:

$$V_{\text{cryst}} = \sum_{p=0}^{\infty} \sum_{q=0}^{p} A_p^q U_p^q(x, y, z). \tag{11.3}$$

If we use the Cartesian coordinate frame for the expressions of spherical harmonics, it will be clear that the polynomials $U_p^q(x, y, z)$ agree within a constant factor to the expressions $r^p \left[Y_p^q(\theta, \phi) + Y_p^{-q}(\theta, \phi)\right]$, and the coefficients A_p^q differ from the coefficients B_p^q by constant factors depending on p and q. The explicit form of the spherical harmonics and homogeneous polynomials for some values of p was cited in Table 10.3. It also contains the expressions for the operators of $\widehat{Q}_p^q$, constructed from the components of the electron angular momentum, and which are equivalent to polynomials of U_p^q, according to 10.3.

Restrictions on the form of the expression for the potential of the crystal field imposed by symmetry. The electrostatic potential, generated by the environment and being a scalar, is invariant with respect to the transformations of the environment symmetry group. Using the terminology of the group theory, we can say that the potential is "transformed" by a single irreducible representation of the group $\mathcal{G}$. However, each member of the sum over p in the expression (11.2)

$$V_p = \sum_{q=-p}^{p} B_p^q r^p Y_p^q(\theta, \phi) = \sum_{q=0}^{p} A_p^q U_p^q, \tag{11.4}$$

being a linear combination of the spherical harmonics of order p, is transformed by the irreducible representation of the rotation the group $\mathcal{K}$ (see Sect. 10.3). The invariance requirement of the group $\mathcal{G}$ imposes additional constraints on the coefficients B_p^q and, consequently, A_p^q. The number of independent combinations and, consequently, the number of independent coefficients B_p^q in the expression (11.4) are equal to the number of individual representations in the expansion of the representation $\mathfrak{D}^{(p)}$ according to the irreducible representations defined by (11.1). Remarkably, the group theory solves completely different physical problems in the same way, regardless the physical meaning that we attach to a vector basis of the irreducible representation.

To simplify the expansion (11.4) we can formulate a few simple rules deriving from the consideration of the effects of individual operations of symmetry:

1. Inversion. If the group $\mathcal{G}$ contains the inversion, the expression for V_{cryst} cannot contain terms with odd values of p, as those terms are odd (i.e., they change their sign) under the inversion (the change of the signs of the coordinates).
2. Reflection σ_h. The associated Legendre function $P_p^q(\cos\theta)$ is the polynomial of degree q with respect to $\sin\theta$ and of degree $p - q$ with respect to $\cos\theta$. Since $\cos\theta$ changes its sign upon plane reflection perpendicular to the polar axis, the expansion (11.4) may contain only terms with even values of $p-q$. Consequently, if the inversion and reflection are present simultaneously, the values of q are also supposed to be even.
3. Rotation around the axis C_n. When turning at some angle ϕ', each term of (11.3) is multiplied by $e^{iq\phi'}$, i.e. the potential remains invariant if $q\phi\prime = 2\pi k$ (k integer). Thus, if $q\phi\prime = 2\pi/n$ (the symmetry transformation), then only $q = kn$ is allowed, i.e. q is supposed to be multiple to the order of the axis n.

Though further simplifications are less obvious, they are very significant. They are related to the fact that the energy levels are determined by the potential (11.2) via its matrix elements. Therefore, in calculations of the energy using the expansion (11.2), we can take account only those terms that lead to non-zero matrix elements. We can show that, according to the matrix element theorem (see 10.3), the matrix element of the $\langle lm|V_p|lm\prime\rangle$ type is non-zero only in the case of $p \leq 2l$. Indeed, the vectors $|lm\rangle$ and $|lm\prime\rangle$ form a basis of the irreducible representation $\mathfrak{D}^{(l)}$, therefore, the product of bases is transformed in accordance with the irreducible representation $\mathfrak{D}^{(l)} \otimes \mathfrak{D}^{(l)} \sim \mathfrak{D}^{(0)} \oplus \mathfrak{D}^{(1)} \oplus, \ldots, \otimes \mathfrak{D}^{(2l)}$. As it was stated above, the term V_p is transformed with the irreducible representation $\mathfrak{D}^{(p)}$. According to the mentioned theorem, the matrix element is non-zero if (and only if) the irreducible representation with the number p is found in the $\mathfrak{D}^{(l)} \otimes \mathfrak{D}^{(l)}$. It is possible only in the case $p \leq 2l$. If, as usual, l is referred to a single electron angular momentum, we are talking about the energy of its single interaction with the crystal field (a strong crystal field). In the case of a weak crystal field the *total momentum of ion* J must be used, while in the case of an intermediate one we should use only the total *orbital* momentum L. Consequently, in the case of an intermediate crystal field, the condition $p \leq 2L$ should be met. However, we should keep in mind that in the final analysis the states $|JM_J\rangle$ or $|LM\rangle$ are a linear combination of products of one-electron functions, and the matrix elements formed with them are expressed in terms of matrix elements, obtained using single-electron functions. Therefore, a non-zero contribution can be made only by the potential terms corresponding to non-zero matrix elements of the type $\langle lm|V_p|lm'\rangle$, i.e. by those terms for which the condition $p \leq 2l$ is met. Finally, we should discard the term with $p = 0$, since it is an additive constant and does not affect the difference between the split energy levels. Thus, the effective potential of the crystal field (V_{cryst}) in the presence of inversion, plane of symmetry and symmetry axis of order n, perpendicular to it, has the form

$$V_{\text{cryst}} = \sum_{p=2}^{p'} \sum_{q=-p}^{p} B_p^q r^p Y_p^q(\theta, \phi), \tag{11.5}$$

where p takes only even values, and the upper limit of summation p' is equal to $\min\{2l, 2J\}$ for a weak crystal field and $\min\{2l, 2L\}$ for a intermediate field. The index q takes only the even values, that are multiple of the order of the symmetry axis n, and zero. We apply the above mentioned theory to construct a general expression of the effective potential for the group of the symmetry of the environment $\mathcal{O}_h$ ("cubic" potential) for $l = 2$ (d-electron), provided that the z axis is the axis C_4. The condition $p \leq 21$ allows us to replace the series (11.2) by a finite sum (11.5) with $p' = 4$. Meaning the group $\mathcal{O}_h$ contains the inversion $\hat{\imath}$, the reflection $\widehat{\sigma_h}$ and the rotations $\widehat{C}_4$ (see Table 10.4 and notes below the Eq. (10.38)), an effective cubic potential can be presented as

$$V_{\text{cryst}} = B_2^0 r^2 Y_2^0(\theta, \phi) + B_4^0 r^4 Y_4^0(\theta, \phi) + B_4^4 r^4 \left[Y_4^4(\theta, \phi) + Y_4^{-4}(\theta, \phi) \right].$$

Expanding the representations $\mathfrak{D}^{(2)}$ and $\mathfrak{D}^{(4)}$ according to the irreducible representations of the group $\mathcal{O}_h$ (or by using the data in Table 11.3), we find that $\mathfrak{D}^{(2)}$ does not contain the identical representation $\mathfrak{A}_{1g}$, and the representation $\mathfrak{D}^{(4)}$ contains it only once. This means that $B_2^0 = 0$, and the coefficients B_4^0 and B_4^4 are not independent. A similar statement can be made about the parameters of the crystal field A_4^0 and A_4^4 in the expression (11.4). According to the formula (11.4) and the data in Table 10.4, the term of the potential expansion with $p = 4$ can be stated as

$$V_4 = A_4^0 \left(35z^4 - 30z^2r^2 + 3r^4\right) + A_4^4 \left(x^4 - 6x^2y^2 + y^4\right).$$

The relation between the coefficients A_4^q can be found from the condition of invariance under transformations of the group $\mathcal{O}_h$. In particular, there must be invariance under the rotations around a threefold axis C_3, specific for the cubic groups: $\widehat{C}_3 V_4 = V_4$. One of the rotations $\widehat{C}_3$ can be obtained making the circular permutation (x, y, z):

$$\widehat{C}_3 V_4 = A_4^0 \left(35x^4 - 30x^2r^2 + 3r^4\right) + A_4^4 \left(y^4 - 6y^2z^2 + z^4\right).$$

Equating the coefficients of x^4 in the transformed and untransformed expressions of the potential, we obtain $3A_4^0 + A_4^4 = 8A_4^0$, where $A_4^4 = 5A_4^0$ comes from. Consequently,

$$V^{\text{cube}} = A_4^0 \left(U_4^0 + 5U_4^4\right). \tag{11.6}$$

In some cases it is more convenient to use the explicit expression of the cubic potential in the Cartesian coordinates, which is easily obtained after substituting the polynomials of the explicit form U_4^0 and U_4^4 from Table 10.3:

$$V^{\text{cube}} = 20A_4^0\left(x^4 + y^4 + z^4 - \frac{3}{5}r^4\right). \tag{11.7}$$

Thus, there are only certain components of the potential of the electrostatic field depending on the coordinates in a certain way which contribute to the interaction of a paramagnetic center with the environment. This dependence is completely determined by the symmetry of the electrostatic field and the magnitude of the angular momentum of the paramagnetic center. Of course, general laws of electromagnetic interactions of multipoles with fields are quite manifested here. A similar situation occurs with a nuclear quadrupole resonance, where only terms of the potential with $p = 2$ (see 8.1) are effective.

Direct calculation of the coefficients of the crystal field potential is possible if we know the charge distribution of surrounding ions or molecules. The simplest model is a point-charge model, which we are going to consider. The expression for the potential

$$V = \frac{q}{4\pi\varepsilon_0|\vec{r} - \vec{R}|},$$

created at the point r by the charge q, whose position is specified by the radius-vector $\vec{R}$ ($r < R$) can be expanded in a series, using a well-known expansion of Legendre polynomials:

$$\frac{1}{|\vec{r} - \vec{R}|} = \frac{1}{R}\sum_{k=1}^{\infty}\left(\frac{r}{R}\right)^k P_k(\cos(\alpha)) \tag{11.8}$$

where $P_k(\cos(\alpha))$ are the Legendre polynomials; α is the angle between vectors $\vec{r}$ and $\vec{R}$. Taking into account the above condition of the vanishing of matrix elements, only even terms in the series with $k \leq 2l$ are of interest. To illustrate the calculation algorithm we shall find crystal-field parameters for d-electrons (i.e., $l \leq 2$) in the cases when charges are located at the vertices of a regular octahedron (i), a cube or a tetrahedron (a cubic crystal field) and (ii) at the vertices of a distorted (elongated or flattened along the axis z) octahedron (a tetragonal crystal field). In the case of cubic symmetry of location of charges surrounding the paramagnetic center, only one coefficient A_4^0 is sufficient to consider in the expression of cubic potential (11.7) for d-electrons. It can be determined by comparing the expressions for the potential on the axis z, obtained from the formulas (11.7) and (11.8). For points on the axis z, the expression (11.8) becomes

$$\frac{1}{|\vec{r} - \vec{R}|} = \frac{1}{R}\sum_{k=1}^{\infty}\left(\frac{z}{R}\right)^k P_k(\cos(\alpha)).$$

Now α is the angle between the radius-vector $\vec{R}$ and the axis z.

In the case of a regular octahedron the superposition of potentials from 6 charges (points in Fig. 11.1a) with $R_1 = R_2$ can be put into the expression (for brevity we set $\frac{q}{4\pi\varepsilon_0} = 1$)

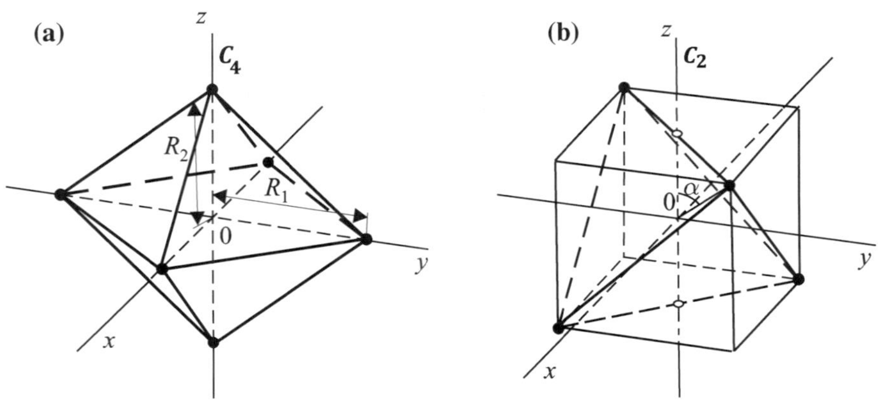

Fig. 11.1 Octahedral (**a**) and tetrahedral (**b**) charge dispositions

$$V^{\rm cube}_{4({\rm oct})}(z) = \frac{z^4}{R^5}\left[P_4(1) + P_4(-1) + 4P_4(0)\right] = \frac{7z^4}{2R^5}. \tag{11.9}$$

At the same time, the substitution $x = 0$, $y = 0$ (the axis z) in (11.7) gives the expression

$$V^{\rm cube} = 8A_4^0 z^4,$$

from which after the comparison with (11.9) we get $8A_4^0 = \frac{7}{2}R^5$, and for an arbitrary charge q we get

$$A^0_{4({\rm oct})} = \frac{7q}{64\pi\varepsilon_0 R^5}. \tag{11.10}$$

Thus, near the origin of coordinates, the effective potential produced by the charges q, located at the vertices of an octahedron, can be calculated using the expression

$$V^{\rm cube}_{4({\rm oct})}(\vec{r}) = \frac{q}{4\pi\varepsilon_0}\frac{35}{4R^5}\left(x^4 + y^4 + z^4 - \frac{3}{5}r^4\right).$$

When the charges are located at the vertices of a cube (Fig. 11.1b) we get $|\cos(\alpha)| = \frac{1}{\sqrt{3}}$ for all charges and

$$V^{\rm cube}_{4({\rm cube})}(z) = \frac{q}{4\pi\varepsilon_0}\frac{8z^4 P_4(1)}{R^5} = -\frac{q}{4\pi\varepsilon_0}\frac{28z^4}{9R^5}. \tag{11.11}$$

The comparison of the expressions (11.9) and (11.11) leads to

$$A^{\rm cube}_{4({\rm cube})} = -\frac{7q}{72\pi\varepsilon_0 R^5}. \tag{11.12}$$

The expression (11.12), but with the coefficient $\frac{1}{2}$, can also be used for calculating potential coefficients of the crystal field in the case of the symmetry of a regular octahedron with the surrounding charges located in its vertices. In this case a regular tetrahedron is oriented so that the vertices coincide with the four corners of a cube, labeled with filled circles in Fig. 11.1b.

If the surrounding charges are disposed in tetragonal symmetry, the potential of the crystal field (11.3) is reduced to

$$V^{\mathrm{t}}_{\mathrm{cryst}}(\vec{r}) = A_2^0 U_2^0 + A_4^0 U_4^0 + A_4^4 U_4^4 \tag{11.13}$$

(the index t shows that this is a tetragonal symmetry). We shall determine the coefficients A_2^0, A_4^0 and A_4^4 for the case when the charges are at the vertices of a stretched or compressed (along one of its axes) octahedron (see Fig. 11.1a, $R_1 \neq R_2$). According to the (11.8), the effective potential on the axis z can be presented as

$$\begin{aligned} V^{\mathrm{t}}_{\mathrm{oct}}(z) &= z^2\left[\frac{P_2(1)+P_2(-1)}{R_1^3} + \frac{4P_2(0)}{R_2^3}\right] \\ &\quad + z^4\left[\frac{P_4(1)+P_4(-1)}{R_1^5} + \frac{4P_4(0)}{R_2^5}\right] \\ &= 2z^2\left(\frac{1}{R_1^3} - \frac{1}{R_2^3}\right) + z^4\left(\frac{2}{R_1^5} + \frac{3}{R_2^5}\right). \end{aligned} \tag{11.14}$$

However, for the points on the axis z, the Eq. (11.13) takes the simple form

$$V^{\mathrm{t}}(z) = 2A_2^0 z^2 + 8A_4^0 z^4. \tag{11.15}$$

Comparing the coefficients at z^2 and z^4 in the Eqs. (11.2) and (11.15), we get for an arbitrary charge q

$$A_2^{0\mathrm{t}} = \frac{q}{4\pi\varepsilon_0}\left(\frac{1}{R_1^3} - \frac{1}{R_2^3}\right), \tag{11.16}$$

$$A_4^{0\mathrm{t}} = \frac{q}{4\pi\varepsilon_0}\left(\frac{1}{4R_1^5} + \frac{3}{16R_2^5}\right). \tag{11.17}$$

To calculate $A_4^{4\mathrm{t}}$ we should compare the expressions for the potential in the form (11.8) and (11.13) on the axis x:

$$
\begin{aligned}
V^{\mathrm{t}}_{\mathrm{oct}}(x) &= x^2\left[\frac{2P_2(0)}{R_1^3}+\frac{P_2(1)+P_2(-1)+2P_2(0)}{R_2^3}\right]\\
&+x^4\left[\frac{2P_4(0)}{R_1^5}+\frac{P_4(1)+P_4(-1)+2P_4(0)}{R_2^5}\right]\\
&= x^2\left(\frac{1}{R_2^3}-\frac{1}{R_1^3}\right)+x^4\left(\frac{3}{4R_1^5}+\frac{11}{4R_2^5}\right),
\end{aligned}
\tag{11.18}
$$

$$
V^{\mathrm{t}}(x) = -A_2^0x^2+\left(3A_4^0+A_4^4\right)x^4. \tag{11.19}
$$

After comparing the coefficients at x^4 in (11.2) and (11.19) we get the relation

$$
3A_4^{0\mathrm{t}}+A_4^{4\mathrm{t}} = \frac{1}{4}\left(\frac{11}{R_1^5}+\frac{3}{R_2^5}\right). \tag{11.20}
$$

Substituting the coefficient $A_4^{0\mathrm{t}}$ from (11.17) from (11.20), we obtain using the Eq. (11.20) the final expression for the coefficient $A_4^{4\mathrm{t}}$:

$$
A_4^{4\mathrm{t}} = \frac{q}{4\pi\varepsilon_0}\left(\frac{3}{16R_1^5}+\frac{2}{R_2^5}\right).
$$

11.3 Energy Levels of Paramagnetic Center in Crystal Field

11.3.1 Case of Strong Field

In this case, when a crystal field is assumed to be strong, an electron state in the approximation of spherically symmetric self-consistent field has to be taken as an unperturbed state. It is known from the qualitative analysis that the energy level of a p-electron is not split by the crystal field of cubic symmetry. It is natural, since three p-orbitals are transformed according to the irreducible representation of $\mathfrak{T}_{1u}$ of the group $\mathcal{O}_h$, i.e. $\mathfrak{D}_u^{(2)} \sim \mathfrak{T}_{1u}$. The conclusion about the absence of splitting also follows from the equivalence of the axes x, y, z in the cubic groups. Therefore, orbitals p_x, p_y and p_z cannot differ in energies (see the explicit form of wave functions (10.13)). Similar reasoning leads to the conclusion that in the case of the tetragonal symmetry of the crystal field a splitting is possible only into two levels, as the axes x and y are equivalent. Indeed, $\mathfrak{D}_u^{(1)} \sim \mathfrak{A}_{1u} \oplus \mathfrak{E}_u$. Similarly, we can conclude that in the field of cubic symmetry orbitals d_{zx}, d_{xy} and d_{yz} have the equal energy. Take into account two circumstances: (1) $\mathfrak{D}_g^{(2)} \sim \mathfrak{A}_{2g} \oplus \mathfrak{E}_g$, (2) in accordance with (10.13) and the data in Table 10.4, the orbitals d_{zx}, d_{xy} and d_{yz} are transformed according to the irreducible representation $\mathfrak{T}_{2g}$ in operations from the symmetry group $\mathcal{O}$. The rest orbitals d_{z^2}

and $d_{x^2-y^2}$ belong to the irreducible representation $\mathfrak{E}_g$. When the symmetry of the crystal field falls up to the tetragonal one we have the expansions $\mathfrak{T}_{2g} \sim \mathfrak{B}^{\text{t}}_{2g} \oplus \mathfrak{E}^{\text{t}}_{g}$, $\mathfrak{E}_g \sim \mathfrak{B}^{\text{t}}_{1g} \oplus \mathfrak{A}^{\text{t}}_{1g}$, i.e. we get four levels. Electron energy levels in the crystal field are traditionally written in small letters, in accordance with corresponding symbols of irreducible representations, e_g, t_{2g}, t_{1g} etc.

As an example, we define the magnitude of d-electron level splitting in the cubic field for two cases: (i) the charges are in the form of a regular octahedron and in the form of a regular tetrahedron with known distances R of charges from the paramagnetic center and (ii) the mean value of r^4 for d-electron: $\langle r^4 \rangle$. Parameters of the crystal field for these cases have been determined earlier, so it is only necessary to express the value of electron energy in the crystal field in terms of these parameters. It makes sense to divide the solution of the problem into two parts: the calculation of the energy levels by the equivalent operator method up to a constant and the definition of this constant by explicit calculation of one of the matrix elements.

According to the perturbation theory the Hamiltonian describing the atomic state in crystal field ($\widehat{\mathcal{H}}$) is the sum of the Hamiltonians for an unperturbed atom ($\widehat{\mathcal{H}}_0$) and the interaction operator ($\widehat{\mathcal{H}}'$) regarded as a perturbation, i.e., $\widehat{\mathcal{H}} = \widehat{\mathcal{H}}_0 + \widehat{\mathcal{H}}'$. In order to apply perturbation theory, one needs to know the state and energy of the unperturbed system and the type of the perturbation operator $\widehat{\mathcal{H}}'$. The vectors for unperturbed states (wave functions) with energies E_i will be denoted by $|i\alpha\rangle$ where index α will be introduced to distinguish between degenerate levels with the same energy E_i ($\alpha = 1, 2, ..., f_i$, with f_i standing for degeneracy multiplicity). (There are different approaches in the perturbation theory to degenerate and non-degenerate levels.) For nondegenerate levels, each value of E_i corresponds to a unique state $|i\rangle$, so that index α can be omitted. In the perturbed state, E_i can be obtained as follows:

$$E_i' = E_i + E_i^{(1)} + E_i^{(2)}, \tag{11.21}$$

where $E^{(1)}$ and $E^{(2)}$ are the corrections to the value of the unperturbed system energy in the state $|i\rangle$ in the first and second approximations respectively. In this case,

$$E_i^{(1)} = \langle i|\widehat{\mathcal{H}}'|i\rangle, \tag{11.22}$$

The correction to the value of energy in the second approximation ($E_i^{(2)}$) in addition takes into account the excited states. For non-degenerated states there is the expression:

$$E_i^{(2)} = -\sum_{k \neq i} \frac{\langle k|\widehat{\mathcal{H}}'|i\rangle\langle i|\widehat{\mathcal{H}}'|k\rangle}{E_i - E_k}. \tag{11.23}$$

In the case of degeneration of the multiplicity f_i, linear combinations of vectors of unperturbed states should be used as perturbed state eigenvectors in the zero approximation $|i\widetilde{\alpha}\rangle$:

$$|i\widetilde{\alpha}\rangle = \sum_{\alpha=1}^{f_i} |i\alpha\rangle\langle i\alpha|i\widetilde{\alpha}\rangle. \tag{11.24}$$

Here, $\langle i\alpha|i\widetilde{\alpha}\rangle$ are numerical factors representing the scalar products of state vectors in space. These factors obey the system of homogeneous linear equations

$$\sum_{\alpha'=1}^{f_i} \left(\langle i\alpha|\widehat{\mathcal{H}}'|i\alpha'\rangle - E_{i\widetilde{\alpha}}^{(1)}\delta_{\alpha\alpha'}\right)\langle i\alpha'|i\widetilde{\alpha}\rangle = 0, \tag{11.25}$$

where α changes from 1 to f_i. The above equation system should be completed by the normalizing condition. The condition for the existence of a non-zero solution of (11.25) leads to the secular equation

$$\mathrm{Det}\left\{\langle i\alpha|\widehat{\mathcal{H}}'|i\alpha'\rangle - E_{i\widetilde{\alpha}}^{(1)}\delta_{\alpha\alpha'}\right\} = 0. \tag{11.26}$$

The equation roots of (11.26) $E_{i\widetilde{\alpha}}^{(1)}$ ($\alpha = 1, 2 \ldots$), some of which may turn out to be identical, serve to adjust values E_i in the first approximation of perturbation theory. The values $\langle i\alpha|i\widetilde{\alpha}\rangle$ can be obtained from system (11.25) after inserting obtained values $E_{\widetilde{\alpha}}^{(1)}$.

Thus, for finding the corrections to the energy value of degenerate levels in the first approximation, we should solve the equation of the fifth degree, since the d-electron initial level is fivefold degenerated. However, the group theory allows us, without solving the secular equation, to identify those combinations of state vectors or wave functions, for which the interaction energy operator with the crystal field is diagonal. According to Wigner theorem (see Sect. 10.3.4), these combinations are the bases of those irreducible representations that relate to energy levels, obtained under the perturbation. After finding basis functions, to calculate the correction to the energy it is sufficient to find the corresponding diagonal matrix element, i.e., the mean value of the perturbation energy (as in the perturbation theory for a degenerate level).

The second approximation for degenerated states will be not required and is not considered.

As it was shown in Sect. 10.3.1, for the cubic group $\mathcal{O}_h$ the states $|2,0\rangle$ and $|2,2_+\rangle$, defined by the Eq. (10.11), form the basis of the irreducible representation $\mathfrak{E}_g$, and the states $|2,2_-\rangle$, $|2,1_+\rangle$ and $|2,1_-\rangle$ are the basis of the representation $\mathfrak{T}_{2g}$. Consequently, these bases can be considered as the bases of eigenvectors for electron energy levels e_g and t_{2g} in the crystal field. In this case it does not matter which function of the basis exactly we choose for calculating the correction to the electron energy due to the effect of the crystal field. For example, we get expressions for the energy of the level e_g using the state $|2,2_+\rangle$, and for the level t_{2g} using the state $|2,2_-\rangle$. In accordance with the general expression for the potential of a crystal field of cubic symmetry (11.6) an equivalent operator $\widehat{Q}^{\mathrm{cube}}$ is supposed to be

$$\widehat{Q}^{\mathrm{cube}} = \widehat{Q}_4^0 + 5\widehat{Q}_4^4. \tag{11.27}$$

Using the data from Table 10.3, we obtain the explicit form of operators $\widehat{Q}_4^0$ and $\widehat{Q}_4^4$:

$$\widehat{Q}_4^0 = 35\hat{L}_z{}^4 - 30\widehat{l^2}\hat{L}_z{}^2 + 25\hat{L}_z{}^2 - 6\widehat{l^2} + 3\widehat{l^4},$$

$$\widehat{Q}_4^4 = \frac{1}{2}\left[(\widehat{l^+})^4 + (\widehat{l^-})^4\right].$$

In the representation of eigenstates $\widehat{l_z}$ the operator $\widehat{Q}_4^0$ is diagonal, so that its matrix elements are simply calculated by substituting its eigenvalues m^2 instead of $\hat{L}_z{}^2$ and $l(l+1)$ instead of $\widehat{l^2}$ into the proper expression. Substituting the values $l = 2$ for d-electron and $m = \pm 2$, we get $\langle 2, \pm 2|\widehat{Q}_4^0|2, \pm 2\rangle = 12$. Matrix elements of the operator $\widehat{Q}_4^4$ are non-zero only if $|m - m'| = 2$, i.e. in the case $l = 2$ (d-electron) only for $m = \pm 2$ and $m' = \mp 2$ respectively. Performing consistently the increase operation (see (1.7)), it is easy to obtain the value of these elements starting from $m = -2$:

$$\langle 2, 2|\widehat{Q}_4^4|2, -2\rangle = 12. \tag{11.28}$$

Now it is easy to define the diagonal matrix elements of the operators (11.27) for the basic states $|2, 2_\pm\rangle$ defined by (10.13). So, for the level e_g

$$\begin{aligned}\langle 2, 2_+|\widehat{Q}^{\mathrm{cube}}|2, 2_+\rangle &= \frac{1}{2}\left[\langle 2, 2|\widehat{Q}_4^0|2, 2\rangle + \langle 2, -2|\widehat{Q}_4^0|2, -2\rangle\right] \\ &+ \frac{5}{2}\left[\langle 2, 2|\widehat{Q}_4^4|2, -2\rangle + \langle 2, -2|\widehat{Q}_4^4|2, 2\rangle\right] = 72,\end{aligned} \tag{11.29}$$

and for the level t_{2g}

$$\begin{aligned}\langle 2, 2_+|\widehat{Q}^{\mathrm{cube}}|2, 2_+\rangle &= \frac{1}{2}\left[\langle 2, 2|\widehat{Q}_4^0|2, 2\rangle + \langle 2, -2|\widehat{Q}_4^0|2, -2\rangle\right] \\ &- \frac{5}{2}\left[\langle 2, 2|\widehat{Q}_4^4|2, -2\rangle + \langle 2, -2|\widehat{Q}_4^4|2, 2\rangle\right] = -48.\end{aligned} \tag{11.30}$$

The values of the matrix elements in the form (11.29) and (11.30) determine the position of energy levels relative initial level of the d-electron in the free ion. If we denote the value of the total splitting of the crystal field as Δ, the resulting levels will be shifted on $\frac{3}{5}\Delta$ and $-\frac{2}{5}\Delta$, i.e. their changes are inversely to their degeneracy order. It is clear that in this case the "center of mass" of the levels is not shifted.

It now remains only to calculate the absolute value of the level shifts. For this purpose, it is convenient to introduce the proportionality factor $\langle l\|\beta\|l\rangle$ between the operator of the crystal field U_p^q and its equivalent operator $\widehat{Q}_p^q$, according to the equation

$$\langle lm|U_p^q|lm'\rangle = \langle l\|\beta\|l\rangle\langle lm|\widehat{Q}_p^q|lm'\rangle. \tag{11.31}$$

As it is follows from the Wigner–Eckart theorem, the factor $\langle l||\beta||l\rangle$ does not dependent on m or q for the given values l and p. (Greek letters in the factor indicate the meaning of the subscript p: symbols α, β and γ are taken respectively for $p = 2$, $p = 4$ and $p = 6$). To find such factors it is enough to calculate one of the matrix elements explicitly, e.g. $\langle 2, 2|U_4^4|2, -2\rangle$, where $U_4^4 = \frac{1}{2}r^4 \sin^4\theta(e^{4i\phi} + e^{-4i\phi})$. Then, taking into account that the state vector $|2, \pm 2\rangle$ can be represented by the function $\sqrt{\frac{5}{4\pi}\frac{3}{8}}R_{n2}(r)\sin^2\theta e^{\pm 2i\phi}$, we can get an explicit expression for the matrix element of the crystal field:

$$\langle 2, 2|U_4^4|2, -2\rangle = \frac{1}{2}\frac{3}{8}\frac{5}{4\pi}\int_0^{2\pi} \mathrm{d}\phi \int_0^{\infty} \mathrm{r}^4|\mathrm{R}_{n2}(\mathrm{r})|^2\mathrm{r}^2\mathrm{dr}\int_0^{\pi} \sin^9\theta \mathrm{d}\theta = \frac{8}{21}\langle \mathrm{r}^4\rangle. \tag{11.32}$$

Equating right hand part of (11.31) and (11.32) and using (11.28), we find for $l = 2$

$$\langle l||\beta||l\rangle = \frac{2}{63}\langle r^4\rangle. \tag{11.33}$$

The values of the level energies e_g and t_{2g} can be obtained by multiplying the values (11.29) or (11.30) by the found factor (11.33) and the coefficient of the crystal field A_4^0. Thus, the energy level of d-electron, which is fivefold degenerate in a spherically symmetric field of the atom, in the crystal field of cubic symmetry is split into two levels: the level e_g, shifted according to (11.29) by the value $-\frac{48}{21}e\langle r^4\rangle A_4^0$, and the level t_{2g}, shifted according to (11.30) by the value $\frac{32}{21}e\langle r^4\rangle A_4^0$. The total splitting ($\Delta$) is calculated by the formula

$$\Delta = \frac{80}{21}e\langle r^4\rangle |A_4^0|. \tag{11.34}$$

It remains to substitute the expression for A_4^0 from the formula (11.10) for the location of the charges at the vertices of a regular octahedron or from the formula (11.12) for the location of the charges at the vertices of a regular tetrahedron, and make calculations for concrete values R. The results of similar calculations for the paramagnetic crystal containing Ti^{3+} will be given below while considering a intermediate crystal field.

Qualitative interpretation of the results. The important new result of the obtained solution is the conclusion about the sequence order of the d-electron energy levels. It follows from the explicit expressions for A_4^0 (11.10) and (11.12) that in the case of octahedral coordination the lower level will be t_{2g} ($A_4^0 < 0$ for $q < 0$), whereas in the case of cubic or tetrahedral coordination the lower level will be e_g or e (Fig. 11.2).

The dependence of the level order on the coordination of the central ion is qualitatively confirmed by considering electrostatic interactions which are schematically presented in Fig. 11.3, which simultaneously shows the spatial distribution of the d-electron density (given by hatching).

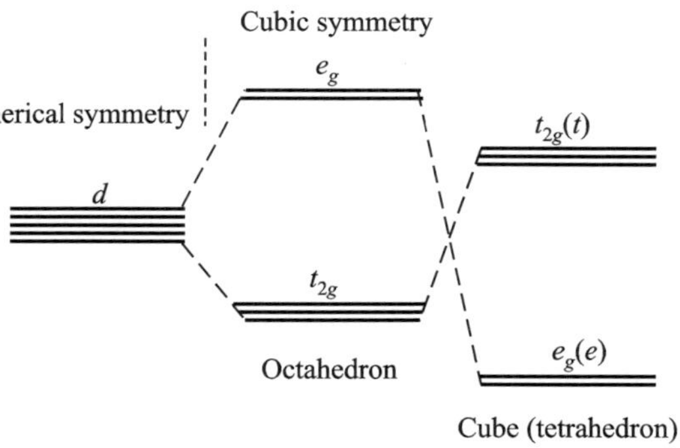

Fig. 11.2 Splitting of energy levels of d-electron (*left*) under effect of the electric field in the cases of octahedral (intermediate) and cubic or tetrahedral (*right*) charge disposition

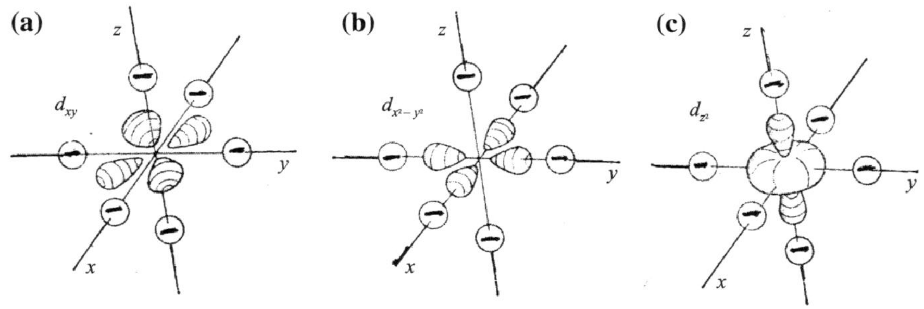

Fig. 11.3 Schematic representation of charge density distribution of d-electrons in the octahedral charge environment

States, relating to the triplet level, have zero density on the coordinate axes, i.e. in the direction to negative charges which create the crystal field. This mutual arrangement of electron clouds and surrounding charges results in the least repulsion between them, so the level t_{2g} is deeper than the level e_g. On the contrary, electron density of states e_g is maximum in the directions to surrounding charges, which leads to a greater interaction energy. In the cases of tetrahedral and cubic coordination of anions the position is reversed, and the ground level is e_g.

In simple cases such visual images help to solve the problem of the splitting patterns and level order.

Let us suppose that it is necessary to determine the result of a slight distortion of an octahedron, for example its shrinkage along the axis z. Obviously, the electron cloud in the state $|2, 0\rangle$ will be mostly close to negative charges on the axis z. As a result the energy of this state will be greatest. Therefore, a double-degenerate (doublet) level e_g will split into two single ones, the lowest of which will be $|2, 2_+\rangle$, i.e. the level $d_{x^2-y^2}$. Of the remaining conditions two of them: $|2, 1_+\rangle$ and $|2, 1_-\rangle|$, i.e. d_{xz}

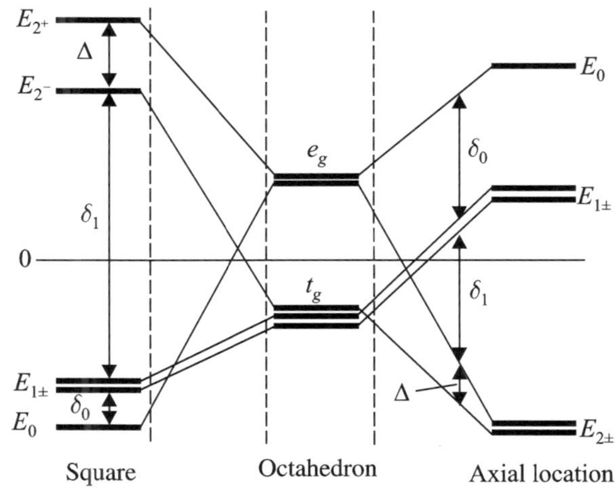

Fig. 11.4 A changing of *d*-electron energy level system in going from *square* to *axial* environment

and d_{yz}, will be of equal energy, but of a greater one than the state $|2, 2_-\rangle$, as the electron density distribution is concentrated in the plane (xy) and the electron cloud is farer from axial charges. In the case of elongating an octahedron along the axis z the level sequence becomes reversed. Figure 11.4 shows how the level system of the d-electron changes while the environment goes from tetragonal (the anions are at the square vertices) to axial (the anions are only the axis z). The square can be considered as the limiting case of an elongated octahedron (the charges on the axis z are so far that their interaction with the central ion can be ignored), and the axial position of the charges can be considered as a limiting case of a flattened octahedron.

As stated above, numerical results, obtained by this simple method, have little accordance with the experimental values, because the adopted model turns out to be too rough for a strong crystal field (see also Chap. 12). However, the qualitative conclusions about the situation and the level sequence turn out to be correct. The solution of a problem of the effect of a crystal field on a single electron will be used below while considering intermediate and weak crystal fields.

11.3.2 Splitting of Terms of Configurations np^k and nd^k in a Intermediate Crystal Field

Now we consider cases with one or more electrons in the unfilled shell.

One electron over the filled shell. Obviously, the solution of the previous problem is directly applicable in the case of a intermediate crystal field as well, if there is one electron over the filled shell, i.e. for the configurations d^1 and d^6. Indeed, in the intermediate crystal field approximation the spin-orbit interaction is the last to be taken into account: the influence of the crystal field leads to the situation that the angular momentum is completely determined by an orbital momentum, which coincides with an orbital angular momentum of a single electron outside closed shells. Therefore free atoms (ions) in the configurations d^1 and d^6 in their ground state have the terms with the same angular momentum as the electron. Consequently, the term splitting in the crystal field is completely determined by a single electron outside closed shells. An example is the splitting of term ^2D of the ion Ti^{3+} in alum (octahedral coordination), where calculation results are in good accordance with the experimental value Δ. Calculations are done by the formula (10.10) with $Z = 5{,}35$. Substitution of values, obtained in X-ray diffraction analysis, in the expression for Δ (11.34) leads to its value equal to $7 \times 10^5\,\text{cm}^{-1}$.

Special consideration is not required for the configurations d^4 and d^9 as well, which can be considered as a hole in closed shells d^5 and d^{10}. It is clear that the additions to energy due to the crystal field differ from the above case only by signs, i.e. the level inversion should be made in the diagrams. Of course, the configuration d^5 is not closed in a literal sense, since the spin is $S \neq 0$. But in the problem of level splitting when the spin-orbit interaction is neglected, as for the case of a intermediate crystal field, the spin does not play any role. In this case, a half-filled shell can be considered as a closed one (i.e., $L = 0$ for it). In this regard, the configuration d^5 is similar to the configuration d^{10}, for which $S = 0$. In the case of the p-electron $L = 0$ for the configurations p^3 and p^6, in the case of f-electrons (rare-earth ions) $L = 0$ for configurations f^7 and f^1.

Ions with several electrons in an unfilled shell. In this case it is necessary to sum up the orbital angular momenta of electrons before to take into account the crystal field in order to get the total orbital momentum of a atom (L). There are no principal differences from the one-electron case in the calculations of matrix elements of equivalent operators: we should only replace l by L. The difference appears in the determination of the proportionality factors $\langle L\|\alpha\|L\rangle$, $\langle L\|\beta\|L\rangle, \ldots$, which in general do not coincide with the previously calculated factors $\langle l\|\alpha\|l\rangle$, $\langle l\|\beta\|l\rangle, \ldots$. As in the case of a single electron, it is necessary and enough to calculate one of the matrix elements. It is sufficiently enough to express new constants via the known one-electron ones. The above difference is due to the fact that, for the calculation of the matrix elements of the crystal field, we should use many-electron wave functions (10.22).

A simple example of a system with several electrons can serve an atom (ion) with two p-electrons in the unfilled shell (configuration np^2). According to Hund rules, the term ^3P is the ground one, which in the crystal field of tetragonal symmetry is split into two levels: $\mathfrak{A}_{2u}$ and $\mathfrak{E}_u$ (see Table 11.3). In this case the crystal field is described by the single member $A_2^0 U_2^0$, where

$$U_2^0 = 3z^2 - r^2.$$

Using the method of equivalent operators, we can write:

$$\langle LM|U_z^0|LM\prime\rangle = \langle L||\alpha||L\rangle\langle LM|\widehat{Q}_2^0|LM\prime\rangle, \tag{11.35}$$

where $\widehat{Q}_2^0 = 3\widehat{L}_z^2 - \widehat{L}^2$. The matrix elements of the equivalent operator $\widehat{Q}_2^0$ are easily calculated using Eqs. (1.4), (1.5) and (1.7).

Thus, the configuration levels p^2 in the tetragonal field are expressed as follows: $2eA_2^0\langle L||\alpha||L\rangle$ (for the level $\mathfrak{A}_{2u}$) and $-eA_2^0\langle L||\alpha||L\rangle$ (for $\mathfrak{E}_u$). As usual, to determine $\langle L||\alpha||L\rangle$ it is necessary to calculate one of the matrix elements in an explicit form. The antisymmetric atomic wave function for the state with $M = L$ of configuration d^2 is

$$|1,1\rangle = \frac{1}{\sqrt{2}}\left(|1,1\rangle^{(1)}|1,0\rangle^{(2)} - |1,0\rangle^{(1)}|1,1\rangle^{(2)}\right).$$

Keeping in mind that one-electron functions are orthonormalized, then a diagonal matrix element of the operator U_2^0 on the functions $|1,1\rangle$ is equal to $-\langle l||\alpha||l\rangle$. However, in accordance with (11.35)

$$\langle 1,1|U_2^0|1,1\rangle = \langle L||\alpha||L\rangle.$$

Comparing these expressions, we obtain

$$\langle L||\alpha||L\rangle = \frac{2}{5}\langle r^2\rangle.$$

The expression for $\langle l||\alpha||l\rangle$ is obtained in the same manner as $\langle l||\beta||l\rangle$ (see Sect. 11.3). Consequently, energy levels are determined by the values $\frac{4}{5}eA_2^0\langle r^2\rangle$ and $-\frac{2}{5}eA_2^0\langle r^2\rangle$ for states $\mathfrak{A}_{2u}$ and $\mathfrak{E}_u$, respectively. This result can also be obtained without calculation: by reversing the configuration level p^1, because the configuration p^2 can be considered as the result of adding one p-hole to half-filled shell p^3. However, our aim was to show on a simple example the procedure of the calculation of the energy of a system with several unpaired electrons.

11.3.3 Paramagnetic Center in a Weak Crystal Field

In this case, the energy values in a crystal field are determined by the matrix elements

$$\langle JM_J|\sum_{i=1}^{N} V_{\text{cryst}}(\vec{r}_i)|JM_J'\rangle,$$

where $|JM_J\rangle$ is the wave function of an atom or ion, which is characterized by the total momentum J, formed by composition of the total orbital and total spin momenta of the atom, and the projection of the total momentum M_J. The wave function $|JM_J\rangle$ is created from a total orbital function of the atom $|LM\rangle$ and total spin function $|SM_S\rangle$, according to the theory of composition of moments:

$$|JM_J\rangle = \sum_{M=-L}^{L} \sum_{M_S=-S}^{S} \langle LSMM_S|JM_J\rangle|LM\rangle|SM_S\rangle, \tag{11.36}$$

where $\langle LSMM_S|JM_J\rangle$ are the Clebsch–Gordan coefficients. Proportionality factors $\langle J||\alpha||J\rangle$, $\langle J||\beta||J\rangle, \ldots$ are introduced by a formula, similar to (11.31):

$$\langle JM_J|\sum_{i=1}^{N} U_p^q(r_i)|JM'_J\rangle = \langle J||\alpha||J\rangle\langle JM_J|\widehat{Q}_p^q|JM'_J\rangle.$$

The qualitative picture of the level splitting is described by the formula (11.1), where J is the meaning of the total electron momentum (see Table 11.3).

Determining the picture of the level splitting in a weak crystal field, we should keep in mind the Kramers theorem, according to which there are always twofold degeneracy at least, if the number of electrons in the atom system is odd, i.e., if the number J is half-integer.

The Eq. (11.1) shows that the level of a free ion in a weak crystal field always splits into twofold degenerate levels, if the symmetry of the environment is below than a cubic one. In this case, a pair of complex conjugate basis vectors corresponds to each level. As to $J \geq 3/2$ in the cubic symmetry field for there is always at least one fourfold degenerate level. However, it appears that degeneracy should be minimal in the ground energy state (i.e., in the above case a twofold one). In another words, a cubic symmetry is inadmissible for a ground state. This phenomenon is quite general and is called the Jahn-Teller effect (or theorem). The Jahn-Teller effect can also be explained by means of the group theory.

The system with an even number of electrons always has at least one non-degenerate level, if the symmetry is below than a cubic one. Due to the Jahn-Teller effect this non-degenerate level is ground. For the same reason, as a rule, strict cubic symmetry is not realized in a weak crystal field. The distance on energy scale from the ground level up to the nearest excited one should also be estimated to find out the possibility of registration of electron paramagnetic resonance.

Energy level calculation for a single d-electron in an unfilled shell in the case of a weak crystal field. This object is selected as an example in order to avoid complicated calculations in our presentation. Although the case of a weak crystal field is only typical for rare-earth ions with the shells being filled with f-electrons, but not d-ones, the course of solution for f-electrons remains the same.

Thus, we express the splitting and the position of the energy levels of the single d-electron paramagnetic center within a weak crystal field in terms of parameters of the crystal field A_p^q and the mean degree values of the electron distance from the center $\langle r^p \rangle$. Let the group symmetry of the crystal field is C_3. It is convenient to divide the calculation into five steps:

(1) the determination of the ground state in the presence of the spin-orbit interaction;
(2) the determination of the form of the effective potential;
(3) the determination of the energy levels up to constant coefficients which are common to the set of states with a given J;
(4) the determination of one of the coefficients by comparing one of the matrix elements of the equivalent operator with the matrix element of a weak crystal field, calculated using one-electron functions;
(5) the obtaining of the expression for energy of levels via $\langle r^p \rangle$ and A_p^q.

Now we shall consider each step separately.

1. The configuration d^1 forms a term with $L = l = 2$ and $S = \frac{1}{2}$, i.e., the term 2^2D. The total momentum takes on values from $J = L + S$ to $J = |L - S|$ i.e., $J = \frac{5}{2}$, $\frac{3}{2}$. In the ground state J takes on the smallest value as the shell is less than half filled (see Sect. 10.2.6), consequently, $J = \frac{3}{2}$. Thus, the ground state of the ion will be the state $^2D_{3/2}$.
2. The value of p in the general expression for the crystal field potential (11.9) must satisfy the inequalities $p \leq 2l$ and $p \leq 2J$, therefore, $p \leq 3$. For even values of p, q and the restriction due to the presence of the trigonal axis: $q = 3k$ (k is an integer), this means that the effective potential should be $V_{\text{cryst}} = A_2^0 U_2^0$.
3. The matrix elements $\langle JM|U_p^q|JM'_J\rangle$ of this potential are calculated by means of the equivalent operator up to a fixed factor:

$$\langle JM_J|U_2^0|JM'_J\rangle = \langle J||\alpha||J\rangle\langle JM_J|\widehat{Q}_2^0|JM'_J\rangle.$$

From this expression for $J = \frac{3}{2}$, we get

$$\langle\frac{3}{2}, \pm\frac{1}{2}|U_2^0|\frac{3}{2}, \pm\frac{1}{2}\rangle = -\langle\frac{3}{2}||\alpha||\frac{3}{2}\rangle\langle\frac{3}{2}, \pm\frac{3}{2}|\widehat{Q}_2^0|\frac{3}{2}, \pm\frac{3}{2}\rangle = -3\langle\frac{3}{2}||\alpha||\frac{3}{2}\rangle.$$

Off-diagonal matrix elements are zero. Consequently, the ground level of ion with $J = 3/2$ in a weak crystal field splits into two symmetrical doublets, each of them characterized by its own value $|M_J|$.
4. To determine $\langle J||\alpha||J\rangle$ it is necessary to calculate explicitly one of the matrix elements. The easiest way to do this is for $M_J = J$. The wave function $|JM_J\rangle$ may be constructed from orbital and spin functions. The values of the corresponding Clebsch–Gordan coefficients are given in Table 11.4. Using them and taking into account the orthogonality and normalization of the spin functions, we find

$$\langle 3/2, 3/2|U_2^0|3/2, 3/2\rangle = \frac{1}{5}\langle 2, 1|U_2^0|2, 1\rangle + \frac{4}{5}\langle 2, 2|U_2^0|2, 2\rangle.$$

Table 11.4 Clebsch–Gordan coefficients for $L = 2$ and 3

M_J	M	M_S	$L = 2$		$L = 3$	
			$J = 5/2$	$J = 3/2$	$J = 7/2$	$J = 5/2$
1/2	1	−1/2	$\sqrt{2/5}$	$\sqrt{3/5}$	$\sqrt{3/7}$	$2/\sqrt{7}$
	0	1/2	$\sqrt{3/5}$	$-\sqrt{2/5}$	$2/\sqrt{7}$	$-\sqrt{3/7}$
3/2	2	−1/2	$1/\sqrt{5}$	$2/\sqrt{5}$	$\sqrt{2/7}$	$\sqrt{5/7}$
	1	1/2	$2/\sqrt{5}$	$-1/\sqrt{5}$	$\sqrt{5/7}$	$-\sqrt{2/7}$
5/2	3	−1/2	–	–	$1/\sqrt{7}$	$\sqrt{6/7}$
	2	1/2	1	0	$\sqrt{6/7}$	$-1/\sqrt{7}$
7/2	3	1/2	–	–	1	0

Applying the method of equivalent operators, we get:

$$\langle 2,1|U_2^0|2,1\rangle = -3\langle L||\alpha||L\rangle, \quad \langle 2,2|U_2^0|2,2\rangle = 6\langle L||\alpha||L\rangle.$$

Since the state $|LM\rangle$ in this example is the one-electron one (there is only one electron in the unfilled shell), then (see (11.32))

$$\langle L||\alpha||L\rangle = -\frac{2}{21}\langle r^2\rangle.$$

After substituting this result in the expression for the matrix element, we get

$$\langle 3/2,3/2|U_2^0|3/2,3/2\rangle = \frac{21}{5}, \quad \langle l||\alpha||l\rangle = -\frac{2}{5}\langle r^2\rangle.$$

Comparing this matrix element with its expression obtained with the equivalent operator for $J = 3/2$, we have:

$$\langle J||\alpha||J\rangle = -\frac{2}{15}\langle r^2\rangle.$$

5. On the basis of these results, the expressions for the energy levels are as follows:

$$E_{1/2} = \frac{2}{5}A_2^0\langle r^2\rangle, \quad E_{3/2} = -\frac{2}{5}A_2^0\langle r^2\rangle.$$

Figure 11.5a shows the energy level diagram of the of the ion in cerium ethyl sulfate ($l = 3$, $J = \frac{5}{2}$) in crystal field and represents the further splitting of the levels in a magnetic field. The structure of the nearest environment of the ion Ce^{3+} in ethyl sulfate is shown in two views: along the axis C_3 (Fig. 11.5b), and in the direction perpendicular to it (Fig. 11.5c). The light and shaded circles represent H_2O molecules in different planes, the ions $C_2H_5SO_4$ are represented as triangles.

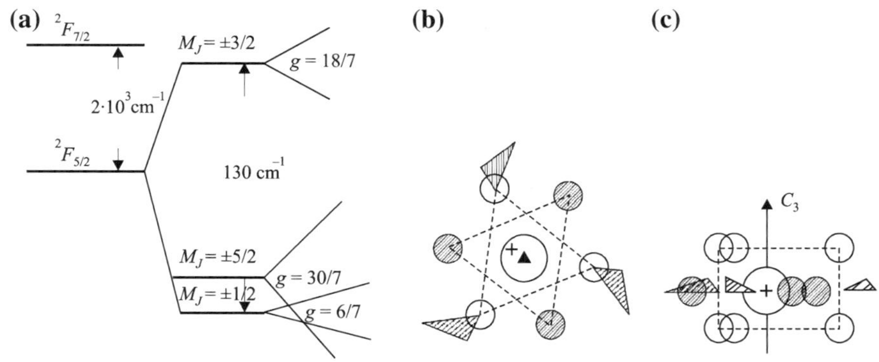

Fig. 11.5 Ion Ce^{3+} energy level diagram in trigonal surroundings (**a**) and the disposition of ions $C_2H_5SO_4$ (the *triangles*) and H_2O (the *circles*) in two projections (**b** and **c**) in cerium ethyl sulfate. Turn the attention to the different scale of splittings

Calculation of energy levels in the case when there are several electrons in an unfilled shell. If there are several electrons in an unfilled shell, then in the calculation of the state energy of the paramagnetic center in a weak crystal field there appears an intermediate step at which the coefficients $\langle L||\alpha||L\rangle$, $\langle L||\beta||L\rangle \ldots$ are expressed in terms of $\langle l||\alpha||l\rangle$, $\langle l||\beta||l\rangle \ldots$. This is done exactly in the same manner as for the calculation in the case of a intermediate crystal field (see the example for the configuration p^2). The rest course of solution is the same as for a single electron in the unfilled shell.

The calculation for the shell which is filled more than a half allows the simplification related to the fact that in the ground state $J = L + S$, so it is not necessary to use the Clebsch–Gordan coefficients. Figure 10.1a shows an example of this situation where the splitting diagram of the level $^4I_{9/2}$ of the ion Nd^{3+} is represented for two kinds of environment symmetry: cubic and trigonal.

In conclusion the following should be mentioned: the method of the equivalent operator allows the easy calculation of the relative values of matrix elements. To calculate their absolute values it is sufficient to perform an explicit calculation of only one matrix element of a crystal field for the same value of the angular momentum. The solution of the secular equation is replaced by the calculation of the diagonal matrix elements in the basis of the irreducible representations of the crystal field symmetry the group. A significant result of the above analysis is the fact that for the $3d$-ions in an intermediate crystal field the degeneracy order of the ground state depends on the number of electrons in the $3d$-shell. In the case of octahedral coordination the ground levels are singlets in configurations $3d^3$ and $3d^8$, doublets for $3d^4$ and $3d^9$, and triplets for $3d^1$ and $3d^2$. For rare-earth ions in a weak crystal field it is required to specially consider each configuration $4f^k$, but, according to the Kramers theorem and the Jahn-Teller effect, it can be stated that for a system with an even number of electrons the ground level is a singlet, but with an odd number of electrons it is a doublet.

Further Readings

Altshuler SA, Kozirev BM (1964) Electron paramagnetic resonance. Academic, New York, 724 pp
Abragam A, Bleaney B (1970) Electron paramagnetic resonance of transition ions. Clarendon Press, Oxford, 911 pp

Chapter 12
Covalent-Coupled Paramagnetic Complexes

We have already met the situations when the crystal field model does not lead to correct results. This is the case of the strong crystal field when the interaction of surrounding ions or molecules with a paramagnetic center is comparable with electrostatic interactions of electrons in a ion. A rigorous approach to the problem of the paramagnetic center state in a crystal involves consideration of the crystal as a single quantum-mechanical system. This approach, however, does not enable to get a solution in a form suitable for a practical use, so we have to resort to all kinds of simplifications. Up to now we have considered a paramagnetic center itself as a quantum mechanical system, but all of its interactions with the rest of a crystal were reduced to the interaction with the electrostatic field created by this remaining part. Further, although not of principle, simplification was to set aside of the consideration the rest of a crystal except for nearest environment of a paramagnetic center. In this way we assumed the existence of purely ionic bonds between the paramagnetic center and the surroundings. These simplifications are not acceptable in many important cases.

12.1 Essentials of Covalent Bond Model

If the interaction of any electron of a paramagnetic center with "foreign" electrons (electrons of surrounding ions or molecules) is comparable in value to the interactions of this electron with other electrons of the paramagnetic center, it is necessary to combine the paramagnetic center and its immediate surroundings into a single "molecule" or complex. The theoretical study of this complex is usually carried out using general methods from the quantum theory of molecules. In other words, this approach involves the formation of chemical covalent bonds between the paramagnetic center and the surrounding ions or molecules (often called as ligands). It is interesting to note that both approaches were developed long before the discovery of the EPR (in 1945). In order to explain the magnetic properties of iron cyanide Linus Pauling, an American physical chemist, used the method of valence bonds in 1931.

V. I. Chizhik et al., *Magnetic Resonance and Its Applications*,
DOI: 10.1007/978-3-319-05299-1_12,

However, since 1932, the school of John Hasbrouck Van Vleck, an American physicist, has been evolving the crystal field theory. Due to its simplicity and elegance, and almost complete success in explaining the paramagnetic properties of crystals it eclipsed the valence bond theory in this area for a long time. Only after the discovery of the electron paramagnetic resonance they revealed limitations of the crystal field model (which was pointed out by J.H. Van Vleck from the very beginning though). In some cases the splitting values in the crystal field turned out to be incorrect, and sometimes even an energy change sign was wrong. In cyanides, for example, the calculations carried out by the crystal field method lead to incorrect values of the effective spin. The field theory cannot explain such effects as the effective reduction of the orbital angular momentum, superhyperfine interaction and superexchange between neighbouring paramagnetic ions (*vide infra*). Finally, attempts to improve the crystal field model by replacing the point charge by the spatial charge distribution, including spatial distribution of the electron density of surrounding ions, failed to reduce the discrepancy between theory and experiment.

The model closest to reality is a model that allows the formation of partially covalent bonds between the center and the ligands. The bond is formed due to the fact that the electrons of the closed shells of the ligands (typically they are s- or p-electrons) shift partially towards the d-orbital of the central ion. A very slight transfer of the ligand electron density to excited s- and p-orbital of the paramagnetic ion occurs too. Thus the chemical bond in paramagnetic complexes is partially ionic and partially covalent. A simplifying factor is a small degree of charge transfer of the ligand electrons to the orbits of the paramagnetic ion. Due to this fact, the connections of the central ion with different ligands can be assumed independent of each other, in the sense that the perturbation of the central ion electronic structure by one of the ligands does not affect the nature of its bonds with the other ligands. Therefore, to understand the nature of bond of the central ion with the ligands we should first consider the relation in a diatomic system MX, where M stands for the paramagnetic ion (metal), and X stands for the ligand.

Molecular orbitals of the complex MX. The states of electrons in the molecule are described by the so-called molecular orbitals, which play in the theory of molecules the same role as the one-electron states s, p, d etc. in the atom in the approximation of a spherically symmetric self-consistent field (also known as atomic orbitals). The molecular orbital (MO) method implies that each molecular orbital is constructed as a linear combination of atomic orbitals (LCAO). The weight, which an atomic orbital possesses in a molecular one, is determined by the condition of minimum energy. Let us scrutinize the calculation of the energy complex and the construction of a molecular orbital by example of a complex MX. Atomic orbitals of M and X are denoted, respectively, $|M\rangle$ and $|X\rangle$, and a molecular orbital formed from them is $|K\rangle$. Let us assume that

$$|K\rangle = \alpha|M\rangle + \beta|X\rangle, \tag{12.1}$$

where the coefficient α determines the contribution to the molecular orbital of a central ion, and the coefficient β determines the contribution of a ligand. The average value of the electron energy is, in general, a functional of $|K\rangle$, and if $|K\rangle$ is chosen in the form (12.1), it is a function of the coefficients α and β , which should vary so that the energy takes the extreme value. The normalization condition

$$\langle K|K\rangle = 1$$

is imposed on α and β. The problem of finding a conditional extremum is solved by the method of Lagrange multipliers. The values of the parameters α and β, which determine the extremum of the energy functional, are found by solving the system of equations

$$\begin{aligned} \frac{\partial \Phi(\alpha,\beta,E)}{\partial \alpha} &= 0, \\ \frac{\partial \Phi(\alpha,\beta,E)}{\partial \beta} &= 0, \\ f(\alpha,\beta) &= 0, \end{aligned} \tag{12.2}$$

where $\Phi(\alpha, \beta, E) \equiv E(\alpha, \beta) - Ef(\alpha, \beta)$ and $f(\alpha, \beta) = \langle K|K\rangle - 1$.

If $\widehat{\mathcal{H}}$ is an electron energy operator in the field of the nucleus and the electrons of the central ion and the ligands, then $E(\alpha, \beta)$ and $f(\alpha, \beta)$ can be expressed as follows:

$$E(\alpha, \beta) = \alpha^2\langle M|\widehat{\mathcal{H}}|M\rangle + \beta^2\langle X|\widehat{\mathcal{H}}|X\rangle + 2\alpha\beta\langle M|\widehat{\mathcal{H}}|X\rangle,$$

$$f(\alpha, \beta) = \alpha^2 + \beta^2 - 2\alpha\beta\langle M|X\rangle - 1$$

(we should keep in mind that the orbitals belonging to different ions are not orthogonal). Introducing the symbols $E_M = \langle M|\widehat{\mathcal{H}}|M\rangle$, $E_X = \langle X|\widehat{\mathcal{H}}|X\rangle$, $E_{MX} = \langle M|\widehat{\mathcal{H}}|X\rangle$ and $S = \langle M|X\rangle$, we transform the system (12.2) to the form

$$\begin{aligned} \alpha E_M + \beta E_{MX} - E(\alpha + \beta S) &= 0, \\ \alpha E_{MX} + \beta E_X - E(\beta + \alpha S) &= 0, \\ \alpha^2 + \beta^2 - 2\alpha\beta S - 1 &= 0. \end{aligned} \tag{12.3}$$

The energy E is found as the solution of the secular equation:

$$(E_M - E)(E_X - E) = (E_{MX} - ES)^2. \tag{12.4}$$

Even without solving the Eq. (12.4), we can conclude about the position of the energy levels relative to the initial levels of the noninteracting paramagnetic ion and ligands. Indeed, the right side of (12.4) is significantly positive, therefore, the root must be either less than the lower value or greater than the upper one of the values E_M, E_X. Since there are only two roots (denoted by E and E^*, $E < E^*$), one of them is less than E_X, and the other one is greater than E_M (ligand orbitals usually correspond to lower energy states):

$$E < E_X, \quad E^* > E_M, \tag{12.5}$$

and one of the new levels (E in Fig. 12.1) is always deeper than the lower one (E_X) by the value $\Delta E = E_X - E$, and the other one (E^*) is higher than the higher one (E_M) by the value $\Delta E^* = E^* - E_M$. The state corresponding to the lower level, is called *binding orbital* (denoted by $|K\rangle$), and the state of the top level is called *antibinding orbital* (denoted by $|K\rangle^*$; the asterisk here does not denote the complex conjugate). Clearly, if the contributions to the orbital of the ligand p-electron and d-electron of paramagnetic center have the same sign, then there is an increase of electron density on the bond line that stabilizes ("binds") the system. If contributions have different signs, then on the bond line there is a node in the density distribution and the ligand repulsion from the paramagnetic center arises.

We can now estimate the relative contributions of the initial states $|M\rangle$ and $|X\rangle$ to the upper and lower levels. To do this we have to express E from (12.3). Multiplying the first and the second equations (12.3), respectively, by α and β, we get

$$\alpha^2 E_M + \alpha\beta E_{MX} - \alpha^2 E - \alpha\beta S = 0,$$

$$\alpha\beta E_{MX} + \beta^2 E_X - \beta^2 E - \alpha\beta S = 0,$$

where after the subtraction a simple relation comes from:

$$\alpha^2 E_M - \beta^2 E_X = (\alpha^2 - \beta^2)E.$$

Now it is easy to express E in a convenient form:

$$E = \frac{\alpha^2 E_M - \beta^2 E_X}{\alpha^2 - \beta^2} = \frac{\alpha^2 E_M - \alpha^2 E_X + \alpha^2 E_X - \beta^2 E_X}{\alpha^2 - \beta^2} = E_X - \Delta E,$$

where

$$\Delta E = \frac{\alpha^2}{\beta^2 - \alpha^2}(E_M - E_X),$$

or

$$E^* = \frac{\alpha^{*2} E_M - \beta^{*2} E_M + \beta^{*2} E_M - \beta^{*2} E_X}{\alpha^2 - \beta^2} = E_M + \Delta E^*,$$

where

$$\Delta E^* = \frac{\beta^{*2}}{\alpha^{*2} - \beta^{*2}}(E_M - E_X). \tag{12.6}$$

Comparing these expressions with inequalities (12.5), we note that ΔE and ΔE^* are always positive, but then the coefficients $\alpha^2/(\beta^2 - \alpha^2)$ and $\beta^{*2}/(\alpha^{*2} - \beta^{*2})$ are also positive, since it was assumed that $E_M \gg E_X$. This yields the inequalities shown in Fig. 12.1 at the corresponding levels: $\beta^2 > \alpha^2$ and $\alpha^{*2} > \beta^{*2}$. These

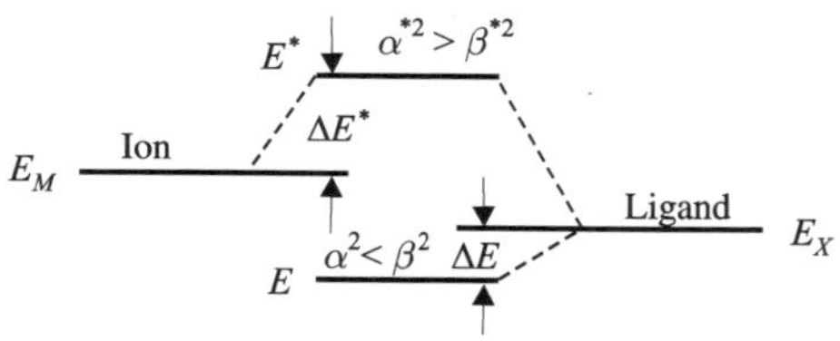

Fig. 12.1 Displacements of the energy levels of paramagnetic ion and ligand in a complex

conditions mean that the ligand wave function contributes greatly to the bonding molecular orbital, while the ion wave function contributes to the antibonding one.

Quantitative characterization of the covalency degree of the complex. It is convenient for complexes to express a molecular orbital in a slightly different way compared with (12.1):

$$|K\rangle = \frac{1}{N}(|X\rangle + \gamma|M\rangle), \tag{12.7}$$

$$|K\rangle^* = \frac{1}{N^*}(|M\rangle - \lambda|X\rangle), \tag{12.8}$$

here $\gamma = \frac{\alpha}{\beta}$; $\lambda = -\frac{\beta^*}{\alpha^*}$; $N^2 = 1 + \gamma^2 - 2\gamma S$; $N^{*2} = 1 + \lambda^2 - 2\lambda S$. The following limiting special cases are possible:

(1) $\lambda = \gamma \simeq 1$, in such a situation, the electron is equally owned by the ion and the ligand, and the bond is largely covalent;

(2) $\lambda, \gamma << 1$, mainly ionic bond; in this case it is expected that $S < 1$ and $(E_{MX} - E_X) << 1$.

If $\gamma = 0$, the electron transfer to the ligands does not occur (see Eq. 12.8). However, the coefficient λ in this case is non-zero, i.e., the ligand electron clouds occupy the central ion. Indeed, the orthogonality condition of $|K\rangle$ and $|K\rangle^*$ implies $\gamma - \lambda + S(1 - \gamma\lambda) \simeq 0$, or, for small values of λ and $\gamma = 0$, $\lambda \simeq S$. This situation is very frequent to occur.

12.2 Formation of Molecular Orbitals in the Paramagnetic Complexes

Since these complexes are quite wide-spread, as an example, which is of great practical importance, we shall consider the case when six identical X ions are at the vertices of a regular octahedron, and a paramagnetic ion with a filling d-shell M in its centre. Between the shell M and the ions X there is a covalent bond caused by overlapping of external valence shell electrons. We introduce an approximation in which the molecular orbitals are the linear combinations of the Hartree-Fock atomic orbitals. The choice of the corresponding linear combinations of the atomic orbitals is determined primarily by the requirements of symmetry: a molecular orbital and,

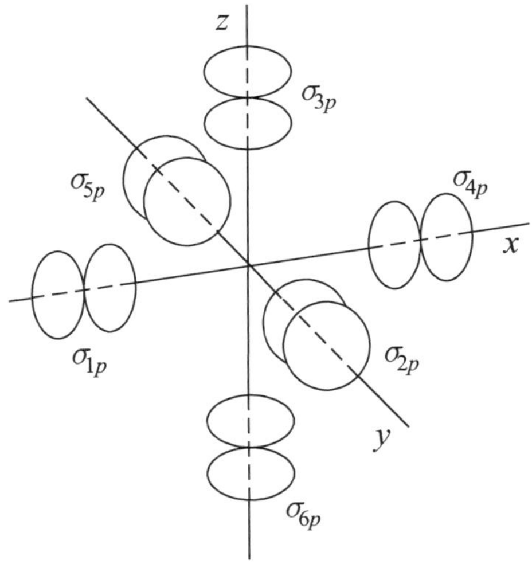

Fig. 12.2 Schematic representation of σ_p-orbitals of octaedral complex

therefore, its constituent linear combinations of atomic orbitals $|M\rangle$ and $|X\rangle$ must be transformed according to irreducible representations of the symmetry group of the complex, in the given case, the $\mathcal{O}_h$ group. In accordance with the distribution of the electron density all molecular orbitals are classified into σ- and π-orbitals. Molecular orbitals with a maximum density along the bond line (in this example they are the coordinate axes), are called *σ-orbitals*. If the electron density is maximal in the directions which do not coincide with a bond line, then *π-orbital* forms. Obviously, s-electrons can only participate in σ-bond, while p- and d-electrons can form both σ- and π-bonds. In this example, the ligand p-electrons contribute to the σ-bond if their density is maximal in the direction of the coordinate axis, where the ligand is located.

12.3 Determination of Combinations of Ligand Orbitals Transformed According to Irreducible Representation of the Symmetry Group of the Complex

Thus, in the interaction of ligand p-electrons with the central ion, a chemical bond of two types may appear.

σ-bond. Designation and schematic distribution of the electron density of p-electrons of 6 ligands, involved in formation of the σ-bond with the central ion, are shown in Fig. 12.2. For short, atomic orbitals of these p-electrons are marked $|\sigma_{ip}\rangle$, where i is the ligand number that determines its position in the Cartesian coordinate system with the paramagnetic center at the origin. A set of atomic orbitals is transformed by the irreducible representation $\mathfrak{G}_\sigma$, which characters are easily found while considering the impact of individual transformations on the set of p-orbitals by

Table 12.1 Irreducible representation characters for $\mathcal{O}_h$ group and these of representations for σ- and π-orbitals for octaedral complex

$\mathcal{O}_h$	$\widehat{E}$	$8\widehat{C}_3$	$3\widehat{C}_2$	$6\widehat{C}_{2d}$	$6\widehat{C}_4$	$\hat{\imath}$	$8\widehat{S}_6$	$3\widehat{\sigma}_h$	$3\widehat{\sigma}_d$	$6\widehat{S}_4$	$n_j^{(\sigma)}$	$n_j^{(\pi)}$	Basis
$\mathfrak{A}_{1g}$	1	1	1	1	1	1	1	1	1	1	1	0	s
$\mathfrak{A}_{2g}$	1	1	1	−1	−1	1	1	1	−1	−1	0	0	–
$\mathfrak{E}_g$	2	−1	2	0	0	2	−1	2	0	0	1	0	$d_{z^2}, d_{x^2-y^2}$
$\mathfrak{T}_{1g}$	3	0	−1	−1	1	3	0	−1	−1	1	0	1	–
$\mathfrak{T}_{2g}$	3	0	−1	1	−1	3	0	−1	1	−1	0	1	d_{xy}, d_{yz}, d_{zx}
$\mathfrak{A}_{1u}$	1	1	1	1	1	−1	−1	−1	−1	−1	0	0	–
$\mathfrak{A}_{2u}$	1	1	1	−1	−1	−1	−1	−1	1	1	0	0	–
$\mathfrak{E}_u$	2	−1	2	0	0	−2	1	−2	0	0	0	0	–
$\mathfrak{T}_{1u}$	3	0	−1	−1	1	−3	0	1	1	−1	0	1	–
$\mathfrak{T}_{2u}$	3	0	−1	1	−1	−3	0	1	−1	1	0	1	p_x, p_y, p_z
$\mathfrak{G}_\sigma$	6	0	2	0	2	0	0	4	2	0	–	–	σ^6
$\mathfrak{G}_\pi$	12	0	−4	0	0	0	0	0	0	0	–	–	π^{12}

counting those orbitals which do not vary or only change the sign under this transformation, i.e., by those that lead to the appearance of $+1$ or -1 on the transformation matrix diagonal. The algebraic sum of these coefficients is the sought-for character of the reducible representation. By Eq. (10.31) we determine the number $n_j^{(\alpha)}$ of the irreducible representations of the group $\mathcal{O}_h$, included in the representation $\mathfrak{G}_\sigma$. As in Sect. 11.1, it is convenient to draw a table to perform these calculations (Table 12.1).

The table contains representation characters of the group $\mathcal{O}_h$, corresponding to operations listed in the heading of the next ten columns as well as the results of calculations $n_j^{(\sigma)}$ and $n_j^{(\pi)}$ by the formula (10.42) for the representations $\mathfrak{G}_\sigma$ and $\mathfrak{G}_\pi$, respectively. The last column shows the bases, which are of our interest, transforming under the relevant irreducible or reducible representations. The results of calculations can be briefly expressed in the expansion:

$$\mathfrak{G}_\sigma \sim \mathfrak{A}_{1g} \oplus \mathfrak{E}_g \oplus \mathfrak{T}_{1u}.$$

The s-orbitals, belonging to the σ-bond and denoted as $|\sigma_{is}\rangle$, are converted likewise. Further consideration applies equally to both p- and s-orbitals, so the indices p and s are omitted. For simplicity, we will also use p and s instead of $|\sigma_{ip}\rangle$ and $|\sigma_{is}\rangle$ for atomic orbitals. To construct a molecular orbital it is necessary to know what linear combinations of the ligand atomic orbitals are transformed according to the irreducible representations $\mathfrak{A}_{1g}$, $\mathfrak{E}_g$, and $\mathfrak{T}_{1u}$. The required combinations are calculated by the formula (10.33), where we should fix $|a\rangle = \sigma_1$. Then

$$|\sigma_{1j}\rangle = \widehat{P}^{(j)}\sigma_1 = \frac{f_j}{g}\sum_G \chi^{(j)}\widehat{G}\sigma_1,$$

where $|\sigma_{1j}\rangle$ denotes the projection of σ_1 onto the molecular orbital transformed by the irreducible representation $\mathfrak{G}_\sigma$. It is convenient to present the calculation as shown in Table 12.2, which shows the results of the transformation from the group $\mathcal{O}_h$ of the

Table 12.2 Transformation of the ligand σ_1 and Y_1 orbitals in $\mathcal{O}_h$ group

Transformation from the group $\mathcal{O}_h$	Transformation result		IP character					
	σ_1	Y_1	$\mathfrak{A}_{1g}$	$\mathfrak{E}_g$	$\mathfrak{T}_{1g}$	$\mathfrak{T}_{2g}$	$\mathfrak{T}_{1u}$	$\mathfrak{T}_{2u}$
$\widehat{E}$	σ_1	Y_1	1	2	3	3	3	3
$\widehat{C}_3^1(x,y,z)$	σ_3	X_3						
$\widehat{C}_3^2(x,y,z)$	σ_2	Z_2						
$\widehat{C}_3^1(x,-y,-z)$	σ_6	$-X_6$						
$\widehat{C}_3^2(x,-y,-z)$	σ_5	Z_5						
$\widehat{C}_3^1(x,y,-z)$	σ_2	$-Z_2$						
$\widehat{C}_3^2(x,y,-z)$	σ_6	X_6						
$\widehat{C}_3^1(x,-y,z)$	σ_5	$-Z_5$						
$\widehat{C}_3^2(x,-y,z)$	σ_3	$-X_3$						
$\sum\left(8\widehat{C}_3\right)$	$2(\sigma_2+\sigma_3+\sigma_5+\sigma_6$	0	1	−1	0	0	0	0
$\widehat{C}_2(z)$	σ_4	$-Y_4$						
$\widehat{C}_2(x)$	σ_1	$-Y_1$						
$\widehat{C}_2(y)$	σ_4	Y_4						
$\sum\left(3\widehat{C}_2\right)$	$\sigma_1+2\sigma_4$	$-Y_1$	1	2	−1	−1	−1	−1
$\widehat{C}_{2d}(x,y)$	σ_2	X_2						
$\widehat{C}_{2d}(-x,y)$	σ_5	$-X_5$						
$\widehat{C}_{2d}(x,z)$	σ_3	$-Y_3$						
$\widehat{C}_{2d}(-x,z)$	σ_6	$-Y_6$						
$\widehat{C}_{2d}(y,z)$	σ_4	Z_4						
$\widehat{C}_{2d}(-y,z)$	σ_4	$-Z_4$						
$\sum\left(6\widehat{C}_{2d}\right)$	$\sigma_2+\sigma_3+\sigma_5+\sigma_6+2\sigma_4$	$X_2-X_5-Y_3-Y_6$	1	0	−1	1	−1	1
$\widehat{C}_4^1(z)$	σ_5	X_5						
$\widehat{C}_4^3(z)$	σ_2	$-X_2$						
$\widehat{C}_4^1(x)$	σ_1	$-Z_1$						
$\widehat{C}_4^3(x)$	σ_1	Z_1						
$\widehat{C}_4^1(y)$	σ_3	Y_3						
$\widehat{C}_4^3(y)$	σ_6	Y_6						
$\sum\left(6\widehat{C}_4\right)$	$2\sigma_1+\sigma_2+\sigma_3+\sigma_5+\sigma_6$	$X_5-X_2+Y_3+Y_6$	1	0	1	−1	1	−1
$\hat{\imath}$	σ_4	$-Y_4$	1	2	3	3	−3	−3
$\sum\left(8\widehat{S}_6\right)=\hat{\imath}\sum\left(8\widehat{C}_3\right)$	$2(\sigma_2+\sigma_3+\sigma_5+\sigma_6)$	0	1	−1	0	0	0	0
$\sum\left(3\widehat{\sigma}_h\right)=\hat{\imath}\sum\left(3\widehat{C}_2\right)$	$2\sigma_1+\sigma_4$	Y_4	1	2	−1	−1	1	1
$\sum\left(6\widehat{\sigma}_d\right)=\hat{\imath}\sum\left(6\widehat{C}_{2d}\right)$	$2\sigma_1+\sigma_2+\sigma_3+\sigma_5+\sigma_6$	$X_2-X_5+Y_3+Y_6$	1	0	−1	1	1	−1
$\sum\left(6\widehat{S}_4\right)=\hat{\imath}\sum\left(6\widehat{C}_4\right)$	$\sigma_2+\sigma_3+\sigma_5+\sigma_6+2\sigma_4$	$X_5-X_2-Y_3-Y_6$	1	0	1	−1	−1	1

orbital σ_1, the resulting totals by classes and necessary characters of the irreducible representation of the group $\mathcal{O}_h$. The totals by classes, containing the inversion $\widehat{i}$, are obtained by inverting the resulting totals of transformations from the corresponding class without inversion. For example, reflections $\widehat{\sigma}_d$ may be represented as the product $\widehat{i}\widehat{C}_{2d}$. For this class $\sum_{i=1}^{N_l}(6\widehat{C}_{2d})\sigma_1 = \sigma_2 + \sigma_3 + \sigma_5 + \sigma_6 + 2\sigma_4$. Inversion for σ_k is reduced to a permutation of the indices, $1 \leftrightarrow 4$, $2 \leftrightarrow 5$, $3 \leftrightarrow 6$. Consequently, $\sum_{i=1}^{N_l}(6\widehat{\sigma}_d)\sigma_1 = 2\sigma_1 + \sigma_2 + \sigma_3 + \sigma_5 + \sigma_6$. After multiplying the received totals by the characters of the irreducible representations contained in the six last columns of Table 12.2, and summarizing in accordance with (10.33), we get one function from every basis (small letters are used to denote the irreducible representation as in the case of an electron in the crystal field):

$$|a_{1g}\rangle = \sigma_1 + \sigma_2 + \sigma_3 + \sigma_4 + \sigma_5 + \sigma_6, \tag{12.9}$$

$$|e'_g\rangle = 2(\sigma_1 + \sigma_4) + (\sigma_2 + \sigma_3 + \sigma_5 + \sigma_6),$$

$$|t'_{1u}\rangle = \sigma_1 - \sigma_4. \tag{12.10}$$

The requirements of orthogonality of the rest basis functions to the obtained functions are used to find the proper combinations. The orbitals of the irreducible representation basis $\mathfrak{T}_{1u}$ can be obtained by cyclic permutation of x, y and z in (12.10). The second molecular orbital of the basis of the irreducible representation $\mathfrak{E}_g$ is found in the form

$$|e''_g\rangle = \alpha_1\sigma_1 + \alpha_2\sigma_2 + \alpha_3\sigma_3 + \alpha_4\sigma_4 + \alpha_5\sigma_5 + \alpha_6\sigma_6,$$

where the coefficients are determined from the orthogonality conditions of the molecular orbital $|e''_g\rangle$ to the molecular orbitals | $|e'_g\rangle$, $|a_{1g}\rangle$, $|t'_{1u}\rangle$, $|t''_{1u}\rangle$ and $|t'''_{1u}\rangle$:

$$\alpha_1 + \alpha_2 + \alpha_3 + \alpha_4 + \alpha_5 + \alpha_6 = 0, \tag{12.11}$$

$$2(\alpha_1 + \alpha_4) - (\alpha_2 + \alpha_3 + \alpha_5 + \alpha_6) = 0,$$

$$\alpha_1 - \alpha_4 = 0, \quad \alpha_2 - \alpha_5 = 0, \quad \alpha_3 - \alpha_6 = 0.$$

From (12.11) we obtain

$$\alpha_1 + \alpha_4 = 0, \quad \alpha_2 + \alpha_5 + \alpha_3 + \alpha_6 = 0.$$

Consequently,

$$|e''_g\rangle = (\sigma_2 + \sigma_5) - (\sigma_3 + \sigma_6).$$

Table 12.3 Sets of the ligand orbital combinations transforming in $\mathcal{O}_h$ group

IR	Orbital of ligands	of paramagnetic ion
$\mathfrak{A}_{1g}$	$\sqrt{1/6}(\sigma_1 + \sigma_2 + \sigma_3 + \sigma_4 + \sigma_5 + \sigma_6)$	s
$\mathfrak{E}_g$	$\sqrt{1/12}\,[2(\sigma_1 + \sigma_4) - (\sigma_2 + \sigma_3 + \sigma_5 + \sigma_6)]$,	$d_{z^2}, d_{x^2-y^2}$
	$(1/2)\,(\sigma_2 + \sigma_5 - \sigma_3 - \sigma_6)$	"
$\mathfrak{T}_{1u}$	$\sqrt{1/2}(\sigma_1 - \sigma_4), \sqrt{1/2}(\sigma_2 - \sigma_5), \sqrt{1/2}(\sigma_3 - \sigma_6)$	p_x, p_y, p_z
$\mathfrak{T}_{1u}$	$(1/2)\,(Y_1 + Y_4 + Y_3 + Y_6), (1/2)\,(Z_2 + Z_5 + Z_1 + Z_4)$,	p_x, p_y, p_z
	$(1/2)\,(X_3 + X_6 + X_2 + X_5)$	"
$\mathfrak{T}_{2u}$	$(1/2)\,(Y_1 + Y_4 - Y_3 - Y_6), (1/2)\,(Z_2 + Z_5 - Z_1 - Z_4)$,	–
	$(1/2)\,(X_3 + X_6 - X_2 - X_5)$	–
$\mathfrak{T}_{1g}$	$(1/2)\,(Y_1 - Y_4 + X_5 - X_2), (1/2)\,(Z_2 - Z_5 + Y_6 - Y_3)$,	–
	$(1/2)\,(X_3 - X_6 + Z_4 - Z_1)$	–
$\mathfrak{T}_{2g}$	$(1/2)\,(Y_1 - Y_4 + X_2 - X_5), (1/2)\,(Z_2 - Z_5 + Y_3 - Y_6)$,	d_{xy}, d_{yz}, d_{zx}
	$(1/2)\,(X_3 - X_6 + Z_1 - Z_4)$	"

Considering the normalization conditions, we get the sets of ligand wave functions, transformed according to the irreducible representations of the group $\mathcal{O}_h$. These sets are shown in Table 12.3 along with the corresponding wave functions of the central ion. It is easy to see that the resulting combinations are quite natural in the geometric aspect. Indeed, the molecular orbital $|a_{1g}\rangle$ is a result of interaction of the s-electron of the paramagnetic center with all the σ_p-electrons of the ligands, and besides all the p-electrons, because of the spherical symmetry of the s-electron density distribution, give the same contribution to the molecular orbital. A specific form of the electronic density distribution of the states d_{z^2} is reflected in the structure of the molecular orbital e'_g: a greater length in the axial direction is reflected in the appearance of the coefficient 2, and different signs of the wave-function lobes lead to the appearance of the minus sign in the linear combination of the ligand atomic orbitals. The formation of the molecular orbitals is shown in Fig. 12.3a, b which indicates relative signs of the wave functions and hatching shows symbolically the electron overlapping areas of the paramagnetic center and the ligands.

π**-bond.** There are twelve electrons, belonging to the ligands, in the π-orbital of the complex MX_6. Their electron density distribution and their symbols are schematically shown in Fig. 12.3c, d. The symbol for the π-electron is the same as the name of the axis along which the electron density is maximal. The positive part of the wave function is directed along the axis positive direction. For example, Z_5 denotes p-electron, located on the negative part of the axis y (see 12.3d), its electron density is extended parallel to the axis z, and its wave function is positive for the positive part of z. Under this condition, the symbols X, Y and Z are transformed in the same way as the coordinates x, y and z.

The procedure of finding the irreducible representations of the group $\mathcal{O}_h$ does not differ from the one just discussed for the σ-orbitals. Making transformations from the group $\mathcal{O}_h$ on the π-orbitals of the ligands, we see that only two characters of the

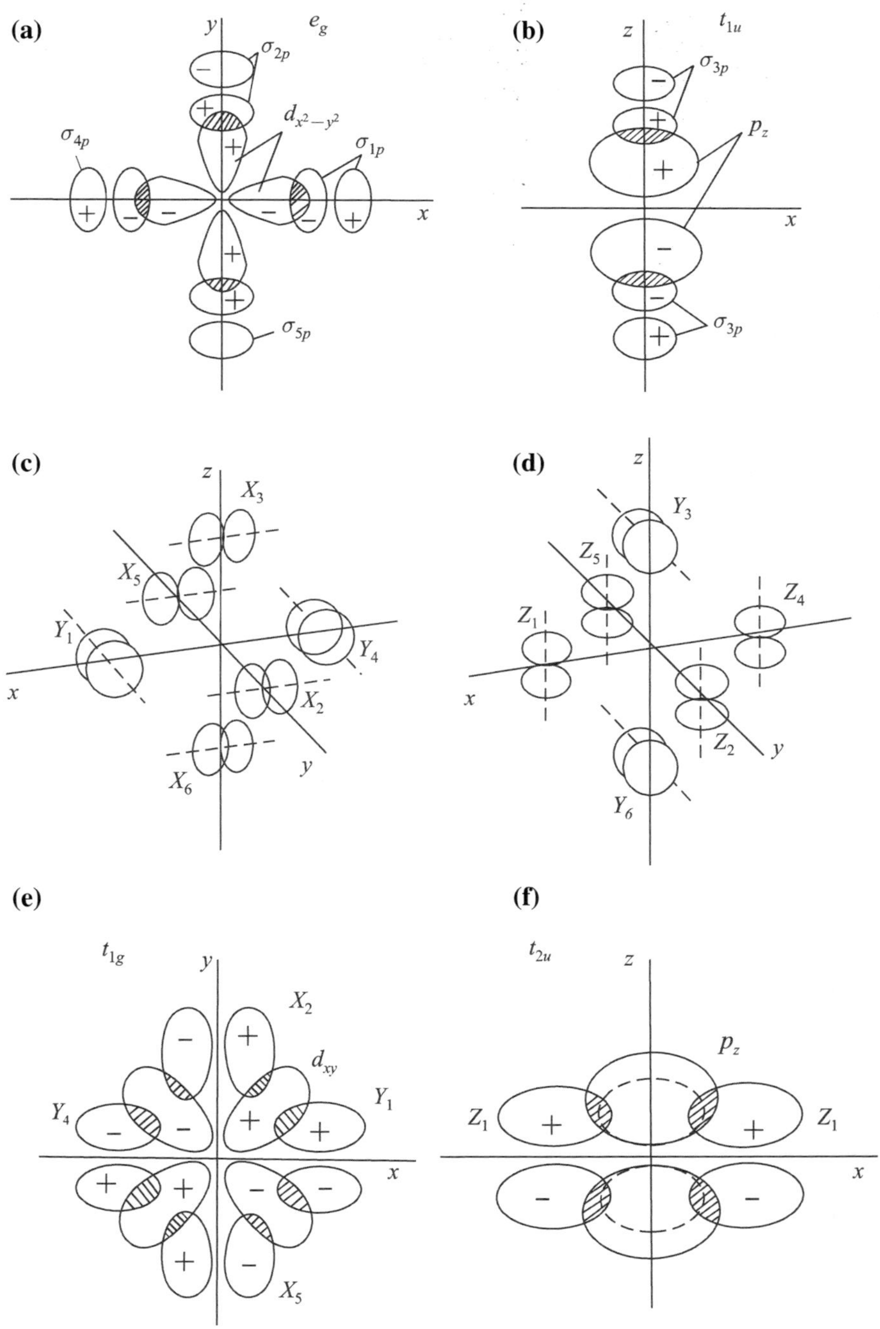

Fig. 12.3 Electron density distribution of π-electrons of ligands (**a**)–(**f**)

reducible representations $\mathfrak{G}_\pi$ are non-zero: $\chi_\pi(\widehat{E}) = 12$ and $\chi_\pi(\widehat{C}_2) = -4$. This implies (see the values of n_j^π in Table 12.1) that

$$\mathfrak{G}_\pi \sim \mathfrak{T}_{1g} \oplus \mathfrak{T}_{2g} \oplus \mathfrak{T}_{1u} \oplus \mathfrak{T}_{2u}. \tag{12.12}$$

Only representations $\mathfrak{T}_{1g}$ and $\mathfrak{T}_{1u}$, included in the reducible representation $\mathfrak{G}_\pi$, are irreducible ones, according to which the atomic orbitals of the central ion (p- and d_{xy}-, d_{yz}-, d_{zx} -electrons) are converted. The results for the π-orbitals are also presented in Table 12.3, and Fig. 12.3e, f schematically shows the formation of the π-bonds in the complex MX_6: in Fig. 12.3e the d_{xy}-electron of the central ion forms the π-bond with four p-electrons of the anions, located in the plane (xy), and Fig. 12.3f shows the formation of the π-bond of p_z-electrons of the central ion with four p-electrons of the ligands, located also in the plane (xy) (the axis y is perpendicular to the plane of the figure).

12.4 Energy Levels of the Octahedral Complex

Combination of the ligand atomic orbitals and the central ion orbital in the same row of Table 12.3 (i.e., transformed according to the same irreducible representation) can form a pair of the molecular orbitals (12.7) and (12.8). Figure 12.4 shows the diagram of the energy level formation of the complex. The central ion levels are shown on the left, the right side shows the ligand levels (the levels for σ- and π-electrons, which have the same energy, are shown separated). In the middle there are levels of the complex, corresponding to the pairs of bonding and antibonding molecular orbitals. Each such pair, as shown in Fig. 12.4, arises from the corresponding pair of levels of the central ion and the ligands belonging to the same irreducible representation. If any of the levels does not have a partner that transforms according to the same irreducible representation, then this level becomes a part of the molecular unchanged orbitals (levels t_{1u}, t_{1g} and t_{2u}). Level labeling consists of labeling the orbital character (σ, σ^*, π or π^*) with a lower index for the original atomic orbital of the central ion (s, p and d), and the irreducible representation labeling. In addition, the degeneracy order of each level without the double spin degeneracy is indicated in brackets. Basically, for each pair we can define covalence parameters similar to the parameter λ, introduced earlier for the complex MX. The fact, that the electron is not completely localized in the paramagnetic center, decreases the effective orbital angular momentum. Formally, the reduction of the orbital momentum can be expressed as a modification of the matrix elements of the orbital angular momentum in the complex, with respect to the corresponding values for the paramagnetic center in the crystal field, for example:

$$\langle t'_{2g}|\hat{\mathbf{L}}|t''_{2g}\rangle = k_{\pi\pi}\langle d_{xy}|\hat{\mathbf{L}}|d_{yz}\rangle.$$

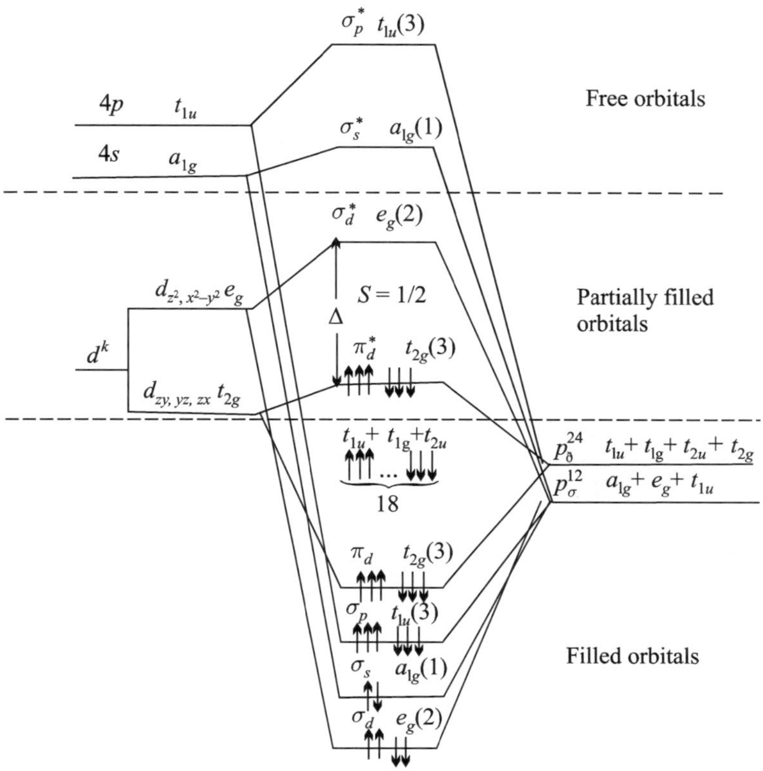

Fig. 12.4 Formation of the energy levels of the octahedral complex MX_6

where $k_{\pi\pi} < 1$ is a constant which determines the reduction degree of the orbital momentum. As an example, we can give an expression, without derivation, for this constant through the covalence parameter λ_t of the orbital πt_{2g}:

$$k_{\pi\pi} = 1 - \frac{1}{2}\lambda_t^2.$$

For other orbitals we usually obtain other expressions which are generally more complex. Obviously, the ligand electrons will occupy all the bonding molecular orbitals and the ligand atomic orbitals, non-interacting with the central ion. In Fig. 12.4 the presence of the electrons on the molecular orbitals is indicated by arrows. Arrow direction indicates the relative orientation of the different electron spins on this orbital. The order of filling of the antibonding orbitals is determined by the Hund rules. This means that at the lower level there will be electrons with the same spin projections until all different orbital states are filled. The first filling antibonding orbital will be π_d^*-orbital (t_{2g}), where there may exist three electrons with the same projections of spin $S = 3/2$, since additional electrons at the same orbital will reduce the total spin (as in atom with shell more than half filled). Thus, the configuration t_{2g}^6 (i.e., there are six electrons at the level t_{2g}) leads to the spin $S = 0$

and, therefore, forms a closed shell. The configuration $t_{2g}^6 e_g^4$ is similar to a closed one, which corresponds to the closed configuration of the free ion d^{10}. The configuration $t_{2g}^6 e_g^2$ corresponds to the configuration d^8 in the free ion, which is not closed ($S = 1$). This example clearly shows the change in the electronic structure caused by the strong interaction of the paramagnetic center with the nearest environment.

12.5 Magnetic Properties of Complexes

First, we consider the possibility to determine the magnetic properties of complexes.

Qualitative determination of the magnetic properties of complexes. The just considered approximate model of the complex leads to the conclusion about its magnetic properties. For example, it is necessary to explain the existence of the diamagnetic and paramagnetic isomer in the compound $[Ni(CN)_4]K_2$. This compound consists of two K^+ ions and the complex ion $[Ni(CN)_4]^{2-}$. The magnetic properties of the compound are determined by the structure of the complex $[Ni(CN)_4]^{2-}$. It is known that the group $(CN)^-$ generally forms complexes with metals. As it can be seen from the chemical formula, each Ni^{2+} ion is surrounded by four $(CN)^-$ groups. Hence, we can suppose the existence of two structures: tetrahedral and tetragonal ones (in the latter case we assume the disposition of $(CN)^-$ groups according to the vertices of the square). To decide whether a substance is paramagnetic, it is necessary to construct the energy level diagrams for the tetrahedral and the square environments.

In the case of tetrahedral environment the level system is constructed similarly to octahedral coordination, except for that the initial levels t_2 and e of the central ion are inverted and that there are no indices g or u in symbols due to the absence of inversion in the symmetry group of a tetrahedron. The resulting level diagram for the tetrahedron complex is shown in Fig. 12.5a where for simplicity we take into account only the σ-bond formation.

In the case of square environment we should base on the level splitting of d-electron in the field of tetragonal symmetry. From Table 11.3 the expansion $\mathfrak{D}^{(2)} \sim \mathfrak{A}_{1g} \oplus \mathfrak{B}_{1g} \oplus \mathfrak{B}_{2g} \oplus \mathfrak{E}_g$ follows. The transformation properties of the d-electron states in the form (10.13) allow us to refer the states to the irreducible representation of the group $\mathcal{D}_{4h}$, as it has been done in Sect. 11.3.1. Qualitative arguments about the relative magnitude of electrostatic interactions make it possible to locate the d-electron levels in the square environment as shown in Fig. 11.4. In addition, it is easy to obtain that $\mathfrak{G}_\sigma \sim \mathfrak{A}_{1g} \oplus \mathfrak{B}_{1g} \oplus \mathfrak{E}_u$ (the characters of the representation $\mathfrak{G}_\sigma$ are non-zero only for the transformations $\widehat{E}$, $2\widehat{C}_2$, $\widehat{\sigma}_h$ and $2\widehat{\sigma}_v$ of the group $\mathcal{D}_{4h}$). Thus, only the states d_{z^2} and $d_{x^2-y^2}$ bind with those of ligands to form the σ-bond. As in the case of octahedron, we must also consider the unpopulated states $4s$ (a_{1g}) and $4p$ (split into a_{2u} and e_u). As a result there are six σ-orbitals, combined from the atomic orbitals of the central ion and the ligands, and five orbitals of the central ion included in the complex without their change (a_{1g}^s, a_{1u} and b_{2g} and the two-fold

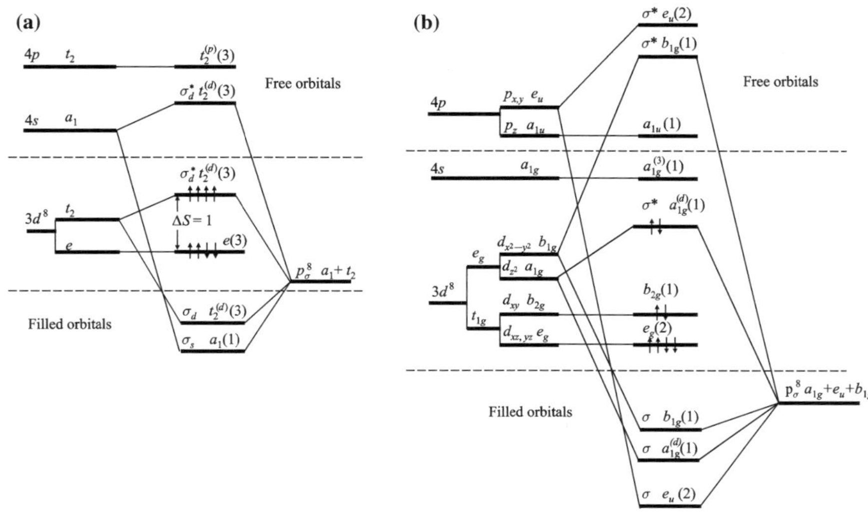

Fig. 12.5 Energy level diagram for tetrahedron (**a**) and square (**b**)

degenerate orbital e_g, see Fig. 12.5b). Account for the π-bond only leads to some level shift without changing the whole picture.

As in the case of octahedral environment, all binding orbitals are filled, because their number is always equal to the number of p-states of the ligands (in this case only p-states are taken into account). If we put eight electrons of the d-shell of the Ni^{2+} ion on the antibinding orbitals, we see (see Fig. 12.5a) that in tetrahedral environment the total spin is $S = 1$, while in square environment all the spins are compensated, therefore, $S = 0$. Thus, without any calculations, considering only the environment symmetry, we can draw a conclusion that there are two isomers of $[Ni(CN)_4]^{2-}$ with different magnetic properties: if the CN^- groups form a square, the complex is diamagnetic, and if they form a tetrahedron, it is paramagnetic.

Low-spin and high-spin systems. In the above given examples the order of filling the electron energy levels depended on the environment symmetry, and the total spin was not determined by the number of electrons in the d-shell of the free ion. The general rule is: the total spin does not reach a maximum value of $\frac{5}{2}$ with the increase of the number of electrons. Therefore, such systems are called *low-spin systems*. Examples of the electron distribution on levels of low-spin systems have already been given in Figs. 12.4 and 12.5. Figure 12.6a shows the electron distribution of d^5 configuration on two levels, that leads to the spin $S = \frac{1}{2}$ (a low-spin system). However, there are cases when the energy difference between the levels of the complex is small, so that the gain in energy due to the parallel arrangement of spins (by Hund rule) exceeds the energy difference between levels. Then, after filling all orbital states with the same projection spins, the next electron, as shown in Fig. 12.6b, occupies the upper level with the same spin orientation. The total spin, as in the case of the intermediate crystal field, is determined by the total number of electrons in the d-shell. Such systems are called *high-spin systems*. They can be distinguished not

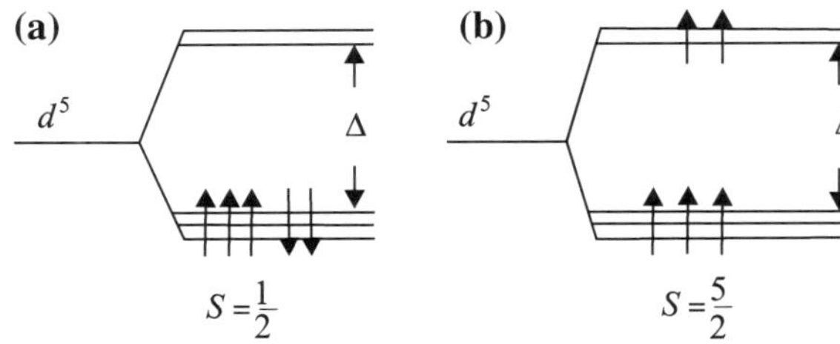

Fig. 12.6 Electron distribution on energy levels in low-spin (**a**) and high-spin (**b**) systems

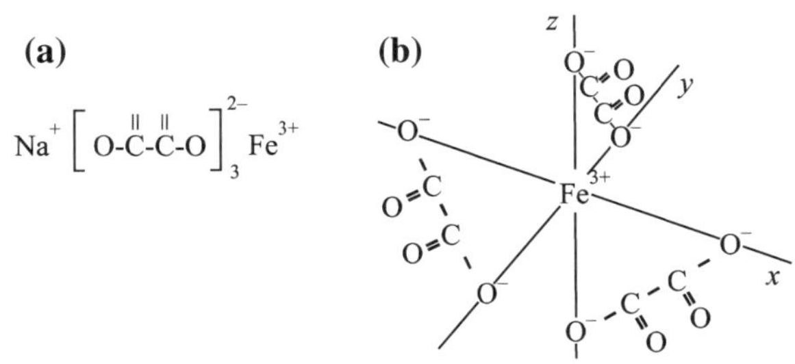

Fig. 12.7 Chemical formula (**a**) and space structure of sodium trihydrate ferro oxalate (**b**)

only by the spectra of electron paramagnetic resonance, but also by magnetochemical techniques, in particular by the measurements of the static magnetic susceptibility (χ_0). It is useful if the conditions for the registration of the electronic paramagnetic resonance are unfavorable (a large width of some lines of the EPR spectrum, a very short spin-lattice relaxation time, a complex and unresolved hyperfine structure of the spectrum, etc.).

Distribution of electrons on orbitals can be obtained from the magnetic susceptibility measurement data (an example). The experimental value of the static susceptibility in sodium trihydrate ferrous oxalate $[NaFe\,(C_2O_4)_3]\cdot 3H_2O$ (12.7a) is equal to $8,35\times 10^{-5}$ (SI units). It is necessary to find the electron distribution along the orbitals of the complex. The structure of the complex ion is that six O^- ions (see Fig. 12.7b) are located at the vertices of octahedron, in the center of which there is the Fe^{3+} ion. Thus, the energy level diagram, shown in Fig. 12.4 is suitable for this example. According to the well-known Curie law, the value χ_0 depends on the spin S as follows:

$$\chi_0 = \frac{g^2 \mu_B n_0 S(S+1)}{3kT} \tag{12.13}$$

(where $g = 2$; n_0 is the number of particles per unit of volume; k is the Boltzmann constant). One can find the value $S(S+1)$ from (12.13) by substituting the value $n_0 = \rho N_A / M$ (N_A is the Avogadro's number; M is the molecular weight, ρ is the matter density):

$$S(S+1) = \frac{3kTM\chi_0}{g^2 \mu_B \rho N_A}.$$

Substituting the numerical values, we obtain $S(S+1) = 9{,}74$, which is close to the value of 8,75 for $S = \frac{5}{2}$. Since the maximum value $S = \frac{3}{2}$ in the octahedral low spin complex, then we can draw a conclusion that this complex is a high-spin one with a short interval (Δ) between the orbitals t_{2g} and e_g (see 12.6b).

In conclusion, we emphasize the following points:

- Covalent bond model leads to a better understanding of the electronic structure of paramagnetic complexes than the crystal field model;
- The symmetry of the complex still plays a decisive role in constructing molecular orbitals;
- One of the important quantitative effects of π-bond formation in a complex is the increase of the energy interval between the levels t_{2g} and e_g.

Further Readings

Altshuler SA, Kozirev BM (1964) Electron paramagnetic resonance. Academic, New York, p. 724

Ballhausen CJ (1980) Molecular electronic structure of transition metal complexes. McGraw-Hill Inc., US. , p. 207. ISBN-13:978-0070034952

Chapter 13
Fine Structure of EPR Spectra in Solids

As it was already mentioned in Chap. 11, a main factor defining the structure of spectra of the electron paramagnetic resonance is the interaction of electrons with the nearest molecular environment and nuclei of atoms and the molecules forming a paramagnetic center. It has been shown in Chap. 11 that interactions with an environment of electrostatic nature lead to big (in comparison with that of Zeeman) splitting of energy levels of a free atom. A consequence of this is that in usual conditions, as a rule, only the bottom level of the formed ones is populated. This level is split, in turn, under influence of the small perturbation caused by a magnetic field on a number "magnetic" sublevel, intervals between which define frequencies of EPR. But unlike the known Zeeman splitting, these intervals, as a rule, are not equal and their dependence on a magnetic field has nonlinear character. Such a structure of EPR spectra is termed as *fine*. Interactions of electron with nuclei are weaker perturbation and serve as the cause of appearance of additional structure of a spectrum which has received the name *hyperfine*. The present chapter is devoted to theoretical aspects of EPR spectroscopy. Principles of registration of EPR spectra and a nuclear magnetic resonance are identical (but there is difference due to another frequency range) and here are not considered. Details of experimental methods are widely presented in the literature (see, for example, Pool C. P. Jr., Electron spin resonance. Intersci. Publ. Inc. New York, 1967).

13.1 Crystal Field Effects on EPR Spectra

13.1.1 Spin Hamiltonian Method for Singlet State in Crystal Field

Let's consider application of the perturbation theory for the description of joint influence of an external magnetic field and spin-orbital interaction on a paramagnetic center in a crystal field or, generally, in a molecular complex. The states of the paramagnetic center in a crystal field (or molecular orbital of paramagnetic complex) will be considered as unperturbed states. The role of perturbation will play the

V. I. Chizhik et al., *Magnetic Resonance and Its Applications*,
DOI: 10.1007/978-3-319-05299-1_13,

Table 13.1 A form of the terms in the expression (13.4)

Product	$\lambda\hat{L} \cdot \hat{S}$	$\mu_B \hat{L} \cdot \vec{B}$
$\lambda\hat{L} \cdot \hat{S}$	$\lambda^2 \hat{S}_\mu \langle \Gamma\vert\hat{L}_\mu\vert 0\rangle\langle 0\vert\hat{L}_\nu\vert\Gamma\rangle \hat{S}_\nu$	$\lambda\mu_B \hat{S}_\mu \langle \Gamma\vert\hat{L}_\mu\vert 0\rangle\langle 0\vert\hat{L}_\nu\vert\Gamma\rangle B_\nu$
$\mu_B \hat{L} \cdot \vec{B}$	$\lambda\mu_B B_\mu \langle \Gamma\vert\hat{L}_\mu\vert 0\rangle\langle 0\vert\hat{L}_\nu\vert\Gamma\rangle \hat{S}_\nu$	$\mu_B^2 B_\mu \langle \Gamma\vert\hat{L}_\mu\vert 0\rangle\langle 0\vert\hat{L}_\nu\vert\Gamma\rangle B_\nu$

interaction of orbital and the spin momenta with an external magnetic field and the spin-orbital interaction, which are described by the operator (see Table 10.1):

$$\widehat{\mathcal{H}} = \mu_B(\hat{L} + g_s\hat{S}) \cdot \vec{B} + \lambda\hat{L} \cdot \hat{S}. \tag{13.1}$$

The peculiarity of the analyzed problem consists that the orbital momentum ($\hat{L}$) is not the integral of motion, i.e. it is not constant in a crystal field whereas the spin momentum ($\hat{S}$) is conserved. Therefore splitting of levels in the magnetic field is actually defined by the spin. In such situation, as we saw from the consideration carried out in Sect. 1.9.1, it is expedient to use the *spin Hamiltonian*. For its construction the Eq. (13.1) should be averaged over orbital movement (i.e. integrated over space using co-ordinate wave functions), but at the same time considering the operators of spin components *as parameters*. After this procedure the Eq. (13.1) will contain spin operators only and will operate only on the spin wave functions. Such an operator $\widehat{\mathcal{H}}_S$ as it was already noticed in Sect. 1.9.1 is called *the spin Hamiltonian*. The averaging can be made at different levels of approximations of the perturbation theory, considering (13.1) as perturbation relatively to the energy of an ion in the crystal field. Thus, the spin Hamiltonian can generally be presented in the form of a polynomial with the spin components $\hat{S}_x$, $\hat{S}_y$, $\hat{S}_z$. Its coefficients are connected with matrix elements of the operator $\hat{L}$ in the basis of the states of a paramagnetic center in a crystal field. Later will the spin Hamiltonian be constructed for a case when the ground state in a crystal field is singlet, that is for a situation with the nondegenerate ground level.

For practical purposes it is quite often used another, semiempirical, approach to construct the spin Hamiltonian. From the general considerations a dependence of the Hamiltonian from the spin component $\hat{S}_x$, $\hat{S}_y$, $\hat{S}_z$ is supposed. Thus the coefficients remain uncertain parameters through which the EPR frequencies are expressed. Using the frequencies defined from an experimental spectrum and the method of the best fitting technique, the values of the parameters of the spin Hamiltonian can be found. Then one can consider these parameters as characteristics of a studied paramagnetic system.

Let us present the derivation of the expression for the spin Hamiltonian in the case of the singlet ground state in the middle crystal field.

Spin Hamiltonian in the first approximation of the perturbation theory. As according to the initial assumption the ground level is nondegenerate, then it is necessary to apply the Eq. (11.22) for the calculation of energy correction in the first approximation of perturbation theory. Let $|\Gamma\rangle$ is a vector of an orbital state in the intermediate crystal field, and let $\Gamma = 0$ corresponds to the ground state. Then

$$E^{(1)} = (\lambda\hat{\vec{S}} + \mu_B\vec{B}) \cdot \langle 0|\hat{\vec{L}}|0\rangle + g_s\mu_B\hat{\vec{S}} \cdot \vec{B}. \tag{13.2}$$

This expression contains spin operators only and represents a spin Hamiltonian as the first approximation of perturbation theory. The matrix element $\langle 0|\hat{\vec{L}}|0\rangle$ is no more any operator, but usual vector quantity. It is possible to show that if $|0\rangle$ is a nondegenerate level, then $\langle 0|\hat{\vec{L}}|0\rangle = 0$. Let us to consider any component of $\hat{\vec{L}}$, for example $\hat{L}_x$. The explicit expression for it is:

$$\hat{L}_x = \mathrm{i}\sum_j \left(z_j\frac{\partial}{\partial y_j} - y_j\frac{\partial}{\partial z_j}\right)$$

Summation is made over all electrons. It follows from the expression that the operator $\hat{L}_x$ is purely imaginary. However it is known that all the operators corresponding to physical quantity are self-conjugate, therefore for them the equality must hold:

$$\langle 0|\hat{L}_x|0\rangle^* = \langle 0|\hat{L}_x|0\rangle,$$

That is compatible with the previous only under the condition

$$\langle 0|\hat{L}_x|0\rangle = 0.$$

As such a reasoning can be applied to each component of the orbital momentum $\hat{\vec{L}}$ it is necessary to draw a conclusion that the average of the orbital momentum is equal zero for a nondegenerate level. This phenomenon is termed by *suppression, blocking* or *freezing of orbital momentum with crystal field.* Really, the degenerate levels always correspond to the non-zero orbital momentum of a free ion (P- or D- states). The crystal field removes this degeneration, and the ion behaves as if it does not have any orbital momentum, and its magnetic properties are defined exclusively by the existence of spin. This effect has been found out in researches of magnetic susceptibility of the crystals containing ions of elements of transition groups, long before the EPR discovery. Experimental values of a magnetic susceptibility did not correspond at all to the values calculated with Curi formula (1.34) if the value of the full momentum of an impulse of an ion was substituted in it. At the same time, in some cases it was possible to achieve satisfactory coincidence if the values only of ion spin were substituted.

The expression for *the spin Hamiltonian in the first approximation of perturbation theory* definitivly becomes

$$\widehat{\mathcal{H}}_S^{(1)} = g_s\mu_B\hat{\vec{S}} \cdot \vec{B}. \tag{13.3}$$

Equation (13.3) predicts $2S+1$ equidistant levels leading to the unique frequency of the electronic paramagnetic resonance corresponding to the free electron g-factor: $g = g_s \simeq 2$. Such a conclusion, however, contradicts to experimental spectra of the electronic paramagnetic resonance of ions of transition elements and other objects to which it would be possible to apply the similar theory. It turns out actually that EPR spectrum contains some lines, and the g-factor value differs from the value g_s. Thus, if a sample is a monocrystal the spectrum depends also on the orientation of a sample relative to the magnetic field vector. For the correct description of an experimental spectrum the higher approximation of the perturbation theory is required.

Spin Hamiltonian in the second-order of the theory of perturbations. According to (11.23) the correction to the energy value for nondegenerate level in the second approximation of the perturbation theory is calculated using the formula

$$E^{(2)} = -\sum_{\Gamma}{}' \frac{\langle \Gamma|\lambda \hat{\vec{L}} \cdot \hat{\vec{S}} + \mu_B(\hat{\vec{L}} + g_s\hat{\vec{S}}) \cdot \vec{B}|0\rangle\langle 0|\lambda \hat{\vec{L}} \cdot \hat{\vec{S}} + \mu_B(\hat{\vec{L}} + g_s\hat{\vec{S}}) \cdot \vec{B}|\Gamma\rangle}{E_\Gamma - E_0}. \tag{13.4}$$

The prime at the summation symbol means that here the summation is carried out over all values Γ, except $\Gamma = 0$. As it is easy to be convinced, the matrix elements of the products $\hat{\vec{S}} \cdot \vec{B}$ become zero because spin operators do not act on vectors of orbital states and can be taken out the symbol of a matrix element:

$$\langle \Gamma|\hat{\vec{S}} \cdot \vec{B}|0\rangle = \langle \Gamma|0\rangle \vec{B} \cdot \hat{\vec{S}}.$$

As $\Gamma \neq 0$, then owing to orthogonality of state vectors $\langle \Gamma|0\rangle = 0$. It leads to the vanishing of 45 summands (for each value Γ) in numerator of the right part of expression (13.4). After disclosing of other scalar products of vectors and their multiplication in numerator it turns out 36 summands for each value Γ which, however, are conveniently grouped (Table 13.1). Designations of lines and columns show the factors defining products on the crossing of corresponding lines and columns. Each of indexes μ and ν can accept values x, y, z. Each element of the table represents nine summands. There are identical summands among theirs.

For the calculation of expression (13.4) which actually is the threefold sum, it is convenient to change a summation order: at first over Γ, and then, having substituted expressions from Table 13.1 in the formula, over μ and ν. Also it is convenient to enter components $\Lambda_{\mu\nu}$ of a tensor Λ by the definition:

$$\Lambda_{\mu\nu} = \sum_{\Gamma}{}' \frac{\langle \Gamma|\hat{L}_\mu|0\rangle\langle 0|\Gamma|\hat{L}_\nu\rangle}{E_\Gamma - E_0}. \tag{13.5}$$

From this expression one can see $\Lambda_{\mu\nu} = \Lambda^*_{\nu\mu}$, i.e. The matrix is Hermitian. Using (13.5), it is possible to present *the spin Hamiltonian in the second-order perturbation theory correction* as follows:

$$\widehat{\mathcal{H}}_S^{(2)} = -\lambda^2 \sum_{\mu,\nu} \Lambda_{\mu\nu} \hat{S}_\mu \hat{S}_\nu - \mu_\mathrm{B}^2 \sum_{\mu,\nu} \Lambda_{\mu\nu} B_\mu B_\nu - \lambda\mu_\mathrm{B} \left(\sum_{\mu,\nu} \Lambda_{\mu\nu} B_\mu \hat{S}_\nu + \sum_{\mu,\nu} \Lambda_{\mu\nu} \hat{S}_\mu B_\nu \right).$$

In the last sum it is possible to change designations of indexes:

$$\sum_{\mu,\nu} \Lambda_{\mu\nu} \hat{S}_\mu B_\nu = \sum_{\mu,\nu} \Lambda_{\nu\mu} \hat{S}_\nu B_\mu.$$

Then taking into account a self-associativity of the matrix Λ, Hamiltonian $\widehat{\mathcal{H}}_S^{(2)}$ takes a form

$$\widehat{\mathcal{H}}_S^{(2)} = -\lambda^2 \sum_{\mu,\nu} \Lambda_{\mu\nu} \hat{S}_\mu \hat{S}_\nu - \mu_\mathrm{B}^2 \sum_{\mu,\nu} \Lambda_{\mu\nu} B_\mu B_\nu - \lambda\mu_\mathrm{B} \sum_{\mu,\nu} \left(\Lambda_{\mu\nu} + \Lambda_{\mu\nu}^* \right) B_\mu \hat{S}_\nu. \tag{13.6}$$

Considering expressions (13.3) and (13.6), it is possible to write down the spin Hamiltonian (up to second-order perturbation theory) in the short form:

$$\widehat{\mathcal{H}}_S = \mu_\mathrm{B} \hat{\vec{S}} \mathsf{g} \vec{B} - \lambda^2 \hat{\vec{S}} \Lambda \hat{\vec{S}} - \mu_\mathrm{B}^2 \vec{B} \Lambda \vec{B}. \tag{13.7}$$

Here the product "vector-tensor-vector" is understood in the usual sense of matrix multiplication: the vector standing on the right is considered as the vector-column, the vector standing at the left–as a vector-string. In (13.7) a new matrix (tensor) g with elements $g_{\mu\nu}$ was entered:

$$g_{\mu\nu} = g_\mathrm{S} \delta_{\mu\nu} - \lambda \left(\Lambda_{\mu\nu} + \Lambda_{\mu\nu}^* \right). \tag{13.8}$$

Further it will be shown that Hamiltonian (13.7) leads to several important consequences:

(1) Energy levels depend on the orientation of a crystal relative to magnetic fields;
(2) Spin degeneration can be partially removed even in the absence of a magnetic field, but only owing to joint action of a crystal field and spin-orbit interaction;
(3) There exists a paramagnetism connected with the deviation of system symmetry from the spherical one;
(4) The g-factor is distinct from the free electron g_S-factor.

It is necessary to notice that application of the spin Hamiltonian in the form (13.7) is not limited by applications to ions of transitive elements. This Hamiltonian can be used for calculation of spectra of any paramagnetic centers interacting with the environment (for example, arising due to an irradiation), and also organic molecules in a triplet state.

13.1.2 Spin Hamiltonian in the Case of Axial Symmetry

In this important special case the axis x and y are equivalent, and the symmetry axis (an axis z) can be accepted as one of the main axes of the tensor Λ. Then this tensor becomes diagonal so there are only elements $\Lambda_{zz} = \Lambda_{\parallel}$ and $\Lambda_{xx} = \Lambda_{yy} = \Lambda_{\perp}$ which are distinct from zero. Hence, g-tensor is completely defined by two elements too: $g_{\perp} = g_s - 2\lambda\Lambda_{\perp}$ and $g_{\parallel} = g_s - 2\lambda\Lambda_{\parallel}$. If to introduce the angle θ between a magnetic field vector $\vec{B}$ and an axis of symmetry and to choose an axis direction x so that $B_y = 0$ then one can write down the first term from (13.7) as $g_{\parallel}\mu_B \hat{S}_z B_0 \cos\theta + g_{\perp}\mu_B \hat{S}_x B_0 \sin\theta$. Here B_0 is a value of the static magnetic field. Expression $\hat{\vec{S}}\Lambda\hat{\vec{S}}$ for this case is reduced to

$$-\Lambda_{\parallel}\hat{S}_z{}^2 - \Lambda_{\perp}\left(\hat{S}_x{}^2 + \hat{S}_y{}^2\right)$$

or in the usually accepted form

$$\begin{aligned}-\Lambda_{\parallel}\hat{S}_z{}^2 - \Lambda_{\perp}\left(\hat{\vec{S}}^2 - \hat{S}_z{}^2\right) &= -\left(\Lambda_{\parallel} - \Lambda_{\perp}\right)\hat{S}_z{}^2 - \Lambda_{\perp}\widehat{S^2} \\ &= -\left(\Lambda_{\parallel} - \Lambda_{\perp}\right)\left(\hat{S}_z{}^2 - \frac{1}{3}\widehat{S^2}\right) - \frac{1}{3}\left(\Lambda_{\parallel} + 2\Lambda_{\perp}\right)\widehat{S^2}. \end{aligned} \quad (13.9)$$

For set of states with the fixed value of spin S the last member in (13.9) effects only on a choice of a reference point of energy, therefore it can be neglected at the definition of EPR frequencies. Hence, the Hamiltonian (13.7) can be replaced by the equivalent one:

$$\widehat{\mathcal{H}}'_S = g_{\parallel}\mu_B \hat{S}_z B_0 \cos\theta + g_{\perp}\mu_B \hat{S}_x B_0 \sin\theta + D\left(\hat{S}_z{}^2 - \frac{1}{3}\widehat{S^2}\right). \quad (13.10)$$

Here the new parameter is introduced

$$D = -\lambda^2\left(\Lambda_{\parallel} - \Lambda_{\perp}\right). \quad (13.11)$$

The last member in (13.7) does also no effect on energy differences and therefore can be rejected.

Zero-field splitting and anisotropy. If a static magnetic field $B_0 = 0$, then from (13.10) we obtain the expression for energy:

$$E^0_{M_S} = D\left[M_S^2 - \frac{1}{3}S(S+1)\right]. \quad (13.12)$$

Here M_S is a value of spin projection on the direction of the symmetry axis. It follows from (13.12) that the parameter D defines the magnitude of the zero-field splitting.

It is easy to estimate the zero-field splitting, for example, for ions of the first transition group (iron group) for which the case of the intermediate crystal field is realized. In this case the interval between levels $\Delta = E_\Gamma - E_0$ is order-of-magnitude $\frac{\Delta}{hc} \simeq 10^4$ cm^{-1} (see Table 10.2), and the constant of spin-orbital interaction is $\frac{\lambda}{hc} \simeq 10^2$ cm^{-1}. Therefore $\frac{D}{hc} \simeq \frac{\lambda^2}{hc\times\Delta} \simeq 1\,\mathrm{cm}^{-1}$. Thus even in the absence of a magnetic field various spin states have different energies as a result of the action of a crystal field. It is pure quantum-mechanical effect of the interaction of spin with a electrostatic crystal field. From positions of nonrelativistic classical physics it is difficult to explain an interaction of the electrostatic crystal field with the electron magnetic momentum.

In the presence of a static magnetic field it is necessary to make and solve the secular equation. The Hamiltonian (13.10) contains in explicit form the angle θ between the magnetic field vector and the symmetry axis. With parallel orientation of the magnetic field vector to the axis of symmetry only a component $g_\parallel$ plays a role, and with perpendicular one $g_\perp$. The difference of $g_\parallel$ and $g_\perp$ from g_s, can be easily estimated. The equation (13.8) gives

$$|g_\parallel - g_\mathrm{s}| \sim |g_\perp - g_\mathrm{s}| \sim 10^{-2}.$$

This value also defines an order of effects of the anisotropy in the phenomenon EPR caused by a crystal field.

Van-Vleck paramagnetism. The last term in the Hamiltonian (13.7) is insignificant for the definition of intervals between energy levels and, hence, does not influence on EPR spectra, however it leads to the consequence representing of the general physical interest. This term, reflecting square-law dependence of energy on a magnetic field, brings the specific contribution to a magnetic susceptibility, namely *Van-Vleck paramagnetism.* Really, if to define the magnetic moment (M_0) of a volume unit via energy as

$$M_0 = -n_0 \frac{\partial E}{\partial B} \tag{13.13}$$

(here n_0 is the number of paramagnetic particles in a volume unit) then the contribution of the above term in the case of parallel orientation of the axis of symmetry and the magnetic field will be:

$$M_0 = 2\mu_\mathrm{B}^2 \Lambda_\parallel B_0 n_0. \tag{13.14}$$

It follows directly from expression (13.6). The magnetic susceptibility, which is defined as a factor of proportionality between M_0 and $\frac{B_0}{\mu_0}$ (μ_0 is the magnetic constant), is expressed as follows:

$$\chi_\mathrm{VV} = 2\mu_\mathrm{B}\mu_0\Lambda_\parallel.$$

The contribution of the Van Vleck susceptibility χ_{VV} is interesting because it is not connected with the presence of a spin, and is defined exclusively by nonzero off-diagonal elements of the orbital momentum. Such elements appear always if symmetry of a paramagnetic center to which belongs the electron is not spherical. Therefore Van Vleck paramagnetism exists for diamagnetic molecules of low symmetry too. Unlike the usual case of paramagnetic susceptibility (*Curi susceptibility*), Van Vleck paramagnetism is not related to populations of energy levels and therefore does not depend on temperature. Its magnitude is comparable with a diamagnetic susceptibility.

13.1.3 Relation Between Characteristics of Crystal Field and Parameters of Spin Hamiltonian

The concrete relation between parameters of the spin Hamiltonian and parameters of a crystal field can be obtained using expressions for $g_{\mu\nu}$ in the form (13.8) and D in the form (13.11) in which it is necessary to substitute elements of tensor Λ, defined as (13.5). Values of splitting are directly connected with parameters of a crystal field (an explicit form of this relationship for some special cases and general methods of calculation have been considered in Sect. 11.3). Let us consider as examples of expression of spin Hamiltonian parameters for ions in P- and D-states in a crystal field of tetragonal symmetry.

P-term. Suppose we want to find the explicit expression of spin Hamiltonian parameters for the paramagnetic center in the P-state formed by one electron in the field of the charges located in tops of the deformed octahedron (see Fig. 11.1*a*). The effective potential of a crystal field in this case is characterized by a single parameter A_2^{0t} which is expressed through geometrical parameters of an octahedron by the Eq. (11.16). The interval (Δ) between the levels, formed as a result of splitting of a P-term, is easily calculated by the method of equivalent operators:

$$\Delta = 3A_2^{0t}\langle l\|\alpha\|l\rangle,$$

where

$$\langle l\|\alpha\|l\rangle = -\frac{2}{5}\langle r^2\rangle, \quad \text{i.e.} \quad \Delta = \frac{6}{5}A_2^{0t}\langle r^2\rangle.$$

Taking into account (11.31), it is possible to express Δ via distances to the nearest point charges R_1 and R_2, creating a crystal field:

$$\Delta = \frac{12}{5}|q|\langle r^2\rangle\left|\frac{1}{R_1^3} - \frac{1}{R_2^3}\right|. \tag{13.15}$$

In the case of an extended octahedron (along an axis z) the state $p_z \equiv |0\rangle$ corresponds to the ground level, and the states $p_x \equiv |1_+\rangle = 2^{-1/2}(|1\rangle + |-1\rangle)$ and $p_y \equiv |1_-\rangle = -\mathrm{i} \cdot 2^{-1/2}(|1\rangle - |-1\rangle)$ — to excited ones (see Eq. (10.11)). According to (13.5)

$$\Lambda_{\|} = \Lambda_{zz} = \frac{\left|\langle 1_+|\hat{L}_z|0\rangle\right|^2 + \left|\langle 1_-|\hat{L}_z|0\rangle\right|^2}{\Delta},$$

$$\Lambda_{\perp} = \Lambda_{xx} = \Lambda_{yy} = \frac{\left|\langle 1_+|\hat{L}_x|0\rangle\right|^2 + \left|\langle 1_-|\hat{L}_x|0\rangle\right|^2}{\Delta}.$$

Thus, it is sufficient to calculate matrix elements of the type $\langle \Gamma|\hat{L}_\mu|0\rangle$ for $\mu = x, z$. As $\hat{L}_z|0\rangle = 0$, then only elements $\langle \Gamma|\hat{L}_x|0\rangle$ can differ from zero. The operator $\hat{L}_x$ is represented in the form $\hat{L}_x = (\frac{1}{2})(\hat{L}^+ + \hat{L}^-)$, and the action of the operator $\hat{L}^\pm$ on the eigenstate of the operator $\hat{L}_z$ is defined by the equation

$$\hat{L}^\pm|0\rangle = \sqrt{2}|\pm 1\rangle.$$

Hence, $\hat{L}_x|0\rangle = \frac{1}{\sqrt{2}}(|1\rangle + |-1\rangle) = |1_+\rangle$, from where we have $\langle 1_+|\hat{L}_x|0\rangle = 1$, and $\langle 1_-|\hat{L}_x|0\rangle = 0$. Finally we have:

$$\Lambda_{\|} = 0, \Lambda_{\perp} = \frac{1}{\Delta}.$$

It follows from expressions (13.8) and (13.11) : $g_{\|} = g_s$, $g_{\perp} = g_s - \frac{\lambda}{\Delta}$ and $D = \frac{\lambda^2}{\Delta}$, that defines the required relation with the geometrical parameters characterizing an environment of the paramagnetic center.

D-Term. The diagram of the energy levels which appeared as a result of splitting of the D-term, formed by one d-electron, in the field of tetragonal symmetry, is presented in Fig. 11.4. The stable configuration of surrounding charges is defined by Jan–Teller effect and represents a slightly oblate octahedron. In this case (which usually is termed as a case (a) the ground state will be $d_{xy} = |2_-\rangle$ (see Eq. 10.21). Another case (b) can be realized in an environment of the charges located in tops of a square (or in an equivalent environment which creates a crystal field of a similar configuration, see Fig. 11.4, the left part). In the case b the ground level is characterized by a state $d_{zz} = |0\rangle$.

For calculation of components tensor $\Lambda_{\|}^{(a)}$ in the case a we shall use the same procedure, as earlier:

$$\Lambda_{\|}^{(a)} = \frac{\left|\langle 2_+|\hat{L}_z|2_-\rangle\right|^2}{\Delta} + \frac{\left|\langle 1_+|\hat{L}_z|2_-\rangle\right|^2 + \left|\langle 1_-|\hat{L}_z|2_-\rangle\right|^2}{\Delta + \delta_1} + \frac{\left|\langle 0|\hat{L}_z|2_-\rangle\right|^2}{\Delta + \delta_1 + \delta_0} \quad (13.16)$$

(Splittings Δ, δ_1 and δ_0 have been defined in Fig. 11.4). The expression similar (13.16) can be obtain for components $\Lambda_{\perp}^{(a)}$ if to replace $\hat{L}_z$ on $\hat{L}_x$. For the case b $\Lambda_{\parallel}^{(b)} = 0$ for the same reasons, as for the P-term, therefore we receive the expression for $\Lambda_{\perp}^{(b)}$

$$\Lambda_{\perp}^{(b)} = \frac{\left|\langle 1_+|\hat{L}_x|0\rangle\right|^2 + \left|\langle 1_-|\hat{L}_x|2_-\rangle\right|^2}{\delta_0} + \frac{\left|\langle 2_-|\hat{L}_x|0\rangle\right|^2}{\delta_0+\delta_1} + \frac{\left|\langle 2_+|\hat{L}_x|0\rangle\right|^2}{\Delta+\delta_1+\delta_0}. \quad (13.17)$$

Among the matrix elements, which are necessary for calculation of (13.16) and (13.17), only elements $\langle 2_+|\hat{L}_x|2_-\rangle = 2$, $\langle 1_+|\hat{L}_x|2_-\rangle = 1$ and $\langle 1_\pm|\hat{L}_x|0\rangle = \sqrt{3}$ are not equal to zero. Hence,

$$\Lambda_{\parallel}^{(a)} = \frac{4}{\Delta}, \quad \Lambda_{\perp}^{(a)} = \frac{1}{\Delta+\delta_1}; \quad \Lambda_{\parallel}^{(b)} = 0, \quad \Lambda_{\perp}^{(b)} = \frac{6}{\delta_0}.$$

As well as for P-term (see (13.15)) the values of splitting, calculated by technique of equivalent operators, are defined by parameters of a crystal field. In turn these are expressed by means of the Eqs. (11.16)–(11.20) via geometrical parameters of the environment R_1 and R_2 (see Fig. 11.1a).

It is necessary to notice that for small distortion of the extended octahedron (see Fig. 11.4 and the comment to it), the ground level remains twice degenerate, and above theory to such case, strictly speaking, is not applicable. Nevertheless it is possible to show that in the case of twice degenerate ground level Hamiltonian (13.7) can be used for calculation of EPR spectra. A leading consideration in support of such a conclusion is the cause that twofold degeneration of an energy level cannot be realized for any integer value of the orbital momentum.

13.1.4 Anisotropy of EPR Spectra

As a result of calculation one determines the frequencies of lines of EPR spectrum and their intensities. In some cases it is enough to determine *selection rules*, in other words to exclude the frequencies having zero intensity (*the forbidden transitions*). In a single crystal the frequencies and the intensities of spectral lines of EPR depend on orientation of a crystal relatively to a magnetic field vector.

In the given section we will analyze a problem of obtaining of an angular dependence of EPR spectra for a case when the environment has a symmetry axis more than the second order. We will also accept the ground energy level in a crystal field is nondegenerate. At first we will consider a simple system with a spin $S = \frac{1}{2}$ on which example we will show an expedient of the definition of frequencies. Then on more difficult example of ion Ni^{2+} ($S = 1$) we will consider all basic features of EPR spectrum of electrons in the environment with a symmetry axis more than the second order, and, at last, we will give an important for practice example of the system of energy levels of the ion Cr^{3+} in a ruby.

The case of $S = \frac{1}{2}$. To calculate an angular dependence of EPR frequency it is necessary for beginning to select the correct form of a spin Hamiltonian. The Hamiltonian in the form (13.10) is appropriate for the problem with a nondegenerate ground level and a symmetry axis more than second order. The last term in (13.10) gives the zero contribution in the case $S = \frac{1}{2}$ so there are only the summands related to the interaction with a magnetic field.

The following stage is the calculation of matrix elements. It is convenient to calculate them using the basis from eigenfunctions of the operators $\hat{S}_z$ and $\hat{S}_x = (\frac{1}{2})\left(\hat{S}^+ + \hat{S}^-\right)$, where $\hat{S}^\pm$ are the increase and decrease operators. In order to the result was expressed in terms of frequency as it is accepted in the calculations related to EPR, let's enter the following labels:

$$\nu_\perp h = \mu_B g_\perp B_0, \quad \nu_\parallel h = \mu_B g_\parallel B_0. \tag{13.18}$$

In these notations the matrix $\langle M_s|\widehat{\mathcal{H}}_S|M_s'\rangle$ (in h units) looks as follows:

$$\begin{bmatrix} \frac{1}{2}\nu_\parallel \cos\theta & \frac{1}{2}\nu_\perp \sin\theta \\ \frac{1}{2}\nu_\perp \sin\theta & \frac{1}{2}\nu_\parallel \cos\theta \end{bmatrix}.$$

The secular equation for definition $\varepsilon \equiv E/h$ follows from above matrix:

$$\begin{vmatrix} \frac{1}{2}\nu_\parallel \cos\theta - \varepsilon & \frac{1}{2}\nu_\perp \sin\theta \\ \frac{1}{2}\nu_\perp \sin\theta & -\frac{1}{2}\nu_\parallel \cos\theta - \varepsilon \end{vmatrix} = 0$$

or

$$\varepsilon^2 - \frac{1}{4}\nu_\parallel^2 \cos^2\theta - \frac{1}{4}\nu_\perp^2 \sin^2\theta = 0.$$

Hence, one obtains

$$\varepsilon_\pm = \pm\frac{1}{2}\sqrt{\nu_\parallel^2 \cos^2\theta + \nu_\perp^2 \sin^2\theta},$$

From here it follows

$$E_\pm = h\varepsilon_\pm = \pm\frac{1}{2}\mu_B B_0\sqrt{\nu_\parallel^2 \cos^2\theta + \nu_\perp^2 \sin^2\theta}, \tag{13.19}$$

and for EPR frequency one gains the dependence on a magnetic field in the form

$$\nu = \frac{\mu_B B_0}{h}\sqrt{\nu_\parallel^2 \cos^2\theta + \nu_\perp^2 \sin^2\theta}. \tag{13.20}$$

The diagram of the energy dependence on the field B_0 (13.19) represents straight lines with the slope determining by the expression $\frac{1}{2}\mu_B B_0\sqrt{\nu_\parallel^2\cos^2\theta+\nu_\perp^2\sin^2\theta}$. As a rule, in an experiment the frequency is set constant and a static magnetic field changes, therefore it is convenient to enter the resonance value of the magnetic field on the basis of (13.20):

$$B_{\mathrm{res}}=\frac{h\nu}{\mu_B g(\theta)}=\frac{g_s B_s}{g(\theta)},$$

Here $g(\theta)=\sqrt{\nu_\parallel^2\cos^2\theta+\nu_\perp^2\sin^2\theta}$; B_s is the resonance value of the field for a free electron.

13.2 Examples of EPR Spectra

13.2.1 EPR Frequencies of the Ion Ni^{2+} in an Intermediate Crystal Field.

In the present item the theoretical dependence of EPR spectrum of the ion Ni^{2+} in compound $NiSiF_6 \cdot 6H_2O$ on the magnetic field and angle θ is calculated.

The obtained dependence will be used to find parameters of the spin Hamiltonian characterizing an experimental spectrum. The ion Ni^{2+} in above compound is disposed in the intermediate crystal field of tetragonal symmetry. The character of splitting of the term ^3F of the ion Ni^{2+} is determined by the configuration of its electron shell d^8 and by the value of S and L for the ground state: $S=1, L=3$. As it follows from the data of Table 11.3, the energy level for $J=3$ is split by the tetragonal crystal field in accordance with the decomposition $D^{(3)}\sim\mathfrak{A}_{2u}\oplus\mathfrak{B}_{1u}\oplus\mathfrak{B}_{2u}\oplus 2\mathfrak{E}_{1u}$, and the ground level is one of nondegenerate levels. The specified symmetry of the problem allows us to apply again the Hamiltonian in the form (13.10). Let's denote $d\equiv\frac{D}{h}$, then the matrix $\frac{1}{h}\langle M_S|\widehat{\mathcal{H}}_S|M_S'\rangle$ becomes

$$\begin{bmatrix}-\nu_\parallel\cos\theta+\frac{1}{3}d & \frac{1}{\sqrt{2}}\nu_\perp\sin\theta & 0\\ \frac{1}{\sqrt{2}}\nu_\perp\sin\theta & -\frac{2}{3}d & \frac{1}{\sqrt{2}}\nu_\perp\sin\theta\\ 0 & \frac{1}{\sqrt{2}}\nu_\perp\sin\theta & \nu_\parallel\cos\theta+\frac{1}{3}d\end{bmatrix}.$$

The secular equation is

$$\left[\left(\frac{1}{3}D-\varepsilon\right)^2-\nu_\parallel^2\cos\theta^2\right]\left(-\frac{2}{3}D-\varepsilon\right)-\nu_\perp^2\sin\theta^2\left(\frac{1}{3}D-\varepsilon\right)=0. \quad (13.21)$$

It is easily solved for special cases for $\theta=0°$ and $\theta=90°$. In the case $\theta=0°$ it has roots

$$\varepsilon_0 = -\frac{2}{3}d \quad \text{and} \quad \varepsilon^{\pm} = \frac{1}{3}d \pm \nu_{\parallel}.$$

For $\theta = 90°$ the equation becomes

$$\left(\frac{1}{3}d - \varepsilon\right)\left[\left(\frac{1}{3}d - \varepsilon\right)\left(-\frac{2}{3}d - \varepsilon\right) - \nu_{\perp}^2\right] = 0$$

and has following roots:

$$\varepsilon_0 = \frac{1}{3}d \quad \text{and} \quad \varepsilon^{\pm} = -\frac{1}{6}d \pm \sqrt{\frac{1}{4}d^2 + \nu_{\perp}^2}.$$

The explicit form of the dependencies on a magnetic field B is obtained after substitution $\nu_{\parallel}$ and $\nu_{\perp}$ from (13.18). For $\theta = 0°$ and for $\theta = 90°$ we have respectively:

$$\begin{aligned} E_0 &= -\frac{2}{3}D, \\ E^{\pm} &= \frac{1}{3}D \pm g_{\parallel}\mu_B B, \\ E_0 &= \frac{1}{3}D, \\ &\text{and} \\ E^{\pm} &= -\frac{1}{6}D \pm \sqrt{\frac{1}{4}D^2 + g_{\perp}^2\mu_B^2 B^2}. \end{aligned} \tag{13.22}$$

These dependencies are presented on the diagram Fig. 13.1a. Considering the extreme cases $\theta = 0°$ and $\theta = 90°$ (the limiting values of angles are specified near corresponding levels in Fig. 13.1a), it is easy to present qualitatively the character of change of the diagram of levels for the intermediate values θ. See below explanations to Fig. 13.1b, c.

13.2.2 Relative Intensities of Lines of EPR Spectrum of Ni^{2+}

To determine of EPR spectrum it is necessary, besides energy values, to calculate relative intensities of the lines corresponding to transitions between Zeeman sublevels $|n\rangle$ and $|n\rangle'$ in a constant magnetic field. A view of the spectrum is determined by the position of spectral lines and their intensities, which are proportional to a square of the module of a matrix element $\langle n|\hat{\mu}_1|n'\rangle$, termed *the dipole moment of transition*. Here $\hat{\mu}_1$ is the operator of the projection of the magnetic dipole moment on the oscillating magnetic field. As $\hat{\mu}_1 = g\mu_B\widehat{S}_1$ ($\widehat{S}_1$ is the projection of a spin similar $\hat{\mu}_1$) instead of an element $\langle n|\hat{\mu}_1|n'\rangle$ it is possible to consider matrix elements $\langle n|\widehat{S}_1|n'\rangle$.

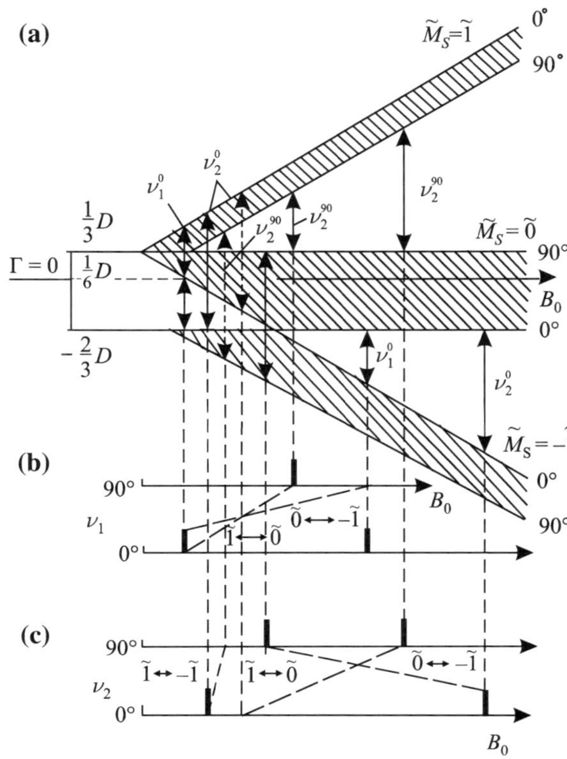

Fig. 13.1 **a** The levels of energy for the ion Ni^{2+} in tetragonal crystal field as a function of magnetic field at different orientations of symmetry axis. Bidirectional *arrows* indicate the EPR transition induced by the fixed frequencies ν_1 (*the short*) and ν_2 (*the long*). **b**, **c** The EPR spectra at the frequencies ν_1 (**b**) and ν_2 (**c**) for $\theta = 0$ and 90°, respectively

At first let us discuss a situation $\theta = 0°$, when the vector of a constant magnetic field is parallel to the symmetry axis of an electrostatic crystal field (an axis z). In this case the spin Hamiltonian $\widehat{\mathcal{H}}_S$ in the form (13.10) commutes with the operator $\hat{S}_z$ and, hence, eigenstates of the operator $\hat{S}_z$ are simultaneously eigenstates of $\widehat{\mathcal{H}}_S$, and it is possible to put $|n\rangle = |M_S\rangle$. If the oscillating magnetic field is perpendicular to a static field (for example, parallel to axis y) then $\widehat{S}_1 = \hat{S}_y$. This implies *the selection rule* coinciding with the known selection rule for magnetic (Zeeman) transitions, namely: non-zero matrix element for $\widehat{S}_1 = \hat{S}_y$ will be only if $M'_S = M_S+1$. Thus, the magnetic resonance in the above situation differs from an usual magnetic resonance only by the shift of the level $M_S = 0$ on the quantity D relative to both other levels. It leads to the appearance of two resonance lines instead of one in a spectrum of a simple magnetic resonance. If the oscillating magnetic field frequency $\nu < \frac{D}{h}$ it is easy to see from Fig. 13.1a that the single transition $M_S = 0 \rightleftharpoons M_S = -1$ takes place for two values of a magnetic field (see spectral lines in Fig. 13.1b, corresponding to the frequency ν_1^0) . If $\nu > \frac{D}{h}$ both transitions are possible: $M_S = 1 \rightleftharpoons M_S = 0$ and $M_S = 0 \rightleftharpoons M_S = -1$, to which there corresponds the determined values of a field (see spectral lines on Fig. 13.1c, corresponding to the frequency ν_2^0). At a parallel alignment of vectors oscillating and static magnetic fields all dipole moments of transition are zero, hence, the magnetic resonance of a paramagnetic center considered at $\theta = 0°$ is impossible.

The appearance of several lines in a spectrum instead of one as a result of the joint effect of a crystal field and spin-orbit interaction is usually called as *thin structure* EPR spectrum.

In the case $\theta \neq 0°$ the situation becomes complicated. In this case the spin-orbit interaction in the presence of a crystal field not only shifts levels, but also mixes up eigenstates of the operator $\hat{S}_z$ which does not commutes now with the spin Hamiltonian. Therefor each eigenstate of the Hamiltonian represents a linear combination of eigenstates of the operator $|M_S\rangle$:

$$|\widetilde{M}_S\rangle = \sum_{M_S} |M_S\rangle\langle M_S|\widetilde{M}_S\rangle. \tag{13.23}$$

Here instead of a general label of the number of a level (n), the more convenient label is used: $\widetilde{M}_S$. In the extreme case of the strong magnetic field when spin-orbit interaction can be neglected in comparison with Zeeman split, levels $\widetilde{M}_S$ approach asymptotically to levels M_S, and states $|\widetilde{M}_S\rangle$ tend to states $|M_S\rangle$. Therefore for the classification of levels it is convenient to use labels $\widetilde{M}_S$ which accept the same values, as M_S. These values are put down at corresponding levels in Fig. 13.1a. One determines the coefficients $\langle M_S|\widetilde{M}_S\rangle$ for the concrete value $\widetilde{M}_S$ from the solution of the system of the linear homogeneous equations:

$$\sum_{M'_S=-S}^{S} \left[\langle M_S|\widehat{\mathcal{H}}_S|M'_S\rangle - E_{\widetilde{M}_S}\delta_{M_S M'_S}\right]\langle \widetilde{M}_S|M'_S\rangle = 0, \tag{13.24}$$

where M_S runs over the values from -S to S. The energy values $E_{\widetilde{M}_S}$ in Eq.(13.24) are the solutions of the secular equation (13.21).

Let's consider the concrete situation: $\theta = 90°$. We will find an explicit form of expansion (13.23) for the ion Ni^{2+} at the mutually perpendicular orientation of a symmetry axis of a crystal field and a magnetic field vector. Equation systems (13.24) for each of three states $|\widetilde{M}_S\rangle$ are convenient for writing down in the matrix form by means of explicit expressions for matrix elements $\langle M_S|\widehat{\mathcal{H}}_S|M'_S\rangle$. Then, using (13.22), we will obtain for $\widetilde{M}_S = 1$:

$$\begin{bmatrix} \frac{1}{2}d + \sqrt{\frac{1}{4}d^2 + \nu_\perp^2} & \frac{1}{2}\nu_\perp & 0 \\ \frac{1}{2}\nu_\perp & -\frac{1}{2}d + \sqrt{\frac{1}{4}d^2 + \nu_\perp^2} & \frac{1}{2}\nu_\perp \\ 0 & \frac{1}{2}\nu_\perp & \frac{1}{2}d + \sqrt{\frac{1}{4}d^2 + \nu_\perp^2} \end{bmatrix} \begin{bmatrix} \langle -1|\widetilde{1}\rangle \\ \langle 0|\widetilde{1}\rangle \\ \langle 1|\widetilde{1}\rangle \end{bmatrix} = 0; \tag{13.25}$$

for $\widetilde{M}_S = -1$:

$$\begin{bmatrix} \frac{1}{2}d-\sqrt{\frac{1}{4}d^2+\nu_\perp^2} & \frac{1}{2}\nu_\perp & 0 \\ \frac{1}{2}\nu_\perp & -\frac{1}{2}d-\sqrt{\frac{1}{4}d^2+\nu_\perp^2} & \frac{1}{2}\nu_\perp \\ 0 & \frac{1}{2}\nu_\perp & \frac{1}{2}d-\sqrt{\frac{1}{4}d^2+\nu_\perp^2} \end{bmatrix} \begin{bmatrix} \langle -1|\widetilde{-1}\rangle \\ \langle 0|\widetilde{-1}\rangle \\ \langle 1|\widetilde{-1}\rangle \end{bmatrix} = 0; \tag{13.26}$$

and for $\widetilde{M}_S = 0$:

$$\begin{bmatrix} 0 & \frac{1}{2}\nu_\perp & 0 \\ \frac{1}{2}\nu_\perp & -d & \frac{1}{2}\nu_\perp \\ 0 & \frac{1}{2}\nu_\perp & 0 \end{bmatrix} \begin{bmatrix} \langle -1|\widetilde{0}\rangle \\ \langle 0|\widetilde{0}\rangle \\ \langle 1|\widetilde{0}\rangle \end{bmatrix} = 0. \tag{13.27}$$

The coefficients $\langle -1|\widetilde{1}\rangle$ and $\langle 1|\widetilde{1}\rangle$ are included into system (13.25) symmetrically, hence, they are equal. From the first equation of the system it follows

$$\frac{\langle 1|\widetilde{1}\rangle}{\langle 0|\widetilde{1}\rangle} = \frac{\nu_\perp}{\sqrt{2}\left(\frac{1}{2}d+\sqrt{\frac{1}{4}d^2+\nu_\perp^2}\right)}.$$

Taking into account the normalization of functions (13.23) it is convenient to put $\langle -1|\widetilde{1}\rangle = 2^{-1/2}\sin\alpha$ and $\langle 0|\widetilde{1}\rangle = \cos\alpha$, therefore the value α is determined by the equation

$$\mathrm{tg}\alpha = \frac{\nu_\perp}{\frac{1}{2}d+\sqrt{\frac{1}{4}d^2+\nu_\perp^2}}.$$

Using the trigonometric formula $\mathrm{tg}\alpha = \frac{\sin 2\alpha}{(1+\cos 2\alpha)}$, we obtain

$$\sin 2\alpha = \frac{\nu_\perp}{\sqrt{\frac{1}{4}d^2+\nu_\perp^2}}. \tag{13.28}$$

From the system (13.26) we obtain the relation

$$\frac{\langle 1|\widetilde{-1}\rangle}{\langle 0|\widetilde{-1}\rangle} = \frac{\nu_\perp}{\sqrt{2}\left(\frac{1}{2}d-\sqrt{\frac{1}{4}d^2+\nu_\perp^2}\right)},$$

whence, taking into account the trigonometric relation $\mathrm{ctg}\alpha = \frac{\sin 2\alpha}{(1-\cos 2\alpha)}$, it follows that $\langle -1|\widetilde{-1}\rangle = \langle 1|\widetilde{-1}\rangle = 2^{-1/2}\cos\alpha$ and $\langle 0|\widetilde{-1}\rangle = -\sin\alpha$. These equalities lead to expression (13.28) too. At last, from system (13.27) it follows $\langle 1|\widetilde{0}\rangle = \langle -1|\widetilde{0}\rangle$ and $\langle 0|\widetilde{0}\rangle = 0$.

Thus, the concrete view of combinations (13.23) for $S = 1$ and $\theta = 90°$ is the following:

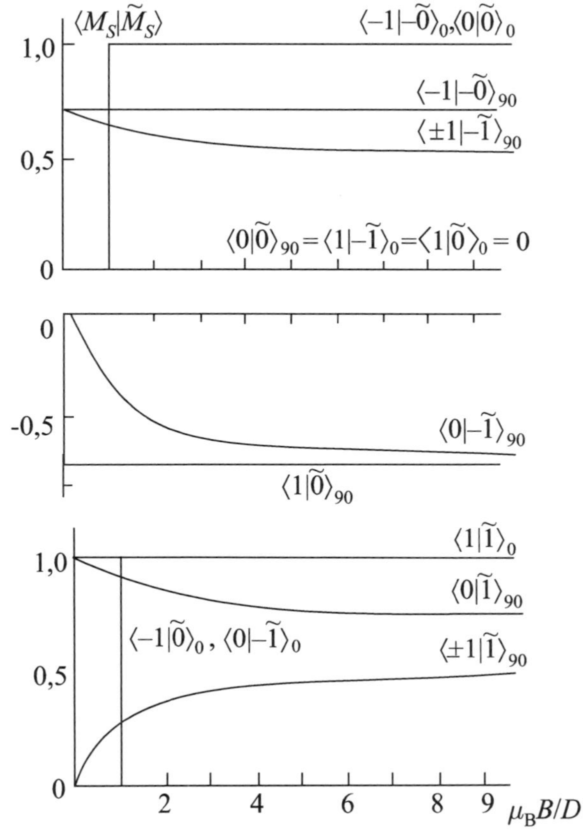

Fig. 13.2 Dependence of coefficients $\langle M_S|\widetilde{M}_S\rangle$ (they are indicated near corresponding graphics) on a magnetic field (relative units)

$$
\begin{aligned}
|\widetilde{-1}\rangle &= \frac{1}{\sqrt{2}}\cos\alpha|-1\rangle - \sin\alpha|0\rangle + \frac{1}{\sqrt{2}}\cos\alpha|1\rangle,\\
|\widetilde{0}\rangle &= \frac{1}{\sqrt{2}}\left(|-1\rangle - |1\rangle\right),\\
|\widetilde{1}\rangle &= \frac{1}{\sqrt{2}}\sin\alpha|-1\rangle + \cos\alpha|0\rangle + \frac{1}{\sqrt{2}}\sin\alpha|1\rangle.
\end{aligned}
\tag{13.29}
$$

In Fig. 13.2 the dependencies of non-zero coefficients on the value of a static magnetic field B in the units of the zero-field splitting $D/(g\mu_B)$ for $\theta = 0°$ and $\theta = 90°$ are presented. As a combined effect of a crystal field and spin-orbit interactions leads to mixing of various states $|M_S\rangle$ the simple selection rule $\widetilde{M}'_S = \widetilde{M}_S \pm 1$ $(\Delta\widetilde{M}_S = 1)$ is not valid. Knowing the states expressed by (13.29), it is easy to calculate the transition probabilities, which are proportional to $\left|\langle\widetilde{M}_S|\widehat{S}_1|\widetilde{M}'_S\rangle\right|^2$.

In the situation $\theta = 90°$ the following variants are possible:

(1) mutually perpendicular orientations of a symmetry axis of a crystal field and vectors of oscillating and static magnetic fields;
(2) parallel orientations of an axis of a crystal field and vector of an oscillating magnetic field;
(3) vectors of oscillating and static magnetic fields are parallel, but the static field vector is directed along an axis x.

At first let's consider the first variant.

Let's find the dipole moments of transitions in the case of mutually perpendicular orientations of a symmetry axis of a crystalline field and oscillating and static magnetic fields ($\widetilde{S}_1 = \hat{S}_y$):

$$\begin{aligned}\langle\widetilde{-1}|\widehat{S}_y|\widetilde{0}\rangle &= \left(\frac{1}{\sqrt{2}}\cos\alpha\langle-1| - \sin\alpha\langle 0| + \frac{1}{\sqrt{2}}\cos\alpha\langle 1|\right)\hat{S}_y\left(\frac{1}{\sqrt{2}}|-1\rangle - \frac{1}{\sqrt{2}}|1\rangle\right)\\ &= -\frac{1}{\sqrt{2}}\sin\alpha\left(\langle 0|\hat{S}_y|-1\rangle - \langle 0|\hat{S}_y|1\rangle\right)\\ &= \frac{1}{2\sqrt{2}}\mathrm{i}\sin\alpha\left(\langle 0|\hat{S}^+|-1\rangle + \langle 0|\hat{S}^-|1\rangle\right) = -\mathrm{i}\sin\alpha.\end{aligned}$$

Other moments of transitions are calculated in a similar way:

$$\langle\widetilde{1}|\widehat{S}_y|\widetilde{0}\rangle = \mathrm{i}\cos\alpha; \quad \langle\widetilde{-1}|\widehat{S}_y|\widetilde{1}\rangle = 0.$$

The second variant: the parallel alignment of a vector of an oscillating field and symmetry axis of a crystal field ($\widetilde{S}_1 = \hat{S}_z$). In this case we obtain following expressions for the dipole moments of transitions:

$$\langle\widetilde{-1}|\widehat{S}_z|\widetilde{0}\rangle = -\cos\alpha; \quad \langle\widetilde{1}|\widehat{S}_z|\widetilde{0}\rangle = \sin\alpha; \quad \langle\widetilde{-1}|\widehat{S}_z|\widetilde{1}\rangle = 0.$$

From these expressions it is evident that in the case of mutually perpendicular orientation of vectors oscillating and static magnetic fields the transitions between the neighboring levels also are allowed similar to the case $\theta = 0°$. However relative intensities of lines depend on parameter α, and hence, according to formulas (13.28) and (13.18), from the relation between values of $\hbar\gamma B_0$ and zero-field splitting D.

In Fig. 13.1b, c the spectra of EPR of the ion Ni^{2+} are given for $\theta = 90°$ and for two fixed frequencies: $\nu_1 < \frac{D}{h}$ (Fig. 13.1b) and $\nu_1 > \frac{D}{h}$ (c). Resonance values of a field are reached when the interval between levels with the allowed transition (Fig. 13.1a) coincides with quantities $h\nu_1$ or $h\nu_2$. Corresponding transitions in Fig. 13.1a are denoted by two-sided arrows, near which the frequency (ν_1 or ν_2) and the angle θ as an upper index at ν_i (0 or 90°) are specified. Dashed arrows designate "forbidden" transitions, i.e. such transitions for which the probability tends to zero if the angle θ reaches one of the limiting values. Long dashed lines in Fig. 13.1b, c join the spectroscopic lines corresponding to the same transition for different θ.

In Fig. 13.3 the angular dependence of the same spectroscopic lines, as it was shown in Fig. 13.1b is presented in a polar frame. The length of the radius vector to the point on the curve, corresponding to transitions $M_S \rightleftharpoons M_S'$ (the dashed terminations of curves indicate

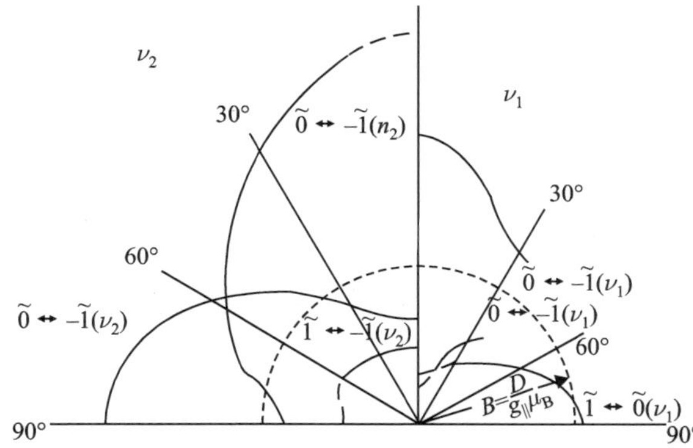

Fig. 13.3 The polar diagram of EPR transitions for the ion Ni^{2+}. A polar angle is the *angle* between the magnetic field and the symmetry axis, and the length of radius-vector is proportional to the resonance value of the magnetic field at the frequencies ν_1 and ν_2 as it Fig. 13.1

an area of the forbidden transitions), is equal to the resonance value of the magnetic field for the frequency specified near corresponding curves. The angle between the vertical and the radius vector is equal to the angle θ between a symmetry axis and a magnetic field. All curves are symmetrical relative to co-ordinate axes, therefore in Fig. 13.3 the curves are presented inside one quadrant only. Values of a field, at which Zeeman splitting $g_{\|}\mu_B B$ is equal to the zero-field one (D), lie on a dashed semicircle with radius $B = \frac{D}{g_{\|}\mu_B}$.

Now let's consider the third variant: the vector of a constant magnetic field is directed along an axis x, but the vector of oscillating field is parallel to a vector static one ($\widehat{S_1} = \hat{S}_x$). Simple calculations show that $\langle\widetilde{-1}|\hat{S}_x|\tilde{1}\rangle = \langle\tilde{1}|\hat{S}_x|\tilde{0}\rangle = 0$, but $\langle\tilde{1}|\hat{S}_x|\widetilde{-1}\rangle = \cos 2\alpha \neq 0$. It means that so-called *double-quantum transition* $|\tilde{1}\rangle \rightleftharpoons |\widetilde{-1}\rangle$, for which $\widetilde{M}'_S = \widetilde{M}_S \pm 2$ and which in a usual magnetic resonance is forbidden, now is allowed (in Fig. 13.1a and Fig. 13.3 the transitions for the case of parallel vectors of fields are not shown).

Let us give an example of the definition of spin Hamiltonian constants from the EPR spectrum of nickel fluor silicate. The experimental spectrum consisted of two lines observed in fields $B_0^{(1)} = 0{,}250$ T and $B_0^{(2)} = 0{,}360$ T for $\theta = 0$ and $B_0^{(3)} = 0.276$ T and $B_0^{(4)} = 0.332$ T for $\theta = 90°$. Measurements were carried out at the wave length $\lambda = 3\,\text{cm}$.

From the equations (13.22) and (13.28) it is easy to express parameters of the spin Hamiltonian via the resonance values of a field:

$$d = \frac{|B_0^{(2)} - B_0^{(1)}|}{B_0^{(2)} + B_0^{(1)}}\nu_0,$$

$$g_\perp = \frac{h}{2\mu_B B_0^{(3)}}\sqrt{\nu_0\left(\nu_0 + \frac{1}{2}d\right)},$$

$$g_\| = \frac{\nu_0 h}{2\mu_B(B_0^{(2)} + B_0^{(1)})} \quad or \quad g_\| = \frac{h}{2\mu_B B_0^{(4)}}\sqrt{\nu_0\left(\nu_0 + \frac{1}{2}d\right)}$$

(Quantity $g_{\parallel}$ can be calculated from two formulas). Here $\nu_0 = \frac{c}{\lambda} = 10$ GHz, the quantity $\frac{2\mu_B}{h}$ should be replaced according to (1.26–1.27) with 2.8×10^{10} Hz/T. Substitution of numerical values gives: $g_{\parallel} = g_{\perp} = 2.34$ and $d = 1.8\,\text{GHz}$ ($0.6\,\text{cm}^{-1}$), i.e. $\nu_0 > d$ that corresponds to frequency ν_2 in Fig. 13.1.

13.2.3 Magnetic Levels of the Ion Cr^{3+} in the Intermediate Crystal Field

Let's consider one more system of energy levels and transitions between them representing practical interest for quantum electronics: the ion Cr^{3+} in an octahedral environment, $S = \frac{3}{2}$. Such a situation is realized for the chromium ions in ruby, at that the value of interactions corresponds to the case of an intermediate crystal field. The matrix $\langle M_S|\widehat{\mathcal{H}}_S|M'_S\rangle$ at the same designations as for $S = 1$ has the following view:

$$\begin{bmatrix} -\frac{3}{2}\nu_{\parallel}\cos\theta + d & \frac{\sqrt{3}}{2}\nu_{\perp}\sin\theta & 0 & 0 \\ \frac{\sqrt{3}}{2}\nu_{\perp}\sin\theta & -\nu_{\parallel}\cos\theta - d & \nu_{\perp}\sin\theta & 0 \\ 0 & \nu_{\perp}\sin\theta & \nu_{\parallel}\cos\theta - d & \frac{\sqrt{3}}{2}\nu_{\perp}\sin\theta \\ 0 & 0 & \frac{\sqrt{3}}{2}\nu_{\perp}\sin\theta & \frac{3}{2}\nu_{\parallel}\cos\theta + d \end{bmatrix}.$$

In the octahedron environment $\nu_{\parallel} = \nu_{\perp} = \nu$. Then the secular equation becomes simpler:

$$\varepsilon^4 - \frac{1}{2}\varepsilon^2\left(5\nu^2 - d^2\right) + 2d\varepsilon d\nu^2\left(1 - 3\cos^2\theta\right) + \frac{9}{16}\nu^4 +$$

$$+\frac{1}{2}d^2\nu^2\left(1 - 9\cos^2\theta\right) + \frac{3}{2}d^3\nu\cos\theta + d^4 = 0.$$

At $\theta = 54°45'$ the coefficient $\left(1 - 3\cos^2\theta\right)$ becomes zero and the corresponding secular equation degenerates in the biquadratic one. The energy dependencies on a magnetic field are symmetrical relative $E = 0$ (Fig. 13.4). Because of mixing of states not only transitions between the neighboring levels $\left(\widetilde{-\frac{1}{2}} \rightleftharpoons \widetilde{\frac{1}{2}}\right)$ with frequency ν_0 are allowed, but also the transitions $\widetilde{M}'_S = \widetilde{M}_S \pm 2$ $\left(\widetilde{-\frac{1}{2}} \rightleftharpoons \widetilde{\frac{3}{2}}\right)$ and $\left(\widetilde{-\frac{3}{2}} \rightleftharpoons \widetilde{\frac{1}{2}}\right)$ with frequencies ν_1 and ν_2, and $\nu_1 = \nu_2$. Such a system of levels appears rather convenient for operation of masers (see Chap. 19). The similar approach is used at examination of EPR spectra of paramagnetic centers of other nature. As an example, EPR spectra of the paramagnetic defect S_2^- arisen in a matrix KI at 4 K are given in Fig. 13.5. Above spectra are presented for four orientations of magnetic field relaive to crystallographic axes. From Fig. 13.5 one can see that EPR spectrum of this object manifests a strong anisotropy relieved in the change the of line position and intensities depending on of the magnetic field orientation.

The analysis of the spectra obtained at special orientations of the monocrystal KI when there is the merging of lines (at 0° and 45°) allows us to get the valuable information on the position and electronic structure of a lattice defect.

Fig. 13.4 The dependence of energy levels of ion Cr^{3+} in ruby on a magnetic field

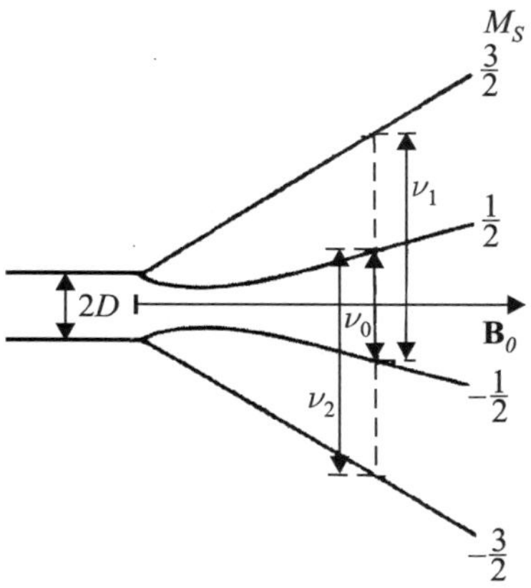

Fig. 13.5 EPR spectra of the crystal defect S_2^- in KI, obtained at different orientations of magnetic field at $T = 4K$. The values of angles between a magnetic field in the plane [100] and the crystallographic axis (100) are specified near each spectrum

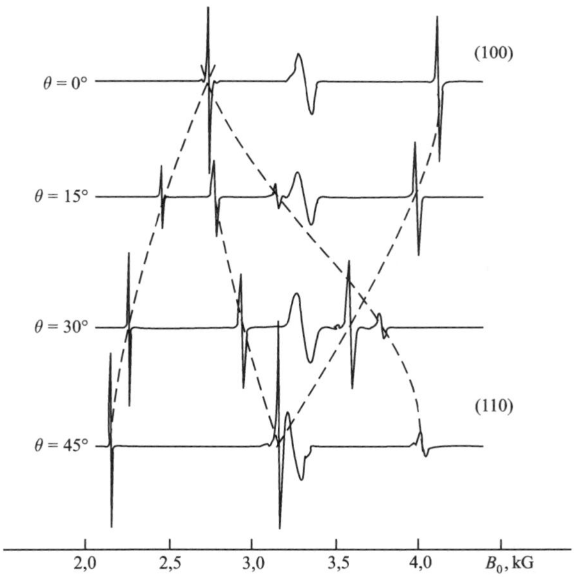

13.2.4 Application of Spin-Hamiltonian Method to Calculation of EPR Spectra of Molecules in Triplet State

As it was already noticed, in the process of the formation of molecules the spins of atoms, entering into molecular composition, tend to compensate each other. As a rule, symmetry of molecules is low enough, and a ground orbital is an isolated orbital singlet, in other words, it is a nondegenerate level occupied by two electrons with opposite orientations of spins: a spin singlet state with $S = 0$. However some molecules (there are not so many such molecules but they play the essential role in the vital biochemical processes) form triplet spin states ($S = 1$) in the ground orbital state. Examples: O_2 and S_2, and also a series of complex organic molecules. Some molecules, though have a zero spin in the ground state, there is a transition of one of electrons in an excited orbital with the selection rule $\Delta M_S = 0$ (without change of a spin state) but under an influence of the ultra-violet irradiation. However in a number of large organic molecules with the conjugate systems of intramolecular bonds the nonradiative change of a spin state takes place. As a result, spins of electrons in the ground and excited orbitals appear to be parallel. Therefore a triplet state with $S = 1$ is formed. The triplet state is a long living (metastable) one because the transition from an excited state in the ground one requires now the change of a spin state ($\Delta M_S = 1$), but such transition is forbidden by selection rules. The molecule in a triplet state also is described by the Hamiltonian in a form (13.9) with $S = 1$. Constants of spin Hamiltonian of a triplet state molecule, unlike paramagnetic centers in a crystal field, are determined mainly by the dipole-dipole interaction between the electrons forming a triplet state, whereas the spin-orbit interaction introduces seldom the appreciable contribution. If there is no axial symmetry the Hamiltonian can be led to the form

$$\widehat{\mathcal{H}}_S = g_{\|}\mu_B \hat{S}_z B_0 cos\theta + g_{\perp}\mu_B \hat{S}_x B_0 \sin\theta + D\left(\hat{S}_z{}^2 - \frac{1}{3}\hat{S}^2\right) + E\left(\hat{S}_y{}^2 - \hat{S}_x{}^2\right),$$

where

$$D = \frac{3}{4}g^2\mu_B^2\left\langle\frac{r^2 - z^2}{r^5}\right\rangle; \quad E = \frac{3}{4}g^2\mu_B^2\left\langle\frac{y^2 - x^2}{r^5}\right\rangle;$$

r is the distance between the electrons forming a triplet state; the averaging is carried out over orbitals of both electrons. This Hamiltonian differs from the Hamiltonian (13.9) for a axial symmetric case with the additional quadratic respectively spin components term with the coefficient E (it is not an energy!), taking into account the deviation from axial symmetry. One more difference is that the g-factor is isotropic and with high precision coincides with one for a free electron.

One of the first registered EPR spectra of molecules in a triplet state was the spectrum of naphthalene in a durol (tetramethylbenzen) matrix. Both molecules have a flat conformation with the symmetry $\mathcal{D}_{2h}$ and form similar schistose structures at crystallization, therefore naphthalene is easily implanted in crystal durol structure. After ultra-violet irradiation strongly anisotropic EPR spectra from naphthalene caused by formation of triplet states are observed. Angular dependencies of the position of spectral lines corresponding to transitions, for which the selection rule $\Delta \tilde{M}_S = \pm 1$ is fulfilled, are presented in Fig. 13.6 for different orientations of the static magnetic field relative to symmetry axes x, y and z of naphthalene molecules. In Fig. 13.6a the magnetic field vector is parallel to a plane (xy) of molecules; in Fig. 13.6b, c the vector is parallel to planes (xz) and (yz), respectively. The spectra look like two doublets which are approximately symmetrical relative to the resonance value of a magnetic field for a free electron (0,331 T at $\nu = 9,272$ GHz). From these spectra the values 0,1 and $-0,14\,\mathrm{cm}^{-1}$ respectively for $\frac{D}{hc}$ and $\frac{E}{hc}$ are obtained. These values are used in calculations of spin density distribution of the electrons forming triplet states.

Observed doubling of number of lines is caused by the split of lines of the thin structure caused by interaction of the electrons with nuclei (it will be considered in the following section). Besides presented in Fig. 13.6, the naphthalene EPR lines are also observed

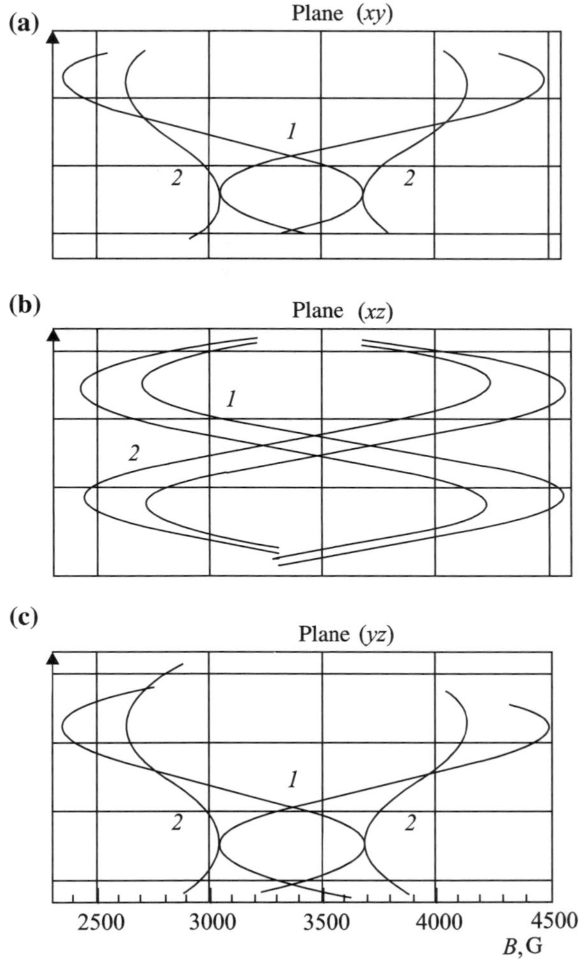

Fig. 13.6 The dependencies of position of lines EPR spectra of naphthalene in durol on the orientation of magnetic field. Magnetic field vector lies: in xy-plane **(a)**, in xz-plane **(b)**, in yz-plane **(c)**

in the range of more weak magnetic fields (0,15–0,17 T). These lines are assigned to transitions $\Delta\widetilde{M}_S = \pm 2$ which are caused by mixing "pure" states $|M_S\rangle$ (see the Sects 13.1.4 and 13.2.1). At sufficient concentration of molecules in the triplet state there can exist additional lines with a small split in EPR spectrum. These lines are assigned to *wandering triplet states*, formed by *different* molecules, in another words by *triplet excitons*. With growth of concentration of molecules the migration of the triplet states plays an essential role, leading to quenching of metastable states, and observation of the EPR becomes impossible.

Thus, the configuration of system of energy levels, frequencies and of transition probabilities do not depend on the nature of Spin Hamiltonian coefficients. Therefore all conclusions, drawn in Sects 13.1.4 and 13.2.1 for the ion d^2, remains valid not only for systems to which the theory of the crystal field is immediately applicable, but also for any system described by the spin Hamiltonian of the kind (13.9). In particular, such approach is suitable for molecules in a triplet state.

13.2.5 The Case of Degeneration of Ground Level for the Intermediate and Weak Crystal Fields

The intermediate crystal field. It is possible to observe EPR practically for all ions having a singlet or doublet ground orbital level (but spin degeneracy remains). Analyzing a situation for intermediate crystal field, it is necessary to apply a perturbation theory for degenerate states demanding the solution of the secular equation. For the description of doublet orbital level it is possible to use a Spin Hamiltonian in the same form as for singlet level, as (and it is possible to show) all nondiagonal matrix elements of the perturbation operator are converted in zero.

In the case of a triplet ground orbital level the perturbation theory already in the first approximation leads to level splitting due to the spin-orbit interactions, and spin degeneracy is took off too. The splitting magnitude has the order of $10^2\,\text{cm}^{-1}$. If the number of electrons is even, a ground level owing to Jan–Teller effect becomes singlet, thus an observation of EPR appears impossible. If the number of electrons is odd then according to the Kramers theorem always there is the twofold degeneration which is removed with magnetic field. The splitting of such a level by magnetic field on two sublevels occurs absolutely similar to how it takes place in a system with a spin $\frac{1}{2}$. Therefore the behavior of such a system is often described as for the effective spin S' (In this case $S' = \frac{1}{2}$). However practically in this case the observation of EPR is quite often inconveniently because of the fast relaxation caused by the affinity of an excited orbital. Sometimes in the above situation it is possible to observe EPR at very low temperatures.

The week crystal field. In this case the even number of electrons represents an adverse situation for observation of EPR. In principle it is possible to observe the resonance if a splitting by crystal field is small so the frequency of transitions between them is comparable with frequency of the oscillated magnetic field. But relaxation processes strongly broaden spectral lines, and for observation of EPR the very low temperatures are required too.

In the case of the odd number of electrons a splitting on twofold degenerate levels (doublets) is typical, the each of them is then splitted by a magnetic field. These levels are characterised by the total ("half-integer") electron momentum J. At closely located doublets EPR is observed at once on several doublets (usually no more than on two). But also in this case relaxation processes complicate resonance observation.

As an example we consider the calculation of the EPR spectrum of the ion Ce^{3+} $(4f^{1/2}F^{5/2})$ in a weak crystal field. The ground level of ion Ce^{3+} is splitted on three doublets with $|M_J| = \frac{1}{2}; \frac{3}{2}; \frac{5}{2}$, two of which are close to each other (see the sect. 11.3.3 and Fig. 11.5). Expressions for the energy of the doublets, obtained by a method of the equivalent operators, are given in Table 13.2. The coefficients of proportionality, which are necessary for the calculation of energy, are expressed here via parameters of a crystal field and average distance of an electron from a nucleus as follows:

$$\langle\frac{5}{2}||\alpha||\frac{5}{2}\rangle = -\frac{2}{35}\langle r^2\rangle, \quad \langle\frac{5}{2}||\beta||\frac{5}{2}\rangle = \frac{1}{140}\langle r^4\rangle.$$

If the perturbation is interaction of a free ion with a magnetic field, the perturbation theory is applied to find the frequency of EPR:

Table 13.2 Expressions for energy of an electron with $J = 5/2$ in weak crystal field and calculated and experimental g-factors

$\|M\|$	E	$g_{\parallel}$		$g_{\perp}$	
		Theory	Exper.	Theory	Exper.
$\frac{1}{2}$	$-8\langle\frac{5}{2}\|\|\alpha\|\|\frac{5}{2}\rangle + 2\langle\frac{5}{2}\|\|\beta\|\|\frac{5}{2}\rangle$	0,857	0,955	2,571	2,185
$\frac{3}{2}$	$-2\langle\frac{5}{2}\|\|\alpha\|\|\frac{5}{2}\rangle - 3\langle\frac{5}{2}\|\|\beta\|\|\frac{5}{2}\rangle$	2,571	–	0	–
$\frac{5}{2}$	$10\langle\frac{5}{2}\|\|\alpha\|\|\frac{5}{2}\rangle + \langle\frac{5}{2}\|\|\beta\|\|\frac{5}{2}\rangle$	4,286	3,72	0	0,2

$$\widehat{\mathcal{H}}' = -g_{\mathrm{L}}\mu_{\mathrm{B}}\left(\hat{\vec{L}} + 2\hat{\vec{S}}\right) \cdot \vec{B},$$

Here g_{L} is the Lande factor known from atomic physics:

$$g_{\mathrm{L}} = 1 + \frac{J(J+1) + S(S+1) - L(L+1)}{2J(J+1)}. \quad (13.30)$$

In this situation the unperturbed vectors (basis) are eigenvectors $|JM_J\rangle$ of the operator of the total momentum $\hat{\vec{J}}$ and its projections J_z on a symmetry axis, belonging to levels of a free ion in a crystal field. Hence, the effective operator of interaction with a magnetic field, similar to the spin operator (13.3) in case of a weak crystal field, can be written down in the first approximation of perturbation theory as

$$\widehat{\mathcal{H}}' = g_{\mathrm{L}}\mu_{\mathrm{B}}B\left[\hat{J}_z \cos\theta + \frac{1}{2}\left(\hat{J}^+ + \hat{J}^-\right)\sin\theta\right]$$

(here θ is the angle between a symmetry axis and a magnetic field vector). Then for the doublet $M_J = \pm\frac{1}{2}$ and $J = \frac{5}{2}$ we can obtain the matrix $\frac{\langle M_J|\widehat{\mathcal{H}}'|M'_J\rangle}{g_{\mathrm{L}}\mu_{\mathrm{B}}B}$:

$$\begin{bmatrix} \frac{1}{2}\cos\theta & \frac{3}{2}\sin\theta \\ \frac{3}{2}\sin\theta & -\frac{1}{2}\cos\theta \end{bmatrix},$$

which leads to the following secular equation for the definition of levels of energy ε' (in units $g_L\mu_{\mathrm{B}}B$):

$$\left(\cos\theta - 2\varepsilon'\right)\left(-\cos\theta - 2\varepsilon'\right) - 9\sin\theta^2 = 0.$$

The roots of this equation are $\varepsilon'_{1,2} = \pm\frac{1}{2}\sqrt{\cos^2 + 9\sin\theta 2}$. Hence, the frequency of EPR will be escribed by expression

$$\nu = \frac{g_{\mathrm{L}}\mu_{\mathrm{B}}B}{h}\sqrt{\cos^2 + 9\sin\theta 2}.$$

According to Eq. (13.30) $g_{\mathrm{L}} = \frac{6}{7}$, therefore the values of g-factor for parallel and perpendicular orientations of a magnetic field vector concerning to a symmetry axis appear equal, accordingly,

$$g_{\parallel} = g_{\mathrm{L}} = \frac{6}{7},\ g_{\perp} = 3g_{\mathrm{L}} = \frac{18}{7}.$$

For other doublets the nondiagonal matrix elements are equal to zero. For $M_J = +\frac{3}{2}$ the factor $g_{\parallel} = 3g_{\mathrm{L}} = 18/7$, $g_{\perp} = 0$, whence

$$\nu = \frac{9\mu_{\mathrm{B}}}{7h} B \cos\theta,$$

and for $M_J = \pm\frac{5}{2}$ the factor $g_{\parallel} = 5g_{\mathrm{L}} = \frac{30}{7}$, $g_{\perp} = 0$, one obtains

$$\nu = \frac{15\mu_{\mathrm{B}}}{7h} B \cos\theta.$$

It is necessary to note that the above theory gives only the qualitatively correct description of the influence of an environment of paramagnetic center on EPR spectrums, but it does not lead to good numerical coincidence to the experiment data.

To calculate intervals between the doublets, it is not enough data on the electronic paramagnetic resonance only, it is necessary to use experimental data on the magnetic permeability. Relations of line intensities in the EPR spectrum and values of g-factor show that the ground level is the doublet with $J = \frac{1}{2}$, and the following one (located on $3\,\mathrm{cm}^{-1}$ above) with the value $J = \frac{5}{2}$. The doublet with $J = \pm\frac{3}{2}$, as show magnetic susceptibility data, will be at a distance approximately on $130\,\mathrm{cm}^{-1}$ from them, and the spectrum of EPR from it is not observed.

In Table 13.2 the results of the theoretical calculation of the g-factor are compared with the experimental data for EPR spectrum in cerium ethylsulfate, crystal structure of which has trigonal symmetry.

The experience has shown that the best results are obtained if linear combinations of states with different values J, corresponding to the ground and first excited levels, are used as basic wave functions. In other words it should be considered an impurity of states with $J = \frac{7}{2}$. Thus in the expression for effective potential of a crystal field it is necessary to consider the terms of higher (sixth) order that appears essential at definition of Zeeman splitting in the presence of a crystal field of trigonal symmetry. It should be expected, because only beginning from $J = \frac{7}{2}$, the expression for effective potential includes the terms reflecting the presence of a symmetry axis of the third order.

Further Readings

Pilbrow JR (1990) Transition ion electron paramagnetic resonance. Clarendon Press, New York

Altshuler SA, Kozirev BM (1964) Electron paramagnetic resonance. Academic, New York

Weil JA, Bolton JR (2007) Electron paramagnetic resonance. Elementary theory and practical applications., 2nd edn. John Wiley & Sons, New York, ISBN 978-0471-75496-1

Eaton GR, Eaton SS, Barr DP, Weber RT (2010) Quantitative EPR: a practitioners guide, 1st edn. Springer, The Netherlands c. ISBN 978-3-211-9294-76

Chapter 14
Electron-Nuclear Interactions and Hyperfine Structure of EPR Spectra

14.1 Spin Hamiltonian in the Presence of Hyperfine Interactions

As it was already mentioned in Sect. 1.11, the interactions between an electron and nuclear magnetic momenta are termed as *hyperfine interactions* (HFI). Particularly, such interactions cause the hyperfine structure of EPR spectra. By analogy with the way how it was done in Sect. 13.1.1, to deduce the spin Hamiltonian for an electron in electrostatic and magnetic field, one can construct the spin Hamiltonian taking into account hyperfine interactions. There are three types of hyperfine interactions of a nucleus: (i) with orbital motion, (ii) with an electron spin and (iii) the contact interaction (see Sect. 1.11). The logic of the deduction does not differ from that what has been applied at the deduction of the expression (13.7), but it is necessary to substitute spin-orbit interaction in (13.1) using the Hamiltonian of electron-nuclear interaction:

$$\widehat{\mathcal{H}}_{\rm hf} = \frac{\mu_0}{2\pi}\hbar\gamma_I\mu_{\rm B}\hat{\vec{I}}^{(i)}\cdot\left(\frac{\hat{\vec{l}}}{r^3} - \frac{\hat{\vec{s}}}{r^3} + 3\frac{\vec{r}(\hat{\vec{s}}\cdot\vec{r})}{r^5} + \frac{8\pi}{3}\delta(\vec{r})\right) \qquad (14.1)$$

The part of a spin Hamiltonian responsible for hyperfine interactions of one unpaired electron after averaging on orbital variables is presented in the form

$$\widehat{\mathcal{H}}_{\rm hf} = h\sum_{i=1}^{N}\sum_{\mu,\nu} A_{\mu\nu}^{(i)}\hat{I}_\mu{}^{(i)}\hat{S}_\nu. \qquad (14.2)$$

Here the first summation is carried out over those N nucleus which interact with an electron, the second one acts on co-ordinates x, y, z, as well as in (13.6); μ, ν accept values x, y, z, i.e. $x_x \equiv x, x_y \equiv y, x_z \equiv z$; $\hat{I}_\mu$ are the components of the nuclear spin; $\hat{S}_\nu$ are the components of the electron spin; $A_{\mu\nu}^{(i)}$ are the components of the tensor of

V. I. Chizhik et al., *Magnetic Resonance and Its Applications*,
DOI: 10.1007/978-3-319-05299-1_14,

the second rank $\mathsf{A}^{(i)}$, looking as

$$A_{\mu\nu}^{(i)} = \frac{\mu_0 \gamma_I^i g_s \mu_B}{2h} \left(-\delta_{\mu\nu} \langle 0 | \frac{1}{r^3} | 0 \rangle + 3 \langle 0 | \frac{x_\mu x_\nu}{r^5} | 0 \rangle + \frac{4}{3} \langle 0 | \delta \left(\vec{r} - \vec{r}_i \right) | 0 \rangle \right).$$

Taking into account properties of δ-functions one obtains

$$A_{\mu\nu}^{(i)} = \frac{1}{2} \mu_0 \gamma_I^i g_s \mu_B \left(-\delta_{\mu\nu} \langle 0 | \frac{1}{r^3} | 0 \rangle + 3 \langle 0 | \frac{x_\mu x_\nu}{r^5} | 0 \rangle + \frac{4}{3} \left| \psi(\vec{r}_i) \right|^2 \right). \qquad (14.3)$$

The equation for $\widehat{\mathcal{H}}_{IS}$ can be written down in the compact form through a tensor $\mathsf{A}^{(i)}$:

$$\widehat{\mathcal{H}}_{\text{hf}} = h \sum_{i=1}^{N} \hat{\vec{I}} \mathsf{A}^{(i)} \hat{\vec{S}}.$$

According to the Eq. (14.3) components $A_{\mu\nu}^{(i)}$ are determined by wave functions of an unpaired electron, therefore their values can be used for obtaining of some information on electronic structure of paramagnetic center.

In single-crystal samples, the hyperfine structure of EPR spectrum depends on the orientation of a crystal relative to the magnetic field. In case of crystalline samples, the wave function of the electronic state in the crystal field should be substituted into (14.3). If the paramagnetic center belongs to a complex with covalent bonds, then it is necessary to use as a wave function the molecular orbital, describing a state of unpaired electron. Expression (14.3) is written in the form suitable for molecules, too. In this case $\vec{r}_i$ is the radius vector of the nucleus position in a molecule.

As an example the angular dependence of EPR spectra of naphthalene in durol matrix are presented in Fig. 13.6. They illustrate the manifestation of anisotropic hyperfine interaction in the naphthalene triplet state ($S = 1$). The crystal field of a durol matrix causes the appearance of the fine structure in the form of two lines, each of which splits, in turn, into two ones because of the hyperfine interaction of unpaired electrons with protons. As well as the position of lines of the fine structure, the value of the hyperfine splitting in the EPR spectrum strongly depends on the orientation of a magnetic field relative to crystal symmetry axes. Orientation dependence of the hyperfine splitting of the EPR spectra contains a considerable quantity of the information on microstructure and dynamics of crystals, but their interpretation quite often happens rather difficult.

An EPR spectrum in polycrystals and amorphous substances is averaged over all directions. This averaging does not necessarily lead to the complete elimination of interactions, but causes their modification. Individual components of the spectrum often merge that leads to the strong inhomogeneous broadening (the Sect 1.7.2 see) of the observable lines of EPR spectrum.

14.2 Isotropic Hyperfine Interactions

14.2.1 Isotropic Hyperfine Spin Hamiltonian

In the aspect of extraction of the information on electronic structure of a substance the situation, when the hyperfine interaction can be considered as isotropic, is optimum. In contrast to solids, there is the complete averaging for the first two terms of (14.1) in solutions. It is caused by the presence of the rapid molecular motion, when the appreciable change of a position of a molecule or its orientation occurs during the time much smaller than a reciprocal value of the width of its spectrum (in terms of frequency) determined by hyperfine interactions.

In this case the isotropic term of the Hamiltonian (14.1) representing the contact interaction is only maintained. This case is realized in solutions containing stable free radicals, molecules or molecular ions having not compensated spin. The free radicals usually represent large molecules of low symmetry, the ground state of which has the minimal degeneration. In this case the level has only double spin degeneration. Besides, as a rule, the unpaired electron belongs to an outside shell, hence, it participates in covalent bonds. The forming of a covalent bond and an augmentation of its degree of covalency lead to the increase of an interval between ground and the nearest excited level (see the Sect. 12.1, Fig. 12.1 and the expression for energy level shift at formation of the molecular orbitals (12.6)). This fact allows us to consider the basic state of an unpaired electron in a free radical as an ideal *isolated doublet*. In this case there is no the fine structure of EPR spectrum, therefore the spin Hamiltonian, including as well the interaction with an external magnetic field, is led to a simple form

$$\widehat{\mathcal{H}}_{\mathrm{hf}} = -g_s\mu_B\hat{\vec{S}}\cdot\vec{B} - \sum_{i=1}^{N}\gamma_I^{(i)}\hbar\vec{B}\cdot\hat{\vec{I}}^{(i)} + \sum_{i=1}^{N} hA_i\hat{\vec{I}}^{(i)}\cdot\hat{\vec{S}}. \tag{14.4}$$

Here A_i is the scalar to which the tensor $\mathsf{A}^{(i)}$ is reduced in the isotropic case and for which it follows from (14.3)

$$A_i = \frac{2}{3}\mu_0\gamma_I^i g_s\mu_B\left|\psi\left(\vec{r}_i\right)\right|^2. \tag{14.5}$$

As one can see from (14.5) that the hyperfine interaction constant A_i is directly proportional to the probability to find an electron with non-compensated spin in the nucleus location, in other words, to *spin density* on a nucleus (a constant sign is defined by the sign of the gyromagnetic ratio for a nucleus). This circumstance allows us to determine a spatial distribution of spin density or *delocalization* of unpaired electron in a molecule using the phenomenon of EPR. Sensitivity of these unique techniques is rather great: for example, for the atom of hydrogen (as an experiment and an estimation under the formula (14.5)) give $A_{\mathrm{H}} = 1,43$ GHz (in terms of a magnetic field: 51 mT) while the width of a EPR line in solutions of free radicals is less on 2–3 orders. Therefore, thanks to EPR technique, the changes of spin density on a nucleus less than 1 % can be detected. EPR is also very sensitive techniques of the detection of configurational interaction, when there is a small impurity of an excited state wave function to the ground state of unpaired electron. Such hyperconjugation consists in transport of electron density through two chemical bonds. The presence of an insignificant impurity of exited s-states to the ground state of paramagnetic ions with unfilled d- and f-shells explains the appearance of the isotropic hyperfine structure of EPR spectra of such ions.

14.2.2 Breit-Raby Formula and Hyperfine Structure of EPR Spectra

Energy levels of the hyperfine interaction are calculated from the secular equation obtained by means of the Hamiltonian (14.4). The full number of states for paramagnetic center with a spin S interacting with N nuclei is equal $(2S+1)\prod_{i=1}^{N}(2I^{(i)}+1)$. Generally the magnetic field removes the degeneration completely but, if the electron interacts with several identical nuclei, some levels can appear degenerated because of the accidental coincidence of values of the energy for different states. The contact interaction mixes up "pure" electron and nuclear states which can be presented with multiplicative basic vectors $|M_S m_{I_1} m_{I_2} \ldots m_{I_N}\rangle$, formed by multiplication of state vectors of a noninteracting electron ($S = \frac{1}{2}$) and nuclei:

$$|M_S m_{I_1} m_{I_2} \ldots m_{I_N}\rangle = |M_S\rangle|m_{I_1}\rangle|m_{I_2}\rangle \ldots |m_{I_N}\rangle.$$

The secular equation for the Hamiltonian (14.4) with single constant A has the following form:

$$\text{Det}\left\{\langle M_S m_I | g_s \mu_B \hat{\vec{S}} \cdot \vec{B} - \gamma_I \hbar \hat{\vec{I}} \cdot \vec{B} + hA\hat{\vec{I}} \cdot \hat{\vec{S}} | M'_S m'_I\rangle - E\delta_{M_S M'_S}\delta_{m_I m'_I}\right\} = 0. \tag{14.6}$$

Similar to how it has been made in case of the spectrum of nickel ion in a crystal field, Hamiltonian eigenstates (14.4) are convenient to denote with quantum numbers $\widetilde{M}_S$ and $\widetilde{m}_I$. All states can be grouped in accordance of values of the full moment F: $I + \frac{1}{2}$ and $I - \frac{1}{2}$. For $S = \frac{1}{2}$ it is possible to gain the following expression for the energy of a state with $\widetilde{M}_S$ and $\widetilde{m}_I$ (the Breit–Rabi formula):

$$E_{I\pm\frac{1}{2},\widetilde{M}_F} = -\frac{hA}{4} - \gamma_I \hbar B \widetilde{M}_F \pm \frac{hA}{4}(2I+1)\sqrt{1 + \frac{4\widetilde{M}_F\kappa}{2I+1} + \kappa^2}. \tag{14.7}$$

Here

$$\kappa = \frac{(\gamma_i - \gamma_s)B}{2\pi A(2I+1)} \simeq -\frac{\gamma_s B}{2\pi A(2I+1)}; \tag{14.8}$$

γ_s and γ_I are the gyromagnetic ratios for an electron and a nucleus, respectively; the parameter κ can be expressed through g-factors of an electron and a nucleus (g_s and g_I), nuclear magneton $\mu_{\rm n}$ and proton mass $m_{\rm p}$(see Sect. 1.1.2):

$$\kappa = \frac{(g_s\mu_B + g_I\mu_{\rm n})B}{hA(2I+1)} \quad \text{or, accordingly,} \quad \kappa \simeq \frac{(g_s + g_I')\mu_B B}{hA(2I+1)}, \quad g_I' = g_I\frac{m_{\rm e}}{m_{\rm p}}.$$

If the approximate expression for κ from the definition (14.8) is used then the second term in Eq. (14.7) can be neglected too. The upper sign in (14.7) is used for $F = I + \frac{1}{2}$, and lower one does for $F = I - \frac{1}{2}$.

14.2.3 *Energy Levels and Selection Rules in Weak Magnetic Field*

The diagram of energy levels constructed according to the Eq. (14.7) for a case $I = \frac{1}{2}$ is given in Fig. 14.1. The standard selection rule $\Delta\widetilde{M}_F = \pm 1$ ($\widetilde{M}_F = \widetilde{M}_S + \widetilde{m}_I$), is strictly fulfilled only in the limiting cases of the strong or weak magnetic fields. In the limiting case of a weak magnetic field when Zeeman energy of $2\mu_B B$ is comparable with hA, the rule $\Delta\widetilde{M}_F = \pm 1$ operates. As an example transitions between states $|F, \widetilde{M}_F\rangle$, namely: $|1,-1\rangle \rightleftharpoons |0,0\rangle$ and $|1,0\rangle \rightleftharpoons |1,-1\rangle$, satisfying to this rule, and a corresponding spectrum on frequency $\nu_1 > \frac{hA}{2}$, are shown in Fig. 14.1 (spectral lines *1* and *2*). In the limit of the strong fields ($2\mu_B B \gg hA$) for EPR transitions only between identical nuclear spin states are resolved: $\Delta m_I = 0$. These transitions and a spectrum corresponding to them also are shown for frequency $\nu_2 \gg A$ in Fig. 14.1 (spectral lines *3* and *4* for $m_I = 1$ and $m_I = -1$, respectively). The value of a constant A of the isotropic hyperfine interaction can be calculated from the experimental spectra on a basis of Eq. (14.7) if the value of a field B is known.

Because of mixing of "pure" states in the intermediate case the transitions with $\Delta\widetilde{M}_F$, being different from the rule ± 1, become possible. What is more, transitions for parallel static and oscillating magnetic fields are allowed with the selection rule $\Delta\widetilde{M}_F = 0$. To distinguish the two sorts of transitions, the terms *π-transitions* for the transitions induced by an oscillating field perpendicular to static one and *σ-transitions* for another case.

14.2.4 *EPR Spectroscopy of Free Radicals*

It is expedient to consider separately the limiting case of the strong field, using the first approximation of perturbation theory (see Eq. (11.22)). Thus the both sums in the hyperfine interaction Hamiltonian (14.4) are the perturbation relative to the Hamiltonian of the interaction of an electron spin moment with an external magnetic

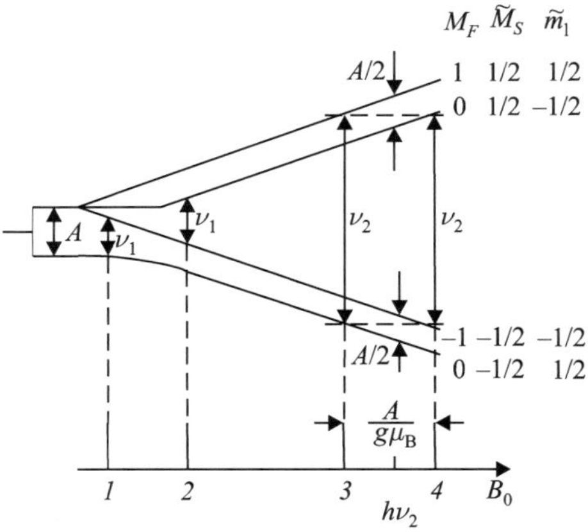

Fig. 14.1 The dependence of the energy levels of an electron interacting with a nucleus ($I = 1/2$) on magnetic field

field (the first term in (14.4). In this case the diagonal elements of the matrix of the secular equation (14.6) correspond to the values of energy:

$$E_{m_I \pm \frac{1}{2}} = \pm \frac{1}{2} g_s \mu_B B \pm \frac{1}{2} h A M_I - \gamma_I \hbar B m_I,$$

Hence the hyperfine interaction constant A is determined from splitting in EPR spectrum (see the interval between lines *3* and *4* in Fig. 14.1, a nuclear spin $I = \frac{1}{2}$).

It is necessary to note that the above conception of "the limiting strong magnetic field" does not correspond to the terminology accepted by experimenters according to which field used for observation of EPR in millimetric and centimetric range are considered as "strong". In many real situations these are not so strong that one could use the perturbation theory as a first approximation. For obtaining exact values of constants of the hyperfine interaction from experimental values it is necessary to use the Breit–Rabi formula. In EPR spectrometers a spectrum is obtained more often by magnetic field change at fix oscillating field frequency, therefore for calculations it is convenient to consider a constant of hyperfine interaction A, measured in magnetic field unities: $a = \frac{A}{g_s \mu_B}$.

The diagrams of levels of an electron interacting with one nucleus with $I = 1$ and its EPR spectrum in a strong magnetic field are presented in Fig. 14.2a, c respectively. If there is an interaction of single electron with N identical nuclei with identical hyperfine interaction constants, then some levels appear degenerated ("accidental degeneracy") that leads to a spectrum with unequal component intensities. In Fig. 14.2 the diagram of levels (a, c) and spectra (b, d) are presented for $I = \frac{1}{2}$ and $N = 2$ (a, b) as well for $I = 1$ and $N = 1$ (c, d). As one can see in the case of interaction of an electron with one nucleus the spectrum consists from $2I + 1$ lines of equal intensity. If the interaction with N identical nuclei with the total spin I_Σ is characterized by the same constant, the spectrum also contains $2I_\Sigma + 1$ lines, but their intensities are different. In particular, if $I = 1/2$ the intensities are proportional to binomial coefficients.

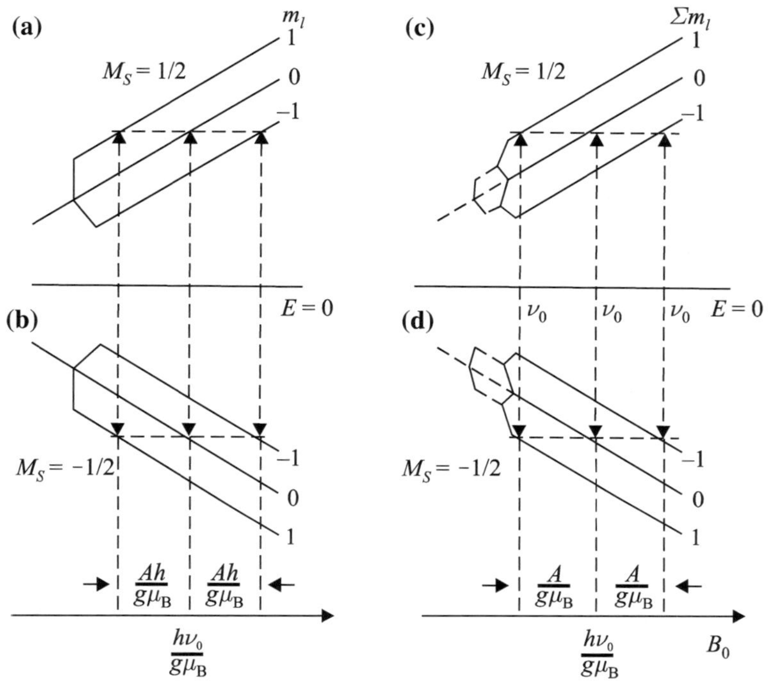

Fig. 14.2 The diagrams of energy levels (**a**, **c**) and spectra (**b**, **d**) of an electron interacting with a nucleus having spin 1 (**a**, **b**) or two nuclei with spin 1/2 (**c**, **d**)

Thus, in simple cases it is possible to estimate the character of distribution of electron density of an unpaired electron on atoms in a molecule by the structure of spectrum without calculation.

The Fig. 14.3a represents the EPR spectrum of the solution of semiquinone. One can see from the spectrum that the unpaired electron equally interacts with all four protons of the molecule, hence, the probability of its staying near both atoms of oxygen is identical, in other words, its spin density is distributed uniformly.

Figure 14.4 illustrates more complicated case of the EPR spectrum of dimethylhidropiren semiquinone. This compound is characterized by very different constants of the hyperfine interaction. As in the last case, the experimental spectrum in Fig. 14.4a was received using differential passage techniques (see Sect. 2.3). In Fig. 14.4b the theoretical spectrum, constructed using values of constants of the hyperfine interaction providing its best fitting to the observed spectrum, is presented. In this case, as well as for semiquinone, the uniform delocalization of spin density of unpaired electron on a few nuclei takes place. Interaction with group of four nearest protons leads to the appearance of five groups of lines (compare with five semiquinone spectral lines). Interaction with the second group of four protons in positions *4*, *5*, *9* and *10* should lead to split each line on five ones, and the interaction with protons of groups CH_3 yields seven, in total 35 lines. However actually each

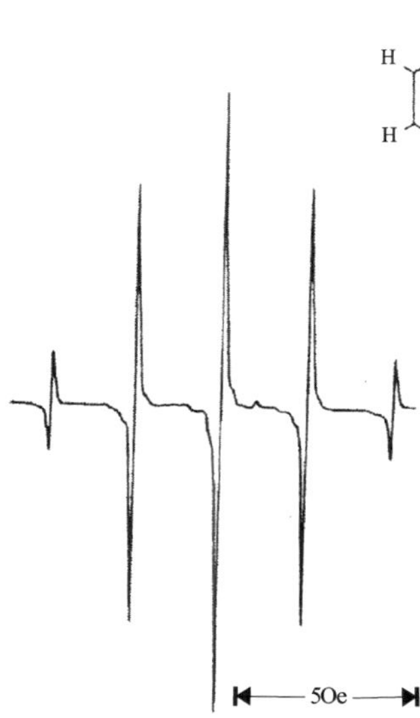

Fig. 14.3 The appearance of the experimental EPR spectrum of aqueous solution of semiquinone, registered using differential passage techniques

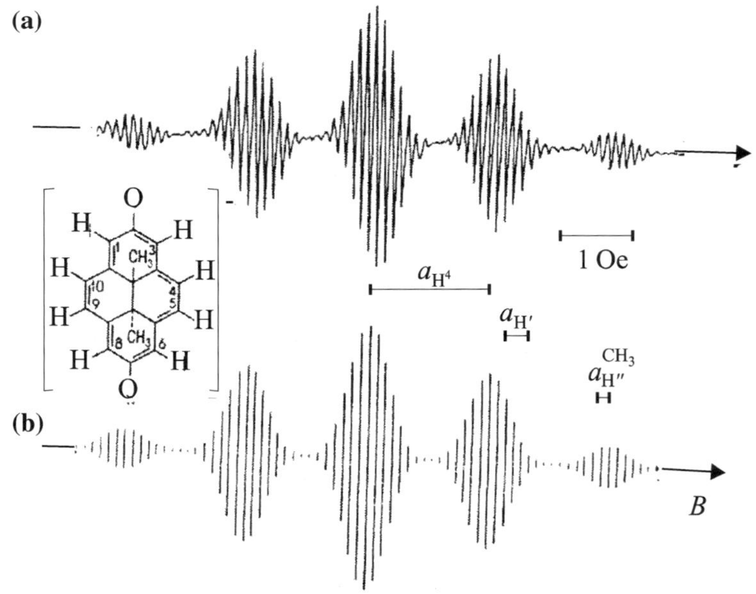

Fig. 14.4 The approximate view of **a** the experimental EPR spectrum and **b** the theoretical one of dimethylhidropiren semiquinon

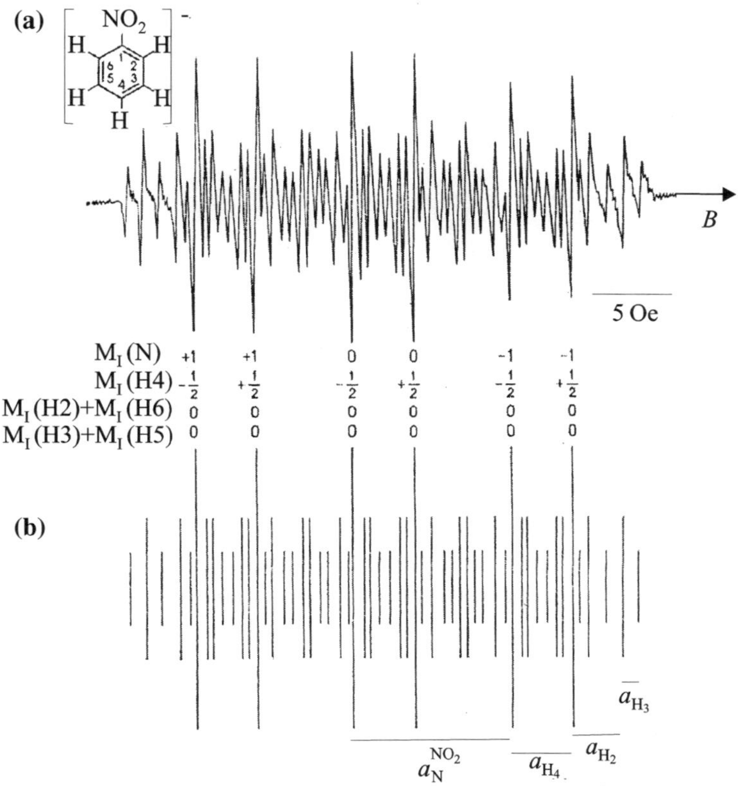

Fig. 14.5 The schematic representation of **a** the experimental EPR spectrum and **b** the theoretical spectrum of solution of nitrobenzene anion radical. The structure formula is at *top* to *left*

group contains only 15 lines, which intensity does not submit to the law of binomial coefficients. Such spectrum can be explained if to admit that the constant of a hyperfine interaction of an electron with methyl protons (A_{CH_3}) because of the accidental reason is twice less than the constant of interaction with protons of the second group (A_H).

It is more difficult to decipher spectra which are characterized by close, but unequal constants which are not in the multiple relations. Figure 14.5 represents the experimental (at top) and theoretical (at foot) spectra of an anion-radical of nitrobenzene. From the analysis of the found values of constants of a hyperfine interaction and the chemical structural formula of the radical it is possible to draw a conclusion that the unpaired electron is not completely localized on nitrogen atom (as it follows from the formula) but it is partially delocalized on an aromatic ring.

The most difficult problem is to refer the found constants of a hyperfine interaction with nuclei of certain atoms in complex molecules. An effective technique of the solution of the problem is an isotopic replacement of nuclei in certain chemical positions (the deuteration in the case of atoms of hydrogen).

As hyperfine interaction constants are directly related to spin density at the nuclei location, the EPR-spectroscopy of the free radicals allows us to acquire information not only on an electronic structure of rather complicated molecules, which are free radicals, but also about processes of the transfer of electron density by various actions (for example, optical). So, this is a basis for the application of EPR for examination of energetic processes in living organisms, for example photosynthesis. The analysis of EPR spectra has shown that under the influence of light in chloroplasts (special components of cells of green parts of plants), algae and photosynthesizing bacteria the free radicals, which are primary intermediates of photosynthesis, are formed.

A sensitivity of constants of hyperfine interaction to small changes of electron structure is the basis of the techniques of investigation of intermolecular interactions termed as *by techniques of a spin probe* and *a spin label*.

The spin probe technique consists in addition of a small amount of the free radicals in the explored medium and studying of dependencies of parameters of EPR spectra (a position of lines, an intensity relation and spectral component widths) on properties of medium and experimental conditions. The technique is based on high sensitivity of parameters of EPR spectra to the minor change in allocation of unpaired electron density. The changes are caused by an intermolecular interaction of the free radical with solvent molecules. On the basis of certain model of the influence of medium on the EPR spectrum of the radicals the theory linking medium properties with spectrum parameters can be build. It gives the possibility to interpret experimental spectra and to draw the conclusions about properties of explored system.

In the *spin label* technique the radicals are used wich interact with the studied molecule, (for example, a protein) but are inert in relation to solvent molecules. Nitroxyl radicals having a structure R_2NO (R is the organic radical) are the most suitable to using them as spin labels. Unpaired $2p_\pi$-orbital, oriented orthogonal to plane of the fragment CCNO, interacts with a nuclear magnetic moment of nitrogen ^{14}N having a spin $I = 1$. The hyperfine interaction leads to EPR spectrum consisting of three lines of equal intensity. The radical attaches to the interesting site of a studied molecule, and as result parameters of EPR spectrum become depending on orientation of this molecule in a magnetic field. As the motion of such large molecule as protein is much slower, then the motion of the free radical in a solution, the hyperfine interaction in a spin label is not averaged on all orientations and it remains anisotropic. It allows us to acquire the information on structure and dynamics of a protein fragment in which the label is inserted. Besides, the fact is used that $2p_\pi$-orbital of the free radical possesses polarity and it is possible to get data on the polarity of an environment from the spectrum of a spin label. The technique of a spin label, for example, has been used to investigate the relation between biological functions of phospholipid bilayer and its structure.

Further Readings

Berliner L, Eaton GR, Eaton SS, Hagen WR (2008) Biomolecular EPR spectroscopy. Wiley, New York ISBN: 1420059572
Gerson F (1970) High resolution E. S. R. spectroscopy. Wiley, New York
Hoff AJ (1989) Advanced EPR: applications in biology and biochemistry. Elsevier, Amsterdam
Saifutdinov RG, Larina LI, Vakul'skaya TI, Voronkov MG (2001) Electron paramagnetic resonance in biochemistry and medicine. Kluwer Academic/Plenum Publishers, New York, p 268 ISBN 0-306-46531-0

Part IV
Double Resonances and Polarization Transfer

Chapter 15
Double Resonances

15.1 Preliminary Remarks

A very important aspect of the physics of magnetic resonance are so-called *double resonances*. The essence of the double resonance experiments consists in the influence on the selected spin resonance transition with continuous or pulsed alternating magnetic field and the registration of the response at another transition in the same or another spin system. The most typical double resonances are those which involve two interacting spin systems (A and B). Spin systems A and B are usually two nuclear systems with different resonance frequencies ($\omega_A \neq \omega_B$) or nuclear and electronic spin systems which frequencies ω_n and ω_e are very different: $\omega_\text{n} \ll \omega_\text{e}$. These are so-called *double nuclear–nuclear* and *electron-nuclear resonances*. At the present time investigators use also *double nuclear quadrupole–nuclear magnetic resonance, spin–photon (optical pumping) and spin–phonon (acoustic) resonances*.

The first double resonance experiments were carried out by R. V. Pound in 1950 on the nuclei ^{23}Na in the crystal $NaNO_3$. In this crystal the nucleus ^{23}Na with the spin $I = 3/2$ are situated in the inhomogeneous electric field with an axially symmetric gradient and due to the quadrupole interaction the NMR line splits into three components corresponding to transitions between levels with different values of the quantum number m_I (see Chap. 8):

$$3/2 \longleftrightarrow 1/2, \quad 1/2 \longleftrightarrow -1/2, \quad -1/2 \longleftrightarrow -3/2.$$

At the saturation of any of these transitions under the influence of the strong rf field the change of the populations of spin states occurs and the intensities of other lines will also changed. For example, in the case of the saturation of the transition $3/2 \longleftrightarrow 1/2$ the intensity of the NMR signal, corresponding to the transition $1/2 \longleftrightarrow -1/2$, increases in 5/3 times.

The interest of researchers to double resonances caused by several factors. First, the double-resonance technique allows in many cases drastically increase the intensity of weak signals of nuclear magnetic resonance; second, the double resonances

V. I. Chizhik et al., *Magnetic Resonance and Its Applications*,
DOI: 10.1007/978-3-319-05299-1_15,

Table 15.1 The normalized (relative to protons) NMR signals (S) in the equal magnetic field B_0 and the time of an experiment (T), which is necessary to spend to achieve the same signal-to-noise ratio

Nucleus	Natural abundance, %	S	T
^{1}H	$\simeq$100	$\equiv 1$	$\equiv 1$
^{2}H	0.016	$6.0 \cdot 10^{-6}$	$3 \cdot 10^{10}$
^{13}C	1.1	$3.5 \cdot 10^{-4}$	$8 \cdot 10^{6}$
^{15}N	0.36	$1.2 \cdot 10^{-5}$	$7 \cdot 10^{9}$
^{19}F	100	0.86	1.4
^{31}P	100	0.10	90
^{39}K	93	$2.2 \cdot 10^{-3}$	$2 \cdot 10^{5}$
^{43}Ca	0.130	$3.0 \cdot 10^{-5}$	$1 \cdot 10^{9}$
^{109}Ag	49	$2.3 \cdot 10^{-4}$	$2 \cdot 10^{7}$

make it possible to simplify the NMR spectra by "decoupling" of interactions which determine the multiplicity of spectra; third, they lead to the appearance of new features in NMR spectra that promotes to a more reliable their interpretation.

The formula for the amplitude of the NMR signal (2.17) and the data presented in Table 15.1 allow the comparison of the relative intensity of signals from different nuclei according to their properties and natural abundance. For simplicity, we assume that radiotechnical conditions of the registration for all nuclei are identical, although in the case of the registration of low-frequency resonances it is possible to use coils with many turns, that, in principle, can improve the sensitivity of devices. Table 15.1 contains examples of normalized relative to protons NMR signals (S) in the equal magnetic field B_0. It is clear from the table that decrease of a signal amplitude dramatically increases the time of an experiment (T), which is necessary to spend to achieve the same signal to noise ratio.

The improvement of sensitivity is very important for the registration of the resonances of nuclei, which gives weak signals due to small gyromagnetic ratios and/or low natural abundance. Regardless of the cause, that determines weak signals, in such cases one speaks about the registration of resonances of *rare nuclei*. So far, there are a number of isotopes which are difficult to observe because they cannot be obtained with large enrichment. In this connection the development of methods for the detection of the magnetization of nuclei with small magnetic moments and with a low natural abundance is the actual problem of modern experimental physics.

The theory predicts an increase of the sensitivity using the double-resonance NMR for two-four or even more orders of magnitude compared to conventional methods. The gain in the sensitivity of the registration of resonances of rare nuclei can be obtained due to the use of "reservoir" of energy of widespread nuclei with strong signals. The latter can be realized in two ways: firstly, using the indirect method, in which the spectra of rare nuclei are detected via the signals from widespread nuclei, and secondly, using the direct method in which for the signal amplification the rare nuclei are polarized by means of the double resonance. The indirect method has

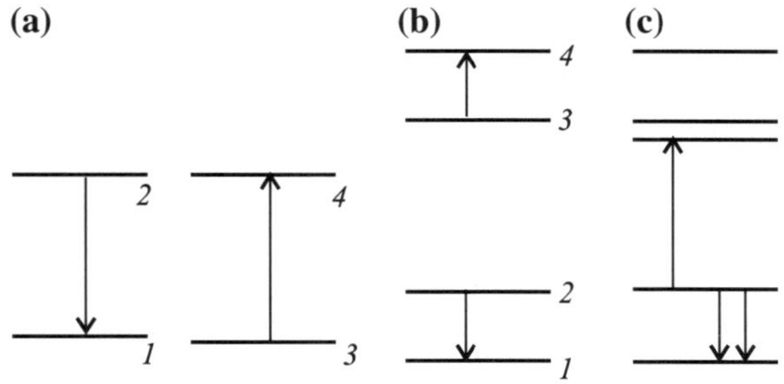

Fig. 15.1 To the explanation of the process of cross-relaxation: transitions in two different spin systems (**a**) and variants of transitions in the single spin system (**b, c**)

higher sensitivity, but the direct one allows us to apply multipulse sequences for the study of high-resolution spectra in solids and a number of other innovative methods for the registration of NMR spectra.

For experiments on double resonances it is very important to consider processes of, so called, *cross-relaxation*, which provides the exchange of energy between two magnetic transitions belonging to different but "mixed" systems of spins or to a multi-level system. These processes were intensively studied in connection with the invention of masers in which the electron paramagnetic resonance in solids is used (see Chap. 19). The peculiarity of the cross-relaxation is that due to this process the thermal equilibrium is established between the different spin transitions (the alignment of spin temperatures). In this case the transitions are realized like it is shown in Fig. 15.1. The number of transitions per time unit is proportional to the probability w_{cr} and for the number of particles in the initial energy levels, i.e., for the population, for example, of the first energy level in Fig. 15.1a or b, we have:

$$\frac{dn_1}{dt} = w_{\mathrm{cr}}(n_2 n_3 - n_1 n_4).$$

Usually the time constant $T_{12} = 1/w_{\mathrm{cr}}$ is introduced for a rough estimation of the time during which the equilibrium between two spin systems or between two transitions in a single system is adopted. The effect of cross relaxation is the most important when $T_2 \ll T_{12} \ll T_1$. For the cross-relaxation is sufficient the approximate (not exact) equality of the transition frequencies but with the increase of difference in the frequencies of two transitions the probability w_{cr} drops sharply (it requires the overlap of spectral lines).

Figure 15.2 shows the effect of the cross-relaxation on the apparent spin-lattice relaxation time T_1 of the nuclei ^{19}F in the solid fluorochlorobenzene ($C_6Cl_3F_3$) in conditions of the equality of the resonance frequencies of the nuclei ^{19}F and ^{35}Cl. The fact is that the resonance frequency of the nuclei ^{19}F ($I = 1/2$) is determined only by the external magnetic field B_0 but the resonance frequency of the nuclei ^{35}Cl depends both on the magnetic field and the energy of quadrupole interaction (see Chap. 8):

Fig. 15.2 The effect of the cross-relaxation on the apparent spin–lattice relaxation time T_1 of the nuclei ^{19}F in the solid fluorochlorobenzene ($C_6Cl_3F_3$)

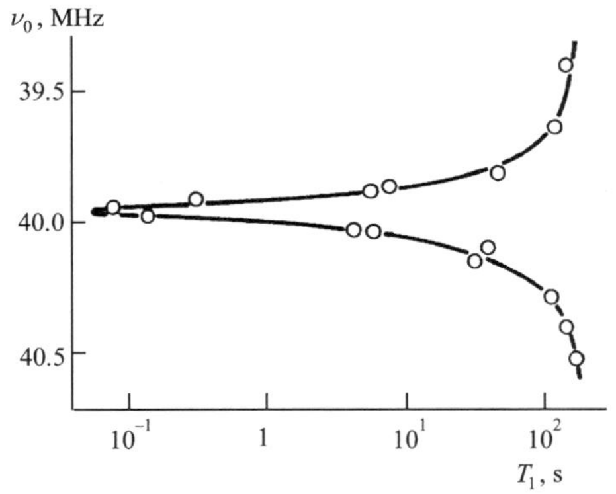

$$\omega_0^{F} = \gamma_F B_0; \qquad \omega_0^{Cl} = \gamma_{Cl} B_0 + \omega_Q.$$

If $\gamma_F > \gamma_{Cl}$ (that is realized in this case) and $B_0 = \omega_Q/(\gamma_F - \gamma_{Cl})$ then the alignment of frequencies is realized (for the example, when $\nu_0 \simeq 40\,\text{MHz}$).

Similar processes occur in the case of multiquantum transitions when the splitting in spectra are characterized by multiple frequencies (see Fig. 15.1c): $\nu_{01} \simeq n\nu_{02}$ (n is the integer), but with the increase of n the probability of cross-relaxation is dramatically reduced.

A variety of the double resonance techniques are currently very high. In this book we will describe only some of them.

15.2 Overhauser Effect

Let us turn to the equation describing the behavior of the longitudinal component magnetization of the spin I under the influence of the interaction with the spins S through the dipole-dipole or scalar mechanisms (see Chap. 3):

$$\frac{d\langle I_z\rangle}{dt} = -\rho_1(\langle I_z\rangle - I_0) - \sigma_1(\langle S_z\rangle - S_0), \tag{15.1}$$

where I_0 and S_0 are the equilibrium values of $\langle I_z\rangle$ and $\langle S_z\rangle$ with

$$\langle I_z\rangle \sim [(n_1 + n_2) - (n_3 + n_4)], \quad \langle S_z\rangle \sim [(n_1 + n_3) - (n_2 + n_4)],$$

and the constant coefficients in (15.1) are given by (see Sects. 3.9–3.11)

$$\rho_1 = w_0 + 2w_1 + w_2; \qquad \sigma_1 = w_2 - w_0. \tag{15.2}$$

The similar equation can be obtained for $\langle S_z \rangle$ by the symmetrical replacing of the symbols with $\rho_1' = w_0 + 2w_1' + w_2$. Mind that w_0, w_1', w_1 and w_2 are the probabilities of relaxation transitions with the change in the total magnetic quantum number $\Delta m = 0, \pm 1, \mp 1$ and ± 2, respectively. We assume that the resonance of the spins S is saturated under the influence of a strong rf field, i.e., there is the alignment of the populations of their upper and lower levels, so $\langle S_z \rangle = 0$. Then from (15.1) we obtain

$$\langle I_z \rangle = I_0 + \frac{\sigma}{\rho} S_0. \tag{15.3}$$

Equation (15.3) predicts a change in the magnetization (polarization) of the system spins I under the effect on the related system of spins S. This result is called *Overhauser Effect* after the name of the scientist, who for the first time considered the possibility of nuclear polarization in metals via electrons (A. W. Overhauser. Phys. Rev., 1953, v. 89, p. 689; v. 91, p. 476; v. 92, p. 411).

The solution of Eq. (15.1) together with a similar equation for $\langle S_z \rangle$ with initial conditions $\langle I_z(0) \rangle - I_0 = 0$, $\langle S_z(0) \rangle - S_0 = 0$ and $\rho = \rho'$ (the condition $\rho \neq \rho'$ leads to more complicated formulas, without any fundamental differences) is

$$\begin{aligned} \langle I_z(t) \rangle - I_0 &= A\left[e^{-(\rho+\sigma)t} - e^{-(\rho-\sigma)t}\right], \\ \langle S_z(t) \rangle - S_0 &= A\left[e^{-(\rho+\sigma)t} + e^{-(\rho-\sigma)t}\right]. \end{aligned} \tag{15.4}$$

Expressions (15.4) describe the so-called *transient Overhauser effect*. The effect is manifested in the increase of the magnetization, if the signs of gyromagnetic ratios are of the same, and in the decrease if the signs are opposite.

In case of the incomplete saturation of the resonance of spins S, formula (15.3) turns to

$$\langle I_z \rangle = I_0 - \frac{\sigma}{\rho}\left[\langle S_z \rangle_\infty - S_0\right] \tag{15.5}$$

($\langle S_z \rangle_\infty$ is the stationary value of $\langle S_z \rangle$). Introducing the coefficient of spin polarization for I: $\alpha = \langle I_z \rangle_\infty / I_0$ and the degree of saturation of the resonance for spins S: $s = (S_0 - \langle S_z \rangle_\infty)/S_0$, rewrite expression (15.5):

$$\alpha = 1 + \frac{\sigma}{\rho} s \frac{S_0}{I_0}.$$

The ratio σ/ρ depends on the type of interaction and can be calculated using formulas in Sects. 3.9–3.11. It follows from them that for two interacting spins ($I = S = 1/2$) through the dipole-dipole mechanism the probabilities of relaxation transitions are proportional to the following values (the coefficients of proportionality in all cases are the same):

$$w_1 \propto \frac{3\tau_c}{1 + \omega_{0I}^2 \tau_c^2}, \quad w_0 \propto \frac{2\tau_c}{1 + (\omega_{0I} - \omega_{0S})^2 \tau_c^2}, \quad w_2 \propto \frac{12\tau_c}{1 + (\omega_{0I} + \omega_{0S})^2 \tau_c^2}.$$

If $\omega_{0I}^2\tau_c^2 \ll 1$, $\omega_{0S}^2\tau_c^2 \ll 1$ we have $w_1 : w_0 : w_2 = 3 : 2 : 12$ and, consequently, $\sigma/\rho = 1/2$. For the scalar relaxation mechanism the only probability w_0 is significant (see Sect. 3.11) and $\sigma/\rho = -1$.

Taking into account the Curie formula for the magnetization it can be written in conditions of the complete saturation of the resonance of the spins S:

$$\alpha = 1 + \frac{\sigma}{\rho}\left|\frac{\gamma_S}{\gamma_I}\right|. \tag{15.6}$$

Equation (15.6) shows that the most powerful Overhauser effect can be achieved with $\gamma_S \gg \gamma_I$.

Especially impressive dynamic polarization of nuclear spins I occurs if the spins S belongs to electrons, but for the realization of this case a substance must contain paramagnetic components. Then $\alpha \gg 1$ and with $\sigma/\rho = -1$ we have

$$\langle I_z\rangle_\infty / I_0 = |\gamma_S/\gamma_I|.$$

This result was first predicted by A. W. Overhauser for metals.

Generally speaking, the Overhauser effect can really be registered only when the nuclear magnetic relaxation processes, which initiate the transitions of spins regardless from each other, cannot suppress (obscure) the relaxation transitions with simultaneous flippings of both spins (probablies w_0 and w_2), because otherwise $\sigma/\rho \to 0$.

15.3 Variants of Pulse Double Resonances

In this section we consider variants of pulse double resonances in the most adapted for the detection of weak signals in solids, when problems of the instrumental sensitivity of the NMR method arise with peculiar acuity.

15.3.1 Double Resonance Spin Echo

One of the first realizations of a pulsed double-resonance technique was proposed in 1958 by D. E. Kaplan and E. L. Hahn who called it as *spin-echo double resonance* (SEDOR). This method uses the standard pulse sequence 90°—τ—180°—τ—echo. If near a nucleus with a spin I, a nucleus with a spin S is located, then a neighboring nucleus creates a local field, that results in the broadening of resonance line like in the case of the inhomogeneity of the static magnetic field B_0. However, the conditions of dephasing and phasing of the nuclear magnetization of the spins I in successive time intervals τ can be changed by applying the 180-degree pulse at the frequency ω_{0S} at the same time as the 180-degree pulse at the frequency ω_{0I}, that results in the change of the direction of the local fields generated by the spin S. Therefore during the second interval τ the dephasing of the nuclear magnetization of the spins I (instead of the usual phase-locking) will continue and the signal echo is reduced in proportion to the concentration of the spin S, which are near the nuclei with spin I. This technique was successfully applied to study the double-resonance ^{59}Co–^{63}Cu in dilute alloys of cobalt in copper.

15.3.2 An Indirect Method for Detecting Nuclear Magnetic Resonance on the Basis of Cross-Relaxation Effects in Double Resonances

It is indicated before that for the cross-relaxation process the approximate equality or the multiplicity of the transition frequencies is required. However, gyromagnetic ratios of different nuclei are such that this condition is almost impossible to fulfill, and therefore it is impossible to organize the effective contact of spins I and S in laboratory coordinate system. An ingenious solution of this problem was suggested by S. R. Hartmann and E. L. Hahn (Bull. Amer. Phys. Soc., 1960, v. 45, p. 498), who showed that one could choose experimental conditions for the contact between the spins I and S in rotating coordinate systems. An alternating magnetic field effects on the nuclei only if its frequency is close to the resonance one. Therefore each of the two fields—B_{1I} at the frequency ω_{0I} and B_{1S} at the frequency ω_{0S}—will affect only on one spin system. Then every type of spins can be considered in its own rotating frame, and for the total set of spins we obtain "a double-rotating coordinate system", in which $\Delta\omega_i = 0$ for each spin system and one can be write in it:

$$\Omega_I = \gamma_I B_{1I}; \qquad \Omega_S = \gamma_S B_{1S}. \tag{15.7}$$

It follows from Eqs. (15.7) that the nutation frequencies in rotating systems coordinates can be made equal changing the magnitudes of the fields B_{1I} and B_{1S}, i.e. we obtain the equality:

$$\Omega_I = \gamma_I B_{1I} = \gamma_S B_{1S} = \Omega_S, \tag{15.8}$$

which is known as the *Hartmann–Hahn condition*. Under this condition the intensive cross-relaxation in the rotating frame is realized between the spins I and S.

It is known that the spectrum (and hence the line broadening) in the case of interaction of different nuclei is determined by the term A of the Hamiltonian describing the dipole-dipole interaction

$$\widehat{\mathcal{H}}_{\text{d}}{}^{A} = \gamma_I \gamma_S r^{-3} \hbar^2 \hat{I}_z \hat{S}_z (1 - \cos^2\theta). \tag{15.9}$$

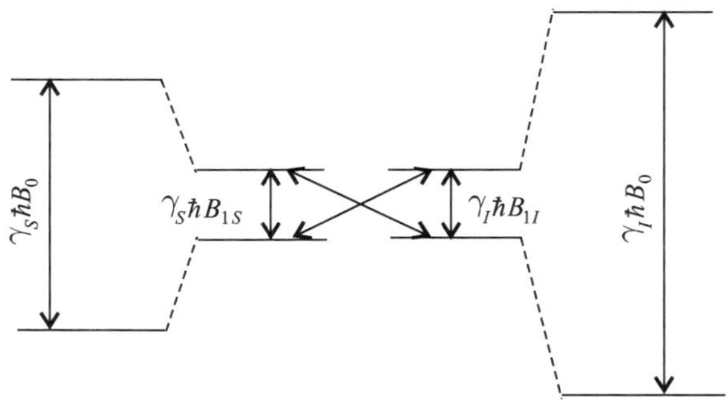

Fig. 15.3 The scheme of energy levels in the laboratory and rotating frames

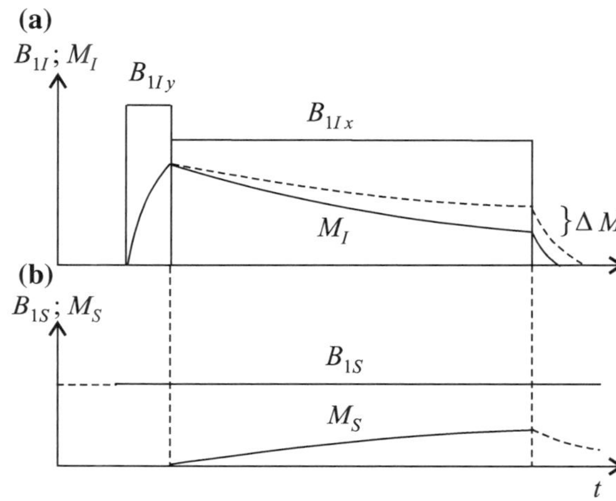

Fig. 15.4 The scheme of the experiments on the realization of the pulsed double resonance (between spins I(**a**) and S(**b**))

Since the axis of the rotating and laboratory coordinate systems coincide, then this Hamiltonian does not change under the transformation of spin variables in rotating frames. External alternating fields force the spins to reorient. Due to the interaction (15.9) the reorientations of the spins I and S are bound and especially in the case if the condition (15.7) is fulfilled. The process of the cross-relaxation in the double-rotating coordinate system is illustrated in Fig. 15.3.

Consider the main stages of the experiments on the realization of the pulsed double resonance (Fig. 15.4). The pulse rf field B_{1Iy} prepares the spin system I so that its magnetization $\vec{M}$, which initially is directed along the axis z, becomes oriented along the axis x of the coordinate system rotating with the frequency $\omega_{0I} = \gamma_I B_0$. Then the conditions of the spin-locking for the spin I are realized (Fig. 15.4a, see also Fig. 3.10). To describe the ongoing processes of the spin system, it is convenient to use the concept of the spin temperature (Sect. 3.2). The value of the equilibrium magnetization is determined by the temperature of the lattice T_L by the formula Curie:

$$M_{0I} = C_I \frac{B_0}{T_L} = \beta_L C_I B_0, \tag{15.10}$$

here $\beta_L = 1/T_L$; C_I is the Curie constant for an ensemble of nuclei with spin I. Since after the "locking" of the magnetization $\vec{M}_{0I}$ by the strong alternating field its vector is along the field $\vec{B}_{1Ix}$ but the value of M_{0I} corresponds to the induction field B_0, the magnetization can be described with inverse spin temperature $\beta_{I\rho}$:

$$M_{0I} = \beta_{I\rho} C_I B_{1Ix}. \tag{15.11}$$

From (15.10) and (15.11) we obtain

$$\beta_{I\rho} = \beta_L \frac{B_0}{B_{1Ix}}, \tag{15.12}$$

i.e. the spin system s I is strongly "cooled" in the rotating frame, because the $B_0/B_{1Ix} \gg 1$. If besides to saturate the resonance transition of the spin S with the help of the strong field B_{1S}, then for this spin system we obtain $\beta_{S\rho} = 0$ (the spin temperature becomes infinite), because the magnetization of the M_S will be absent (see the initial part of Fig. 15.4b). When the Hartmann–Hahn condition (15.7) "switches on", the contact between the spin systems is realized: the system of the spin I, which was at a "low" temperature, will "heated" by the "hot" spins S. Figure 15.4a schematically shows the decrease of the magnetization of the spin system I without (dotted line) and with the interaction with the spin system S. At the same time, the system of spins S "cools" that leads to the appearance of the magnetization M_S (see in Fig. 15.4b). In other words this process leads to leveling the spin temperatures of the systems. The decrease of the magnetization of the spins I is a measure of the magnetization of spins S. The magnitude of ΔM_I gives the information on the spectrum of the spin system S. This is the *indirect method of the detection of the magnetization and spectrum of rare spins*.

The quantity ΔM_I is determined by measuring the amplitude of free induction signals after switching off the long pulse B_{1Ix} (see Fig. 15.4a). The effect after one cycle can be expressed by the relation

$$M_I(\tau) = M_{0I}(1 - R)\exp[-\tau/T_{1\rho I}],$$

where R is the coefficient taking into account the effect of the double resonance. At low temperatures the time $T_{1\rho I}$ is large and $M_I \simeq M_{0I}(1 - R)$. The value R is of the order of the ratio of spin concentrations N_S/N_I, because

$$R = \frac{\gamma_S^2 S(S+1) N_S}{\gamma_I^2 I(I+1) N_I}.$$

It is interesting to note that the two main causes leading to low sensitivity of steady-state nuclear magnetic resonance in solids—great values of T_1 and small values of T_2—are positive factors in the method of pulsed nuclear magnetic double resonance.

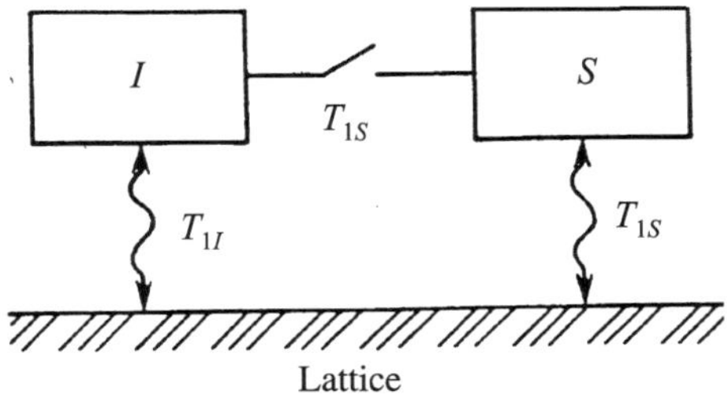

Fig. 15.5 The schematic representation of the "reservoir" of the widespread spins I and the "reservoir" of the rare spins S, and the relation between the "reservoirs"

The heat capacity of rare spins S is very small and it is necessary to prevent its "cooling". To maintain its high spin temperature one can invert the phase of the field B_{1S} (this technique does not allow the spin system to come to the equilibrium state).

Applying short pulses on the nuclear spins S more effectively "heats" them and it allows ones to obtain a high sensitivity with a smaller number of cycles. In this modification, instead of changing the phase of the field B_{1S} at 180° one uses switching on and off the field B_{1S}. During the pulse of the field B_{1S} the exchange of energy between the spin system exists (the Hartmann–Hahn condition is realized) and the magnetization I due to the interaction is reduced. After switching off the field B_{1S} the magnetization of the system S relaxes rapidly with the time constant of T_2 and the magnetization M_I continues to relax with the time constant T_ρ. During the next pulse of the field B_{1S} the decrease of the magnetization M_I accelerates again due to the interaction with spin system S and so on. After k pulses the magnetization M_I reduces according to the law $M_I(k) = M_I \exp(-kR)$.

Figure 15.5 shows the schematic representation of (i) the "reservoir" of widespread spins I and "reservoir" of the rare spins S, connected with the lattice by interactions, characterized by relaxation times T_{1I} and T_{1S}, respectively, (ii) the relation between the two "reservoirs", which may be varied in accordance with experimentalist wishes by changing the amplitude and frequency of rf fields ("switching" of the Hartmann–Hahn condition).

15.3.3 A Direct Method for Recording the NMR Signals.

The described modification of the double resonance allows us also to register the NMR signals of rare nuclei. For the direct observation of the magnetization of the spins S at their resonance frequency it is necessary to switch "heating field" off during the observation time. The information is contained in free induction signals, the Fourier transform of which is the spectrum of nuclear magnetic resonance of spin S. The scheme of the experiment, proposed by A. Pines, M. G. Gibby and J. S. Waugh (J. Chem. Phys., 1972, v. 56, p. 1776) is shown in Fig. 15.6.

The equilibrium magnetization in the field of $\vec{B}_0$ of the spin system I is exposed, as before, by the spin-locking for "cooling" the spins in the rotating frame. The spin system S is irradiated by a series of k pulses of the field B_{1S}, which duration t_S is

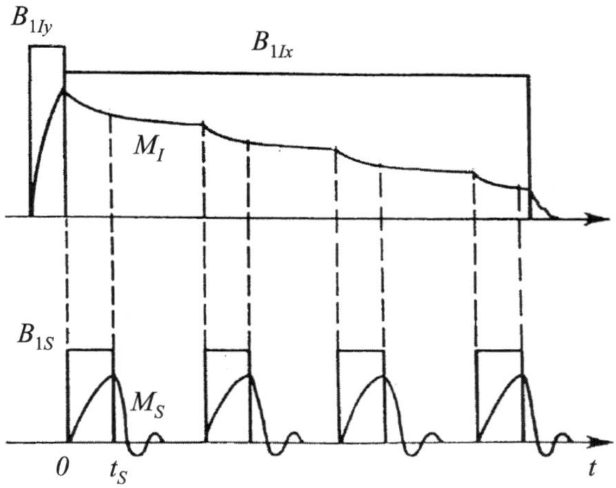

Fig. 15.6 The scheme of the experiment of A. Pines, M. G. Gibby and J. S. Waugh

sufficient to establish the equilibrium of spin temperatures in the systems I and S. The picture is made under the assumption that the time $T_{1\rho I}$ is enough big, so that in the absence of a pulse B_{1S} the magnetization M_I practically does not change. The decrease M_I during the t_S is due to cross-relaxation. As to the of magnetization M_S, it increases during the each pulse of the field B_{1S} and relaxes after it.

In the direct method the free induction signals of the spins S are detected after each pulse of the field B_{1S}. Since the increase in the spin polarization after a single pulse is not yet sufficient for the registration of NMR signals, one must use their accumulation. The number of accumulated signals within a single cycle is limited by (i) the relaxation time $T_{1\rho I}$, which does not allow us to increase the time experiment, (ii) the cross-relaxation time T_{IS}, which does not allow us to apply very short pulses of the field B_{1S}, and (iii) the time of the observation of free precession, i.e. T_{2S}. The accumulation of signals during different cycles is practically difficult due to the large time of the achievement of the thermal equilibrium in the spin system I (the spin–lattice relaxation time in solids can reach several hours).

To remove the last restriction one uses different methods: the temperature modulation of the spin–lattice process, the polarization in a relatively weak field, the addition of paramagnetic impurities, the dynamic polarization and so on. However, the most effective method is the polarization transfer from a system of quadrupole nuclei X, if they are present in a sample with sufficient concentration, to the nuclei I.

Methods for obtaining the thermal contact between the nuclei I and X is well known. One of the examples is described earlier in this chapter: it is enough to get match the Larmor frequencies of spins I with one of frequencies of the quadrupole nuclear resonance of the spins X, i.e. to obtain the *level crossing* for I and X. Practically one uses additional modulation of the magnetic field B_0, that provides the more reliable periodic level crossing. Since the cross-relaxation time T_{IX} is much smaller than the spin-lattice relaxation time T_{1I} and the spin-lattice relaxation of quadrupole nuclei is very fast (T_{1X} is much smaller than T_{IX}), then the thermal

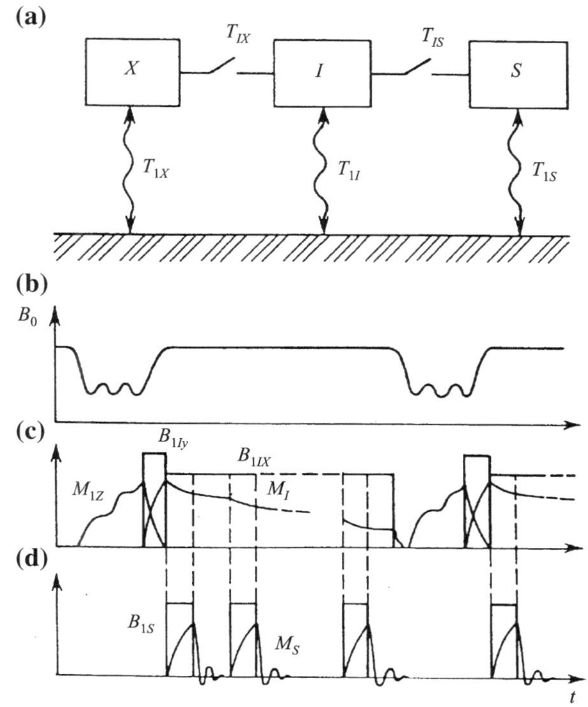

Fig. 15.7 The combination of the double pulse resonance with the "quadrupole polarization" by M. Schwab and E. Hahn: **a** the thermodynamic scheme of the experiment; **b** the magnetic field modulation; **c, d** the behaviour of the magnetizations of spins I and S

equilibrium of the spins I with the lattice, carried out through "the channel X", is achieved during the order of several T_{IX} (tens of seconds).

The combination of the double pulse resonance with the "quadrupole polarization" was realized by M. Schwab and E. Hahn in 1970 (J. Chem. Phys., 1970, v. 52, p. 3153), who for the first time registered the nuclear magnetic resonance ^{13}C at natural abundance in the solid phase. The thermodynamic scheme of the experiment is shown in Fig. 15.7a. The "switching of thermal contacts" is shown in Fig. 15.7b. The difference from scheme, shown in Fig. 15.6, is in the appearance of one more phase: the rapid polarization of the spins I through the interaction with the spins X (the contact T_{IX} in Fig. 15.7a will be closed due to a specific choice of the field strength B_0, that ensures the level crossing for spins I and X). The rest phases of the experiment are similar to ones earlier discussed (see Fig. 15.7c, d). The described method is called as *cascade amplification of the NMR spectrum of spins in small concentrations*.

As already mentioned, the most promising and interesting methods are the direct detection of signals from the spin S, since they open the way to high-resolution spectroscopy in solids. It is obvious that instead of the registration of the FID from the spins S after the pulse B_{1S} it is possible to observe and accumulate signals during pulse sequences, narrowing of NMR lines in solids. Figure 15.8b shows the spectrum of ^{13}C in p-dichlorobenzene (Fig. 15.8a), obtained via the Fourier transform of signals accumulated in the result of 100 cycles of the polarization during 15 min with the

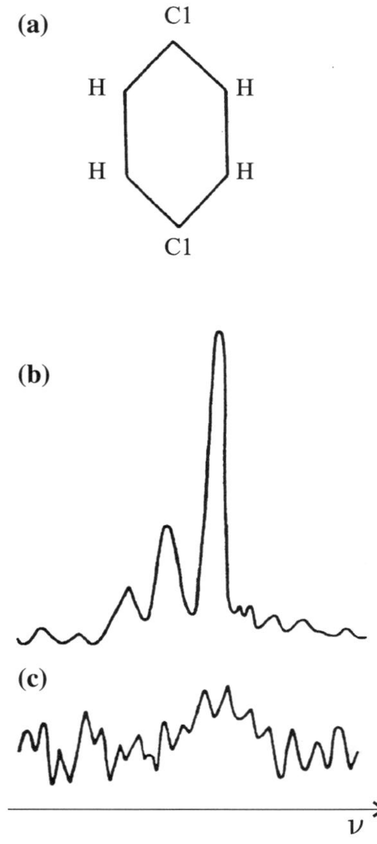

Fig. 15.8 The spectrum of ^{13}C in *p*-dichlorobenzene **a**, obtained in different experiments **b, c** (see the details in the text)

level crossing for the I and X systems and 20 contacts I—S per cycle (mixing time 20 μs). For comparison, Fig. 15.8c shows the same spectrum obtained during 5-h accumulation but without contacts with the quadrupole nucleus X (with the same 20 contacts I—S).

Thus, at present there is a real opportunity for development NMR spectroscopy of solids with improved resolution even for very low concentrations of nuclei, since substances, containing two, three, and even more kinds of spins, are widespread.

15.4 Double Nuclear Quadrupole Resonances

The sensitivity of direct NQR methods decreases with frequency decreasing, and even at frequencies of about 1 MHz observation of NQR signals is not a trivial task. Therefore, for a number of nuclei with small quadrupole moments the direct NQR study is impossible. In addition, very often the studied objects contain very low concentration of quadrupole nuclei and even if they have a rather high quadrupole moment as soon as the signal intensity is determined by the number of nuclei in the selected probe the sensitivity of direct methods is not sufficient enough. So in both cases the more adequate approach is applying indirect methods to register NQR signals. Indirect NQR methods is an intensive developing and very perspective area of magnetic resonance methods, as in a number of cases it is a unique tool for a close and subtle analysis of high-technology mixtures. A detail description of indirect methods can be found in several review papers.[1]

Double nuclear quadrupole resonance (DNQR) is one of indirect NQR methods. This method is not so widely used as it asks that the studied compound contains quadrupole nuclei of two different species: I—nuclei with a high resonance signal and S— nuclei under study, and there are dipole–dipole interactions between them. The main idea is that the resonance energy adsorption by one system (S) should affect the other (I). If the distances between spins of the both spin systems is such as $r_{II} > r_{IS}$, then exiting nuclei of the I species, the signal decay is mainly defined by the interaction between the I and S spin systems. Continuous irradiation on the resonance frequency of the S nuclei, ν_S, results in averaging of the interaction Hamiltonian and decreases the spin–echo decay of the I nuclei. The difference in the amplitude of the spin–echo signal of the I nuclei under irradiation at the frequency ν_S and without it provides information about the energy adsorption by the S nuclei. An example of such a system is $NaClO_3$, where ^{35}Cl and ^{23}Na nuclei belong to the I and S species, respectively.

DNQR methods differ by the choice of reference spin system, from which the NQR signal is recorded, and by means used to realize the energy exchange between the I and S spin systems. As in NMR, DNQR techniques can be visually described by a thermodynamic analogy. Figure 15.5 represents schematically the thermal exchange between "reservoirs" containing the I and S spin species and the lattice. Regardless of the nature of interactions, which result in energy level splitting of the system, it is possible to mark out three main stages of any DNQR experiment:

- to cool the I spin system (see Sect. 15.3.2);
- to heat the S spin system (for example, by a rf field) and make the I and S systems interact between each other;
- to measure the changes in the I spin system.

Theoretical conception used for DNQR description is based on two conceptions: rotating frame and spin temperature.

The rotating frame was introduced in Sect. 1.5 when deriving the Bloch equations describing the magnetization vector behavior within the classical representation (1.55)–(1.64).

In the absence of external magnetic field the Hamiltonian of the quadrupole spin system is

$$\widehat{\mathcal{H}} = \widehat{\mathcal{H}}_Q + \widehat{\mathcal{H}}_1 + \widehat{\mathcal{H}}_d, \tag{15.13}$$

[1] D. Schmid. Magnetic double resonance principles and application in solid state physics. Springer Tracs in Modern Physics (Springer-Verlag, 1973), vol. 68, pp. 1-75 D.T. Edmonds, Phys. Report C 29, 233 (1977). V.S. Grechishkin, A.A. Shpilevoi, Phys. Usp. 39, 713 (1996)

where $\widehat{\mathcal{H}}_Q$ is the quadrupole interaction Hamiltonian in the form of (8.2) (for an axial EFG tensor), $\widehat{\mathcal{H}}_1$ is the Hamiltonian describing the interaction with the rf field $\vec{B}_1 \cos(\omega t)$, which is

$$\widehat{\mathcal{H}}_1 = -\gamma\hbar\hat{I}_x B_{1x}\cos(\omega t),$$

if the field $\vec{B}_1$ is along the x–axis, $\widehat{\mathcal{H}}_d$ is the dipole–dipole interaction Hamiltonian (1.131). Transformation from the laboratory into rotating frame can be made by using the unitary operator

$$\hat{U} = \exp(\mathrm{i}\hat{A}t), \quad \text{at} \quad \hat{A} = \frac{1}{2}\omega\hbar\sum_i\left[(\hat{I}_z{}^i)^2 - \frac{1}{3}\hat{\vec{I}}(\hat{\vec{I}}+1)\right]. \tag{15.14}$$

The Hamiltonian (15.13) modified by (15.14) is

$$\widehat{\mathcal{H}}' = \hat{U}\widehat{\mathcal{H}}\hat{U}^* = \frac{1}{2}\hbar(\omega_Q - \omega)\sum_i\left[(\hat{I}_z{}^i)^2 - \frac{1}{3}\hat{\vec{I}}(\hat{\vec{I}}+1)\right] + \hbar\omega_1\hat{I}_x + \widehat{\mathcal{H}}'_d,$$

where $\widehat{\mathcal{H}}'_d$ is the secular part of the dipole–dipole interaction Hamiltonian that commutes with $\widehat{\mathcal{H}}_Q$. Such a transformation into the rotating frame effectively "reduces" the nuclear quadrupole interactions and increases the role of the $\omega_1\hat{I}_x$ and $\widehat{\mathcal{H}}_d$ terms. As opposed to NMR, this transformation does not correspond to a real mechanical rotation of the sample about an axis.

The spin temperature conception was introduced earlier in Sect. 3.2 to describe relaxation processes in a two-level system in a constant magnetic field $\vec{B}_0$. Another notion important in double magnetic resonance methods is the *adiabatic demagnetization.* If the B_0 value is reduced to B'_0 within the time shorter than T_1, the entropy of the spin system does not change. However, the spin temperature decreases, as the entropy is a function of the B_0/θ_S ratio, and the new spin temperature value is $\theta'_S = (B'_0/B_0)\theta_S$. If the B'_0 value is about internal local fields (i.e. 10^{-3} T), then for a while the spin temperature θ'_S can be lower the lattice temperature. In a zero magnetic field, the spin system comes to an equilibrium within the time always less than T'_1 in high magnetic fields (magnetic fields, in which the spin system is embedded, correspond to the internal local fields values due to spin dipole moments). Such a process to reduce the spin temperature is the adiabatic demagnetization. This process is completely reversible and the magnetic field B_0 switching returns the system in the initial state.

15.4.1 Spin–Echo Double Nuclear Quadrupole Resonance

Spin–echo double nuclear quadrupole resonance is one of the first realization of DNQR. The experiment diagram is similar to those represented in Fig. 16.2. The studied sample must contain quadrupole nuclei of two species: I and S, and one applies to it three pulses: 90° and 180° pulses at the NQR frequency of the I spin system, and simultaneously with the second pulse another 180° pulse at the NQR frequency of the S spin system, which can be unknown. If the current frequency of the latter pulse coincides with the NQR frequency of the S spin system ω_S, the spin–echo signal amplitude observed at I nuclei, decreases with respect to the usual two-pulse echo. The matter is that the second 180° S-pulse inverts the dipole magnetic field that changes the precession rate of individual spin packets, and therefore, the echo amplitude (before application of the 180° S-pulse the precession rate of spin packets was determined by the sum of the local and dipole magnetic fields, whereas after the pulse it is determined by their difference). Thereby, the signal reducing evidences the resonance energy adsorption by the S spin system. The sensitivity gain at such an indirect observation of the resonance adsorption by the "weak" nuclear system as compared to the direct method can be defined by the following expression:

$$\frac{\Delta V_I}{V_S} = \frac{1}{e} \frac{\mu_I \omega_I \gamma_I^2}{\mu_S \omega_S \gamma_S^2} \frac{S_{2,IS}}{S_{2,II}},$$

where ΔV_I is the decreasing of the echo signal amplitude for the I nuclei; V_S is the signal amplitude for the S nuclei; μ_I and μ_S are magnetic moments of the I and S nuclei, respectively; $S_{2,II}$ is the second moment of the adsorption line for I nuclei, $S_{2,IS}$ is the second moment of the adsorption line for I nuclei due to their interaction with S nuclei. As it is seen the sensitivity is higher the greater the difference between the resonance frequency of the I and S nuclei is.

It is to be noted that the nuclei S must have a half-integer spin number, or an integer one but the asymmetry parameter of the EFG tensor η must be close to zero. If not their dipole–dipole interactions with the system of I spins are too weak and $S_{2,IS}$ comes to zero within the first-order approximation of the perturbation theory. In other words one deals with a dumping of dipole–dipole interactions between half-integer and integer spins in solids. The interaction can be reinforced by applying the external magnetic field B_0, but in polycrystals it results in the resonance line broadening. However, that is polycrystals with $\eta \neq 0$ that are of practical interest. That is why practical applications of the spin-echo DNQR are rather restricted. Nevertheless it was successfully used to study ^{23}Na, ^{39}K, ^{41}K, ^{133}Cs, ^{85}Rb, ^{87}Rb NQR in various chlorates.[2]

15.4.2 Stationary Double Nuclear Quadrupole Resonance in the Rotating Frame

This method is based on the stationary irradiation of a sample and firstly was applied by Jones and Hartmann.[3] The main idea is that one applies to the studied spin system two radiofrequency fields: the field $B_{1I} \cos \omega t$ with a frequency that differs from the

[2] M. Emshwiller, E.L. Hahn, D. Kaplan, Phys. Rev. 118, 414 (1960).

[3] E.P. Jones, S.R. Hartmann, Phys. Rev. Lett. 22, 867 (1969). E.P. Jones, S.R. Hartmann, Phys. Rev. B. 6, 757 (1972).

resonance frequency of the I nuclei by a frequency mismatch $\Delta\omega = \omega_I - \omega$, and the field $B_{1S} \cos \omega_S t$ with the resonance frequency of the S nuclei. The field B_{1I} provides a magnetization equilibration in the I system, whereas the field B_{1S} affects the equilibrium system inducing thermal exchange between I and S spins if the Harmann–Hahn condition is fulfilled (15.8). In the NQR case the Hartmann–Hahn matching can be written as

$$\alpha_I \gamma_I B_{1I} = \alpha_S \gamma_S B_{1S}, \tag{15.15}$$

where $\alpha = \sqrt{I(I+1) - m(m-1)}$. Duration of the thermal exchange between two systems is restricted by the cross–relaxation time, and for more effective heating of the I spins the phase of the field B_{1S} is periodically switched by 180°. If the frequency ω_S coincides with one of the frequencies of quadrupole transitions of the S nuclei the magnetization of the I system (in the rotating frame) decreases that results it changing of the NQR adsorption signal.

As compared with direct pulse NQR, sensitivity of stationary DNQR methods can be expressed as

$$\sigma = \sigma_\mathrm{p} \frac{B_{1I}}{2\sqrt{B_{1I}^2 + B_\mathrm{loc}^2}} \frac{T_{1I}}{T_{1I} + T_{IS}},$$

where T_{IS} is the cross–relaxation time between the systems I and S; B_loc is the total induction of the local magnetic field.

Initially this method was applied to study NQR of ^{39}K, ^{40}K, ^{41}K, and other rare or dilute spins.[4] However, it has a number of disadvantages, the most important are following: (i) it is rather difficult to realize in practice high stability and rate in switching of high-power rf fields; (ii) it can be applied to a limited number of compounds as the studied material must contain two quadrupole spin systems, one with a high resonance signal. Currently more frequent are double NMR–NQR methods, in which a weak quadrupole signal from S nuclei is registered by changing in a strong NMR signal of I nuclei.

Depending on contact conditions between I and S systems, the double NMR–NQR methods can be divided into three categories:

- double resonance in high magnetic field (in the rotation frame);
- double resonance in zero magnetic field (in the laboratory frame);
- double resonance in low magnetic field.

[4] D.E. Demco, S. Kaplan, S. Pausak, J.S. Waugh, Chem. Phys. Lett. 30, 77 (1975).

15.4.3 Double NMR–NQR in High Magnetic Field

This method was initially developed by Hartmann and Hahn[5] and is based on the crosse–relaxation effects. Its description, as it is used in NMR, is given in Sect. 15.3.2. In NQR the splitting between energy levels in the S system is determined by quadrupole interactions, and the amplitude of the $\vec{B}_{1S}$ field is estimated from the Hartmann–Hahn condition:

$$\gamma_I B_{1I} = \sqrt{I(I+1) - m(m-1)}\gamma_S B_{1S}.$$

The method sensitivity is defined by

$$\sigma = \frac{\sigma_I}{e}\left[1 - \exp\left(-\frac{T_{1I}}{T_{IS}}\right)\right], \tag{15.16}$$

where σ_I is the signal-to-noise ratio for I-NMR signal recording, T_{1I} is the spin–lattice relaxation time for the I nuclei species in the rotating frame, T_{IS} is the cross–relaxation time between the I and S systems; e is the electron charge.

As it follows from Eq. (15.16) the double resonance sensitivity in the rotating frame does not depend on the NQR frequency and is greater the spin–echo NQR one. It was successfully applied to NQR studies of weak nuclei in samples with natural isotope abundance.[6]

15.4.4 Double NMR–NQR in Zero Magnetic Field

The experiment scheme is shown in Fig. 15.9. In this method originally developed by Redfield, Slusher and Hahn[7] the sample is embedded into a field $\vec{B}_0$ to equilibrate magnetization of the spin system I. Then, one adiabatically reduces the field $\vec{B}_0$, and after that applies a rf field $\vec{B}_{1S}$ for thermal contact between the I and S systems. Then one applies the stationary field $\vec{B}_0$ once again and measures the rest magnetization of the I nuclear system, for example, by applying a 90° pulse $\vec{B}_{1I}$ at the NMR frequency.

Thermal contact between two systems can be effected either by thermal mixing in zero external field,[8] or by solid–effect mechanism,[9] or by irradiating at frequency connecting two multiplets (double resonance with coupled multiplets techniques[10]).

---

[5] S.R. Hartmann, E.L. Hahn, Phys. Rev. 128, 2042 (1962).

[6] R.E. Walstedt, D.A. McArthur, E.L. Hahn, Phys. Lett. 15, 7 (1965). J.L. Ragle, K.L. Sherk, J. Chem. Phys. 50, 3553 (1969). D.T. Edmonds, Phys. Report C 29, 233 (1977).

[7] A.G. Redfield, Phys. Rev. 130, 589 (1963). R.E. Slusher, E.L. Hahn, Phys. Rev. Lett. 12, 246 (1964).

[8] M. Goldman, A. Landesman, Phys. Rev. 132, 610 (1963).

[9] J. Seliger, R. Bling, M. Mali, R. Osredkar, A. Prelesnik, Phys. Stat. Sol. (a) 25, K121 (1974).

[10] S.G.P. Brosnan, D.T. Edmonds, J. Mag. Res. 45, 440 (1981).

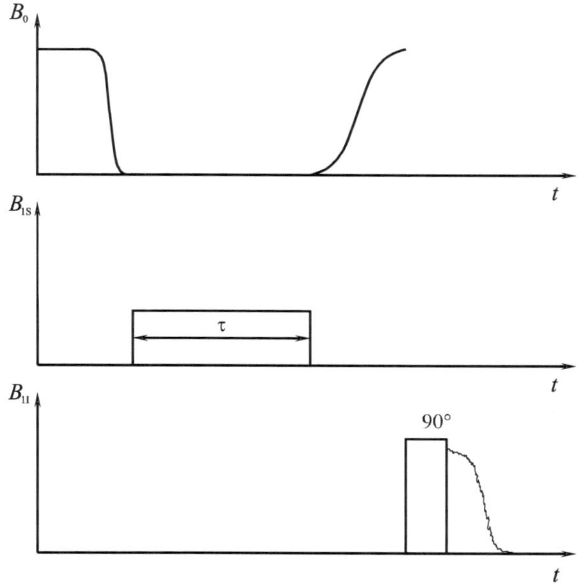

Fig. 15.9 Double NMR-NQR in zero magnetic field

15.4.5 Double NMR–NQR in Low Magnetic Field

This method is more universal and can be applied to study both integer and half-integer spin systems. As opposed to the methods described above, in this method to make the coupling between I and S systems more effective one applies an intermediate field much lower than the local fields, whereas the quadrupole system is irradiated in zero magnetic field. It allows us to avoid broadening the resonance lines for the S nucleus species.

During adiabatic demagnetization, which is achieved by cycling the magnetic field $\vec{B}_0$, (see Fig. 15.10), when magnetic field $\vec{B}_0$ is switching off one has a moment when energy level splitting in the NMR system, due to Zeeman interactions, is equal to one of the quadrupole frequencies: $B_0' = \omega_S'/\gamma_I$. Due to spin–flip transitions induced by dipole–dipole interactions the energy levels of both systems become equal that makes possible cross–relaxation processes. And it is the same when the magnetic field $\vec{B}_0$ is switching on. This method is also called by the *double resonance with level crossing*. After the magnetic field is switched off one applies the rf field $\vec{B}_{1S}$ and if its frequency is equal to the resonance frequency of one of quadrupole transitions of the S system, its level populations are equalized. When switching $\vec{B}_0$ on there is an energy transfer from the system S to I, which decreases the polarizability of the latter. If the frequency of $\vec{B}_{1S}$ is not equal to any NQR frequency, the state of the S spin system does not change. Similar to the methods described above, the S spin

Fig. 15.10 Double NMR–NQR in low magnetic field

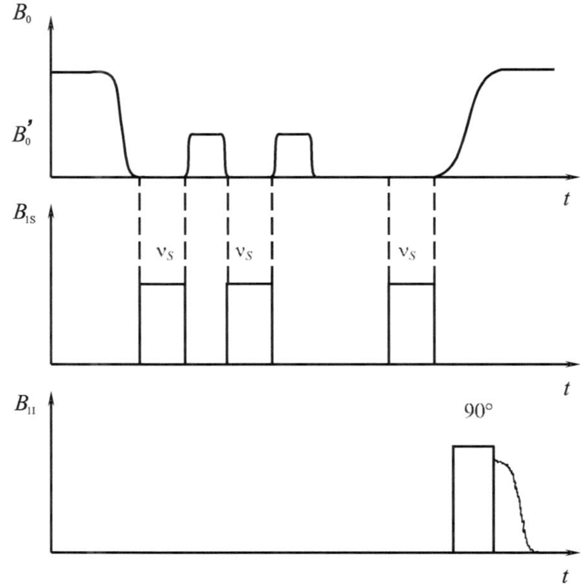

system state is tested by changing in the free induction decay signal of the I spin system after applying the 90° pulse.[11]

The main advantage of this method is that its sensitivity does not decreases with detected frequency decreasing. It allows us to use it to study light nuclei with small quadrupole moments, such as ^{2}H, ^{7}Li, ^{10}B, ^{14}N, ^{17}O, ^{23}Na, and not only single crystals, but polycrystalline samples as well.

Of cause the above-listed methods methods do not exhaust all double nuclear quadrupole methods actually used. This area has received a new impetus due to the need in high-sensitive remote detectors of ^{14}N nuclei belonging to explosives and narcotics. This subject is briefly discussed in Sect. 9.7.

[11] D.T. Edmonds, J.P.G. Mailer, J. Mag. Res. 26, 93 (1977).

Chapter 16
Two-Dimensional NMR Fourier Spectroscopy

16.1 Essence of Multidimensional NMR-Spectroscopy

Two-dimensional Fourier-transform spectroscopy of nuclear magnetic resonance is one of the capital and most extensive research concepts in the NMR spectroscopy that allows the investigation of complex systems of coupled nuclear spins. Usual one-dimensional spectra are failed to be decoded because of inevitable lineoverlap for large molecules. The idea of two-dimensional NMR spectroscopy is that an action of radio-frequency pulses on one of spins causes the perturbation of a state of another. As a result the changes in signals of NMR of the subsystem are produced. A two-dimensional spectrum is just a visual form of representation of interaction between the subsystems.

In Chap. 4 it has been shown that the high resolution NMR-spectroscopy is extremely informative research technique of molecule structure in a liquid state. In principle, the NMR-spectroscopy allows us to study in details stereochemical structure of molecules, their geometrical parameters up to definition internuclear distances and the angles between chemical bonds. Besides, the NMR spectroscopy represents absolutely irreplaceable tool for examination of dynamic effects, such, as chemical and spin exchanges, an intramolecular motion etc (see Sect. 4.3). There are two modes of the registration of a spectrum: stationary (see Sect. 2.3) and Fourier transform of a signal of NMR after pulse action (Sect. 2.4). The use of Fourier transform has yielded revolution in the area of a NMR-spectroscopy. Besides purely technical advantages of Fourier-transform spectroscopy described in Sect. 2.4, there is a possibility to use manifold combinations of pulse procedures. A combination of pulse or stationary irradiation at two or more frequencies has made Fourier-transform spectroscopy much more flexible technique than stationary spectroscopy.

Though the theory of spectra, considering a different magnetic shielding of nuclei in dependence on its positions in a molecule and scalar spin-spin interactions between

V. I. Chizhik et al., *Magnetic Resonance and Its Applications*,
DOI: 10.1007/978-3-319-05299-1_16,

them, is well developed and allows the extraction of molecular parameters from spectra, there is a variety of restrictions for practical use of the NMR-spectroscopy. First of all, they are related to a limit of spectral resolution: hardware restrictions (inhomogeneity of a magnetic field, a quick signal attenuation of a nuclear induction in pulse techniques or insufficiently slow passage and excessively high power of irradiation in stationary technique) and physical cause (natural width of a line of NMR, overlapping of separate spectroscopic lines). The last causes are especially essential for the molecular systems with restricted mobility of molecules: viscous fluids, heterogeneous and anisotropic mediums, such as mesomorphous systems and solids. In these cases the natural width of a line appears rather considerable, and the spectrum becomes complicated under the influence of the dipole-dipole interactions which are not averaged. As a result, it is impossible to separate individual lines and to carry out their reference and interpretation. NMR spectra can appear unsuitable for the analysis if they contain too many lines with the close resonance frequencies. It takes place in NMR spectra of complex molecules, such as polymers, biological molecules etc. Special difficulty for the analysis is represented by spectra of aqueous solutions as proton spectra of diluted compounds are masked by wings of the very intensive resonance line of water.

In the Sect. 15.3 the series of the techniques allowing to suppress some interactions leading to occurrence of a great number of lines has been viewed. As a rule, they use an irradiation with two or more frequencies and Overhauser effect. Such techniques extended possibilities of NMR spectroscopy for study of large molecules, however their advantages have been revealed to the full after realization of two-dimensional Fourier-transform spectroscopy (2D-spectroscopy) in 1971. This technique was based on Jeener's experiments on double resonance. 2D-spectroscopy has later been extended to more than two dimensions. There are several types of 2D-spectroscopy. By using 2D-spectroscopy one identifies groups of the nuclei coupled by scalar interactions (*two-dimensional correlation spectroscopy*), separates chemical shifts and scalar interactions (*J-spectroscopy*), studies a coherence of the states bounded by forbidden multiquantum transitions (*multiquantum spectroscopy*). In spite of the fact that the direct a dipole-dipole interaction between nuclei spins is averaged in the mobile phases and in usual NMR spectra is not manifested, in 2D-spectra in the combination with the transient Overhauser effect allows us to determine internuclear distances with sufficient accuracy. In many cases multidimensional NMR Fourier-transform spectroscopy is uniquely expedient for obtaining the detailed information about chemical (primary) and spatial (secondary and tertiary) structure of very complex molecules, including biopolymers. It is one of standard experimental techniques of the molecular biology, having advantage in comparison with a X-ray diffraction analysis, that allows the investigation of structure and dynamics of molecules not only in solid, but in a state, close to the natural one. In particular, the molecular mechanisms of gene expresson are studied with its help.

16.2 Two-Dimensional Spectroscopy: Obtaining and Interpretation of 2D-Spectra

The 2D spectrum is obtained as a result of double Fourier transform of the response of a spin system on the action of radio-frequency pulses, which contains *two independent time variables* serving as integration variables for double Fourier transform. One of variables is always the real time (t_2) during which the acquisition of data has been done. The other one is an interval (t_1) in sequence of radio-frequency pulses. The obtained spectrum (2D-spectrum) represents the function of two frequency variables, ω_1 and ω_2, which are conjugated to times t_1 and t_2. This function can be presented as a graphic in the form of a relief or intensity isolines on a plane.

Experiment on obtaining the 2D-spectrum, in general, consists in four items:

(1) *preparation stage* which consists in creating of a nonequilibrium state of spins, at least, of one isotope by action, at least, one radio-frequency pulse. As a result, a precessing transversal nuclear magnetization is created;
(2) *evolution stage* during which there exists the free precession in the plane xy of the magnetization of nuclei exposed to action of radio-frequency pulses in the preparatory stage; for the performance of two-dimensional experiment the duration of the evolution (t_1) is incremented with the fixed step at each repetition of pulse sequence and represents one of time scales for the subsequent two-dimensional Fourier transform;
(3) *mixing stage* to which some pulse action precedes; during mixing there is a magnetization or coherence transfer between spin subsystems which are coupled by dipole-dipole or scalar interactions or between which there is a chemical exchange; in some types of two-dimensional experiments the pulse action can coincide with the mixing stage;
(4) last stage of experiment represents *the measurement of transversal magnetization*, as well as in the one-dimensional experiments; current time of data acquisition (t_2) is used as the second time scale at Fourier transform. Conjugated to times t_1 and t_2 frequencies ω_1 and ω_2 represent scales of chemical shifts and (or) the spin-spin splittings. Characteristic feature for a two-dimensional spectrum is the presence of the extrema on a main diagonal ("diagonal peaks"). Besides, there are extrema out of a diagonal ("cross-peaks") from a position of which, as a rule, the important information on interactions in spin system is taken. Depending on used pulse sequence it is possible to assign different physical sense to the new frequency variable ω_1 in the 2D-spectrum. Cross-peaks correspond to projections of the signals appearing when subsystems of spins are related by the dipole-dipole or scalar interaction. Thus, the dipole-dipole interaction leads to transfer of magnetization from one subsystem to another (Overhauser effect), and in the case of the scalar interaction the pulse excitation of one group of spins leads to the fact that further evolution of coupled spins goes in interconnection. This phenomenon is termed as *coherence transfer*.

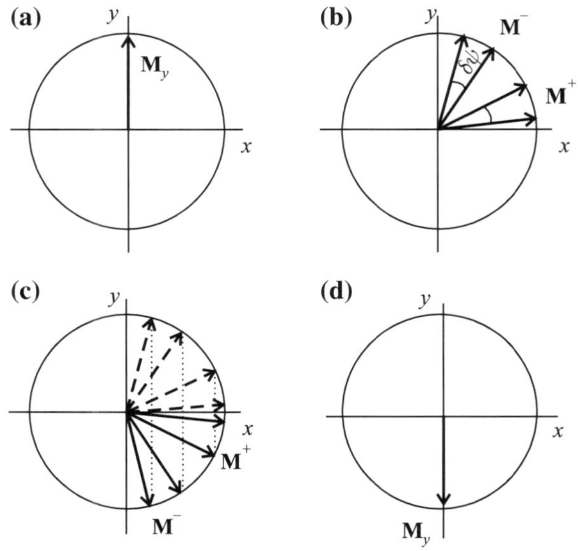

Fig. 16.1 The action of the 90–180 pulse sequence on traversal magnetisation of nuclei A only: **a** Just after 90-pulse; **b** During an evolution t_1 between pulses; **c** After 180-pulse; **d** At the moment of echo

16.3 J-Spectroscopy: Separation of Electronic Shielding and Scalar Interaction Effects

Let us show the advantage of the introduction of second-time variable on the example of obtaining the two-dimensional, so-called, J-spectrum of nuclear spin system of type AX ($I_A = I_X = \frac{1}{2}$). If nuclei A and X are coupled by the scalar interaction, the Hamiltonian of spin system $\widehat{\mathcal{H}}$ has a form:

$$\widehat{\mathcal{H}} = \gamma_A(1 - \sigma_A)\hat{\vec{I}}_A\vec{B} + \gamma_X(1 - \sigma_X)\hat{\vec{I}}_X\vec{B} + J\hat{\vec{I}}_A\hat{\vec{I}}_X, \tag{16.1}$$

here J is the constant of scalar (spin-spin) interaction of nuclei A and X in units of cyclic frequency; γ_A, γ_X, σ_A and σ_X are the gyromagnetic ratios and constants of shielding of nuclei A and X; $\hat{\vec{I}}_A$, $\hat{\vec{I}}_X$ are operators of spins of nuclei A and X.

It is well-known the spectrum of the system with the Hamiltonian (16.1), consists of two doublets with the splittings is equal to J, aligned on the resonance frequencies of nuclei A and X. Such a spectrum can be obtained in experiment by Fourier transform of a signal FID of one of nuclei, for example after a 90° pulse. A shortcoming of the above technique is that components of the obtained spectrum are widened by always existing heterogeneity of a static magnetic field which can be characterised by constant T_2^*. If $T_2^* J < 1$ then a spectrum is not resolved and one cannot measure a constant J. It is possible to obtain also a spectrum using a spin echo signal. This case is illustrated in Fig. 16.1 where the motion of magnetization of nuclei A in the plane xy of the rotating frame is shown. Immediately after the first pulse B_1 that is directed along the axis x (a preparation stage) all spins of nuclei A are completely in

Fig. 16.2 The pulse sequence to obtain 2D J-spectrum: **a** Pulses for nuclei A; **b** Pulse for nuclei X

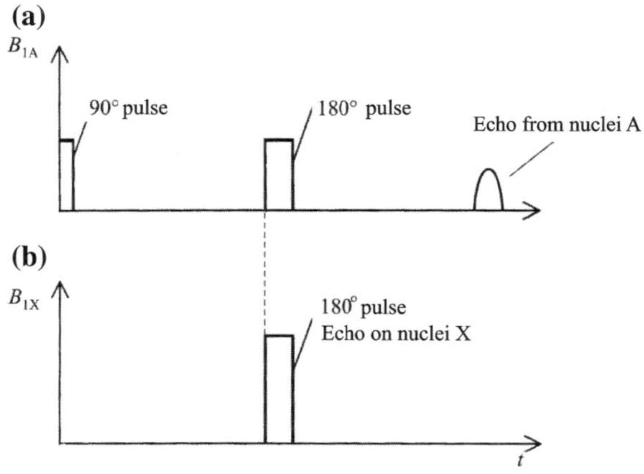

phase and form a vector of the total magnetization with the greatest possible value of a projection M_y (see Fig. 16.1a). In the ideally homogeneous field the spins would be separated on two groups: $\vec{M}^+$ and $\vec{M}^-$ with frequencies of the precession $\Omega + \frac{1}{2}J$ and $\Omega - \frac{1}{2}J$ where Ω is the frequency mismatch relative a frame rotation frequency (the evolution stage, see Fig. 16.1b). In fact the inhomogeneity of the static magnetic field causes additional distribution of the frequencies $\delta\Omega$ and, hence, additional dephasing $\delta\psi = \delta\Omega\, t_1$. In Fig. 16.1b the positions of the vectors $\vec{M}^+$ and $\vec{M}^-$ just before of a 180° pulse at $\frac{t_1}{2}$ are shown. After the 180° pulse the vectors will occupy symmetrical in relation to the axis x positions so the group of spins lagging behind to an pulse will appear ahead (a mixing stage, Fig. 16.1c). As a result in the time t_1 all vectors are again phased irrespective of the cause of their dephasing, in the axis direction $-y$ (see Fig. 16.1d). It was the known experiment on the obtaining of a Hahn spin echo (see Sect. 2.5.2).

Fourier transform of the spin echo with t_2 as time-variable, as well as in case of the free induction, results in a spectrum in which electronic shielding and scalar interactions are manifested. Besides, spectral lines are broadened because of inhomogeneity of the static magnetic field. If to perform Fourier transform not on real time t_2, but on time of the occurrence of a peak of an echo t_1 (for this purpose it is necessary to repeat the experiment, varying t_1 step-by-step) the single Lorentz line which width is determined by value of T_2, instead of T_2^*, will be obtained. But such a "spectrum" will not contain any more information neither on chemical shifts, nor on scalar coupling constants.

The information on scalar interactions can be obtained if to modify the experiment so that 180° pulses operate both on nuclei A, and on a nuclei X, as it is shown in Fig. 16.2. Then the situation will change. Preparation and evolution stages (see Fig. 16.3a, b) will not differ from the previous experiment, but the mixing stage will be another. The 180° pulse influencing the nucleus X, inverts its spin states, that it is possible to interpret the situation as the change of signs of local fields near the nuclei A. As the result of a combined effect of 180° pulses on nuclei A and X,

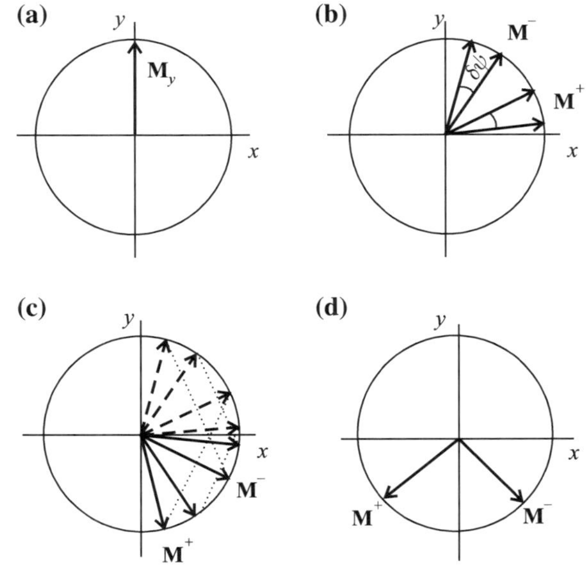

Fig. 16.3 The action of the 90°–180° sequence on traversal magnetization of nuclei A and 180°-pulse on B: **a** Just after 90°-pulse; **b** During an evolution t_1 between pulses; **c** After 180°-pulse; **d** At moment of "echo"

the components $\vec{M}^+$ and $\vec{M}^-$ not only will be reflected in an axis x, but also will exchange in places (see Fig. 16.3c). Owing to that, the lagging group remains the lagging one, and at the moment of the occurrence of the signal peak of the echo (t_1) there will be no any more complete phasing (see Fig. 16.3d). The amplitude of the echo, viewed as function t_1, will appear modulated with function $\cos Jt_1$. The spectrum of such a function represents now a doublet with frequencies $+\frac{1}{2}J$ and $-\frac{1}{2}J$. Remarkable property of such a spectrum is that neither field inhomogeneity nor chemical shift does not affect on the width and position of lines. The spectroscopic width of everyone components is spotted by a time of the transversal relaxation T_2. Such an experiment shows possibility to single out the influence of scalar interactions on a view of NMR spectrum irrespectively of other interactions.

If there are some spin systems with different chemical shifts or gyromagnetic ratios, the obtained spectrum will represent the superposition of doublets (in the general cases, multiplets) relating to each of systems. Therefore one cannot determine, to which group of nuclei characterising by certain chemical shift, a multiplet belongs.

The relationship between chemical shifts and scalar multiplets is carried out by means of two-dimensional spectroscopy. Thus, for the obtaining of a two-dimensional spectrum one uses not only the envelope of echo maximums, how it takes place in the first Jeener's experiments, but whole its contour, as it was done in one-dimensional Fourier-transform spectroscopy. As a result of a series of experiments with step-by-step change t_1 we obtain a two-dimensional data file, which is exposed to double Fourier transform. If to neglect the signal attenuation, caused by inhomogeneity of magnetic field and relaxation, it is possible to present the

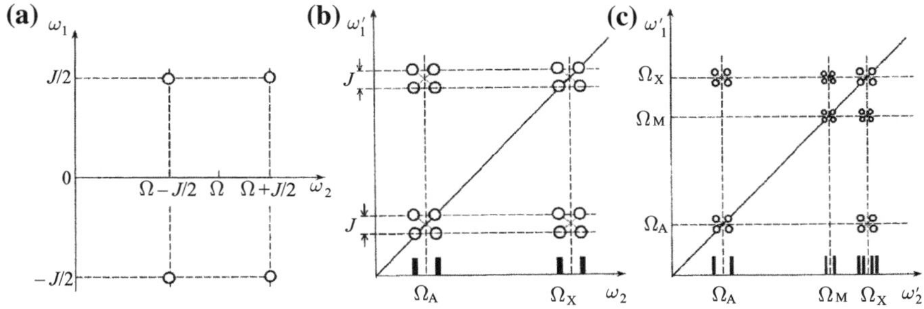

Fig. 16.4 Schematic representatios of 2D-spectra: **a** J-spectrum of nucleus A in a system AB; **b** Correlation spectrum of a system AX; **c** Correlation spectrum of a system AMX

2D-spectrum $S(\omega_1, \omega_2)$ of the nuclei A with the help δ—functions of Dirac:

$$S(\omega_1, \omega_2) = \delta\left(\Omega_1 - \frac{1}{2}J, \Omega_2 - \Omega - \frac{1}{2}J\right) + \delta\left(\omega_1 + \frac{1}{2}J, \omega_2 - \Omega + \frac{1}{2}J\right). \tag{16.2}$$

Here Ω is the cyclic frequency relative some reference frequency (usually it is reference frequency of the phase lock detector).

Figure 16.4a represents schematically the spectrum of nucleus A in accordance with Eq. (16.2), the maximums of intensity (peaks) are designated by circles, and their position are clear from the frequency intervals specified on axes. The projection of a 2D-spectrum on the axis ω_2 is an usual (one-dimensional) spectrum of NMR whereas its projection on the axis ω_1 represents the superposition of above mentioned multiplets with centres at $\omega_1 = 0$. The analysis of arrangement of peaks on a plane $\omega_1\omega_2$ allows us to ascertain rather easily conformity between chemical shifts and observable constants of scalar interactions. As the line broadening caused by inhomogeneity of the magnitostatic field is excluded from a spectrum on the axis ω_1, the technique allows the measurements of coupling constants with high accuracy.

16.3.1 Correlation 2D-Spectroscopy (COSY): Identification of Groups of the Nuclei Connected by Interactions

To know which group of spins is connected with a certain subsystem in the presence of several spin subsystems, ***CO****rrelation* ***S****pectroscop****y*** (COSY) appeared the most suitable. The COSY was proposed by J. Jeener and developed then by R. Ernst with colleagues. The elementary experiment with system of type AX (A and X are the different nuclei), in accordance with this version of spectroscopy, can also be

Table 16.1 The energy values and equilibrium relative energy level populations in the system AX

m_{IA}	m_{IB}	M_I	E_{m_I}	Population
1/2	1/2	1	$-(1/2)hB_0(\gamma_A+\gamma_X)$	$1+(k+1)q$
1/2	−1/2	0	$-(1/2)hB_0(\gamma_A-\gamma_X)$	$1+(k-1)q$
−1/2	1/2	0	$(1/2)hB_0(\gamma_A-\gamma_X)$	$1-(k-1)q$
−1/2	−1/2	−1	$(1/2)hB_0(\gamma_A+\gamma_X)$	$1-(k+1)q$

considered from positions of the classical phenomenological theory. The idea consists in that the changes occurring in the subsystem A at the stage of evolution have been reflected in a signal from the nuclei X. For obtaining of a correlation spectrum the pulse sequence similar presented in Fig. 16.2a, b is used, but the second pulse acting on the nucleus A is not 180° but 90°. Besides, instead of acquisition of an echo from the nuclei A a signal FID from the nuclei X is registered after a 90° pulse too. For the convenience one assumes that the 90° pulse acts on X just after the second pulse on A though both pulses operate simultaneously in real experiment as it is shown in Fig. 16.2. It should be noted that the calculation by quantum mechanic using of density matrix apparatus leads to the same result in both situations.

Correlation between the spin subsystems leads to the situation that after a 90° pulse z-magnetization of the nuclei X turns out modulated by the precession of transversal magnetization of the nuclei A. As a result by the time t_1 the vectors of transversal magnetization $\mathbf{M}^{\pm}$, as well as in the first experiment of Jeener, will rotate on angles $\Omega_A \pm \frac{1}{2}J$ accordingly (see Fig. 16.3). The second 90° pulse will transmute y-components in z-components of magnetization of the nuclei A. The change of z-components of magnetization of the nuclei A is equivalent to change of population of levels of system AX that will lead also to change a z-magnetization of the nuclei X. To calculate this change, it is necessary to keep in mind that the total populations of the levels related, by transitions of the nuclei A, is constant, as there was not any effect of RF field on the nuclei X. The values of energy and equilibrium relative population of levels of the system AX are given in Table 16.1 where $k = \frac{\gamma_A}{\gamma_B}$ and $q = \frac{\gamma_B \hbar B_0}{2kT}$.

Differences of the obtained population determine initial amplitudes of precessing magnetization $\vec{M}_X^{\pm}$, arising after the following immediately behind the second pulse on nuclei A the 90° pulse on the frequency of the nuclei X.

Thus, initial amplitude of the free induction of each group of the nuclei X is a function of t_1, being superposition of periodic functions with frequencies $\Omega_A \pm \frac{1}{2}J$. A signal of the free induction of the nuclei X, considered as function of t_2, has two spectroscopic components: $\Omega_X \pm \frac{J}{2}$, and as a function of t_1 and t_2 has ten components (see Fig. 16.4b): $(0, \Omega_X \pm \frac{J}{2})$; $(\Omega_A \pm \frac{J}{2}, \Omega_X \pm \frac{J}{2})$; $(\Omega_X \pm \frac{J}{2}, \Omega_A \pm \frac{J}{2})$.

In Fig. 16.4b, two-dimensional correlation spectra of system AMX are schematically presented. It is seen that is impossible to choose from the one-dimensional spectrum, which nuclei from the system AMX interact among themselves: A–M, A–X, X–M or everyone with all. From the two-dimensional spectrum presented in Fig. 16.4c it is seen that the common cross-peaks belong to nuclei A and X (large circles), but simultaneously here there are also cross-peaks from nuclei of M and X (small circles). Hence, interactions of nuclei A–X and M–X take place, but the interaction between nuclei A and M is absent.

It is necessary to note that if a system of nuclei does not participate in scalar interactions its peaks settle down only on a main diagonal. This circumstance is used in the analysis of 2D-spectra of NMR and for simplification of the complicated spectra. For example, for inhibition of single intensive lines one uses various angles of rotation of magnetization by the second pulse (see Fig. 16.2). Depending on this angle (α) the phase relationships between precessing spins change that allows us to obtain the additional information on spin system from 2D-spectra. If $\alpha = \pi$ the function of modulation of an echo $s_{\pi}(t_1)$ is

$$s_{\pi}(t_1) = \cos Jt_1. \tag{16.3}$$

If $\alpha = \frac{\pi}{2}$,

$$s_{\pi/2}(t_1) = \frac{1}{4}\left[1 + \cos Jt_1 + (1 \cos Jt_1) \cos \Delta\Omega t_1\right] \tag{16.4}$$

($\Delta\Omega$ is the difference of chemical shifts of the nuclei A and X in terms of cyclic frequency).

From comparison of the relations (16.3) and (16.4) the technique of the obtaining of the spectrum only of coupled spins and of the exception of isolated ones can be derived. Indeed, if $J = 0$ the intensity of an echo just after the sequence with $\alpha = \pi$ (and only in this case) is twice more than in the case $\alpha = \frac{\pi}{2}$. Then, subtracting from the first spectrum the doubled second one, it is possible to exclude the contribution of isolated spins completely. It is very useful when isolated spins are much more abundant than the interacting among themselves, for example in the diluted aqueous solutions. In this case, if not to use special techniques, single intensive lines can mask the week multiplets completely.

If nuclei A and X are nuclei *of the same isotope* and differ only by magnetic shielding constants (*the homonuclear spectroscopy*), it is enough to use the pulse sequence presented in Fig. 16.2a. In this case both 90 and 180° pulses carry out a rotation of equilibrium magnetizations at the given angle both nuclei A and nuclei X in the total scale of their chemical shifts. The FID in this version of 2D-spectroscopy is registered from both types of nuclei. However the analysis based on phenomenological approach, similar performed above, does not allow us to explain this and more complicated experiments on 2D-spectroscopy. Usually the density matrix apparatus

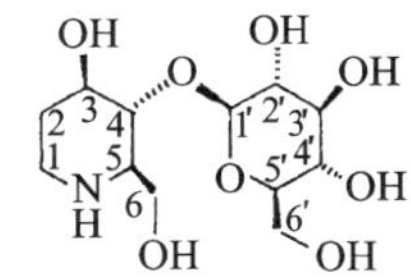

Fig. 16.5 The structure formula of inhibitor of glycosidase

(see 1.9) is used for this purpose. It is known from the quantum statistics that if to neglect interactions with the lattice, the reduced density operator ($\hat{\rho}(t)$), satisfies, so-called, the main quantum equation which is analogue to the Liouville-von Neuman equation for non-reduced density operator. The requirement of the isolation of a spin system from a lattice in pulse NMR experiments is realized selecting the experiment duration less than the spin-lattice relaxation time. In this case the solution looks like complex exponential with the combinations of spin projections in the exponent. A time dependence of the density operator follows Eq. (1.125). An action of radio-frequency pulses is represented by the operator of rotation $\hat{R}$ (see Eqs. (1.126) and (1.128)) whereas an evolution during intervals between the pulses is described by Eq. (1.125) with the operator $\widehat{\mathcal{H}}$ taking into account interactions in spin system.

In other words, at some k-th stage of the formation of a two-dimensional spectrum the initial density operator will be transformed in accordance with the formula

$$\hat{\rho}(t) = \exp(-\mathrm{i}\hat{Q})\hat{\rho}(k)\exp(\mathrm{i}\hat{Q}), \tag{16.5}$$

where $\hat{Q}$ is the operators $\hat{R}$ or $\widehat{\mathcal{H}}$ depending on the needs, and $\hat{\rho}(k)$ is the density operator at the initial moment of k-th stage. Then, for example, an action of the sequence shown on Fig. 16.2a, it is described by the scheme:

$$\hat{\rho}(0)\xrightarrow{\frac{\pi}{2}\left(\hat{I}_{\mathrm{A}x}+\hat{I}_{\mathrm{B}x}\right)}\hat{\rho}(1)\xrightarrow{\Omega_{\mathrm{A}}t_1\hat{I}_{\mathrm{A}z}}\xrightarrow{\Omega_{\mathrm{B}}t_1\hat{I}_{\mathrm{B}z}}\xrightarrow{2\pi J\hat{I}_{\mathrm{A}z}\hat{I}_{\mathrm{B}z}}\hat{\rho}(2)\xrightarrow{\alpha\left(\hat{I}_{\mathrm{A}x}+\hat{I}_{\mathrm{B}x}\right)}\hat{\rho}(3). \tag{16.6}$$

As an example of an application of the homonuclear 2D-spectroscopy to examination of chemical structure we will regard the spectrum of the natural compound which chemical formula is given in Fig. 16.5. The structure of the compound is close to inhibitors of glycosidase (biologically important enzyme to degrade glucose). Its one-dimensional spectrum is given in Fig. 16.6. Though it is possible to refer the groups of signals, associated into multiplets, to protons in certain positions, more complete analysis is impeded by overlapping of parts of the spectrum relevant to different rings. The one-dimensional spectrum nevertheless is expedient for using for primary orientation in the two-dimensional spectrum given in Fig. 16.7 (the scales on abscissa and ordinate are identical). The reference point is the doublet from proton H_1', it is termed *anomeric* and very well stand out the one-dimensional spectrum. In Fig. 16.8 the part of the spectrum arranged in more week field is given in a wider scale.

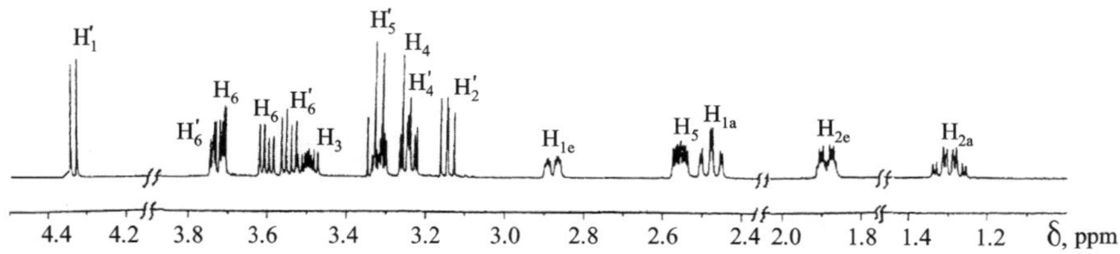

Fig. 16.6 Diagram of the one-dimensional proton spectrum of inhibitor of glycosidase

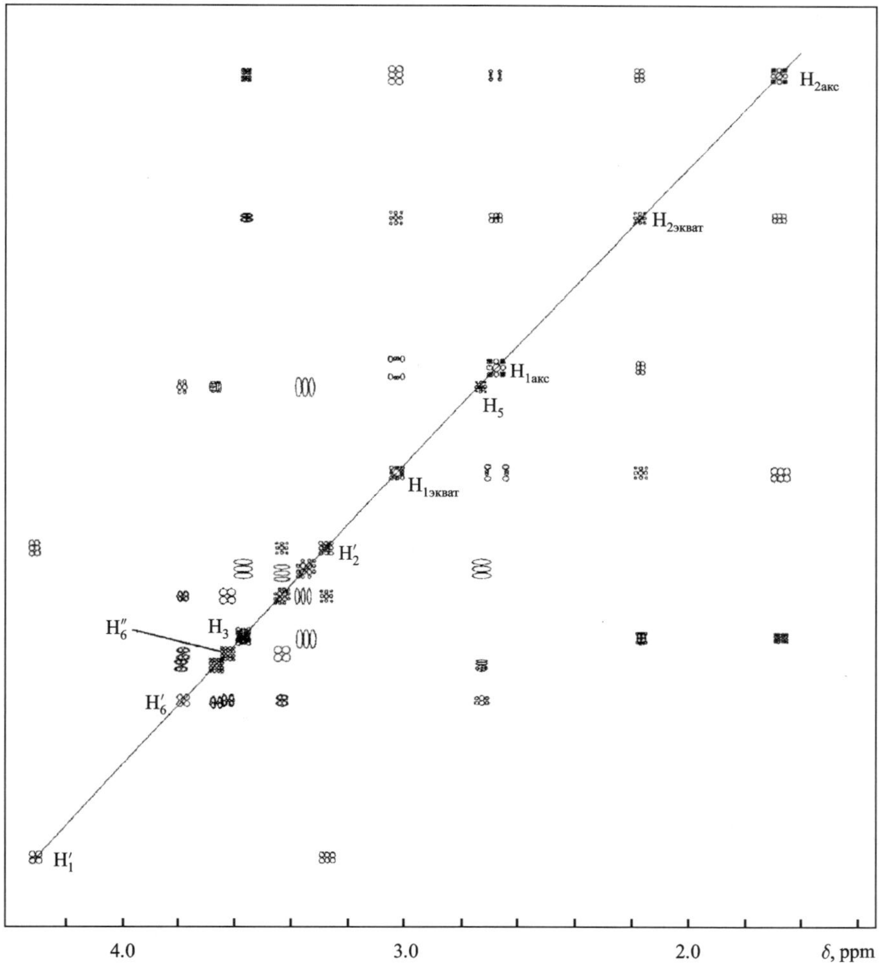

Fig. 16.7 The schematic representation of the correlation spectrum of inhibitor of glycosidase

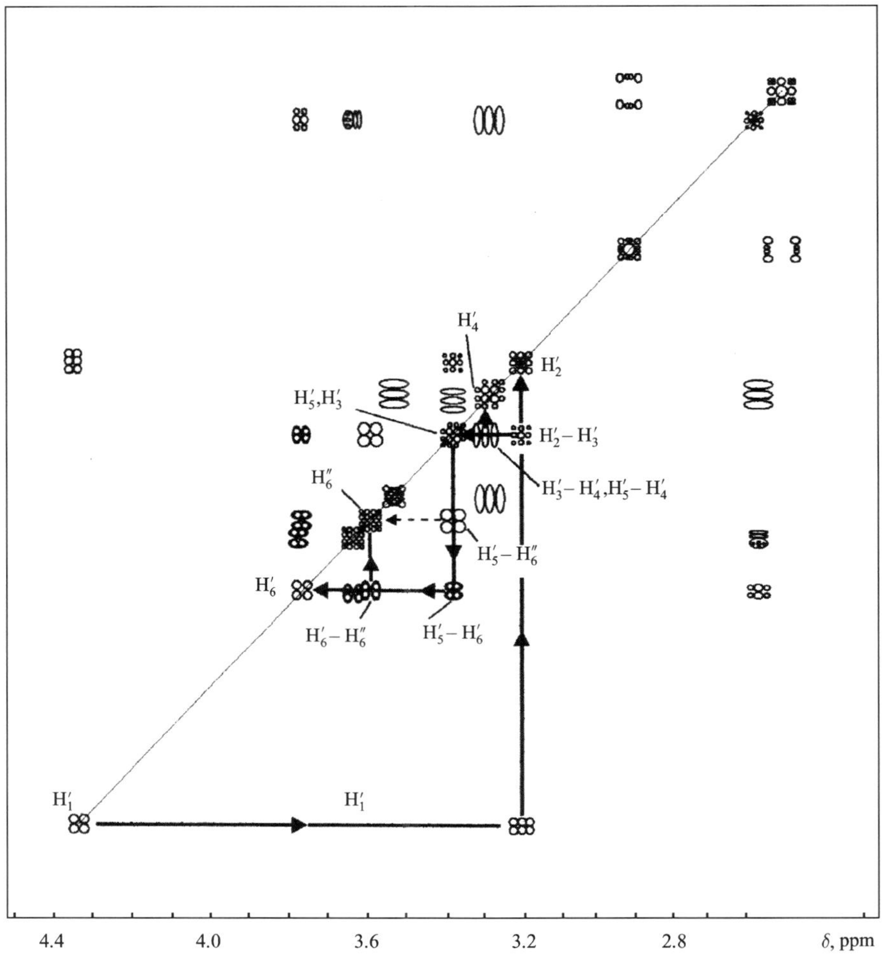

Fig. 16.8 The schematic representation of the low-frequency part of correlation spectrum of inhibitor of glycosidase

Here the procedure of consecutive reference of protons in a ring of glucose which is identified by the presence of groups OH (straight lines with arrows in Fig. 16.8) is demonstrated. The correlation is possible thanks to that interacting protons are shown in the form of off-diagonal peaks (cross-peaks). As the spectrum is two-dimensional, cross-peaks are easily discriminated. Corresponding lines in the one-dimensional spectrum, for example, H_3' and H_5', are overlapped, despite higher frequency resolution. The resolution in 2D-spectrum is possible because the position of a cross-peak is determined by two chemical shifts. Other parts of a spectrum are similarly analyzed. By selecting special conditions of obtaining of the spectra it is possible to find out indirect spin-spin interactions of the protons removed from each other.

The obtained information allows us to draw the conclusion about structure of molecules. In particular, the existence of cross-peaks of protons H_4 and H_1' indicate that in this place there is a link of two rings in one molecule.

Heteronuclear (A and X are nuclei of different types) correlation spectra are widely applied to studying of a structure of the complicated organic compounds (Hydrogen-1 and Carbon-13), and also organophosphorus compounds. Their spectra are computed using expressions (16.5) and (16.6).

16.3.2 2D-Spectroscopy on the Basis of Nuclear Overhauser Effect (NOESY)

In this version of spectroscopy *the transient Overhauser effect* caused by direct dipole-dipole interaction is used (see Sect. 15.2) and Eq. (15.4). The Overhauser effect is the cause of the change of the longitudinal magnetization of the interacting nuclei. The longitudinal component arises in an experiment on correlation spectroscopy after the second pulse too, but it is not reflected in the COSY spectrum. It is possible to observe its change, if after the mixing time τ_m we add one more ("sounding") 90° pulse, i.e. apply the sequence $(\frac{\pi}{2})$ - t_1 - $(\frac{\pi}{2})$ - τ_m - $(\frac{\pi}{2})$ - t_2. The amplitude of FID after the third pulse is proportional to the longitudinal magnetization component just before the pulse and hence, it is a measure of Overhauser effect. The consideration, similar to that explains the occurrence of a two-dimensional spectrum in the correlation spectroscopy, leads to the conclusion that in this case there should be the cross-peaks in a spectrum corresponding to interacting nuclei. However in this case the question under consideration is about a dipole-dipole interaction but not about scalar interaction like in COSY. The advantage of technique NOESY is that quantity of the effect depends on the distances between nuclei. For molecules with low mobility (proteins) the intensity of cross-peak is proportional to $\frac{\tau_c}{\tau_m d^6}$, where d is the distance between interacting nuclei, τ_c is the correlation time for the thermal rotary motion of a molecule. It is possible to estimate the distance between nuclei having compared intensity of cross-peaks for unknown and known distances choosing the optimum value τ_m (at small τ_m the Overhauser effect will not be noticeable yet, and at too long one the relaxation process will dominate).

16.3.3 Two-Dimensional Exchange Spectroscopy (EXSY and ROESY)

This is an extension of the 2D-spectroscopy of known in one-dimensional spectroscopy experiments for systems with chemical or configurational exchange. In pulse sequence EXSY (***EX***change ***S***pectroscop***Y***) the same pulses, as in NOESY, are used, but optimization τ_m is performed on the greatest sensitivity of a spectrum to an exchange time. The first 90° pulse, as well as in other two-dimensional experiments, directs vectors of equilibrium z-magnetization of each exchanging subsystems along an axis x of a rotating frame. During the evolution time t_1 the transversal magnetizations precess with everyone's own frequency. The 90° pulse, which turns x—magnetization into z-magnetization, precedes the mixing stage. During the mixing time there is a mutual transfer of magnetization of subsystems owing to a chemical or configurational exchange, that is fixed by the registration of FID. After double Fourier transform there are the cross-peaks caused by an exchange in a two-dimensional spectrum. Their height in relation to corresponding diagonal peak is proportional to $\frac{k\tau_m}{1-k\tau_m}$, where k is the constant of an exchange rate. At the same time with this the peaks caused by Overhauser effect can be observed (simultaneously). Various switching patterns of the relative phases of 90° pulses and a reference signal of a phase-lock detector with the subsequent summation of output signals are used to their separation.

If in the mixing stage we include a radio-frequency field "capched" spins (the spin-locking, see §3.14) then there is *the magnetization transfer in the rotating frame* (ROESY—***R***otation Frame ***O***verhauser ***E***ffect ***S***pectroscop***Y***). In the one-dimensional experiment the 90° pulse preceeds the selective 180° inverting magnetization of one of subsystems. Data acquisition occurs after the stage of spin-locking during which there is an exchange of transversal magnetizations (transversal cross-relaxation). For the obtaining of a two-dimensional spectrum the pulse sequence is used that differs from NOESY by the application of the spin-loking at the mixing stage. To eliminate the off-resonance effects and simplify the spectrum the switching of phases of a 90° pulse, a locking field and a reference signal are also used. The technique allows us to explore the very slow exchange (similarly how relaxation rates measured in a rotating frame allow us to investigate slow molecular motions, see Sect. 3.14).

16.3.4 Multiquantum Spectroscopy: The Basic Concepts

This term, multiquantum spectroscopy, refers to the variety of the techniques of 2D-spectroscopy using connection between the pair of states with arbitrary difference of magnetic quantum numbers. As it was already specified, the variety of techniques of 2D-spectroscopy cannot be described without the quantum-mechanical approach using the apparatus of density matrix (see Sect. 1.9). Even the experiments, reviewed earlier in this section, cannot be correctly explained by means of the semiclassical approach if the angle of rotation for the mixing pulse (see Fig. 16.2) is arbitrary.

For the description of processes in the bound spin systems the concept *coherent states* or, simple, *coherences* have been entered. This concept is an expansion of semiclassical concept *transitions between energy states*, to which the occurrence of transversal precessing magnetization corresponds. The concept of *transitions* is valid only for a pair of energy levels between which transitions are permited, i.e. for which magnetic quantum numbers can differ only an unity, whereas the coherence can be exist for any pair of the states connected by an interaction.

The concept "coherence" is used in theories of vibration and wave processes and means the existence of correlation between phases of two wave processes. As to quantum systems this concept means a correlation between phases of wave functions of two connected states. The coherence is estimated quantitatively by a value of nondiagonal element of the density matrix linking that states (see Sect. 1.9.2). Equality to zero of a nondiagonal element of a density matrix means the lack of mutual correlation between phases of wave functions of states. In this case states mean incoherent. The difference of magnetic quantum numbers of states determines *the order of coherence*: One regards *the one-quantum coherence* corresponding to usual induced transition, *zero-coherence* corresponding to z-magnetizations and *a multiquantum coherence of the order* $p \geq 2$ corresponding to the forbidden transitions.

One of the major advantages of NMR Fourier-transform spectroscopy is, in particular, the possibility to study the coupling of spin systems, based on the fact of presence of a coherence of states with any difference of their quantum numbers. The multiquantum coherence cannot be observed immediately, but, it is possible to carry out *the coherence transfer* of the multiquantum- or zero- coherence into the observable one-quantum coherence using proper pulse sequences (the simplest experiment on observation FID after a 90° pulse can be treated as a trivial example of the transfer of the zero-quantum coherence in the one-quantum coherence).

2D-spectra on the basis of making and recording of a multiquantum coherence are used, first of all, for the simplification of complicated spectra (when other versions of 2D-spectroscopy do not lead to satisfactory results), for the selective registration of the specified order coherence, and also for the increase of the volume of information related to special features of relaxation processes.

Simplification of spectra is related, first of all, with a reduction of number of transitions with increasing of the order of a coherence. The technique is very effective for the examination of structure of mesomorphous systems when there is incomplete averaging of the dipole-dipole interactions, and the NMR spectrum becomes very complicated.

The selective observation of a coherence of the certain order allows us to select signals from interesting types of spin systems by suppression of signals from the others. As an example we consider the study of scalar coupling $^{13}C - {}^{13}C$ in non-enriched samples (selective enrichment with carbon-13 is very complicated and expensive). The natural content of magnetic nuclei of carbon is only 1 % and the probability to meet interacting pair is equal 0,01 %. In these conditions the splitting caused by indirect spin-spin interactions are masked with very intense signals from the isolated nuclei carbon-13. At the same time these splittings are rather informative and easily interpreted: because of the small content of the isotope, in a single molecule we can practically meet no more than one interacting pair. The problem is solved by creating of a double-quantum coherence and its subsequent registration by means of pulse sequence $(\frac{\pi}{2})_x$ - τ - π_y - τ - $(\frac{\pi}{2})_x$. Here τ means pulse-to-pulse spacing, and the indices x, y specify a phase of radio-frequency pulses (or, that is equivalent, the orientation of rotating components of B_1 in a rotating frame). Thus, if to choose $\tau = \frac{1}{4J}$ and to change cyclicly a phase ϕ of the last pulse then a summed signal will import the contribution only from interacting spins, and contributions of the isolated spins will be compensated. The obtained pulse sequence $(\frac{\pi}{2})_x$ - $\frac{1}{4J}$ - π_y - $\frac{1}{4J}$ - $(\frac{\pi}{2})_x$ - t_1 - $(\frac{\pi}{2})_\phi$ was called as name INADEQUATE (Incredible Natural Abundance Double Quantum Transfer Experiment).

The lifetime of a multiquantum coherence is determined by a "true" relaxation time T_2 and does not depend on the inhomogeneity of the magnetic field. This circumstance allows us to adapt multiquantum spectroscopy for measuring long-range (hence, small) constants of scalar interaction. In the case of usual techniques (for example COSY) a transversal magnetization damps because of inhomogeneity of the magnetic field before the effect of a weak coupling is revealed.

The double transfer of one-quantum coherence in heteronuclear experiments is used for a study of biomolecules. At a preparatory stage the coherence is transmitted from nuclei I with a great gyromagnetic ratio γ to nuclei S with small γ, and after an evolution stage the reverse transfer of the coherence is carried out. As a result it is possible to obtain the high sensitivity, which is inherent for nuclei I, and to obtain an information on interactions, in which nuclei S take part. The particular interest represents the version of the technique using cross-correlation between the dipole-dipole mechanism of relaxation and the relaxation due to anisotropy of magnetic shielding of nuclei ^{15}N. Cross-correlation leads to considerable narrowing of resonance lines and, accordingly, to an increase of peak intensity. For example, in the field of 20 T, spectral line width of one doublet component ^{15}N – ^{1}H from the amide groups of peptide bond in protein molecules is reduced approximately in 20 times. The effect is reached by application to nuclei I of the pulse sequence $(\frac{\pi}{2})_x$ - τ_1 - π_y - τ_1 - $(\frac{\pi}{2})_x$ - t_1 - $\frac{\pi}{2}_x$ - t_2 - $(\frac{\pi}{2})_x$ - τ_2 - π_y - τ_2 - *acquisition of data*—and, after an interval τ_1, to nuclei S of the sequence π_y - τ_1 - $(\frac{\pi}{2})_x$ - t_1 - $(\frac{\pi}{2})_x$ - t_2 - $(\frac{\pi}{2})_x$ - τ_2 - π_y.

Sequence abbreviation is TROSY (***T***ransversal ***R***elaxation ***O***ptimized ***S***pectro scop***y***), i.e. the spectroscopy optimized with respect to transversal relaxation time.

16.3.5 Conclusive Remarks

Physicists have offered for researchers and experts of different specialities (chemists, biologists, geologists, physicians) the set of manifold techniques for the obtaining of 2D-spectra and their use for the solution of problems related to structure and dynamics of molecules. Here only some of them have been described. Techniques of the analysis of much more complex molecules, than given above in examples, such, as proteins and DNA are developed. Such analysis is possible after accumulation of the wide experience of interpretation of 2D-spectra of various molecules. Well developed standard procedures are carried out using the computer by means of special, frequently rather complicated, programs.

The two-dimensional spectroscopy has great importance for study of solids, however it is absolutely special area that stays out of our consideration. The special techniques using the specific features of NMR in solids are developed for obtaining of two-dimensional spectra. In particular, the combination of techniques of 2D-spectroscopy and the magic angle spinning is used.

Further Readings

Bates RD (1993) Dynamic nuclear polarization. Magn Reson Rev 16:237

Bax A (1985) A simple description of two-dimensional NMR spectroscopy. Bull Magn Reson 7:167

Bax A (1985) Two-dimensional nuclear magnetic resonance in liquids. Delft University Press, pp 200 (ISBN-10: 9027714126 / ISBN-13: 9789027714121)

Billeter M, Orekhov V (eds) (2012) Novel sampling approaches in higher dimensional NMR. Springer Verlag, NY

Blümich B, Spiess HW (1988) Two-dimensional solid-state NMR spectroscopy: new possibilities for the investigation of the structure and dynamics of solid polymers. Angew Chem Int Ed Engl 27:1655

Bremser W, Fachinger W (1985) Multidimensional spectroscopy. Magn Reson Chem 23:1056

Derome AE (1987) Modern NMR techniques for chemistry research. Pergamon Press. pp 415 (ISBN 0-08-032513-0)

Ernst RR, Bodenhausen G, Wokaun A (1987) Principles of NMR in one and two dimensions. Oxford University Press, NY

Ernst RR (1987) Two-dimensional NMR spectroscopy: a powerful tool for the investigation of molecular structure and dynamics. Chimia 41:323

Freebolin H (2011) Basic one- and two-dimensional spectroscopy. Wiley-VCH, pp 411 (ISBN 978-3-527-32782-9)

Freeman R (2003) Magnetic resonance in chemistry and medicine. Oxford University Press, pp 336

Griesinger C, Bennati M, Vieth HM, Luchinat C, Parigi G, HÃűfer P, Engelke F, Glaser SJ, Denysenkov V, Prisner TF (2012) Dynamic nuclear polarization at high magnetic fields in liquids. Prog Nucl Magn Reson Spectrosc 64:4

Günther H (1995) NMR Spectroscopy: Basic Principles, Concepts, and Applications in Chemistry, 2nd edn. Wiley, NY

Harris RK, Wasylishen RE (eds) Encyclopedia of NMR. Wiley, NY (ten volumes; ISBN 978-0-470-05821-3)

Keeler J (2010) Understanding NMR Spectroscopy, 2nd edn. Wiley, NY (ISBN 978-0-470-74609-7)

Kessler H, Gehrke M, Griesinger C (1988) Two-dimensional NMR spectroscopy: background and overview of the experiments. Angew Chem Int Ed Engl 27:490

Munowitz M, Pines A (1987) Principles and applications of multiple-quantum NMR. Adv Chem Phys 66:1

Rahman A.-U (1989) One- and two-dimensional NMR spectroscopy. Elsevier, NY

Schraml J, Bellama JM (1988) Two-dimensional NMR spectroscopy. Wiley, NY

Suryaprakash N (2011) Two-dimensional higher quantum correlation and resolved techniques for the spectral analysis of complex 1H spectra of scalar-coupled spins. Annu Rep NMR Spectrosc 72:61

Part V
Quantum Radiofrequency Electronics (Radioelectronics)

Chapter 17
Basic Physical Ideas of Quantum Radioelectronics

In this and follow-up chapters we will describe the physical principles for the development of devices, in which an element that determines their properties is an ensemble of micro-particles (nuclei, atoms, molecules, and so on) in special state. In the classical radioelectronics, there are numerous devices that use the so-called *narrow-band elements*. Their particular feature is the specifical reaction to radio signals in a more or less narrow band of frequencies. Examples are a resonant circuit, consisting of a capacitor and inductor, or a combination of resistors and capacitors (Wien bridge, double T-shaped bridge, and so on). These narrow-band elements are used to improve characteristics of amplifiers and generators of the radio diapason. In the middle of 1950s a new branch of radioelectronics was born, in which the *spectral lines as narrow-band elements are used*. The emission and absorption spectra are the manifestation of purely quantum properties of matter. Hence the name is *quantum radioelectronics*. This branch has all the basic features of classical electronics, but there is one peculiarity: quantum narrow-band elements can be sources of electromagnetic radiation.

For the denomination of quantum generators and amplifiers one uses several terms. The most popular term is *maser*. The word is formed by abbreviation of the expression "Microwave Amplification by Stimulated Emission of Radiation". Optical quantum generators were called *lasers* (the word "microwave" is replaced with the word "light"). Sometimes, there are also such combinations as "optical or infrared maser". In this case one keeps in mind the decoding of the word "maser", when the first letters stands for "molecular amplification".

17.1 Quantum Amplification of Electromagnetic Fields

Let us consider a particle which possesses two energy states with values $E_1 < E_2$. In the following the term "particle" means a molecule, atom, ion, electron or nucleus. The transition from one state to another is accompanied by radiation or absorption of electromagnetic energy at a frequency $\nu = (E_2 - E_1)/h$. If the particle is at upper

V. I. Chizhik et al., *Magnetic Resonance and Its Applications*,
DOI: 10.1007/978-3-319-05299-1_17,

energy level, there is some probability of its *spontaneous* transition to the lower level with the emission of a quantum of energy. Using the approach developed by A. Einstein, we can write the expression for the number of particles dn_2, performing a spontaneous transition in the time interval dt:

$$\mathrm{d}n_2 = -A_{21}n_2\mathrm{d}t,$$

where A_{21} is the probability of spontaneous transitions per time unit; n_2 is the level population (the minus sign in this expression means the decrease of the number of particles in the state E_2). Then

$$n_2(t) = n_2(0)\exp(-A_{21}t)$$

and the value of $1/A_{21} = \tau$ is the average lifetime of a particle in the excited state.

The second opportunity for particles to make the transition from the state of E_2 is the *induced or stimulated radiation*, i.e. the transition under the influence of external radiation. A peculiar feature of this process is the fact that the emission of particles is coherent with the external radiation.

For a particle at the lower level, there is only one possibility: the transition with the absorption of external electromagnetic energy. The second and third processes are called *induced radiation and absorption*. The intensity of the second and third processes is directly proportional to the spectral density of external radiation ρ_ν and we can write:

$$\mathrm{d}n_2 = -B_{21}\rho_\nu n_2\mathrm{d}t;\ \mathrm{d}n_1 = -B_{12}\rho_\nu n_1\mathrm{d}t,$$

where $B_{21}\rho_\nu$ and $B_{12}\rho_\nu$ are the probabilities of induced transitions per time unit.

The coefficients A_{21}, B_{21} and B_{12} are called *Einstein coefficients*. A. Einstein showed in the analysis of radiation of an absolutely black body that

(1) the probability of stimulated emission and absorption are equal, i.e. $B_{12} = B_{21}$ (more precisely, this ratio should be written in the form of $g_1B_{12} = g_2B_{21}$, where g_1 and g_2 are the multiplicities of degenerated states 1 and 2, but for simplicity we assume that states are nondegenerated);
(2) $A_{21} = B_{21}8\pi h\nu^3/c^3$ and therefore in the radio diapason the stimulated emission prevails over the spontaneous one even at very low densities of radiation. This is a very favorable factor because the spontaneous emission of particles is random (noises) and not related to the phase of a signal.

We do not mention the relaxation processes, which are considered elsewhere and which will be taken into account if necessary.

It is easy to write an expression for the power, radiated by an ensemble of particles due to induced transitions. Taking into account that the absorbed power is $P_1 = B_{12}\rho_\nu n_1h\nu$ and the radiated one is $P_2 = B_{21}\rho_\nu n_2h\nu$, the resulting power is given by

$$P = P_2 - P_1 = (n_2 - n_1)B_{21}\rho_\nu h\nu. \tag{17.1}$$

Equation (17.1) shows that to amplify the power of electromagnetic fields due to the particle radiation it is necessary to realize

$$n_2 > n_1, \tag{17.2}$$

i.e. the population of the upper energy level must be greater than the lower one. The difference $n_2 - n_1$ is called *the number of active particles*. For the realization of the condition (17.2) for populations of the levels it is necessary to make the significant deviation from the thermodynamic equilibrium, for which the number of particles n_i^0 at E_i is described by

$$n_i^0 = Ae^{-E_i/(kT)}; \quad n_2^0/n_1^0 = e^{-(E_2-E_1)/(kT)} \tag{17.3}$$

(here A is a constant). According to (17.3) at the thermodynamic equilibrium the number of particles is always smaller in a state with higher energy. In the absence of the thermodynamic equilibrium the distribution of particles over states can be arbitrary. However, formally an arbitrary distribution can be described by the formula (17.3) introducing an effective temperature (the analog of the spin temperature, see Chap. 3):

$$n_2/n_1 = e^{-(E_2-E_1)/(kT_{\text{ef}})}.$$

We are primarily interested in the situation when $n_2 > n_1$. For this case $T_{\text{ef}} < 0$ and it is accepted to speak about the state with the "negative" temperature.

Let us return to formula (17.1), which implies that the resulting radiated power of an ensemble of particles is directly proportional to the number of active particles (method of obtaining this difference can be any). The radiated power is also proportional to the transition frequency (ν). Consequently, the amplification of electromagnetic waves with the stimulated emission is easier to obtain at high frequencies. This is a very important circumstance, because in the classical (usual) radioengineering the transition to the millimeter or submillimeter diapasons is bounded with increasing experimental difficulties.

The expression (17.1) is written under the assumption that the energy of both possible states are precisely defined. However, in reality there is always a "smearing" of energy levels that leads to the emission of electromagnetic waves in some interval $\Delta\nu$. For the description of this effect one introduces the spectral line shape function $g(\nu)$. Taking this into account, the above formula (17.1) can be rewritten as

$$P = (n_2 - n_1)B_{21}\rho_\nu h\nu g(\nu). \tag{17.4}$$

It should be noted that the values P and ρ_ν in this approximation are linearly related to each other. Since the value of ρ_ν is directly proportional to the power of the external radiation, then the formula (17.4) describes the process of quantum amplification of electromagnetic waves. The Einstein coefficient B_{21} determines the quantum amplification. (Recall that the Einstein coefficients were introduced phenomenologically, but they can be calculated on the basis of the theory of quantum transitions, see Sect. 17.2).

Up to now it was assumed that the emission of a quantum system is used to amplify an external signal. If the radiation power of the particles according to the formula (17.4) exceeds the power of losses in a radio receiver (for example, in a resonator containing an excited substance), then continuous oscillations can arise and sustain at the frequency of the transition even in the absence of external signals, i.e. in this case a system will be a quantum generator.

17.2 Calculation of Probabilities of Quantum Transitions

The calculation of transition probabilities is one of the main problems of quantum mechanics. Wave functions describing the states of a system of microparticles are solutions of the Schrödinger equation:

$$\widehat{\mathcal{H}}_0\Psi(x,t) = \mathrm{i}\hbar\partial\Psi(x,t)/\partial t, \tag{17.5}$$

where $\widehat{\mathcal{H}}_0$ is the energy operator; $\Psi(x,t)$ is the wave function (for brevity it is indicated only one spatial coordinate). For stationary states the Eq. (17.4) has the form

$$\widehat{\mathcal{H}}_0\psi(x) = E_n\psi_n(x),$$

here $\psi_n(x)$ and E_n are the eigenfunctions and eigenvalues of the energy operator. In the following we assumed that the quantum task with the Hamiltonian $\widehat{\mathcal{H}}_0$ is already solved.

If there is an additional interaction, the wave function of the system $\Psi'(x,t)$ must satisfy the equation

$$(\widehat{\mathcal{H}}_0 + \widehat{\mathcal{H}}'(t))\Psi'(x,t) = \mathrm{i}\hbar\partial\Psi'(x,t)/\partial t, \tag{17.6}$$

where $\widehat{\mathcal{H}}'(t)$ is the perturbation operator. The function Ψ' can be found in the form of series in the functions of the unperturbed system:

$$\Psi'(x,t) = \sum_n a_n(t)\Psi_n(x,t) = \sum_n a_n(t)\psi_n(x)e^{-\mathrm{i}E_nt/\hbar}. \tag{17.7}$$

Let us substitute (17.7) into Eq. (17.6), multiply on the left by $\psi_m^*\exp(\mathrm{i}E_m\hbar^{-1}t)$ and integrate over the spatial variables. Then, taking into account that $\int\psi_m^*\psi_n\mathrm{d}x = \delta_{mn}$ ($\delta_{nm} = 1$ if $m = n$ and $\delta_{mn} = 0$ if $m \neq n$), we obtain

$$\mathrm{i}\hbar\frac{\mathrm{d}a_m}{\mathrm{d}t} = i\hbar\dot{a}_m(t) = \sum_n a_n(t)\mathcal{H}'_{mn}(t)e^{-\mathrm{i}(E_n-E_m)t/\hbar}, \tag{17.8}$$

$$\mathcal{H}'_{mn}(t) = \int \psi_m^* \widehat{\mathcal{H}}'(x,t)\psi_n \mathrm{d}x.$$

Now the task is to find the coefficients of $a_m(t)$ from the equation system (17.8). Let at time $t = 0$ the system is in the state with the energy E_l. Then

$$a_n(0) = 0, \ \text{if } n \neq l, \qquad a_n(0) = 1, \ \text{if } n = l \tag{17.9}$$

and the probability to find the system in the state with the energy E_n at the end of time t is

$$W_{ln}(t) = |a_n^{(l)}(t)|^2, \tag{17.10}$$

where $a_n^{(l)}(t)$ is the solution of Eq. (17.8) with initial conditions (17.9).

If the perturbation is small, then for not too large time intervals one can expect that the coefficients $a_n(t)$ for $n \neq l$ will be small. Therefore, for obtaining a_m in the first approximation we substitute to the right side of Eq. (17.8) $a_n^{(l)}$ in the form (17.9), i.e. $a_n^{(l)} = \delta_{nl}$. After the integration of Eq. (17.8) we obtain, taking into account the initial conditions:

$$a_m^{(l)}(t) = -\frac{i}{\hbar}\int_0^t \mathcal{H}'_{ml}(\tau)e^{\mathrm{i}\omega_{ml}\tau}\mathrm{d}\tau + \delta_{ml},$$

$$\omega_{ml} = (E_m - E_l)/\hbar.$$

Define the probability of transition under the influence of harmonic perturbations:

$$\vec{A} = 2\vec{A}_1 \cos\omega t, \tag{17.11}$$

assuming that the $\vec{A}$ is one of the components of the electromagnetic field (i.e. $\vec{\mathcal{E}}$ or $\vec{B}$). Since all the calculations for electric and magnetic interactions are formally the same, then we will substitute $\vec{\mathcal{E}}$ and $\vec{B}$ in the final formulas when particular examples are considered. The components of the electromagnetic field $\vec{\mathcal{E}}$ and $\vec{B}$ interact, respectively, with the electric or magnetic dipole, which we denote by $\vec{\mu}$ in both cases. Thus the interaction of the field and dipole is described by the formula

$$\widehat{\mathcal{H}}' = -(\hat{\vec{\mu}}, \vec{A}) = -\mu_x A,$$

if the vector $\vec{A}$ is directed along the axis x (this does not reduce the generality of results, and the subscript x will be omitted). Take into account that

$$\vec{A} = 2\vec{A}_1 \cos\omega t = \vec{A}_1[\exp \mathrm{i}\omega t + \exp(-\mathrm{i}\omega t)].$$

It appears that only the second term is significant:

$$\vec{A}(t) = \vec{A}_1 e^{-\mathrm{i}\omega t}.$$

The final expression for the probability of transitions from the state l to the state m for the time t is of the form

$$W'_{lm}(t) = \frac{A_1^2}{4\pi^2}|\mu_{ml}|^2 \frac{\sin^2(\omega_{ml} - \omega)t/2}{(\omega_{ml} - \omega)^2/4}. \tag{17.12}$$

It is obvious that for the first term in the expansion of the cosine in the exponents the denominator of the formula similar to (17.12) would contain a very large number $(\omega_{ml} + \omega)$ and the corresponding probability would be extremely small. For small values of the sine argument (for small times) we obtain from (17.12)

$$W_{lm}{}'(t) = \hbar^{-2} A_1^2 |\mu_{ml}|^2 t^2. \tag{17.13}$$

The expression (17.12) was obtained under the assumption of discrete energy spectrum system, but the real energy levels are broadened. As it has already been said the function of this distribution $g(\nu_{ml})$ are usually introduced, which characterizes the shape spectral line of a transition. Integrating (17.12) over all frequencies ν_{ml} with the weight $g(\nu_{ml})$, we obtain the total probability for the transition:

$$W_{lm}(t) = \hbar^{-2} A_1^2 |\mu_{ml}|^2 g(\nu) t.$$

It was assumed in the calculations that the function $g(\nu_{ml})$ varies smoothly compared to $(\sin^2 x)/x^2$.

Let us introduce the transition probability per unit time:

$$W_{lm} = W_{lm}(t)/t = \hbar^{-2} A_1^2 |\mu_{ml}|^2 g(\nu). \tag{17.14}$$

Taking into account that the spectral density of energy for the electromagnetic radiation is $\rho_\nu = (2A_1)^2/(8\pi)$, then using formula (17.14) we can obtain the expressions for the Einstein coefficients.

The results obtained can be used to describe the probability transitions in the theory of magnetic resonance. Consider magnetic dipoles in a constant magnetic field B_0 along z which are irradiated by the alternating field with the amplitude of the magnetic component B_1. If $\vec{\mu} = \gamma\hbar\vec{J}$, $J = 1/2$ (in units of $\hbar$) and the basis for the description states is the set of wave functions belonging to the eigenvalues operator $\hat{J}_z$, i.e. $J_z = \pm 1/2$, then from (17.14) we have

$$W_{1/2,-1/2} = \frac{1}{4}\gamma^2 B_1^2 g(\nu). \tag{17.15}$$

In conclusion, we will show that Eq. (17.8) can be solved exactly for a two-level system in the presence of perturbations of the form $\widehat{\mathcal{H}}'(t) = -\vec{\mu}\vec{A}_1 \exp(-\mathrm{i}\omega t)$ and with initial conditions (17.9). We assume that in our analysis only two energy levels with the indexes m and n are important and the perturbation occurs at frequencies very close to ω_{mn}, i.e. $\omega_{mn} = \omega + \Omega$, $\Omega \ll \omega_{mn}$. In our case, the wave function is described by the formula (see expression (17.7): $\Psi(x,t) = a_m \Psi_m(x,t) + a_n \Psi(x,t)$. Then the system of Eq. (17.8) can be rewritten as follows:

$$\begin{aligned} \mathrm{i}\hbar\dot{a}_m &= a_n F_{mn} \exp[\mathrm{i}(\omega_{mn} - \omega)t], \\ \mathrm{i}\hbar\dot{a}_n &= a_m F_{nm} \exp[\mathrm{i}(\omega_{nm} + \omega)t], \end{aligned} \tag{17.16}$$

here $F = -\vec{\mu}\vec{A}$. Using (17.16) and the equality $F_{mn} = F^*_{nm}$, we obtain

$$\mathrm{i}\hbar\dot{a}_m = a_n F_{mn} \exp \mathrm{i}\Omega t,$$

$$\mathrm{i}\hbar\dot{a}_n = a_m F^*_{mn} \exp(-\mathrm{i}\Omega t).$$

Introduce the denotation $a_n \exp \mathrm{i}\Omega t = b_n$, then

$$\mathrm{i}\hbar\dot{a}_m = b_n F_{mn}, \quad \mathrm{i}\hbar(\dot{b}_n - \mathrm{i}\Omega b_n) = F^*_{mn} a_m.$$

Excluding the a_m, we have

$$\ddot{b}_n - i\Omega\dot{b}_n + \frac{|F_{mn}|^2}{\hbar^2} b_n = 0. \tag{17.17}$$

The solution of Eq. (17.17) is sought, as usual, in the form of $b_n = c' \exp \mathrm{i}\beta t$, where c' is a constant, and $\beta_{1,2} = \Omega/2 \pm \sqrt{\Omega^2/4 + |F_{mn}|^2/\hbar^2} \equiv \Omega/2 \pm \Delta$. As a result, we obtain

$$\begin{aligned} a_n &= C e^{\mathrm{i}\alpha_1 t} + D e^{-\mathrm{i}\alpha_2 t}, \\ a_m &= -C \frac{\hbar\alpha_1}{F^*_{mn}} e^{\mathrm{i}\alpha_2} t + D \frac{\hbar\alpha_2}{F^*_{mn}} e^{-\mathrm{i}\alpha_1} t, \\ \alpha_1 &= -\Omega/2 + \Delta, \quad \alpha_2 = \Omega/2 + \Delta. \end{aligned} \tag{17.18}$$

Let us assume that at the initial time $t = 0$ the system is in a state Ψ_m. The evolution of the system at following times is described with the function $\Psi(x,t)$. The coefficients C and D must be determined from the initial conditions

$$|a_n(0)|^2 = 0, \quad |a_m(0)|^2 = 1. \tag{17.19}$$

Using (17.18) and (17.19), we have $C = -D$; $D = F^*_{mn}/(2\hbar\Delta)$. Then the explicit form of $a_n^{(m)}$ is as follows:

$$a_n^{(m)} = -\mathrm{i}\frac{F_{mn}^*}{\hbar\Delta} e^{-\mathrm{i}\Omega t/2} \sin \Delta t.$$

The expression for the transition probability W_{mn} according to (17.10) has the form

$$W_{mn}(t) = \left|a_n^{(m)}\right|^2 = \frac{(\mu_{mn}A_1)^2/\hbar^2}{((\omega_{mn}-\omega)/2)^2 + (\mu_{mn}A_1/\hbar)^2} \sin^2\left[t\sqrt{\left(\frac{\omega_{mn}-\omega}{2}\right)^2 + \left(\frac{\mu_{mn}A_1}{\hbar}\right)^2}\right]. \tag{17.20}$$

This result was obtained by I. Rabi in 1937. It is easy to see that the formulas of the theory perturbation (17.12), (17.13) and the exact formula (17.20) is approximately the same for small perturbations (A_1).

Considering the interaction of electromagnetic fields with quantum systems we used the classical expression (17.11) for the components of external electromagnetic fields. Why can we use the classical expressions for the components of electromagnetic fields to describe their interactions with quantum particles? It appears that even for very low energy densities the electromagnetic fields we approximate them with the huge number of photons. For example, the power of 10^{-15} W (the approximate noise level in the radio diapason) for the frequency of 10^{10} Hz ($\lambda \simeq 3$ cm) corresponds to the passage of 10^8 photons per second. It is known from the theory and experiment that, if the energy of a system is composed of a large number of quanta, that system is adequately described with equations of classical physics.

17.3 General Causes of Broadening of Spectral Lines

Many of the causes of the broadening of spectral lines were already considered in the book along the description of the magnetic resonances. However, in quantum radioelectronics there are some peculiarities. Let us return to formula (17.4), which describes the radiation power of molecular systems. There is the factor $g(\nu)$ in this formula, which characterizes the frequency distribution of the radiation, i.e. the shape of a spectral line. Normally, this function is normalized according to the relation

$$\int_{-\infty}^{\infty} g(\nu)\mathrm{d}\nu \equiv 1. \tag{17.21}$$

To characterize the distribution one introduces the concept of *width of spectral line* $\Delta\nu_l$, which is defined as the frequency interval between the points at which $g(\nu)$ is reduced to a half of its the maximum value. We proceed from the uncertainty relation, represented in form $\Delta E \cdot \Delta t \sim \hbar$, which indicates that the energy uncertainty of a microparticle is ΔE, if the duration of the measurement is Δt. The information about the energy state of a particle can be obtained only if the particle will make a transition from this state in some others with the absorption or emission of a photon, thus "announcing" about its existence and its previous state. Consequently, the lifetime (τ) of a particle at an energy level is that limiting time interval, which is available to a researcher to obtain an information. The stronger effect on a particle, the faster it will change its state and, therefore, the greater the uncertainty in the knowledge of its energy. Since $\Delta E = h \cdot \Delta \nu_l$, then taking into account the relation of uncertainties we obtain

$$\Delta \nu_l \sim 1/(2\pi\tau), \tag{17.22}$$

i.e. the less the lifetime of a particle at energy levels, the more the spectral width of the transition connecting the levels. The evaluation of the line width from the ratio (17.22) is valid for all processes that cause transitions between energy levels.

In the previous chapters many causes of the broadening of spectral lines have been considered. In this chapter we will discuss the general features of two processes broadening spectral lines that are always present but have different manifestations in various devices of quantum radioelectronics.

17.3.1 The Natural Line Width

Each spectral line has natural width, which is due to the effect of spontaneous radiation. The possibility of the spontaneous transition of a quantum system from a certain excited state to lower energy states leads to the fact that the excited states cannot longer be considered as strictly stationary. If the probability of transitions to lower states is still small, these states are called *quasi-stationary*.

Let us consider a system with two states with energies E_1 and E_2. The quasi-stationary state is characterized by the law of decay:

$$\exp(-A_{21}t), \tag{17.23}$$

where A_{21} is the probability the spontaneous transition from the upper state to the lower one, and the value of $1/A_{21} = \tau$ is the average lifetime of a particle in excited (upper) state. For time intervals, which are comparable with the average time life of the particle in the upper state, previously used approximations are not enough. The initial conditions corresponding to this situation: $a_2(0) = 1$; $a_1(0) = 0$. In contrast to previous approaches, these values cannot be substituted into the right-hand side of Eq. (17.8), because the values a_1 and a_2 are essentially time-dependent (square of the modulus of the coefficient a_2 characterizes the probability of realization of the initial state). Taking into account expression (17.23) we can try to solve the system (17.8) substituting a_2 in the form $\exp(-A_{21}t/2)$ (such a representation of the coefficient a_2 satisfies the initial conditions). Now we have to assume that the initial state corresponds to an atom with energy E_2 and the final state corresponds to an atom with energy E_1 and a photon $\hbar\omega$. Using the expansion of the wave function on stationary states in the form (17.7) and the decay law (17.23) for the coefficient a_2, one can obtain

$$a_1(t) = \frac{1}{\hbar}\mathcal{H}'_{12}\frac{e^{-\mathrm{i}(\omega-\omega_{12})t - A_{21}t/2} - 1}{(\omega-\omega_{12})^2 - \mathrm{i}A_{21}/4}.$$

The distribution of the radiation is determined by the dependence of the probability of the transition from ω. For the time $t \gg \tau$, when all the atoms have moved into the lower state, the probability of the emission of a photon is

$$|a_1|^2 = \frac{|\mathcal{H}'_{12}|^2}{\hbar^2}\frac{1}{(\omega-\omega_{12})^2 + A_{21}^2/4}.$$

In the resonance approximation, which is usually used in the radiation theory, it is assumed that the range $\omega \simeq \omega_{12}$ is of great importance. Then the we may assume that $\mathcal{H}'_{12}$ is independent on the frequency, and the radiation intensity is determined, taking into account the normalization (17.21) of the function $g(\nu)$, as follows:

$$g(\nu) = \frac{A_{21}}{(\omega-\omega_{12})^2 + A_{21}^2/4}. \qquad (17.24)$$

This frequency distribution is called the *Lorentzian line shape*. The line width of (17.24) is defined as

$$\Delta\nu_l = A_{21}/(2\pi).$$

The natural line width is strongly dependent on the frequency (it is proportional to ν^3). For the radiofrequency diapason up to far millimeter range the natural linewidth is completely masked by other mechanisms of broadening. For example, in the centimeter range $\Delta\nu_l \simeq 10^{-8}$ Hz for a electric dipole and $\Delta\nu_l \simeq 10^{-13}$ Hz for a magnetic one. These values are vanishingly small compared to the width of the lines which are realized in the ordinary conditions under the influence of relaxation. However, in the optical range the width of spectral lines is determined by the process of spontaneous emission.

17.3.2 The Broadening of Spectral Lines Due to the Interaction of Microparticles with Electromagnetic Fields

Since the electromagnetic fields cause transitions between energy states, this process leads to the decrease of the lifetime of particles in energy states. It is possible to obtain the dependence of $g(\nu)$ in the presence of external radiation using Rabi's formula (17.20) for the induced transitions of a dipole.

The time of the interaction of a dipole with the electromagnetic field will be limited due to one or more relaxation processes. We assume that the interaction with the electromagnetic field is interrupted with an average of the time interval τ. If the population of the state is N, then Ndt/τ jumps out of this state will be performed during dt. The probability, that the interaction of particles with the radiation field will be interrupted after the time t due to the relaxation process, we assume in the form $f(t) = C\exp(-t/\tau)$, here C is a constant. Note that the condition $\int_0^\infty f(t)dt = 1$ must be satisfied, because it is known that the jump is necessary ever to happen. Thus, the change of the population with the emission of a photon at a frequency $\nu = \omega/(2\pi)$ during the time interval dt is

$$\mathrm{d}N(t,\nu) = N\frac{\mathrm{d}t}{\tau}f(t)\,W(t) = C_1 e^{-t/\tau}\frac{\sin^2 t\sqrt{k^2+(\omega-\omega_{12})^2/4}}{k^2+(\omega-\omega_{12})^2/4}\mathrm{d}t,$$

where $W(t)$ is the probability of the induced transitions, defined by (17.20); $k = \mu A_1/\hbar$ with $\mu \equiv \mu_{12}$; C_1 is a constant. Since the total distribution of the particles, which emit quanta at the frequency ν, is the result of the integration of $\mathrm{d}N(t,\nu)$ with respect to time in the interval (0, ∞), then for $g(\nu) \sim N(\nu) = \int_0^\infty \mathrm{d}N(t,\nu)$ we obtain the expression

$$g(\nu) = C_2\int_0^\infty e^{-t/\tau}\frac{\sin^2 t\sqrt{k^2+(\omega-\omega_{12})^2/4}}{k^2+(\omega-\omega_{12})^2/4}\mathrm{d}t$$

(C_2 is a constant). Taking into account the normalizing we finally obtain

$$g(\nu) = \frac{\sqrt{1/(4\tau^2)+(\mu A_1/\hbar)^2}}{((\omega-\omega_{12})/2)^2+1/(4\tau^2)+(\mu A_1/\hbar)^2}. \tag{17.25}$$

Equation (17.25) shows that the width of a spectral line under the influence of the radiation field is

$$\Delta\nu_l = (2/\pi)\sqrt{1/(4\tau^2)+(\mu A_1/\hbar)^2}. \tag{17.26}$$

With the increase of the amplitude of the acting field (A_1) line width (17.26) can be infinitely big. At the same time the alignment of the populations of the states, between which transitions occur, is realized and this process is called *the saturation of a spectral line*.

17.4 Frequency Stabilization with a Quantum Generator

The basis of all modern frequency standards is quantum generators or filters which use spectral lines belonging to the super-high range of radio waves. However, in the vast majority of experiments it is required to measure time intervals, using relatively

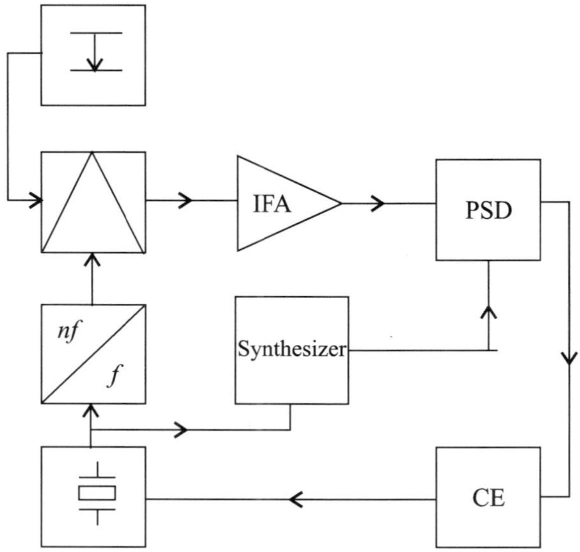

Fig. 17.1 Frequency stabilization with a quantum generator (*IFA*—intermediate–frequency amplifier, *PSD*—phase–sensitive detector, *CE* control element)

low-frequency oscillations, for example in the areas of radionavigation (beacon service) and universal time service, in systems of precise phase and frequency measurements, and so on. It often requires to use signals at certain frequencies, whose values are integers with lots of zeros after the significant digits. Since frequencies of reference quantum generators are not multiples of "round" numbers, a synthesizer should be included in a frequency standard, which converts the reference oscillator frequency in the row of standard frequencies. Moreover, the low output power of quantum generators also prevents their direct use in measurement systems. These difficulties can be avoided using automatic frequency control of a quartz generator.

The most general block-scheme of a frequency standard with a quantum generator is shown in Fig. 17.1. The signal at the intermediate frequency is created by mixing the signals of the quantum generator and the frequency-multiplied one from the auxiliary quartz generator which operates at a frequency 5 to 10 MHz. The main element is the phase-sensitive detector (PSD), which output voltage is proportional to the phase difference of the signals from the quantum generator and auxiliary quartz generator. This voltage is applied to a control element (CE) of the frequency of a quartz generator.

The modern radio devices make it easy to get any "round" frequency at the output of the standard, if the coefficient of the frequency multiplying of a quartz oscillator is properly chosen.

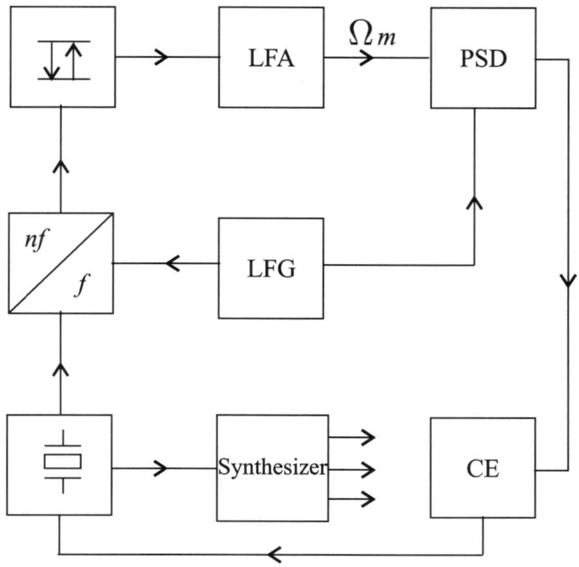

Fig. 17.2 Frequency stabilization with a quantum discriminator (*LFA*—low-frequency amplifier, *PSD*—phase-sensitive detector, *LFG*—low-frequency generator, *CE*—control element)

17.5 Frequency Standards with a Quantum Discriminator

Besides the *active quantum-mechanical frequency standards*, in which the generation occurs due to the interactions of excited atoms or molecules with a resonator (i.e. the energy is stored in a molecular system), there are *passive quantum-mechanical frequency standards* or, as they are sometimes called, *quantum-mechanical frequency standards with a quantum discriminator*, in which the molecular system is not the source of energy. In these devices a spectral line is used as a narrow-band filter, the frequency characteristic of which determines the quality of the frequency stabilization of an auxiliary quartz generator. The block-scheme of a frequency standard is shown in Fig. 17.2.

In passive quantum-mechanical frequency standards the frequency of quartz generator (f) is transformed with the synthesizer or multiplier in the frequency corresponding to the spectral transition of the working substance of the discriminator. The output signal from the quantum discriminator carries the information about the deviation of the frequency of the converted signal of a quartz generator from the frequency of the chosen spectral transition. Since frequencies of spectral lines are multidigit numbers, then for the obtaining of "round" values of frequencies of output signals the frequency synthesizers are used (see Fig. 17.2).

Let us consider the principle of adjustment of frequency of an external oscillator to the frequency of a spectral line. To determine the sign of the frequency deviation from the center of a spectral line the phase modulation in frequency multiplication stages is realized. Since the time derivative of the current phase is a frequency, this

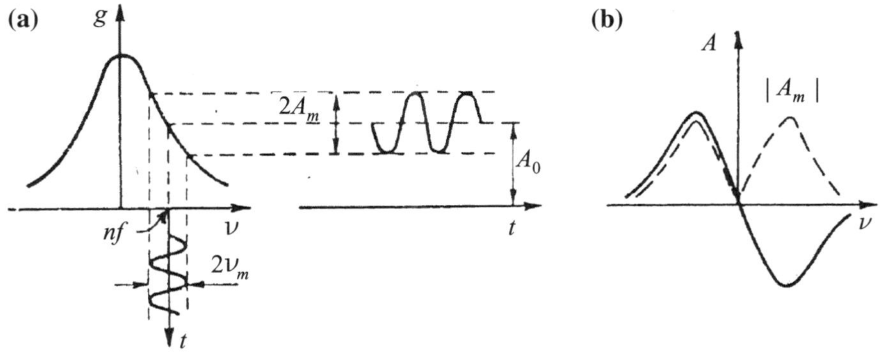

Fig. 17.3 The effect of a frequency modulated signal on a spectral line (**a**) and the signal (**b**) from PSD (Fig. 17.2)

procedure means what we have obtained the frequency modulation of the signal from the quartz generator. As a result the frequency of this signal (ν) will be

$$\nu = nf + \nu_m \cos \Omega_m t,$$

where f is the frequency the quartz generator; n is the multiplicity of the multiplier; ν_m and Ω_m are the amplitude and frequency of modulation, respectively (Fig. 17.3a).

Thus, the amplitude of induced absorption (or emission) will be characterized by the dependence

$$A(t) = A_0 + A_m \cos(\Omega_m t + \phi), \tag{17.27}$$

where A_0, A_m and ϕ are the certain quantities which are independent of time. If the modulation amplitude ν_m of frequency is small compared with the line width, i.e. $\Delta\nu_l \gg \nu_m$, then in the range of $2\nu_m$ the line shape function $g(\nu)$ can be approximated by a straight line and the value A_m is defined by the relation

$$|A_m(\nu)| = K|g'(\nu)|,$$

where $g'(\nu) \equiv dg(\nu)/d\nu$; K is the conversion factor of the detector of a spectral line. If the signal described by the expression (17.27) is applied to a phase-sensitive detector, the output signal A_{pd} will be proportional to the derivative $g'(\nu)$, because the phase of the signal $A(t)$ for $nf > \nu_0$ and $nf < \nu_0$ is shifted by 180° (see Fig. 17.3b). The middle part of the diagram is an almost linear dependence, i.e. the output signal of the phase-sensitive detector is proportional to the frequency deviation of nf from the central frequency of a spectral line. Locking the feedback loop through the control stage (see in Fig. 17.2), we easily get the automatic congruence between the quartz generator frequency and the frequency of a spectral line. The accuracy of regulation is typically 10^{-12} to 10^{-15} if the relative frequency shift of a quartz generator is about 10^{-7} during a month.

Further Readings

Dudkin VI, Pakhomov LN (2006) Quantum electronics. Devices and their applications. Tekhnosphera, Moscow (in Russian)
Marcuse D (1980) Principles of quantum electronics, 1st edn. Academic Press, New York. ISBN-10: 0124710506; ISBN-13: 978–0124710504
Siegman AE (1971) An introduction to lasers and masers. McGraw-Hill, New York
Singer JR (1959) Masers. Wiley, Newyork
Troup G (1959) MASERs microwave amplification and oscillation by stimulated emission. Methuen & Co, London
Vanier J, Audoin C (1989) The quantum physics of atomic frequency standards. Adam Hilger, Bristol
Vuylsteke AA (1960) Elements of maser theory. D. Van Nostrand Co Inc, New York
Yariv A (1988) Quantum electronics, 3rd edn. Wiley, New York

Chapter 18
Quantum Generators with Molecular and Atomic Beams

18.1 Selection Over Energy States of Molecules and Atoms Moving in Inhomogeneous Fields

For the realization of the process of quantum amplification of electromagnetic waves it is necessary to create an anomalous difference the populations of energy levels: the population of the upper level must be greater than the lower one (see Sect. 17.1). More over, it was shown in Sect. 3.2, that it is impossible to create such stationary state in two-level system. In certain devices of quantum electronics it does not require to reach this condition, but one need to obtain the difference of populations which is very different from the equilibrium. Both these tasks can be solved in the same way, if it is possible to direct particles through a special sorting system, in which the deviation of particle trajectories is realized due to the interaction of the particle dipole with the inhomogeneous field. Since calculations are almost identical for the electric and magnetic dipoles, we introduce the vector $\vec{A}$, which will denote an electric field component $\vec{\mathcal{E}}$ in the case of an electric dipole and magnetic field components $\vec{B}$ in the case of a magnetic dipole.

Let us consider a particle which energy can be either E_1 or E_2 and, for definiteness, we assume that $E_2 > E_1$. States of the particle are described with wave functions $\psi_{1,2}$, determined from the Schrödinger equation:

$$\widehat{\mathcal{H}}_0\psi_{1,2} = E_{1,2}\psi_{1,2},$$

where $\widehat{\mathcal{H}}_0$ is the Hamiltonian of the particle in the absence of a field $\vec{A}$. If the particle has a dipole moment $\vec{\mu}$ and is placed in an electric or magnetic field $\vec{A}$, then the new Hamiltonian of the system will be

$$\widehat{\mathcal{H}} = \widehat{\mathcal{H}}_0 - \hat{\vec{\mu}}\vec{A}. \tag{18.1}$$

It is useful to seek the wave function, describing the new state of the particle, in the form:

V. I. Chizhik et al., *Magnetic Resonance and Its Applications*,
DOI: 10.1007/978-3-319-05299-1_18,

$$\psi = a\psi_1 + b\psi_2.$$

If the field $\vec{A}$ is time-independent, then a and b are constants.

Substitute the wave function of ψ in the Schrödinger equation with the Hamiltonian (18.1):

$$[\widehat{\mathcal{H}}_0 - \hat{\vec{\mu}}\vec{A}](a\psi_1 + b\psi_2) = E(a\psi_1 + b\psi_2). \tag{18.2}$$

Here E is the energy of the particles in the presence of the field $\vec{A}$. To identify new energy values, we will multiply (18.2) consecutively on ψ_1^* and ψ_2^* and integrate over all space variables taking into account that the field $\vec{A}$ can be considered as homogeneous within a molecular system. Assuming for simplicity that $\vec{\mu}||\vec{A}$, we obtain:

$$\begin{aligned} a(E_1 - E) - b\mu_{12}\vec{A} &= 0, \\ -a\mu_{21}\vec{A} + b(E_2 - E) &= 0, \end{aligned} \tag{18.3}$$

where $\mu_{ik} = \int \psi_i^* \hat{\mu}\psi_k \mathrm{d}v$ are the matrix elements dipole moment between the states i and k (in the following we will restrict the analysis, for simplicity, by the case of non-degenerated states for which $\mu_{ii} = 0$). The system of Eq. (18.3) has a solution if its determinant equals zero:

$$E^2 - E(E_1 + E_2) + E_1E_2 - |\mu_{12}|^2A^2 = 0,$$

that allows us to find new values of the particle energy ($h\nu = E_2 - E_1$):

$$E^{(1,2)} = (E_1 + E_2)/2 \pm \sqrt{(h\nu)^2/4 + |\mu_{12}|^2A^2}.$$

We introduce the notation $E^{(1,2)}$ so that for $\vec{A} \to 0$ the new the energy $E^{(1)}$ will pass to E_1 and $E^{(2)}$ to E_2 (Fig. 18.1). Then

$$\begin{aligned} E^{(1)} &= (E_1 + E_2)/2 - \sqrt{(h\nu)^2/4 + |\mu_{12}|^2A^2}, \\ E^{(2)} &= (E_1 + E_2)/2 + \sqrt{(h\nu)^2/4 + |\mu_{12}|^2A^2}. \end{aligned} \tag{18.4}$$

The potential energy of the dipole $\vec{\mu}$ in the field $\vec{A}$ is the difference $E^{(1,2)} - E_{1,2}$. By combining the expressions (18.4) we obtain

$$E_{\mathrm{pot}}^{(1,2)} = \mp(h\nu/2)\left[\sqrt{1 + (4|\mu_{12}|^2A^2)/(h\nu)^2} - 1\right]. \tag{18.5}$$

The force acting on a particle is defined as

$$\vec{F} = -\mathrm{grad}\,E_{\mathrm{pot}}.$$

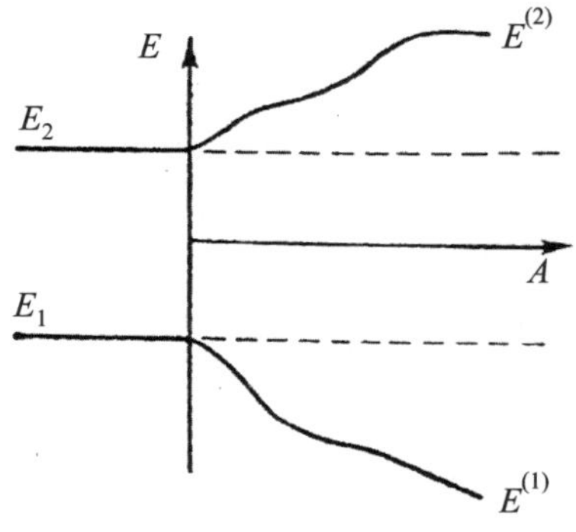

Fig. 18.1 The scheme of energy states for a particle with two energy levels

Therefore, the force exits ($\vec{F} \neq 0$) if the field $\vec{A}$ is inhomogeneous.

The particle in the upper energy state E_2 is subjected by the force directed towards the reducing of the field and, in contrary, the particle with energy E_1 is subjected by the force directed towards the increasing of the field. This effect is the basis for the selection of flying particles with static inhomogeneous electric or magnetic fields.

18.2 Rotational-Inversional Spectra of Molecules

The first quantum generator was created on the basis of the ammonia (NH_3) spectrum which is not belong, strictly speaking, to the phenomenon of magnetic resonance but due to the historical reasons this generator is worth to be described. Besides we will apply the theoretical developments to the description of other devices.

The majority of the spectra in the microwave region is due to transitions between rotational energy levels of molecules in gaseous phase. Usually, the rotation of a molecule (as well as any body) can be described by an ellipsoid of inertia, the axis of which coincide with the direction of the principal axes of inertia and can be determined by the principal moments of inertia (I_x, I_y, I_z). Many properties of rotational motion of molecules can be described by classical expressions, attracting the principle of correspondence. In the classical mechanics the rotational energy of a body in the principal axes of inertia is expressed by the following relation:

$$E = \frac{1}{2} I_x \omega_x^2 + \frac{1}{2} I_y \omega_y^2 + \frac{1}{2} I_z \omega_z^2 \tag{18.6}$$

or, taking into account the equalities $I_i \omega_i = J_i$ (i corresponds to the axis x, y, z; J_i is the component of the angular momentum $\vec{J}$), by the ratio

$$E = \frac{J_x^2}{2I_x} + \frac{J_y^2}{2I_y} + \frac{J_z^2}{2I_z}.$$

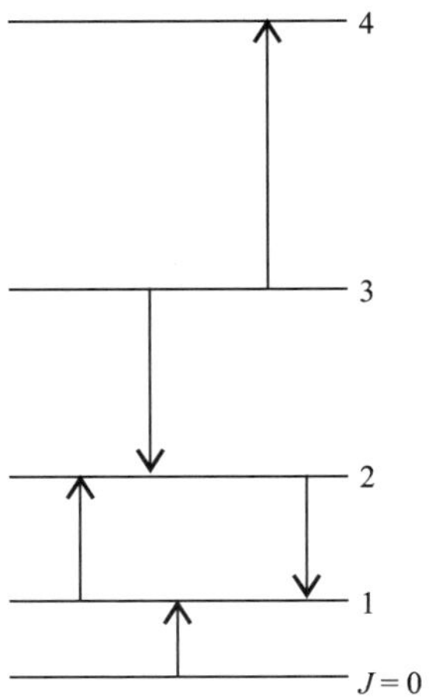

Fig. 18.2 The scheme of energy states for a linear molecule

For a linear molecule $I_x = I_y = I_A$, $I_z = 0$ and the expression 18.6 reduces to the following:

$$E = \frac{J_x^2 + J_y^2}{2I_A} = \frac{\vec{J}^2}{2I_A}.$$

Since the total angular momentum is quantized according to the rule $\vec{J}^2 \rightarrow \hbar J(J+1)$, then the quantum mechanical expression for energy of a linear molecule takes the form (we have used the principle of correspondence):

$$E = BJ(J+1), \tag{18.7}$$

where B is the so-called rotational constant associated with the moment of inertia molecule; J is the rotational quantum number, which takes values of 0, 1, 2, 3, …Thus, for linear molecules one obtains a system of divergent energy levels (Fig. 18.2). Allowed quantum transitions obey the rule $\Delta J = \pm 1$.

It follows from (18.7) that the lines in a rotational spectrum are located at equal distances on the frequency scale. The values of the parameter B, expressed in units of frequency ($\Delta E = h\nu$), lies typically in the microwave range (eg, for molecules OSC $B = 6.081$ GHz, for the HCN $B = 44.316$ GHz, and so on). The view of the spectrum of linear molecules is very simple, and it is immediately clear which information on the properties of a molecule can be derived from it. First of all, one can determine the moment inertia of the molecule and, since the masses of nuclei are determined independently (for example, by mass-spectrometry), then internuclear distances in the molecule are accurately calculated.

Let us consider now a symmetric-top molecule (the symmetry axis z), which is characterized by the relation $I_x = I_y = I_B \neq I_z = I_C$. It can be obtained from the expression (18.6)

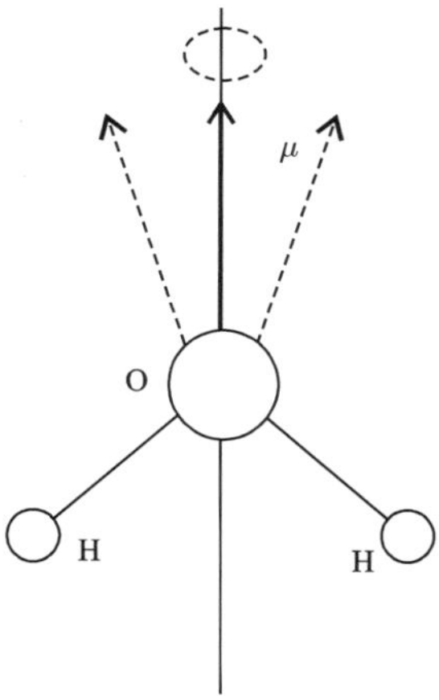

Fig. 18.3 The example of a symmetrical molecule

$$E = \frac{\vec{J}^2}{2I_B} + J_z^2\left(\frac{1}{2I_C} - \frac{1}{2I_B}\right). \tag{18.8}$$

The projection of $\vec{J}$ at z axis is quantized by the rule $J_z \to \hbar K$, where K is an integer varying from J to $-J$, i.e. $|K| \leq J$. Then expression (18.8) can be rewritten as

$$E = B(J+1)J + (C-B)K^2. \tag{18.9}$$

where B and C are constants, which include the universal constants and inertia moments about the respective axes. Equation (18.9) describes the law of the distribution of energy levels of molecules such as symmetric top. If the top is "oblate", then $(C-B) < 0$, for a "prolate" top $(C-B) > 0$.

The concrete location of the levels we consider on the example the ammonia molecule. It is convenient to draw the scheme of levels taking into account the rules of transitions: $\Delta J = \pm 1$; $\Delta K = 0$. The latter selection rule can be explained as follows. Figure 18.3 shows a water molecule, which has the axis of twofold symmetry (all arguments are valid also for cases of the symmetry axes of higher orders). The rotation of such a molecule at 180° transforms it into a physically equivalent position. If an electric dipole of the molecule is directed at the certain angle to the axis of symmetry, then after turning it will take a new position (dashed line in Fig. 18.3) that contradicts the condition of symmetry of the molecule. There would be no contradictions, if we assume the orientation of the dipole along the axis of symmetry. But in this case components of the dipole moment, which are perpendicular to the axis of symmetry, are equal to zero. The moment of forces acting on the dipole due to interaction with electric field $\vec{E}$, is the vector product $[\vec{\mu}, \vec{E}]$. Therefore, the electric field component, directed along the axis of symmetry of the molecule does not affect on the rotation of the molecule, and perpendicular components cause changes in molecular rotation around the directions which do not coincide with the axis of symmetry. Consequently,

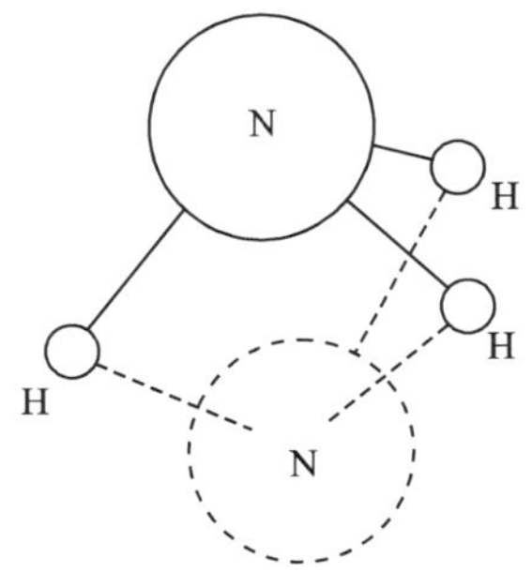

Fig. 18.4 The inversion of a symmetrical molecule

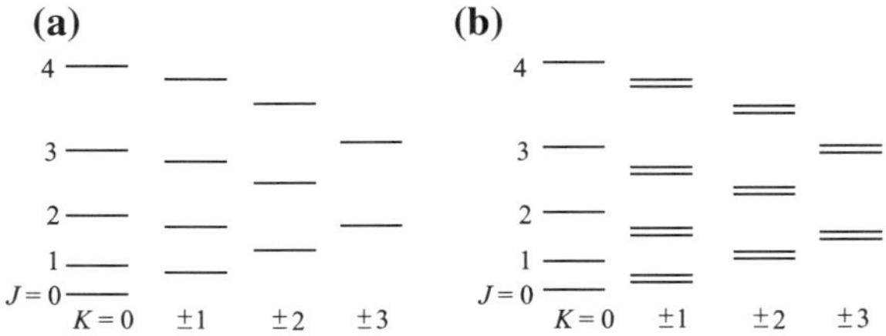

Fig. 18.5 The scheme of energy states for a symmetrical-top molecule without (**a**) and with (**b**) regard for the inversion process

the radiation field cannot change the angular momentum about the axis symmetry of the molecule. From the correspondence principle, we obtain the selection rule: $\Delta K = 0$.

Therefore it is convenient to construct the system of levels of a symmetric top in the form of columns for different values of K (transitions between the columns are forbidden). Ammonia molecule is oblate like a top (Fig. 18.4), and for it the relation $(C - B) < 0$ is valid. As a result, we obtain the scheme of energy levels shown in Fig. 18.5a. In the construction of the columns it is taken into account that $|K| \leq J$ (i.e. the projection of a vector cannot be greater than its length). It follows from expression (18.9) that all states, except those for which $K = 0$, possess double-degeneration with respect to K, because this number is included in the formula in the square. One can see the coincidence of the energy levels of the first column with the energy spectrum of the linear molecule (Fig. 18.2).

In the absence of external influences the vector $\vec{J}$ in the space remains unvariable orientation. The molecule performs a complex motion consisting of rotation around the axis of symmetry and the precession of the axis of the vector around the full angular momentum.

The scheme of energy states, shown in Fig. 18.5a, is really complicated because of the special vibrations in the ammonia molecule. The nitrogen atom can be located above the plane of the hydrogen atoms and below (see the dotted image in Fig. 18.4), and there is some probability of the transition from one state to another. This transition is called *inversion*. Both structures are physically equivalent, but for the transition between them it is necessary to overcome the finite energy barrier. The nitrogen atom makes rapid oscillations on one side of the plane of the hydrogen atoms and after a large number of such oscillations can jump to the other side, and begin to oscillate in its new position. The frequency of fast oscillations corresponds to the infrared

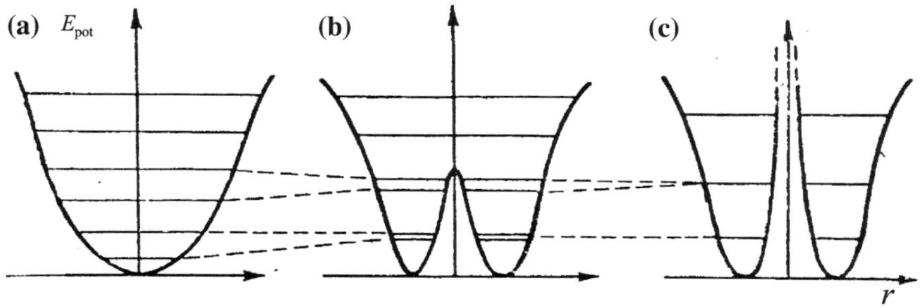

Fig. 18.6 The scheme of energy states for the vibrational degree of freedom of a molecule NH_3: "flat" (**a**), real (**b**), and "super rigid" (**c**) molecule

spectrum and the frequency of jumps (inversion) belongs to the microwave range of radio waves.

Besides the atom of the ammonia molecule can participate in different types of oscillations, but to describe the process of inversion one usually simplifies the picture of the movement. Let us consider pseudo-vibrations of a molecule, which involving a rigid frame of the three hydrogen atoms and a nitrogen atom (such a representation preserves the basic features of the process). If the molecule of ammonia was plane, the potential energy of the nitrogen atom would have a dependence shown in Fig. 18.6a. The diagram also shows the vibrational energy levels corresponding to approximating the potential curve of a harmonic oscillator. However, if in the real location of atoms in the ammonium molecule it would be required to overcome the infinite potential barrier for the inversion transition, the assignment of vibrational levels would be consistent with Fig. 18.6c (since the potential curves are approximately in two times steeper, the distance between vibrational levels would be twice as large). For the finite value of the potential barrier the picture is intermediate (see Fig. 18.6b): the energy levels above the barrier are roughly the same as for the case shown in Fig. 18.6a, but below the barrier there is the doublet splitting energy states. This is the *inversion splitting*. Since $\omega_{\rm vibr} \gg \omega_{\rm rot}$, then for each vibrational state we have a system of rotational states (see Fig. 18.5a). As a result, the lower vibrational-rotational levels will be split (the scheme of arrangement is shown in Fig. 18.5b). The magnitude of the inversion splitting depends strongly on the height of the potential barrier (see Table 18.1). The proximity of levels to the "top" of the potential barrier is also influenced on the value of the inversion splitting. For example, for the second vibrational state the ammonia molecule (Fig. 18.6b) the splitting is of 1080 GHz. However, in experiments, these transitions are usually not observed as $\hbar\omega_{\rm vibr} \gg KT$ and the second levels of the vibrational states are not populated.

It follows from above qualitative considerations that the magnitude of the inversion splittings is the same for different J and K. But in reality this is not so. Due to the centrifugal perturbation of the structure of a molecule the frequencies of inversion transition for different J and K are different, and this leads to a rich and well-resolved rotation-inversion spectrum of ammonia which is located close to frequency of 24 GHz.

Table 18.1 The invertion frequencies for different hydrides

Molecule	Potential barrier, cm^{-1}	Frequency of an invertion transition
NH_3	2077	24 GHz
PH_3	6075	0.14 MHz
AsH_3	11220	0.5 oscilation per year

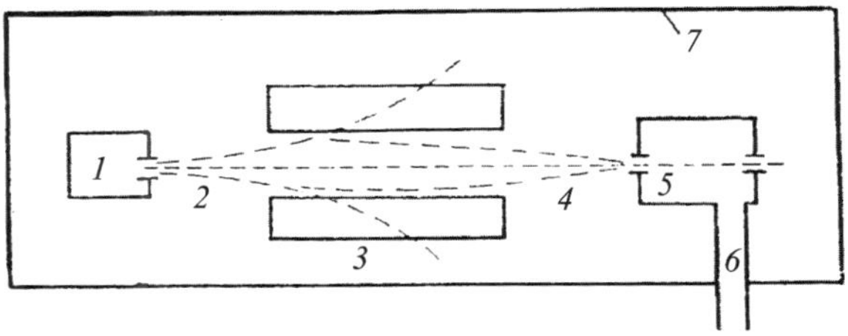

Fig. 18.7 The scheme of the generator with a beam of the ammonia molecules

If molecule is an asymmetric top ($I_x \neq I_y \neq I_z$), then the system of levels is very complex. Since such molecules are not used in quantum electronics devices, they will not consider.

18.3 Generator with a Beam of the Ammonia Molecules

The schematic representation of the oscillator with a beam of ammonia molecules is shown in Fig. 18.7. The source of the molecular beam is a chamber *1*, which contains the ammonia gas under a pressure of approximately 10 Pa at temperature of about 300 K. The camera contains one or more holes forming the molecular beam. To form a molecular beam in which there are no collisions (rather than hydrodynamic flow of gas) the size of the holes must not exceed the mean free path of molecules in the chamber. Since the probability of molecules to go out through the hole is proportional to their speed, distribution of molecular velocities in the beam is slightly different from the Maxwell one for a gas.

Ammonia molecules, flying zone *2* (see Fig. 18.7), are distributed among energy states (see Figs. 18.5 and 18.6) in accordance with the Boltzmann law, i.e. the higher the energy level, the smaller amount of molecules is found in it. Naturally, such a beam of molecules is not suitable for the purposes of quantum amplification. It was shown in Sect. 18.1, that in the case of two-level system one can to select molecules in the upper and lower energy states using the movement in inhomogeneous static electric or magnetic fields. The molecule of ammonia possesses an electric dipole, so in this case for sorting the electric field should be applied. Location of the energy levels of the ammonia molecules is such that the inversion splitting are a set of isolated pairs of levels (see Fig. 18.5b). Consequently, we can use the results obtained in Sect. 18.1 for the ammonia molecules.

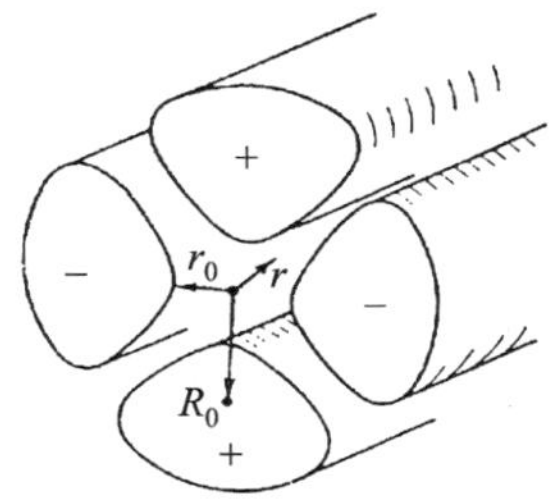

Fig. 18.8 The quadrupole capacitor

At the present time there are many different designs of sorting systems. In the first devices for this purpose the *quadrupole capacitor* was used, which intensity of electric field is calculated by the formula

$$|\mathcal{E}| = \frac{V}{R_0}\frac{r}{r_0},$$

where V is the difference of potential between adjacent electrodes; R_0 is the distance to the focus of a hyperbolic surface of the electrode, and the remaining notation is clear from Fig. 18.8.

To calculate the trajectory of the molecule through the sorting system *3* (see in Fig. 18.7), we will use the results of Sect. 18.1. For the NH_3 molecules the relation $\mu\mathcal{E} \ll h\nu_{\text{inv}}$ is realized for any real fields. Therefore, the formula (18.5) becomes

$$E_{\text{pot}}^{(1,2)} \simeq \mp|\mu|^2\mathcal{E}^2/(h\nu_{\text{inv}}). \tag{18.10}$$

The trajectory of molecules in a nonuniform field is described by the equation

$$m\frac{\mathrm{d}^2\vec{r}}{\mathrm{d}t^2} = -\mathrm{grad}\,E_{\text{pot}}^{(1,2)},$$

which in view of (18.10) can be put in the form

$$\mathrm{d}^2\vec{r}/\mathrm{d}t^2 \mp \Omega^2\vec{r} = 0, \tag{18.11}$$

here $\Omega = (\mu V/(R_0 r_0))\sqrt{2/(h\nu_{\text{inv}})}$.

The solution of Eq. (18.11), taken with a minus sign (lower inversion level), describes the trajectory of the molecules according to the formula

$$\vec{r}(t) = \vec{r}(0)\exp\Omega t, \tag{18.12}$$

where $\vec{r}(0)$ is the initial value of $\vec{r}(t)$. Thus, the molecules deviate from the axis and migrate into the peripheral region of the sorting system. These molecules do not clutter the working volume because the cavity of the electrodes of the quadrupole capacitor are filled with liquid nitrogen, and ammonia freezes at the electrodes.

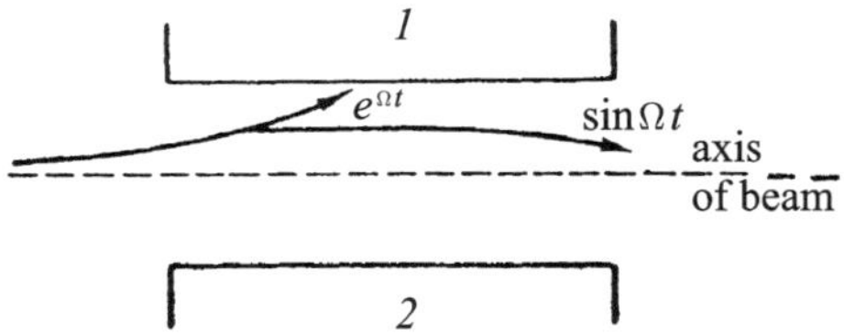

Fig. 18.9 The movement of beam molecules in the sorting system

For the "active" molecules at the upper inversion level, the solution of Eq. (18.11) can be represented as a periodic function:

$$\vec{r}(t) = \vec{A} \sin (\Omega t + \phi), \tag{18.13}$$

where the constants $\vec{A}$ and ϕ are determined from the initial conditions of the molecule flight within the sorting system. Figure 18.9 shows the molecule trajectories according to (18.12) and (18.13) in the space between electrodes, *1* and *2*. Of course, if the active molecule flies too fast or at a large angle to the beam axis, the sorting system cannot decline it in the right direction. Thus, we come to the conclusion that there is a certain angle of the capture of molecules. As a result, only small part of the active molecules forms the beam at the output of the sorting system. To ensure the adequate performance of the quadrupole capacitor the potential difference of about 20–30 kV is applied to its electrodes. The distance r_0 (see Fig. 18.8) is usually chosen about several mm, and the length sorting system is about 100–150 mm. Such a sorting system allows ones to get the output (the zone of *4* in Fig. 18.7) approximately 10^{14} molecules per second, if the input beam intensity is of 10^{18} molecules per second.

To prevent the scattering of a beam of ammonia molecules in the air, all considered elements and the resonator *5* (see Fig. 18.7) are placed in a sealed cover *7*, from which gaseous medium pumped to a pressure of 10^{-4} Pa. The radiofrequency power output is taken with a waveguide *6*.

Thus, the beam of active molecules enters in the cavity *5* (see Fig. 18.7). As a result of the interaction with the cavity field the molecules radiate electromagnetic energy. It follows from the formulas of Sect. 17.2 that the radiation of molecule would be effective if the frequency of the acting field is matched with the frequency spectral transition. Then such a signal is amplified in the cavity.

Our task is to find the conditions when the self-excitation of the system "a resonant circuite–a beam of excited molecules" is realized. It is natural that self-excitation of such a system requires that the power radiated by an ensemble of molecules would be more than the power of losses in a resonator, i.e.

$$P_{\text{rad}} > P_{\text{los}}. \tag{18.14}$$

The process of the self-excitation is as follows. If the resonator is tuned approximately on the frequency of the spectral transition, then its pass-band comprises a noise component at the transition frequency, that induces the radiation of molecules. This leads to the increase of the spectral density of electromagnetic field at the radiation frequency. The process is developing like an avalanche before the saturation effect of the spectral transition (see Sect. 17.3). As a result, the stationary amplitude of oscillations appears in the resonator.

Equation (18.14) can be concretized expressing P_{rad} and P_{los} via parameters of the beam and resonator. The certain selection of molecules in respect of their velocities is performed during the formation of the beam and in the sorting system. Therefore, we can assume that molecules are flying with the equal speed and have passed the cavity during the τ. Then, the power radiated by the beam of molecules is

$$P_{\text{rad}} = N_0 h\nu W(\tau), \tag{18.15}$$

where N_0 is the number of molecules, entering in the resonator per 1 s; $W(\tau)$ is the probability of the emission of a photon $h\nu$ during the time interval τ. Since molecules fly with high speed ($\bar{v} \simeq 5 \cdot 10^4$ cm/s) and, therefire τ is very small (the cavity length is usually 10–15 cm), then $W(\tau)$ one can use the expression (17.13), i.e.

$$W(\tau) = (\mu \mathcal{E}_1 \tau / \hbar)^2,$$

where μ is the matrix element for the inversion transition; $\mathcal{E}_1$ is the amplitude of electric field acting in the resonator.

Let us express the power of losses in terms of the quality of a resonator. The general definition of the quality factor (Q) of an oscillating system is given by

$$Q = 2\pi E_{\text{st}} / E_T, \tag{18.16}$$

where E_{st} is the energy stored in the system; E_T is the energy which is spent in the system for a period (T) of oscillations. It follows from the formula (18.16) that the total quality factor of the system (Q_L) (load system) can be expressed through partial quality factors characterizing the different channels of energy dissipation:

$$\frac{1}{Q_L} = \sum_i \frac{1}{Q_i} = \sum \frac{E_{Ti}}{2\pi E_{\text{st}}}.$$

For example, the quality of the resonator *5* (see Fig. 18.7) is defined by the quality of non-loaded resonator (Q_0) and the quality of connection with a waveguide (Q_1): $1/Q_L = 1/Q_0 + 1/Q_1$. With the help of the formula (18.16) one can obtain that

$$P_{\text{los}} = \omega E_{\text{st}} / Q_L. \tag{18.17}$$

If we express the E_{st} in (18.17) via the density of energy, the condition of the self-excitation (18.14) can be written as

$$N_0 > \frac{h}{16\pi^2} \frac{1}{\mu^2 \tau^2} \frac{V}{Q_L} \tag{18.18}$$

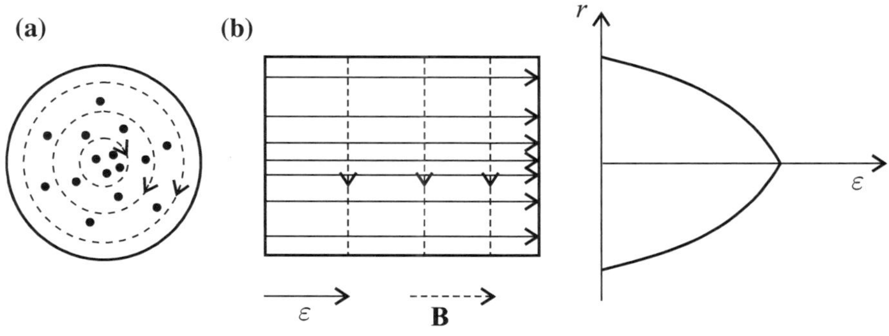

Fig. 18.10 The structure of electromagnetic fields in a cylindrical resonators for the TM_{010} mode

(V is the volume of a resonator). In the analysis of the expression (18.18) should be mind that N_0 is the number of emitting molecules at a certain frequency.

It was indicated in Sect. 18.2 that the inversion spectrum of the ammonia consists of many lines due to different centrifugal perturbations of the structure of the molecule. The most intensive line belongs to the transition between levels with $J = 3$, $K = \pm 3$ (see in Fig. 18.5) of the ground vibrational state (see Fig. 18.6b). The frequency of this transition is 23,870.13 MHz ($\lambda \simeq 1.25$ cm). Under normal conditions only 6 % of all molecules belong to this state. As it can be seen from the foregoing, one of the most significant sites molecular generator is a resonator. The natural frequencies of oscillations and Q-quality are the main characteristics of the resonator. In quantum generators on the beams *cylindrical resonators* are often used. The generator with the beam of ammonia molecules contains a cylindrical resonator with the excitation of oscillations of the TM_{010} (Fig. 18.10a), which naturally consistent with a beam of flying molecules. The inlet and outlet holes are located in the center of the ends of the resonator, and molecules pass through the region of maximum electric field (which dependence is shown in Fig. 18.10b), that ensures good connection a resonator and an excited beam. The natural frequency of the oscillations in a resonator for the TM_{010} mode depends only on the radius of a resonator, Q-factor takes very large values of ($\sim 10^4$) and it increases with the length of a resonator.

18.4 The Equation Describing the Work of a Molecular Generator

The simplified consideration of the condition of self-excitation of a molecular generator, which was done in the previous section, yields the condition (18.18) for the number of molecules, that must enter in a resonator. However, the approach used does not allow the estimation of the frequency or amplitude of the generation in a resonator. This information can be obtained by solving the equation for the electric field $\mathcal{E}$ in a resonator:

$$\frac{d^2\mathcal{E}}{dt^2} + \frac{\omega_0}{Q}\frac{d\mathcal{E}}{dt} + \frac{\omega_0^2}{\varepsilon}\mathcal{E} = 0, \tag{18.19}$$

in which the complex dielectric constant (ε) takes into account properties of the excited beam, and ω_0 and Q are the frequency and Q factor of a resonator.

The first step should be to establish the explicit form of the permittivity of media, which can be expressed via the polarizability of a molecule:

$$\varepsilon = \varepsilon' - i\varepsilon'' = 1 + 4\pi N(\varkappa' - i\varkappa''),$$

where N is the number of molecules in 1 cm^3; $\varkappa = \varkappa' - i\varkappa''$ is the polarizability of a molecule. The polarizability, in turn, is associated with molecular dipole moment by the ratio

$$\mu = \varkappa\mathcal{E}$$

or for the electric field of the form (17.11)

$$\mu = \varkappa 2\mathcal{E}_1 \cos\omega t. \tag{18.20}$$

To obtain expression for the polarizability the classical Eq. (18.20) should be compered with the analogical one calculated with the quantum-mechanical approach:

$$<\mu>= \int \Psi^* \hat{\mu} \Psi dv.$$

The distribution function of molecules over the time, spent in a resonator, can be expressed using the following approximate expression:

$$dN = \frac{N}{\tau} e^{-t/\tau} dt. \tag{18.21}$$

Distribution of (18.21) leads to a Lorentzian spectral line shape. In spite of the fact that in the molecular generator there is a partial selection of molecules on velocities, this approximation is quite acceptable for the evaluations of working characteristics of the generator. Calculations lead to expressions:

$$\varepsilon' = 1 - A\xi \frac{(\omega_{mn} - \omega)\tau}{(\omega_{mn} - \omega)^2 + (1/\tau)^2 + \xi\mathcal{E}_1^2},$$

$$\varepsilon'' = -A\xi \frac{1}{(\omega_{mn} - \omega)^2 + (1/\tau)^2 + \xi\mathcal{E}_1^2} \tag{18.22}$$

with $A = 4\pi N_0\hbar/V$; $\xi = |\mu_{mn}|^2/\hbar^2$; V is the volume of a resonator.

We can now proceed to the solution of Eq. (18.19). Let us seek the solutions in the form

$$\mathcal{E} = \mathcal{E}_1 e^{i\omega t}. \tag{18.23}$$

Substituting (18.23) into (18.19), one can obtain

$$-\omega^2 + i\omega_0 \omega Q + \frac{\omega_0^2}{\varepsilon} = 0,$$

from which we easily have (equating to zero separately the real and imaginary parts):

$$\begin{aligned} \frac{\varepsilon'}{\varepsilon'^2 + \varepsilon''^2} &= \left(\frac{\omega}{\omega_0}\right)^2, \\ \frac{\varepsilon''}{\varepsilon'^2 + \varepsilon''^2} &= \frac{-\omega}{Q\omega_0}. \end{aligned} \tag{18.24}$$

The expressions (18.22) for ε' and ε'' contain two unknown values: an amplitude $\mathcal{E}_1$ and a frequency ω. Then there are two algebraic equations for the two unknowns. It is easy to see from Eq. (18.24) that

$$\xi\tau^2\mathcal{E}_1^2 = AQ\xi\tau^2\omega/\omega_0 - 1 + \tau^2(\omega_{mn} - \omega)^2 + A\xi(\omega_{mn} - \omega)\tau^3.$$

Naturally, the frequency of oscillations will be close to the frequency of the spectral transition. Besides, we assume that the resonator is fine-tuned to the frequency of the transition. Then $\omega \simeq \omega_{mn} \simeq \omega_0$ and

$$\xi\tau^2\mathcal{E}_1^2 \simeq AQ\xi\tau^2 - 1. \tag{18.25}$$

Equation (18.25) determines the amplitude ($\mathcal{E}$) of the steady-state oscillations. The condition of the self-excitation can be obtained from the expression for the amplitude with the demand of its reality (i.e. that $\mathcal{E}_1^2 > 0$):

$$AQ\xi\tau^2 > 1. \tag{18.26}$$

Solving (18.26) with respect to N_0, we find that

$$N_0 > \frac{h}{8\pi^2}\frac{1}{|\mu_{mn}|^2\tau^2}\frac{V}{Q},$$

and this agrees with (18.18) up to a numerical factor. The difference in the numerical coefficients is explained by different approximations of the distribution of molecules in respect of the flight time through the resonator.

If the condition of the self-excitation is strongly fulfilled, then the unit on the right side of the Eq. (18.25) can be neglected, and the amplitude of oscillations of in the resonator is determined simple expression:

$$\mathcal{E}_1^2 = AQ, \tag{18.27}$$

which implies that the amplitude of steady-state oscillations depends only on resonator parameters and beam intensity.

Using the expression (18.27) one can calculate the maximum power of a quantum generator with a beam of molecules:

$$P = \frac{\omega E_{\mathrm{st}}}{Q} = \frac{1}{2} N_0 h\nu.$$

It follows from this expression that the the maximum output power of the generator is equal to half of the power that could radiate by all molecules which enter in a resonator. This can be explained by the fact that the radiated molecules begin to absorb electromagnetic energy during the further passage through a resonator. If we take the $N_0 = 10^{14}$ molecules/s (the beam intensity frequently occurring in the experiment), and $\nu = 24$ GHz, we obtain $P \simeq 10^{-2}$ erg/s $= 10^{-9}$ W, i.e. the power that is easy to register with a conventional microwave receiver. Eliminating from Eq. (18.24) $\mathcal{E}_1^2$, one can derive the equation for the steady-state oscillation frequency (ω) in a resonator:

$$\omega^3 + \omega\omega_0^2\left(\frac{\omega_0\tau}{Q} + \frac{1}{Q^2} - 1\right) - \frac{\omega_0^3\omega_{nm}\tau}{Q} = 0,$$

the approximate solution of which is:

$$\omega \simeq \omega_{mn}\left(1 + \frac{2Q}{\omega_{mn}\tau}\frac{\omega_0 - \omega_{mn}}{\omega_0} - \frac{1}{Q\omega_{mn}\tau}\right). \tag{18.28}$$

We will later return to the analysis of the expression (18.28).

18.5 The Stability of the Frequency of the Quantum Generator with a Beam of Ammonia Molecules

Since quantum generators are of great interest as high quality standards of frequency and time, the question about stability and reproducibility of their frequencies requires special consideration.

18.5.1 The Influence of the Spectral Line Width

It was shown in Sect. 17.3 that the lifetime of a particle at an energy level determines the monochromaticity of radiation for the transition from one energy state to another (i.e. the width of spectral lines). This law keeps for all processes that reduce the time of observation of a particle in an energy state. The expression (18.28) contains a parameter τ, which characterizes the mean interaction time of a molecule with the resonator field. Therefore a researcher has the time τ for the detection of radiation of molecules. In this regard, it is obvious that in order to reduce a spectral line width it is desirable to increase the flight time of molecules through a resonator. This can be achieved *by increasing the resonator length* or *by using "slow" molecules*. To reduce the average velocity of a beam, one can use special fields which slow down molecules or remove "fast" molecules from a beam. However, methods for the reduction of the beam velocity is very complicated. The possibility of increasing the resonator length is limited by influence of the Doppler effect on a spectral line. For compromising τ the value of the last term in (18.28) would be approximately 10^{-11}, and we must remember that the value of the second term in brackets depends on the accuracy of the tuning of a resonator to the frequency of a spectral line. If the resonator tuning is produced for the maximum output power, then the second term in brackets takes the value of the order of 10^{-9} and defines the basic deviation from the frequency ω_{mn}. To maintain the resonator setting in the time, one uses the thermocompensation of its dimensions or thermostabilization.

To characterize the spectral line width the parameter *quality factor of line* (Q_l) is introduced, which is defined as

$$Q_l = (\omega_{mn})/2. \tag{18.29}$$

The expedience of introducing the line quality factor Q_l in the form (18.29) can be explained as follows. Transform (18.29), using (17.22):

$$Q_l = \frac{2\pi\nu_{mn}}{2}\frac{1}{2\pi\Delta\nu_l} = \frac{\nu_{mn}}{2\Delta\nu_l}.$$

The last expression coincides with the common definition of the quality factor of tuned circuit in electronics. The small difference in the numerical factors is due to different definition of the spectral distribution width (bandwidth) of resonant systems. Taking into account the smallness of the last term in expression (18.28) one can rewrite it as

$$\omega - \omega_{mn} \simeq \frac{Q}{Q_l}(\omega_0 - \omega_{mn}). \tag{18.30}$$

For the generator with the beam of ammonia molecules ratio Q/Q_l is approximately 10^{-3}.

The phenomenon of the change of a generator frequency due to inaccurate tuning of the resonator on the frequency of a spectral line is so-called *frequency pulling* which holds for all generators operating in the autooscillating regime, regardless of whether the molecular system is a source of energy for generation or serves as a narrow-band filter in the feedback loop (as it is usually done in NMR-generators). The frequency pulling is one of the main causes of instability or poor reproducibility of the frequency of generators.

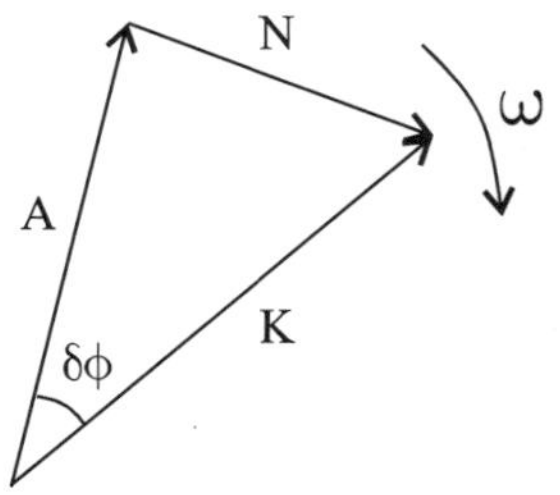

Fig. 18.11 The vector diagram of the oscillating fundamental mode ($\vec{A}$), the noise signal ($\vec{N}$) and the resulting the vector ($\vec{K}$) at any point in time

18.5.2 The Influence of Thermal Noise

The general cause of frequency fluctuations is the existence of thermal noises in a resonator. The noise power for any passive circuit at the temperature T in the radio diapason is estimated by the formula

$$P_{\mathrm{N}} = kT \cdot \Delta\nu,$$

where k is the Boltzmann constant; $\Delta\nu$ is the bandwidth of a system. Naturally, the noise signals are added to the generated one with random phases, and this causes fluctuations in the frequency of the output signal.

Figure 18.11 shows the vector diagram of the oscillating fundamental mode ($\vec{A}$), the noise signal ($\vec{N}$) and the resulting the vector ($\vec{K}$) at any point in time. The oscillation amplitude A is proportional to the square root of the output power of the generator P_0 and $N \propto \sqrt{P_{\mathrm{N}}}$. To estimate the maximum effect of noises on the phase fluctuations, consider the perpendicular location of the vectors $\vec{A}$ and $\vec{N}$. The additional phase shift of oscillations due to a single fluctuation will be

$$\delta\phi = N/A = \sqrt{kT \cdot \Delta\nu/P_0}.$$

The number of possible fluctuations (n) during the time t is determined by the bandwidth: $n = t \cdot \Delta\nu$, and the total additional phase shift is proportional to the root the square of the number of fluctuations, i.e. $\phi_N = \delta\phi\sqrt{n}$. Taking into account that the total phase of the oscillation is $\phi = \omega t$, we obtain the expression for the relative instability:

$$\frac{\phi_N}{\phi} = \frac{1}{2\pi}\frac{\Delta\nu}{\nu}\sqrt{\frac{kT}{P_0}}\frac{1}{\sqrt{t}}.$$

Thus, the relative error due to thermal noises can be reduced by averaging over time (but it should be kept in mind that the gain is proportional to $\sqrt{t}$). This error is the smaller, the more output generator power and the less relative bandwidth of the system ($\Delta\nu/\nu$).

18.5.3 The Influence of the Doppler Effect

Significant contribution to the broadening of spectral characteristics of molecular system makes the Doppler effect. The expansion the spectral lines is associated with the shift of a radiation frequency (ν) of a moving object:

$$\nu = \nu_{mn} \frac{1 - \vec{v}\vec{s}/c}{\sqrt{1 - v^2/c^2}}, \tag{18.31}$$

where $\vec{v}$ is the velocity of the emitter; $\vec{s}$ is the unit vector in the direction of propagation of electromagnetic waves; c is the light velocity. For ordinary thermal velocities $v^2/c^2 \simeq 10^{-11}$–10^{-12}. Replacing the denominator in the formula (18.31) by 1, we obtain

$$(\nu - \nu_{mn})/\nu_{mn} = -\vec{v}\vec{s}/c = \pm v/c_{\text{ph}},$$

here c_{ph} is the phase velocity of propagation of the electromagnetic wave in the direction of travel (plus or minus signs depend on the direction of relative motion the emitter and the wave front). If the emitting molecules located in a gas with an isotropic velocity distribution, the Doppler effect only leads to the broadening of spectral lines, but in molecular beams the line shift can appear. Assuming that the distribution of velocities obeys the Maxwell law, we obtain for the width of spectral lines:

$$\Delta\nu/\nu_{mn} \simeq 7 \cdot 10^{-7}\sqrt{T/M}$$

(T is the absolute temperature; M is the molecular weight). In most cases, the value of $\sqrt{T/M} > 1$ and the relative contribution of the Doppler broadening will be more than 10^{-6} and exceed all other contributions to the broadening of spectral lines. However, for directed beams there are ways to reduce this effect.

In the generator with a beam of ammonia molecules the cylindrical resonator is used, in which for the oscillations of the TM_{0n0} the phase velocity of propagation electromagnetic waves along the resonator axis $c_{\text{ph}} = \infty$. Hence the implementation of oscillations of the TM $_{0n0}$ type in a resonator should exclude the broadening and frequency shift of spectral lines due to the Doppler effect, if the motion of molecules is directed along the axis of the resonator. However the divergence of the beam does not get rid of the influence of the Doppler effect. There is another significant cause of the increasing role of the Doppler effect: the emergence of a traveling wave in the resonator due to the non-uniformity of the molecule radiation along the resonator axis and the presence of energy flow due to power output to the external recording circuit. The interaction of molecules with a wave, which carries the energy along the resonator axis with finite velocity, causes the Doppler broadening of spectral lines. To neutralize this effect, contrary beams are used: two beams are directed in one resonator from opposite sides.

18.5.4 The Influence of Intramolecular Interactions

Until now, the consideration of the molecular generator was carried out under the assumption that in the process of radiation only one spectral transition is involved. As to the rotation–inversion spectrum of ammonia molecules, this approximation is valid, because the inversion lines for different J and K are separated strongly enough. However, the spectrum of the ammonia with the basic nitrogen isotope ^{14}N possesses a well developed hyperfine structure, which is caused by different intramolecular interactions. The strongest of them is the quadrupole interaction of the electric quadrupole moment of the nitrogen nucleus and intramolecular inhomogeneous electric field (see Sect. 1.12). As a result the state of the molecule are described by the total momentum $\vec{F}_1 = \vec{I}_\mathrm{N} + \vec{J}$, were $\vec{I}_\mathrm{N}$ is the nuclear spin (the nitrogen ^{14} N spin is equal to 1); $\vec{J}$ is the full angular momentum of the molecule. The quantum number F_1 can range from $I_\mathrm{N} + J$ to $|I_\mathrm{N} - J|$ depending on the magnitude and mutual orientation of the vectors $\vec{I}_N$ and $\vec{J}$. For the isotope ^{14}N we obtain (with $J = 3$): F_1= 4, 3, 2. Thus, each energy level, which is characterized by the quantum numbers J and K, under the influence of the quadrupole interaction is split into $2J+1$ (with $J < I_N$) or $2I_N+1$ (with $I_N < J$) levels. The value of quadrupole splitting is determined by the relation

$$\Delta E_\mathrm{Q} = eqQ\frac{3C(C+1) - 4I_\mathrm{N}(I_\mathrm{N}+1)J(J+1)}{8I_\mathrm{N}(2I_\mathrm{N}-1)(2J-1)(2J+3)}\left[\frac{3K^2}{J(J+1)} - 1\right], \quad (18.32)$$

where q is the electric field gradient at the location of the nucleus; eqQ is the quadrupole coupling constant; $C = F_1(F_1+1) - I_\mathrm{N}(I_\mathrm{N}+1) - J(J+1)$.

The selection rules for transitions with a change of the quantum number F_1 are as follows: $\Delta F_1 = 0, \ \pm 1$. Therefore, the spectral line of the inversion transition is split into five components (Fig. 18.12a). The interval from the central line to the others of a few MHz, and the lateral lines are not make a significant contribution to the radiation at the generation mode. However, it appears that central line has a fine structure, which is due to differences values of the quadrupole coupling constants in the upper and lower inversion states. The splitting of the central line ($\Delta F_1 = 0$) is shown in enlarged scale in Fig. 18.12b. Intensity of three components are proportional to the number of possible projections of the vector $\vec{F}_1$ in the final state, i.e. $2F_1+1$. It means that for ^{14}N the ratio holds 7 : 9 : 5. Because of the broadening of the spectral lines due to the Doppler effect and other causes, the radiation has a continuous spectrum in the band of 2–2.5, kHz (shown with the dashed line in Fig. 18.12b). The asymmetry of the line would be insignificant if all components, that define its contours, are identically sorted in the inhomogeneous electric field. However, after the passage of the sorting system the radiation at the lines with $\Delta F_1 = 0$ is in accordance with the ratio of 1 : 0.92 : 0.5 (see Fig. 18.12c). If the sorting process is incomplete, it will be obtained another relationship between the line intensities, that is the generation frequency may depend on the voltage applied to the electrodes of a sorting system. Thus, the quadrupole interaction leads to the broadening of the spectral line and

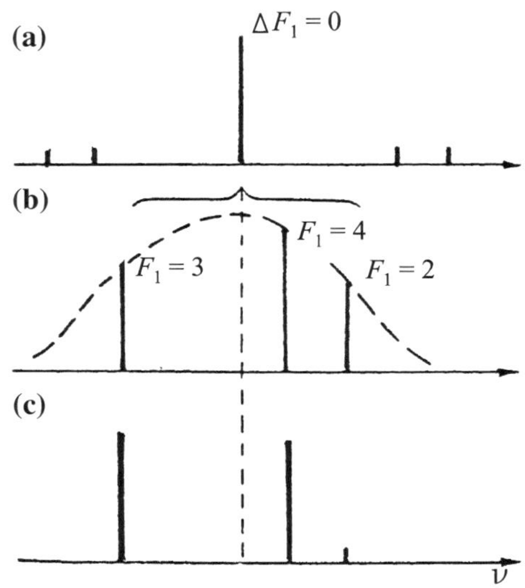

Fig. 18.12 The splitting of the spectral line for the state with $J = K = 3$: **a** the general view of the splitting; **b** and **c** the thin structure of the central line before and after the sorting, respectively

to the shift of its central part. Both of these factors are negative for the generator as the standard time or frequency. Therefore, one uses two methods, excluding the quadrupole splitting in the spectrum.

a. From (18.32) it is clear that for the transition $J = 3$, $K = 2$ the quadrupole splitting is absent (the expression in brackets vanishes). The benefits is not possible to realize completely, since that line is approximately of an order weaker than one with $J = 3$, $K = 3$, that leads to the worse signal to noise ratio.
b. If one uses ammonia isotope ^{15}N, then there is no quadrupole interaction, since the nuclear spin of ^{15}N is equal to 1/2 and such nuclei have no the electric quadrupole moment.

In addition to electrical interactions in the ammonia molecule there is an interaction of nucleous magnetic moments with the magnetic field caused by the fast rotation of the molecule. The contribution in the hyperfine structure of the spectrum comes from the interaction of the total spin of hydrogen atoms (protons) and the nitrogen nuclear spin with the magnetic field arising due to the molecule rotation (spin-rotational interaction). Magnetic splittings are much smaller than quadrupole ones, thus the stability of generators, based on emission lines without quadrupole splitting is higher on several orders.

18.6 The Spectral Transition Used in the Hydrogen Maser

The hydrogen atom is the simplest atomic system and its energy spectrum was calculated the most precisely. There will not be considered excited Coulomb energy levels because under usual conditions all hydrogen atoms located in the lower ("ground") state $1S_{1/2}$ and the other states have no effect on the processes described below. The ground state of the hydrogen atom has a fine structure due to the fact that the nucleus (proton) has a spin $\hat{\vec{I}}$ and the associated magnetic moment $\vec{\mu}$. The magnetic interaction between the electron and nucleus is described Hamiltonian (see Sect. 1.11)

$$\widehat{\mathcal{H}} = hA\hat{\vec{I}} \cdot \hat{\vec{J}}, \tag{18.33}$$

where $\hat{\vec{J}}$ is the total angular momentum of the electron; A is the coupling constant. As a result the atom state is described by the total angular momentum of the atom: $\hat{\vec{F}} = \hat{\vec{J}} + \hat{\vec{I}}$. The amendments to the energy level due interaction (18.33) are calculated from the expression

$$E_F = \frac{hA}{2}[F(F+1) - I(I+1) - J(J+1)]. \tag{18.34}$$

The number of possible states is determined by the quantum number F, which assumes as usual, the values from $I + J$ to $|I - J|$, i.e. $2I + 1$ or $2J + 1$ values (if $I < J$ or $I > J$, respectively). Thus, for the hydrogen atom in the ground state we obtain $F = 1,\ 0$ because $J = 1/2$ and for the proton $I = 1/2$. From (18.34) we have $E_1 = hA/4$, $E_0 = -3hA/4$ (see Fig. 14.1). The frequency of the transition between these states is

$$\nu_0 = (E_1 - E_0)/h = A.$$

The frequency of the transition between the sublevels of the hyperfine structure of the ground state of a hydrogen-like atom with a nucleus having spin I, is given by

$$\nu_0 = A\frac{2I+1}{2}.$$

The value of ν_0 depends on the properties of nuclei, and even for one chemical element can vary greatly for different isotopes, for example, for the three isotopes of hydrogen it takes the values $\nu_0^{\mathrm{H}} = 1420.4$ MHz; $\nu_0^{\mathrm{D}} = 327.4$ MHz; $\nu_0^{\mathrm{T}} = 1516.7$ MHz.

Due to the fact that the transitions between the hyperfine levels are observed almost always in the presence of an external magnetic field (which partially remains even inside magnetic screens), it is necessary to determine its effect on frequency transitions. The Hamiltonian of the magnetic interactions of an atom is given by

$$\widehat{\mathcal{H}} = hA\hat{\vec{I}}\hat{\vec{J}} - \frac{\mu_J}{J}\hat{\vec{J}}\vec{B}_0 - \frac{\mu_I}{I}\hat{\vec{I}}\vec{B}_0 \tag{18.35}$$

($\vec{B}_0$ is the external static magnetic field). If in (18.35) the interaction of electrons and nuclei with the magnetic field is smaller than their interaction with each other, one can use the Breit-Rabi formula (see Chap. 14)

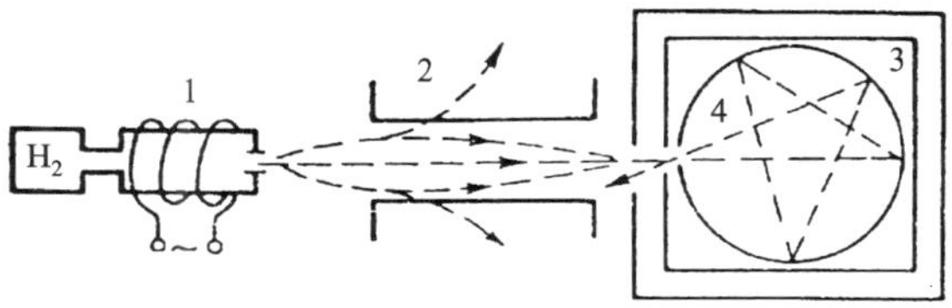

Fig. 18.13 The scheme of the generator with a beam of the hydrogen atoms

$$E_F^{(1,2)}(B_0) = -\frac{\Delta E}{2(2I+1)} - \frac{\mu_I m_F B_0}{I} \pm \frac{\Delta E}{2}\sqrt{1 + \frac{4m_F}{2I+1}x + x^2}, \qquad (18.36)$$

where $x = (\mu_I/I - \mu_J/J)(B_0/\Delta E)$; $\Delta E = hA$; indices 1, 2 correspond to the levels with quantum number $F = I \pm 1/2$; m_F is the quantum number characterizing the projection of the total angular momentum of the atom $\hat{\vec{F}}$ on the direction of the vector field $\vec{B}_0$. The scheme of the energy levels in the presence of a magnetic field is shown in Fig. 14.1. The transition between the states of $F = 1,\ m_F = 0$ and $F = m_F = 0$ possesses the weakest frequency dependence on a magnetic field. This transition is used in the generator with the beam of hydrogen atoms. In conclusion, we note that the initial (i.e. in the absence of a magnetic field) splitting of $1S_{1/2}$ in the hydrogen atom is $1, 420, 405, 751.768 \pm 0.005$ Hz.

18.7 The Arrangement of the Generator with the Beam of Hydrogen Atoms

The hydrogen generator was created in 1960. Its block-scheme is similar to ammonia-beam maser and consists of three main elements: source of the beam of hydrogen atoms, sorting system and resonator (Fig. 18.13). The source of the beam of atoms is a gas-discharge tube, in which molecular hydrogen is decomposed into atoms. The dissociation of molecular hydrogen occurs upon heating to temperatures above 3000 °C, but in practice usually one use the high-frequency discharge (at frequencies of about 100 MHz) directly in the beam source. The most effective dissociation of hydrogen is at pressures of (1–5)$\cdot 10^2$ Pa. The beam is formed by the system channel which number, diameter and length are determined by the required intensity beam and its optimum angle divergence (see Sect. 18.3).

The formed beam is directed to the sorting system (*2*), which in this case should create a strong inhomogeneous magnetic field because the hydrogen atom possesses only a magnetic moment. In the most generators with beams of hydrogen atoms a six-pole magnet, creating an axial symmetric magnetic field with a minimum value along the direction of beam propagation, is used. Comparing the results of Sect. 18.1 with the data on the behavior of energy levels of hydrogen atoms in a magnetic field, represented in Fig. 14.1, we can conclude that the atoms in the states, characterized by the quantum numbers of $F = 1, m_F = 1$ and $F = 1, m_F = 0$, will move in the lower field (i.e. to the beam axis). The atoms in the states, characterized by quantum numbers $F = 1, m_F = -1$ and $F = 0, m_F = 0$, will be declined toward the larger field, i.e. out from the beam. Usually the direction of the magnetic field in a resonator is not related to the configuration of the field in a sorting system, and in a resonator atoms have an isotropic orientation, i.e. they can have all three momentum projection on the magnetic field: $m_F = 1,\ 0,\ -1$. Then one ninth of the ammount of hydrogen atoms will belong to the transition $F = 1, m_F = 0 \longleftrightarrow F = 0,\ m_F = 0$. These estimations are valid for an ideal sorting system. In reality, if 10^{16}–10^{17} atoms per second are emitted from a source, the resonator gets only 10^{12}–10^{13} atoms in the state $F = 1, m_F = 0$.

To define *the conditions of self-excitation of the generator with a beam of hydrogen atoms* we can used (18.18), if under the parameter μ we mean the magnetic dipole of the hydrogen atom (the other symbols keep their same meaning):

$$N_0 > \frac{h}{8\pi^2}\frac{1}{\mu^2\tau^2}\frac{V}{Q}.$$

That understanding is possible because of the complete analogy between the formulas, describing the behavior of magnetic and electric dipoles. Energy interaction of the magnetic moment of the hydrogen atom with the resonator field in two orders of magnitude smaller than in the case of the electric dipole of the ammonia molecule. Therefore for the self-excitation of the hydrogen generator under equal other conditions it is required to increase the threshold particle number N_0 in 10^4 times. In addition, the volume of the resonator of the hydrogen maser is much higher than in the generator with the beam of ammonia molecules due to the higher wavelength (about 20 cm instead of 1.25 cm).

With the help of real sorting systems it is impossible to increase the intensity of the beam entering the resonator in such degree. This difficulty was overcome by using a special "accumulative" bulb *4* (see Fig. 18.13), placed in the resonator (*3*). The accumulative bulb is a quartz vessel with a single inlet. The hydrogen atoms fly into the bulb and, after about 10^4–10^5 random collisions with walls, fly out through the inlet. Therefore, the residence time of atoms in the resonator (τ) increases so that the condition self-excitation can be fulfilled. The bulb is made of quartz to reduce the dielectric losses in the resonator (to maintain its high quality factor). However, the collision of an atom with a quartz wall can lead to transitions from the excited state ($F = 1$) to the bottom ($F = 0$). This random process leads to a rapid decay of the excited state and, consequently, to a sharp reduction of the time $\tau_{(\mathrm{i})}$ of the interaction of the hydrogen atom with the resonator field, i.e. the conditions for the self-excitation of the generator are deteriorated. The value of τ_{i} can be done close to τ, if the walls of the quartz bulb cover with inert substance. The best results are achieved by the application of paraffins or Teflon. The considered cause of the reduction of the atom lifetime in the excited state is not unique. Other reasons are listed in the next subsection but here we note that the value achieved in the experiments is $\tau_{\mathrm{i}} \simeq 0,\ 5$–$2$ c, which is sufficient for the reliable condition of self-excitation of the generator.

Since the frequency of the transition $F = 1, m_F = 0 \longleftrightarrow F = 0,\ m_F = 0$ is still dependent on the magnetic field (see Fig. 14.1), the resonator is placed in a multilayer magnetic screen (5 in Fig. 18.13) which prevents the penetration of magnetic fields of a sorting system, surrounding subjects and the Earth field in the space of the radiation of atoms. However, it is necessary to keep a small static magnetic field in the resonator for the splitting of states with different m_F to ensure the tuning of the resonator to the desirable spectral transition. It is sufficient to get a residual magnetic field of the order of 10^{-6} T.

18.8 The Width of Spectral Lines of Hydrogen Atoms in the Accumulative Bulb and Causes of the Instability of the Hydrogen Generator Frequency

The advantage of the generator with a beam of hydrogen atoms is a very high stability of its frequency, which is determined primarily by the narrowness of the spectral lines in a accumulation bulb. Let us consider the factors that influence on the line width of the hyperfine splitting ($F = 1, m_F = 0 \longleftrightarrow F = 0,\ m_F = 0$) and the frequency shift of this transition.

The presence of a storage bulb determines the time interaction of hydrogen atoms with the resonator field. The time of the interaction is limited by the time of the existence of hydrogen atoms in a bulb (τ), which depends on the size of the inlet (it is also the output), the size of a bulb and the mean flying time of atoms from a wall to a wall (τ_{fl}):

$$\tau = \tau_{\text{fl}} S_{\text{bulb}}/S_0,$$

where S_0 and S_{bulb} is, respectively, the areas of the inlet and the surface of a bulb (the value of S_0/S_{bulb} determines the probability of the departure of atoms out of a bulb).

The collision of atoms with walls of a bulb can lead to three different effects.

a. The magnetic fields created by walls of a bulb can cause transitions of atoms from one state to another, just as it was considered in Sect. 17.2. But in this case, the amplitude and frequency of the alternating field will be random and the problem is reduced to the calculation of probabilities of transitions under the influence of fluctuating field, as is done for the probability of relaxation transitions (see Sects. 3.6 and 3.7).

b. In the case of the special "mild" covering of walls the probability of the change of a state of an atom, when it strikes a wall, may be negligible. However, when the elastic collision with a wall occurs there is a deformation of the electron shell of an atom, and, therefore, there is the change of the interaction energy of the nucleus and electron in an atom. The phase shift during a collision is equal to

$$\phi = \frac{1}{\hbar}\int\limits_{\Delta t}(\delta E_1 - \delta E_0)\mathrm{d}t = \frac{1}{\hbar}\int\limits_{\Delta t}\delta(E_1 - E_0)\mathrm{d}t, \qquad (18.37)$$

where δE_1 and δE_0 are the changes in the energy of the upper and lower levels; Δt is the time of collision. Integral in (18.37) can be calculated if the explicit form of the interaction potential of an atom with surface is known. For hydrocarbonate surfaces and hydrogen atom at room temperature of $\phi \simeq 10^{-4}$ rad per a collision is realized. The full phase shift after n collisions is proportional to $\sqrt{n}$ and $\phi_\Sigma = \phi\sqrt{n}$, i.e. we can assume that a complete loss of the radiation coherence ($\phi \simeq 1$) happen only after the 10^8 collisions.

c. It is also possible to notice a chemical reaction of an atomic hydrogen with the wall surface, resulting in the irretrievably loss of free hydrogen atoms. The residence time of an atom in a bulb due to the reaction on walls is usually limited to a few seconds.

Collisions between hydrogen atoms can cause the formation of molecular hydrogen or the process of spin exchange (simultaneous change of the orientation of magnetic moments of atoms). These processes lead to the loss of the radiation coherence, but their contribution to the line width up to the densities of the order 10^{17}–10^{18} atoms in the 1 cm^3 is very small. The similar effects can arise due to collisions of hydrogen atoms and molecules of residual gas in a bulb. Due to the big residence time of atoms in a storage bulb the requirements to the vacuum system in the hydrogen generator is significantly higher than in the generator with a beam of ammonia molecules: the residual gas pressure must not exceed 10^{-5} Pa.

The residual inhomogeneous magnetic field in the bulk of a storage bulb also leads to the broadening of spectral lines. Because hydrogen atoms move randomly in a bulb, the magnetic field, acting on it, fluctuates (Fig. 18.14). One can distinguish a constant component of the field B_{av} and a variable component with an amplitude of δB: $\vec{B} = \vec{B}_{\mathrm{av}} + \delta\vec{B}$. The effect of the field $\vec{B}_{\mathrm{av}}$ on atoms leads to a change in the transition frequency, which can be obtained from the formula (18.36). In the case of the hydrogen atom it is (in Hz)

$$\nu = \nu_0 + 2.75 \cdot 10^{11} B_{\mathrm{av}}^2 \qquad (18.38)$$

(B_{av} is the value of the field, expressed in units of 10^4 G or Tesla). If the B_{av} changes with time on the value of ΔB, then the stability and reproducibility of the generator frequency at $\Delta B \ll B_{\mathrm{av}}$ can be estimated using (18.38) from the formula:

$$\Delta\nu/\nu_0 \simeq 4 \cdot 10^2 B_{\mathrm{cp}}^2 (\Delta B / B_{\mathrm{cp}}),$$

which implies that the fluctuations of the field within a few percent of $B_{\mathrm{av}} = 10^{-2}$ G (10^{-6} T) lead to the relative frequency instability of 10^{-11}.

If the variable component of the residual field δB contains the transition frequency ν in its spectrum, then it can cause transitions of atoms from the excited state. These transitions are similar to the relaxation ones (see Chap. 3), and the effective relaxation

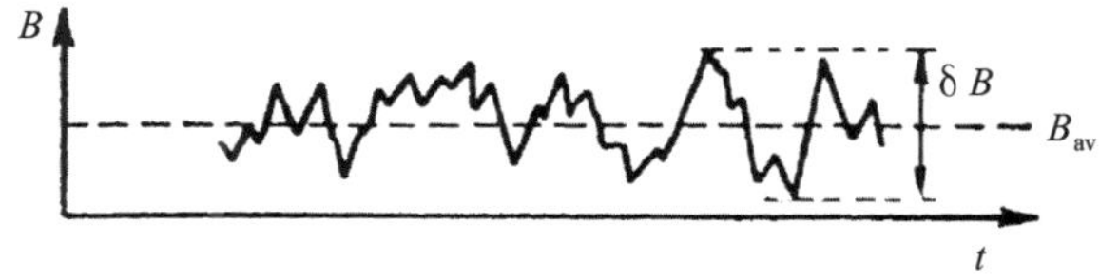

Fig. 18.14 The magnetic field acting on an atom in an accumulative bulb

time is given by

$$\tau_B = \frac{\hbar^2}{\mu^2(\delta B)^2} \frac{(2\pi \nu \tau_{\text{bulb}})^2 + 1}{\tau_{\text{bulb}}},$$

where τ_{bulb} is the mean time of flight of the atom through the bulb. For a bulb with a diameter of 16 cm $\tau_{\text{bulb}} \simeq 3 \cdot 10^{-5}$ c. If $\delta B \simeq 10^{-2}$ G (10^{-6} T), the value of τ_B is greater than 10^4 s, i.e. the effect is negligible.

In the generator with a beam of hydrogen atoms there is almost *no the broadening or shift of the spectral line due to the longitudinal Doppler effect (of the first order)* because of the isotropic character of the fast motion of hydrogen atoms in the accumulative bulb. *Transverse Doppler effect (of the second order)* leads to relative frequency deviations 10^{-14}–10^{-15} at $t \simeq 20$ °C.

The total effect of the broadening of the spectral line $F = 1, m_F = 0 \longleftrightarrow F = 0, m_F = 0$ of hydrogen atoms in the accumulative bulb can be described by an effective time of decay of the excited state $\tau_\Sigma \simeq 2$ s, which corresponds to $\Delta\nu_l \simeq 0.15$ Hz. The extremely small width of the line determines the highest quality factor of a line: $Q_l \simeq 10^9$ (the quality factor of lines in the ammonia generator is about 10^6). This leads to the fact that in spite of the higher Q of the resonator ($5 \cdot 10^4$, compared to 10^3–$5 \cdot 10^3$ in the ammonia generator) the influence of the inaccurate tuning of a resonator to the frequency of a spectral transition is relatively small (see (18.30)). Because of the narrow bandwidth the thermal noises also produce a small effect on the frequency stability of the generator ($\sim 10^{-15}$). As a result, the generator with a beam of hydrogen atoms showed a record frequency stability and reproducibility, that can be illustrated by the specifications of one of the first production frequency standards (see Sect. 18.10).

18.9 Cesium Standard of Frequency and Time

A schematic representation of the device of the quantum discriminator with a beam of atoms Cs is shown in Fig. 18.15.

The main isotope of cesium ^{133}Cs (natural abundance of 100 %) has a spin $I = 7/2$. In the cesium atom the electron shell consists of five filled layers and one electron in the outer shell 6*s*. Therefore, the ground state of the cesium atom is $6S_{1/2}$. The interaction of the unpaired electron and nucleus is described by (18.33) and leads to the splitting of the energy level of the ground state. The total angular momentum of the atom is characterized by the quantum numbers of $F = 3; 4$, since $I + J = 4$, and

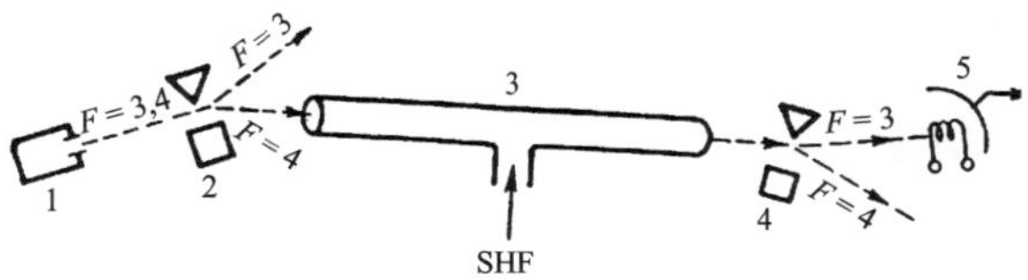

Fig. 18.15 The scheme of the quantum standard with a beam of cesium atoms

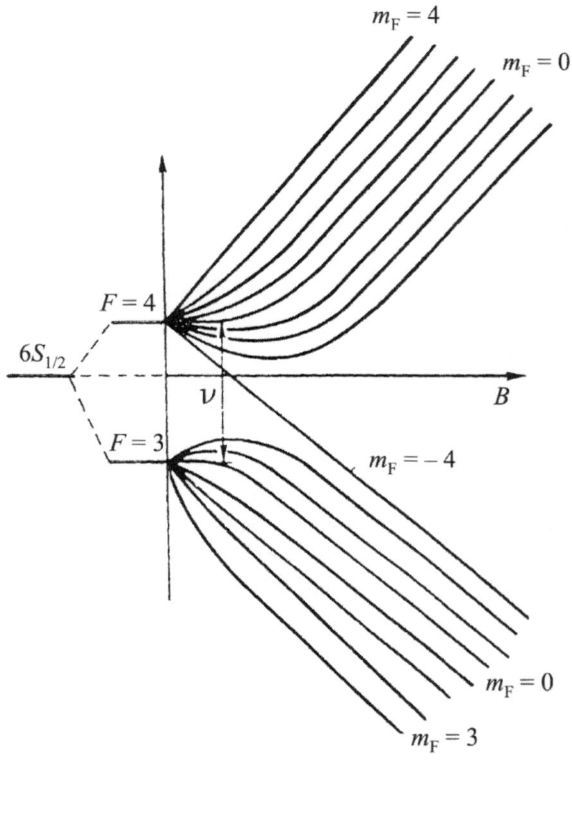

Fig. 18.16 The scheme of the energy levels for the ground state of the cesium(133) atoms

$|I - J| = 3$. The energy of the state with $F = 3$ and 4 in the absence and presence of a magnetic field is described by formulas (18.34) and (18.36).

The dependence of energy levels on the magnitude of the field is shown in Fig. 18.16. The initial splitting of the ground state $6S_{1/2}$ for the ^{133}Cs is 9192631770 Hz. The transition $F = 3$, $m_F = 0 \longleftrightarrow F = 4$, $m_F = 0$, which is used in the standard, is characterized by the smallest dependence on the magnetic field.

It is possible to sort the beam of cesium atoms in the same way as in the hydrogen generator using the inhomogeneous magnetic field. In the inhomogeneous field of special type (flat gradient) the beam of cesium atoms can be split into two rays (see Fig. 18.15). As a result, the Boltzmann distribution along energetic states for atoms emitted from the source is drastically disturbed, since after the passing through the magnet gap (2) atoms with $F = 3$ and $F = 4$ are spatially separated. Each of the obtained beams is equivalent for the further use. Figure 18.15 shows the case when one uses atoms with $F = 4$. These atoms are directed in the atomic-beam tube (*3*), in which they are affected by the radiosignal (SHF) at a frequency close to the frequency of the spectral transition $F = 3$, $m_F = 0 \longleftrightarrow F = 4$, $m_F = 0$. If the coincidence of that frequencies is fairly good (within the line width), then the induced transitions from the state $F = 4$ to the state $F = 3$ take place, i.e. at the output of the tube we again obtain two kinds of atoms. They can be split again in the space with the help of inhomogeneous field (magnet *4*). The presence of the atoms with $F = 3$ is determined with a special detector (*5*), consisting of a heated filament and collector. For the detection of cesium atomic beams one uses the effect of the surface ionization, when, being in contact with the surface of the heated metal, an atom gives its valence electron and leaves the surface as an ion, which can be easily detected. This process holds with high probability, if the work of electron exit out a metal excesses the ionization energy of an atom. For cesium the ionization energy is of 3.87 eV (for rubidium −4.16 eV, for potassium −4.3 eV). In this case, the filament of a detector is made of tungsten, for which the work of exit is 4.5 eV, and therefore the probability of the cesium ionization is close to unity. Cesium ions are trapped with a collector, which is usually an input electrode of an amplifier with secondary emission of electrons. High sensitivity of the detection of cesium atoms, changing their state, ensured due to the strong deviation from the Boltzmann distribution (there are no opposite transitions).

Thus, the detector of a beam of cesium atoms produces a current which is proportional to the intensity of the spectral line at the frequency of a microwave signal supplied to the atomic-ray tube. Therefore, applying the ideas of Sect. 17.5, it is possible to create the frequency standard using the transition $F = 3$, $m_F = 0 \longleftrightarrow m_F = 4$, $m_F = 0$. The dependence of the frequency of this transition on the magnetic field can be represented by the formula

$$\nu = \nu_0 + 4,27 \cdot 10^{10} B^2,$$

which differs from (18.38) only the numerical values of the constants. It follows from the formula that the space of interaction of atoms with a microwave field should be well shielded from magnetic fields (all arguments, given previously about the effect of magnetic fields on the operation of the hydrogen generator, are valid). The line-shape function $g(\nu)$ for the atomic-beam tube, shown in Fig. 18.15, has a bell-shaped form with a line width

$$\Delta\nu_l = 1,4\bar{v}/L,$$

where $\bar{v}$ is the average velocity of the atoms; L is the length of the space of the interaction of atoms with a microwave field. On the basis of this formula, we can conclude that it is possible to decrease the line width, increasing the length of a tube. However, the inhomogeneity of the magnetic field within the interaction space and some other effects limit the possibility of the realization of very narrow lines. Nevertheless in the unique devices the tube length was up to several meters and the line width was 10 Hz, which gives a quality factor of a line is equal to 10^9.

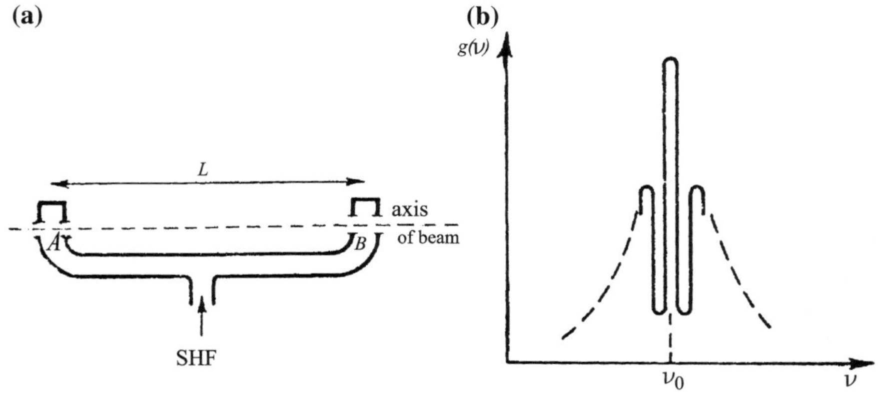

Fig. 18.17 The microwave system for the realization of the Ramsey-resonance (**a**) and the shape of spectral line (**b**)

The registration of a spectral line using a cylindrical atomic-beam tube is called *the Rabi-resonance* (named after its author). N. F. Ramsey showed that the effective line width would be less if one applied the method of separated oscillating fields: *Ramsey-resonance*. The scheme of a tube modified for this purpose is shown in Fig. 18.17a, and the shape of the spectral line of atoms interacting in regions A and B with a microwave field is presented in Fig. 18.17b. The line shape is described by a rather complicated function, however, for the properly tuned device is significant only the central part of the line. Therefore, the effective width of the line is determined by the central peak and has a value

$$\Delta\nu_l = 0,65\bar{v}/L,$$

where L is the distance between areas A and B. The application of the U-shaped atomic-beam tube greatly reduces requirements for the homogeneity of the magnetic field in the of interaction zone between areas A and B: it requires only a coincidence of magnetic fields in these areas.

18.10 A Comparative Analysis of Different Quantum Standards of Time and Frequency with Beams of Atoms and Molecules

The first standards of time and frequency based on the achievements of quantum radioelectronics were created in 1960–1970s. Their accuracy significantly surpassed the astronomical possibilities of the measurements of current time (the system of universal time UT-2 showed the accuracy about 10^{-8}). We will illustrate the question by the comparative analysis of the first quantum standards which were produced in USSR (Russia).

The first quantum mechanical frequency standard was made on the basis of *the generator with the beam of ammonia molecules*. It was a unique measuring system with the mass of 700 kg and required the special service. This device was characterized by the following parameters (ammonia $^{14}NH_3$ with $J = 3$, $K = 3$): the frequency deviation from the nominal $\pm 5 \cdot 10^{-9}$, the relative instability for 1 ms: $\pm 3 \cdot 10^{-9}$, for 10 ms: $\pm 1 \cdot 10^{-10}$, for 0.1 s: $\pm 1 \cdot 10^{-10}$, for 1 s: $\pm 3 \cdot 10^{-10}$.

Note that the relative instability of the standard decreases with observation time, as it follows from the foregoing analysis. However, over longer periods some circumstances can influence on the generator frequency: nonuniformity of the beam intensities, foulness of the sorting system due to the "hoar-frost" of freezing ammonia molecules, the temperature drifts of the resonator tuning, and so on. This led to the deterioration of long-time frequency stability the standard: for 1 h $\pm 1 \cdot 10^{-9}$).

Then frequency standards were created using the molecular beam of $^{14}NH_3$ with $J = 3$, $K = 2$ and of $^{15}NH_3$ with $J = 3$, $K = 3$, but their characteristics were significantly exceeded by devices on the basis of generators with the beams of hydrogen atoms.

The technical characteristics **hydrogen frequency standard** were the following: the relative frequency deviation from the nominal $\pm 4 \cdot 10^{-11}$; the relative error frequency reproduction (from switching to switching) $\pm 3 \cdot 10^{-13}$, the relative random variation of the frequency for 1 s: $\pm 1.5 \cdot 10^{-12}$, for 10 s: $\pm 3 \cdot 10^{-13}$, for 1 h: $\pm 5 \cdot 10^{-14}$, for a day: $\pm 7 \cdot 10^{-14}$. The hydrogen frequency standard provided its characteristics in four days after switching on or in 2 h if the temperature control system was preheated beforehand. The standard included two hydrogen generator (the second unit is needed for the fine tuning of the resonator frequency to the spectral line).The standard was a stationary device (mass of one set was of 330 kg).

The more compact **standard time and frequency with the beam of cesium atoms** operates in a passive mode. Its technical characteristics: the relative error of the mean frequency of $\pm 5 \cdot 10^{-11}$; the relative random variation of the frequency for 1 s: $\pm 5 \cdot 10^{-11}$, for 100 s: $\pm 2 \cdot 10^{-11}$, for a day: $\pm 2 \cdot 10^{-11}$; the mass of the device is 36 kg.

The standard with the beam of cesium atoms, like the hydrogen maser, is *self-calibrating, i.e. a primary standard of time and frequency*. It is characterized by high long-time stability and small time establishing ($\simeq$2 h). Its disadvantages are the sensitivity to rotations and limited life of a sealed vacuum atomic beam tube.

Generally speaking, frequency standards like the cesium one can be created with a beam of any alkali metal atoms or some other elements of the periodic table. The preference, given to the standard with a beam of cesium atoms, is caused by the several reasons. The main one is the lowest frequency dependence of the transition between the hyperfine energy levels on the magnitude and inhomogeneity of external magnetic field. It was pointed out in Sects. 18.6 and 18.9, that this dependence for low fields is defined by the formula

$$\nu = \nu_0 + \beta B^2, \tag{18.39}$$

Table 18.2 The parameters in formula (18.39) for different atoms

Isotope	ν_0, MHz	$\beta \cdot 10^8$, Hz/T^2
^{1}H	1402.5	2750
^{23}Na	1771.6	2170
^{87}Rb	6834.6	570
^{133}Cs	9192.3	427
^{205}Tl	21310.8	20

where ν_0 is the transition frequency in the absence of a magnetic field; β is the constant factor. The relative variation of frequency ν in the presence of the magnetic field change is calculated as

$$\Delta\nu/\nu_0 = (2\beta/\nu_0)B^2(\Delta B/B).$$

Thus, the lowest frequency dependence on the magnitude of the field B and its variation $(\Delta B/B)$ will be realized for a standard, built with a beam of atoms with a low coefficient of β and a high frequency of the transition of the initial splitting ν_0.

Table 18.2 lists the parameters in formula (18.39) for different atoms. The table shows that the cesium-133 has a very low ratio β/ν_0. This indicator is better only for thallium-205. However, the accuracy frequency reproduction achieved in systems with beams of thallium until does not exceed the accuracy of cesium frequency standards. The use of thallium beams is difficult mainly due to complexities in the detection of atoms, as the work of exit for tungsten is less than the ionization potential of thallium. Consequently, the pure tungsten cannot give effect for the detector with surface ionization, and it is not suggested a reliable design of the detector of thallium atom beams. The lack of effective indicators of the beam is partly compensated by a better use of the beam because the thallium-205 has spin $I = 1/2$, and more atoms participate in transitions than in the case of alkali metals, for which $I \geq 3/2$ (the energy levels are split to a bigger number of sub-levels).

Further Readings

Townes CH, Schawlow AL (1955) Microwave Spectroscopy, McGraw-Hill Publishing Company LTD, NY-London-Toronto (reissued in 2012)

Vanier J, Audoin C (1989) The quantum physics of atomic frequency standards. Adam Hilger, Bristol

Vuylsteke AA (1960) Elements of maser theory. D. Van Nostrand Co Inc, NY-Toronto-London

Yariv A (1988) Quantum electronics, 3rd edn. Wiley, New York

Chapter 19
Quantum Amplifiers Based on Electron Paramagnetic Resonance

Already in the first papers on the ammonia maser, it was stated that if the self-excitation condition is not satisfied, the device can serve as an amplifier of microwave power with low intrinsic noises. This is due to the fact that the probability of spontaneous emission in the radio diapason is very low and noises of an amplifier are determined only by the thermal electrical fluctuations in a resonator (actually the working medium is a combination of neutral particles). In spite of the intrinsic noises of the ammonia maser are very small, these devices (and the like) have not got any significant spread because of the extremely narrow bandwidth and the inability of the change of a working frequency, i.e. because of those peculiarities which have made it possible to use them as standards of time and frequency.

The greatest successes in the elaboration of quantum amplifiers are associated with the use of the phenomenon of electron paramagnetic resonance in solids. Positions of EPR spectral lines are determined by the external magnetic field and, in addition, the lines are almost always characterized by a large width. In Chap. 18 the quantum devices were considered, in which the creation of anomalous level populations takes place in the certain part of the space and its use to generate electromagnetic waves in another one. If the working substance is solid, such a technique cannot be used.

Two-level quantum devices on EPR with a motionless sample have not got the expansion because of the fact that continuous operation is not possible (see Sect. 3.2), but at their example it is easier and more convenient to consider some questions that are of the general importance. In particular, such a question is the description of a working substance of a quantum amplifier using the conception of *magnetic quality factor* Q_{m}.

19.1 Magnetic Quality Factor

In Sect. 18.3 the general definition of the quality factor of a system (Q_L) was introduced, according to which one can write:

V. I. Chizhik et al., *Magnetic Resonance and Its Applications*,
DOI: 10.1007/978-3-319-05299-1_19,

$$1/Q_L = 1/Q_0 + 1/Q_1 + 1/Q_2 + 1/Q_{\mathrm{m}}, \qquad (19.1)$$

where Q_0 is the quality factor of an unloaded resonator; Q_1 and Q_2 are the quality factors, which characterize the energy leakage to an inlet and outlet waveguides; Q_{m} is the *magnetic quality factor* characterizing the change energy under the interaction with a working substance. If the working substance is excited, then under the interaction with the electromagnetic field it will not absorb but emit the energy. Then Q_{m} will be negative and the expression (19.1) can be rewritten as

$$1/Q_L = 1/Q_0 + 1/Q_1 + 1/Q_2 - 1/|Q_{\mathrm{m}}|.$$

The condition of the self-excitation can be expressed in terms of quality factors as follows:

$$0 \geq Q_L \text{ or } 1/|Q_{\mathrm{m}}| \geq 1/Q_0 + 1/Q_1 + 1/Q_2.$$

To realize the regime of the amplification is sufficient to satisfy the inequalities

$$1/Q_0 + 1/Q_1 + 1/Q_2 > 1/|Q_{\mathrm{m}}| > 1/Q_0 + 1/Q_1, \qquad (19.2)$$

if the index 2 in (19.2) refers to an outlet waveguide. It is easy to determine the explicit form of Q_{m} starting from the expressions for the macroscopic magnetization of the sample and for the power absorbed (or radiated) by the magnetization under the interaction with an alternating field. Taking into account the expression for the energy stored in a resonator and formula (18.17), we obtain

$$|Q_{\mathrm{m}}| = 1/(4\pi\chi''), \qquad (19.3)$$

where χ'' is the component of the dynamic susceptibility of the sample (see Chap. 1). Thus the magnetic quality factor is completely determined by the properties of a sample. For weak signals (no saturation of the transition) $\chi'' = (1/2)\chi_0\omega_0 T_2$ (T_2 is the transverse relaxation time; χ_0 is the static susceptibility of the sample which is determined by the formula Curie). For electronic resonance $\chi_0 = N\mu^2/(kT)$ (N is the number of paramagnetic particles in the 1 cm^3; μ is the magnetic moment paramagnetic particles). Then

$$|Q_{\mathrm{m}}| = \frac{kT}{2\pi\omega_0 T_2 N\mu^2}. \qquad (19.4)$$

From (19.2)–(19.4) one can conclude that it is advantageous to have a low value $|Q_{\mathrm{m}}|$. Therefore, to improve the quantum amplification it is necessary to use the low temperature of the working medium (T) and to increase the number paramagnetic particles in the 1 cm^3(N). However, the increase of the number of paramagnetic particles can lead to the broadening of a spectral line (to the decrease of T_2), and therefore one usually chooses some compromise value N. An important result is that

Fig. 19.1 The scheme of the transitions between three energy levels

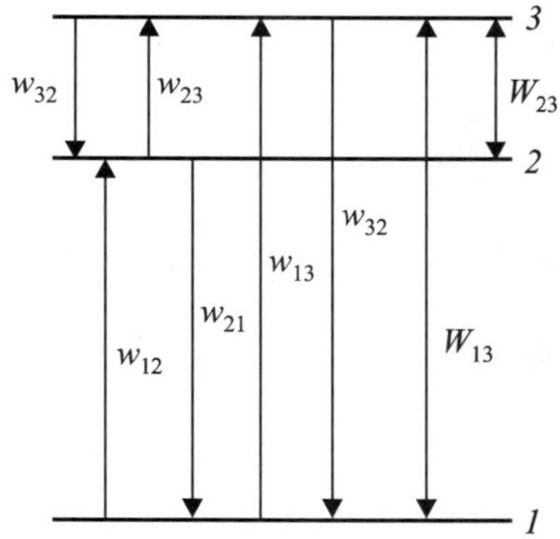

the magnetic quality factor decreases with the increase of a frequency, i.e. conditions of the quantum amplification are improved with the working frequency (in classical radioelectronics one can see an opposite trend.)

19.2 The Use of Auxiliary Radiation

For practical applications *the three-level maser* is the most perspective one, in which the anomalous difference of populations between two any levels is achieved by saturation of the transition, binding other pair of levels.

Consider three levels with energies E_1, E_2 and E_3 (Fig. 19.1). Let the number of particles on them will be n_1, n_2 and n_3. The particles can move between these states under the influence of relaxation processes and external radiations. Introduce the probability per time unit: w_{ik}—for the relaxation and W_{ik}—for the external radiation (recall that $w_{ik} \neq w_{ki}$; $W_{ik} = W_{ki}$).

Consider the case of the existence two external radiation: the first one at the transition frequency ν_{13} and the second one at the frequency ν_{23}. Then the kinetic equations for the number of particles at the levels will be (the third equation is the condition of a closed system):

$$\begin{aligned} dn_1/dt &= -(w_{12} + w_{13})n_1 + w_{21}n_2 + w_{31}n_3 - (n_1 - n_3)W_{13}, \\ dn_2/dt &= -(w_{21} + w_{23})n_2 + w_{12}n_1 + w_{32}n_3 - (n_2 - n_3)W_{23}, \\ N &= n_1 + n_2 + n_3. \end{aligned} \tag{19.5}$$

We assume that the external radiation at the transition frequency ν_{13} is of large power, i.e. $W_{13} \gg w_{13}, w_{31}, w_{23}, w_{32}, W_{23}$. For the stationary state of a system, $dn_i/dt = 0$ and the transition $1 \to 3$ is saturated ($n_1 = n_3$). Then it follows from (19.5) that

$$
\begin{aligned}
n_1 &= N\frac{w_{21}+w_{23}+W_{23}}{(2w_{21}+w_{12})+(2w_{23}+w_{32})+3W_{23}},\\
n_2 &= N\frac{w_{12}+w_{32}+W_{23}}{(2w_{21}+w_{12})+(2w_{23}+w_{32})+3W_{23}}.
\end{aligned}
\tag{19.6}
$$

To estimate the possibility of the realization of an anomalous population consider the difference $n_1 - n_2 = n_3 - n_2$. At thermodynamic equilibrium in the absence of an external influence, for any pair of energy states one can write: $w_{ij} - w_{ji} \simeq w_{ji}h\nu_{ji}/(kT)$ for $i > j$. Then

$$
n_1 - n_2 = n_3 - n_2 \simeq \frac{N}{3}\frac{h}{kT}\frac{w_{21}\nu_{12} - w_{32}\nu_{23}}{w_{21}+w_{32}+W_{23}}. \tag{19.7}
$$

We introduce the following denotation: ν_s is the frequency of the transition, which will be used for the amplification; ν_p is the frequency of auxiliary pump radiation; ν_i is the frequency of the idle transition. Taking into account that $\nu_p/\nu_s = 1+\nu_i/\nu_s$, one can write the condition of the inversion of the level populations for the transitions $1 \leftrightarrow 2$ or $2 \leftrightarrow 3$ in short form:

$$
\nu_p/\nu_s > 1 + w_s/w_i,
$$

where w_s and w_i are the probabilities of relaxation transitions at frequencies ν_s and ν_i. If in a substance $w_s \simeq w_i$, then the necessary condition is $\nu_p/\nu_s > 2$, i.e. the pump frequency should be much higher frequency of the "signal" transition. It is interesting that in the first quantum amplifier the condition $\nu_p \simeq 2\nu_s$ was realized but the ratio of relaxation times for transitions was artificially changed using the cross relaxation with paramagnetic impurity (see Sect. 15.1).

Equation (19.7) can be written in the following form:

$$
\Delta n_{32} = \frac{N}{3}\frac{h\nu_{23}}{kT}I = |\Delta n_{32}^0|I,
$$

where $|\Delta n_{32}^0| \equiv |n_3^0 - n_2^0|$ is the population difference in the thermodynamic equilibrium; I is, so called, inversion coefficient, which shows how the equilibrium difference Δn_{32}^0 can be changed under the influence of pumping radiation. The inversion coefficient depends on the ratio of the frequencies transitions and probabilities of relaxation processes:

$$
I = \frac{|\Delta n_{32}|}{|\Delta n_{32}^0|} = \frac{\nu_p/\nu_s - (1+w_s/w_i)}{1+w_s/w_i}.
$$

If $w_s/w_i \to 0$ (optimal case), then $I \to (\nu_p/\nu_s) - 1$; if $w_s \simeq w_i$, then $I = \nu_p/(2\nu_s) - 1$. Thus, if the transition frequencies ν_p and ν_s differ less than in ten times (this case is always realized in practice) then the inversion coefficient takes values of the order of a few units. Therefore $|\Delta n_{32}| \sim |\Delta n_{32}^0|$, and to increase the Δn_{32}^0 it is useful to reduce the temperature of a working substance.

19.3 Resonator Quantum Amplifiers

The power, which can be taken from a molecular system at the transition between levels of i and k, is equal to (see 17.4)

$$P = (n_i - n_k)h\nu_{ik}W_{ik}.$$

For the three-level maser one obtains, according to (19.7):

$$P_{\text{output}} = \frac{N}{3}\frac{h^2\nu_{23}}{kT}\frac{w_{21}\nu_{12} - w_{32}\nu_{23}}{w_{21} + w_{32} + W_{23}}W_{23}. \tag{19.8}$$

Equation (19.8) gives for weak amplified signals (i.e. $w_{32} \gg W_{23}$):

$$P_{\text{output}} = \text{const} \cdot P_{\text{input}}.$$

It was taken into account that $W_{23} = \text{const} \cdot P_{\text{input}}$ (see Sect. 17.2). Consequently, the molecular system provides the linear amplification of weak input radiation. In the case of strong signals $w_{32}, w_{21} \ll W_{23}$ and P_{output} does not depend on the value of the input signal (the saturation of the transition $2 \leftrightarrow 3$). Under these conditions one can estimate from (19.8) the maximum power which can be obtained from the molecular system. For the electron paramagnetic resonance, which frequencies belong to centimeter region of radio diapason, P_{max} is approximately 10 μW. Despite the relatively low power output the dynamic range of quantum amplifiers is very large, because they have very low intrinsic noises (see Sect. 19.5).

Quantum amplifiers can be *transit* (Fig. 19.2a) and *reflective* (Fig. 19.2b) types. The picture shows the resonator (1), the working substance (2), and inlet (3) and outlet (4) waveguides. It is not shown the pump waveguide, magnet required to obtain the EPR effect, and other accessories. In the reflective resonator amplifier the inlet and output waveguides, as well as resonator with the active substance, are connected with the circulator (5), which provides a one-way propagation radio waves.

To find the amplification coefficient of the transit resonator amplifier (G_{tr}) one can use the theory of the electrodynamics that describes the propagation of radio waves in hollow systems. In particular, there is the expression which connects the power of the incident and transmitted waves:

$$G_{\text{tr}} = \frac{P_{\text{output}}}{P_{\text{input}}} = \frac{Q_1^{-1}Q_2^{-1}}{(2Q_L)^{-2} + (\nu/\nu_0 - 1)^2},$$

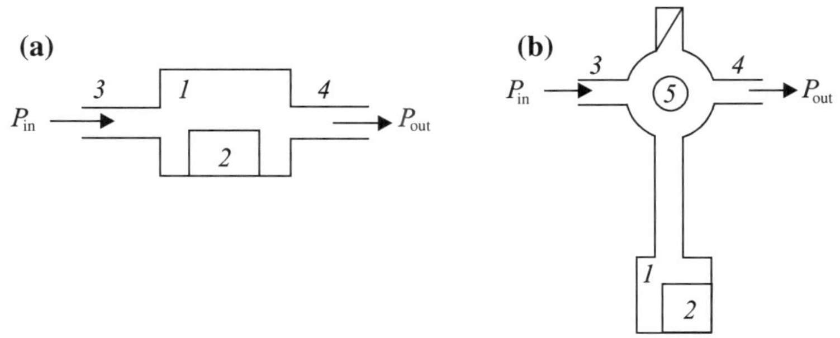

Fig. 19.2 The schemes of the resonator quantum amplifier of the transit (**a**) and reflective (**b**) types

where Q_1 and Q_2 is the Q-factors of the linkage of a resonator and waveguides 3 and 4; Q_L is the quality factor of a loaded resonator (see 19.1); ν_0 is the resonator frequency. If for simplicity we put $Q_1 = Q_2$, then for $Q_0 \gg Q_1, Q_2$ and $\nu \simeq \nu_0$ one can obtain $G_{pr} = 4Q_L^2/Q_1^2$. The maximum amplification is realized near the threshold of the self-excitation, i.e. with $|Q_m| \simeq Q_1/2$ (see Eq. 19.2). Let us construct the product:

$$\Delta\nu G_{tr}^{1/2} \equiv S_{tr} \simeq \nu_0/|Q_m|, \tag{19.9}$$

where $\Delta\nu$ is the bandwidth of an amplifier; S_{pr} is, so-called, the amplification area for a device of this type.

For the device of reflective type the amplification area is

$$S_{refl} \simeq 2\nu_0/|Q_m|. \tag{19.10}$$

It follows from (19.9) and (19.10) that the value of the amplification area depends only on the level of operating frequency and the properties of the working substance.

Thus, it is impossible to change arbitrarily the amplification factor and bandwidth of the quantum amplifier (general property of regenerative receivers). For example, when one needs to increase the gain factor in four times it would be necessary to accept the decrease of the bands in a half. The amplification area of a reflective amplifier is in two times more than for a transit one. This is due to the fact that in a reflective amplifier there is a single waveguide and in a transit amplifier the power is uselessly dissipated in the input waveguide (the waveguide for a more high-frequency pump signal is made as a limited one and the power of the amplified signal is not dissipated in it).

19.4 Quantum Amplifiers with Traveling Wave

As follows from (19.9) and (19.10), it is impossible in resonator quantum amplifiers to obtaine a large bandwidth without, at the same time, the significant decrease of an amplification coefficient. The increase the gain, while maintaining the required level a bandwidth, is possible either including a few quantum devices one after another or improving the efficiency of the interaction of the signal field and paramagnetic crystal. The second way, radically different from those already discussed, is to place the working substance in a periodic structure.

The efficiency of the interaction of microwave fields with a paramagnetic substances is reached by slowing down the propagation of radio waves along a periodic structure. An excited paramagnetic substance leads to negative losses and, therefore, an exponential increase of the power is observed. The unidirectional amplification (this is the condition stability of an amplifier) is possible if there are areas in the structure with the circularly polarized electromagnetic field, and if the transitions at the signal frequency are allowed only for circular polarization (see Sect. 17.2).

Devices with an extended interaction space in which a signal propagates along the active medium, is called *quantum amplifiers with traveling wave*. Let us calculate the transfer coefficient of such quantum amplifier. In waveguide the energy density of the incident radiation (u_0) related to the power (P) by the relation:

$$P = u_0 S v_{\mathrm{gr}},$$

where S is the sectional area of the waveguide; v_{g} is the group velocity of the propagation for this type of waves. The change of the power in the layer dz will be:

$$\mathrm{d}P = P(\alpha_{\mathrm{m}} - \alpha_{\mathrm{d}})\mathrm{d}z, \tag{19.11}$$

where the coefficient α_{d} characterizes the attenuation (decrease) of a wave in the structure; α_{m} characterizes the increase of its power under the interaction with the active substance. Naturally, we are interested in the case when $\alpha_{\mathrm{m}} \gg \alpha_{\mathrm{d}}$. Then

$$P(l) = P(0)\exp(\alpha_{\mathrm{m}} l)$$

(l is the length of a sample placed in the periodic structure). Consequently, the amplification factor of the traveling wave maser is

$$G_{\mathrm{tr.w}} = \exp(\alpha_{\mathrm{m}} l).$$

Express α_{m} in terms of the parameters of the active medium. For this we represent the power increment in another form:

$$\mathrm{d}P = N_{\mathrm{e}} h\nu W, \tag{19.12}$$

where N_{e} is the number of active particles in the layer dz; W is the probability of induced transitions, which for a two-level system is given by (17.15). The number of active particles can be related to the spin density N_0: $N_{\mathrm{e}} \simeq (N_0/2)(h\nu/(kT))S\mathrm{d}z$. Comparing expressions (19.11) and (19.12), one obtains

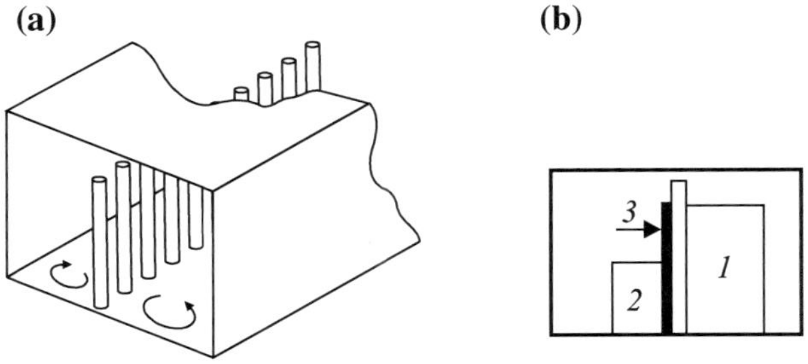

Fig. 19.3 The schemes of the quantum amplifier with traveling wave: **a** the comb structure, **b** the cross-section of the amplifier

$$\alpha_{\mathrm{m}} = \frac{\pi N_0 (h\nu)^2 \gamma^2 g(\nu)}{4kT v_{\mathrm{gr}}}. \tag{19.13}$$

It can be seen from (19.13) that in addition to the previous conditions of the optimization of the properties of a working substance we have obtained another one: for a maximum of amplification it is desirable to reduce the group velocity of the wave propagation along a crystal.

There are three types of slow-wave structures: (1) dielectric, (2) geometrical, (3) resonant. The greatest deceleration factors (10^2–10^3) can be obtained using the resonant structures. They possess a definite bandwidth limited by the upper and lower critical frequencies and than less their bandwidth, the greater the deceleration. The example of such systems is a comb structure (Fig. 19.3a), in which the fields of the circular polarization are located at both sides of the rods forming the structure.

The bandwidth of slow-wave structures is almost always much greater than the width of EPR spectral line. Therefore, unlike the resonator quantum amplifiers, the bandwidth of the traveling-wave amplifier ($\Delta\nu$) is commensurate with a line width of the EPR ($\Delta\nu_l$). If we approximate the function of a spectral line by the Lorentzian, the amplification coefficient, expressed in decibels, is given by the formula:

$$G_{\mathrm{dB}} = G_{\mathrm{dB}}(\nu_0)\frac{1}{1 + (\Delta\nu)^2/(\Delta\nu_l)^2}.$$

We define the bandwidth at 3 dB:

$$\Delta\nu = \Delta\nu_l \left(\frac{3}{G_{\mathrm{dB}}(\nu_0) - 3}\right)^{1/2}. \tag{19.14}$$

It follows from (19.14) that the amplification area for the traveling-wave amplifier increases with the gain. For $G_{\mathrm{dB}}(\nu_0) \simeq 30$–$40$ dB the bandwidth of the amplifier $\Delta\nu \simeq 0.3\Delta\nu_l$ and can reach several MHz. The change of the average operating frequency is performed by varying the frequency of electron paramagnetic resonance, i.e. by variation of the external magnetic field. The limits of the change are determined by the bandwidth of a slow-wave structure and usually reaches a few hundred megahertz.

The traveling-wave amplifiers are characterized by very high stability due to the automatic decoupling input and output. Figure 19.3b shows the scheme of the arrangement of main elements of the amplifier in the comb-structure. If the active working substance (1) is located in of "proper" polarization zone (see Sect. 17.2), then for a backward wave this polarization will be "wrong" and, therefore, a backward wave will not be amplified. To attenuate the backward wave an absorbing substance (2) is placed at the other side of rods of a comb structure. The absorbing material may be a maser substance with a high concentration of paramagnetic or ferrite. In the first case, the greater concentration of paramagnetic centers leads to the increase of the probability of relaxation transitions and the conditions for the polarization of a substance are not fulfilled (see Sect. 19.2). Therefore, the population inversion for the signal transition is not realized and this substance absorbs electromagnetic power. Because in the zone of the absorbing substance the polarization is "wrong" for the signal wave, then it is not absorbed. For better separation of forward and backward waves the special layer (3) from aluminium oxide is placed near the rods. The use of ferrites for absorption of a backward wave is advantageous because they do not absorb energy at the pump frequency. However in this case the change of the working frequency of an amplifier is complicated, because the frequency of electron paramagnetic resonance and the absorption frequency of a ferrite differently depend on the magnetic field.

19.5 Noises of Quantum Amplifiers

As it was already mentioned, the main advantage of quantum amplifiers is the extremely low noise level. First of all, let us consider briefly the general questions associated with the description of noises in electrical circuits.

The power of thermal electrical noises in a passive circuit, consisting of resistances, is determined by only the temperature (T) of these resistances (Nyquist theorem):

$$P_{\mathrm{n}}(\nu, t)\mathrm{d}\nu = kT\mathrm{d}\nu, \tag{19.15}$$

and the mean square of an electromotive force, generated by the resistance of R, is described by the formula

$$\langle e_{\mathrm{n}}^2(\nu, T)\rangle\mathrm{d}\nu = 4kTR\mathrm{d}\nu. \tag{19.16}$$

Formulas (19.15) and (19.16) show that the thermal noise power of the radio frequency diapason in passive circuits are independent of the operating frequency. It should be kept in mind that these formulas are derived under the assumption $h\nu \ll kT$ (this relation is valid in the radio range for most experiments). In general, the power of the noises can be written as follows:

$$P_{\mathrm{n}}(\nu,t)\mathrm{d}\nu = \frac{h\nu}{\exp(h\nu/(kT)) - 1}\mathrm{d}\nu. \tag{19.17}$$

The noise power at the output of a circuit, consisting of N elements, is determined by

$$P_{\mathrm{n}} = \sum_{i=1}^{N} p_i P_{\mathrm{n}\,i} \quad \text{with} \quad \sum_{i=1}^{N} p_i = 1, \tag{19.18}$$

where $P_{\mathrm{n}\,i}$ is the noise power generated by i-th element of the scheme; p_i is the coefficient, which characterizes the absorption power in the i-th element if at output of the circuit the power of 1 W is provided. When the coefficients p_i are determined in nonreciprocal systems the direction of the signal propagation is changed to the opposite one (eg, in a circulator).

To characterize the noise of amplifiers one usually uses such parameters as the *noise factor* (F) and *noise temperature* (T_{N}).

The noise factor is the ratio of the noise power at the output of the real and the ideal amplifiers

$$F = 1 + \frac{P_{\mathrm{ng}}}{kT\int_0^\infty G(\nu)\mathrm{d}\nu} \tag{19.19}$$

where P_{ng} is the noise power generated by real the amplifier; integral expression is the amplified input noise (input impedance is at the temperature T). In accordance with the international agreement the noise factor of an amplifier should be calculated, taking the temperature of the noise source $T_0 = 290\,\mathrm{K}$.

The noise temperature is defined by the formula

$$T_{\mathrm{N}} = \frac{P_{\mathrm{ng}}}{k\int_0^\infty G(\nu)\mathrm{d}\nu}. \tag{19.20}$$

Taking into account (19.19) and (19.20), the relationship between T_{N} and F can be expressed as:

$$F = 1 + \frac{T_{\mathrm{N}}}{T_0}. \tag{19.21}$$

The expression for the total output power of the amplifier allows us to draw the important conclusions. If the frequency response of the amplifier has a rectangular shape with a bandwidth of $\Delta\nu$, then using (19.20), one obtains

$$P_{\mathrm{output}} = G[P_{\mathrm{input}} + k(T + T_{\mathrm{N}})\Delta\nu],$$

from which it follows, first, that the sensitivity of the receiver is determined by the sum of temperature of the noise source and the noise temperature of the amplifier, and secondly, that in the case of high temperature of the input resistance it is not expedient to apply the low-noise receivers.

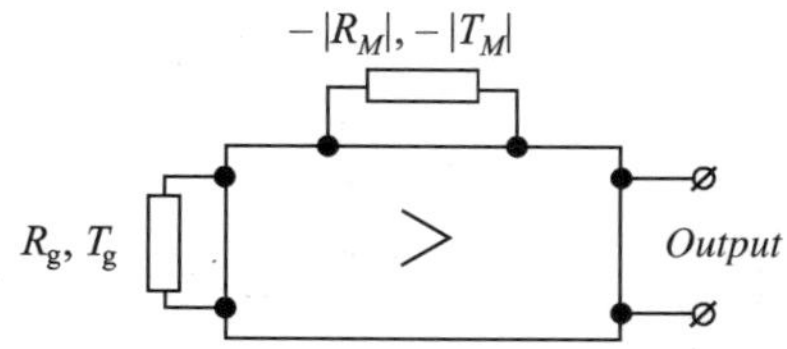

Fig. 19.4 The equivalent schemes of the quantum amplifier with traveling wave

Consider the equivalent scheme of the quantum amplifier (Fig. 19.4). Let its input impedance (R_g) is at temperature T_g. The interaction of electromagnetic waves with the exited substance can be described by an effective negative resistance R_m, which characterizes the negative losses. The question raises: what temperature should be attributed to this negative resistance? The noise E.M.F., which is generated by the spin system, can be described (see 19.16) as

$$\langle e_n^2 \rangle = 4kT_m R_m \Delta\nu$$

(T_m is the effective temperature (see Sect. 19.1)). One should use namely the temperature T_m, because this quantity characterizes the difference between the populations of states and, therefore, the ability of the spin system to absorb or radiate energy. Relations (19.15) and (19.16) were derived from the thermodynamics laws. Thus these formulas remain valid regardless if the absorption or emission of energy occur in the spin system, that results in an unexpected result:

$$\langle e_n^2 \rangle = 4k(-|T_m|)(-|R_m|)\Delta\nu,$$

i.e. the product of the negative resistance and negative spin temperature leads to a real (positive) noise E.M.F. Therefore, any statements, concerning the thermal noise in a ordinary system with losses at the temperature T, can be extended to a radiating quantum system, using the effective temperature. Applying (19.18) to the quantum amplifier we obtain

$$P_n = p_{gR} P_n(R_g, T_g) + p_m P_m(-|R_m|, -|T_m|).$$

Obviously, $p_g = G$, i.e. equal to the amplification factor, since with the change of the direction of propagation of electromagnetic waves in the amplifier (nonreciprocal system) it will amplify signals from the output to the input. We can realize this situation in the traveling-wave amplifiers changing the magnetic field to the opposite direction. Then

$$p_m = (1 - G),$$

$$P_n = G P_{nR}(R_g, T_g) + (1 - G) P_{ng}(-|R_m|, -|T_m|). \qquad (19.22)$$

The second term in (19.22) describes the amplifier noises, therefore

$$P_{ng} = (G - 1)k|T_m|\Delta\nu.$$

Using the definition of the noise temperature of the amplifier (19.20) we obtain $T_N \simeq |T_m|$, if $G \gg 1$. The noise factor of the quantum amplifier found from the expressions (19.19) and (19.21):

$$F = 1 + |T_m|/T_0.$$

Relations (19.6) and (19.7) allow us to calculate $|T_m|$ for the three-level maser:

$$|T_m| = T\nu_{23}\frac{w_{21} + w_{32}}{w_{21}\nu_{12} - w_{32}\nu_{23}}.$$

If $\nu_{12} \simeq 2\nu_{23}$ and w_{ik} are approximately equal, $|T_m| \simeq 2T$. Hence, to obtain the small noise level it is necessary to work at low temperatures. In the previous sections we indicated other reasons of the desirable reduction of the temperature of the working substance. These requirements lead to the fact that quantum amplifiers on EPR operate at liquid helium temperatures.

Thus, we have obtained for a quantum amplifier $T_N \leq 10$ K, $F \simeq 1.03$, that is better than it is required in the vast majority of experiments.

It seems that reducing the $|T_m|$, one can arbitrarily reduce the T_N. But the reduction of $|T_m|$ leads to the violation of the condition $h\nu \ll k|T_m|$, which was used for the derivation of (19.15) and (19.16). Applying the general formula (19.13), we obtain

$$P_{ng} = (1 - G)\frac{h\nu}{\exp(-h\nu/(k|T_m|)) - 1}\Delta\nu,$$

i.e. P_{ng} tends to the value of $(G - 1)h\nu$, if $h\nu \gg k|T_m|$. Therefore, the noise temperature limit of the quantum amplifier is determined by the process of the spontaneous emission and is equal to $T_{n\,min} = h\nu/k = 0.5$ K for the frequency of 10 GHz.

19.6 Some Design Features of Quantum Amplifiers on EPR

In the majority of quantum amplifiers, in which effect electron spin resonance is used, as an active substance one can employ crystals of ruby (crystal of corundum Al_2O_3 with addition of paramagnetic chromium ions), rutile (TiO_2), crystals doped with paramagnetic ions of rare earth elements. The active development of radio physics and other related areas of science and technology, such as radio astronomy and ultra long-ranged radiolocation, requires a highly sensitive receiving systems of millimeter diapason. The use of quantum amplifiers based on EPR can significantly increase the sensitivity of the receiving systems of this range. The progress is related to the solution of fundamental problems in the physics of paramagnetic crystals and relaxation processes. Examples of perspective working substances for quantum

Table 19.1 The comparative characteristics of some quantum amplifiers

Amplifier type	Characteristics							
	Active substance	λ_s, cm	λ_p, cm	B_0, 10^4 G	G, dB	$\Delta\nu$, MHz	$\Delta\nu_{chage}$, MHz	T_n, K
Single resonator	Ruby	21	2.65	0.2	20	1	–	10
	Andalusite	0.7	0.3	3	20	100	–	25
Two resonators	Ruby	3.2	1.27	0.43	25	13	–	10
Traveling-wave	Ruby	13.2	2.36	0.25	26	16	150	13
	Ruby	7.2	1.0	0.33	40	25	140	5
	Rutile	10–15	0.57	0.4	25–40	5–20	1000	10

amplifiers in the millimeter range are rutile, emerald, andalusite. All of them are characterized by a big primary splitting of energy levels of paramagnetic ions in the crystal electrostatic field.

The special requirements to the resonator of quantum amplifiers are claimed because it is necessary have the biggest possible quality factor at both the signal frequency and the frequency of the pump transition. When one deals with electromagnetic waves with $\lambda = 3$ cm and shorter, cavity resonators are used in which oscillations are excited at the lowest type for the signal frequency and at the high one for the pump frequency.

For comparison, Table 19.1 shows the characteristics of some amplifiers. As it can be seen from this table, the traveling-wave amplifiers have better characteristics. At the present almost all quantum paramagnetic amplifiers of centimeter and millimeter bands are based on the principle of a traveling wave.

Summarizing the ideas of previous sections, we can identify a clear trend: for the best operation of an amplifier it is favorable to cool a working active substance. It is determined by several factors: (1) the magnetic quality factor decreases with the temperature drop (see Sect. 19.1), (2) the saturation of a pump transition in a three-level maser is better at low temperatures when relaxation transition probabilities are small, (3) the number of active particles and, hence, the output power of an amplifier increases with decreasing temperature (see 19.8), (4) the quantum amplifier noise is less at low temperatures. The result is that it is necessary to work with liquid helium temperatures. The increase in the operating temperature yields the deterioration of amplifier characteristics (gain and noise factor). There are a few reports about the realization of the quantum amplification at temperatures of liquid hydrogen ($\simeq$10 K), nitrogen ($\simeq$70 K) and oxygen ($\simeq$90 K), but these devices were not used in the practice of experiments due to bad their characteristics. The necessity to work at liquid helium temperature results in additional experimental difficulties. The cryostat design is much more complicated and its volume increases. Therefore magnets sometimes become very cumbersome. The size and weight of the magnet are determined by many factors (level magnetic field, requirements of its homogeneity within the dimensions of the working substances, magnetic properties of used materials, and so on), but in some cases masers have such a compact design that they can be placed in the focus of antenna systems. Now magnets are replaced by superconducting solenoids which are put inside cryostat. This way is perspectives because of the decrease of the size of equipment, increase of the level and stability of magnetic fields (increase of operating frequencies).

Naturally, on the basis of substances, which spectra have the splitting in zero magnetic field, a quantum amplifier can be created without the use of magnetic systems, but such a maser will operate at the frequency determined by the intracrystalline electric field of a chosen substance and the operating frequency can not be changed.

Further Readings

Dudkin VI, Pakhomov LN (2006) Quantum Electronics. Devices and their Applications. Tekhnosphera, Moscow (in Russian)

Singer JR (1959) Masers. John Whiley and Sons Inc., New York

Vuylsteke AA (1960) Elements of Maser Theory. D. Van Nostrand Co Inc, NY-Toronto-London

Yariv A (1988) Quantum electronics, 3d edn. Wiley, New York

Chapter 20
The Use of Optical Radiation in Quantum Radioelectronics Devices

At present, there are both active and passive devices of the quantum radioelectronics, in which to increase the intensity of a spectral line in the radio diapason one uses an auxiliary radiation of optical range. This method is called *optical pumping*. The peculiarity of optical pumping is that, in contrast to all previously discussed methods, the auxiliary radiation affects on transitions at very high frequencies and in this case $h\nu \gg kT$. Devices with the optical pumping possess some certain design features. Working substances of these devices can be vapours of atoms of different groups of the periodic system, but in the most vapours of alkali metals are used.

20.1 The Use of Optical Pumping to Increase the Intensity of a Spectral Line at Radiofrequency

The principle of obtaining an optical pumping for abnormal population levels, transitions between which leads to the emission in the radio range, can be illustrated by simple calculations.

Consider the scheme of three levels, as it is shown in Fig. 20.1 (here we use the same notation as in Fig. 19.1). We assume that $h\nu_{13}, h\nu_{23} \gg kT$ (optical range) and $h\nu_{12} \ll kT$ (radio range). It follows from these relations: $w_{12} \simeq w_{21}[1 - (h\nu_{12}/(kT)]$; $w_{13}, w_{23} \simeq 0$. Therefore, the calculations, although like those in Sect. 19.2, but they differ in some details. Kinetic equations for the changes of populations have the form

$$\begin{aligned} \frac{\mathrm{d}n_2}{\mathrm{d}t} &= n_1\left(w_{21} - w_{21}\frac{h\nu_{12}}{kT} + W_{12}\right) - n_2(w_{21} + W_{21}) + n_3 w_{32}, \\ \frac{\mathrm{d}n_3}{\mathrm{d}t} &= n_1 W_{13} - n_3(w_{31} + w_{32} + W_{13}). \end{aligned} \tag{20.1}$$

V. I. Chizhik et al., *Magnetic Resonance and Its Applications*,
DOI: 10.1007/978-3-319-05299-1_20,

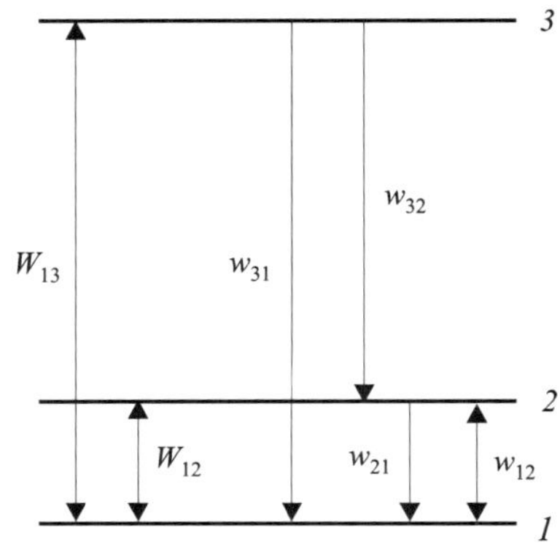

Fig. 20.1 The scheme of the transitions between three energy levels for the optical pumping

For optical transitions it is practically impossible to realize the conditions of the saturation of transitions, because of $W_{13} \ll w_{31}, w_{32}$ and, therefore, $n_1 \neq n_3$. The solvation of (20.1) for the stationary case gives

$$\frac{n_2 - n_1}{n_1} = \left(1 + \frac{W_{12}}{w_{21}}\right)^{-1} \left[\frac{w_{32}}{w_{21}} \frac{W_{13}}{w_{31} + w_{32} + W_{13}} - \frac{h\nu_{12}}{kT}\right]. \tag{20.2}$$

As it is seen from (20.2), for the inversion of populations of the first and second levels we need to formulate:

$$\frac{w_{32}}{w_{21}} \frac{W_{13}}{w_{31} + w_{32} + W_{13}} > \frac{h\nu_{12}}{kT},$$

i.e. there is a maximum operating frequency ν_{12} which cannot be exceeded to get the effect of quantum amplification. At the initial moment (after switch light on) particles move from the state 1 to the state 3 with the absorption of light quanta. Therefore, the intensity of the passed light must decrease. Then, after spontaneous transitions from the third level (the probability of spontaneous transitions in the optical range is very high) particles will gradually move to state 2, and the substance becomes transparent to light. If one applies the electromagnetic radiation at the transition $1 \leftrightarrow 2$, then the level 1 will again be populated. This will again entail transitions $1 \leftrightarrow 3$, i.e. new absorption of light quanta. Thus, the intensity of the light passed through the working substance depends on the number of transitions between the first and second levels. Therefore, we have obtained a method of registration of transitions in radio frequency range (transition $1 \leftrightarrow 2$) via the change of the light intensity (transition $1 \leftrightarrow 3$). This method possesses very high sensitivity because the the energy of an optical quantum in 10^5to10^6 times more than the quantum of electromagnetic radiofrequency fields. For example, the vapor of alkali metals (the pressure is of 10^{-5} Pa) change the intensity of the light for about 30 %. The expression (20.2) shows that the efficiency of the optical pumping is higher, the lower the relaxation probability w_{21}. In other words, for the successful optical pumping it is

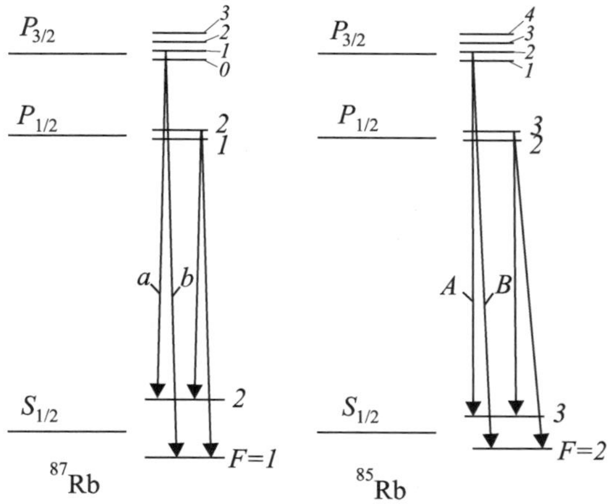

Fig. 20.2 The scheme of the transitions between the energy levels of two isotopes of rubidium

necessary to reduce the probability of random mixing of particles between the first and second levels.

If the optical radiation was applied to the transition between levels 2 and 3 (see Fig. 20.1), the deviation from Boltzmann distribution would be in the other direction, and then it was possible to get a the situation when $n_1 \gg n_2$, i.e. we can strongly increase the intensity of a spectral absorption line of atoms or molecules in the radio range.

20.2 The Quantum Standard of Time and Frequency With the Rubidium Vapor

The optical pumping for the polarization of atoms with magnetic moments we will consider on the example of rubidium, which has two main isotopes: ^{85}Rb and ^{87}Rb with natural abundance 73 and 27 %. The splitting of the $5S_{1/2}$, $5P_{1/2}$ and $5P_{3/2}$ is due to the presence of the hyperfine interaction of the nucleus with a single valence electron of the atoms (in the rubidium atom this electron is in the state with a principal quantum number $n = 5$). The physics of this interaction described in Sects. 1.11 and 14.2, the formulas for the calculation of the splitting is given in Sect. 18.6. The complexity of the energy spectrum (see Fig. 20.2) leads to the fact that each of lines of well-known alkali doublets is split into two (the real number of components more but most of them are unresolved in ordinary conditions).

As a result of the overlaping of lines the two isotopes doubles have the form shown in Fig. 20.3, that is the lines a and A are almost overlaped but the lines b and B are separated.

Figure 20.4 shows the main elements of the optical part of the device with the working substance of vapor ^{87}Rb, which is enclosed in a bulb (*3*). The light source

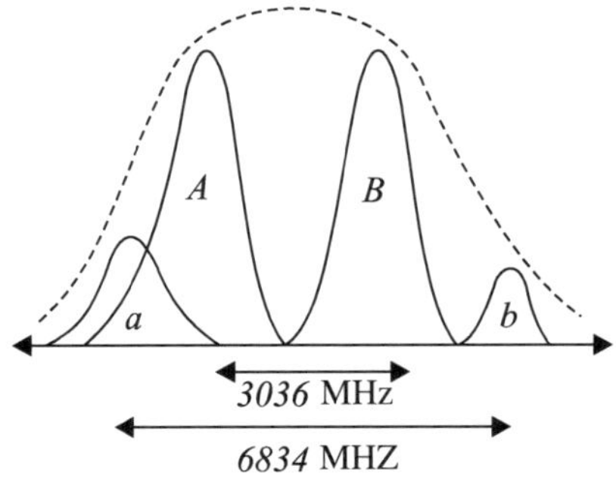

Fig. 20.3 The scheme of the disposition of the spectral lines of two isotope of rubidium

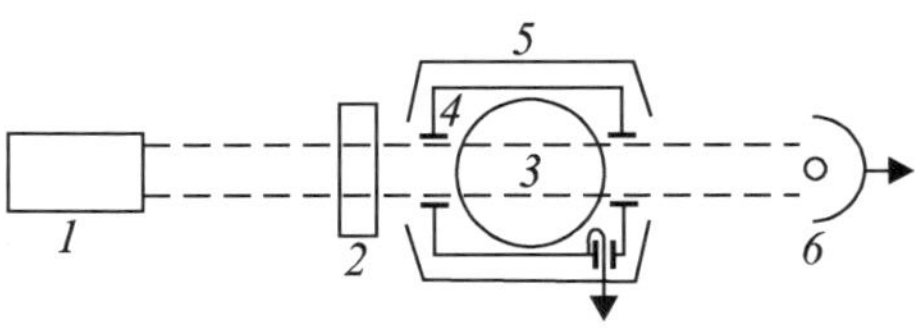

Fig. 20.4 The optical part of the rubidium standard

for the optical pumping is a discharge lamp (*1*) containing rubidium with the natural ratio of isotopes or only isotope ^{87}Rb. Because of the spectral lines broadening due to the processes in a discharge lamp its emission occurs in wide band (dashed line in Fig. 20.3). To achieve the effect of the optical pumping in the bulb, the lamp radiation is directed through a special filter with ^{85}Rb or ^{87}Rb, depending on, the population of which level one wants to change.

If one applies a filter (*2*) with ^{85}Rb, it is obvious that the atoms ^{87}Rb from the state with $F = 1$ (see Fig. 20.2) will be transferred (via state P) to a level with $F = 2$, because the line A in Fig. 20.3 produces the absorption of light by atoms in the same range as the line a; the existence of the B line has practically no effect on the absorption of light at the frequency of the line b. Thus, at the upper level ($F = 2$) the number of atoms will be greater than at the lower one and their interaction with the resonator (*4*) can lead to the self-excitation at frequency spectral transition $\nu_0 \simeq 6834$ MHz. The obtained generator can be used to build a frequency standard in accordance with the scheme shown in Fig. 17.1. Since the energy spectrum of rubidium becomes more complicated in the presence magnetic field (similar to the spectrum of cesium, see Fig. 18.16), it is necessary to use a magnetic screen (*5*). To reduce the influence of the residual magnetic field, the resonator is tuned to the frequency of transition $F = 2, m_F = 0 \longleftrightarrow F = 1, m_F = 0$.

If the filter (*2*) contains the isotope ^{87}Rb, then rubidium atoms in the bulb (*3*) are pumped from the state with $F = 2$ to the state with $F = 1$, i.e. the population difference $\Delta n_{12} = n_1 - n_2$ is abnormally increased, that leads to the drastic increase of the intensity absorption line at a frequency ν_0. Such a substance can be used in the scheme of passive frequency standard with a quantum discriminator (see Fig. 17.2). The device should be complemented by a photodetector (*6*), which provides a high signal/noise ratio at the input of the low frequency amplifier (advantages of optical methods of the registration of transitions between levels associated with radiofrequency radiation are described in Sect. 20.1). A commercially available quantum frequency standard was created on the principle described. This standard of the low mass (less than 30 kg) had high operational characteristics: the relative error of the frequency of $\pm 1 \cdot 10^{-10}$; the deviation from the nominal $\pm 3 \cdot 10^{-11}$ for a month; the random variations of frequency over 1 s: $\pm 2 \cdot 10^{-11}$; during a day: $\pm 5 \cdot 10^{-12}$.

The advantages of the passive quantum frequency standard with rubidium include long-time stability, small size and insensitivity to rotation.

The main disadvantage of the rubidium frequency standard is the fact that it does not involve a self-calibrating element. It would seem that this standard should be similar to devices with beams of hydrogen or cesium atoms. But in contrast to them, the rubidium vapor cell is filled with a buffer gas at a relatively high pressure (about 10^4 Pa) to reduce the width the spectral lines of atoms (the probability of w_{21}), which greatly increases due to atomic collisions with the walls of the bulb. Besides, the buffer gas reduces the contribution of the Doppler effect in the width and shift of spectral lines, and reduces the dependence of the transition frequency on the intensity of optical pumping. However the hyperfine splitting depends on the buffer gas pressure, and thus, the monochromatization of a spectral line leads to the loss of the internal calibration. As a result the rubidium standard is secondary and its frequency should be checked with primary standard. Usually the diapason of possible corrections is in limits $\pm 2 \cdot 10^{-10}$.

Further Readings

Dudkin VI, Pakhomov LN (2006) Quantum electronics. Devices and their applications. Tekhnosphera, Moscow (in Russian) (in Russian)

Vanier J, Audoin C (1989) The quantum physics of atomic frequency standards. Adam Hilger, Bristol

Chapter 21
Magnetic Resonance Quantum Magnetometry

One of the applications of Quantum Radiophysics is quantum magnetometry, based on the phenomenon of magnetic resonance and optical orientation atomic and nuclear moments. Specialists in many fields of physics and engineering face the necessity of the magnetic measurements, but the long-known classical methods do not provide modern requirements to accuracy and reliability of measurements. In the quantum magnetometry the measurement of field magnitude is performed by measuring the frequency of transitions between Zeeman sublevels of the spin system, that can be done with high accuracy by means of radioelectronics. The chapter is devoted to a brief exposition of the physical principles and guidelines of construction of nuclear and atomic quantum magnetometers. The most bright applications of the quantum magnetometry are realized for the measurement of the magnetic field of the Earth (especially its variations in space and time).

21.1 The Method of Free Nuclear Induction in the Magnetic Field of the Earth

From the expression for the Larmor frequency of the nuclear magnetic resonance $\nu_0 = \gamma B_0/(2\pi)$ it follows that there is the possibility of measuring a magnetic field $\vec{B}_0$ via resonance frequency. However, under the transition from strong fields to the Earth field the frequency of nuclear magnetic resonance decreases on four orders of magnitude or more, and the quality factor of receiving circuits also decreases considerably. It can be shown that due to this causes the signal-to-noise ratio should decrease by million times, and signals cannot be registered. To compensate the decrease of the NMR signals in weak magnetic fields B_0, including Earth's magnetic field, the following techniques are used: (i) the increase of a sample volume, (ii) the preliminary polarization of nuclei with an additional magnetic field, (iii) the Overhauser effect, (iv) the use of coils of complex geometry which allow us to increase the ratio of a

V. I. Chizhik et al., *Magnetic Resonance and Its Applications*,
DOI: 10.1007/978-3-319-05299-1_21,

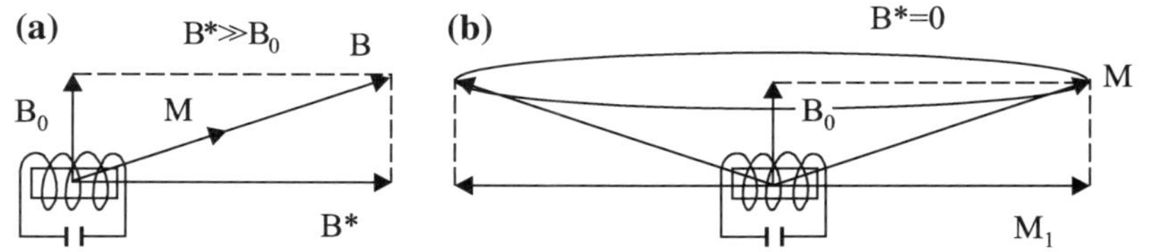

Fig. 21.1 The scheme of the Packard–Varian method

signal to external electromagnetic hindrances, (v) the application of signal digital processing, (vi) the accumulation of NMR signals.

First time the solution of the problem was suggested by M. Packard and R. Varian (Phys. Rev., 1954, v. 93, p. 941). *The method of Packard–Varian* in principle is very simple. A sample is placed, as usual, in the coil of a input receiver circuit tuned to the resonant frequency. The pre-polarization of a sample is produced with an auxiliary strong field ($\vec{B}^*$) which is created by passing a current through a polarizing solenoid (it can be used a receiving coil). The axes of a receiving coil and a vector of the auxiliary field $\vec{B}^*$ must be approximately perpendicular to the Earth magnetic field $\vec{B}_0$ (Fig. 21.1a).

After the time $(3-5)T_1$ the magnetization vector $\vec{M}$ is directed along the vector of the field $\vec{B} = \vec{B}_0 + \vec{B}^*$, i.e. almost perpendicular to $\vec{B}_0$, and the magnetization value is proportional to B in accordance with Curie law. The the field $\vec{B}^*$ must be switched off so quick that the vector $\vec{M}$ do not strongly change both its magnitude and direction. As a result, the non-stationary state arises in which the total nuclear magnetization $\vec{M}$ precesses around the vector field $\vec{B}_0$ (see Fig. 21.1b) and it is possible to detect a signal free nuclear induction in the receiving coil. The initial amplitude of the signal is proportional to the strong field B^* and its frequency ν_0 is determined by the low field B_0. It should be mentioned that the coefficient γ in the Larmor formula is a fundamental physical constant which is independent of external conditions. Therefore the free nuclear induction method provides a high accuracy of magnetic measurements.

21.2 Parameters of the Free Induction Signal in Weak Fields

21.2.1 The Signal-to-Noise Ratio

Because of the low signal intensity of the nuclear magnetic resonance in the Earth field it is very important the right choice of equipment parameters affecting on the signal-to-noise ratio: $E/\sqrt{\overline{U}^2}$. To obtain this expression in an analytic form we will use some simplifications. First of all, we assume that the complete sample magnetization has been achieved and the disappearance of the field $\vec{B}^*$ occurs rapidly, i.e. there is no a signal loss.

According to the law of electromagnetic induction the emf in the coil of a receiver circuit is described by

$$E = -n\frac{\mathrm{d}F}{\mathrm{d}t}, \tag{21.1}$$

where n is the number of turns of the coil; $F = BS$ is the induction flux ($B = 4\pi M_x$) through a cross section area S; $M_x = M\cos 2\pi\nu_0 t$; $M \simeq \chi_0 B^*$ (the coil axis and the vector field $\vec{B}^*$ are directed along the x); χ_0 is the static nuclear magnetic susceptibility of a working substance. Substituting these values in (21.1), one can then obtain the signal power:

$$P = \frac{2}{9}I^2(I+1)^2\frac{\gamma^4 h^4 B^{*2} n^2 S^2 \nu_0^2}{R}\left(\frac{N}{kT}\right)^2, \tag{21.2}$$

where R is the coil resistance; I is the nuclear spin; N is the number of nuclei per unit volume; k is the Boltzmann constant; T is the absolute temperature of a sample; h is the Planck constant. Taking into account that $S = V/(2l)$ (V is the coil volume, $2l$ is its length) and using the well-known expression for the inductance of a long coil : $L = 4\pi n^2 S/(2l)$ we obtain (21.2) in the form

$$P = \frac{1}{9}I^2(I+1)^2\gamma^2 h^4\left(\frac{N}{kT}\right)^2 QV\nu_0^3\left(\frac{B^*}{B_0}\right)^2.$$

The noise power is expressed by Nyquist formula: $P_\mathrm{n} = kT\Delta\nu F$, where T is the coil temperature; $\Delta\nu$ and F are the bandwidth and noise factor of the device. Substituting P and P_n in the expression $E/\sqrt{\overline{U}^2} = \sqrt{P}/\sqrt{P_\mathrm{n}}$, we obtain

$$\frac{E}{\sqrt{\overline{U}^2}} = \frac{Nh^2\gamma\eta I(I+1)}{6}\left(\frac{\nu_0^3}{k^3T^3}\,\frac{QV}{\Delta\nu F}\right)^{1/2}\frac{B^*}{B_0}. \tag{21.3}$$

In the expression (21.3) *the filling factor* (η) is phenomenologically introduced to take into account the degree of the filling of the coil by a sample. The factor B^*/B_0 has also appeared reflecting the increase of the signal intensity in the weak field B_0 due to pre-polarization.

21.2.2 The Shape of the Free Induction Signal in a Inhomogeneous Field

The Bloch equations (1.56) predict an exponential decay of the transverse component of the freely precessing magnetization: $\exp(-t/T_2)$. In addition to the participation in relaxation processes, magnetic moments are also subjected to external influences.

These include the influence of inhomogeneity of an external magnetic field, the reaction of a receiver circuit, the effect of fluid motion.

Consider the effect of the field inhomogeneity (δB), which leads to the distribution of resonance frequencies in a sample. If there is a Lorentzian distribution of frequencies, the signal shape is exponential with the index: $1/T_2^* = 1/T_2 + \gamma\delta B/2$. Such a case, for example, is realized for a fluid containing ferromagnetic particles. For other frequency distributions the attenuation of free nuclear induction will not be exponential. Knowing the dependence of the magnetic field on coordinates and summarizing the signals from separate elements of a working substance one can obtain the shape of a signal, which will depend on the geometry of a sample:

$$M_\perp(t) = M_\perp(0)\exp(-t/T_2)\int\limits_{V_0} \sin\big([\gamma B_0 + \gamma\delta B(x, y, z)]t\big)\,\mathrm{d}x\,\mathrm{d}y\,\mathrm{d}z. \qquad (21.4)$$

In general case these calculations are complex, however, if the small inhomogeneity of the external field can be considered as the constant field gradient (G) in a sample, the calculations are simplified. Let the vector field $\vec{B}_0$ is directed along the axis of z, $B_z = B_0 - G_1x - G_2y$, and the axis x is the axis of the cylindrical sample. Introducing the notation $\gamma G_1 l = \beta_1$, $\gamma G_2 a_0 = \beta_2$, we obtain from (21.4) to the first order in the values $G_1 l/B_0$ and $G_2 a_0/B_0$ ($2l$ and $2a$ are the length and diameter of a sample, consequently):

$$M_\perp(t) = 2M_\perp(0)\exp\left(-\frac{t}{T_2}\right)\left|\frac{\sin\beta_1 t}{\beta_1 t}\right|\left|\frac{J_1(\beta_2 t)}{\beta_2 t}\right|, \qquad (21.5)$$

where $J_1(\beta_2 t)$ is the Bessel function of the first order.

If the gradient field vector is perpendicular to the axis of the sample ($G_1 = 0$), then the expression (21.5) is simplified:

$$M_\perp(t) = 2M_\perp(0)\exp\left(-\frac{t}{T_2}\right)\left|\frac{J_1(\beta_2 t)}{\beta_2 t}\right|, \qquad (21.6)$$

If the vector of the field gradient is directed along the axis of the sample ($G_2 = 0$), then

$$M_\perp(t) = M_\perp(0)\exp\left(-\frac{t}{T_2}\right)\left|\frac{\sin\beta_1 t}{\beta_1 t}\right|. \qquad (21.7)$$

It is seen from formulae (21.6) and (21.7) that in both cases there are beats which frequencies are caused by changes in the value of the static magnetic field within a sample: $\Omega_1 = 2\gamma G_1 l$ and $\Omega_2 = 2\gamma G_2 a_0$. This phenomenon can be used to measure the field gradient (see also Sect. 2.6 on the question).

21.2.3 The Reaction of a Resonant Receiver Circuit (Coil Reaction or Radiation Damping)

One of the important aspects of the theory of free nuclear precession in the Earth magnetic field is the study of the reaction of a resonant circuit (see Sect. 2.9), i.e. the feedback effect on the nuclear magnetization via electromagnetic oscillations in the receiver coil arising due to the nuclear precession. The quantitative analysis of the coil reaction can be carried out based on classical concepts, because a spin system can be described by the macroscopic magnetization. The theory of the phenomenon is based on differential equations which describe the coupled system consisting of the precessing magnetization and tuned receiver coil. However, the equations obtained are non-linear and their general solution is difficult. Nevertheless, the solution can be obtained using well-reasonable assumptions about the smallness of the time constant of the receiver circuit in the comparison with T_1 and T_2. In this case, the current in the rf circuit will be described by relations which are valid for stationary regime. In this case, the system of equations was solved by the method of "slowly varying amplitudes", assuming $T_1 \simeq T_2 = T$. The solution is sought in the form $M_x = M_\perp(t)\cos[\omega t + \phi(t)]$, $M_y = M_\perp(t)\sin[\omega t + \phi(t)]$, $M_z = M_z(t)$. Here $M_\perp(t)$, $M_z(t)$, $\phi(t)$ are the slowly varying functions. The alternating field $\vec{B}$ of the coil reaction is formed by the current caused due to the induced emf (see (21.1)). Omitting the rather cumbersome intermediate steps, we write the final expressions for the components of the magnetization and the phase of the nuclear precession in the presence of the coil reaction:

$$M_\perp(t) = M(0)\exp\frac{-t}{T}\frac{1}{\mathrm{ch}(\xi - \xi_0)}, \tag{21.8}$$

$$M_z(t) = M(0)\exp\frac{-t}{T}\mathrm{th}(\xi - \xi_0), \tag{21.9}$$

$$\phi(t) = \ln\frac{\mathrm{ch}(\xi - \xi_0)}{\mathrm{ch}\xi_0}\mathrm{tg}\delta. \tag{21.10}$$

Here $M(0) \simeq \chi_0 B^*$; $\xi = (T/\tau)[1 - \exp(-t/T)]$; $\xi_0 = \ln \mathrm{tg}[\theta(0)/2]$; and $\tau^{-1} = 2\pi\eta\gamma QM(0)\cos\delta$; $\theta(0)$ is the angle between the component M_z and magnetization $\vec{M}$ when $t = 0$; $\delta = \mathrm{arctg}(2\Delta\omega Q/\omega_{\mathrm{rf}})$ is the phase shift between current and voltage in the resonant circuit due to the difference $\Delta\omega$ between the frequency of the free nuclear induction ω_0 and the frequency of the resonant circuit ω_{rf}; Q is the quality factor.

Figure 21.2 shows the behavior of the normalized functions $M_\perp(t)$ and $M_z(t)$ for $\theta(0) = \pi/2$. The main characteristic that determines the effect of the coil reaction is the parameter T/τ. When $T/\tau \gg 1$ the damping of the component $M_\perp(t)$ accelerates and it is significantly different from the exponent. If we compare the expression for $E/\sqrt{\overline{U}^2}$ in the form (21.3) and τ, one can see that signal intensity and the effect

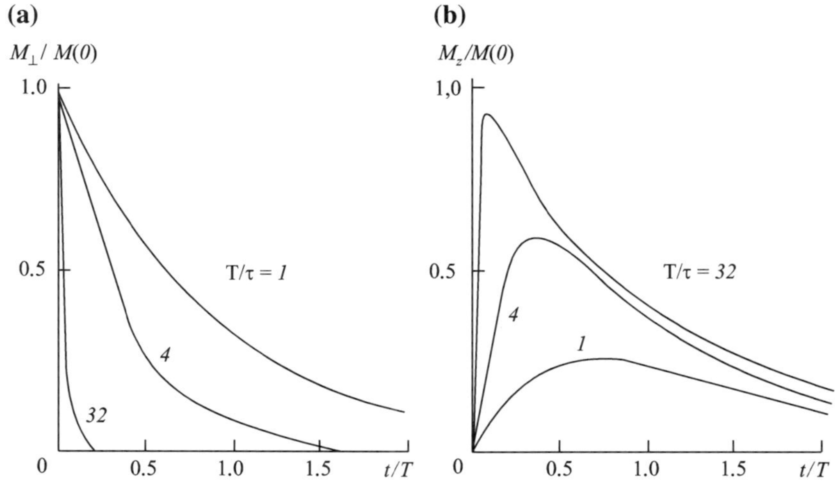

Fig. 21.2 The behavior of the components of nuclear magnetization (**a**–for $M_\perp$ and **b**–for M_z) under the influence of the coil reaction

of the reaction circuit are determined by the same the parameters η, γ, Q, $M(0)$. Striving to increase the ratio of $E/\sqrt{\overline{U}^2}$, we thus reinforce the coil reaction.

Under the influence of the coil reaction at $\theta(0) = \pi/2$ the free nuclear precession only decays. In the case of $\theta(0) > \pi/2$, in accordance with the shape of the function $[\mathrm{ch}(\xi - \xi_0)]^{-1}$, the component $M_\perp$, first, (at $\xi - \xi_0 < 0$) increases, at $\xi - \xi_0 = 0$ reaches a maximum, and then (for $\xi - \xi_0 > 0$) decreases (see also Sect. 2.9). Here it is appropriate to emphasize that the coil reaction leads only to the alignment of the nuclear macroscopic magnetization along the vector $\vec{M}_0$. The final establishment of the equilibrium occurs as a result of nuclear magnetic relaxation. The change of the free induction frequency under the influence of the coil reaction is easy to determine from (21.10):

$$\delta\omega = \frac{\mathrm{d}\phi}{\mathrm{d}t} = \mathrm{tg}\frac{\delta}{\tau}\exp\left(-\frac{t}{T}\right)\mathrm{th}(\xi - \xi_0).$$

This expression shows that $\delta\omega$ depends on time as well as $M_z(t)$. When $\Delta\omega = 0$, i.e. there is exact tuning the rf circuit, then $\delta\omega = 0$.

Let us estimate value of the parameter T/τ, that may be realized in practice. If $B^* = (100 - 400)$ G ((1 – 4) $\cdot\, 10^{-2}$ T), $\eta = 0.25 - 0.5$; $Q = 30 - 60$, then for water ($T \simeq 3$ s for protons) it is obtained $T/\tau = 0.1 - 2.0$ and for benzene ($T \simeq 19\,\mathrm{s}$)$-T/\tau = 0.5 - 8.0$ (see Fig. 21.2).

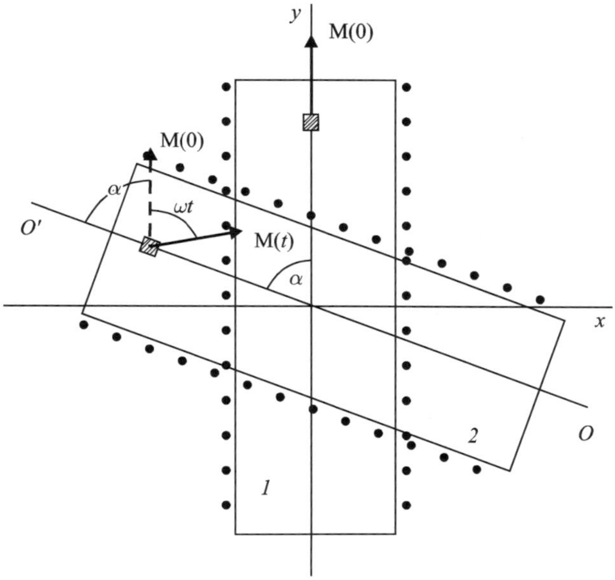

Fig. 21.3 The influence of the coil rotation on the registration of the precession of the nuclear magnetization

21.3 The Influence of Fluid Flow on the NMR Signals

Already in the first experiments on the observation of nuclear magnetic resonance in the Earth magnetic field it was found that the rotation of the cylindrical sensor in a plane perpendicular to the vector field $\vec{B}_0$ after switching off the polarizing field $\vec{B}^*$ causes a change in the resonance frequency of the registered signal. The effect leads to errors in magnetometrical measurements but it can be used for the investigations in the area of hydrodynamics, for example, for the measurements of velocity of liquid flows. This effect is due to the change of spatial positions of the elementary volumes of fluid, that results in the change of the relative orientation of the vector nuclear magnetization $\vec{M}$ and the axis of a sensor coil.

Figure 21.3 shows that if the sensor rotates around the vector $\vec{B}_0 \| z$ in the plane (xy) during the observation time t at an angle α, there will be, first, phase run $\omega_0 t$ due to the precession of the spins in a magnetic field, and, second, an additional phase shift due to the rotation of a sensor. Indeed, after switching off the polarizing field $\vec{B}^*$ the nuclear magnetization "remember" its initial orientation, which is created by the field $\vec{B}^*$. Let the initial orientation of the vector $\vec{M}(0)$ is along the axis y at the coil position 1 (see Fig. 21.3). Then turn the sensor at an angle α from the initial position to the position 2. Therefore, the sensor in the position 2 will accept the nuclear precession from all volume elements as the signal with a phase modified by the rotation of the sensor (i.e. α). When the constant rotational speed of the sensor ($\Omega = \mathrm{d}\alpha/\mathrm{d}t = \text{const}$) is realized the frequency of the observed signal is equal to

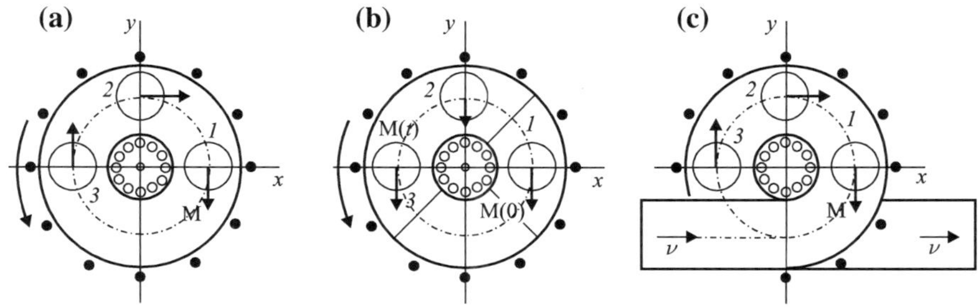

Fig. 21.4 The use of a toroidal coil for the registration of the precession of the nuclear magnetization: **a**–immobile sample, **b**–sample with baffles, **c**–mobile sample

$$\omega = \omega_0 \pm \Omega,$$

here $\omega_0 = \gamma B_0$, and plus and minus signs determine the direction of the sensor rotation. If $\Omega \neq$ const, then there will be frequency modulation of the signal. The rotation of cylindrical NMR sensors causes a large dynamic error of measurements by absolute proton magnetometers.

Figure 21.4 illustrates the manifestation of the additional phase shift when the free precession of nuclei in the Earth magnetic field is registered using a toroidal sensor. If one uses an ordinary toroidal sensor with an internal sample (Fig. 21.4a), then under the rotation or swings of a sensor in the plane (xy) around the vector external field $\vec{B}_0 \| z$ all volume elements (ΔV_i) of fluid maintain its position in space unchanged due to inertia effect and weak adhesion to the walls of the sensor. Therefore, the vectors $\vec{M}_i$ for the numbered in Fig. 21.4 elements $\Delta V_1, \Delta V_2, \Delta V_3$ will have the same initial phase $(\alpha = 0)$ of their precession around the vector field $\vec{B}_0$ during the time of signal observation. In this case, α is the angle between the magnetization vector $\vec{M}$ and the a circular axis of the toroid, along which the polarizing field $\vec{B}^*$ is created. Since for all ΔV_i elements $\alpha_i =$ const, then $d\alpha_i/dt = 0$ and the frequency of the observed signal is

$$\omega = \omega_0 = \gamma B_0. \tag{21.11}$$

If in the toroidal sensor there are baffles (see Fig. 21.4b), then the inner liquid will rotate together with the coil at the same angular velocity $\Omega = d\alpha/dt$. The frequency of the observed signal, like the case of the cylindrical sensor, will be proportional to the angular velocity sensor:

$$\omega = \omega_0 \pm \Omega,$$

i.e. the coils, shown in Figs. 21.3 and 21.4b, are the angular velocity sensors.

If the liquid flows through a toroidal or curved sensor (see Fig. 21.4c), then the volume elements ΔV_i moving along the axis cause the change of the phase of the precessing vector $\vec{M}_i$, and hence the frequency the observed signal. Thus, in the flowing toroidal or curved sensors the additional phase of the free precession of nuclei are directly proportional to $d\alpha/dt = \Omega$. The induced signal will be:

$$\mathcal{E}_1 = \mathcal{E}(0)\sin\left[\omega_0 t + \alpha(t)\right]. \tag{21.12}$$

From (21.12) one can conclude that the frequency of the observed signal ω will differ from ω_0 on the value of Ω. Sign of Ω will depend on the flow direction, i.e. $\omega = \omega_0 \pm \Omega$. If we express Ω via the velocity of fluid flow v and the toroid radius R, then the registered frequency is

$$\omega = \gamma B_0 \pm v/R.$$

Thus, the change of the frequency of the observed signal in flowing sensors is proportional to the velocity of flow and inversely proportional to the radius of the sensor curvature. The effect can be used for the measurements of fluid velocity or outlay.

21.4 Dynamic Nuclear Polarization in a Weak Magnetic Field

Besides the polarization with the auxiliary magnetic field $\vec{B}^*$ in practical applications one uses also *dynamic nuclear polarization* with the Overhauser effect (see Sect. 15.2). As it was described the dynamic polarization is observed in systems with two interacting spin varieties and its essence is to change the degree of orientation of the spins selected affecting on the spins of a different sort. The dynamic nuclear polarization can successfully be performed for a wide range of substances, in particular, for the solutions of paramagnetic ions or radicals in hydrogencontaining liquids. When one needs to create a device with a continuous nuclear precession (spin generators and filters) the dynamic nuclear polarization method is much convenient than the method used the auxiliary magnetic field. In the combined use of both types of polarization one can obtain much higher nuclear magnetization than in an every individual case. The process of the dynamic nuclear polarization can be described phenomenologically by means of two coupled equations for the longitudinal components of the nuclear and electron magnetization (see Sect. 15.2). The equation for the nuclear magnetization can be written as

$$\frac{dM_z}{dt} + \frac{M_z - M_0^*}{T_1} = 0,$$

where

$$M_0^* = \chi_0 B_0 \left(1 + s\frac{\gamma_S}{\gamma_I}\frac{T_1}{T_{IS}}\right). \tag{21.13}$$

Here the indexes S and I belong to the electron and nuclear systems, respectively; M_0^* is the new equilibrium value of the nuclear magnetization in the presence of the deviation of the electron magnetization M_S from the equilibrium value M_S^0 according to the factor $s = (M_S^0 - M_S)/M_S^0$; T_{IS} is the relaxation time, which characterizes the relationship electronic and nuclear systems. The presence of the second term in expression (21.13) indicates the possibility of the increase of the nuclear magnetization.

The rate of proton spin–lattice relaxation in a solution of paramagnetic ions can be represented as

$$1/T_1 = 1/T_1^0 + 1/T_1'. \tag{21.14}$$

Here $1/T_1^0$ is the relaxation rate of the nuclear spins in the absence of spins S; $1/T_1'$ is the nuclear spin relaxation rate due to the influence of S. We denote $\xi = T_1'/T_{IS}$. Then, according to (21.14)

$$T_1/T_{IS} = \xi f. \tag{21.15}$$

Here $f = 1 - T_1/T_1^0 = T_1/T_1'$ is *the leakage coefficient* ($f < 1$) which characterizes the incomplete electronic contribution $1/T_1'$ to the nuclear relaxation rate $1/T_1$ (see (21.14)). Using (21.15), rewrite (21.13) as follows:

$$M_0^* = \chi_0 B_0 \left(1 + s \frac{\gamma_S}{\gamma_I} \xi f \right). \tag{21.16}$$

If the normal Zeeman splitting of electronic energy levels and their full saturation are realized then $s_{\max} = 1$. The coefficient ξ depends on the type of interaction between spins: it can be shown (see Sect. 15.2), that in the case of the scalar interaction of electron and nuclear spins $\xi = -1$, and for the direct dipole–dipole interaction $\xi = 1/2$. Then for $s_{\max}$ and $f = 1$ and for the scalar interaction the increase of the magnetization as a result of the dynamic polarization will be as follow: $\eta = M_0^*/M_0 \simeq |\gamma_S|/\gamma_I = 660$. In the case of the dipole interaction in the same conditions it is possible to get the negative polarization with $\eta \simeq -330$ (it is taken into counts different signs of the gyromagnetic ratio protons and electrons). In real cases, in the result of incomplete saturation of electron paramagnetic resonance and the fact that $f < 1$, the value of η dramatically reduced, and the effect is not sufficient for the practical use in geomagnitometers.

Much higher degree of dynamic nuclear polarization is obtained in the case when an unpaired electron spin S belongs to an atom with a nuclear spin I between which the static interaction $\hat{\vec{I}} \cdot \hat{\vec{S}}$ is realized. As a result, the electronic energy levels possess a fine structure (see Chap. 14) in which there are two groups of Zeeman sublevels, separated by the interval in ten times greater than the Zeeman splitting for the free electron spins. The dynamic nuclear polarization of this type can be observed in a number of solutions of stable free radicals, such as the radicals disulfonate peroksilamina $(KSO_3)_2NO$ (Fremy salt). This is a stable free radical, which molecule contains an odd number of electrons. The unpaired electron is localized in at the nucleus of nitrogen-14 with the spin $I = 1$. In the solid state the Fremy salt is diamagnetic, that can be explained by the formation of dimers in which the electron spins are compensated. In solution the dimers dissociate into diamagnetic ions K^+ and ion-radicals $[(SO_3)_2NO]^{2-}$.

Let us consider the system of energy levels for the ion-radical of the Fremy salts. The most difficult case is when the binding energy of the total angular momentum of an electron shell $\vec{J}$ and a nuclear spin $\vec{I}$ is comparable with the energy their interaction with the external field (the case of intermediate magnetic fields). The projection of the magnetic moment of the atom $(\mu_F)_z$ can only be determined from the general dependence of the energy of the atom $E(F, m_F, B_0)$ on the external magnetic field according to the expression

$$(\mu_F)_z = -\frac{\partial E(F, m_F, B_0)}{\partial B_0},$$

where F is the total angular momentum of an atom; m_F is the magnetic quantum numbers with the values from F to $-F$ ($2F + 1$ values). To determine the energy of an atom it is necessary to find the eigenvalues of the Hamiltonian included all types of interactions:

$$\widehat{\mathcal{H}} = hA\widehat{I}\widehat{J} + \mu_B |g_J| \widehat{J} B_0 - g_I \mu_n \widehat{I} B_0.$$

Here A is the hyperfine interaction constant ($A = 35.4$ MHz for the Fremy salt); μ_B is the Bohr magneton; μ_n is the nuclear magneton; g_I is the g-factor of a nucleus; g_J is the g-factor of an electron shell (factor Lande). The problem is reduced to the solution of the secular equation. For the energy levels of our interest the solution is given by the Breit-Rabi formula (see Sect. 14.2.2). From this formula it is possible to find the energy difference between neighboring levels (for $J = 1/2$):

$$\Delta E^{\pm} = E(I \pm 1/2, m_F, \chi) - E(I \pm 1/2, m_F - 1, \chi), \tag{21.17}$$

where

$$\chi = bB_0; \quad b = 2\mu_B \frac{|g_J| + g_I'}{hA(2I + 1)}; \quad g_I' = g_I \frac{m_e}{m_p};$$

m_e and m_p are the masses of an electron and a proton. In the case of weak magnetic fields ($\chi \ll 1$), we obtain the expression

$$\Delta E^{\pm} = \Gamma^{\pm} B_0 \mp (2m_F - 1) G B_0^2, \tag{21.18}$$

where

$$\Gamma^{\pm} = \mu_B \left[-g_I' \left(1 \pm \frac{1}{2I + 1} \right) \pm \frac{|g_J|}{2I + 1} \right];$$

$$G = \left(\mu_B \frac{|g_J| + g_I'}{2I + 1} \right)^2 \frac{1}{\Delta\nu_0}; \quad \Delta\nu_0 = A \frac{2I + 1}{2}.$$

It is seen from (21.18) that in this case the levels are not equidistant and the interval between adjacent levels with m_F and $m_F \pm 1$ is a quadratic function of the magnetic

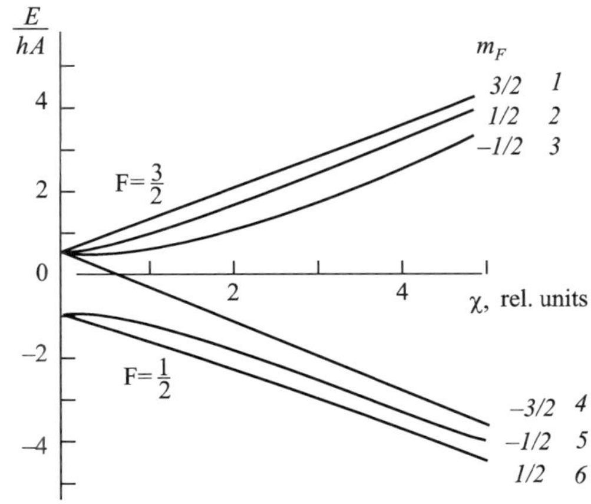

Fig. 21.5 The dependence of the energy levels on the magnetic field for the Fremy salt

field magnitude. The dependence of the energy levels on χ for the salt Fremy is shown in Fig. 21.5.

The effect of the dynamic nuclear polarization at the condition of full saturation electron transitions in the case of weak magnetic fields ($2\pi A \gg \gamma_S B_0$) can be estimated as follows. The equilibrium populations $n(F, m_F)$ for each of the six sublevels of the hyperfine structure in the high-temperature approximation are given by the expression

$$n(F, m_F) = \frac{n}{6}\left[1 - \frac{E(F, m_F, B_0)}{kT}\right], \tag{21.19}$$

where n is the number of nuclei per volume unit. Substituting in (21.19) the level energies found by the formula (21.17), one obtains a set of values of the equilibrium populations. Then find the equilibrium electron magnetization by summarizing the magnetization on each level with the population, defined by (21.19):

$$M_S^0 = \sum n(F, m_F) m_F g_F \mu_B. \tag{21.20}$$

Here g_F is the g-factor of the atom, which is determined from the expression

$$g_F = g_J \frac{F(F+1) + J(J+1) - I(I+1)}{2F(F+1)} + g_I' \frac{F(F+1) + I(I+1) - J(J+1)}{2F(F+1)}.$$

Performing the summation, one can find that $M_S^0 = 11\mu_B^2 n B_0/(27kT)$. Under the influence of the RF field at the transition frequency, the level populations are aligned. The new values of the populations of these levels are calculated using the expression

$$n^S(F, m_F) = n^S(F', m_F') = \frac{n(F, m_F) + n(F', m_F')}{2}. \tag{21.21}$$

Assuming that the populations of the other levels are in equilibrium (i.e. neglecting the relaxation transitions between these levels and all others) one can calculate the resulting electron magnetization M_S by the formula (21.20) using (21.21).

The RF field $\vec{B}_1$, vector of which is perpendicular to the constant field $\vec{B}_0$, stimulates only the transitions with the change $m_F \pm 1$. Among of the seven such transitions there are the most interesting two, which lead to the maximum of the dynamic nuclear polarization: the transitions 1–6 and 4–5 (see Fig. 21.5).

Using the formula (21.16), it is easy to determine the achievable gain of the nuclear magnetization: $\eta = M_0^*/(\chi_0 B_0)$. In aqueous solutions the interaction of the radicals of Fremy salt and water protons is dipole–dipole. If $f = 1$, one can get $\eta_{1\text{-}6} = -6750$ and $\eta_{4\text{-}5} = 6500$. Experimentally for the Earth magnetic field it was obtained $\eta = 600$. The low (compared with the calculated) value is due to the lower value of the coefficient of leakage ($f < 1$), the incomplete saturation of the transition, partially overlapping lines of the transitions 1–6 and 4–5, giving the dynamic nuclear polarization of opposite signs. The overlapping of lines can easily be removed using the auxiliary constant magnetic field during the polarization. When the value of this field was 3 G ($3 \cdot 10^{-4}$ T) the nuclear magnetization became in 2600 times bigger than the equilibrium one in the Earth field.

21.5 Generators on the Nuclear Magnetic Resonance

The free nuclear induction method is not suitable for continuous measurements of a field, so in some cases it is necessary to obtain an undamped nuclear precession. Mind that at low fields the sharp decrease of the intensity of the nuclear magnetic resonance occurs and it is necessary to use the additional artificial polarization (static or dynamic) of a nuclear spin system. Taking into account this requirement, NMR-generators of several types have been developed. A detailed examination of the cumbersome theory of the NMR-generators is out of the scope of this book, and we will discuss here only two types of such generators.

21.5.1 Schmelzer Generator

Let us turn to the Bloch spin detector with crossed coils (see Sect. 2.3). When mutually perpendicular arrangement of the coils is realized in the receiver coil only the nuclear induction signal will be registered (see (2.8)). If the condition $\Delta\omega = 0$ is fulfilled then the dispersion component becomes equal to zero, and the induced signal depends on the absorption components v. For the calculation of this component we can use the formula for the nuclear susceptibility χ'' (see Sect. 1.5). Express the amplitude of the alternating field as $B_1 = k_1 U_{0\,\max}$, where $U_{0\,\max}$ is the voltage in the transmission coil; k_1 is the coefficient of proportionality(usually $k_1 \simeq (0.02 - 0.05)$ G/V or

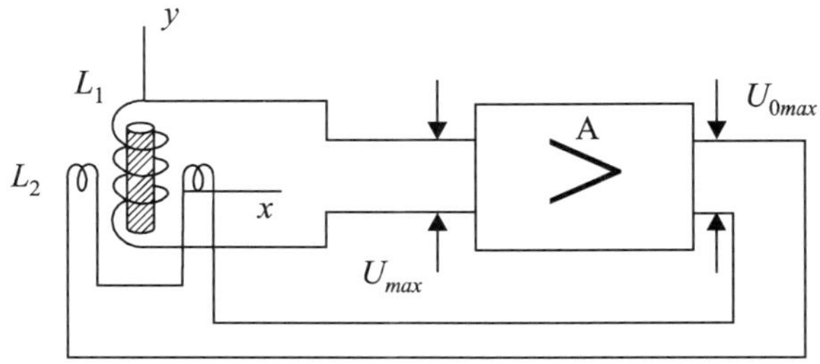

Fig. 21.6 The simplified scheme of the Schmelzer generator

$(0.02 - 0.05) \cdot 10^{-4}$ T/V). Then the amplitude of the signal in the receiving coil is calculated as

$$U_{\max} = aU_{0\max}/(1 + bU_{0\max}^2), \tag{21.22}$$

where $a = Ak_1$; A is the experimental constant; $b = \gamma^2 T_1 T_2 k_1^2$. Suppose that the transmitter coil is powered by the output stage of the receiver (Fig. 21.6). Under certain conditions such a device can operate as an autogenerator, and it is called *the Schmelzer generator*. Let us consider the operation of the scheme in more detail.

A sample is placed in a magnetic field $\vec{B}_0$, the vector of which is perpendicular to the plane of the figure. Between the coils L_1 and L_2 there is no direct inductive (and capacitive) coupling. The coils L_1 and L_2 are connected to the input and the output of the amplifier (▷), respectively. The nuclear magnetization $\vec{M}_0$ is oriented along the vector field $\vec{B}_0$. However due to the existence of electric fluctuations (noises) in the receiver coil the alternating field $\vec{B}_1$ arises and it declines $\vec{M}$ from the $\vec{B}_0$ direction. As a result the transverse component of the nuclear magnetization appears, which precesses in the field B_0 with the Larmor frequency ω_0 and induces E.M.F. ($U_{\max}$) in the coil L_1. After amplification, the voltage is applied to the coil L_2, in which the secondary alternating field, in turn, acts on the sample, and so on. Thus, due to this process and if the conditions of amplitude and phase balances in the feedback are fulfilled the undamped oscillations can appear. The voltage in the the transmitter coil is

$$U_{0\max} = KU_{\max} \tag{21.23}$$

(K is the gain of the amplifier). Substituting (21.23) in (21.22), one obtains

$$U_{0\max} = \sqrt{aK/b - 1/b}.$$

This expression describes the real signal if

$$K_{\min} > 1/a. \tag{21.24}$$

Fig. 21.7 The choice of the working points for the Schmelzer generator

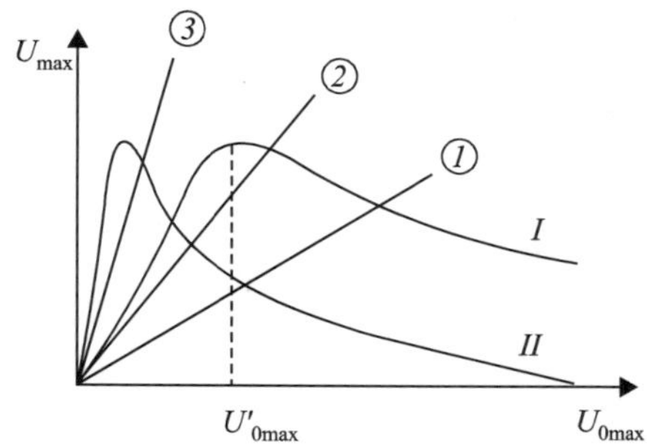

Inequality (21.24) expresses the condition of the self-excitation of the generator. The condition of self-excitation can be illustrated graphically as it is shown in Fig. 21.7: curves I and II were calculated using the formula (21.22) with $(T_2)_{\mathrm{I}} < (T_2)_{\mathrm{II}}$ and direct lines 1–3 reflects the formula (21.23) with $K_1 > K_2 > K_3$. It can be seen that the steady state generation is realized at the points of intersection of the curves I, II and direct lines 1–3, and for each pair of dependences there is only one point of intersection.

For small values $U_{0\,\max}$ equation (21.22) can be rewritten as $U_{\max} \simeq aU_{0\,\max}$. Thus, for $K < 1/a$ the curve and straight line will never intersect, i.e. the condition of self-excitation will not be fulfilled. In addition, the optimal operation of the generator corresponds to the condition $U_{0\,\max} = U'_{0\,\max}$ (see Fig. 21.7). It follows from (21.22) that for this case $bU_{0\,\max} = 1$. Therefore, for optimal regime the relation $K = 2/a = 2K_{\min}$ is realized.

It is necessary to point out that in the Schmelzer generator, like to any self-oscillating device, the phase shift in the feedback yields the shift the oscillating frequency from the resonance frequency ω_0 (see also Sect. 18.4). This drift is determined by the relaxation time T_2 and the magnitude of the phase shifts: the wider a resonance line and greater phase shifts, the greater the frequency deviation.

21.5.2 NMR Generator with a Moving Fluid

In 1957 F. I. Skripov developed the method for the registration of the continuous nuclear induction signal in the Earth magnetic field. The signal was obtained from the moving liquid using a static (!) polarization field. The method was realized at the laboratory in the country (laboratory of the Saint-Petersburg (ex Leningrad) State University) where it was possible to avoid distortions of magnetic field and industrial hindrances. The scheme of the spin detector with flowing liquid is shown in Fig. 21.8. When the flow goes through the magnetizing coil (P), the nuclear magnetization $\vec{M}$ is installed which is proportional to the sum of fields: a strong $\vec{B}^*$ and weak (the Earth

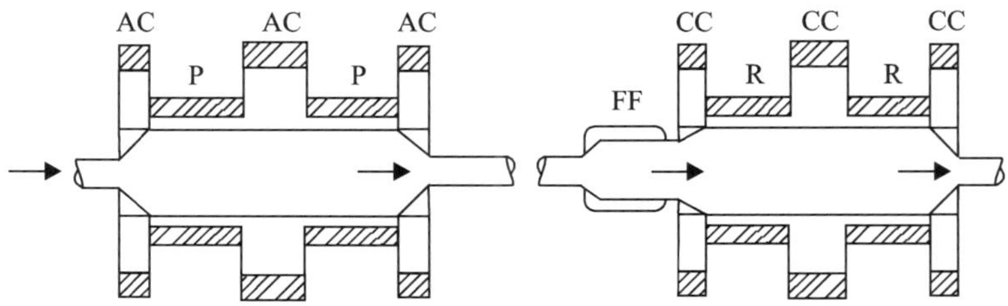

Fig. 21.8 The design of coils in the generator of Skripov

field) $\vec{B}_0$. Then, via a long thin tube the liquid reaches the registration area with the resonant circuit, where the precession of the nuclei in the Earth field $\vec{B}_0$ is registered. The velocity of flow is big enough to maintain the value $\vec{M}$ without the significant decrease.

As the field, acting on a volume element of the moving sample at any real velocity, varies adiabatically, at the end of the narrow tube the nuclear magnetization vector is along the vector of the weak field $\vec{B}_0$. Then the liquid flows between two phasing coils (FF) which create an alternating magnetic field with a frequency of oscillation being equal or very close to the resonance one. If the field amplitude in the phasing coils is properly chosen the nuclear magnetization vector nutates at an angle of approximately 90°, that is similar to the 90° rf pulse. As a result, the arising precession of the vector $\vec{M}$ is lasting in time due to renewing the working substance in the receiving coil (R), i. e. we have obtained a continuous signal.

If the phasing coils are powered by an external oscillator, the regime of the spin filter is realized. However, when the free induction signal after amplification will put at the phasing coils, it is possible to realize the regime of a generator. If between the phasing coils and receiver resonant circuit there is no direct inductive (and capacitive) coupling, then the feedback is implemented through the precession of the nuclear magnetization. The frequency of the generator changes within a wide range following automatically the change of Earth field.

The neutralization of the strong field $\vec{B}^*$ in the site of the receiving coil is achieved with three additional coils (AC) connected in series between themselves and the main magnetizing coils (P). The system of coils is calculated in such a way that, with an admissible weakening of the field of the main coils not more than 20 %, its dipole, octupole and 32-pole moments are "invisible" in the receiving coil site. Since the magnetic quadrupole and 64-dipole moments are absent because of the symmetry, the first uncompensated moment is a 128-pole one. Using a simple current adjustment it was possible to reduced the effect of the field $\vec{B}^*$ in $2 \cdot 10^6$ times at a distance of 100 cm from the center of the magnetizing coils.

It is worth to be especially underlined that the receiving circuit system has similar compensation coils (CC), that weaken the impact of variable fields, which arise as a result of industrial interferences.

21.6 Principles of the Design of Proton Geomagnitometers

21.6.1 Spin Detectors of Free Nuclear Induction

Spin detectors consist of a sensor and devices for the polarization of a sample and stimulation of the precessing magnetization. The most important element of the spin detector, which influences on the value of $E/\sqrt{\overline{U}^2}$, is the sensor of the signal of free nuclear induction comprising a coil (or a system of coils) and a working substance. The working substances are usually different proton-containing fluids selected in accordance with specified temperature range and relaxation times (water, alcohol, oil, and some others). The most simple sensor is a coil with the cylindrical winding. Relatively simple calculations show that for optimal value of $E/\sqrt{\overline{U}^2}$ the ratio of an average radius of the winding to its thickness and the ratio of a thickness of the winding to its length should be equal to about 0.4 and 0.2, respectively.

However, a simple cylindrical sensor is very sensitive to alternating fields of industrial interferences. Therefore one usually uses more complex designs of sensors that enhances noise immunity. These include, for example, two cylindrical coils connected in the circuit so that the NMR signals are added up but interferences are mutually compensated. In addition, coils protected with an electrostatic screen. Sensors with a toroidal winding have the very high noise immunity. Such sensors are also characterized by a weak dependence of the amplitude of the free nuclear induction signal on the orientations of the toroid and the field $\vec{B}_0$. If the field $\vec{B}_0$ is in the plane of the toroidal sensor, then the signal is reduced by half, and in the cylindrical sensor the signal disappears completely.

21.6.2 The Pre-polarization with an Auxiliary Field

Due to the low level of the NMR signal in weak fields of special importance is how to minimize losses of the nuclear magnetization created during the pre-polarization. In this connection consider the behavior of the magnetization during the shutdown of the polarizing field.

The behavior of the macroscopic nuclear magnetization is described with the system of Bloch equations (1.56). Even if we ignore the influence of relaxation processes during the shutdown then their solution can be obtained only in the form of slowly convergent series, difficult for applications. Some important practical recommendations can be obtained using a simplified model (here we also neglect the relaxation processes).

As it is seen from the Fig. 21.9a at the beginning of the process the summary field $\vec{B}$ mainly changes in the magnitude and not in the direction (this is fulfilled very precisely because usually the field B^* is in a hundred times more than the field B_0 and the angle α is very small). Later, when the polarizing field B^* becomes the

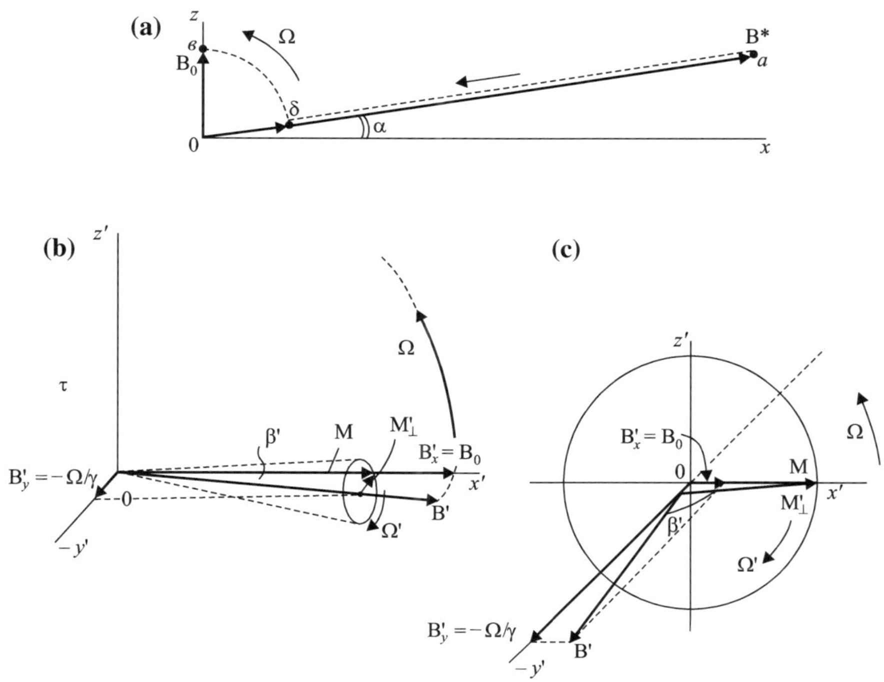

Fig. 21.9 The process of the shutdown of the polarizing field: the behaviour of magnetic field (**a**) and magnetization (**b, c**)

similar magnitude as the field B_0, the angular velocity of the turn of the field vector $\vec{B}$ is greatly increased, but its magnitude varies relatively little.

The initial adiabatic process. The field $\vec{B}$ changes only in the magnitude ($\alpha =$ const) until the moment when it becomes $B = B_0$. The vector of the total field $\vec{B}$ in Fig. 21.9a goes from the point a to the point δ. Since the relaxation is absent (our supposition), the magnetization of a sample will not change either in magnitude or in direction, i.e. in this time interval no requirements to a method of the shutdown are established.

The final step shutdown of the field. The vector $\vec{B}$ only changes in direction with a constant angular velocity Ω and $B = B_0$ (the tip of vector $\vec{B}$ at Fig. 21.9a goes from the point δ to the point b). Because of the smallness of the angle α it can be neglected compared with the angle of the rotation of the vector $\vec{B}$. The further analysis can be made in a coordinate system x', y', z', moving together with the vector of the field $\vec{B}$: the axis of z' coincides with the direction of $\vec{B}$; the axis of y' coincides with the axis of y. The system x', y', z' the precession of the vector $\vec{M}$ occurs around the vector of the effective field $\vec{B}' = \vec{B} + \vec{\Omega}/\gamma$ (see Fig. 21.9b, c).

Suppose that the situation of slow removal of the field $\vec{B}$ is realized: $\Omega/(\gamma B_0) \ll 1$, i.e. the rate of the rotation of the field $\vec{B}$ is much less than the angular velocity of the precession of the nuclear magnetization in the field $\vec{B}_0$. Besides the vector $\vec{M}$ precesses around the field $\vec{B}'$ with the frequency $\Omega' = \gamma B'$ being at a small angle of a precession cone β' (see Fig. 21.9b). Thus, in the laboratory coordinate system x, y, z vector $\vec{M}$ rotates together with the field $\vec{B}$. At the end of the process ($t = t_{\text{off}}$), when $\Omega t_{\text{off}} = \pi/2$, the vector $\vec{M}$ is under the small angle to the vector of the field $\vec{B}_0$. Due to the smallness of the angle β' the precession signal just after the slow shutdown is very small and cannot be observed in practice. Usually such slow shutdown (*adiabatic*) is performed before the 90° pulse.

Now, let $\Omega/(\gamma B_0) \gg 1$, i.e. we have a situation of quick shutdown. Performing the similar qualitative consideration, we can see that now the angle β' is close to $\pi/2$ (see Fig. 21.9c). In this case, $B' = \sqrt{B_0^2 + (\Omega/\gamma)^2} \simeq \Omega/\gamma$ and $\Omega' = -\gamma B' \simeq -\Omega$. Consequently, in the laboratory coordinate system x, y, z the vector $\vec{M}$ is almost fixed, i.e. "come off" from the vector of the field $\vec{B}$. This shutdown is *non-adiabatic*. After the time $t_{\text{off}} = (\pi/2)/\Omega$, i.e. after the turn of the field $\vec{B}$ to the point b (see Fig. 21.9a), the non-stationary state of the vector $\vec{M}$ arises, when this vector is almost perpendicular to the the field $\vec{B}_0$, and the process of the free precession of the magnetization around the field $\vec{B}_0$ is realized. Thus, the condition non-adiabatic shutdown is the requirement

$$\Omega/\omega_0 \gg 1 \quad \text{or} \quad t_{\text{off}}/T_0 \ll 1, \tag{21.25}$$

where T_0 is the period of the free precession in the B_0. If the conditions (21.25) are not fulfilled, there is a decrease of the initial signal amplitude.

After the slow shutdown of the field $\vec{B}$ it is necessary to apply the 90° rf pulse on the sample for the obtaining the free nuclear induction. It is important that the initial phase of the free precession after the 90° pulse is strictly bounded with the phase of the alternating magnetic field of this pulse. This fact is essential in experiments in which the phase relationships between the frequency of the external impact and the frequency of precession play an important role. These experiments include the method of spin echo, as well as measurements of the magnetic Earth field using phase methods.

21.6.3 Registration of Nuclear Induction Signal

Measurements of the Earth magnetic field using the proton magnetometers is performed via measurements of the frequency of nuclear precession (ν_0). Consider the methods of the minimization of errors in magnetometers.

First, it is necessary to consider the lowest limit of an error of frequency measurements, i.e. the influence of noises. We consider the following case for the signal-to-noise ratio $E/\sqrt{\overline{U}^2} \equiv \rho \gg 1$ and $\vec{E} \perp \vec{U}$. Then the change of the phase of the signal due to noise ($\delta\phi$) is small and one can write: $\delta\phi \simeq \text{tn}\delta\phi = \rho^{-1}$.

Because of noises at the beginning and end of the measurement interval (t_{reg}) are not correlated, the standard error of measurements of the nuclear induction frequency is expressed by the obvious relation:

$$\delta\omega_0 = \sqrt{(\delta\phi)^2_{\text{beg}} + (\delta\phi)^2_{\text{end}}}/t_{\text{reg}} = \sqrt{\rho^{-2}_{\text{beg}} + \rho^{-2}_{\text{end}}}/t_{\text{reg}}. \tag{21.26}$$

In the case of an exponential shape of the signal of free nuclear induction, one can write:

$$\rho_{\text{end}} = \rho_{\text{beg}} \exp(-t_{\text{reg}}/T_2),$$

and substituting this expression in (21.26), one obtains

$$\delta\omega_0 = (\rho_{\text{beg}} T_2)^{-1}\alpha^{-1}\sqrt{1 + \exp 2\alpha},$$

where $\alpha = t_{\text{reg}}/T_2$. The mean square error for the magnetic field is equal to

$$\delta B = (\gamma\rho_{\text{beg}} T_2)^{-1}\alpha^{-1}\sqrt{1 + \exp 2\alpha}.$$

It is easy to show that the minimum error of measurements is $\alpha_{\text{opt}} = 1.1$.

The most widespread magnetometers are with an electronic frequency meter. They are convenient and easy to use, have a digital output, their accuracy is determined by stability of the quartz oscillator. There are two methods of measuring frequency.

1. It is possible to measure the number (N_{st}) of periods (T_{st}) of a signal of a standard frequency (ν_{st}) during the time of the fixed number (N_0) of periods (T_0) of the NMR signal. The frequency measured is given by

   $$T_0 N_0 = T_c N_c.$$

   Then it is necessary to use the conversion formula

   $$\nu = \nu_c N_0/N_c, \tag{21.27}$$

   that is a shortcoming of the method. It follows from (21.27) that the relative measurement error is

   $$\frac{\delta\nu_0}{\nu_0} = \frac{\delta\nu_c}{\nu_c} + \frac{\delta N_0}{N_0} + \frac{\delta N_c}{N_c}, \tag{21.28}$$

   where $\delta\nu_c/\nu_c$ is the stability of a standard generator; $\delta N_0/N_0$ is the error due to noises; $\delta N_{\text{st}}/N_{\text{st}} = 1/N_{\text{st}}$, since N_{st} is determined within ± 1. In this case, the main factor is the noise error, because the others can be made to be small compared with it.
2. The task is to determine the number of periods (N_0) of the measured frequency (ν_0) during the time interval defined by a fixed number (N_{st}) of periods of the standard frequency (ν_{st}). The number N_0 is directly proportional to the measured frequency (and, consequently, to the value of the field B_0), see (21.27). The error is calculated according also to formula (21.28). Since the accuracy of the reference frequency meter is a unit, i.e. one period of the signal frequency, and the number of periods N_0, e.g. for 1 s, is equal to about $2 \cdot 10^3$ then $\delta N_0/N_0 \simeq 0.5 \cdot 10^{-3}$,

that is an unacceptably large error in magnetometry. To improve the accuracy of relative measurements, the signal requency must be multiplied in many times to get the situation when the error is determined by noises. Creating a signal frequency multipliers (for example, in $10^2 - 10^3$ times) is very difficult, that is a disadvantage of this method.

21.7 Optical Polarization in Quantum Magnetometry

The practical use of proton magnetic resonance in the Earth field is associated with certain technical difficulties. One cause of the difficulties is the low frequency resonance (in average 2 kHz), and to fix, for example, the change of the field at $1 \cdot 10^{-5}$ G (1 nT), it is necessary to measure the frequency with an accuracy of 0.05 Hz. The other difficulty consists in the requirement of a rather high power source to make a nuclear pre-polarization. To avoid both of these problems it is possible to use magnetic resonance of atoms with optical polarization (optical pumping). The optical orientation of atoms is usually realized in the vapor of rubidium or cesium. The physical picture of the process of optical pumping already described in Sect. 20.1. In Sect. 18.9 the field dependence of the energy of levels for atoms ^{133}Cs in the ground state ($I = 7/2$, $J = 1/2$, $g_I > 0$) is presented. In Sect. 20.2 we have described the energy spectrum atoms of ^{87}Rb in the ground and optically excited states. The ground state for the alkali metal atoms is the state $ns\ ^2S_{1/2}(L = 0, S = 1/2)$, where $n = 5$ and $n = 6$ for rubidium and cesium, respectively. Due to the hyperfine interaction the level $ns\ ^2S_{1/2}$ is split into two sub-levels according to the possible values of the quantum numbers: $F = I \pm 1/2$. The nearest excited state is $np\ ^2P$-state ($L = 1$, $S = 1/2$). The transition between $np\ ^2P$- and $ns\ ^2S_{1/2}$-state corresponds to the spectral D-line in the visible or infrared regions of the spectrum. Excited $np\ ^2P$-state is also split into sublevels $np\ ^2P_{1/2}$ and $np\ ^2P_{3/2}$ due to spin–orbit interactions in accordance with two possible values of the quantum numbers: $J = L \pm 1/2$. Therefore the spectral D-line is also split into two close lines: D_1 and D_2. Due to the hyperfine interaction each of the $np\ ^2P$-levels is split for a number of sub-levels in according with different values of the quantum number F. It is worth noting that, in contrast to the nuclear magnetic resonance, the transition frequency depends on the field B_0 in the nonlinear manner: $\nu^{\pm} = [\Gamma^{\pm} B_0 \mp (2m_F - 1) G B_0^2]/h$ (see (21.18)). This is one of the markable shortcomings of devices with optical pumping.

When a system of atoms is exposed with radiation at the spectral frequency of the D-line, atoms absorb light quanta and move from i-th sublevel S-state to k-th sublevel of exited P-states with a probability of W_{ik}, which is proportional to the radiation intensity. Due to the spontaneous radiation the lifetime of atoms in excited states is very short: about $10^{-6} - 10^{-8}$ s. After this time, an atom emits a photon of light and moves to one of sublevels with probability A_{kl} (Einstein coefficient).

In the method of optical pumping, there are shifts of the frequency in rf range under the influence of optical radiation. The shift of the first type depends on the

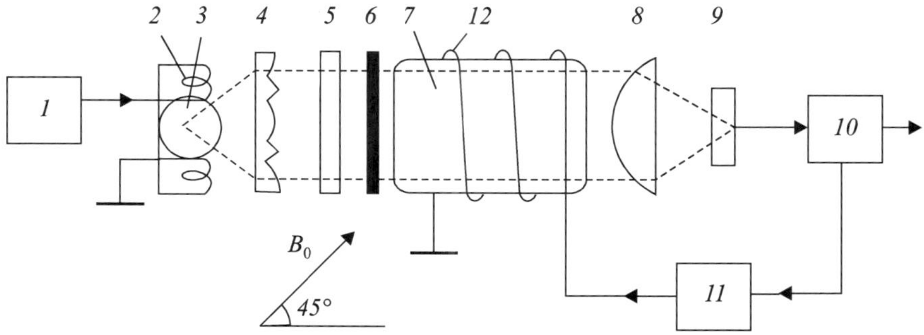

Fig. 21.10 The cesium magnetometer with the optical pumping

difference between frequencies of the incident light radiation and atomic absorption line. Various levels are shifted under the influence of light radiation at different values, that results in a change of the precession frequency, determined by the energy difference of adjacent levels. The shift of the second type is due to the fact that, during the optical pumping, atoms are in the excited state for a certain time. In this case, the resulting g-factor is determined by shift of g-factors of the ground and excited states in some proportion. Besides, this shift depends on the magnitude of the measured field. Both considered types of the shifts depend on the intensity of the pump radiation. The study of nature of the frequency shift and the development of their neutralization are important tasks of the quantum magnetometry.

Optical detection of the resonance signal is either via the change in intensity of the light passing through the sample and creating the orientation of the atoms along the vector field $\vec{B}_0$ or registering the modulation of the intensity of a weak auxiliary light beam (which is perpendicular to the magnetic field). In the latter case the two beams (polarization and registration) can be replaced by a single one oriented at a certain angle to the vector of magnetic field. Optical method of the detection of the magnetic resonance is much more sensitive than the method based on the direct detection of E.M.F. of electromagnetic induction, because the energy of an optical photon is about $10^6 - 10^9$ greater than energy of a photon of radio frequency range. Modulation of the light, passing through the sample, can be used for self-excitation of the system. The signal from the output of a photodetector, which records the light passing through the sample, will be modulated by the Larmor frequency. If this signal after the amplification feeds a coil surrounding the sample so that it creates the rf field which is perpendicular to the vector of constant field $\vec{B}_0$, the feedback loop will be closed and the self-excitation a system can be realized.

Figure 21.10 shows the scheme of a typical spin generator used in the cesium magnetometer, based on the described method of a feedback through a light beam. The light source is a cesium spectral lamp (3) excited by a coil (2) of a high-frequency generator (1). On the way to the cell (7) with the working substance (vapour of ^{133}Cs) the light passes a Fresnel lens (4) to form a beam, a polaroid (5) for linearly polarized light, and a quarter-wave plate (6) to convert linearly polarized light into circularly polarized light. Passed through a cell light is focused by a lens (8) and falls on the photodetector (9). The modulated signal from a photodetector is directed in a feedback loop consisting of an amplifier (10), phase shifter (11), and the feedback coils (12), which create radio frequency field to maintain the precession. The phase shifter provides the change of the signal phase at $-90°$, required due to the fact that the radiofrequency modulation of light is shifted in a phase relatively the rf field for $+90°$. The use of the light beam as a signal carrier eliminates the direct connection of the input of an amplifier with its output, because there is no a receiving coil in which together with an induced atomic E.M.F. spurious signals could also be imposed from the output of the amplifier. The magnitude of the signal (U) of the described spin generator depends on the angle (α) between the axis of the optical system and the field $\vec{B}_0$ in accordance with the law $U = U_m \sin\alpha \cos\alpha$. The maximum value of the signal can be obtained with the optimal angle $\alpha_{\text{opt}} = 45°$, and if $\alpha = 0$ and $\alpha = 90°$ then zones with no signal are formed.

It was earlier considered the phenomenon of optical polarization of atoms in the ground state. But it is possible to realize the polarization atoms in the metastable state, which is pseudoground. The use for this purpose the metastable ^{4}He eliminates the major shortcomings of "alkaline" magnetometers.

Let us consider the scheme of energy levels of atoms ^{4}He, given in Fig. 21.11. According to the Pauli principle, two existing electrons of the atom ^{4}He can be in a lower energy state ($l = 0$, $n = 1$), when their spin moments are in the opposite directions, so that their summary spin is $S = 0$. In this $1\,^1S_0$-state ^{4}He is named as *parahelium*. A nuclear spin of ^{4}He is also absent. In this state the atom is diamagnetic. The nearest excited state of parahelium has a configuration $1s_1 2s_1$ and it is indicated as $2\,^1S_0$.

But there is another "metastable" state of the helium atom, when spin moments are in one direction, forming the resulting spin $S = 1$. According to the Pauli principle, one of the electrons must be in the state different from that of another electron by number l or n. In this case the lowest state is the state with $n = 2, l = 0$. The configuration of the atom is the same as in the first excited state parahelium ($1s_1 2s_1$) and it is referred to as $2\,^3S_1$. In this state the helium is paramagnetic, it is called *orthohelium*. In the zero approximation the energy of the first excited state of $2\,^1S_0$ parahelium and the ground state of $2\,^3S_1$ orthohelium are of the same value. Shown in Fig. 21.11 the difference between energy levels $2\,^1S_0$ and $2\,^3S_1$ is to due the interaction between the electrons and the correction to the energy comprises two contributions corresponding to the Coulomb interaction and exchange energy. The transitions between the para- and ortho-states have a very small probability. Therefore, even taking into account the effect of collisions with other atoms, electrons, and cell walls, the lifetime of ^{4}He in the metastable state $2\,^3S_1$ is sufficient to carry out its optical polarization.

Used for the optical pumping the nearest excited $2P$-level orthohelium ($L = 1$) is split into three sub-levels due to the spin–orbit interaction: $2\,^3P_2$, $2\,^3P_1$, $2\,^3P_0$ according to the three possible values of the quantum number $J = S+L$. The ground level $2\,^3S_1$ of the metastable atom ^{4}He is split into three well resolved equidistant Zeeman sublevels. Frequency of transitions between these sublevels are used for the

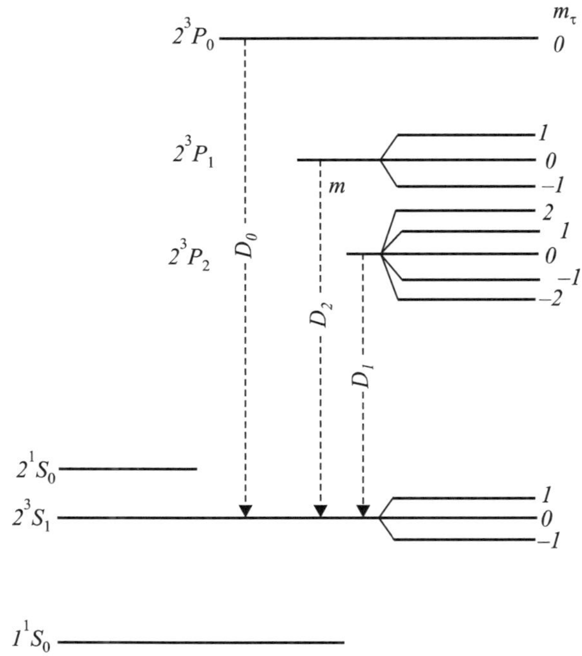

Fig. 21.11 The sceme of energy ststes of the helium ^{4}He atom

measurements of a magnetic field: $\nu_0 = B_0\gamma/(2\pi)$, where $\gamma/(2\pi) = 2.8 \cdot 10^6$ Hz/G (28 Hz/nT).

In comparison with the previously described "alkaline" devices, the helium magnetometer has the following advantages: a linear dependence of the frequency of the magnetic resonance on the magnetic field and the absence of frequency shifts because of unresolved structure of a spectrum. The shortcoming of the helium magnetometer is the presence of the previously described shift of the resonance frequency under the influence of light.

Among other objects, for which the atom optical orientation can be realized, we describe the nuclear paramagnet ^{3}He. The energy level scheme for ^{3}He is shown in Fig. 21.12. In the ground $1\,^1S_0$-parastate, the spin moments of the two electrons of the atom are compensated and therefore the total electron spin $S = 0$ that results in a closed shell configuration $(1s)^2$. In the ground metastable $2\,^3S_1$-orthostate the spin moments of the electrons are in the same direction, giving the total spin $S = 1$ and the configuration $1s2s$—the same as in the first excited state of parahelium.

If one does not take into account the interaction between the electrons, the energy of the first excited state of the parahelium and the ground state orthohelium are of an equal magnitude. The consideration of the electron interaction leads to the splitting of the ortho- and parastates. The nearest excited state of the orthohelium is $2\,^3P$-level ($L = 1$), which is split due to the spin–orbit interaction on three sub-levels corresponding to the values J ($|L - S| \leq J \leq L + S$).The D_1- and D_2-transitions

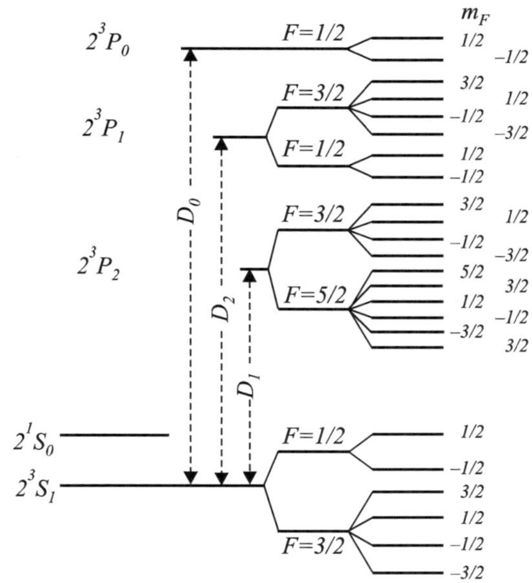

Fig. 21.12 The sceme of energy ststes of the helium ^{3}He atom

between $2\,^3S_1$- and $2\,^3P$-levels are used for the optical pumping. The nucleus ^{3}He possesses the spin $I = 1/2$ and therefore the parastates are split in a magnetic field into two Zeeman sublevels and ortostates have a hyperfine structure.

Because in the ground $1\,^1S_0$-parastate the closed electron shell of the atom ^{3}He screens the nuclear spin from the external disturbances, the relaxation time of nuclear moments of ^{3}He can be longer in several orders of magnitude than the relaxation times of other nuclei.

To decrease the broadening of NMR-line caused by the metastability exchange *a spatial separation of places of optical pumping and NMR signal detection* was suggested. In this method the gas cell takes the form of a dumb-bell and consists of a camera for the optical pumping and a "resonance" camera (for the NMR detection) connected by a diffusion pipe and this pipe contain a diaphragm with a small hole. In the pumping camera one excites a gas discharge and circularly polarized light from a lamp is focussed at a plasma. The pumping light drives ^{3}He atoms in the metastable $2\,^3S_1$-state, existing in the gas discharge. During diffusion along the pipe, atoms ^{3}He from the metastable state pass in the ground one. Thus the resonance camera comprises only atoms mostly in $1\,^1S_0$-state. The resonance camera is placed in a coil of a receiver resonant circuit in which the NMR signal is registered.

In the non-discharged helium cell the nuclear relaxation times can reach values up to hundreds of seconds. By modifying the design of the dual-chamber cell and gas pressure one can control the value of T_1 and gets a line quality factor for NMR-lines of $10^7 - 10^8$. As a result, self-generating devices such as a maser or spin generator, based on the optical orientation of atoms ^{3}He in the two-camera cell, have a frequency shift much smaller than in other self-generating magnetometers (see Sect. 18.5 about frequency shifts in masers). The potential sensitivity of these magnetometers is unprecedentedly high, but the practical implementation of the limit parameters is difficult because of noises of receiving equipment.

21.8 Some Applications of Quantum Magnetometers

High sensitivity and accuracy of quantum magnetometers based on magnetic resonance have led to their almost exclusive use in areas of science and technology where it is necessary to measure very weak magnetic effects. It is well-known the widespread use of quantum magnetometers in the mineral prospecting. The use of proton magnetometer allowed the precise determination of the value of the proton gyromagnetic ratio, which is the fundamental physical constant. Quantum magnetometers are used in the investigation of the possibility of earthquake prediction and implementation of magnetic imaging of buried archaeological objects. Even this, far from complete, recapitulation shows that many application of quantum magnetometry occur at the intersection with other fields of science, including humane areas. Let us briefly comment on two the last-mentioned applications of quantum magnetometry.

One of the most important practical applications of quantum magnetometry are attempts *to predict earthquakes in seismic areas*. For this purpose over the past decades the large range of experimental and theoretical studies of the magnetic properties of rocks, subjected to mechanical stress, was carried out. Empirical laws were explained on the basis of the theory of magnetism. Received theoretical formulas satisfactorily describe experimental dependencies of the magnetic susceptibility and residual magnetization of rock from the weak uniaxial compression or tension. The theoretical predictions of the possible change of the magnetic field of the Earth ΔB are in limits of 10 nT. Practically ΔB can vary from 0.1 to 10 nT, less probable, to 30 nT. A very small lower limit of expected seismomagnetic earthquake precursors and a clear need to exclude cases of failure forecasting system or false alarms impose very strict requirements to the sensitivity and long-term stability of the magnetometers. The indicated requirements can be satisfied only using modern proton magnetometers which sensitivity and stability are about $0.1 - 0.2$ nT that explains their almost exclusive use for the solvation of this problem. Systematic studies of the seismomagnetic effect in order to predict earthquakes held in many countries, including the U.S.A., Japan, Russia, Turkey, China. But it was found that the forecast of earthquakes depends not only on the quality of magnetometers, but also on the level magnetic noises in a particular region. Possible disturbances include, for example, the natural variations of the magnetic field of the Earth, as well as noises of industrial and geological origin. Natural variations are eliminated using differential magnetometers or measurements by two (more) absolute magnetometers.

Significant advances of *application of physical methods in archeology* has recently given a rise to a new branch of science. Physical methods have been successfully used in all stages of research of archaeological objects: from prospecting of areas to laboratory study of findings. Among the existing physical methods of "field research" the magnetic prospecting demonstrates the best universality, speed, simplicity of the processing and interpretation of the data. The most important prerequisite of the applicability of magnetic prospecting in archeology is the existence of the sufficient contrast of magnetic properties of objects and the host environment. The contrast is sufficient if it provides the possibility to detect the magnetic anomaly over an object. It is convenient to carry out a brief examination of the question on a specific example of rural settlements of the ancient period in the Northern Black Sea. Typical objects for these settlements are remains of walls, which were built with blocks of limestone and located in argillaceous (clayey) soil (loam). Such, practically non-magnetic, walls give negative anomalies on the background of weakly magnetic loam. The model calculations and experimental data leads to values of expected anomalies from the walls of about $2 - 10\,\text{nT}$. The positive anomalies of the same order of magnitude are observed at other archaeological sites (for example, brick-works). Therefore, the investigation of buried archaeological objects requires the application of highly sensitive magnetometers. Processing and presentation of the magnetic survey data are carried out using a computer. The picture of the configuration of limited in amplitude negative anomalies is a ready plan and introscopic image of buried walls. Using introscopic images of rural settlements it is possible before the excavation to determine their type and key planning elements, that are required to identify the object investigated how the antique monument or local barbarian settlement.

Further Readings

Bates RD (1993) Dynamic nuclear polarization. Magn. Reson. Rev. 16:237

Lambert JB, Shawl CE, Stearns JA (2000) Chem. Soc. Rev. NMR in Archaeology 29:175

Mohoric A, Stepisnik J (2009) NMR in the Earth's magnetic field. Prog. Nucl. Magn. Reson. Spectrosc. 54:166

Pomerantsev NM, Ryzhkov VM, Skrotskii GV (1972) Physical foundations of quantum magnetometry. Publ. House "Nauka", Moscow (in Russian)

Appendix A
Symbols and Abbreviations

Abbreviations

ADC	Analog-to-digital converter
AFR	Antiferromagnetic resonance
COSY	Correlation spectroscopY (two-dimensional NMR experiment)
DM	Dzyaloshinskii–Moriya (interaction)
DNQR	Double nuclear quadrupole resonance
EFG	Electric field gradient
emf	Electromotive force
EPR	Electron paramagnetic resonance
ESR	Electron spin resonance
EXSY	Exchange spectroscopY (NOESY experiment applied to detect chemical exchange and conformational motion)
FID	Free induction decay
FMR	Ferromagnetic resonance
fMRI	Functional magnetic resonance imaging
GTO	Gaussian type orbital
HFI	Hyperfine interactions
hff	Hyperfine field
HW-8	Haeberlen–Waugh (pulse sequence)
INADEQUATE	Incredible natural abundance double quantum transfer experiment
IR	Inversion–recovery magnetic resonance imaging technique
LC	Liquid crystal, liquid crystalline
LC-circuit	Resonance circuit
LCAO	Linear combination of atomic orbitals
MO LCAO	Molecular orbital as linear combination of atomic orbitals (method)

V. I. Chizhik et al., *Magnetic Resonance and Its Applications*,
DOI: 10.1007/978-3-319-05299-1,

vMREV-8	Mansfield–Rhi–Elleman–Vaughan (pulse sequence)
MRI	Magnetic resonance imaging
vMW-2, MW-4	Mansfield–Waugh (pulse sequence)
MWH	Mulliken-Wolfsberg-Helmholz (method)
NMR	Nuclear magnetic resonance
NOESY	Nuclear overhauser effect SpectroscopY
NQR	Nuclear quadrupole resonance
OW	Ostroff–Waugh (pulse sequence)
PAPS	Phase alternated pulse sequence
PDLF	Proton detected local fields
PSD	Phase-sensitive detector
rf, RF	Radio frequency
RFG	Radio frequency generator
RIS	Rotational isomeric states (molecular conformations)
RKKY	Ruderman–Kittel–Kasuya–Yosida (interaction)
rms	Root mean square
ROESY	Rotation frame overhauser effect spectroscopY
SCF	Self-consistent field
SD	Synchronous detector
SLF	Spectroscopy of local fields
S/N	Signal-to-noise (ratio)
STO	Slater type orbital
TROSY	Transversal relaxation optimized spectroscopY
TZV	Triple-Zeta-Valence (basis set)
WHH-4 (or WAHUHA)	Waugh–Huber–Haeberlen (pulse sequence)

Latin Symbols

A	hyperfine interaction constant
A	spin–orbit interaction constant
$\vec{A}$	magnetic field vector potential
$\vec{B}$	magnetic field
$\vec{B}_0$	static external magnetic field (produced by a magnet of a spectrometer), directed along Z-axis of laboratory frame
$\vec{B}_1$	weak harmonic (radio frequency) magnetic field, directed in XY-plane (orthogonal to $\vec{B}_0$)
$\vec{B}_{\text{rf}}$	the same as $\vec{B}_1$
C_n	rotation axis of n-fold symmetry (symmetry groups)
$\mathfrak{D}^{(l)}$	l-th irreducible representation of the rotation group (group theory)
e	elementary charge (charge of electron) ($e = -1.602176565(35) \times 10^{-19}$ C)
$\vec{\mathcal{E}}$	electric field

E_n	energy level, eigenvalue of Hamiltonian
eq, eQ	see Q and q
f_i	dimension of i-th irreducible representation (group theory)
g	Landé g-factor
g	symbol for the spectroscopic term with $l = 4$
g_s	g-factor of free electron
$\vec{G}$	magnetic field gradient
h	Planck's constant ($h \approx 6.62606957(29) \times 10^{-34}$ J s or J Hz^{-1})
$\hbar$	Planck's constant divided by 2π ($\hbar \approx 1.0546 \times 10^{-34}$ J s rad^{-1})
$\widehat{\mathcal{H}}$	Hamiltonian, energy operator
$\hat{\vec{I}}, \hat{I}_z, \hat{I}_x, \hat{I}_y, \hat{I}_+, \hat{I}_-$	spin-operator (of one particle or total of all particles in a system) and its components
$\vec{J}$	total atom (ion) moment
J_{ij}	constant of isotropic J-coupling (or indirect spin–spin coupling) between spins i and j
$\mathsf{J}_{ij}, \mathsf{J}_{xy,ij}, \ldots$	tensor of J-coupling (matrix 3×3) and its components
$\vec{k}$	vector of conjugate space
k_B	Boltzmann constant ($k_B \approx 1.3806488(13) \times 10^{-23}$ J K^{-1})
l	orbital quantum number
$\vec{l}$	electron orbital moment
$\vec{L}$	atom (ion) orbital moment
m_e	electron rest mass ($m_e = 9.10938215(45) \times 10^{31}$ kg)
m_l	magnetic quantum number
m_s	spin quantum number
$\vec{M}, M_z, M_x, M_y$	nuclear magnetization and its projections onto the axes of laboratory frame
n	principal quantum number
$\hat{\vec{p}}$	operator of linear momentum
P_i	probability of i-th state of the system
$P_n(x), P_n^l(x)$	Legendre polynomial of order n (also known as Legendre functions of the first kind) and associated Legendre polynomial of order n and degree l; useful as polynomial of $\cos\Theta$ of colatitude angle in spherical coordinates
Q	in the expressions $e\mathsf{Q}^*$, $e\mathsf{Q}^*_{xy}$, …a tensor of the electric quadrupole moment of nucleus (matrix 3×3) and its components
q	in expression eq principal component of EFG tensor $eq = V_{cc}$
q	electric charge
R	Universal Gas Constant ($R \approx 8.314$ J K^{-1} mol^{-1})
$\vec{s}$	spin of electron
$\vec{S}$	total atom (ion) spin

$S_{\text{C-D}}$	parameter of orientational order of chemical bond C–D (C–H)
$S = S_{zz}$, $S_{xx} - S_{yy}$	orientational order parameter and biaxiality parameter of liquid crystal
$\hat{\vec{S}}$	spin-operator for spin $S = \frac{1}{2}$ (in NMR)
S_n	rotoreflection axis of n-fold symmetry (also known as "rotation-reflection axis" or "improper rotation axis") (symmetry groups)
T	temperature (in units of K)
T_1	longitude relaxation time, or spin–lattice relaxation time
T_2	transverse relaxation time, or spin–spin relaxation time
T_2^*	characteristic time of free induction decay, fictitious transverse relaxation time which accounts inhomogeneity of external magnetic field
T_E	time interval between 90-degree pulse and the echo maximum
U_l^m	uniform spherical polinomial of degree l and order m
V	potential of static electric field
V_{cc}, V_{aa}, V_{bb}	principal components of EFG tensor (in the principal frame of EFG tensor abc)
Y_l^m	spherical harmonics of degree l and order m
Z	nuclear charge
Z	partition function $Z = \sum_i \exp(-E_i/k_B T)$

Greek Symbols

$\vert\alpha\rangle$	eigenfunction of $\hat{I}_z$ for $I = \frac{1}{2}$: $\hat{I}_z$ $1\alpha\rangle = +\frac{1}{2}\vert\alpha\rangle$
$\vert\beta\rangle$	eigenfunction of $\hat{I}_z$ for $I = \frac{1}{2}$: $\hat{I}_z$ $1\beta\rangle = -\frac{1}{2}\vert\beta\rangle$
γ	gyromagnetic ratio
$\delta(x)$	Dirac delta function
δ_{ij}	Kronecker delta or Kronecker's delta
ϵ_0	Vacuum permittivity ($\epsilon_0 = 8.85418782 \times 10^{-12}\ \text{m}^{-3}\,\text{kg}^{-1}\,\text{s}^4\,\text{A}^2$)
η	biaxiality parameter of the EFG tensor, also know as asymmetry parameter of the EFG tensor
Θ	colatitude angle (angle between vector $\vec{B}_0$ and other vector in a system)
λ	wave length
μ_B	Bohr magneton ($\mu_\text{B} = 927.400968(20) \times 10^{-26}\,\text{J}\,\text{T}^{-1}$)
μ_0	magnetic constant, also know as vacuum permeability or permeability of free space ($\mu_0 = 4\pi \times 10^{-7}\ \text{V}\,\text{s}\,\text{A}^{-1}\,\text{m}^{-1}$)
ν	frequency (in units of Hz)

ρ	density of matter; spin density
σ	reflection plane (symmetry groups)
σ	isotropic chemical shift (NMR)
$\boldsymbol{\sigma}_i, \boldsymbol{\sigma}_{xz,i}, \ldots$	tensor of chemical shift of spin i (matrix 3×3) and its components
χ	diamagnetic susceptibility
χ_{VV}	Van Vleck paramagnetic susceptibility
ω	angular frequency (in units of rad s^{-1})

Index

V. I. Chizhik et al., *Magnetic Resonance and Its Applications*,
DOI: 10.1007/978-3-319-05299-1,

Printed by Printforce, the Netherlands